Energy Harvesting Technologies and Applications for the Internet of Things and Wireless Sensor Networks

Energy Harvesting Technologies and Applications for the Internet of Things and Wireless Sensor Networks

Guest Editors

Slim Naifar
Olfa Kanoun
Carlo Trigona

MDPI

Basel • Beijing • Wuhan • Barcelona • Belgrade • Novi Sad • Cluj • Manchester

Guest Editors

Slim Naifar
National Engineering School of Sfax
Sfax
Tunisia

Olfa Kanoun
Chemnitz University of Technology
Chemnitz
Germany

Carlo Trigona
University of Catania
Catania
Italy

Editorial Office
MDPI AG
Grosspeteranlage 5
4052 Basel, Switzerland

This is a reprint of the Special Issue, published open access by the journal *Sensors* (ISSN 1424-8220), freely accessible at: https://www.mdpi.com/journal/sensors/special_issues/EHTAIoTWSN.

For citation purposes, cite each article independently as indicated on the article page online and as indicated below:

Lastname, A.A.; Lastname, B.B. Article Title. *Journal Name* **Year**, *Volume Number*, Page Range.

ISBN 978-3-7258-2509-7 (Hbk)
ISBN 978-3-7258-2510-3 (PDF)
https://doi.org/10.3390/books978-3-7258-2510-3

Contents

About the Editors

Slim Naifar

Slim Naifar, Ph.D., is an associate professor at the Higher Institute of Applied Sciences and Technology of Gabès, Tunisia. He received his engineering degree in electromechanics from the National Engineering School of Sfax, Tunisia, in 2012. He later obtained a master's degree in robotics from the National Engineering School of Sfax in collaboration with Rice University, Texas, USA. In 2018, he completed a joint Ph.D. in Electrical Engineering from Chemnitz University of Technology, Germany, and Mechanical Engineering from the University of Sfax, Tunisia. Slim Naifar has been actively involved in various scientific and industrial projects, contributing to significant advances in his field. In 2022, he was awarded the prestigious "Richard Hartmann" sponsorship award for his outstanding industry-related scientific and technical achievements to the economy of Saxony, Germany. His current research interests include vibration energy-harvesting converters, magnetostrictive and piezoelectric transducers, pendulum-based converters, and bistable structures. He has co-authored more than 50 research papers.

Olfa Kanoun

Olfa Kanoun is a Full Professor of measurement and sensor technology at Chemnitz University of Technology, Germany. She received her diploma in electrical engineering and information technology from the Technical University of Munich in 1995 and her Ph.D. from the University of Bundeswehr Munich in 2001. Prof. Kanoun is a world-renowned expert in sensors and sensor system design with over two decades of experience. Her research focuses on impedance spectroscopy, impedimetric sensors, energy harvesting, wireless sensors, and micro- and nanosensors. Her expertise extends to self-powered wireless sensors, energy transfer, and flexible polymer sensors based on nanomaterials for physical and chemical measurements. She has made several significant contributions to the development of intelligent field sensors based on impedance spectroscopy, wireless sensors, and sensor systems based on nanotechnology. She has been recognized as one of the world's top 2% best scientists since 2019, according to an analysis by Ioannidis et al. In 2022, she was recognized with the IEEE IM Society Technical Award for pioneering the evolution of impedance spectroscopy from laboratory scale to field sensors.

Throughout her career, Prof. Kanoun has led numerous fundamental research and collaborative projects. She has supervised more than 40 dissertations and has published over 600 papers in peer-reviewed scientific journals and at international conferences.

Carlo Trigona

Professor Trigona received his M.S. in Automation Engineering and Control of Complex Systems (cum laude) in 2006, followed by a Ph.D. in Electronic, Automation, and Control of Complex Systems in 2010 from the same university. His post-doctoral research included positions at the University of Montpellier in France (2010–2011), the University of Catania in Italy (2011–2017), and Chemnitz University of Technology in Germany (2017–2018). He then served as an Assistant Professor (2018–2020) and a tenure-track Professor (2020–2023) at DIEEI, University of Catania.

Professor Trigona has received several awards for his research endeavors, notably the 2020 IEEE-I&M Outstanding Young Engineer Award for his exceptional contributions to advancing I&M concepts in sensors and transducers for energy harvesting and the 2023 IEEE-I&M Best Application Award for his innovative sensors that incorporate the response of plants to a wide range of environmental factors, becoming low, environmentally friendly, biodegradable instruments. Regarding his research profile, he has co-authored more than 250 scientific publications with over 2900 citations. These contributions span chapters in books, papers in international journals, proceedings of international conferences, trademarks, and patents. His research interests encompass sensors, transducers, MEMS, fluxgate magnetometers, energy harvesting, and innovative areas such as green and biodegradable sensors and transducers based on living organisms.

Preface

This reprint explores transformative advances in energy harvesting technologies reshaping the Internet of Things (IoT) and Wireless Sensor Networks (WSNs). Synthesizing cutting-edge research, it examines innovative methodologies and materials driving the evolution of self-powered electronic systems. From novel piezoelectric materials to advanced thermal harvesting techniques, this collection provides researchers, industry professionals, and graduate students with comprehensive insights into current capabilities and future possibilities. The work offers in-depth analysis of state-of-the-art technologies, implementation case studies, and critical evaluations of emerging trends in miniaturization and efficiency optimization. As we approach widespread IoT deployment, these advances in energy harvesting become crucial for creating sustainable, autonomous systems. Through this compilation of international research efforts, we aim to catalyze further innovation in self-sustaining IoT ecosystems.

Slim Naifar, Olfa Kanoun, and Carlo Trigona
Guest Editors

sensors

MDPI

Editorial

Energy Harvesting Technologies and Applications for the Internet of Things and Wireless Sensor Networks

Slim Naifar [1,2], Olfa Kanoun [1,*] and Carlo Trigona [3]

1 Professorship of Measurement and Sensor Technology, Department of Electrical Engineering and Information Technology, Chemnitz University of Technology, 09126 Chemnitz, Germany
2 Laboratory of Electromechanical Systems, National Engineering School, University of Sfax, Sfax 3038, Tunisia
3 Dipartimento di Ingegneria Elettrica, Elettronica e Informatica, University of Catania, Viale Andrea Doria 6, 95125 Catania, Italy; carlo.trigona@dieei.unict.it
* Correspondence: olfa.kanoun@etit.tu-chemnitz.de

The field of Internet of Things (IoT) technologies is advancing rapidly, driven by the critical need for autonomous and sustainable wireless sensor networks [1–6]. This growth has been accompanied by increased emphasis on energy harvesting (EH) techniques, which aim to develop technologies for capturing and converting ambient energy into usable electrical power. These methods leverage various environmental energy sources, converting them into electrical energy through appropriate transduction mechanisms and energy management circuits. The use of ambient energy sources such as solar, thermal, vibrational, and radio frequency energy has led to innovations in energy harvesting systems, incorporating advanced materials and novel transduction methods, significantly contributing to the development of efficient and sustainable wireless sensor nodes [7–14]. Research in this field also emphasizes minimizing the energy consumption of wireless sensor nodes, thereby enhancing the efficiency and sustainability of Wireless Sensor Networks (WSNs) [15–18].

The following Special Issue compiles 31 papers that offer a comprehensive overview of recent advancements in energy harvesting technologies and their integration into IoT systems. The collection highlights key findings and their implications for future research and applications. Included are studies covering energy harvesting devices, their optimization and potential applications (Contributions 1–9), piezoelectric and pyroelectric nanogenerators (Contributions 10–11), as well as investigations into thermoelectric generators and solar harvester modules (Contributions 12–14). Additionally, this Special Issue features research on the design and applications of radio frequency harvesters (Contributions 15–19), wireless power transfer technologies (Contributions 20–21), and the design, implementation, and evaluation of wireless-powered communication networks (Contributions 22–30). A comprehensive review paper is also included, focusing on aging mechanisms and their impact on the electrical performance of e-textiles (Contribution 31).

In Contribution 1, research is conducted on a vibration-based cantilever beam of composite-laminated piezoelectric patches through an experimental study of its characteristics and a modeling study of energy harvesting. The experimental study investigates the harvesting capacity and electromechanical characteristics of the cantilever harvester. In Contribution 2, a pendulum transducer prototype is tested with an online hydrometric measurement station. The system is successfully validated through experimental studies in a river. In Contribution 3, process optimization of key parameters such as beam spacing, flux density, and optimal impedance load matching of magnetic coupled piezoelectric harvesters is presented. The analysis of calculated voltage outputs based on theoretical and finite element models indicates that optimized parameters significantly enhance system efficiency. In Contribution 4, an energy harvesting system is proposed to collect downward airflow from helicopters or multi-axis unmanned rotary-wing aircraft. This wind force drives a magnet to rotate, generating a repulsive force that causes a double elastic steel

Citation: Naifar, S.; Kanoun, O.; Trigona, C. Energy Harvesting Technologies and Applications for the Internet of Things and Wireless Sensor Networks. *Sensors* **2024**, *24*, 4688. https://doi.org/10.3390/s24144688

Received: 27 June 2024
Revised: 9 July 2024
Accepted: 16 July 2024
Published: 19 July 2024

system to slap and vibrate periodically, generating more electricity than traditional systems. In Contribution 5, a parameter identification method for the multiparameter identification study of a linear-arch composite beam piezoelectric energy harvester is proposed. An experimental platform verifies that the method has high accuracy and practicability. In Contribution 6, optimized cantilever geometries developed using the design of experiments approach are analyzed and combined with frequency up-conversion excitation to convert random kinetic ambient motion into periodic excitation of the harvester. In Contribution 7, piezoelectric energy harvesting is studied in the context of a low-velocity impact of a rigid mass on a composite beam. The methodology includes modeling the open-circuit impact response in a finite element package, formulating a lumped parameter model for the piezoelectric transducer connected to the harvesting circuit, and experimentally verifying the impact using a custom portable configuration with impactor motion control. In Contribution 8, a quantitative comparison of three passive rectifiers in a low-power, low-voltage electromagnetic energy harvesting subsystem is presented. In Contribution 9, a novel method to increase the bandwidth of a cantilever beam by using an embedded transverse out-of-plane movable mass is presented. The concept is investigated through experimentation with a movable mass, in the form of a solid sphere, embedded within a stationary proof mass with hollow cylindrical chambers.

In Contribution 10, flexible generators based on lead-free barium titanate ($BaTiO_3$) and a polydimethylsiloxane (PDMS) polymer were developed. Through a comparative study, the authors investigate the impact of multi-walled carbon nanotubes (MWCNTs) through structural, morphological, electrical, and electromechanical measurements. In Contribution 11, the authors report for the first time a composite of ferroelectric antimony sulfoiodide (SbSI) nanowires and non-ferroelectric titanium dioxide (TiO2) nanoparticles as a pyroelectric nanogenerator. In Contribution 12, the authors propose a novel flexible triboelectric nanogenerator (FTENG) using a flexible micro-needle-structured polydimethylsiloxane (MN-PDMS) in combination with comfortable, commercially available 2D-polyester fibers and electroless nickel-plated cotton cloth, both of which are widely used in daily life.

In Contribution 13, the authors present a simulation of an environment monitoring device powered by a thermoelectric generator (TEG) that harvests energy from the temperature difference between air and soil. In Contribution 14, an energy harvesting system for solar energy is presented, featuring a flexible battery, a semi-flexible solar harvester module, and a Bluetooth Low Energy (BLE) microprocessor module.

In Contribution 15, the authors model and optimize various components of a Radiofrequency (RF) Energy Harvesting (EH)-assisted, QB-enabled Internet of Things (IoT) system. In Contribution 16, a new RF Energy Harvesting (RF-EH) system designed for Wireless Sensor Network (WSN) feeding is proposed. This system utilizes two different monitored architectures controlled by switch circuits based on input power levels. In Contribution 17, the authors propose a self-threshold voltage compensated Radio Frequency to Direct Current (RF-DC) converter that operates at 900 MHz and 2.4 GHz for RF energy harvesting applications. In Contribution 18, the authors present a multiband ambient RF energy harvester equipped with a high-gain wideband circularly polarized antenna, aimed at enabling self-powered wireless sensors. In Contribution 19, the authors introduce a metamaterial (MTM)-integrated high-gain rectenna for RF sensing and energy harvesting applications, operating at the 2.45 GHz ISM band. The novel MTM superstrate approach with a three-layered integration method is presented for the first time for rectenna applications.

In Contribution 20, the authors introduce a multiple concurrent slotframe scheduling (MCSS) protocol for wireless power transfer (WPT)-enabled wireless sensor networks. Simulation results from their study indicate that MCSS outperforms the traditional TSCH medium access control protocol and TSCH multiple slot frame scheduling (TMSS) in terms of average end-to-end delay, aggregate throughput, and average harvested energy. In Contribution 21, the authors discuss an Internet of Things (IoT)-enabled smart meter that utilizes energy-efficient simultaneous wireless information and power transfer (SWIPT) for a wireless-powered smart grid communication network. In Contribution 22, the detection problem is tackled using an optimal stopping framework, dealing with highly unstable likelihoods due to noisy measurements. In Contribution 23, the Bluetooth Low Energy (BLE) specification introduces support for Low Power Nodes (LPNs) via the friendship feature, where an LPN pairs with a neighboring friend node. In Contribution 24, the authors explore the LIoT concept by designing, implementing, and evaluating an LIoT node's communication and energy harvesting performance. In Contribution 25, an innovative wireless monitoring system for brake system diagnostics is presented. Due to the difficulty of applying a wired monitoring system to a freight convoy, a low-power system architecture focusing on energy harvesting and wireless communication was developed. In Contribution 26, a residual energy estimation-based medium access control (REE-MAC) protocol for wireless powered sensor networks (WPSNs) is presented. In Contribution 27, the energy consumption of popular IoT wireless technologies such as Sigfox, LoRaWAN, NB-IoT, Wi-Fi, and BLE is analyzed. Smart meters' energy consumption is broken down into metering, standby, and communication processes. In Contribution 28, the authors discuss how wireless-powered communication networks (WPCNs) will enable massive machine-type communications (MTCs), a significant service domain for 5G and beyond systems. In Contribution 29, the focus is on distributed transmit power control for energy-efficient wireless-powered secure communications, improving sum secrecy energy efficiency through distributed algorithms. In Contribution 30, the authors examine energy-efficient wireless-powered secure communications involving multiple sets of transmitters, receivers, and energy harvesting nodes.

In Contribution 31, possible aging mechanisms are reviewed, elaborating on the effect of aging on the electrical performance of e-textiles. The review provides an overview of various laboratory methods for investigating accelerated functional aging. Finally, a model of cumulative fatigue damage theory is proposed to model the changes in e-textile properties over their lifetime.

The above studies collectively contribute to advancing the fields of energy harvesting and IoT technologies, showcasing innovative approaches, practical applications, and theoretical insights that pave the way for sustainable and autonomous sensor networks. We extend our gratitude to all of the authors and reviewers whose contributions have enriched this Special Issue.

Conflicts of Interest: The authors declare no conflict of interest.

List of Contributions

1. Xue, X.; Sun, Q.; Ma, Q.; Wang, J. A Versatile Model for Describing Energy Harvesting Characteristics of Composite-Laminated Piezoelectric Cantilever Patches. *Sensors* **2022**, *22*, 4457. https://doi.org/10.3390/s22124457.
2. Naifar, S.; Grimmeisen, F.; Viehweger, C.; Hu, Z.; Bauer, A.; Hörschelmann, P.; Kanoun, O. Pendulum-Based River Current Energy Converter for Hydrometric Monitoring Systems. *Sensors* **2022**, *22*, 4246. https://doi.org/10.3390/s22114246.
3. Koszewnik, A.; Ołdziej, D.; Amaro, M. Parameter Optimization of a Magnetic Coupled Piezoelectric Energy Harvester with the Homogenized Material—Numerical Approach and Experimental Study. *Sensors* **2022**, *22*, 4073. https://doi.org/10.3390/s22114073.
4. Wang, Y.; Chu, M. Analysis of Double Elastic Steel Wind Driven Magneto-Electric Vibration Energy Harvesting System. *Sensors* **2021**, *21*, 7364. https://doi.org/10.3390/s21217364.
5. Zhang, X.; Zhang, C.; Wang, L.; Chen, L.; Chen, X.; Xu, D.; Fan, H.; Zhu, F. A Method for Parameter Identification of Composite Beam Piezoelectric Energy Harvester. *Sensors* **2021**, *21*, 7213. https://doi.org/10.3390/s21217213.
6. Gljušćić, P.; Zelenika, S. Experimental Characterization of Optimized Piezoelectric Energy Harvesters for Wearable Sensor Networks. *Sensors* **2021**, *21*, 7042. https://doi.org/10.3390/s21217042.
7. Margelis, N.; Plagianakos, T.; Karydis-Karandreas, P.; Papadopoulos, E. Assessment of Impact Energy Harvesting in Composite Beams with Piezoelectric Transducers. *Sensors* **2021**, *21*, 7445. https://doi.org/10.3390/s21227445.

8. Xu, Y.; Bader, S.; Magno, M.; Mayer, P.; Oelmann, B. System Implementation Trade-Offs for Low-Speed Rotational Variable Reluctance Energy Harvesters. *Sensors* **2021**, *21*, 6317. https://doi.org/10.3390/s21186317.
9. Jackson, N.; Rodriguez, L.; Adhikari, R. Wide Bandwidth Vibration Energy Harvester with Embedded Transverse Movable Mass. *Sensors* **2021**, *21*, 5517. https://doi.org/10.3390/s21165517.
10. Bouhamed, A.; Jöhrmann, N.; Naifar, S.; Böhm, B.; Hellwig, O.; Wunderle, B.; Kanoun, O. Collaborative Filler Network for Enhancing the Performance of $BaTiO_3$/PDMS Flexible Piezoelectric Polymer Composite Nanogenerators. *Sensors* **2022**, *22*, 4181. https://doi.org/10.3390/s22114181.
11. Mistewicz, K. Pyroelectric Nanogenerator Based on an SbSI–TiO_2 Nanocomposite. *Sensors* **2022**, *22*, 69. https://doi.org/10.3390/s22010069.
12. Chung, C.; Huang, Y.; Wang, T.; Lo, Y. Fiber-Based Triboelectric Nanogenerator for Mechanical Energy Harvesting and Its Application to a Human–Machine Interface. *Sensors* **2022**, *22*, 9632. https://doi.org/10.3390/s22249632.
13. Paterova, T.; Prauzek, M.; Konecny, J.; Ozana, S.; Zmij, P.; Stankus, M.; Weise, D.; Pierer, A. Environment-Monitoring IoT Devices Powered by a TEG Which Converts Thermal Flux between Air and Near-Surface Soil into Electrical Energy. *Sensors* **2021**, *21*, 8098. https://doi.org/10.3390/s21238098.
14. Páez-Montoro, A.; García-Valderas, M.; Olías-Ruíz, E.; López-Ongil, C. Solar Energy Harvesting to Improve Capabilities of Wearable Devices. *Sensors* **2022**, *22*, 3950. https://doi.org/10.3390/s22103950.
15. Gautam, S.; Solanki, S.; Sharma, S.; Chatzinotas, S.; Ottersten, B. Boosting Quantum Battery-Based IoT Gadgets via RF-Enabled Energy Harvesting. *Sensors* **2022**, *22*, 5385. https://doi.org/10.3390/s22145385.
16. Benkalfate, C.; Ouslimani, A.; Kasbari, A.; Feham, M. A New RF Energy Harvesting System Based on Two Architectures to Enhance the DC Output Voltage for WSN Feeding. *Sensors* **2022**, *22*, 3576. https://doi.org/10.3390/s22093576.
17. Basim, M.; Khan, D.; Ain, Q.; Shehzad, K.; Shah, S.; Jang, B.; Pu, Y.; Yoo, J.; Kim, J.; Lee, K. A Highly Efficient RF-DC Converter for Energy Harvesting Applications Using a Threshold Voltage Cancellation Scheme. *Sensors* **2022**, *22*, 2659. https://doi.org/10.3390/s22072659.
18. Nguyen, H.; Le, M. Multiband Ambient RF Energy Harvester with High Gain Wideband Circularly Polarized Antenna toward Self-Powered Wireless Sensors. *Sensors* **2021**, *21*, 7411. https://doi.org/10.3390/s21217411.
19. Lee, W.; Choi, S.; Kim, H.; Hwang, S.; Jeon, S.; Yoon, Y. Metamaterial-Integrated High-Gain Rectenna for RF Sensing and Energy Harvesting Applications. *Sensors* **2021**, *21*, 6580. https://doi.org/10.3390/s21196580.
20. Lee, S.; Nguyen-Xuan, S.; Kwon, J.; Kim, E. Multiple Concurrent Slotframe Scheduling for Wireless Power Transfer-Enabled Wireless Sensor Networks. *Sensors* **2022**, *22*, 4520. https://doi.org/10.3390/s22124520.
21. Masood, Z.; Ardiansyah; Choi, Y. Energy-Efficient Optimal Power Allocation for SWIPT Based IoT-Enabled Smart Meter. *Sensors* **2021**, *21*, 7857. https://doi.org/10.3390/s21237857.
22. Koutsioumpos, M.; Zervas, E.; Hadjiefthymiades, E.; Merakos, L. A Comprehensive Study of Event Detection in WPCN Networks with Noisy Measurements. *Sensors* **2022**, *22*, 2163. https://doi.org/10.3390/s22062163.
23. Sultania, A.; Delgado, C.; Blondia, C.; Famaey, J. Downlink Performance Modeling and Evaluation of Batteryless Low Power BLE Node. *Sensors* **2022**, *22*, 2841. https://doi.org/10.3390/s22082841.
24. Perera, M.; Katz, M.; Häkkinen, J.; Godaliyadda, R. Light-Based IoT: Developing a Full-Duplex Energy Autonomous IoT Node Using Printed Electronics Technology. *Sensors* **2021**, *21*, 8024. https://doi.org/10.3390/s21238024.
25. Zanelli, F.; Mauri, M.; Castelli-Dezza, F.; Sabbioni, E.; Tarsitano, D.; Debattisti, N. Energy Autonomous Wireless Sensor Nodes for Freight Train Braking Systems Monitoring. *Sensors* **2022**, *22*, 1876. https://doi.org/10.3390/s22051876.
26. Lee, S.; Kwon, J.; Kim, E. Residual Energy Estimation-Based MAC Protocol for Wireless Powered Sensor Networks. *Sensors* **2021**, *21*, 7617. https://doi.org/10.3390/s21227617.

27. Saavedra, E.; Mascaraque, L.; Calderon, G.; del Campo, G.; Santamaria, A. The Smart Meter Challenge: Feasibility of Autonomous Indoor IoT Devices Depending on Its Energy Harvesting Source and IoT Wireless Technology. *Sensors* **2021**, *21*, 7433. https://doi.org/10.3390/s21227433.
28. Iqbal, M.; Sadi, Y.; Coleri, S. Minimum Length Scheduling for Multi-Cell Full Duplex Wireless Powered Communication Networks. *Sensors* **2021**, *21*, 6599. https://doi.org/10.3390/s21196599.
29. Gautam, S.; Solanki, S.; Sharma, S.; Chatzinotas, S.; Ottersten, B. Hybrid Active-and-Passive Relaying Model for 6G-IoT Greencom Networks with SWIPT. *Sensors* **2021**, *21*, 6013. https://doi.org/10.3390/s21186013.
30. Lee, K. Distributed Transmit Power Control for Energy-Efficient Wireless-Powered Secure Communications. *Sensors* **2021**, *21*, 5861. https://doi.org/10.3390/s21175861.
31. Biermaier, C.; Bechtold, T.; Pham, T. Towards the Functional Ageing of Electrically Conductive and Sensing Textiles: A Review. *Sensors* **2021**, *21*, 5944. https://doi.org/10.3390/s21175944.

References

1. Kanoun, O.; Bradai, S.; Khriji, S.; Bouattour, G.; El Houssaini, D.; Ben Ammar, M.; Naifar, S.; Bouhamed, A.; Derbel, F.; Viehweger, C. Energy-Aware System Design for Autonomous Wireless Sensor Nodes: A Comprehensive Review. *Sensors* **2021**, *21*, 548. [CrossRef]
2. Sun, Y.; Li, Y.Z.; Yuan, M. Requirements, challenges, and novel ideas for wearables on power supply and energy harvesting. *Nano Energy* **2023**, *115*, 108715. [CrossRef]
3. Fernando, X.; Lăzăroiu, G. Spectrum sensing, clustering algorithms, and energy-harvesting technology for cognitive-radio-based internet-of-things networks. *Sensors* **2023**, *23*, 7792. [CrossRef] [PubMed]
4. Ali, A.; Shaukat, H.; Bibi, S.; Altabey, W.A.; Noori, M.; Kouritem, S.A. Recent progress in energy harvesting systems for wearable technology. *Energy Strategy Rev.* **2023**, *49*, 101124. [CrossRef]
5. Shaukat, H.; Ali, A.; Ali, S.; Altabey, W.A.; Noori, M.; Kouritem, S.A. Applications of sustainable hybrid energy harvesting: A review. *J. Low Power Electron. Appl.* **2023**, *13*, 62. [CrossRef]
6. Dziadak, B.; Makowski, Ł.; Kucharek, M.; Jóśko, A. Energy harvesting for wearable sensors and body area network nodes. *Energies* **2023**, *16*, 1681. [CrossRef]
7. Zhao, B.; Qian, F.; Hatfield, A.; Zuo, L.; Xu, T.B. A review of piezoelectric footwear energy harvesters: Principles, methods, and applications. *Sensors* **2023**, *23*, 5841. [CrossRef]
8. Giusa, F.; Maiorca, F.; Noto, A.; Trigona, C.; Andò, B.; Baglio, S. A Diode-less Mechanical Voltage Multiplier: A Novel Transducer for Vibration Energy Harvesting. *Sens. Actuators A Phys.* **2014**, *212*, 34–41. [CrossRef]
9. Shaukat, H.; Ali, A.; Bibi, S.; Altabey, W.A.; Noori, M.; Kouritem, S.A. A Review of the recent advances in piezoelectric materials, energy harvester structures, and their applications in analytical chemistry. *Appl. Sci.* **2023**, *13*, 1300. [CrossRef]
10. Zhao, L.C.; Zou, H.X.; Wei, K.X.; Zhou, S.X.; Meng, G.; Zhang, W.M. Mechanical intelligent energy harvesting: From methodology to applications. *Adv. Energy Mater.* **2023**, *13*, 2300557. [CrossRef]
11. Caban, J.; Vrabel, J.; Górnicka, D.; Nowak, R.; Jankiewicz, M.; Matijošius, J.; Palka, M. Overview of energy harvesting technologies used in road vehicles. *Energies* **2023**, *16*, 3787. [CrossRef]
12. Wang, T. Pendulum-based vibration energy harvesting: Mechanisms, transducer integration, and applications. *Energy Convers. Manag.* **2023**, *276*, 116469. [CrossRef]
13. Zuo, J.; Dong, L.; Yang, F.; Guo, Z.; Wang, T.; Zuo, L. Energy harvesting solutions for railway transportation: A comprehensive review. *Renew. Energy* **2023**, *202*, 56–87. [CrossRef]
14. Panda, S.; Hajra, S.; Oh, Y.; Oh, W.; Lee, J.; Shin, H.; Vivekananthan, V.; Yang, Y.; Mishra, Y.K.; Kim, H.J. Hybrid nanogenerators for ocean energy harvesting: Mechanisms, designs, and applications. *Small* **2023**, *19*, 2300847. [CrossRef]
15. Chéour, R.; Jmal, M.W.; Khriji, S.; El Houssaini, D.; Trigona, C.; Abid, M.; Kanoun, O. Towards Hybrid Energy-efficient Power Management in Wireless Sensor Networks. *Sensors* **2021**, *22*, 301. [CrossRef]
16. Shaikh, F.K.; Zeadally, S. Energy harvesting in wireless sensor networks: A comprehensive review. *Renew. Sustain. Energy Rev.* **2016**, *55*, 1041–1054. [CrossRef]
17. Seah, W.K.; Eu, Z.A.; Tan, H.P. Wireless sensor networks powered by ambient energy harvesting (WSN-HEAP)-Survey and challenges. In Proceedings of the 2009 1st International Conference on Wireless Communication, Vehicular Technology, Information Theory and Aerospace & Electronic Systems Technology, Aalborg, Denmark, 17–20 May 2009; pp. 1–5. [CrossRef]
18. Elshrkawey, M.; Elsherif, S.M.; Wahed, M.E. An enhancement approach for reducing the energy consumption in wireless sensor networks. *J. King Saud Univ.-Comput. Inf. Sci.* **2018**, *30*, 259–267. [CrossRef]

sensors

MDPI

Article

A Versatile Model for Describing Energy Harvesting Characteristics of Composite-Laminated Piezoelectric Cantilever Patches

Xiaomin Xue *, Qing Sun, Qiangli Ma and Jiajia Wang

Department of Civil Engineering, Xi'an Jiaotong University, Xi'an 710049, China; sunq@xjtu.edu.cn (Q.S.); 3120322062@stu.xjtu.edu.cn (Q.M.); wjj970226@stu.xjtu.edu.cn (J.W.)

* Correspondence: xuexm@xjtu.edu.cn

Abstract: Vibration-based energy harvesters consisting of a laminated piezoelectric cantilever have recently attracted attention for their potential applications. Current studies have mostly focused on the harvesting capacity of piezoelectric harvesters under various conditions, and have given less attention to the electromechanical characteristics that are, in fact, crucial to a deeper understanding of the intrinsic mechanism of piezoelectric harvesting. In addition, the current related models have mostly been suitable for harvesting systems with very specific parameters and have not been applicable if the parameters were vague or unknown. Drawing on the available background information, in this study, we conduct research on a vibration-based cantilever beam of composite-laminated piezoelectric patches through an experimental study of its characteristics as well as a modeling study of energy harvesting. In the experimental study, we set out to investigate the harvesting capacity of the system, as well as the electromechanical (voltage/current-strain and power-strain relationships) characteristics of the cantilever harvester. In addition, we summarize some pivotal rules with regard to several design variables, which provide configuration design suggestions for maximizing energy conversion of this type of harvesting system. In the modeling study, we propose a coupled electromechanical model with a set of updated parameters by using an optimization program. The preceding experimental data are used to verify the superiority of the model for accurately predicting the amount of harvested energy, while effectively imitating the characteristics of a cantilever harvesters. The model also has merit since it is suitable for diversified harvesters with vague or even unknown parameters, which cannot be dealt with by using traditional modeling methods. Overall, the experimental study provides information on a comprehensive way to enhance the harvesting capacity of piezoelectric cantilever transducers, and the modeling study provides a wide scope of applications for cantilever harvesters even if precise information is lacking.

Keywords: vibration-based harvesting; composite-laminated piezoelectric patch; cantilever harvester; electromechanical model; optimized parameters

Citation: Xue, X.; Sun, Q.; Ma, Q.; Wang, J. A Versatile Model for Describing Energy Harvesting Characteristics of Composite-Laminated Piezoelectric Cantilever Patches. *Sensors* **2022**, *22*, 4457. https://doi.org/10.3390/s22124457

Academic Editors: Carlo Trigona, Olfa Kanoun and Slim Naifar

Received: 27 April 2022
Accepted: 2 June 2022
Published: 13 June 2022

1. Introduction

In recent years, vibration-based energy harvesting has attracted attention for its potential use in self-energized, autonomous, and wireless monitoring systems [1,2]. The harvesters are usually classified as piezoelectric [3] electrostatic [4,5], electromagnetic [6,7], magnetostrictive [8,9], electroactive polymer [10], etc. Among the transducers, piezoelectric energy harvesters have received significant attention due to their high energy density, simple fabrication, and their compatibility with MEMS technology [11].

One of most popular harvesters is the piezoelectric cantilever beam harvester using laminated patches that usually consist of one or two piezoelectric layers (unimorph or bimorph) with some other substructures [1,3]. For example, Liu et al. [12] fabricated a MEMS-based piezoelectric laminated cantilever for energy harvesting application integrated with silicon proof mass material, which could realize a quite low resonant frequency.

Montanini et al. [13] designed a cantilever-type piezoelectric generator by gluing a PZT patch onto an FRP cantilever beam, and investigated its effect on vibration frequency, vibration amplitude, resistance load, etc. Aramaki et al. [14] fabricated a 2-DOF system using a piezoelectric MEMS vibrational energy harvester with 1-DOF and an Al cantilever, which reached a higher level of output power by 3.4 μW. Usharani et al. [15] presented an energy harvester associated with a piezoelectric-patched cantilever beam that had a step section. It could achieve higher power in a broadband operating frequency. Morimoto et al. [16] fabricated a piezoelectric energy harvester composed of PZT films transferred onto stainless steel cantilevers for improved energy conversion efficiency and toughness.

In addition to the fabrication research on possible applications of such harvesters, researchers have proposed various corresponding mathematical models. Initially, SDOF electromagnetic models [17,18] were presented based on the assumption of a cantilever beam mass-spring-damper system. These modelsprovided simple expressions and were very convenient to use; however, the models were inaccurate because they lacked important aspects of a physical system such as strain distribution and dynamic model shape, as well as their influence on electrical response [19]. As an alternative approach, approximated distributed parameter models [20–22] were developed by using Hamilton's principle and the Rayleigh–Ritz method on the assumption of the Euler–Bernoulli beam theory, which were more accurate approximations as compared with SDOF models. The models mainly considered one single vibration mode and oversimplified the backward coupling in the mechanical domain, and some incorrect modeling attempts easily resulted in inexact solutions [23]. Additionally, Erturk and Inman [19,24] proposed an advanced distributed model to solve its closed-form analytical solution. Its effectiveness was experimentally tested for predicting vibration responses and voltage output. Previous models have targeted energy prediction for harvesters with explicit configurations, and have not been applicable for harvesters with ambiguous or complicated configurations.

Fabrication and modeling studies in the related literature have both basically focused on harvesters' energy conversion capacities by means of experimental observation and prediction of approximate harvested energy amounts by means of appropriate modeling. For example, most current models have been deemed to be valid through the predictive accuracy of the gross energy amount (such as peak or average electric output). In fact, models should be further tested through electromechanical characteristics that are more pivotal to have a deeper understanding of the piezoelectric harvesting mechanism. Unfortunately, so far, there has not been sufficient model validations. Thus, further detailed model validations are urgently needed. In addition, the current models are mainly applicable for regular or simple piezoelectric-laminated cantilever harvesters (such as unimorph or bimorph) with explicit configurations. However, most harvesting products are encapsulated and complicated (such as MFC and PPA introduced in Section 2), with configurations that are not that simple and explicit. Thus, it is also indispensable to present a model that can versatilely fit for piezoelectric-laminated cantilever harvesters with diversified configurations.

On the basis of the abovementioned background, in this paper, we conduct a series of studies on energy harvesting by using two popular composite-laminated piezoelectric cantilever patches (MFC and PPA) in terms of experimental and modeling studies. In the experimental study, the vibration-based energy harvesting system is set up by using the two types of cantilever patches. Their mechanical and electric output are recorded under different conditions through tuning frequency in exciter, changing resistance in the circuit, and replacing mass weight at the free end of harvesters. On the basis of the experimental results, some crucial parameter settings are summarized for the configuration design of harvesters that maximize energy conversion. Additionally, a full investigation is also executed with regard to the harvesting characteristic relationships including voltage/current-strain and power-strain relationships, which provides a detailed information about the intrinsic electromechanical mechanism. Based on the preceding experimental study, we propose a coupled electromechanical model with set of updated parameters by using an optimization program. The model is verified to be superior through test data, and it is accurate for predicting harvested energy and effective for imitating the characteristic behaviors of the piezoelectric cantilever harvesters. Furthermore, the model can be applied for piezoelectric cantilever harvesters with diversified configurations (even vague or unknown configurations) that cannot be easily dealt with by using traditional modeling methods.

2. Energy Harvesting Experiment

For this study, we used two popular commercialized products, i.e., macro fiber composite (MFC) and piezo protection advantage (PPA), as the cantilever beam harvesters and conducted an experiment to investigate their vibration-based harvesting properties. Herein, we briefly introduce the products below.

2.1. Harvester Samples

(1) MFC patch

One composite piezoelectric patch, i.e., macro fiber composite (MFC), is a product developed by the NASA Langley Center. It is a layered, planar transducer that employs unidirectional piezoceramic fibers embedded in a thermosetting polymer matrix. The matrix is externally laminated with two polyamide films that are printed interdigited electrodes. The schematic diagram for the construction of the P1 type MFC is shown in Figure 1a, where its capacitance is 1.83 nF, tensile modulus is 30.33 GPa, block force is 162 N, and free strain is 1500.

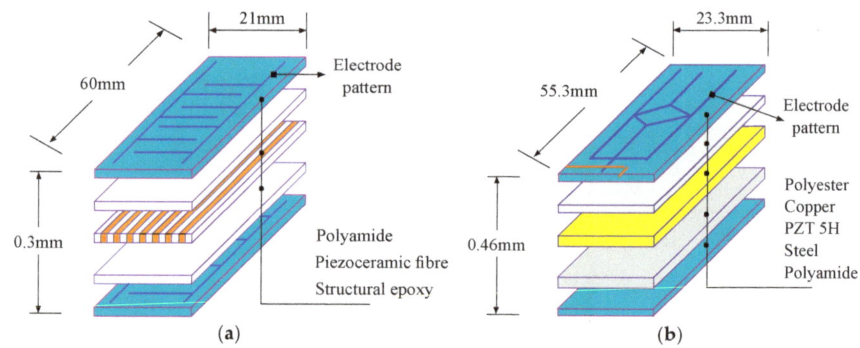

Figure 1. Schematic diagram for the construction of the piezoelectric laminated products. (**a**) MFC patch; (**b**) PPA patch.

(2) PPA patch

The other composite piezoelectric patch, i.e., piezo protection advantage (PPA), is one of the piezoelectric unimorph patches produced by the Mide's company. It is also a laminated transducer but employs a whole layer of piezoceramic in the middle. In addition to the PZT layer, it also has four layers including polyester, copper, steel, and polyamide films. A schematic diagram of its construction is also shown in Figure 1b, where its capacitance is 100 nF, tensile modulus is 30.33 GPa, block force is 0.2 N, and free deflect is 0.8 mm.

2.2. Experimental Procedure

As previously mentioned, piezoelectric laminated patches are commonly utilized for cantilever harvesters. Therefore, we used MFC and PPA patches to construct a cantilever harvester with an electric system, which is described below.

As shown in Figure 2, the test system apparatus included the following: a vibration exciter (Bruel & Kjaer Type 4832), a variable resistor (Fumi ZX21 resistance box), data acquisition (QuantumX MX840B module), a signal generator (Bruel & Kjaer Type 2732), and a power amplifier (AWG2005). The vibration exciter was used to vibrate the cantilever

MFC/PPA patch; one end of the patch was clamped and connected to the exciter by a harness (see Figure 3). The variable resistor was used to assemble a circuit for carrying an electric current. Data acquisition was adopted to realize multifunctional measuring missions, which simultaneously measured and recorded the voltage (V), current (I), and strain (S) data from the MFC/PPA. Note that the strain was recorded through a strain gauge pasted at the fixed-center end of the sample. Based on the data collected, the electric power P was calculated by following the electric power relation $P = V \cdot I$. In Figure 2, MFCs are shown as the testing sample and compensator in the system. In the same way, they were replaced by PPAs when we were about to measure PPA.

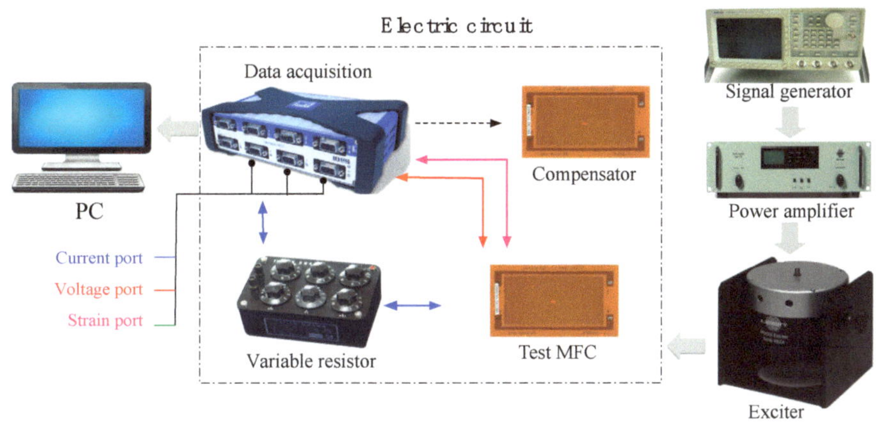

Figure 2. Flowchart for the energy harvesting system with MFC cantilever patch.

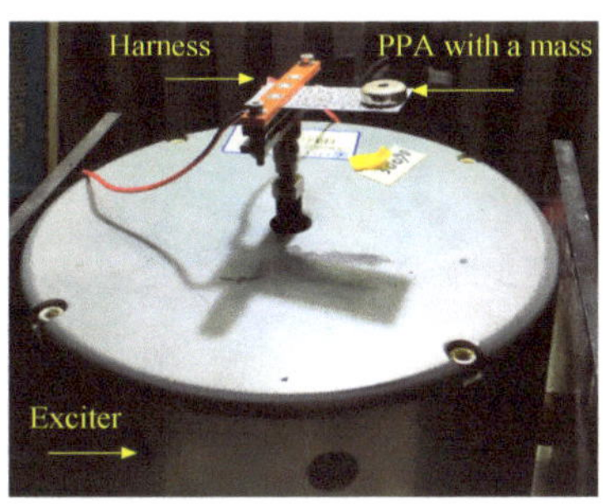

Figure 3. PPA clamped to the exciter.

2.3. Experimental Investigation

By using the setup, series of preliminary tests were carried out to measure the mechanical and electric responses from MFC and PPA under harmonic excitations. Excitation frequency from the exciter was tuned from 1 to 150 Hz with a 5 Hz interval, and the resistance load was changed from 0 to 90 kOhm with a 5 kOhm interval. In the experiment, PPA was also added by a mass weight that could lower the structural dominant frequency. The 4 mass weights chosen were 0 g, 12.5 g, 18.75 g, and 25 g.

2.3.1. Resonant Frequency Determination

It is important to know the natural frequency of a harvester because it dramatically affects the harvesting capability. Thus, a set of sweep frequency tests was executed to determine the natural frequencies of the MFC and PPA structures. Prior to the test, the resistance was maintained at a constant value of 20 kOhm.

First, we increased the excitation frequency slowly from 1 to 150 Hz, and then recorded the strain response. Next, we decomposed the strain data into frequency domain by using Fourier transform, and finally, obtained the single-sided amplitude spectrum sets, as shown in Figure 4.

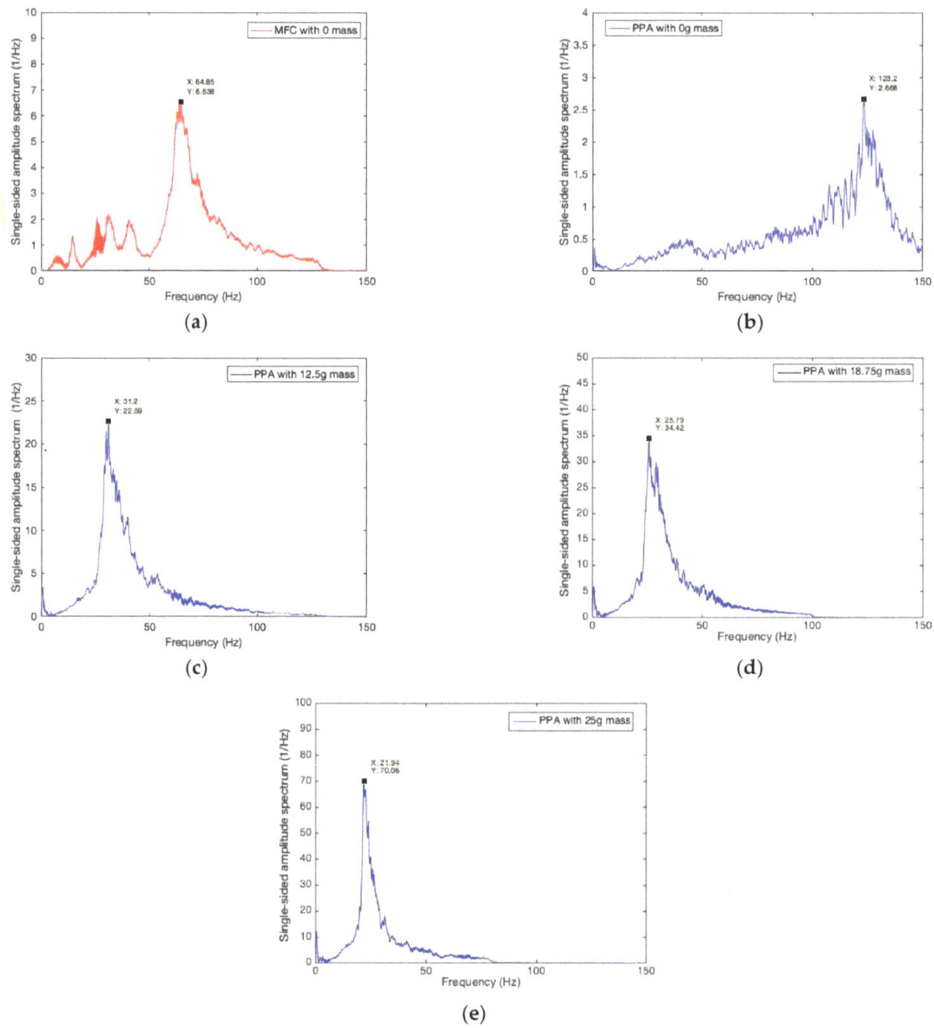

Figure 4. Amplitude spectrum of frequency sweep for the strain response of PPA and MFC. (**a**) Amplitude spectrum for MFC with 0 g mass; (**b**) Amplitude spectrum for PPA with 0 g mass; (**c**) Amplitude spectrum for PPA with 12.5 g mass; (**d**) Amplitude spectrum for PPA with 18.75 g mass; (**e**) Amplitude spectrum for PPA with 25 g mass.

In Figure 4a, the strain amplitude spectrum for MFC without mass approaches a dramatic peak of 64.85 Hz where resonance is aroused. This indicates that 64.85 Hz is the value of its structural dominant frequency. Similarly, PPA was tested using the four different mass weights. When the PPA cantilever is set without mass, its natural frequency is 123.2 Hz (see Figure 4b). When the mass weight is 12.5 g, the natural frequency changes to a lower value of 31.2 Hz, and the strain amplitude increases drastically to a higher value of 22.59 Hz^{-1}. Next, the same pattern is repeated with mass weights up to 18.75 g and 25 g, the natural frequencies are further decreased to 25.79 Hz and 21.94 Hz, respectively, while the strain amplitude is further magnified to 34.42 Hz^{-1} and 70.06 Hz^{-1}.

On the one hand, we found that the maximization of strain and electric power could be aroused, while excitation frequency sweeps are close to the structure's natural frequency. On the other hand, the natural frequency could be changed, while the tip end had a different weight mass. This indicates there should be an effective way to adopt a suitable structure to accommodate the environment, and thereupon, more energy conversion would be excavated.

2.3.2. Harvesting Capability Comparison

As illustrated in Section 2.3.1, maximization of electric power (harvesting capability) is stimulated when several excitation frequencies approach the dominant frequencies of MFC and PPA cantilever patches, and, in this section, we compare their harvesting capabilities.

Figure 5 shows the results from MFC under excitation with its approximated dominant frequency (60 Hz). Figure 6 shows the corresponding results from PPA under excitation with its approximated dominant frequency (120 Hz). As displayed in Figure 5a, strain response from the fixed-center end of the patch is measured as an amplitude of 1.96 $\times 10^{-3}$. Under this stimulation, the voltage response is then obtained with 1.019 V amplitude, as shown in Figure 5b. Likewise, the current response is obtained as 12.22 μA, as shown in Figure 5c. Based on the voltage and current results, electric power can be calculated based on a simple electric power equation, which is the most important variable associated with vibrational-electric energy conversion. We found that the corresponding peak power was 13.54 μW, as shown in Figure 5d.

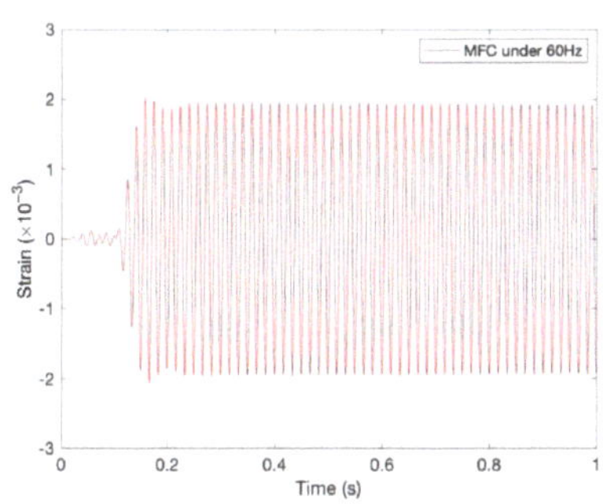

(a)

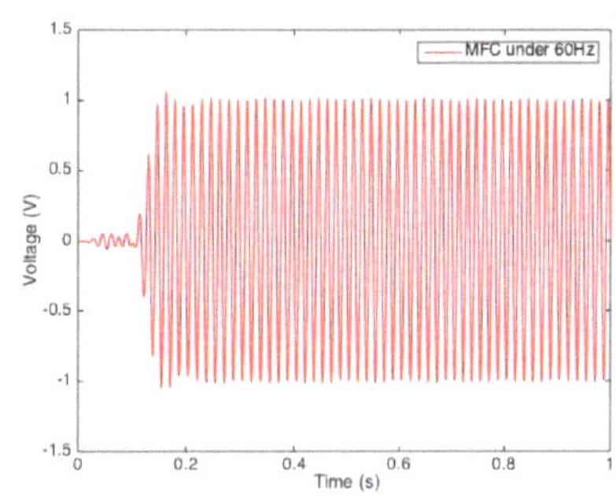

(b)

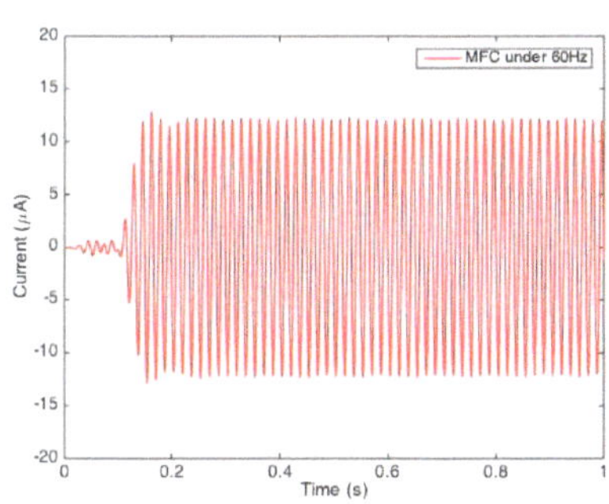

(c)

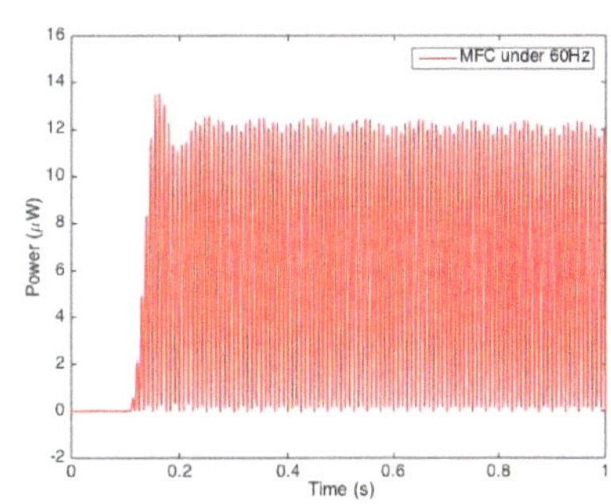

(d)

Figure 5. Mechanical and electric responses from MFC under resonant excitation. (**a**) Strain response for MFC under 60 Hz; (**b**) Voltage response for MFC under 60 Hz; (**c**) Current response for MFC under 60 Hz; (**d**) Power response for MFC under 60 Hz.

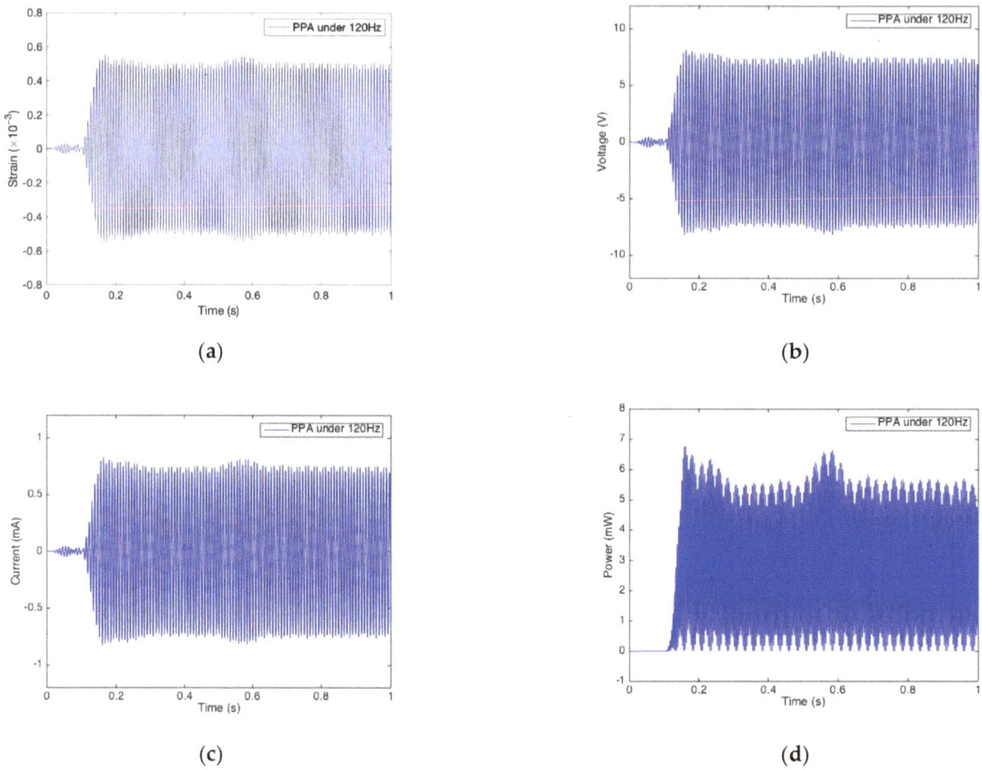

Figure 6. Mechanical and electric responses from PPA under resonant excitation. (**a**) Strain response for PPA under 120 Hz; (**b**) Voltage response for PPA under 120 Hz; (**c**) Current response for PPA under 120 Hz; (**d**) Power response for PPA under 120 Hz.

Likewise, Figure 6 shows the results from PPA under it dominant frequency excitation (120 Hz). As illustrated in Figure 6b–d, we found the responses' amplitudes of strain, voltage, current, and power were 0.5442×10^{-3}, 8.083 V, 0.8137 mA, and 6.98 mW, respectively.

It should be noted that the exciter is an open-loop system, and the strain data are not at the same level from MFC (in Figure 5a) and PPA (in Figure 6a). For the sake of comparability between these two patches, we standardized the electric responses under the unit of strain amplitude (1×10^{-3}). For example, the strain amplitude from MFC is 1.96×10^{-3} in Figure 5a, and the corresponding voltage amplitude is 1.019 V in Figure 5b. Through standardization, the comparable voltage is standardized to 0.52 V (1.019 V/1.96 V) amplitude. At the same time, the comparable current/power is standardized to 6.23 μA/6.9 μW. For PPA, the comparable voltage, current, and power are 14.85 V, 1.50 mA, and 12.83 mW, respectively. By contrast, it is obvious that the harvesting capability of PPA is much more superior as compared with that of MFC, where voltage/current collection from the PPA is 28/241 times larger than MFC, and amazingly, power collection is 1986 times larger than that of MFC.

2.3.3. Harvesting Characteristic Investigation

In addition to the electric harvesting capacity, we further investigate their electromechanical characteristics, which is pivotal to explore the energy conversion mechanism. As shown in Figure 7a,c, it is quite clear that the voltage/current-strain curve is in the rotational symmetric shape of ellipses suggesting their hysteretic nonlinear relationship. In addition, the power-strain curve is axial symmetry and it is quite unique in that it looks like a butterfly shape, as shown in Figure 7b,d. We previously discussed that a larger peak power indicated a stronger harvesting capability of a harvester. In fact, we found the fullness of the butterfly shape also indicated the harvesting capability of the harvester. For example, the power-strain butterfly in MFC, shown in Figure 7d, is much thinner than that in PPA, as shown in Figure 7c, which demonstrates the same conclusion (as proven in Section 2.3.2) that PPA has a much greater capability of energy harvesting than MFC.

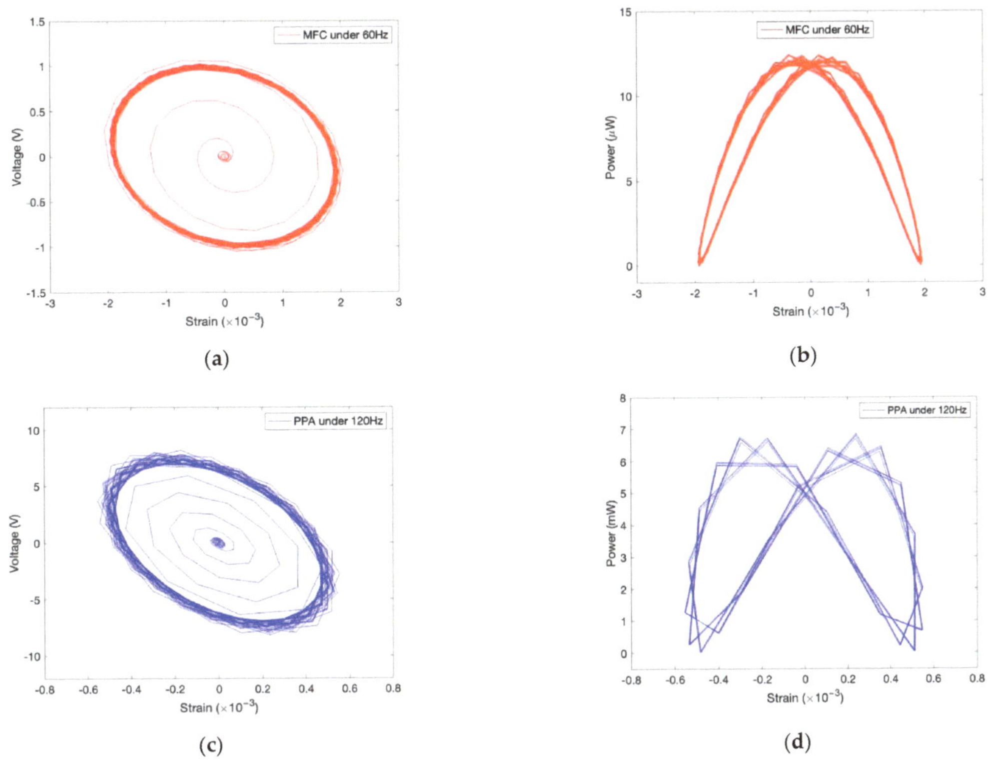

Figure 7. Electromechanical characteristics from MFC and PPA under resonant excitations. (**a**) Voltage vs. strain for MFC under 60 Hz; (**b**) Power vs. strain for MFC under 60 Hz; (**c**) Voltage vs. strain for PPA under 120 Hz; (**d**) Power vs. strain for PPA under 120 Hz.

2.3.4. Rational Resistance Determination

Herein, we observe how the resistance load affects the harvested electric power, given that the excitation frequency is set as its approximate resonant frequency. In Figure 8, for comparison, all the generated powers of the cases are normalized within the unit range (0–1). The normalized powers of the MFC are listed with different resistances from 0 to 90 kOhm, as shown in Figure 8a. We find the electric power of the MFC almost increases linearly along with the resistant load. Unlike the MFC, the power of the PPA increases to a peak power, and then decreases, with an increase in resistance load. Take PPA without mass as example, the peak power is stimulated when resistance is changed to 10 kOhm. This value of resistance can be deemed to be a rational setting in the circuit for this case. Likewise, the rational resistances are all determined to be, respectively, 30 kOhm, 35 kOhm, and 50 kOhm in the cases of 12.5 g, 18.5 g, and 25 g levels for cantilever PPA.

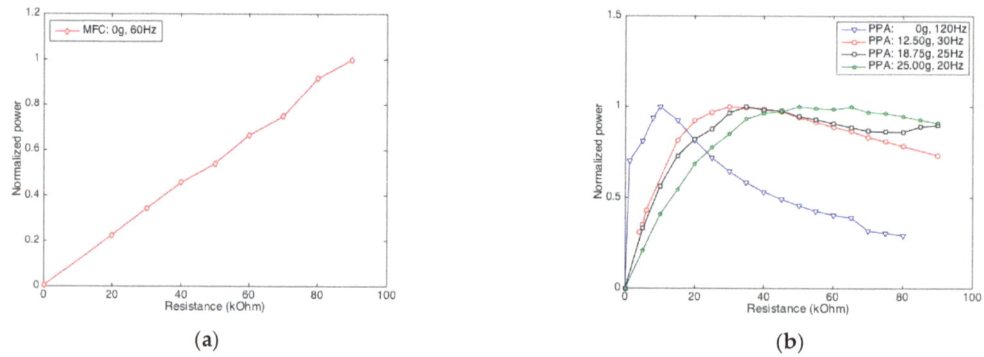

Figure 8. Normalized-harvested powers by varying load resistances for MFC and PPA. (**a**) Power vs. resistance for MFC; (**b**) Power vs. resistance for PPA.

2.3.5. Appropriate Excitation Amplitude Determination

We also observe how the excitation amplitude load affects the generated power. Figure 9 displays the relationship of the generated peak/average power with the excitation amplitude from the PPA harvester under two typical cases. In Figure 9a for the case of 100 Hz frequency, 10 kOhm resistance without mass, the results demonstrate that a larger excitation amplitude stimulates more electric power. In addition, the power increases more dramatically when the harvester is excited under a larger amplitude of excitation. Likewise, Figure 9b shows the results of another typical case that represent similar changes in the generated electric power with the excitation amplitude.

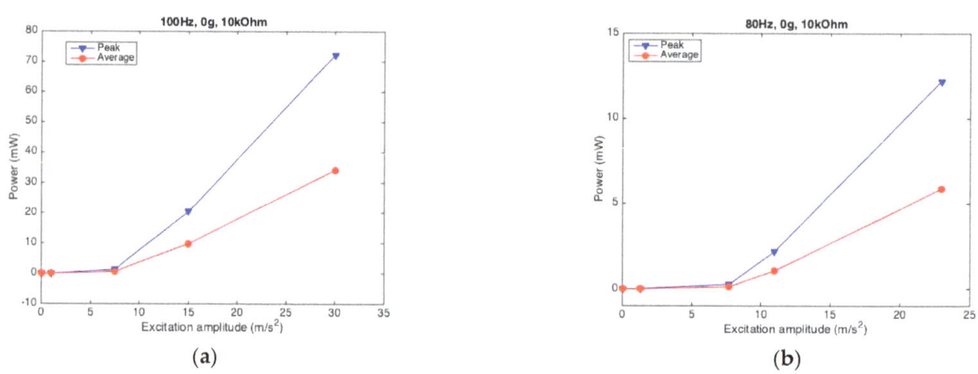

Figure 9. Normalized-harvested powers by varying excitation amplitudes for PPA. (**a**) under 100 Hz excitation; (**b**) under 80 Hz excitation.

The above phenomena illustrate that a more intense vibrational environment would arouse much more electric energy, which is preferable for harvesting more electric power converted from mechanical energy. However, one concern is that the structure and its harvester should stay safe with respect to their mechanical strength.

3. Coupled Electromechanical Modeling

3.1. *Piezoelectric Basic Equation*

When a poled piezoelectric material is mechanically strained, on the one hand, it becomes electrically polarized and electric charge is generated on the material's surface. The phenomenon is known as the "direct piezoelectric effect" and is the basis on which piezoelectric materials are utilized as a sensor or harvester. If electrodes are connected to the piezoelectric surfaces, on the other hand, the generated electric charge can be gathered and used. This phenomenon is known as the "inverse piezoelectric effect", and it is utilized especially as an actuator in vibration control applications.

The above phenomena are usually described by using a set of constitutive equations. On the assumption that the total strain in the transducer is the sum of mechanical strain induced by the mechanical stress and the controllable actuation strain caused by the applied electric voltage, their strain-charge form is given by:

$$S_i = s_{ij}^{E} \cdot T_j + e_{mi} \cdot E_m \tag{1}$$

$$D_m = e_{mi} \cdot T_i + \varepsilon_{ik}^{T} \cdot E_k \tag{2}$$

where T is the stress vector; S is the strain vector; E is the vector of applied electric field; D is the vector of electric displacement; s is the matrix of compliance coefficients; e is the matrix of piezoelectric coupling coefficients; ε is the matrix of electric permittivity; and the indexes i, j = 1, 2, . . . , 6 and m, k = 1, 2, 3 refer to different directions within the material coordinate systems, as illustrated in Figure 10.

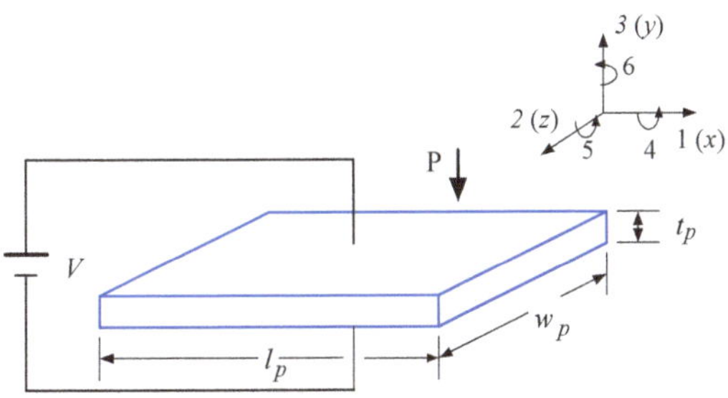

Direction	Axis
1	x
2	z
3	y
4	shear around x
5	shear around z
6	shear around y

Figure 10. Axis nomenclature for the piezoelectric constitutive equation.

The constitutive Equation (2) describes the "direct piezoelectric effect" about how electric displacement is aroused in the stress field and electric field. Based on this, the collected electric charge can be determined from the electric displacement when the piezoelectric transducer is subject to a non-zero stress field, that is:

$$q = \iint [D_1 \; D_2 \; D_3] \begin{bmatrix} dA_1 \\ dA_2 \\ dA_3 \end{bmatrix} \tag{3}$$

where dA_1, dA_2, and dA_3 indicate the differential electrode areas in y-z, x-y, and x-z planes, respectively, as shown in Figure 10. Then, the aroused inner voltage V_p can be obtained via $V_p = \frac{q}{C_p}$, where C_p is the capacitance of the piezoelectric layer.

Combined with the piezoelectric basic equations, in this paper, we adopt a coupled electromechanical model in order to effectively describe the complicated vibration-to-electric behaviors of the laminated piezoelectric cantilever patch. Note that a piezoelectric ceramic material is commonly adopted due to its high-power density and simple structure, however, it demonstrates worse tenacious, flexible, and fatigue performance. Therefore, substructures such as some polymer and metallic materials are usually laminated with piezoelectric ceramic for improving those defects. On the basis of this, we present a suitable model for laminated piezoelectric structures.

3.2. Coupled Electromechanical Model

3.2.1. Coupled Mechanical Equations

In terms of the mechanical problem, the configuration is connected to a resistor with resistance load R, as shown in Figure 11. Suppose the harvester is a laminated cantilever patch consisting of N piezoelectric layers and K substructure layers. In view of Euler–Bernoulli beam assumptions, the partial differential equation of the cantilever beam with a tip mass under an excitation at fixed end (displacement y_b) can be expressed by:

$$\frac{\partial^2 M(x,t)}{\partial x^2} + c_s \Lambda \frac{\partial^5 y(x,t)}{\partial x^4 \partial t} + c_a \frac{\partial y(x,t)}{\partial t} + \overline{m} \frac{\partial^2 y(x,t)}{\partial t^2} = -[\overline{m} + m\delta(x-l)] \frac{\partial^2 y_b(x,t)}{\partial t^2} \tag{4}$$

where c_s/c_a is the damping coefficient of strain rate/viscous air, Λ is the inertia moment of the cross section, $\overline{m}$ is the mass per unit length l, m is the tip mass, δ is the Dirac delta function, y_b is the y-direction displacement at base position, $y(x,t)$ is the relative deflection to its base at position x and time t, and M is the internal bending moment, i.e.,:

$$M(x,t) = -b \cdot \left(\sum_{pi=1}^{K} \int_{h_{pi-1}}^{h_{pi}} T_1^{pi} y dy + \sum_{sj=1}^{N} \int_{h_{sj-1}}^{h_{sj}} T_1^{sj} y dy \right) \tag{5}$$

where b and h are the width and the thickness of the laminated beam, respectively; h_{pi} is the thickness of the pi-th piezoceramic layer and there are altogether K piezoceramic layers; h_{si} is the thicknesses of the sj-th substructure layers and there are altogether N substructure layers; the axial stress components in the sj-th substructure at the 1-direction are presented by T_1^{sj}, and the pi-th piezoceramic layer at the 1-direction is presented by T_1^{pi}.

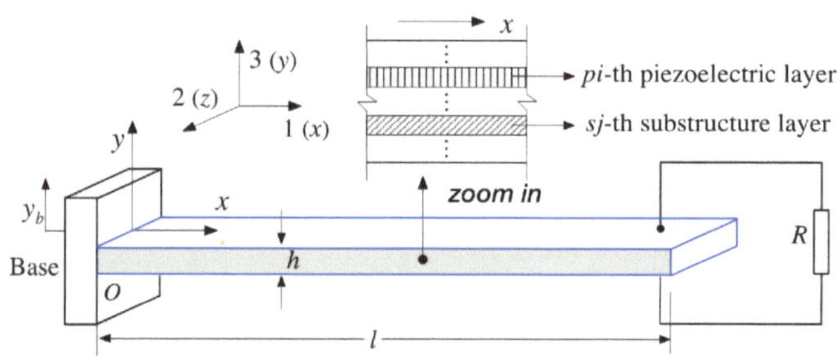

Figure 11. Schematic diagram of the universal composite-laminated piezoelectric cantilever harvester.

The piezoceramic part abides by constitutive equations (see Equations (1) and (2)) and the substructure parts abide by Hooke's law in substructures. Then, the axial strain at 1-direction from neutral axis of the composite beam S_1 is proportional to the curvature of the beam at x position, that is:

$$S_1(x,\ y,\ t) = -y \frac{\partial^2 y(x,t)}{\partial x^2} \tag{6}$$

The electric field component E_3 should be expressed in terms of the respective voltage term in each piezoelectric layer, as shown in Figure 11. It should be noted that different (series or parallel) connections of the piezoelectric layers differ from each other resulting in different mechanical equations. Herein, if the piezoelectric layers are assumed to be identical, the voltage generated from the piezoceramic layers is V_p. In the series connection, the voltage across the electrodes of each layer should be divided by K, which is equal to $V_p(t)/K$. For the case of parallel connection, the voltage is equal to $V_p(t)$. Then, the instantaneous electric field E_3 in series connection can be equally distracted, that is:

$$E_3 = V_p(t) / \sum_{pi=1}^{K} h_{pi} \tag{7}$$

Likewise, for the configuration with parallel connection, the electric fields in each layer are:

$$E_3 = V_p(t) / \left(\frac{1}{K} \cdot \sum_{pi=1}^{K} h_{pi} \right) \tag{8}$$

where h_{pi} is the assumed center-to-center interdigitated electode spacing of the pi-th piezoelectric layer.

Combined with the relationships from bending moment, curvature to the axial strain in the beam in Equation (6), Equation (5) can be replaced by:

$$M(x,t) = -EI \frac{\partial^2 y(x,t)}{\partial x^2} + \varsigma \cdot V_p(t)[H(x) - H(x-l)] \tag{9}$$

where EI is the bending stiffness of the composite cross section of the laminated beam, $H(x)$ is the Heaviside function, and ς is a coupling coefficient.

On the assumption of the overall stiffness equal to a sum of that from every layer, the bending stiffness term EI is usually divided into two parts (one refers to the piezoelectric layers, and the other refers to the substructural layers). It is given by:

$$EI = \sum_{pi=1}^{K} c_{11}^{E} I_{pi} + \sum_{sj=1}^{N} E_{sj} I_{sj} \tag{10}$$

where E_{sj} is the Young modulus of sj-th substructure layer, I_{sj} are the relative substructural inertia moments of the cross section, and c_{11}^{E} is the elastic stiffness component at constant electric field.

In Equation (9), the coupling coefficient ς can be expressed by:

$$\varsigma = e_{31} b \cdot \vartheta(h_{pi}, h_{si}) \tag{11}$$

where the coupling geometric coefficient ϑ is dependent on the dimensional and combinatorial relationship of each substructure and piezoelectric layers (h_{pi} and h_{si}). It is also dependent on the circuit connection method.

Based on the above, the coupled mechanical Equation (4) can be substituted with:

$$EI \frac{\partial^4 y(x,t)}{\partial x^4} + c_s I \frac{\partial^5 y(x,t)}{\partial x^4 \partial t} + c_a \frac{\partial y(x,t)}{\partial t} + \overline{m} \frac{\partial^2 y(x,t)}{\partial t^2} - \varsigma v_p(t) \left[\frac{d\delta(x)}{dx} - \frac{d\delta(x-l)}{dx} \right] = -[\overline{m} + m\delta(x-l)] \frac{\partial^2 y_b(x,t)}{\partial t^2} \tag{12}$$

Equation (4) is a partial differential equation and its analytical solution is difficult, and even impossible, to solve. Herein, we employ the mode superposition method to solve it effectively. The model expression is as follows:

$$y(x,t) = \sum_{k=1}^{\infty} \varphi_k(x) \cdot q_k(t) \tag{13}$$

where $q_k(t)$ is the modal mechanical response, and $\varphi_k(x)$ is the mass normalized eigenfunction of the k-th vibration mode whose expression is:

$$\varphi_k(x) = C_k \left[\cos\text{h} \frac{\beta_k x}{L} - \cos \frac{\beta_k x}{L} - \frac{\sin \beta_k - \sin\text{h}\beta_k + \beta_k \frac{m}{\overline{m}L} (\cos \beta_k - \cos\text{h}\beta_k)}{\cos \beta_k + \cos\text{h}\beta_k - \beta_k \frac{m}{\overline{m}L} (\sin \beta_k - \sin\text{h}\beta_k)} \cdot \left(\sin\text{h} \frac{\beta_k x}{L} - \sin \beta_k \right) \right] \tag{14}$$

where C_k is a modal amplitude constant that can be evaluated by normalizing the eigenfunctions according to the orthogonality conditions, and β_k is the eigenvalues for k mode that can also be evaluated by the short circuit conditions [22].

By adopting the mode superposition method, Equation (12) is replaced by the mechanical equation of motion in modal coordinates, that is:

$$\frac{d^2 q_k(t)}{dt^2} + 2\xi_k \omega_k \frac{dq_k(t)}{dt} + \omega_k^2 q_k(t) - \lambda_k v_p(t) = f_k(t) \tag{15}$$

where ω_k is the undamped natural frequency of the kth vibration mode, ξ_k is the modal mechanical damping ratio derived from the proportional damping assumption. The modal electromechanical coupling term $\lambda_k = \varsigma \frac{d\varphi_k(x)}{dx}\Big|_{x=L}$ and modal mechanical forcing function is:

$$f_k(t) = -\overline{m}\frac{d^2 y_b(t)}{dt^2}\int_0^L \varphi_k(x)dx - m\varphi_k(l)\frac{d^2 y_b(t)}{dt^2} \tag{16}$$

3.2.2. Coupled Electric Equations

In the case of a laminated cantilever, the mechanical strain can be assumed to be the axial strain in the case of curvature. Then, the tensorial expression of the piezoelectric constitutive (Equations (1) and (2)) can be simplified to scalar equations as:

$$T_1 = c_{11}^E S_1 - e_{31} E_3 \tag{17}$$

$$D_3 = e_{31} S_1 + \varepsilon_{33}^s E_3 \tag{18}$$

where c_{11}^E is the elastic stiffness of the piezoelectric layer at constant electric field, e_{31} is the piezoelectric constant, and ε_{33}^s is the permittivity component at constant strain with the plane-stress assumption. Note the subscripts 1, 2, and 3 represent longitudinal, lateral, and vertical direction of the cantilever beam, respectively, as shown in Figure 11.

The axial strain components of piezoelectric derived from Equation (17) and substructure layers in the laminated patch is proportional to its curvature at the position x. Based on this, the stimulated electric field E_3 can be solved.

Because the circuit admittance across the electrodes is $1/R$, the electric response is obtained based on Gauss's law in Equation (3) as:

$$\frac{V_p(t)}{R} = \frac{d}{dt}\left(\int_A D_3 dA_3\right) \tag{19}$$

where D_3 is the 3-directional electric displacement in the piezoelectric layer and A_3 is the electrode area in the x-z plane.

Due to the average strain in the piezoelectric layer subjected by the bending, substitute Equation (20) into Equation (21), and then yield:

$$-e_{31} h_{pc} b \int_0^L \frac{\partial^3 y(x,t)}{\partial x^2 \partial t} dx = C_p \frac{dV_p(t)}{dt} + \frac{V_p(t)}{R} \tag{20}$$

where C_p is capacitance equal to $\frac{bl\varepsilon_{33}^S}{h_p}$ in the electric circuit, and h_{pc} is the equivalent distance between the neutral axis and the center of piezoceramic layer.

Based on Kirchhoff law, the current source I_p can be derived as:

$$I_p(t) = C_p \frac{dV_p(t)}{dt} + \frac{V_p(t)}{R} \tag{21}$$

It is easy to observe that Equations (20) and (21) have the same terms on the right side. Thus, current I_p can be deduced directly by the agreement with the left term between them.

In fact, the measured electric results in the experiment are received from the electric circuit through data acquisition, not the generated ones (I_p and V_p) directly from the harvester. It is easy to understand the voltage across the resistance should be the same as the generated one, i.e., $V(t) = V_p(t)$, because they have a parallel relationship. Therefore, the measured current $I(t)$ from the experiment can be calculated based on Ohm's Law shown below as:

$$I(t) = V_p(t)/R \tag{22}$$

Next, substitute the modal expansion to Equation (20), and then achieve:

$$\sum_{k=1}^{\infty} \theta_k \frac{dq_k(t)}{dt} = C_p \cdot \frac{dV_p(t)}{dt} + \frac{V_p(t)}{R} \tag{23}$$

where the modal coupling term $\theta_k = -e_{31} h_{pc} b \int_0^l \frac{d\varphi_k(x)}{dt} dx$.

3.2.3. Coupled Electromechanical Equations

The mechanical equation (Equation (15)), and electric circuit equation (Equation (23)), present coupled mechanical and electric relations of the variables modal mechanical response $q_k(t)$ and voltage response $V(t)$ or $V_p(t)$. Based on them, the variables $q_k(t)$ and $V(t)/V_p(t)$ can be theoretically calculated. Moreover, the electric circuit equations (Equations (21) and (23)), are combined to calculate the results of source current $I_p(t)$ and measured current $I(t)$, successively.

3.3. Coupled Equations' Solutions

To derive a steady-state solution from the coupled electromechanical equations, we employ frequency response functions (FRFs) to solve their mechanical and electrical results. On the assumption that the 3rd directional displacement (y_b) in Equation (4) is harmonic, its modal forcing function can be expressed as $f_k(t) = F_k e^{\omega t j}$. Derived from Equation (16), amplitude F_k can be then calculated as:

$$F_k = \omega^2 \left[\overline{m} Y_0 \int_0^L \varphi_k(x) dx + m \varphi_k(l) \right] \tag{24}$$

where Y_0 represents the deformation amplitude at the 3rd direction from the base, ω represents the excitation frequency, and j represents the imaginary part numeric values.

In this paper, model mechanical $q_k(t) = Q_k e^{\omega t j}$ and voltage response $V(t) = V_0 e^{\omega t j}$ are deemed to be single frequency harmonic whose amplitudes Q_k and V_0 are complex valued based on the hypothesis of linear system. Therefore, the solution of the model (Equations (20) and (23)) can be replaced as an equation set:

$$\lambda_k V_0 + \left(\omega_k^2 - \omega^2 + 2\xi_k \omega_k \omega \cdot j \right) Q_k = F_k \tag{25}$$

$$\left(\frac{1}{R} + \frac{1}{2} C_p \omega \cdot j \right) V_0 - \omega \sum_{k=1}^{\infty} \theta_k Q_k \cdot j = 0 \tag{26}$$

As soon as Q_k and V_0 are determined based on the above, the mechanical and electric solutions are obtained. The presentational electric expressions are listed below as:

$$V(t) = e^{\omega t j} \cdot \frac{\sum_{k=1}^{\infty} \frac{\omega \theta_k F_k \cdot j}{\omega_k^2 - \omega^2 + 2\xi_k \omega_k \omega \cdot j}}{\frac{1}{R} + \frac{1}{2} C_p \omega \cdot j + \sum_{k=1}^{\infty} \frac{\omega \theta_k F_k \cdot j}{\omega_k^2 - \omega^2 + 2\xi_k \omega_k \omega \cdot j}} \tag{27}$$

$$I(t) = e^{\omega t j} \cdot \frac{\sum_{k=1}^{\infty} \frac{\omega \theta_k F_k \cdot j}{\omega_k^2 - \omega^2 + 2\xi_k \omega_k \omega \cdot j}}{1 + \frac{1}{2} C_p \omega R \cdot j + \sum_{k=1}^{\infty} \frac{\omega \theta_k F_k \cdot j}{\omega_k^2 - \omega^2 + 2\xi_k \omega_k \omega \cdot j} R} \tag{28}$$

The complex voltage amplitude V_0 can be substituted into Equation (27) to achieve the steady mechanical response expression as follows:

$$\eta(t) = \left(F_r - \lambda_k \frac{\sum_{k=1}^{\infty} \frac{\omega\theta_k F_k \cdot j}{\omega_k^2 - \omega^2 + 2\xi_k\omega_k\omega \cdot j}}{\frac{1}{R} + \frac{1}{2}C_p\omega \cdot j + \sum_{k=1}^{\infty} \frac{\omega\theta_k F_k \cdot j}{\omega_k^2 - \omega^2 + 2\xi_k\omega_k\omega \cdot j}} \right) \times \frac{e^{\omega t j}}{\omega_k^2 - \omega^2 + 2\xi_k\omega_k\omega \cdot j} \quad (29)$$

Then, the transverse deformation response at point x can be achieved by substituting Equation (29) into Equation (13) as:

$$y(x,t) = \sum_{k=1}^{\infty} \left[\left(F_r - \lambda_k \frac{\sum_{k=1}^{\infty} \frac{\omega\theta_k F_k \cdot j}{\omega_k^2 - \omega^2 + 2\xi_k\omega_k\omega \cdot j}}{\frac{1}{R} + \frac{1}{2}C_p\omega \cdot j + \sum_{k=1}^{\infty} \frac{\omega\theta_k F_k \cdot j}{\omega_k^2 - \omega^2 + 2\xi_k\omega_k\omega \cdot j}} \right) \times \frac{e^{\omega t j}}{\omega_k^2 - \omega^2 + 2\xi_k\omega_k\omega \cdot j} \right] + y_b(x,t) \quad (30)$$

4. Coupled Electromechanical Modeling by Using Identified Parameters

4.1. Model Parameters

In the coupled electromechanical model, there are many parameters that must be assigned accurate values, otherwise the model would demonstrate inaccurate, even invalid, results. In fact, it is quite challenging, if not impossible, to determine accurate values for all these parameters. This could be attributed to the following three reasons: One reason is that some parameters are difficult to clearly defined, and thereby, they are often simplified based on sets of certain assumptions. The simplification easily causes the model to provide inaccurate results. The second reason is that most laminated harvesting products are encapsulated, and the configuration, geometrical, mechanical, etc. information for every layer is not very specific. The third reason is that the design of some laminated harvesters is complicated, such as MFC, in which the "piezoelectric layer" is not made of a single piezoelectric material but fibers that are embedded in a polymer matrix.

In view of the above, we propose one strategy to deal with the crucial issues. First, we outline which parameters are not specifically defined and are difficult to assign in the model.

(1) Parameter h_{pi} represents the center-to-center interdigitated electrode spacing (assumed as the thickness) of the pi-th piezoelectric layer. It is simplified on the assumption that identical and electric fields at each layer are equally distracted, which may cause the model's result to be not accurate enough.
(2) Parameter h_{pc} represents (equivalent) the distance from the neutral axis of the structural cross section to the center of the piezoelectric layer. Similar to h_{pi}, it is simplified on the assumptions that identical electric fields at each layer are equally distracted. In addition, it could be even harder to assign when the harvester has multiple piezoelectric layers, and how to propose its proper equivalent value is another issue.
(3) Parameter EI represents the equivalent bending stiffness of the composite cross section for the constant electric field condition. In order to obtain its value, simplified relations (referred in Equation (10)) are commonly proposed based on some inevitable assumptions by ignoring non-uniformity field and coupling effect within layers.
(4) Parameter ϑ represents the coupling geometric coefficient that is comprehensively dependent on the thickness of every layer in the harvester, and also dependents on the circuit connection method. In general, it is a complicated coefficient that is not direct and easy to obtain accurately.

It takes little effort to discover that the above parameters have comprehensively equivalent definitions, and it is quite difficult to assign them accurate values. The simplification of hypothetical principles also produces more difficulties resulting in inaccurate, or even invalid, evaluations of them. Thus, determining effective values for the abovementioned parameters is pivotal for determining the model's validity. In this paper, we set out to propose a method of parameter identification for those four parameters by using a genetic algorithm (GA) procedure, which is introduced in the next subsections.

4.2. Parameter Identification Procedure

Taking the example of a laminated harvester with a single piezoelectric layer, there are four parameters to identify which are h_{pc}, h_{p1}, EI, and ϑ. We propose Equation (31) with these four parameters (instead of electromechanical expressions (Equations (15), (21) and (23)) to describe the mechanical-electric behavior in a cantilever harvester. It is given by:

$$[V, I]_{sim} = f\left(S_1, \widetilde{\Theta}\right) \tag{31}$$

where Θ denotes the vectors of the identified parameters (h_{pc}, h_{p1}, EI, and ϑ) that represent the intrinsic attributes of the harvesting systems. In order to determine the vector Θ, a unified optimization strategy can be qualified through measuring the discrepancy between model predictions and experimental data. The smaller the discrepancy, the closer the identified parameters $\widetilde{\Theta}$ is to the intrinsic ones Θ. Herein, we employ a genetic algorithm (GA) to carry out the said minimization due to its powerful computational capability of multi-agent parallelized stochastic search.

A flow chart on the GA for parameter estimation is shown in Figure 12. First, experimental data were collected, which included the mechanical data (strain $S_{1,exp}$) and electric data (voltage V_{exp} and current I_{exp}). Secondly, we initialized a set of parameters $\widetilde{\Theta}$ and incorporated the strain data $S_{1,exp}$ into the model. Then, the simulated voltage V_{sim} and current I_{sim} were calculated. During the phase of the GA, the fitness function "Fit" is defined as the discrepancy between the experimental data and simulated result from the model. The lower the fitness function value, the better the individual chromosome. Through appropriate iterative circulation, the optimal individual (identified parameters) is eventually computed out. Note that the above variables with subscript "sim/exp" indicate simulation/experimental data.

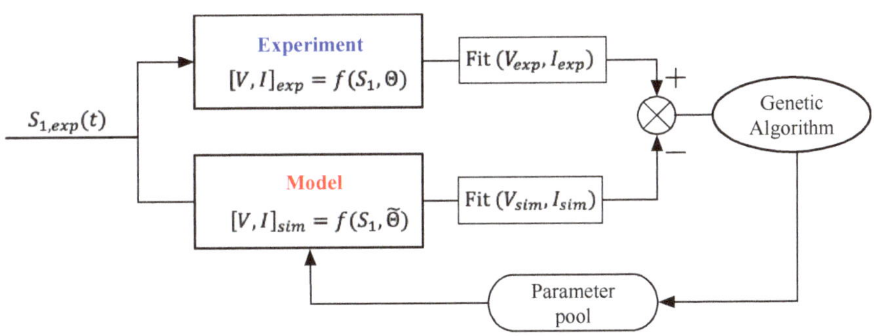

Figure 12. GA flowchart of parameter estimation for the electromechanical model.

4.3. Genetic Algorithm (GA) Detail

A GA involves several basic working parts including chromosome representation, fitness function, genetic operators, and termination criteria.

(1) Chromosome representation

In GA terminology, chromosomes are simply the variations in the set of unknown variables of the problem that the GA is going to solve. In this study, the model tackled has four variables that are represented simply by Θ. Thus, the chromosome representation is expressed as:

$$\Theta_u = \left\{h_{pc,u}\ ,\ h_{p1,u},\ EI_u,\ \vartheta_u\ \right\},\ u = 1, \ldots, U \tag{32}$$

where u denotes the u-th individual in a chromosome set, whose total number is U.

(2) Fitness function

A fitness function measures the status of one chromosome achieving the preset objective. Each chromosome needs to be awarded an evaluation that indicates how close it came to meeting the objective. Here, we define the fitness function ("Fit") by using a combination weight function including the root-mean-square error (*RMSE*) for each output (voltage and current responses) that is designed between the simulated and experimental results, which are:

$$RMSE_{V,u} = \frac{\sqrt{\frac{1}{W}\left\{\sum_{w=1}^{W}\left[V_{exp,w}(S,\Theta) - V_{sim,w}\left(S,\widetilde{\Theta}\right)\right]^2\right\}}}{\sqrt{\frac{1}{W}\left[\sum_{w=1}^{W} V_{exp,w}(S,\Theta)\right]^2}} \quad (33)$$

$$RMSE_{I,u} = \frac{\sqrt{\frac{1}{W}\left\{\sum_{w=1}^{W}\left[I_{exp,w}(S,\Theta) - I_{sim,w}\left(S,\widetilde{\Theta}\right)\right]^2\right\}}}{\sqrt{\frac{1}{W}\left[\sum_{w=1}^{W} I_{exp,w}(S,\Theta)\right]^2}} \quad (34)$$

$$\text{Fit} = \alpha \cdot RMSE_{V,min} + (1-\alpha) \cdot RMSE_{I,min} \quad (35)$$

where $RMSE_{V,u}/RMSE_{I,u}$ denotes the voltage/current's discrepancy between simulation and experimental results at the u-th individual. $V_{exp,w}$ and $V_{sim,w}$ denote the voltage, respectively, from experimental and simulation results (as subscribed with "*exp*" and "*sim*") at the w-th time point whose total number is W. The fitness function consists of voltage and current discrepancies. A weight coefficient "α" is also employed for balancing the importance among voltage and current predictions, and then yields the fitness function in Equation (35).

(3) Termination strategy

The traditional GA empirically adopts a fixed number of generations, which either lead to excessive computations or solutions with inadequate accuracy. To avoid this issue, a set of termination rules is designed on the aspects of global accuracy and average accuracy. The rules can comprehensively guarantee the best and the average individuals of every generation resulting in sufficient accurate solutions with appropriate computation loading. The rules are set as follows:

Rule 1: Define a minimum threshold $e_{1,min}$. As soon as the finest fitness value meets $e_{1,min}$ at the v-th iteration, automatically terminate the procedure and extract the best solution, that is:

$$\text{Fit}_v \leq e_{1,min} \quad (36)$$

where $e_{1,min}$ is empirically set as 10% in the present study.

Rule 2: Define a calibrated threshold $e_{2,min}$. As soon as the difference of the finest fitness values between the kth iteration and $(v+1)$-th iteration meets the $e_{2,min}$, automatically terminate the procedure and extract the best solution. It is given by:

$$|\text{Fit}_v - \text{Fit}_{v+1}| \leq e_{2,min} \quad (37)$$

where $e_{2,min}$ is empirically set as 10% of the $e_{1,min}$ in the present study.

In general, Rule 1 is responsible for guaranteeing the accuracy of the final solution, as well as Rule 2 is responsible for avoiding excessive and inefficient computation.

Other than the above, there are some other basic settings needed to complete the GA procedure. Here, the population is set as 60 individuals with 20 binary genes, the roulette wheel selection with elitism strategy is used, and the crossover/mutation operator is set as 0.85/0.01.

5. Model Validation with Experimental Data

In this study, we introduce a coupled electromechanical model with a set of identified parameters that are obtained by using GA procedure. In this section, we verify the validity of the proposed method.

5.1. Comparison of Model and Experimental Data

5.1.1. Original Model (without Original Parameters)

Based on related simplified equations, the parameters are evaluated, referred to as the original parameters in this study.

Then, taking PPA for example, the original parameters are deduced and evaluated as follows:

(1) Parameter h_{p1} is the equivalent thickness of the piezoelectric layer and is equal to 50 μm. (2) Parameter h_{pc} is the distance from the center of the piezoelectric layer to the structural neutral axis. Then, the relation is deduced as $h_{pc} = |h_{s4} + h_{s3} - h_{s2} - h_{s1}|/2$ and is equal to 150 μm. (3) Parameter EI is deduced based on Equation (10) as $EI = \sum_{sj=1}^{4} E_{s1} \cdot I_{s1} + c_{11}^{E} \cdot I_p$ and is equal to 107 N·cm^2. (4) Parameter ϑ is dependent on the thickness of every layer and the circuit connection method. It is deduced as $\vartheta = |h_{s4} + h_{s3} - h_{s2} - h_{s1}|$ and is equal to 10 mm. The valuations are listed in Table 1.

Table 1. Optimal settings of energy harvesters of MFC and PPA.

Structural Type	Harmonic Frequency (Hz)	Optimal Resistance (kOhm)	Normalized Max/Mean Power (mW)
MFC	64.85	-	6.9 μW/2.76 μW
PPA with 0 g mass	123.20	10	12.83/5.31
PPA with 12.5 g mass	31.20	30	7.16/2.94
PPA with 18.75 g mass	25.79	35	3.69/1.70
PPA with 25 g mass	21.94	50	2.32/0.98

By using the model with original parameters, representative results (under the case of 120 Hz excitation, 10 kOhm resistance, and 0 g tip mass) from the PPA harvester are displayed in Figure 13. It is easy to observe that the input strain (see Figure 13a) is nearly the same between the model simulation and experimental data in terms of the sinusoidal amplitude (around 0.56 $\times 10^{-3}$) and frequency (exactly 120 Hz). After substituting the strain data into the model, we observe the mechanical response (tip deformation of y direction), which represents the exact same frequency as the experimental data and an amplitude around 1.6 mm that is close to the experimental data, as illustrated in Figure 13b. Next, the electric response (current and voltage) is slightly smaller, around 12%, as compared with the experimental data, as shown in Figure 13b,c. For the comparison of characteristic relationships, the strain-voltage characteristic loops are illustrated; the elliptical figures in simulation are slightly rotated anticlockwise versus those from the experimental data, as shown in Figure 13e. The strain-power characteristic loops are shown in Figure 13f; the butterfly shapes in simulation are a vertically compressed version of those in experimental mode. Meanwhile, the maximum power prediction from the model simulation is much smaller, around 30%, as compared with that of the experimental data.

Next, taking MFC for example, we find that its piezoelectric layer is assembled of piezoelectric fibers with polymer matrix and interdigited electrodes, and it is impossible to obtain the parameters' values by using conventional methods.

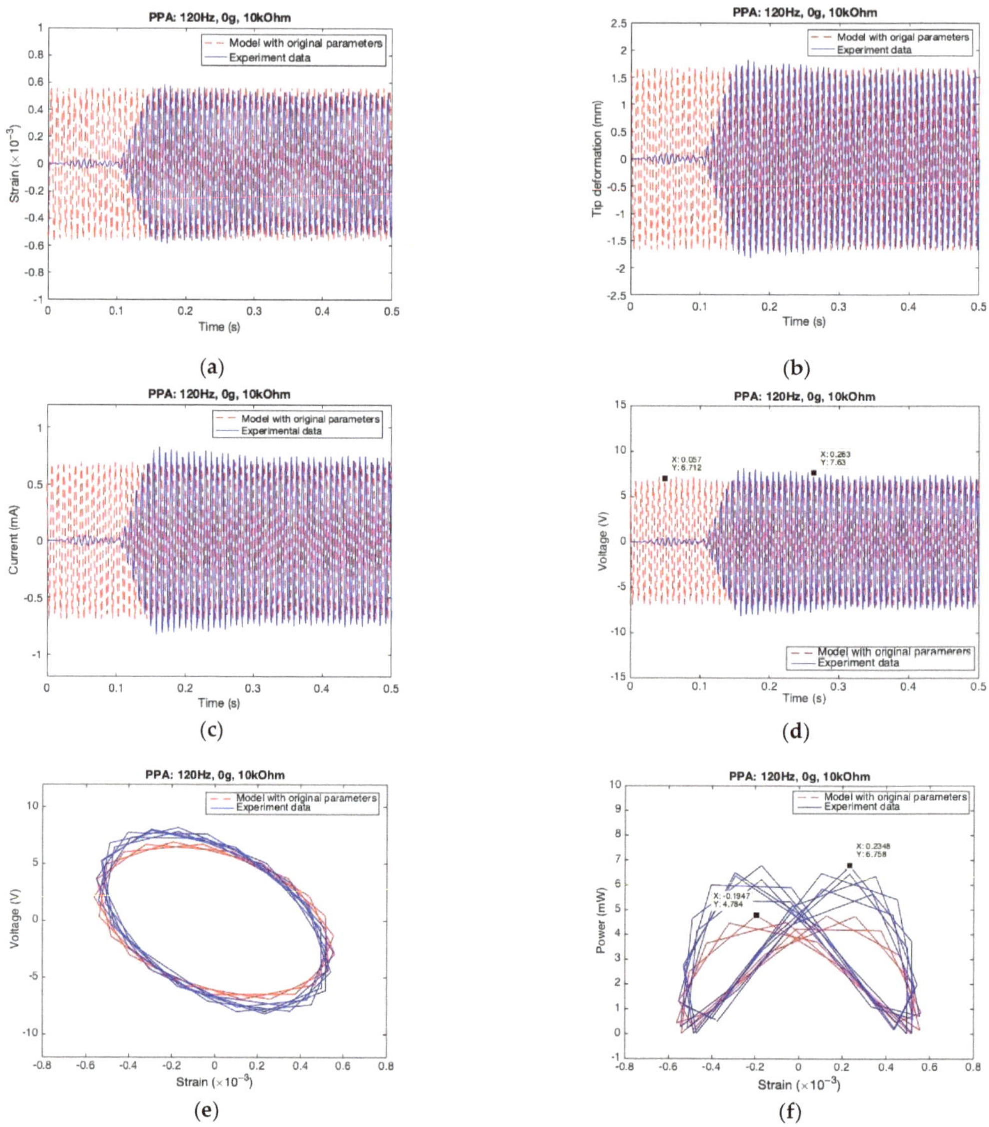

Figure 13. Comparison between the experimental and the model simulation results with original parameters in the PPA. (**a**) Strain vs. time; (**b**) Deformation vs. time; (**c**) Current vs. time; (**d**) Voltage vs. time; (**e**) Voltage vs. strain; (**f**) Power vs. strain.

5.1.2. Proposed Model (with Updated Parameters)

Using the GA procedure, we identified the parameter values which are listed in Table 2. Then, we substituted the updated parameter values into the model again. The mechanical responses (strain and deformation response) using the updated parameter values in the model are basically the same as the experimental data, shown in Figure 13a,b. However, the electric responses (current and voltage) have better agreement with the experiment data, shown in Figure 14a,b. For The comparison of characteristic relationships showed that the strain-voltage loops between simulation and experiment data were even more identical. The strain-power characteristic loops, as shown in Figure 13d, are butterfly shaped in simulation and just slightly vertically compressed as compared with those in the experimental results. The maximum power prediction from the model simulation is slightly lower, around 13.9%, than that in the experimental data. Apparently, the prediction is much more accurate when the model is employed with identified parameters using the GA procedure.

Table 2. Original and optimized parameters of PPA and MFC for the electromechanical model.

Parameter	Unit	PPA		MFC	
		Original	**Optimized**	**Original**	**Optimized**
h_{pc}	μm	50	49.63	-	23.32
h_p	μm	150	129.74	-	0.5458
EI	N·cm^2	107	106.2	-	1.558
ϑ	mm	10	11.88	-	0. 8994

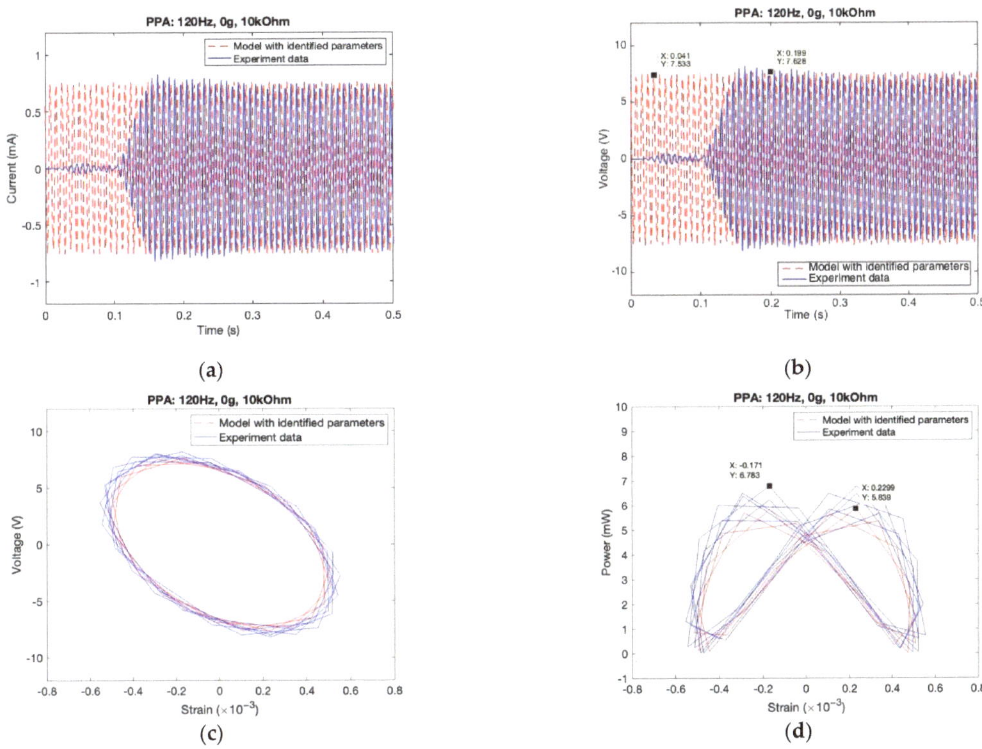

Figure 14. Comparison between experimental and the model simulation data with optimized parameters in PPA. (**a**) Current vs. time; (**b**) Voltage vs. time; (**c**) Voltage vs. strain; (**d**) Power vs. strain.

According to an MFC harvester, the parameters in the model are not known because of its complicated laminated structure. Given this, we use the GA procedure to determine its potential vague parameters. Here, dataset of one random case (60 Hz excitation and 90 kOhm resistance) is utilized for the GA procedure, and its identified parameters are obtained and listed in Table 1.

Then, we can substitute the parameters into the model to predict the harvesting behavior for other cases. Figure 15 illustrates one of the simulation results that are in a good agreement with the experimental results, which proves the proposed method is efficient and accurate.

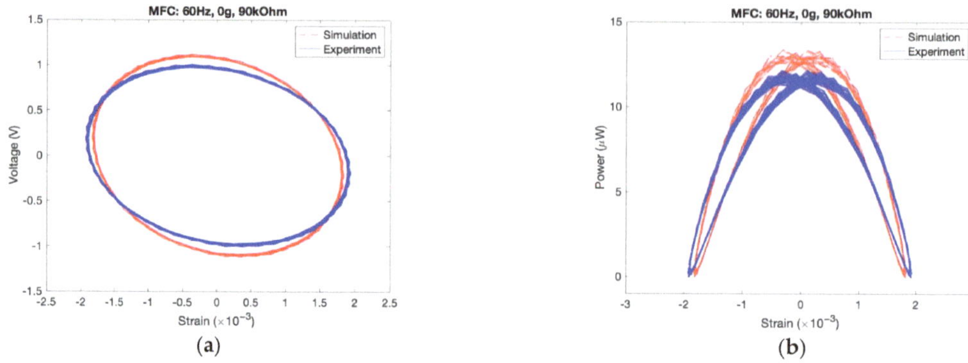

Figure 15. Comparison of experimental and simulation results in MFC. (**a**) Voltage vs. strain; (**b**) Power vs. strain.

5.2. Error Analysis

"Fit" (from Equation (35)) is the error between the simulation and experiment results, which represents the model's accuracy for a certain case (under a certain frequency, resistance, and mass weight). We found that the Fit values varied according to different cases. In order to test the effectiveness of the proposed model, an error analysis should be carried out. Four groups of cases in PPA, as previously mentioned in Section 2, were classified according to four weight levels at the tip (0 g, 12.5 g, 18.75 g, and 25 g). Each group consisted of cases with different excitation frequencies, while resistances were set as the optimal resistance.

Figure 16 shows the "Fit" values vs. excitation frequency for the four groups. As displayed in Figure 16a, we found the "Fit" value for the original model is lowest at 34% when the excitation frequency approaches the resonant frequency (120 Hz). When the excitation frequency is changed to a value that is farther away from the resonant frequency, the error increases further until the ultimate error of 82%. In contrast, the "Fit" values for the model with updated parameters are all reduced with a minimum of 14% and a maximum of 49%. Figure 16b,d illustrate the results of the other three groups, in the same manner. The "Fit" values for the model with updated parameters are all reduced from the values of the original model, which proves the updated model with optimized parameters is superior.

In general, electric power prediction is our most important goal. From this perspective, the simulation error should also consider the proportion of harvested power in different cases. As proven before, the harvested power is significantly dominant around the case of excitation frequency at resonance, where the model is quite accurate. In contrast, negligible harvesting energy might not matter if the model is not sufficiently accurate. In view of this, the "Fit" value cannot indicate authentic effectiveness of one modeling method for predicting generated power. It, in fact, should also consider the harvested energy percentage of the whole power among the group cases, which is called the "energy ratio (ER)". It is given by:

$$\mathrm{ER}(r) = \frac{P(r)}{\sum_{r=1}^{N} P(r)} \times 100\% \tag{38}$$

where $P(r)$ indicates the harvested power from the r-th (ordinal number) case, and $\sum_{r=1}^{N} P(r)$ indicates the gross power summed in each group where N is the max-ordinal number. ER represents harvesting capacities when exaction frequencies are varied.

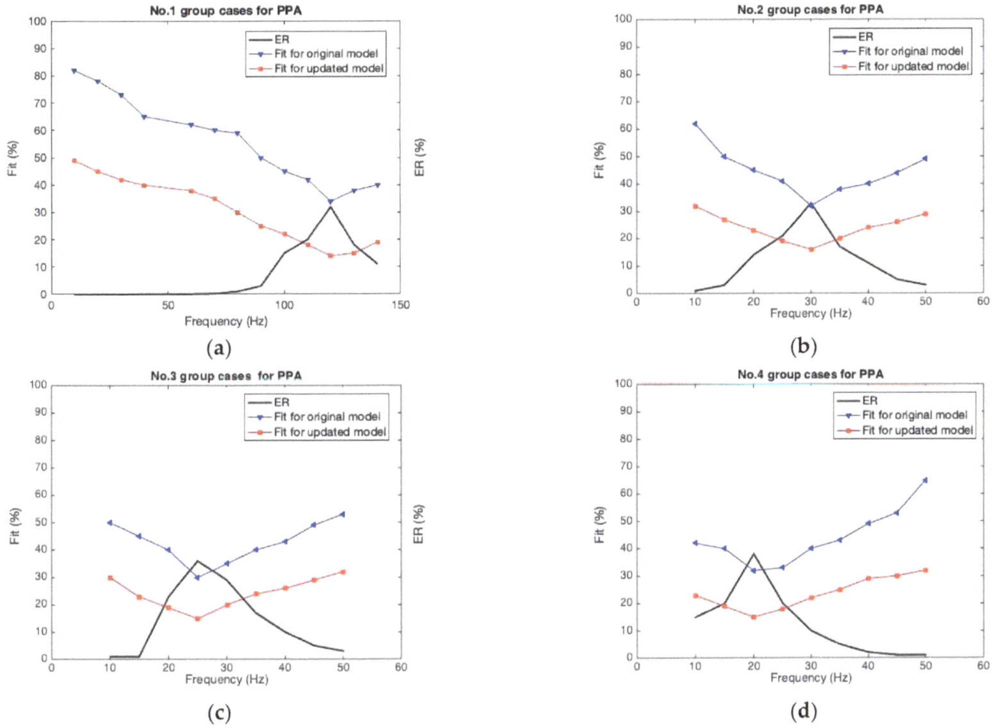

Figure 16. Error analysis of models with original and optimized parameters. (**a**) Error analysis for No. 1 group; (**b**) Error analysis for No. 2 group; (**c**) Error analysis for No. 3 group; (**d**) Error analysis for No. 4 group.

Then, another measurable indicator called the equivalent error (EE), within each group of cases, is devised as:

$$\mathrm{EE} = \left\{ \sum_{r=1}^{N} [\mathrm{ER}(r) \cdot \mathrm{Fit}(r)] \right\} \times 100\% \tag{39}$$

where $\mathrm{Fit}(r)$ is from Equation (35), and $\mathrm{ER}(r)$ is from Equation (38).

Then, as shown in Table 3, it is as expected, the "EE" from the original method is also reduced significantly when the proposed method is use. Regarding the No. 1 group of examples, simulation accuracy by using the identified parameters is remarkably improved by 56.36%. Regarding the No. 2 group, the "EE" is improved by 48.17%. The other groups show that the simulation using the proposed method is more accurate. It is obvious that the model with optimized parameters gains much better simulations for imitating the characteristic loops of a cantilever harvester owing to its full consideration of the electric energy.

Table 3. Equivalent error analysis of the models with original and optimized parameters.

	Equivalent Error (EE) (%)				
	No. 1 Group	**No. 2 Group**	**No. 3 Group**	**No. 4 Group**	**Average**
Model with original parameters	38.77	42.12	46.24	41.37	42.04
Model with optimized parameters	16.92	21.83	25.19	21.2	21.85
EE improved by	56.36	48.17	45.52	48.76	48.03

5.3. Change Analysis of Responses with Vibration Frequency

The proposed model was validated as effective and sufficiently accurate, as described in Section 5.2. Here, we combine the experimental results and the simulation results and discuss the capacities of mechanical responses and electric responses. An example of the PPA cantilever harvester with 0 g mass under 10 kOhm resistance is shown in Figure 16, which illustrates the relationships of different responses' amplitudes with different vibration frequencies.

In Figure 17a, the strain amplitude in the experimental results reaches a peak of 0.547×10^{-6} at 120 Hz vibration frequency, which closely approaches the dominant frequency (123 Hz). Under the same conditions, the predicted strain amplitude by model simulation is 0.556×10^{-6}. As the vibration frequency changes farther away from the dominant frequency, the strain amplitude decreases rapidly until the minimum (approaching zero). In general, all the simulations for strain response are very close to the experimental results, with an error of 5.6% on average.

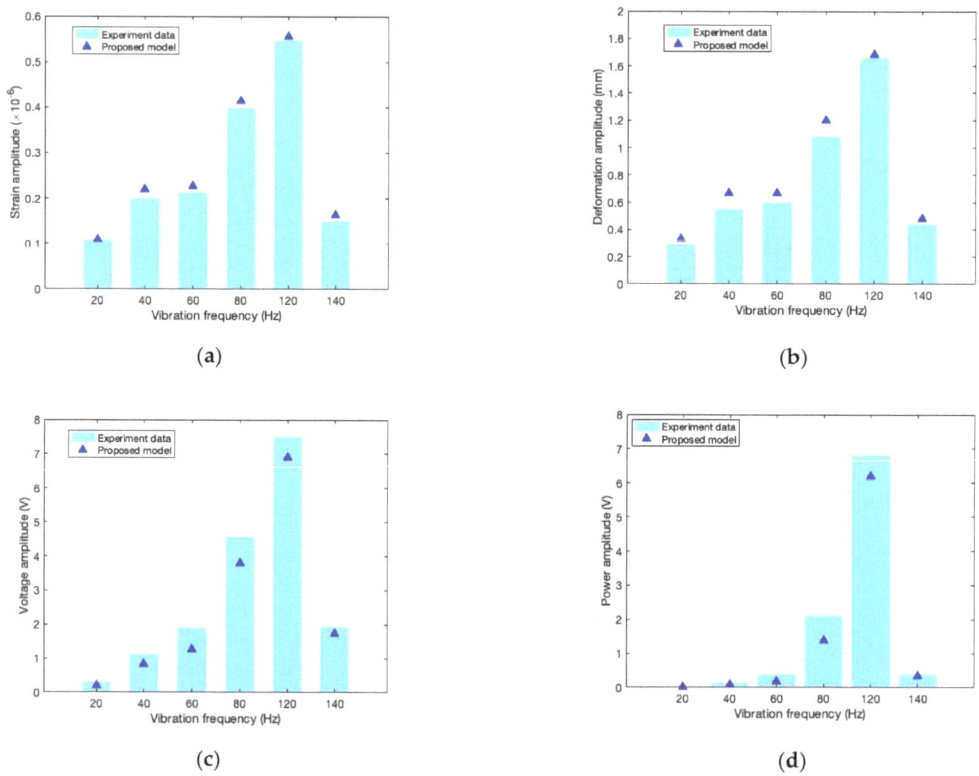

Figure 17. Changing law of responses' amplitudes with vibration frequency. (**a**) Strain amplitude vs. vibration frequency; (**b**) Deformation amplitude vs. vibration frequency; (**c**) Voltage amplitude vs. vibration frequency; (**d**) Power amplitude vs. vibration frequency.

In Figure 17b, the deformation amplitude of the y-direction at the free tip of the cantilever also reaches a peak (1.66 mm) at a vibration frequency of 120 Hz. It also decreases rapidly towards zero when the vibration frequency changes farther from the dominant frequency. The simulations for deformation also agree well with the experimental results, with an error of 11.3% on average.

In Figure 17c, the electric voltage amplitude demonstrates a similar pattern to the vibration frequency as compared with mechanical responses. The simulation results are basically similar to the experimental results, with an average error increased to 20%.

In Figure 17d, harvested power illustrates a significant peak of 6.787 mW at 120 Hz vibration, and it decreases very sharply, approaching zero farther away from the dominate frequency. The predictions for gross power by the model are basically in good agreement with the test results within an error of 25%.

In general, mechanical responses and electric responses of cantilever harvesters all represent a very similar pattern that a resonance phenomenon triggers more easily when vibration frequency changes closer to the structural dominant frequency. On the contrary, the responses decrease rapidly approaching zero when vibration frequency changes farther from the dominant frequency. Regarding the simulation effect of the proposed model, we found that the mechanical responses are predicted quite accurately, and the electric responses are still predicted effectively, with an error that remains below 25%.

6. Conclusions

In this study, we investigated a vibration-based cantilever beam from two popular piezoelectric laminated patches (MFC and PPA) in terms of an experimental study of its characteristic behaviors and a modeling study of energy harvesting. In the experiment, we investigated the harvesting capacity of the system as well as the electromechanical behavior of the cantilever harvester. Furthermore, some pivotal rules with regard to several design variables were summarized, which should be very useful for maximizing energy conversion. On the basis of identified parameters, a coupled electromechanical model for a laminated piezoelectric cantilever harvester was proposed with a set of optimized parameters determined by an optimization program. The model was tested and found to be superior for accurately estimating harvested energy amount, and describing the characteristic behavior effectively. The model was also demonstrated to be more versatile and applicable for diversified harvesters with vague or unknown parameters, which is an impossible task for traditional modeling methods.

In summary, this study conducted a characteristic experiment and harvesting modeling. The combined results should support realizing more efficient conversion of vibration energy to electrical energy, and should have great application potential in piezoelectric cantilever harvesters.

Author Contributions: Conceptualization, X.X. and Q.S.; methodology, X.X.; software, X.X.; validation, X.X., Q.M. and J.W.; formal analysis, X.X.; investigation, X.X.; resources, X.X.; data curation, Q.M.; writing—original draft preparation, J.W.; writing—review and editing, Q.M.; visualization, J.W.; supervision, Q.S.; project administration, X.X.; funding acquisition, X.X. All authors have read and agreed to the published version of the manuscript.

Funding: This research is financially supported by the National Natural Science Foundation of People's Republic of China, under grants No. 11502188. This support is gratefully acknowledged.

Institutional Review Board Statement: Not applicable.

Informed Consent Statement: Not applicable.

Data Availability Statement: The data presented in this study are available on request from the corresponding author. The data are not publicly available due to privacy.

Conflicts of Interest: The authors declare no conflict of interest.

References

1. Aridogan, U.; Basdogan, I.; Erturk, A. Analytical modeling and experimental validation of a structurally integrated piezoelectric energy harvester on a thin plate. *Smart Mater. Struct.* **2014**, *23*, 045039. [CrossRef]
2. Safaei, M.; Sodano, H.A.; Anton, S.R. A review of energy harvesting using piezoelectric materials: State-of-the-art a decade later (2008–2018). *Smart Mater. Struct.* **2019**, *28*, 113001. [CrossRef]

3. Anton, S.R.; Sodano, H.A. A review of power harvesting using piezoelectric materials (2003–2006). *Smart Mater. Struct.* **2007**, *16*, R1. [CrossRef]
4. Khan, F.U.; Qadir, M.U. State-of-the-art in vibration-based electrostatic energy harvesting. *J. Micromechanics Microengineer.* **2016**, *26*, 103001. [CrossRef]
5. Boisseau, S.; Despesse, G.; Seddik, B.A. Electrostatic conversion for vibration energy harvesting. *Small-Scale Energy Harvest.* **2012**, *31*, 1–39.
6. Moss, S.D.; Payne, O.R.; Hart, G.A.; Ung, C. Scaling and power density metrics of electromagnetic vibration energy harvesting devices. *Smart Mater. Struct.* **2015**, *24*, 023001. [CrossRef]
7. Spreemann, D.; Manoli, Y. *Electromagnetic Vibration Energy Harvesting Devices: Architectures, Design, Modeling and Optimization*; Springer: Berlin, Germany, 2012; Volume 35.
8. Mohammadi, S.; Esfandiari, A. Magnetostrictive vibration energy harvesting using strain energy method. *Energy* **2015**, *81*, 519–525. [CrossRef]
9. Narita, F.; Fox, M. A review on piezoelectric, magnetostrictive, and magnetoelectric materials and device technologies for energy harvesting applications. *Adv. Eng. Mater.* **2018**, *20*, 1700743. [CrossRef]
10. Yuan, X.; Changgeng, S.; Yan, G.; Zhenghong, Z. Application review of dielectric electroactive polymers (DEAPs) and piezoelectric materials for vibration energy harvesting. *J. Phys. Conf. Ser.* **2016**, *744*, 012077. [CrossRef]
11. Chandrasekaran, S.; Bowen, C.; Roscow, J.; Zhang, Y.; Dang, D.K.; Kim, E.J.; Misra, R.D.; Deng, L.; Chung, J.S.; Hur, S.H. Micro-scale to nano-scale generators for energy harvesting: Self powered piezoelectric, triboelectric and hybrid devices. *Phys. Rep.* **2019**, *792*, 1–33. [CrossRef]
12. Liu, H.; Quan, C.; Tay, C.J.; Kobayashi, T.; Lee, C. A MEMS-based piezoelectric cantilever patterned with PZT thin film array for harvesting energy from low frequency vibrations. *Phys. Procedia* **2011**, *19*, 129–133. [CrossRef]
13. Montanini, R.; Quattrocchi, A. Experimental characterization of cantilever-type piezoelectric generator operating at resonance for vibration energy harvesting. In Proceedings of the AIP Conference, Ancona, Italy, 29 June–1 July 2016; Volume 1740, p. 060003.
14. Aramaki, M.; Yoshimura, T.; Murakami, S.; Kanda, K.; Fujimura, N. Electromechanical characteristics of piezoelectric vibration energy harvester with 2-degree-of-freedom system. *Appl. Phys. Lett.* **2019**, *114*, 133902. [CrossRef]
15. Usharani, R.; Uma, G.; Umapathy, M.; Choi, S.B. A new piezoelectric-patched cantilever beam with a step section for high performance of energy harvesting. *Sens. Actuators A Phys.* **2017**, *265*, 47–61. [CrossRef]
16. Morimoto, K.; Kanno, I.; Wasa, K.; Kotera, H. High-efficiency piezoelectric energy harvesters of c-axis-oriented epitaxial PZT films transferred onto stainless steel cantilevers. *Sens. Actuators A Phys.* **2010**, *163*, 428–432. [CrossRef]
17. Roundy, S.; Wright, P.K.; Rabaey, J. A Study of Low Level Vibrations as a Power Source for Wireless Sensor Nodes. *Comput. Commun.* **2003**, *26*, 1131–1144. [CrossRef]
18. Dutoit, N.E.; Wardle, B.L.; Kim, S.G. Design considerations for MEMS-scale piezoelectric mechanical vibration energy harvesters. *Integr. Ferroelectr.* **2005**, *71*, 121–160. [CrossRef]
19. Erturk, A.; Inman, D.J. A Distributed Parameter Electromechanical Model for Cantilevered Piezoelectric Energy Harvesters. *J. Vib. Acoust.* **2008**, *130*, 1257–1261. [CrossRef]
20. Sodano, H.A.; Park, G.; Inman, D.J. Estimation of Electric Charge Output for Piezoelectric Energy Harvesting. *Strain* **2004**, *40*, 49–58. [CrossRef]
21. Chen, S.-N.; Wang, G.-J.; Chien, M.-C. Analytical modeling of piezoelectric vibration-induced micro power generator. *Mechatronics* **2006**, *16*, 379–387. [CrossRef]
22. Hagood, N.W.; Chung, W.H.; Von Flotow, A. Modelling of piezoelectric actuator dynamics for active structural control. *J. Intell. Mater. Syst. Struct.* **1990**, *1*, 327–354. [CrossRef]
23. Ajitsaria, J.; Choe, S.Y.; Shen, D.; Kim, D.J. Modeling and analysis of a bimorph piezoelectric cantilever beam for voltage generation. *Smart Mater. Struct.* **2007**, *16*, 447. [CrossRef]
24. Erturk, A.; Inman, D.J. An experimentally validated bimorph cantilever model for piezoelectric energy harvesting from base excitations. *Smart Mater. Struct.* **2009**, *18*, 25009–25018. [CrossRef]

sensors MDPI

Article

Pendulum-Based River Current Energy Converter for Hydrometric Monitoring Systems

Slim Naifar [1,*], Felix Grimmeisen [2,3], Christian Viehweger [1], Zheng Hu [1], Arthur Bauer [2], Peter Hörschelmann [4] and Olfa Kanoun [1]

[1] Chair for Measurement and Sensor Technology, Technische Universität Chemnitz, Reichenhainerstraße 70, 09126 Chemnitz, Germany; christian.viehweger@etit.tu-chemnitz.de (C.V.); zheng.hu@etit.tu-chemnitz.de (Z.H.); olfa.kanoun@etit.tu-chemnitz.de (O.K.)

[2] SEBA Hydrometrie GmbH & Co. KG (SEBA), Gewerbestraße 61a, 87600 Kaufbeuren, Germany; grimmeisen@seba.de (F.G.); a.bauer@seba.de (A.B.)

[3] Institute of Applied Geosciences (AGW), Karsruhe Institute of Technology (KIT), Kaiserstraße 12, 76131 Karlsruhe, Germany

[4] JuB-Creative Product GmbH, Industriestraße 12, 99846 Seebach Germany; p.hoerschelmann@jub-cp.de

* Correspondence: slim.naifar@etit.tu-chemnitz.de

Abstract: Energy harvesting from flowing water is important for supplying hydrometric monitoring systems. Nevertheless, it is challenging due to the chaotic water flow in only one main direction and the relatively weak energy profile. In this paper, a novel energy harvester has been proposed, designed, and validated. The converter consists of a pendulum, a gearbox, two overrunning clutches, a spiral spring, and a generator. By coupling the kinetic energy via an oscillating mass equipped with a magnetic spring, it is possible to accommodate the power supply, electronics, and sensors with data transmission in a completely closed, encapsulated, stable housing without an interface to the outside. In addition, an energy management circuit and a battery charging circuit were developed that could be housed in the sealed enclosure. The pendulum transducer prototype was tested with the developed online hydrometric measurement station, which consists of a multi-channel data logger with a cellular modem and a tipping bucket rain gauge sensor. The overall system was successfully validated by experimental studies in a river.

Keywords: energy harvesting; kinetic converter; hydrometric monitoring system; autonomous systems; water flow

Citation: Naifar, S.; Grimmeisen, F.; Viehweger, C.; Hu, Z.; Bauer, A.; Hörschelmann, P.; Kanoun, O. Pendulum-Based River Current Energy Converter for Hydrometric Monitoring Systems. *Sensors* **2022**, *22*, 4246. https://doi.org/10.3390/s22114246

Academic Editor: Alfio Dario Grasso

Received: 29 April 2022
Accepted: 31 May 2022
Published: 2 June 2022

Publisher's Note: MDPI stays neutral with regard to jurisdictional claims in published maps and institutional affiliations.

1. Introduction

With the increase in demand for distributed sensor networks, great focus has been devoted to harvesting environmental energy in order to replace batteries and extend the sensor nodes' lifetime [1–4] in wireless sensor networks. Considering the cost and the environmental issues, ambient energy provides key benefits to be considered as one of the major sources of energy in the future [5,6]. Furthermore, environmental energy harvesters can increase the self-powered micro device's adaptability. So far, numerous harvesting devices have been designed to convert environmental energy like solar [7] or thermal power [8], vibration and kinetic energy [9] into electrical power. Specifically, fluid kinetic energy is considered one of the most easily reachable and renewable energy sources in the environment [10]. In this context, hydrokinetic energy of flowing water in rivers presents a potential source of energy for powering wireless sensor systems such as hydrometric monitoring systems [11].

Fluid kinetic energy is mostly harvested by traditional hydro-turbine structures, which are able to harness the energy of moving or falling water [12]. Hydroelectric systems are mainly used at large scales for generating electrical power [13,14]. Water has also received great attention for small-scale energy harvesting to produce electrical power [15]. For example, in [16], a commercial hydrogenerator developed and patented by Vulcano is reported. The harvester was installed in an irrigation pipe and is able to generate almost constant output power in the range of 18 mW.

An alternative solution is to use fluid dynamics structures to induce a structure vibration, through which the fluid kinetic energy can be converted into vibration energy, and then, use a different method, including piezoelectricity and electromagnetism, to finally convert vibration energy into electrical energy. As a result, an eel-like hydraulic power harvester has been designed to convert ocean current into electrical power [17]. Piezoelectric eel energy harvesters are composite devices made of piezoelectric polymers deposited on structural supports. The eels are generally placed in river flows behind a bluff body used to generate vortexes [18].

Shan et al. presented a piezoelectric energy harvester (PEH) made from macro fiber composites (MFC) that is immersed in water vortexes caused by cylinders upstream [19]. The PEH consists of an MFC (M8514-P2) beam and a polyvinyl chloride (PVC) layer. The results show that the output power increases with the increase of the flow velocity, which varied from 0.05 m/s to 0.5 m/s. A maximum power output of 1.32 μW corresponding to a power density 1.1 mW/m^2 was obtained at the water velocity of 0.5 m/s and a cylinder diameter of 30 mm. In [20], authors reported on an upright vortex-induced piezoelectric energy harvester fabricated by a piezoelectric cantilever beam and a cylinder to harvest energy from water flow. The results show that the maximum output power of the harvester is 84.49 μW with an energy density of 60.35 mW/m^2 at a velocity of 0.35 m/s. In [10], harvest water flow induced vibration with a tandem arrangement of two piezoelectric energy harvesters (PEHs) in the direction of flowing water. Results indicate that the maximum output power of the downstream piezoelectric energy harvester (DPEH) reaches 371.7 μW, which is approximately 2.56 times of that of the upper piezoelectric energy harvester (UPEH), at a specific spacing between the upstream and the downstream piezoelectric energy harvesters.

However, most of reported water flow harvesters were investigated under laboratory conditions. Further, all of them have an active element in contact with the flowing water, which may significantly affect the reliability of such solutions when considering real implementation.

This paper addresses this issue by eliminating the contact between the flowing water and the converter while maintaining its ability to harvest a sufficient amount of energy to power the wireless sensor system. This paper presents an energy-autonomous measuring system that can use the naturally available kinetic energy in a hydrometric measuring station's environment, thereby enabling low-maintenance measurements and modern remote data transmission. Contrary to traditional water flow harvesters, the developed converter is able to power the measuring station in a completely closed, encapsulated, stable housing without an interface to the outside. This paper is structured as follows: As part of Section 2, the design configuration including river flow analysis, converter working principle and an equivalent physical model are presented. The developed energy management and battery charging circuits are described in Section 3. Additionally, the hydrometric monitoring system and its current consumption are assessed. The results of laboratory characterization and field measurements conducted in a river are discussed in Section 4.

2. Device Configuration

The kinetic converter consists of a rotating pendulum, which is connected to a bi-directional mechanism, a spiral torsion spring, gears, and a DC generator having a working principle similar to that of kinetic wristwatches. The converter is placed close to the center of gravity of a rectangular base, which is floating on the surface of the river and attached from one of its corners to a fixed frame. Figure 1a shows the schematic of a rotating pendulum along with the coordinate axes where α, β, and ϕ represents respectively the roll, pitch, and yaw angles. In the configuration under consideration, the pendulum, which is presented by a weight suspended from a pivot, is at horizontal position. Under roll and pitch oscillations, the pendulum is able to rotate in the XY plane.

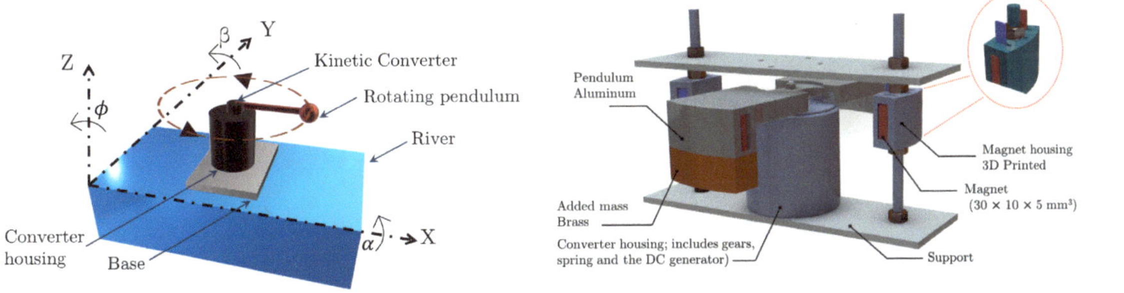

(a) (b)

Figure 1. (**a**) Schematic of a pendulum-based kinetic converter on the river, (**b**) improved pendulum design for river current energy harvesting.

On the other side, as power consumption of hydrological measurement systems significantly exceeds that in the case of wristwatches, a significantly heavier weight has to be associated with the kinetic converter to be able to harvest a valuable amount of energy from river water flow. The challenge thereby is, as the river water flow excites the kinetic harvester, the heavy pendulum must be quickly stabilized in a static position parallel to the direction of the applied excitation and will then require a large excitation in a different direction to turn again.

2.1. River Flow Analysis

It is particularly important to consider the position of the converter in the river. Random placement of the converter directly in the river will require continuous adjustment of its position based on different parameters, such as the river flow rate. Ideally, the kinetic converter should be located where the flow oscillates in time. In order to determine potential locations of the converter, we propose to analyse river water flow analysis. River flows can be considered as quasi-one-dimensional channels with variable and rough cross-sections. Water in the river past a fixed obstacle is of particular interest. The Reynolds number largely determines the qualitative behavior of fluid flow over an obstacle. Natural flows are usually turbulent due to the large Reynolds number R_e, which can be determined for a river by the following formula:

$$R_e = \frac{Q}{\mu l} \tag{1}$$

where Q is the volume flow rate, l is the length of the river, and μ is the kinematic viscosity.

Nevertheless, a comparative finite element analysis study of flows passing through an obstacle for different R_e values is presented using Computational Fluid Dynamics (CFD) module of Comsol Multiphysics version 5.4. The study is performed under similar conditions in terms of the size of the canal and that of the obstacle. Specifically, the study is realized in a 6 m length and 4 m width canal with a rectangular obstacle having width and length of 0.2 and 0.4 m (see Figure 2a). Both laminar flow and $k - \epsilon$ interfaces in Comsol were employed. For low Reynolds R_e values (below 100), the flow is steady. A 2D model is developed to simulate the flow at relatively low R_e values using laminar flow interface, and results are presented in Figure 2a. On the left boundary, a constant inlet velocity is ramped up over time with a mean inflow velocity of 0.1 m/s. In addition, a constant outlet pressure is defined on the right boundary.

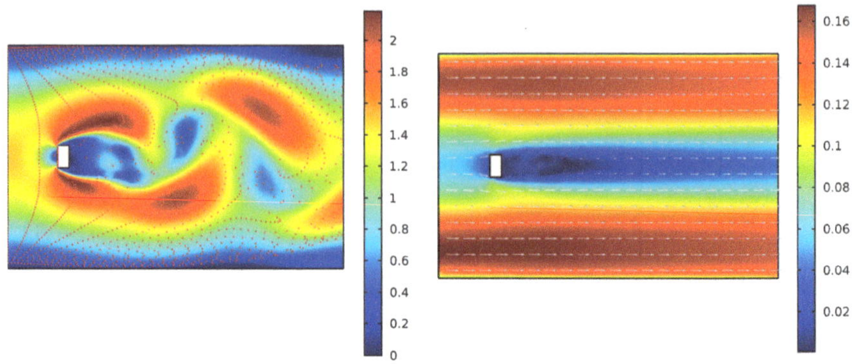

(a) (b)

Figure 2. Comparative finite element analysis of laminar and turbulent flows: (**a**) model built with laminar flow interface where $R_e \approx 50$; velocity magnitude in m/s—red points present particle tracing with mass; drag-driven particle movement; (**b**) model built with turbulent flow $k - \epsilon$ interface where $R_e \approx 5 \times 10^5$; velocity field in m/s—arrows represent the fluid flow vectors.

In contrast, a 3D model is developed to simulate the flow at higher R_e values ($3 \times 10^5 < R_e < 3.5 \times 10^6$) in a canal of a 6 m length, 4 m width, and 1 m height with water flowing through it at a velocity of 0.1 m/s. In this case, the underlying Navier–Stokes equations for flows are solved with a RANS turbulence model using the Comsol module "Turbulent flow, $k - \epsilon$ interface" (see Figure 2b).

Results are depicted in Figure 2a,b for the models built using the laminar and the turbulent flow interfaces, respectively. Figure 2a shows alternating vortex street forms where the vortices and streamlines are laminar (Kármán vortex street), where the maximum velocity magnitude is approximately 2 m/s. In Figure 2b, there are no clear vortex streets to be observed. The flow is laminar with an unsteady and turbulent wake where the maximum velocity magnitude is approximately 0.16 m/s. These results indicate that harvesting energy from a flow using a standard pendulum structure is difficult, as the oscillation in time in the flow depends on several parameters, including the velocity of inflow and the kinematic viscosity values, which are additionally affected by the temperature.

This paper presents a new pendulum design based on restoring forces as a method of harvesting flow energy even in the absence of vortex streets. Additionally, the proposed solution can increase the converter's sensitivity to unidirectional and small excitations. The magnetic pendulum design is based on a heavy weight having a specific geometry and is made of non-magnetic material with drag wings. A hole is drilled in one wing, and an additional mass is added to the end of the second wing to create an unbalanced structure. Four rectangular magnets are placed near the ends of the lateral wings. Four additional magnets are attached to two outer magnet housings so that a repulsive force is created between the magnets. The magnet holder is attached to the kinetic transducer through the housing, as shown in Figure 1b.

2.2. Working Principle of the Kinetic Converter

Figure 3 shows the energy conversion of the input river flow into electricity through the designed kinetic converter. The converter is formed of several mechanical components to store the input kinetic energy at a determined level before transmitting it to a DC generator. It is mainly composed of a pendulum, gears including a bi-directional mechanism, a spiral torsion spring with a magnetic casing, and a DC generator.

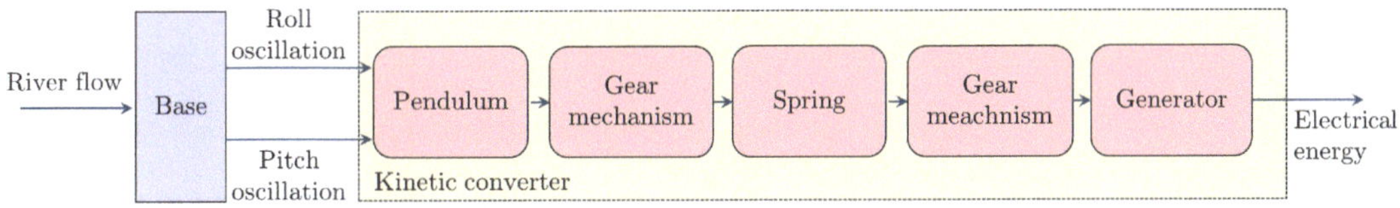

Figure 3. Block diagram representing the river flow kinetic energy converter.

Based on the bi-directional mechanism, the rotating pendulum is able to transmit its rotary motion via its gear wheel to one drive gear wheel of each of two directionally actuated couplings. One coupling transmits the torque to its output shaft when the direction of rotation is clockwise, the other transmits the torque to its associated output shaft when the direction of rotation is counterclockwise. If the directional clutches are each driven in the opposite direction of rotation—and this also applies when rotary motion is initiated from the output side—they run freely, i.e., they do not transmit torque in these opposite directions, neither from the input side to the output side nor vice versa from the output side to the input side. The permanently excited DC generator builds up a counter-torque through its own cogging torque, which keeps the output wheel of the spring mechanism in the idle state until a specific torque is exceeded. Only when the spring tension exceeds a specific value as a result of the winding and there is thus a slightly higher torque on the spring accumulator output wheel compared to the generator detent torque acting directly or indirectly is the movement to drive the generator is triggered. The harvested energy is stored in the spring, which releases when reaching a predefined holding torque, which is controlled through the magnetic casing acting on the spiral spring. This structure makes the obtained output from the DC generator regular with a fixed amplitude.

2.3. Physical Modelling

In order to replicate the behaviour of the kinetic converter, an equivalent physical model of the kinetic converter is developed in MATLAB version R2019a, Simulink version 9.3, and Simscape version 4.6. Both Simulink and Simscape library blocks are used to reproduce the behavior of the converter.

The model is composed of four main blocks, which are presented in Figure 4 as follows:

(a) Pendulum:
The input of the pendulum model is the acceleration in m/s^2; the output is the angular velocity in rad/s. The pendulum design can be developed directly in the model or imported as a CAD file. In the model under consideration, the pendulum is represented by a rigid body formed by a string and a bob block. The bi-directional mechanism is included in the pendulum model. A graphical representation of the resulting pendulum rotation is included in order to count the number of pendulum turns as a function of the applied acceleration. In this case, the number of pendulum turns refers to a 360-degree rotation of the pendulum without considering the magnetic solution presented in Figure 1b.

(b) Torsion spiral spring:
A nonlinear spiral torsion spring is configured by setting the parameters of the spring constant. For the nonlinear torsion spring, the deformation vector and the torque vector are configured depending on the properties of the spring. The selected spring is made of rectangular section austenitic stainless steel, number 1.4310 grade X10CrNi18-8, having 0.2 mm thickness and 10 mm width. The inner and outer diameters of the wound spring are, respectively, 20 mm and 48 mm. To reproduce the behavior of the magnetic force and the kinetic friction between the spring and the magnet housing, a fundamental friction clutch coupling is employed. Inputs and outputs of this block are coupled with kinetic and static force values. The clutch remains locked unless the torque transmitted through the clutch falls outside the static friction limits, which are determined as a function of the DC generator holding torque,

which in turn depends on the coupled resistance load value. As soon as the spring winding torque overcomes the static friction of the clutch, it begins to unwind.

(c) Gear set:
The model of the engaged gear is based on the transmission ratio. In the considered model, it is equal to 0.0211.

(d) DC generator:
A DC machine block is selected to model the DC generator. Several parameters such as the coil resistance, inductance, and torque constant are included to build the model of the DC generator. In addition, the load characteristics of the DC generator and their impacts on other parameters were examined. The DC generator used in this work is a commercially available DC motor model FF-130 SH-11340 having an operating range between 3 to 12 V. At no load, the motor has a speed of 6100 r/min and a current of approximately 0.033 A.

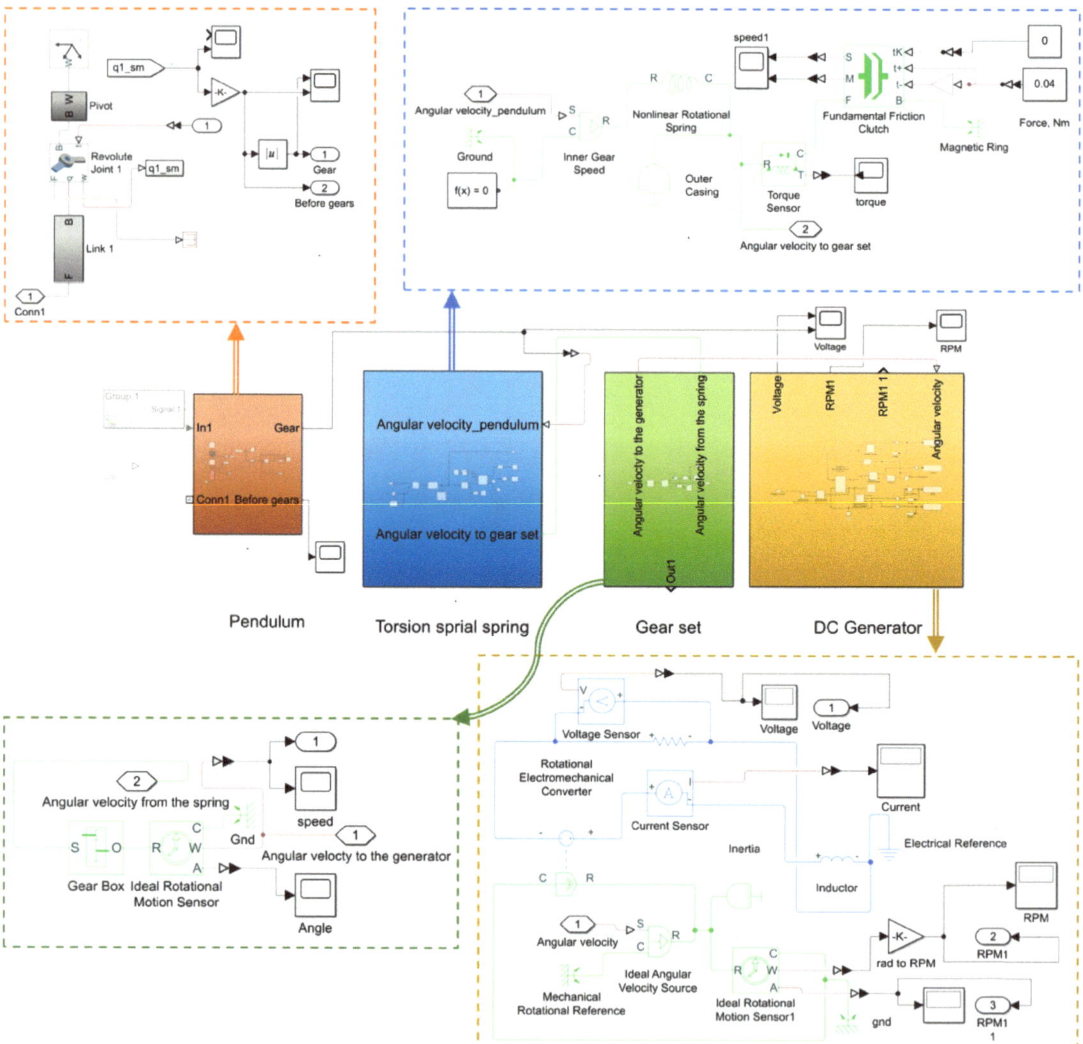

Figure 4. Developed Matlab/Simulink model of the kinetic energy converter.

For the simulation model, it is worth noting that no mechanical losses due to air resistance and friction were considered. Table 1 presents different parameters that were used in the developed physical model.

Table 1. Parameters used in the physical model.

Component	Parameter	Value
Pendulum	Weight	850 g
	Diameter	25 cm
	Weight of the added mass	490 g
Gears	Transmission ratio spring-generator	0.0211
	Transmission ratio pendulum-spring	1
Spring	Material	X10CrNi18-8
	Spring rate	0–3.5 N·m
	Plate thickness	0.2 mm
	Plate width	10 mm
	Outer diameter	48 mm
	Inner diameter	20 mm
DC generator	Rotor Inertia	1.68×10^{-7} kg·m^2
	Armature resistance	13 Ω
	Armature inductance	0.009 H
	Torque constant	0.00963 N·m/A

The complete system is studied under pulse excitations acting on the pendulum. By defining the generator holding torque, we can study the magnitude of the generated pulses in addition to the pulses repetition frequency. Once the device parameters, such as the spiral spring properties, the magnetic casing, the DC generators, and the gear ratio, are fixed, the only parameter that can affect the holding torque is the load resistance across the generator terminals. Thus, by varying the holding torque as a function of the load resistance, the amount of energy stored in the spring before release changes, which needs to be analyzed to find a balance between converter efficiency and the requirements of the energy management and storage circuits, which will be presented in the next section. To that end, we propose to analyze the converter generated pulses as a function of several load resistances while calculating the required number of pendulum turns required to spring release.

For instance, the simulated generator output voltage under 82 Ω and 4.7 kΩ load resistances is shown in Figure 5. For the case of 82 Ω load resistance, a maximum voltage output of approximately 4 V is obtained. The pulse is observed after complete turns of the pendulum. For the case of 4.7 kΩ load resistance, the output voltage has a high amplitude level of approximately 12 V, and it requires 11 pendulum turns. Figure 5c shows the pendulum number of turns required to release as function of several load resistances from 20 Ω to 22 kΩ. In the next section, the power consumption of the hydrometric monitoring system is investigated, which will allow us to select the suitable converter parameter.

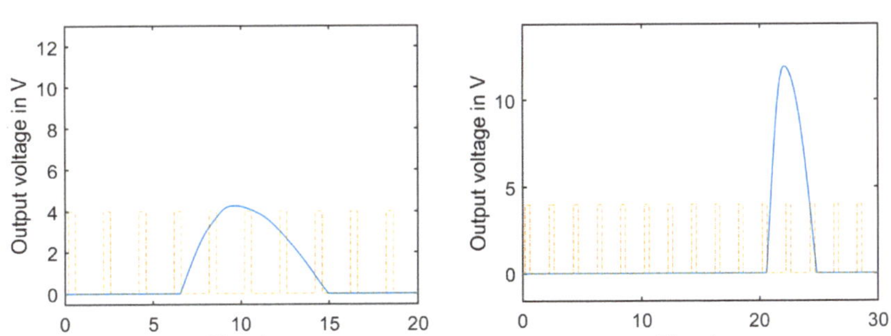

(**a**) (**b**)

Figure 5. *Cont.*

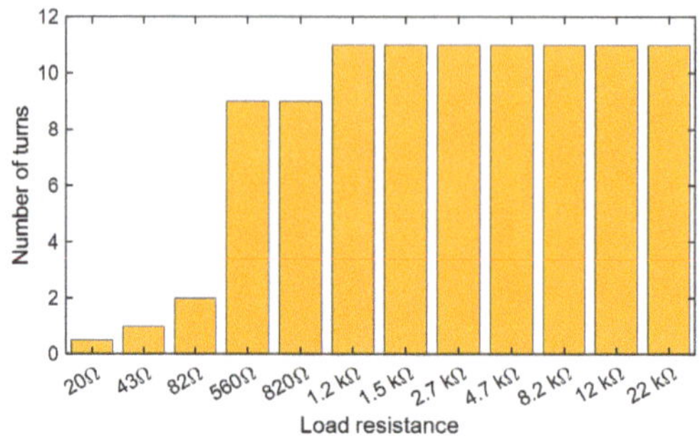

(c)

Figure 5. Generated pulses from the DC generator for different load resistances: (**a**) 82 Ω, (**b**) 4.7 KΩ. Rectangular signal presents the pendulum turns. (**c**) Pendulum number of turns required to release as function of several load resistances.

3. Hydrometric Monitoring System

A block diagram of the autonomous monitoring system is shown in Figure 6, which can be powered by the kinetic converter. In the following section, the hydrometric monitoring system is presented with a detailed analysis of its daily current consumption based on experimental investigations.

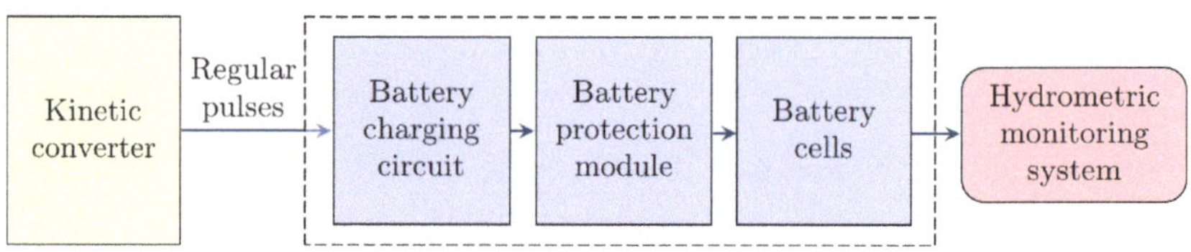

Figure 6. Block diagram of the autonomous hydrometric monitoring system.

3.1. System Design and Power Consumption

The used hydrometric monitoring system enables data transmission based on 3G and 4G standards. For this purpose, an electronic platform of the data logger device with new modem hardware and firmware was developed (SEBA types UniLogCom 4G and SlimLogCom 4G). The development work on the hardware and firmware for integrating the 4G modem finally enabled decisive and increased flexibility with even lower power consumption. Figure 7 shows the developed electronic platform UniLogCom with integrated 4G modem.

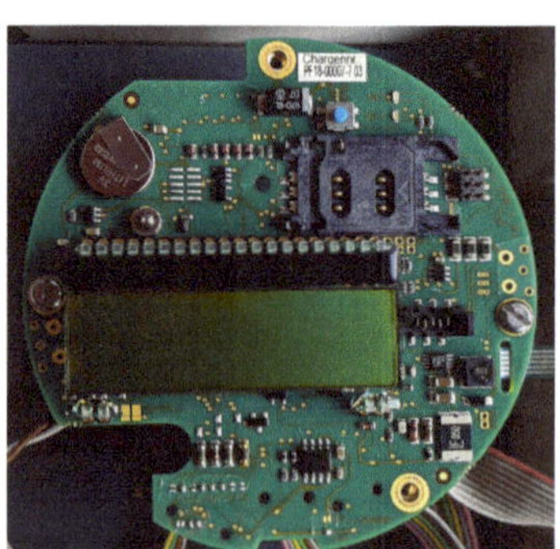

Figure 7. Electronic platform UniLogCom.

In order to know the power consumption of the electronic platform, extensive power consumption tests were undertaken. For this purpose, the UniLogCom data logger was connected to the tipping bucket precipitation sensor (SEBA type RG50).

The power consumption test series was performed using the UniLogCom according to the following procedure with three measurements each:

- Rain pulse measurement (RG50);
- Connection setup with RS232 interface;
- FTP push remote data transfer;
- 8-min data retrieval from modem.

The UniLogCom is able to operate with a voltage level between 6 and 15 V. To that end, we performed the current consumption investigations with 6 V and 12 V supply per test series, and the modem functions were activated or deactivated. Figure 8 shows the setup used for the experimental power consumption investigations.

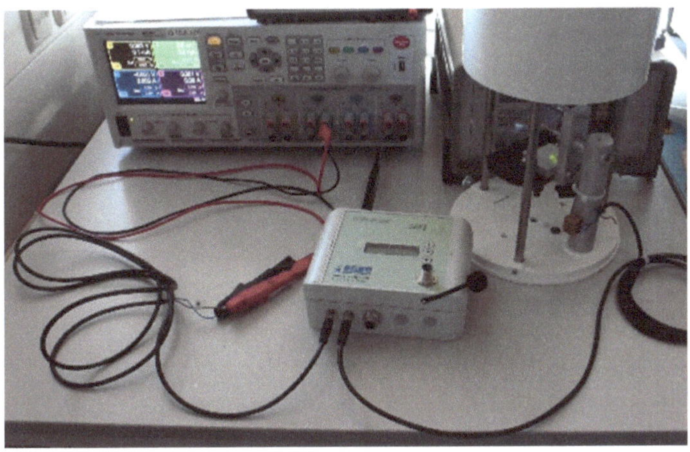

Figure 8. Test setup for the accurate recording of power consumption.

A precipitation sensor RG50 tipping bucket was used in the power consumption tests. The RG50 is a highly accurate precipitation gauge with pulse output for data collector systems. It has a plastic rocker with ball bearings on one side and a bubble level and adjustment screw. The precipitation gauge can optionally be equipped with a heater so that it can also be used at temperatures below freezing. The RG50 precipitation gauge is used exclusively for recording precipitation quantities in the form of pulse discharges.

(a) Current consumption measurement rain pulse (RG50):
The UniLogCom was supplied with 6 V/12 V and the modem functions were deactivated. The rain rocker was pressed once to record the current pulse. Measured signal at 6V is shown in Figure 9a.

(b) Power consumption measurement 8 minutes modem data retrieval:
The UniLogCom was supplied with 6 V/12 V and the time slot for data retrieval was activated. During the active 8 minute time slot, a manual data retrieval was triggered. (CH01, CH2, and CH3 data size < 1kB). Measured signal at 6V is shown in Figure 9b.

(c) Power consumption test logging interval:
The UniLogCom was supplied with 6 V/12 V and the modem functions were deactivated. The logging interval, which is to be expected daily at 00:00, was mapped below. During the measurement CH03—rain sum and CH32—supply are recorded. Measured signal at 6 V is shown in Figure 9c.

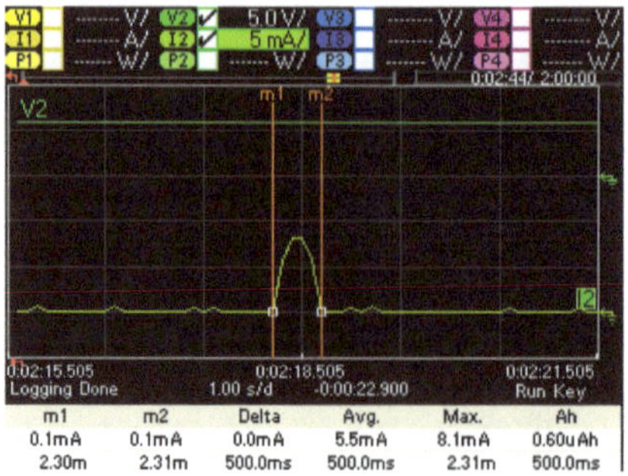

(**a**)

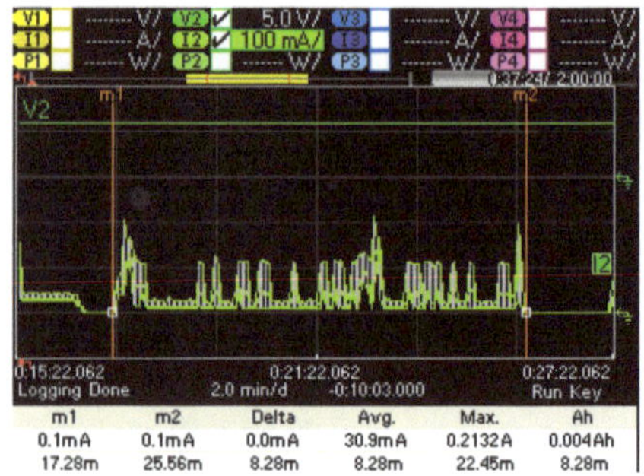

(**b**)

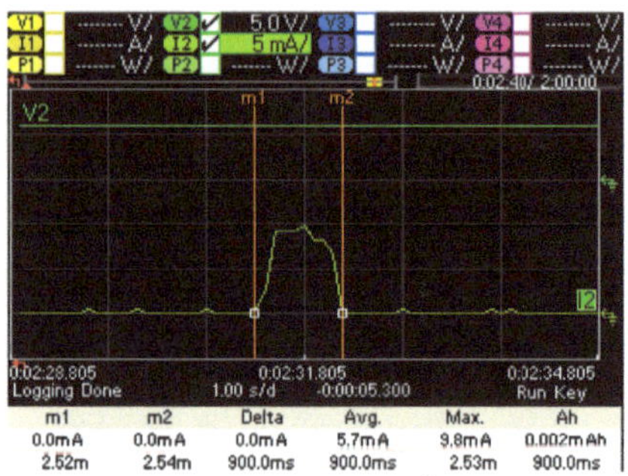

(**c**)

Figure 9. (**a**) Current consumption rain pulse—measurement 1 at 6 V, (**b**) current consumption minutes time slot + data retrieval—measurement 1 at 6 V, (**c**) current consumption logging interval 00:00 h—measurement 1 at 6 V.

Results of the daily current consumption of the UniLogCom with RG50 sensor with 6 V and 12 V supplies are shown in Tables 2 and 3, respectively.

Table 2. Daily current consumption of the UniLogCom with RG50 sensor with a 6 V supply.

Type of Load	Number of Events	Current Consumption per Event in mAh	Total Current Consumption in mAh
FTP push	1	1.28330	1.28330
Logging	2	0.00143	0.00285
Time slot	0.143	4.79140	0.68517
Rain pulses	100.000	0.00069	0.06900
Quiescent current	-	-	2.16000
Daily current consumption	-	-	4.200

Table 3. Daily current consumption of the UniLogCom with RG50 sensor with a 12 V supply.

Type of Load	Number of Events	Current Consumption per Event in mAh	Total Current Consumption in mAh
FTP push	1	0.83140	0.83140
Logging	2	0.00085	0.00170
Time slot	0.143	2.20955	0.31597
Rain pulses	100.000	0.00034	0.03400
Quiescent current	-	-	2.04000
Daily current consumption	-	-	3.223

3.2. Energy Management and Battery Charging Circuits

In order to link the energy generation and the measurement task in a defined way, an intelligent energy management is of central importance. This part is particularly challenging due to the influence of the solution on the behavior of the transducer and the challenge of selecting a storage element suitable for the intended application in terms of intermittent energy extracted from the ambient source and intermittent use of the stored energy depending on the hydrological measurement system. Based on the current consumption investigation, a storage element consisting of two series-connected Li-ion batteries of 3.7 V each is used to power the hydrological measurement system. To protect the battery from being overcharged, a commercial voltage protection circuit based on HY2120 is inserted between the interfaces of the battery charger to the batteries [21]. The state of the battery group can be captured by this IC. Together with power MOSFET with a parallel body diode, this IC can decide the current flowing orientation through its over-charging and discharging controlling logic: When the overcharging state is triggered, the power generated by the battery charger cannot be injected into the battery group; when the dis-overcharging state is trigged, the battery will not deliver its power to the measurement device. Table 4 shows an overview of the main components used for the realization of the energy management and storage element.

Table 4. Components used in the battery charging circuit.

Component	Number	Details
L-ion battery	2	3.7 V 3000 mAh Samsung ICR18650-30B
Voltage protection module	1	Overcharge voltage: 4.25–4.35 V ± 0.05 V Over discharge voltage: 2.3–3.0 V ± 0.05 V Maximum continuous current: 3 A Maximum peak current: 5 A
Constant-current/ constant-voltage battery charger	1	Based on LT1512 Output voltage: 8.4 V Minimum input voltage: 2 V

The charging circuit is shown in Figure 10, which is mainly based on LT1512. The LT1512 is an IC battery charging chip specifically optimized to use the single-ended primary inductance converter (SEPIC) topology. This special topology has unique advantages for battery charging: it operates at input voltages above, equal to, or below the battery voltage; it has no off-state battery discharge path, and eliminates the snubber losses of flyback designs. This IC also has a current measurement point that is ground referenced and does not need to be connected directly to the battery.

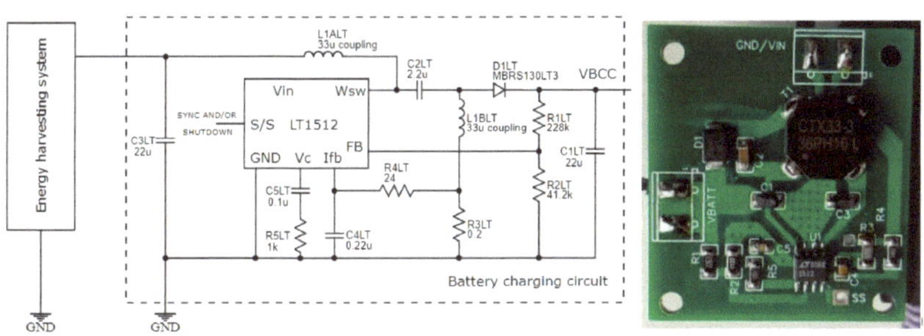

(a) (b)

Figure 10. Constant current/constant voltage battery charger: (**a**) schematic of the circuit, (**b**) designed PCB.

4. Experimental Validation

4.1. Laboratory Characterization

The characterization and evaluation of the energy management circuit is performed through laboratory tests with the developed converter and hydrometric measurement system, including a rain gauge sensor, as shown in Figure 11. Because the load resistance connected to the converter output can change the characteristics of the voltage output in terms of pulse amplitude and width, the output response under several load resistances was first investigated. Figure 12a,b, respectively, show the open-circuit voltage output of the converter and the charging current to the batteries for load resistances from 560 Ω to 22 kΩ. Results are shown in Figure 12. For load resistances below 560 Ω, the open-circuit output voltage of the converter is below 2 V, which is below the minimum voltage level required for battery charging circuit operation.

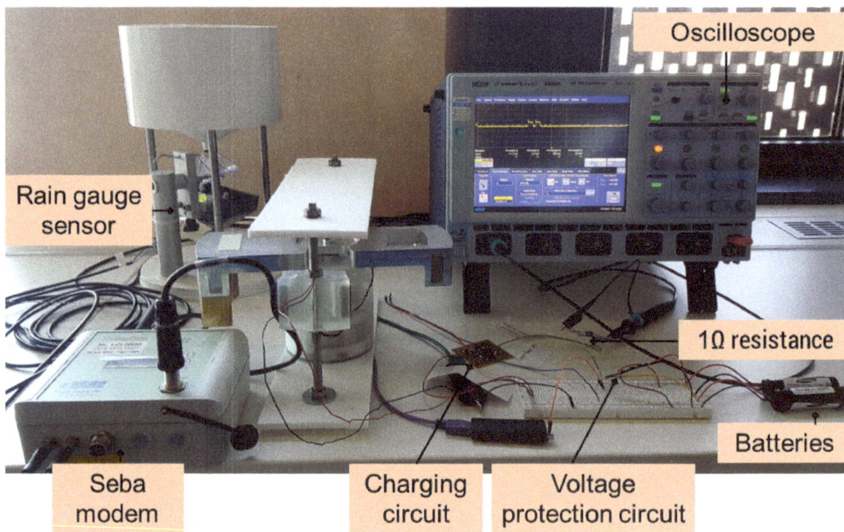

Figure 11. Experimental setup for laboratory characterization of the energy management and storage circuits.

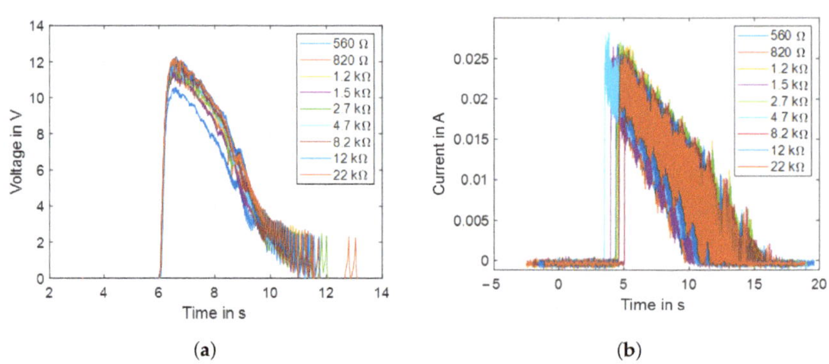

Figure 12. (**a**) Open circuit voltage output, (**b**) charging current to the batteries.

To determine the optimum load resistance, both the efficiency of the kinetic converter and the efficiency of the charging circuit must be considered. For this purpose, the generated power per trip was evaluated under different load resistances and the generated power per pendulum revolutions was determined (see Table 5). The results show that the optimum load resistance is approximately 4.7 kΩ, with the generated power in the generator being released at 0.0303 mAh at 8.4 V, which requires 11 complete pendulum turns.

Table 5. Characterization of the converter voltage output under different load resistances.

Load Resistance	Required Turns of the Pendulum before Release	Harvested Power per Pulse at 8.4 V in mAh	Harvested Power per Pendulum Turn at 8.4 V in mAh
20 Ω	not working	0	0
43 Ω	working continuously	0.0012	0
82 Ω	1	0.0058	0.0058
560 Ω	9	0.0228	0.0025
820 Ω	9	0.0206	0.0023
1.2 kΩ	11	0.0287	0.0026
1.5 kΩ	11	0.0241	0.0021
2.7 kΩ	11	0.0294	0.0026
4.7 kΩ	11	0.0303	0.0027
8.2 kΩ	11	0.0250	0.0022
12 kΩ	11	0.0249	0.0022
22 kΩ	11	0.0265	0.0024

4.2. Field Measurement

In addition to laboratory prototyping of the pendulum transducer, charging circuit for the batteries, and hydrometric sensors, field measurements were conducted on a river in Chemnitz, Germany. For this purpose, the system with magnetic pendulum transducers first had to be robustly integrated into a watertight enclosure of dimensions 500 × 280 × 250 mm, as shown in Figure 13a. An illustration of the converter and the energy management and storage circuits are illustrated in Figure 13b. The pendulum converter prototype was finally equipped with a SlimlogCom 4G platform and fitted with the two sensors for precipitation RG50 and water level and temperature sensor DST-22. For the energy supply, the system was also expanded with buffer batteries. The electronic platform is compactly integrated in the housing of the RG50 (see Figure 13c).

Figure 14a shows the field measurement position. By considering a kinematic viscosity of approximately 1 mm^2/s and $Q = 0.15\ m^3/s$ (see Figure 14b), the Reynolds number of the river is typically $R_e > 10^6$.

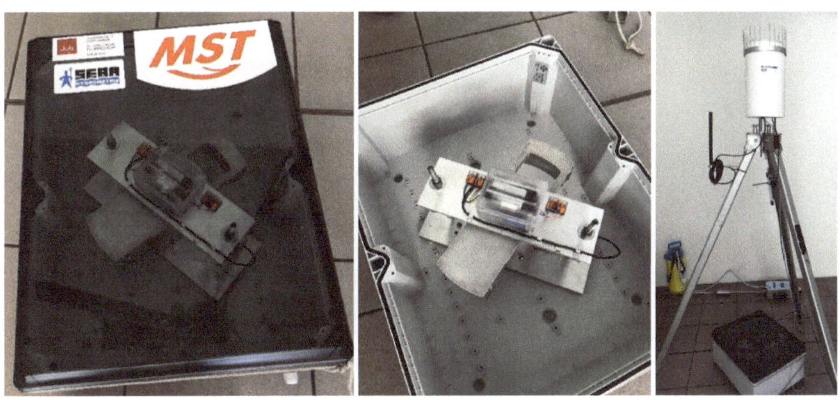

(**a**) (**b**) (**c**)

Figure 13. (**a**) Housing of the final pendulum converter system, (**b**) pendulum based converter and energy management and storage circuits integrated in the housing, (**c**) components of the final prototype with precipitation sensor, electronic platform, and temperature sensor.

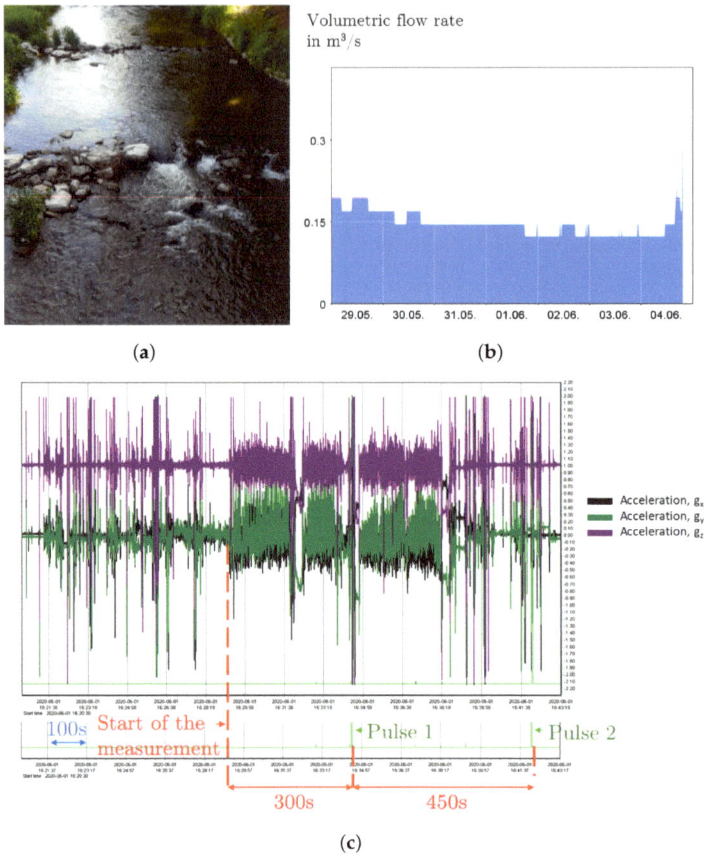

Figure 14. (**a**) Field measurement location, (**b**) volumetric flow rate of the river during the measurement period according to data provided by the Saxon State Office for Environment, Agriculture and Geology, (**c**) acceleration values and output voltage from the MSR 160 data logger.

The housing containing the converter was attached with a cord to a static frame fixed to the selected obstacle in the river. The length of the cord is equal to the determined recirculation zone based on the finite element analysis. Further, a data logger (MSR 160) was attached to the housing, and the acceleration along the three axes and the voltage output from the converter were measured to determine the time between pulses and therefore the harvested energy.

Recorded signals from the data logger are shown in Figure 14c, which include the measured acceleration in the three axes and the output voltage form the kinetic converter. It was found that the pulse repetition rate ranged from approximately 5 to 7 min. It should be noted that theoretically and based on the daily current consumption of the load determined in the previous section, which is approximately 4–4.3 mAh at 7 V, the lifetime of the autonomous sensor system can be determined. Table 6 shows the estimated lifetime of the system as function of the time between pulses that can be generated by the kinetic converter. Theoretically, the system has an infinite lifetime if the generator is able to generate an average of one pulse/10 min or more. It is worth noting that these results are calculated without considering self-discharge of the batteries and for a daily electric charge consumption of the sensor system of 4.2 mAh at 7 V. Further, according to measurements done at different locations, the system's performance was similar and confirms the reproducibility of the obtained values.

Table 6. Estimated system lifetime as function of the pulse frequency.

Time between Pulses in Minutes	Harvested Electric Charge at 7 V in mAh in Year	Estimated System Lifetime in Year
5	3153	unlimited
6	2628	unlimited
7	2253	unlimited
8	1971	unlimited
9	1752	unlimited
10	1576	unlimited
20	788	6.5
30	525	4.7
0	0	3.14

5. Discussion

Compared to the state of the art, the proposed conversion principle is characterized by the fact that even low water flow excitations can be converted. A high specific converter power density can be achieved due to the spring energy storage. In addition, the pendulum with magnetic spring allows the converter to operate even under unidirectional excitation. Without the magnetic spring principle, this would not be possible because the pendulum remains in a stationary state. Based on the novel approach, redundant battery-operated storage concepts can be optimized in the future by energy-autonomous conceptions of measuring stations. Likewise, a complete abandonment of redundant storage concepts is conceivable, which would allow for more cost-efficient maintenance intervals. The system represents an interesting solution for seamless monitoring of dangerous environmental phenomena such as extreme flood events with threats to humans, animals, and property or water quality issues (e.g., overturning of water bodies).

6. Conclusions

The present paper proposes a novel energy harvesting approach for harvesting energy from water flow to supply hydrometric monitoring systems. The converter consists of a rotating pendulum connected to a bi-directional mechanical system, spiral torsion spring, gears, and a DC generator having a similar working principle to that of kinetic wristwatches. The converter performance was evaluated for different load resistances by a model based approach. For this, a physical model is developed for the whole converter in Matlab Simscape. Taking into account all events and components, the daily consumption of the hydrometric monitoring system was experimentally determined. It was found that the daily current consumption of the electronic platform with RG 50 with a 6 V supply is approximately 4.2 mAh.

In addition, corresponding energy management and storage circuits were designed and fabricated. A prototype of the complete system consisting of the novel converter, energy management and storage circuits, and a data logger platform with a modem (SlimLogCom 4G) and a precipitation sensor RG50 was fabricated. For the characterization of the overall system, laboratory tests were conducted, in addition to testing in a river. Laboratory characterization shows that the converter can harvest approximately 7.2 mAh and 4.3 mAh electric charge per day at 7 V when considering 6 min and 10 min separation times between two successive releases, respectively. In addition, measurements conducted in a river demonstrated that the harvester could generate at least one pulse in five to seven minutes. Thus, it has been demonstrated that the harvester can fully supply the hydrometric monitoring system, thereby avoiding the need for external power sources.

In comparison to previous energy harvesting systems for rivers, the proposed energy harvesting solution could enable the use of hydrometric monitoring systems under various conditions and avoid the need to replace batteries by preserving and utilizing the collected energy. This solution is advantageous in that the harvesting system remains enclosed within a stable housing with no external interface to the flowing water, resulting in a high level of reliability. The system under consideration has the potential to allow the use of hydrometric monitoring systems under a variety of conditions and even with additional sensors as well as a higher number of communication events, while eliminating the need for battery replacement.

7. Patents

A part of this work has been patented.

Author Contributions: Conceptualization, S.N., F.G., O.K. and C.V.; methodology, S.N.; software, A.B.; validation, S.N., Z.H. and F.G.; investigation, S.N., A.B. and F.G.; writing—original draft preparation, S.N.; writing—review and editing, S.N., F.G., Z.H. and O.K.; supervision, O.K.; project administration, O.K., F.G. and P.H.; funding acquisition, O.K., F.G. and P.H. All authors have read and agreed to the published version of the manuscript.

Funding: This research was supported by the Federal Ministry of Education and Research (BMBF), Germany, within the project number #02WQ1384A.

Conflicts of Interest: The authors declare no conflict of interest.

Nomenclature

Symbol/Abbreviation	Description/Meaning
α	Roll angle [°]
β	Pitch angle [°]
ϕ	Yaw angle [°]
R_e	Reynolds number
Q	Volume flow rate [m^3/s]
l	Length of the river [m]
μ	Kinematic viscosity [m^2/s]
k	Turbulent kinetic energy [J/kg] = [m^2s^{-2}]
ϵ	Rate of dissipation of the turbulent kinetic energy [m^2/s^{-3}]
CAD	Computer-Aided Design
CFD	Computational Fluid Dynamics
DC	Direct Current
DPEH	Downstream Piezoelectric Energy Harvester
FTP	File Transfer Protocol
MFC	Micro Fiber Composite
PEH	Piezoelectric Energy Harvester
PVC	Polyvinyl Chloride
RANS	Reynolds-Averaged Navier–Stokes
RG50	Precipitation Sensor
SEPIC	Single-Ended Primary Inductance Converter
UPEH	Upstream Piezoelectric Energy Harvester

References

1. Rahimzadeh, M.; Samadi, H.; Mohammadi, N. Analysis of Energy Harvesting Enhancement in Piezoelectric Unimorph Cantilevers. *Sensors* **2021**, *21*, 8463. [CrossRef] [PubMed]
2. Kargar, S.; Hao, G. An Atlas of Piezoelectric Energy Harvesters in Oceanic Applications. *Sensors* **2022**, *22*, 1949. [CrossRef] [PubMed]
3. Bradai, S.; Naifar, S.; Kanoun, O. Development of a hybrid vibration converter for real vibration source/Entwicklung eines Hybrid-Vibrationswandlers für eine echte Schwingungsquelle. *Tech. Mess.* **2019**, *86*, 57–61. [CrossRef]
4. Bradai, S.; Naifar, S.; Trigona, C.; Baglio, S.; Kanoun, O. An electromagnetic/magnetoelectric transducer based on nonlinear RMSHI circuit for energy harvesting and sensing. *Measurement* **2021**, *177*, 109307. [CrossRef]
5. Communication from the Commission and Communication from the Commission. In *Energy for the Future: Renewable Sources of Energy. White Paper for a Community Strategy and Action Plan*; European Union: Maastricht, The Netherlands, 1997; p. 599.
6. Pop-Vadean, A.; Pop, P.; Latinovic, T.; Barz, C.; Lung, C. Harvesting energy an sustainable power source, replace batteries for powering WSN and devices on the IoT. *IOP Conf. Ser. Mater. Sci. Eng.* **2017**, *200*, 012043. [CrossRef]
7. Satharasinghe, A.; Hughes-Riley, T.; Dias, T. A Review of Solar Energy Harvesting Electronic Textiles. *Sensors* **2020**, *20*, 5938. [CrossRef] [PubMed]
8. Kishore, R.; Priya, S. A Review on Low-Grade Thermal Energy Harvesting: Materials, Methods and Devices. *Materials* **2018**, *1*, 1433. [CrossRef] [PubMed]

9. Naifar, S.; Bradai, S.; Viehweger, C.; Kanoun, O. Survey of electromagnetic and magnetoelectric vibration energy harvesters for low frequency excitation. *Measurement* **2017**, *106*, 251–263. [CrossRef]
10. Song, R.; Hou, C.; Yang, C.; Yang, X.; Guo, Q.; Shan, X. Modeling, Validation, and Performance of Two Tandem Cylinder Piezoelectric Energy Harvesters in Water Flow. *Micromachines* **2021**, *12*, 872. [CrossRef] [PubMed]
11. Kong, H.; Roussinova, V.; Stoilov, V. Renewable energy harvesting from water flow. *Int. J. Environ. Stud.* **2018**, *76*, 84–101. [CrossRef]
12. Shaikh, F.; Zeadally, S. Energy harvesting in wireless sensor networks: A comprehensive review. *Renew. Sustain. Energy Rev.* **2016**, *55*, 1041–1054. [CrossRef]
13. Raja Singh, R.; Raj Chelliah, T.; Agarwal, P. Power electronics in hydro electric energy systems—A review. *Renew. Sustain. Energy Rev.* **2014**, *32*, 944–959. [CrossRef]
14. Ferrarese, G.; Malavasi, S. *Energy Harvesting: Technology, Methods and Applications*, 3rd ed.; Williams, R., Ed.; Nova Science Pub Inc.: New York, NY, USA, 2015.
15. Azevedo, J.; Lopes, J. Energy harvesting from hydroelectric systems for remote sensors. *AIMS Energy* **2016**, *4*, 876–893. [CrossRef]
16. Morais, R.; Matos, S.; Fernandes, M.; Valente, A.; Soares, S.; Ferreira, P.; Reis, M. Sun, wind and water flow as energy supply for small stationary data acquisition platforms. *Comput. Electron. Agric.* **2008**, *64*, 120–132. [CrossRef]
17. Taylor, G.; Burns, J.; Kammann, S.; Powers, W.; Welsh, T. The Energy Harvesting Eel: A small subsurface ocean/river power generator. *IEEE J. Ocean. Eng.* **2001**, *26*, 539–547. [CrossRef]
18. Kamenar, E.; Zelenika, S.; Blazevic, D.; Samanic, I. River flow energy harvesting by employing piezoelectric eels. In Proceedings of the 14th EUSPEN International Conference, Dubrovnik, Croatia, 2–6 June 2014; Volume 1, pp. 63–66.
19. Shan, X.; Song, R.; Liu, B.; Xie, T. Novel energy harvesting: A macro fiber composite piezoelectric energy harvester in the water vortex. *Ceram. Int.* **2015**, *41*, S763–S767. [CrossRef]
20. Song, R.; Shan, X.; Lv, F.; Xie, T. A study of vortex-induced energy harvesting from water using PZT piezoelectric cantilever with cylindrical extension. *Ceram. Int.* **2015**, *41*, S768–S773. [CrossRef]
21. HY2120 Data Sheet 2-Cell Lithium-ion/ Polymer Battery Packs Protection ICs. Available online: https://www.hycontek.com/wp-content/uploads/DS-HY2120_EN.pdf (accessed on 22 May 2022).

sensors

MDPI

Article

Parameter Optimization of a Magnetic Coupled Piezoelectric Energy Harvester with the Homogenized Material—Numerical Approach and Experimental Study

Andrzej Koszewnik [1,*], Daniel Ołdziej [1] and Mário B. Amaro [2]

[1] Department of Robotics Control and Mechatronics, Faculty of Mechanical Engineering, Bialystok University of Technology, Wiejska 45C, 15-351 Bialystok, Poland; d.oldziej@pb.edu.pl
[2] Instituto Superior Técnico, 1049-001 Lisbon, Portugal; marioamaro@tecnico.ulisboa.pt
* Correspondence: a.koszewnik@pb.edu.pl; Tel.: +48-571-443-052

Abstract: This paper presents the process optimization of some key parameters, such as beam spacing, flux density and optimal impedance load matching of magnetic coupled piezoelectric harvesters. In order to do this, the distributed parameters model of this structure, containing macro-fiber components (MFC) with homogenous material in the piezoelectric fiber layer, was determined. Next, the computational model of this structure was designed on the basis of the first-order shear theory (FOST). The performed analysis of the calculated voltage outputs on the basis of the theoretical approach and finite element model by various beam spacing allowed us to indicate that optimized parameters play an important role in enhancing the efficiency of the system. Experiments carried out in a laboratory stand for this structure, allowed for the verification of the numerical results. In the effect, it can be noted that magnetic coupled harvesters will be relevant for a wide range of application sectors, as well as useful for the evolving composite industry.

Keywords: magnetic coupled systems; piezoelectric; energy harvesting; macro-fiber composites (MFC); parameter optimization; homogenous material

Citation: Koszewnik, A.; Ołdziej, D.; Amaro, M.B. Parameter Optimization of a Magnetic Coupled Piezoelectric Energy Harvester with the Homogenized Material—Numerical Approach and Experimental Study. *Sensors* **2022**, *22*, 4073. https://doi.org/10.3390/s22114073

Academic Editors: Slim Naifar, Olfa Kanoun and Carlo Trigona

Received: 28 April 2022
Accepted: 26 May 2022
Published: 27 May 2022

1. Introduction

In recent years, energy harvesting technology has received extensive attention and the development of self-sufficient systems has been promoted. It is known that energy harvesting systems can convert various energy sources in the environment into electrical energy, providing energy for sensors with low power demand [1–3]. This leads to decreasing the maintenance costs of inspections by structural health monitoring (SHM) systems, and also reducing battery disposal [4–6]. The aforementioned ambient energy harvesting technology can bring about a revolutionary development to wireless sensor networks as well as expand the development area for intelligent self-powering sensors used in the Industrial Revolution 4.0 [7–9].

In numerous energy harvesting technologies, vibration is a promising source of high power density [10]. As a common physical phenomenon, vibration is widely found in production and living environments, such as home appliances, vehicles, natural environments [11], and railways or bridges [12]. It also exists in life phenomena, such as the heartbeat [13] and limb activities [14]. At present, the commonly used electromechanical conversion methods include the piezoelectric method [15–17], the electromagnetic method [18,19], the electrostatic method [20], etc. Among them, the piezoelectric method, which is based on a cantilever structure, has gained great attention due to its high energy conversation efficiency and simple processing design [21–23].

The aforementioned above method of harvesting energy, using cantilever beam structures, faces many difficulties. One of them is a mismatch resonance frequency band to the excitation frequency for mechanical systems; another one is that there is no single frequency-stable excitation. As a result, regardless of the type of obstacles, each of them leads to a significant decrease in electromechanical system efficiency and, finally, to a reduction in their implementation in many applications. A similar problem has been observed by professor Litak and his group in [24], where the performed analysis of a complex structure allowed to indicate that sources of vibrations in those structures are generally broadband. As a result, this led to the conclusion that the use of a conventional single-beam harvester, especially in complex structures, requires a proper matching of their properties to consider the mechanical structure.

Taking this fact into account, many researchers have now decided to consider the complex problem of the real source of excitation and propose several optimization techniques in order to increase the amplitude of the generated piezo-voltage, as well as expand the bandwidth by using linear and nonlinear techniques [25].

Among the linear optimization methods, a plurality of resonance frequencies is mostly generated in the form of an array. This method utilizes a plurality of individual cantilever beams, which is simple to design. However, the volume and the matching circuit of the system are relatively large [26]. Another linear method is the multi-degree of freedom technique, which is relatively complex but more robust, and it generally establishes double or more degrees of freedom in composite forms. As a result, such a method, due to the existing coupling between multi modalities, allows for reducing the individual response amplitude for each degree of freedom [27].

In the case of nonlinear optimization methods, there are at least three methods that introduce nonlinear stiffness and enhance the efficiency of the system. One of them is a technique called the "Stop Blocking Method", which allows for introducing the piecewise stiffness to the system [28,29]. As a result, the bandwidth of this system is extended; however, this occurs at the expense of decreasing the peak value of the generated voltage. Unfortunately, the necessity of applying a vibration signal with an amplitude greater than the stopper spacing, which does not work in the case of small structures, is a drawback of this technique.

Another nonlinear method is the multi-stable technique (bi-stable or tri-stable), which also introduces nonlinear stiffness into the system by a magnetic force [30,31]. One solution to this method is a cantilever beam with two axially opposite magnets corresponding to the bi-stable system. Then, this kind of structure allows for generating a higher amplitude of voltage from a piezo-patch harvester, especially in regions of multiple potential wells, as well as in the wide frequency range. Unfortunately, as it was mentioned above, also in this case, the need to make a structure with high precision as well as the necessity of generating an excitation signal with high amplitude are drawbacks of this method.

The last nonlinear method is a technique that utilizes vertical auxiliary magnets. The aforementioned method is the most studied one by many researchers in the world, due to its numerous benefits. Paper [32] is an example of the research conducted in this field. Its authors tested an energy harvesting system with bi-directional resonance frequency tunability. Another example can be found in the paper written by Zhang, who analyzed the resonance response of a single fixed auxiliary magnet structure. He concluded that the considered structure had a certain effect on widening the frequency band [33]. Yet another example is also included in the paper published by Firoozy and his group, who established a distributed reduced-order model based on the Galerkin method, with consideration of the angle between the magnets during motion [34]. As a result, they indicated that the process of parameter optimization of the structure with vertical auxiliary magnets should be analyzed further in order to enhance the efficiency of this kind of system.

The instance of research where the aforementioned problem was further analyzed is in a paper written by Abdelmoula et al. [35], who performed a comparative study on a broadband piezoelectric energy harvester with single and dual magnetic forces. The obtained results by them showed that the inclusion of a second attractive magnetic force with the same spacing distance leads to a changed magnetic coupled system, from a softening behavior (single magnetic) to a hardening behavior (dual magnetic forces). Moreover, they indicated that decreasing the spacing distance results in an enhancement in the hysteresis region and, hence, broadband resonance regions take a place. Another example in this field is a paper published by Rui, who studied four different harvester modes containing low/high-frequency beams with attraction/repulsion [36]. He concluded that an energy harvesting system with conventional MFC patches and with a repulsion mode gives better results than an attractive mode, due to generating high output power in a wide operating frequency range. In addition, on the basis of the obtained results, he indicated that internal resonance phenomena are such that magnetically coupled systems can occur due to a nonlinear configuration using magnets. Yet another example is a paper written by Shih and Su, who analyzed a magnet-induced nonlinear U-shaped bi-directional piezoelectric harvester with an attraction and repulsion mode to overcome the shortcomings of the conventional harvester [37]. The obtained results by them, indicate that the proposed nonlinear U-shaped harvester demonstrates the capability of bi-directional harvesting and improved performance with the aid of nonlinear magnetic force.

While the potential of magnetically coupled harvesters is established in principle, as it is in the review paper [38], its implementation in specific engineering sectors, such as SHM, requires different views, especially in the field of nonlinearities of the piezo-harvester. An instance of these considerations is paper [39], where the influence of nonlinear geometric responses of a piezoelectric composite plate, considering von Karman's large strain theories into the classic plate solution, has been investigated by using a 3D element model. Then, on the basis of the obtained results, the authors indicated that this problem cannot be omitted, especially when the correct prediction of the stress-strain over the PEH is analyzed. Other scientific works in this field focused on introducing piezoelectric coupling to the shell element [40,41] and the results indicate the influence of nonlinearity on the piezoelectric laminated shell is significant to achieving better performances. As a result, this led to developing investigations and modeling of the piezo structure on the basis of high order theories as a higher-order layer-wise plate finite element [42]. In summary, despite many scientific contributions related to the formulations of plates and shells for piezoelectric laminated elements, there is a gap in the verification of numerical results considering the shell finite elements of the piezoelectric element in the magnetically coupled system. This paper addresses this knowledge gap by carrying out numerical analyses and subsequently validating them against experimental results.

In this context, this paper is focused on the analysis of the performance of a magnetic coupled piezoelectric energy harvester in detail, through simulations and experiments. In contrast to [43,44], the piezo-harvester with a three-dimensional material in the piezoelectric fiber layer in this paper is modeled using homogenization techniques, such as the representative volume element (RVE) [45,46]. In the effect, it allows for improvement of the electromechanical properties of this composite and increases the harvesting effect. In addition, a theoretical distributed parameter model was established; simulations based on an FEM model were performed, as well as test rigs for different parameter configurations, which were carried out in a laboratory stand. The obtained results show that the magnetic coupled piezoelectric energy harvester allows for enhancing the effect of energy harvesting from vibrations and improves the performance of the conventional harvester by choosing optimal parameters.

This paper is organized as follows: Section 2 describes the methodology of modeling a coupled magnetic smart structure by taking into account a laminated structure of the piezo-composite. Section 3 presents simulation results calculated on the basis of the theoretical distributed parameter model with consideration of different configurations of the beam space. Section 4 presents the computational model of the structure with a homogenous model of macro-fiber composite (MFC), which is also a core novelty of this manuscript. In Section 5, experimental investigations were carried out for a coupled magnetic beam structure by taking into account different beam spaces to verify numerical results. The recorded voltage output signals from both piezo-composite harvesters allow for determining the optimized parameters of the tested structure. Section 6 concludes the main findings of this work.

2. Mathematical Model of the Coupled Magnetic Smart Structure with a Laminated Model of the MFC Element

The magnetic coupled piezoelectric energy harvesters studied in this manuscript are shown in Figure 1. This structure is composed of two monomorphic cantilever beams with piezo-harvester stripes with a three-dimensional homogenized material and two magnets used to enhance the efficiency of the system.

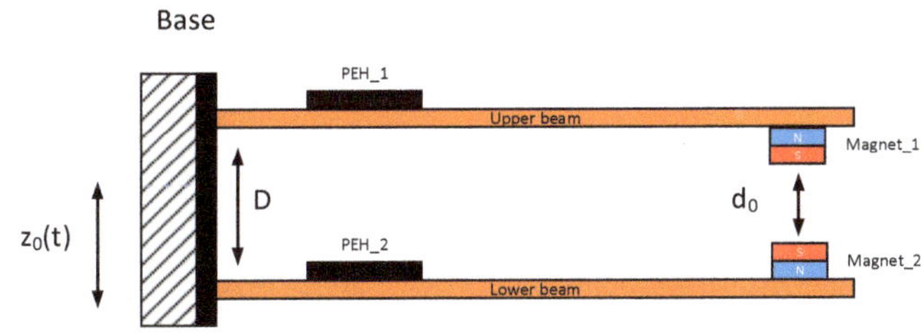

Figure 1. Schematic diagram of the magnetic coupled piezoelectric energy harvester.

As a result, two degrees of freedom of the nonlinear piezoelectric dynamics model with consideration of the nonlinear magnetic force can be expressed as:

$$\begin{aligned} M_1\ddot{z}_1(t) + C_1\dot{z}_1(t) + K_1z_1(t) + \Theta V_{p\_1}(t) + QF_{mag}(t) &= \mu_1 M_1\ddot{z}_0(t) \\ M_2\ddot{z}_2(t) + C_2\dot{z}_2(t) + K_2z_2(t) + \Theta V_{p\_2}(t) - QF_{mag}(t) &= \mu_2 M_2\ddot{z}_0(t) \\ C_p\dot{V}_{p1}(t) - \Theta\dot{z}_1(t) &= -\frac{V_{p1}(t)}{R_L} \\ C_p\dot{V}_{p2}(t) - \Theta\dot{z}_2(t) &= -\frac{V_{p2}(t)}{R_L} \end{aligned} \quad (1)$$

where:

$z(t)$—the relative displacement of the cantilever beam.
$z_0(t)$—the base excitation.
μ—the correction factor of the cantilever beam.
Q—the system factor, ($Q = 1$ for the attracted effect, $Q = -1$ for the repulsed effect).
$F_{mag}(t)$—the magnetic force.
M_1, M_2—the equivalent mass per unit of the upper and lower beams, respectively.
K_1, K_2—the equivalent stiffness of the upper and lower beams, respectively.
C_1, C_2—the equivalent damping of the upper and lower beams, respectively.

Taking into account [4,47,48], it is known that the macro-fiber composite shown in Figure 2 can be considered as five-layer elements, including a single active layer, two Kapton layers and two electrode layers. As a result, the equivalent values of masses M_1 and M_2 can be expressed in the following form:

$$\begin{cases} M_1 = 0.23b_{peh}\left(\sum_{i=1}^{n_i} n_i\rho_i t_i + \sum_{i=1}^{n_j} n_j\rho_j t_j + \sum_{i=1}^{n_k} n_k\rho_k t_k\right) + 0.23\rho b_{up}L_{up}t_{up} + m_{mag} \\ M_2 = 0.23b_{peh}\left(\sum_{i=1}^{n_i} n_i\rho_i t_i + \sum_{i=1}^{n_j} n_j\rho_j t_j + \sum_{i=1}^{n_k} n_k\rho_k t_k\right) + 0.23\rho b_{lo}L_{lo}t_{lo} + m_{mag} \end{cases} \quad (2)$$

where:

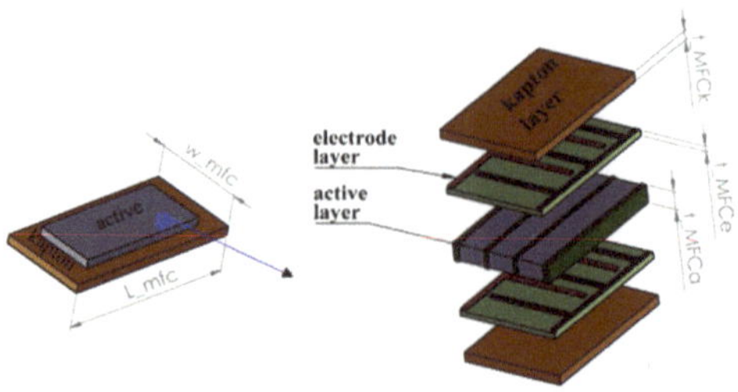

Figure 2. The structure of the macro-fiber composite.

n_i, n_j, n_k—the total amount of the active, Kapton and electrode layers, respectively.
ρ_i, ρ_j, ρ_k—the density of the active, Kapton and electrode layers, respectively.
t_i, t_j, t_k—the thickness of the active, Kapton and electrode layers, respectively.
ρ—the density of the host structures.
b_{up}, b_{low}—the width of the upper and lower beams, respectively.
L_{up}, L_{low}—the length of the upper and lower beams, respectively.
t_{up}, t_{low}—the thickness of the upper and lower beams, respectively.
m_{mag}—the mass of the magnet.

The contribution of the magnetic force in the system leads to an appearing additional magnetic stiffness K_{mag} that can be considered as a function of the distance d between both magnets. This leads to expressing the equivalent stiffnesses of both beams K_1 and K_2 in the following forms, respectively:

$$\begin{cases} K_1 = K_{up\_b} + K_{mag} \\ K_2 = K_{low\_b} + K_{mag} \end{cases} \quad (3)$$

where:

K_{up_b}—the stiffness of the upper beam without the impact of the magnetic force can be expressed as:

$$K_{up\_b} = \frac{3E_1I_1}{L_1{}^3} = \frac{b_{peh}}{4L_{peh}{}^3}\left(\sum_{i=1}^{n_i} n_iE_ih_i{}^3 + \sum_{j=1}^{n_j} n_jE_jh_j{}^3 + \sum_{k=1}^{n_k} n_kE_kh_k{}^3\right) + \frac{3E_{up}I_{up}}{L_{up}{}^3} \quad (4)$$

K_{low_b}—the stiffness of the lower beam without the impact of the magnetic force can be expressed as:

$$K_{low\_b} = \frac{3E_2I_2}{L_2{}^3} = \frac{b_{peh}}{4L_{peh}{}^3}\left(\sum_{i=1}^{n_i} n_iE_ih_i{}^3 + \sum_{j=1}^{n_j} n_jE_jh_j{}^3 + \sum_{k=1}^{n_k} n_kE_kh_k{}^3\right) + \frac{3E_{lo}I_{lo}}{L_{lo}{}^3} \quad (5)$$

K_{mag}—the magnetic stiffness ($K_{mag} = \left|\frac{\delta F_{mag}}{\delta d}\right|$)

On the other hand, the equivalent damping C_1 and C_2 of the lower and upper beams are obtained by using the logarithmic decay method.

The aforementioned magnetic force, according to the "Vertical Magnetic Method", is introduced to the system by two cylindrical magnets of the same size located axially in the vicinity of the free ends of both beams. Then, supposing that d_0 is the static spacing of the magnets, while $\Delta d(t) = z_1(t) - z_2(t)$ is the change of the spacing between the magnets over time, this nonlinear force can be expressed as:

$$F_{mag}(d) = \frac{6\pi B_r{}^2 r^4 t_{mag}{}^2}{\mu_{air}(\Delta d(t) + d_0)^4} \tag{6}$$

Then, the deflections equations of the upper and lower beams in the simplified form of the multilayer piezo-composite can be established as:

$$E_1 I_1 \ddot{w}_1(x) = (m_{mag} - F_{mag})(L_{up} - x) + \frac{m_{peh}\rho_{peh} + \rho_{up} b_{up} L_{up} t_{up} g}{2L_{up}} x^2 \tag{7}$$

$$E_2 I_2 \ddot{w}_2(x) = (m_{mag} - F_{mag})(L_{lo} - x) + \frac{m_{peh}\rho_{peh} + \rho b_{lo} L_{lo} t_{lo} g}{2L_{lo}} x^2 \tag{8}$$

where:

E_1, E_2 denote the Young modulus of the lower and upper beams with the macro-fiber piezo-composite, respectively ($E_1 = E_2 = \frac{t_b E_b + t_{peh} E_{peh}}{t_b + t_{peh}}$).

The calculation of the moment of inertia I_1 and I_2 of the lower and upper beams with piezo-composites attached to their top surface requires determining the position of the neutral plane for each of them. Then, taking into account the considered structure, the location of this plane can be calculated in the following form:

$$t_{np} = \frac{0.5 t_{up}{}^2 + p t_{peh}\left(t_{up} + 0.5 t_{peh}\right)}{t_{up} + p t_{peh}} \tag{9}$$

where $p = \frac{E_{peh}}{E_b}$ is the ratio of the modulus, while the moment inertia of both beams, by assuming that the thickness of both beams is the same ($t_b = t_{up} = t_{low}$), can be written as:

$$I_1 = I_2 = \frac{1}{12} b_b t_b{}^3 + b_b t_b (t_n - 0.5 t_b)^2 + \frac{n}{12} b_{peh} t_{peh}{}^3 \left(H\left(x - x_{peh2}\right) - H\left(x - x_{peh1}\right)\right) + p \cdot b_{peh} t_{peh} \left(t_b - t_n + 0.5 t_{peh}\right)^2 \tag{10}$$

As a result, solving Equations (7) and (8) by considering the boundary conditions of both beams given by Equation (11) leads to determining the static deflection of the upper $w_1(L)$ and the lower beam $w_2(L)$, and next, to expressing the static magnetic force in the form:

$$w_1(0) = 0; \dot{w}_1(0) = 0;\ w_2(0) = 0;\ \dot{w}_2(0) = 0 \tag{11}$$

$$F_{mag} = \frac{6\pi B_r{}^2 r^4 t_{mag}{}^2}{\mu_{air}} \frac{1}{(D - w_1(L_1) + w_2(L_2) - 2t_{mag})^4} \tag{12}$$

Similarly, taking into account the fact that the $\Delta d(t)$ given in Equation (6) is the change of the spacing between the magnets over time, the final equation of the dynamic magnetic force can be expressed in the following form:

$$F_{mag}(t) = \frac{6\pi B_r{}^2 r^4 t_{mag}{}^2}{\mu_{air}} \frac{1}{(\Delta d(t) + D - w_1(L_1) + w_2(L_2) - 2t_{mag})^4} \tag{13}$$

The magnetic force given by Equation (13) leads to assessing the parameter of the coupled magnetic harvesters model. Using the Runge–Kutta algorithm and ode45 solver in Matlab, first the ordinary differential equation of two degrees of freedom of the nonlinear piezoelectric dynamics model have been solved, and next, the output voltages at different excitation frequencies can be obtained under simulation conditions.

3. Simulation Analysis of Different Modes (Finite Element Model)

The process of the assessment of the nonlinear system optimization parameters is described in this section. Firstly, taking into account two degrees of freedom of the nonlinear piezoelectric dynamics model given by Equation (2), the eigenvalue problem of this structure has been solved. Then, the calculated first two natural frequencies (6.58 Hz, and 43.35 Hz) on the basis of Equation (2) allows the indication that the best optimal parameters of the energy harvesting system should be obtained for the system working in the low-frequency range up to 15 Hz. Taking into account the value of the first natural frequency, the optimal load resistance for this system has been looked for by assuming that the base acceleration z_0 is constant (0.6 m/s^2), while the beam space between the upper and lower beams as well as a resistive load R_L are different. Considering this strategy, the simulations were performed for three different beams space (10, 22, and 30 mm) and twenty resistive loads in the range of 20–400 kΩ with the interval of 20 kΩ, respectively. The obtained results, included in Figure 3, show that the highest power output of this system is achieved for the structure with the beam space of 10 mm and the piezo-patch connected to the resistive load of R_L = 180 kΩ. However, the lowest power was recorded for the beam space D of 30 mm and the energy harvesters connected to the resistive load of R_L = 220 kΩ. As a result, it can be concluded that increasing the beam space for coupled magnetic piezo-harvesters leads to a slight decrease in the power output with a simultaneous slight shift in the optimal resistive load.

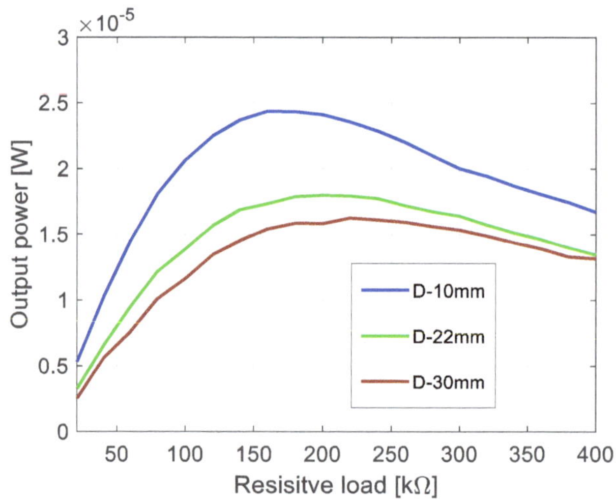

Figure 3. The comparison of the power output of the system for various distances between the beams.

Next, the performance of the energy harvesting system was analysed in the frequency domain for three different quasi-optimal impedance loads. For this purpose, simulations of voltage generated by both piezo-composites were performed by assuming that the excitation signal was a sinusoidal signal with the frequency changing in the range of 5–22 Hz with the interval of 0.25 Hz. The results in Figure 4 indicate that the highest voltage output peaks from both piezo-harvesters were obtained for the frequency of 6.58 Hz, the system with the beam space of 10 mm and the piezo-composite connected to the resistive load of 220 kΩ. Further analysis of the diagrams in Figure 4 showed that the beam space and the impact of the magnetic force with the repulsion effect lead to shifting the frequency range where energy harvesting from vibration is enhanced. As a result, their highest impact is observed for the structure with the beam space of 10 mm in the vicinity of the frequency range of 11–13 Hz, where the voltage from the upper piezo-harvester suddenly increased with a simultaneous decrease in the voltage generated from the bottom piezo-composite. A similar effect is also visible for the structures with the beam spaces of 22 mm and 30 mm, however, in these cases, the aforementioned nonlinearity of the system occurs faster in the frequency ranges of 9–11 Hz and 8–9.5 Hz, respectively. Finally, it can be concluded that decreasing the beam space in the magnetic coupled structure leads to generating higher voltage output at resonance and over it, as well as to changing the frequency where the magnetic forces have the strongest effect on the process.

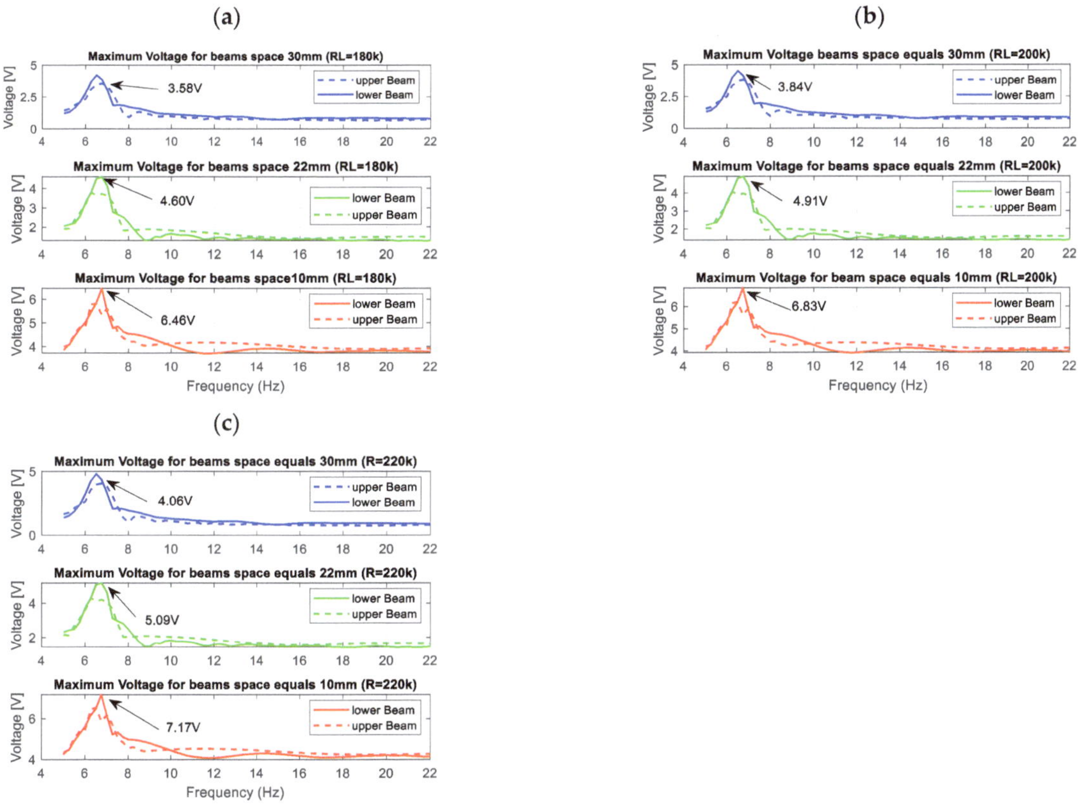

Figure 4. The comparison of the maximum voltage from both piezo-harvesters by various beams spaces for different optimal resistive loads (**a**) R_L = 180 kΩ, (**b**) R_L = 200 kΩ, (**c**) R_L = 220 kΩ.

Another parameter that could have an influence on the behavior of the magnetically coupled piezo-harvesters with a homogenized material is the magnetic remanence of the magnets located on the free ends of both beams. In order to check how this parameter affects the efficiency of this system, a simulation was performed again for three various beam spaces, but in this case, for only one impedance load of 220 kΩ. The analysis of the results presented in Figure 5 indicated again a high efficiency of this system for the magnetic remanence of the magnet close to 1.2 T by the lowest beam space. In the case of further increasing the value of the magnetic remanence, it can be noticed that the efficiency of this system has been reduced due to a decreasing voltage output generated from the piezo as well as the impact of the magnetic force on the structure.

The last step in this section was the analysis of phase portraits of the lower and upper beams for three different models of the structures. In order to do this, the numerical calculations were performed for the three different frequency ranges: 11–13 Hz, 9–11 Hz and 8.5–9.5 Hz, indicated above, for the structure with the beam space of 10 mm, 22 mm, and 30 mm, respectively. Observing portrait phases presented in Figure 6, it can be noticed that the impact of the magnetic force on the structure, by simultaneously decreasing beam space, leads to appearing chaotic trajectories located around particular equilibrium points.

This effect is especially visible in the diagram related to the model of the structure with the beam space of 10 mm, where the amplitude of the vertical displacements of both beams for the frequency of excitation close to 11.5 Hz are the highest, and the generated trajectories of both beams are strongly crooked. A similar however weaker effect can also be observed for the diagrams related to the structure with the beam space of 22 mm for the frequency close to 10 Hz. On the other hand, in the case of the structure with the beam space of 30 mm, the differences in portrait phases of both beams for the chosen frequency are insignificant, so they can be omitted for further analysis. Finally, taking into account all simulation results, it can be concluded that the efficiency of the coupled magnetic beams with harvesters can be significantly enhanced for the structure with the beam space of 10 mm, which is exciting to vibration with the first natural frequency as well as the frequency close to 11.5 Hz.

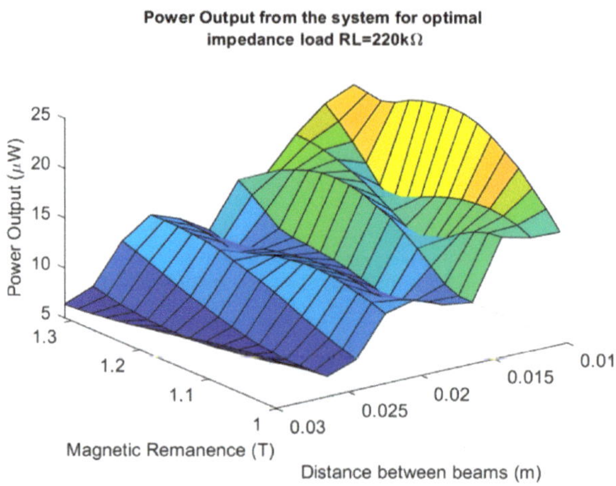

Figure 5. The power output generated by the EH system excited to vibration with the first natural frequency and connected to the optimal resistive load R_L = 220 kΩ.

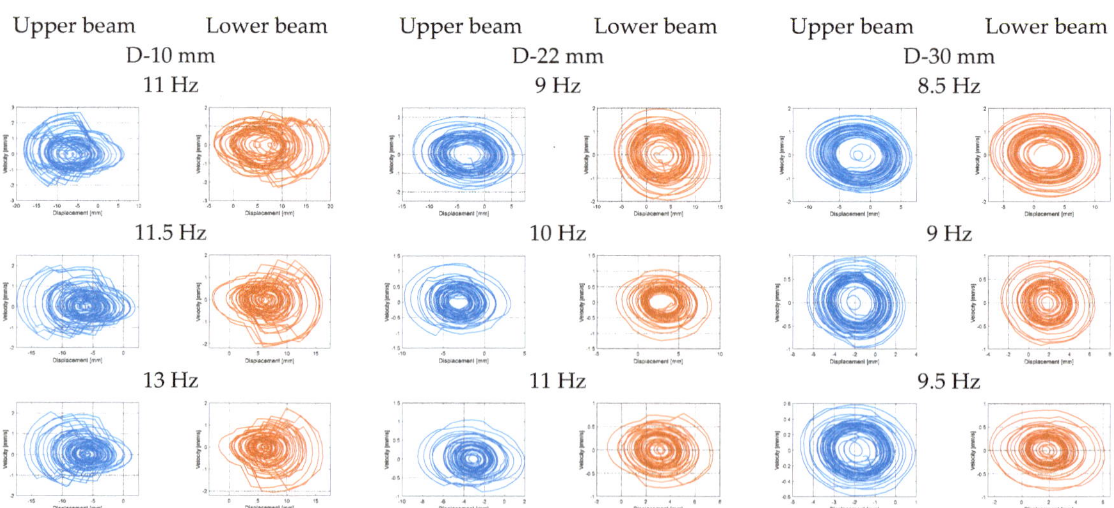

Figure 6. Comparison of the phase portraits for the model of the coupled magnetic piezo-harvester by different beams space and frequency excitation (upper beam—blue line, lower beam—orange line).

4. The Computational Model of the Coupled Magnetic Beam Structure

In order to verify the values of the optimization parameters calculated in the previous section, the numerical calculations of the magnetic coupled piezoelectric harvesters with a homogenized material are described in this section. Taking into account the above fact, in the first step, the FE model of piezo-composite composed of a single active layer, two electrode layers, and two Kapton layers, as it is shown in Figure 2, is designed by using the homogenization technique, such as the representative volume element (RVE), in order to obtain better mechanical and electrical properties of this element (see Table 1). Overall, it is achieved by considering a stress-strain effect in the laminate structure as well as the fact, that the normal plane section of the reference surface of the laminated shell remains plane after the deformation, but is not necessarily normal to the deformed reference surface that is the main feature of the first-order shear theory (FOST). Then, according to this theory described in detail in [49] only displacements, forces applied to the structure and electrical potentials are enough to increase the computational efficiency model as well as obtain a higher value of voltage generated by the piezo-element.

Table 1. Material properties of homogenized MFC layer of MFC8514 P2 [4].

			Mechanical Parameters			
Young's Modulus (GPa)	**Poisson's Ratio (-)**	**Shear Modulus (GPa)**	**Piezoelectric Charge Coefficient (pC/N)**	**Relative Permittivity (-)**		
E_x 31.6	v_{xy} 0.4	G_{xy} 4.9	d_{31} −173	$\varepsilon_r{}^T$ 2253		
E_y 17.1	v_{yz} 0.2	G_{yz} 2.5	d_{32} −150			
E_z 9.5	v_{xz} 0.4	G_{xz} 2.4	d_{33} 325			
			Geometrical parameters			
overall length [mm]	**overall width [mm]**	**active length [mm]**	**active width [mm]**	**thickness of fiber layer [μm]**	**thickness of electrode layer [μm]**	**thickness of Kaption layer [μm]**
103	17	85	14	180	25	30

In the next step, the process of discretization of the considered structure is performed using the RVE technique. Then, according to this method, the piezoelectric fiber layer with a homogenized material is modeled using a Solid 226-node coupled brick element, while the electrode and Kapton layers of the harvester are modeled using an 8-node coupled-brick element, Solid186. On the other hand, other passive elements, such as the upper beam, lower beam and magnets are modeled using an 8-node coupled-brick element, Solid186. In the effect, it leads to determining the computational model of the considered structure, which is shown in Figure 7, where the thickness of the adhesive layer (less than 15 μm) is omitted.

The obtained results of this model allow us to solve the eigenvalue problem by using the Ansys software and a modal analysis toolbox. For this purpose, the behavior of the FE model of the considered structure is analysed in the selected frequency range of 1–200 Hz, and the obtained eigenvalues are listed in Table 2.

Taking into account those results, it can be noticed that the obtained values of the first three lowest natural frequencies of the considered structure are close to those calculated on the basis of Equation (2). It allows for concluding that the results in the theoretical approach with the considered proper location of a piezo-composite with a homogenous material on the structure are calculated properly.

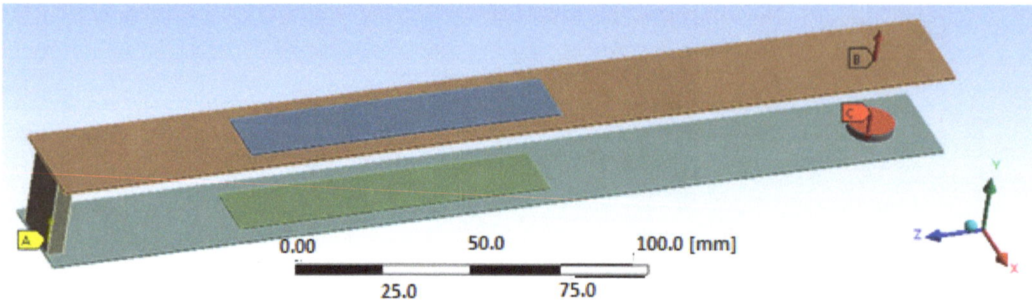

Figure 7. The numerical model of the smart intact structure with both piezo-composites attached to the host structure: (A) fixed support, (B) magnetic force—upper magnet, (C) magnetic force—lower magnet.

Table 2. Values of natural frequencies of the intact structure and damage structures.

Mode of Vibration	Eigenvalue [Hz]	
	Upper Beam	Lower Beam
First	6.64	6.64
Second	43.45	43.76
Third	147.40	147.7

Next, harmonic analysis of the considered structure is performed in order to assess the efficiency of magnetically coupled energy harvesters by different beam spaces. In order to do this, first, the computational models of the structure included in the lower and upper beams with shell models of the MFC elements attached to their top surfaces were excited to vibration by considering a sinusoidal excitation with the frequency changing in the range of 5–22 Hz. Then, the determined vertical displacements of nodes of the upper magnets and the lower magnet allowed to calculate the magnitude of the magnetic force according to Equation (13), and next, to check the behavior of the FE model by considering the impact of the magnetic forces in the repulsion mode on the structure. The results obtained (see Figure 8) from the "Piezo and Mems toolbox" of the Ansys software showed that the numerical values of the output voltage generated from both piezo-composites are close to the theoretical results. It is especially visible in the frequency regions over the first natural frequency, where the repulsion effect of the magnetic force leads to decreasing the voltage from the upper piezo and a simultaneous increase of the voltage from the lower piezo. Further analysis of these frequency regions showed that the highest amplitudes of voltage from both piezo-harvesters were generated for the computational model with a beam space of 10 mm. It allowed for concluding that the highest efficiency of the coupled magnetic system for a real structure should be also achieved in the same frequency range of 11–13 Hz.

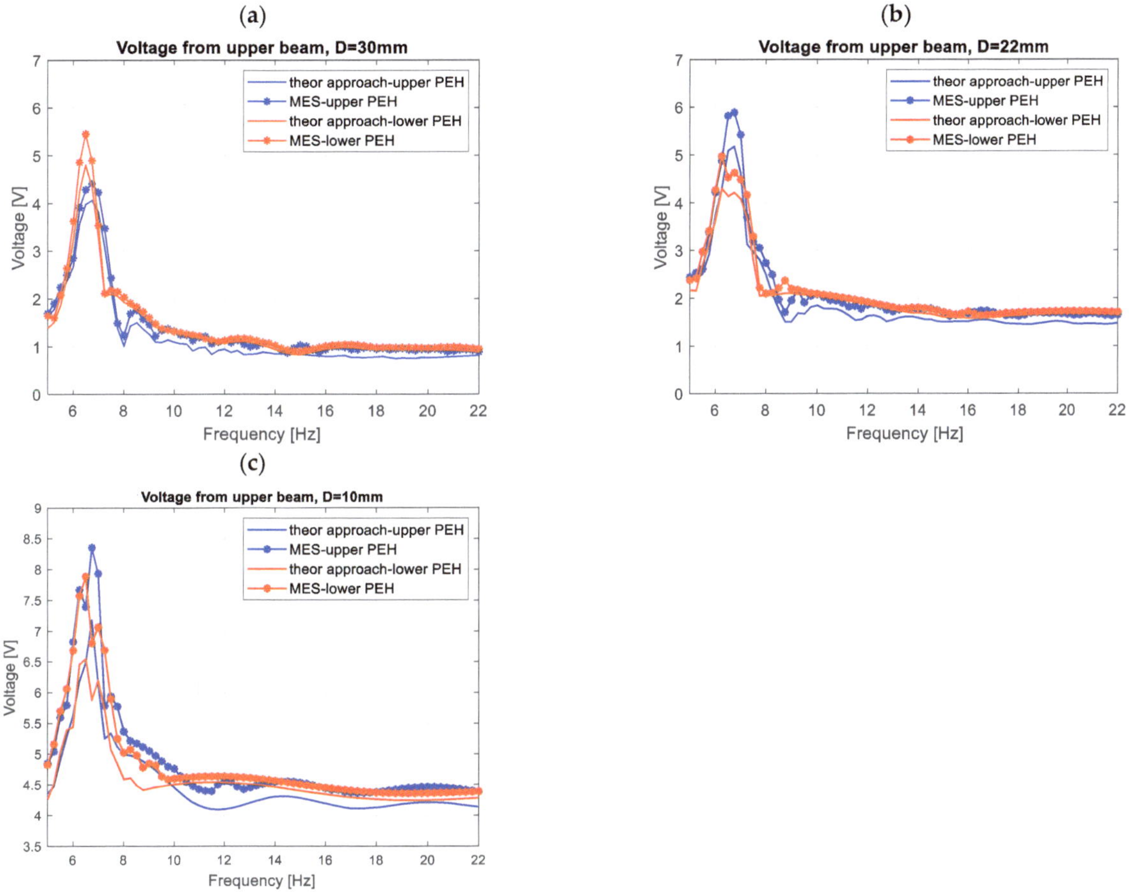

Figure 8. The comparison of theoretical and numerical results of the voltage output generated from piezo-stripe harvesters by different beam spaces (**a**) D = 30 mm, (**b**) D = 22 mm, (**c**) D = 10 mm.

5. Experimental Verification

In this section, the performance of the magnetic coupled piezoelectric energy harvesting system, located on two beams, is tested in a laboratory stand (see Figure 9). In order to assess these performances, both glass fiber beams with chosen beam spaces were equipped with a piezo-patch sensor MFC8514 P2, developed by the Smart Materials company, that was located at a distance of 10 mm from the fixed ends of those beams. Apart from this element, both beams were retrofitted into a ring neodymium magnet MP 14 × 8/4 × 3 that is used to introduce the magnetic effect to the system. Regardless of the beam space, this structure is excited to vibration, in the same way, each time, by a harmonic signal generated from the Signal generator developed by Agilent, and then it is applied to the vibration shaker, TIRA S-51110 M. On the other hand, from the measurement point of view, both voltages from piezo-stripe harvesters were measured and recorded by the data acquisition module, PXI 4499, connected to the vibrating structure, while the measurement of the excitation acceleration, as well as displacements of the free ends of those beams, were realized by the laser displacement sensor, LQ10A65PUQ.

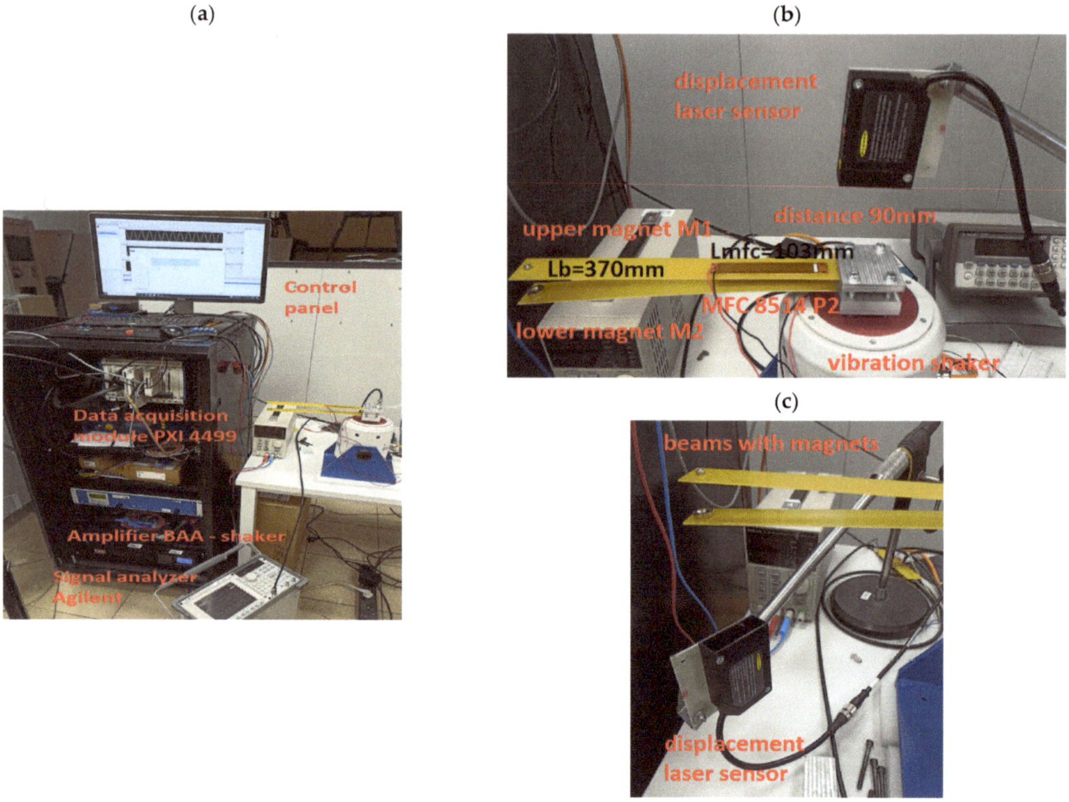

Figure 9. The view of a test rig during a lab test (**a**) the view of whole lab stand, (**b**) the view of lab stand with displacement laser sensor to measure a base vibration, (**c**) the view of lab stand with displacement laser sensor to measure the lower beam vibration.

Taking into account the scope of simulations described in the previous section, the first step of the experimental tests was related to determining the frequency response functions of the coupled magnetic piezo-harvesters for the structure with three different beam spaces. In order to do this, the harmonic signal in the form of a chirp signal u(t) = 2sin(ωt), which corresponds to the excitation acceleration 0.6 m/s^2, was applied to the vibration shaker by a TIRA signal amplifier in the frequency range of 5–55 Hz, while the voltage signal from both piezo-harvesters was recorded by using a measurement card.

Observing the diagrams in Figure 10, it can be noticed that the generated frequency response functions properly verified the simulation amplitude plots. Especially, it is shown in the vicinity of the first natural frequency, where decreasing the beam space leads to generating the highest amplitude peak of voltage. Further analysis of these amplitude plots also confirmed the fact that the magnetic force acts as the strongest on the structure in a different frequency range versus the beam space. As a result, the repulsed effect of the magnetic force for a real structure with the beam space of 10 mm again leads to the increase of the output voltage from the lower harvester and to a simultaneous decrease in the voltage from the upper harvester in the frequency range of 11–13 Hz, while for a real structure with higher beam spaces, this effect was achieved for the frequencies closer to the first natural frequency.

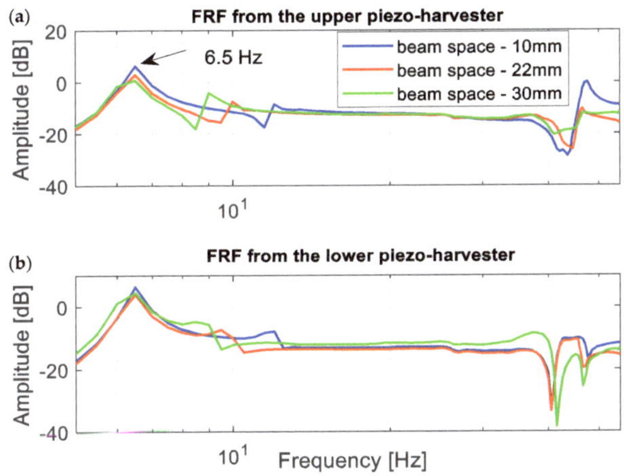

Figure 10. The comparison of the amplitude plot of the coupled magnetic beam structure with different beam spaces by using (**a**) upper harvester, (**b**) lower harvester.

In the next step, the process of impedance matching for the magnetic coupled piezo-harvesters located on the top and bottom beams were carried out in a laboratory stand. For this purpose, a series of measurements of the output voltages from both piezo-harvesters were conducted in the time domain at various beam spaces D and various impedance loads R_L. Again, as it was described in the previous section, the structure was excited to vibration by using a chirp signal with the same amplitude and the same frequency range (5–55 Hz). The recorded voltages from both piezo-composites firstly allowed to calculate the power output from the system according to Equation (14), and next to indicate the best matching impedance load.

$$P = \frac{1}{nR_L{\sigma_a}^2}\sum_{i=1}^{n} {U_i}^2 \tag{14}$$

where:

σ_a—the standard deviation of excitation acceleration calculated for n steps.
R_L—the impedance load.
n—the width of the window.

Observing the diagrams in Figure 11, it can be noticed that experimental tests properly verified the output voltages calculated from the theoretical model. As a result, again the highest efficiency of the system is achieved for the structure with the lowest beam space of 10 mm and the piezo-composite connected to the impedance load of 200 kΩ, while the lowest efficiency for the structure with the beam space of 30 mm. In addition, the lower convergence between the experimental and the numerical results for chosen impedance loads could be the effect of heterogenous adhesion between the sensor and the host structure, as well as the nonlinearity of the piezo-sensor. However, omitting some discrepancies in these diagrams, it can be finally concluded that the optimal impedance for this system is in the range of 200–210 kΩ.

In the next step, the behavior of the coupled magnetic piezo-harvester with the repulsed effect of the magnetic force was assessed for various beam spaces by simultaneously connecting the system to the optimal impedance load. In order to do this, measurements of voltages from both piezo-composites were carried out in the time domain in the narrow frequency range of 5–14 Hz with a sampling time of 10 ms. Then, the obtained maximum amplitudes of these voltages for a chosen frequency excitation allow to draw characteristics presented in Figure 11. Observing these diagrams, it can be noticed again that the beam space in the structure leads to shifting the frequency where the magnetic effect is the most visible. In the case of the structure with a beam space of 30 mm, this effect was achieved between 8.5–9 Hz. Then, the magnetic force introduced to the system leads to reducing the amplitude of the voltage generated by the bottom piezo-composite from 0.9 V to 0.2 V, and, at the same time, increases the amplitude of voltage from the top harvester. A similar effect was achieved for the structure with a beam space of 22 mm. Then, the highest impact of the magnetic force was located in the frequency range of 9.5–10.5 Hz, where the difference between amplitudes of voltage generated from the top and bottom harvesters was close to 0.9 V (see Figure 12). On the other hand, in the last case (10 mm), decreasing the beam space leads to an appearance of the magnetic effect only in the frequency range of 12–12.5 Hz, and generating the highest difference between particular amplitudes of voltage from both piezo-composites (close to 1 V).

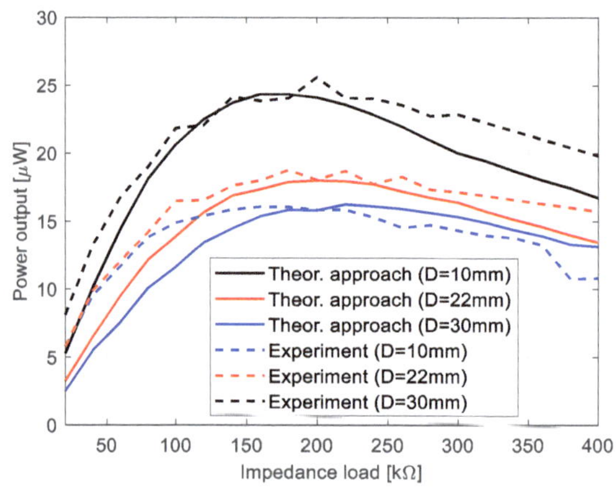

Figure 11. The comparison of a power output from the piezo-harvester located on the top of the structure, generated and calculated from the theoretical model.

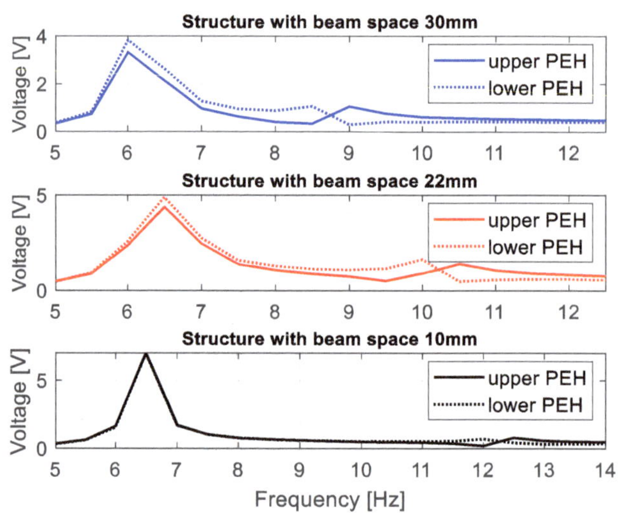

Figure 12. The comparison of the voltage generated from the top and bottom piezo-composites by various beam spaces in the structure.

The obtained output voltages from both piezo-harvesters allow for calculating the efficiency of this system in two steps. In the first one, the root mean square values for these signals and also harmonic signal excitation were calculated, while in the second one, the power of these signals was calculated, respectively. In the effect, the efficiency of the real magnetic coupled structure can be expressed as:

$$\eta = \frac{P_{out}}{P_{in}} \cdot 100\% \tag{15}$$

where:

$P_{out} = RMS(V_{out})^2 = \frac{1}{N}\sum_{i=1}^{N} V_{out}{}^2$ is the power of the output voltage signal from the piezo-harvester,

$P_{in} = RMS(u)^2 = \frac{1}{N}\sum_{i=1}^{N} u^2$ is the power of the excitation harmonic signal.

The obtained results presented in Figure 13 indicate that the beam space between magnets in the magnetic coupled harvester structure plays an important role in enhancing the efficiency of this system. Especially, it is shown for a system that works at resonance, where decreasing the beam space leads to four times the increasing efficiency of this system (from 3% up to 12% for the lower beam) as well as for a system that works with frequencies over a frequency resonance where the decreasing of this parameter leads to a temporary increase of this system efficiency up to 0.5%. In the effect, on the basis of the obtained results, it can be concluded that the behavior of the system with the lowest beam space connecting with vibrating structures can be useful to power more amounts of small electrical sensors with low power demand. Moreover, it allows the support of the SHM system, where the fast detection of some changes in the early stage of the mechanical structure performances is required.

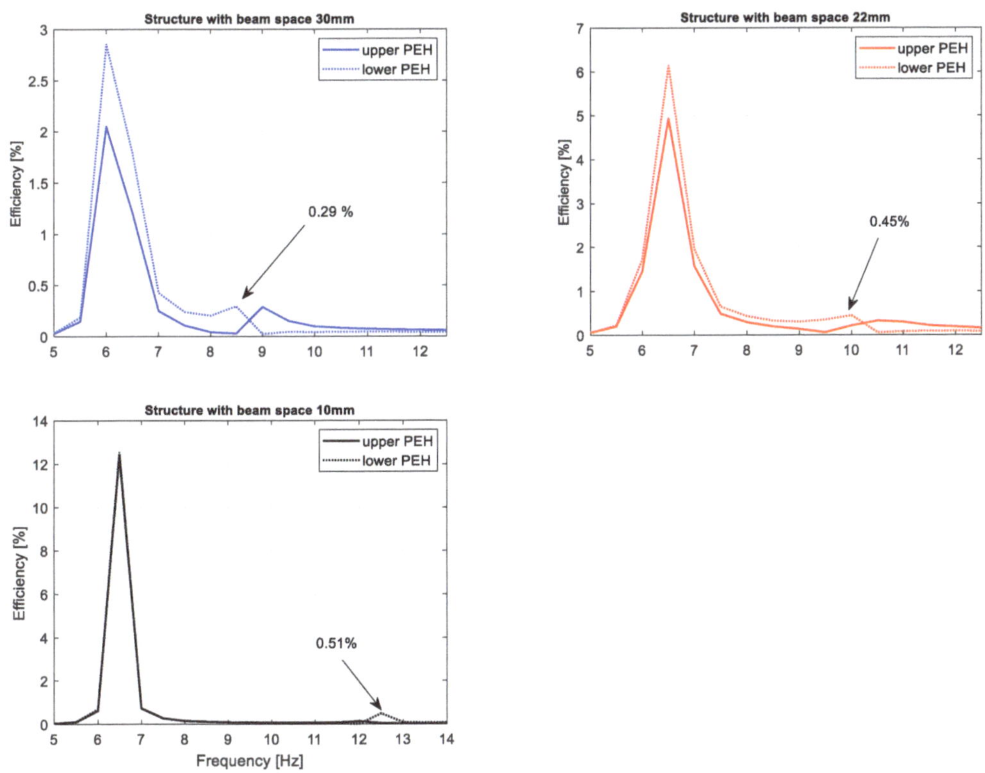

Figure 13. The comparison of the efficiency of real magnetic coupled harvesters by various beam spaces in the structure.

The last step of the experimental test was related to determining the portrait phase of a real structure with considering the repulsion effect of the magnetic force for three different configurations of beam spaces. Taking into account the results presented in Figure 14, the measurements of displacements of both beams in the vertical direction were carried out in three different frequency ranges with an increment of 0.2 Hz. In the case of the structure with the beam space of 10 mm, it was in the frequency range of 11.6–13.2 Hz; for the structure with the beam space of 22 mm, it was in the range of 9–11 Hz; while for the structure with the beam space of 30 mm, it was in the range of 8.4–10 Hz. For each time, both the upper and lower laser displacement sensors were located at a distance of 90 mm from the upper surface of the upper beam and the underside surface of the lower beam, respectively (see Figure 9). In addition, in order to perform precise tests, both measurement points on the lower and upper beams were located at a distance of 20 mm from the free ends of those beams. Observing the diagrams in Figure 14, it can be seen that the magnetic force has the strongest influence on a coupled system with a beam space of 10 mm in a frequency close to 11.8–12 Hz. Then, the trajectories of particular portrait phases of the upper and lower beams are the farthest versus their equilibrium points, as well as they are more chaotic than the other ones. Similar behavior can also be observed for the two other real structures with a beam space of 22 mm and 30 mm. Then, both trajectories of both beams are located, respectively, in the frequency range close to 10.2–10.4 Hz, as well as 9.2 Hz, and are less chaotic in comparison to the previous one. Finally, taking into account all experimental results, it can be concluded that the highest efficiency for the considered structure was achieved for the real structure with the beam space of 10 mm working with the first natural frequency as well as the frequency close to 11.8 Hz.

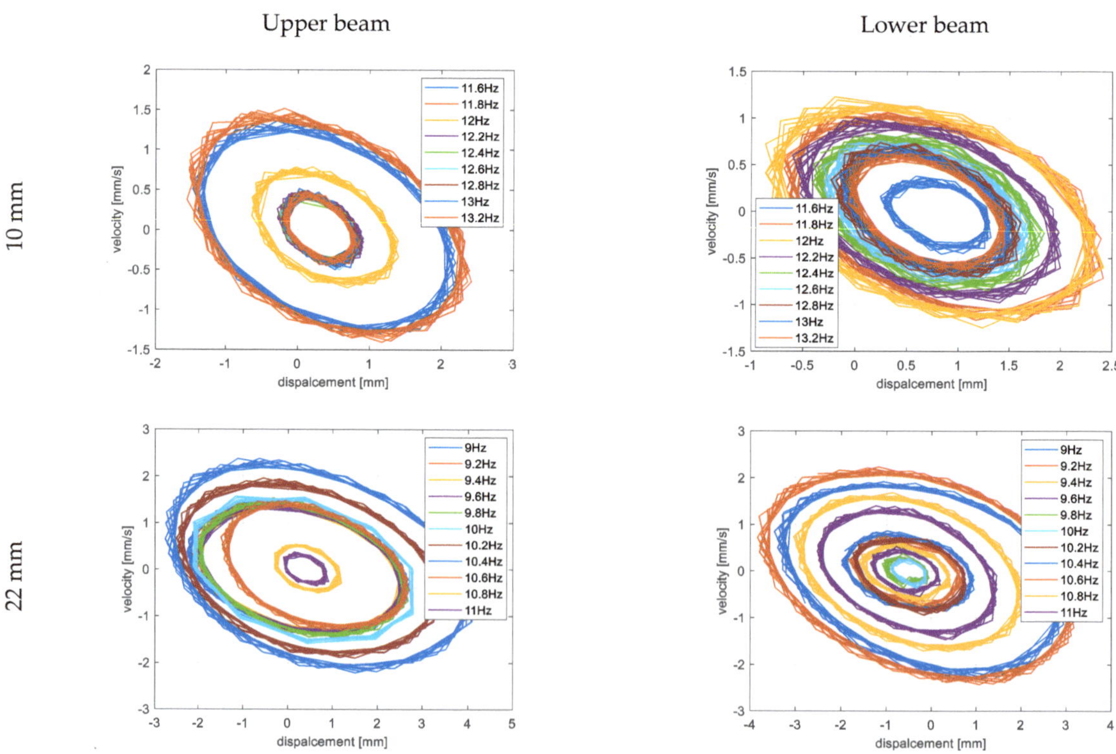

Figure 14. *Cont.*

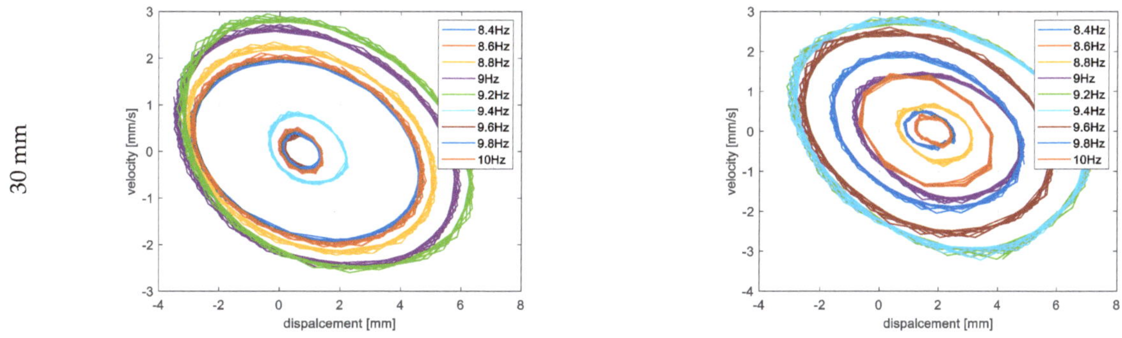

Figure 14. Comparison of portrait phases of the upper and lower beams of the structure in three different beam spaces.

6. Summary and Conclusions

The use of piezoelectric patches has expanded the possibilities of the use of energy harvesting in recent times. Nonlinearity in the piezo-patches, as well as the nonlinearity of magnetic forces introduced to the system, has led to investigations in this paper on structures composed of thin piezo-stripes by creating computational models for them. With the current focus on using traditional and modern sensors to aid digital twinning and model updating, such a focus on the behavior of the sensors becomes even more important. Composite structures are being introduced into a range of sectors, including renewable energy, and so this example is also relevant for future expansion in terms of the sustainability of such solutions.

Taking the magnetic effect introduced to the system into account, as well as a laminate structure of the macro-fiber composite, the mathematical model of the considered structure was first determined. The obtained model firstly allows for checking the behavior of the system for three different spaces. The results presented in Figure 3 indicated that a gradual increase in the beam space affects the value of the optimal impedance load in the range of 180–220 kΩ. Further analysis of voltage generated from both macro-fiber composites by a chosen impedance load indicates that the highest peaks were obtained by the system with the resistive load of 220 kΩ excited to vibration with the first natural frequency. In addition, the analysis of diagrams in Figure 4 leads to the conclusion that the increase in the distance between both magnets in the coupled magnetic harvester systems results in the shift frequency range, where the efficiency of the energy harvesting system is additionally enhanced. As a result, taking into account the magnetic force introduced to this model, it can be concluded that the highest efficiency of the system was achieved for a system with a beam space of 10 mm at the frequency range of 11–13 Hz.

Next, the coupled magnetic harvester system was analyzed for various values of the magnetic remanence of ring magnets in the range of 1–1.3 T. The results presented in Figure 5 show that the increase of this parameter only up to 1.2 T, for only the model of the structure with the beam space of 10 mm, allows to obtain a significant value of the power output. In the case of other models, the calculated power output is too short and, finally, the efficiency of these systems can be also very short. The portrait phase of the considered models, presented in Figure 6, is the confirmation of these conclusions.

Next, the computation model of the coupled magnetic harvester with a homogenized material in the active layer was created in the Ansys software. The voltages calculated from the finite element models of both harvesters showed that these values are close to the theoretical results, especially in the frequency regions over the first natural frequency, where the repulsion effect of the magnetic force is the most active. As a result, it allows concluding that the modeling of the coupled magnetic structure was performed properly.

Experimental investigations carried out in the laboratory on both time and frequency domains allowed us to verify the simulation results. Especially, it is observable in Figure 11, where the experimental results are close to the simulation results. In addition, the analysis of diagrams presented in Figures 13 and 14 again indicates that the beam space affects the efficiency of the real coupled magnetic harvester system (increasing from 3% to 12%) as well as the frequency range where it is enhanced. Finally, it can be concluded that the highest efficiency of a real system is obtained for the structure with a beam space of 10 mm, which is excited to vibration with a harmonic signal in the frequency range of 11–13 Hz.

Author Contributions: Conceptualization, A.K.; Methodology, A.K.; Software, D.O. and M.B.A.; Validation, A.K. and D.O.; Writing—original draft, A.K.; Writing—review & editing, D.O. and M.B.A. All authors have read and agreed to the published version of the manuscript.

Funding: This work is supported by the University Work No. WZ/WM-IIM/2/2022 of the Faculty of Mechanical Engineering, Bialystok University of Technology.

Institutional Review Board Statement: Not applicable.

Informed Consent Statement: Not applicable.

Data Availability Statement: Data for the experiments are available from the authors on request.

Conflicts of Interest: The authors declare no conflict of interest. The funders had no role in the design of the study; in the collection, analyses, or interpretation of data; in the writing of the manuscript, or in the decision to publish the results.

References

1. Ambrożkiewicz, B.; Litak, G.; Wolszczak, P. Modelling of Electromagnetic Energy Harvester with Rotational Pendulum Using Mechanical Vibrations to Scavenge Electrical Energy. *Appl. Sci.* **2020**, *10*, 671. [CrossRef]
2. Müller, F.; Krack, M. Explanation of the self-adaptive dynamics of a harmonically forced beam with a sliding mass. *Arch. Appl. Mech.* **2020**, *90*, 1569–1582. [CrossRef]
3. Qiu, Z.; Li, C.; Zhang, X. Experimental study of active vibration control for a kind of two-link flexible manipulator. *Mech. Syst. Signal Process.* **2019**, *118*, 623–644. [CrossRef]
4. Koszewnik, A.; Lesniewski, K.; Pakrashi, V. Numerical Analysis and Experimental Verification of Damage Identification Metrics for Smart Beam with MFC Elements to Support Structural Health Monitoring. *Sensors* **2021**, *21*, 6796. [CrossRef] [PubMed]
5. O'Leary, K.; Pakrashi, V.; Kelliher, D. Optimization of composite material tower for offshore wind turbine structures. *Renew. Energy* **2019**, *140*, 928–942. [CrossRef]
6. Cahill, P.; Hazra, B.; Karoumi, R.; Mathewson, A.; Pakrashi, V. Vibration energy harvesting based monitoring of an operational bridge undergoing forced vibration and train passage. *Mech. Syst. Signal Process.* **2018**, *106*, 265–283. [CrossRef]
7. Yadav, A.; Goonewardena, M.; Ajib, W.; Dobre, O.; Elbiaze, H. Energy Management for Energy Harvesting Wireless Sensors With Adaptive Retransmission. *IEEE Trans. Commun.* **2017**, *65*, 5487–5498. [CrossRef]
8. Hadas, Z.; Rubes, O.; Ksica, F.; Chalupa, J. Kinetic Electromagnetic Energy Harvester for Railway Applications—Development and Test with Wireless Sensor. *Sensors* **2022**, *22*, 905. [CrossRef]
9. Kanoun, O. *Energy Harvesting for Wireless Sensor Networks*; Walter De Gruyter Inc.: Berlin, Germany, 2018.
10. Koszewnik, A. The influence of a slider gap in the beam–slider structure with an MFC element on energy harvesting from the system: Experimental case. *Acta Mech.* **2021**, *232*, 819–833. [CrossRef]
11. Orrego, S.; Shoele, K.; Ruas, A.; Doran, K.; Caggiano, B.; Mittal, R.; Kang, S. Harvesting ambient wind energy with an inverted piezoelectric flag. *Appl. Energy* **2017**, *194*, 212–222. [CrossRef]
12. Wang, H.; Jasim, A.; Chen, X. Energy harvesting technologies in roadway and bridge for different applications—A comprehensive review. *Appl. Energy* **2018**, *212*, 1083–1094. [CrossRef]
13. Karami, M.A.; Inman, D.J. Powering pacemakers from heartbeat vibrations using linear and nonlinear energy harvesters. *Appl. Phys. Lett.* **2012**, *100*, 042901. [CrossRef]
14. Bassani, G.; Filippeschi, A.; Ruffaldi, E. Nonresonant Kinetic Energy Harvesting Using Macrofiber Composite Patch. *IEEE Sens. J.* **2018**, *18*, 2068–2076. [CrossRef]
15. Koszewnik, A.; Oldziej, D. Performance assessment of an energy harvesting system located on a copter. *Eur. Phys. J. Spéc. Top.* **2019**, *228*, 1677–1692. [CrossRef]
16. Koszewnik, A. Analytical Modeling and Experimental Validation of an Energy Harvesting System for the Smart Plate with an Integrated Piezo-Harvester. *Sensors* **2019**, *19*, 812. [CrossRef]
17. Koszewnik, A.; Grześ, P.; Walendziuk, W. Mechanical and electrical impedance matching in a piezoelectric beam for Energy Harvesting. *Eur. Phys. J. Spéc. Top.* **2015**, *224*, 2719–2731. [CrossRef]
18. Litak, G.; Margielewicz, J.; Gąska, D.; Wolszczak, P.; Zhou, S. Multiple Solutions of the Tristable Energy Harvester. *Energies* **2021**, *14*, 1284. [CrossRef]

19. Kumar, A.; Balpande, S.; Anjankar, S. Electromagnetic Energy Harvester for Low Frequency Vibrations Using MEMS. *Procedia Comput. Sci.* **2016**, *79*, 785–792. [CrossRef]
20. Feng, Y.; Yu, Z.; Han, Y. High-performance gap-closing vibrational energy harvesting using electret-polarized dielectric oscillators. *Appl. Phys. Lett.* **2018**, *112*, 032901. [CrossRef]
21. Ksica, F.; Hadas, Z.; Hlinka, J. Integration and test of piezocomposite sensors for structure health monitoring in aerospace. *Measurement* **2019**, *147*, 106861. [CrossRef]
22. Okosun, F.; Guerin, S.; Celikin, M.; Pakrashi, V. Flexible amino acid-based energy harvesting for structural health monitoring of water pipes. *Cell Rep. Phys. Sci.* **2021**, *2*, 100434. [CrossRef]
23. Koszewnik, A. Assessment parameters of fractional order capacitor of piezo-patch harvester located on a multicopter. *Eur. Phys. J. Spéc. Top.* **2022**, 1–12. [CrossRef]
24. Margielewicz, J.; Gąska, D.; Litak, G.; Wolszczak, P.; Yurchenko, D. Nonlinear dynamics of a new energy harvesting system with quasi-zero stiffness. *Appl. Energy* **2022**, *307*, 118159. [CrossRef]
25. Zhou, S.; Cao, J.; Litak, G. Bifurcation, chaotic and hysteresis phenomena of broadband tristable energy harvesters. *MATEC Web Conf.* **2018**, *241*, 01025. [CrossRef]
26. Dechant, E.; Fedulov, F.; Fetisov, L.Y.; Shamonin, M. Bandwidth Widening of Piezoelectric Cantilever Beam Arrays by Mass-Tip Tuning for Low-Frequency Vibration Energy Harvesting. *Appl. Sci.* **2017**, *7*, 1324. [CrossRef]
27. Usharani, R.; Uma, G.; Umapathy, M.; Choi, S.-B. A Novel Piezoelectric Energy Harvester Using a Multi-Stepped Beam with Rectangular Cavities. *Appl. Sci.* **2018**, *8*, 2091. [CrossRef]
28. Wu, Y.; Badel, A.; Formosa, F.; Liu, W.; Agbossou, A. Nonlinear vibration energy harvesting device integrating mechanical stoppers used as synchronous mechanical switches. *J. Intell. Mater. Syst. Struct.* **2014**, *25*, 1658–1663. [CrossRef]
29. Liu, S.; Cheng, Q.; Zhao, D.; Feng, L. Theoretical modeling and analysis of two-degree-of-freedom piezoelectric energy harvester with stopper. *Sens. Actuators A Phys.* **2016**, *245*, 97–105. [CrossRef]
30. Wang, W.; Cao, J.; Litak, C.R.B.G. Multiple Solutions of asymmetric potential bistable energy harvesters: Numerical simulations and experimental validation. *Eur. Phys. J. B* **2018**, *91*, 254. [CrossRef]
31. Yan, B.; Zhou, S.; Litak, G. Nonlinear analysis of the tristable energy harvester with a resonant circuit for performance enhancement. *Int. J. Bifurc. Chaos* **2018**, *28*, 1850092. [CrossRef]
32. Challa, V.R.; Prasad, M.G.; Shi, Y.; Fisher, F.T. A vibration energy harvesting device with bidirectional resonance frequency tunability. *Smart Mater. Struct.* **2008**, *17*, 015035. [CrossRef]
33. Zhang, G.-C.; Chen, L.-Q.; Li, C.-P.; Ding, H. Primary resonance of coupled cantilevers subjected to magnetic interaction. *Meccanica* **2017**, *52*, 807–823. [CrossRef]
34. Firoozy, P.; Khadem, S.; Pourkiae, S. Broadband energy harvesting using nonlinear vibrations of a magnetopiezoelastic cantilever beam. *Int. J. Eng. Sci.* **2017**, *111*, 113–133. [CrossRef]
35. Abdelmoula, H.; Zimmerman, S.; Abdelkefi, A. Accurate modeling, comparative analysis, and performance enhancement of broadband piezoelectric energy harvesters with single and dual magnetic forces. *Int. J. Non-Linear Mech.* **2017**, *95*, 355–363. [CrossRef]
36. Rui, X.; Li, Y.; Liu, Y.; Zheng, X.; Zeng, Z. Experimental Study and Parameter Optimization of a Magnetic Coupled Piezoelectric Energy Harvester. *Appl. Sci.* **2018**, *8*, 2609. [CrossRef]
37. Shih, H.-A.; Su, W.-J. Theoretical analysis and experimental study of a nonlinear U-shaped bi-directional piezoelectric energy harvester. *Smart Mater. Struct.* **2019**, *28*, 015017. [CrossRef]
38. Junxiang, J.; Shagang, L.; Lifeng, F.; Zhao, D. A review of piezoelectric Vibration Energy Harvesting with magnetic coupling based on different structural characteristics. *Micromachines* **2021**, *12*, 436. [CrossRef]
39. Rabinovitch, O. Geometrically nonlinear behavior of piezoelectric laminated plates. *Smart Mater. Struct.* **2005**, *14*, 785–798. [CrossRef]
40. Neto, M.A.; Leal, R.P.; Yu, W. A triangular finite element with drilling degrees of freedom for static and dynamic analysis of smart laminated structures. *Comput. Struct.* **2012**, *108–109*, 61–74. [CrossRef]
41. Nestrovic, T.; Marcinkovic, D.; Chandrashekar, Z.; Trajkov, M. Implementation of a user defined piezoelectric shell element for analysis of active structures. *Finite Elem. Anal. Des.* **2012**, *52*, 11–22. [CrossRef]
42. Nath, J.K.; Kapuria, S. Coupled efficient layer wise and smeared third order theories for vibration of smart piezo-laminated cylindrical shell. *Compos. Struct.* **2012**, *94*, 1886–1899. [CrossRef]
43. Margielewicz, J.; Gąska, D.; Litak, G.; Wolszczak, P.; Trigona, C. Nonlinear Dynamics of a Star-Shaped Structure and Variable Configuration of Elastic Elements for Energy Harvesting Applications. *Sensors* **2022**, *22*, 2518. [CrossRef] [PubMed]
44. Tang, L.; Yang, Y. A nonlinear piezoelectric energy harvester with magnetic oscillator. *Appl. Phys. Lett.* **2012**, *101*, 094102. [CrossRef]
45. Deraemaeker, A.; Benelechi, S.; Benjeddou, A.; Preumont, A. Analytical and numerical computation of homogenized properties of MFCs: Application to a composite boom with MFC actuators and sensors. In Proceedings of the III ECCOMAS thematic conference on Smart Structures and Materials, Gdansk, Poland, 9–11 July 2007.
46. Deraemaeker, A.; Nasser, H.; Benjeddou, A.; Preumont, A. Mixing Rules for the Piezoelectric Properties of Macro Fiber Composites. *J. Intell. Mater. Syst. Struct.* **2009**, *20*, 1475–1482. [CrossRef]

47. Rubes, O.; Tofel, P.; Macku, R.; Skarvada, P.; Ksica, F.; Hadas, Z. Piezoelectric Micro-fiber Composite Structure for Sensing and Energy Harvesting Applications. In Proceedings of the 18th International Conference on Mechatronics—Mechatronika (ME), Brno, Czech Republic, 5–7 December 2018; pp. 1–6.
48. Hu, K.-M.; Li, H. Large deformation mechanical modeling with bilinear stiffness for Macro-Fiber Composite bimorph based on extending mixing rules. *J. Intell. Mater. Syst. Struct.* **2020**, *32*, 127–139. [CrossRef]
49. Saratoro, M.; Medeiros, R.; Vandepitte, D.; Tita, V. Computational model for supporting SHM system design: Damage identification via numerical analysis. *Mech. Syst. Signal Process.* **2017**, *84*, 445–461. [CrossRef]

sensors

MDPI

Article

Analysis of Double Elastic Steel Wind Driven Magneto-Electric Vibration Energy Harvesting System

Yi-Ren Wang * and Ming-Ching Chu

Department of Aerospace Engineering, Tamkang University, Tamsui, New Taipei City 25137, Taiwan; 609430045@gms.tku.edu.tw
* Correspondence: 090730@mail.tku.edu.tw

Abstract: This research proposes an energy harvesting system that collects the downward airflow from a helicopter or a multi-axis unmanned rotary-wing aircraft and uses this wind force to drive the magnet to rotate, generating repulsive force, which causes the double elastic steel system to slap each other and vibrate periodically in order to generate more electricity than the traditional energy harvesting system. The design concept of the vibration mechanism in this study is to allow the elastic steel carrying the magnet to slap another elastic steel carrying the piezoelectric patch to form a set of double elastic steel vibration energy harvesting (DES VEH) systems. The theoretical DES VEH mechanism of this research is composed of a pair of cantilever beams, with magnets attached to the free end of one beam, and PZT attached to the other beam. This study analyzes the single beam system first. The MOMS method is applied to analyze the frequency response of this nonlinear system theoretically, then combines the piezoelectric patch and the magneto-electric coupling device with this nonlinear elastic beam to analyze the benefits of the system's converted electrical energy. In the theoretical study of the DES VEH system, the slapping force between the two elastic beams was considered as a concentrated load on each of the beams. Furthermore, both SES and DES VEH systems are studied and correlated. Finally, the experimental data and theoretical results are compared to verify the feasibility and correctness of the theory. It is proven that this DES VEH system can not only obtain the electric energy from the traditional SES VEH system but also obtain the extra electric energy of the steel vibration subjected to the slapping force, which generates optimal power to the greatest extent.

Keywords: vibration energy harvesting system; piezoelectric patch; nonlinear vibration; frequency response

Citation: Wang, Y.-R.; Chu, M.-C. Analysis of Double Elastic Steel Wind Driven Magneto-Electric Vibration Energy Harvesting System. *Sensors* **2021**, *21*, 7364. https://doi.org/10.3390/s21217364

Academic Editors: Slim Naifar, Olfa Kanoun and Carlo Trigona

Received: 19 September 2021
Accepted: 1 November 2021
Published: 5 November 2021

1. Introduction

The development of green energy technology has always been an important research topic, and vibration often occurs in life, therefore, the research on vibration energy harvester (VEH) that uses vibration to generate electricity is a very popular field. Roundy et al. [1] proposed the use of micro electromechanical system (MEMS) manufacturing technology to design electro-vibration energy converters. Roundy et al. [2] also found that piezoelectric converters can convert more power than capacitive converters. In addition, piezoelectric converters do not need to provide an additional voltage source, which is one of the reasons why piezoelectric converters are more advantageous than capacitive converters. Roundy and Wright [3] also pointed out that the model of the VEH system based on the principle of piezoelectricity is increasing, and most of them have been verified in practice. Therefore, this technology is not limited to theoretical predictions. In the review work of Iqbal et al. [4], they concluded that due to the large amount of vibration-based energy available in the environment, the VEH has been identified as the energy harvester with the highest potential. Recently, Jiang et al. [5] also reviewed the piezoelectric vibration energy harvesting technologies with magnetic coupling and determine the potential benefits of the magnetic force on energy-harvesting techniques. From this, we can see the rise of mechanical vibration into green energy in the field of energy.

In the aerospace field, Li et al. [6] reviewed the design of using fluid-induced vibration as an energy harvesting device and suggested that nonlinear coupling phenomena should be considered in the design of energy harvesters because these phenomena are the core key to causing vibration. It can be seen from the foregoing that the importance of nonlinear vibration in the VEH system. Nayfeh and Mook [7] proposed that when the natural frequencies of the same degree-of-freedom are multiples, the high-frequency modes are excited but the low-frequency modes produce higher amplitudes. This is a characteristic of typical nonlinearity, which is called internal resonance. They also mentioned a variety of analytical methods for analyzing nonlinear systems, and the method of multiple scales (MOMS) is more conducive to the analysis of non-conservative systems than other methods. Therefore, the MOMS method is very useful for the theoretical analysis of our system.

In the VEH study, Erturk and Inman [8] analyzed the frequency response, the voltage, current, and power output generated by the vibration of the fixed-free Euler–Bernoulli beam with piezoelectric patch. Rajora et al. [9] used Hooke's Law to calculate the electrical energy conversion performance of piezoelectric patches and learned that piezoelectric patches have different electrical energy conversion benefits in different material thicknesses. Masana and Daqaq [10] applied axial force to a fixed-fixed Euler–Bernoulli beam to generate transversal vibration. In their work, the method of multiple scales (MOMS) was used for analysis. In addition, they also used fixed points plot to analyze the transverse vibration of the beam and proved that the axial disturbance can control the frequency and amplitude of the nonlinear elastic beam in the transverse direction.

Bistable energy harvester (BEH) is another research focus of the non-linear VEH system. Most BEH systems use a cantilever beam as the main mechanism, attach a magnet to its free end, and place a magnet that repels it opposite the free end and use the repulsive force of the magnet to excite the beam. The deformation of the piezoelectric patch is increased, and thus a greater power generation benefit is produced. Andó et al. [11] proposed a 2D BEH, which makes one cantilever beam deflects in the Z direction, and the other cantilever beam deflects in the Y direction. This device can be excited in two different directions at the same time, in order to achieve better power generation. Harne and Wang [12] reviewed the relevant theories about the BEH system and pointed out that the design of the BEH system in various studies largely relies on magnetic attraction, repulsion, and mechanical loading to generate the bistable vibration. Gao et al. [13] connected a spring to a magnet and placed it on an elastic structure and found that the device connected to the spring and the magnet has better electrical energy conversion than traditional BEH. Zhu and Zu [14] designed a magnetic induction buckled beam piezoelectric energy harvester. The beam is fixed at one end, and the other end is movable and subjected to magnetic induction pressure. Their results show that magnetic induction can be used to generate buckled phenomena and can effectively trigger inter-well motion. Wang and Zhu [15] proposed an energy harvesting system with an oscillation horizontal cantilever beam facing the direction of airflow. The effect of airflow velocities on the system's output power was investigated. Their experimental results show that if the magnet interaction is properly integrated, the power output can be increased by 30%. Wang et al. [16] proposed a rotating magnetic excitation VEH system. They showed that the magnetic force can excite the elastic steel to vibrate nonlinearly to generate electrical energy. These magneto-PZT concepts proved the feasibility of the energy harvesting system. However, only a simplified theoretical model was studied, and the main issue of the slap between steels is not considered, which may enhance the output power.

In addition, in recent years, an energy harvesting system based on the slapping force has been proposed. Fu et al. [17] used three parallel cantilever beams to collide with each other to generate vibration and generated power through the "collision". They proved that the power generation from the slap between steels is possible. Wang et al. [18] designed two parallel cantilever beams to form a double elastic steel VEH (DES VEH) system. In order to achieve the perfect performance of the slap, the placement of the PZT patch is very important. They obtained the peak and node of the elastic steel of this DES VEH system and placed the PZT at the node to obtain the maximum power generation benefit of this DES VEH system. Wang et al. [19] used the nonlinear model of transverse vibration fixed-fixed beam to simulate the DES VEH system. The position of the PZT and the position of the external force were located to produce the maximum electric power output. The lateral DES VEH model proposed in their research proved to be feasible.

In this research, a double elastic steel vibration energy harvester system is presented. This system uses the wind-driven magnetic field to excite the elastic steel to vibrate and generate electricity. This system collects the downward airflow from a helicopter or a multi-axis unmanned rotary-wing aircraft and uses this wind force to drive the magnet to rotate, generating repulsive force, which makes the DES VEH system vibrate periodically. The design concept of the vibration mechanism in this study is to allow the elastic steel carrying the magnet to slap another elastic steel carrying the piezoelectric patch to form a set of DES VEH systems. The results of this study show that this device can not only obtain the electric energy from the traditional energy harvesting system but also obtain the electric energy of the steel vibration subjected to the slapping force. Instead of just considering the deformation of the PZT on a traditional single elastic beam vibration energy harvester, this study implements another characteristic of the PZT electric power generation by subjecting to the clapping force. The present work applies the deformation and mechanical stress on the PZT simultaneously, which perfectly makes the power generation to the greatest extent. The concept of the wind-driven magneto-electric vibration energy harvesting system is shown in Figure 1.

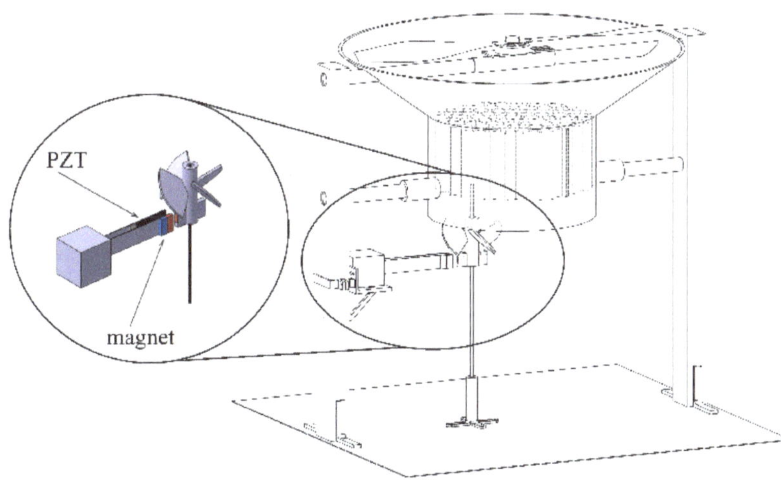

Figure 1. Concept of the wind driven magneto electric vibration energy harvesting system.

The theoretical DES VEH mechanism of this research is composed of a pair of cantilever beams, with magnets attached to the free end of one beam, and PZT attached to the other beam as shown in Figure 2. The present work first analyzes the single-beam system. The detailed study of the nonlinear single-beam vibration equation of motion, frequency response, and its power generation benefit are presented in Sections 2–4. We obtained the equation of motion of the nonlinear elastic beam by Newton's 2nd Law, Taylor series expansion, and Euler's angle transformation. Secondly, we used the MOMS method to analyze the frequency response of this nonlinear system. Then combined the piezoelectric patch and the magneto-electric coupling device with this nonlinear elastic beam to analyze the benefits of the coupling system's vibration energy converted into electrical energy. The slapping force of the DES VEH system can be considered as a concentrated load on the SES VEH system. A detailed study of the DES VEH system and its power generation benefit are introduced in Section 5. Furthermore, both single elastic steel and double elastic steel VEH systems were studied and correlated. In Section 6, the experiments are performed. The experimental data and theoretical results are compared to verify the feasibility and correctness of the theory and are presented in Section 7. This research also explores the differences between the SES VEH and DES VEH systems in the experiment and analyzes the power generation benefits of the overall nonlinear system.

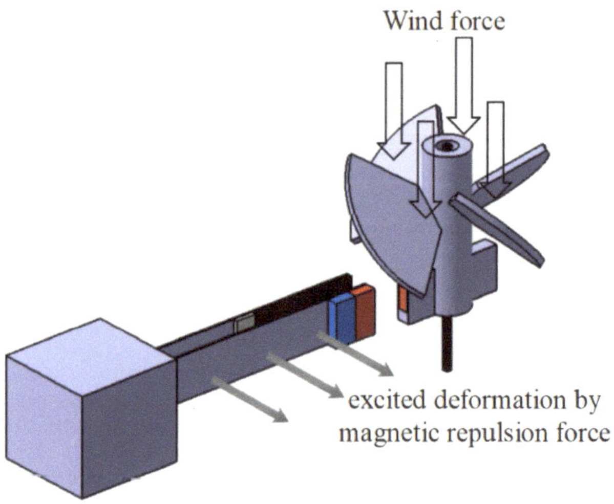

Figure 2. Schematic of the DES VEH.

2. Theoretical Model

2.1. Beam Equations of Motion

The initial status of the beam was considered straight, and each cross-section was a plane that follows stress-strain laws. Using Newton's 2nd Law, Euler's angle transformation, and Taylor series expansion, this study derives the basic equations of motion for the nonlinear beam, but excludes any rotations in the beam; i.e., limiting it to planar motions. The schematic diagram and coordinates of each symbol are shown in Figure 3. According to the nonlinear 2D Euler–Bernoulli beam theory [20], the equations of motion for the 2D beam are as follows:

$$\overline{m}\ddot{\overline{u}} - EA\overline{u}'' = EA\left(\frac{1}{2}\overline{W}'^2 - \overline{u}'\overline{W}'^2\right)' + EI_A\left(\overline{W}'\left(\overline{W}''' - \overline{u}'''\overline{W}' - 2\overline{u}''\overline{W}'' - 3\overline{u}'\overline{W}'''\right)\right)' \quad (1)$$

$$\begin{aligned}&\overline{m}\ddot{\overline{W}} + EI_A\overline{W}^{iv} - EA\left(\overline{u}'\overline{W}' - \overline{u}'^2\overline{W}' + \tfrac{1}{2}\overline{W}'^3\right)' + \overline{q} - EI_A\left(\overline{u}'\overline{W}''' + (\overline{u}'\overline{W}')''\right.\\&\left.-\left(\overline{u}'^2 - \overline{W}'^2\right)\overline{W}''' - \overline{u}'(\overline{u}'\overline{W}')'' - \left(\overline{u}'^2\overline{W}' - \tfrac{1}{3}\overline{W}'^3\right)''\right)' = 0\end{aligned} \quad (2)$$

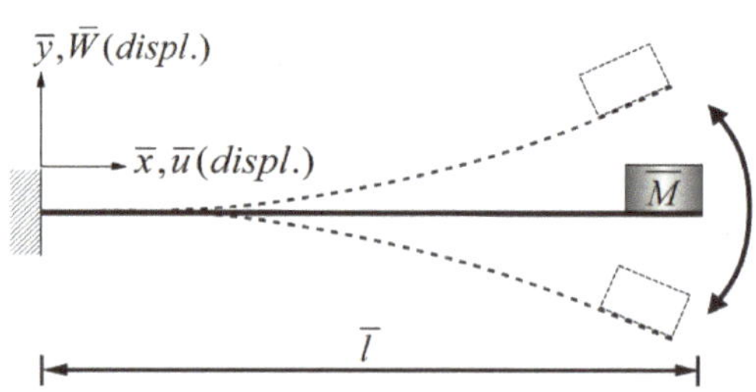

Figure 3. Coordinate of the fixed-free beam with tip mass.

In Equations (1) and (2), $\overline{m}$ is the mass of the beam, E is Young's modulus, A represents the cross-section area of the beam, I_A is the moment of inertia of the beam, $\overline{q}$ denotes the uniform distributed force on the beam, $()'$ is $(d/d\overline{x})$, $\dot{()}$ represents $(d/d\overline{t})$, and the $\overline{u}$ and $\overline{W}$ represent the displacements on $\overline{x}$ and $\overline{y}$ directions, respectively. $\overline{M}$ is the tip mass (of the magnet). It is noted that the magnet mass is treated as the boundary condition of the beam-free end and is not shown in the equation of motion of the beam. The boundary conditions are:

$$\begin{cases} \overline{W}(0,\overline{t}) = 0 \\ \overline{W}'(0,\overline{t}) = 0 \\ EI_A\overline{W}''(\overline{l},\overline{t}) = I_m\overline{\omega}^2\overline{W}'(\overline{l},\overline{t}) \\ EI_A\overline{W}'''(\overline{l},\overline{t}) = -\overline{M}\overline{\omega}^2\overline{W}(\overline{l},\overline{t}) \end{cases} \tag{3}$$

where $\overline{\omega}$ denotes $\sqrt{EI_A/\overline{m}\overline{l}^4}$.

The axial force on the free end in the x direction is zero, therefore, this equation of motion can be simplified to a simple $\overline{y}$ direction equation, which is expressed as follows:

$$\begin{aligned} &\overline{m}\ddot{\overline{W}} + \overline{\mu}\dot{\overline{W}} + EI_A(\overline{W}''' + \overline{W}'\overline{W}''^2 + \overline{W}'''\overline{W}'^2)' \\ &= -\tfrac{1}{2}\overline{m}(\overline{W}'\textstyle\int_{\overline{l}}^{\overline{x}}(\int_0^{\overline{x}}\ddot{\overline{W}'^2}\,d\overline{x})d\overline{x})' + \overline{q} \end{aligned} \tag{4}$$

It is noted that we add a damping term in Equation (4), where $\overline{\mu}$ denotes the damping coefficient of the beam. The dimensionless equation of Equation (4) can be expressed as follows:

$$W^{**} + \mu W^{*} + W^{iv} + W''^3 + W^{iv}W'^2 + 4W'W'''W'' = -\frac{1}{2}(W'\int_1^x(\int_0^x W'^2dx)^{**}dx)' + q \tag{5}$$

where $l = \frac{\overline{l}}{\overline{l}} = 1$, $\tau = \overline{\omega}\overline{t}$, $W = \frac{\overline{W}}{\overline{l}}$, $\tau = \sqrt{\frac{EI}{\overline{m}\overline{l}^4}}\overline{t}$, $\mu = \frac{\overline{l}^2}{\sqrt{\overline{m}EI_A}}\overline{\mu}$, $q = \frac{\overline{q}}{\overline{m}\overline{l}\overline{\omega}^2}$, $\overline{\omega} = \sqrt{\frac{EI_A}{\overline{m}\overline{l}^4}}$, $()'$ represents d/dx, $()^*$ denotes $d/d\tau$.

The dimensionless boundary conditions are expressed as follows:

$$\begin{cases} W(0,\tau) = 0\,, \\ W'(0,\tau) = 0, \\ W''(l,\tau) = 0, \\ W'''(l,\tau) = -MW(l,\tau) \end{cases} \tag{6}$$

where $M = \frac{\overline{M}}{\overline{l}\overline{m}}$.

2.2. Theoretical Model of Piezoelectric Patch

This study uses the piezoelectric equation of the piezoelectric patch (PZT patch) of Rajora et al. [9] as follows:

$$C_p\dot{V} + \frac{1}{\overline{R}_p}V + \int_{\overline{a}}^{\overline{b}} eh_pt_h\dot{\overline{W}}''\,d\overline{x} = 0 \tag{7}$$

where $\overline{R}_P$ is resistance, e is the piezoelectric coupling coefficient, h_p denotes the thickness of the piezo patch, t_h represents the width of the piezo patch, C_p is the capacitance of the piezoelectric patch. The Coulomb force of this PZT patch acting on the nonlinear elastic beam is:

$$\text{Coulomb force} = \int_{\overline{a}}^{\overline{b}} eh_pt_h\dot{\overline{W}}''\,d\overline{x}\frac{V}{h_p} = (et_h\int_{\overline{a}}^{\overline{b}}\overline{W}''\,d\overline{x})V = C_f(\int_{\overline{a}}^{\overline{b}}\overline{W}''\,d\overline{x})V \tag{8}$$

where $\bar{a}$ and $\bar{b}$ are the positions of the piezo patch from one end to the other end, C_f is the Piezoelectric coupling coefficient, The dimensionless PZT equation can be written as:

$$\nu^* + R_p\nu + \hat{k}\int_a^b (W'')^* dx = 0 \quad (9)$$

where $v = \frac{V}{C_f}$, $R_P = \frac{1}{R_p C_p \bar{\omega}}$, $\hat{k} = \frac{e h_p t_h}{C_p C_f}$, $()^* = d/d\tau$, $()' = d/dx$. The dimensionless voltage (v) can be found from Equation (9):

$$\nu = -\frac{\hat{k}}{e^{R_P\tau}}\int_0^\tau (\int_a^b (W'')^* dx) e^{R_P\tau} d\tau \quad (10)$$

From Equation (10), we can obtain the dimensionless Coulomb force of the PZT acting on the nonlinear beam. Let $\eta^2 = \frac{C_f^2}{\bar{l}m\bar{\omega}^2}$, we can obtain:

$$\frac{C_f^2(\int_a^b W'' dx)\nu}{\bar{l}m\bar{\omega}^2} = \eta^2(\int_a^b W'' dx)\nu = -\frac{\hat{k}\eta^2}{e^{R_P\tau}}\int_a^b W'' dx \int_0^\tau (\int_a^b (W'')^* dx) e^{R_P\tau} d\tau \quad (11)$$

We combine the PZT force (Equation (11)) with Equation (5) to obtain the dimensionless beam equation with PZT as:

$$\begin{aligned} &W^{**} + W^{iv} + \mu W^* + W''^3 + W^{iv}W'^2 + 4W'W'''W'' \\ &-\tfrac{\hat{k}\eta^2}{e^{R_P\tau}}\int_a^b W'' dx \int_0^\tau (\int_a^b (W'')^* dx) e^{R_P\tau} d\tau = -\tfrac{1}{2}(W'\int_1^x (\int_0^x W'^2 dx)^{**} dx)' + q \end{aligned} \quad (12)$$

2.3. Theoretical Model of Magneto-Electric Equation

The external force exerted by the magnet's magneto-electric effect on the nonlinear elastic beam can be expressed by Lorentz Force. The magnet's magneto-electric equation is:

$$\overline{C}_M\dot{\bar{I}} + \overline{R}_M\bar{I} + \overline{C}_G\dot{\overline{W}} = 0 \quad (13)$$

where $\overline{C}_G$ is the electromagnetic coupling coefficient, $\overline{C}_M$ is inductance, $\bar{I}$ is current, $\overline{R}_M$ is resistance. The magneto-electric Lorentz force acting on the nonlinear elastic beam is:

$$\text{Lorentz force } = \overline{C}_G\bar{I} = B\bar{L}\sin\theta \quad (14)$$

where B is the magnetic field and $\bar{L}$ is the length of the beam. The dimensionless equation is:

$$i^* + R_M i + W^* = 0 \quad (15)$$

The dimensionless current is:

$$i = -\frac{1}{e^{R_M\tau}}\int_0^\tau W^* e^{R_M\tau} d\tau \quad (16)$$

where $R_M = \frac{\overline{R}_M}{\overline{C}_M\bar{\omega}}$. Substituting Equation (16) into Equation (12), the equation of motion integrating magnetoelectricity and piezoelectricity can be obtained as follows:

$$\begin{aligned} &W^{**} + W^{iv} + \mu W^* + W''^3 + W^{iv}W'^2 + 4W'W'''W'' \\ &-\tfrac{\hat{k}\eta^2}{e^{R_P\tau}}\int_a^b W'' dx \int_0^\tau (\int_a^b (W'')^* dx) e^{R_P\tau} d\tau + C_G\tfrac{1}{e^{R_M\tau}}\int_0^\tau W^* e^{R_M\tau} d\tau \\ &= -\tfrac{1}{2}(W'\int_1^x (\int_0^x W'^2 dx)^{**} dx)' + q \end{aligned} \quad (17)$$

where $C_G = \frac{\overline{C}_G^2}{\overline{C}_M\bar{m}\bar{\omega}^2}$.

2.4. Method of Multiple Scales, MOMS

We use the method of multiple scales (MOMS) of Nayfeh and Mook [7] to analyze this nonlinear problem. The MOMS method divides time into two time scales, namely, the fast and the slow time scales. Let the $T_0 = \tau$ be the fast-time term and $T_1 = \varepsilon^2\tau$ be the slow-time term and $W(x,\tau,\varepsilon) = \varepsilon^1 W_0(T_0,T_1,T_2,\ldots.) + \varepsilon^3 W_1(T_0,T_1,T_2,\ldots.)$. Equation (17) can be expressed as:

$$\begin{aligned}
&\left(\varepsilon\tfrac{\partial^2 W_0}{\partial {T_0}^2} + \varepsilon^3\,\tfrac{\partial^2 W_1}{\partial {T_0}^2} + 2\varepsilon^3\,\tfrac{\partial^2 W_0}{\partial T_0 \partial T_1} + \cdots\right) + \left(\varepsilon {W_0}^{iv} + \varepsilon^3 {W_1}^{iv}\right)\\
&+\varepsilon^2\mu\left(\varepsilon\tfrac{\partial W_0}{\partial T_0} + \varepsilon^3\tfrac{\partial W_1}{\partial T_0} + \varepsilon^3\tfrac{\partial W_0}{\partial T_1} + \cdots\right) + \left(\varepsilon^3 {W''_0}^3 + 2\varepsilon^5 {W''_0}^2 W''_1 + \cdots\right)\\
&+\left(\varepsilon^3 W_0^{\mathrm{iv}} {W'_0}^2 + 2\varepsilon^5 W_0^{\mathrm{iv}} W'_0 W'_1 + \varepsilon^5 W_1^{iv} {W'}_0^2 + \cdots\right)\\
&+4\left(\varepsilon^3 W'_0\, W'''_0\, W''_0 + \varepsilon^5 W'_0\, W'''_0\, W''_1 + \varepsilon^5 W'_0\, W'''_1\, W''_0 + \varepsilon^5 W'_1\, W'''_0\, W''_0 + \cdots\right)\\
&-\tfrac{\hat{k}\eta^2}{e^{R\tau}}\int_a^b (\varepsilon W''_0 + \varepsilon^3 W''_1)dx \int_0^\tau \left(\int_a^b \left(\tfrac{\partial^3(\varepsilon W''_0 + \varepsilon^3 W''_1)}{\partial T_0 \partial x^2}\right)dx e^{R\tau}\right)d\tau\\
&+C_G\tfrac{1}{e^{R\tau}}\int_0^\tau \left(\varepsilon\tfrac{\partial W_0}{\partial T_0} + \varepsilon^3\tfrac{\partial W_1}{\partial T_0} + \varepsilon^3\tfrac{\partial W_0}{\partial T_1}\right)e^{R\tau}d\tau\\
&= -\tfrac{1}{2}\varepsilon^3\left(W'_0\int_1^x \left(\int_0^x {W'_0}^2 dx\right)^{**} dx\right)' + \varepsilon^3 q
\end{aligned} \tag{18}$$

where ε is the time scale of small disturbances and is a minimum value, the effect of high order terms $\varepsilon^5, \varepsilon^7 \ldots$ on the system is neglected. The term composed of the order of ε^1 in Equation (18) is:

$$\frac{\partial^2 W_0}{\partial {T_0}^2} + {W_0}^{iv} = 0 \tag{19}$$

The term composed of the order of ε^3 is:

$$\begin{aligned}
&\tfrac{\partial^2 W_1}{\partial {T_0}^2} + {W_1}^{iv} = -2\,\tfrac{\partial^2 W_0}{\partial T_0 \partial T_1} - \mu\tfrac{\partial W_0}{\partial T_0} - {W''_0}^3 - W_0^{\mathrm{iv}} {W'_0}^2 - 4W'_0\, W'''_0\, W''_0\\
&-\tfrac{1}{2}\left(W'_0\int_1^x \left(\int_0^x {W'_0}^2 dx\right)^{**} dx\right)' - C_G\tfrac{1}{e^{R\tau}}\int_0^\tau \left(\tfrac{\partial W_0}{\partial T_0}\right)e^{R\tau}d\tau\\
&+\tfrac{\hat{k}\eta^2}{e^{R\tau}}\int_a^b W''_0 dx \int_0^\tau \left(\int_a^b \left(\tfrac{\partial^3 W_0}{\partial T_0 \partial x^2}\right)dx e^{R\tau}\right)d\tau + q
\end{aligned} \tag{20}$$

3. Frequency Response

3.1. Eigen Value and Frequency

The W_0 is defined as $W_0 = X(x)Y(t)$ by using the separation of variables. Equation (19) can be written as:

$$X(x)\ddot{Y}(\tau) + X^{iv}(x)Y(\tau) = 0 \tag{21}$$

With the boundary conditions, the characteristic equation can be found as:

$$\alpha^3(-1 - \cos\alpha l \cosh\alpha l) + M(-\sin\alpha l \cosh\alpha l + \cos\alpha l \sinh\alpha l) = 0 \tag{22}$$

The mode shape of the nth harmonic is:

$$\phi_n(x) = (-\sin\alpha_n x - \sinh\alpha_n x) + \frac{(\sin\alpha_n + \sinh\alpha_n)}{(-\cos\alpha_n - \cosh\alpha_n)}(-\cos\alpha_n x - \cosh\alpha_n x) \tag{23}$$

The general solution of W_0 can be expressed as:

$$W_0(x,\tau) = \sum_{n=1}^{\infty} \xi_{0n}(\tau)\phi_n(x) \tag{24}$$

In which $\xi_{0n}(\tau)$ is expressed as:

$$\xi_{0n}(\tau) = B_n(T_1)e^{-i\zeta_n}e^{i\omega_n T_0} + \overline{B}_n(\mathrm{T}_1)e^{i\zeta_n}e^{-i\omega_n T_0} \tag{25}$$

where ζ_n is phase angle, B_n represents the amplitude of the nth mode.

3.2. System Frequency Response

Equations (19) and (20) are multiplied by ϕ_m using the mode shapes' orthogonal properties, integrated by 0–ℓ (dimensionless $\ell = 1$). The integral value is zero when $\phi_m \neq \phi_n$ and is not zero when $\phi_m = \phi_n$. The equations, based on the order of ε^1, can be expressed as follows:

$$\zeta_{0m}^{**} + \alpha_m^4 \zeta_{0m} = 0 \tag{26}$$

Order of ε^3:

$$\begin{aligned}
&\zeta_{1m}^{**} + \alpha_m^4 \zeta_{1m} = -2\tfrac{\partial}{\partial T_1}\zeta_{0m}^{*} - \mu\zeta_{0m}^{*} \\
&-\tfrac{1}{Q_m}\zeta_{0i}\zeta_{0j}\zeta_{0k}(\int_0^1 \phi''_i\phi''_j\phi''_k\phi_m dx + \int_0^1 \alpha_m^4\phi_m\phi'_j\phi'_k\phi_m dx + 4\int_0^1 \phi'_i\phi''_j\phi'''_k\phi_m dx) \\
&-\tfrac{1}{2Q_m}\zeta_{0i}(\zeta_{0j}^{**}\zeta_{0k} + 2\zeta_{0j}^{*}\zeta_{0k}^{*} + \zeta_{0j}\zeta_{0k}^{**})(\int_0^1 \phi_m(\phi''_i\int_1^x\int_0^x \phi'_j\phi'_k dxdx + \phi'_i(\int_1^x\int_0^x \phi'_j\phi'_k dxdx)')dx) \\
&+\tfrac{1}{Q_m}\tfrac{k\hat{\eta}^2}{e^{R\tau}}\int_0^1 \phi_m(\int_a^b \phi''_i\zeta_{0i}dx\int_0^\tau (\int_a^b \phi''_j\zeta_{0j}^{*}dx)e^{R\tau}d\tau)dx - C_G\tfrac{1}{e^{R\tau}}\int_0^\tau \zeta_{om}^{*}e^{R\tau}d\tau + q_m
\end{aligned} \tag{27}$$

where $Q_m = \int_0^l \phi_m^2 dx$. In order to consider the frequency response of the system, we assume that the system is subjected to a simple harmonic external force as:

$$q_m = \widehat{q}_m e^{i\Omega T_0} = \widehat{q}_m e^{i(\omega_m + \varepsilon^2\sigma)T_0} = \widehat{q}_m e^{i\omega_m T_0}e^{i\sigma T_1} \tag{28}$$

where $\Omega = \omega_m + \varepsilon\sigma$, σ denotes the tuned frequency near the linear natural frequency (ω_m). In order to find the response of the 1st mode (m = 1), the secular terms need to select harmonic with the ω_1 term. Substituting Equations (25) and (28) into Equation (27), and exciting the 1st mode (m = 1), the secular terms of Equation (27) are:

$$\begin{aligned}
&-2(i\omega_1 B_1 e^{i\zeta_1}e^{i\omega_1 T_0} - \omega_1\varsigma_1^{*}B_1 e^{-i\zeta_1}e^{i\omega_1 T_0}) - \mu(i\omega_1 B_1 e^{-i\varsigma_1}e^{i\omega_1 T_0}) + \widehat{q}_1 e^{i\omega_1 T_0}e^{i\sigma T_1} \\
&-\tfrac{1}{Q_1}(3B_1\overline{B_1}B_1 e^{-i\varsigma_1}e^{i\omega_1 T_0}(\int_0^1 \phi_1\phi''^3_1 dx + \int_0^1 \phi_1\phi_1^{iv}\phi'^2_1 dx + \int_0^1 4\phi_1\phi'_1\phi''_1\phi'''_1 dx) \\
&+2B_1\overline{B_2}B_2 e^{-i\varsigma_1}e^{i\omega_1 T_0}(\int_0^1 3\phi''_1\phi''^2_2\phi_1 dx + \int_0^1 \phi_1\phi_1^{iv}\phi'^2_2 dx + \int_0^1 2\phi_1\phi_2^{iv}\phi'_1\phi'_2 dx \\
&+\int_0^1 4\phi'_1\phi''_2\phi'''_2\phi_1 dx + \int_0^1 4\phi'_2\phi''_2\phi'''_1\phi_1 dx + \int_0^1 4\phi'_2\phi''_1\phi'''_2\phi_1 dx) \\
&-2B_1\overline{B_3}B_3 e^{-i\varsigma_1}e^{i\omega_1 T_0}(\int_0^1 3\phi''_1\phi''^2_3\phi_1 dx + \int_0^1 \phi_1^{iv}\phi_1\phi'^2_3 dx + \int_0^1 2\phi_1\phi_3^{iv}\phi'_1\phi'_3 dx \\
&+\int_0^1 4\phi'_1\phi''_3\phi'''_3\phi_1 dx + \int_0^1 4\phi'_3\phi''_3\phi'''_1\phi_1 dx + \int_0^1 4\phi'_3\phi''_1\phi'''_3\phi_1 dx)) \\
&-\tfrac{1}{2Q_1}(-4\omega_1^2 B_1\overline{B_1}B_1 e^{-i\varsigma_1}e^{i\omega_1 T_0}(\int_0^1 \phi''_1\phi_1(\int_1^x\int_0^x \phi'_1\phi'_1 dxdx)dx \\
&+\int_0^1 \phi'_1\phi_1(\int_1^x\int_0^x \phi'_1\phi'_1 dxdx)'dx) \\
&+4(-\omega_1^2 - \omega_2^2)B_1\overline{B_2}B_2 e^{-i\varsigma_1}e^{i\omega_1 T_0}(\int_0^1 \phi''_2\phi_1(\int_1^x\int_0^x \phi'_1\phi'_2 dxdx)dx \\
&+\int_0^1 \phi'_2\phi_1(\int_1^x\int_0^x \phi'_1\phi'_2 dxdx)'dx) \\
&+4(-\omega_1^2 - \omega_3^2)B_1\overline{B_3}B_3 e^{-i\varsigma_1}e^{i\omega_1 T_0}(\int_0^1 \phi''_3\phi_1(\int_1^x\int_0^x \phi'_1\phi'_3 dxdx)dx \\
&+\int_0^1 \phi'_3\phi_1(\int_1^x\int_0^x \phi'_1\phi'_3 dxdx)'dx)) - C_G\tfrac{1}{e^{R\tau}}\int_0^\tau (i\omega_1)(B_1 e^{-i\zeta_1}e^{i\omega_1 T_0})e^{R\tau}d\tau \\
&-C_G\tfrac{1}{e^{R\tau}}\int_0^\tau (i\omega_1)(B_1 e^{-i\zeta_1}e^{i\omega_1 T_0})e^{R\tau}d\tau
\end{aligned} \tag{29}$$

where $Q_1 = \int_0^l \phi_1^2 dx$. Similarly, for the 2nd mode, the secular terms need to select harmonic with the ω_2 term. For the 3rd mode, the secular terms need to select harmonic with the ω_3 term, and these secular terms will no longer be listed for the sake of simplicity. After the secular terms are selected, they are determined to be zero; therefore, a solvability condition can be obtained.

Taking the external disturbance of the first mode as an example, the solvability condition of the 1st mode is multiplied by $e^{i\zeta_1}$, and let $\Gamma = \sigma T_1 + \varsigma_1$. If the frequency response is a steady state, $\Gamma^* = \sigma + \varsigma_1^* = 0$, we can know $\varsigma_1^* = -\sigma$. The real part of the 1st mode's solvability condition can be expressed as:

$$
\begin{aligned}
&2\omega_1\sigma B_1 - \widehat{q}_1\cos\Gamma - \tfrac{1}{Q_1}\Big(3B_1\overline{B_1}B_1\Big(\textstyle\int_0^1 \phi_1\phi''^3_1 dx + \int_0^1 \phi_1\phi_1^{iv}\phi'_1{}^2 dx + \int_0^1 4\phi_1\phi'_1\phi''_1\phi'''_1 dx\Big)\\
&+2B_1\overline{B_2}B_2\Big(\textstyle\int_0^1 3\phi''_1\phi''^2_2\phi_1 dx + \int_0^1 \phi_1\phi_1^{iv}\phi'_2{}^2 dx + \int_0^1 2\phi_1\phi_2^{iv}\phi'_1\phi'_2 dx + \int_0^1 4\phi'_1\phi''_2\phi'''_2\phi_1 dx\\
&+\textstyle\int_0^1 4\phi'_2\phi''_2\phi'''_1\phi_1 dx + \int_0^1 4\phi'_2\phi''_1\phi'''_2\phi_1 dx\Big)\\
&+2B_1\overline{B_3}B_3\Big(\textstyle\int_0^1 3\phi''_1\phi''^2_3\phi_1 dx + \int_0^1 \phi_1^{iv}\phi_1\phi'_3{}^2 dx + \int_0^1 2\phi_1\phi_3^{iv}\phi'_1\phi'_3 dx + \int_0^1 4\phi'_1\phi''_3\phi'''_3\phi_1 dx\\
&+\textstyle\int_0^1 4\phi'_3\phi''_3\phi'''_1\phi_1 dx + \int_0^1 4\phi'_3\phi''_1\phi'''_3\phi_1 dx))\\
&-\tfrac{1}{2Q_1}\Big(-4\omega_1^2 B_1\overline{B_1}B_1\Big(\textstyle\int_0^1 \phi''_1\phi_1(\int_1^x\int_0^x \phi'_1\phi'_1 dxdx)dx + \int_0^1 \phi'_1\phi_1(\int_1^x\int_0^x \phi'_1\phi'_1 dxdx)'dx)\\
&+4(-\omega_1^2-\omega_2^2)B_1\overline{B_2}B_2\Big(\textstyle\int_0^1 \phi''_2\phi_1(\int_1^x\int_0^x \phi'_1\phi'_2 dxdx)dx)+\int_0^1 \phi'_2\phi_1(\int_1^x\int_0^x \phi'_1\phi'_2 dxdx)'dx)\\
&+4(-\omega_1^2-\omega_3^2)B_1\overline{B_3}B_3(\textstyle\int_0^1 \phi''_3\phi_1(\int_1^x\int_0^x \phi'_1\phi'_3 dxdx)dx\\
&+\textstyle\int_0^1 \phi'_3\phi_1(\int_1^x\int_0^x \phi'_1\phi'_3 dxdx)'dx))
\end{aligned}
\tag{30}
$$

The imaginary part is:

$$
-\mu\omega_1 B_1 - \widehat{q}_1\sin\Gamma - C_G\frac{1}{e^{R\tau}}\int_0^\tau \omega_1 B_1 e^{R\tau}d\tau \tag{31}
$$

In the same way, multiply the 2nd and 3rd modes' secular terms by $e^{i\varsigma_2}$ and $e^{i\varsigma_3}$, respectively to obtain their solvability conditions. In order to eliminate the time correlated term (Γ), the squares of Equations (30) and (31) are added up. These solvability conditions can be solved numerically by using IMSL© subroutine—NEQNF nonlinear solver and the Levenberg-Marquardt algorithm. The response plot of amplitudes B_1, B_2, and B_3 vs. the tuned frequency (σ) of this system can be drawn as fixed points plots. Figures 4–6 are the frequency responses for the 1st, 2nd, and 3rd modes, respectively. The maximum amplitude happens near the linear natural frequency (σ = 0) for each mode. Those values are used for the perturbation method in the next section.

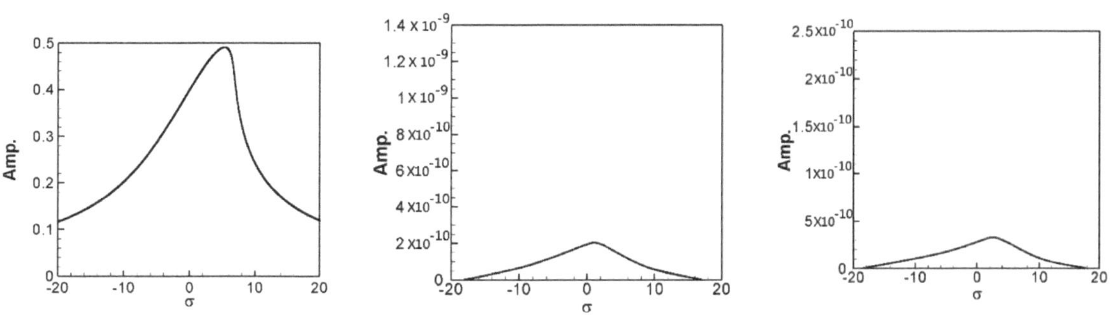

(a) (b) (c)

Figure 4. SES system fixed points plots of the 1st mode being excited. (**a**) 1st mode response, (**b**) 2nd mode response, and (**c**) 3rd mode response.

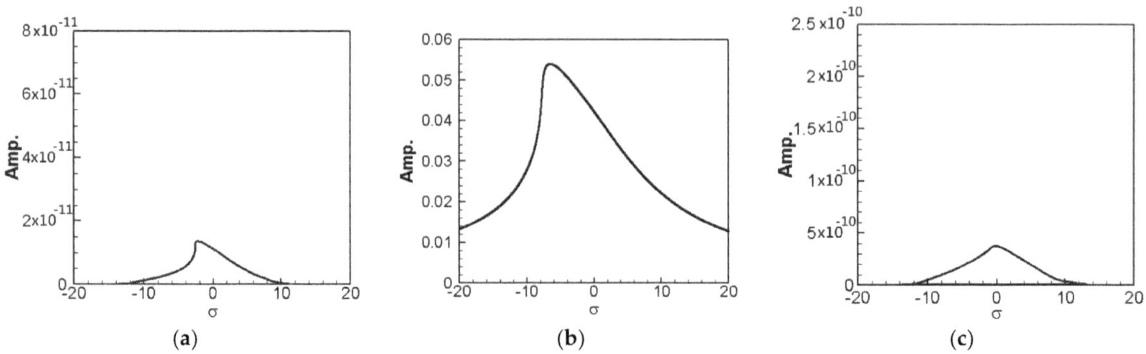

Figure 5. SES system fixed points plots of the 2nd mode being excited. (**a**) 1st mode response, (**b**) 2nd mode response, and (**c**) 3rd mode response.

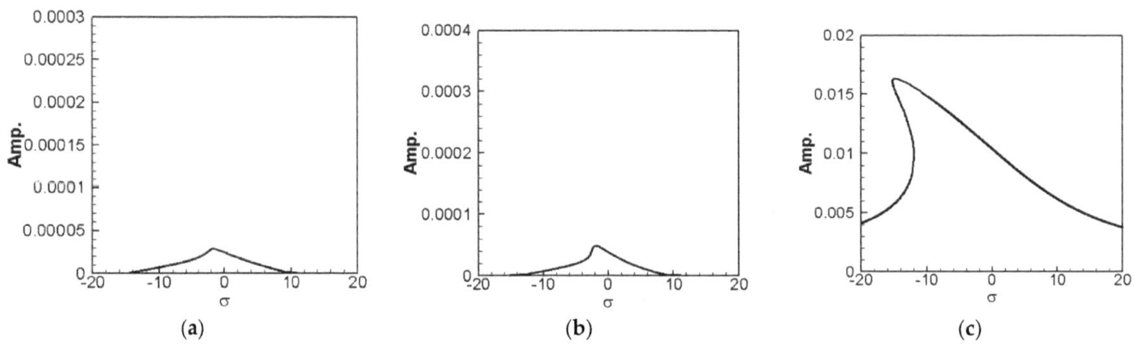

Figure 6. SES system fixed points plots of the 3rd mode being excited. (**a**) 1st mode response, (**b**) 2nd mode response, and (**c**) 3rd mode response.

4. SES VEH Power Generation Benefit Analysis

This study uses the perturbation technique to analyze the power benefit of the nonlinear SES VEH system. The ξ_n of this beam can be expressed as: $\xi_n = \overline{\xi}_n + \widetilde{\xi}_n$, where $\overline{\xi}_n$ is the equilibrium term and is obtained from Section 3.2 (the steady-state response B_n), $\widetilde{\xi}_n$ is the disturbance term. It is noted that the equilibrium term $\overline{\xi}_n$ is chosen as the maximum amplitude form the fixed points plot from Figures 4–6 for each mode to show the best effect of the theoretical model. $W = \sum_{n=1}^{3} (\overline{\xi}_n + \widetilde{\xi}_n)\phi_n$ is then substituted into Equation (17) and orthogonalized as:

$$
\begin{aligned}
&\overline{\xi}_m^{**} + \widetilde{\xi}_m^{**} + \overline{\xi}_m \phi_m{}^{iv} + \widetilde{\xi}_m \phi_m{}^{iv} + \mu \overline{\xi}_m^{*} + \mu \widetilde{\xi}_m^{*} \\
&+\tfrac{1}{Q_m}\big(\textstyle\int_0^1 \phi_m (\overline{\xi}_m + \widetilde{\xi}_m)^3 \phi''{}_m{}^3 dx + \int_0^1 \phi_m (\overline{\xi}_m + \widetilde{\xi}_m) \phi_m{}^{iv} \phi_m \phi'{}_m{}^2 dx \\
&+4\textstyle\int_0^1 \phi_m (\overline{\xi}_m + \widetilde{\xi}_m)^3 \phi_m' \phi_m'' \phi_m''' dx\big) + \tfrac{1}{2Q_m}\big((\overline{\xi}_m + \widetilde{\xi}_m)((\overline{\xi}_m + \widetilde{\xi}_m)**(\overline{\xi}_m + \widetilde{\xi}_m) \\
&+2(\overline{\xi}_m + \widetilde{\xi}_m)*(\overline{\xi}_m + \widetilde{\xi}_m)* \\
&+(\overline{\xi}_m + \widetilde{\xi}_m)(\overline{\xi}_m + \widetilde{\xi}_m)**)\big(\textstyle\int_0^1 \phi_m \phi_m'' \int_1^x \int_0^x \phi'{}_m{}^2 dxdx + \int_0^1 \phi_m \phi_m' (\int_1^x \int_0^x \phi'{}_m{}^2 dxdx)'\big) \\
&-\tfrac{1}{Q_m}\big(\tfrac{\hat{k}\eta^2}{e^{R\tau}}(\overline{\xi}_m + \widetilde{\xi}_m)\textstyle\int_0^1 \phi_m (\int_0^1 \phi''{}_m dx)^2 (\int_0^\tau (\overline{\xi}_m + \widetilde{\xi}_m)^* e^{R\tau} d\tau) dx\big) \\
&\qquad + C_G \frac{1}{e^{R\tau}} \int_0^\tau (\overline{\xi}_m + \widetilde{\xi}_m)^* e^{R\tau} d\tau - \frac{1}{Q_m} \int_0^l \phi_m q_m dx = 0
\end{aligned}
\tag{32}
$$

where m = 1–3. The solution of Equation (32) is carried out using the RK-4 method, and the response of voltage v and time t of the system is made as a time response of a nonlinear SES

VEH system. Figure 7 gives an example of the mutual verification diagrams of fixed-point plots, phase plots, and time response plots of the 1st mode's forced response, which prove the correctness of the theoretical model. We learned the amplitude of each mode through the fixed points plots in Section 3.2, and substituted it into the piezoelectric equation, and obtained the voltage and time of the 1st, 2nd, and 3rd mode by the RK-4 method (Figure 8). The results show that as the vibration frequency increases, the power generation volts and benefits also increase.

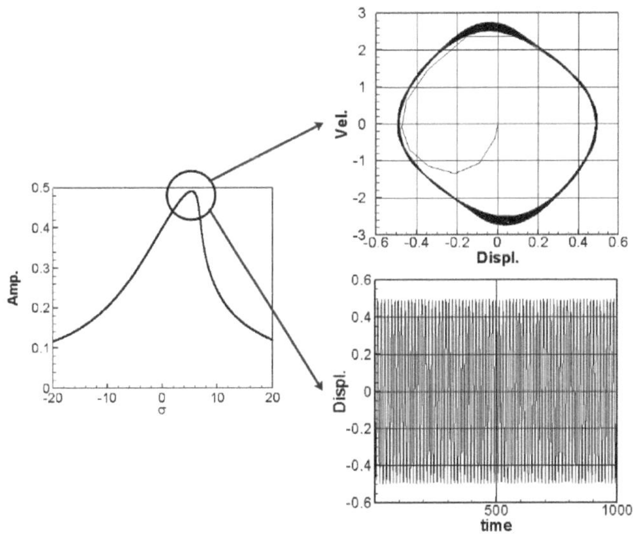

Figure 7. SES verification diagrams of fixed-point plots, phase plots, and time response plots of the 1st mode's forced response.

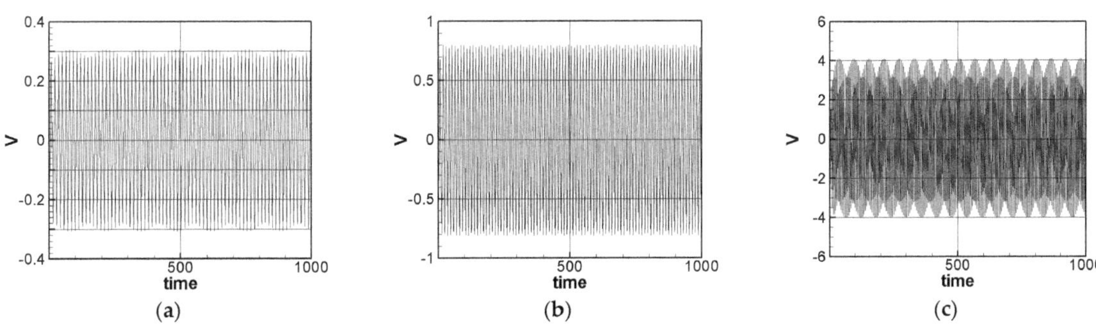

Figure 8. SES system theoretical voltage output (dimensionless). (**a**) 1st mode, (**b**) 2nd mode, and (**c**) 3rd mode.

5. DES VEH Power Generation Benefit Analysis

5.1. DES Frequency Response

This study assumes that the piezo patch is placed at the free end of the beam. The flapping force on the piezo patch from the other elastic steel sheet can be expressed in the dimensionless form as $F_{TP} = W^{**}$, as shown in Figure 9.

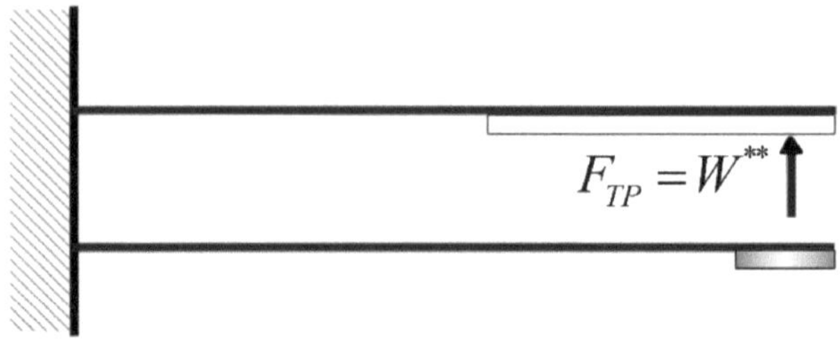

Figure 9. Schematic of the double elastic beam.

Since $\xi = A_P e^{i\omega t}$, $F_{TP} = -A_P\omega^2 \sin\omega t \phi(x)|_{x=1}$. The concentrated load can be considered the external force of SES, in which A_P is the amplitude of the other elastic steel on the tip of the beam (x = 1), which can be obtained by MOMS from Section 3.2. Finally, the equation of motion for the nonlinear DES system using the flapping external force of the piezo patch is expressed as follows:

$$\begin{aligned}
&\frac{\partial^2 W_1}{\partial T_0{}^2} + W_1{}^{iv} = -2\frac{\partial^2 W_0}{\partial T_0 \partial T_1} - \mu\frac{\partial W_0}{\partial T_0} - W''_0{}^3 - W_0^{iv} W'_0{}^2 - 4W'_0 W'''_0 W''_0 \\
&-\tfrac{1}{2}\left(W'_0 \int_1^x \left(\int_0^x W'_0{}^2 dx\right)^{**} dx\right)' - C_G \frac{1}{e^{R\tau}} \int_0^\tau \left(\frac{\partial W_0}{\partial T_0}\right) e^{R\tau} d\tau \\
&+\frac{\hat{k}\eta^2}{e^{R\tau}} \int_a^b W''_0 dx \int_0^\tau \left(\int_a^b \left(\frac{\partial^3 W_0}{\partial T_0 \partial x^2}\right) dx e^{R\tau}\right) d\tau + q + W^{**}
\end{aligned} \tag{33}$$

By using orthogonal properties, the 1st mode's secular terms can be found and the solvability condition is written as:

$$\begin{aligned}
&-2(i\omega_1 B_1 e^{i\zeta_1} e^{i\omega_1 T_0} - \omega_1 \varsigma_1^* B_1 e^{-i\zeta_1} e^{i\omega_1 T_0}) - \mu(i\omega_1 B_1 e^{-i\varsigma_1} e^{i\omega_1 T_0}) + \widehat{q}_1 e^{i\omega_1 T_0} e^{i\sigma T_1} \\
&-\tfrac{1}{Q_1}(3B_1\overline{B_1}B_1 e^{-i\varsigma_1} e^{i\omega_1 T_0}(\int_0^1 \phi_1 \phi''^3_1 dx + \int_0^1 \phi_1 \phi_1^{iv} \phi'_1{}^2 dx + \int_0^1 4\phi_1 \phi'_1 \phi''_1 \phi'''_1 dx) \\
&+2B_1\overline{B_2}B_2 e^{-i\varsigma_1} e^{i\omega_1 T_0}(\int_0^1 3\phi''_1 \phi''^2_2 \phi_1 dx + \int_0^1 \phi_1 \phi_1^{iv} \phi'_2{}^2 dx + \int_0^1 2\phi_1 \phi_2^{iv} \phi'_1 \phi'_2 dx \\
&+\int_0^1 4\phi'_1 \phi''_2 \phi'''_2 \phi_1 dx + \int_0^1 4\phi'_2 \phi''_2 \phi'''_1 \phi_1 dx + \int_0^1 4\phi'_2 \phi''_1 \phi'''_2 \phi_1 dx) \\
&+2B_1\overline{B_3}B_3 e^{-i\varsigma_1} e^{i\omega_1 T_0}\Big(\int_0^1 3\phi''_1 \phi''^2_3 \phi_1 dx + \int_0^1 \phi_1^{iv} \phi_1 \phi'_3{}^2 dx + \int_0^1 2\phi_1 \phi_3^{iv} \phi'_1 \phi'_3 dx \\
&+\int_0^1 4\phi'_1 \phi''_3 \phi'''_3 \phi_1 dx + \int_0^1 4\phi'_3 \phi''_3 \phi'''_1 \phi_1 dx + \int_0^1 4\phi'_3 \phi''_1 \phi'''_3 \phi_1 dx\Big)) \\
&-\tfrac{1}{2Q_1}\Big(-4\omega_1^2 B_1\overline{B_1}B_1 e^{-i\varsigma_1} e^{i\omega_1 T_0}\Big(\int_0^1 \phi''_1 \phi_1 (\int_1^x \int_0^x \phi'_1 \phi'_1 dxdx) dx \\
&+\int_0^1 \phi'_1 \phi_1 (\int_1^x \int_0^x \phi'_1 \phi'_1 dxdx\ )' dx\Big) \\
&+4(-\omega_1^2 - \omega_2^2) B_1\overline{B_2}B_2 e^{-i\varsigma_1} e^{i\omega_1 T_0}\Big(\int_0^1 \phi''_2 \phi_1 (\int_1^x \int_0^x \phi'_1 \phi'_2 dxdx) dx \\
&+\int_0^1 \phi'_2 \phi_1 (\int_1^x \int_0^x \phi'_1 \phi'_2 dxdx\ )' dx\Big) \\
&+4(-\omega_1^2 - \omega_3^2) B_1\overline{B_3}B_3 e^{-i\varsigma_1} e^{i\omega_1 T_0}\Big(\int_0^1 \phi''_3 \phi_1 (\int_1^x \int_0^x \phi'_1 \phi'_3 dxdx) dx \\
&+\int_0^1 \phi'_3 \phi_1 (\int_1^x \int_0^x \phi'_1 \phi'_3 dxdx\ )' dx)) - C_G \frac{1}{e^{R\tau}} \int_0^\tau (i\omega_1)(B_1 e^{-i\zeta_1} e^{i\omega_1 T_0}) e^{R\tau} d\tau \\
&+\tfrac{1}{Q_1}(-\omega_1^2 B_1 e^{-i\zeta_1} e^{i\omega_1 T_0}) \int_0^{l_p} \phi_1 \phi_1(x)|_{x=1} dx = 0
\end{aligned} \tag{34}$$

Similarly, the 2nd and 3rd modes' solvability conditions can be found. The fixed points plot and the frequency response for each mode can be made. The fixed-point plots are not shown for the sake of simplicity.

5.2. DES VEH Power Generation Benefit Analysis

Substituting the slapping force (from Section 5.1) into Equation (32), we can obtain the nonlinear DES VEH equation as:

$$\begin{aligned}
&\overline{\xi}_m^{**} + \widetilde{\xi}_m^{**} + \overline{\xi}_m {\phi_m}^{iv} + \widetilde{\xi}_m {\phi_m}^{iv} + \mu \overline{\xi}_m^{*} + \mu \widetilde{\xi}_m^{*} \\
&+ \tfrac{1}{Q_m}\Big(\textstyle\int_0^1 \phi_m (\overline{\xi}_m + \widetilde{\xi}_m)^3 {\phi''_m}^3 dx + \int_0^1 \phi_m (\overline{\xi}_m + \widetilde{\xi}_m) {\phi_m}^{iv} \phi_m {\phi'_m}^2 dx \\
&+ 4\textstyle\int_0^1 \phi_m (\overline{\xi}_m + \widetilde{\xi}_m)^3 \phi_m' \phi_m'' \phi_m''' dx\Big) + \tfrac{1}{2Q_m}\Big((\overline{\xi}_m + \widetilde{\xi}_m)\big((\overline{\xi}_m + \widetilde{\xi}_m)^{**}(\overline{\xi}_m + \widetilde{\xi}_m) \\
&+ 2(\overline{\xi}_m + \widetilde{\xi}_m)^{*}(\overline{\xi}_m + \widetilde{\xi}_m)^{*} + (\overline{\xi}_m + \widetilde{\xi}_m)(\overline{\xi}_m + \widetilde{\xi}_m)^{**}\big)\Big)\Big(\textstyle\int_0^1 \phi_m \phi_m'' \int_1^x \int_0^x {\phi'_m}^2 dx dx \\
&+ \textstyle\int_0^1 \phi_m \phi_m' \big(\int_1^x \int_0^x {\phi'_m}^2 dx dx\big)'\Big) - \tfrac{1}{Q_m}\Big(\tfrac{\hat{k}\eta^2}{e^{R\tau}}(\overline{\xi}_m + \widetilde{\xi}_m)\textstyle\int_0^1 \phi_m \big(\int_0^1 {\phi''}_m dx\big)^2 \\
&\big(\textstyle\int_0^\tau (\overline{\xi}_m + \widetilde{\xi}_m)^{*} e^{R\tau} d\tau\big) dx\Big) + C_G \tfrac{1}{e^{R\tau}} \textstyle\int_0^\tau (\overline{\xi}_m + \widetilde{\xi}_m)^{*} e^{R\tau} d\tau \\
&- \tfrac{1}{Q_m}\textstyle\int_0^1 \phi_m q_m dx + \tfrac{1}{Q_m}(\overline{\xi}_m + \widetilde{\xi}_m)^{**}\int_0^{l_p} \phi_m \phi_m(x)\big|_{x=l_p} dx = 0
\end{aligned} \tag{35}$$

The solution of Equation (35) is carried out using the RK-4 method, and the response of voltage v and time t of the system is made as a time response of a nonlinear DES VEH system. Figures 10–12 are the mutual verification diagrams of the fixed-point plots, phase plots, and time response plots of the 1st, 2nd, and 3rd modes' forced response, which prove the correctness of the theoretical model. In the same way, the perturbation equation of the 1st, 2nd, and 3rd modes can be obtained by substituting the amplitude from the fixed points plots (in Section 3.2) into the piezoelectric equation. The voltage output can be obtained by using the RK-4 method and the results are shown in Figure 13. The results show that as the vibration frequency increases, the power generation volts and benefits also increase.

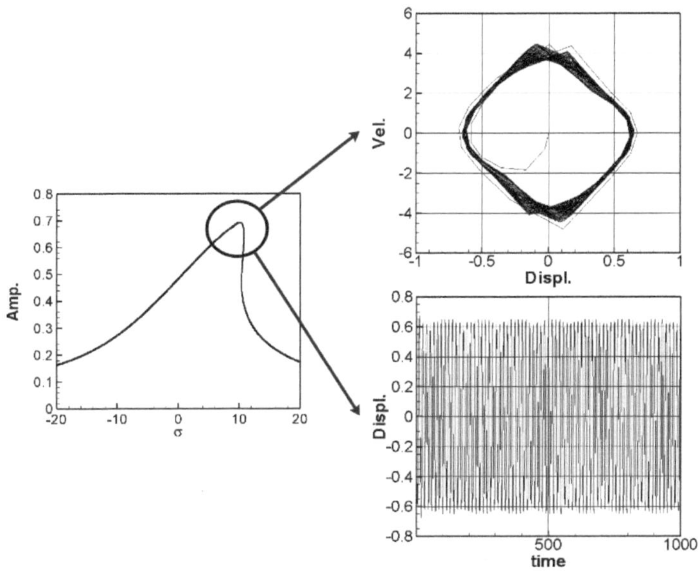

Figure 10. DES verification diagrams of fixed-point plots, phase plots, and time response plots of the 1st mode's forced response.

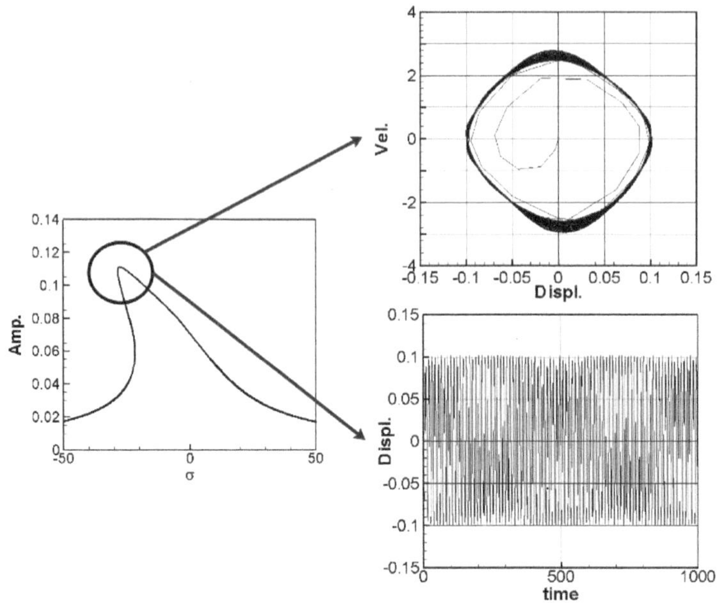

Figure 11. DES verification diagrams of fixed-point plots, phase plots, and time response plots of the 2nd mode's forced response.

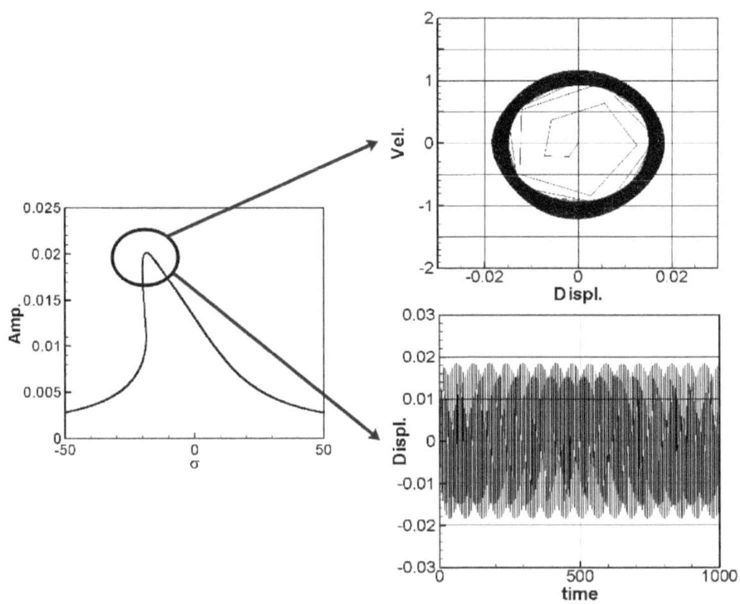

Figure 12. DES verification diagrams of fixed-point plots, phase plots, and time response plots of the 3rd mode's forced response.

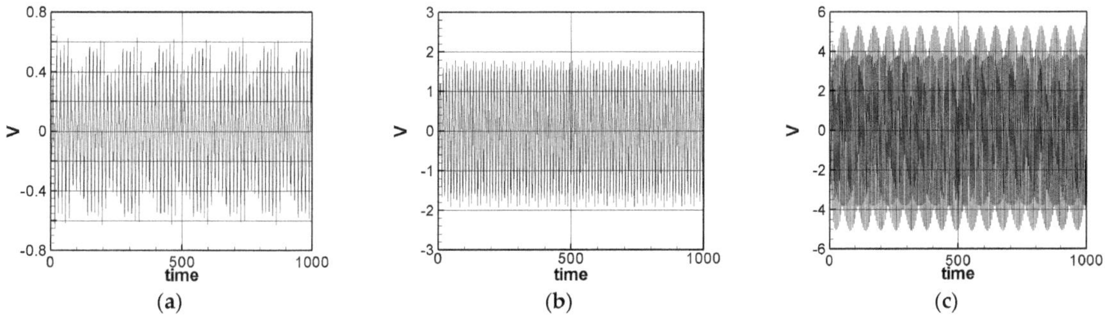

Figure 13. DES VEH system theoretical voltage output (dimensionless). (**a**) 1st mode, (**b**) 2nd mode, and (**c**) 3rd mode.

6. Experiment Setup

We construct a simple experiment with the schematic diagram of the 3D theoretical model shown in Figure 1. This mechanism is composed of three parts, namely a brushless motor with a two-bladed rotor, a wind collecting cover, and the energy harvesting system. The design principle of the energy harvesting system in this study is to collect the airflow under the rotor and turn the windmill to drive the magnets of the system; then, the magnetic field is periodically transformed to form a repulsive force that excites the elastic steel, and the two elastic steels slap and vibrate each other, and generate electrical energy through the PZT. This study only discusses the situation of windmill blades at a pitch angle of 45°. The complete experimental setup is shown in Figure 14. Piezoelectricity is the electricity generated by piezo element by an effect known as the piezoelectric effect. It is the ability of certain materials to generate an AC (alternating current) voltage when subjected to mechanical stress or vibration, or to vibrate when subjected to an AC voltage, or both. If we take the example of PZT, the maximum current can vary from micro-Amperes to several Amperes, and the voltage generated in 1–100 V, depending upon the size of *PZT*. The experimental Piezo patch is the Bi-morph Type PZT (produced by Superex Technology^{TM}). This PZT is one of the best of Superex Technology's company. The dimension of the PZT is 60 mm × 20 mm. The distance between the two elastic beams is 1/13 of the elastic beam. This distance was chosen randomly, to ensure that the elastic beams can flap freely and slap each other. The purpose of this research is to prove that the concept of the DES VEH has better power generation than the traditional SES VEH system. The distance between the two elastic beams selected in this study has shown its capability in generating power. The other different dimensions of the PZT and the different distance between the two elastic beams may result in different power generation. This will be recommended for future work in this study.

As shown in Figure 14a,c, the windmill is installed directly under the middle of the wind collector. Most of the airflow enters the windmill. The only external force exerted on the beam is the repulsive force between the magnets. The beams are installed parallel to the airflow and keep a little distance from the middle of the air collector. This will allow the beams to withstand less airflow impact and less damage to the PZT sensor. This will also avoid the random vibration caused by external loads other than the magnetic force. In this study, a honeycomb wind collector was designed, as shown in Figure 14a, to collect the airflow from the downstream of the rotor blades. This honeycomb structure also rectifies the airflow to make it move smoothly, thereby exciting the windmill. The rotor speed used in the experimental setup was set to a maximum of 2000 RPM which almost reaches the maximum speed as the same type of rotor installed on a multi-axes drone. Since the wind collector was installed and the beams are installed parallel to the airflow and keep a little distance from the middle of the air collector. This somehow reduces the direct impact of the airflow on the beams and the magnet. No changes in the beams and the magnet were observed during the entire experiment. The only factor that excites the beams was the magnetic repulsive force as designed in this study.

(**a**)

(**b**) (**c**)

Figure 14. The assembly of the wind driven magneto-electric vibration energy harvester system. (**a**) brushless motor with a two-bladed rotor, (**b**) wind collecting cover, (**c**) windmill, and the magneto-electric VEH system.

6.1. Natural Frequency and Internal Resistance

We use impact hammer, accelerometer, and fast Fourier transform (from the IMC© system) to measure the first three natural frequencies of this system, which are 3.98 Hz, 19.53 Hz, and 54.35 Hz. From Equation (22), substituting the length of the beam and the mass ratio of the magnet to the beam, the first three eigen values can be obtained as: 2.1264, 4.7142, and 7.8590. The theoretical natural frequency of the first three modes can be calculated as: 3.980 Hz, 19.562 Hz, and 54.367 Hz. These values are consistent with the frequencies from the experimental analysis. Before analyzing the power of the wind-driven magneto-electric energy harvester system, it is necessary to add resistance to the system as the system load. In order to obtain the best power of the system, the closer the additional resistance value is to the internal resistance of the system, the best electrical power can be obtained. The following describes the estimation of internal resistance. First, the electric power equation is as follows:

$$P = \bar{I}V = \bar{I}^2 R = \frac{V^2}{R} \tag{36}$$

where P is power, $\bar{I}$ denotes current, V represents voltage, and R is resistance.

We use Thevenin's Theorem to find the internal resistance. Thevenin's Theorem gives,

$$V_L = \frac{R_L}{R_T + R_L} V_T \tag{37}$$

where R_T is internal resistance, R_L denotes the loading resistance, V_T is the open circuit voltage, and V_L is the loading voltage. From Equations (36) and (37) we can obtain:

$$P = \frac{V_L^2}{R_L} = \frac{\frac{R_L^2}{(R_T+R_L)^2} V_T^2}{R_L} = \frac{R_L V_T^2}{(R_T - R_L)^2 + 4R_T R_L} = \frac{V_T^2}{\frac{(R_T-R_L)^2}{R_L} + 4R_T} \tag{38}$$

When the loading resistance is equal to the internal resistance of the system, the best electric power will be obtained.

We used the IMC© system to obtain the open-circuit voltage of the system and used low-pass filtering to filter out high-frequency noise, and then use the Butterworth filter to obtain the voltage. Excite the 1st mode of the VEH system, and its output voltage is shown in Table 1. Take the root mean square and the average voltage is 4.3454 volts. In the next step, we choose 18 K ohm as the load of the system, and the average value of the output voltage is 1.4377 volts. Substituting this result into Equation (38), the theoretical internal resistance value can be calculated as 36,404 ohms (about 36 K ohms). We also tested the load resistance from 30 K to 62 K and obtained the output voltage and power as shown in Table 1. Through Table 1, the ohm-volt diagram and the ohm-power diagram can be made as shown in Figure 15. It can be observed that when the load resistance is 36 K ohms, the highest output power is 0.194 mW. Therefore, in the following Sections, when measuring the power generation efficiency of the system, 36 K ohms will be used as the experimental load resistance.

Table 1. Experimental measurement of system resistance.

Resistance (ohm)	Voltage (V)					Average (V)	mW
	A	B	C	D	E		
18,000	1.4340	1.4284	1.4706	1.4326	1.4232	1.4377	0.1148
30,000	2.0682	2.2663	2.2344	2.4732	2.4556	2.2995	0.1763
33,000	2.2042	2.3711	2.4706	2.5141	2.6418	2.4404	0.1805
36,000	2.5687	2.6546	2.6215	2.7136	2.6590	2.6435	0.1941
39,000	2.6012	2.6111	2.7039	2.6727	2.6950	2.6568	0.1810
43,000	2.6584	2.8030	2.8323	2.7838	2.6848	2.7525	0.1762
51,000	3.0149	2.9188	2.9670	3.0128	2.9255	2.9678	0.1727
Open circuit (V)	4.5053	4.1695	4.2287	4.4984	4.3251	4.3454	

6.2. SES and DES System Output Voltage

In the experiment, we used an inverter motor to drive the RPM of the windmill. When the 1st mode, 2nd mode, and 3rd mode were excited, the windmill RPM was about 120, 480, and 1650. We tested two VEH systems as shown in Figure 16. Figure 16 shows the traditional SES VEH with the PZT patch placed at the root of the elastic steel, and the DES VEH system with the PZT patch placed on the free end of one of the elastic steels and place the magnet on the free end of the other elastic steel. The PZT patch is deformed and slapped to generate a voltage in the DES system. The experimental results of the voltage output of the SES and DES systems are shown in Tables 2 and 3, respectively.

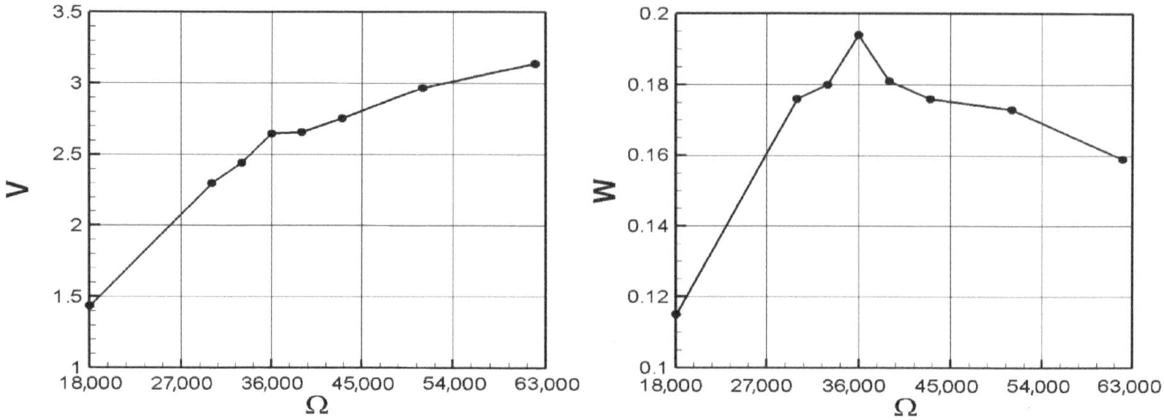

(a) (b)

Figure 15. Diagram of ohm-volt and ohm-power, (**a**) ohm-volt diagram, and (**b**) ohm-power diagram.

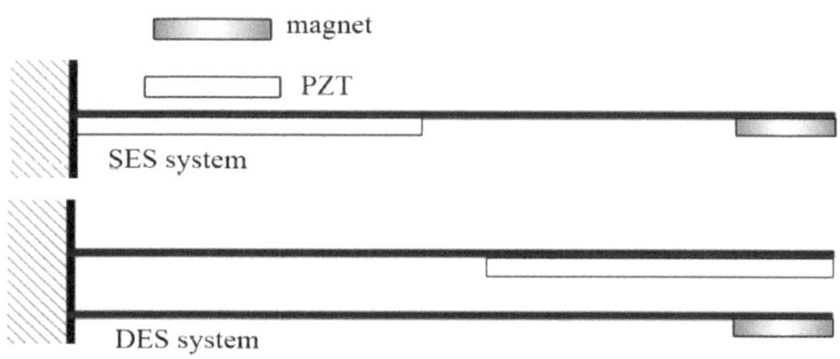

Figure 16. Schematic of SES and DES systems.

Table 2. Experimental results of the voltage and power of the SES system.

Case	SES Mode 1			SES Mode 2			SES Mode 3		
	Voltage (V)	mW	Max V	Voltage (V)	mW	Max V	Voltage (V)	mW	Max V
1	2.0563	0.1174	3.8102	3.2891	0.3005	10.3193	3.3664	0.3148	13.6078
2	2.0993	0.1224	3.8217	3.0991	0.2668	9.5380	3.2767	0.2982	10.3314
3	2.1410	0.1273	3.8929	3.2660	0.2963	11.2396	3.2085	0.2860	11.3319
4	2.1901	0.1332	3.7929	3.1179	0.2701	10.5006	3.5251	0.3452	13.3956
5	2.1886	0.1331	3.7948	3.1323	0.2725	9.3771	3.5416	0.3484	13.6456
Ave.	2.1351	0.1267	3.8225	3.1809	0.2812	10.1949	3.3836	0.3185	12.4625

Table 3. Experimental results of the voltage and power of the DES system.

Case	DES Mode 1			DES Mode 2			DES Mode 3		
	Voltage (V)	mW	Max V	Voltage (V)	mW	Max V	Voltage (V)	mW	Max V
1	2.2058	0.1352	14.11527	3.5632	0.3527	14.46878	3.4834	0.3371	14.9354
2	2.4851	0.1715	14.69722	3.2502	0.2934	13.64842	3.5404	0.3482	14.2494
3	2.3477	0.1531	14.05054	3.3983	0.3208	14.23503	3.3271	0.3075	14.5385
4	2.2659	0.1426	13.8839	3.7611	0.3929	14.05283	3.6828	0.3767	14.9435
5	2.1075	0.1234	13.99478	3.2616	0.2955	13.83062	3.556	0.3512	14.4558
Ave.	2.2824	0.1452	14.1483	3.4469	0.3311	14.0471	3.5179	0.3441	14.6245

We have observed in the experiments that whether it is SES or DES system, the higher the mode, the higher the volts of power generation, which is consistent with the trend of the theoretical prediction (Figures 8 and 13). In addition, it is also found on the SES system that if the vibration frequency is too large, the repulsive force of the magnet and the restoring force of the elastic steel cannot be fully matched, resulting in the elastic steel sometimes having a large swing amplitude, sometimes it only vibrates at a small amplitude at the original equilibrium position, which is repeated iteratively. Figure 17 shows the voltage output diagram of the first three modes of the SES system. It can be observed that the frequency of the 3rd mode of the SES system is too fast at 15 to 20 s, resulting in the elastic steel vibration being unable to match the frequency generated by the magnet repulsion. The elastic steel has irregular amplitude. This phenomenon is unfavorable for power generation. In the experiment of the DES system (see Figure 18), in addition to the original repulsion force and the impact of the slap force, the elastic steel has a larger amplitude Moreover, because of the impact of the slap force and at high frequencies, the repulsive force of the magnet cannot match the elastic force of the steel, just like the SES system. As a result, sometimes the voltage output is not good. Therefore, if the repulsive force of the magnet can be effectively adjusted or the slap force can be reduced, in theory, a better slap situation can be achieved. Therefore, the DES system can meet the theoretical prediction.

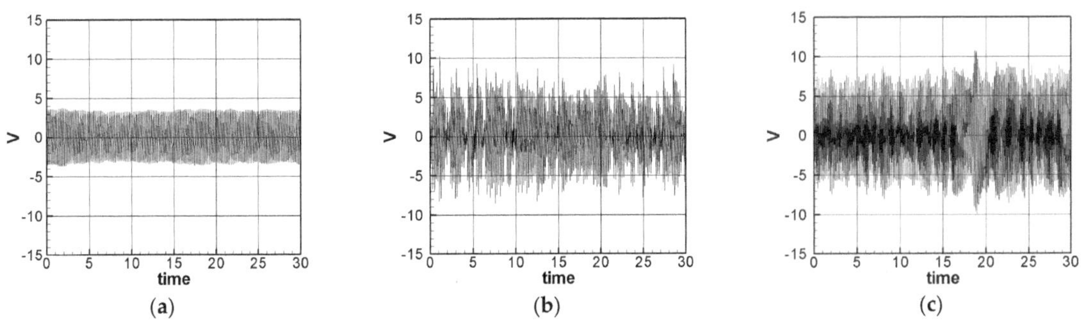

Figure 17. SES system experimental voltage output (with dimension (volt)). (**a**) 1st mode, (**b**) 2nd mode, and (**c**) 3rd mode.

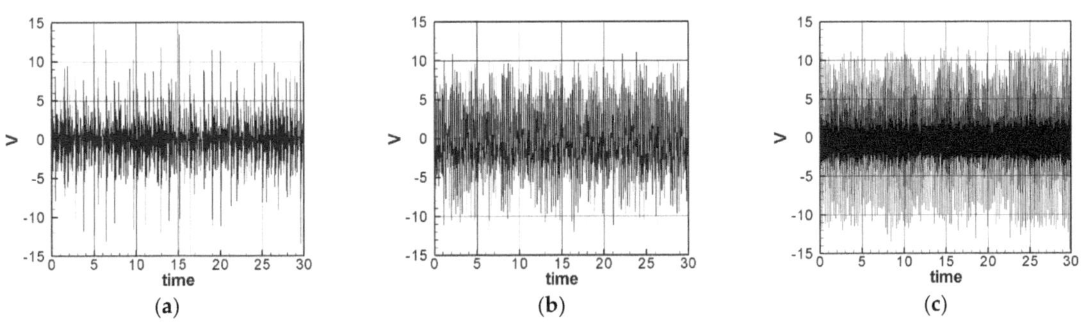

Figure 18. DES system experimental voltage output (with dimension (volt)). (**a**) 1st mode, (**b**) 2nd mode, and (**c**) 3rd mode.

7. Theoretical and Experimental Verification

This Section will verify the theoretical and experimental results of the SES and DES system. We took the beam mass, length, magnet mass, etc., used in the experiment, and put these parameters into the theoretical functions, and got the dimensional output voltages. The root mean squares (RMS) of theoretical output voltages of the SES and DES system are detailed in Table 4. It can be observed that the voltages of either the SES or DES system increase with an increase in the modes, and the comparison with the experimental results also has the same trend. Through experiments, it is found that the SES system or the DES system is relatively stable in the first two modes. When the third mode is excited, the frequency is too fast, resulting in an irregular vibration of the elastic steel. Especially when the 3rd mode is excited in the DES system, the impact of the flapping force destroys the periodic vibration of the elastic steel. The repulsive force of the magnet and the restoring force of the elastic steel on both of the SES and DES systems cannot be fully matched, resulting in the elastic steel sometimes having a large swing amplitude, sometimes it only vibrates at a small amplitude at the original equilibrium position, which is repeated iteratively. This phenomenon is unfavorable for power generation. In the experiment of the DES system, in addition to the original repulsion force and the impact of the slap force, the elastic steel has a larger amplitude. Moreover, because of the impact of the slap force and at high frequencies, the repulsive force of the magnet cannot match the elastic force of the steel, just like the SES system. As a result, sometimes the voltage output is not good. Due to these factors, the experiments of the SES system or the DES system in the 3rd mode cannot agree with the theoretical results. In future engineering, if the repulsive force of a magnet can be designed and adjusted to match the flapping force of the elastic steel so that it can be more idealized, we believe it will have a better effect.

Table 4. SES and DES system theoretical output voltage (with dimension (volt)).

		SES	DES
RMS voltage (V)	Mode1	2.2660	2.4484
	Mode2	3.4741	3.7988
	Mode3	6.0436	6.7436

Table 2 shows the experimental results of the first three modes of the SES system. The first three modes root mean square (RMS) voltage is 2.1351, 3.1809, and 3.3836 volts, respectively. The ratio is 1:1.4898:1.5848. The theoretically estimated SES and DES system's 1st, 2nd and 3rd mode's voltage outputs with dimension are shown in Figures 19 and 20. Figure 19 is for the SES system and Figure 20 is for the DES system, respectively. The theoretical root mean square (RMS) output voltages with dimension are shown in Table 4.

From Table 4, the root mean square voltages for the first three modes of the SES system are 2.2660, 3.4741, and 6.0436 respectively. The ratio is 1:1.5331:2.6671. According to the experimental results, it can be shown that the output voltage agrees with the theoretical prediction very well. The 1st mode output voltage relative error is 6.1308%, the 2nd mode output voltage relative error is 9.2175%, and the 3rd mode voltage relative error is 78.6145%. Then, compare the voltage ratio of the 1st and 2nd modes. The experimental voltage ratio of the first two modes is 1:1.4898, and the theoretical voltage ratio is 1:1.5331; the relative error of the ratio of the two modes is about 2.9064%. It is confirmed that the experimental values of the first two modes conform to the theoretical predictions.

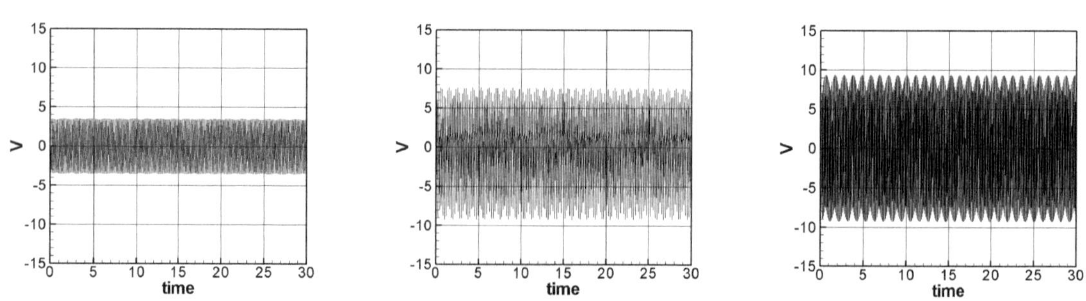

Figure 19. SES system theoretical voltage output (with dimension (volt)). (**a**) 1st mode, (**b**) 2nd mode, and (**c**) 3rd mode.

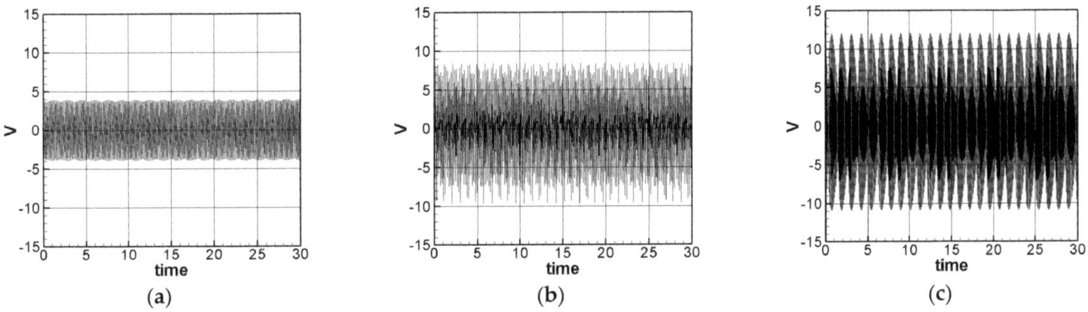

Figure 20. DES system theoretical voltage output (with dimension (volt)). (**a**) 1st mode, (**b**) 2nd mode, and (**c**) 3rd mode.

Similarly, for the DES system, the first three modes of the experimental RMS voltage are 2.2824, 3.4469, and 3.5179 volts, respectively. The ratio is 1:1.5102:1.5413. The theoretically estimated DES system's voltages are 2.4484, 3.7988, and 6.7436, respectively. The ratio is 1:1.5515:2.7543. The 1st mode output voltage relative error is 7.2730%, the 2nd mode output voltage relative error is 10.2092%, and the 3rd mode voltage relative error is 91.6939%. The experimental voltage ratio of the first two modes is 1:1.5102, and the theoretical voltage ratio is 1:1.5515; the relative error of the ratio of the two modes is about 2.7347%. It is confirmed that the experimental values of the first two modes conform to the theoretical predictions.

The objective of this work is to propose a DES VEH system and to show that this novel concept can provide a better power generation than the traditional SES VEH system. This study proves its characteristics theoretically. This research just carried out a simple experimental setup to demonstrate the working principle of the theoretical model. Future work may suggest more complex and precise experiments. However, compared with the absolute values of the theoretical and experimental voltage outputs, for the SES system, the 1st mode output voltage relative error is 6.1308%, the 2nd mode output voltage relative error is 9.2175%, and the 3rd mode voltage relative error is 78.6145%. For the DES system, the 1st mode output voltage relative error is 7.2730%, the 2nd mode output voltage relative error is 10.2092%, and the 3rd mode voltage relative error is 91.6939%. The errors of the first two modes are almost less 10% and can be considered acceptable for engineering applications. The 3rd mode is difficult to achieve its ideal flapping condition and it is hard to match the magnetic repulsive force as mentioned in this work. This may require future improvements. It can be seen that the theories and experiments of the first two modes of the SES system and the DES system can be mutually verified. In future engineering applications, proper adjustment of the repulsive force of the magnet will reduce the occurrence of irregular vibration in the third mode of the SES and DES system, in order to obtain a higher voltage output.

8. Conclusions

This research is based on the main structure of a fixed-free nonlinear elastic beam with a tip mass. The frequency response and amplitude are analyzed by the MOMS method and fixed points plots. The SES and DES VEH systems are excited by wind and magnetic force to convert the vibration energy to electric power. Through experiments to verify the correctness of the theory of this study. The nonlinear beam model also supports the results from the experiments of the double elastic steel energy harvesting system. The proposed DES system has better power generation than the traditional single elastic steel system, which was used in refs. [15,16]. The use of wind and magnetic force applied on the DES system also has a wider range of practical applications than a simple external load from a shaker, which was used in refs. [17–19]. The novel concept and practical application have been proved feasible in this work. The conclusions of this research are as follows:

1. Regardless of theory or experiment, the voltage outputs of the SES or DES system increase as the excitation modes increase and the voltage output of the DES system in each mode is also greater than that of the SES system.
2. The experimental voltage values of the SES and DES systems in the first and second modes agree with theoretical predictions. The reason for the large difference between the theoretical value and the experimental value of the 3rd mode is due to the irregular motion of the third mode. The frequency generated by the repulsive force of the magnet driven by the rotation of the windmill cannot be matched with the elastic force of the elastic steel, resulting in the 3rd mode not being able to produce the same voltage output as the theoretical value.
3. Regardless of whether it is an SES system or a DES system, the theoretical prediction in each mode is greater than the experimental result. According to the theoretical estimate, the voltage output is the 3rd mode > the 2nd mode > the 1st mode. The voltage output effect of the DES system is also better than that of the SES system.

In summary, we recommend using the DES energy harvesting system. However, when the 3rd mode is excited, the periodicity of the elastic steel is destroyed. In future engineering applications, if the repulsive force of the magnet can be adjusted to control the slap force on the elastic steel so that it can maintain a periodic state after slap, which will enable the third mode of the DES system to approach the theoretical condition. In the future, we hope that through this piezoelectric energy harvesting system driven by wind and magnetic force, we can collect the downward airflow generated by helicopters and multi-axis unmanned rotorcrafts and recycle this energy to maximize the effectiveness of DES VEH.

Author Contributions: Y.-R.W. organized the content; Y.-R.W. wrote this article; M.-C.C. performed the experiment. All authors have read and agreed to the published version of the manuscript.

Funding: This research was supported by the Ministry of Science and Technology of Taiwan, Republic of China (grant number: MOST 110-2221-E-032-026).

Institutional Review Board Statement: Not applicable.

Informed Consent Statement: Not applicable.

Data Availability Statement: Not applicable.

Acknowledgments: The authors thank J.K. Hsiao for his help in the experiment of system internal resistance.

Conflicts of Interest: The authors declare no conflict of interest.

References

1. Roundy, S.; Wright, P.K.; Pister, K.S.J. Micro-electrostatic vibration-to-electricity converters. In Proceedings of the IMECE2002, ASME International Mechanical Engineering Congress & Exposition, New Orleans, LA, USA, 17–22 November 2002.
2. Roundy, S.; Wright, P.K.; Rabaey, J. A study of low level vibrations as a power source for wireless sensor nodes. *Comput. Commun.* **2003**, *26*, 1131–1144.
3. Roundy, S.; Wright, P.K. A piezoelectric vibration based generator for wireless electronics. *Smart Mater. Struct.* **2004**, *13*, 1131–1142.
4. Iqbal, M.; Nauman, M.M.; Khan, F.U.; Abas, P.E.; Cheok, Q.; Iqbal, A.; Aissa, B. Vibration based piezoelectric, electromagnetic, and hybrid energy harvesters for microsystems applications: A contributed review. *Int. J. Energy Res.* **2021**, *45*, 65–102.
5. Jiang, J.; Liu, S.; Feng, L.; Zhao, D. A review of piezoelectric vibration energy harvesting with magnetic coupling based on different structural characteristics. *Micromachines* **2021**, *12*, 436.
6. Li, D.; Wu, Y.; Da Ronch, A.; Xiang, J. Energy harvesting by means of flow-induced vibrations on aerospace vehicles. *Prog. Aerosp. Sci.* **2016**, *86*, 28–62. [CrossRef]
7. Nayfeh, A.H.; Mook, D.T. *Nonlinear Oscillations*; Wiley-Interscience: New York, NY, USA, 1979.
8. Erturk, A.; Inman, D.J. A distributed parameter electro-mechanical model for cantilevered piezoelectric energy harvesters. *J. Vib. Acoust.* **2008**, *130*, 041002. [CrossRef]
9. Rajora, A.; Dwivedi, A.; Vyas, A.; Gupta, S.; Tyagi, A. Energy harvesting estimation from the vibration of a simply supported beam. *Int. J. Acoust. Vib.* **2017**, *22*, 2186–2193. [CrossRef]
10. Masana, R.; Daqaq, M.F. Electromechanical modeling and nonlinear analysis of axially loaded energy harvesters. *J. Vib. Acoust.* **2011**, *133*, 011007.

11. Andò, B.; Baglio, S.; Maiorca, F.; Trigona, C. Two dimensional bistable vibration energy harvester. *Procedia Eng.* **2012**, *47*, 1061–1064. [CrossRef]
12. Harne, R.L.; Wang, K.W. A review of the recent research on vibration energy harvesting via bistable systems. *Smart Mater. Struct.* **2013**, *22*, 023001. [CrossRef]
13. Gao, Y.-J.; Leng, Y.-G.; Fan, S.-B.; Lai, Z.-H. Studies on vibration response and energy harvesting of elastic-supported bistable piezoelectric cantilever beams. *Acta Phys. Sin.* **2014**, *63*, 090501.
14. Zhu, Y.; Zu, J. A magnet-induced buckled-beam piezoelectric generator for wideband vibration-based energy harvesting. *J. Intell. Mater. Syst. Struct.* **2004**, *25*, 1890–1901.
15. Wang, L.; Zhu, D. Enhancing output power of a cantilever-based flapping airflow energy harvester using external mechanical interventions. *Sensors* **2019**, *19*, 1499. [CrossRef] [PubMed]
16. Wang, Z.; He, L.; Zhang, Z.; Zhou, Z.; Zhou, J.; Cheng, G. Research on a piezoelectric energy harvester with rotating magnetic excitation. *J. Electron. Mater.* **2021**, *50*, 3228–3240.
17. Fu, Y.; Ouyang, H.; Davis, R.B. Triboelectric energy harvesting from the vibro-impact of three cantilevered beams. *Mech. Syst. Signal Process.* **2019**, *121*, 509–531. [CrossRef]
18. Wang, Y.-R.; Wong, M.-S.; Chen, B.-Y. Analytical and experimental studies of double elastic steel sheet (DESS) vibration energy harvester system. *Energies* **2020**, *13*, 1793.
19. Wang, Y.-R.; Hung, C.-C.; Tseng, J.-T. Transverse vibration energy harvesting of double elastic steel. *Int. J. Struct. Stab. Dyn.* **2021**, *21*, 2150113.
20. Nayfeh, A.H.; Pai, P.F. *Linear and Nonlinear Structural Mechanics*; Wiley-Interscience: New York, NY, USA, 2004; Chapters 4 and 5.

sensors

MDPI

Article

A Method for Parameter Identification of Composite Beam Piezoelectric Energy Harvester

Xuhui Zhang [1,2,*], Chao Zhang [1], Lin Wang [1], Luyang Chen [1], Xiaoyu Chen [1], Dongmei Xu [1,2], Hongwei Fan [1,2] and Fulin Zhu [1]

1 College of Mechanical Engineering, Xi'an University of Science and Technology, Xi'an 710054, China; 17205201064@stu.xust.edu.cn (C.Z.); 18205018017@stu.xust.edu.cn (L.W.); chenluyang@stu.xust.edu.cn (L.C.); 19105016004@stu.xust.edu.cn (X.C.); dongmeixu@xust.edu.cn (D.X.); hw_fan@xust.edu.cn (H.F.); 20205224060@stu.xust.edu.cn (F.Z.)

2 Shaanxi Key Laboratory of Mine Electromechanical Equipment Intelligent Monitoring, Xi'an 710054, China

* Correspondence: zhangxh@xust.edu.cn; Tel.: +86-133-6397-5328

Abstract: This paper proposes a parameter identification method for the multiparameter identification study of the linear–arch composite beam piezoelectric energy harvester. According to the voltage response characteristics of the system under short-circuit conditions, the mechanical equation is solved by transient excitation, combined with the backbone curve theory and logarithmic attenuation method, to obtain the system's linear damping, linear stiffness, and nonlinear stiffness. According to the voltage response characteristics of the system under open-circuit conditions, combined with the electrical equations, the system electromechanical coupling coefficient and equivalent capacitance coefficient are obtained; numerical simulation results show that the identification parameters have good accuracy. Finally, an experimental platform was built for verification, and the results show that the method has high accuracy and practicability.

Keywords: piezoelectric energy harvester; stiffness; damping; parameter identification

Citation: Zhang, X.; Zhang, C.; Wang, L.; Chen, L.; Chen, X.; Xu, D.; Fan, H.; Zhu, F. A Method for Parameter Identification of Composite Beam Piezoelectric Energy Harvester. *Sensors* **2021**, *21*, 7213. https://doi.org/10.3390/s21217213

Academic Editors: Slim Naifar, Olfa Kanoun and Carlo Trigona

Received: 26 September 2021
Accepted: 28 October 2021
Published: 29 October 2021

1. Introduction

With the rapid development of wireless sensor network technology, low-power wireless sensors have been widely used in large-scale equipment state detection [1]. Vibration energy harvesting technology has attracted much attention due to its application prospects in microelectronic equipment power supply and equipment condition monitoring [2–7]. The use of piezoelectric self-capture energy-powered wireless monitoring technology to replace traditional wired monitoring is expected to greatly alleviate the wiring difficulties and power supply difficulties of online monitoring of complex environment equipment in coal mines [8]. For this reason, some experts and scholars have proposed a variety of structural energy harvesters and performed dynamic modeling and response characteristics analysis [9–13]. However, it is difficult to accurately express the mechanical and electrical parameters in the model through traditional mathematical modeling methods, which results in certain errors in the system dynamic response analysis and experimental results. Therefore, a lot of attention has been paid to the parameter identification technology of piezoelectric energy harvesting systems. Feldman et al. [14] proposed two accurate identification methods; the stiffness and damping parameters of the nonlinear vibration system are obtained based on the free vibration of the classical nonlinear system, and the feasibility of the method is verified numerically.

Lu et al. [15] proposed steady-state excitation parameter identification and transient excitation parameter identification methods for a vibration system containing both nonlinear stiffness and nonlinear damping. Taking the vibration isolation system with nonlinear stiffness and nonlinear damping as an example, the two types of parameter identification methods given are verified by numerical simulation. The results show that the results of the two identification methods are relatively consistent, and the steady-state parameter identification method is more accurate and convenient than the two, but an experimental platform has not been built to verify the identification algorithm. Yuan et al. [16–18] used the restoring force surface method and the Hilbert transform method to study the nonlinear parameter identification of piezoelectric twin-crystal thin plates. According to the experimental results, the accuracy of the displacement–stiffness function is proved and the nonlinear characteristics of the system are accurately displayed. Zhou [19] proposed a tri-stable piezoelectric energy harvester and identified the damping and electromechanical coupling coefficients of the system through the genetic algorithm (GA) method. The results show that the algorithm is very accurate in identifying parameters. Chen et al. [20] proposed an electrical parameter identification method for the laminated disc piezoelectric energy harvester to identify the electromechanical coupling coefficient and equivalent capacitance coefficient of the system. The result is highly consistent with the theoretical value and verifies the accuracy of the algorithm.

Alper Erturk et al. [21] used a nonlinear least-squares optimization algorithm for a cantilever beam piezoelectric energy harvester to identify the electromechanical coupling coefficient and damping coefficient of the system. The comparison of theoretical and experimental results proved the effectiveness of the algorithm, and it was found that adding coupling nonlinearity will shift the nonlinear response of the system to the right. Adnan Kefal et al. [22] proposed a new parameter identification algorithm based on Lyapunov's theory, and by analyzing the time-domain response of the array piezoelectric bimorph under the random vibration of the low-frequency structure, the array parameter values were identified, and the accuracy of the algorithm was proved. Porfiri et al. [23] proposed two techniques for estimating the electromechanical coupling coefficient and piezoelectric modal capacitance; the effectiveness of the algorithm was proved by the comparison of bi-crystal piezoelectric cantilever beam experiment and finite element simulation analysis. Delpero et al. [24] proposed a method of measuring the electromechanical coupling coefficient and applied it to experiments on piezoelectric energy harvesters of different sizes; the experimental results are in good agreement with the analysis and prediction. Binh Duc Truong et al. [25] proposed a parameter identification method based on least-squares minimization, which estimates the damping and stiffness coefficients of the system according to the frequency response function of the system dynamics equation, and the algorithm is suitable for linear and nonlinear systems.

Excitation in the environment is usually low-frequency and multidirectional; traditional linear beams can only respond to excitation in a single direction, which is difficult to achieve in practical applications. In order to realize the self-powering of wireless monitoring nodes in coal mines. Zhang [26] designed two linear–arch composite beam multidirectional piezoelectric energy harvesters, as shown in Figure 1; the energy harvester can collect multidirectional vibration energy. The introduction of the arch structure makes the system dynamics model complicated, and it is difficult to achieve accurate modeling with the traditional modeling method. Therefore, this paper proposes a method to identify the system mechanical and electrical parameters of the linear–arch composite beam piezoelectric energy harvester. According to the voltage response characteristics of the system under short-circuit conditions, the transient excitation of the mechanical equations is solved. Combining the ridge theory and logarithmic attenuation method, the system's linear damping, linear stiffness, nonlinear stiffness, and other mechanical parameters are obtained with an identification procedure, according to the voltage response characteristics of the system under open-circuit conditions; combined with the electrical equations, electrical parameters such as the electromechanical coupling coefficient and equivalent capacitance coefficient of the system are obtained. The identified parameters are finally compared with the experimental results to verify the correctness of the procedure. The parameter identification method proposed in this paper is of great significance for guiding the modeling under complex nonlinear conditions. It is helpful in improving the accuracy of theoretical modeling and providing theoretical guidance for optimizing system parameters.

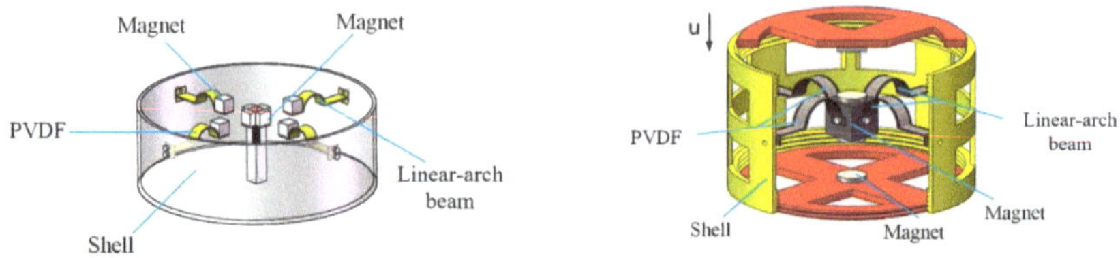

Figure 1. Multidirectional piezoelectric energy harvester based on linear–arch composite beam.

2. The Structure and Theoretical Model of the Piezoelectric Energy Harvester

The piezoelectric vibration energy harvester based on a composite beam (PEH-C) is shown in Figure 2. The system is mainly composed of a base, a load resistor, a composite beam, a piezoelectric film, and a mass. In order to enhance the directional sensitivity of the system to vibration sources in the real environment, the composite beam is made of a combination of linear and arch. The horizontal length of the composite beam is L. Under the excitation of external vibration along the Z-axis, the composite beam vibrates up and down, and the piezoelectric material pasted on the beam deforms at the same time. In this way, the positive piezoelectric effect of the piezoelectric material is used to convert the vibration energy in the actual environment into electrical energy.

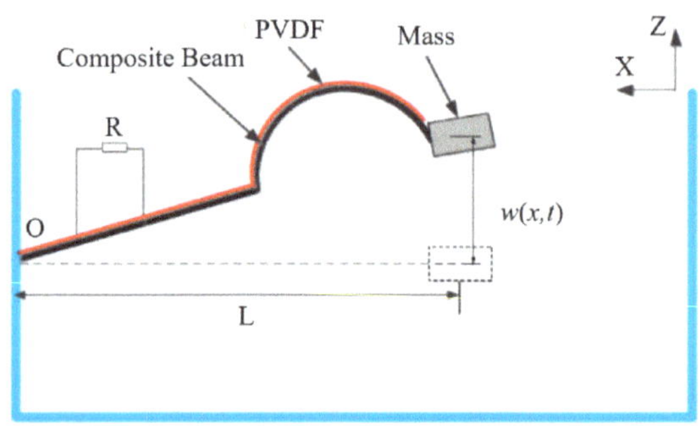

Figure 2. Schematic diagram of the structure of the PEH-C.

$w(x,t)$ is the displacement of a point on the linear–arch composite beam along the Z-axis at time t, According to the Rayleigh–Ritz theory, the relative displacement $w(x,t)$ of linear–arch composite beam vibration can be discretized as a combination of modes

$$w(x,t) = \sum_{i=1}^{n} \varphi_i(x)x(t) \tag{1}$$

where $\varphi_i(x)$ represents the mode shape of the ith mode of the linear–arch composite beam and $x(t)$ represents the generalized modal coordinates.

For the linear–arch composite beam in this paper, one end is clamped and fixed on the base, and the other end is free. The mode of the linear–arch composite beam is difficult to obtain; therefore, the allowable function is adopted to simplify it. The allowable function can be expressed as [27]

$$\varphi_i(x) = 1 - \cos\left[\frac{(2i-1)\pi x}{2L}\right] \tag{2}$$

Since the excitation of piezoelectric energy harvester is mainly low-frequency, the first-order modal bending vibration of the linear–arch composite beam plays a leading role. This paper only considers the first-order mode of the linear–arch composite beam; therefore, $i = 1$, $\varphi_1(L) = 1$, and $w(L,t) = x(t)$.

For the piezoelectric vibration energy harvester based on composite beam (PEH-C) system, considering the structural nonlinearity, the generalized Hamilton principle and Lagrange function are used to establish the system parameter model, and the dynamic equation of the PEH-C system is obtained [28]:

$$\begin{gathered} m\ddot{x}(t) + c\dot{x}(t) + \beta x(t)^3 + \alpha x(t) - \vartheta v(t) = -H_s\ddot{z}(t) \\ \vartheta \dot{r_1}(t) + c_p\dot{v}(t) + v(t)/R = 0 \end{gathered} \tag{3}$$

where m is the composite beam mass, c is the composite beam first-order damping, α is the linear stiffness coefficient, β is the nonlinear stiffness coefficient, ϑ is the electromechanical coupling coefficient, $x(t)$ is the response displacement, H_s is the system foundation excitation coefficient, $v(t)$ is the output voltage, $z(t)$ is the external basic incentive, c_p is the system equivalent capacitance coefficient, and R is the resistance value of external load.

3. Theoretical Modeling of Parameter Identification Method

From Equation (3), when the system is in a short-circuit condition, the output voltage of the system is 0, so the mechanical equation in the system dynamics equation can be rewritten in the following form:

$$m\ddot{x}(t) + c\dot{x}(t) + \alpha x(t) + \beta x^3(t) = f(t) \tag{4}$$

where $f(t)$ is the external excitation. $x(t)$ is the end displacement, and the positive direction of the Z-axis shown in Figure 2 is the positive direction of the displacement.

We select external excitation as unit pulse excitation; that is, $f(t) = \delta(t)$. Then, Equation (4) can be written as

$$m\ddot{x}(t) + c\dot{x}(t) + \alpha x(t) + \beta x^3(t) = 0 \tag{5}$$

Then the solution of Equation (5) can be set as

$$x(t) = A(t)\cos[\varphi(t)] \tag{6}$$

In the formula, $A(t)$ is the amplitude of the system over time, and $\varphi(t)$ is the phase angle. When the system moves along the main vibration curve under the condition of free vibration, the instantaneous frequency and instantaneous phase angle have the following relationship:

$$\varphi(t) = \int_0^{\tau} \omega_d(\tau)d\tau \tag{7}$$

In the formula, ω_d is the instantaneous frequency.

For the system described by the Equation (5), the instantaneous frequency analytic formula that changes with time is [29]

$$\omega^2{}_d(t) = \omega^2{}_{dd}(t) + \frac{3}{4}\frac{\beta}{m}A^2(t) \tag{8}$$

In the formula, the damped natural angular frequency $\omega_{dd} = \omega_n\sqrt{1-\zeta^2}$, the undamped natural angular frequency $\omega_n = \sqrt{\alpha/m}$, and damping ratio $\zeta = c/(2\sqrt{\alpha \times m})$; we assume $A(t) = A_0 e^{-\zeta\omega_n t}$ according to the theory of multiscale method and combine Equations (7) and (8) to obtain

$$\varphi(t) \approx \sqrt{1-\zeta^2}\omega_n t + \frac{3\beta}{16\alpha\sqrt{1-\zeta^2}}{A_0}^2\left(1-e^{-2\zeta\omega_n t}\right) \tag{9}$$

$$\omega_d \approx \sqrt{1-\zeta^2}\omega_n + \frac{3\beta\zeta\omega_n}{8\alpha\sqrt{1-\zeta^2}} {A_0}^2 e^{-2\zeta\omega_n t} \tag{10}$$

Substituting Equation (9) into (6), we can obtain

$$x(t) \approx A_0 e^{-\zeta\omega_n t} \cos\left[\sqrt{1-\zeta^2}\omega_n t + \frac{3\beta}{16\alpha\sqrt{1-\zeta^2}} {A_0}^2 (1-e^{-2\zeta\omega_n t})\right] \tag{11}$$

According to the time-domain attenuation curve of the free vibration of the system and the Hilbert signal processing method, the system envelope curve is obtained and its backbone curve is drawn. The least-squares method is used to linearly fit the backbone curve to determine the system stiffness coefficient, and then the system damping coefficient is identified by the logarithmic attenuation method.

For identification of system stiffness coefficient, according to Equation (8), we draw the backbone curve of the system and fit the backbone curve to a straight line by the least-squares method, which is recorded as

$$y \approx b_0 x + b_1 \tag{12}$$

Then, we combine Equations (8) and (12) and simplify the combination to obtain

$$\alpha \approx b_1 m / \left(1-\zeta^2\right) \tag{13}$$

$$\beta \approx 4 b_0 m / 3 \tag{14}$$

For identification of system damping coefficient, for the PEH-C, the envelope under free vibration attenuation can be fitted with an exponential function, and its equation is set as

$$A(t) \approx A e^{at} \tag{15}$$

Combining Equations (6), (11) and (15), we can obtain

$$\zeta \approx -a / \sqrt{\alpha / m} \tag{16}$$

$$c = 2\zeta\sqrt{\alpha m} \approx -2ma \tag{17}$$

Combining Equations (13) and (16), we can obtain

$$\alpha \approx \frac{b_1 m}{1 - ma^2/\alpha} \tag{18}$$

Equation (18) can be solved in parallel with vertical Equations (16) and (17) to obtain the linear stiffness coefficient α, damping ratio coefficient ζ, and damping coefficient c of the system:

$$\alpha \approx b_1 m + ma^2 \tag{19}$$

$$\zeta \approx -a / \sqrt{b_1 + a^2} \tag{20}$$

$$c = 2\zeta\sqrt{\alpha m} = -2ma \tag{21}$$

For the identification of the system electromechanical coupling coefficient and equivalent capacitance coefficient of the system, we make the system response output of the electrical equation in the dynamic equation in an open-circuit state, take $R = 10^7\,\Omega$, and add a sinusoidal excitation to make the system output stable at a fixed frequency. At this time, the output voltage, voltage change rate, and velocity signal response output can be regarded as the first-order harmonic state, namely

$$\dot{r_1}(t) = p_0 \sin(\omega t) + q_0 \cos(\omega t) \tag{22}$$

$$\dot{v}(t) = p_1 \sin(\omega t) + q_1 \cos(\omega t) \tag{23}$$

$$v(t) = p_2 \sin(\omega t) + q_2 \cos(\omega t) \tag{24}$$

In the equation, p_0, q_0, p_1, q_1, p_2 and q_2 are coefficients, which are measured by experiment. Substituting Equations (22)–(24) into Equation (3) and solving them, we can obtain

$$\vartheta[p_0 \sin(\omega t) + q_0 \cos(\omega t)] + c_p[p_1 \sin(\omega t) + q_1 \cos(\omega t)] + \frac{[p_2 \sin(\omega t) + q_2 \cos(\omega t)]}{R} = 0 \tag{25}$$

Since Equation (25) is constant, the coefficients of the sin term and cos term of the equation can be set to 0 at this time, and Equation (25) can be solved to obtain the system electromechanical coupling coefficient ϑ and the equivalent capacitance coefficient c_p expression:

$$\vartheta = \frac{p_2 q_1 - p_1 q_2}{R(p_1 q_0 - p_0 q_1)} \tag{26}$$

$$c_p = \frac{q_2 p_0 - p_2 q_0}{R(p_1 q_0 - p_0 q_1)} \tag{27}$$

4. The Verification of Parameter Identification Value

4.1. Numerical Verification of Mechanical Parameter Identification

In order to verify the correctness of the system's mechanical and electrical parameter modeling, the relevant parameters in the system dynamics equation are selected for numerical simulation. The parameter values are shown in Table 1.

Table 1. The parameter value table of the energy capture system.

Parameter	Name	Value	Unit
m	composite beam mass	0.0042	Kg
c	the first-order damping of composite beam	0.02	Ns/m
α	linear stiffness coefficient	22.1	N/m
β	nonlinear stiffness coefficient	41,963	N/m^3
H_s	basic excitation coefficient	0.004016	Kg
c_p	equivalent capacitance coefficient of piezoelectric film	2×10^{-7}	C
ϑ	electromechanical coupling coefficient	5.192×10^{-4}	C/m
R	load resistance	1×10^7	
$\ddot{z}(t)$	vibration acceleration	$8 \times \sin(72.5t)$	m/s^2

For the system of Equation (4), the unit pulse excitation $\delta(t)$ is selected for simulation analysis, and the Runge–Kutta algorithm is used to solve the dynamic system. Subsequently, the Hilbert transform method is used to process the time-domain attenuation signal; the system vibration attenuation curve, envelope, and backbone curve are drawn; and we use the least-squares method to fit the envelope and backbone curve data. The result is shown in Figure 3.

As shown in Figure 3b, the fitting equation of the envelope of the system displacement attenuation curve can be expressed as

$$y \approx 0.1268e^{-2.27t} \tag{28}$$

As shown in Figure 3c, the system backbone curve fitting equation can be expressed as

$$y \approx 7.391 \times 10^6 t + 5259.042 \tag{29}$$

From Equations (28) and (29), $a = -2.27$, $b_0 = 7.391 \times 10^6$, and $b_1 = 5259.042$. Substituting the values of a, b_0, and b_1 into Equations (19), (14) and (21), the system linear stiffness coefficient α, nonlinear stiffness coefficient β, and linear damping coefficient c can be obtained: $\alpha \approx 22.11$, $\beta \approx 41{,}389.6$, and $c \approx 0.0191$.

In order to analyze the accuracy of the identification algorithm, an error analysis of the identification coefficient is performed, as shown in Table 2.

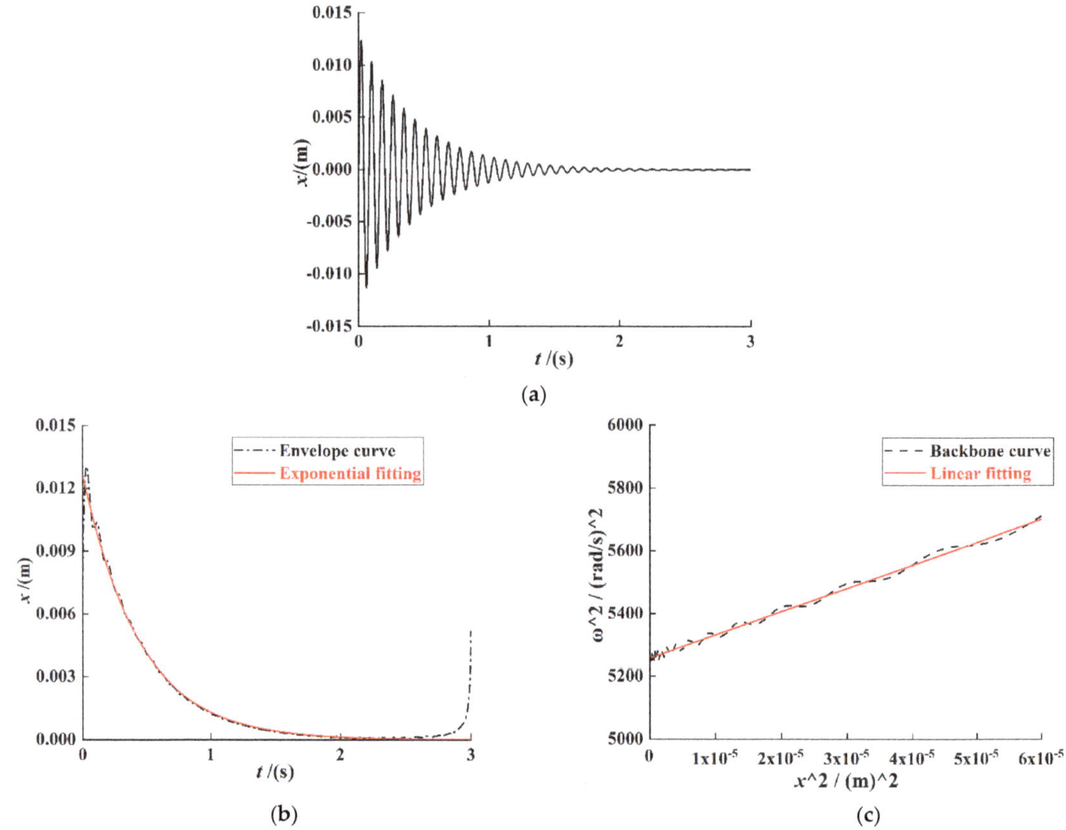

Figure 3. System response curve under free vibration attenuation. (**a**) Attenuated signal curve. (**b**) Signal envelope curve. (**c**) Signal backbone curve.

Table 2. Identification results of mechanical parameters.

Parameter Type	c	α	β
Select value	0.02	22.1	41,963
Identification value	0.0191	22.1096	41,389.6
Error	4.5%	0.43%	1.37%

It can be seen from Table 2 that the identification error of the damping coefficient based on the logarithmic attenuation method is only 4.5%. According to the backbone curve of the free attenuation signal, the errors of the linear stiffness and the nonlinear stiffness are obtained by the identification; they are 0.43% and 1.37%, respectively. The errors are less than 5%, and the errors of the last two are even less than 2%, which is within an acceptable range. Therefore, the identification method has good accuracy.

4.2. Numerical Verification of Electrical Parameter Identification

According to the above-given conditions, Equation (3) is dynamically solved and analyzed by the Runge–Kutta numerical method, and the stable time period data are selected to obtain the system response characteristic curve. The simulation result is shown in Figure 4.

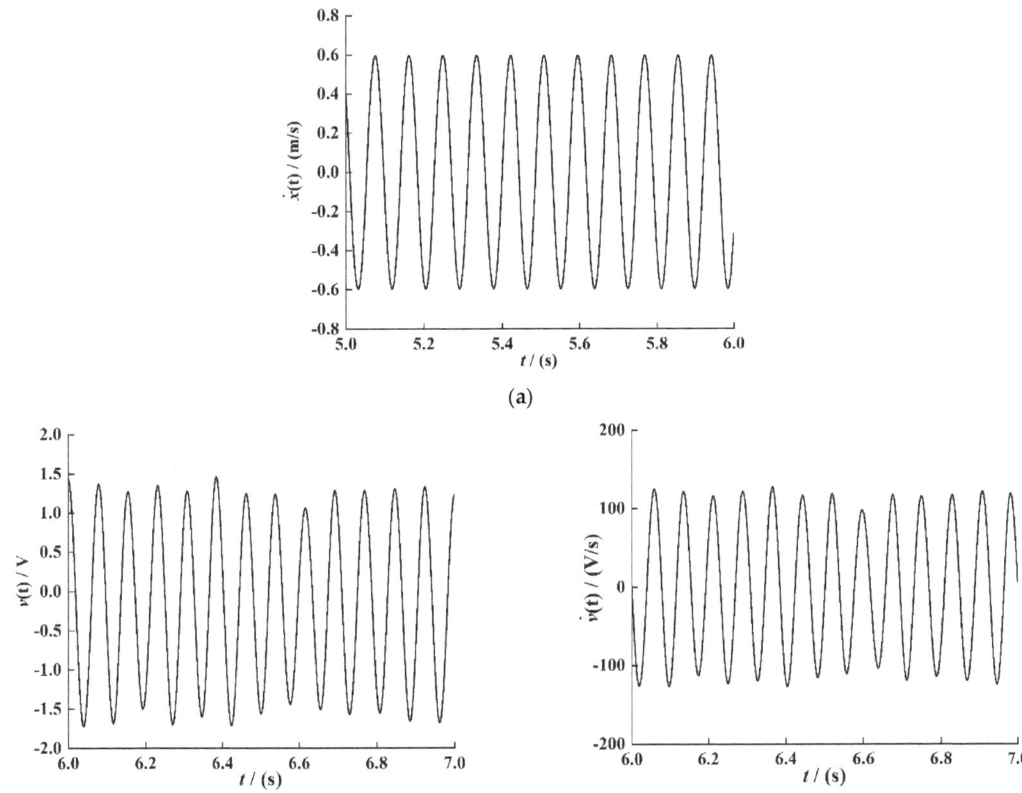

(b) (c)

Figure 4. Response characteristic curve under system stability (t = 5–6 s). (**a**) Velocity signal. (**b**) Voltage signal. (**c**) Voltage change ratio signal.

The least-squares method is used to curve-fit the time–velocity signal, time–voltage signal, and time–voltage change ratio signal in Figure 4; the fitting results are as follows:

$$\dot{r_1}(t) = -0.2084 \times \sin(72.5 \times t) - 0.5671 \cos(72.5 \times t) \tag{30}$$

$$v(t) = 20.08 \times \sin(72.5 \times t) - 8.056 \times \cos(72.5 \times t) \tag{31}$$

$$\dot{v}(t) = 530.9 \times \sin(72.5 \times t) + 1476 \times \cos(72.5 \times t) \tag{32}$$

According to Equations (30)–(32), $p_0 = -0.2084$, $p_1 = 530.9$, $p_2 = 20.08$, $q_0 = -0.5671$, $q_1 = 1476.3$, and $q_2 = -8.06$. Substituting these values into the Equations (26) and (27), the system electromechanical coupling coefficient ϑ and the equivalent capacitance coefficient c_p value can be obtained:

$$\vartheta = 5.1977 \times 10^{-4} \tag{33}$$

$$c_p = 2.0025 \times 10^{-7} \tag{34}$$

In order to analyze the accuracy of the identification algorithm, the electromechanical coupling coefficient ϑ and the equivalent capacitance coefficient c_p are obtained by the identification and analyzed for error, and the results are shown in Table 3.

It can be seen from Table 3 that the errors of the electromechanical coupling coefficient and equivalent capacitance coefficient of the system are 0.11% and 0.13%, respectively, which is mainly due to a certain error in the first harmonic fitting of the system response signal. The error of both is less than 0.5%; therefore, the identification method has good accuracy.

Table 3. Results of electrical parameter identification.

Parameter Type	ϑ	c_p
Select value	5.192×10^{-4}	2×10^{-7}
Identification value	5.1977×10^{-4}	2.0025×10^{-7}
Error	0.11%	0.13%

5. Experimental Analysis

In order to obtain the actual dynamic model parameter values of the piezoelectric energy harvester and verify the correctness of the parameter identification method, an experimental platform was built for experimental verification, as shown in Figures 5 and 6. The experimental platform is composed of laser vibrometer (LV-S01), vibrometer controller (LV-S01), oscilloscope (DSOX3024T), shaker (E-JZK-5T), handheld vibrometer (CoCo-80), computer (Lenovo), controller (VT-9008), amplifier (E5871A) and acceleration sensor; the resolution of laser vibrometer was 1 μm/s. During the experiment, we used the laser vibrometer to collect the response output velocity signal of the PEH-C and used the handheld vibrometer for data collection and storage; next, under the excitation conditions, the output terminal of PEH-C was directly connected to the oscilloscope probe, and the voltage signal was obtained through the oscilloscope. The length of the composite beam in the *X*-axis direction was 40 mm, the width of the composite beam was 8 mm, the thickness was 0.2 mm, the length of the linear beam was 20 mm, and the radius and chord length of the arched part were 10 mm and 20 mm. The width of the PVDF pasted on the composite beam was the same as that of the composite beam; the thickness was 0.11 mm.

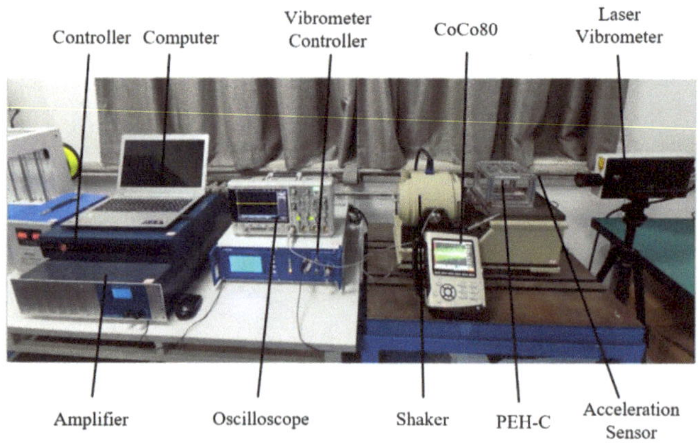

Figure 5. Experimental platform for vibration test of the PEH-C.

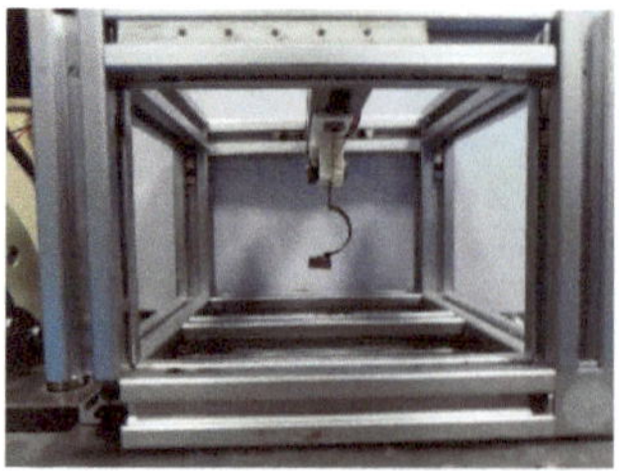

Figure 6. PEH-C.

We set the excitation amplitude of the vibration platform to 10 m/s^2 and the excitation frequency to 13 Hz, and we conducted a fixed frequency experiment under simple harmonic excitation. The result is shown in Figure 7. Figure 7a shows the original velocity data signal collected by CoCo80. The signal has an obvious external interference signal, which is mainly caused by the external environment. We intercepted the signal of the attenuation part and conducted variational mode (VMD) processing to filter out the interference signal, and we obtained the velocity signal under the open-circuit condition of the composite beam piezoelectric vibrator, as shown in Figure 7b. As shown in Figure 7c, the displacement signal was obtained by integrating the velocity signal. The image has obvious distortion in 2–4 s, indicating that the envelope of the displacement signal is no longer smooth at this time, which may be caused by the VMD decomposition filtering of the signal and the signal curve not being smooth. We intercepted the 1-s signal, and the result is shown in Figure 7d.

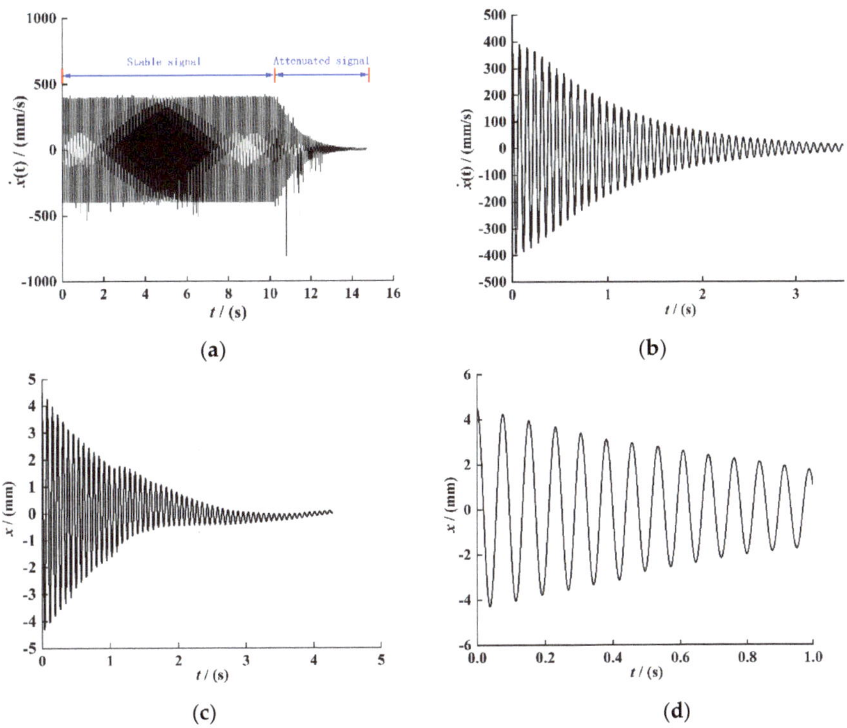

Figure 7. The response to time-domain signals of the PEH-C. (**a**) Original Velocity signal. (**b**) Velocity signal after VMD filtering. (**c**) Displacement attenuation signal. (**d**) Displacement attenuation sinal.

For the signal data shown in Figure 7d, the Hilbert transform method was used to process the displacement time-domain attenuation signal and draw its envelope and backbone curve; the results are shown in Figure 8a,b.

It can be seen from Figure 8a that the expression of the fitting equation for the envelope of the displacement attenuation curve is

$$y \approx 4.5244e^{-0.9484t} \tag{35}$$

It can be seen from Figure 8b that the expression of the system backbone curve fitting equation is

$$y \approx 3.809 \times 10^6 t + 6812.33 \tag{36}$$

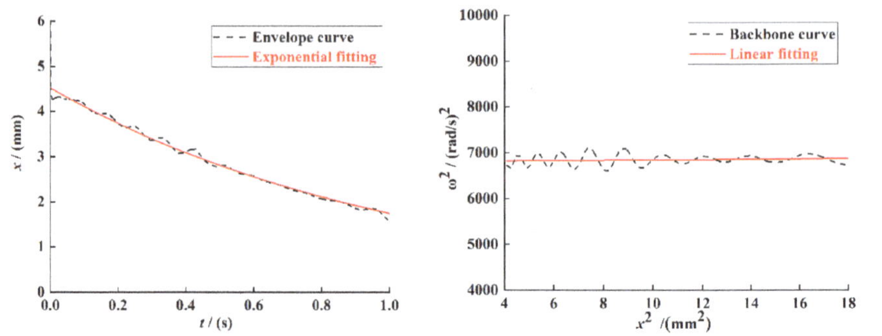

(**a**) (**b**)

Figure 8. System response curve under free vibration attenuation. (**a**) Signal envelope curve. (**b**) Signal backbone curve.

From Equations (35) and (36), $a = -0.9484$, $b_0 = 3.809$, and $b_1 = 6812.33$. Substituting the values of a, b_0, and b_1 into Equations (19), (14) and (21), the actual linear stiffness coefficient α, the nonlinear stiffness coefficient β, and the linear damping coefficient c in the dynamic model of the PEH-C can be obtained:

$$\alpha \approx b_1 m + m a^2 = 4.2 \times 10^{-3} \times \left(6812.33 + (-0.9484)^2\right) = 28.62 \tag{37}$$

$$\beta \approx \frac{4 b_0 m}{3} = \frac{4 \times 3.809 \times 4.2 \times 10^3}{3} = 21330.6 \tag{38}$$

$$c \approx -2ma = 2 \times 4.2 \times 10^{-3} \times 0.9484 = 0.008 \tag{39}$$

In order to ensure the accuracy of the damping coefficient of the PEH-C, the velocity attenuation signal in Figure 8b was processed by Hilbert transform to obtain its envelope; finally, the curve fitting was performed based on the least-squares method. The result is shown in Figure 9.

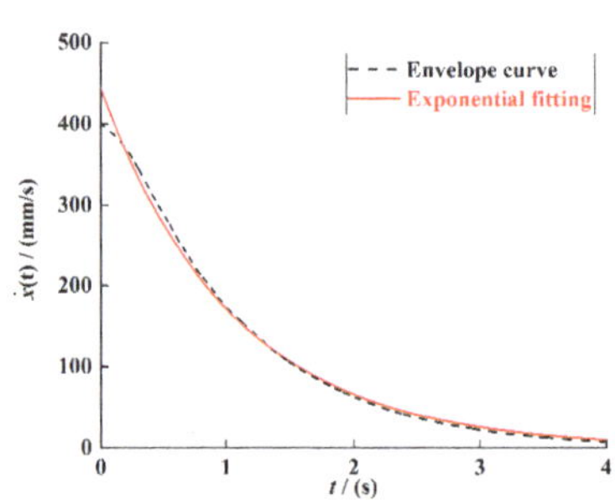

Figure 9. Velocity attenuation signal envelope and its fitting curve.

The envelope curve fitting equation is

$$y \approx 443.8 e^{-0.9483t} \tag{40}$$

Thus, we can obtain the size of the damping coefficient as 0.008; the comparison shows that the damping coefficient identified based on the velocity attenuation curve is consistent with the identification result based on the displacement attenuation signal.

Figure 10a shows the output voltage diagram of the PEH-C under the same conditions. In order to eliminate the interference of hybrid wave on the signal, the experimental data were filtered, and the steady-state output voltage was intercepted within 6–7 s. The result is shown in Figure 10b; the time interval in which the voltage is negative at 6–8 s is significantly longer than the time interval in which the voltage is positive. The voltage signal in the figure is obviously asymmetric, and the voltage is not completely stable, which may be caused by the manufacturing error of the PEH-C and the arched section of the composite beam. Figure 10c shows the voltage change rate curve at the same time. In order to avoid the problem of asynchrony when collecting voltage and velocity signals, according to the theoretical simulation results, the velocity signal change trend should be 180° out of phase with the voltage change rate trend. The result is shown in Figure 10d.

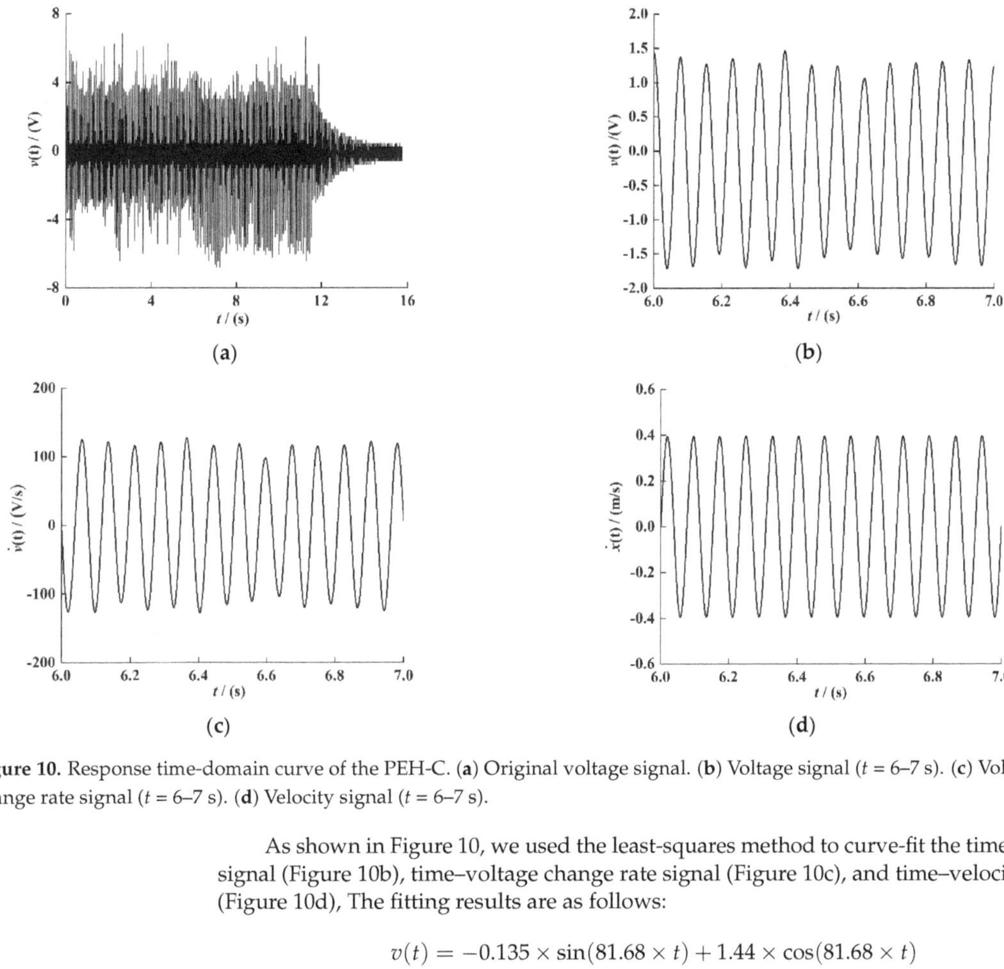

Figure 10. Response time-domain curve of the PEH-C. (**a**) Original voltage signal. (**b**) Voltage signal (t = 6–7 s). (**c**) Voltage change rate signal (t = 6–7 s). (**d**) Velocity signal (t = 6–7 s).

As shown in Figure 10, we used the least-squares method to curve-fit the time–voltage signal (Figure 10b), time–voltage change rate signal (Figure 10c), and time–velocity signal (Figure 10d), The fitting results are as follows:

$$v(t) = -0.135 \times \sin(81.68 \times t) + 1.44 \times \cos(81.68 \times t) \tag{41}$$

$$\dot{v}(t) = -118 \times \sin(81.68 \times t) - 5.86 \times \cos(81.68 \times t) \tag{42}$$

$$\dot{r_1}(t) = 0.394 \times \sin(81.68 \times t) + 0.00265 \times \cos(81.68 \times t) \tag{43}$$

From Equations (41)–(43), we can see that $p_0 = 0.394$, $p_1 = -118$, $p_2 = -0.135$, $q_0 = 0.00265$, $q_1 = -5.86$, and $q_2 = 1.44$. Substituting these values into Equations (26) and (27), the system electromechanical coupling coefficient ϑ and the equivalent capacitance coefficient c_p value can be obtained:

$$\vartheta = 8.57 \times 10^{-5} \tag{44}$$

$$c_p = 2.85 \times 10^{-7} \quad (45)$$

According to Equation (41), the system resonance frequency $\omega_n = 81.68$ rad/s and the simultaneous equation $\omega_n = \sqrt{\alpha/m}$; we bring the mechanical parameter α into it, and therefore, $\alpha = 28.02$, Table 4 shows the identification results of system mechanics and electrical parameters.

Table 4. The identification results of the mechanical and electrical parameters of the PEH-C.

Parameter	α	β	c	ϑ	c_p
Identification value	28.02	21,330.6	0.008	8.57×10^{-5}	2.85×10^{-7}

We substitute the identification result into (3) for a solution and give the system external excitation as $\sin(81.68 \times t)$; the simulation result is shown in Figure 11.

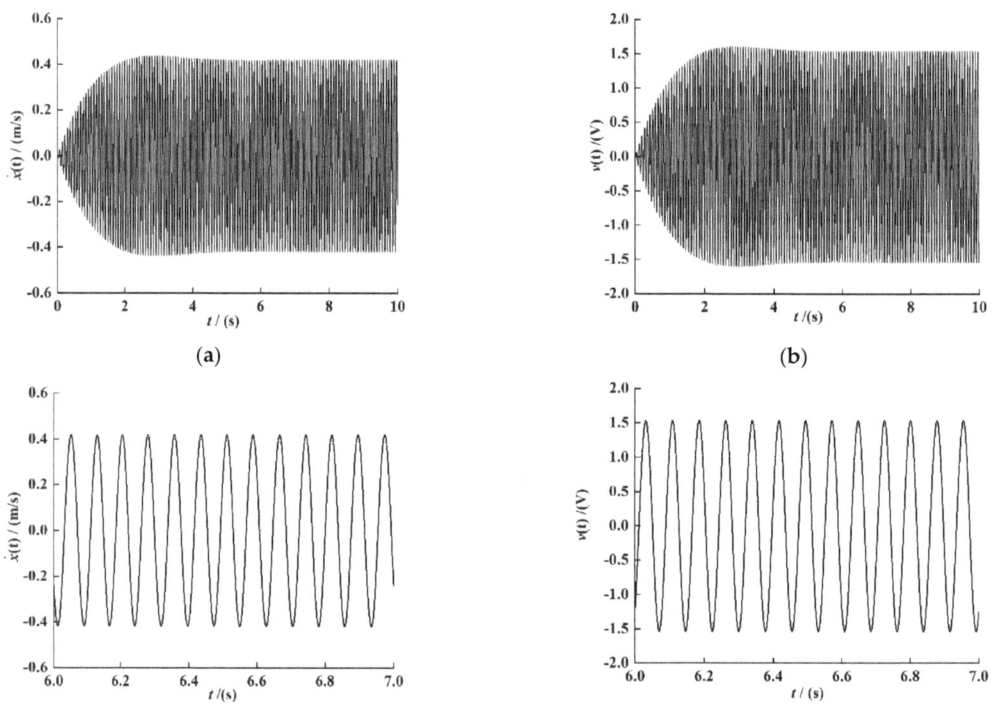

Figure 11. System numerical simulation response time-domain signal. (**a**) Velocity signal. (**b**) Voltage signal. (**c**) Velocity interception signal under stability. (**d**) Voltage interception signal under stability.

Comparing Figures 7, 10 and 11, it can be seen that the theoretical analysis is consistent with the experimental results, and it is not difficult to find that the identification method has good accuracy.

6. Conclusions

Based on the unique short-circuit and open-circuit output characteristics of the PEH-C, this paper proposes a high-precision identification method for the damping coefficient, stiffness coefficient, electromechanical coupling coefficient, and equivalent capacitance coefficient of the PEH-C. The analytical expressions of the parameters are obtained through theoretical modeling, and then the Runge–Kutta numerical method is used to verify the accuracy of the algorithm. The results show that the error of the damping coefficient identified by the algorithm is 4.5%; the errors of linear stiffness and nonlinear stiffness are 0.43% and 1.37%, respectively; and the errors of electromechanical coupling coefficient and equivalent capacitance coefficient are 0.11% and 0.13%, respectively. An experimental platform was built for verification, and the experimental results show that the parameters obtained through parameter identification have good accuracy and applicability. The error in this study is less than 5%, which is within an acceptable range when compared with the error analysis of related studies in the reference literature. We can use this parameter identification method to obtain the mechanical and electrical parameters of the system, solve the system dynamics model more accurately, and better guide the design of piezoelectric energy harvesters.

Author Contributions: Conceptualization, X.Z., C.Z. and L.W.; methodology, L.C. and X.C.; software, L.C. and L.W.; validation, F.Z. and D.X.; writing—original draft preparation, C.Z.; writing—review and editing, X.Z.; visualization, F.Z.; supervision, X.Z.; project administration, X.Z. and D.X.; funding acquisition, X.Z. and H.F.; All authors have read and agreed to the published version of the manuscript.

Funding: This research was funded by the National Natural Science Founds of China (Grant No. 51974228), Shaanxi Innovative Talent Plan Project (No. 2018TD-032), and the Key R&D project in Shaanxi (Grant No. 2018ZDCXL-GY-06-04).

Institutional Review Board Statement: Not applicable.

Informed Consent Statement: Not applicable.

Data Availability Statement: Not applicable.

Acknowledgments: Authors acknowledge the support from the students Meng Zuo and Yan Guo for their assistance in the modeling and experiments.

Conflicts of Interest: The authors declare no conflict of interest.

References

1. Wu, Y.; Qiu, J.; Zhou, S.; Ji, H.; Chen, Y.; Li, S. A piezoelectric spring pendulum oscillator used for multi-directional and ultra-low frequency vibration energy harvesting. *Appl. Energy* **2018**, *231*, 600–614. [CrossRef]
2. Yang, Z.; Zhou, S.; Zu, J.; Inman, D. High-Performance Piezoelectric Energy Harvesters and Their Applications. *Joule* **2018**, *2*, 642–697. [CrossRef]
3. Tran, N.; Ghayesh, M.H.; Arjomandi, M. Ambient vibration energy harvesters: A review on nonlinear techniques for performance enhancement. *Int. J. Eng. Sci.* **2018**, *127*, 162–185. [CrossRef]
4. Liu, W.; Badel, A.; Formosa, F.; Zhu, Q.; Zhao, C.; Hu, G.-D. A Comprehensive Analysis and Modeling of the Self-Powered Synchronous Switching Harvesting Circuit With Electronic Breakers. *IEEE Trans. Ind. Electron.* **2018**, *65*, 3899–3909. [CrossRef]
5. Lee, J.; Oh, J.; Kim, H.; Choi, B. Strain-based piezoelectric energy harvesting for wireless sensor systems in a tire. *J. Intell. Mater. Syst. Struct.* **2015**, *26*, 1404–1416. [CrossRef]
6. Yang, Z.; Zu, J. Toward Harvesting Vibration Energy from Multiple Directions by a Nonlinear Compressive-Mode Piezoelectric Transducer. *IEEE/ASME Trans. Mechatron.* **2015**, *21*, 1787–1791. [CrossRef]
7. Peigney, M.; Siegert, D. Piezoelectric energy harvesting from traffic-induced bridge vibrations. *Smart Mater. Struct.* **2013**, *22*, 095019. [CrossRef]
8. He, Q.; Dong, C.; Li, K.; Wang, J.; Xu, D.; Li, X. A multiple energy-harvester combination for pattern-recognizable power-free wireless sensing to vibration event. *Sens. Actuators A Phys.* **2018**, *279*, 229–239. [CrossRef]
9. Chen, Z.; Yang, Y.; Lu, Z.; Luo, Y. Broadband characteristics of vibration energy harvesting using one-dimensional phononic piezoelectric cantilever beams. *Phys. B Condens. Matter* **2013**, *410*, 5–12. [CrossRef]
10. Zhang, X.H.; Yang, W.J.; Zuo, M.; Tan, H.; Fan, H.; Mao, Q.; Wan, X. An Arc-shaped Piezoelectric Bistable Vibration Energy Harvester: Modeling and Experiments. *Sensors* **2018**, *18*, 4472. [CrossRef] [PubMed]
11. Zhang, X.; Zuo, M.; Tan, H.; Wu, Z.; Lai, Z.; Fan, H. Dynamics analysis of multi-field coupled piezoelectric energy harvester under random exci-tation. In *IOP Conference Series: Materials Science and Engineering*; IOP Publishing: Bristol, UK, 2019; Volume 531, p. 012038.
12. Sun, S.; Tse, P.W. Modeling of a horizontal asymmetric U-shaped vibration-based piezoelectric energy harvester (U-VPEH). *Mech. Syst. Signal Process.* **2019**, *114*, 467–485. [CrossRef]
13. Cao, J.; Zhou, S.; Wang, W.; Lin, J. Influence of potential well depth on nonlinear tristable energy harvesting. *Appl. Phys. Lett.* **2015**, *106*, 173903. [CrossRef]

14. Feldman, M.; Braun, S. Nonlinear vibrating system identification via Hilbert decomposition. *Mech. Syst. Signal Process.* **2017**, *84*, 65–96. [CrossRef]
15. Lu, Z.-Q.; Hu, G.S.; Ding, H.; Chen, L.-Q. Nonlinear jump based parameter estimation for both stiffness and damping nonlinearities. *J. Vib. Control.* **2018**, *25*, 325–335. [CrossRef]
16. Yuan, T.-C.; Yang, J.; Chen, L.-Q. Nonlinear dynamics of a circular piezoelectric plate for vibratory energy har-vesting. *Commun. Nonlinear Sci. Numer. Simul.* **2018**, *59*, 651–656. [CrossRef]
17. Chen, L.-Q.; Yuan, T. Harmonic Balance in the Dynamic Analysis of Circular Composite Plate Harvester. *Procedia IUTAM* **2017**, *22*, 200–207. [CrossRef]
18. Chen, T.-C.; Yang, J.; Chen, L.-Q. Restoring force surface method and Hilbert transform one for nonlinear system identification. *J. Vib. Shock.* **2019**, *38*, 73–78.
19. Zhou, S.; Cao, J.; Inman, D.J.; Lin, J.; Liu, S.; Wang, Z. Broadband tristable energy harvester: Modeling and experiment verification. *Appl. Energy* **2014**, *133*, 33–39. [CrossRef]
20. Yuan, T.; Yang, J.; Chen, L.Q. Nonlinear characteristic of a circular composite plate energy harvester: Experiments and simula-tions. *Nonlinear Dyn.* **2017**, *90*, 2495–2506. [CrossRef]
21. Stanton, S.C.; Erturk, A.; Mann, B.P.; Inman, D.J. Nonlinear piezoelectricity in electroelastic energy harvesters: Modeling and experi-mental identification. *J. Appl. Phys.* **2010**, *108*, 074903. [CrossRef]
22. Kefal, A.; Maruccio, C.; Quaranta, G.; Oterkus, E. Modelling and parameter identification of electromechanical systems for energy harvesting and sensing. *Mech. Syst. Signal Process.* **2019**, *121*, 890–912. [CrossRef]
23. Porfiri, M.; Maurini, C.; Pouget, J. Identification of electromechanical modal parameters of linear piezoelectric structures. *Smart Mater. Struct.* **2007**, *16*, 323–331. [CrossRef]
24. Delpero, T.; E Bergamini, A.; Ermanni, P. Identification of electromechanical parameters in piezoelectric shunt damping and loss factor prediction. *J. Intell. Mater. Syst. Struct.* **2012**, *24*, 287–298. [CrossRef]
25. Truong, B.D.; Le, C.P.; Halvorsen, E. Parameter identification from frequency response of MEMS energy harvesters Smart Sensors, Actuators, and MEMS VIII. *Int. Soc. Opt. Photonics* **2017**, *10246*, 102460U.
26. Zhang, X.H.; Lai, Z.P.; Wu, Z.H.; Tan, H.Z.; Zuo, M.; Fan, H.W. Theoretical modeling and experimental study of a new bistable piezoelectric vibration energy harvesting system. *J. Vib. Eng.* **2019**, *32*, 87–96.
27. Erturk, A.; Inman, D.J. *Piezoelectric Energy Harvesting*; John Wiley & Sons: Hoboken, NJ, USA, 2011.
28. Zhang, X.H.; Zuo, M.; Yang, W.J.; Wan, X. A Tri-Stable Piezoelectric Vibration Energy Harvester for Composite Shape Beam: Nonlinear Modeling and Analysis. *Sensors* **2020**, *20*, 1370. [CrossRef] [PubMed]
29. Kovacic, I.; Brennan, M.J. *The Duffing Equation: Nonlinear Oscillators and Their Behaviour*; Wiley: Chichester, UK, 2011; pp. 180–181.

sensors

MDPI

Article

Experimental Characterization of Optimized Piezoelectric Energy Harvesters for Wearable Sensor Networks

Petar Gljušćić [1,2] and Saša Zelenika [1,2,*]

1 University of Rijeka, Faculty of Engineering, Vukovarska 58, 51000 Rijeka, Croatia; pgljuscic@riteh.hr
2 University of Rijeka, Centre for Micro- and Nanosciences and Technologies, Radmile Matejčić 2, 51000 Rijeka, Croatia
* Correspondence: szelenika@uniri.hr; Tel.: +385-(0)51-406517

Abstract: The development of wearable devices and remote sensor networks progressively relies on their increased power autonomy, which can be further expanded by replacing conventional power sources, characterized by limited lifetimes, with energy harvesting systems. Due to its pervasiveness, kinetic energy is considered as one of the most promising energy forms, especially when combined with the simple and scalable piezoelectric approach. The integration of piezoelectric energy harvesters, generally in the form of bimorph cantilevers, with wearable and remote sensors, highlighted a drawback of such a configuration, i.e., their narrow operating bandwidth. In order to overcome this disadvantage while maximizing power outputs, optimized cantilever geometries, developed using the design of experiments approach, are analysed and combined in this work with frequency up-conversion excitation that allows converting random kinetic ambient motion into a periodical excitation of the harvester. The developed optimised designs, all with the same harvesters' footprint area of 23 × 15 mm, are thoroughly analysed via coupled harmonic and transient numerical analyses, along with the mostly neglected strength analyses. The models are validated experimentally via innovative experimental setups. The thus-proposed ϕ = 50 mm watch-like prototype allows, by using a rotating flywheel, the collection of low-frequency (ca. 1 to 3 Hz) human kinetic energy, and the periodic excitation of the optimized harvesters that, oscillating at their eigenfrequencies (~325 to ~930 Hz), display specific power outputs improved by up to 5.5 times, when compared to a conventional rectangular form, with maximal power outputs of up to >130 mW and average power outputs of up to >3 mW. These power levels should amply satisfy the requirements of factual wearable medical systems, while providing also an adaptability to accommodate several diverse sensors. All of this creates the preconditions for the development of novel autonomous wearable devices aimed not only at sensor networks for remote patient monitoring and telemedicine, but, potentially, also for IoT and structural health monitoring.

Keywords: piezoelectric energy harvesters; medical sensor networks; DoE; frequency up-conversion; optimized geometry; FE numerical modelling; experimental assessment

Citation: Gljušćić, P.; Zelenika, S. Experimental Characterization of Optimized Piezoelectric Energy Harvesters for Wearable Sensor Networks. *Sensors* **2021**, *21*, 7042. https://doi.org/10.3390/s21217042

Academic Editor: Francesc Pozo

Received: 1 October 2021
Accepted: 20 October 2021
Published: 24 October 2021

1. Introduction

Innovative wearable devices worn by the users as accessories, e.g., as a watch or a clothing element, are the result of the ongoing development of a new class of electronic gadgets characterized by rapidly decreasing size and power requirements, increasingly used in biomedical applications [1,2]. In order to operate, such devices require a reliable power source. When in the form of a conventional or a rechargeable battery, these sources have, however, limited lifetimes and a negative environmental impact. They are, hence, increasingly and efficiently replaced with energy harvesting (EH) systems, thus enabling the development of a new class of autonomous wearable devices and resulting sensor networks [3,4]. In fact, EH involves the collection of low-level energy from the environment and its conversion into usable electrical energy.

The most common ambient energy sources are kinetic—converted via the piezoelectric effect or by using the electromagnetic or triboelectric principles, thermal (waste heat)—collected by using thermoelectric (Seebeck effect) devices, solar or light energy—converted by using photovoltaics, as well as radio frequency (RF)—which can be harvested by using specialized antennas [5,6]. The resulting autonomous EH-powered wearable devices can include various sensors, e.g., heart rate, blood pressure, glucose level, temperature sensors or oximeters, as well as an energy storage element and data processing and communication components, making them suitable for creating autonomous sensor networks aimed at remote patient monitoring and telemedicine, professional athletics, work safety in high risk professions and various IoT or structural health monitoring (SHM) applications [3,4,7,8]. The concept of EH integration in autonomous sensor networks is outlined in Figure 1.

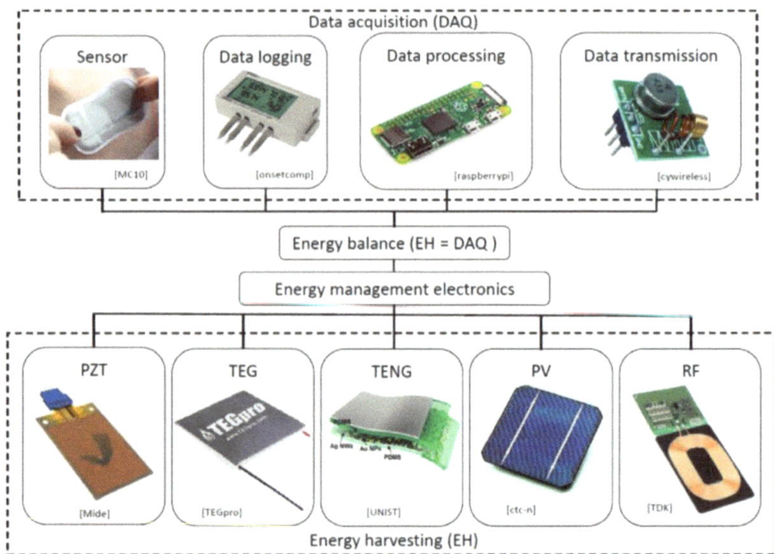

Figure 1. EH integration in autonomous devices with sensors and the respective data processing and communication modules, enabling the creation of sensor networks [8].

In this framework, kinetic energy, present in all moving systems, also induced by human motion, is a very reliable and copious source of ambient energy. The piezoelectric principle, due to its simplicity, versatility, miniaturization potential and energy density, represents, in turn, a favourable approach to the collection and conversion of ambient kinetic energy [5,6].

Based on a seminal work of Priya in 2005 [9], numerous solutions aimed at human motion EH are recently suggested in literature. The device suggested by Pozzi et al. [10], based on an inverted piezoelectric windmill to utilize the relative movement at the human knee, represents in this frame one of the early solutions. Despite its rather hefty size, the system generated a relatively low power output that, even with some improvements proposed subsequently by the same research group [11], remained limited to the mW range. The introduction of a flywheel, which collects kinetic energy at the human wrist and periodically excites a piezoelectric energy harvester (PEH), resulted then in a new type of rotational EH device [12]. As shown in Figure 2, the most common form of a PEH device used in this framework is a bimorph cantilever of rectangular shape, comprising a metallic substrate with two layers of piezoelectric materials deposited on it. In fact, the bimorph configuration was found to be more efficient compared to the unimorph one [13]. The device is then clamped at the excited end, allowing the bimorph to oscillate—thus deforming the piezoelectric material and generating charge (i.e., voltage). In some cases, a tip mass is attached to the free end of the cantilever in order to increase the displacements and tune the response of the device to the excitation frequency, resulting in higher power outputs [5,6].

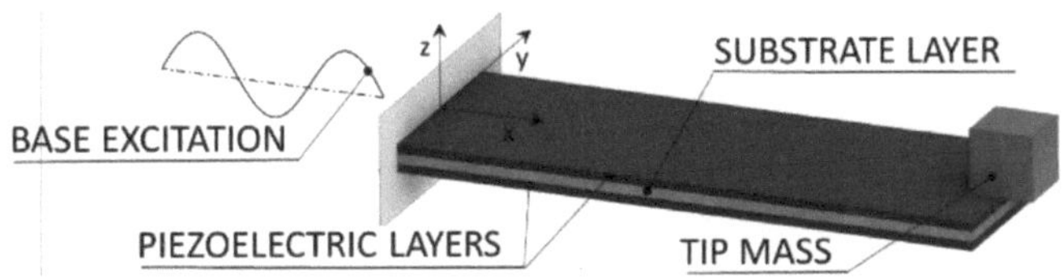

Figure 2. Bimorph piezoelectric energy harvester [3].

In [14] a miniaturized wearable device is then proposed, comprising a flywheel that magnetically excites a single rectangular PEH, whereas in [15] a device, utilizing a flywheel with magnetic plucking, is studied from the standpoint of the relative positions of the excitation magnets. The latter solution is somewhat improved by introducing several thin-film PEHs [16]. Although the performances of the above devices are increased compared to conventional PEHs exposed to human motion excitation, their power output levels are still rather limited, generating merely ~40 μW.

Various innovative approaches have thus been proposed in recent literature. The double pendulum mechanism for leg movement EH, albeit providing a viable approach and showing promising performances (~2 mW), uses a relatively complex motion mechanism [17]. A handheld device, consisting of a rotating flywheel equipped with magnets, interacting with stationary magnets attached to radially placed piezoelectric beams fixed on both ends, results again in a rather bulky and complex device, generating a small amount of power, i.e., 0.18 μW [18].

What is more, a very small number of these studies considers the dynamical stresses present in the piezoelectric materials, and thus the durability of the developed devices [18,19]. This would indicate that a notable number of proposed devices would, in practice, be unable to perform, in the long run, as they were meant to.

Another recent research direction is based on improving PEHs' performances by changing the geometry of the harvesters. It has thus been shown that the performances of a PEH bimorph can be improved using a trapezoidal shape instead of a rectangular one of comparable width; by inverting the trapezoid, the performances can be enhanced even further [20]. Significant power gains compared to the reference shape, although in the μW power range only, were, in turn, achieved by using complex optimized geometries, with the best performing variation being the one without area constraints [21]. Although the static stress was considered in this case, the overall performances of the PEH are, again, uncertain when the effects of the dynamical stresses in long-term operation, coupled to the narrow cross-section near the fixture of the harvester, are considered. As in the case of the trapezoidal and inverse PEHs, a large portion of the reference rectangular PEH, whose use could potentially lead to further gains in performances, is removed in the optimization process, here, as well.

The process of topology optimization has also been applied to PEHs, both to the thickness of the layers as well as to the surface itself, but the resulting geometries, although providing output voltage improvements, are overly complex, while strength considerations are again generally omitted [22,23].

Miniaturized EH devices are also being studied as means of human motion utilization. Various approaches are being considered in this frame. If compared to a conventional MEMS device, the recently suggested harmonically excited one displays then notable power gains and a reduction of the needed excitation frequency, but its operational frequency of 40 Hz and the low power output (<0.5 μW), limits its potential in wearable applications [24]. A miniaturized rotational harvester using a flywheel and mechanically plucked MEMS cantilevers, enables, in turn, generating 11 μW, decreasing, however, by half with a variation in plucking frequency [25]. Despite representing a promising mean of human motion EH, the low power outputs of such design configurations strongly limit their application field.

Another innovative approach to human motion PEH is the utilization of piezoelectric textiles [26]; while the output powers in this case are again low, the possibility of increasing the amount of worn EH material in the form of clothing could provide means to improve performances. Due to the commonplace treatment of clothing elements (crumpling, washing etc.), the durability of such devices is, however, still an issue that strongly limits their potential applications.

In any case, the proposed solutions clearly show that the collection of the high amounts of kinetic energy due to arm and leg motion is a particularly efficient way of harvesting human motion energy [27]. Numerous potential solutions aimed at using this energy via wearable PEHs are then presented in literature, the most promising ones relying on changing the boundary conditions at the cantilever free end via active tuning or damping control, using the above mentioned approach of optimising the bimorphs' shape, extending the viable frequency bandwidth by employing several harvesters or complex geometries with bi-stable or nonlinear responses, or, finally, by using a frequency up-conversion (FUC) approach, wherein the cantilever's free end is plucked and the bimorph is let oscillating at its eigenfrequency [3]. After a careful consideration of these approaches, with particular attention to reliability, technological simplicity and ease of operation, a viable and promising path towards a functional wearable device that utilizes kinetic energy from human motion could be an improved wrist-worn device, using a flywheel to excite a number of optimized PEHs.

This work will, thus, provide an enhanced approach to the employment of such a path, where the whole reference rectangular shape of the PEH is used and segmented into optimized sections. The optimization itself is performed by using design of experiments (DoE) methods. The resulting shapes, combined with a suitable miniaturized plucking mechanism allowing the bimorphs to oscillate at their eigenfrequencies, will provide the basis of an innovative highly efficient wrist-worn EH device aimed at powering autonomous wearable sensors. A detailed study of the strength issues will also be performed to ensure the needed resilience of the proposed class of devices.

An overview of the viability of PEH devices as a power source for wearable applications is therefore given in Section 2 of this work, where the major drawback of this EH principle, as well as a potential solution, are pointed out. Section 3 summarises the influence of the PEH devices' geometry on their electro-mechanical response, allowing innovative designs to be proposed. The considered PEH shapes are then optimized and their responses are analysed. A detailed description of the experimental setups used to validate the numerical results is given in Section 4. Due to the dynamical nature of PEHs' operation, stress analyses are performed in Section 5 to determine the maximum excitation levels, while not exceeding the fatigue limit of the piezoelectric material. The results obtained via the described experimental setups are analysed and compared in Section 6. The conclusions of the work are finally drawn in Section 7, where an outlook for future work is provided as well.

2. Energy from Human Motion and Frequency Up-Conversion

Kinetic energy generated by human motion is generally in the form of vibrations or impacts, and, not unlike in several IoT or SHM settings, rather than being periodical, it is characterized by random and chaotic movements. In previous art it was, in fact, established that the conventional activities of daily living (ADL) result in non-periodical excitations of varying amplitudes, with frequencies between ~0.5 and 10 Hz (with a predominant occurrence of lower ones) [28] and with respective maximal acceleration values of up to ~1 g [29]. As shown in Figure 3, for human walking, these ranges of values have been confirmed, also, by studies performed at our premises, in which dominant frequencies of around 1 and 1.8 Hz, with accelerations of ca. 0.1 g, were recorded. In the case of running, values of ~1.3 and ~2.6 Hz, with accelerations of ~0.7 g were, in turn, recorded [30].

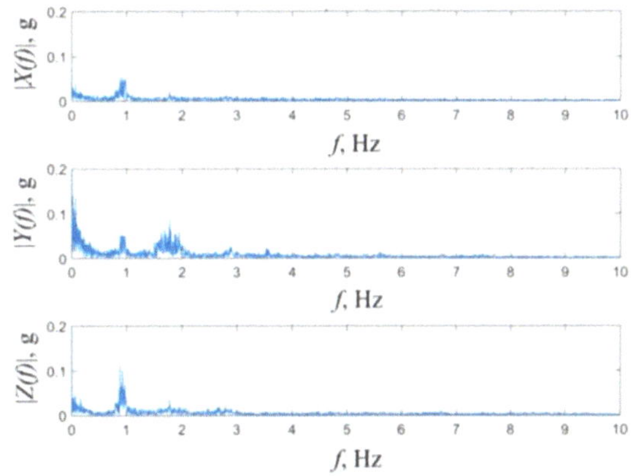

Figure 3. Accelerations during human walking in the frequency domain [30].

On the other hand, it was established that all cantilever-based PEHs are characterized by the same major drawback, i.e., the rapid decrease in conversion efficiency, along with a corresponding decrease of voltage and power outputs, when the excitation frequency moves even slightly away from the eigenfrequency of a specific device [3,5,6]. By introducing PEHs into wearable technologies, this issue, considering the outlined nature of human motion, becomes especially relevant. One of the most promising ways to overcome it is to use the frequency up-conversion (FUC) mechanism [12,31,32], i.e., by converting random human motion into a periodical excitation of the harvesting device by impacting (plucking) the free end of the PEH and letting it oscillate at its eigenfrequency. The FUC approach ensures, thus, that the PEH transduction operates always at the maximum conversion efficiency, despite the random nature of the excitation [12,31,32], significantly facilitating the successful integration of PEH principles into wearables and other applications characterised by random kinetic excitation spectra [3,8,32].

3. Geometry Optimization and Influence on the PEH Response

The most commonly used and commercially readily available PEH type is a cantilever with a rectangular planar layout. A considerable amount of studies, investigating the influence of geometry on PEHs' response is available in literature [3,5,6,12,20–23], mainly focusing on the increase of PEH power density, particularly important for wearable applications, and the broadening of the excitation bandwidth, resulting in a wider applications domain. In order to expand the thus experimentally validated findings on the possibility of replacing a rectangular PEH with trapezoidal or inverse trapezoidal shapes, resulting in a substantial increase in specific power outputs [20], a numerical model is developed using ANSYS® [3]. The damping coefficients required in the carried finite element (FE) analyses are determined here via uncoupled modal analyses, while, due to the complex interactions induced by the forward and backwards electromechanical coupling effects, the optimal load resistance values are determined via the well-established practice of multiple harmonic analyses [3,33–35]. In fact, the approximate equation for determining the optimal resistance [36,37] provides this value for a single set of working conditions, i.e., a single working point.

An initial study comprising numerous FE analyses shows hence that, when compared to a conventional rectangular PEH of the same overall surface area, the design configuration segmented in two trapezoidal (A) and an inverted trapezoidal (B) shape (Figure 4a), allows attaining a higher specific power output, since it allows a more efficient utilization of the limited volume available in wearable devices [3,12]. What is more, the maximal power output of each segment can be specifically matched to an electrical load equivalent to that of a diverse wearable sensor [3]. A further performance-enhancing design variant, comprising a triangular notch at the clamped end (Figure 4b), thus increasing the compliance of the harvester and inducing stress concentration effects, is also analysed [12].

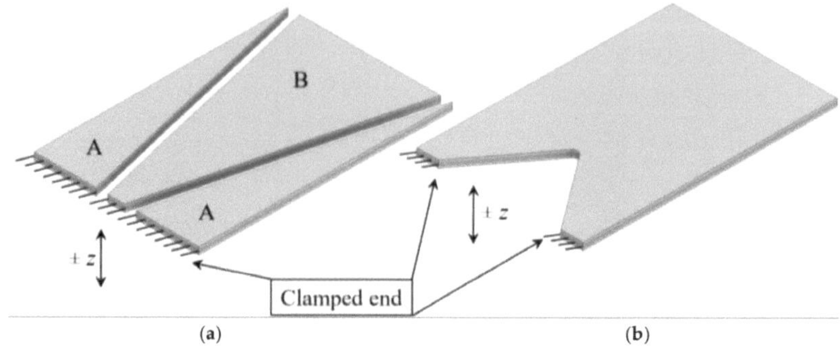

Figure 4. Segmented PEH (**a**) and PEH with a triangular notch at the clamped end (**b**), with the excitation displacement of the clamped end denoted by z.

The results of the performed study show thus the viability of geometry alteration as means to increase the specific power output of PEHs as well as, especially when coupled to FUC-based loading, to broaden the respective useful excitation spectrum. To determine the ideal PEH dimensions, resulting in the highest power outputs, a more in-depth analysis, comprising a structured optimization process, as described in our previously disclosed studies [12,38] is, however, needed [12]. The bimorph harvesters considered in the optimization process, comprising a stainless steel substrate layer and two PZT-5A piezoelectric layers, are commercially available ones [39]. It has an overall surface area of length (l) × width (w) = 23 × 15 mm, whereas the thickness of the stainless steel substrate is t_s = 0.15 mm and of the two PZT-5A piezoelectric layers is t_{pzt} = 0.254 mm, making the overall thickness of the bimorph t~0.7 mm. The relevant basic electromechanical properties of the used materials are reported in Table 1, while the strength limit of the piezoelectric material will be quantified in Section 5 of this work, where it is used as the relevant criterion for the performed stress analyses. In the performed simulation, a variable resistive load is then connected to the PEHs subjected to a 1 g harmonic excitation [38]. The central composite design of experiments (DoE) algorithm is, hence, employed to generate random combinations of the characteristic dimensions of the segmented and notched harvesters, while taking into account also the technological and practical aspects, i.e., the space needed for clamping and soldering the connections [40]. Via modal and coupled 3D FE harmonic analyses and suitable optimization criteria, the optimal combination of the considered dimensions is therefore determined. The inverted trapezoidal shape has thus an optimal width at its narrow end of 3 mm, and, at its broader end, of 14 mm, with the width of the two trapezoidal half-portions of Figure 4a adjusted accordingly to use the whole available rectangular footprint. The width and height of the notch in the corresponding design configuration are, in turn, respectively 11 and 6.5 mm.

In Table 2, are, then, listed the correspondingly obtained maximal output powers, as well as the respective optimal load resistance values; the comparison of the performances of the optimized PEHs to that of a conventional rectangular PEH, having an overall surface area on which the optimal shapes are based, is also provided. It is, thus, initially established that the maximal output power values (P_{max}) are obtained using the notched PEH design variant. When the combined power outputs of all the three segments are taken into account, the segmented PEH shows also far better results than the conventional rectangular version. When the specific power output P_{Smax}, i.e., the maximal power output normalized to the PEH surface area, is considered, the highest values are achieved using the inverted trapezoidal configuration, with the second highest values obtained with the notched PEH. The specific power output of the two trapezoidal segments is lower than that of the rectangular shape, but it should be viewed as a further addition to the already high specific power attained via the inverted trapezoidal shape [38].

Table 1. Properties of PZT-5A and stainless steel [39].

	PZT-5A (3195HD)		
Elastic modulus	E_{PZT}	52	GPa
Poisson ratio	ν_{PZT}	0.31	-
Density	ρ_{PZT}	7800	kg/m^3
Piezoelectric strain	d_{31}	390	pC/N
coefficients	d_{33}	−190	pC/N
	Stainless Steel		
Elastic modulus	E_s	193	GPa
Poisson ratio	ν_s	0.29	-
Density	ρ_s	7800	kg/m^3

Table 2. Comparison of optimized PEH parameters [38].

	$R_{L\ opt}$, kΩ	P_{max}, µW	P_{Smax}, µW/m^2
Trapezoidal	7	26.3	175.5
Inverted	12	131.5	672.7
Notched	7	168.9	545.9
Rectangular	5	141.3	409.5

When the free end of each studied PEH device is subject to an identical plucking deflection δ_z and their resulting free oscillations at the respective eigenfrequencies are preliminarily calculated via FE coupled transient analyses, it is, in turn, concluded that the maximum peak-to-peak voltage generated by the inverted trapezoidal segment alone at its optimal load resistance $R_{L\ opt}$ is comparable to that generated by the conventional rectangular PEH device. When the surplus voltage generated by the two trapezoidal segments is also considered, the segmented PEH clearly outperforms the conventional one. The maximum peak-to-peak voltage values generated by the notched PEH is comparable to that of the inverted trapezoidal shape. This data supports, therefore, fully the practicability of the integration of the optimized PEH shapes with the FUC excitation approach [38], which will be investigated in more detail in the below treatise.

The illustrated approach is then at the basis of a prototype ϕ = 50 mm watch-like wearable device shown in Figure 5, that the Precision Engineering Laboratory of the Faculty of Engineering of the University of Rijeka, Croatia [41] is developing in collaboration with medical institutions, where FUC is achieved via a rotating flywheel equipped with multiple plectra. The prototype being developed includes optimized segmented bimorph harvesters, as determined in the above analyses.

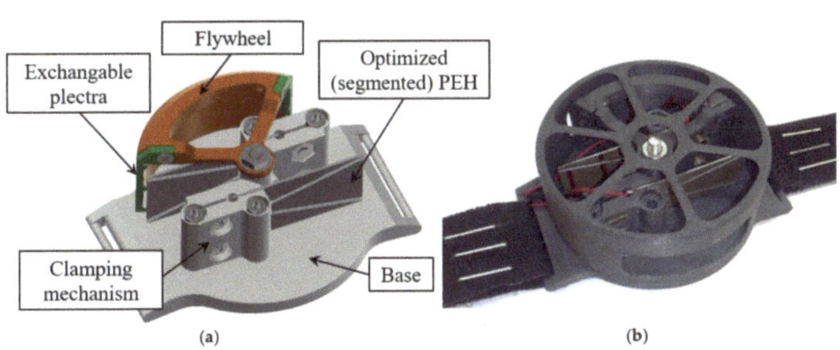

Figure 5. Schema of the devised watch-like wearable PEH device with two optimized segmented bimorph cantilevers (**a**) and the first prototype of the device (**b**).

4. Experimental Setups

The performed extensive static, modal, harmonic, and transient numerical FE analyses have to be validated by suitable experimental measurements. Two different measurement setups are thus developed at the Precision Engineering Laboratory [41], i.e., a harmonic setup and the FUC setup. The performed experiments will then provide means to fully characterised the optimized PEHs to be used in the suggested wearable prototype.

The first setup, schematically shown in Figure 6a, is employed to induce a harmonic excitation to the clamped PEH bimorph over a predefined frequency range. It is based on the Brüel and Kjær® LDS V201 electrodynamic permanent magnet shaker (indicated in the Figure 6b with 1) coupled with the LDS LPA100 power amplifier. The excitation parameters are controlled via an NI LabVIEW® virtual instrument operating on the NI MyRIO 1900 device (2), which is used for data acquisition (DAQ) as well. The acceleration of the shaker and of the fixture of the PEHs, is measured via the Vernier® 3D-BTA accelerometer (3) connected to the Vernier® BT-MDAQ adapter (4). The tested $l \times w \times t = 23 \times 15 \times 0.7$ mm PEH (5), is finally connected to the DAQ unit via the variable resistance box (6) [38].

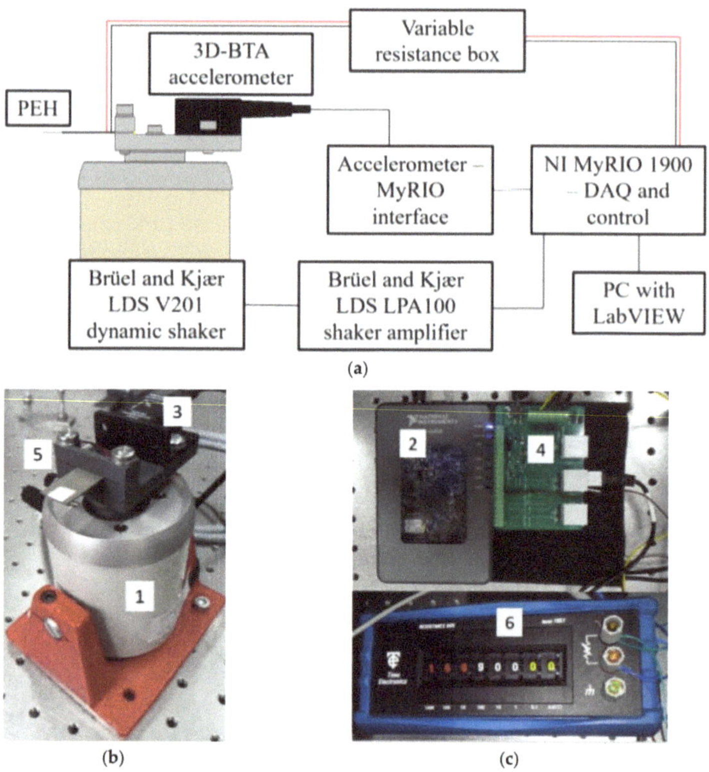

Figure 6. Schema of the experimental setup for harmonic analyses (**a**) with the actual execution showing the Brüel and Kjær® shaker with the accelerometer and the bimorph harvester [38] (**b**), as well as the control, DAQ and variable resistor units [38] (**c**).

The setup aimed at emulating the FUC excitation approach, shown schematically in Figure 7a, comprises, in turn, the 3D-printed PEH clamping mechanism (indicated in Figure 7b with 1) and the 3D-printed rotating exchangeable plectra mounted on the shaft of a DC motor (2). A purely mechanical (i.e., not magnetic) plucking mechanism is used, here, to prevent the possible damping effects of magnets on the oscillating PEHs [15,38]. The voltage output generated by the tested PEH (5) is measured by employing the Agilent®

DSO-X 2012A oscilloscope (3), while the displacement of the PEH's free end is acquired by using a Metrolaser® Vibromet 500V laser doppler vibrometer (4). The PEH itself, having the same overall dimensions as in the case of the harmonic analysis setup, is connected to the oscilloscope via the variable resistance box (6) [38].

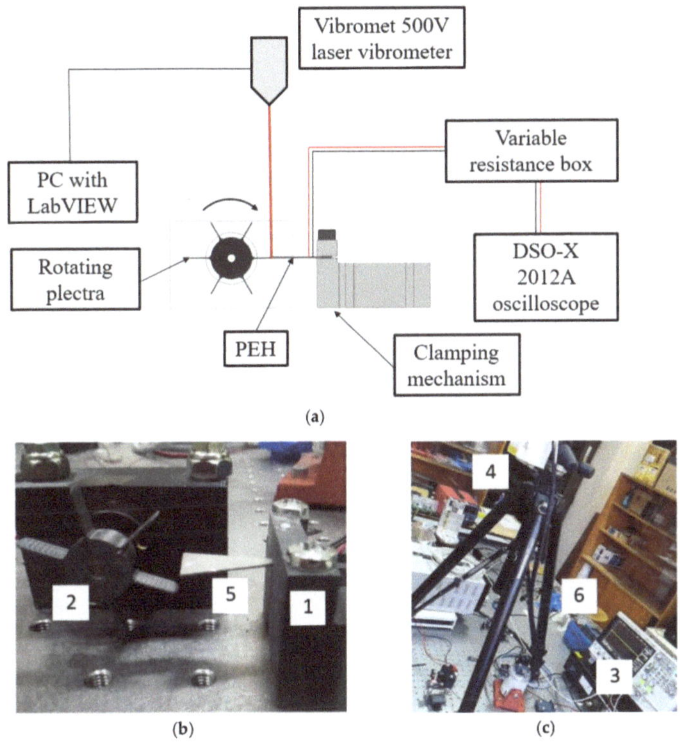

Figure 7. Schema of the FUC experimental setup (**a**) with the actual execution showing the plucking mechanism [38] (**b**) and the DAQ system [38] (**c**).

5. Strength Analysis of Optimized PEH Shapes

During regular operation, a kinetic PEH is subjected to dynamical working conditions. These conditions induce dynamical stresses of the used brittle piezoelectric material. Special attention is thus to be devoted to the fatigue lifetime of the considered PEHs, which is almost completely neglected in prior art. In fact, the criteria of maximum power or voltage output alone, most commonly used in literature [21,28,32], can be relevant only when a small number of operational cycles suffices for a particular application. When, however, a bimorph PEH device is intended for wearable or other devices with common dynamical excitation conditions, the maximum power or voltage outputs can be viable criteria only when the stress levels are lower that the fatigue limit of the used piezoelectric material, i.e., for PZT-5A, R_d = 55 MPa [42]. FE stress analyses are thus performed in ANSYS® for the conventional rectangular PEH shape, used as a reference, and for the considered optimized PEH shapes.

In this frame, a variant of the segment harvester shape, aimed at further increasing the power output by introducing wavy contours (notches) along the edges of the segments, as shown in Figure 8, is considered, as well. The assumption is that in this configuration the bigger stress in the piezoelectric material will result in a boost in charge generation, and thus in an increase in voltage and power outputs. In all the performed FE analyses one end of the PEH is then fixed, i.e., clamped, while the free end, without tip mass, is statically bent, with deflections δ_z ranging from 0.05 to 1 mm, allowing the resulting stresses to be obtained (Figure 9). As expected, the highest stresses occur at the fixture, with a more uniform stress distribution along the trapezoidal shapes (Figure 9b, e). The effect of the wavy edges, i.e., of the stress concentrators on the segmented shape, can, in turn, be clearly observed in the areas of increased stress levels, particularly in Figure 9e. The stresses are redistributed, in this case, towards the concentrators and away from the fixture, making the overall stress distribution more uniform. These considerations will have a significant impact on the power output of the optimized PEHs, as elaborated in Section 6 below.

The thus attained maximum bending stress values σ_{max} in the piezoelectric layers, as determined via the performed FE analyses, are reported in Figure 10a versus the respective tip displacements δ_z. To determine the maximum allowable deflection of the considered PEH shapes, the fatigue bending limit of the used PZT-5A piezoelectric material R_d is also marked on the graph. From the stress data analysis, it can be concluded that, in order to keep the stress levels below the dynamical strength limit, the tip deflection δ_z of around 0.5 mm can be applied to the notched shape, while δ_z = 0.6 mm is applicable to the inverted trapezoidal shape with and without stress concentrators. The trapezoidal shape with stress concentrators and the conventional rectangular shape can be subjected to δ_z = 1 mm, while an even larger deflection can be applied to the trapezoidal shape with a straight edge. By limiting the plucking deflection of the optimized PEHs to these values, the fatigue safety of the piezoelectric layers can be assured, and thus long-lasting operation of the device can be achieved.

Figure 8. Trapezoidal (**a**) and inverted trapezoidal (**b**) PEH segments with wavy edges (added stress concentrators).

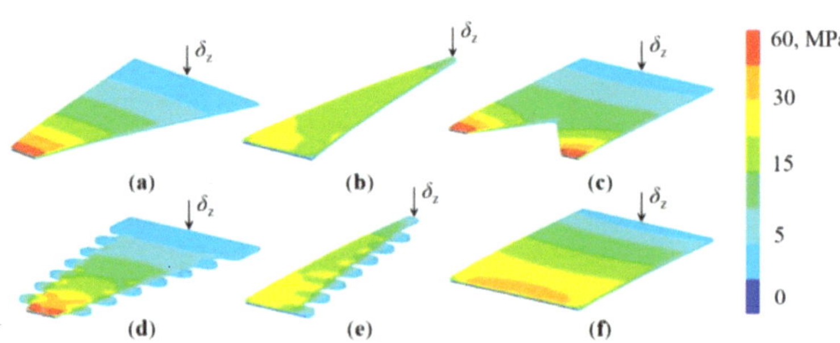

Figure 9. FE calculated stresses in the piezoelectric layers of the PEHs for a δ_z = 0.5 mm free-end deflection: inverted (**a**), trapezoidal (**b**), notched (**c**), wavy inverted (**d**), wavy trapezoidal (**e**) and conventional rectangular shape (**f**).

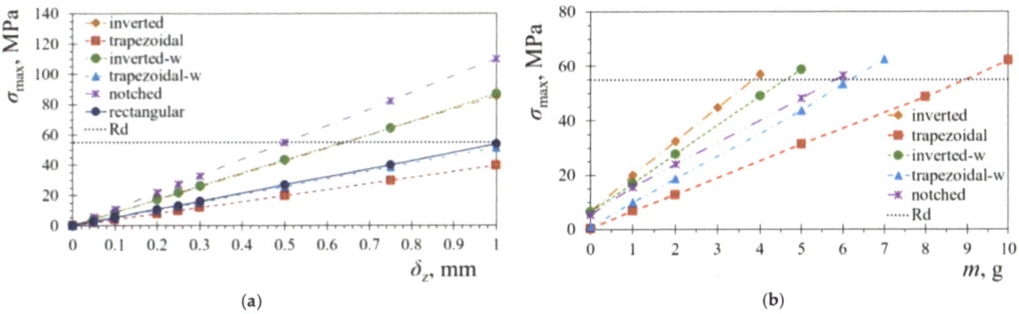

Figure 10. Piezoelectric material bending stresses of the considered PEH shapes: for various tip displacements δ_z (**a**) and for various tip masses m (**b**) when the PEHs are subject to harmonic excitations.

Further numerical analyses can be performed in ANSYS® when the tip masses m are introduced as well, while each of the considered PEH shapes is subjected to a 1 g harmonic excitation. The optimal tip-mass values, corresponding to the fatigue-strength limit, can hence be determined. The thus obtained results are shown in Figure 10b, where it can be observed that, as could have been expected, the largest tip mass of m = 9 g can be safely attached to the trapezoidal PEH, while only m = 4 g can be used with the inverted trapezoidal shape. The rectangular shape, not shown in Figure 10b for reasons of clarity, can securely withstand m = 25 g that cannot, however, be packed in a volume suitable for a practical use in wearable (wrist-worn) applications, even when high-density materials, such as tungsten, would be used to obtain it.

Since the behaviour of the bimorph PEHs changes significantly with every variation of the tip mass value, coupled harmonic analyses are performed next for the considered geometries with an optimal tip mass attached to the free end. A sweep through load-resistance values R_L is then conducted, again, to determine the optimal resistances and the respective maximal power outputs P_{max}. The thus obtained results are shown in Figure 11a. Figure 11b,c show, in turn, the specific power values P_{Smax} normalized, respectively, by the tip-mass and the PEH surface-area values.

In Figure 11 it can, hence, be observed that, when only the maximal power output levels are considered, the best performances are attained with the rectangular PEH loaded with m = 25 g. As already pointed out, if such a device would be employed in wearables, the size of the harvester would be unsuitable due to the large volume of the tip mass. When the power outputs are normalized by the tip masses (Figure 11b), the sum of the powers of the two trapezoidal segments is the highest, with the output of a single trapezoidal segment being comparable with that of the rectangular PEH. A similar behaviour can be observed if the maximal output powers are normalized by the surface area of the PEHs (Figure 11c). It can also be noted, here, that in all the cases of Figure 11 the performances are reduced for most of the optimized shapes, i.e., the inverted trapezoid, the notched shape as well as the shapes with stress concentrators along the edges. This is due to the limited ability of these shapes to withstand larger tip masses, while ensuring a continuous operation in dynamical working conditions. If, however, the specific requirements of PEH devices for wearable applications, in terms of high specific power outputs and low masses and volumes, as well as limited tip deflections are taken into account, the combination of two trapezoidal and one inverted segment represents surely a very practicable choice. What is more, the absolute power levels generated at the optimal loads by the developed optimised PEHs with tip masses (Figure 11a), i.e., 0.5 mW for the inverted trapezoidal shape, 0.9 mW for the notched one and 2.2 mW for a single trapezoidal shape, taking into account the respective duty cycles, are more than enough to power a combination of wearable sensors, data-processing and communication components, as established in [3], thus, enabling fully operational and long-lasting autonomous wearable device.

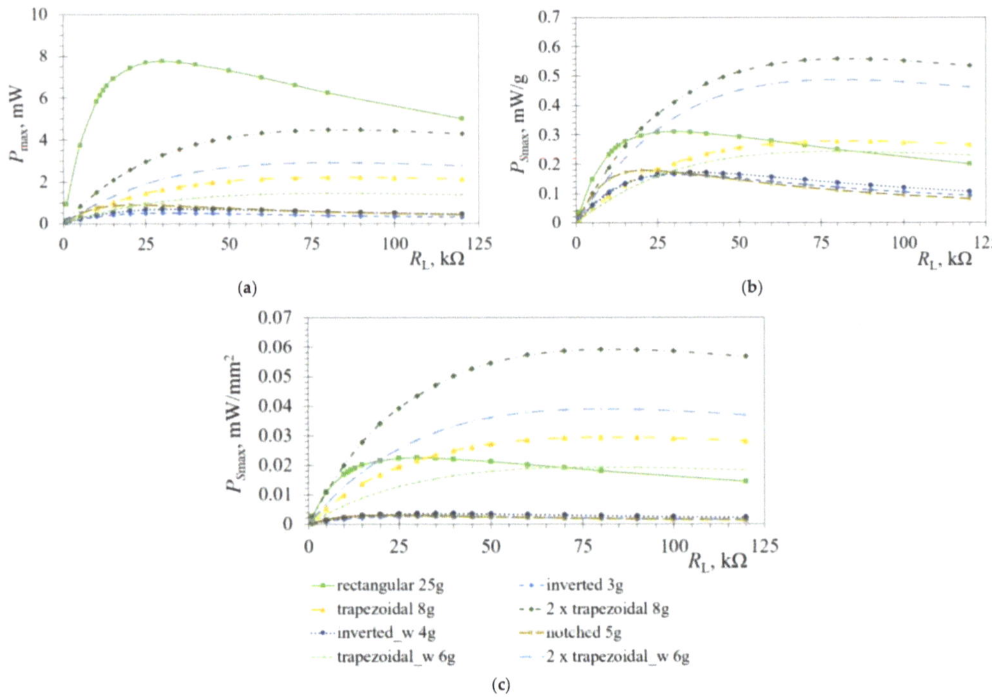

Figure 11. Coupled harmonic responses of the various PEH shapes with optimal tip masses m: P_{max} (**a**), P_{Smax} normalised by m (**b**), and P_{Smax} normalised by the PEH surface area (**c**) vs. R_L.

6. Results and Discussion

The described experimental setups were used to measure the performances of the rectangular and of the optimized PEHs undergoing both harmonic and plucking excitations. The thus obtained responses are then graphically depicted and compared to the numerical results.

6.1. Damping Ratio

The first step in validating the numerical models is determining the damping ratio of the harvesters, by measuring the pure mechanical response of the bimorph PEHs. Two excitation approaches were, therefore, used to acquire the data. Firstly, the harvester is excited by plucking its free end, and letting it oscillate at its eigenfrequency. The second approach, used to have a better control of the excitation process, is to excite the PEH fixture by using the electrodynamic shaker operating at the eigenfrequency of a specific PEH, stopping the shaker, and letting the harvester oscillate freely. In both cases, the displacement amplitude of the PEHs' free end is measured via a Metrolaser® vibrometer. A typical mechanical response attained in one of the measurements is shown in Figure 12. Based on the thus acquired data, the amplitudes of two consecutive response peaks y_n and y_{n+1} are quantified, allowing the logarithmic decrement δ to be calculated as:

$$\delta = \ln\left(\frac{y_n}{y_{n+1}}\right) \tag{1}$$

The damping ratio ζ is, in turn, determined as:

$$\zeta = \frac{\delta}{\sqrt{4 \cdot \pi^2 + \delta^2}} \tag{2}$$

When a commercially available rectangular PEH [39] was considered, a slight difference of 3% was found when comparing the damping ratios obtained via the two approaches, i.e., $\zeta = 0.03$ in the case of plucking, and $\zeta = 0.031$ in the case of shaker excitation. The thus determined damping ratios were then used to calculate the Rayleigh damping coefficients, essential for the harmonic and transient FE analyses [43].

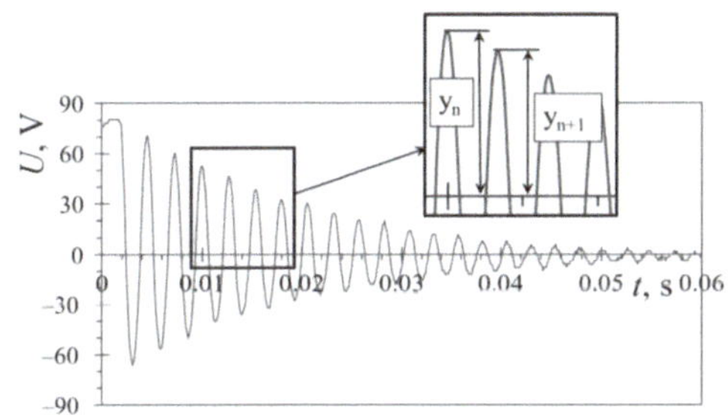

Figure 12. Typical mechanical response obtained by plucking the PEH's free end.

6.2. Model Validation

If the numerical models are to be considered as a suitable tool for the development of innovative PEH design configurations, they need to be validated first. Before the validation itself, detailed measurements of the PEH layer thicknesses were made. The validation itself comprised, then, an FE mesh sensitivity analysis, followed by the validation of the harmonic and transient results.

6.2.1. Bimorph PEH Layer Thickness Measurement

To be able to accurately model the bimorph PEH devices, the thicknesses of the respective constituent layers have to be precisely measured. Measurements were hence carried on at the Precision Engineering Laboratory [41] by employing the Olympus® SZX16 optical stereomicroscope equipped with a digital camera. A set of images, such as the one shown in Figure 13, was thus captured under 180× magnification, and the individual layers of the commercial harvester [39] were measured using the calibrated image-analysis software. Following a simple statistical analysis, it was, hence, concluded that the bimorph PEH could be modelled with the metallic substrate layer thickness $t_s = 0.163$ mm (with the respective deviation $\sigma = \pm\ 0.0043$ mm) and the piezoelectric layer thicknesses $t_{PZT} = 0.251$ mm ($\sigma = \pm\ 0.0045$ mm). A slight difference, with respect to the nominal thickness values as reported in Section 3, was, therefore, observed, which, in turn, due to exponential correlation between the thickness of the harvester and the respective area moment of inertia of its cross-section, had a significant impact on the dynamical response of the device [5,6].

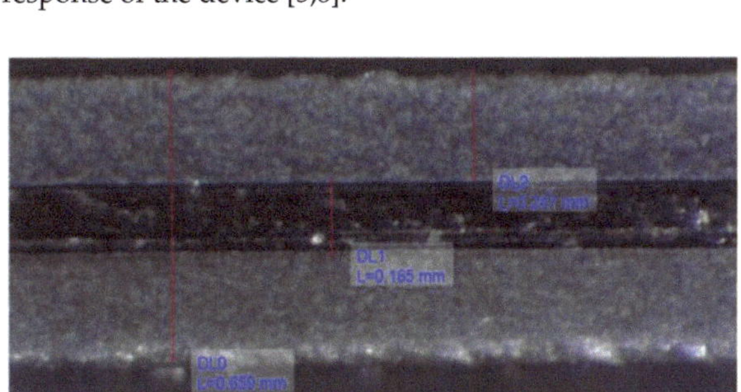

Figure 13. An example of the layer thickness measurement.

6.2.2. Mesh Sensitivity Analysis

Since the accuracy of the FE model can be significantly influenced by the type of elements used in modelling, as well as the element size (mesh density) [44], a sensitivity analysis was, therefore, performed for two different element types (i.e., hexahedral (brick or cube) and tetrahedral elements) of varying densities. The latter was implemented by varying the edge-element length, set between 0.25 and 2 mm. What is more, in the case of hexahedral elements, two different methods of volume bonding (using the ANSYS® "glue volumes" function and the merging of neighbouring nodes between the substrate and piezoelectric layers) were studied. The relative errors of the various models were then assessed by comparing the solutions with different meshes, in terms of the values of the first eigenfrequency f_1 obtained via the FE models for the rectangular PEH, with the experimental values of $f_{1\text{base}}$ = 539.38 Hz ($\sigma = \pm 1.47$ Hz), obtained in the case of base excitation, and $f_{1\text{FreeEnd}}$ = 535.15 Hz ($\sigma = \pm 1.04$ Hz), attained for the free-end excitation. The slight 0.79% difference of these two approaches is mainly due to PEHs' clamping implementation, affecting the results. The relative errors e_r of the modal numerical results with respect to experimental data for both excitation types, are then shown in the following Tables. The mesh-sensitivity validation, using the merged nodes and the hexahedral elements, is reported in Table 3, while the same, using glued volumes with hexahedral and tetrahedral elements, is given in Tables 4 and 5.

Table 3. Mesh sensitivity validation for hexahedral (brick) elements with merged nodes.

Element Length, mm	f_1, **Hz**	e_{r_base}, %	e_{r_FreeEnd}, %
2	503.32	6.91	6.13
1.75	532.69	1.25	0.46
1.5	540.11	0.14	0.93
1.25	544.05	0.86	1.65
1	525.13	2.68	1.89
0.75	534.79	0.85	0.06
0.5	534.28	0.95	0.16
0.25	539.45	0.01	0.80

Table 4. Mesh sensitivity analysis for hexahedral (brick) elements with glued volumes.

Element Length, mm	f_1, **Hz**	e_{r_base}, %	e_{r_FreeEnd}, %
2	503.62	6,86	6.07
1.75	532.5	1.28	0.49
1.5	540.11	0.14	0.93
1.25	544.05	0.86	1.65
1	525.13	2.68	1.89
0.75	534.79	0.85	0.06
0.5	534.28	0.95	0.16
0.25	539.45	0.01	0.80

Table 5. Mesh sensitivity analysis for tetrahedral elements with glued volumes.

Element Length, mm	f_1, **Hz**	e_{r_base}, %	e_{r_FreeEnd}, %
2	567.97	5.16	5.95
1.75	554.15	2.70	3.49
1.5	542.6	0.60	1.39
1.25	538.63	0.14	0.65
1	544.35	0.92	1.71
0.75	543.65	0.79	1.58
0.5	542.04	0.49	1.28
0.25	543.66	0.79	1.58

Based on the reported data, it can be concluded that the most accurate results were achieved by using elements with an edge length of around 0.5 to 0.75 mm. It is also important to note that the smaller element sizes, i.e., those <0.5 mm, resulted in significantly extended computational times. The hexahedral elements then better matched the experimental data ($e_{r\_FreeEnd}$ = 0.06% for the 0.75 mm edge length) as compared with the tetrahedral ones of the same size ($e_{r\_FreeEnd}$ = 1.58%). On the other hand, the layer-bonding method did not seem to significantly affect the results for element sizes < 1.75 mm.

Given these considerations, hexahedral elements with a 0.75 mm edge length were used in the subsequent harmonic and transient analyses where possible, while tetrahedral elements of a comparable size were used when a specific geometry necessitates it. The layers were, in turn, bonded using the "glue volumes" function, since it is a much quicker and simpler bonding method that does not significantly affect the results.

6.2.3. Harmonic Response Validation

Numerical FE harmonic analyses were conducted next, by sweeping through an excitation spectrum around the previously determined first eigenfrequency of the rectangular bimorph PEH. The model boundary conditions (excitation, clamping distance and parallel electrical coupling) were set as close as possible to the experimental ones, with the electrical connections and electrodes being considered ideal, i.e., having zero resistance. To determine the maximal power outputs, and thus assess the optimal load resistance, the harmonic analyses were then conducted with varying load resistances. As it can be observed in Figure 14, $R_{L\ opt}$ = 5 kΩ was thus determined for the rectangular PEH.

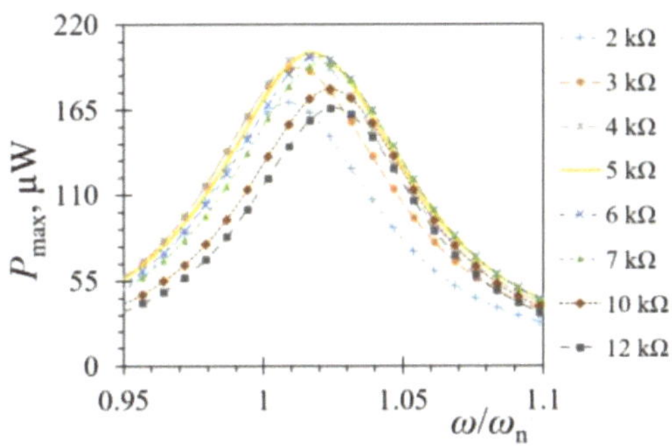

Figure 14. FE obtained powers at various load resistance values.

The experimental data, acquired by using the measurement setup described in Section 4, is, in turn, shown in Figure 15, where it is compared to the FE numerical results, displaying a close match. The maximum voltage output of 1.41 V was achieved experimentally at 544 Hz, while the highest voltage at the same frequency obtained via the harmonic FE model was 1.39 V, resulting in a difference of merely ~1.4%. Despite the good match in the obtained eigenfrequencies and the maximal voltages, there are, however, some discrepancies in the overall shape of the response curves, particularly away from the coupled electromechanical eigenfrequency. This could be attributed to slight experimental excitation control errors as well as clamping inaccuracies (a small difference in cantilevers' length can have a noteworthy influence on PEHs' responses). The evaluation of the damping ratio, and the subsequent calculation of the Rayleigh damping coefficients, as well as some limitations in the ANSYS® modelling of the coupled electromechanical piezoelectric effects evidenced in [45], could each, in turn, have had an effect on the numerical results.

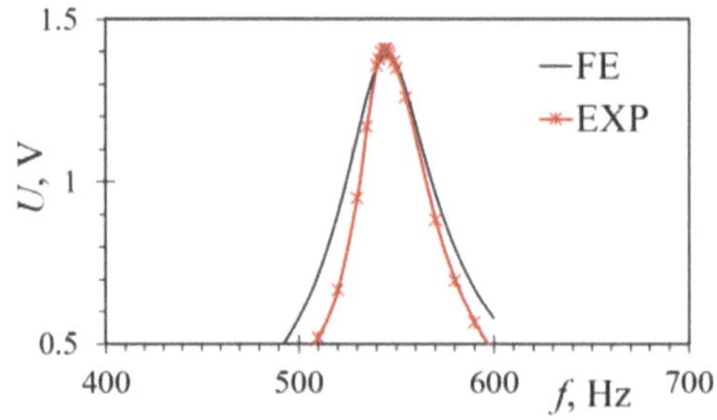

Figure 15. Comparison of experimental and FE voltage output values of a harmonically excited rectangular PEH device.

In addition to the experimental measurements at the optimal load resistance value of 5 kΩ, the power outputs were measured and determined numerically too, as shown in Figure 16, for load resistances varying in the range from 1 up to 150 kΩ. The thus obtained results show a close match, with a ~3.5% difference in the maximum power value and a shift of the peak determined experimentally towards $R_L = 6$ kΩ. These small differences can possibly be attributed to the additional resistances and other minor inaccuracies present in the non-ideal experimental system, neglected in the FE model, due e.g., to the clamping system, excitation control, electrical connections and similar effects.

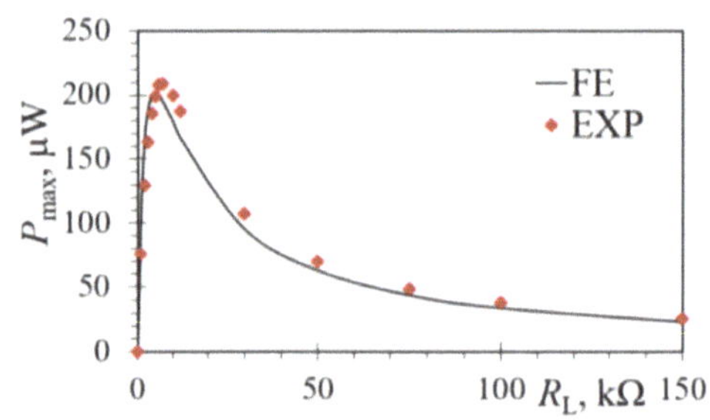

Figure 16. Comparison of experimental and FE power outputs vs. R_L for a harmonically excited rectangular bimorph PEH device.

6.2.4. Transient FUC Response Validation

To validate the FE model from the FUC perspective, transient analyses of a rectangular bimorph PEH, excited by plucking its free end, were performed next. The boundary conditions remained the same as in the harmonic model, while the excitation was introduced as a displacement δ_z of the free end of the harvester. To exclude the possibility of overstressing the PEHs during the extensive experimental tuning of the FUC setup, $\delta_z = 0.6$ mm was, initially, chosen. The numerical and experimental tests were then carried on at the determined optimal load resistance $R_{L\ opt} = 5$ kΩ. The obtained results are shown in Figure 17. A close match of the experimental and the FE responses can thus be seen, particularly in the first five cycles, where the highest voltages, i.e., the majority of the power, is generated. The maximum measured peak-to-peak voltage was $U_{p\text{-}p\_EXP} = 20.1$ V, whereas the corresponding FE value was $U_{p\text{-}p\_FE} = 21.26$ V, giving a difference of ~5%. The differences between the FE and experimental results in the next four cycles ranged from ~1% to ~9%. A better matching in the subsequent cycles could be achieved by tuning the damping ratio and the respective damping coefficients, but this would cause a significant mismatch in the foremost section of the coupled electromechanical response.

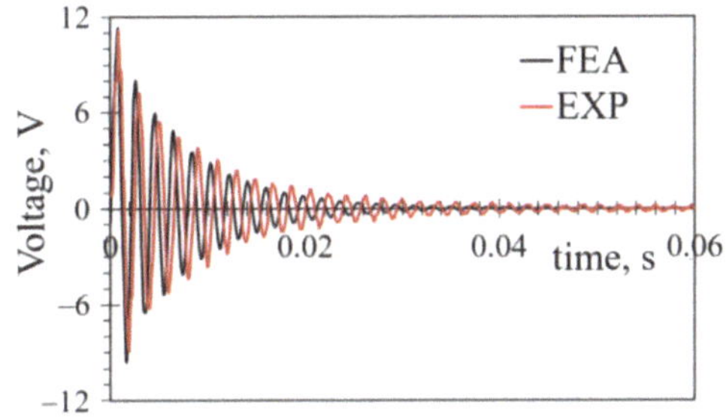

Figure 17. Comparison of experimental and transient FE FUC responses for a rectangular bimorph PEH excited by plucking its free end.

The power generated via the FUC-based excitation at $R_{L\ opt}$ = 5 kΩ was calculated from the thus-obtained voltages and is displayed in Figure 18. The hence obtained average power values over a 0.05 s time interval were P_{av_FE} = 0.454 mW and P_{av_EXP} = 0.468 mW, with a slight difference of ~3%. For the deflection of δ_z = 0.6 mm, the attained maximal powers were, in turn, P_{max_FE} = 12.53 mW and P_{max_EXP} = 12.42 mW, with a difference of merely 0.88%.

Based on these results, it can be concluded that the used FE models match well the experimental responses in the cases of both excitation approaches, and they can, therefore, be unhesitatingly employed to develop optimized PEH designs.

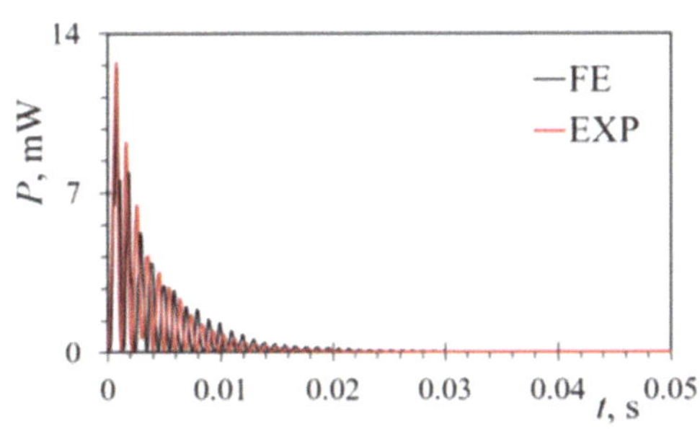

Figure 18. Experimental and transient FE FUC power outputs for a rectangular bimorph PEH excited by plucking its free end.

6.3. Optimized PEH Responses

Based on the described optimization process, five different PEH bimorph device designs with optimized shapes were made by cutting commercially available rectangular PEHs [39] with water-jet technology [46]. The responses of the harvesters were then assessed by plucking their free ends, using the FUC experimental setup. The output voltages were measured at the optimal load resistances of each individual device, and the resulting power outputs calculated.

The first eigenfrequencies f_1 of all the optimized PEHs were assessed first, and compared with the respective FE modal analysis results (Table 6). The measured eigenfrequency values (depending on the shape, being between ~325 and ~930 Hz) match closely the FE results. In fact, in most cases the difference is <1%. Only in case of the trapezoidal PEH with wavy edges, there is a slightly increase in this difference, which is still <5%; this could be attributed to clamping inaccuracies induced by the small size and the geometrical complexity of this particular harvester.

Table 6. Experimentally assessed mechanical eigenfrequency values compared with the modal FEA results.

	f_{1_FEA}, Hz	f_{1_EXP}, Hz	Diff.
Trapezoidal	930.2	934.1 ($\sigma = \pm 1.63$)	0.44%
Inverted	324.5	324.7 ($\sigma = \pm 0.46$)	0.065%
Notched	374.1	374.6 ($\sigma = \pm 0.79$)	0.14%
Trap. with wavy edges	762.5	793.7 ($\sigma = \pm 2.98$)	4.09%
Inv. with wavy edges	333.5	333.6 ($\sigma = \pm 0.6$)	0.033%
Rectangular	534.8	535.1 ($\sigma = \pm 1.04$)	0.056%

The FUC responses for each individual PEH were experimentally assessed next, and the resulting voltages compared to the transient FE results. The initial displacement of the free end of the harvesters, obtained in all the cases by plucking the harvesters with the same rectangular 3D-printed plectrum, while the DC actuator rotated at the same speed, was measured during each experiment, and the obtained results for the analysed PEHs, considerably improved in accuracy and consistency with respect to the preliminary ones reported in [38], are depicted in Figure 19, exhibiting, in all cases, a close match. The initial free end displacement values were, again, limited due to strength considerations (cf. Section 5).

In Figure 19a are shown the responses for the inverted trapezoidal bimorph harvester. The displacement of the free end was $\delta_z = 0.27$ mm, while the optimal load resistance was $R_{L\ opt} = 13$ kΩ. The largest measured peak-to-peak voltage was $U_{p\text{-}p\_EXP} = 9.66$ V, while the corresponding FE value was $U_{p\text{-}p\_FE} = 10.04$ V (the difference is ~3.8%). The oscillation period of the experimental and FE results was similar: t~0.06 s.

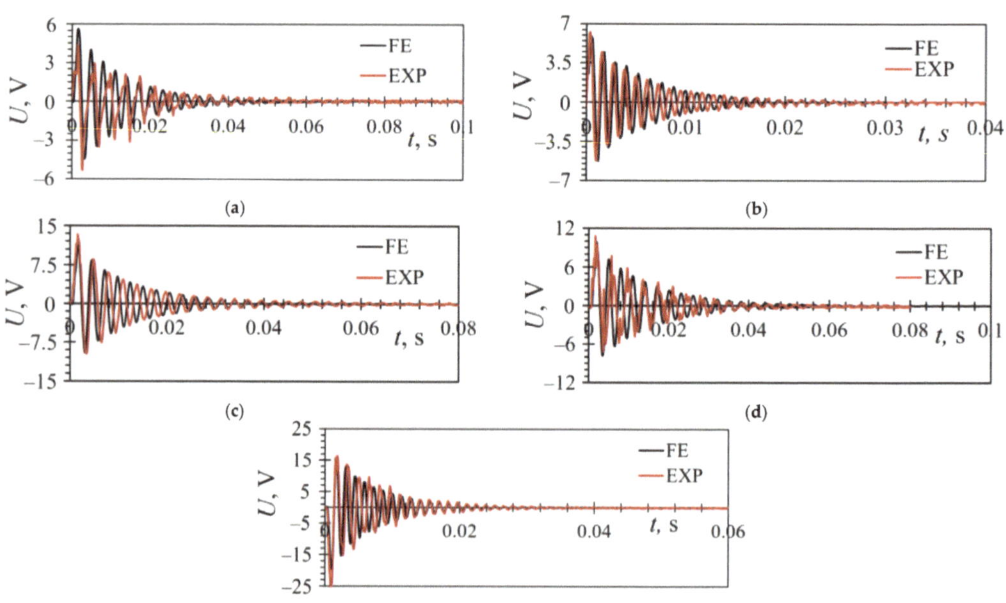

Figure 19. Comparison of experimental and transient FE responses of the PEHs excited by plucking for: an inverted trapezoidal (**a**), a trapezoidal (**b**), a notched (**c**), an inverted wavy trapezoidal (**d**) and a trapezoidal wavy (**e**) bimorph PEH.

The responses for the trapezoidal PEH is, in turn, displayed in Figure 19b. In this case $\delta_z = 0.2$ mm, $R_{L\ opt} = 7$ kΩ, $U_{p\text{-}p\_EXP} = 11.46$ V, $U_{p\text{-}p\_FE} = 11.07$ V (difference of ~3.5%), and the oscillation period was t~0.03 s.

The response for the PEH device with a triangular notch at the clamped end is displayed in Figure 19c. The characteristic performances were: δ_z = 0.47 mm, $R_{L\ opt}$ = 7 kΩ, $U_{p\text{-}p\_EXP}$ = 23.1 V, $U_{p\text{-}p\_FE}$ = 21.4 V (difference of ~7.5%), and t~0.08 s.

The responses for the inverted trapezoidal harvester with stress concentrators along the edge of the bimorph and the corresponding trapezoidal PEH cantilever are, finally, given in Figure 19d, e, whose respective characteristic values were: δ_z = 0.48 mm, $R_{L\ opt}$ = 12 kΩ, $U_{p\text{-}p\_EXP}$ = 17.9 V, $U_{p\text{-}p\_FE}$ = 17.61 V (~1.6% difference), and t~0.06 s (Figure 19d), and δ_z = 0.38 mm, $R_{L\ opt}$ = 13 kΩ, $U_{p\text{-}p\_EXP}$ = 41.62 V, $U_{p\text{-}p\_FE}$ = 40.98 V (1.5% difference), t~0.03 s (Figure 19e).

The thus obtained maximal peak-to-peak voltages $U_{max\_p\text{-}p}$, as well as the average, the maximal and the maximal specific powers P_{av}, P_{max}, and P_{s_max}, are listed in Table 7. The average powers were calculated, here, over the reported oscillation periods for the respective bimorph PEHs, while all the reported data were obtained for the stated initial deflections δ_z, as obtained experimentally in the described conditions.

Table 7. Experimentally obtained voltage and power values.

	$U_{max\_p\text{-}p}$, V	P_{av}, µW	P_{max}, mW	P_{s_max}, µW/mm^2
Trapezoidal	11.46	71	2.79	37
Inverted	9.66	25	1.1	6
Notched	23.12	411	12.66	42
Trapezoidal with wavy edges	41.62	442	31.95	43
Inverted with wavy edges	17.9	148	4.74	24
Rectangular	20.1	468	12.42	36

6.4. Discussion

As stated, in the employed experimental setup the displacements of the free end were attained by plucking all the used bimorph harvesters with the same rectangular 3D-printed plectrum rotating at the same speed. This resulted in a different δ_z for each PEH shape, thus making comprehensive comparison difficult. Shown in Table 8, the obtained voltages and powers were, hence, normalized by δ_z, resulting in a dimensionality ($U_{n\_max\_p\text{-}p}$, V/mm; P_{n_xx}, mW/mm) that, as evidenced also in the respective graphical representation of Figure 20, provided a better means of comparing responses between the used piezoelectric harvesting devices.

Table 8. Experimentally obtained voltages and powers normalized by the respective δ_z values.

	$U_{n\_max\_p\text{-}p}$, V/mm	P_{n_ave}, µW/mm	P_{n_max}, mW/mm	$P_{n\_s\_max}$, µW/mm^2/mm
Trapezoidal	45.83	1130	44.57	590
Inverted	36.46	360	15.43	80
Notched	49.18	2087	57.33	188
Trapezoidal with wavy edges	52.03	690	49.92	670
Inverted with wavy edges	37.3	641	20.57	106
Rectangular	33.4	2490	34.8	101

It can thus be observed that, in terms of the normalised output voltages, all optimized bimorph PEH devices, if compared with the conventional rectangular shape, generated larger outputs, the highest value of which was obtained from the notched PEH. An increase in output voltages can also be noted in the PEHs with wavy edges, compared with their counterparts, without added stress concentrators (Figure 20a).

When the normalised average power is considered, as shown in Table 8, the highest output was obtained from the rectangular PEH, while the optimized bimorph PEHs generated less power. When, however, the average power is normalized over the respective PEH surface area (Figure 20b), the specific normalized average power output of both trapezoidal

PEHs is noticeably higher than that of the rectangular PEH, while the values for the notched PEH are slightly lower but comparable (~5%) to those of the rectangular one. The specific normalized average power of both inverted trapezoidal PEH shapes was lower than that of the rectangular one but, as already pointed out, this output is, in practice, an addition to that of the respective trapezoidal segments.

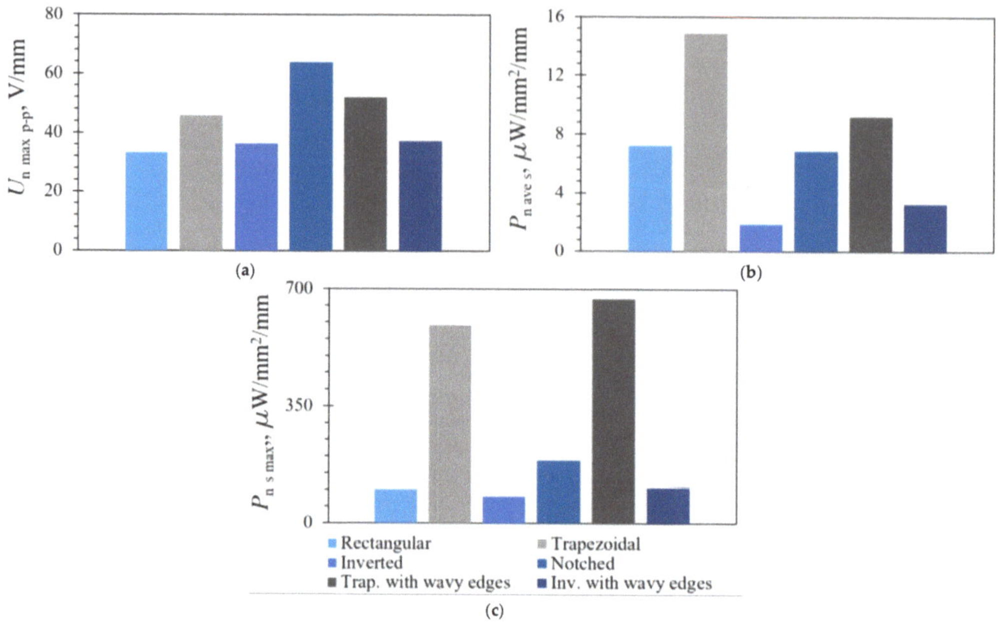

Figure 20. Normalized maximal voltages (**a**), as well as respectively calculated specific average (**b**) and specific maximal (**c**) power outputs for the considered bimorph PEH shapes.

When, finally, the maximum power outputs, calculated from the normalized voltages and divided over the surface area of the respective PEHs, were observed, as shown in Figure 20c, it could be noted that, especially, both the trapezoidal PEHs, but also (less pronouncedly) the notched PEH, significantly outperformed the conventional rectangular bimorph. Both optimized inverted segments, by themselves, generated a specific normalised maximum power output comparable to that of the rectangular shape, with the inverted PEH with wavy edges having even slightly outperformed it. Moreover, as with the maximal voltages, a noticeable increase in the normalised specific maximal powers was obtained for both PEHs with added stress concentrators, compared with those without them.

If the determined optimal bimorphs' design configurations would then be excited by the respective maximal allowable initial free end displacement δ_{z_max}, as determined in Figure 10a in Section 5, the corresponding maximal power outputs and the average power outputs during the matching oscillation periods would be those reported in Table 9.

When the thus-obtained power outputs of the segments with straight edges, excited by the respective maximum allowable displacements, were summed, and the resulting performances observed for the segmented device as a whole, the maximum and average power output values were, respectively, $P_{max\_\delta zmax}$ = 134.06 mW and $P_{ave\_\delta zmax}$ = 3.38 mW, which, when compared with the performance of the rectangular PEH reported in Table 9, represents a significant improvement. Due to the increased stress levels and, therefore, the resulting limited allowable initial free-end displacements, the summed performances of the segments with stress concentrators, in terms of average power output, were slightly lower than those of the rectangular bimorph, i.e., $P_{ave\_\delta zmax}$ = 1.76 mW, but, even in this case, the maximal power output exceeded that of the rectangular PEH, since it summed to $P_{max\_\delta zmax}$ = 112.1 mW.

Table 9. Maximal and average power outputs achievable by the optimized PEHs at the maximal allowable initial free-end displacements δ_{zmax}, as determined in Section 5.

	$P_{max\_\delta zmax}$, mW	$P_{ave\_\delta zmax}$, mW
Trapezoidal	62.4	1.58
Inverted	9.26	0.22
Notched	28.7	1.04
Trapezoidal with wavy edges	49.9	0.69
Inverted with wavy edges	12.3	0.38
Rectangular	34.8	2.49

7. Conclusions and Outlook

The possibility of using energy harvesting, especially the respective piezoelectric approach, in wearable sensor networks, was analysed in this work, along with a detailed review of the state-of-the-art. Given the random nature of human motion, the promising frequency up-conversion (FUC) approach was applied to overcome the issues of PEHs related to their optimal performances at the respective eigenfrequencies only. Based on the initial studies available in literature, the influence of bimorph PEHs' geometry on their outputs was systematically analysed and a DoE-based approach to geometry optimization was applied. Initial FE modal, harmonic and transient analyses allowed, then, establishing a clear indication of an increase in performances of the optimized bimorph PEHs as compared with the conventional one with a rectangular planar layout. Two optimized PEH designs (segmented and notched PEH) were, hence, proposed, both utilizing a larger portion of the available volume defined by the reference rectangular cantilever, compared with the solutions proposed in prior art.

By using the developed experimental setups, a thorough validation of the used numerical models was performed next, which proved the suitability of the used FE models. In fact, the measurement results, compared with FE data, prove not only the viability of the FE approach, but also the clearly improved performances of the optimized bimorph PEHs subject to excitation via plucking of their free ends, in terms of the normalised specific maximal output voltages and powers. In terms of the obtained specific maximal power outputs, a single trapezoidal segment of a three-segment device, displayed, thus, a 484% (in terms of the absolute maximal power: 28%) improvement in performance relative to the power output obtained via the reference rectangular bimorph, while the specific power output of the notched PEH increases by 86% (64.7% in terms of the maximal power). The addition of stress concentrators on the edges of the segments showed further improvements, i.e., an increase in specific power output of 14% (12% in maximal power) for the inverted trapezoidal shape, and a 32.5% (33.3% in maximal power) improvement for the trapezoidal segment. This, in turn, resulted in an increase in performances of 563% (43.5% when maximal power is considered), when a single trapezoidal segment with added stress concentrators is compared with the conventional rectangular bimorph. The inverted segments did display specific power-output levels that were slightly lower, although still comparable to the reference rectangular device, which can be attributed to the narrow cross-section near the clamped end that limited the allowable free-end plucking displacement. As repeatedly evidenced above, the latter power output should also be viewed as only a portion of the device's overall power output that is to be added to that of the two trapezoidal segments, further confirming the viability of the segmented-design configuration approach.

The absolute power outputs potentially achievable by the optimized PEHs excited at the maximal allowable displacements of the free end, ranged, in turn and in terms of the maximal power outputs, from $P_{max\_\delta zmax}$ = 9.26 mW, for the inverted trapezoidal PEH, to $P_{max\_\delta zmax}$ = 134.06 mW, when the straight-edged segmented PEH is considered. In terms of the average power outputs, in the oscillation periods, characteristic of the considered shapes, the power outputs ranged, then, from $P_{ave\_\delta zmax}$ = 0.22 mW, for the inverted trapezoidal PEH, to $P_{ave\_\delta zmax}$ = 3.38 mW, for the straight-edged segmented bimorph PEH. The obtained overall performances of the straight-edged segmented design configuration are, therefore, significantly better than those of the conventional rectangular PEH bimorph with the same footprint ($P_{max\_\delta zmax}$ = 34.8 mW and $P_{ave\_\delta zmax}$ = 2.49 mW).

It should also be noted, here, that, when the results attained while keeping the excitation magnitude the same for all the PEH shapes, without considering stress limitations—as done in the initial studies and in most of the available literature—are compared to the results where the limitations caused by strength considerations are duly taken into account, a noteworthy difference in trends and performances is obtained. In fact, when the strength considerations are neglected, the inverted trapezoidal PEH clearly outperforms the trapezoidal one, but when the initial plucking deflection is limited due to excessive stress levels in the narrowest cross-section of the inverted shape, the trapezoidal bimorph shows much better performances. This demonstrates that the performances of the bimorph PEHs reported in literature, where strength considerations were neglected, should be considered with a fair degree of attentiveness.

The performed strength analyses then permit clearly establishing the allowable free-end-deflection and tip-mass values for each of the considered bimorph design configurations that still ensure a long and safe operation of the PEH transducers. The performances of the herein proposed PEHs, compared to a notable portion of the available literature, represent, therefore practically achievable values that can be expected for a long-lasting and optimally operating wearable devices used in real-life operating conditions. What is more, the proposed segmented PEH not only permits optimally using the available design volume—thus, amply meeting the power requirements needed for the wearable sensor nodes constituted by medical sensors with the respective data elaboration and communication modules as established in [3]; it also introduces the possibility of matching the response of each harvester segment with a different resistive load (i.e., sensor with the respective add-on modules), further improving the overall efficiency of the suggested wearable devices.

A kinetic EH approach, aimed at powering wearable devices and using an innovative combination of the FUC excitation modality with PEH geometry optimization, all with the goal of maximizing the resulting efficiency and energy density, along with a careful consideration of the dynamic strength and durability of the devices, are, therefore, all significant contributions of the herein performed work.

Given the influence of the initial free-end displacements on the outputs of the optimised harvesters, a thorough study of the influence of the stiffness (dependent also on the 3D printing parameters) and of the excitation frequency of the used plucking plectra on the FUC responses is under way, which will extend the herein presented treatise. From the design point of view, an improved fixture base is also being developed to facilitate a more precise clamping of the harvesters. In future work, additional harmonic excitation experiments will also be performed, along with those varying load resistances, so as to understand better the influence of the latter on the coupled electromechanical performances of bimorph PEH devices. A suitable power-management system is also being developed, which should enable matching different loads with each PEH segment, thus further optimizing the overall performances of the wearable devices. A detailed study of the damping effects is also planned.

All this will lead to the design of an integrated wearable device, comprising optimized bimorph PEHs excited by human and other random motion via the FUC approach, using a flywheel with several plectra, which will result in an autonomous apparatus that could be used in the nodes of sensor networks in telemedicine, in remote patient monitoring and in autonomous drug delivery. Such a device could also be implemented as a worker-safety monitor in dangerous working environments, in industry 4.0, in IoT settings and in structural health-monitoring (SHM) systems for e.g., airplanes or for civil engineering structures.

Author Contributions: Conceptualization, P.G. and S.Z.; methodology, S.Z.; software, P.G.; validation, P.G.; formal analysis, P.G.; investigation, P.G.; resources, P.G. and S.Z.; data curation, P.G.; writing—original draft preparation, P.G.; writing—review and editing, S.Z.; visualization, P.G.; supervision, S.Z.; project administration, S.Z.; funding acquisition, S.Z. All authors have read and agreed to the published version of the manuscript.

Funding: Not applicable.

Institutional Review Board Statement: Not applicable.

Informed Consent Statement: Not applicable.

Data Availability Statement: The following data files are available online at https://repository.riteh.uniri.hr/islandora/object/riteh:2274: Figure 3: ADL_accel_measurement, Figure 9: FEA_bending_stress_0.5mm, Figure 10: FEA_bending_stress, Figure 11: FEA_max_power_mass, Figure 12: EXP_log_decrement, Figure 13: thickness_meas, Tables 3–5: FEA_mesh_sensitivity, Figure 14: FEA_power_at_load, Figure 15, Figure 16, Figure 17, Figure 18 and Table 6: model_validation, Figure 19: EXP_voltage_FUC, Table 7, Table 8 and Figure 20: EXP_powers.

Acknowledgments: Work enabled by using the equipment funded via the EU ERDF project RC.2.2.06-0001 "Research Infrastructure for Campus-based Laboratories at the University of Rijeka", supported by the University of Rijeka grant uniri-tehnic-18-32 "Advanced mechatronics devices for smart technological solutions", and partially performed in the frame of the EU COST Action CA18203 "Optimizing Design for Inspection".

Conflicts of Interest: The authors declare no conflict of interest. The funders had no role in the design of the study; in the collection, analyses, or interpretation of data; in the writing of the manuscript, or in the decision to publish the results.

References

1. Kamišalić, A.; Fister, I.; Turkanović, M.; Karakatič, S. Sensors and Functionalities of Non-Invasive Wrist-Wearable Devices: A Review. *Sensors* **2018**, *18*, 1714. [CrossRef]
2. Majumder, S.; Mondal, T.; Deen, M. Wearable Sensors for Remote Health Monitoring. *Sensors* **2017**, *17*, 130. [CrossRef]
3. Gljušćić, P.; Zelenika, S.; Blažević, D.; Kamenar, E. Kinetic Energy Harvesting for Wearable Medical Sensors. *Sensors* **2019**, *19*, 4922. [CrossRef]
4. Tan, Y.K. *Energy Harvesting Autonomous Sensor Systems: Design, Analysis, and Practical Implementation*; CRC Press: Boca Raton, FL, USA, 2013; ISBN 978-1-4398-9273-2.
5. Priya, S.; Inman, D.J. (Eds.) *Energy Harvesting Technologies*; Springer: New York, NY, USA, 2009; ISBN 978-0-387-76463-4.
6. Kaźmierski, T.J. (Ed.) *Energy Harvesting Systems: Principles Modeling and Applications*; Springer: New York, NY, USA, 2011; ISBN 978-1-4419-7565-2.
7. Du, S.; Jia, Y.; Arroyo, E.; Fernandez, S.; Riches, S.T.; Seshia, A.A. MEMS Piezoelectric Energy Harvester Powered Wireless Sensor Module Driven by Noisy Base Excitation. In Proceedings of the 20th International Conference on Solid-State Sensors, Actuators and Microsystems & Eurosensors XXXIII (TRANSDUCERS & EUROSENSORS XXXIII), Berlin, Germany, 23–27 June 2019; pp. 350–353. [CrossRef]
8. Zelenika, S.; Hadas, Z.; Bader, S.; Becker, T.; Gljušćić, P.; Hlinka, J.; Janak, L.; Kamenar, E.; Ksica, F.; Kyratsi, T.; et al. Energy Harvesting Technologies for Structural Health Monitoring of Airplane Components—A Review. *Sensors* **2020**, *20*, 6685. [CrossRef]
9. Priya, S. Modeling of Electric Energy Harvesting Using Piezoelectric Windmill. *Appl. Phys. Lett.* **2005**, *87*, 184101. [CrossRef]
10. Pozzi, M.; Aung, M.S.H.; Zhu, M.; Jones, R.K.; Goulermas, J.Y. The Pizzicato Knee-Joint Energy Harvester: Characterization with Biomechanical Data and the Effect of Backpack Load. *Smart Mater. Struct.* **2012**, *21*, 075023. [CrossRef]
11. Pozzi, M. Magnetic Plucking of Piezoelectric Bimorphs for a Wearable Energy Harvester. *Smart Mater. Struct.* **2016**, *25*, 045008. [CrossRef]
12. Gljušćić, P.; Zelenika, S. Assessment of Performances of Optimized Piezoelectric Energy Harvesters for Wearables. In Proceedings of the 20th EUSPEN International Conference, Geneva, Switzerland, 8–12 June 2020; pp. 49–52.
13. Yoshikawa, S.; Saarmaa, E.; Kottenstette, N.; Jacques, R. Unimorph and Bimorph Actuators. *J. Acoust. Soc. Am.* **1998**, *104*, 1828. [CrossRef]
14. Pillatsch, P.; Yeatman, E.M.; Holmes, A.S. A Piezoelectric Frequency Up-Converting Energy Harvester with Rotating Proof Mass for Human Body Applications. *Sens. Actuat. A Phys.* **2014**, *206*, 178–185. [CrossRef]

15. Xue, T. On Magnetic Plucking Configurations for Frequency Up-Converting Mechanical Energy Harvesters. *Sens. Actuat. A Phys.* **2017**, *253*, 101–111. [CrossRef]
16. Xue, T. Wearable Inertial Energy Harvester with Sputtered Bimorph Lead Zirconate Titanate (PZT) Thin-Film Beams. *Smart Mater. Struct.* **2018**, *13*, 085026. [CrossRef]
17. Izadgoshasb, I.; Lim, Y.Y.; Tang, L.; Padilla, R.V.; Tang, Z.S.; Sedighi, M. Improving Efficiency of Piezoelectric Based Energy Harvesting from Human Motions Using Double Pendulum System. *Energ. Convers. Manag.* **2019**, *184*, 559–570. [CrossRef]
18. Bao, B.; Wang, Q.; Wu, N.; Zhou, S. Hand-Held Piezoelectric Energy Harvesting Structure: Design, Dynamic Analysis, and Experimental Validation. *Measurement* **2021**, *174*, 109011. [CrossRef]
19. Moro, L.; Benasciutti, D. Harvested Power and Sensitivity Analysis of Vibrating Shoe-Mounted Piezoelectric Cantilevers. *Smart Mater. Struct.* **2010**, *19*, 115011. [CrossRef]
20. Benasciutti, D.; Moro, L.; Zelenika, S.; Brusa, E. Vibration Energy Scavenging via Piezoelectric Bimorphs of Optimized Shapes. *Microsyst. Technol.* **2010**, *16*, 657–668. [CrossRef]
21. Park, J.; Lee, S.; Kwak, B.M. Design Optimization of Piezoelectric Energy Harvester Subject to Tip Excitation. *J. Mech. Sci. Technol.* **2012**, *26*, 137–143. [CrossRef]
22. de Almeida, B.; Pavanello, R. Topology Optimization of the Thickness Profile of Bimorph Piezoelectric Energy Harvesting Devices. *J. Appl. Comput. Mech.* **2018**, *5*, 113–127. [CrossRef]
23. Homayouni-Amlashi, A.; Mohand-Ousaid, A.; Rakotondrabe, M. Multi Directional Piezoelectric Plate Energy Harvesters Designed By Topology Optimization Algorithm. *IEEE Robot. Autom. Lett.* **2020**, *5*, 462–469. [CrossRef]
24. Huang, M.; Hou, C.; Li, Y.; Liu, H.; Wang, F.; Chen, T.; Yang, Z.; Tang, G.; Sun, L. A Low-Frequency MEMS Piezoelectric Energy Harvesting System Based on Frequency Up-Conversion Mechanism. *Micromachines* **2019**, *10*, 639. [CrossRef]
25. Lockhart, R.; Janphuang, P.; Briand, D.; de Rooij, N.F. A Wearable System of Micromachined Piezoelectric Cantilevers Coupled to a Rotational Oscillating Mass for On-Body Energy Harvesting. In Proceedings of the 27th International Conference on Micro Electro Mechanical Systems (MEMS), San Francisco, CA, USA, 26–30 January 2014; pp. 370–373.
26. Mokhtari, F.; Spinks, G.M.; Fay, C.; Cheng, Z.; Raad, R.; Xi, J.; Foroughi, J. Wearable Electronic Textiles from Nanostructured Piezoelectric Fibers. *Adv. Mater. Technol.* **2020**, *5*, 1900900. [CrossRef]
27. In, V.; Palacios, A. Energy Harvesting. In *Symmetry in Complex Network Systems: Understanding Complex Systems*; Springer: Berlin/Heidelberg, Germany, 2018; pp. 295–316, ISBN 978-3-662-55543-9.
28. Bai, Y.; Tofel, P.; Hadas, Z.; Smilek, J.; Losak, P.; Skarvada, P.; Macku, R. Investigation of a Cantilever Structured Piezoelectric Energy Harvester Used for Wearable Devices with Random Vibration Input. *Mech. Syst. Signal Pract.* **2018**, *106*, 303–318. [CrossRef]
29. EU EnABLES (European Infrastructure Powering the Internet of Things) H2020-INFRAIA-2014-2015 Grant No. 730957. Available online: www.enables-project.eu (accessed on 21 March 2021).
30. Malvić, M. *Low Level Ambient Energy Harvesting Systems*; Student project report paper; University of Rijeka, Faculty of Engineering: Rijeka, Croatia, 2021.
31. Kathpalia, B.; Tan, D.; Stern, I.; Erturk, A. An Experimentally Validated Model for Geometrically Nonlinear Plucking-Based Frequency up-Conversion in Energy Harvesting. *Smart Mater. Struct.* **2018**, *27*, 015024. [CrossRef]
32. Kuang, Y.; Zhu, M. Design Study of a Mechanically Plucked Piezoelectric Energy Harvester Using Validated Finite Element Modelling. *Sensor. Actuat. A Phys.* **2017**, *263*, 510–520. [CrossRef]
33. Bao, C.; Dai, Y.; Wang, P.; Tang, G. A Piezoelectric Energy Harvesting Scheme Based on Stall Flutter of Airfoil Section. *Eur. J. Mech. B-Fluid.* **2019**, *75*, 119–132. [CrossRef]
34. Han, C.S.; Liu, K.; Kim, J.H.; Kang, D.H.; Cho, Y.S. Internal-Field-Dependent Low-Frequency Piezoelectric Energy Harvesting Characteristics of in Situ Processed Nb-Doped Pb(Zr,Ti)O3 Thin-Film Cantilevers. *J. Alloys Compd.* **2019**, *781*, 898–903. [CrossRef]
35. Moon, K.; Choe, J.; Kim, H.; Ahn, D.; Jeong, J. A Method of Broadening the Bandwidth by Tuning the Proof Mass in a Piezoelectric Energy Harvesting Cantilever. *Sensor. Actuat. A-Phys.* **2018**, *276*, 17–25. [CrossRef]
36. Zouari, M.; Naifar, S.; Bouattour, G.; Derbel, N.; Kanoun, O. Energy Management Based on Fractional Open Circuit and P-SSHI Techniques for Piezoelectric Energy Harvesting. *Tm-Tech. Mess.* **2019**, *86*, 14–24. [CrossRef]
37. Noh, J.; Kim, P.; Yoon, Y.-J. Load Resistance Optimization of a Magnetically Coupled Two-Degree-of-Freedom Bistable Energy Harvester Considering Third-Harmonic Distortion in Forced Oscillation. *Sensors* **2021**, *21*, 2668. [CrossRef]
38. Gljušćić, P.; Zelenika, S.; Perčić, M.; Kamenar, E. Experimental Characterisation of Performances of Optimized Piezoelectric Energy Harvesters. In Proceedings of the 21st International Conference & Exhibition, Virtual, 7–10 June 2021; pp. 131–134.
39. Piezo.com. Available online: https://piezo.com/ (accessed on 20 February 2021).
40. Dean, A.; Morris, M.; Stufken, J.; Bingham, D. *Handbook of Design and Analysis of Experiments*; CRC Press: Boca Raton, FL, USA, 2015; ISBN 978-1-4665-0434-9.
41. University of Rijeka, Croatia, Faculty of Engineering, Precision Engineering Laboratory Equipment. Available online: http://precenglab.riteh.uniri.hr/category/lab-equipment (accessed on 19 July 2021).
42. Okayasu, M.; Ozeki, G.; Mizuno, M. Fatigue Failure Characteristics of Lead Zirconate Titanate Piezoelectric Ceramics. *J. Eur. Ceram. Soc.* **2010**, *30*, 713–725. [CrossRef]
43. Meirovitch, L. *Fundamentals of Vibrations*; McGraw-Hill: Boston, MA, USA, 2001; ISBN 978-0-07-041345-0.
44. Nicholson, D.W. *Finite Element Analysis: Thermomechanics of Solids*; CRC Press: Boca Raton, FL, USA, 2003; ISBN 978-0-429-21444-8.

45. Yang, Y.; Tang, L. Equivalent Circuit Modeling of Piezoelectric Energy Harvesters. *J. Intel. Mat. Syst. Struct.* **2009**, *20*, 2223–2235. [CrossRef]
46. RWS Waterjet Service GmbH. Available online: https://www.stm-waterjet.com/en/ (accessed on 21 February 2021).

sensors

MDPI

Article

Assessment of Impact Energy Harvesting in Composite Beams with Piezoelectric Transducers

Nikolaos Margelis, Theofanis S. Plagianakos *, Panagiotis Karydis-Karandreas and Evangelos G. Papadopoulos

Control Systems Lab, School of Mechanical Engineering, National Technical University of Athens, 15780 Athens, Greece; nmargelis@gmail.com (N.M.); panosgary@gmail.com (P.K.-K.); egpapado@central.ntua.gr (E.G.P.)

* Correspondence: fanplag@gmail.com

Abstract: Piezoelectric energy harvesting (PEH) is studied in the case of a low-velocity impact of a rigid mass on a composite beam. A methodology is outlined, encompassing modelling of the open-circuit impact response in a finite element (FE) package, formulation of a lumped parameter (LP) model for the piezoelectric transducer connected with the harvesting circuit, and experimental verification of the impact using a custom portable configuration with impactor motion control. The subcircuit capacitor charging effect, the impactor mass and velocity on the harvesting subcircuit response, and the obtained output power are quantified. The results indicate that the current methodology can be used as a design tool for the structure and the harvesting circuit to achieve power output from composite beams with piezoelectric patches under impact conditions.

Keywords: composites; piezoelectric energy harvesting; impact response; impact testing machine

Citation: Margelis, N.; Plagianakos, T.S.; Karydis-Karandreas, P.; Papadopoulos, E.G. Assessment of Impact Energy Harvesting in Composite Beams with Piezoelectric Transducers. *Sensors* **2021**, *21*, 7445. https://doi.org/10.3390/s21227445

Academic Editors: Slim Naifar, Olfa Kanoun and Carlo Trigona

Received: 29 September 2021
Accepted: 5 November 2021
Published: 9 November 2021

1. Introduction

Piezoelectric energy harvesting (PEH) is defined as the principle of converting mechanical to electric energy via vibrating piezoelectric transducers, and storing that energy by means of an appropriately designed electric circuit. This process evolves among two main stages: (i) conversion of mechanical strain, stemming from vibration to electric voltage in the piezoelectric (PE) transducer; and (ii) driving the produced current to an electric circuit feeding a battery. Based on this process, pioneering autonomous systems including piezoelectric sensors and appropriate harvesting circuitries have been designed since the late 1990s [1,2]. Most PEH configurations developed since then have been designed by considering as input continuous forced harmonic excitation [3,4]. This led to a variety of devices using ambient or human-motion-induced energy for IoT applications and wireless sensor networks [5,6]. The energy conversion process physics has been studied extensively, employing lumped-parameter models or models based on analytical continuum mechanics for capturing the response of the vibrating structure in the linear [7–9] and nonlinear [10–14] regime. A key part of a harvesting device is the circuit. Great effort has been placed on the design of electric circuits maximizing the power output of the PEH device, as in the pioneering works of Ottman et al. [15,16], Lefeuvre et al. [17,18] and others [19–23], leading to various options for the design of a PEH device working under predefined conditions [24,25]. The basic architecture of a PEH circuit has been commercialized [26,27].

An interesting application of PEH is low-energy impact. Neglecting energy loss during an impact, the initial kinetic energy of the impactor is partially converted to potential energy of impactor–target contact, kinetic and elastic energy in the mechanical substructure, while the rest is instantly stored as electric energy in the piezoelectric transducer. The temporal variation of this energy equilibrium may be quantified by means of finite element methods [28], used to model the impacted structure as a continuum solid and encompass the interaction between impactor and structure by means of a contact law. In initial PEH systems, impact was exploited as a frequency up-conversion mechanism, enabling power harvesting at low excitation frequencies. Umeda et al. [29] studied the impact of a steel ball on a piezoelectric vibrator by developing an electrical equivalent model with constants obtained from the admittance characteristics of the system measurements. A plucking-based up-conversion strategy for knee-joint energy harvesting has been developed by Pozzi and Zhu, who modelled a resistively shunted piezoelectric bimorph with finite elements and validated their model experimentally [30]. In several works, the use of an impact as a source of power has been studied by connecting the piezoelectric transducer to a resistance. Renaud et al. [31] developed a lumped-parameter model, assumed a single-mode impulse excitation based on a method reported by Lee [32] and provided a closed expression of electric resistance for optimizing power output. Gu and Livermore studied an impact-driven coupled vibration harvester consisting of a composite and a piezoelectric bimorph beam, by assuming Euler–Bernoulli-type kinematics and displacement at the contact point induced by inelastic collision [33]. The effect of impactor mass and velocity has been studied by Basari et al., based on fundamental energy equilibrium equations and impact tests [34]. Doria et al. [35] studied the effect of the shape and duration of a finite-width impulse of base acceleration on a cantilever beam at open-circuit configuration using analytical, experimental and numerical methods. The contact stiffness between the impactor and the target has been considered by Jaquelin et al. [36] and Martinez-Ayuso et al. [37], who developed analytical solutions based on Euler–Bernoulli beams. Fu and Liao implemented a Hunt–Crossley [38] contact model for predicting the power output of a resistively shunted piezoelectric beam subjected to impact [39]. Circuits including rectifiers have been implemented by Ferrari et al. [40] and Wong and Dahari [41], who studied impacted piezoelectric beams experimentally, in the case of impact induced by a vibrating beam with tip mass and raindrops, respectively. PEH harvesting systems based on impact for powering devices have been developed by Feng et al. [42], who introduced the impact event as an electric pulse excitation in the case of a self-monitoring helmet, and by Chen et al. [43] in the case of road speed-bumps.

In most of the above-mentioned works on PEH, impact has been studied as a frequency up-conversion mechanism, induced as an impulse excitation. However, impulse excitation—mostly induced to the model as a half sine force or voltage time signal—should fit the induced signal amplitude to a measurement. From a design perspective, the velocity of the impactor would be easier to estimate as an analysis input parameter. Moreover, impulse excitation does not encompass the physics of the impact, which is introduced by the contact law between the impactor and the structure hosting the PEH circuit, and mainly depends on the stiffness and mass of the impactor and target. Thus, important parameters such as the impact-force–time profile determining the stress distribution in the impacted structure are not captured accurately. Concerning the circuit, mainly resistive circuits are connected to the structure under impact, leading to a qualitative and not quantitative prediction of the harvested energy. This work aims to complement the current state of the art by presenting a finite element (FE)-based methodology capable of predicting the impact dynamics, including force–time history, and the harvested energy in customized commercial PEH circuits comprised of rectification, charging and discharging phases. The methodology is validated against measurements performed in an in-house low-cost impact testing machine and processed using real-time hardware/software. The results validate the expected harvested power, and the effect of impact parameters such as impactor mass and velocity on the coupled electromechanical response.

2. Description of the Method

The objective of the experimental-numerical method developed here is the prediction of the electric power that can be harvested during impacts which start vibrations on composite beams with piezoelectric patches connected to PEH circuits. The methodology evolves in two stages, indicated by the orange and the blue dotted lines in the flowchart shown in Figure 1.

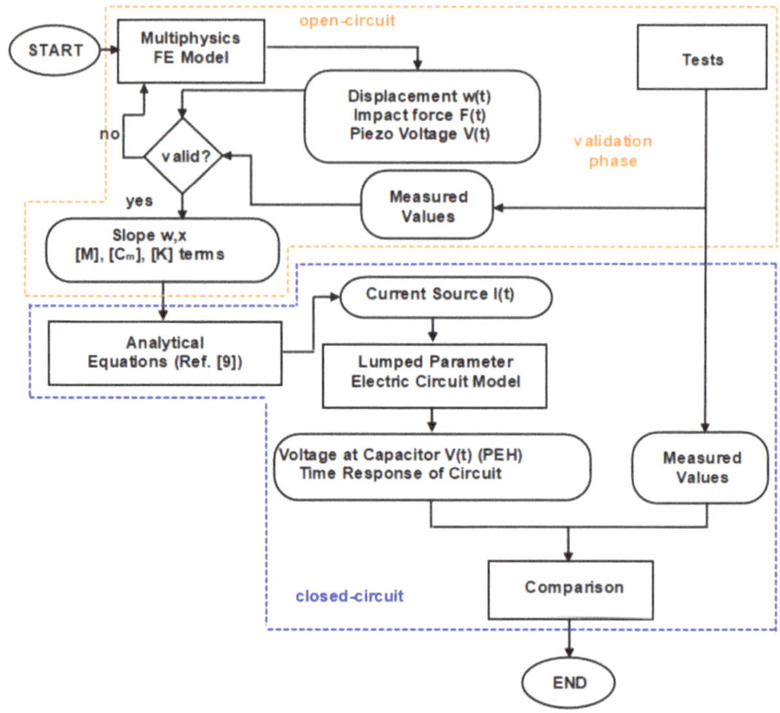

Figure 1. Flowchart of the experimental-numerical methodology.

In the first stage, a finite element (FE) model of a composite beam with a piezoelectric patch is developed in the multiphysics commercial software Abaqus, where the patch is in open-circuit mode [44]. This stage serves to validate the numerical model of the beam subject to impact tests. The validation leads to fine-tuning of the FE model in terms of contact stiffness and energy dissipation on the composite material and its support. In this context, the methodology is described as experimental-numerical, since some of the model parameters (e.g., contact stiffness and damping) are experimentally determined. Next, the selected terms of modal matrices and the predicted slope of the FE model are used as input to the analytical equations for predicting the current in closed-circuit mode [9]. In this context, the piezoelectric patch is modelled as an electromechanically uncoupled current source. Then, the structure of the PEH circuit is programmed in a commercial circuit-design software [45] to yield the transient and steady-state responses of the electromechanical system. The predictions of the closed electric circuit model are compared with measurements.

In the next sections, the basic equations of motion of the coupled electromechanical system are formulated, and each stage of the methodology is described explicitly.

2.1. Coupled Electromechanical System in Open-Circuit Conditions

The beam geometry is shown in Figure 2, and the relevant geometric parameters are listed in Table 1. The PE patch, consisting of a piezoceramic material embedded in polyimide, is bonded with epoxy on the upper surface of the beam and is assumed to be tightly connected with the beam. The beam laminate consists of orthotropic composite plies with arbitrary orientation and an isotropic piezoelectric transducer. The composite material is assumed to exhibit linear elastic behavior. Through-thickness polarization and linear piezoelectric response are assumed for the PE patch transducer.

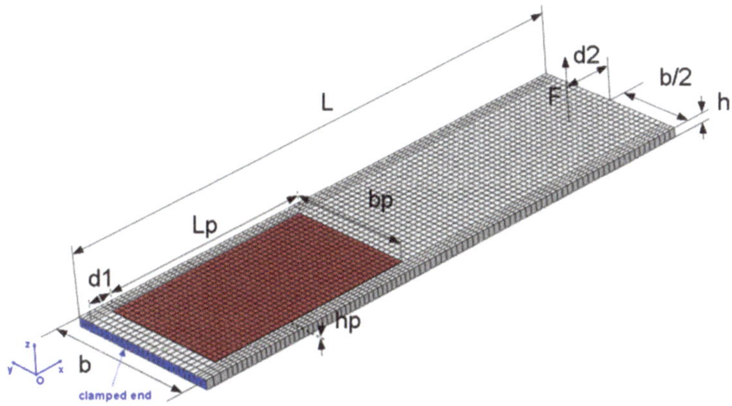

Figure 2. Composite beam with PE patch: FE mesh and geometric parameters.

Table 1. Beam and PE patch geometric parameters.

Parameter	Value (mm)	Parameter	Value (mm)
Beam length L	128	PE patch length L_p	50
Beam width b	37	PE patch width b_p	30
Beam thickness h	2.15	PE patch thickness h_p	0.2
Distance from support d_1	5.5	Distance from free end d_2	10

The ply constitutive equations in the natural coordinate system Oxyz (Figure 2) are [46]:

$$\begin{aligned} \sigma_i &= C_{ij}^{E} S_j - (e_{3i})^T E_3 \\ D_3 &= e_{3j} S_j + \varepsilon_{33}^{S} E_3 \end{aligned} \quad (1)$$

where i, j = 1, ... ,6; σ_i and S_j are the mechanical stress and engineering strain in vectorial notation; E_3 is the component of electric field vector along the thickness direction; D_3 is the electric displacement vector component; C_{ij} is the elastic stiffness tensor; e_{3j} is the piezoelectric tensor arising from the piezoelectric charge tensor and the stiffness tensor and ε_{33} is the material electric permittivity. Superscripts E and S indicate constant electric field and strain conditions, respectively, and superscript T indicates matrix transposition. Equation (1) is formulated in a general manner to encompass the behavior of both a piezoelectric and a passive composite ply (e_{mj} = 0). The electric field vector component E_3 is the gradient of the electric potential V along the thickness of the PE patch:

$$E_3 = -\partial V / \partial z \quad (2)$$

The through-thickness mechanical displacements of the composite laminate are described employing the first-order shear laminate theory:

$$\begin{aligned} u(x,y,z) &= u^0(x,y) + \beta_x(x,y)z \\ v(x,y,z) &= v^0(x,y) + \beta_y(x,y)z \\ w(x,y,z) &= w^0(x,y) \end{aligned} \quad (3)$$

where u^0, v^0, w^0 and β_x, β_y are displacements and rotations of the midsurface, respectively. As explicitly described in Section 2.2, a higher-order layerwise approximation of the through-thickness displacement field [47] is utilized at a specific stage of the method to directly provide the slope of transverse displacement along the longitudinal axis of the beam. The PE patch is modelled as a 3D solid continuum, perfectly bonded on the composite substrate, with electric potential as nodal DOF in addition to mechanical displacements.

The equations of motion of the beam subjected to impact are expressed in variational form using Hamilton's principle as [28]:

$$\int_{t_0}^{t_0+dt}\left(\int_V \delta K dV - \int_V \delta H dV - \int_V \delta W_d dV + \int_\Gamma \delta u^T \tau d\Gamma\right)dt = 0 \quad (4)$$

where index V denotes beam volume; δu is the vector of mechanical displacements arising from the kinematic assumptions (3); τ are the tractions at the boundary surface Γ and δH, δK and δW_d are the variations of the electromechanical, kinetic, and dissipated energy of the beam, respectively. The energy terms in Equation (4) may be expressed as a function of stress, strain and electric field, which are substituted using Equations (1) and (2) and the strain–displacement relations. The degrees of freedom of the beam are approximated along the midsurface (mechanical displacements) and within the transducer volume (electric voltage) by means of quadratic (8 nodes/FE) and linear (4 nodes/FE) shape functions, respectively [44]. Numerical integration of Equation (4) yields the coupled electromechanical discrete system in the time domain:

$$\begin{bmatrix} [M_{uu}] & 0 \\ 0 & 0 \end{bmatrix}\begin{Bmatrix} \ddot{u} \\ \ddot{V} \end{Bmatrix} + \begin{bmatrix} [C_{uu}] & 0 \\ 0 & 0 \end{bmatrix}\begin{Bmatrix} \dot{u} \\ \dot{V} \end{Bmatrix} + \begin{bmatrix} [K_{uu}] & [K_{u\upsilon}] \\ [K_{\upsilon u}] & [K_{\upsilon\upsilon}] \end{bmatrix}\begin{Bmatrix} \bar{u} \\ V \end{Bmatrix} = \begin{Bmatrix} F(t) \\ Q(t) \end{Bmatrix} \quad (5)$$

where [M], [C] and [K] are the system mass, damping and stiffness matrices, divided into elastic (index u) and electric (index υ) parts; F is the external force vector and Q denotes the external electric charge. In the case of an open-circuited piezoelectric transducer, as in the first stage of the methodology illustrated in Figure 1, the external electric charge is assumed to be zero.

The external force appearing in Equation (5) is attributed to the contact between a rigid impactor, modelled as a point mass with initial velocity, and the beam. This impact force is expressed as a function of transverse displacement using a linear contact law:

$$F(t) = \begin{Bmatrix} k_y\left(w^i(t) - w^0(x_0, y_0, t)\right), & w^i(t) > w^0(x_0, y_0, t) \\ 0 & w^i(t) \le w^0(x_0, y_0, t) \end{Bmatrix} \quad (6)$$

where the index i denotes the impactor, (x_0, y_0) are the coordinates of impact on the midsurface and k_y is the contact stiffness derived by material and geometric properties [48]. The contact law of Equation (6) is represented schematically by interposition of a linear spring with stiffness k_y between beam and impactor, as shown in Figure 3.

Figure 3. Modeling of impactor–target contact during impact.

Note that the present formulation accounts for low-velocity impacts causing no material damage. It was implemented using the linear spring connection element SPRINGA [44]. More sophisticated models describing the contact between impactor and composite beam [49–51] could be applied, but the linear contact law was used for the sake of simplicity.

Solution of Equation (5) provides the time profile of displacements of the composite beam and the electric potential at the terminals of the open-circuited piezoelectric transducer, which in this case acts as a sensor. The time profile of impact force may be predicted using Equation (6) and used as input for predicting electric current in a harvesting circuit, as explicitly described in the following section.

2.2. Determination of the PEH Patch Model

The response of the PEH patch is predicted by developing a model with degrees of freedom depending on the vibration modes considered, which in the single-mode case is described as a lumped parameter (LP) model with parameters determined by the FE simulation. The model is based on the formulation of Erturk and Inman [9]—the piezoelectric patch is modelled as a current source in parallel to an electric impedance (Figure 4) and a capacitor, which is practically the capacitive part of the piezoelectric transducer.

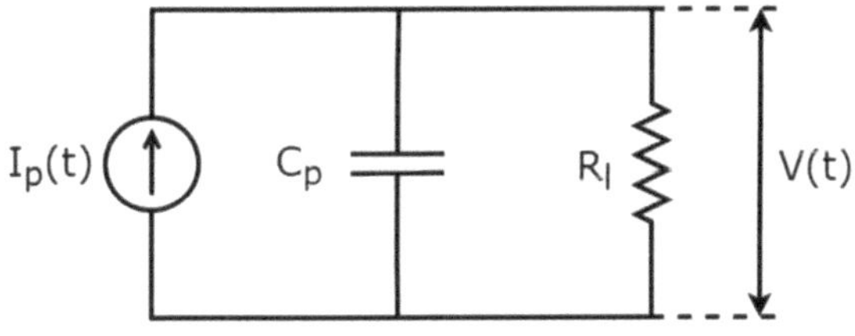

Figure 4. Modeling of piezoelectric patch in closed-circuit configuration.

The closed-circuit response is described by the ordinary differential equation [9]:

$$\frac{\overline{\varepsilon}_{33}^{S} \cdot b \cdot L}{h_p} \cdot \frac{dV}{dt} + \frac{V(t)}{R_l} + \sum_{j=1}^{\infty} \kappa_j \frac{d\eta_j(t)}{dt} = 0 \tag{7}$$

The capacity in the circuit of Figure 4 is:

$$C_p = \frac{\varepsilon_{33} bL}{h_p} \tag{8}$$

where L, b and h_p are the length, width and thickness of the transducer, respectively.

Considering that an exclusively bending vibration along the length of the beam is excited by the impact event and assuming proportional damping, the transverse displacement is separated to spatial and time components:

$$w(x,t) = \sum_{j=1}^{n} \varphi_j(x)\eta_j(t) \tag{9}$$

where j denotes an eigenmode, n is the number of the eigenmodes considered and φ is an eigenvector. The circuit current source is derived by comparing the analytical bending equation of motion with the dynamics of a resistor–capacitor circuit, as:

$$I(t) = -\sum_{j=1}^{n} \kappa_j \frac{d\eta_j(t)}{dt} \tag{10}$$

where j denotes an eigenmode and η is the transient vector of modal coordinates in the space-time decomposition of the transverse displacement. κ is the electromechanical coupling coefficient, provided as:

$$\kappa_j = e_{31} h_p b \int_{x_s}^{x_f} \frac{d^2\varphi_j(x)}{dx^2} dx = e_{31} h_p b \left(\left. \frac{d\varphi_j(x)}{dx} \right|_{x_f} - \left. \frac{d\varphi_j(x)}{dx} \right|_{x_s} \right) \tag{11}$$

with x_s and x_f denoting the start and end of the patch in terms of length coordinate x.

The methodology followed for determining stiffness, mass, damping, piezoelectric and permittivity terms of the LP model, including all necessary terms appearing in Equations (9)–(11), may be briefly summarized as follows.

(a) Solving of Equation (5) for the free-vibration response of the open-circuit system in the frequency domain provides the eigenvectors, which are used for the formulation of the modal mass, damping, stiffness and force transformation matrices, respectively [52]:

$$\begin{aligned}[M_{ffmod}] &= [\varphi]^T[M][\varphi] \\ [C_{ffmod}] &= [\varphi]^T[C][\varphi] \\ [K_{ffmod}] &= [\varphi]^T[K][\varphi] \\ [F_{sfmod}] &= [\varphi]^T[F]\end{aligned} \tag{12}$$

(b) The modal matrices are used for determining η:

$$[M_{ffmod}]\{\ddot{\eta}\} + [C_{ffmod}]\{\dot{\eta}\} + [K_{ffmod}]\{\eta\} = \{F_{sfmod}\}u \tag{13}$$

where u is a modal force amplitude scaling parameter that can be predicted by Equations (5) and (6) or obtained by measurement of the impact force. The linear system of Equation (13) consists of n uncoupled equations.

(c) The electromechanical coupling coefficient κ is provided by Equation (11) using PE properties, geometrical parameters and the slope of the modal transverse displacement. The latter is directly calculated in modal space, in addition to the eigenvectors and modal matrices, by implementing a C^1-continuous 2D higher-order layerwise FE [47], which encompasses this slope as a nodal DOF.
(d) The electric current flowing through the PEH circuit is calculated using Equation (10). As an alternative to points (b–c), η may be derived at the point of excitation by means of a high-speed camera.
(e) The PE transducer is modelled in the PEH circuit as a current source and a capacitor with C_p appearing in Equation (8).
(f) The harvested power may be calculated as:

$$P(t) = \frac{\int_{t_1}^{t_2} V_{out}(t)I_{out}(t)dt}{t_2 - t_1} \tag{14}$$

where V_{out} and I_{out} are respectively the voltage and current in the discharging branch of the PEH circuit described in the following section.

2.3. PEH Circuit

The PEH circuit studied here is the commercial PI E-821.00 [27], shown in Figure 5a. The circuit was modelled in Ltspice software [45], as presented in Figure 5b. The model includes a charging branch, consisting of a diode rectifier and a capacitor that charges due to the alternative current produced by the PE transducer. When the voltage of the capacitor reaches a reference value (nominal value 12 V), an automatic switch is activated, allowing part of the electric energy in the capacitor to be released into the discharging branch. The automatic switch is turned off when the voltage of the capacitor falls below a low threshold (nominal value 6 V). The switch is implemented by a MOSFET controlled by an operational amplifier which compares the voltage of the capacitor to the reference voltage. This response is modelled with the use of a Schmitt trigger supplied by external voltage sources for simplicity. The discharging branch consists of a buck converter tuning the output to be a pulse of constant voltage.

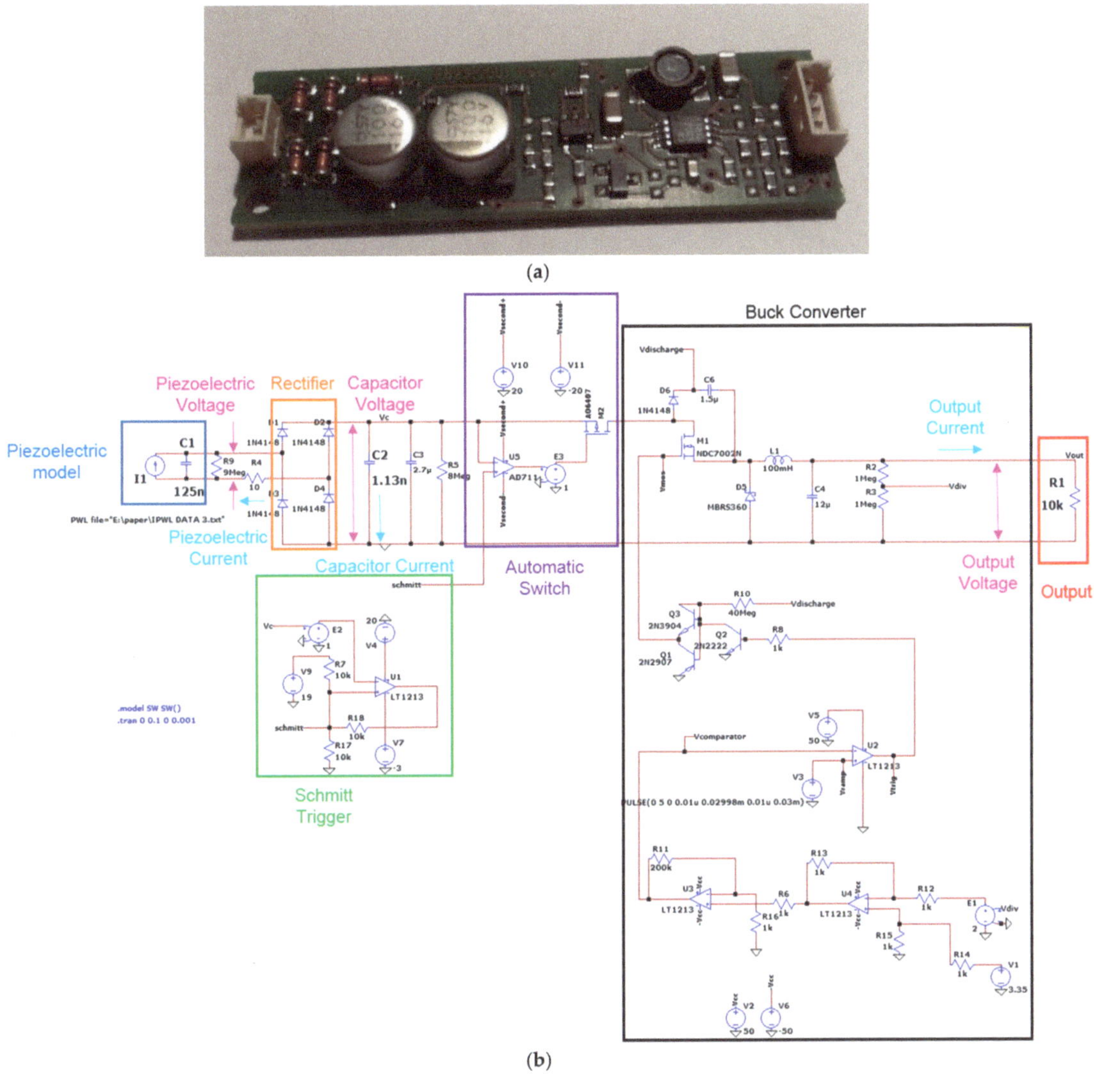

Figure 5. PEH circuit: (**a**) physical item, (**b**) modelling approach in Ltspice.

3. Experimental Configuration for Impact Testing

3.1. Custom Impact Frame for Controlled Impact Velocity

Impact tests were performed using a custom portable experimental configuration [53], which was complemented in terms of repeatability and data acquisition capabilities (Figure 6).

Its basic version includes a light aluminum frame, prestressed by steel cables for providing extra rigidity, and an impactor instrumented for impact force measurement. In its current version, the impactor is moved by a DC servomotor [54] in controllable configuration by means of encoder feedback (see Appendix A). The perpendicularity of impact is ensured by means of two parallel bearing sets. Impact force and impactor velocity are measured by a custom force sensor [55] and the encoder, respectively.

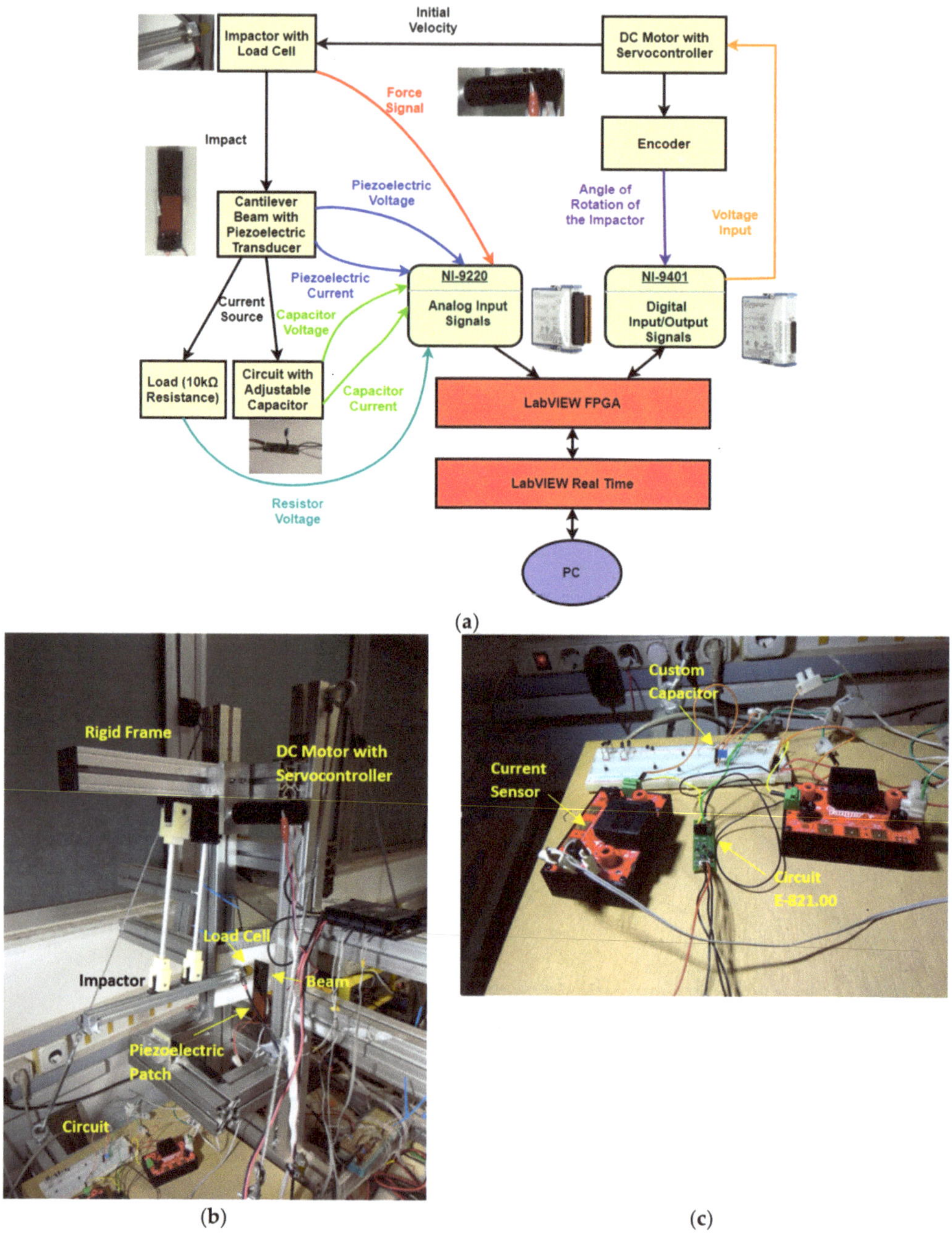

Figure 6. Experimental configuration for impact tests: (**a**) schematic representation with measurement system, (**b**) photo of impact frame with beam specimen, (**c**) PEH circuit including an E-821.00 module.

The impact test takes place in two successive steps: (i) the impactor is raised up to the desired height, and (ii) the velocity of the impactor is provided by the user to initiate its trajectory (see Appendix A). The motion of the impactor in both stages is controlled by implementing a modified state feedback controller [56] with gravity compensation. The controller is based on a lumped-parameter model of the impactor, including constants derived by CAD models and static, constant velocity and constant acceleration tests (see Appendix A). The appropriate gains were determined by the root-locus method. All signals (load cell, encoder feedback, circuit voltage and current) are obtained using high-speed data acquisition modular equipment (NI CRIO 9074 [57] including an analog input module (AI–NI 9220) with 16 channels being capable to operate at a maximum rate of 10^5 samples/s per channel, and a digital I/O module (NI 9401) with a maximum sampling rate of 10^7 samples/s, while a rate of 10^4 samples/s is used for the control of the impactor. The software implemented for acquisition, control and storage of the signals is Labview FPGA [58]. The developed experimental configuration [59] enables impactor tip angle and velocity to be set by the test engineer to perform impact tests of various types [60] (see Appendix A). The displacement at the load application point is measured by processing images [61] acquired by a high-speed camera. A polyurethane tip was used in the impact tests. The positions of impactor and piezoelectric patch on the tested clamped composite beam are shown in Figure 2.

3.2. Materials and Specimens

A composite beam of graphite/epoxy material with a P.876-A12 DuraAct® [62] (PI Ceramic GmbH, Lederhose, Germany) surface attached piezoelectric patch and lamination $[-45/45/90_2/0/90]_S$ was studied. The patch consists of PIC255 piezoceramic embedded in polyimide. For the sake of simplicity, the piezoceramic part of the patch was modelled, and its properties are listed in Table 2. The elastic properties of the composite material considered were obtained from static tests [63] performed by the manufacturer (Hellenic Aerospace Industry S.A., Athens, Greeece). They were validated additionally by performing modal tests on the cantilever beam leading to measurement of the fundamental eigenfrequency at 89.7 Hz. A relatively large modal loss factor of η = 12.5% was assumed in the LP system, which includes the contribution of clamped support, viscoelastic material and intensive usage of the specimen in forced harmonic response tests [64]. The electromechanical properties of all materials considered are listed in Table 2.

Table 2. Electromechanical properties of materials considered (* under constant stress).

Property	Composite Material	Piezoelectric Material
Density (kg/m^3)	1554	7800
	Elastic Properties	
E_{11} (GPa)	138.40	62.10
E_{22} (GPa)	8.50	62.10
E_{33} (GPa)	8.50	48.30
G_{12} (GPa)	4.30	23.20
G_{13} (GPa)	4.30	21.30
G_{23} (GPa)	4.30	21.30
ν_{12}	0.31	0.33
ν_{13}	0.31	0.43
ν_{23}	0.31	0.43
	Piezoelectric Properties	
d_{31} (10^{-12} m/V)	-	−191
d_{32} (10^{-12} m/V)	-	−191
d_{33} (10^{-12} m/V)	-	409
	Dielectric Properties (ε^0 = 8.85 × 10^{-12} F/m)	
$\varepsilon_{33}/\varepsilon^0$	3.5	1832 *

4. Results and Discussion

According to the methodology shown in Figure 1, the validation stage included the study of the composite beam impact response in open-circuit electric conditions; therefore, the piezoelectric transducer acted as a sensor. The contact stiffness k_y was tuned by correlating model prediction with measured impact force and piezoelectric voltage time profiles, yielding a value of 500 N/m. In the prediction stage, the closed-circuit response was studied in three cases, each with a different electric component connected to the PE terminals: (i) a resistance of 10 kΩ; and (ii) the E-821.00 harvesting module twice, each time with a different charging subcircuit capacitor. The latter configuration was achieved by replacing the standard 200 μF capacitor with either of two other custom ones having a capacity of 100 μF and 1 nF, respectively.

4.1. Open-Circuit Impact Response

Figure 7 illustrates FE-predicted and measured time history of impact force, transverse displacement, and piezoelectric voltage at an impact with initial velocity of 0.75 m/s. The measured force signal shown in Figure 7a indicates a globally dominated response [53], as derived from the long duration of the impact and the single fundamental vibration mode participating in the response. Chattering was observed in the beginning of the measured force signal, assumed to be mainly attributable to the dimensional clearance of the force sensor in the impactor assembly. The FE model successfully captured the impact duration, while it overestimated the impact force amplitude. This deviation may be attributed to modelling the impactor as a concentrated rigid mass, whereas in the test the polyurethane tip became deformed during impact. A better comparison during the initial impact phase was observed between prediction and measurement in the time history of the displacement, shown in Figure 7b. The measured piezoelectric voltage (Figure 7c) had a similar time profile with the displacement, as expected in an open-circuit PE response, while it was more sensitive to the initial chattering observed in the force–time history. The end of the impact event was determined from the force sensor signal, considering the initial noise, and is denoted with a dashed vertical line.

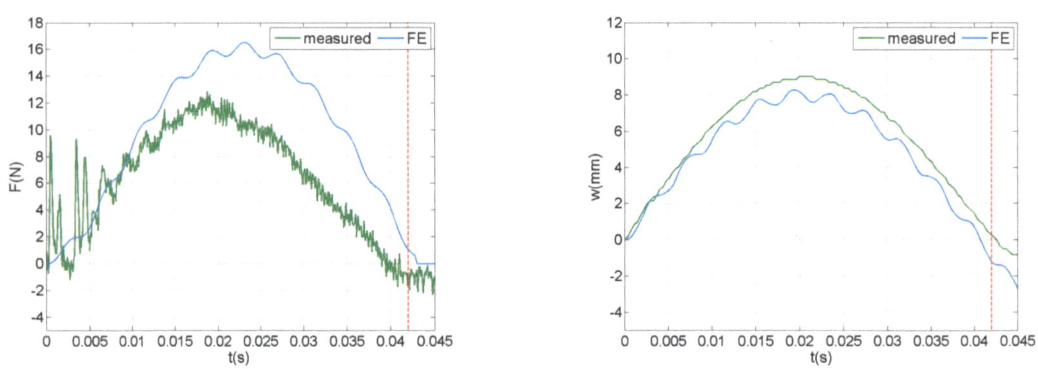

(**a**) (**b**)

Figure 7. *Cont.*

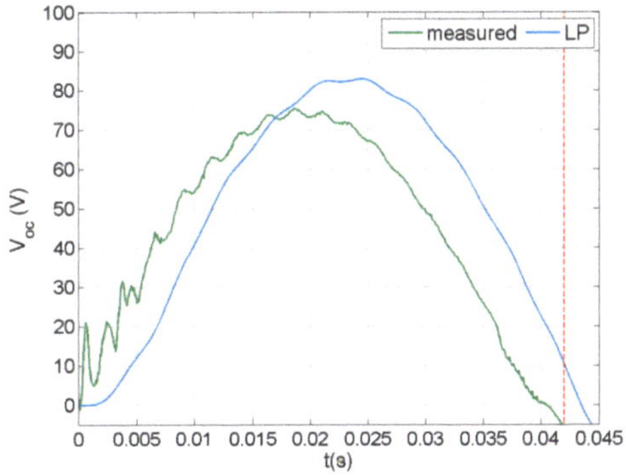

(c)

Figure 7. Time history of predicted and measured impact response (v^i = 0.75 m/s) in open-circuit configuration: (**a**) impact force, (**b**) tip displacement and (**c**) electric potential at piezoelectric terminals.

4.2. Closed-Circuit Resistive Impact Response

Predictions of the lumped parameter (LP) model for the resistor voltage and circuit current are compared with measurements in Figure 8a,b, respectively. The LP model predictions followed the measured trends for both the resistor voltage and circuit current, but missed the fluctuation measured during the initial impact phase. The change of sign in both graphs during impact indicates that the impactor moved backwards when it lost contact with the beam. The free vibration which followed the impact event, provided a smoother signal, which contained practically the first bending mode.

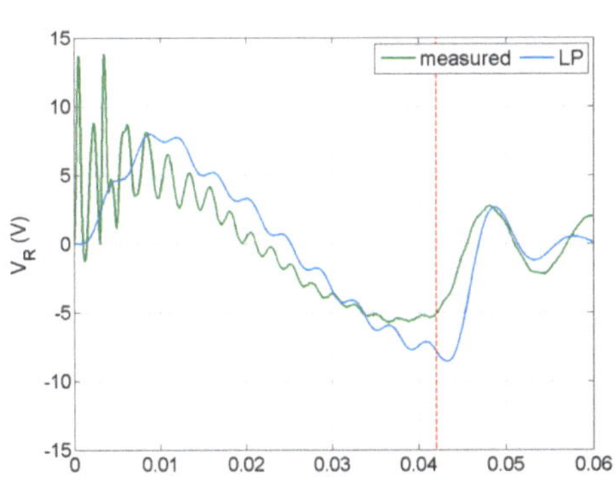

(a)

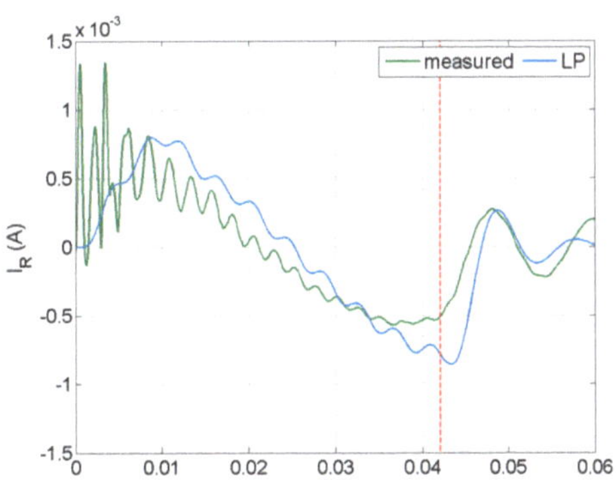

(b)

Figure 8. Impact response in a resistive circuit with R = 10 kΩ: (**a**) resistor voltage, (**b**) circuit current.

4.3. Closed-Circuit PEH Response

The dynamic response of the PEH circuit connected to the piezoelectric composite beam under an impact with velocity of 0.75 m/s is illustrated in Figures 9 and 10. A 100 μF capacitor was employed in the circuit. Voltage and current were measured at the piezoelectric terminals and at the capacitor, as illustrated in Figure 5b. The current predictions and measurements seemed to be more sensitive to the impact event and the induced vibration, while the voltage appeared to have a smoother time response. This trend, also observed in the forced frequency response of a similar system [64], is more evident in the measurements presented in Figure 10. This may be attributed to the rectification in the charging subcircuit, which mainly affected the voltage signal as designated by the manufacturer. As shown in Figure 10, the capacitor voltage eventually reached a value of

0.22 V, which is lower than the discharging threshold of 12 V, thus there was no current output in the discharging subcircuit.

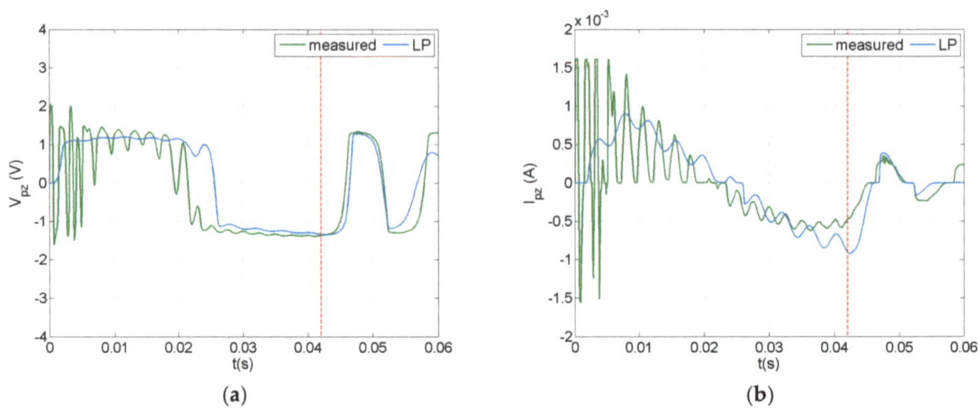

Figure 9. Predicted and measured signals at the piezoelectric terminals of the PEH circuit with a 100 µF capacitor in the case of impact: (**a**) voltage, (**b**) current.

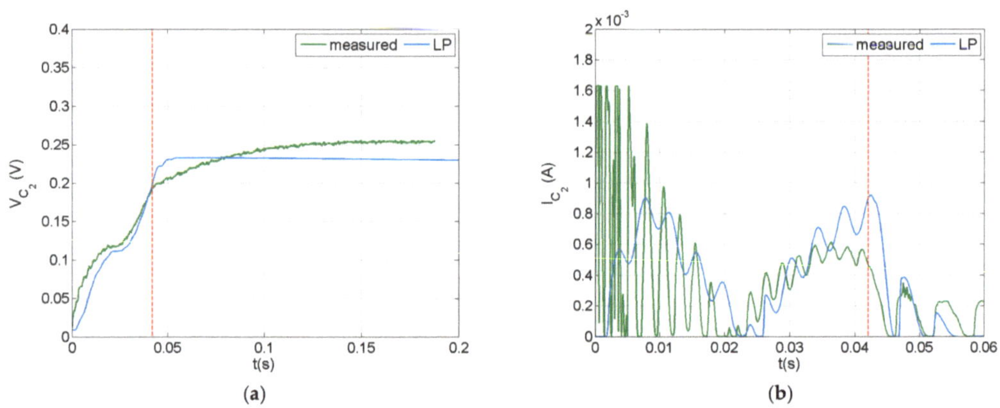

Figure 10. Predicted and measured signals at the capacitor of the PEH circuit with a 100 µF capacitor in the case of impact: (**a**) voltage, (**b**) current.

The LP model could predict the general signal trends, but missed the initial fluctuation during impact. This trend was more evident before rectification (Figure 9). As already observed in the open-circuit case (Section 4.1), parameters not included in the impact modelling approach, such as the deformability of impactor and clearance in the force sensor assembly, strongly affected the fluctuation in the dynamic response during the initial phase of the impact event. Nevertheless, this methodology could predict the limit values of impact-induced signals and therefore could be used as a design tool for PEH circuits in applications related to low-velocity impact events.

The design capabilities of the developed methodology were studied by implementing a 1 nF capacitor in a customized version of the E-821.00 PEH circuit for achieving voltage output. This capacity is smaller than the 2.7 µF capacity of the MOSFETs, so practically the circuit operated with the MOSFETs also acting as capacitors.

Predictions and measurements of output voltage in the case of an impact velocity of 1.25 m/s are shown in Figure 11. A very good match between predicted and measured signal was observed, indicating the enhanced capabilities of the methodology. The low value of capacitance required for obtaining harvested power indicates that the cantilever beam configuration is not the most appropriate for impact harvesting, since the strain in the piezoelectric transducer was not sufficient to charge capacitors of a practical application range in realistic implementation at macro-scale. To this end, other configurations should be further studied, focusing on the design of the composite structure hosting the transducer.

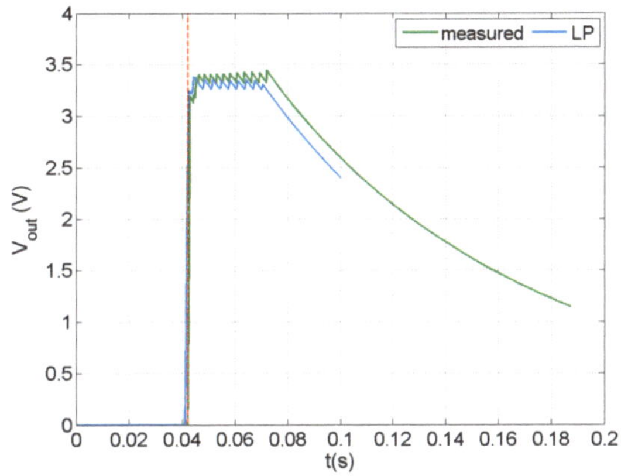

Figure 11. Predicted and measured voltage in the output of the discharging subcircuit in the case of a PEH circuit with capacity of 2.7 μF.

The current methodology could be used as a design tool for harvesting circuits in the case of structures subjected to impact. In case of known impactor mass and velocity, a custom PEH circuit can be designed to harvest power during impact and subsequent vibration. Conversely, for the fixed PEH with a 2.7 μF capacitor studied herein, the mass and velocity of the impactor can be modified, representing different impact conditions.

Figure 12 illustrates the effect of initial impactor velocity on the harvested power. The blue stars indicate experimental data, while the green circles are predictions of the present LP model for an impactor mass of 0.3 kg. For velocities exceeding 1 m/s, the beam acted as a harvester. Power increases had two discrete slopes. However, a precise prediction of the harvested power at high velocities requires a thorough study of the mechanical response of the beam in terms of geometric linearity and mechanical strength, which is beyond the scope of the current work.

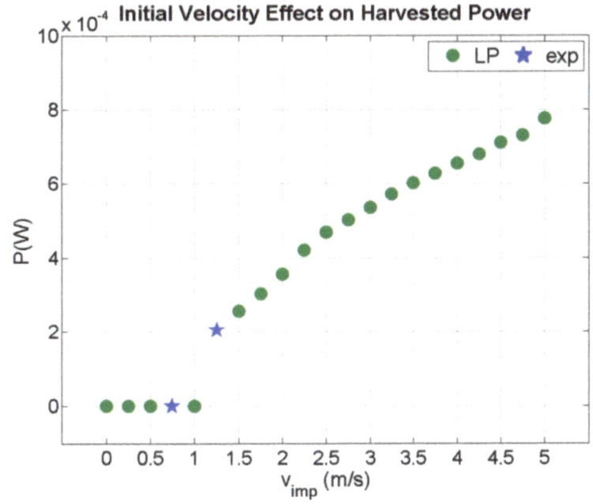

Figure 12. Predicted output power as a function of impact velocity (m^i = 0.3 kg).

In Figure 13, the effect of a mass impacting the beam with initial velocity 0.75 m/s is shown. The threshold for harvesting power in the E.821.00 PEH circuit with a modified capacitor was 0.8 kg. The harvested power increased with mass, as expected due to the increased strain induced in the piezoelectric patch by the larger impulse of the impactor.

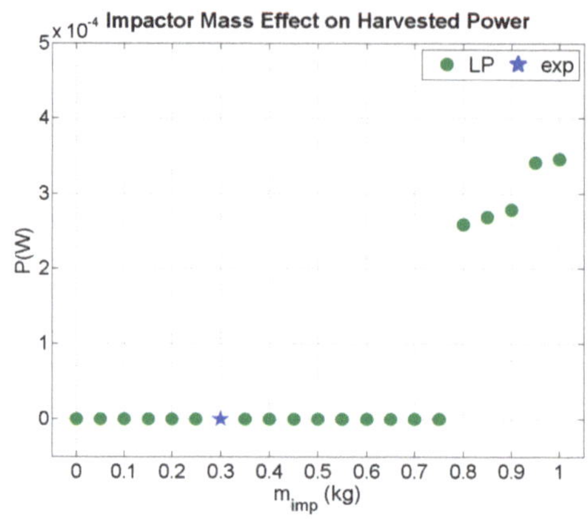

Figure 13. Predicted output power as a function of impactor mass (v^i = 0.75 m/s).

In addition to the prediction of harvested power, the current methodology can be used as a design tool for prediction of the structural integrity of a structure subjected to impact. The prediction of the impact force–time profile, as presented in Figure 7, provides an overview of the impact event and the timestamp of maximum impact force. Figure 14 illustrates the longitudinal stress distribution at ply 3 of 12 at maximum impact force in the case of an impact of a 0.3 kg mass with a velocity of 4 m/s. This ply has a 90° fiber orientation with respect to the coordinate system presented in Figure 2. In specific regions, the compressive stress exceeded the compressive strength of the epoxy matrix (158 MPa) [63]. This prediction dictates that an additional design loop is required to ensure the structural integrity of the harvesting system and tune its harvesting capability by properly setting the capacity of the output subcircuit.

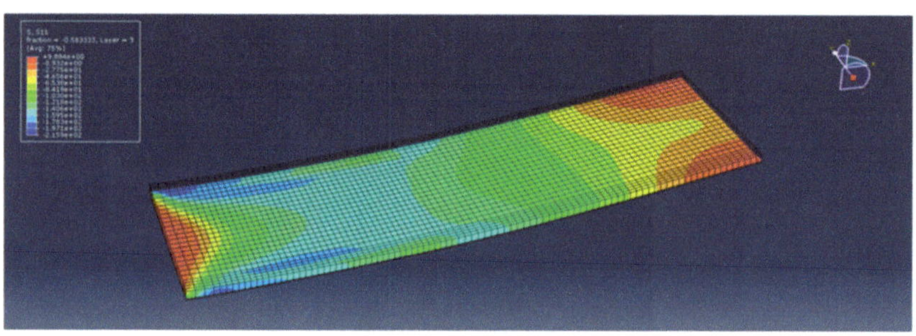

Figure 14. Longitudinal stress distribution in ply 3 of a composite beam at maximum impact force during the impact of a 0.3 kg mass hitting with 4 m/s initial velocity.

5. Summary and Conclusions

A FE-based methodology was developed for studying PEH from composite beams with piezoelectric patches subjected to low-velocity impact. Equivalent LP models were implemented for predicting the post-impact free vibration, electromechanical response of the beam, and the impact-induced electric response of the harvesting circuit. A commercial PEH circuit was modified to enable harvested power output under the specific impact conditions achieved using a custom portable experimental configuration, which involves programming in real-time software for controlling impactor velocity and acquiring sensory signals. The main conclusions derived are listed below:

1. Predictions of the current methodology compared well with measurements of the electric response of the harvesting circuit, indicating its applicability in the design of PEH systems subjected to impact.
2. Deviations between predictions and measurements were observed for the impact force due to the geometric clearance of the force sensor, approximation of impactor–beam contact stiffness and impactor deformability.
3. Modification of the commercial PEH circuit in terms of capacity in the charging subcircuit led to harvesting of power in the impact conditions studied experimentally. The design of this modification was possible via the development of a circuit model equivalent to the commercial circuit in LTspice software, and respective verification.
4. The harvested power increased with impactor mass and velocity beyond a threshold. Up to that threshold, no electric power produced due to the impact event and subsequent beam vibration could be extracted.
5. A major capability of this methodology is the prediction of impact-force–time profile and stress distribution during impact events in composite beams with arbitrary lamination. In this context it can be used for the design of composite impact harvesters with piezoelectric patches, enabling the prediction of harvested power within applicable structural integrity limits.

The current methodology includes the modelling of PEH in composite beams by developing an LP electromechanical model based on FE and solving for the electric response of the harvesting circuit. To this end, the extension of the methodology to plate and shell structures is straightforward and will be studied in future work, aiming at the design of appropriate PEH circuits for various impact conditions.

Author Contributions: Conceptualization, T.S.P. and E.G.P.; methodology, all; software, N.M. and P.K.-K.; validation, N.M. and P.K.-K.; formal analysis, N.M.; investigation, N.M., T.S.P. and P.K.-K.; resources, T.S.P. and E.G.P.; data curation, N.M.; writing—original draft preparation, T.S.P., N.M. and P.K.-K.; writing—review and editing, all; visualization, N.M., P.K.-K. and T.S.P.; supervision, T.S.P. and E.G.P.; project administration, T.S.P. and E.G.P.; funding acquisition, T.S.P. and E.G.P. All authors have read and agreed to the published version of the manuscript.

Funding: This research was co-financed by the European Regional Development Fund of the European Union and Greek national funds through the Operational Program Competitiveness, Entrepreneurship, and Innovation, under the call RESEARCH–CREATE–INNOVATE (project code: T1EDK-01533).

Institutional Review Board Statement: Not applicable.

Informed Consent Statement: Not applicable.

Data Availability Statement: Data available upon request from the authors.

Acknowledgments: The authors would like to specially thank Eduardo Barrera from the Telematics and Electronics Engineering Department of Universidad Politécnica de Madrid for his valuable advice regarding the FPGA programming for impact data acquisition.

Conflicts of Interest: The authors declare no conflict of interest.

Appendix A

Appendix A.1. Parameters of Impactor Mechanism

The model of the moving mechanism for impact testing is shown in Figure A1.

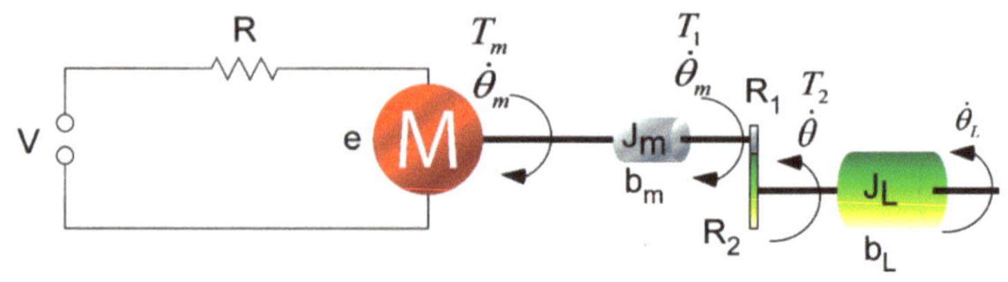

Figure A1. Model of motor–gearbox–load.

The differential equation of motion of the impactor is described in Equation (A1):

$$n \cdot K_T \cdot i = \left(n^2 \cdot J_m + J_L\right) \cdot \ddot{\theta} + \left(n^2 \cdot b_m + b_L\right) \cdot \dot{\theta} + T_C \cdot \text{sgn}\dot{\theta} + T_W \tag{A1}$$

where T_m is the motor's torque, $\dot{\theta}_m$ is the motor's rotational speed, J_m is the rotor's inertia of the motor, b_m is the viscous friction coefficient of the motor, $\dot{\theta}$ is the rotational speed of the impact pendulum (load), J_L is the inertia of the load, b_L is the viscous friction coefficient of the load, $n = R_2/R_1$ is the planetary gearhead's reduction, i is the current that passes through the motor, K_T is the torque constant, T_c is the torque produced by the Coulomb friction, u is the output torque after the gearhead, T_W is the torque produced by the gravitational forces (weight) of the impact pendulum and b_e is the equivalent coefficient of the viscous friction $b_e = n^2 \cdot b_m + b_L$.

The parameters of the model were experimentally determined (Table A1).

Table A1. Parameters of the impactor mechanism.

T_C (Nm)	b (Nms)	M_{tot} (kg)	J_m (kgm^2)	J_L (kgm^2)	n	k_T (Nm/A)
0.068	0.46	0.367	3.47E-6	0.02	28	0.054

Appendix A.2. Control of Impactor Mechanism

The block diagram of the closed-loop system with a PV controller is illustrated in Figure A2.

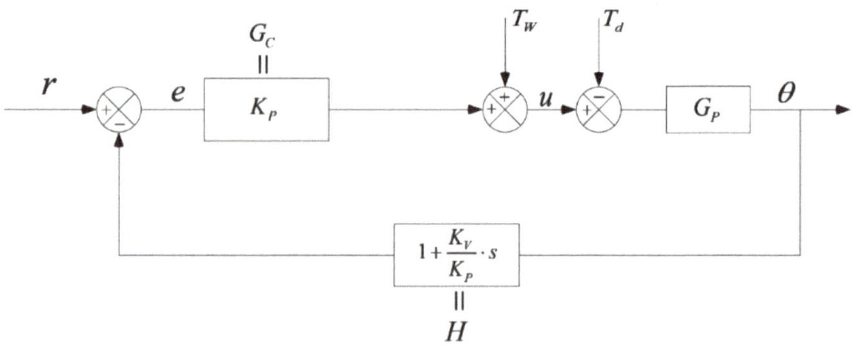

Figure A2. Simplified closed-loop block diagram for a non-linear PV (or PD *) controller with weight compensation T_w, where $r(t)$ is the reference input signal, $G_c(s)$, $G_p(s)$ and $H(s)$ are transfer functions of controller, plant and feedback loop, respectively, and T_d is a term for the disturbances.

A non-linear PV (or PD *) controller is an alternative form of a simple PD controller with an extra term to compensate weight. More specifically, instead of using the first derivative of the error, the first derivative of the output is considered in order to avoid overshoot:

$$u(t) = K_P \cdot e(t) - K_V \cdot \frac{dy}{dt} + T_W = K_P \cdot (\theta_d - \theta) - K_V \cdot \dot{\theta} + T_W \quad \text{(A2)}$$

where $u(t)$ is the input, $e(t)$ is the measured error, $y(t)$ is the output, K_P and K_V are the gains of the controller, θ and θ_D are actual and desired position of the impact pendulum, respectively, $\dot{\theta}$ is its rotational speed and T_W is a term to compensate the torque produced by its weight.

Appendix A.3. Real-Time Implementation of the Control System

CRIO–9074 hardware was used for measuring the velocity of the impactor by means of an encoder, for implementing the state feedback controller, and for generating the PWM input to the plant via the motor controller (Figure A3).

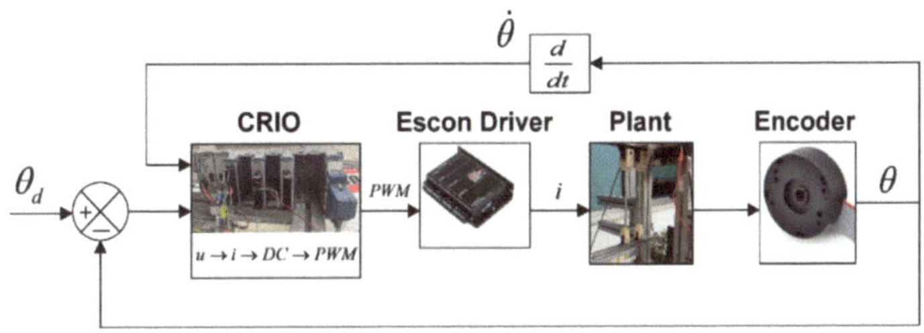

Figure A3. Real-time block diagram representation for the closed loop.

The flowchart of the consolidated automatic control and data acquisition system for the control of the impact pendulum and the sampling of measurements from sensors is represented in Figure A4. The process was performed in parallel at real-time and FPGA levels. The FPGA level ensures the efficiency of analog and digital measurements and output signals, while the real-time (RT) level is exclusively focused on implementation of the PV controller, data storage and visualization. The red dashed line in Figure A4 illustrates the flow of analog data from the FPGA level to the RT level, while the green dashed lines indicate the flow of digital signals between the FPGA and the RT level.

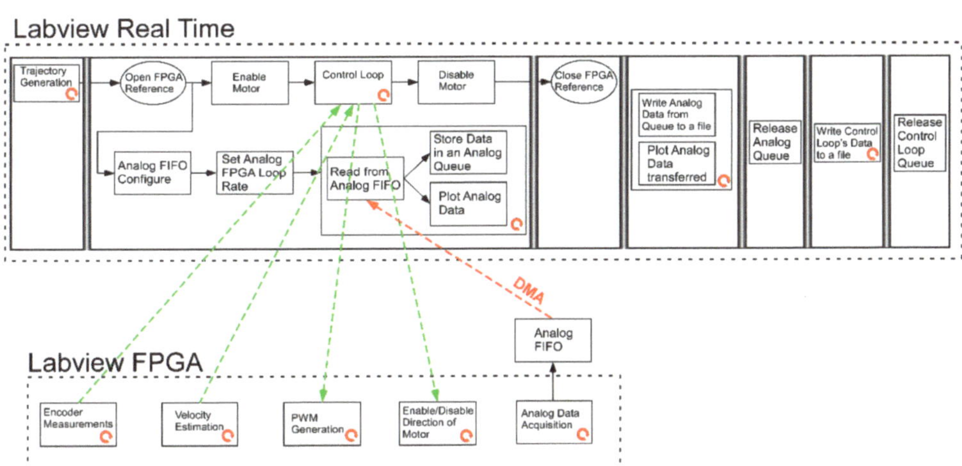

Figure A4. Flow chart of the integrated control and acquisition system, where DMA and FIFO denote direct memory access and first in, first out sequence, respectively.

The execution frequency of each loop at real-time and FPGA levels is presented in Table A2.

Table A2. Loop rates of main processes.

Labview Real-Time Level	
Control Loop	$f_{control\ loop} = 1$ kHz
Labview FPGA Level	
Encoder Measurements	$f_{digital\ loop} = 10$ kHz
Analog Data Acquisition	$f_{analog\ loop} = 16$ kHz

Position and velocity profiles of the impactor in a typical impact test are illustrated in Figure A5.

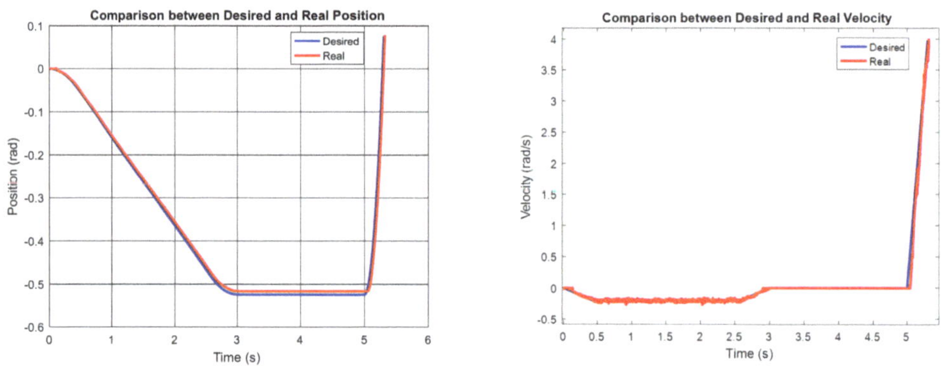

(**a**) (**b**)

Figure A5. Implementation evidence of the developed real-time control system: (**a**) position profile and (**b**) velocity profile of the impactor.

The tip angle of the impactor in a typical impact test is illustrated in Figure A6. The tip angle a3 is low to justify the assumption of vertical impact.

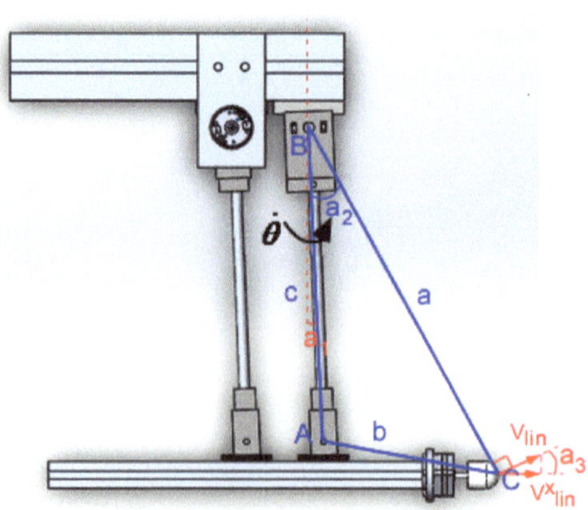

Figure A6. Schematic representation of impactor tip angle, where v_{lin} is linear velocity; v^{x}_{lin} is its x-component; p is the angle between x-axis and linear velocity vector; a is the distance between the tip of the impactor and point B, measured before testing and a1, a2 and a3 are rotation angles.

References

1. Kymissis, J.; Kendall, C.; Paradiso, J.; Gershenfeld, N. Parasitic power harvesting in shoes. In Proceedings of the Digest of Papers Second International Symposium on Wearable Computers (Cat. No.98EX215), Institute of Electrical and Electronics Engineers (IEEE), Pittsburgh, PA, USA, 6 August 2002.
2. Elvin, N.G.; Elvin, A.; Spector, M. A self-powered mechanical strain energy sensor. *Smart Mater. Struct.* **2001**, *10*, 293–299. [CrossRef]
3. Cook-Chennault, K.; Thambi, N.; Sastry, A.M. Powering MEMS portable devices—A review of non-regenerative and regenerative power supply systems with special emphasis on piezoelectric energy harvesting systems. *Smart Mater. Struct.* **2008**, *17*. [CrossRef]
4. Kim, H.S.; Kim, J.-H.; Kim, J. A review of piezoelectric energy harvesting based on vibration. *Int. J. Precis. Eng. Manuf.* **2011**, *12*, 1129–1141. [CrossRef]
5. Mitcheson, P.D.; Yeatman, E.; Rao, G.K.; Holmes, A.S.; Green, T. Energy harvesting from human and machine motion for wireless electronic devices. *Proc. IEEE* **2008**, *96*, 1457–1486. [CrossRef]
6. Safaei, M.; Sodano, H.A.; Anton, S.R. A review of energy harvesting using piezoelectric materials: State-of-the-art a decade later (2008–2018). *Smart Mater. Struct.* **2019**, *28*, 113001. [CrossRef]
7. Roundy, S.; Wright, P.K. A piezoelectric vibration based generator for wireless electronics. *Smart Mater. Struct.* **2004**, *13*, 1131–1142. [CrossRef]
8. Lu, F.; Lee, H.P.; Lim, S.P. Modeling and analysis of micro piezoelectric power generators for micro-electromechanical-systems applications. *Smart Mater. Struct.* **2003**, *13*, 57–63. [CrossRef]
9. Erturk, A.; Inman, D.J. *Piezoelectric Energy Harvesting*; John Wiley and Sons Ltd.: Oxford, UK, 2011.
10. Erturk, A.; Hoffmann, J.; Inman, D.J. A piezomagnetoelastic structure for broadband vibration energy harvesting. *Appl. Phys. Lett.* **2009**, *94*, 254102. [CrossRef]
11. Wickenheiser, A.M.; García, E. Broadband vibration-based energy harvesting improvement through frequency up-conversion by magnetic excitation. *Smart Mater. Struct.* **2010**, *19*, 065020. [CrossRef]
12. Masana, R.; Daqaq, M.F. Electromechanical modeling and nonlinear analysis of axially loaded energy harvesters. *J. Vib. Acoust.* **2010**, *133*, 011007. [CrossRef]
13. Harne, R.; Wang, K.W. A review of the recent research on vibration energy harvesting via bistable systems. *Smart Mater. Struct.* **2013**, *22*. [CrossRef]
14. Jia, Y. Review of nonlinear vibration energy harvesting: Duffing, bistability, parametric, stochastic and others. *J. Intell. Mater. Syst. Struct.* **2020**, *31*, 921–944. [CrossRef]
15. Ottman, G.; Hofmann, H.; Bhatt, A.; Lesieutre, G. Adaptive piezoelectric energy harvesting circuit for wireless remote power supply. *IEEE Trans. Power Electron.* **2002**, *17*, 669–676. [CrossRef]
16. Ottman, G.K.; Hofmann, H.F.; Lesieutre, G. Optimized piezoelectric energy harvesting circuit using step-down converter in discontinuous conduction mode. *IEEE Trans. Power Electron.* **2003**, *18*, 696–703. [CrossRef]
17. Lefeuvre, E.; Badel, A.; Richard, C.; Petit, L.; Guyomar, D. A comparison between several vibration-powered piezoelectric generators for standalone systems. *Sens. Actuators A Phys.* **2006**, *126*, 405–416. [CrossRef]
18. Lefeuvre, E.; Audigier, D.; Richard, C.; Guyomar, D. Buck-boost converter for sensorless power optimization of piezoelectric energy harvester. *IEEE Trans. Power Electron.* **2007**, *22*, 2018–2025. [CrossRef]
19. Shen, H.; Qiu, J.; Ji, H.; Zhu, K.; Balsi, M. Enhanced synchronized switch harvesting: A new energy harvesting scheme for efficient energy extraction. *Smart Mater. Struct.* **2010**, *19*, 115017. [CrossRef]
20. Ramadass, Y.K.; Chandrakasan, A.P. An efficient piezoelectric energy harvesting interface circuit using a bias-flip rectifier and shared inductor. *IEEE J. Solid State Circuits* **2010**, *45*, 189–204. [CrossRef]
21. Tabesh, A.; Fréchette, L.G. A low-power stand-alone adaptive circuit for harvesting energy from a piezoelectric micropower generator. *IEEE Trans. Ind. Electron.* **2010**, *57*, 840–849. [CrossRef]
22. Liang, J.; Liao, W.-H. Impedance modeling and analysis for piezoelectric energy harvesting systems. *IEEE ASME Trans. Mechatron.* **2011**, *17*, 1145–1157. [CrossRef]
23. Huguet, T.; Lallart, M.; Badel, A. Bistable vibration energy harvester and SECE circuit: Exploring their mutual influence. *Nonlinear Dyn.* **2019**, *97*, 485–501. [CrossRef]
24. Guyomar, D.; Lallart, M. Recent progress in piezoelectric conversion and energy harvesting using nonlinear electronic interfaces and issues in small scale implementation. *Micromachines* **2011**, *2*, 274–294. [CrossRef]
25. Caliò, R.; Rongala, U.B.; Camboni, D.; Milazzo, M.; Stefanini, C.; De Petris, G.; Oddo, C.M. Piezoelectric energy harvesting solutions. *Sensors* **2014**, *14*, 4755–4790. [CrossRef] [PubMed]
26. Linear Technology Corporation. *Piezoelectric Energy Harvesting Power Supply*; LTC-3588; Linear Technology Corporation: Norwood, MA, USA, 2010.
27. PI Motion/Positioning. E-821 Electronic Module for Energy Harvesting Using Piezo Actuators for Energy Generation. 2014. Available online: https://www.physikinstrumente.com/en/products/controllers-and-drivers/piezo-drivers-for-open-loop-operation-of-piezo-actuators/e-821-electronic-module-for-energy-harvesting-608000/ (accessed on 28 September 2021).
28. Plagianakos, T.S.; Papadopoulos, E.G. Low-energy impact response of composite and sandwich composite plates with piezoelectric sensory layers. *Int. J. Solids Struct.* **2014**, *51*, 2713–2727. [CrossRef]

29. Umeda, M.; Nakamura, K.; Ueha, S. Analysis of the transformation of mechanical impact energy to electric energy using piezoelectric vibrator. *Jpn. J. Appl. Phys.* **1996**, *35*, 3267. [CrossRef]
30. Pozzi, M.; Zhu, M. Plucked piezoelectric bimorphs for knee-joint energy harvesting: Modelling and experimental validation. *Smart Mater. Struct.* **2011**, *20*, 055007. [CrossRef]
31. Renaud, M.; Fiorini, P.; van Schaijk, R.; van Hoof, C. Harvesting energy from the motion of human limbs: The design and analysis of an impact-based piezoelectric generator. *Smart Mater. Struct.* **2009**, *18*, 35001. [CrossRef]
32. Lee, E.H. The impact of a mass striking a beam. *J. Appl. Mech.* **1940**, *7*, A129–A138. [CrossRef]
33. Gu, L.; Livermore, C. Impact-driven, frequency up-converting coupled vibration energy harvesting device for low frequency operation. *Smart Mater. Struct.* **2011**, *20*, 045004. [CrossRef]
34. Basari, A.A.; Awaji, S.; Sakamoto, S.; Hashimoto, S.; Homma, B.; Suto, K.; Okada, H.; Okuno, H.; Kobayashi, K.; Kumagai, S. Evaluation on mechanical impact parameters in piezoelectric power generation. In Proceedings of the 10th Asian Control Conference (ASCC), Sabah, Malaysia, 31 May–3 June 2015; pp. 1–6.
35. Doria, A.; Medè, C.; Desideri, D.; Maschio, A.; Codecasa, L.; Moro, F. On the performance of piezoelectric harvesters loaded by finite width impulses. *Mech. Syst. Signal. Process.* **2018**, *100*, 28–42. [CrossRef]
36. Jacquelin, E.; Adhikari, S.; Friswell, M. A piezoelectric device for impact energy harvesting. *Smart Mater. Struct.* **2011**, *20*, 105008. [CrossRef]
37. Martinez-Ayuso, G.; Friswell, M.I.; Adhikari, S.; Khodaparast, H.H.; Featherston, C.A. Energy harvesting using porous piezoelectric beam with impacts. *Procedia Eng.* **2017**, *199*, 3468–3473. [CrossRef]
38. Hunt, K.; Crossley, F.R.E. Coefficient of restitution interpreted as damping in vibroimpact. *J. Appl. Mech.* **1975**, *42*, 440–445. [CrossRef]
39. Fu, X.; Liao, W.-H. A dimensionless model of impact piezoelectric energy harvesting with dissipation. *Act. Passiv. Smart Struct. Integr. Syst.* **2016**, *9799*, 97991. [CrossRef]
40. Ferrari, M.; Cerini, F.; Ferrari, V. Autonomous sensor module powered by impact-enhanced energy harvester from broadband low-frequency vibrations. In Proceedings of the Transducers & Eurosensors XXVII: The 17th International Conference on Solid-State Sensors, Actuators and Microsystems, Barcelona, Spain, 16–20 June 2013; pp. 2249–2252. [CrossRef]
41. Wong, C.H.; Dahari, Z. Development of vibration-based piezoelectric raindrop energy harvesting system. *J. Electron. Mater.* **2017**, *46*, 1869–1882. [CrossRef]
42. Feng, T.; Aono, K.; Covassin, T.; Chakrabartty, S. Self-powered monitoring of repeated head impacts using time-dilation energy measurement circuit. *IEEE Trans. Biomed. Circuits Syst.* **2015**, *9*, 217–226. [CrossRef] [PubMed]
43. Chen, N.; Jung, H.; Jabbar, H.; Sung, T.H.; Wei, T. A piezoelectric impact-induced vibration cantilever energy harvester from speed bump with a low-power power management circuit. *Sens. Actuators A Phys.* **2017**, *254*, 134–144. [CrossRef]
44. Dassault Systems. *Abaqus/CAE User's Guide, Version 6.14*; Dassault Systems: Vienna, Austria, 2014.
45. Available online: https://www.analog.com/en/analog-dialogue/articles/get-up-and-running-with-ltspice.html (accessed on 23 September 2021).
46. Tiersten, H.F. *Linear Piezoelectric Plate Vibrations*; Springer Science: New York, NY, USA, 1969.
47. Plagianakos, T.S.; Saravanos, D.A. High-order layerwise finite element for the damped free-vibration response of thick composite and sandwich composite plates. *Int. J. Numer. Methods Eng.* **2009**, *77*, 1593–1626. [CrossRef]
48. Yigit, A.; Christoforou, A. Impact dynamics of composite beams. *Compos. Struct.* **1995**, *32*, 187–195. [CrossRef]
49. Skrinjar, L.; Slavič, J.; Boltežar, M. A review of continuous contact-force models in multibody dynamics. *Int. J. Mech. Sci.* **2018**, *145*, 171–187. [CrossRef]
50. Yigit, A.S.; Christoforou, A.P.; Majeed, M.A. A nonlinear visco-elastoplastic impact model and the coefficient of restitution. *Nonlinear Dyn.* **2011**, *66*, 509–521. [CrossRef]
51. Siorikis, D.; Rekatsinas, C.; Christoforou, A.; Saravanos, D. Experimental and numerical investigation of contact laws for the rapid simulation of low-energy impacts on laminated composite plates. *Compos. Struct.* **2017**, *168*, 646–656. [CrossRef]
52. Walgrave, S.C.; Ehlbeck, J.M. Understanding modal analysis. *SAE Tech. Pap. Ser.* **1978**, *87*. [CrossRef]
53. Plagianakos, T.; Lika, K.; Papadopoulos, E.G. Low-velocity impact response of smart sandwich composite plates with piezoelectric transducers: Modeling and experiments. *J. Intell. Mater. Syst. Struct.* **2016**, *27*, 774–785. [CrossRef]
54. Maxon Motor. *ESCON Module 50/5 Servo Controller-Hardware Reference*; Maxon Motor: Sachseln, Switzerland, 2014.
55. Available online: https://www.botasys.com/ (accessed on 9 October 2020).
56. Nise, N.S. *Control System Engineering*, 9th ed.; John Wiley & Sons: Oxford, UK, 2019.
57. Available online: https://www.ni.com/en-us/support/model.crio-9074.html (accessed on 18 September 2021).
58. National Instruments Co. *LabVIEW FPGA Course Manual*; National Instruments Co.: Austin, TX, USA, 2009.
59. Karydis-Karandreas, P. Modeling and Control of an Impact Pendulum for Impact Testing of Plates with Pi-Ezoelectric Transducers. Master's Thesis, National Technical University of Athens, Athens, Greece, 2019.
60. Christoforou, A.P.; Yigit, A. Characterization of impact in composite plates. *Compos. Struct.* **1998**, *43*, 15–24. [CrossRef]
61. Komitopoulos, S. Active Vibration Control of Composite Beams Using Piezoelectric Transducers and Real-Time Software. Master's Thesis, National Technical University of Athens, Athens, Greece, 2021.
62. Available online: https://www.piceramic.com/en/ (accessed on 27 September 2020).

63. Karachalios, E.; Muñoz, K.; Jimenez, M.; Prentzias, V.; Goossens, S.; Geernaert, T.; Plagianakos, T.S. LRI-fabricated composite demonstrators for an aircraft fuselage on the basis of a Building Block design approach. *Compos. Part. C Open Access* **2021**, *6*, 100178. [CrossRef]
64. Plagianakos, T.S.; Margelis, N.; Leventakis, N.; Bolanakis, G.; Vartholomeos, P.; Papadopoulos, E.G. Finite element-based assessment of energy harvesting in composite beams with piezoelectric transducers. *Proc. Inst. Mech. Eng. Part. L J. Mater. Des. Appl.* **2021**. [CrossRef]

sensors

MDPI

Article

System Implementation Trade-Offs for Low-Speed Rotational Variable Reluctance Energy Harvesters

Ye Xu [1], Sebastian Bader [1,*], Michele Magno [1,2], Philipp Mayer [2] and Bengt Oelmann [1]

1 Department of Electronics Design, Mid Sweden University, 85170 Sundsvall, Sweden; ye.xu@miun.se (Y.X.); michele.magno@pbl.ee.ethz.ch (M.M.); bengt.oelmann@miun.se (B.O.)
2 Department of Information Technology and Electrical Engineering, ETH Zurich, 8092 Zurich, Switzerland; mayerph@iis.ee.ethz.ch
* Correspondence: sebastian.bader@miun.se

Abstract: Low-power energy harvesting has been demonstrated as a feasible alternative for the power supply of next-generation smart sensors and IoT end devices. In many cases, the output of kinetic energy harvesters is an alternating current (AC) requiring rectification in order to supply the electronic load. The rectifier design and selection can have a considerable influence on the energy harvesting system performance in terms of extracted output power and conversion losses. This paper presents a quantitative comparison of three passive rectifiers in a low-power, low-voltage electromagnetic energy harvesting sub-system, namely the full-wave bridge rectifier (FWR), the voltage doubler (VD), and the negative voltage converter rectifier (NVC). Based on a variable reluctance energy harvesting system, we investigate each of the rectifiers with respect to their performance and their effect on the overall energy extraction. We conduct experiments under the conditions of a low-speed rotational energy harvesting application with rotational speeds of 5 rpm to 20 rpm, and verify the experiments in an end-to-end energy harvesting evaluation. Two performance metrics—power conversion efficiency (PCE) and power extraction efficiency (PEE)—are obtained from the measurements to evaluate the performance of the system implementation adopting each of the rectifiers. The results show that the FWR with PEEs of 20 % at 5 rpm to 40 % at 20 rpm has a low performance in comparison to the VD (40–60 %) and NVC (20–70 %) rectifiers. The VD-based interface circuit demonstrates the best performance under low rotational speeds, whereas the NVC outperforms the VD at higher speeds (>18 rpm). Finally, the end-to-end system evaluation is conducted with a self-powered rpm sensing system, which demonstrates an improved performance with the VD rectifier implementation reaching the system's maximum sampling rate (40 Hz) at a rotational speed of approximately 15.5 rpm.

Keywords: energy harvesting; rotational energy harvesting; kinetic energy harvesting; variable reluctance; self-powered sensors; internet of things; smart sensors

Citation: Xu, Y.; Bader, S.; Magno, M.; Mayer, P.; Oelmann, B. System Implementation Trade-Offs for Low-Speed Rotational Variable Reluctance Energy Harvesters. *Sensors* **2021**, *21*, 6317. https://doi.org/10.3390/s21186317

Academic Editors: Carlo Trigona, Olfa Kanoun and Slim Naifar

Received: 1 September 2021
Accepted: 18 September 2021
Published: 21 September 2021

1. Introduction

Energy harvesting has drawn much interest due to its capability of extending battery lifetime, or even enabling fully energy-autonomous operation of new generations of smart sensors and internet of things edge devices [1–3]. It can be employed in a number of different application domains, ranging from biomedical engineering, via industrial applications, to structural health monitoring [4–6]. A typical energy harvesting system includes an energy transducer, an interface circuit, and a load that uses the accumulated energy to perform a specific task [3,7–9]. The energy transducer converts energy from the physical domain to the electrical domain, generating a source-dependent output signal [10,11]. For kinetic energy harvesting, electromagnetic and piezoelectric energy transducers are common choices, which create an alternating current (AC) output [1,7,12–14]. The intended system loads are commonly low-power electronic systems, such as Internet of Things (IoT) nodes or sensing systems, requiring a direct current (DC) input [9]. In order to effectively supply power to these loads, the interface circuit needs to perform AC-DC conversion, impedance matching, voltage regulation, and potentially energy buffering. A typical implementation combines a rectifier circuit with a power management integrated circuit (PMIC) [12,15–18].

Variable reluctance energy harvesting (VREH) has been identified to be a suitable technique for rotational kinetic energy harvesting with low rotational speeds [19–23]. VREHs are a type of electromagnetic energy harvesters, which generate an AC output as the result of magnetic flux changes inside a pickup coil [11,20]. As opposed to many other electromagnetic transducers, VREHs do not require a relative translation between coil and magnet, which makes them suitable for applications with large diameter rotating objects, as well as low rotational speeds [20,21]. However, the output voltage and frequency of VREH transducers is low at low rotational speeds [21,23], making the system design susceptible to interface circuit implementation. In [22], a self-powered rpm sensing system based on an VREH implementation was proposed. The system uses a Cockroft-Walton voltage multiplier in combination with a commercial LTC3588-1 PMIC as interface circuit. In the system analysis, it was shown that the power delivered to the load was below 20 % of the power that can be obtained with a matched impedance load. Consequently, significant improvements on the system's interface circuit are possible in order to optimize the electrical circuit for the power conversion.

Other studies have investigated and compared interface circuit designs for energy harvesting systems [3,12,16,24–30], as well as self-powered sensor systems incorporating a number of these interface circuits [17,18,31–34]. Many of these studies are generic without specific application constraints [3,12,29], whereas others are focused on piezoelectric transducers with relative high output voltages [15,18,27,30,35,36]. Interface circuits for electromagnetic transducers with relative low voltages have been addressed in [17,28,33,34,37]. In [28], Ulusan et al. proposed and evaluated an integrated interface circuit, including AC-DC and DC-DC conversion, based on an active voltage doubler (VD). However, active rectifier solutions may create cold-start challenges and are more difficult to be implemented effectively with discrete components, which is desirable for small-scale applications. Discrete interface circuit solutions have been implemented based on a passive VD and a discrete PMIC in [17,33], and based on a negative voltage converter rectifier (NVC) and PMIC in [34]. The difference in transducers being used in these studies makes a direct comparison between the respective interface circuit implementation difficult. Moreover, none of the studies related the converted power to that of a matched impedance load, which has been argued to be an important metric for energy harvesting interface circuits [24,27,30,32,35,36,38,39].

In this paper, we conduct a systematic comparison of interface circuits based on three passive rectifier implementations for a VREH system as depicted in Figure 1. The rectifiers under evaluation are a full-wave bridge rectifier (FWR) [18,29,40–42], a Delon full-wave voltage doubler (VD) [33,40,43–45], and a negative voltage converter (NVC) [29,34,42,46,47]. System implementations using these rectifiers in combination with a state-of-the-art DC-DC PMIC (TI BQ25570) are evaluated experimentally under the conditions of a low-speed rotational energy harvesting application. In order to evaluate the interface circuits, power conversion efficiency (PCE) and power extraction efficiency (PEE) are used as metrics. While PCE is limited to the losses within the interface circuits [48], PEE includes effects on the transducer by relating the obtained output power to the output power with a matched impedance load [24,30,32,38]. Moreover, the achievable sampling rates of a self-powered rpm sensing system based on the respective interface circuits are investigated.

The key contributions of this study are summarized as follows: (i) a systematical and experimental evaluation of interface circuits based on different rectifiers, including their losses and effects on the extracted power. This particularly enables a direct comparison between VD and NVC; (ii) a demonstration that the optimum rectifier selection can change over the range of operating conditions in a VREH system implementation; (iii) a self-powered, wireless RPM sensing system operating at low rotational speeds with an PEE of up to 70 %.

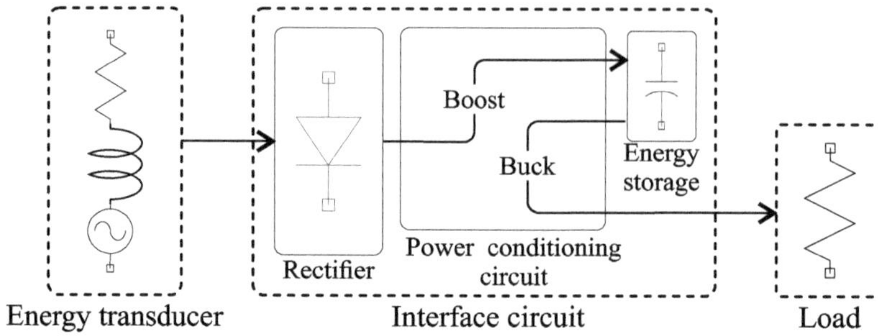

Figure 1. Overview of the Variable Reluctance Energy Harvesting system under evaluation.

2. Variable Reluctance Energy Harvesting System

A high-level architectural overview of the targeted variable reluctance energy harvesting system is given in Figure 1. It can be divided into three core modules, namely the variable reluctance energy transducer, the interface circuit including AC-DC and DC-DC conversion, as well as the electrical load.

2.1. *Variable Reluctance Energy Harvester*

The variable reluctance energy harvester is a type of electromagnetic energy harvester, which operates based on Faraday's law of induction. A VREH generates an electromotive force due to changes in magnetic flux, and it can for example be used in rotational kinetic energy harvesting applications [21]. In a VREH, both the magnet and the pickup coil are stationary, whereas the reluctance variation, and thus the changes in magnetic flux, are a result of the relative motion between a ferromagnetic structure and the pickup unit.

In this work, an m-shaped, on-rotor VREH structure is used [22]. This structure is illustrated in Figure 2. The magnetic flux distribution varies periodically between the two extreme positions in the core of the pickup coil as the shaft rotates. As a result, an alternating voltage is induced across the pickup coil, with a frequency of

$$f_c = \omega N_t, \tag{1}$$

where ω is the rotational speed in rad/s, and N_t is the number of teeth of the tooth wheel.

To maximize energy extraction from the VREH, a complex conjugate matched impedance is required as load to the pickup coil. As the VREH transducer is inductive, its complex conjugate matched impedance can be implemented with a resistor R_{Load} and a capacitor C_{Load}, such that

$$R_{\text{Load}} = R_{\text{Coil}} \tag{2}$$

$$C_{\text{Load}} = \frac{1}{(2\pi f_c)^2 \cdot L_{\text{Coil}}}, \tag{3}$$

where R_{Coil} and L_{Coil} are the DC resistance (DCR) and internal inductance of the pickup coil. By means of the maximum power transfer theorem, connecting an impedance matched load to the pickup coil can determine the maximum amount of power that is dissipated in the load, presenting the power extraction ability of the VREH transducer. This is used to derive one of the key performance metrics, power extraction efficiency, further discussed in Section 3.3. The maximum amount of power is a function of multiple system parameters and operating conditions and can be estimated as [21]

$$P_{\max} = \frac{N_{\text{Coil}}^2}{8R_{\text{Coil}}} \Delta_\phi^2 \left[\frac{\omega N_{\text{tooth}}}{2\pi} \right]^2. \tag{4}$$

where N_{Coil} is the number of coil turns, Δ_ϕ is the magnetic flux difference between aligned and unaligned positions shown in Figure 2, ω is the angular velocity, and N_{tooth} refers to the number of teeth in the toothed wheel. Neglecting the inductive part of the coil's impedance, an accurate estimation is only achieved at low operation frequencies of f_c.

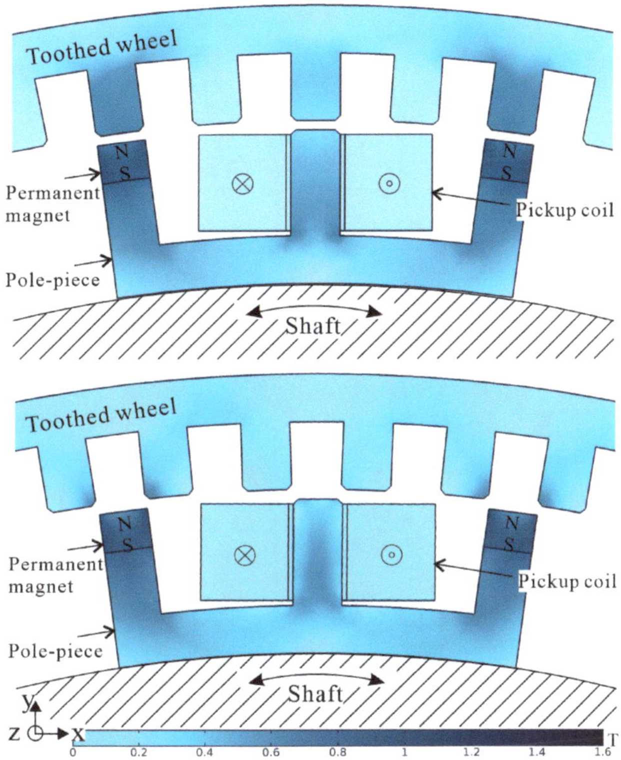

Figure 2. Magnetic flux density distribution of the VREH at the two extreme positions aligned position (**top**) and unaligned position (**bottom**).

2.2. Interface Circuit

The interface circuit is composed of two units, a rectifier circuit (AC-DC) and a power conditioning circuit (DC-DC). For the power conditioning circuit, an off-the-shelf DC-DC PMIC of type Texas Instruments BQ25570 is used. This PMIC is an inductive switching regulator that boosts the input voltage to charge an intermediate storage device (e.g., a supercapacitor or battery). The energy that is stored is then made available to the load via an additional integrated buck converter.

For the rectifier circuit, three implementations commonly used in low-power energy harvesting systems have been chosen, which are evaluated separately and compared to each other. The three rectifier implementations are a full-wave diode bridge rectifier (FWR), a Delon full-wave voltage doubler (VD), and a negative voltage converter (NVC). Circuit implementations for these three rectifiers are depicted in Figure 3, together with their ideal input and output voltages. The terminals *IN+* and *IN−* of a rectifier are connected to the pickup coil to maintain the differential input. A rectified voltage is produced at the terminal *OUT* with a reference voltage on the terminal *GND*.

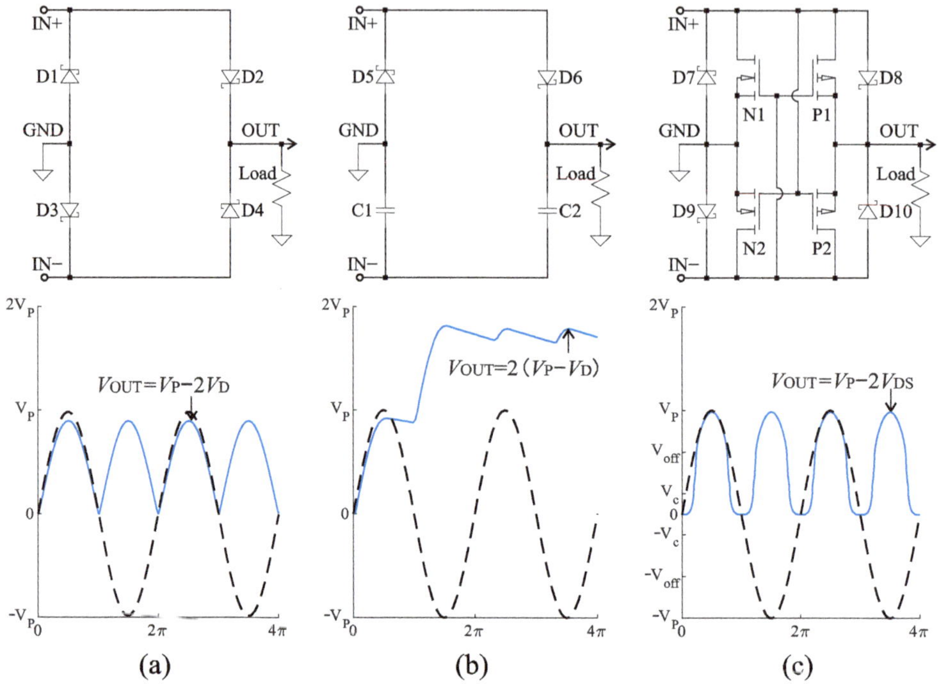

Figure 3. Schematic diagrams of the three rectification circuits, (**a**) full-wave diode bridge rectifier (FWR), (**b**) Delon full-wave voltage doubler (VD), and (**c**) negative voltage converter (NVC), and associated qualitative waveforms of input and output voltage. V_P is the amplitude of the input voltage across *IN+* and *IN−*.

The FWR consists of four diodes arranged in the form of a bridge. This circuit topology is the most common circuit used in energy harvesting applications due to its simple operation and implementation. During the positive half-cycle of the input voltage across *IN+* and *IN−*, *D2* and *D3* are forward-biased and conduct current. Conversely, during the negative half-cycle of the input voltage, *D1* and *D4* are forward-biased and conduct current, but *D2* and *D3* are reverse-biased. As a result, a full-wave rectified voltage is produced at its output. However, the output voltage level is reduced by the voltage drops of the two diodes. In most cases, a filter solution, such as a smoothing capacitor, is connected in parallel with the load across the output terminals reducing the ripple of the output DC voltage.

The VD (or single stage voltage multiplier) consists of diodes and capacitors to rectify and boost the input voltage across *IN+* and *IN−*. Its output voltage is twice that of the input AC peak-voltage V_P. During the positive half cycle of the input, the diode *D6* is forward biased, and the current flows though the diode and charges the capacitor *C2* to a voltage of V_P. Similarly, during the negative cycle *C1* is charged through *D5* to V_P. The output voltage across both the capacitor thus is $2V_P$. For a real implementation, the charge on each capacitor is reduced by the forward voltage drop of the diodes (V_D).

Based on a full gate cross-coupled topology, an NVC uses MOSFETs instead of diodes, and converts the negative half of the input voltage across *IN+* and *IN−* to positive ones for voltage rectification. Due to the low on-resistance of the MOSFETs, the NVC typically incurs a lower forward voltage drop (V_{DS}), defined by the the sum of drain-source voltages of two conducting MOSFETs. As shown in Figure 3c, during the positive half-cycle, *P1* and *N2* are turned on, and *N1* and *P2* are turned off. The current flows from *IN+* to *OUT* via *P1* and returns to *IN−* over *N2*. Therefore, the forward voltage drop is defined by *P1* and *N2* during the positive half-cycle. Moreover, the influence of the on-resistance causes an increase in V_{OUT} when the input voltage V_p surpasses a critical value V_c [49]. When V_p

reaches V_{off}, the NVC operates correctly and provides a low on-resistance conduction path for current to flow, where V_{off} depends on the highest threshold voltage $V_{GS(th)}$ between *N*- and *P*-MOSFETs. In this study, Schottky diodes are used in parallel with each MOSFET to adapt to a low-voltage rectification, because the forward voltage of the Schottky diodes is lower than $V_{GS(th)}$ of the MOSFETs [34,47].

The effect of this can be observed in Figure 4. With input voltages greater than V_{off} (approximately 1.2 V), the input voltage is approximately equal to the output voltage, resulting in a high conversion efficiency. With help of the parallel connected Schottky diodes, the NVC acts as an equivalent FWR at low input voltages of 0.1 V to 0.8 V. In contrast, the NVC without the additional diodes has significantly lower rectification performance, and is practically unusable at voltages below 0.4 V.

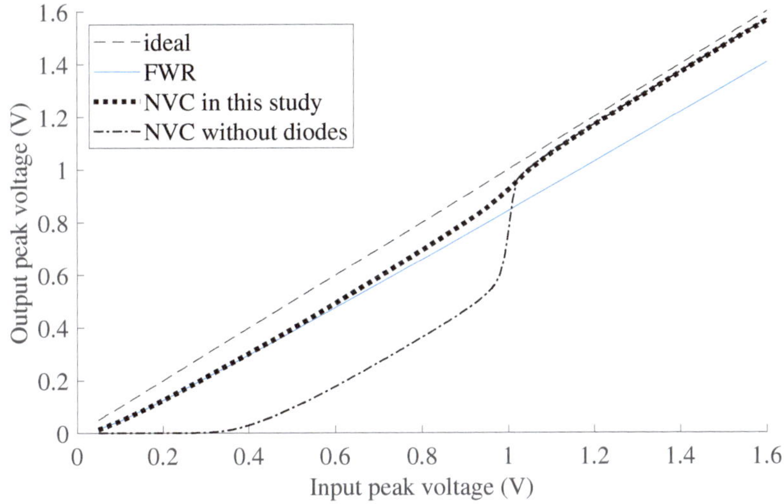

Figure 4. Simulated input and output voltage of the rectifiers using the circuits shown in Figure 3, with an AC voltage source of 7.55 Hz and a load of 10 kΩ.

2.3. *Wireless RPM Sensing System*

The intended electrical load in this study is a wireless RPM sensing system, as illustrated in Figure 5. The system's main components are a MEMS gyroscope and a Bluetooth Low Energy (BLE) System-on-Chip (SoC). The rotational speed of the host structure is thus obtained by measuring the angular velocity using the gyroscope. The BLE SoC transmits the measurements, together with ambient temperature readings, to a Bluetooth host device, where the data will be further processed and used. The MEMS gyroscope is a Bosch BMG-250 with low current draws of 3 µA and 850 µA at 1.8 V in idle and data acquisition mode, respectively.

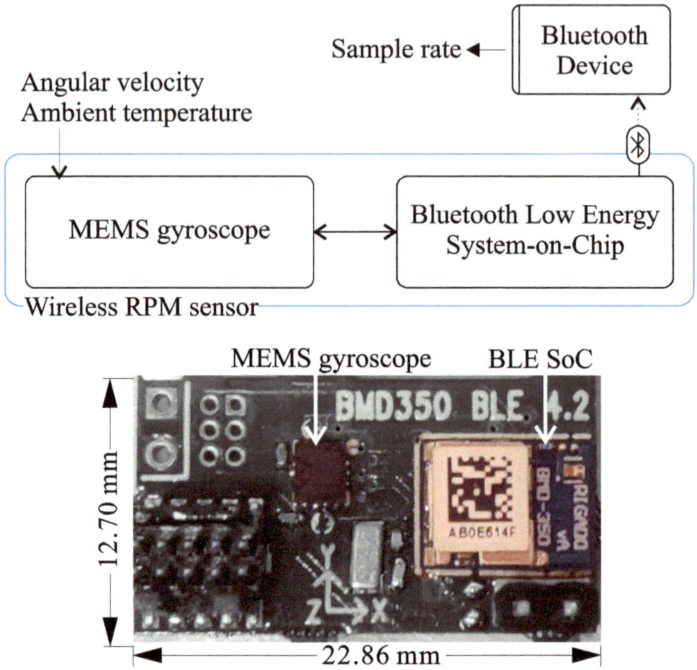

Figure 5. Block diagram of the wireless RPM sensing system (**top**), and the sensor node implementation (**bottom**), including a Bosch BMG250 MEMS gyroscope and a Rigado BMD350 module based on a Nordic Semiconductor nRF52832 SoC.

The sensing system can be operated in different operating modes, based on the duty-cycling of its sub-systems. These operating modes affect the system's achievable sample rate and its average power consumption, as shown in Table 1 and Figure 6. In mode 1, both the sensing and communication sub-systems are duty-cycled. This results in the lowest power consumption when the system is idle with a current draw of 5 µA, but requires the MEMS sensor (BMG-250) to be reinitialized for every sample. Due to the relatively long initialization time of the gyroscope (59 ms), the maximum achievable sample rate in this mode of operation is limited to 15.625 Hz and the contribution to the total energy consumption is high (63.67 µJ). In order to further increase the sample rate, in mode 2, the MEMS gyroscope remains always on avoiding initialization, whereas the BLE continues to be duty-cycled. The sample rate in this mode is limited to a maximum of 40 Hz due to the minimum advertisement interval of the BLE unit of 25 ms. An example of the current consumption profile for operating mode 1 is depicted in Figure 7.

Table 1. Operation modes of the wireless RPM sensor, including supported sample rate ranges and component activity level.

	Mode 1	Mode 2
Gyroscope	duty-cycled	active
BLE SoC	duty-cycled	duty-cycled
Sample rate f_s (Hz)	(0–15.625)	(15.625–40)

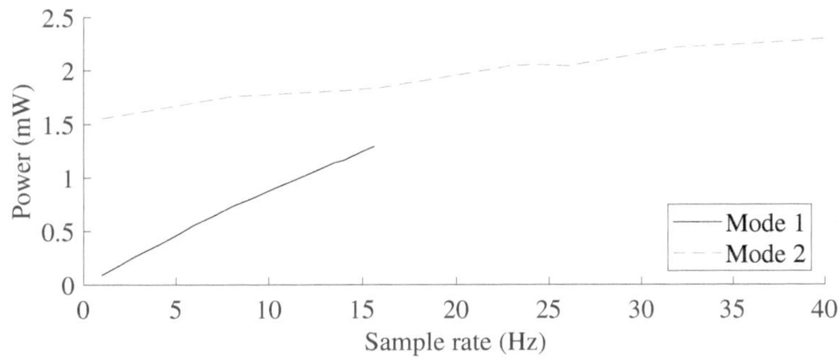

Figure 6. The average power requirement as a function of sample rate for the wireless sensor working in the two operating modes with a constant supply voltage of 1.8 V.

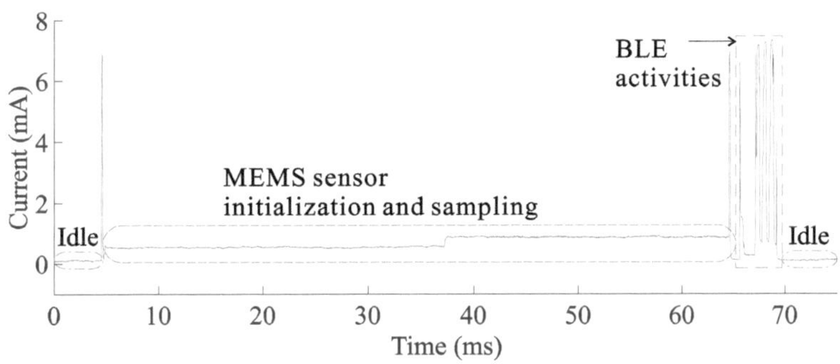

Figure 7. The current profile for the wireless sensor working in operating mode 1. The sensor is operated with a constant supply voltage of 1.8 V.

The targeted applications of the sensing system are angular speed measurements for rotating objects with low-speed and large shaft diameters. The VREH is a suitable technique for such applications, making the sensing system energy autonomous. Specific constraints of a machine health monitoring system for a hydraulic motor [50] have been considered, leading to an investigation of the proposed system in a laboratory test bench with rotational speeds of 5 rpm to 20 rpm and a shaft diameter of 270 mm. More details on the experimental setup are presented in Section 3.

3. Experimental Setup

To evaluate the effect of the rectifier selection on the system performance, three sets of experiments are conducted. In the first set of experiments, the maximum output power of the VREH transducer is determined. For this, the output power is acquired under different loads for given excitation conditions. In the second set of experiments, the circuit losses and the DC output power of the system are investigated. This allows to obtain general performance metrics of the individual implementations, and compare them to each other. In the third set of experiments, the wireless RPM sensor is attached to the VREH as a load. These experiments provide information on the effects of the gained performance improvements on the final system operation. Schematic overviews of the experimental setup for each set of experiments are provided in Figure 8, and more details are provided in Section 3.2.

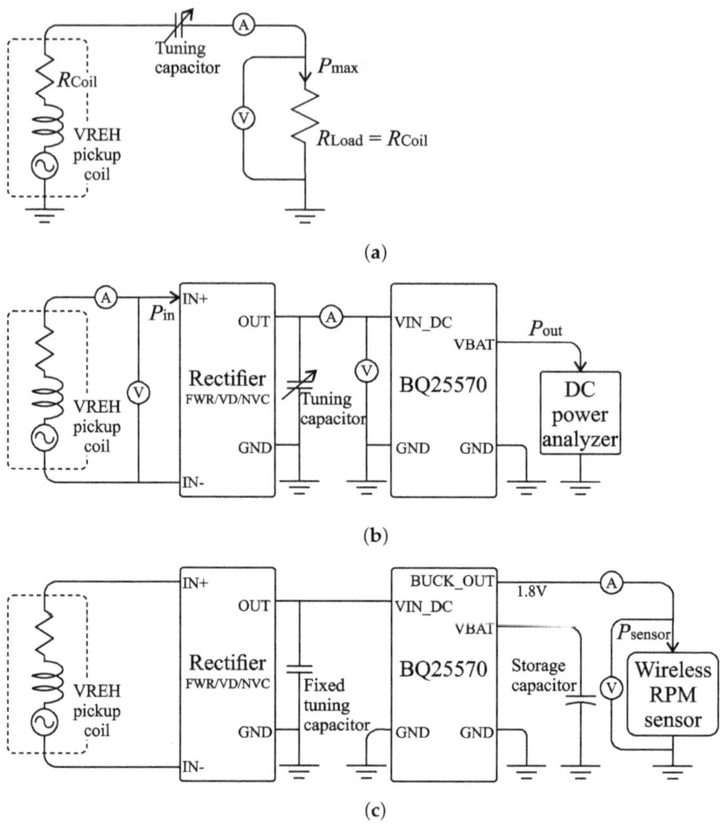

Figure 8. Schematic diagrams of the experimental setups for (**a**) evaluating maximum output power under impedance matched load; (**b**) evaluation of the interface circuit implementations; and (**c**) the full system evaluation with a wireless RPM sensor as the electrical load.

For the interface circuit implementation, a printed circuit board (PCB) containing the three rectifier circuits has been produced. Component selection has a significant influence on the rectifier performance, and care has been taken to select diodes and MOSFETs with beneficial properties (e.g., low forward voltage drop). Table 2 provides a list of the selected components, as well as their characteristics. A tuning capacitor is provided at the output of the rectifier circuit. In order to simplify the evaluation of different capacitance values, it is provided in form of a capacitor decade box with capacitance ranges of 1 µF to 4000 µF.

Table 2. Specification of key components for the rectifier circuit implementations.

Component	Specification	Characteristics
C1, C2	Ceramic capacitor	16 V, 100 µF
D1 – D10	Schottky diode CTS05S40-L3F, SOD882	$V_F = 0.31$ V @ $I_F = 0.1$ A $I_R = -13$ µA @ $V_R = -10$ V $I_R = -50$ µA @ $V_{BR} = -40$ V
N1, N2	N-MOSFET BSS138PW-115, SOT323	$C_{iss} = 40$ pF, $C_{oss} = 8.5$ pF, $C_{rss} = 4.5$ pF @ $V_{DS} = 5$ V $C_{iss} = 42$ pF, $C_{oss} = 14$ pF, $C_{rss} = 7$ pF @ $V_{DS} = 1$ V $R_{DSon} = 1\,\Omega$, $V_{GS(th)} = 1.2$ V $V_{SD} = 0.75$ V *
P1, P2	P-MOSFET PMN27XPE-115, SOT457	$C_{iss} = 1700$ pF, $C_{oss} = 300$ pF, $C_{rss} = 204$ pF @ $V_{DS} = -5$ V $C_{iss} = 1800$ pF, $C_{oss} = 500$ pF, $C_{rss} = 305$ pF @ $V_{DS} = -1$ V $R_{DSon} = 27$ mΩ, $V_{GS(th)} = -1.0$ V $V_{SD} = -0.7$ V *

* Body diode forward voltage.

3.1. Mechanical Setup and Test Conditions

All experiments are conducted on a setup for rotational energy harvesting, primarily targeting industrial applications with low rotational speeds. The shaft diameter of the setup is 270 mm, and experiments are conducted under low rotational speeds in the range of 5 rpm to 20 rpm. A photo of the mechanical setup is provided in Figure 9, outlining the key components involved.

The VREH pickup unit is mounted on a stationary POM cylinder, whereas the tooth wheel is mounted on a rotary mounting disc. The mounting disc is driven via a steel shaft by a controllable DC motor providing the relative rotary motion between pickup unit and tooth wheel with a constant air gap of 1 mm. The tooth wheel consists of 90 equally spaced teeth, and other geometrical parameters of the prototype are shown in Figure 10.

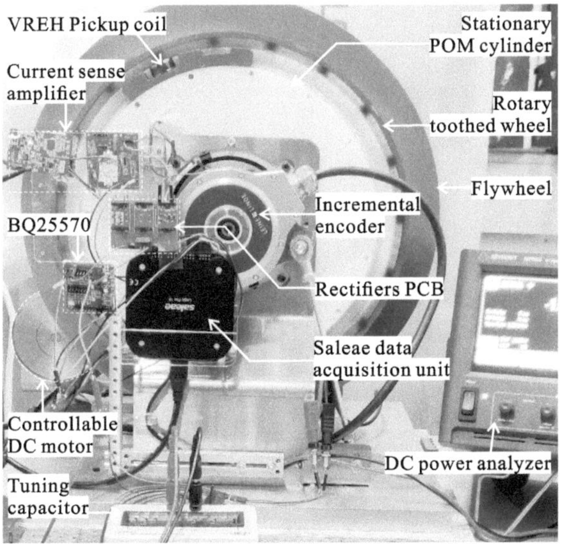

(a)

Figure 9. *Cont.*

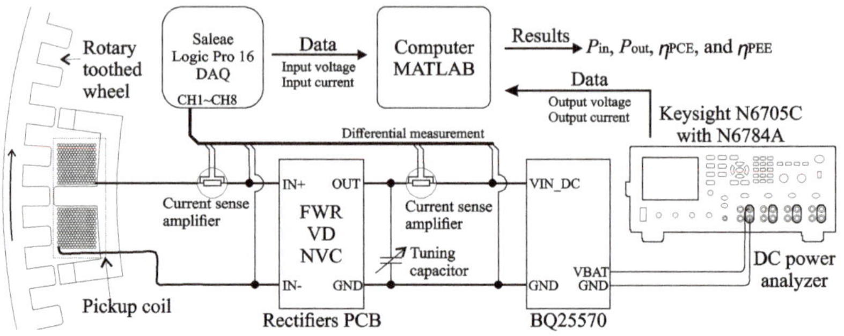

(b)

Figure 9. Overview of the experimental setup with the rotating shaft, the VREH transducer, the interface circuit implementations, a reference encoder, and data acquisition instruments in view of (**a**) the practical implementation and (**b**) a schematic diagram.

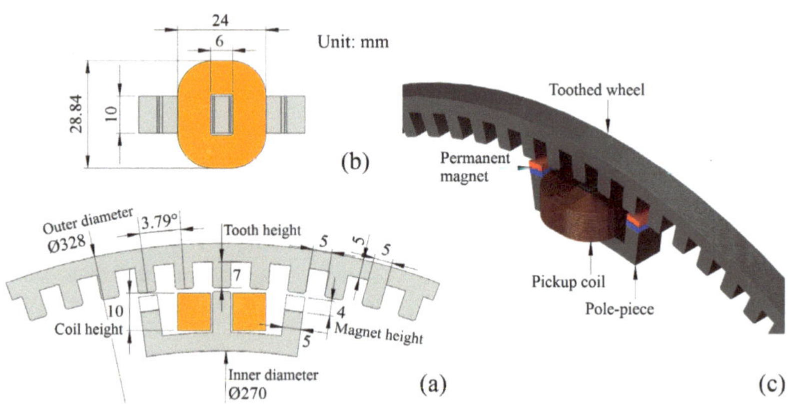

Figure 10. Geometric parameters in the VREH prototype (**a**) cross-sectional view, (**b**) top view (only showing the pickup unit) and (**c**) perspective.

An incremental encoder provides a reference pulsed output signal under rotation and is used to monitor the rotational speeds, as well as count mechanical cycles of the toothed wheel. A flywheel is mounted on the setup to provide a stable rotational speed.

3.2. Measurement Setup

To characterize the system performance, input and output power of the interface circuits are measured. For the input power, the AC input current and input voltage are digitized using a Saleae Logic Pro 16 data acquisition unit (DAQ). The DAQ provides 16 channels with inbuilt 12-bit analog-to-digital converters (ADC), allowing input signals in the range of $-10\,\text{V}$ to $10\,\text{V}$. Prior to digitization, the input current is converted into a voltage using a current sense amplifier (uCurrent Gold). Both analog input signals are sampled at a frequency of $1\,\text{kHz}$.

For output power measurements, a DC power analyzer (Keysight N6705C with N6784A module) is used. The power analyzer is operated as a constant-voltage (CV) current sink with a voltage of $2.5\,\text{V}$, and monitors the current drain from the *VBAT* terminal of the BQ25570. The same instrument is also used to measure the intermediate power at the output of the rectifier (i.e., before the PMIC).

Input power variations occur due to small air gap differences in the setup, and are further influenced by the low update rate of the MPPT circuit ($0.0625\,\text{Hz}$) of the BQ25570. In order to minimize the effects of these variations, all measurements are conducted over a number of full rotations (200 mechanical rotations). An example for this averaging process is given in Figure 11. All acquired data is transferred to a host computer for further processing in MATLAB. As a results, P_{in} and P_{out} are obtained according to

$$\begin{aligned} P_{\text{in}} &= \frac{1}{T}\int_{t_0}^{t_0+T}[V_{\text{in}}(t)I_{\text{in}}(t)]dt \\ P_{\text{out}} &= V_{\text{out}}\frac{1}{T}\int_{t_0}^{t_0+T}I_{\text{out}}(t)dt\,, \end{aligned} \tag{5}$$

where V_{in}, I_{in}, V_{out}, and I_{out}, are the input voltage, input current, output voltage, and output current, respectively.

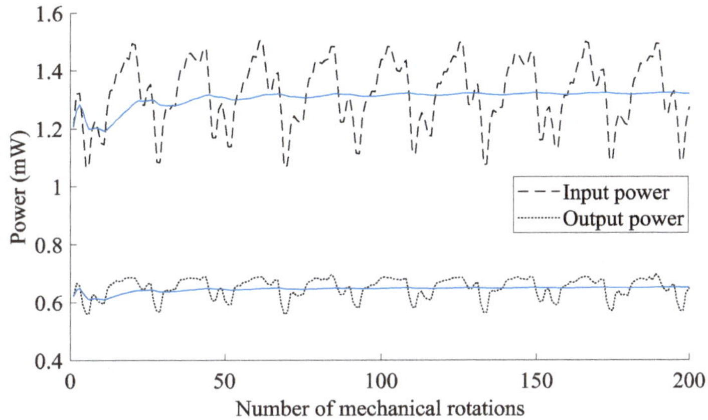

Figure 11. Input and output power variations caused by the mechanical setup and MPPT update rate (0.0625 Hz). The average power (blue solid lines) are obtained by taking multiple mechanical rotations into count. The example is for an interface circuit with the FWR and a tuning capacitance of 1600 µF at a rotational speed of 10 rpm.

3.3. Performance Metrics

Two metrics are used to evaluate the performance of the interface circuits and their effects on the energy harvesting system performance, respectively. These metrics are the power conversion efficiency (PCE) and the power extraction efficiency (PEE).

The PCE describes the conversion efficiency of the interface circuit. It takes the efficiencies of the rectifier circuit (η_{Rec}) and of the power conditioning circuit ($\eta_{\text{DC-DC}}$) into account, such that

$$\text{PCE} = \eta_{\text{Rec}} \cdot \eta_{\text{DC-DC}} = \frac{P_{\text{out}}}{P_{\text{in}}}\,. \tag{6}$$

Herein, P_{in} and P_{out} are the input and output power levels of the interface circuit, respectively (see Figure 8b). η_{Rec} is expressed by $1 - P_{\text{loss}}/P_{\text{in}}$, where P_{loss} is the power dissipated by a rectifier circuit. For diode-based rectifiers, it mainly depends on the voltage V_{D} and the current I_{D}. For MOSFET-based rectifiers, in addition to the conduction loss due to the drain-source on-resistance R_{DSon}, the power loss is also affected by the drain-to-source switching loss [51]. It therefore depends on the input voltage (V_{in}), voltage frequency (f_{e}), threshold voltage ($V_{\text{GS(th)}}$), load current (I_{Load}), and the output capacitance (C_{oss}) [29,51,52]. In this work, the switching loss can be neglected due to the low frequencies generated by the energy transducer provided in Table 3. $\eta_{\text{DC-DC}}$ is the conversion efficiency of the commercial PMIC (BQ25570), which is a function of input and output parameters as defined in its data sheet [53].

Table 3. Electrical performance of the VREH transducer at different rotational speeds under matched load and open circuit conditions.

RPM	Matched Load					Open Circuit	
	P_{max} (mW)	RMS Voltage (V)	Peak Voltage (V)	R_{Load} (Ω)	C_{Tuning} (μF)	Peak Voltage (V)	Frequency (Hz)
5	0.423	0.276	0.390	180.5	600	0.798	7.55
10	1.808	0.571	0.808	180.5	160	1.619	15.15
15	4.112	0.861	1.218	180.5	70	2.447	22.51
20	6.966	1.121	1.585	180.5	35	3.221	30.02

The PEE replaces the actual input power with the achievable power level under the current excitation, given a matched load (P_{max}). As such, it takes into account how the transducer output is affected by the interface circuit. Consequently, PEE is a good measure of the energy harvesting system performance, similar to a normalized output power, with

$$\mathrm{PEE} = \frac{P_{out}}{P_{max}} = \frac{P_{in}}{P_{max}} \cdot \mathrm{PCE}\,. \tag{7}$$

4. Results and Analysis

In this section, we present and analyze the results obtained from the three sets of experiments. First, in Section 4.1, the output power of the transducer under matched loads is provided. We then present and analyze results of the interface circuit performance (PCE and PEE) in Section 4.2. Finally, in Section 4.3, we evaluate effects of the different implementations on the operating conditions of the wireless RPM sensing system as an electrical load.

4.1. Output Power at Matched Load

Based on the experimental setup shown in Figure 8a, the maximum output power of the VREH transducer was evaluated for a matched load condition. The load resistance was selected to equal the DC resistance of the pickup coil, with $R_{Load} = R_{Coil} = 180.5\,\Omega$. Moreover, a tuning capacitor was used to compensate for the internal inductance of the pickup coil ($L_{Coil} \approx 690\,\mathrm{mH}$, number of turns: 3000).

Figure 12 illustrates an example output of the current and voltage measurement for the case of a rotational speed of approximately 10 rpm ($\approx$1.06 rad s^{-1}). According to (1), the measurement demonstrates an AC output with a frequency $f_c = \omega N_t = 1.06\,\mathrm{rad\,s^{-1}} \cdot 90 = 15.15\,\mathrm{Hz}$. The measurement shows that the output current and voltage are approximately in phase, which demonstrates an effective impedance matching. Moreover, Table 3 lists the maximum output power of the VREH transducer as a function of rotational speed under matched load conditions, as well as the corresponding voltage and frequency performance parameters. These output power values (i.e., P_{max}) are used as a reference for the performance evaluation of the individual interface circuit implementations.

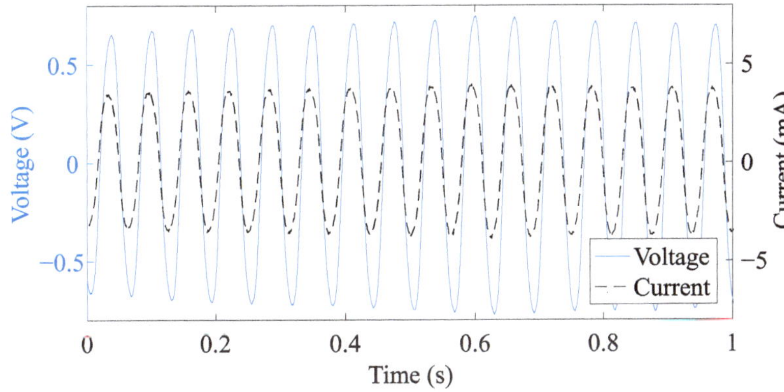

Figure 12. Measured waveforms of transient voltage and current, exemplified for a matched load condition, $R_{Load} = 180.5\,\Omega$ and $C_{Tuning} = 160\,\mu F$, for the VREH operating at 10 rpm.

4.2. PCE and PEE Evaluation

Figure 13 shows the PCE and PEE estimations for the interface circuit implementations at four different rotational speeds (i.e., 5 rpm, 10 rpm, 15 rpm, 20 rpm). The results show that the circuit performance (PCE and PEE) is dependent on the used rectifier circuit and the rotational speed. It can moreover be tuned using the capacitance between the rectifier circuit and the PMIC.

It can be observed that for each condition an optimal tuning capacitance exists maximizing PCE and PEE, respectively. However, for optimization of the PCE and PEE respectively, different capacitance values are required, with optimizations of the PEE requiring a larger capacitance than optimizations of the PCE. As the PEE increases even with decreasing PCE, the results demonstrate that the tuning capacitance has an effect on the operating point of the transducer, increasing its power output. For example, at the rotational speed of 10 rpm the NVC-based system reaches its maximum PCE of 72 % with a tuning capacitance of 5 µF. However, the PEE under this condition is only 36 %, whereas the maximum PEE of 47 % is achieved with a tuning capacitance of 1600 µF. As the PEE-metric incorporates the PCE, this shows that under this condition the increased output power of the VREH transducer outweighs the higher losses in the interface circuit.

This becomes even more obvious when extracting the maximum PCE and PEE values for each condition, as depicted in Figure 14. Figure 14a shows PCE and PEE values for an PCE-optimized capacitance, whereas Figure 14b shows those for an PEE-optimized capacitance. It can be observed that a higher rotational speed increases PCE and PEE. This is due to the higher output voltages and currents obtained from the transducer at higher rotational speeds, which improves conversion efficiencies. Moreover, the results show that the capacitance should be tuned for a maximized PEE value, as a higher PEE is equivalent to a higher output power.

Table 4 lists the input and output conditions, as well as the individual conversion efficiencies of AC-DC and DC-DC conversion for each of the interface circuit implementations. Comparing the three rectifier circuits, the FWR demonstrates the lowest performance under all of the operating conditions. This is to be expected due to the relative low output voltages of the VREH transducer, combined with the higher forward voltage drop of the FWR circuit. As a result, the FWR provides low conversion efficiencies, particularly at low rotational speed (i.e., low input voltage), and its output levels lead to relatively low efficiencies of the PMIC as well [53]. The VD, in contrast, shows significantly higher performance values. This is a result of three effects: The VD itself provides a higher conversion efficiency due to its reduced voltage drop in comparison to the FWR; its output levels lead to a higher PMIC efficiency; and input power levels demonstrate that it extracts more power from the VREH transducer. Finally, the NVC shows a performance that is highly condition dependent. At low rotational speed (5 rpm), the NVC performs similarly to the FWR, but its performance increases rapidly with increasing rotational speed, outperforming the VD at the highest input conditions. This difference can be explained based on the characteristics of the MOSFETs in the NVC circuit. At a low rotational speed, even with the parallel connected Schottky diodes, the input voltage is far below the voltages required for the switching of the transistors, effectively converting the NVC into an FWR (see Figure 3). With increasing input voltages at higher rotational speeds the on-resistances are reduced, and thus a lower voltage drop is achieved. Although the PMIC efficiency remains below that of the VD solution, the improved rectifier efficiency, together with an increased power extraction from the VREH transducer, results in an overall increased output power.

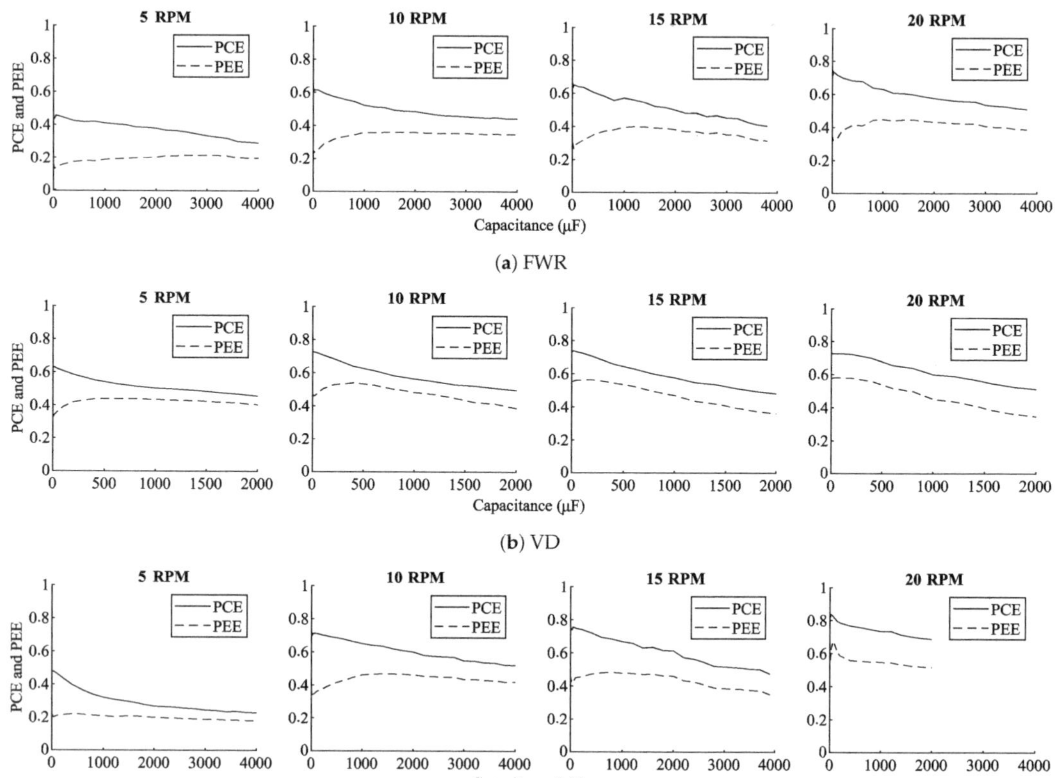

(a) FWR

(b) VD

(c) NVC

Figure 13. PCE and PEE development for different tuning capacitances for interface circuits based on (**a**) FWR, (**b**) VD, and (**c**) NVC, at the rotational speeds of 5 rpm, 10 rpm, 15 rpm, and 20 rpm.

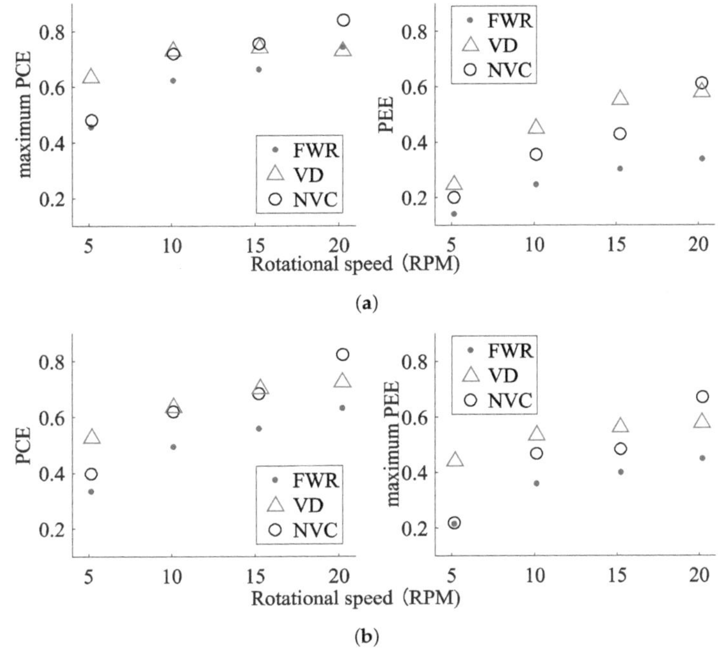

Figure 14. Optimized PCE and PEE values at different rotational speeds for (**a**) PCE-optimized tuning capacitance and (**b**) PEE-optimized tuning capacitance.

Table 4. Input and output conditions of the PEE-optimized interface circuits based on FWR, VD, and NVC.

		Rectifier Input			Rectifier Output			Efficiencies	
	RPM	**Power (mW)**	**Voltage (V)**	**Current (mA)**	**Power (mW)**	**Voltage (V)**	**Current (mA)**	**Rectifier**	**PMIC**
FWR	5	0.276	0.480	0.575	0.145	0.327	0.443	0.525	0.630
	10	1.310	0.822	1.593	0.811	0.619	1.310	0.619	0.801
	15	2.964	1.185	2.501	1.948	0.960	2.029	0.657	0.844
	20	4.926	1.556	3.165	3.670	1.319	2.783	0.745	0.850
VD	5	0.310	0.433	0.809	0.226	0.672	0.337	0.729	0.822
	10	1.420	0.770	1.917	1.130	1.371	0.824	0.796	0.863
	15	3.219	1.139	2.827	2.704	2.125	1.273	0.840	0.909
	20	5.477	1.414	3.952	4.697	2.617	1.795	0.858	0.912
NVC	5	0.232	0.496	0.468	0.120	0.442	0.271	0.516	0.773
	10	1.383	0.779	1.776	1.005	0.691	1.455	0.726	0.842
	15	2.906	1.214	2.393	2.281	1.149	1.985	0.785	0.872
	20	5.657	1.491	3.794	5.258	1.431	3.674	0.929	0.889

4.3. End-to-End System Evaluation

For the end-to-end energy harvesting system evaluation, the VREH system is used to supply the wireless RPM sensor described in Section 2.3, and as depicted in Figure 8c. In this scenario, the interface circuit provides a regulated 1.8 V output through the PMIC's *BUCK_OUT* terminal. An additional power conversion through the PMIC's internal buck converter thus needs to be taken into account. The efficiency of this conversion mainly depends on the output current defined by the sensing system's sample rate (see Table 1), resulting in efficiencies of approximately 85 % to 93 % for a supply voltage of 1.8 V.

Four 220 µF tantalum capacitors connected in parallel provide an energy buffer at the *VBAT* terminal. Tantalum capacitors offer a high performance in terms of volumetric efficiency, electrical reliability, and temperature range, contributing to the minimized and robust wireless sensing system. This capacitance can accumulate sufficient energy for the wireless sensor's initialization and duty cycling operation. For the current system design, a fixed tuning capacitance is provided for each rectifier implementation. The tuning capacitance has been selected to provide a near-optimal PEE under the range of rotational speeds investigated. Based on the measurement results presented in Section 4.2, the three tuning capacitors have been selected as 2000 µF, 1000 µF, and 400 µF, for the FWR, NVC, and VD, respectively.

Figure 15 shows the obtained output power levels and achieved sample rates of the wireless RPM sensor under different operational speeds and with the respective rectifier implementations. In accordance with the results presented in Section 4.2, the FWR implementation demonstrates the worst performance, whereas the VD implementation provides the highest performance at low rotational speeds. At 10 rpm, for example, the achievable sample rates are approximately 10 Hz, 9 Hz, and 7 Hz, for the VD, NVC, and FWR implementations, respectively.

At about 15.5 rpm, both the VD and NVC implementations reach the maximum sample rate of 40 Hz for the current system implementation. For the same sample rate, the FWR requires a rotational speed of approximately 18 rpm.

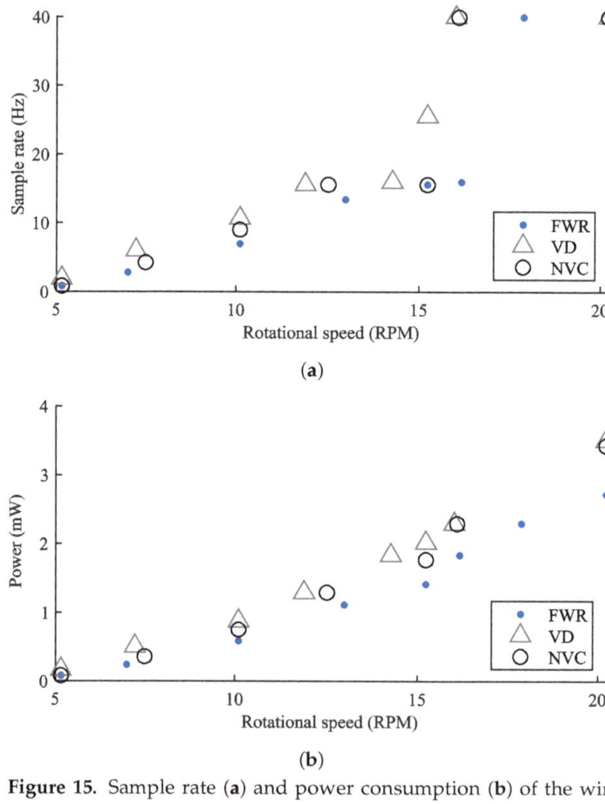

Figure 15. Sample rate (**a**) and power consumption (**b**) of the wireless RPM sensor operated at different rotational speeds, and supplied through the interface circuits with the three rectifiers.

5. Conclusions

In this paper, three interface circuits based on passive rectifiers, namely FWR, VD, and NVC, have been compared with respect to their influences on the performance of a low-speed rotational kinetic energy harvesting system. The rectifier circuits were individually integrated with a commercial PMIC in order to experimentally evaluate the interface circuits' conversion efficiency (PCE) and their influence on the energy harvesting system performance (PEE).

Based on a tuning capacitor, it was shown that a trade-off exists between the optimization of PCE and PEE. Although a larger tuning capacitance was shown to negatively effect the conversion efficiency of the interface circuit, the PEE and thus the total output power continues to increase up to a specific point (PEE optimum). This shows that the tuning capacitance has an influence on the operating point of the VREH transducer, which can outweigh conversion losses in the circuits.

In the comparison of the individual interface circuit implementations, significant performance differences were observed, thus indicating the importance of proper interface circuit selection based on the application conditions. At low rotational speeds, a solution based on a VD rectifier was demonstrated to outperform the other rectifiers, doubling the PEE at a rotational speed of 5 rpm. This is a result of the VD's higher conversion efficiency, its output levels leading to a higher PMIC efficiency, as well as a higher extracted power from the VREH transducer. However, the losses in the NVC rectifier are significantly reduced at higher rotational speeds, leading to an increased output power.

The general results of the interface circuits were verified in an end-to-end energy harvesting system evaluation based on a self-powered wireless RPM sensing system. The results of this evaluation demonstrated significant differences of achievable sample rates based on the rectifier circuit selection. The highest sample rates were obtained with an VD-based interface circuit at a low rotational speed, whereas the VD and NVC performed similarly at higher rotational speeds. For the energy harvesting system implementation, a fixed tuning capacitor was selected for each interface circuit solution, leaving further performance improvement opportunities. A dynamic tuning capacitance, for example, may lead to an increased PEE, but effective implementations of this dynamic adjustment would need to be investigated. Furthermore, other solutions to improve the operating point of the the VREH transducer without negatively affecting the PCE are opportunities for further investigation.

By applying a similar optimization procedure as presented in [22], optimal dimensions of the magnet height, coil height, and tooth height can be identified to result in an VREH with highest volumetric power density for a given size. Alternatively, the size of the VREH device can be minimized for an expected output power requirement. Thus, using the optimized structure is essential to further improve the sensor performance, as well as to minimize the entire self-powered system.

In conclusion, the selection and tuning of interface circuits for low-speed electromagnetic energy harvesting systems is shown to have significant effects on the end-to-end energy harvesting system performance, demonstrating that an optimal solutions is condition-sensitive. This knowledge can be used as a design guideline for energy harvesting systems with low input voltages, and for further investigations on hybrid or adjustable solutions.

Author Contributions: Conceptualization, Y.X., S.B. and B.O.; methodology, Y.X., S.B., M.M., P.M. and B.O.; simulations, Y.X. and P.M.; experimental setup and measurements Y.X.; data analysis Y.X., S.B. and M.M.; writing—original draft preparation, Y.X.; writing—review and editing, S.B., M.M., P.M. and B.O.; supervision, S.B. and B.O. All authors have read and agreed to the published version of the manuscript.

Funding: This research was funded by the Knowledge Foundation, under grant NIIT 20180170, as well as by the SNF Bridge Discovery project AEROSENSE (ID 187087).

Institutional Review Board Statement: Not applicable.

Informed Consent Statement: Not applicable.

Data Availability Statement: Data is contained within the article.

Conflicts of Interest: The authors declare no conflict of interest. The funders had no role in the design of the study; in the collection, analyses, or interpretation of data; in the writing of the manuscript, or in the decision to publish the results.

References

1. Khaligh, A.; Zeng, P.; Zheng, C. Kinetic Energy Harvesting Using Piezoelectric and Electromagnetic Technologies—State of the Art. *IEEE Trans. Ind. Electron.* **2010**, *57*, 850–860. [CrossRef]
2. Soyata, T.; Copeland, L.; Heinzelman, W. RF Energy Harvesting for Embedded Systems: A Survey of Tradeoffs and Methodology. *IEEE Circuits Syst. Mag.* **2016**, *16*, 22–57. [CrossRef]
3. Newell, D.; Duffy, M. Review of Power Conversion and Energy Management for Low-Power, Low-Voltage Energy Harvesting Powered Wireless Sensors. *IEEE Trans. Power Electron.* **2019**, *34*, 9794–9805. [CrossRef]
4. Zelenika, S.; Hadas, Z.; Bader, S.; Becker, T.; Gljušćić, P.; Hlinka, J.; Janak, L.; Kamenar, E.; Ksica, F.; Kyratsi, T.; et al. Energy Harvesting Technologies for Structural Health Monitoring of Airplane Components—A Review. *Sensors* **2020**, *20*, 6685. [CrossRef]
5. Surmenev, R.A.; Chernozem, R.V.; Pariy, I.O.; Surmeneva, M.A. A review on piezo- and pyroelectric responses of flexible nano- and micropatterned polymer surfaces for biomedical sensing and energy harvesting applications. *Nano Energy* **2021**, *79*, 105442, doi:10.1016/j.nanoen.2020.105442. [CrossRef]
6. Aranda, J.J.; Bader, S.; Oelmann, B. Self-Powered Wireless Sensor Using a Pressure Fluctuation Energy Harvester. *Sensors* **2021**, *21*, 1546. [CrossRef]
7. Siddique, A.R.M.; Mahmud, S.; Heyst, B.V. A comprehensive review on vibration based micro power generators using electromagnetic and piezoelectric transducer mechanisms. *Energy Convers. Manag.* **2015**, *106*, 728–747. [CrossRef]
8. Prauzek, M.; Konecny, J.; Borova, M.; Janosova, K.; Hlavica, J.; Musilek, P. Energy Harvesting Sources, Storage Devices and System Topologies for Environmental Wireless Sensor Networks: A Review. *Sensors* **2018**, *18*, 2446. [CrossRef] [PubMed]
9. Sun, H.; Yin, M.; Wei, W.; Li, J.; Wang, H.; Jin, X. MEMS based energy harvesting for the Internet of Things: A survey. *Microsyst. Technol.* **2018**, *24*, 2853–2869. [CrossRef]
10. Narita, F.; Fox, M. A Review on Piezoelectric, Magnetostrictive, and Magnetoelectric Materials and Device Technologies for Energy Harvesting Applications. *Adv. Eng. Mater.* **2018**, *20*, 1700743. [CrossRef]
11. Díez, P.L.; Gabilondo, I.; Alarcón, E.; Moll, F. A Comprehensive Method to Taxonomize Mechanical Energy Harvesting Technologies. In Proceedings of the 2018 IEEE International Symposium on Circuits and Systems (ISCAS), Florence, Italy, 27–30 May 2018; pp. 1–5. [CrossRef]
12. Szarka, G.D.; Stark, B.H.; Burrow, S.G. Review of Power Conditioning for Kinetic Energy Harvesting Systems. *IEEE Trans. Power Electron.* **2012**, *27*, 803–815. [CrossRef]
13. Zhou, M.; Al-Furjan, M.S.H.; Zou, J.; Liu, W. A review on heat and mechanical energy harvesting from human—Principles, prototypes and perspectives. *Renew. Sustain. Energy Rev.* **2018**, *82*, 3582–3609. [CrossRef]
14. Iqbal, M.; Nauman, M.M.; Khan, F.U.; Abas, P.E.; Cheok, Q.; Iqbal, A.; Aissa, B. Vibration-based piezoelectric, electromagnetic, and hybrid energy harvesters for microsystems applications: A contributed review. *Int. J. Energy Res.* **2021**, *45*, 65–102. [CrossRef]
15. Tabesh, A.; Frechette, L.G. A Low-Power Stand-Alone Adaptive Circuit for Harvesting Energy From a Piezoelectric Micropower Generator. *IEEE Trans. Ind. Electron.* **2010**, *57*, 840–849. [CrossRef]
16. Dayal, R.; Dwari, S.; Parsa, L. Design and Implementation of a Direct AC–DC Boost Converter for Low-Voltage Energy Harvesting. *IEEE Trans. Ind. Electron.* **2011**, *58*, 2387–2396. [CrossRef]
17. Chen, M.; Wang, M.; Yu, H.; He, G.; Zhu, Y.; Liu, Y.; Wang, G. A Self-Powered 3.26-μW 70-m Wireless Temperature Sensor Node for Power Grid Monitoring. *IEEE Trans. Ind. Electron.* **2018**, *65*, 8956–8965. [CrossRef]
18. Chew, Z.J.; Zhu, M. Adaptive Self-Configurable Rectifier for Extended Operating Range of Piezoelectric Energy Harvesting. *IEEE Trans. Ind. Electron.* **2020**, *67*, 3267–3276. [CrossRef]
19. Kroener, M.; Ravindran, S.K.T.; Woias, P. Variable reluctance harvester for applications in railroad monitoring. *J. Phys. Conf. Ser.* **2013**, *476*, 012091. [CrossRef]
20. Häggström, F.; Gustafsson, J.; Delsing, J. Energy Harvesting Technologies for Wireless Sensors in Rotating Environments. In Proceedings of the IEEE Emerging Technology and Factory Automation (ETFA), Barcelona, Spain, 16–19 September 2014; pp. 1–4. [CrossRef]
21. Xu, Y.; Bader, S.; Oelmann, B. A Survey on Variable Reluctance Energy Harvesters in Low-Speed Rotating Applications. *IEEE Sens. J.* **2018**, *18*, 3426–3435. [CrossRef]
22. Xu, Y.; Bader, S.; Magno, M.; Mayer, P.; Oelmann, B. Energy-autonomous On-rotor RPM Sensor Using Variable Reluctance Energy Harvesting. In Proceedings of the 2019 IEEE 8th International Workshop on Advances in Sensors and Interfaces (IWASI), Otranto, Italy, 13–14 June 2019; pp. 175–180. [CrossRef]
23. Xu, Y.; Bader, S.; Oelmann, B.. Design, modeling and optimization of an m-shaped variable reluctance energy harvester for rotating applications. *Energy Convers. Manag.* **2019**, *195*, 1280–1294. [CrossRef]
24. Ramadass, Y.K.; Chandrakasan, A.P. An Efficient Piezoelectric Energy Harvesting Interface Circuit Using a Bias-Flip Rectifier and Shared Inductor. *IEEE J. Solid-State Circuits* **2010**, *45*, 189–204. [CrossRef]
25. Peters, C.; Handwerker, J.; Maurath, D.; Manoli, Y. A Sub-500 mV Highly Efficient Active Rectifier for Energy Harvesting Applications. *IEEE Trans. Circuits Syst. Regul. Pap.* **2011**, *58*, 1542–1550. [CrossRef]

26. Cardoso, A.J.; de Carli, L.G.; Galup-Montoro, C.; Schneider, M.C. Analysis of the Rectifier Circuit Valid Down to Its Low-Voltage Limit. *IEEE Trans. Circuits Syst. Regul. Pap.* **2012**, *59*, 106–112. [CrossRef]
27. Hehn, T.; Hagedorn, F.; Maurath, D.; Marinkovic, D.; Kuehne, I.; Frey, A.; Manoli, Y. A Fully Autonomous Integrated Interface Circuit for Piezoelectric Harvesters. *IEEE J. Solid-State Circuits* **2012**, *47*, 2185–2198. [CrossRef]
28. Uluşan, H.; Gharehbaghi, K.; Zorlu, Ö.; Muhtaroğlu, A.; Külah, H. A Fully Integrated and Battery-Free Interface for Low-Voltage Electromagnetic Energy Harvesters. *IEEE Trans. Power Electron.* **2015**, *30*, 3712–3719. [CrossRef]
29. Gomez-Casseres, E.A.; Arbulú, S.M.; Franco, R.J.; Contreras, R.; Martínez, J. Comparison of passive rectifier circuits for energy harvesting applications. In Proceedings of the 2016 IEEE Canadian Conference on Electrical and Computer Engineering (CCECE), Vancouver, BC, Canada, 15–18 May 2016; pp. 1–6. [CrossRef]
30. Dell'Anna, F.; Dong, T.; Li, P.; Wen, Y.; Yang, Z.; Casu, M.R.; Azadmehr, M.; Berg, Y. State-of-the-Art Power Management Circuits for Piezoelectric Energy Harvesters. *IEEE Circuits Syst. Mag.* **2018**, *18*, 27–48. [CrossRef]
31. Parthasarathy, D.; Enoksson, P.; Johansson, R. Prototype energy harvesting wheel speed sensor for anti-lock braking. In Proceedings of the 2012 IV International Congress on Ultra Modern Telecommunications and Control Systems, St. Petersburg, Russia, 3–5 October 2012; pp. 364–369. [CrossRef]
32. Aktakka, E.E.; Najafi, K. A Micro Inertial Energy Harvesting Platform With Self-Supplied Power Management Circuit for Autonomous Wireless Sensor Nodes. *IEEE J. Solid-State Circuits* **2014**, *49*, 2017–2029. [CrossRef]
33. Buccolini, L.; Conti, M. An Energy Harvester Interface for Self-Powered Wireless Speed Sensor. *IEEE Sens. J.* **2017**, *17*, 1097–1104. [CrossRef]
34. Magno, M.; Kneubühler, D.; Mayer, P.; Benini, L. Micro Kinetic Energy Harvesting for Autonomous Wearable Devices. In Proceedings of the 2018 International Symposium on Power Electronics, Electrical Drives, Automation and Motion (SPEEDAM), Amalfi, Italy, 20–22 June 2018; pp. 105–110. [CrossRef]
35. Lu, C.; Tsui, C.; Ki, W. A Batteryless Vibration-based Energy Harvesting System for Ultra Low Power Ubiquitous Applications. In Proceedings of the 2007 IEEE International Symposium on Circuits and Systems, New Orleans, LA, USA, 27–30 May 2007; pp. 1349–1352. [CrossRef]
36. Kong, N.; Ha, D.S. Low-Power Design of a Self-powered Piezoelectric Energy Harvesting System With Maximum Power Point Tracking. *IEEE Trans. Power Electron.* **2012**, *27*, 2298–2308. [CrossRef]
37. Roscoe, N.M.; Judd, M.D. Optimization of Voltage Doublers for Energy Harvesting Applications. *IEEE Sensors J.* **2013**, *13*, 4904–4911. [CrossRef]
38. Ngo, K.D.; Phipps, A.; Nishida, T.; Lin, J.; Xu, S. Power converters for piezoelectric energy extraction. In Proceedings of the ASME International Mechanical Engineering Congress and Exposition, Chicago, IL, USA, 5–10 November 2006; Volume 47659, pp. 597–602. [CrossRef]
39. Abouzied, M.A.; Osman, H.; Vaidya, V.; Ravichandran, K.; Sánchez-Sinencio, E. An Integrated Concurrent Multiple-Input Self-Startup Energy Harvesting Capacitive-Based DC Adder Combiner. *IEEE Trans. Ind. Electron.* **2018**, *65*, 6281–6290. [CrossRef]
40. Magno, M.; Spadaro, L.; Singh, J.; Benini, L. Kinetic energy harvesting: Toward autonomous wearable sensing for Internet of Things. In Proceedings of the 2016 International Symposium on Power Electronics, Electrical Drives, Automation and Motion (SPEEDAM), Anacapri, Italy, 22–24 June 2016; pp. 248–254. [CrossRef]
41. Asthana, P.; Khanna, G. Power amplification interface circuit for broadband piezoelectric energy harvester. *Microelectron. J.* **2020**, *98*, 104734. [CrossRef]
42. Edla, M.; Lim, Y.Y.; Deguchi, M.; Padilla, R.V.; Izadgoshasb, I. An Improved Self-Powered H-Bridge Circuit for Voltage Rectification of Piezoelectric Energy Harvesting System. *IEEE J. Electron Devices Soc.* **2020**, *8*, 1050–1062. [CrossRef]
43. Hosain, M.K.; Kouzani, A.Z. Design and analysis of efficient rectifiers for wireless power harvesting in DBS devices. In Proceedings of the 2013 IEEE 8th Conference on Industrial Electronics and Applications (ICIEA), Melbourne, Australia, 19–21 June 2013; pp. 651–655. [CrossRef]
44. Tissier, J.; Latrach, M. A 900/1800 MHz dual-band high-efficiency rectenna. *Microw. Opt. Technol. Lett.* **2019**, *61*, 1278–1283. [CrossRef]
45. Mouapi, A.; Hakem, N.; Kandil, N. Performances Comparison of Shottky Voltage Doubler Rectifier to support RF Energy Harvesting. In Proceedings of the 2020 IEEE International Conference on Environment and Electrical Engineering and 2020 IEEE Industrial and Commercial Power Systems Europe (EEEIC/I CPS Europe), Madrid, Spain, 9–12 June 2020; pp. 1–5. [CrossRef]
46. Morel, A.; Quelen, A.; Berlitz, C.A.; Gibus, D.; Gasnier, P.; Badel, A.; Pillonnet, G. 32.2 Self-Tunable Phase-Shifted SECE Piezoelectric Energy-Harvesting IC with a 30nW MPPT Achieving 446% Energy-Bandwidth Improvement and 94% Efficiency. In Proceedings of the 2020 IEEE International Solid- State Circuits Conference—(ISSCC), San Francisco, CA, USA, 16–20 February 2020; pp. 488–490. [CrossRef]
47. Mayer, P.; Magno, M.; Benini, L. Smart Power Unit—mW-to-nW Power Management and Control for Self-Sustainable IoT Devices. *IEEE Trans. Power Electron.* **2021**, *36*, 5700–5710. [CrossRef]
48. Wu, L.; Do, X.; Lee, S.; Ha, D.S. A Self-Powered and Optimal SSHI Circuit Integrated With an Active Rectifier for Piezoelectric Energy Harvesting. *IEEE Trans. Circuits Syst. Regul. Pap.* **2017**, *64*, 537–549. [CrossRef]
49. Herbawi, A.S.; Paul, O.; Galchev, T. An ultra-low-power active AC-DC CMOS converter for sub-1V integrated energy harvesting applications. Sensors **2013**, *13*, 883.

50. Bosch Rexroth. Radial Piston Motors Hägglunds CBm. Available online: http://wwww.boschrexroth.com (accessed on 23 February 2021).
51. Jadli, U.; Mohd-Yasin, F.; Moghadam, H.A.; Pande, P.; Nicholls, J.R.; Dimitrijev, S. Measurement of Power Dissipation Due to Parasitic Capacitances of Power MOSFETs. *IEEE Access* **2020**, *8*, 187043–187051. [CrossRef]
52. Ren, C.; Xu, M.; Zhou, J.; Lee, F.C. Analytical loss model of power MOSFET. *IEEE Trans. Power Electron.* **2006**, *21*, 310–319. [CrossRef]
53. Texas Instruments. BQ25570 Datasheet (Rev. G). Available online: http://www.ti.com/product/BQ25570 (accessed on 19 November 2020).

sensors

MDPI

Article

Wide Bandwidth Vibration Energy Harvester with Embedded Transverse Movable Mass

Nathan Jackson *, Luis A. Rodriguez and Rahul Adhikari

Center for High Technology Materials and Mechanical Engineering Department, University of New Mexico, Albuquerque, NM 87106, USA; luis.a.rod.mil@gmail.com (L.A.R.); rahulvinda@unm.edu (R.A.)
* Correspondence: njack@unm.edu; Tel.: +1-505-272-7095

Abstract: One of the biggest challenges associated with vibration energy harvesters is their limited bandwidth, which reduces their effectiveness when utilized for Internet of Things applications. This paper presents a novel method of increasing the bandwidth of a cantilever beam by using an embedded transverse out-of-plane movable mass, which continuously changes the resonant frequency due to mass change and non-linear dynamic impact forces. The concept was investigated through experimentation of a movable mass, in the form of a solid sphere, that was embedded within a stationary proof mass with hollow cylindrical chambers. As the cantilever oscillated, it caused the movable mass to move out-of-plane, thus effectively altering the overall effective mass of the system during operation. This concept combined high bandwidth non-linear dynamics from the movable mass with the high power linear dynamics from the stationary proof mass. This paper experimentally investigated the frequency and power effects of acceleration, the amount of movable mass, the density of the mass, and the size of the movable mass. The results demonstrated that the bandwidth can be significantly increased from 1.5 Hz to >40 Hz with a transverse movable mass, while maintaining high power output. Dense movable masses are better for high acceleration, low frequency applications, whereas lower density masses are better for low acceleration applications.

Keywords: bandwidth; energy harvester; piezoelectric material; cantilever; MEMS

Citation: Jackson, N.; Rodriguez, L.A.; Adhikari, R. Wide Bandwidth Vibration Energy Harvester with Embedded Transverse Movable Mass. *Sensors* **2021**, *21*, 5517. https://doi.org/10.3390/s21165517

Academic Editors: Slim Naifar, Olfa Kanoun and Carlo Trigona

Received: 30 July 2021
Accepted: 14 August 2021
Published: 17 August 2021

1. Introduction

With the increase in the demand for the Internet of Things, cyber-physical systems, and smart buildings, there continues to be a high interest in vibration energy harvesting to create self-sustaining systems through the harvesting and conversion of mechanical energy from the ambient environment into usable electrical energy. Most vibration energy harvesters consist of a cantilever beam as a mechanical system that oscillates due to an applied vibration from the environment. The different kinetic energy converting mechanisms include piezoelectrics, electromagnetics, electrostatics, or triboelectrics. The energy converting mechanisms are different for each of these, but all these methods rely on a similar mechanical oscillating system. Typically, these cantilevers are linear systems with high Q-factors or narrow bandwidths of <2 Hz (~1% of resonant frequency) operating at low frequencies of <250 Hz [1–6]. The narrow bandwidths allow them to have high power density, but it limits their use in real-life applications, as even a 1% change in resonant frequency will significantly reduce the amount of power harvested. Large changes in resonant frequency can occur due to small changes in the vibration source or even due to manufacturing non-uniformities of the cantilever. Therefore, wider bandwidth devices are necessary for practical applications.

Since bandwidth is one of the major challenges associated with vibration energy harvesting, it was extensively investigated. Previous attempts to solve this issue have included developing non-linear cantilevers and spring designs based on duffing resonators [7–10], impact driven mechanical stoppers [11–13], additional magnetic forces [14–16], bistable non-linear devices [17–19], and the designing of an array of devices with varying frequencies [4]. However, these methods have numerous disadvantages, such as hysteresis effects, which depend on frequency sweep testing protocol, low power density, as decreasing the Q-factor reduces the peak power harvested, a larger footprint, thus decreasing the overall power density, complex manufacturing, especially at the micro-scale, and the need for external power, which reduces the overall efficiency of the system. Impact-based systems causing non-linear dynamics were also recently investigated and have demonstrated an increase in bandwidth [20,21], but the methods investigated to date consist of impact methods that are not scalable down to the micro-scale needed for IoT applications.

Recently, there were numerous attempts to increase the bandwidth by using a lateral sliding mass, which changes the center of gravity of the proof mass during oscillation, resulting in a continuously changing resonant frequency in a process referred to as dynamic tuning [22]. This dynamic tuning has the effect of widening the bandwidth. Previous attempts to create a sliding mass include using rolling cylinders [23–26], sloshing liquids [22,27–30], and a combination of the two [31]. These wide bandwidth methods involve altering the resonant frequency during oscillation by changing the effective location of the proof mass. The various methods demonstrated limited success as they rely on a large change in the center of gravity to significantly increase the bandwidth. The large lateral displacements required to widen the bandwidth are potentially feasible in macro-scale devices, but scaling down to the micro-scale would limit lateral displacement of the movable mass. In addition, these devices increase the bandwidth when compared to a traditional linear (stationary proof mass) system, but altering the center of gravity has only a minor effect on the overall effective mass, which limits the bandwidth enhancement capabilities.

This paper investigated the experimental validation of a novel movable mass energy harvesting system aimed at widening the bandwidth. Instead of a lateral moving mass, this paper investigated a vertical or transverse movable mass component. The transverse movable mass concept combined: (i) a continuous alteration of the effective proof mass to change the resonant frequency during operation, and (ii) the non-linear dynamics based on the impact of the movable mass on the cantilever substrate, in order to increase the bandwidth without a significant decrease in power. This is the first time a vertical movable mass was embedded into the proof mass of an energy harvesting system. The development of a transverse movable mass has the potential to significantly alter the effective mass of the system, resulting in the continuous altering of the resonant frequency during operation. The transverse movable mass is potentially a preferred option over a lateral sliding mass because it provides a larger change in the overall mass of the system, which correlates to an increase in bandwidth. This method also has the potential to be applied to both macro- and micro-scale devices, as embedded powders in silicon cantilevers were previously developed [32], which could act as a movable mass for MEMS cantilevers. This paper uses a piezoelectric cantilever as the energy harvesting system, but the method of widening the bandwidth can be applied to any cantilever device. This paper investigated the power and bandwidth effects of a transverse movable mass cantilever-system by performing experimental testing associated with varying the acceleration, the amount of movable mass, the size of the movable mass, and varying the material density of the movable mass. The movable mass method under investigation was expected to be utilized for applications requiring large acceleration and low frequency, such as aerospace, automotive, or pacemaker applications [33,34]

2. Materials and Methods

2.1. Concept

The simplistic equation for determining the resonant frequency of a rectangular cantilever with stationary proof mass is given by the equation [35]:

$$f = \left(\frac{1}{2\pi}\sqrt{\frac{E}{4m}}\right)\sqrt{\frac{wt^3}{L^3}} \tag{1}$$

where E is the elastic modulus of the beam, m is the mass, and w, t, and L are the width, thickness, and length of the beam, respectively. The resonant frequency can be altered by changing the physical dimensions of the cantilever, the elastic modulus, or the mass. Changing the elastic modulus of the beam was previously attempted as a tuning method [36], but it is difficult to implement in practical applications. Changing the dimensions of the beam during operation is not feasible. Therefore, changing the effective mass to alter the resonant frequency is the most feasible option. A vertical moving mass will not change the center of gravity, but instead works by altering the overall effective mass during operation as the movable mass can be in free fall, which effectively eliminates its contribution to the overall mass at that specific time. Therefore, the cantilever will effectively have two resonant frequencies: (1) when the movable mass is on the surface contributing to the effective mass, and (2) when the movable mass is in free fall, this will result in two separate resonant frequencies. If the weight percent (wt.%) of the movable mass is large, then the two frequencies will have a wide range. If the system has multiple movable masses, then it will have multiple resonant frequencies, thus resulting in a wide bandwidth effect. Numerically, the change in resonant frequency from the movable mass is given by the following equation:

$$\frac{f'}{f} = \frac{\left(\frac{1}{2\pi}\sqrt{\frac{E}{4m'}}\right)\sqrt{\frac{wt^3}{L^3}}}{\left(\frac{1}{2\pi}\sqrt{\frac{E}{4m}}\right)\sqrt{\frac{wt^3}{L^3}}} \tag{2}$$

where m′ is the mass of the stationary proof mass and m is total mass of the stationary proof mass plus the movable mass. If all other properties except the mass remain constant, the normalized increase in resonant frequency, by altering the effective proof mass while the movable mass is in free fall, can be simplified to:

$$\frac{f'}{f} = \frac{\sqrt{\frac{1}{m'}}}{\sqrt{\frac{1}{m}}} \tag{3}$$

where f′ represents the frequency when the movable mass is in free fall and f represents the frequency when both masses are stationary. Thus, a 50% reduction in mass during free fall would result in, approximately, a 1.4× increase in the resonant frequency. Therefore, having higher wt.% of movable mass will further increase the frequency range, resulting in a wider bandwidth. Having multiple moving masses would then create multiple resonant frequencies, which would effectively increase the bandwidth, and more movable mass would generate an increased change in resonant frequency, thus resulting in a larger bandwidth. However, for the movable mass to contribute to the change in frequency, the acceleration force needs to be large enough to propel the movable mass from the surface.

The concept of creating a vertical movable mass was briefly described in our conference paper [37]. The experimental setup for validating the concept included a stationary proof mass with single or multiple chamber(s)/cavities that could be partially filled with a solid movable mass. The chamber was designed to promote vertical movement and prevent lateral movement of the solid mass. As the cantilever beam oscillates with a large tip displacement, the movable mass will be propelled off the cantilever substrate and thus be in free fall. While the movable mass is in free fall, the effective mass of the system is significantly reduced depending on the wt.% of the movable mass. Therefore, during operation, the cantilever is continuously altering its resonant frequency and, thus, effectively widening the bandwidth. Figure 1 demonstrates the concept by having a solid movable sphere mass within a stationary proof mass. While at rest (Figure 1a) the overall mass of the system consists of: (i) a stationary proof mass chamber, and (ii) the mass of the solid sphere. The mass of the cantilever is considered negligible. The second mode of operation is when the cantilever displacement is large, causing the movable mass to displace as shown in Figure 1b. In this mode the effective mass only consists of the stationary mass, as the solid movable mass is in free fall. This concept also involves an impact mode (not shown in Figure 1), which creates the non-linear dynamic effect. The impact is caused from the solid movable mass falling, due to gravity, and impacting the cantilever substrate. The effect of the impact on the resonant frequency will depend on the force of the impact, which is determined by the vertical displacement of the movable mass, the density of the mass, and the stiffness of the substrate. This concept was designed for low frequency, high acceleration applications, as the concept requires a large displacement of the cantilever to generate enough force to move the solid mass. The energy harvesting operation uses a piezoelectric cantilever that generates stress and strain during oscillation, which then converts mechanical energy into electrical energy.

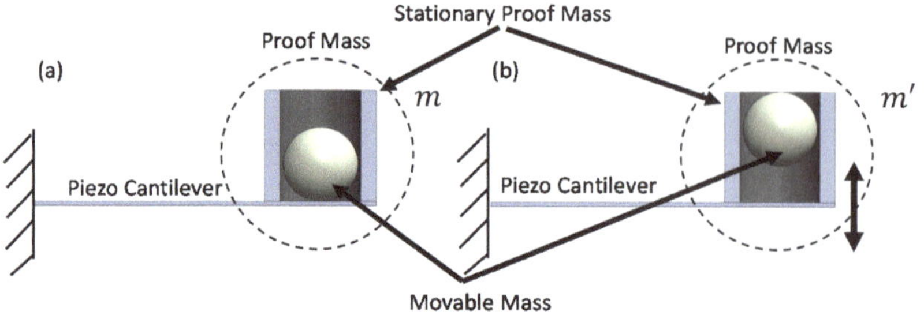

Figure 1. Schematic of the concept for transverse vertical movable mass demonstrating the various modes of operation: (**a**) demonstrates when the cantilever is at rest and the overall proof mass consists of the stationary proof mass and the movable mass, and (**b**) demonstrates when the cantilever has a large enough displacement to propel the movable mass off the substrate so the effective proof mass is represented only by the stationary proof mass.

The system involves non-linear dynamics due to both the impact as well as the continuous change in the effective mass of the system. In addition, if multiple movable masses are in a chamber, then there is an influence from the impact forces between the movable masses as well as when they impact the substrate. When the movable mass is stationary or at rest, the effective mass of the system will be significantly higher than when the movable mass is in free fall (depending on the density and wt.% of the movable mass). With the addition of the movable mass component, the resonant frequency of the system at rest will be significantly lower than the resonant frequency while the movable mass is in free fall (dependent on the wt.% of movable mass). Thus, the system will begin to oscillate at a lower frequency, which will cause the movable mass to move out of plane. Since the time for the movable mass to fall and the time for the cantilever to complete a cycle are different, the system will be going in and out of resonant frequency, and additional frequencies generated from the impact of mass will cause the system to be non-linear. Modelling the complete dynamics of the entire system are beyond the focus of this paper, which focuses instead on the experimental validation of the concept.

2.2. Cantilever Structure

In this paper, a commercial piezoelectric energy harvester (Volture V25W, Mide), with dimensions of 4.6 × 3.8 × 0.06 cm (l, w, and t), was used as the cantilever structure to validate the concept at the macro-scale. The stationary proof mass, which was 1.5 × 3.8 × 2.5 cm (l, w, and t) in size, was custom designed and 3D printed (Ultimaker S5) using PLA. The lightweight PLA was used to reduce the mass of the stationary proof mass to ensure that the wt.% of the movable mass was high. The dimensions of the proof mass and cantilever are scaled up versions of typical microsystem energy harvester devices [1,15,38]. The 3D printed proof mass consisted of three 11 mm diameter hollow chambers along the width of the cantilever, designed to restrict the 10 mm diameter spheres from rolling in the lateral direction. These chambers had a height of 20 mm and a base thickness of 5 mm to allow the movable mass to impact the proof mass instead of directly impacting the cantilever, which could lead to mechanical failure. The chambers were symmetrically aligned on the cantilever. A lid was 3D printed to prevent the movable mass from exiting the chamber. The overall mass of the 3D printed stationary proof mass was ~10 g. Figure 2 is a schematic and image of the cantilever with the custom proof mass. The proof mass was attached to the cantilever using double-sided adhesive tape.

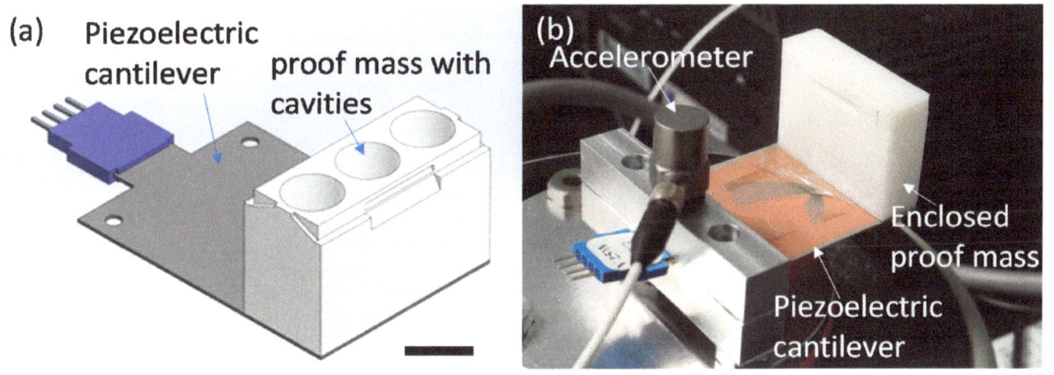

Figure 2. Image of the cantilever system: (**a**) schematic of the piezoelectric cantilever and custom designed proof mass with three chambers for the movable mass (the top lid is not shown), and (**b**) image of the cantilever with proof mass and cap with accelerometer mounted on vibration shaker. Scale bar is 1 cm.

The movable mass in our investigation consisted of solid spheres that were 10 mm in diameter, 5 mm diameter balls were also investigated to determine the effect of varying ball dimensions. The balls, which had a density of 14.95 g cm $^{-3}$, were made from tungsten carbide (labelled W in the paper; MSE Supplies), the main type of material investigated because of its large density since, theoretically, the bandwidth should be dependent on the wt.% of the movable mass. According to Equation (3), the higher percentage of movable mass should generate a wider range of resonant frequencies between the stationary resonant frequency and the free fall resonant frequency, which should correlate to a larger increase in bandwidth. Other materials, including stainless steel 316L (SS, 8 g cm^{-3}; GoodFellow), Aluminum oxide (Al_2O_3, 3.9 g cm^{-3}), and Teflon (PTFE, 2.2 g cm^{-3}; United States Plastic), were investigated to determine the effects of changing the wt.% of the movable mass by changing the density of the balls. The outer surfaces of the balls were polished to reduce friction between the balls and the surface of the proof mass.

2.3. Experimental Methodology

The cantilever device with proof mass was mounted on a vibration shaker (ET 126, Labworks) via a custom-made mounting chuck with an accelerometer to provide acceleration feedback control. The system from Labworks has an integrated accelerometer and vibration control software. The cantilever was connected to an oscilloscope (Tektronix 4-channel 100 MHz with 5 GS/s sampling rate) with a variable load resistor to provide impedance matching for power measurements. Power was calculated based on the V_{rms} and impedance. The optimal power was determined by matching the electrical impedance with the impedance of the piezoelectric cantilever. Discrete manual frequency sweeps were performed in 0.5 Hz steps. A sweep up and down test was performed to determine if sweeping had any impact. In addition, random frequency values were tested and compared to the sweep data to determine if sweeping had any impact on the power and bandwidth. Once the force of the cantilever was large enough to cause vertical displacement of the balls within the chambers, the oscilloscope readings became non-linear due to the impact force, meaning the voltage and power amplitude varied with time and were no longer a simple sinusoidal waveform. The average V_{rms} value, over the course of 1 min at a specific frequency, was measured during the non-linear dynamic range. The frequency range of the impact from the movable mass is called non-linear bandwidth (NLBW), as it is essentially bandwidth at which the movable mass operates, resulting in non-linear dynamics from the impact. The start and end frequency of the movable mass, which is the frequency range at which the impact from the movable mass was observed quantitatively through data from the oscilloscope, as voltage output was no longer sinusoidal, but instead consisted of multiple frequency outputs as demonstrated in Figure 3b. The frequency range of the movable mass could also be determined through hearing the impact of the ball. The NLBW was essentially the frequency range at which the movable mass had significant displacement, causing the mass to be propelled off the surface and impact the substrate.

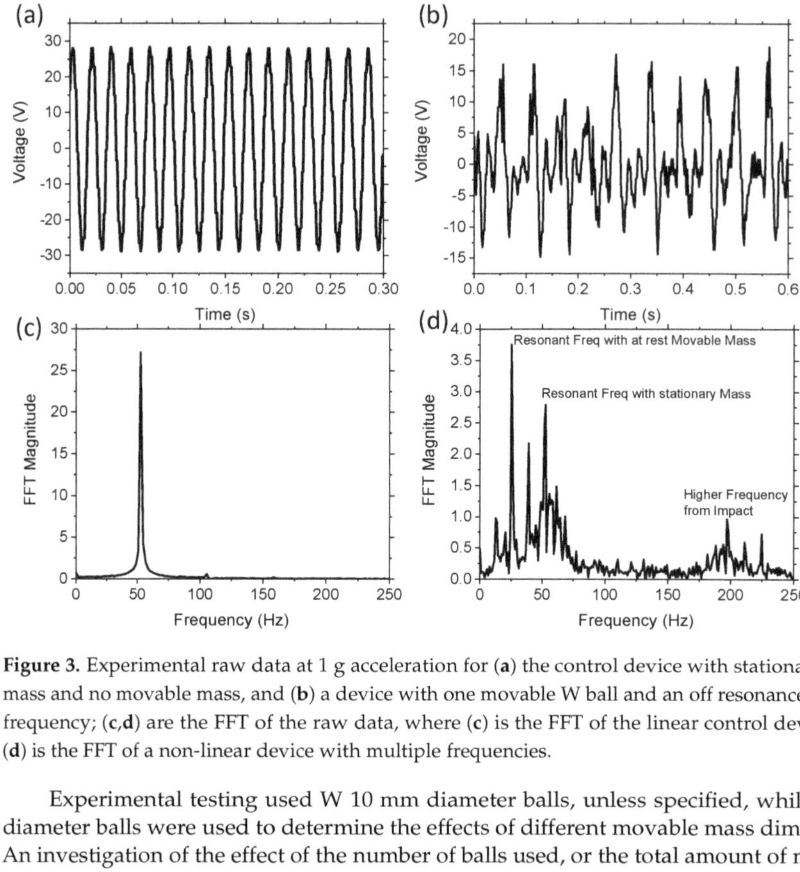

Figure 3. Experimental raw data at 1 g acceleration for (**a**) the control device with stationary proof mass and no movable mass, and (**b**) a device with one movable W ball and an off resonance applied frequency; (**c**,**d**) are the FFT of the raw data, where (**c**) is the FFT of the linear control device, and (**d**) is the FFT of a non-linear device with multiple frequencies.

Experimental testing used W 10 mm diameter balls, unless specified, while 5 mm diameter balls were used to determine the effects of different movable mass dimensions. An investigation of the effect of the number of balls used, or the total amount of movable mass, consisted of comparing power vs. frequency measurements for a device with three balls (one in each chamber) and a device with a single ball in the middle chamber. The effects of acceleration on the bandwidth and power magnitude were determined using accelerations of 0.1, 0.5, 1, and 1.5 g. To determine the effects of changing the wt.% of movable mass, different materials were investigated with varying accelerations as shown in Table 1. The wt.% of the movable mass was calculated as the amount of movable mass divided by the total mass, which included the mass of the balls and the stationary proof mass. The amount of movable mass varied from 22.5% to 70.2%. To determine the effect of the size of the balls, we compared systems with one 10 mm W sphere vs. eight 5 mm W spheres, as they had nearly the same movable mass component of 7.87 g and 7.84 g, with a moveable mass percentage of 44% and 43.9%, respectively, as shown in Table 1, which was considered equivalent.

Table 1. Various scenarios of movable mass and amount of movable mass used in each test.

Material	Mass of One Ball (Grams)	Quantity of Balls	Movable Mass (Grams)	Movable Mass (%)
Tungsten Carbide (W)	7.87	3	23.61	70.2
Tungsten Carbide (W)	7.87	1	7.87	44.0
Tungsten Carbide 5 mm ball	0.98	8	7.84	43.9
SS 316L (SS)	4.168	3	12.50	55.6
Al_2O_3	1.76	3	5.28	34.6
PTFE	0.97	3	2.91	22.5
PTFE	0.97	6	5.82	36.8

3. Results and Discussion

The initial validation of the concept was investigated experimentally by comparing the output voltage as a function of time for two devices operating at 1 g acceleration. The first device was the control, which consisted of the cantilever beam and a stationary 3D printed proof mass without any movable balls in the chambers. The second device consisted of a cantilever beam, stationary proof mass, and one W ball in the middle chamber. In both cases, the vibration source was applied at the resonant frequency of the cantilever with proof mass. The control proof-mass was lighter, having a resonant frequency of ~53 Hz, whereas, when we added the extra mass of the movable ball, the resonant frequency was ~25 Hz.

Figure 3a is a section of the raw data from the oscilloscope while operating near the resonant frequency of the control device with no movable component. The movable mass was restricted from moving in the lateral direction due to the size of the chamber, and thus allowed the ball to only move in the vertical transverse direction. The results of the control system with no movable mass demonstrated a typical linear device with a sinusoidal output. The FFT of these results demonstrated a single resonant frequency at 53 Hz, as shown in Figure 3c. The experimental raw data for the device with movable mass is shown in Figure 3b. The results demonstrate that the movable mass generated an output voltage consisting of multiple frequencies, with a peak resonant frequency of approximately 25 Hz. The reduction in resonant frequency was due to the increased overall proof mass. The large voltage peaks consisted of linear frequencies when the movable mass was in contact with the cantilever and when the system was in free fall. Visual confirmation of the ball moving was seen through the chamber and the ball displacement was approximately 1 cm in height. The FFT of the movable mass device is demonstrated in Figure 3d, which shows multiple frequencies, including a peak at 53 Hz. The 53 Hz peak is demonstrated in both Figure 3c,d, as this represents the resonant frequency of the cantilever with the stationary 3D printed mass. Figure 3d also has a large peak at 25 Hz, which is the resonant frequency of the cantilever with the additional W mass while at rest on the surface. Numerical analysis using Equation (1) estimates the resonant frequency of the two masses to be 26 Hz and the free fall state resonant frequency to be 52 Hz, which is in good agreement with the experimental results. The difference was likely to be a result of a slight difference in the elastic modulus of the cantilever. The FFT data demonstrated multiple frequencies in the 10–75 Hz range, even though the excitation frequency was at 25 Hz, which was due to the non-linear impact of balls and the changing resonant frequency of the cantilever. There were also higher frequency peaks at around 175–225 Hz, which was due to the impact of the ball hitting the chamber. The results validate that there was a wider range of frequencies occurring in the system with the movable mass. The addition of these multiple frequencies should have the effect of increasing the bandwidth, as significant voltage output can be measured through a wide range of frequencies compared to the linear control system in Figure 3c, which only has a single frequency output.

Next, we varied the number of balls in each chamber to determine if the wt.% of movable mass had a significant impact on bandwidth. The power, as a function of applied frequency, was measured for the (i) control (no movable mass), (ii) one middle 10 mm diameter W ball, and (iii) three W balls (one in each chamber), which were experimentally investigated at 1 g acceleration. The power was measured by connecting the output voltage of the cantilever to a variable load resistor to match the impedance for maximum power. The results are demonstrated in Figure 4. The control system had a typical linear dynamic response, with a sharp peak and full width half maximum (FWHM) of 1.5 Hz, and a peak power of 6.43 mW at 53 Hz. The movable mass device (a single W ball), with a 44% movable mass, had a wider bandwidth with a peak frequency shifted to the left (lower frequency) due to the additional mass in the system. The average power was calculated for each frequency over a 1 min duration, and the standard deviation was represented by error bars. The single ball system had a much wider bandwidth compared to the control, which had no movable mass component, and a FWHM (~30.5 Hz compared to 1.5 Hz) and no distinguishable peak, but the average power measured was 1.4 mW. The results of adding a single ball with a 44 wt.% movable mass significantly increased the bandwidth, but decreased the power. The three-ball system, with a 70.2 wt.% movable mass, demonstrated a high bandwidth and a high peak power of 3.8 mW at 15 Hz. The area under the curve, which represents the power over a range of frequencies (mW-Hz), was increased by 117% and 252% for the one-ball and three-ball system, respectively, compared to the control. The three-ball system had a NLBW range of 34 Hz, which demonstrated average power values of >0.7 mW during that frequency range, with the exception of one point around 10 Hz. This validated the concept that a vertical movable mass can widen the bandwidth and increase power generation. The peak power of the movable mass system was reduced compared to the control, but the bandwidth was significantly increased. Increasing the percent of the movable mass further increased the peak power, while still maintaining a high bandwidth as the power over a frequency range was significantly increased. Balls placed on the outside chambers caused high frequency twisting modes to decrease their resonant frequency as well, but the research was interested in low frequency first mode operation to increase the power density.

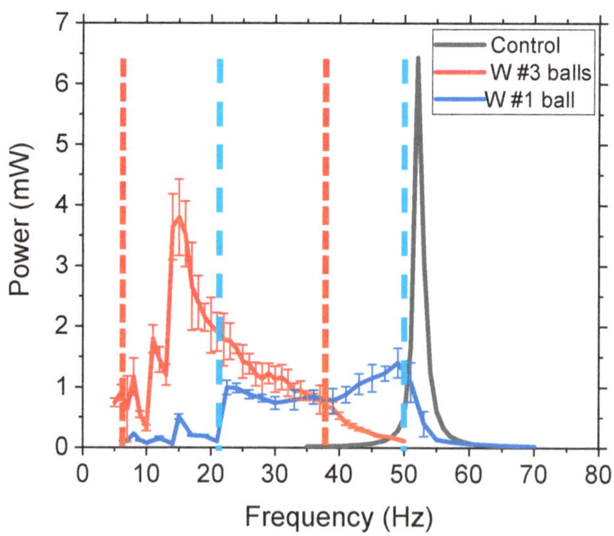

Figure 4. Experimental data of power as a function of frequency for 10 mm W balls at 1 g acceleration for systems with zero (control), one, and three movable balls. The error bars represent the standard deviation of the power measurements during the 1 min testing duration. The dashed lines represent the start and stop frequencies for the movable mass representing the NLBW.

The non-linear energy harvester's bandwidth and power measurements are commonly dependent on the frequency sweep characteristics of the applied excitation where there is a hysteresis effect; thus the operation of the device is dependent on the testing parameters of sweeping the frequency either up or down [39–41]. However, this is not practical as most applications have randomly changing discreet frequencies instead of a continuous sweep. To investigate if the frequency sweep direction had an impact on the power or bandwidth, the frequency was swept in both directions at a rate of 0.05 Hz/s from 5–50 Hz at 0.5 g with a three-ball system of W. The results (Figure 5) demonstrated no significant difference in amplitude or bandwidth due to sweeping the frequency. The results were further validated by randomly selecting frequencies and measuring their output power, which demonstrated that, within the standard deviation, the results were the same. Therefore, this method has potential impact in real-life applications as the power and bandwidth can be accurately predicted regardless of testing parameters. Comparing the results of the three-ball system in Figure 4 (1 g) and Figure 5 (0.5 g) the general shape of the output power vs. frequency was the same, but the 1 g acceleration had higher power, as expected, and higher bandwidth due to the increased impact force. Increasing the acceleration to 1 g and 1.5 g obtained similar results, demonstrating no significant hysteresis effect.

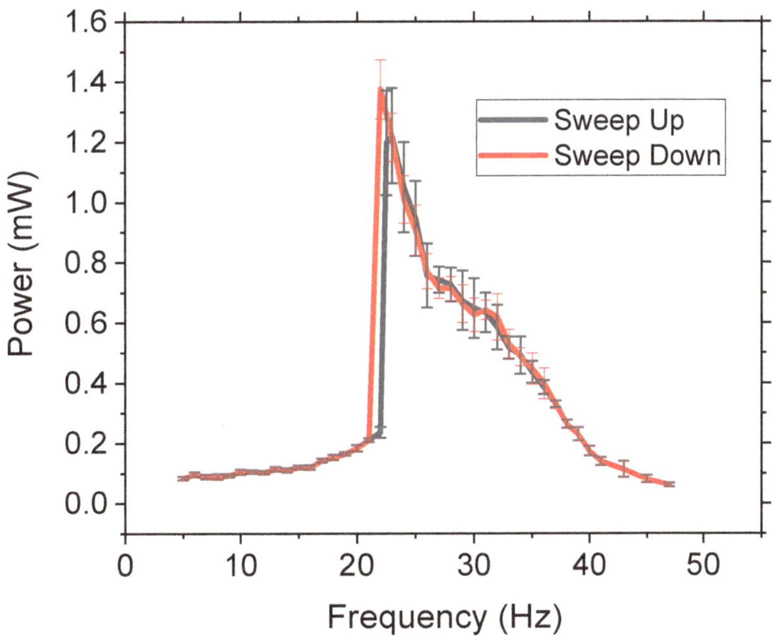

Figure 5. Comparison of sweep up and sweep down frequency testing on the system with three W balls of 10 mm at 0.5 g. Error bars are equal to one standard deviation.

The power and bandwidth are dependent on the acceleration that is applied, and it was believed that the vertical movable mass method required high acceleration because low acceleration would not have a large enough tip displacement to propel the mass within the chamber, thereby limiting the non-linear dynamic effects. To investigate this, various accelerations from 0.1, 0.5, 1, and 1.5 g were experimentally tested with the three-ball system using W balls. The results are presented in Figure 6. At 0.1 g, the system behaved like a linear system with a slight increase in the FWHM of 1.73 Hz as compared to the control's 1.2 Hz. As the acceleration increased, there was a shift in the peak power amplitude towards a lower frequency. This occurred because the higher acceleration resulted in higher force, which increased the tip displacement and caused the movable mass to propel off the surface at lower frequencies. This, in turn, increased the NLBW as the frequency range of the movable mass was increased, thus generating a larger non-linear region due to the impact force. Once the mass started moving, the peak power increased along with the bandwidth. At 1.5 g, the peak power occurred at 6 Hz because the high acceleration and low frequency produced a large enough force to propel the movable mass off of the substrate. This resulted in a peak power of 10.52 mW. Higher acceleration also caused the bandwidth to increase as the area under the curve increased from 1.73, 13.99, 53.91, and 147.23 mW-Hz for 0.1, 0.5, 1, and 1.5 g, respectively. The area under curve represents the total amount of power that is harvested over a range of frequencies. The experimental results demonstrated that this method is efficient for low frequency, high acceleration applications.

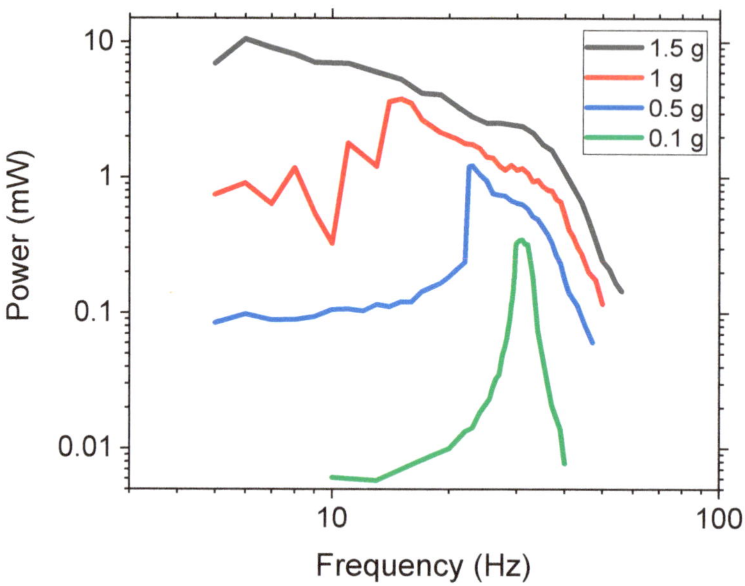

Figure 6. Log scale power as a function of frequency for varying acceleration with three W balls.

Various density balls, as highlighted in Table 1, were used to determine the effects varying the wt.% of movable mass had on the bandwidth and power. Three 10 mm diameter balls, one in each chamber, were used in the experiment with varying accelerations from 0.1, 0.5, and 1 g. The results are shown in Figure 7. At a low acceleration of 0.1 g (Figure 7a), all of the materials used for the balls (W, SS, Al_2O_3, and PTFE) had an output that more closely represented a linear system. As the density of the balls increased, there was a peak power shift towards lower frequencies, which was expected since the overall weight of the proof mass increased. The peak power was dependent on the density of the balls and the overall mass of the system, as expected, demonstrating an increase in the power with an increase in the percentage of movable mass. However, the FWHM value decreased as the amount of movable mass increased, with values of 2.61, 3.01, 3.7, and 5.2 Hz for W, SS, Al_2O_3, and PTFE, respectively, compared to the control's FWHM of 1.2 Hz. The PTFE balls had a wider bandwidth at a low acceleration because they required less force to propel as compared to the denser masses, such as W, which required more force. Therefore, low-density movable masses provided wider bandwidths than dense materials at low acceleration because the higher density balls did not easily propel off the substrate, resulting in limited impact force.

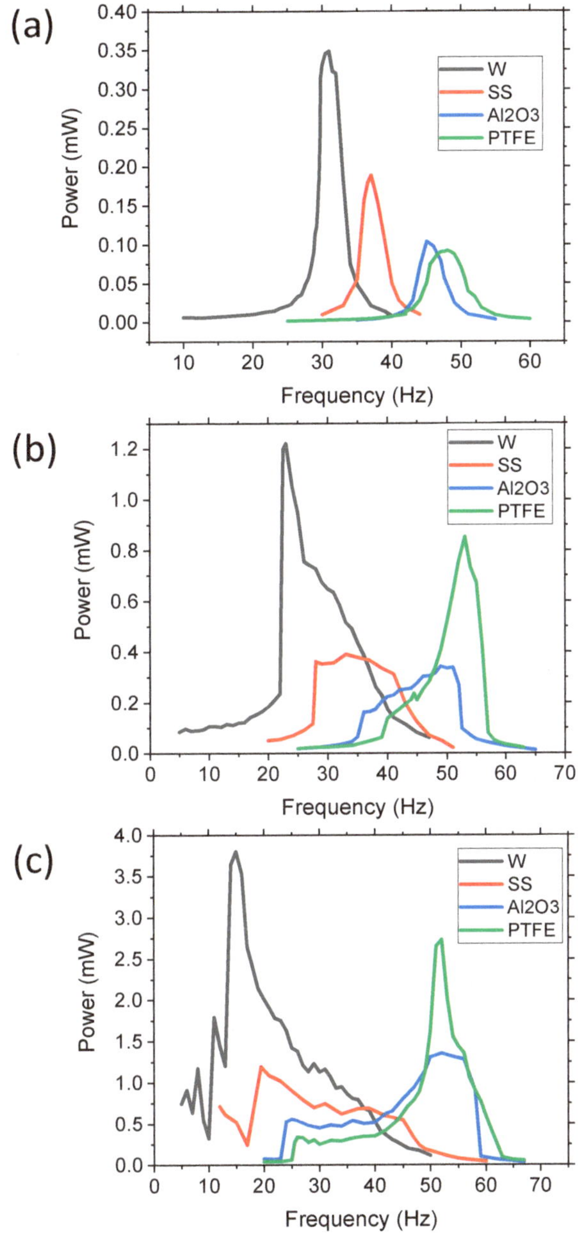

Figure 7. Experimental results of varying ball materials (densities) with varying accelerations: (**a**) 0.1 g, (**b**) 0.5 g, and (**c**) 1 g. Each test consisted of three balls of 10 mm diameter of W (tungsten), SS (stainless steel 316L), Al_2O_3 (aluminum oxide), or PTFE (Teflon).

As the acceleration increased, the cantilevers with denser masses demonstrated an increase in bandwidth and all of the systems had a larger NLBW, as shown in Figure 7b,c, because the acceleration produced a large enough force to propel the movable mass. For example, at 0.5 g, the W balls had a high intensity peak (1.22 mW) at a low frequency and a FWHM of 9.29 Hz. However, it started to develop a broader peak at higher frequencies, thus increasing the bandwidth of usable energy. Usable energy, as defined by the authors, is power >0.7 mW, which was determined based on the typical power requirements for wireless sensor networks, wherein 0.7 mW is adequate for most IoT applications. However, at 0.5 g, the SS balls lacked a high-intensity, narrow peak, instead consisting of a broad peak with an average power of 0.38 mW and a bandwidth of 15.4 Hz. The Al_2O_3, on the other hand, had a broad peak at lower frequencies and started to develop a high-intensity peak at the resonant frequency of the stationary system. This resulted in an average power of 0.25 mW and a bandwidth of 14.52 Hz. The PTFE movable mass system had a high-intensity peak at the resonant frequency of the stationary proof mass, with a peak power of 0.85 mW and a FWHM of 6.81 Hz, but it also consisted of a wider peak at lower frequencies due to the non-linear movable masses. The reason for the high peak exhibited by PTFE was attributed to two issues: (1) the low density of the balls caused a lower impact as compared to W, so the non-linear effects were less, and (2) the less-dense balls had a larger displacement within the chamber, thus the time required for the balls to impact the substrate was longer, resulting in a larger linear dynamic region for a longer amount of time. For example, if the PTFE balls had a 1.5 cm vertical displacement within the chamber the free fall, the time it would take for the balls to fall would be ~55 ms, whereas the time for the cantilever to complete one cycle is ~19 ms, thus the cantilever would be in linear mode for nearly three oscillation cycles before the impact. The W ball system had a large narrow peak at low acceleration due to the cantilever's limited amount of force and displacement, which resulted in the low displacement of the balls and a low impact force. As acceleration increased, the balls had a larger displacement, resulting in a larger NLBW region. The denser balls had a larger impact force when they were propelled. Therefore, the low-density movable masses resulted in a more linear system, but at high acceleration, the system had both linear and nonlinear components.

The power, as a function of the frequency results, of applying a 1 g acceleration (Figure 7c) demonstrated a similar trend as the results in Figure 7b. The Al_2O_3 ball appeared to be shifting in the same manner as the PTFE samples to form a peak at the linear resonant frequency. At 1 g, however, a significant narrow peak was still absent, but increasing the acceleration is likely to cause a rise in the peak at the stationary resonant frequency. Both PTFE and Al_2O_3 had two distinguished modes: (i) a wide plateau mode at lower frequencies of 20–45 Hz approximately (non-linear region), and (ii) a higher intensity peak region (linear). This resulted in the two systems having both a narrow high-power density section (the linear component) and a lower power density wider bandwidth section (the non-linear component). The total power over the usable frequency range resulted in an area under curve value of 28.07 and 28.8 mW-Hz for Al_2O_3 and PTFE, respectively. SS and W also increased the bandwidth and power, with the SS balls having an average power of 0.7 MW and a FWHM of 29.9 Hz. The W balls had a large peak power of 3.79 mW and a wider NLBW region, with an overall area under the curve of 54.07 mW-Hz. Therefore, we determined that a higher wt.% movable mass provided a wider bandwidth and higher power at increased acceleration. Applying a higher acceleration to the W ball system would likely cause the output to shift, similar to the outputs of PTFE and Al_2O_3, but we maintained the acceleration limit at 1 g as most practical applications operate at <1 g acceleration.

Figure 8 demonstrates the bandwidth effects of the varying materials from 0.1 g to 1 g. The results of the 0.1 g bandwidth investigation (Figure 8a) were measured by the FWHM (as these devices behaved more like a linear system) and the NLBW, as the devices still had some movable mass components. For low acceleration, the FWHM values were similar to the NLBW range, with the NLBW being slightly reduced in all cases. However, the difference between the FWHM and NLBW increased as the density of the balls increased, which made sense since the denser balls had a larger impact force on the cantilever. In all cases, the bandwidths were significantly larger than the control with no movable mass, thus validating the concept. The bandwidth also increased as the density of the balls or total movable mass decreased because the lower density balls had larger displacements at the lower acceleration force, whereas the low acceleration force was unable to propel the higher density balls from the surface. However, as the acceleration was increased to 1 g (Figure 8b), the NLBW approached a steady state, with all materials having similar bandwidth of 34–36 Hz. The FWHM values did not apply to the non-linear systems as they have large peaks and wide band regions, thus the term usable bandwidth was employed to represent bandwidths with a power of >0.7 mW. The usable bandwidth increased as the amount of movable mass increased. As the 1.5 g W balls demonstrated an NLBW of 43 Hz, therefore demonstrating that the saturated NLBW was dependent on the acceleration applied, it was determined that the NLBW could be further increased by increasing the acceleration.

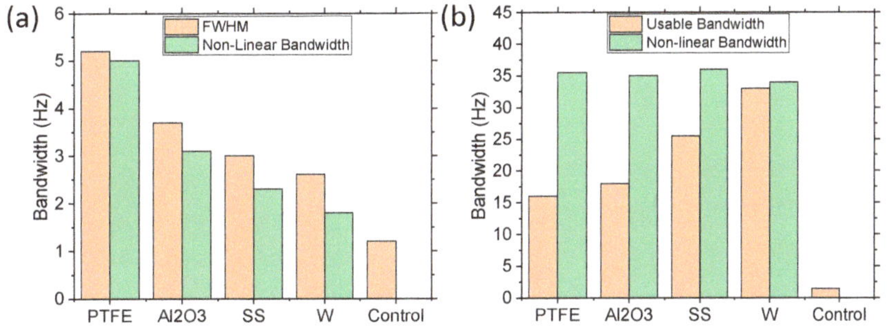

Figure 8. Bandwidth comparison of varying ball materials at (**a**) 0.1 g and (**b**) 1 g, demonstrating the FWHM bandwidth and the NLBW (range of frequency of movable masses).

The effect of the physical dimensions of the movable masses on power and bandwidth were investigated by comparing devices with (i) one 10 mm diameter W sphere with mass of 7.87 g and (ii) eight 5 mm W spheres with mass of 7.84 g. The density of the balls was the same and only the diameter changed (5 mm vs. 10 mm). By increasing the number of 5 mm diameter balls, the two devices had approximately the same wt.% of movable mass, differing only by 0.1%. The effects with varying acceleration from 0.1, 0.5, and 1 g are demonstrated in Figure 9. In all three accelerations, the larger diameter movable mass had higher power harvesting capabilities, but the bandwidth of the two devices was the same for an applied acceleration. For 0.1 g, the devices acted like a linear system with peaks of 0.122 and 0.09 mW and a FWHM of 3.56 and 3.85 Hz, compared to the control's 1.2 Hz FWHM. At 0.5 g, the devices had more non-linear dynamic properties and had peak power values of 0.43 mW and 0.17 mW, with a bandwidth of 16.24 and 17.3 Hz for the 10 mm and 5 mm balls, respectively. At 1 g, the peak power increased to 1.40 mW and 0.51 mW and the bandwidth increased to 38.1 Hz and 41.2 Hz. For each acceleration, there was a slight increase in bandwidth for the smaller balls, but it was not statistically different. However, the power difference between the various accelerations was significant. Therefore, the size of the movable mass did significantly impact the amount of power harvested, but it did not significantly affect the bandwidth. The bandwidth was, thus, dependent on the percent of movable mass as well as the applied acceleration.

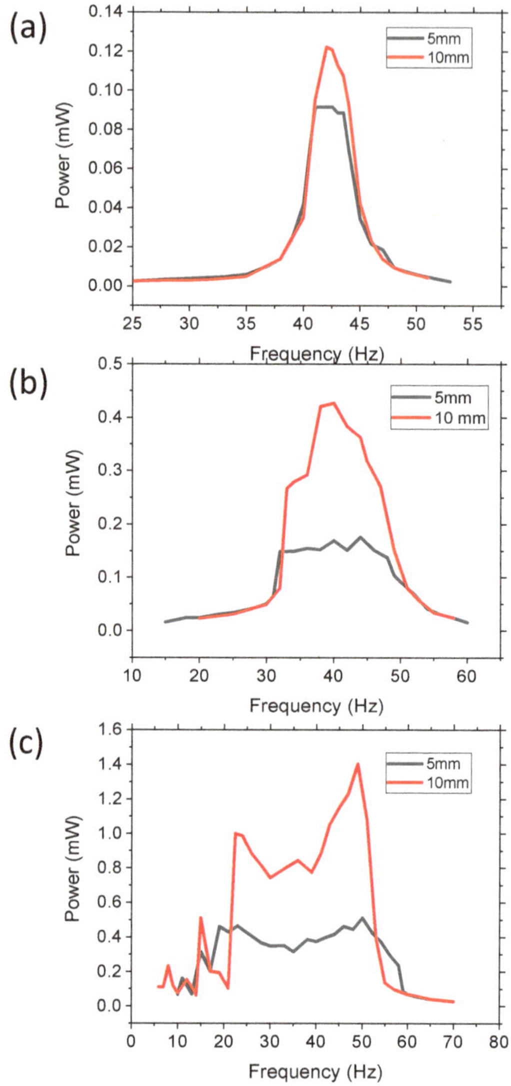

Figure 9. Demonstrating the effect of varying the size of the movable mass, while keeping movable mass percentage the same, by using one 10 mm W ball and eight 5 mm W balls and varying the acceleration from (**a**) 0.1 g, (**b**) 0.5 g, and (**c**) 1 g.

4. Conclusions

In conclusion, we successfully validated a novel concept of widening the bandwidth of a cantilever device by creating an embedded vertical or transverse movable mass system. The results demonstrated that the bandwidth and power effects were dependent on the acceleration and the amount, size, and density of the movable mass. The size of the movable mass directly affected the amount of power that could be harvested, but it did not have a significant effect on bandwidth. However, the density of the movable mass, which affected the overall wt.%, significantly affected both the power and bandwidth, with lower density masses preferred for low acceleration applications.

In the end, all of the abovementioned parameters influenced the bandwidth and amount of power, and further investigation is required to optimize the parameters for a specific application. By modifying these parameters, the end user may fine tune the bandwidth and power specifications to match the specifications of the application. The concept of a vertical movable mass system demonstrated a significant increase in bandwidth for high acceleration, low frequency applications. However, the concept can be implemented in low acceleration applications by using less dense movable masses. The power magnitude was dependent on the size, density, and overall percentage of the movable mass component. The dynamics of the system consisted of both linear and non-linear components, which requires further investigation in the future to enhance device performance. The vertical movable mass method described in this paper has potential use in both macro- and micro-scale cantilevers, but the concept requires further investigation at the micro-scale level to demonstrate validity.

This novel method provides researchers with an alternative means of widening the bandwidth without significantly decreasing power, and it has potential to be scaled down to the micro-scale to power IoT applications. The concept was experimentally validated using a piezoelectric energy harvesting cantilever, but the concept could also be applied to other energy harvesting systems, such as electromagnetics and electrostatics as well as other cantilever applications that require a wide bandwidth. This method is unique as it provides increased bandwidth, while also providing high peak power, demonstrating it can potentially provide both high power and high bandwidth. However, the results are dependent on various parameters and would need to be optimized to the specific application.

5. Patents

An invention titled "Enhanced Frequency Bandwidth of Cantilever Beam using Transverse Movable Mass," was filed through the USPTO.

Author Contributions: R.A., assisted in obtaining experimental data and developing the 3D printed proof masses. L.A.R. aided in the initial analysis and experimentation. N.J. had the initial idea, analyzed the data, and helped write the manuscript. All authors have read and agreed to the published version of the manuscript.

Funding: This research was funded by the National Science Foundation EPSCoR Program under award no. OIA-1757207.

Institutional Review Board Statement: Not applicable.

Informed Consent Statement: Not applicable.

Data Availability Statement: Data is contained within the article.

Acknowledgments: The authors would like to thank all members of the SMART Laboratory group at University of New Mexico. This material is based upon work supported by the National Science Foundation EPSCoR Program under award no. OIA-1757207.

Conflicts of Interest: The authors declare no conflict of interest.

References

1. Andosca, R.; McDonald, T.; Genova, V.; Rosenberg, S.; Keating, J.; Benedixen, C.; Wu, J. Experimental and theoretical studies on MEMS piezoelectric energy harvesters with mass loading. *Sens. Actuators A Phys.* **2012**, *178*, 76. [CrossRef]
2. Anton, S.; Sodano, H. A review of power harvesting using piezoelectric materials (2003–2006). *Smart Mater. Struct.* **2007**, *16*, R1. [CrossRef]
3. Elfrink, R.; Matova, S.; Nooijer, C.; Jambunathan, M.; Goedbloed, M.; Molengraft, J.; Vullers, R.; Renaud, M.; Van Schaijk, R. Shock induced energy harvesting with a MEMS harvester for automotive applications. In Proceedings of the 2011 International Electron Devices Meeting, Washington, DC, USA, 5–7 December 2011.
4. Jackson, N.; O'Keeffe, R.; Waldron, F.; O'Neill, M.; Mathewson, A. Evaluation of low-acceleration MEMS piezoelectric energy harvesting devices. *Microsyst. Technol.* **2014**, *20*, 671–680. [CrossRef]
5. Shen, D.; Park, J.; Ajitsaria, J.; Choe, S.; Wikkle, H.; Kim, D. The design fabrication and evaluation of a MEMS PZT cantilever with an integrated Si proof mass for vibration energy harvesting. *J. Micromech. Microeng.* **2008**, *18*, 055017. [CrossRef]
6. Zhang, G.; Gao, S.; Liu, H.; Niu, S. A low frequency piezoelectric energy harvester with trapezoidal cantilever beam: Theory and experiment. *Microsyst. Technol.* **2017**, *23*, 3457–3466. [CrossRef]
7. Cottone, F.; Vocca, H.; Gammaitoni, L. Nonlinear energy harvesting. *Phys. Rev. Lett.* **2009**, *2012*, 080601. [CrossRef]

8. Hajati, A.; Kim, S. Ultra-wide bandwidth piezoelectric energy harvesting. *Appl. Phys. Lett.* **2011**, *99*, 083105. [CrossRef]
9. Marinkovic, B.; Koser, H. Demonstration of wide bandwidth energy harvesting from vibrations. *Smart Mater. Struct.* **2012**, *21*, 065006. [CrossRef]
10. Sebald, G.; Kuwano, H.; Guyomar, D.; Ducharne, B. Experimental Duffing oscillator for broadband piezoelectric energy harvesting. *Smart Mater. Struct.* **2011**, *20*, 102001. [CrossRef]
11. Liu, H.; Lee, C.; Kobayashi, T.; Tay, C.J.; Quan, C. Investigation of a MEMS piezoelectric energy harvester system with a frequency-widened-bandwidth mechanism introduced by mechanical stoppers. *Smart Mater. Struct.* **2012**, *21*, 035005. [CrossRef]
12. Olszewski, O.; Houlihan, R.; Blake, A.; Mathewson, A.; Jackson, N. Evalulation of vibrational PiezoMEMS harvester that scavenges energy from a magnetic field surronding an AC current-carrying wire. *J. Microelectromech. Syst.* **2017**, *26*, 1298–1305. [CrossRef]
13. Zhao, L.; Yang, Y. An impact-based broadband aeroelastic energy harvester for concurrent wind and base vibration energy harvesting. *Appl. Energy* **2018**, *212*, 233–243. [CrossRef]
14. Challa, V.; Prasad, M.; Shi, Y.; Fisher, F. A vibration energy harvesting device with bidirectional resonance frequency tunability. *Smart Mater. Struct.* **2008**, *17*, 015035. [CrossRef]
15. Jackson, N. PiezoMEMS Nonlinear Low Acceleration Energy Harvester with an Embedded Permanent Magnet. *Micromachines* **2020**, *11*, 500. [CrossRef]
16. Yang, W.; Towfighian, S. A hybrid nonlinear vibration energy harvester. *Mech. Syst. Signal Process.* **2017**, *90*, 317–333. [CrossRef]
17. Pellegrini, S.P.; Tolou, N.; Schenk, M.; Herder, J.L. Bistable vibration energy harvesters: A review. *J. Intell. Mater. Syst. Struct.* **2013**, *24*, 1303–1312. [CrossRef]
18. Podder, P.; Mallick, D.; Amann, A.; Roy, S. Influence of combined fundamental potentials in a nonlinear vibration energy harvester. *Sci. Rep.* **2016**, *6*, 37292. [CrossRef] [PubMed]
19. Yan, B.; Yu, N.; Zhang, L.; Ma, H.; Wu, C.; Wang, K.; Zhou, S. Scavenging vibrational energy with a novel bistable electromagnetic energy harvester. *Smart Mater. Struct.* **2020**, *29*, 025022. [CrossRef]
20. Jacquelin, E.; Adhikari, S.; Friswell, M.I. A piezoelectric device for impact energy harvesting. *Smart Mater. Struct.* **2011**, *20*, 105008. [CrossRef]
21. Ju, S.; Ji, C.-H. Impact-based piezoelectric vibration energy harvester. *Appl. Energy* **2018**, *214*, 139–151. [CrossRef]
22. Jackson, N. Tuning and widening the bandwidth of vibration energy harvesters using a ferrofluid embedded mass. *Microsyst. Technol.* **2020**, *26*, 2043–2051. [CrossRef]
23. Chandwani, J.; Somkuwar, R.; Deshmukh, R. Multi-band piezoelectric vibration energy harvester for low-frequency applications. *Microsyst. Technol.* **2019**, *25*, 3867–3877. [CrossRef]
24. Lihua, C.; Jiangtao, X.; Shiqing, P.; Liqi, C. Study on cantilever piezoelectric energy harvester with tunable function. *Smart Mater. Struct.* **2020**, *29*, 075001. [CrossRef]
25. Porcar-Climent, A.; Jackson, N. Rolling mass for wide bandwidth vibration energy harvesting. In Proceedings of the 2019 19th International Conference on Micro and Nanotechnology for Power Generation and Energy Conversion Applications (PowerMEMS), Krakow, Poland, 2–6 December 2019; pp. 1–4.
26. Somkuwar, R.; Chandwani, J.; Deshmukh, R. Wideband auto-tunable vibration energy harvester using change in centre of gravity. *Microsyst. Technol.* **2018**, *24*, 3033–30440. [CrossRef]
27. Jackson, N.; Stam, F. Sloshing liquid-metal mass for widening the bandwidth of a vibration energy harvester. *Sens. Actuators A Phys.* **2018**, *284*, 17–21. [CrossRef]
28. Jackson, N.; Stam, F.; Olszewski, O.; Doyle, H.; Quinn, A.; Mathewson, A. Widening the bandwidth of vibration energy harvesters using a liquid-based non-uniform load distribution. *Sens. Actuators A Phys.* **2016**, *246*, 170–179. [CrossRef]
29. Jackson, N.; Stam, F.; Olszewski, O.; Houlihan, R.; Mathewson, A. Broadening the bandwidth of piezoelectric energy harvesters using liquid-filled mass. *Procedia Eng.* **2015**, *120*, 328–332. [CrossRef]
30. Liu, D.; Li, H.; Feng, H.; Yalkun, T.; Hajj, M.R. A multi-frequency piezoelectric vibration energy harvester with liquid filled container as the proof mass. *Appl. Phys. Lett.* **2019**, *114*, 213902. [CrossRef]
31. Somkuwar, R.; Chandwani, J.; Deshmukh, R. Bandwidth widening of piezoelectric energy harvester by free moving cylinders in liquid medium. *Microsyst. Technol.* **2020**, *27*, 1959–1970. [CrossRef]
32. Jackson, N.; Pedrosa, F.J.; Bollero, A.; Mathewson, A.; Olszewski, O.Z. Integration of thick-film permanent magnets for MEMS applications. *J. Microelectromech. Syst.* **2016**, *25*, 716–724. [CrossRef]
33. Jackson, N.; Olszewski, O.Z.; O'Murchu, C.; Mathewson, A. Shock-induced aluminum nitride based MEMS energy harvester to power a leadless pacemaker. *Sens. Actuators A Phys.* **2017**, *264*, 212–218. [CrossRef]
34. Tang, G.; Yang, B.; Hou, C.; Li, G.; Liu, J.; Chen, X.; Yang, C. A piezoelectric micro generator worked at low frequency and high acceleration based on PZT and phosphor bronze bonding. *Sci. Rep.* **2016**, *6*, 1–10. [CrossRef] [PubMed]
35. Blevins, R.; Plunkett, R. Formulas for natural frequency and mode shape. *J. Appl. Mech.* **1980**, *47*, 461. [CrossRef]
36. Jackson, N.; Kumar, K.; Olszewski, O.; Schenning, A.P.; Debije, M.G. Tuning MEMS cantilever devices using photoresponsive polymers. *Smart Mater. Struct.* **2019**, *28*, 085024. [CrossRef]
37. Rodriguez, L.; Jackson, N. Wide Bandwidth Piezoelectric Energy Harvester with Transverse Movable Mass. In Proceedings of the International Mechanical Engineering Congress and Exposition (IMECE2020), Portlad, OR, USA, 16–19 November 2020.

38. Elfrink, R.; Kamel, T.; Goedbloed, M.; Matova, S.; Hohlfeld, D.; Van Andel, Y.; Van Schaijk, R. Vibration energy harvesting with aluminum nitride-based piezoelectric devices. *J. Micromech. Microeng.* **2009**, *19*, 094005. [CrossRef]
39. Le, C.P.; Halvorsen, E.; Søråsen, O.; Yeatman, E.M. Microscale electrostatic energy harvester using internal impacts. *J. Intell. Mater. Syst. Struct.* **2012**, *23*, 1409–1421. [CrossRef]
40. Soliman, M.; Abdel-Rahman, E.; El-Saadany, E.; Mansour, R. A wideband vibration-based energy harvester. *J. Micromech. Microeng.* **2008**, *18*, 115021. [CrossRef]
41. Tvedt, L.G.W.; Nguyen, D.S.; Halvorsen, E. Nonlinear behavior of an electrostatic energy harvester under wide-and narrowband excitation. *J. Microelectromech. Syst.* **2010**, *19*, 305–316. [CrossRef]

sensors

MDPI

Article

Collaborative Filler Network for Enhancing the Performance of $BaTiO_3$/PDMS Flexible Piezoelectric Polymer Composite Nanogenerators

Ayda Bouhamed [1,*], Nathanael Jöhrmann [2], Slim Naifar [1], Benny Böhm [3], Olav Hellwig [3], Bernhard Wunderle [2] and Olfa Kanoun [1]

1 Measurement and Sensor Technology, Faculty of Electrical Engineering and Information Technology, Technische Universität Chemnitz, Reichenhainer Straße 70, 09126 Chemnitz, Germany; slim.naifar@etit.tu-chemnitz.de (S.N.); olfa.kanoun@etit.tu-chemnitz.de (O.K.)
2 Materials and Reliability of Microsystems, Faculty of Electrical Engineering and Information Technology, Technische Universität Chemnitz, Reichenhainer Straße 70, 09126 Chemnitz, Germany; nathanael.joehrmann@s2002.tu-chemnitz.de (N.J.); bernhard.wunderle@etit.tu-chemnitz.de (B.W.)
3 Functional Magnetic Materials, Faculty of Natural Sciences, Technische Universität Chemnitz, Reichenhainer Straße 70, 09126 Chemnitz, Germany; benny.boehm@physik.tu-chemnitz.de (B.B.); olav.hellwig@physik.tu-chemnitz.de (O.H.)
* Correspondence: ayda.bouhamed@etit.tu-chemnitz.de

Abstract: Wearable sensors are gaining attention in human health monitoring applications, even if their usability is limited due to battery need. Flexible nanogenerators (NGs) converting biomechanical energy into electrical energy offer an interesting solution, as they can supply the sensors or extend the battery lifetime. Herein, flexible generators based on lead-free barium titanate ($BaTiO_3$) and a polydimethylsiloxane (PDMS) polymer have been developed. A comparative study was performed to investigate the impact of multiwalled carbon nanotubes (MWCNTs) via structural, morphological, electrical, and electromechanical measurements. This study demonstrated that MWCNTs boosts the performance of the NG at the percolation threshold. This enhancement is attributed to the enhanced conductivity that promotes charge transfer and enhanced mechanical property and piezoceramics particles distribution. The nanogenerator delivers a maximum open-circuit voltage (V_{OC}) up to 1.5 V and output power of 40 nW, which is two times higher than NG without MWCNTs. Additionally, the performance can be tuned by controlling the composite thickness and the applied frequency. Thicker NG shows a better performance, which enlarges their potential use for harvesting biomechanical energy efficiently up to 11.22 V under palm striking. The voltage output dependency on temperature was also investigated. The results show that the output voltage changes enormously with the temperature.

Keywords: PDMS/$BaTiO_3$ nanocomposite; MWCNTs; flexible piezoelectric nanogenerators; biomechanical energy harvesting; temperature dependency

Citation: Bouhamed, A.; Jöhrmann, N.; Naifar, S.; Böhm, B.; Hellwig, O.; Wunderle, B.; Kanoun, O. Collaborative Filler Network for Enhancing the Performance of $BaTiO_3$/PDMS Flexible Piezoelectric Polymer Composite Nanogenerators. *Sensors* **2022**, *22*, 4181. https://doi.org/10.3390/s22114181

Academic Editor: Giovanna Marrazza

Received: 8 May 2022
Accepted: 28 May 2022
Published: 31 May 2022

Publisher's Note: MDPI stays neutral with regard to jurisdictional claims in published maps and institutional affiliations.

1. Introduction

Nowadays, energy harvesting technologies are facing an immense breakthrough owing to the increase of sophistication of our society using Internet of Things (IoT) technology. Therefore, millions of sensors are used that require continuous charging [1–3].

To this aim, the replacement of batteries with a continuous power source is becoming inevitable and indispensable in providing continuous environment and human health monitoring.

Additionally, the environmental concern has also pushed for efficient and eco-friendly generating energy sources. Therefore, the demand for energy harvesting technologies that can collect ambient energy such as wind energy, solar energy, and mechanical energy is growing [1–3]. Among all these sources, the ambient energy of mechanical vibrations is one of the most abundant sources that can be converted to usable electrical energy [4–6]. Various principles exist to harvest wasted mechanical energy as electricity, including the piezoelectric and triboelectric principles [7,8]. Both principles can efficiently harvest energy in low-frequency environments and can be lightweight. However, a triboelectric energy harvester has some limitations, such as lower durability caused by structural changes over time and post-stress conditions [9]. For this reason, piezoelectric energy harvesters are more preferred.

Among piezoelectric energy harvesters, polymer nanocomposite based piezoelectric energy harvesters attract great attention because of the possibility to convert vibrational and mechanical energy sources from human activities, such as pressure, bending, and stretching motions into electrical energy. Recently, different piezoelectric nanomaterials have been used to develop flexible nanogenerators (NGs) such as lead zirconate titanate (PZT), zinc oxide (ZnO), barium titanate ($BaTiO_3$), or in poly (vinylidene fluoride) (PVDF) [10–12]. Between all these materials, PZTs have been frequently used, owing to their high piezoelectric properties and thermal stability [13]. However, the toxicity of lead led researchers to consider and use lead-free piezoceramic with comparable piezoelectric properties to PZT. Recently, $BaTiO_3$ is explored for the realization of nanogenerators for different applications, including medical [14].

In fact, $BaTiO_3$ is one of the promising lead-free piezoceramic due to its very high piezoelectric constant $d_{33} > 200$ pC/N, their perovskite crystal structure that leads to a high dielectric constant (100–11,000), and biocompatibility. $BaTiO_3$ piezoelectric nanoparticles are preferred to be embedded into a soft polymer matrix to fabricate simple, scalable, and wearable piezoelectric nanogenerators [14].

For example, Shin et al. [15] reported about a highly performant flexible piezoelectric composite composed of $BaTiO_3$ and poly-(vinylidene fluoride-co-hexafluoropropylene) (PVDF-HFP), which could achieve a maximum output voltage and current under cyclic bending in the range of 5 V to 750 nA, respectively. In another work, Yan and Jeong [16] developed flexible composites with different orientations of $BaTiO_3$ nanofibers in a polydimethylsiloxane (PDMS) polymer. The nanogenerator with vertically aligned $BaTiO_3$ nanofibers exhibited a high piezoelectric performance with an output power of 0.1841 μW under very low mechanical stress around 0.002 MPa.

Several works have illustrated that the performances of nanogenerators can be significantly improved by keeping them under a high electric voltage field of several kV for a very long period of time in order to orient the crystal domains of the piezoelectric material and to align the piezoelectric dipoles into an identical direction. Lin et al. [17] fabricated a stretchable piezoelectric nanogenerator using $BaTiO_3$ /PDMS, which was polarized at an ambient temperature by applying an electric field of 80 kV/cm for 12 h. The nanogenerator delivered an open-circuit voltage and short-circuit current of 5.5 V and 350 nA, respectively. However, this process is long and unsafe. Intensive efforts have been made to improve the performance by introducing different carbon materials, such as carbon black, graphene, and carbon nanotubes (CNTs) [18–21].

Luo et al. [18] developed high-performance flexible piezoelectric nanogenerators containing a 30 wt.% $BaTiO_3$/PDMS/3.2 wt.% carbon black composite film. The performance was enhanced by 143% compared to the nanocomposite without carbon black.

Additionally, many other studies have doped CNTs within the $BaTiO_3$/PDMS composite. In general, the use of CNTs as filler leads to improving the output voltage, as they serve as nano-electrical bridges. Yan et al. [22] reported that the addition of 2 wt.% multiwalled carbon nanotubes (MWCNTs) to a nanogenerator with 40 wt.% $BaTiO_3$ leads to an enhanced output performance of the nanogenerators due to the improvement of both the electrical and dielectric properties. Park et al. [12] demonstrated that the addition of single-walled carbon nanotubes (SW-CNTs) to PDMS/$BaTiO_3$ can lead to a higher output voltage of ~3.2 V compared to the one containing both SW-CNTs and reduced graphene (rGO) oxide. This can be explained with the enhanced electrical properties of SW-CNT that can significantly influence the output performance. Furthermore, the NG with rGO shows less performance caused by the difference in the degree of mixing of both kinds of reinforcements, which leads to geometrical difference between the SW-CNT networks and the laminated rGO structures.

Importantly, the highest performance of nanogenerator filled with conductive particles were found to be at a critical loading of the filler known by the percolation threshold, where there was a sudden increase on the electrical conductivity in the nanocomposites. The percolation threshold depends greatly on the filler geometry and state of the filler dispersion in the polymer matrix. For a polymer composite containing spherical conductive particles, e.g., carbon black, the percolation thresholds were usually found to be very high, which enormously minimized the flexibility and increased the cost of the final composite. In order to tackle this problem, several approaches have been proposed by including conductive fillers with a higher dimension, such as one-dimensional (1D) fibers as MWCNTs or two-dimensional (2D) plates as graphene nanoplates were employed to ensure lower percolation thresholds [23].

A low percolation threshold can be achieved, usually with use of one-dimensional (1D) carbonaceous fibers such as MWCNTs. However, the main challenge is to avoid their agglomeration caused by their high surface area and Van der Wall attractive forces.

Therefore, efforts were devoted towards minimizing the percolation threshold by optimizing the processing conditions.

Several approaches have been followed by researchers to optimize the electrical conductivity and reduce the percolation threshold in polymer composites, including optimization of the processing conditions such as mixing temperature and speed or by involving powerful processes that are able to unbundle the agglomeration of nanoparticles, such as sonication or the calendaring process [24,25].

In this work, we explored the potential of polymer composites based on lead-free ceramics $BaTiO_3$ to realize a wearable and flexible piezoelectric nanogenerator with less piezoceramics concentrations and enhanced performance without any additional poling processes. To achieve a cost-effective NG with improved performance, composite films containing different ratios of $BaTiO_3$ piezoelectric nanoparticles in a polydimethylsiloxane (PDMS) polymer matrix were investigated through electromechanical measurements using a vibration shaker. Then, MWCNTs were dispersed into the $BaTiO_3$/PDMS composite with a low concentration ranging from 0.3 wt.% to 1 wt.% to boost its performance. The distribution of nanoparticles was also examined via scanning electron microscope (SEM) and X-ray diffraction (XRD). The impact of the addition of MWCNTs was also addressed by evaluating and comparing the piezoelectric performance, as well the electrical and mechanical properties of the nanocomposites. The effects of layer thickness on the performance of the nanogenerator were also examined, and the potential of the realized NG for biomechanical energy harvesting was verified in ambient conditions and at different temperatures.

2. Materials and Methods

2.1. Materials

In this study, two different reinforcements were used: $BaTiO_3$ and MWCNTs. The $BaTiO_3$ particles have a high purity and 99.5% trace metals basis. They were purchased from Sigma Aldrich with a dimension less than 2 µm. As well, MWCNTs were purchased from Sigma Aldrich with an outer diameter of 6–9 nm and length of 1 µm. Concerning the polymer matrix, it consists of a soft polymer polydimethylsiloxane (PDMS—Sylgard 184) that was purchased from Dow Corning, GmbH.

2.2. Fabrication Process of the Piezoelectric Nanogenerator

To realize a flexible piezoelectric nanogenerator based on the $BaTiO_3$/PDMS composite, two layers of Kapton polyimide (PI) flexible substrate were used, in which the copper layer was coated in it to serve as an electrode. A laminate flexible composite was placed between the two sides. This laminate composite consisted of three layers composed of two dielectric layers of PDMS and a composite layer, which was deposited in between to form a sandwiched structure, as shown in Figure 1.

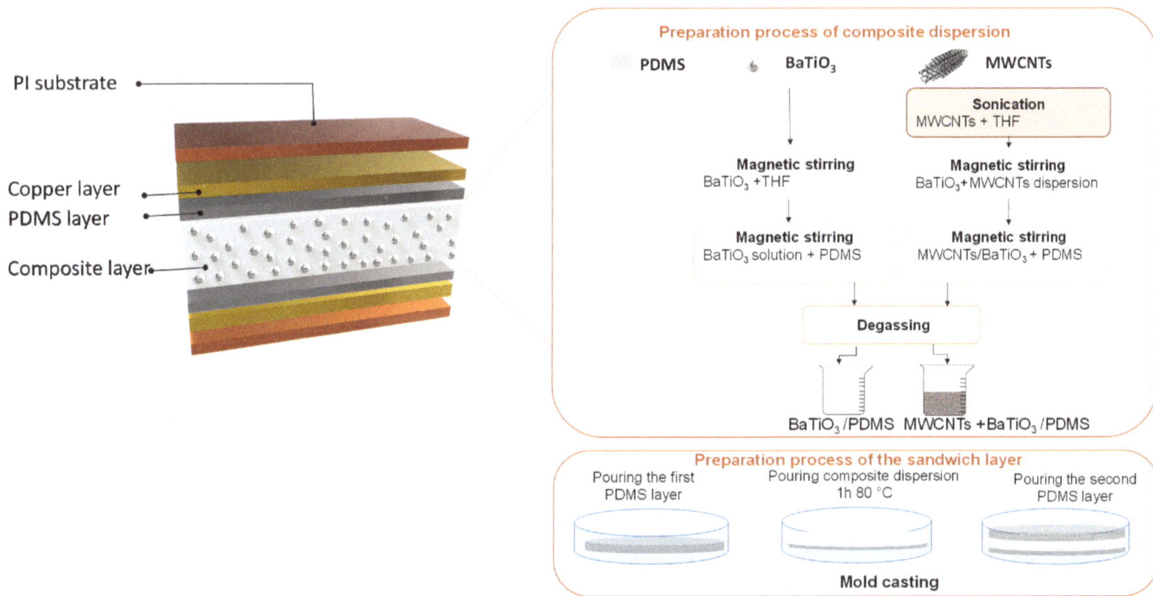

Figure 1. Flowchart of the fabrication process of different types of nanocomposites with an illustration about the structure of the final nanogenerator.

To produce the sandwiched structure, we start first by the fabrication of piezoelectric composites. Different composite compositions were made in this work, as summarized in Table 1. To prepare the composites, $BaTiO_3$ was mixed in different ratios from 10 wt.% to 40 wt.% with tetrahydrofuran (THF) using a magnetic stirrer for 1 h, as shown in Figure 1. Then, the required amount of soft polymer polydimethylsiloxane (PDMS—Sylgard 184) was added and mixed for 2 h at 70 °C.

Table 1. Summary of the composites compositions and dimensions realized in this work.

Type of Composite	$BaTiO_3$ Concentration	MWCNTs Concentration	Composite Thickness
Without MWCNTs	10 wt.%	-	500 µm
	15 wt.%	-	
	20 wt.%	-	
	30 wt.%	-	
	40 wt.%	-	
With MWCNTs	15 wt.%	0.3 wt.%	500 µm
		0.5 wt.%	300, 500, 800 µm
		0.75 wt.%	500 µm
		1 wt.%	500 µm

As the effect of the addition of MWCNTs on the nanogenerator performance is also addressed in this work. The preparation of the hybrid nanocomposite was approximately similar to the process used before, as shown in Figure 1. Due to their small size and large surface area, MWCNTs was firstly dispersed in THF using a horn sonicator Sonoplus HD 7300 for 15 min at 30% amplitude to unbundle the MWCNT agglomerations. To form the hybrid composite, the process presented in Figure 1 was followed.

After mixing, all dispersions were kept for the degassing process in the vacuum chamber before deposition. In meanwhile, the first PDMS layer was prepared and deposited in a glass mold to be cured for 10 min at 150 °C. Subsequently, nanocomposite material was deposited and cured at 80 °C for 1 h, followed by the deposition of the second PDMS layer.

In fact, these two dielectric layers are made to avoid charging in the electrodes, as well as to protect the composite layer during peeling from the mold to be damaged.

2.3. Nanocomposites and Nanogenerators Characterizations

The performances of the nanogenerators were investigated by means of an experimental setup consisting of a voltage generator, a shaker, and a digital oscilloscope for the acquisition of the output signal.

A fixed and harmonically distorted mechanical load of a 30 Hz frequency and having an RMS value of around 0.1 N were used to determine the optimal material composition.

To determine the generated output power, variable load resistances ranging from kilo Ohms to Mega Ohms were used to characterize the output power of the nanogenerator. To that end, the power output (P) of the piezoceramic composite was determined by measuring the voltage across a variable resistance placed in parallel to the nanogenerator, where the consumed power could be calculated as the result of the square of the output voltage divided by the value of the load resistance, as demonstrated in Equation (1):

$$P = \frac{V_{out}^2}{R_{ext}} \quad (1)$$

To investigate deeply the uniformity of piezoceramics particles distribution and the influence of the addition of carbon nanotubes on the quality of the nanocomposite layers, scanning electron microscopy measurements were conducted on the cross-section using Zeiss Auriga 40. To prepare the samples for cross-sectional SEM observation, small samples of 0.5 × 0.5 cm were cut using a lever cutter tool. Then, all the samples were coated with a very thin Au layer in order to avoid the charging effect. The images were taken at 30 kV with a working distance of 2 mm using the InLens SE-Detector (secondary electrons).

In parallel, tensile tests were conducted for some samples using the universal tensile testing machine Instron ElectroPuls E10000 at a rate of 3 mm/min in order to determine the mechanical properties of the composites. To this aim, the specimens were prepared with a standard dog bone of ISO37 and had a very small size with an overall length of 35 mm, gauge length l_0 of 10 mm, and width of 2 mm.

As well, the crystalline structure of the different prepared composites was examined using X-ray diffractometer Rigaku SmartLab with Cu-K_α radiation (wavelength 0.154 nm) operated at 45 kV and 200 mA in Bragg-Brentano geometry using a K_β filter for Cu.

The scanning was in the 2θ range of 10° to 90°, with a step interval of 0.25° and scan speed 10°/min.

As well, the influence of a conductive nanofillers addition on the nanocomposite electrical properties was investigated using a Keithley 2636 source meter.

3. Results and Discussion

3.1. Influence of $BaTiO_3$ Concentration on the Nanogenerator Performance

Upon a repetitive cyclic load, impacting and releasing the NG using a vibration mechanical shaker at 30 Hz, an open output voltage was detected, as illustrated in Figure 2a. The composite material made out of $BaTiO_3$/PDMS exhibits typical piezoelectric signals. In fact, when the NG is compressed, a positive voltage pulse is detected where a flow of electrons from the bottom electrode to the upper electrode will be created. Then, when the NG is released, an opposite pulse signal is observed, owing to the recovering of electrons to the original state.

In fact, the piezoelectric voltage output can be expressed by Equation (2):

$$V = \frac{d_{33}}{\varepsilon_{33}}\sigma t \quad (2)$$

where d_{33}, ε_{33}, σ, and t are the piezoelectric coefficient, effective electrical permittivity, mechanical stress change, and film thickness, respectively.

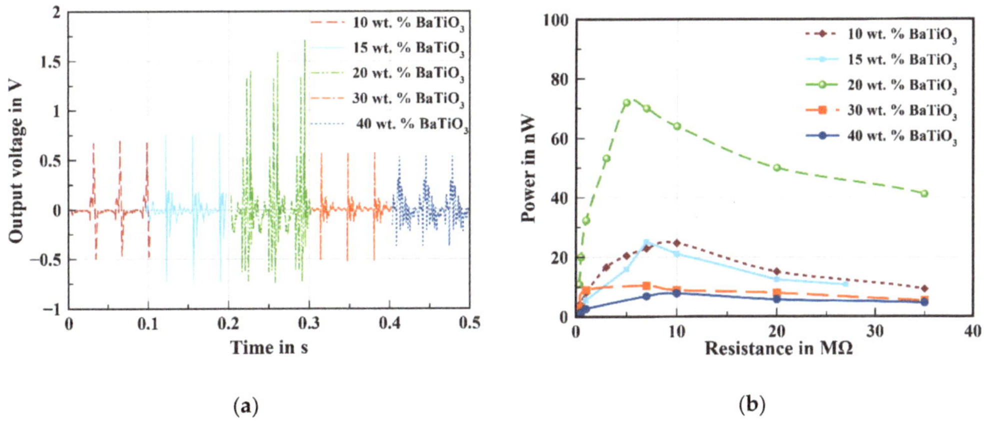

Figure 2. Performance of nanogenerators depending on the concentration of $BaTiO_3$: (**a**) output voltage and (**b**) output power.

According to Figure 2, the content of $BaTiO_3$ greatly affects the output voltage. The nanogenerators containing 20 wt.% $BaTiO_3$ exhibit a high output voltage around 1.7 V, as well as a high output power of 72 nW, which is around three times higher than the sample prepared with 10 wt.%.

This can be attributed to the enhanced piezoelectric coefficient of the composite material due to the increased content of piezoelectric particles within the polymer matrix.

Additionally, this enhancement can be also related to the improved young modulus of the composite materials owing to the integration of more ceramic particles.

Kim et al. [26] found that the electric power output of a piezoelectric nanogenerator depends on various physical parameters of the constituent materials, including the piezoelectric coefficient, Young's modulus, and dielectric constant, where the Young's modulus plays a crucial role.

In addition, it was noted that, by increasing the amount of $BaTiO_3$, the output performance increased first and then reduced after 20 wt.% $BaTiO_3$. This reduction is mainly related to the agglomeration of $BaTiO_3$ within the polymer matrix and ineffective contact surface area leading to a decreased material mechanical property, as well as piezoelectric property, as demonstrated in Table 2. In fact, the Young's modulus was increased by doping $BaTiO_3$ with PDMS; then, it was reduced at 40 wt.% $BaTiO_3$, caused by the inhomogeneous particle's distribution within the polymer matrix.

Table 2. Young's modulus for different composites.

Composite Composition	Young's Modulus (MPa)
15 wt.%	0.08
20 wt.%	0.18
30 wt.%	0.24
40 wt.%	0.19
15 wt.% + 0.3 wt.% MWCNTs	0.09
15 wt.% + 0.75 wt.% MWCNTs	0.23

Figure 3a–c show the cross-sectional morphology images of the 15 wt.% $BaTiO_3$/PDMS, 20 wt.% $BaTiO_3$/PDMS, and 40 wt.% $BaTiO_3$/PDMS, respectively. Those images were taken using SEM to verify the uniformity of particles distribution. According to these figures, $BaTiO_3$ particles are not well-distributed within the PDMS matrix and are sedimented at

the bottom, leading to the formation of a lot of free space in the upper part, especially for samples prepared with 15 wt.% $BaTiO_3$, as demonstrated in Figure 3g. By increasing the amount of $BaTiO_3$ within the polymer matrix, less free space between neighboring particles was seen for the sample containing 20 wt.% $BaTiO_3$. However, the sedimentation was more pronounced in the 40 wt.% $BaTiO_3$ case.

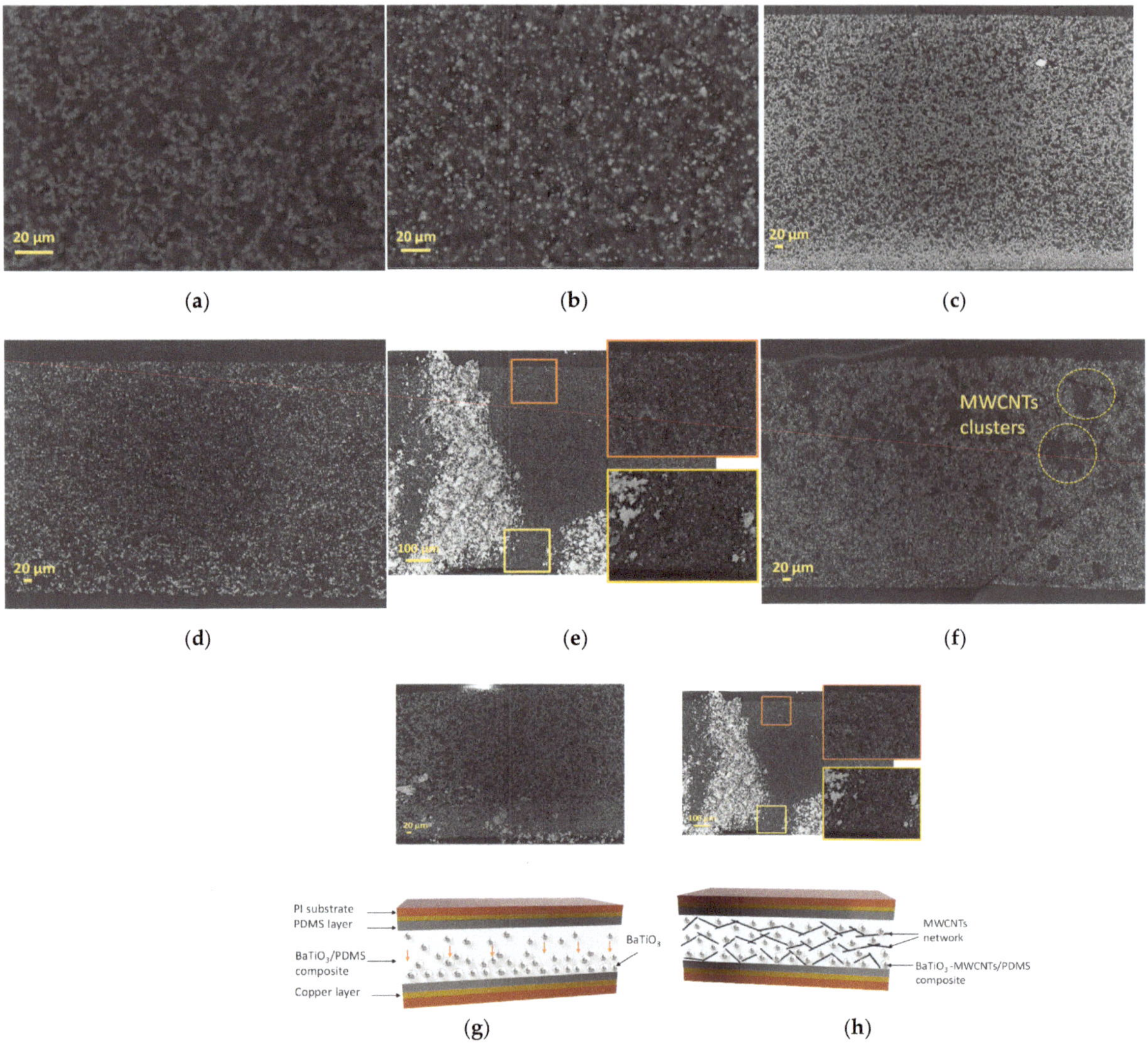

Figure 3. SEM images done in the cross-section for different composites and showing the effects of the addition of MWCNTs: (**a**–**c**) composites with 15 wt.% $BaTiO_3$, 20 wt.% $BaTiO_3$, and 40 wt.% $BaTiO_3$, respectively; (**d**–**f**) 15 wt.% $BaTiO_3$/PDMS composite containing 0.3 wt.% MWCNTs, 0.5 wt.% MWCNTs, and 0.75 wt.% MWCNTs, respectively; and (**g**,**h**) illustration of the distribution of $BaTiO_3$ in two different cases without and with MWCNTs, respectively.

3.2. Influence of Addition of MWCNTs on the Nanogenerator Performance

To minimize the free spaces existing in the sample with 15 wt.% $BaTiO_3$, the sedimentation of the particles, and to achieve a better performance at a low concentration, MWCNTs were doped within 15 wt.% $BaTiO_3$/PDMS with different concentrations.

By the addition of MWCNTs within the composite, enhancement on the $BaTiO_3$ particles distribution was remarqued at 0.5 wt.% MWCNTs. In fact, MWCNTs were acting as niches for $BaTiO_3$ particles, prompting them to not settle down, as shown in Figure 3d,e,h. In spite of that, increasing the amount of MWCNTs leads to the formation of multiple MWCNTs clusters within the polymer matrix, as shown in Figure 3f.

The enhancement of piezoceramics particles distribution at low MWCNT concentrations leads to the improvement of the piezoelectric performance of the nanogenerators, as demonstrated in Figure 4a,b.

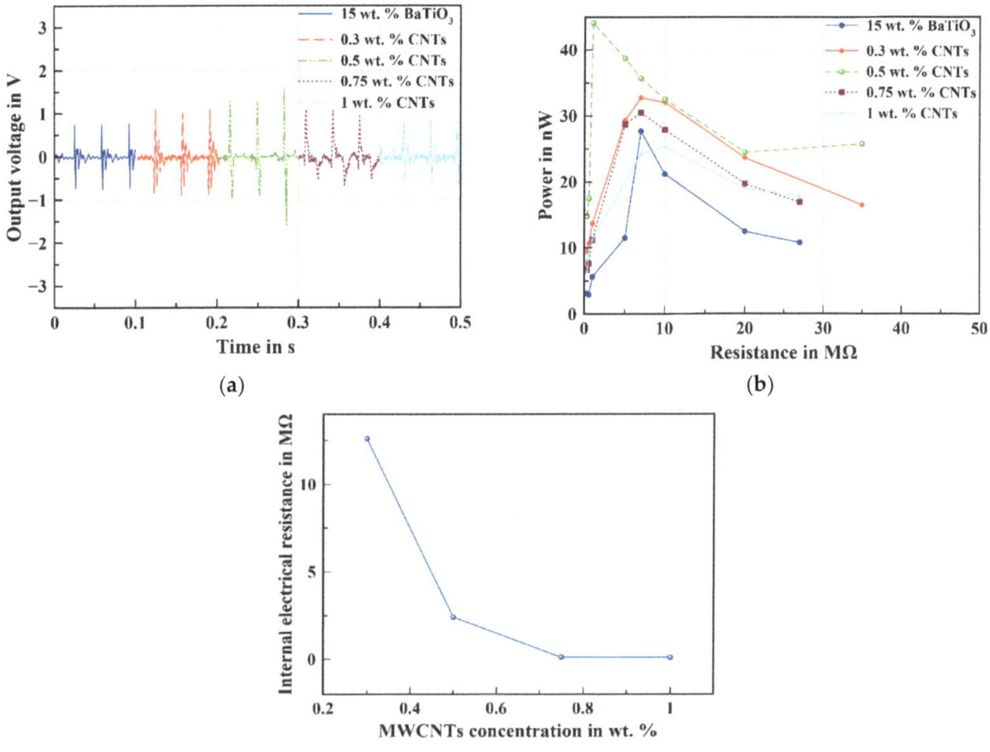

Figure 4. Effect of MWCNT concentration on the performance of the nanogenerators: (**a**) output voltage, (**b**) output power, and (**c**) internal electrical resistance.

By evaluating the performance of the nanogenerators containing different amounts of MWCNTs, the important role of MWCNTs was obvious to see. For the composite containing 0.5 wt.% MWCNTs, it was noted that the output voltage was two times higher, as well as the output power. The high output voltage of the NG with MWCNTs compared to the one with only $BaTiO_3$ is not only related to the uniform distribution of nanoparticles within the polymer matrix, which is one of the key factors in obtaining a high output voltage, but it is also due to their high electrical conductivity that acts as bridges between piezoceramics nanoparticles and leads at the end to transmit efficiently the electrical charge from the top electrode to the bottom electrode generated during pressing. MWCNTs then act as nanobridges between $BaTiO_3$. This effect was suppressed by the presence of only nonconductive PDMS polymers between $BaTiO_3$. Similar results were found by Sun et al. [27], the addition of MWCNTs within the ZnO/PDMS composite leads to efficiently enhancing the export of the charge generated by ZnO NPs. Therefore, increasing the amount of CNTs within the composite helps to ensure the self-polarization process within the material owing to the improved electron transport by the formation of conductive pathways within the matrix. According to Figure 4c, the addition of MWCNTs leads to reducing the internal electrical resistance of the nanocomposite where the resistance is sharply minimized at 0.5 wt.% MWCNTs, indicating the formation of the conductive network. By increasing the amount of MWCNTs, the electrical resistance shows a minor reduction, indicating that the 0.5 wt.% presents the percolation threshold of this composite. In fact, many studies show that the dielectric property of composites filled with a conductive filler shows a significant increase near the percolation threshold [28,29]. Additionally, Banerjee et al. [30] illustrated that increasing the MWCNT volume fraction from 1 to 4% within the PZT/epoxy matrix increases the strain coefficient, d_{33}, from 0.06 to 0.45 pC/N. This increase can be attributed to the increase in polarization of the composite due to the increased conductivity by the MWCNTs inclusions. In addition, the introduction of MWCNTs can also help to improve the composite mechanical properties, as depicted in Table 2, and ensure a better load transfer within the nanogenerator.

In Reference [31], they demonstrated the synergistic effect between isotropic and anisotropic fillers in constituting a collaborative network that led to enhanced mechanical properties of the composite due to the different aspect ratio and the different surface characteristics, which avoids the filler flocculation phenomenon.

It can be seen from the XRD graph presented in Figure 5 that the film prepared with $BaTiO_3$ has no other phases except the ones of the $BaTiO_3$ phases, indicating that no other impurities are created during processing. In addition, the graph demonstrates sharp peaks, which is a sign of good crystallinity.

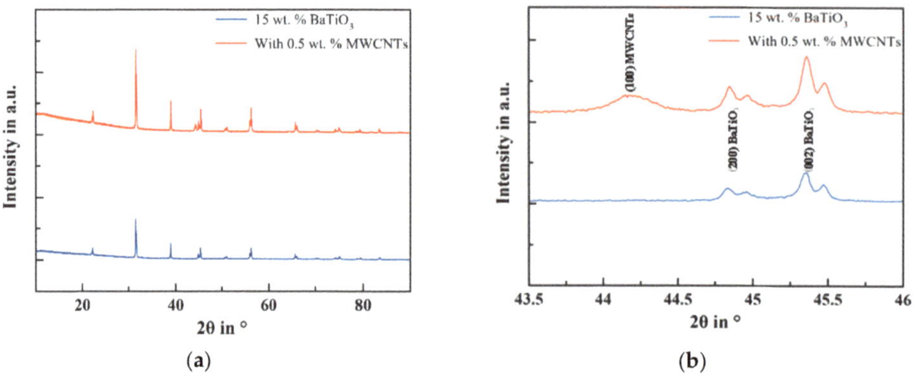

Figure 5. X-ray spectra for different composites with and without MWCNTs: (**a**) spectrum from 2θ = 10° to 90° and (**b**) zoom view in the range of 2θ = 43.5–46°.

The appearance of the peaks (002/200) near 45° and peaks (103/310) near 75° indicate the presence of both crystal phases, which are the tetragonal and cubic phases. As the peaks of PDMS were not interfering with peaks of $BaTiO_3$, the Scherrer method for the calculation of particle sizes for single crystallite peaks was adopted as shown in Equation (3) to calculate the particle sizes for samples containing only $BaTiO_3$ and the samples with $BaTiO_3$ and MWCNTs.

$$D = \frac{K\lambda}{(\beta cos\theta)} \tag{3}$$

where λ is the X-ray wavelength (0.154051 nm), β is the full width at half-maximum of (111), and θ is the scattering angle. The measurements show that the particles' diameters are 1007.36 Å and 955 Å for samples without and with MWCNTs, respectively.

The comparison of the XRD spectrum of composites containing MWCNTs to the spectrum of the sample without MWCNTs shows that the peak intensities of $BaTiO_3$ were higher, reflecting the important role of MWCNTs to boost the crystallinity. In fact, the addition of MWCNTs leads to avoid the formation of $BaTiO_3$ aggregates and sedimentation that may hinder the nucleation and crystal growth.

3.3. Nanogenerator Performance under Simulated Environments

The performance of the nanogenerator can be tuned as a function of the thickness and applied frequency, as shown in Figure 6a,b. Therefore, three different thicknesses: A, B, and C have been investigated for the nanogenerator under a vibration shaker of 30 Hz, which are 300 μm, 500 μm, and 800 μm, respectively. The study demonstrates that a thicker film shows the highest performance. Additionally, the performance of the optimal nanogenerator geometry was investigated under different force-frequency ranging from 10 Hz to 45 Hz. The study illustrates that, as the external force frequency was rising, the output voltage gradually increased from 0.6 V to 2.93 V at 29 Hz, as depicted in Figure 6b. However, the output voltage was gradually reduced to 0.9 V at 45 Hz. This reduction in the output voltage could be explained by the short time of the compressive force that led to inhibiting the material to recover and returning to its original position before the next force impact.

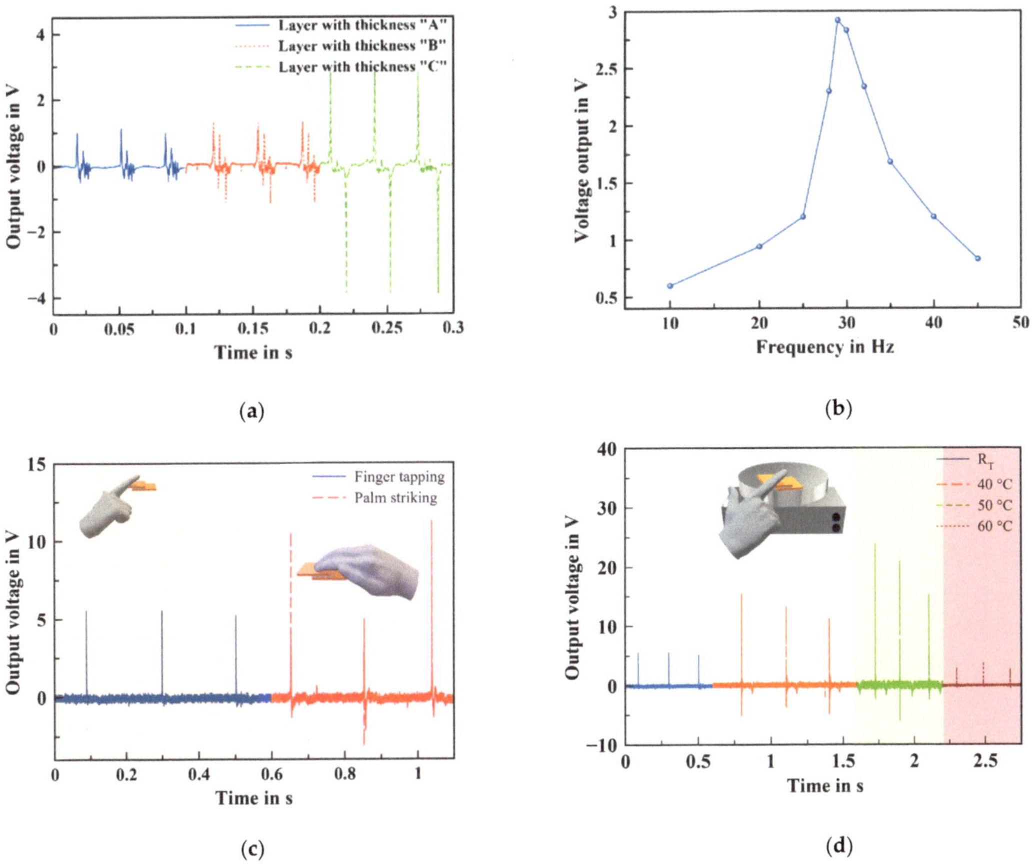

Figure 6. Nanogenerators performances under different conditions: (**a**) output voltage depending on the thickness, (**b**) test under different frequencies from 10 Hz to 45 Hz, (**c**) tapping and striking tests, and (**d**) test under finger tapping and different temperatures.

In order to validate the potential use of the NG for practical applications, the NG was tested under different scenarios, such as finger tapping and palm striking, as illustrated in Figure 6c. The nanogenerator demonstrates the ability to harvest biomechanical energy efficiently up to 5.56 V and 11.22 V under finger and palm striking, respectively.

One of the important things that needs to be addressed is the effect of the environmental changes on the performances of the nanogenerator. Herein, the effect of a temperature change on the NG performance was examined from room temperature up to 60 °C. The measurements were conducted by placing the nanogenerator on top of the hotplate for 10 min. Then, a compressive force was applied by finger tapping. Figure 6d shows the output voltages at different temperatures. It is noted that the output voltage increased until 50 °C from 5.56 V to 23.86 V. Then, the NG exhibited a decrease in the output voltage to reach 3.74 V. The increase can be explained by the enhanced electrical properties of the nanocomposite. In fact, MWCNTs exhibit a negative temperature coefficient (NTC) behavior, due to their semi-conductive behavior. By increasing the temperature, the electrons' mobility will be improved, which facilitates the electron tunneling effect, resulting in an increase in conductivity. Additionally, the mobility of polymer molecular chains will be increased, leading to expansion of the nanocomposite, which can boost the distribution of piezoceramics within the polymer matrix. In this way, the nanogenerator electromechanical coupling coefficient will be increased with the temperature. However, the decreasing at 60 °C may be due to the destruction of the conductive network caused by the thermal expansion of the polymer, which inhibited the electron transfer. In fact, the expansion of the polymer will increase the tunneling resistance between neighboring MWCNTs, leading to a sharp reduction of the conductivity, as well to creation of free spaces between ceramic particles.

4. Conclusions

In this work, eco-friendly and cost-effective flexible nanogenerators were developed using a solution mixing method followed by mold casting. The prepared nanogenerators are based on lead-free piezoelectric $BaTiO_3$ nanoparticles known by their high piezoelectric coefficient. The results show that the performance of nanogenerators containing only $BaTiO_3$ particles depends greatly on the concentration of $BaTiO_3$. In fact, increasing the $BaTiO_3$ concentration improves the output voltage and power, owing to the enhanced piezoelectric coefficient of the composite material. However, excessive $BaTiO_3$ leads to the reduction of the composite performance, caused by the lack of homogeneity and the sedimentation of the particles at the bottom.

To boost the performance of the $BaTiO_3$/PDMS composite, MWCNTs as conductive elements were incorporated within the composite.

Thereby, the $BaTiO_3$ particles were better and homogeneously dispersed in the PDMS matrix, and the sedimentation of the particles was avoided. Additionally, the addition of MWCNTs contributes to the improvement of the electrical and mechanical properties of the composite, especially at the percolation threshold. This later significantly enhanced the performance of the NG, which became two times higher compared to the NG without MWCNTs.

The study has also shown the importance of geometrical parameters such as the composite thickness and the test conditions on the output performance.

Increasing the composite thickness leads to enhancing the output voltage, as well as the applied frequency.

The optimized NG structure favors the scavenging of biomechanical energy without any additional poling process around 5.56 V under gentle finger tapping and 11.22 V under palm striking. Therefore, the realized NG offers great opportunities for achieving wearable energy harvesters for self-powered electronics.

This work demonstrates the importance of addressing the temperature aspect on the NG response. The results showed that the output voltage increased enormously by increasing the temperature. As a consequence, several aspects should be deeply investigated in the future in regard to the environmental effects.

Author Contributions: Conceptualization, methodology, validation, and formal analysis, A.B.; investigation, A.B., N.J., S.N. and B.B.; writing—original draft preparation, A.B.; writing—review and editing, A.B., N.J., S.N., B.B., O.H., B.W. and O.K.; and visualization, and supervision, O.K. All authors have read and agreed to the published version of the manuscript.

Funding: This research received no external funding.

Institutional Review Board Statement: Not applicable.

Informed Consent Statement: Not applicable.

Data Availability Statement: Not applicable.

Conflicts of Interest: The authors declare no conflict of interest.

References

1. Kanoun, O.; Bradai, S.; Khriji, S.; Bouattour, G.; El Houssaini, D.; Ben Ammar, M.; Naifar, S.; Bouhamed, A.; Derbel, F.; Viehweger, C. Energy-aware system design for autonomous wireless sensor nodes: A comprehensive review. *Sensors* **2021**, *21*, 548. [CrossRef] [PubMed]
2. Kanoun, O.; Khriji, S.; Naifar, S.; Bradai, S.; Bouattour, G.; Bouhamed, A.; El Houssaini, D.; Viehweger, C. Prospects of Wireless Energy-Aware Sensors for Smart Factories in the Industry 4.0 Era. *Electronics* **2021**, *10*, 2929. [CrossRef]
3. Kanoun, O.; Keutel, T.; Viehweger, C.; Zhao, X.; Bradai, S.; Naifar, S.; Trigona, C.; Kallel, B.; Chaour, I.; Bouattour, G.; et al. Next generation wireless energy aware sensors for internet of things: A review. In Proceedings of the 2018 15th International Multi-Conference on Systems, Signals & Devices (SSD), Hammamet, Tunisia, 19–22 March 2018.
4. Naifar, S.; Bradai, S.; Viehweger, C.; Kanoun, O.; Choura, S. Investigation of the magnetostrictive effect in a terfenol-D plate under a non-uniform magnetic field by atomic force microscopy. *Mater. Des.* **2016**, *97*, 147–154. [CrossRef]
5. Viehweger, C.; Keutel, T.; Kanoun, O. Energy harvesting for wireless sensor nodes in factory environments. In Proceedings of the 2014 IEEE 11th International Multi-Conference on Systems, Signals & Devices (SSD14), Barcelona, Spain, 11–14 February 2014.
6. Naifar, S.; Bradai, S.; Keutel, T.; Kanoun, O. Design of a vibration energy harvester by twin lateral magnetoelectric transducers. In Proceedings of the IEEE International Instrumentation and Measurement Technology Conference (I2MTC), Montevideo, Uruguay, 12–15 May 2014.
7. Zheng, Q.; Shi, B.; Li, Z.; Wang, Z.L. Recent progress on piezoelectric and triboelectric energy harvesters in biomedical systems. *Adv. Sci.* **2017**, *4*, 1700029. [CrossRef] [PubMed]
8. Pan, S.; Zhang, Z. Fundamental theories and basic principles of triboelectric effect: A review. *Friction* **2019**, *7*, 2–17. [CrossRef]
9. Godwinraj, D.; George, S.C. Recent advancement in TENG polymer structures and energy efficient charge control circuits. *Adv. Ind. Eng. Polym. Res.* **2021**, *4*, 1–8. [CrossRef]
10. Lee, E.J.; Kim, T.Y.; Kim, S.W.; Jeong, S.; Choi, Y.; Lee, S.Y. High-performance piezoelectric nanogenerators based on chemically-reinforced composites. *Energy Environ. Sci.* **2018**, *11*, 1425–1430. [CrossRef]
11. Fakhri, P.; Amini, B.; Bagherzadeh, R.; Kashfi, M.; Latifi, M.; Yavari, N.; Kani, S.A.; Kong, L. Flexible hybrid structure piezoelectric nanogenerator based on ZnO nanorod/PVDF nanofibers with improved output. *RSC Adv.* **2019**, *9*, 10117–10123. [CrossRef]
12. Park, K.I.; Lee, M.; Liu, Y.; Moon, S.; Hwang, G.T.; Zhu, G.; Kim, J.E.; Kim, S.O.; Kim, D.K.; Wang, Z.L.; et al. Flexible nanocomposite generator made of $BaTiO_3$ nanoparticles and graphitic carbons. *Adv. Mater.* **2012**, *24*, 2999–3004. [CrossRef]
13. Villafuerte-Castrejón, M.E.; Morán, E.; Reyes-Montero, A.; Vivar-Ocampo, R.; Peña-Jiménez, J.A.; Rea-López, S.O.; Pardo, L. Towards lead-free piezoceramics: Facing a synthesis challenge. *Materials* **2016**, *9*, 21. [CrossRef]
14. Bouhamed, A.; Binyu, Q.; Böhm, B.; Jöhrmann, N.; Behme, N.; Goedel, W.A.; Wunderle, B.; Hellwig, O.; Kanoun, O. A hybrid piezoelectric composite flexible film based on PVDF-HFP for boosting power generation. *Compos. Sci. Technol.* **2021**, *208*, 108769. [CrossRef]
15. Shin, S.H.; Kim, Y.H.; Lee, M.H.; Jung, J.Y.; Nah, J. Hemispherically aggregated $BaTiO_3$ nanoparticle composite thin film for high-performance flexible piezoelectric nanogenerator. *ACS Nano* **2014**, *8*, 2766–2773. [CrossRef] [PubMed]
16. Yan, J.; Jeong, Y.G. High performance flexible piezoelectric nanogenerators based on $BaTiO_3$ nanofibers in different alignment modes. *ACS Appl. Mater. Interfaces* **2016**, *8*, 15700–15709. [CrossRef] [PubMed]
17. Lin, Z.H.; Yang, Y.; Wu, J.M.; Liu, Y.; Zhang, F.; Wang, Z.L. $BaTiO_3$ nanotubes-based flexible and transparent nanogenerators. *J. Phys. Chem. Lett.* **2012**, *3*, 3599–3604. [CrossRef]
18. Luo, C.; Hu, S.; Xia, M.; Li, P.; Hu, J.; Li, G.; Jiang, H.; Zhang, W. A Flexible Lead-Free $BaTiO_3$/PDMS/C Composite Nanogenerator as a Piezoelectric Energy Harvester. *Energy Technol.* **2018**, *6*, 922–927. [CrossRef]
19. Yaqoob, U.; Chung, G.S. Effect of reduced graphene oxide on the energy harvesting performance of P (VDF-TrFE)-$BaTiO_3$ nanocomposite devices. *Smart Mater. Struct.* **2017**, *26*, 095060. [CrossRef]
20. Park, K.I.; Jeong, C.K.; Ryu, J.; Hwang, G.T.; Lee, K.J. Nanocomposites: Flexible and Large-Area Nanocomposite Generators Based on Lead Zirconate Titanate Particles and Carbon Nanotubes. *Adv. Energy Mater.* **2013**, *3*, 1530. [CrossRef]

21. Bouhamed, A.; Shi, Y.; Naifar, S.; Bautista-Quijano, J.R.; Kanoun, O. Carbon nanotubes for high performance flexible piezoelectric polymer composite nanogenerators. In Proceedings of the 2019 5th International Conference on Nanotechnology for Instrumentation and Measurement (NanofIM), Sfax, Tunisia, 29–30 October 2019.
22. Yan, J.; Jeong, Y.G. Roles of carbon nanotube and $BaTiO_3$ nanofiber in the electrical, dielectric and piezoelectric properties of flexible nanocomposite generators. *Compos Sci Technol.* **2017**, *144*, 1–10. [CrossRef]
23. Dong, S.; Wu, X.; Wang, E.; Wang, X. Reduced percolation threshold of multi-walled carbon nanotubes/polymer composites by filling aligned ferromagnetic particles. *J. Intell. Mater. Syst. Struct.* **2020**, *31*, 187–197. [CrossRef]
24. Kanoun, O.; Bouhamed, A.; Ramalingame, R.; Bautista-Quijano, J.R.; Rajendran, D.; Al-Hamry, A. Review on conductive polymer/CNTs nanocomposites based flexible and stretchable strain and pressure sensors. *J. Sens.* **2021**, *21*, 341. [CrossRef]
25. Bouhamed, A.; Müller, C.; Choura, S.; Kanoun, O. Processing and characterization of MWCNTs/epoxy nanocomposites thin films for strain sensing applications. *Sens. Actuator A Phys.* **2017**, *257*, 65–72. [CrossRef]
26. Kim, H.S.; Lee, D.W.; Kim, D.H.; Kong, D.S.; Choi, J.; Lee, M.; Murillo, G.; Jung, J.H. Dominant role of Young's modulus for electric power generation in PVDF–$BaTiO_3$ composite-based piezoelectric nanogenerator. *J. Nanomater.* **2018**, *8*, 777. [CrossRef]
27. Sun, H.; Tian, H.; Yang, Y.; Xie, D.; Zhang, Y.C.; Liu, X.; Ma, S.; Zhao, H.M.; Ren, T.L. A novel flexible nanogenerator made of ZnO nanoparticles and multiwall carbon nanotube. *Nanoscale* **2013**, *5*, 6117–6123. [CrossRef]
28. Han, R.; Liu, Y.; Shi, J.; Chen, G.X.; Li, Q. Enhanced dielectric properties and breakdown strength of polymer/carbon nanotube composites by coating an $SrTiO_3$ layer. *e-Polymers* **2021**, *21*, 272–278. [CrossRef]
29. Shen, Y.; Lin, Y.H.; Nan, C.W. Interfacial effect on dielectric properties of polymer nanocomposites filled with core/shell-structured particles. *Adv. Funct. Mater.* **2007**, *17*, 2405–2410. [CrossRef]
30. Banerjee, S.; Du, W.; Sundar, U.; Cook-Chennault, K.A. Piezoelectric and dielectric characterization of mwcnt-based nanocomposite flexible films. *J. Nanomater.* **2018**, *2018*, 6939621. [CrossRef]
31. Tagliaro, I.; Cobani, E.; Carignani, E.; Conzatti, L.; D'Arienzo, M.; Giannini, L.; Martini, F.; Nardelli, F.; Scotti, R.; Stagnaro, P.; et al. The self-assembly of sepiolite and silica fillers for advanced rubber materials: The role of collaborative filler network. *Appl. Clay Sci.* **2022**, *218*, 106383. [CrossRef]

sensors

MDPI

Article

Pyroelectric Nanogenerator Based on an SbSI–TiO_2 Nanocomposite

Krystian Mistewicz

Institute of Physics—Center for Science and Education, Silesian University of Technology, Krasińskiego 8, 40-019 Katowice, Poland; krystian.mistewicz@polsl.pl; Tel.: +48-32-603-41-56

Abstract: For the first time, a composite of ferroelectric antimony sulfoiodide (SbSI) nanowires and non-ferroelectric titanium dioxide (TiO_2) nanoparticles was applied as a pyroelectric nanogenerator. SbSI nanowires were fabricated under ultrasonic treatment. Sonochemical synthesis was performed in the presence of TiO_2 nanoparticles. The mean lateral dimension d_a = 68(2) nm and the length L_a = 2.52(7) μm of the SbSI nanowires were determined. TiO_2 nanoparticles served as binders in the synthesized nanocomposite, which allowed for the preparation of dense films via the simple drop-casting method. The SbSI–TiO_2 nanocomposite film was sandwiched between gold and indium tin oxide (ITO) electrodes. The Curie temperature of T_C = 294(2) K was evaluated and confirmed to be consistent with the data reported in the literature for ferroelectric SbSI. The SbSI–TiO_2 device was subjected to periodic thermal fluctuations. The measured pyroelectric signals were highly correlated with the temperature change waveforms. The magnitude of the pyroelectric current was found to be a linear function of the temperature change rate. The high value of the pyroelectric coefficient p = 264(7) nC/(cm^2·K) was determined for the SbSI–TiO_2 nanocomposite. When the rate of temperature change was equal dT/dt = 62.5 mK/s, the maximum and average surface power densities of the SbSI–TiO_2 nanogenerator reached 8.39(2) and 2.57(2) μW/m^2, respectively.

Keywords: antimony sulfoiodide; titanium dioxide; nanowires; pyroelectric effect; nanogenerator; renewable energy

Citation: Mistewicz, K. Pyroelectric Nanogenerator Based on an SbSI–TiO_2 Nanocomposite. *Sensors* **2022**, *22*, 69. https://doi.org/10.3390/s22010069

Academic Editors: Slim Naifar, Olfa Kanoun and Carlo Trigona

Received: 4 November 2021
Accepted: 21 December 2021
Published: 23 December 2021

Publisher's Note: MDPI stays neutral with regard to jurisdictional claims in published maps and institutional affiliations.

1. Introduction

Efficiency in thermal energy harvesting is an important challenge to produce green energy for sustainable development. Most of the waste heat generated in industry is available at low temperatures, generally below 373–503 K [1]. Low-temperature waste heat is especially difficult to recover successfully using currently available technologies [2]. Thermal energy can be converted into electric energy by applying electrochemical systems [3], thermogalvanic cells [4,5], thermoelectric [6–8], thermomagnetic [9,10], and pyroelectric generators [11–13]. However, all mentioned technologies suffer from low efficiency and limited reliability. Therefore, new materials and engineering concepts must be proposed and developed in the field of low-temperature waste heat recovery to ensure their future large-scale and commercial application.

Inorganic ferroelectric crystals or ceramics [14,15], ferroelectric polymers [16], and non-ferroelectric compounds [17] are three main groups of materials commonly used to construct pyroelectric generators. Material morphology is a key factor affecting the pyroelectric performance of the device. The efficiency of thermal energy conversion to electrical power can be enhanced through quantum confinement in pyroelectric nanomaterials [18]. Thus, there is growing interest in the development and examination of pyroelectric nanogenerators based on thin films [19], nanowires [20,21], or nanoparticles [22].

Antimony sulfoiodide (SbSI) is a ternary compound that possesses ferroelectric properties with a Curie temperature of 295 K [23]. The first report on the pyroelectric properties of SbSI single crystals was provided by Imai et al. [24]. The large pyroelectric coefficient (1.2 μC/(cm^2·K)) of bulk single crystals of SbSI was measured over a temperature range encompassing the ferroelectric phase transition [25]. The significantly much lower pyroelectric coefficients of 8.06 $pC/(cm^2{\cdot}K)$ and 180 $nC/(cm^2{\cdot}K)$ were determined for SbSI films grown by flash evaporation [26] and physical vapor deposition [27], respectively. However, the pyroelectric properties of SbSI nanowires have not been investigated so far. Recently, nanowires of SbSI and other compounds that belong to the chalcohalide family of materials have received a great amount of attention due to the fact of their possible applications in piezoelectric nanogenerators for mechanical energy harvesting [28,29], underwater ultrasounds sensors [30], photovoltaic devices [31,32], piezo- and photocatalysis [33,34].

Many different techniques have been proposed till now to fabricate antimony sulfoiodide films including electron beam evaporation [35], flash evaporation [36,37], physical vapor deposition [27], pulsed laser deposition [38,39], molecular beam epitaxy [40], solution processing [41], and combined solution and vapor processes [42,43]. The majority of the aforementioned techniques possess serious drawbacks such as complexity or the need to use high-temperature treatments. A typical way to fabricate continuous films of SbSI nanowires is based on the application of a binding polymer matrix, e.g., polyacrylonitrile (PAN) [32]. As a result, the electrical properties of such a prepared composite can be worse in comparison to the electrical properties of pristine SbSI nanowires. Ye and co-workers [44] presented sol-gel fabrication of thin films and bulk solids of SbSI micro-crystallite-doped organically modified TiO_2. The optical absorption spectrum of this material exhibited an evident quantum confinement effect [44], whereas the electrical properties of the material were not examined. It should be noted that the application of TiO_2 nanoparticles in composites for pyroelectric nanogenerators has been reported in the literature as a factor improving pyroelectric performance [45].

In this paper, a facile fabrication of the SbSI–TiO_2 nanocomposite in one step is presented for the first time. TiO_2 nanoparticles were used as binders in the synthesized nanocomposite. The dense film of the SbSI–TiO_2 nanocomposite was prepared using a simple drop-casting method and sandwiched between the electrodes. Such a device was applied as a pyroelectric nanogenerator.

2. Materials and Methods

2.1. Material Synthesis and Nanogenerator Fabrication

Figure 1 presents the preparation of the pyroelectric nanogenerator. In the first stage, the chemical elements (i.e., antimony, sulfur, and iodine) were weighted in a stoichiometric ratio. The reagents (i.e., 868 mg of Sb, 227 mg of S, and 921 mg of I_2) were mixed with titanium dioxide (TiO_2) nanoparticles purchased from Merck KGaA. According to the data provided by the material manufacturer, the average size of the TiO_2 nanoparticles was 21 nm. The mass of TiO_2 (860 mg) was adjusted to constitute approximately 30% of the total mass of the nanocomposite. The mixture was added to the beaker containing 10 mL of pure ethanol (Figure 1a). Such a prepared suspension of the reagents was subjected to ultrasonic irradiation, generated by the VCX-750 ultrasonic reactor (Sonics & Materials, Inc., Newtown, CT, USA) as shown in Figure 1b. The material synthesis was carried out at a temperature of 323 K over 4 h. After the synthesis was completed, the orange gel of the SbSI–TiO_2 nanocomposite was drop-casted onto an indium tin oxide (ITO) coated polyethylene terephthalate (PET) substrate as depicted in the Figure 1c. The drop-casting deposition of the material was performed in several steps using an electronic pipette. The sample was then placed on a hot plate and dried at a temperature of 333 K (Figure 1d). When the sample drying was finished, the 200 nm thick gold electrode was sputtered on the top side of the SbSI–TiO_2 nanocomposite (Figure 1e) by applying a Q150R ES rotary pumped coater (Quorum Technologies Ltd., Laughton, United Kingdom). Finally, the prepared SbSI–TiO_2 nanogenerator was connected to the external measurement circuit (Figure 1f).

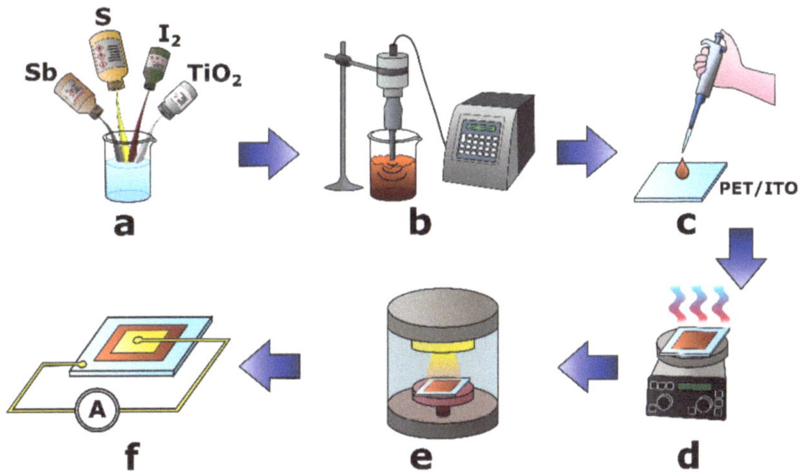

Figure 1. Schematic fabrication process flow of the SbSI–TiO_2 pyroelectric nanogenerator. A detailed description is given in the text.

2.2. Structural and Chemical Characterization of the SbSI–TiO_2 Nanocomposite

The morphology and chemical composition of the SbSI–TiO_2 nanocomposite were investigated using scanning electron microscopy (SEM) and energy-dispersive X-ray spectroscopy (EDS), respectively. The experiments were performed using a Phenom Pro X microscope (Thermo Fisher Scientific, Waltham, MA, USA) integrated with an EDS spectrometer. The SEM studies were carried out at the acceleration voltage of 10 kV, whereas the EDS survey was accomplished at 15 kV. The EDS data were analyzed using the ProSuite Element Identification computer program (Thermo Fisher Scientific).

2.3. Electrical Measurements

The SbSI–TiO_2 nanocomposite was inserted into the SH-242 environmental test chamber (Espec) to examine its electrical properties. All experiments were carried out at a constant relative humidity (*RH*) of 50%. The sample was tested under dark conditions to eliminate the influence of excess carries photogeneration on the measured electric signals. The change in temperature was achieved via air flow inside the chamber. The measurements were performed at a standard atmospheric pressure (1 atm). It was confirmed for the SbSI nanowires that their electrical properties are sensitive to the various gases including oxygen and hydrogen [46]. However, the air circulation inside the chamber was closed. Thus, the gas composition of the air remained the same. As a result, the effect of the gas on the electrical properties of the sample was eliminated. Before pyroelectric investigations, the SbSI–TiO_2 nanocomposite was poled to align the electric dipoles in the SbSI nanowires and enhance the pyroelectric output of the nanogenerator. It was conducted by cooling the device below the Curie temperature and applying an external electric field of 12 kV/m, which is higher than the coercive field of SbSI [23]. The pyroelectric response of the SbSI–TiO_2 nanocomposite was registered for different rates of temperature change (dT/dt) in the range from 1.8 to 62.5 mK/s. The low values of dT/dt were chosen to avoid possible interference caused by the thermoelectric effect [47]. The electric output of the SbSI–TiO_2 nanocomposite was measured with the 6430 Sub-Femtoamp Remote SourceMeter (Tektronix, Beaverton, OR, USA). Data acquisition was performed using a PC computer with a GPIB bus and an appropriate program in LabView (National Instruments, Austin, TX, USA).

3. Results and Discussion

Figure 2a presents the prepared pyroelectric nanogenerator based on the SbSI–TiO_2 nanocomposite sandwiched between the Au electrode and ITO-coated PET. A typical SEM micrograph of the SbSI–TiO_2 nanocomposite is shown in Figure 1b. This material consisted of randomly distributed one-dimensional nanocrystals of SbSI and agglomerations of TiO_2 nanoparticles. The concentrations of the chemical elements were averaged over the sample. The EDS survey proved that titanium dioxide constituted 29.9% of the total mass of the nanocomposite, which is in good agreement with the initial amount of TiO_2 (30%) used for material preparation. TiO_2 nanoparticles served as fillers in the synthesized nanocomposite, which allowed for preparing dense films via a facile drop-casting method. Many attempts have been made to fabricate films containing only SbSI nanowires. However, they were not successful. The presence of voids in the layers of pure SbSI nanowires (without TiO_2 nanoparticles) resulted in the formation of the short-circuited samples after sputtering of the gold electrodes. Another way to fabricate a continuous film of SbSI nanowires is by applying a polymer as a binder [32]. Such a method possesses a significant drawback. It was found in [32] that the increase in polymer concentration in the SbSI–PAN nanocomposite led to an undesirable reduction in its electrical conductivity and photovoltaic performance.

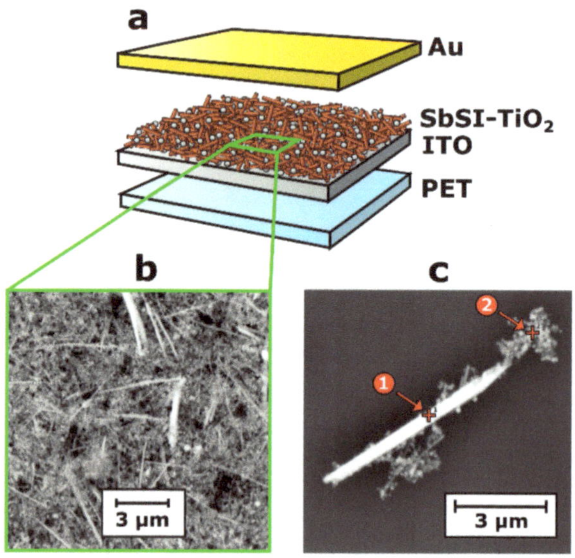

Figure 2. The layer structure of the pyroelectric nanogenerator (**a**), SEM micrograph of the SbSI–TiO_2 film (**b**), and a selected nanowire (**c**).

The elemental atomic ratio of 0.37:0.36:0.27 for Sb, S, and I averaged over nanocomposite volume was close to the stoichiometric composition of SbSI (Table 1). The theoretical values should be 0.33 for each of the elements in this ternary compound. No other elements were detected in the sample. The small amount of material was deposited on the Si substrate. An individual SbSI nanowire was separated from the nanocomposite to examine its chemical composition (Figure 2c). The EDS signal for Si was subtracted from the atomic concentrations evaluated for other components. The elemental atomic ratio registered for spot 1 in Figure 2c corresponded to the chemical composition expected for antimony sulfoiodide (Table 1). The values of atomic concentrations determined for spot 2 in Figure 2c indicate the presence of agglomerated TiO_2 nanoparticles and slight traces of antimony. It should be noted that an excess amount of Sb is frequently reported in the literature for nanowires [32,34], nanocrystals [33], and thin films [27] of SbSI. This effect can be attributed to the fact that the surface of crystalline SbSI nanowires is surrounded by fuzzy shells, which have a chemical composition that may be different from the concentrations of elements in the core of the nanowire [32]. Atomic concentrations of titanium and oxygen are in agreement with the theoretical values for stoichiometric TiO_2 (Table 1). Since titanium dioxide exhibits excellent chemical stability [48,49], it should be resistant to ultrasonic treatment.

Table 1. The chemical composition determined for the areas of the sample shown in Figure 2b,c.

Chemical Element	Atomic Concentration, %		
	Area in Figure 2b	Spot 1 in Figure 2c	Spot 2 in Figure 2c
Sb	15.6	36.8	13.8
S	14.9	30.4	0
I	11.1	32.8	0
Ti	18.5	0	28.8
O	39.9	0	57.4

The prepared material was characterized with scanning electron microscopy to determine the average diameter and length of the SbSI nanowires as presented in Figure 3. Manual image analysis was performed on 250 randomly selected nanowires. The diameters and lengths vary in the wide ranges, reaching up to 220 nm and 8.2 μm, respectively. One can see that the distribution of nanowires sizes followed a log-normal function [50,51]:

$$f(x) = \frac{A}{\sqrt{2\pi}\cdot\sigma\cdot x}\exp\left[-\frac{\left[\ln\left(\frac{x}{x_a}\right)\right]^2}{2\sigma^2}\right], \quad (1)$$

where A is a constant, x means the size of the nanowires (diameter or length), x_a denotes the average value of the nanowire size, and σ is its standard deviation. It should be underlined that the log-normal function is frequently reported in the literature as a relation describing size distributions of nanowires [50–52] and nanoparticles [53–55]. The mean diameter d_a = 68(2) nm and the length L_a = 2.52(7) μm of the SbSI nanowires were determined. The obtained value of d_a is equal within the measurement uncertainty to the average diameter (69(3) nm) of the SbSI nanowires reported in [32].

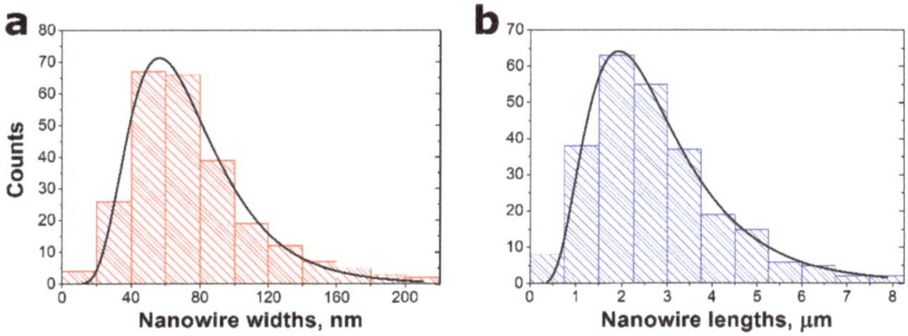

Figure 3. Distributions widths (**a**) and lengths (**b**) of SbSI nanowires in the SbSI–TiO_2 nanocomposite. The black lines represent log-normal functions expected for the nanoscale object size distribution described by Equation (1). The values of the fitted parameters of Equation (1) are given in the text.

The influence of temperature on the electric current flowing through the SbSI–TiO_2 nanocomposite under constant bias voltage is shown in Figure 4a. The rise in temperature leads to a sharp enhancement of the electric current. It is a typical behavior for a semiconducting material. Moreover, when the temperature increases, the reduction in the grain boundary resistance can result in the reduction in the barrier for the mobility of charge carriers participating in grain boundary conduction [56]. Taking into account the geometrical dimensions of the investigated sample, the electric conductance of the SbSI–TiO_2 nanocomposite was calculated (Figure 4b). It is well known that in the case of ferroelectric material, the Arrhenius plot of an electric conductance consists of two linear curves with different slopes corresponding to the paraelectric and ferroelectric phases [56,57]. Thus, a change in activation energy is observed near Curie temperature (T_C) [58–60]. This effect was also documented in the literature for thin films of SbSI [35,61]. Two different theoretical curves were least squares fitted to the experimental results, shown in Figure 4b, using a well-known relation [56,57]:

$$\sigma(T) = \sigma_0 \cdot \exp\left(-\frac{E_A}{k_B T}\right), \tag{2}$$

where σ_0 is the pre-exponential coefficient, E_A means an activation energy, and k_B denotes the Boltzmann constant. The values of activation energies of E_A = 0.537(1) eV and E_A = 0.271(2) eV were determined for the paraelectric and ferroelectric phases, respectively. They could not be compared with the literature data for SbSI due to the fact that these parameters are thickness dependent [61]. Finding the intersection of the linear dependencies in the Arrhenius plot of an electric conductance (Figure 4b) allowed for the estimation of the Curie temperature of the examined material (T_C = 294(2) K). The obtained value was consistent with the data from the literature for ferroelectric SbSI listed in Table 2.

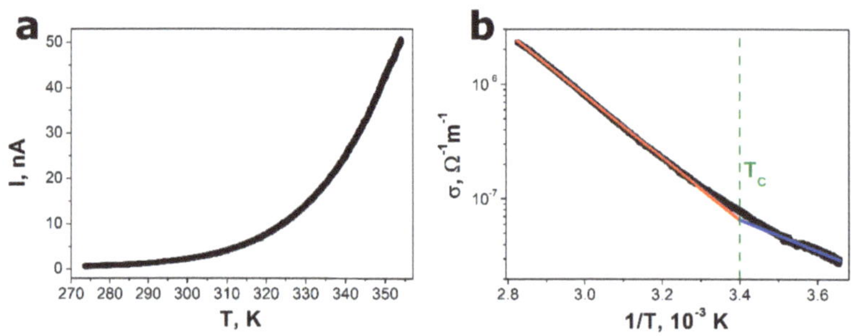

Figure 4. The temperature dependence of the electric current registered under a constant bias voltage of 0.1 V (**a**) and the Arrhenius plot electrical conductance of the SbSI–TiO_2 nanocomposite (**b**). The solid red and blue lines in (**b**) represent the best fitted theoretical dependence (2) in the paraelectric and ferroelectric phases, respectively. The green, dashed line indicates the reciprocal value of the determined Curie temperature. The values of the fitted parameters of Equation (2) are given in the text.

Table 2. A comparison of the determined Curie temperature of the SbSI–TiO_2 nanocomposite with literature data for SbSI.

Material	Preparation Method	T_C, K	Reference
bulk crystal of SbSI		295	[23]
bulk crystal of SbSI	vapor phase growth	292–293	[24]
bulk crystal of SbSI	Bridgman method and vapor transport technique	293	[25]
SbSI film	physical vapor deposition	295	[27]
SbSI film	electron beam evaporation	294	[35]
SbSI film	flash evaporation	293	[37]
SbSI film	pulsed laser deposition	292	[38]
SbSI film	pulsed laser deposition	290–294	[39]
SbSI nanowires	sonochemical synthesis	291(2)	[62]
SbSI–TiO_2 nanocomposite	sonochemical synthesis	294(2)	this paper

The sample of the SbSI–TiO_2 nanocomposite was subjected to periodic temperature fluctuations as presented in Figure 5a,c. The experiments were performed for different amplitudes of temperature changes. A rectangular-shaped transient characteristic of pyroelectric current was observed (Figure 5b,d) when the triangular temperature waveform was applied to the SbSI–TiO_2 nanogenerator. Such behavior is well documented in the literature for other pyroelectric materials, e.g., PVDF–TiO_2 composite [45], PVDF–ZnO nanocomposite [63], $Bi_{0.5}Na_{0.5}TiO_3$–P(VDF–TrFE) nanocomposite [64], P(VDF–TrFE) [65], $PbTiO_3$–P(VDF–TrFE) [66], $LiNbO_3$–polypropylene [67], CeO_2-doped $Na_{0.5}Bi_{0.5}TiO_3$ ceramics [68], $BaZr_{0.2}Ti_{0.8}O_3/Ba_{0.7}Ca_{0.3}TiO_3$ heterostructures [69], and CdS nanorods [47]. This proves that the measured electric response of the SbSI–TiO_2 nanocomposite was due to the true pyroelectric effect. It should be noted that only SbSI nanowires contributed to the pyroelectric response of the nanocomposite. The non-ferroelectric inclusion of TiO_2 cannot be polarized, since its net dipole moment is almost zero [45]. A slight decrease in the amplitude of the pyroelectric response, shown in Figure 5d, can be attributed to the so-called "aging effect" [70].

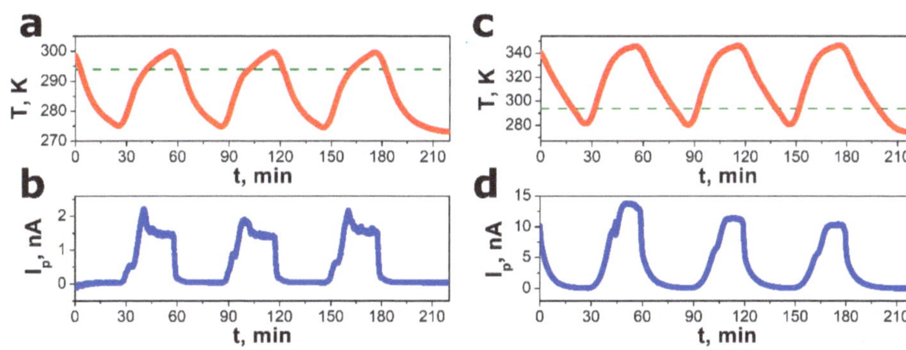

Figure 5. Cyclic changes in temperature (**a**,**c**) and corresponding output current (**b**,**d**) of the SbSI–TiO_2 pyroelectric nanogenerator. The green horizontal line represents the value of the Curie temperature T_C = 294(2) K.

The only positive pyroelectric response of the SbSI–TiO_2 nanogenerator is observed (Figure 5b,d) due to the fact that the rectifying *p–n* heterojunction was formed between the SbSI nanowires and the ITO electrode as well between the SbSI nanowires and the TiO_2 nanoparticles. Usually, SbSI is regarded as *p*-type semiconductor [31,37,61,71,72], whereas both ITO [73] and TiO_2 [74] exhibit *n*-type electrical conductivity. The work functions of ITO, TiO_2, and SbSI were 4.4–4.5 eV [75], 4.2–4.7 eV [76], and 5 eV [31,71,72], respectively. Accordingly, since the work function of *p*-type SbSI is higher than that of n-type ITO and *n*-type TiO_2, a built-in electric field was formed in the ITO/SbSI and TiO_2/SbSI interfaces. When the temperature increased ($dT/dt > 0$), a positive pyroelectric potential was formed. Thus, the ITO/SbSI and TiO_2/SbSI interfaces were forward-biased pn junctions. In such a case, the electrons could flow freely across the interfaces, and the pyroelectric current was measured in the external circuit. When the device cooled down ($dT/dt < 0$), the ITO/SbSI and TiO_2/SbSI interfaces were reversely biased barriers, leading to almost zero electric output of the nanogenerator. This is the process of creating, separating, preserving, and accumulating charges [77]. As a result, the device generated a rectified electric output. The same effect is observed in the case of direct-current generators based on *p–n* [78–82] or Schottky [77,83–85] junctions. Morozovska et al. [20] proposed that the rectification effect of the junction barrier allows the application of a ferroelectric nanowire array fixed between flat electrodes as the direct current generator. Moreover, one should remember that the built-in electrical field of a *p–n* junction barrier can create a polar axis in the solid [86]. The pyroelectric effect can be enhanced, as the electric dipole moment, due to the charge separation at the junction, depending on the temperature mainly through the temperature dependence of the dielectric constant [86].

The magnitude of the pyroelectric current depends on the rate of temperature change (dT/dt). It is described by the following equation [11–13]:

$$I_p = \frac{dQ}{dt} = pA\frac{dT}{dt}, \tag{3}$$

where Q represents a pyroelectric charge, p denotes the pyroelectric coefficient, and A means the electrode area. The pyroelectric current density ($J_p = I_p/A$) is presented as a function of the rate of temperature change in Figure 6a. One can see that the experimental results followed a linear dependence. According to Equation (3), the slope of J_p versus dT/dt was equal to the pyroelectric coefficient. The value of the pyroelectric coefficient of p = 264(7) nC/(cm^2·K) was determined for the SbSI–TiO_2 nanocomposite by fitting the theoretical Formula (3) to the experimental data in the graph in Figure 6a.

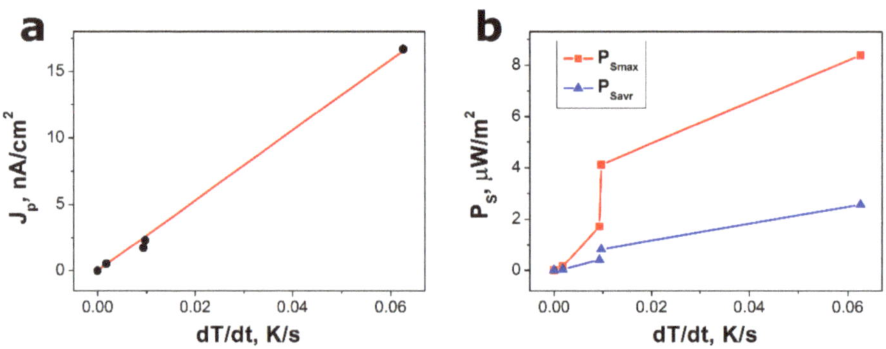

Figure 6. The output pyroelectric current density (**a**) and the surface power density (**b**) of the SbSI–TiO_2 pyroelectric nanogenerator as a function of the temperature change rate. The red line in Figure (**a**) represents the best fit dependence (3). The red and blue experimental points in Figure (**b**) show the values of the maximum and average surface power density, respectively.

The maximum power density (P_{Smax}) of the SbSI–TiO_2 nanogenerator was calculated using the peak values of the output pyroelectric current ($I_{p\ max}$), voltage (U_{max}), and electrode area of the device (A):

$$P_{Smax} = \frac{U_{max} \cdot I_{p\ max}}{A}. \tag{4}$$

One can see that the maximum surface power density is proportional to the maximum value of the pyroelectric current density. The average values of the surface power density (P_{Savr}) were determined using the following equation:

$$P_{Savr} = \frac{P_{avr}}{A} = \frac{1}{A \cdot \Delta t} \cdot \int_0^{\Delta t} U \cdot I_p dt, \tag{5}$$

where Δt represents the period time of a heating–cooling cycle. Figure 6b shows that the increase in the temperature change rate resulted in the obvious rise of P_{Smax} and P_{Savr}. The pyroelectric coefficient, the maximum and average power densities, calculated for the SbSI–TiO_2 nanocomposite, were compared with the values of these parameters reported in the literature for other pyroelectric materials (Table 3). They were divided into five main groups, i.e., non-ferroelectric materials, inorganic ferroelectric bulk crystals or ceramics, pure ferroelectric polymers, ferroelectric thin films or nanomaterials, and ferroelectric composites. It should be underlined that the SbSI–TiO_2 nanocomposite had the highest values for the pyroelectric coefficient and surface power density among all the ferroelectric composites as listed in Table 3. Furthermore, the pyroelectric performance of the SbSI–TiO_2 nanocomposite was much better than that achieved for the thin films of SbSI [26,27] and nanowires of antimony selenoiodide (SbSeI) [87], which is an isostructural material for SbSI [88]. It is expected that the incomplete coverage of the TiO_2 on the SbSI nanorods should lead to the formation of pores between these two different materials. A matrix–void composite with low permittivity is desired for pyroelectric applications, exhibiting a high figure of merit [89,90]. Therefore, the high pyroelectric performance of the SbSI–TiO_2 nanocomposite can be attributed to the good pyroelectric properties of SbSI as well as to the incomplete coverage of the TiO_2 on the SbSI nanorods.

Table 3. Comparison of the pyroelectric performance achieved for the SbSI–TiO_2 nanocomposite and other materials (BCs—bulk ceramics; BNT—$Bi_{0.5}Na_{0.5}TiO_3$; NPs—nanoparticles; NRs—nanorods; NWs—nanowires; *p*—pyroelectric coefficient; P_S—surface power density; PVC—poly(vinyl chloride); PVDF polyvinylidene difluoride; P(VDF–TrFE)—poly(vinylidenefluoride-co-trifluoroethylene); PZT—lead zirconate titanate; SC—single crystal; TFs—thin films). The abbreviations "max" and "avr" refer to the maximum and average surface power densities, respectively.

Group of Materials	Material	p, nC/(cm^2·K)	P_S, μW/m^2	Reference
non-ferroelectric materials	ZnO NWs	1.5		[91]
	ZnO TFs	1.0–1.4		[92]
	CdS NRs	470		[47]
inorganic ferroelectric bulk crystals or ceramics	PZT BCs	53.3	3700 ^{max}	[93]
	PZT BCs	20	13.6 ^{avr}	[94]
	$BaTiO_3$ BCs	10	2240 ^{max}	[95]
	$BaTiO_3$ BCs	16		[96]
	$LiNbO_3$ SC	5–8	219 ^{max}	[97]
	SbSI SC	1200		[25]
pure ferroelectric polymers	PVDF	1.94		[45]
	PVDF	4	108 ^{max}	[98]
	PVDF		0.13 ^{max}	[99]
	P(VDF–TrFE)	2.4		[65]
	P(VDF–TrFE)	4.39	128 ^{max}	[100]
ferroelectric thin films or nanomaterials	$Ba_{0.8}Sr_{0.2}TiO_3$ TFs	25		[101]
	$KNbO_3$ NWs	0.8		[102]
	SbSI TFs	0.008		[26]
	SbSI TFs	180		[27]
	SbSeI NWs	44(5)	0.59(4) ^{max}	[87]
ferroelectric composites	$BaTiO_3$–PVC	10.6		[103]
	PVDF–diamond NPs	8.7		[104]
	PVDF–TiO_2	2.45		[45]
	PVDF–ZnO NPs	~2.9		[63]
	PVDF–$CH_3NH_3PbI_3$	0.004	1.75 ^{max}	[105]
	P(VDF–TrFE)–BNT NPs	5		[64]
	P(VDF–TrFE)–$PbTiO_3$ NPs	4		[66]
	SbSI NWs–TiO_2 NPs	264(7)	8.39(2) ^{max} 2.57(2) ^{avr}	this paper

Electrical energy (ΔE_V) generated by a unit of the volume (V) of the SbSI–TiO_2 nanocomposite during the time interval (t) was evaluated applying a formula given below:

$$\Delta E_V = \frac{1}{V} \cdot \int_0^t U \cdot I_p \mathrm{d}t. \quad (6)$$

Figure 7 presents a time dependence of the volume energy density generated by the SbSI–TiO_2 device during three periodic thermal input cycles. An increase in the rate of temperature change leads to an enhancement of the ΔE_V. When the temperature difference (ΔT) was equal to 25 K, the electrical energy produced per unit volume of the

SbSI–TiO_2 nanocomposite per cycle amounted to 15.2 μJ/cm^3. This value was lower than ΔE_V = 230 μJ/cm^3 achieved for the PVDF film per single thermal cycle at the temperature change ΔT = 23 K [93].

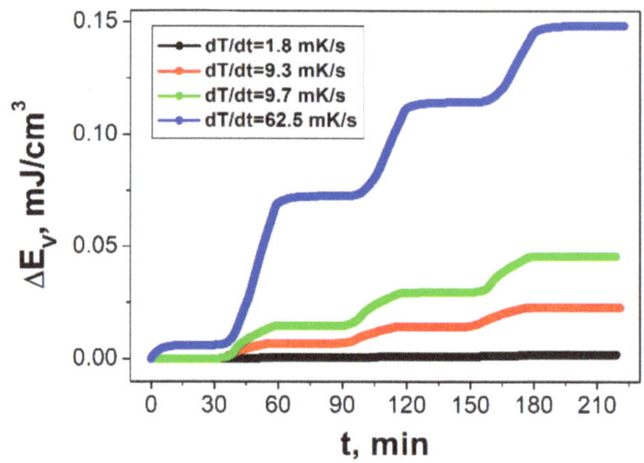

Figure 7. Volume energy density generated by the SbSI–TiO_2 device during three periodic thermal input cycles measured for different rates of temperature change (• 1.8 mK/s; • 9.3 mK/s; • 9.7 mK/s; • 62.5 mK/s).

The probable impact of the size of the TiO_2 nanoparticles on the pyroelectric efficiency of the SbSI–TiO_2 nanocomposite can be explained as follows. A decrease in a TiO_2 nanoparticle's size leads to a reduction in the electrical conductivity of this nanomaterial [106]. Wei and co-workers [107] presented that lower conductivity of the ceramic inclusion in the ferroelectric composite results in higher polarization, which is favorable to the poling of the composite. Therefore, the enhancement of the pyroelectric performance of the SbSI–TiO_2 nanocomposite should be observed with the decrease of the TiO_2 nanoparticle's size.

4. Conclusions

For the first time, a method of fabrication of the SbSI–TiO_2 pyroelectric nanogenerator was described. It involved the sonochemical preparation of the SbSI–TiO_2 nanocomposite in a single step and its drop-casting deposition on ITO-coated PET. TiO_2 nanoparticles served as fillers in the synthesized nanocomposite, which were crucial to obtain dense films with the desired morphology. It is a new alternative approach to fabricate devices based on nanowires. This method eliminates the need for application of a polymer matrix as a binder. The proposed technology is simple and universal. It can be used for preparation of nanogenerators from other one-dimensional chalcohalide nanostructures.

The SEM and EDS studies of the SbSI–TiO_2 nanocomposite confirmed the formation of stoichiometric SbSI nanowires with an average diameter and length of 68(2) nm and 2.52(7) μm, respectively. The Curie temperature of 294(2) K was determined for the examined material, which is in good agreement with the data reported in the literature for ferroelectric bulk crystals, thin films, and pristine nanowires of SbSI.

The SbSI–TiO_2 nanocomposite sandwiched between the ITO and Au electrodes was subjected to periodic temperature fluctuations. The registered pyroelectric response was well correlated to the temperature changes. The measured signal originated from the pyroelectric properties of SbSI nanowires. Non-ferroelectric TiO_2 nanoparticles were unable to contribute to the electric output of the nanogenerator, since they cannot be polarized. The pyroelectric current density was found to be a linear function of the rate of temperature change. The large pyroelectric coefficient of 264(7) nC/(cm^2·K) was determined for the SbSI–TiO_2 nanocomposite. This value was higher than the pyroelectric coefficient reported in the literature for many different pyroelectric materials including ZnO nanowires, PZT bulk ceramics, $BaTiO_3$ bulk ceramics, $LiNbO_3$ single crystal, PVDF polymers, SbSI thin films, and nanowires of SbSeI, which is an isostructural material to the SbSI. The maximum and average surface power densities of the SbSI–TiO_2 nanogenerator reached 8.39(2) and 2.57(2) $\mu W/m^2$, respectively. The electrical energy produced per unit volume of 15.2 $\mu J/cm^3$ was obtained per one thermal cycle for which the temperature change was equal to 25 K. It is expected that the values of the aforementioned parameters can be enhanced by adjusting the weight concentration of the titanium dioxide in the nanocomposite. Such investigations will be performed in the near future. The results, presented in this paper, demonstrate that the ferroelectric SbSI–TiO_2 nanocomposite has great potential for use in pyroelectric sensors and thermal energy harvesters.

Funding: This paper was partially supported by the Silesian University of Technology (Gliwice, Poland) through the Rector's pro-quality grant No. 14/010/RGJ21/0006 and Rector's habilitation grant No. 14/010/RGH21/0008 in the area of scientific research and development.

Institutional Review Board Statement: Not applicable.

Informed Consent Statement: Not applicable.

Conflicts of Interest: The author declares no conflict of interest. The funders had no role in the design of the study; in the collection, analyses, or interpretation of data; in the writing of the manuscript or in the decision to publish the results.

References

1. Pandya, S.; Velarde, G.; Zhang, L.; Wilbur, J.D.; Smith, A.; Hanrahan, B.; Dames, D.; Martin, L.W. New approach to waste-heat energy harvesting: Pyroelectric energy conversion. *NPG Asia Mater.* **2019**, *11*, 26. [CrossRef]
2. Xu, Z.Y.; Wang, R.Z.; Yang, C. Perspectives for low-temperature waste heat recovery. *Energy* **2019**, *176*, 1037–1043. [CrossRef]
3. Rahimi, M.; Straub, A.P.; Zhang, F.; Zhu, X.; Elimelech, M.; Gorski, C.A.; Logan, B.E. Emerging electrochemical and membrane-based systems to convert low-grade heat to electricity. *Energy Environ. Sci.* **2018**, *11*, 276–285. [CrossRef]
4. Duan, J.; Feng, G.; Yu, B.; Li, J.; Chen, M.; Yang, P.; Feng, J.; Liu, K.; Zhou, J. Aqueous thermogalvanic cells with a high Seebeck coefficient for low-grade heat harvest. *Nat. Commun.* **2018**, *9*, 5146. [CrossRef]
5. Owusu, M.; Mensah-Darkwa, K.; Andrews, A.; Davis, F.; Phelan, P. Effect of transient low-grade solar heat on liquid thermogalvanic cells. *Mater. Today Proc.* **2021**, *38*, 767–772. [CrossRef]
6. Jaziri, N.; Boughamoura, A.; Müller, J.; Mezghani, B.; Tounsi, F.; Ismail, M. A comprehensive review of Thermoelectric Generators: Technologies and common applications. *Energy Rep.* **2020**, *6*, 264–287. [CrossRef]
7. Zulkepli, N.; Yunas, J.; Ambri Mohamed, M.; Hamzah, A.A. Review of Thermoelectric Generators at Low Operating Temperatures: Working Principles and Materials. *Micromachines* **2021**, *12*, 734. [CrossRef]
8. Zoui, M.A.; Bentouba, S.; Stocholm, J.G.; Bourouis, M. A Review on Thermoelectric Generators: Progress and Applications. *Energies* **2020**, *13*, 3606. [CrossRef]
9. Kishore, R.A.; Priya, S. A Review on Low-Grade Thermal Energy Harvesting: Materials, Methods and Devices. *Materials* **2018**, *11*, 1433. [CrossRef]
10. Kishore, R.A.; Davis, B.; Greathouse, J.; Hannon, A.; Kennedy, D.E.; Millar, A.; Mittel, D.; Nozariasbmarz, A.; Kang, M.G.; Kang, H.B.; et al. Energy scavenging from ultra-low temperature gradients. *Energy Environ. Sci.* **2019**, *12*, 1008–1018. [CrossRef]
11. Ryu, H.; Kim, S.-W. Emerging Pyroelectric Nanogenerators to Convert Thermal Energy into Electrical Energy. *Small* **2019**, *17*, 1903469. [CrossRef]
12. Bowen, C.R.; Taylor, J.; LeBoulbar, E.; Zabek, D.; Chauhan, A.; Vaish, R. Pyroelectric materials and devices for energy harvesting applications. *Energy Environ. Sci.* **2014**, *7*, 3836–3856. [CrossRef]
13. Lingam, D.; Parikh, A.R.; Huang, J.; Jain, A.; Minary-Jolandan, M. Nano/microscale pyroelectric energy harvesting: Challenges and opportunities. *Int. J. Smart Nano Mater.* **2013**, *4*, 229–245. [CrossRef]
14. Lee, F.Y.; Jo, H.R.; Lynch, C.S.; Pilon, L. Pyroelectric energy conversion using PLZT ceramics and the ferroelectric–ergodic relaxor phase transition. *Smart Mater. Struct.* **2013**, *22*, 025038. [CrossRef]
15. Zhang, Y.; Xie, M.; Roscow, J.; Bao, Y.; Zhou, K.; Zhang, D.; Bowen, C.R. Enhanced pyroelectric and piezoelectric properties of PZT with aligned porosity for energy harvesting applications. *J. Mater. Chem. A* **2017**, *5*, 6569–6580. [CrossRef] [PubMed]
16. Qiu, X. Patterned piezo-, pyro-, and ferroelectricity of poled polymer electrets. *J. Appl. Phys.* **2010**, *108*, 011101. [CrossRef]
17. Hsiao, C.-C.; Yu, S.-Y. Improved Response of ZnO Films for Pyroelectric Devices. *Sensors* **2012**, *12*, 17007–17022. [CrossRef] [PubMed]

18. Morozovska, A.N.; Eliseev, E.A.; Glinchuk, M.D.; Shevliakova, H.V.; Svechnikov, G.S.; Silibin, M.V.; Sysa, A.V.; Yaremkevich, A.D.; Morozovsky, N.V.; Shvartsman, V.V. Analytical description of the size effect on pyroelectric and electrocaloric properties of ferroelectric nanoparticles. *Phys. Rev. Mater.* **2019**, *3*, 104414. [CrossRef]
19. Velarde, G.; Pandya, S.; Karthik, J.; Pesquera, D.; Martin, L.W. Pyroelectric thin films—Past, present, and future. *APL Mater.* **2021**, *9*, 010702. [CrossRef]
20. Morozovska, A.N.; Eliseev, E.A.; Svechnikov, G.S.; Kalinin, S.V. Pyroelectric response of ferroelectric nanowires: Size effect and electric energy harvesting. *J. Appl. Phys.* **2010**, *108*, 042009. [CrossRef]
21. Chen, X.; Shao, J.; Li, X.; Tian, H. A Flexible Piezoelectric-Pyroelectric Hybrid Nanogenerator Based on P(VDF-TrFE) Nanowire Array. *IEEE Trans. Nanotechnol.* **2016**, *15*, 295–302. [CrossRef]
22. Meneses-Franco, A.; Trujillo-Rojo, V.H.; Soto-Bustamante, E.A. Synthesis and characterization of pyroelectric nanocomposite formed of $BaTiO_3$ nanoparticles and a smectic liquid crystal matrix. *Phase Transit.* **2010**, *83*, 1037–1047. [CrossRef]
23. Fatuzzo, E.; Harbeke, G.; Merz, W.J.; Nitsche, R.; Roetschi, H.; Ruppel, W. Ferroelectricity in SbSI. *Phys. Rev.* **1962**, *127*, 2036. [CrossRef]
24. Imai, K.; Kawada, S.; Ida, M. Anomalous Pyroelectric Properties of SbSI Single Crystals. *J. Phys. Soc. Jpn.* **1966**, *21*, 1855–1860. [CrossRef]
25. Bhalla, A.S.; Newnham, R.E.; Cross, L.E.; Dougherty, J.P.; Smith, W.A. Pyroelectricity in SbSI. *Ferroelectrics* **1981**, *33*, 3–7. [CrossRef]
26. Rao, T.S.; Mansingh, A. Electrical and Optical Properties of SbSI Films. *Jpn. J. Appl. Phys.* **1985**, *24*, 422. [CrossRef]
27. Narayanan, S.; Pandey, R.K. Physical vapor deposition of antimony sulpho-iodide (SbSI) thin films and their properties. In Proceedings of the 1994 IEEE International Symposium on Applications of Ferroelectrics, State College, PA, USA, 12–16 May 2014; IEEE: Piscataway, NJ, USA, 1994; pp. 309–311. [CrossRef]
28. Purusothaman, Y.; Alluri, N.R.; Chandrasekhar, A.; Kim, S.-J. Photoactive piezoelectric energy harvester driven by antimony sulfoiodide (SbSI): A $A^{V}B^{VI}C^{VII}$ class ferroelectric-semiconductor compound. *Nano Energy* **2018**, *50*, 256–265. [CrossRef]
29. Toroń, B.; Mistewicz, K.; Jesionek, M.; Kozioł, M.; Stróż, D.; Zubko, M. Nanogenerator for dynamic stimuli detection and mechanical energy harvesting based on compressed SbSeI nanowires. *Energy* **2020**, *212*, 118717. [CrossRef]
30. Mistewicz, K.; Jesionek, M.; Kim, H.J.; Hajra, S.; Kozioł, M.; Chrobok, Ł.; Wang, X. Nanogenerator for determination of acoustic power in ultrasonic reactors. *Ultrason. Sonochem.* **2021**, *78*, 105718. [CrossRef]
31. Nie, R.; Yun, H.-S.; Paik, M.-J.; Mehta, A.; Park, B.-W.; Choi, Y.C.; Seok, S.I. Efficient Solar Cells Based on Light-Harvesting Antimony Sulfoiodide. *Adv. Energy Mater.* **2018**, *8*, 1701901. [CrossRef]
32. Mistewicz, K.; Matysiak, W.; Jesionek, M.; Jarka, P.; Kępińska, M.; Nowak, M.; Tański, T.; Stróż, D.; Szade, J.; Balin, K.; et al. A simple route for manufacture of photovoltaic devices based on chalcohalide nanowires. *Appl. Surf. Sci.* **2020**, *517*, 146138. [CrossRef]
33. Wang, C.; Zhang, M.; Fang, Y.; Chen, G.; Li, Q.; Sheng, X.; Xu, X.; Hui, J.; Lan, Y.; Fang, M.; et al. SbSI Nanocrystals: An Excellent Visible Light Photocatalyst with Efficient Generation of Singlet Oxygen. *ACS Sustain. Chem. Eng.* **2018**, *6*, 12166–12175. [CrossRef]
34. Mistewicz, K.; Kępińska, M.; Nowak, M.; Sasiela, A.; Zubko, M.; Stróż, D. Fast and Efficient Piezo/Photocatalytic Removal of Methyl Orange Using SbSI Nanowires. *Materials* **2020**, *13*, 4803. [CrossRef] [PubMed]
35. Yoshida, M.; Yamanaka, K.; Hamakawa, Y. Semiconducting and Dielectric Properties of C-Axia Oriented SbSI Thin Film. *Jpn. J. Appl. Phys.* **1973**, *12*, 1699. [CrossRef]
36. Mansingh, A.; Sudersena Rao, T. Growth and characterization of flash evaporated ferroelectric antimony sulphoiodide thin films. *J. Appl. Phys.* **1985**, *58*, 3530–3535. [CrossRef]
37. Mansingh, A.; Rao, T.S. I-V and C-V characteristics of ferroelectric SbSI (film)-Si-metal. *Ferroelectrics* **1983**, *50*, 263–264. [CrossRef]
38. Surthi, S.; Kotru, S.; Pandey, R.K. Preparation and electrical properties of ferroelectric SbSI films by pulsed laser deposition. *J. Mater. Sci. Lett.* **2003**, *22*, 591–593. [CrossRef]
39. Surthi, S.; Kotru, S.; Pandey, R.K. SbSI Films for Ferroelectric Memory Applications. *Integr. Ferroelectr.* **2002**, *48*, 263–269. [CrossRef]
40. Inagaki, S.; Nakamura, M.; Hatada, H.; Nishino, R.; Kagawa, F.; Tokura, Y.; Kawasaki, M. Growth of visible-light-responsive ferroelectric SbSI thin films by molecular beam epitaxy. *Appl. Phys. Lett.* **2020**, *116*, 072902. [CrossRef]
41. Choi, Y.C.; Hwang, E.; Kim, D.-H. Controlled growth of SbSI thin films from amorphous Sb_2S_3 for low-temperature solution processed chalcohalide solar cells. *APL Mater.* **2018**, *6*, 121108. [CrossRef]
42. Nie, R.; Seok, S.I. Efficient Antimony-Based Solar Cells by Enhanced Charge Transfer. *Small Methods* **2020**, *4*, 1900698. [CrossRef]
43. Gödel, K.C.; Steiner, U. Thin film synthesis of SbSI micro-crystals for self-powered photodetectors with rapid time response. *Nanoscale* **2016**, *8*, 15920–15925. [CrossRef]
44. Ye, H.; Xu, Y.; Mackenzie, J.D. Semiconducting ferroelectric SbSI quantum dots in organically modified TiO_2 matrix. In *Sol-Gel Optics V*; International Society for Optics and Photonics: Bellingham, WA, USA, 2000; Volume 3943, pp. 95–101. [CrossRef]
45. Gan, W.C.; Abd Majid, W.H. Effect of TiO2 on enhanced pyroelectric activity of PVDF composite. *Smart Mater. Struct.* **2014**, *23*, 045026. [CrossRef]
46. Mistewicz, K.; Nowak, M.; Starczewska, A.; Jesionek, M.; Rzychoń, T.; Wrzalik, R.; Guiseppi-Elie, A. Determination of electrical conductivity type of SbSI nanowires. *Mater. Lett.* **2016**, *182*, 78–80. [CrossRef]
47. Zhang, M.; Hu, Q.; Ma, K.; Ding, Y.; Li, C. Pyroelectric effect in CdS nanorods decorated with a molecular Co-catalyst for hydrogen evolution. *Nano Energy* **2020**, *73*, 104810. [CrossRef]

48. Hu, J.; Liu, Q.; Zhang, H.; Chen, C.-D.; Liang, Y.; Du, R.-G.; Lin, C.-J. Facile ultrasonic deposition of SnO_2 nanoparticles on TiO_2 nanotube films for enhanced photoelectrochemical performances. *J. Mater. Chem. A* **2015**, *3*, 22605–22613. [CrossRef]
49. Zhang, Q.; Li, C. High Temperature Stable Anatase Phase Titanium Dioxide Films Synthesized by Mist Chemical Vapor Deposition. *Nanomaterials* **2020**, *10*, 911. [CrossRef] [PubMed]
50. Borchert, J.W.; Stewart, I.E.; Ye, S.; Rathmell, A.R.; Wiley, B.J.; Winey, K.I. Effects of length dispersity and film fabrication on the sheet resistance of copper nanowire transparent conductors. *Nanoscale* **2015**, *7*, 14496–14504. [CrossRef]
51. Khanarian, G.; Joo, J.; Liu, X.-Q.; Eastman, P.; Werner, D.; O'Connell, K.; Trefonas, P. The optical and electrical properties of silver nanowire mesh films. *J. Appl. Phys.* **2013**, *114*, 024302. [CrossRef]
52. Lu, C. On the bending strength of ZnO nanowires. *Phys. Lett. A* **2008**, *372*, 6113–6115. [CrossRef]
53. Yu, S.-P.; Liu, Y.-H.; Sun, A.-C.; Hsu, J.-H. Determining the size distribution of magnetic nanoparticles based on analysis of magnetization curves. *J. Appl. Phys.* **2009**, *106*, 103905. [CrossRef]
54. Pauw, B.R.; Kastner, C.; Thunemann, A.F. Nanoparticle size distribution quantification: Results of a small-angle X-ray scattering inter-laboratory comparison. *J. Appl. Cryst.* **2017**, *50*, 1280–1288. [CrossRef]
55. Wu, B.-K.; Chern, M.-Y.; Lee, H.-Y. Size-controllable synthesis and bandgap modulation of single-layered RF-sputtered bismuth nanoparticles. *Nanoscale Res. Lett.* **2014**, *9*, 249. [CrossRef] [PubMed]
56. Bochenek, D.; Niemiec, P.; Skulski, R.; Adamczyk, M.; Brzezińska, D. Electrophysical properties of the multicomponent PBZT-type ceramics doped by Sn^{4+}. *J. Electroceram.* **2019**, *42*, 17–30. [CrossRef]
57. Dhananjay, D.; Nagaraju, N.; Krupanidhi, S.B. Off-centered polarization and ferroelectric phase transition in Li-doped ZnO thin films grown by pulsed-laser ablation. *J. Appl. Phys.* **2007**, *101*, 104104. [CrossRef]
58. Mansingh, A.; Srivastava, K.N.; Singh, B. Effect of surface capacitance on the dielectric behavior of ferroelectric lead germanate. *J. Appl. Phys.* **1979**, *50*, 4319. [CrossRef]
59. Połomska, M.; Hilczer, B.; Michalczyk, M. Electric conductivity of some ferroelectric crystals. *Ferroelectrics* **1981**, *39*, 1217–1220. [CrossRef]
60. Goto, Y.; Sawaguchi, E. Electric Conductivity of Ferroelectric $Pb_5Ge_3O_{11}$. *J. Phys. Soc. Jpn.* **1979**, *46*, 1580–1582. [CrossRef]
61. Rao, T.S.; Mansingh, A. Electrical properties of antimony sulphoiodide (SbSI) thin films. *Ferroelectrics* **1989**, *93*, 53. [CrossRef]
62. Mistewicz, K.; Nowak, M.; Stróż, D. A Ferroelectric-Photovoltaic Effect in SbSI Nanowires. *Nanomaterials* **2019**, *9*, 580. [CrossRef] [PubMed]
63. Tan, K.S.; Gan, W.C.; Velayutham, T.S.; Abd Majid, W.H. Pyroelectricity enhancement of PVDF nanocomposite thin films doped with ZnO nanoparticles. *Smart Mater. Struct.* **2014**, *23*, 125006. [CrossRef]
64. Mahdi, R.I.; Gan, W.C.; Halim, N.A.; Velayutham, T.S.; Abd Majid, W.H. Ferroelectric and pyroelectric properties of novel lead-free polyvinylidene fluoride-trifluoroethylene–$Bi_{0.5}Na_{0.5}TiO_3$ nanocomposite thin films for sensing applications. *Ceram. Int.* **2015**, *41*, 13836–13843. [CrossRef]
65. Mahdi, R.I.; Gan, W.C.; Abd Majid, W.H. Hot Plate Annealing at a Low Temperature of a Thin Ferroelectric P(VDF-TrFE) Film with an Improved Crystalline Structure for Sensors and Actuators. *Sensors* **2014**, *14*, 19115–19127. [CrossRef] [PubMed]
66. Krause, M.; Graz, I.; Bauer-Gogonea, S.; Bauer, S.; Ploss, B.; Zirkl, M.; Stadlober, B.; Helbig, U. $PbTiO_3$—P(VDF-TrFE)—Nanocomposites for Pressure and Temperature Sensitive Skin. *Ferroelectrics* **2011**, *419*, 23–27. [CrossRef]
67. Zhang, F.; Kang, H.; Lin, Y.; Guan, L.; Aslan, H.; Zhang, M.; Niu, L.; Dong, M. Studying the Pyroelectric Effects of $LiNbO_3$ Modified Composites. *Nanoscale Res. Lett.* **2020**, *15*, 106. [CrossRef] [PubMed]
68. Halim, N.A.; Abd Majid, W.H.; Velayutham, T.S. Ferroelectric, pyroelectric and piezoelectric properties of CeO_2-doped $Na_{0.5}Bi_{0.5}TiO_3$ ceramics. *SN Appl. Sci.* **2019**, *1*, 582. [CrossRef]
69. Sharma, A.P.; Behera, M.K.; Pradhan, D.K.; Pradhan, S.K.; Bonner, C.E.; Bahoura, M. Lead-free relaxor-ferroelectric thin films for energy harvesting from low-grade waste-heat. *Sci. Rep.* **2021**, *11*, 111. [CrossRef]
70. Xue, D.; Gao, J.; Zhang, L.; Bao, H.; Liu, W.; Zhou, C.; Ren, X. Aging effect in paraelectric state of ferroelectrics: Implication for a microscopic explanation of ferroelectric deaging. *Appl. Phys. Lett.* **2009**, *94*, 082902. [CrossRef]
71. Nakamura, M.; Hatada, H.; Kaneko, Y.; Ogawa, N.; Sotome, M.; Tokura, Y.; Kawasaki, M. Non-local photocurrent in a ferroelectric semiconductor SbSI under local photoexcitation. *Appl. Phys. Lett.* **2020**, *116*, 122902. [CrossRef]
72. Nakamura, M.; Hatada, H.; Kaneko, Y.; Ogawa, N.; Tokura, Y.; Kawasaki, M. Impact of electrodes on the extraction of shift current from a ferroelectric semiconductor SbSI. *Appl. Phys. Lett.* **2018**, *113*, 232901. [CrossRef]
73. Yu, Z.; Perera, I.R.; Daeneke, T.; Makuta, S.; Tachibana, Y.; Jasieniak, J.J.; Mishra, A.; Bäuerle, P.; Spiccia, L.; Bach, U. Indium tin oxide as a semiconductor material in efficient p-type dye-sensitized solar cells. *NPG Asia Mater.* **2016**, *8*, e305. [CrossRef]
74. Anitha, V.C.; Banerjee, A.N.; Joo, S.W. Recent developments in TiO_2 as n- and p-type transparent semiconductors: Synthesis, modification, properties, and energy related applications. *J. Mater. Sci.* **2015**, *50*, 7495–7536. [CrossRef]
75. Park, Y.; Choong, V.; Gao, Y.; Hsieh, B.R.; Tang, C.W. Work function of indium tin oxide transparent conductor measured by photoelectron spectroscopy. *Appl. Phys. Lett.* **1996**, *68*, 2699. [CrossRef]
76. Mansfeldova, V.; Zlamalova, M.; Tarabkova, H.; Janda, P.; Vorokhta, M.; Piliai, L.; Kavan, L. Work Function of TiO_2 (Anatase, Rutile, and Brookite) Single Crystals: Effects of the Environment. *J. Phys. Chem. C* **2021**, *125*, 1902–1912. [CrossRef]
77. Wang, X.; Song, J.; Liu, J.; Wang, Z.L. Direct-Current Nanogenerator Driven by Ultrasonic Waves. *Science* **2007**, *316*, 102–105. [CrossRef]

78. Yang, R.; Xu, R.; Dou, W.; Benner, M.; Zhang, Q.; Liu, J. Semiconductor-based dynamic heterojunctions as an emerging strategy for high direct-current mechanical energy harvesting. *Nano Energy* **2021**, *83*, 105849. [CrossRef]
79. Lu, Y.; Hao, Z.; Feng, S.; Shen, R.; Yan, Y.; Lin, S. Direct-Current Generator Based on Dynamic PN Junctions with the Designed Voltage Output. *iScience* **2019**, *22*, 58–69. [CrossRef]
80. Liu, J.; Liu, F.; Bao, R.; Jiang, K.; Khan, F.; Li, Z.; Peng, H.; Chen, J.; Alodhayb, A.; Thundat, T. Scaled-up Direct-Current Generation in MoS_2 Multilayer-Based Moving Heterojunctions. *ACS Appl. Mater. Interfaces* **2019**, *11*, 35404–35409. [CrossRef]
81. Lu, Y.; Yan, Y.; Yu, X.; Zhou, X.; Feng, S.; Xu, C.; Zheng, H.; Yang, Z.; Li, L.; Liu, K.; et al. Polarized Water Driven Dynamic PN Junction-Based Direct-Current Generator. *Research* **2021**, *2021*, 7505638. [CrossRef]
82. Meng, J.; Guo, Z.H.; Pan, C.; Wang, L.; Chang, C.; Li, L.; Pu, X.; Wang, Z.L. Flexible Textile Direct-Current Generator Based on the Tribovoltaic Effect at Dynamic Metal-Semiconducting Polymer Interfaces. *ACS Energy Lett.* **2021**, *6*, 2442–2450. [CrossRef]
83. Wang, W.; Han, X.; Niu, J.; Jin, X.; Wang, H.; Shao, H.; Lin, T. Direct-current energy generators from polypyrrole-coated fabric/metal Schottky diodes with considerably improved output. *J. Mater. Chem. A* **2020**, *8*, 24166–24174. [CrossRef]
84. Ding, X.; Shao, H.; Wang, H.; Yang, W.; Fang, J.; Zhang, D.; Lin, T. Schottky DC generators with considerable enhanced power output and energy conversion efficiency based on polypyrrole-TiO_2 nanocomposite. *Nano Energy* **2021**, *89*, 106367. [CrossRef]
85. Shao, H.; Fang, J.; Wang, H.; Zhou, H.; Lin, T. Direct current energy generators from a conducting polymer–inorganic oxide junction. *J. Mater. Chem. A* **2017**, *5*, 8267–8273. [CrossRef]
86. Sandomirsky, V.; Schlesinger, Y.; Dashevsky, Z. Pyroelectric effect of a p –n junction in a nonpolar solid. *Appl. Phys. Lett.* **2005**, *86*, 133501. [CrossRef]
87. Mistewicz, K.; Jesionek, M.; Nowak, M.; Kozioł, M. SbSeI pyroelectric nanogenerator for a low temperature waste heat recovery. *Nano Energy* **2019**, *64*, 103906. [CrossRef]
88. Grigas, J.; Talik, E.; Adamiec, M.; Lazauskas, V. X-ray photoelectron spectra and electronic structure of quasi-one-dimensional SbSeI crystals. *Condens. Matter Phys.* **2007**, *10*, 101–110. [CrossRef]
89. Suyal, G.; Seifert, A.; Setter, N. Pyroelectric nanoporous films: Synthesis and properties. *Appl. Phys. Lett.* **2002**, *81*, 1059. [CrossRef]
90. Seifert, A.; Sagalowicz, L.; Muralt, P.; Setter, N. Microstructural evolution of dense and porous pyroelectric $Pb_{1-ax}Ca_xTiO_3$ thin films. *J. Mater. Res.* **1999**, *14*, 2012–2022. [CrossRef]
91. Yang, Y.; Guo, W.; Pradel, K.C.; Zhu, G.; Zhou, Y.; Zhang, Y.; Hu, Y.; Lin, L.; Wang, Z.L. Pyroelectric Nanogenerators for Harvesting Thermoelectric Energy. *Nano Lett.* **2012**, *12*, 2833–2838. [CrossRef]
92. Bukowski, T.J.; McCarthy, K.; McCarthy, F.; Teowee, G.; Alexander, T.P.; Uhlmann, D.R.; Dawley, J.T.; Zelinski, B.J.J. Piezoelectric properties of sol-gel derived ZnO thin films. *Integr. Ferroelectr.* **1997**, *17*, 339–347. [CrossRef]
93. Zhang, K.; Wang, Y.; Wang, Z.L.; Yang, Y. Standard and figure-of-merit for quantifying the performance of pyroelectric nanogenerators. *Nano Energy* **2019**, *55*, 534–540. [CrossRef]
94. Saito, T.; Kawai, Y.; Ono, T. Thermoelectric power generator based on a ferroelectric material combined with a bimetal thermostat. In *Transducers & Eurosensors XXVII, Proceedings of the 17th International Conference on Solid-State Sensors, Actuators and Microsystems, Barcelona, Spain, 16–20 June 2013*; IEEE: Piscataway, NJ, USA, 2013; pp. 2280–2283. [CrossRef]
95. Narita, F.; Fox, M.; Mori, K.; Takeuchi, H.; Kobayashi, T.; Omote, K. Potential of energy harvesting in barium titanate based laminates from room temperature to cryogenic/high temperatures: Measurements and linking phase field and finite element simulations. *Smart Mater. Struct.* **2017**, *26*, 115027. [CrossRef]
96. Zhao, K.; Ouyang, B.; Yang, Y. Enhancing Photocurrent of Radially Polarized Ferroelectric $BaTiO_3$ Materials by Ferro-Pyro-Phototronic Effect. *iScience* **2018**, *3*, 208–216. [CrossRef] [PubMed]
97. Karim, H.; Sarker, M.R.H.; Shahriar, S.; Shuvo, M.A.I.; Delfin, D.; Hodges, D.; Tseng, T.-L.; Roberson, D.; Love, N.; Lin, Y. Feasibility study of thermal energy harvesting using lead free pyroelectrics. *Smart Mater. Struct.* **2016**, *25*, 055022. [CrossRef]
98. Zhao, T.; Jiang, W.; Liu, H.; Niu, D.; Li, X.; Liu, W.; Li, X.; Chen, B.; Shi, Y.; Yin, L.; et al. An infrared-driven flexible pyroelectric generator for non-contact energy harvester. *Nanoscale* **2016**, *8*, 8111–8117. [CrossRef] [PubMed]
99. You, M.-H.; Wang, X.-X.; Yan, X.; Zhang, J.; Song, W.-Z.; Yu, M.; Fan, Z.-Y.; Ramakrishna, S.; Long, Y.-Z. A self-powered flexible hybrid piezoelectric–pyroelectric nanogenerator based on non-woven nanofiber membranes. *J. Mater. Chem. A* **2018**, *6*, 3500–3509. [CrossRef]
100. Kim, J.; Lee, J.H.; Ryu, H.; Lee, J.-H.; Khan, U.; Kim, H.; Kwak, S.S.; Kim, S.-W. High-Performance Piezoelectric, Pyroelectric, and Triboelectric Nanogenerators Based on P(VDF-TrFE) with Controlled Crystallinity and Dipole Alignment. *Adv. Funct. Mater.* **2017**, *27*, 1700702. [CrossRef]
101. Mascot, M.; Fasquelle, D.; Velu, G.; Ferri, A.; Desfeux, R.; Courcot, L.; Carru, J.C. Pyro, Ferro and Dielectric Properties of $Ba_{0.8}Sr_{0.2}TiO_3$ Films Deposited by Sol-Gel on Platinized Silicon Substrates. *Ferroelectrics* **2008**, *362*, 79–86. [CrossRef]
102. Yang, Y.; Jung, J.H.; Yun, B.K.; Zhang, F.; Pradel, K.C.; Guo, W.; Wang, Z.L. Flexible Pyroelectric Nanogenerators using a Composite Structure of Lead-Free $KNbO_3$ Nanowires. *Adv. Mater.* **2012**, *24*, 5357–5362. [CrossRef]
103. Olszowy, M.; Pawlaczyk, C.; Markiewicz, E.; Kułek, J. Dielectric and pyroelectric properties of $BaTiO_3$–PVC composites. *Phys. Stat. Sol. A* **2005**, *202*, 1848–1853. [CrossRef]
104. Li, W.; Zheng, A.; Lin, Y.; Liu, P.; Shen, M.; Zhou, L.; Liu, H.; Yuan, J.; Qin, S.; Zhang, X.; et al. High pyroelectric effect in poly(vinylidene fluoride) composites cooperated with diamond nanoparticles. *Mater. Lett.* **2020**, *267*, 127514. [CrossRef]

105. Sultana, A.; Ghosh, S.K.; Alam, M.M.; Sadhukhan, P.; Roy, K.; Xie, M.; Bowen, C.R.; Sarkar, S.; Das, S.; Middya, T.R.; et al. Methylammonium Lead Iodide Incorporated Poly(vinylidene fluoride) Nanofibers for Flexible Piezoelectric–Pyroelectric Nanogenerator. *ACS Appl. Mater. Interfaces* **2019**, *11*, 27279–27287. [CrossRef] [PubMed]
106. Avinash, B.S.; Chaturmukha, V.S.; Jayanna, H.S.; Naveen, C.S.; Rajeeva, M.P.; Harish, B.M.; Suresh, S.; Lamani, A.R. Effect of particle size on band gap and DC electrical conductivity of TiO_2 nanomaterial. In *AIP Conference Proceedings*; AIP Publishing LLC.: Melville, NY, USA, 2016; Volume 1728, p. 020426. [CrossRef]
107. Wei, N.; Zhang, D.-M.; Yang, F.-X.; Han, X.-Y.; Zhong, Z.-C.; Zheng, K.-Y. Effect of electrical conductivity on the polarization behaviour and pyroelectric, piezoelectric property prediction of 0–3 ferroelectric composites. *J. Phys. Appl. Phys.* **2007**, *40*, 2716–2722. [CrossRef]

sensors

Article

Fiber-Based Triboelectric Nanogenerator for Mechanical Energy Harvesting and Its Application to a Human–Machine Interface

Chen-Kuei Chung *, You-Jun Huang, Tun-Kai Wang and Yu-Lung Lo

Department of Mechanical Engineering, National Cheng Kung University, Tainan 701, Taiwan
* Correspondence: ckchung@mail.ncku.edu.tw

Abstract: Mechanical energy harvesters including piezoelectric nanogenerators, electromagnetic generators and triboelectric nanogenerators (**TENG**) used to convert the mechanical motion into electricity are more and more important in the recent decades. Specifically, the fiber-based TENG (**FTENG**) has gained considerable favors due to its flexibility, light weight, and high environmental tolerance for the wearable devices. The traditional FTENGs made of Teflon result in better performance but are not suitable for long-term wear in person. Here, we propose a novel FTENG using a flexible micro-needle-structured polydimethylsiloxane (**MN-PDMS**) together with the comfortable commercially available **2D-polyester fibers**, and **electroless nickel-plated cotton cloth** of which two are widely used in human daily life. The MN-PDMS is formed by a laser engraved mold for improving its output performance of FTENG compared to the flat-PDMS. The open-circuit voltage (Voc) and the short-circuit current (Isc) of MN-FTENG increased to 73.6 V and 36 μA, respectively, which are 34% and 37% higher than the flat-FTENG. In terms of power, the performance of MN-FTENG reaches 1.296 mW which is 89% higher than that of flat-TENG and it can also light up 90 LEDs. For application, human motion at the joints can be detected and collected with various signals that are used for the human–machine interface (HMI) through the cooperation of components for the Internet of Things (IoT). It can light up the LED bulb through MN-FTENG to potentially develop IoT HMI systems for human motion control of robot in the future.

Keywords: mechanical energy harvester; triboelectric nanogenerators; TENG; fiber; polyester; polydimethylsiloxane; PDMS; sensor; human–machine interface; HMI

Citation: Chung, C.-K.; Huang, Y.-J.; Wang, T.-K.; Lo, Y.-L. Fiber-Based Triboelectric Nanogenerator for Mechanical Energy Harvesting and Its Application to a Human–Machine Interface. *Sensors* **2022**, *22*, 9632. https://doi.org/10.3390/s22249632

Academic Editors: Carlo Trigona, Olfa Kanoun and Slim Naifar

Received: 7 November 2022
Accepted: 5 December 2022
Published: 8 December 2022

1. Introduction

Mechanical energy harvesters [1,2] can convert mechanical motion into electricity during movement. The application of mechanical energy harvesters can be used for wearable self-powered sensing devices and to act as generators, which makes them have great potential as battery-free HMIs in the IoT. Typical mechanical energy harvesters including piezoelectric nanogenerators (PENG) [3,4], electromagnetic generators (EMG) [5,6]. Among them, the Triboelectric Nano-generator (TENG) [7] is a self-powered sensor [8,9] with high mechanical sensitivity and power generation efficiency. TENG has four different modes of power generation mechanisms [10], these are: vertical contact-separation mode, lateral sliding mode, single-electrode mode, and freestanding triboelectric-layer mode. These four modes of TENGs can almost correspond to all types of motion in human daily life. Compared with other mechanical energy harvesters, which often require the use of metal materials or ceramic materials, TENG can be composed of a variety of materials, including a variety of ductile materials. This makes TENG useful for wearable devices through material selection and creates a lot of convenience for a human's daily life in the future.

The rise of flexible devices in recent years makes a new revolution in human–machine interfaces (HMIs) and internet of things (IoT) applications, which can meet people's growing pursuit of a more convenient life [11–13]. However, traditional bulky and rigid electronic devices obstruct the compliant interfacing with human skin [14,15]. Therefore, conventional devices affect the comfort level of users to a large extent. To overcome these drawbacks, it is essential to develop flexible sensors with excellent mechanical flexibility and stretchability. These sensors attached to curved and dynamic surfaces, such as human skin, can continuously monitor physiological and environmental indicators in real time [16,17]. TENGs composed of flexible and stretchable materials have the potential to be used as sensors in wearable devices.

Among many studies [18,19], the fiber-based TENG (FTENG) has gained considerable favors among many teams in recent years due to its flexibility [20], light weight [21], high force sensitivity, and high environmental tolerance for inserting into clothing, insoles and other products as a wearable device and integrate into human beings without any sense of violation. FTENG can also be divided into 1D [22], 2D [23], and 3D [24] dimensions according to different weaving methods. This makes FTENG more suitable to use to collect mechanical energy and signals of human motion than other types of mechanical energy harvesters. Higher woven dimension provides better flexibility, stretchability, and effective contact area, but meanwhile, it also increases the process complexity and the cost of triboelectric layers. In our daily life, FTENG can be used as an HMI to operate mechanical operations. FTENG may collect mechanical energy during sensing the motion force and be used as a power supply system for wearable devices or as a human health monitoring system [25], and for the purpose of sensing or self-powering, better TENG output performance is necessary, and it is also necessary that it can only be used in practical applications if it meets the characteristics of durability and washability. However, in most of the wearable TENG studies, FTENG is assembled into TENG by using Teflon [26,27] or other materials that are not flexible and stretchable to be drawn and woven into textile-like triboelectric layers using various weaving techniques. This type of FTENG is not only complicated to manufacture, but also lacks high enough flexibility and stretchability. Therefore, this kind of FTENG is not suitable for a wearable device to be worn close to the body for a long time.

Here, we propose a novel FTENG using a flexible micro-needle-structured polydimethylsiloxane (MN-PDMS) together with the comfortable commercially available 2D-polyester fibers. We uniformly coat the flexible and extensible polydimethylsiloxane (PDMS) on the polyester fibers (PET) as the main material of FTENG. Additionally, an electroless nickel-plated cotton cloth is used as the conductive fiber here for the electrode which is widely used in human daily life. These ductile materials can greatly improve the flexibility and stretchability of FTENG. It is noted that the MN-PDMS may enhance the contact surface and energy harvesting performance of Al-PDMS TENGs [28,29] and is formed by a laser-engraved mold for improving its output performance of TENG compared to the flat-PDMS [30,31]. The FTENG with MN-PDMS here is named **MN-FTENG**. The MN-FTENG can be used for high-sensitivity motion tests such as human joint moment and applied on HMI with an Arduino controller. The open-circuit voltage (Voc) and the short-circuit current (Isc) of MN-FTENG increase to 73.6 V and 36 μA, which are 34% and 37% higher than the original FTENG because of the increase in effective surface area. Additionally, the output power of MN-FTENG reaches 1.296 mW with 89% higher than FTENG without microstructure. In addition, MN-FTENG can also light up 90 LEDs, it is 1.3 times of flat-FTENG. The MN-FTENG is used for detecting the human motion at joints to collect various signals those are used for the human–machine interface (HMI) through the cooperation of components for the potential Internet of Things (IoT). The LED bulb is lit up and controlled through MN-FTENG which implies the potential to develop HMI systems for the human motion control of robots in the future.

2. Experimental Procedures

2.1. The Material Selection and Assembly of MN-FTENG

Polyester fiber is a kind of linear polymer obtained by polycondensation reaction of saturated dibasic acid and dihydric alcohol. It is associated with an ester group "-COO-", so it is commonly called polyester. Polyester fiber has high strength and elastic recovery ability, so it is flexible and stretchable. Unlike most of the studies, where the fiber material mostly acts as a shape-maintaining material, the fiber base in this study is also part of the triboelectric layer. Therefore, the effect of fiber materials on the mechanical energy harvesting performance must also be considered here. According to the Triboelectric Series [10], PET is the material which gains electrons easily and stays in the negative zone. When we use it as a triboelectric layer to make a wearable device and work with the human skin that easily lose electrons, it can produce a larger surface electrical difference than other cheap and common fiber material such as Nylon and cotton. This is helpful for greater mechanical energy harvesting. The polyester fiber used in this experiment is a kind of common commercially available twill weave polyester fiber. Its anti-wrinkle and high-strength properties can stabilize the film thickness in the subsequent coating process and improve the tensile resistance and dimensional stability of the triboelectric layer. The electrode used in this experiment is a commercially available conductive cloth, which is the composition of the metal nickel and the fiber by electroless plating. It is a flexible fiber material with conductivity similar to a metal material. By forming a metal film on each fiber, the characteristics of the metal can be imparted under the characteristics of the fiber. The composite process of attaching metal thin films to fibers by electroless plating can achieve the same conductivity as metal foils while maintaining flexibility and stretchability. The polymethylmethacrylate (PMMA) used in this experiment is the polymer with the best optical properties, with more than 90% light transmittance for light wavelengths from 340 nm to 1100 nm. PMMA can directly absorb a laser of specified wavelength to achieve a good etching effect. In this experiment, PMMA was processed by a laser-engraving machine as a mold, and PDMS was transferred over the mold. Since there is no mutual pollution and no serious loss, the PMMA mold can be reused several times. The polydimethylsioxane (PDMS) is a silicone elastomer material with a hydrophobic surface and a contact angle of about 100–120°. The polymer is thermally and chemically stable due to its Si-O bond strength and is easy to prepare. The PDMS used in this study is for the production of thin film rapid prototyping of TENG by mixing elastic agent and curing agent in a ratio of 10:1 by weight.

Figure 1a–f shows the manufacturing steps and process of MN-FTENG. First, the PMMA mold was processed by a laser engraving machine and ablated to form continuous microstructure grooves (Figure 1a). The PDMS and curing agent were mixed well and poured onto the PMMA mold (Figure 1b) and degassed for 10 min to fill the holes (Figure 1c). Then, the PET (90 × 90 mm) and conductive cloth (70 × 70 mm) were laid flat on the mold (Figure 1d) and kept some time for the PDMS stuck to the fiber structure (Figure 1e) and followed by putting it into an 80 °C oven for curing. After curing, the polyester fiber triboelectric generator with PDMS microneedle structure can be easily peeled from the mold and cut into 80 × 80 mm size to remove irregular burrs (Figure 1f). The completed MN-FTENG is shown in Figure 1g. With the stretchability and flexibility of the materials used, the device can be freely bent and recovered and can be used in a bent state, as shown in Figure 1h).

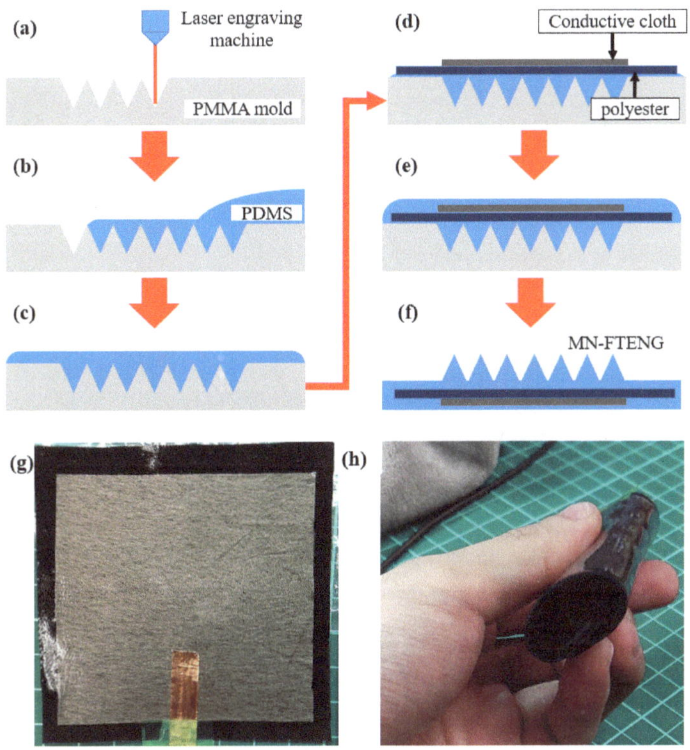

Figure 1. The process of preparing MN-FTENG and the appearance of MN-FTENG in this experiment: (**a**) Microstructures were fabricated on the PMMA mold with a laser engraving machine, (**b**) PDMS was poured into PMMA mold, (**c**) air bubbles was evacuated from PDMS in a vacuum chamber, (**d**) Lay polyester and conductive cloth on PDMS, (**e**) let PDMS fully penetrate into fiber in the vacuum chamber, (**f**) after baking and demolding, MN-FTENG is formed, (**g**) the appearance of MN-FTENG and (**h**) MN-FTENG can still be used after bending 360 degrees.

2.2. The Experiment and Measurement of MN-FTENG

In order to simulate the motion form of TENG touching and separating with the skin, we use a TENG pneumatic cylinder test platform to achieve vertical contact-separation mode. The TENG pneumatic cylinder test platform uses the pneumatic cylinder (FESTO D:S-PAZ-DW20-100PPV, Esslingen, Germany). Figure 2 shows the schematic electrical system of MN-FTENG performance test. The two triboelectric layers of MN-FTENG, i.e., aluminum and MN-PDMS are stickled to the drive shaft and the stage, respectively. The two electrodes of MN-FTENG are aluminum and the conductive fiber inside the MN-FTENG. The test platform is actuated by air pressure (the input air pressure is 2 kg/cm^2), and carried out under the conditions of 8 Hz, 16 N, and the distance between the test piece and the low-purity aluminum electrode is 25 mm. During the test, the aluminum electrode and the test piece are connected to the transient waveform recorder with copper wires to measure the open-circuit voltage signal and short-circuit current needs to be measured in parallel with a high resistance load. In addition, various DC performances of flat-FTENG and MN-FTENG can also be tested by simply connecting capacitors and Light Emitting Diodes (LEDs) through rectifying flat-FTENG or MN-FTENG's signals by a bridge rectifier. Due to the high stability of this platform, the platform can also maintain performance without degradation during high cycle beating experiments to test the stability of MN-FTENG.

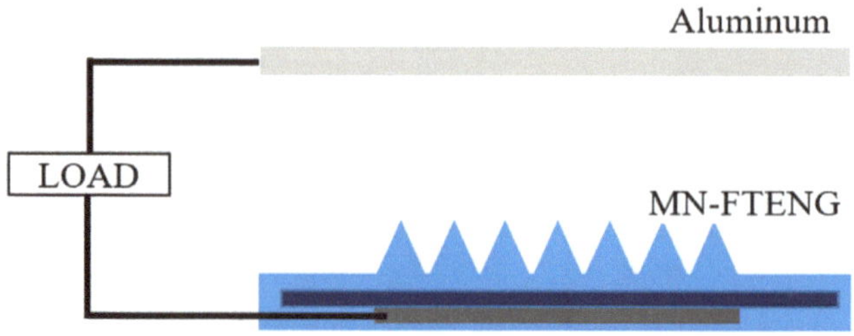

Figure 2. The electrical system of MN-FTENG. The two triboelectric layers of MN-FTENG are aluminum and MN-PDMS. Additionally, the two electrodes of MN-FTENG are aluminum and the conductive fiber inside the MN-FTENG. Since human skin has the positive triboelectric series, in the wearable device, the aluminum is replaced by human skin which becomes the other friction layer of MN-FTENG and turns the MN-FTENG into Single-electrode mode.

For the actual testing the performance of MN-FTENG as a wearable device in the HMI system, the aluminum electrode is replaced by human skin which becomes the other friction layer of MN-FTENG since human skin has the positive triboelectric series. Additionally, this turns the MN-FTENG into Single-electrode mode. Additionally, the MN-FTENG is cut into 30 mm $\times$ 80 mm for better wearability. The force sensitivity of MN-FTENG is confirmed through a dropping test of the weight drop from the fixed height of 3, 5, 7 and 10 cm. In this way, the MN-FTENG and the load cell will be impacted by the same force, and this force will be measured out through the load cell. The measured force ranges from approximately 0.5 to 10 N. The characterization of the microstructure of MN-FTENG was recorded by optical microscopy (OM, Olympus BX 51 M, Tokyo, Japan). The open-circuit voltage (Voc) and the short-circuit current (Isc) of MN-FTENG were recorded by an oscilloscope (HIOKI Memory HiCorder MR8870-20, Nagano, Japan). The electric output power of flat-FTENG and MN-FTENG can be calculated by the output voltage and output current under load by W = IV. Additionally, the development boards in HMI-related experiments are Arduino UNO.

3. Results and Discussions

3.1. The Output Performance of FTENG and MN-FTENG

The microstructures fabricated in this study are regulated by laser parameters of power, speed and density. The height and width of the microneedles of the PDMS mold after processing were measured by OM. As shown in Figure 3a,b, the surface of the flat and unstructured test piece is flat, and the microneedle test piece can see that its structure height is 374 μm. Both of their PDMS layers are about 800 μm thick and the whole triboelectric layers are about 4 mm thick. At the same time, by increasing the number of microneedles per unit area, a structure adjacent to the microneedles is fabricated to increase the contact area. The performance difference is compared between the microneedle fiber triboelectric nanogenerator (MN-FTENG) and the unstructured planar fiber triboelectric nanogenerator (flat-FTENG) in its open circuit voltage and short circuit current. From the open-circuit voltage and short-circuit current of Figure 4a,b, it shows that the voltage and current between FTENG and MN-FTENG have a clear trend of increasing. Compared with the flat-FTENG, the open circuit voltage of MN-FTENG increased from 54.6 V to 73.6 V. While the short-circuit current increased from 26.16 μA to 36 μA. It means that the MN-PDMS brings 34% higher open-circuit voltage signal and 37% higher short circuit signal to the MN-FTENG. From the above data, we can find that under the same mechanical force, the microstructure array can indeed increase the power generation performance. In addition, in the calculation of the current density, the MN-FTENG reaches a maximum of 7.34 mA/m^2. As listed in Table 1, MN-FTENG with the microstructure can achieve better performance than some other FTENGs. In addition, FTENG without the microstructure also has a good enough performance in the field of fiber-based TENG. The higher output performance of MN-FTENG is attributed to the enhanced contact area and high triboelectric polarity difference between the MN-PDMS and Al tribo-layers, as well that both PDMS and PET are highly flexible to make more electron transfer and conduction during continuous contact-and-separate deformation compared to some materials.

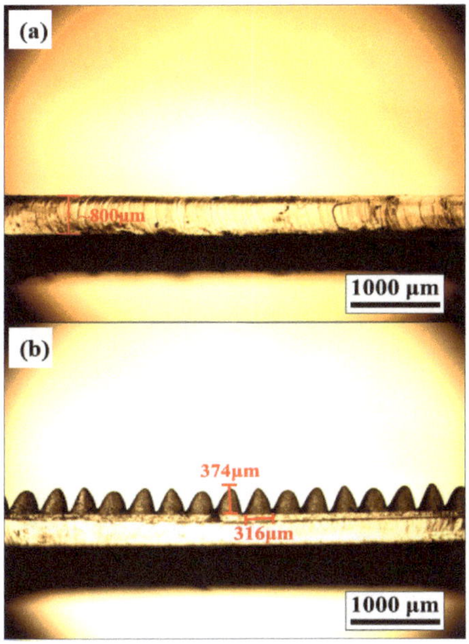

Figure 3. (**a**,**b**) OM diagram of smooth surface and 374 μm microneedle structure on PDMS. Both of these two samples are ~800 μm thick.

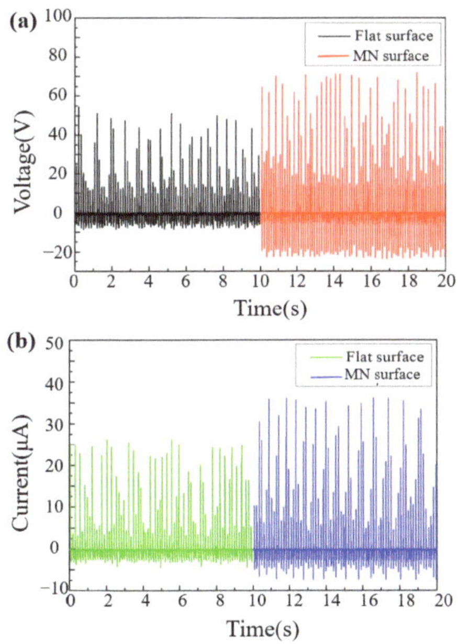

Figure 4. (**a**) Open circuit voltage and (**b**) current diagram of flat surface FTENG and 374 μm microneedled structure MN-FTENG.

Table 1. Comparison of fiber based TENG.

Material	Voc	Isc	Dimension	Ref.
Silicone-PVA	42.9 V	0.51 μA	1D	[22]
Polyester-Ni-Parylene	50 V	4 μA	2D	[23]
PTFE-Ag	50.6 V	0.29 μA	3D	[24]
PET-PDMS-Al	54.6 V	26.16 μA	2D	This research
PET-MNPDMS-Al	73.6 V	36.00 μA	Enhanced 2D	This research

As shown in Figure 5a,c, the voltage increases with the increase in the resistance until it is close to the open-circuit voltage; the current decreases with the increase in the resistance due to the ohmic loss. In the high resistance region, the current reduction becomes slow. After calculation, the maximum power of unstructured FTENG also appears at 1MΩ, which is 0.68 mW. Additionally, the power density is 0.139 W/m^2 in Figure 5b. Additionally, Figure 5d shows the maximum power of MN-FTENG appears at 1 MΩ, which is 1.296 mW. Additionally, the power density is 0.264 W/m^2. It can be seen that the microstructure increases the effective friction area, which can allow more PDMS area to rub against the Al electrode to generate more triboelectric charges, increase the charge density, increase the voltage and current, and finally increase the power by 189% and power density to 0.264 W/m$^{2.}$ Higher output power enables MN-FTENG to drive more electrical appliances, which can reduce battery consumption and energy waste when applied in wearable HMIs in the future.

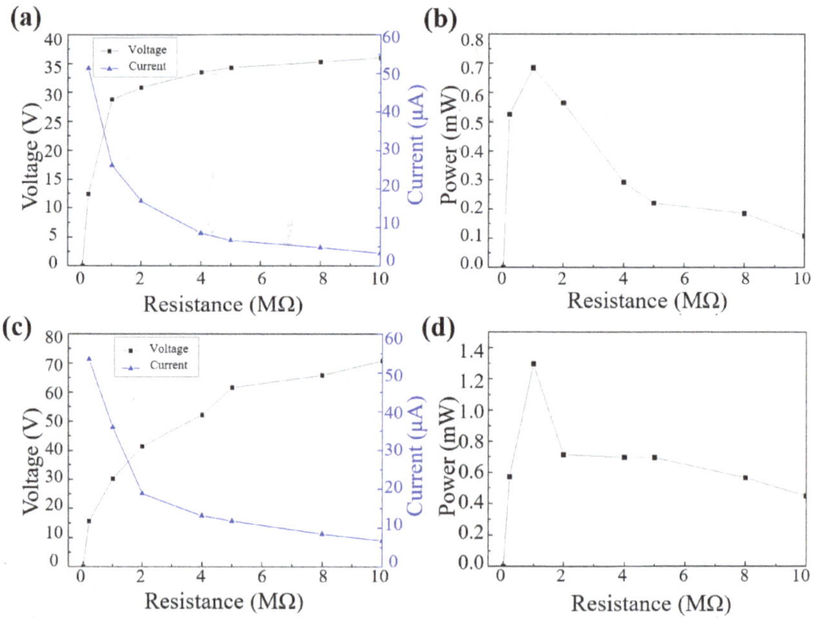

Figure 5. (**a**) Voltage–, current–resistance curve and (**b**) power–resistance curve of unstructured FTENG; (**c**) voltage–, current–resistance curve and (**d**) power–resistance curve of MN-FTENG. These four pictures show that the maximum power of MN-FTENG and FTENG appear at 1 MΩ, which are 0.68 mW and 1.296 mW.

3.2. Durability and Washing Test

TENG is essentially a green energy technology that converts mechanical energy into electrical energy, so its structural durability and output stability under repeated mechanical force operation will be one of its performance indicators. Therefore, in this section, MN-FTENG will be tested for long-term durability and the stability of its output performance will be observed. Continuous recording of its output voltage waveform in 5000 cycles of beating test, it can be seen that MN-FTENG has quite stable output performance, and there is no phenomenon of output performance degradation due to structural damage, as shown in Figure 6a. This indicates that the MN-FTENG proposed in this study has excellent structural stability in continuous long-term durability tests and has the advantage of stable output in practical use, which can be used as a self-powered system or sensor application.

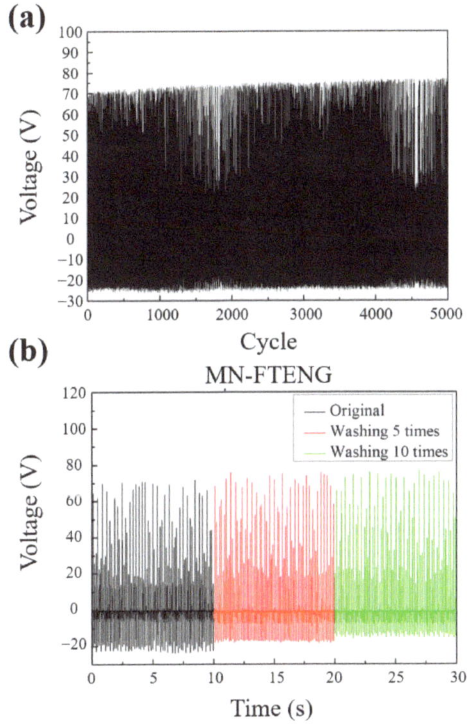

Figure 6. (**a**) The electrical performance of MN-FTENG is stable after 5000 cycles of beating test, and the inset is the enlarged waveform; (**b**) MN-FTENG has quite stable power generation performance in 0 times, 5 times and 10 times of cleaning tests.

The fiber-based TENG has the characteristic of flexibility and is often used in human body wear in many applications. However, if it is to be practically applied to general clothing, its performance stability after cleaning will be a key indicator whether it has the feasibility of physical application. Therefore, this section will discuss whether the power generation efficiency of MN-FTENG declines after cleaning to check its stability. In the cleaning process of this experiment, first rinse the test piece with clean water and rub the surface with detergent, put it into a beaker containing a detergent aqueous solution and ultrasonically shake it for 5 min, then rinse the excess foam with clean water, blow off excess water with an air gun. Finally, the test piece is dried at 55 °C for 10 min to remove residual moisture. The voltage data of the test piece cleaned 0 times, cleaned 5 times, and cleaned 10 times were used as the index to observe the performance difference. From Figure 6b, it can be seen that the voltage output of MN-FTENG is quite stable after 5 and

10 times of cleaning, and the voltage difference is within ±5 V, which shows the cleaning stability of MN-FTENG and its potential of daily wearable sensors which must be easy to clean as the clothing is frequently worn by humans and soiled.

3.3. Energy Storage Characteristics and LED Lighting

To supply the output voltage of TENG to small electronic devices, TENG can be used to charge and stabilize the capacitor and then drive the small components with a stable DC voltage. Therefore, studying the efficiency of MN-FTENG in capacitor charging and discharging is also an indicator for confirming the practical applicability of MN-FTENG. As shown in Figure 7, the unstructured FTENG can charge 3.3 μF to 0.7 V in 20 s, while the 374 μm MN-FTENG can charge 3.3 μF to 3.3 V in 15 s. A higher capacitor charging voltage can make the mechanical energy harvester drive more electronic components while sensing, as shown in Figure 7. While the unstructured FTENG takes 14 s from saturation to full discharge, and the MN-FTENG takes more than 21 s from saturation to full discharge. A longer discharge time can make the electronic components have a longer operating time.

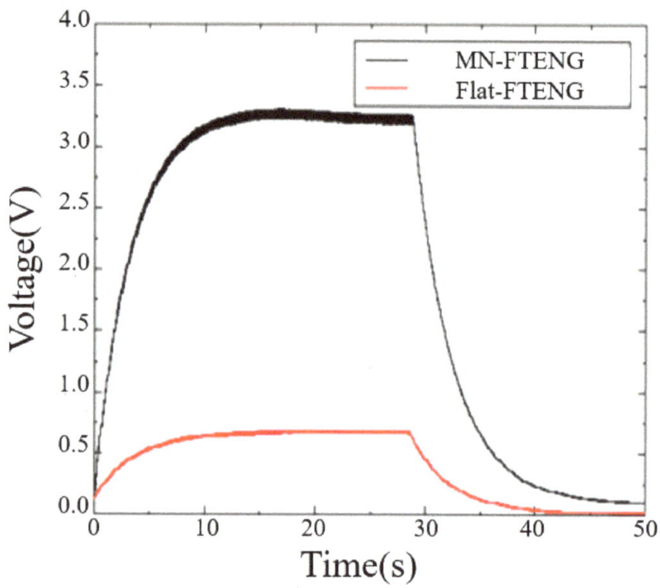

Figure 7. Charge–discharge curves of MN-FTENG and FTENG with 3.3 μF capacitors, it shows that the unstructured FTENG can charge 3.3 μF to 0.7 V in 20 s, while the 374 μm MN-FTENG can charge 3.3 μF to 3.3 V in 15 s. Additionally, the unstructured FTENG takes 14 s from saturation to full discharge, and the MN-FTENG takes more than 21 **s** from saturation to full discharge.

The MN-FTENG fabricated in this study can be used to drive LEDs as lighting or indicator lights, and the Voc and Isc of the TENG will directly affect the number and brightness of LEDs. In this experiment, the green LEDs are connected in series, and the TENG is rectified by a bridge rectifier and then directly output to drive the LEDs. The contribution of microstructures to the output performance can be directly judged through the number of emitting LEDs. It can be seen in Figure 8 that the MN-FTENG can light up 90 LEDs, while the unstructured FTENG can light up 70 LEDs. Therefore, it can be judged that the excellent performance output of MN-FTENG proposed in this study makes it have significant advantages in capacitor charging or LED driving, which is sufficient for application in various fields. The experiments in this chapter prove that MN-FTENG has the ability to generate electricity and drive electrical appliances by itself. In the future, when developing the HMI system, this characteristic can drive the small electrical appliances in the system to achieve the purpose of battery-free driving.

Figure 8. (**a**) Unstructured FTENG can drive 70 LEDs and (**b**) MN-FTENG can drive 90 LEDs.

3.4. Force Sensitivity Test

The MN-FTENG fabricated in this study has quite high mechanical sensitivity due to the deformation and recovery of the PDMS microstructure. As a sensor, it can have different outputs and responses according to the magnitude of the external force. The output voltage of MN-FTENG under various forces is shown in Figure 9. The force range is between 0.5 N to 10 N and the output voltage range is between 5 V to 60 V. It reveals that its output and force have a very high mechanical sensitivity 7 V/N, which is much higher than other mechanical energy harvesters, as listed in Table 2 since FTENG with high deformation can convert more mechanical energy into electric energy. Additionally, the coefficient of determination of $R^2 = 0.97$. Based on this calculation, the 16 N will have an output of approximately 80 V, which is in line with the measurement results in Section 3.1. This shows that MN-FTENG can be used as a sensor to measure the force differences by the output signal of MN-FTENG. Higher mechanical sensitivity indicates better sensing ability for tiny motions, which also means that MN-FTENG is more suitable for capturing tiny human motions, such as arm bending and knee bending, compared to other mechanical energy harvesters. Additionally, a coefficient of determination of 0.97 means a sufficiently stable sensing capability.

Table 2. Comparison of fiber-based MN-TENG with fiber EMG and PENG.

Mechanical Energy Harvester	Voc	Isc	Force Sensitivity	Material	Ref.
EMG	9 mV	10 mA	0.56 V/Hz	Polyurethane/iron	[32]
PENG	4.0 V	2.6 μA	0.27 V/N	P(VDF-TrFE) micropillar array	[33]
Fiber-TENG	73.6 V	36.0 μA	7 V/N	PET + PDMS/Al	This research

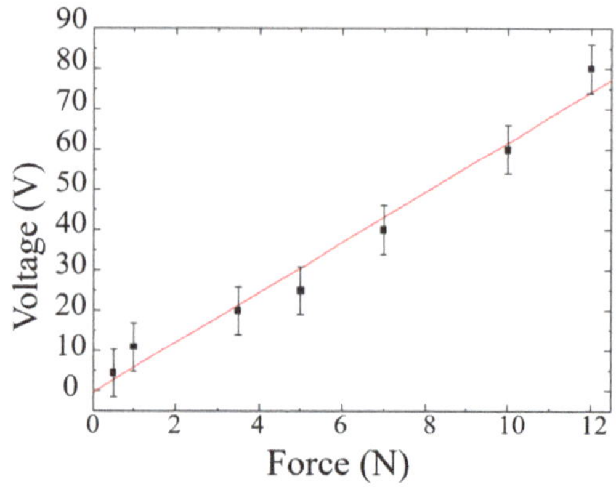

Figure 9. The output voltage of MN-FTENG under various forces, the slope of the sensitivity is 7 V/N, which is significantly higher than other Mechanical energy harvester. Additionally, the coefficient of determination of this experiment is 0.97, which shows high reliability of MN-FTENG's force sensitivity. The force comes from the weight dropping from a specified height (3 cm, 5 cm, 7 cm, and 10 cm) and is measured by the load cell. The same force acts on MN-FTENG and the output voltage is measured by HIOKI. The measured force range is 0.5 N to 10 N, and the voltage range is 5 V to 80 V.

Based on the output performance in Sections 3.1 and 3.3–3.4, MN-FTENG is sufficient to be a mechanical energy harvester for daily use and applied to wearable HMI. In the following sections, we will actually test the benefits of MN-TENG for human motion capture and whether it can be applied to HMI.

3.5. Human Motion Detection

In this section, the motion detection of large joints such as elbows and knees will be carried out. As shown in Figure 10a,c, MN-FTENG of 30 mm × 80 mm is used to make patches and fixed to the elbow and knee joints. The highest voltage of the elbow motion captured by MN-FTENG reaches 0.6 V, and its voltage output will change correspondingly with the speed of hand movement, while Figure 10b,d illustrate that there will be slight differences between the waves and the spacing due to the instability of the motion frequency. The motion of the knee joint and its power generation waveform, it is worth noting that the large area of the knee joint also increases the voltage to 0.8 V. With Figure 9, it can be known by extrapolation that the force of squeezing MN-FTENG when the elbow bending is about 0.05 N, and the knee is 0.1 N. Since the elbows and knees have little external resistance when flexed, this output is in line with the force that MN-FTENG should be subjected to.

The height and density of the waveform reflect the force, speed and frequency of human movement. In the future, the user's force, speed or frequency can be judged through different waveforms, and then the user's current movement situation can be understood. To sum up, the MN-FTENG developed in this research has good electrical properties, flexible properties and durable structure. It can reliably collect motion waveforms at joints in human motion detection and has the potential to be applied to wearable devices. It can be used as a self-powered sensor for human movement to meet the development of multi-sensor technology in the IoT era, or as an activity detection for post-injury rehabilitation, or to develop a HMI system for human movement control equipment. Next, we will develop the control system of MN-FTENG for wearable HMI.

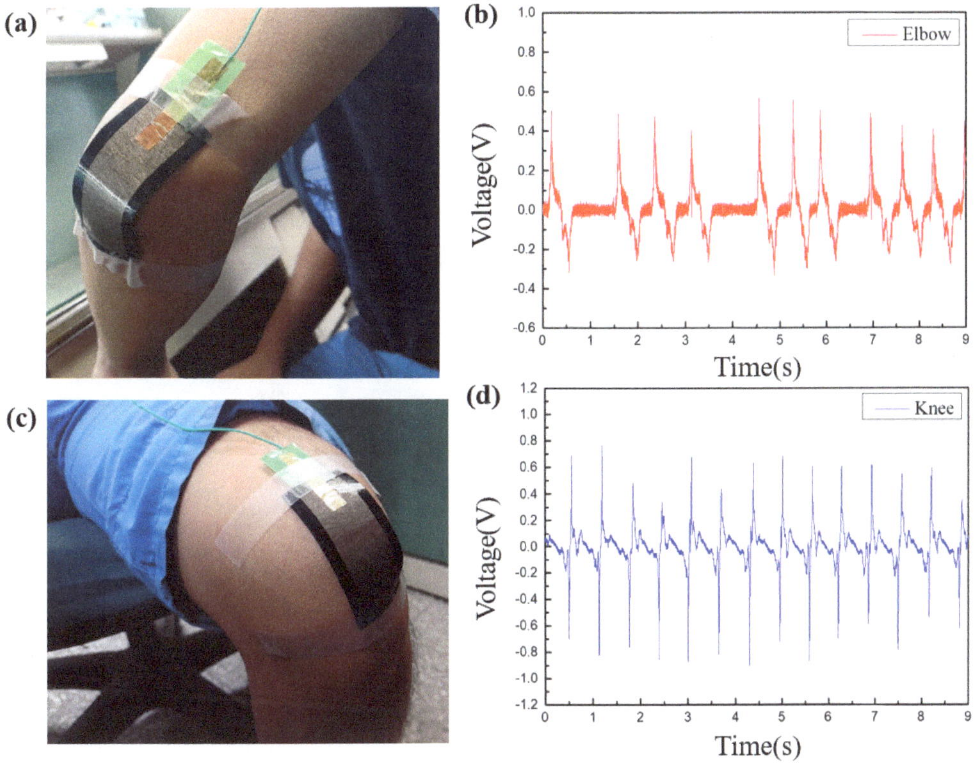

Figure 10. The electric signal of human body motion captured by MN-FTENG, (**a**,**b**)the voltage of the elbow motion signal captured by MN-FTENG reaches 0.6 V; (**c**,**d**) The knee joint reaches 0.8 V. With the result of Figure 9, it also shows the force of squeezing MN-FTENG during elbow bending and knee bending is about 0.05 N and 0.1 N by extrapolation.

3.6. HMI and Its Applications

With the feature of MN-FTENG above, through the simple connection between MN-FTENG, control panel and the mechanical device, it is possible to control the lighting of lamps, start the motor, reverse rotation, stop and other commands by clicking, touching of other human motion while MN-FTENG converts the motion signal of the human body into electrical signal and inputs it into the control panel, so that it controls the connected device, as shown in Figure 11a. In this experiment, we combine the MN-FTENG, Arduino UNO and a light bulb as a control system. Whenever the user slightly touches the MN-FTENG, the MN-FTENG will convert the user's action into an electrical signal and send it to the Arduino UNO to control the light bulb to turn on or off, as shown in Figure 11b,c. Because the mechanical harvester has the characteristics of generating and sensing by itself, it can use this feature to generate signals and provide power for control components. In this system, MN-FTENG does not need to be powered on all the time to operate like most other high-sensitivity HMIs, and can start or drive other devices by relying on the power it captures. This is a very energy efficient way to save energy in systems that require a lot of HMIs.

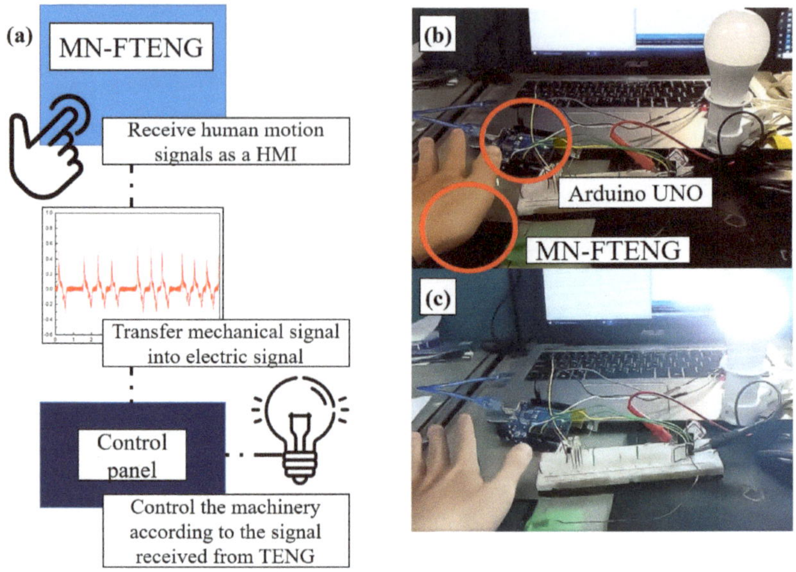

Figure 11. MN-FTENG is used as a self-generating sensor for HMI to drive machinery. (**a**) MN-FTENG converts the motion signal of the human body into electrical signal and inputs it into the control panel, so that it controls the connected device. (**b**) After touching MN-FTENG, MN-TENG will output signal to Arduino UNO, and the Arduino UNO will (**c**) turn on or off the light switch after receiving the signal.

4. Conclusions

In this study, we develop the FTENG as a mechanical energy harvester for HMI sensors. We use a coated with PDMS using commercially available PET and conductive cloth. FTENG with high mechanical sensitivity is used as a self-powered sensor with high flexibility and stretchability, and the microstructure array PMMA mold is fabricated by laser engraving, and the fiber-base triboelectric nanogenerator with microneedle structure (MN-FTENG) is fabricated by demolding. The open-circuit voltage (Voc) and the short-circuit current (Isc) of MN-FTENG with microstructure increase from 54.6 V and 26.16μA to 73.6 V and 36 μA, respectively, which are 34% and 37% higher than FTENG without microneedle structure. Additionally, the output power of MN-FTENG reaches 1.296 mW with 89% higher than that of flat FTENG. In the durability test, the MN-FTENG still maintains a good output performance in 5000 tapping cycles. In the structural cleaning test, good performance stability was also maintained, and the voltage difference was within ±5 V. In the capacitor charging test, MN-FTENG can charge 3.3 μF to 3.3 V within 15 s and MN-FTENG can also light up 90 LEDs. The MN-FTENG has a high and stable mechanical sensitivity of 7 V/N with a coefficient of determination of 0.97 which is higher than most other mechanical harvesters. The above performance is high enough for MN-TENG to be used as a mechanical energy harvester for HMI. As a human body joint motion mechanical energy harvester, MN-FTENG can harvest the motion mechanical energy of the elbow and knee joints as a voltage signal to sense human motion. MN-PET-TENG acts as a self-powered HMI, which can harvest the energy during human motion and store it in capacitors to drive device actions. Additionally, through the control device, the movement of the user's body is used to operate the machine and turns on or off the LED light through the MN-TENG for potential IoT application.

Author Contributions: C.-K.C. proposed the original idea; C.-K.C., Y.-J.H. and T.-K.W. worked together on the conceptualization, organization, discussion and the draft writing; Y.-J.H. and T.-K.W. made the experiments, figure drawing, information survey. C.-K.C. and Y.-L.L. discussed on the concept, methodology and writing of the mechanical sensing, and application. C.-K.C. made the final review, editing, and supervision. All authors have read and agreed to the published version of the manuscript.

Funding: This research was partially sponsored by the Ministry of Science and Technology (MOST), Taiwan, under no. MOST108-2221-E-006-187 and 110-2221-E-006-177.

Data Availability Statement: Data are the coauthors' research results and drawing.

Acknowledgments: We also would like to thank the Core Facility Center in the National Cheng Kung University for analysis instrument support.

Conflicts of Interest: The authors declare no conflict of interest.

References

1. Vallem, V.; Sargolzaeiaval, Y.; Ozturk, M.; Lai, Y.C.; Dickey, M.D. Energy harvesting and storage with soft and stretchable materials. *Adv. Mater.* **2021**, *33*, 2004832. [CrossRef] [PubMed]
2. Liu, L.; Guo, X.; Lee, C. Promoting smart cities into the 5G era with multi-field Internet of Things (IoT) applications powered with advanced mechanical energy harvesters. *Nano Energy* **2021**, *88*, 106304.
3. Xu, Q.; Wen, J.; Qin, Y. Development and outlook of high output piezoelectric nanogenerators. *Nano Energy* **2021**, *86*, 106080. [CrossRef]
4. Hanani, Z.; Izanzar, I.; Amjoud, M.B.; Mezzane, D.; Lahcini, M.; Uršič, H.; Prah, U.; Saadoune, I.; Marssi, M.E.; Luk'yanchuk, I.A.; et al. Lead-free nanocomposite piezoelectric nanogenerator film for biomechanical energy harvesting. *Nano Energy* **2021**, *81*, 105661. [CrossRef]
5. Moradian, K.; Raghebi, M.; Fanaei, T. Fabrication and Investigation of a Millimeter-Scale Electromagnetic Generator for Large-Amplitude Impact Motions. *FME Trans.* **2022**, *50*, 65. [CrossRef]
6. Wang, L.; Wang, H.; Fu, M.; Xie, Z.; Liang, J. Three-Port Power Electronic Interface with Decoupled Voltage Regulation and MPPT in Electromagnetic Energy Harvesting Systems. *IEEE Trans. Ind. Appl.* **2022**, *58*, 2144–2154. [CrossRef]
7. Wu, C.; Wang, A.C.; Ding, W.; Guo, H.; Wang, Z.L. Triboelectric nanogenerator: A foundation of the energy for the new era. *Adv. Energy Mater.* **2019**, *9*, 1802906. [CrossRef]
8. Wang, Z.L. Triboelectric nanogenerator (TENG)—Sparking an energy and sensor revolution. *Adv. Energy Mater.* **2020**, *10*, 2000137. [CrossRef]
9. Lin, L.; Chung, C.K. PDMS Microfabrication and Design for Microfluidics and Sustainable Energy Application: Review. *Micromachines* **2021**, *12*, 1350. [CrossRef]
10. Wang, Z.L. Triboelectric nanogenerators as new energy technology for self-powered systems and as active mechanical and chemical sensors. *ACS Nano* **2013**, *7*, 9533–9557. [CrossRef]
11. Wong, T.H.; Liu, Y.; Li, J.; Yao, K.; Liu, S.; Yiu, C.K.; Huang, X.; Wu, M.; Park, W.; Zhou, J.; et al. Triboelectric Nanogenerator Tattoos Enabled by Epidermal Electronic Technologies. *Adv. Funct. Mater.* **2022**, *32*, 2111269. [CrossRef]
12. Ke, K.H.; Lin, L.; Chung, C.K. Low-cost micro-graphite doped polydimethylsiloxane composite film for enhancement of mechanical-to-electrical energy conversion with aluminum and its application. *J. Taiwan Inst. Chem. Eng.* **2022**, *135*, 104388. [CrossRef]
13. Yin, R.; Wang, D.; Zhao, S.; Lou, Z.; Shen, G. Wearable sensors-enabled human–machine interaction systems: From design to application. *Adv. Funct. Mater.* **2021**, *31*, 2008936. [CrossRef]
14. Gogurla, N.; Kim, Y.; Cho, S.; Kim, J.; Kim, S. Multifunctional and Ultrathin Electronic Tattoo for On-Skin Diagnostic and Therapeutic Applications. *Adv. Mater.* **2021**, *33*, 2008308. [CrossRef]
15. Singh, K.R.; Nayak, V.; Singh, J.; Singh, R.P. Nano-enabled wearable sensors for the Internet of Things (IoT). *Mater. Lett.* **2021**, *304*, 130614. [CrossRef]
16. Yao, S.; Ren, P.; Song, R.; Liu, Y.; Huang, Q.; Dong, J.; O'Connor, B.T.; Zhu, Y. Nanomaterial-enabled flexible and stretchable sensing systems: Processing, integration, and applications. *Adv. Mater.* **2020**, *32*, 1902343. [CrossRef]
17. Wu, W.; Haick, H. Materials and wearable devices for autonomous monitoring of physiological markers. *Adv. Mater.* **2018**, *30*, 1705024. [CrossRef]
18. Liu, J.; Gu, L.; Cui, N.; Xu, Q.; Qin, Y.; Yang, R. Fabric-based triboelectric nanogenerators. *Research* **2019**, *2019*, 1091632. [CrossRef]
19. Paosangthong, W.; Torah, R.; Beeby, S. Recent progress on textile-based triboelectric nanogenerators. *Nano Energy* **2019**, *55*, 401–423. [CrossRef]
20. Sim, H.J.; Choi, C.; Kim, S.H.; Kim, K.M.; Lee, C.J.; Kim, Y.T.; Lepro, X.; Baughman, R.H.; Kim, S.J. Stretchable triboelectric fiber for self-powered kinematic sensing textile. *Sci. Rep.* **2016**, *6*, 35153. [CrossRef]
21. Song, W.; Yin, X.; Liu, D.; Ma, W.; Zhang, M.; Li, X.; Cheng, P.; Zhang, C.; Wang, J.; Wang, Z.L. A highly elastic self-charging power system for simultaneously harvesting solar and mechanical energy. *Nano Energy* **2019**, *65*, 103997. [CrossRef]

22. Yang, Y.; Xie, L.; Wen, Z.; Chen, C.; Chen, X.; Wei, A.; Cheng, P.; Xie, X.; Sun, X. Coaxial triboelectric nanogenerator and supercapacitor fiber-based self-charging power fabric. *ACS Appl. Mater. Interfaces* **2018**, *10*, 42356–42362. [CrossRef] [PubMed]
23. Pu, X.; Li, L.; Song, H.; Du, C.; Zhao, Z.; Jiang, C.; Cao, G.; Hu, W.; Wang, Z.L. A self-charging power unit by integration of a textile triboelectric nanogenerator and a flexible lithium-ion battery for wearable electronics. *Adv. Mater.* **2015**, *27*, 2472–2478. [CrossRef] [PubMed]
24. Kwak, S.S.; Kim, H.; Seung, W.; Kim, J.; Hinchet, R.; Kim, S.W. Fully stretchable textile triboelectric nanogenerator with knitted fabric structures. *ACS Nano* **2017**, *11*, 10733–10741. [CrossRef] [PubMed]
25. Yi, F.; Zhang, Z.; Kang, Z.; Liao, Q.; Zhang, Y. Recent advances in triboelectric nanogenerator-based health monitoring. *Adv. Funct. Mater.* **2019**, *29*, 1808849. [CrossRef]
26. Wang, W.; Yu, A.; Liu, X.; Liu, Y.; Zhang, Y.; Zhu, Y.; Lei, Y.; Jia, M.; Zhai, J.; Wang, Z.L. Large-scale fabrication of robust textile triboelectric nanogenerators. *Nano Energy* **2020**, *71*, 104605. [CrossRef]
27. Cheng, R.; Dong, K.; Liu, L.; Ning, C.; Chen, P.; Peng, X.; Liu, D.; Wang, Z.L. Flame-retardant textile-based triboelectric nanogenerators for fire protection applications. *ACS Nano* **2020**, *14*, 15853–15863. [CrossRef]
28. Chung, C.K.; Ke, K.H. High contact surface area enhanced Al/PDMS triboelectric nanogenerator using novel overlapped microneedle arrays and its application to lighting and self-powered devices. *Appl. Surf. Sci.* **2020**, *508*, 145310. [CrossRef]
29. Ke, K.H.; Chung, C.K. High-performance Al/PDMS TENG with novel complex morphology of two-height microneedles array for high-sensitivity force-sensor and self-powered application. *Small* **2020**, *16*, 2001209. [CrossRef]
30. Trinh, V.L.; Chung, C.K. A Facile Method and Novel Mechanism Using Microneedle-Structured PDMS for Triboelectric Generator Applications. *Small* **2017**, *13*, 1700373. [CrossRef]
31. Trinh, V.L.; Chung, C.K. Harvesting mechanical energy, storage, and lighting using a novel PDMS based triboelectric generator with inclined wall arrays and micro-topping structure. *Appl. Energy* **2018**, *213*, 353–365. [CrossRef]
32. Askari, H.; Saadatnia, Z.; Asadi, E.; Khajepour, A.; Khamesee, M.B.; Zu, J. A flexible hybridized electromagnetic-triboelectric multi-purpose self-powered sensor. *Nano Energy* **2018**, *45*, 319–329. [CrossRef]
33. Chen, X.; Tian, H.; Li, X.; Shao, J.; Ding, Y.; An, N.; Zhou, Y. A high performance P (VDF-TrFE) nanogenerator with self-connected and vertically integrated fibers by patterned EHD pulling. *Nanoscale* **2015**, *7*, 11536–11544. [CrossRef] [PubMed]

sensors

MDPI

Article

Environment-Monitoring IoT Devices Powered by a TEG Which Converts Thermal Flux between Air and Near-Surface Soil into Electrical Energy

Tereza Paterova [1], Michal Prauzek [1,*], Jaromir Konecny [1], Stepan Ozana [1], Petr Zmij [2], Martin Stankus [1], Dieter Weise [3] and Alexander Pierer [3]

1 Department of Cybernetics and Biomedical Engineering, VSB—Technical University of Ostrava, 708 00 Ostrava, Czech Republic; tereza.paterova@vsb.cz (T.P.); jaromir.konecny@vsb.cz (J.K.); stepan.ozana@vsb.cz (S.O.); martin.stankus@vsb.cz (M.S.)

2 Brose CZ Spol. s r.o., 742 21 Koprivnice, Czech Republic; petr.zmij@brose.com

3 Fraunhofer Institute for Machine Tools and Forming Technology IWU, 09126 Chemnitz, Germany; Dieter.Weise@iwu.fraunhofer.de (D.W.); Alexander.Pierer@iwu.fraunhofer.de (A.P.)

* Correspondence: michal.prauzek@vsb.cz; Tel.: +420-59-732-5857

Abstract: Energy harvesting has an essential role in the development of reliable devices for environmental wireless sensor networks (EWSN) in the Internet of Things (IoT), without considering the need to replace discharged batteries. Thermoelectric energy is a renewable energy source that can be exploited in order to efficiently charge a battery. The paper presents a simulation of an environment monitoring device powered by a thermoelectric generator (TEG) that harvests energy from the temperature difference between air and soil. The simulation represents a mathematical description of an EWSN, which consists of a sensor model powered by a DC/DC boost converter via a TEG and a load, which simulates data transmission, a control algorithm and data collection. The results section provides a detailed description of the harvested energy parameters and properties and their possibilities for use. The harvested energy allows supplying the load with an average power of 129.04 μW and maximum power of 752.27 μW. The first part of the results section examines the process of temperature differences and the daily amount of harvested energy. The second part of the results section provides a comprehensive analysis of various settings for the EWSN device's operational period and sleep consumption. The study investigates the device's number of operational cycles, quantity of energy used, discharge time, failures and overheads.

Keywords: thermoelectric generator; energy harvesting; IoT; WSN; DC/DC boost converter; LoRaWAN

Citation: Paterova, T.; Prauzek, M.; Konecny, J.; Ozana, S.; Zmij, P.; Stankus, M.; Weise, D.; Pierer, A. Environment-Monitoring IoT Device Powered by a TEG Which Converts Thermal Flux between Air and Near-Surface Soil into Electrical Energy. *Sensors* **2021**, *21*, 8098. https://doi.org/10.3390/s21238098

Academic Editors: Slim Naifar, Olfa Kanoun and Carlo Trigona

Received: 2 November 2021
Accepted: 1 December 2021
Published: 3 December 2021

Publisher's Note: MDPI stays neutral with regard to jurisdictional claims in published maps and institutional affiliations.

1. Introduction

Interest in energy harvesting has been accelerating as the number of applications deployed long term in wireless sensor networks (WSN) and the Internet of Things (IoT) grows and the economic advantages over battery-based energy sources become more prevalent [1–3]. The topic of energy harvesting also relates to a subgroup of WSN called environmental WSN (EWSN) [4]. WSNs are typically used to obtain information to monitor the chemical, biological or population-related parameters of the environments under surveillance [5,6]. EWSNs are often deployed in remote areas, obstructing power supply from power grids and impeding regular battery subsystem maintenance [7]. Sources that provide energy through energy harvesting to operate wireless sensor nodes in environmental applications include solar radiation, temperature differences, flow-based systems (e.g., wind power), kinetic energy and radio frequency (RF) [8–11].

This contribution investigates thermal energy that can be converted into electrical energy by a thermoelectric generator (TEG) using spatial variations in temperature [12]. Thermoelectric transducers and generators are based on the Seebeck effect [13] and are composed of several pairs of p-type and n-type semiconductor blocks ordered in parallel and connected electrically in series [12]. The open circuit voltage of a thermoelectric element depends on the temperature difference (ΔT) between a hot and a cold surface and also on material properties called Seebeck coefficients [14,15]. The authors in [16] verified that TEG is a suitable solution for powering energy harvesting nodes. TEG can be also used to extend the battery life of devices by generating power from waste heat [17–19].

Thermal energy harvesting subsystems have been applied in many other outdoor applications; these are summarized in Table 1, which describes the state-of-the-art. The presented article explores TEG applications, which can directly convert temperature differences ΔT between soil and near-surface air. This TEG energy harvesting method, which exploits air and soil, has been studied in state-of-the-art publications listed in Table 2; these publications were used as references for the experiment proposed in the current study.

Table 1. Summary of TEG used in outdoor applications.

Author, Source	TEG/TEG Material	Purpose
Datta et al. [20]	TXL-287-03Z	LED lights for signage, illumination of roadways
Tahami et al. [21]	SP1848	LED traffic lights and wireless sensors
Lan et al. [22]	GM250-127-28-10	Wireless sensor node
Wang et al. [23]	Bi_2Te_3	Temperature sensor
Priya et al. [24]	TGM-127-1.4-2.5	Wearable biomedical IoT node
Praveena et al. [25]	$BhTe_3$, Sb_2Te_3	IoT location and temperature sensors
Seyoum et al. [26]	926-1192-ND	Temperature sensors and low power RF chips

Table 2. Summary of TEG reference solutions.

Label	Author, Source	TEG	DC/DC Convertor	Soil Depth (m)	ΔT (°C)	Power (μW)
R1	Ikeda et al. [27]	KTGM 199-2	LTC3109	0.3	2–35	94–369
R2	Huang et al. [28]	TG12-6-02	–	0.3–3.0	0–26.5	76–335
R3	Wang et al. [29]	TG12-8	LTC4071	2.5	3–25	200–324

The presented paper provides and investigates a simulation of an environment-monitoring IoT device powered by a TEG that produces energy through the temperature difference between air and soil. The TEG power source is connected to a boost converter, which links the generated power to the system's load, an energy storage unit (supercapacitor) and an embedded sensor (e.g., a temperature sensor to measure the temperature of an environment) that measures a certain parameter in the environment. A novelty of the current paper is presented in the use of the finite element method (FEM) to simulate the temperature difference between the lower and upper surfaces of the TEG in connection with the hardware consumption model. The obtained temperature gradient is used to drive simulation, including energy conversion and consumption model.

A major motivation of this study is the design of a rapid prototyping procedure that can evaluate advanced designs of TEG powered environmental sensors without any necessary infield deployment. The first step of the rapid prototyping FEM method provides very fast paced evaluation about the efficiency of the proposed hardware solution containing TEG energy harvesting element based on historical data from particular deployment location. Next, the energy converter can optimize hardware configuration capacitors and converter circuits. The last model of the EWSN node is powered by incoming harvested energy, and it can implement advanced energy management strategies based on machine learning in the future. The authors now present a simulation study that evaluates a rapid prototyping procedure; in the future, this approach allows the effective design of TEG-powered EWSN devices.

The paper is organized as follows: Section 2 describes related studies to this particular contribution. Section 3 provides the theoretical background to TEGs and boost converters and a description of temperature differences between air and soil. Section 4 describes an experiment with a module, e.g., sensor simulation and each part of the presented module and historical input data. Section 5 contains the results of the proposed experiments. Section 6 provides a discussion of the final results. Section 7 provides conclusions and a summary of the overall contributions of the presented paper and outlines potential challenges in future studies.

2. Related Work

Thermal energy is used in many outdoor applications as a reliable power source since it offers a less environmentally degrading means of obtaining power by reusing or recycling waste thermal energy. Another reason for the abundant application of TEGs is their capability of powering low-power electronics by exploiting even very small (only a few degrees Celsius) temperature gradients. Table 1 lists some applications and types of TEG in the state of the art.

Datta et al. [20] developed a device prototype for harvesting thermoelectric energy from heat on asphalt road and pavement surfaces. The energy was transferred to a TEG (TXL-287-03Z) embedded at the edge of the road surface. The results showed that the produced energy is sufficient for powering pavement health monitoring and roadway communications devices in off-grid areas [20]. The device can be also used as an alternative source for self-powered electronic road signs and markings and LED street illumination in remote areas.

The authors of [21] explored a novel approach for harvesting energy from roads and pavements, demonstrating the possibility of powering LED traffic lights and wireless sensors embedded into pavement structures from a TEG (SP1848) [21].

Rudolph et al. [22] developed a simulation that evaluated the possibility of powering a wireless sensor node using TEG technology. The authors connected a heat exchanger to a TEG to use the waste heat produced by a heavy-duty truck [22]. Yang et al. [23] proposed and developed a thermal management system that harvests and recycles the thermal waste of high-power light-emitting diodes (HP-LED) using a TEG that contains Bi_2Te_3 [23]. A voltage boost converter (LTC3108) was used to aid in powering a temperature sensor that monitored the surface temperature of the HP-LED.

Priya et al. [24] developed a thermal energy harvesting regulator driven by human body heat. The device contains a TEG and is suitable for powering compact, wearable biomedical IoT nodes with many features. It is ideal for sensors that can be worn on the body or low-power IoT systems that require more power for enhanced features yet have practical limitations in employing intermediate power storage between charging and regulating cycles [24].

Praveena et al. [25] used a TEG module containing $BhTe_3$ and Sb_2Te_3 to generate electrical energy from the heat dissipated by a vehicle's engine and exhaust silencer and subsequently powered IoT sensors. A DC/DC converter was used to boost the voltage to the required level since the initial produced voltage was insufficient for powering applications.

Seyoum et al. [26] presented a design for an ambient-powered wireless bolt for use in high-end electro-mechanical systems. The bolt is equipped with a temperature sensor and a low-power RF chip powered from a TEG. The bolt includes a DC/DC converter for raising the low TEG voltage and ensuring continuous wireless monitoring of these critical fasteners.

3. Background

This section provides a principle explanation of the individual model components: a TEG and a boost converter. This overview also describes the general regularities of temperature differences between underground soil and air.

3.1. Thermoelectric Generator

A TEG is a solid-state device that is able to directly convert heat flux into electrical energy through a phenomenon called the Seebeck effect [13]. In order to generate electrical energy, the TEG consists of many n-type and p-type semiconductors sandwiched between two electrically insulating materials that possess thermoelectric properties. When one side of a TEG is heated and the other side remains cool, voltage is generated through the Seebeck effect. The TEG principle is illustrated in Figure 1.

The p-type components are doped to provide a greater number of positively charged carriers or holes, thereby providing a positive Seebeck coefficient (α). In a similar manner, n-type components are doped to provide more negatively charged carriers, thereby providing a negative α. Table 3 lists the Seebeck coefficient values for chemical elements used in TEGs.

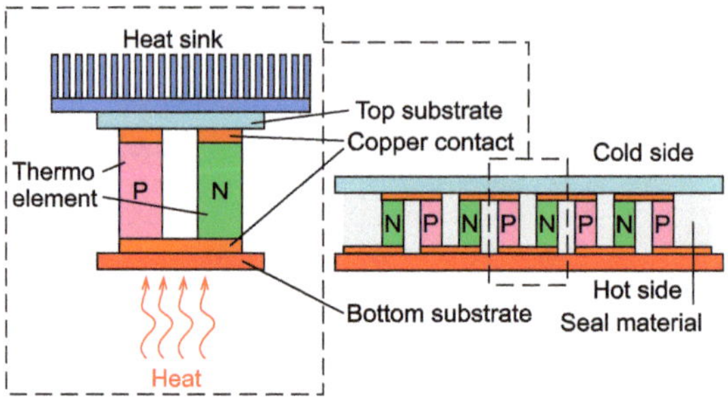

Figure 1. Principle scheme of a thermoelectric generator. The Seebeck effect is exploited to produce electrical energy.

As an electrical connection forms between the two junctions, every positively charged carrier moves to the n-junction and every negatively charged carrier moves to the p-junction. The most used element in thermoelectric generators is lead telluride (PbTe). For the flow of electrons, a suitable material with high electrical conductivity and low thermal conductivity was used. Compounds such as bismuth sulphide (Bi_2S_3), tin telluride ($SnTe$), bismuth telluride (Bi_2Te_3), indium arsenide ($InAs$) and germanium telluride ($GeTe$) are, therefore, frequently used for this purpose [30].

Table 3. Seebeck coefficient α values for various chemical elements (at 300 K) [31].

Element	Pb	In	Sn	Al	Au	Na	Ni	Ag
α ($\mu V/K^{-1}$)	−1.05	1.68	−1	−1.66	1.94	−6.3	−19.5	1.51

Using the ΔT from the TEG and the coefficient α, the induced thermoelectric voltage V_{TEG} can be expressed by the following formula.

$$V_{TEG} = \alpha \times \Delta T. \tag{1}$$

A TEG has many advantages over other types of energy harvester. For example, the TEG can be used to harvest a variable amount of power according to the ΔT of the TEG. A TEG also contains no moving parts, is able to function under extreme and zero gravitational forces, exhibits high fault tolerance and enables noiseless operation.

3.2. Boost Converter

A boost converter (also known as step-up converter) is one of the simplest types of switch-mode converters that can increase an input voltage. It raises the level of DC voltage from low to high while also decreasing the current from high to low. A boost converter consists of an inductor, a semiconductor switch, a diode, a capacitor and a load.

The principle of the boost converter is based on an inductor that resists changes in a current by either increasing or decreasing the energy stored in the inductor's magnetic field. When the switch is closed, current flows through the inductor, which stores some energy by generating a magnetic field. When the switch is opened, a reduction in the output current occurs since load impedance is higher. This causes a reduction in the magnetic field to maintain the current for the load and a reversal in polarity, which increases the voltage to charge the capacitor through the diode. If the switch is cycled quickly enough, the inductor is not fully discharged between charging stages and creates a greater output voltage than input voltage [32,33].

The boost converter can function in either continuous mode or discontinuous mode. In continuous mode, the current through the inductor never falls to zero, whereas in the discontinuous mode, the current can fall to zero and, hence, affect output voltage. A major advantage of a boost converter is its high efficiency of up to 50% [34].

3.3. *Temperature Difference between Soil and Air*

Soil temperature varies depending on whether it is surface soil (depth = 0 cm) or a deeper layer of soil (depth > 0 cm), called underground soil. Underground soil temperature is affected by temperature changes in the near-surface air, which varies cyclically both in daily (short cycle) and yearly (long cycle) conditions [35]. The temperature of underground soil at a depth of a few tenths of a metre is affected mainly by the annual long cycle of air temperature changes and not the daily short cycle. The greater the soil depth, the greater the temperature difference between the air and the soil. Ikeda et al. [27] concluded that the temperature difference between near-surface air and shallow underground soil at a depth of 0.3 m is suitable for power generation. At a depth of 0.5 m, the daily temperature fluctuations are very small, and at a depth of 1 m, the temperature does not fluctuate over the course of a day [35].

4. Experiment

This section describes an experimental sensor module composed of a TEG, sensor, boost converter and a load. The remainder of this section provides input data and reference solution descriptions to evaluate the model.

A scheme of the setup is shown in Figure 2. Each block represents a mathematical description of EWSN. These blocks are described below in the following subsections.

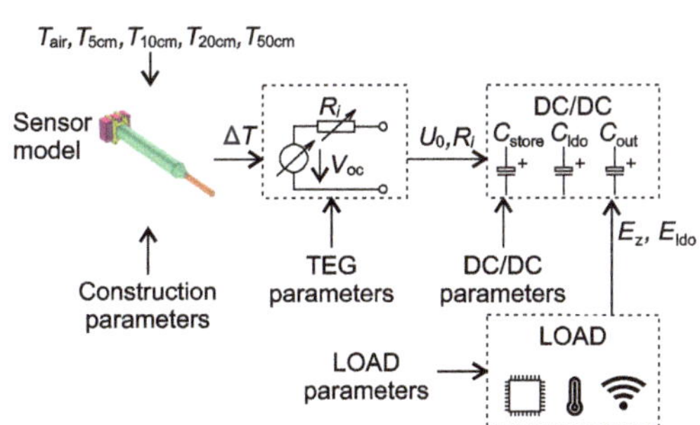

Figure 2. Block diagram of the simulation model. The sensor module supplies a DC/DC converter through the TEG module. The load simulates data transmission, control algorithm and data collection.

4.1. *Input Data*

The input data used in the experiment were obtained from the Czech Hydrometeorological Institute [36], which provides certified data. The founder of this organization is the Ministry of the Environment of Czech Republic. The data contained values of air temperature and near-surface soil temperature collected over a period of one year (2016). The air temperature values were measured at 10 min intervals in the Churanov area in Czech Republic. The soil temperature values were measured at the same location and times at depths of 0.05 m, 0.1 m, 0.2 m and 0.5 m. The Churanov Monitoring station in the Czech Republic is located at the coordinates 49.0683° latitude, 13.615° longitude and 1117.8 m elevation.

4.2. Sensor Simulation Model

This section describes the sensor simulation setup and sensor composition. For the purposes of implementation, the finite element method (FEM) was applied to simulate the temperature difference between the lower and upper surfaces of the TEG. The simulation model was created in Comsol Multiphysics 5.6 software with the Heat Transfer Module.

Figure 3 shows the composition of the sensor divided into eleven domains. The purple area depicts the heat sink, the yellow area represents a holder with the TEG and copper base, the green area represents the isolation tube and the orange area represents a copper rod.

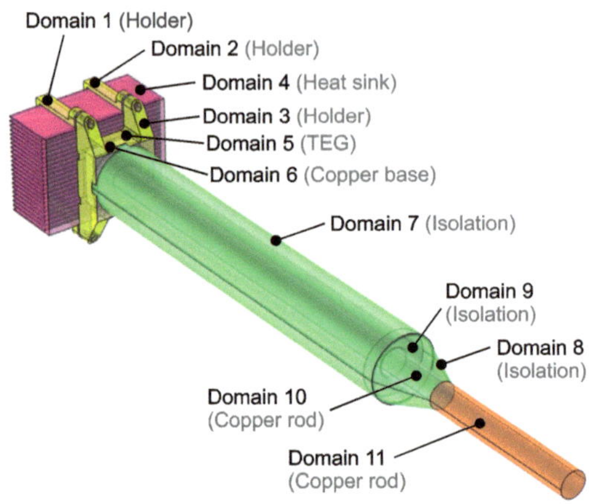

Figure 3. Sensor composition: Domain 1—Holder, left part; Domain 2—Holder, right part; Domain 3—Holder, lower part; Domain 4—Heat sink; Domain 5—TEG; Domain 6—Copper base; Domain 7—Isolation tube, main part; Domain 8—Isolation tube, lower conical part; Domain 9—Inner insulation fill; Domain 10—Copper rod, middle part; Domain 11—Copper rod, lower active part.

The simulation procedure comprised several stages:

1. Addition of a 3D component: When a new 3D component is added to the Comsol model builder, the geometry must be defined. In this case, the geometry was imported from the original 3D model created in Autodesk Inventor Professional 2022. The transfer of geometry was performed using a STEP file. Some post-import actions were performed in order to adjust the geometry for simulation purposes. In particular, because of the different boundary conditions, the copper rod was divided into two sections with a workplane to form two separate domains (Domain 10; Domain 11). Suitable boundaries denoted "selections" in the Comsol notation for several of the domains were selected and grouped into four categories for further processing, and different boundary conditions were applied.

2. Global definitions: In this section of the Comsol Model Builder, additional functions were set for required purposes, i.e., an interpolation function to define the temperature profile in the soil, a function table containing measured air temperatures and a function table containing measured soil temperatures at several predefined depths (i.e., 5, 10, 20 and 50 cm).

3. Addition of physics: Due to the fact that the nature of the explored phenomenon is thermally spread in a solid mass, the Heat Transfer in Solids interface (ht) was used for this part of the simulation. This interface is generally used to model heat transfer in solids by conduction, convection or radiation.

4. Definition of Materials: The model includes several types of material that had to be defined for the purposes of the simulation. Table 4 lists the materials applied in the simulation from the built-in Comsol material library.

Table 4. Materials applied from the Comsol material library to the various domains.

Material	Selection
Aluminium	Domain 4
Bismuth telluride	Domain 5
Copper	Domains 6,10,11
Polypropylene	Domains 7,8
Polyurethane	Domain 9
Polylactic acid, PLA	Domains 1,2,3

5. Definition of the Heat Transfer problem: In order to launch the simulation, the initial conditions had to be defined. In this case, by using the "Initial Values" context menu of the Heat Transfer interface, the initial value for the entire model (all domains) was set to 15 °C. Using the "Temperature" context menu of the Heat Transfer interface, the next step defined the Dirichlet for the boundary conditions. Table 5 contains details of the conditions of the outer surface temperatures on the sensor model.

Table 5. Outer surface temperatures on the sensor model.

Category (Selections)	Description
heat_sink_surfaces	Outer surfaces of the heat sink (Domain 4) exposed to ambient temperature (air temperature)
holder_surfaces	Outer surfaces of the holder (Domains 1, 2, 3) exposed to ambient temperature (air temperature)
isolation_tube_surfaces	Outer surfaces of the isolation tube (Domains 7, 8) exposed to ambient temperature (soil thermal heat source approximated by linear interpolation of the temperatures at 5, 10, 20, 50 cm depth)
copper_active_area_surfaces	Outer surfaces of the active copper part (Domain 11) exposed to ambient temperature (soil thermal heat sources at 50 cm depth)

All outer surfaces above the ground were exposed to ambient air temperature. The outer surfaces of the isolation tube were exposed to the soil heat source according the interpolated temperature profile. Finally, the uncovered active end of the copper rod was exposed to the soil heat sources corresponding to a depth of 50 cm.

6. Addition of a study: Using "Add Study" from the main menu, a time-dependent study was added to reflect the dynamics of the explored process. Since the definition of the problem incorporated only one type of physical phenomenon (heat transfer in solids), specification of the vector of the output times where the solution would be sought was necessary. In this case, a range in seconds reflecting the entire period of the solution (1 year) was selected, the time intervals being the period between two air temperature measurements (soil temperature was measured once every hour).

7. Post-processing (results) Under the "Datasets" node in the Model Builder, the objects of interest (points and surfaces) where the results would be displayed were defined. Under the "Results" node in Model Builder, a 3D, 2D or 1D plot group according to the particular objects of interest can be added.

Figure 4 illustrates the temperature distribution in the sensor, showing temperature changes in the soil from the surface to a certain depth.

The relevance of the simulation is ensured by the fact that no user component, such as equation modification, and no custom parameters in physics and materials were used. In particular, in this case, used materials have great impacts on simulation relevance. In this simulation, all materials are defined in the COMSOL Material Library and exactly agrees with the materials that are designed for prototype devices. Therefore, the properties of the materials and, thus, the results of the simulation are guaranteed by the COMSOL supplier. More detailed information about COMSOL simulation and materials can be found in [37–39].

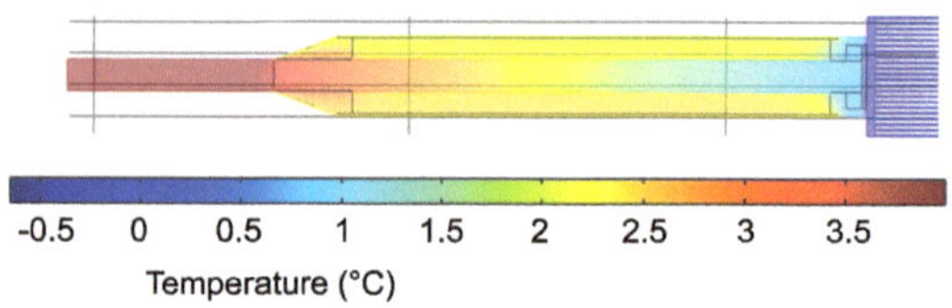

Figure 4. Model of the temperature distribution in the sensor. Red represents higher temperature, and blue represents lower temperature.

4.3. Thermoelectric Generator Block

A TEC1-12706 [40] module specified by the manufacturer for cooling was used for the experiment. This TEG can be exposed to various temperature differences to generate a suitable amount of electrical energy. The TEG's properties were obtained by experimental measurement of current and voltage characteristics for each temperature difference. The experimental setup equipped by TEG is shown in Figure 5. This setup contains TEG itself, heating by water and cooling by an air. The hot water is pumped by an electrical pump from a small tank to one side of the TEG. On the second side, there is a heat sink with a fan in order to ensure heat dissipation. Moreover, two temperature sensors are placed on both sides of the TEG.

Figure 5. Experimental setup equipped by TEC1-12706. The fan cools the TEG from one side, and hot water is pumped onto the second side of the TEG.

The open-circuit voltage V_{oc} and internal resistance R_i were calculated for each temperature difference from these characteristics. The characteristics of the TEG model can be expressed according to Equation (2):

$$V_{oc},\ R_i = f(\Delta T), \tag{2}$$

where V_{oc} is the TEG's open-circuit voltage, R_i is the internal resistance and ΔT is temperature difference between the hot and a cold sides.

Figure 6 graphs the $V_{oc}(\Delta T)$ and $R_i(\Delta T)$ of the module according to temperature change. The open-circuit voltage demonstrates an approximately linear function, while the internal resistor value follows a mostly nonlinear function. The $V_{oc}(\Delta T)$ and $R_i(\Delta T)$ values were used to determine a Thévenin-equivalent series circuit.

4.4. Boost Converter Block

TEG produces an open circuit voltage from tens to hundreds of millivolts, which is insufficient for the direct supply of electrical devices such microcontrollers or transmission modules. The voltage can be boosted by a DC/DC converter, for example, an LTC3109 module, which is a device dedicated to converting electrical energy from extremely low input voltage sources such as TEGs [41]. It allows operation with a voltage input as low as 30 mV and both input polarities. The LTC3109 module provides an output of 2.2 V for very low power systems and one extra output of 2.35–5 V [41].

The mathematical model of the DC/DC converter was designed with respect to the main physical aspects and basic functionality of the LTC3109. The functionality of the DC/DC converter was evaluated by measurement using experimental printed circuit layout boards, which are shown in Figure 7. This setup allows measuring voltage and current on the input and output pins and V_{STORE} as well. By using these parameters, the efficiency of DC/DC conversion could be evaluated. Moreover, the efficiency of charging and discharging of the supercapacitor could be measured.

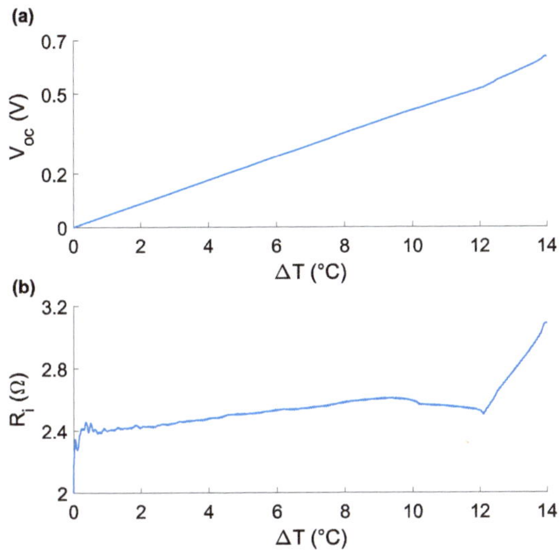

Figure 6. (**a**) Graph of open-circuit voltage V_{oc} according temperature difference ΔT; (**b**) graph of internal resistance R_i according temperature difference ΔT.

Figure 7. Experimental printed circuit board equipped by LTC3109 DC/DC converter.

A typical application circuit described in the LTC3109 datasheet uses three additional capacitors as energy sources in the case that input energy is in short supply. The capacitor C_{LDO} (typically 2.2 µF) stores the V_{LDO} output, and the capacitor C_{OUT} (typically 470 µF) is connected to the V_{OUT} pin. The supercapacitor C_{STORE} (1 F, 5.25 V) is able to supply output when input provides insufficient power. In the presented study, the DC/DC converter model was used to produce $V_{OUT} = 2.35$ V output voltage.

The functionality of the boost DC/DC block is shown in Figure 8 as a demonstration of $V_{OUT} = 3.3$ V output voltage. This figure illustrates the dependency of the output voltages according to whether input power is present or absent. It also includes nine distinguishing indicators that are important in understanding the functionality of the LTC3109 module.

From the first indicator (*1), the input voltage (V_{TEG}) proceeds from zero to 0.3 V and C_{LDO} begins to charge, with a consequent rise in V_{LDO}. When V_{LDO} reaches 2.2 V, the output capacitor C_{OUT} begins to charge (indicator *2). The output capacitor C_{OUT} is directly connected to the V_{OUT} pin; therefore, the output voltage V_{OUT} corresponds to the C_{OUT} voltage. When V_{OUT} reaches 92.5% of nominal voltage (indicator *3), the power good (PGOOD) pin is activated. The storage capacitor C_{STORE} begins to charge once V_{OUT} reaches 100% of nominal voltage (indicator *4) and is fully charged when V_{STORE} is 5.25 V (indicator *5). The sequence tagged with indicator *6 shows how the LTC3109 functions with charged capacitors and no input power ($V_{TEG} = 0$). C_{STORE} is discharged by a load, and V_{OUT} holds nominal voltage until V_{STORE} drops (indicator *7) below V_{OUT} nominal voltage (3.3 V), and then V_{OUT} is equal to V_{STORE}. When V_{OUT} drops below 91% of nominal voltage, the power good (PGOOD) pin drops (indicator *8). LDO output V_{LDO} maintains 2.2 V until V_{STORE} drops below 2.2 V, when V_{LDO} also starts falling (indicator *9).

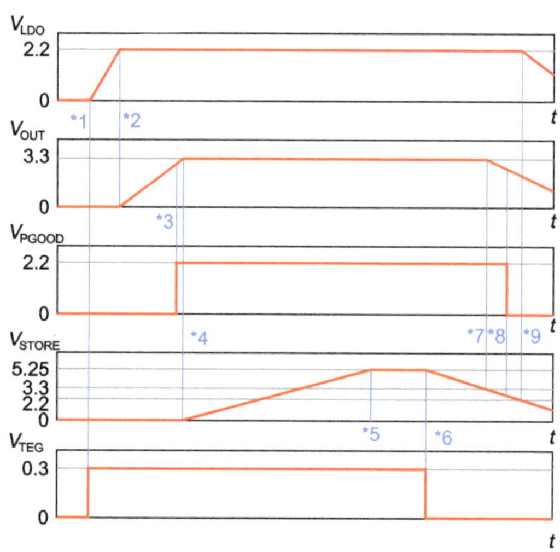

Figure 8. Output voltage sequences for input (V_{TEG}), storage (V_{STORE}), outputs (V_{LDO}, V_{OUT}) and status (V_{PGOOD}).

A mathematical model of the TEG was used as an input DC/DC converter to provide V_{oc} and R_i values. LTC3109 is able to optimize the power transfer from TEG; therefore, the ideal power transfer is assumed in this experiment. The input current to the DC/DC converter is calculated according to the following equation.

$$I_{in} = \frac{V_{oc}}{2 \cdot R_i} \tag{3}$$

The V_{TEG} voltage is calculated according to the following equation.

$$V_{TEG} = R_i \cdot I_{in} \tag{4}$$

Figure 9 charts the efficiency of voltage conversion in the LTC3109 module. The harvested energy can be calculated with the equation:

$$E_{Harvested} = V_{TEG} \cdot I_{in} \cdot \eta(V_{TEG}) \cdot \Delta t, \tag{5}$$

where $E_{Harvested}$ is the harvested energy, $\eta(V_{TEG})$ is the efficiency depending on V_{TEG} and Δt is the time between two simulation steps.

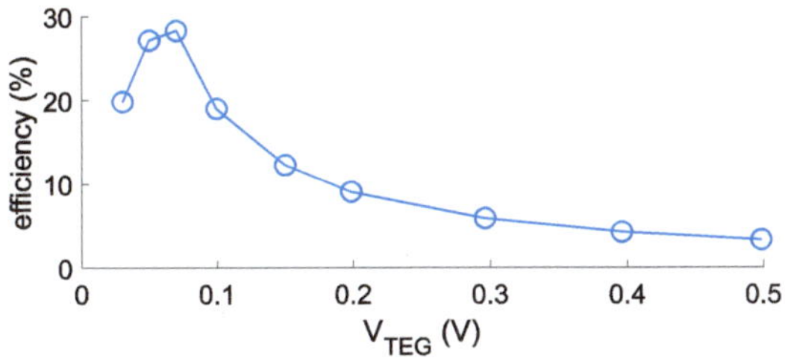

Figure 9. LTC3109 efficiency according to input voltage V_{TEG}.

The harvested energy $E_{Harvested}$ is used to supply a load and to charge the capacitors. The general equation that describes the relationship between capacitor voltage and energy is as follows:

$$E = \frac{1}{2} \cdot C \cdot V^2, \tag{6}$$

where E is the energy stored in the capacitor, C is the capacitance and V is the capacitor voltage. Equation (6) allows the calculation of the capacitor voltage according the current amount of energy stored in the capacitor.

The self-discharging model is based on the parameters of the supercapacitor KW-5R5C105-R [42]. The supercapacitor's datasheet specifies that the initial voltage drops to 70% after 2000 h; therefore, in one simulation step, the supercapacitor voltage decreases by approximately 0.002972% and can be described as follows:

$$E(k) = E(k-1) \cdot 29.72 \cdot 10^{-6}, \tag{7}$$

where E is supercapacitor energy, and k is simulation step.

The simulation of boost converter block also includes the charging and discharging efficiency of the supercapacitor. The calculation of efficiency is based on power dissipation on serial resistor R_{ESR} = 30 Ω [42]. The efficiency can be calculated as follows:

$$\eta = \frac{P}{P + R_{ESR} \cdot I_C^2}, \tag{8}$$

where P is charge/discharge power, I_C is capacitor current and η is the efficiency.

4.5. Load Block

The theoretical model device contains a power source, a microcontroller, an environmental sensor, non-volatile memory and a wireless communications interface. The microcontroller was selected specifically for low power consumption. Moreover, a modern, 32-bit device with a wide range of integrated peripherals is desirable for use. An NXP KL25Z [43] device based on an ARM Cortex-M0+ based microcontroller was also selected because it fulfills the requirements of the architecture and provides a number of low power modes that allow finetuning the energy profile.

An environmental sensor serves as a data source. It should be noted that no particular sensor is prescribed for the defined task, but a sensor of the type that is expected in an advanced design is still desirable; a Bosch BME688 4-in-1 sensor [44] was, therefore, selected. This device was able to measure ambient temperature, air humidity and atmospheric pressure, and the integrated gas sensor was able to detect volatile organic compounds (VOCs), volatile sulphur compounds (VSCs) and various other gases. The BME688 was connected to the microcontroller via the I2C bus.

Generally speaking, data measurement and data transmission are two asynchronous operations. The data transmission channel may not be available immediately when measurement is complete, although it may become ready independently of the measurement. A memory buffer is required to synchronize these two operations; a 24CW1280 EEPROM [45] was, therefore, selected, although FRAM technology was also considered. FRAM memory has many advantages, especially a fast write time and a large number of write cycles, but it was unsuitable for the model device since it runs with a voltage as low as 1.8 V. Unfortunately, no FRAM devices were available that could function with a voltage this low.

A LoRaWAN was considered a viable communications solution since it is one of the most popular communications solutions for IoT devices today; a Semtech SX1261 [46] LoRa transceiver was, therefore, selected as a communications link. The SX1261 was connected to the microcontroller via the SPI bus. LoRaWAN technology offers three communication classes (indicated as class A, B and C) that cover various use cases. Class A offers the best energy saving mode but only very limited downstream capabilities; class C provides a continuous downstream channel but is the most energy demanding since the modem is permanently powered.

Table 6 lists the parameters for power consumption of the simulated peripherals and their modes. The parameters were evaluated by using a LoRaWAN experimental setup depicted in Figure 10. The setup consist of semtech LoRa module, development board for LoRaWAN measurement, base board with MCU and power supply board. All boards include several test points that allows measuring voltage and current.

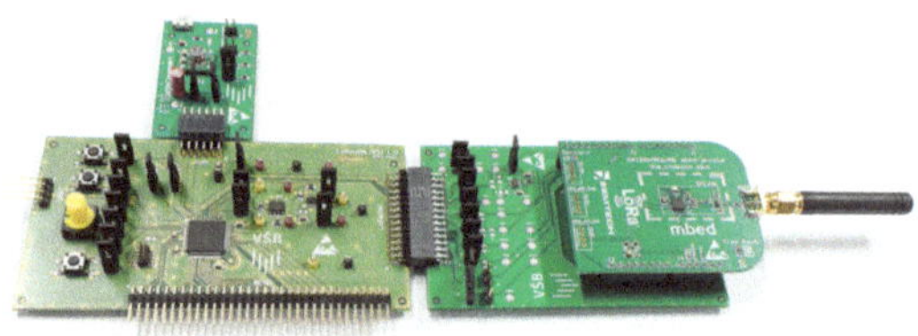

Figure 10. Experimental setup with MCU, LoRaWAN module and energy source.

Table 6. Load parameters of the system modes.

V_{LDO}	
MCU sleep mode (case 1)	$P_{SLEEP} = 3.63\ \mu W$
MCU sleep mode (case 2)	$P_{SLEEP} = 33.7\ \mu W$
MCU run	$P_{MCU} = 1.18\ \mu W$
MCU write to memory	$E_{MEM} = 11\ \mu J$
V_{OUT}	
Measurement	$E_{MEA} = 99\ mJ$
LoRaWAN transmission	$E_{TX} = 68.5\ mJ$
Total time for measurement and transmission	$t_{CYCLE} = 12.3\ s$

The experiment is based on switching two modes represented by sleep and run. If the device is in sleep mode, unused peripherals are deactivated, and the MCU is in VLPS. The required energy is calculated according to Equation (9).

$$E_{SLEEP} = P_{SLEEP} \cdot \Delta t \tag{9}$$

If run mode is active, the total consumption is composed of peripheral consumption (Equation (10)).

$$E_{RUN} = P_{MCU} \cdot t_{CYCLE} + E_{MEM} + E_{MEA} + E_{TX} + P_{SLEEP} \cdot (\Delta t - t_{CYCLE}) \tag{10}$$

From the technical documentation describing individual components, the power profile of the model device was then estimated. This estimation may not always be entirely accurate: For example, data caching may not be necessary because the data transmission channel is either immediately available or a measurement operation may require more time than expected. The estimation considered a variety of possible scenarios but discarded rare events such as data measurement failure or network provisioning operations performed on the wireless interface.

The power profile of the model device has two fundamental states: idle or operational. When the model device is idle, the remainder of the model device powers off. The only powered component in this mode is the MCU, but it does not execute instructions (in the MCU's technical documentation, this mode is called Very Low Leakage Sleep 1, VLLS1). VLLS1 mode is the lowest power mode in which the real-time clock (RTC) circuit is operational. This is an important factor since the RTC must wake up the CPU at a preset time. The current during VLLS1 mode (including the current of the RTC) is estimated to be in the range of 0.973 to 16.08 μA [43]. The relatively wide margin is caused by the current's heavy dependence on the ambient temperature. When the MCU exits sleep mode, a program is executed. Due to the fact that the model device has a low power character, the lowest power execution mode is selected. In the KL25Z MCU, this mode is called a Very Low Power Run, VLPR. Power consumption is greatly reduced in the VLPR mode, but the current may vary greatly. Different on-chip MCU peripherals can be enabled or disabled according to the needs of the application, but the individual machine instructions comprising the application draw different amounts of power. These are major factors that cause current variability in VLPR mode; the current is estimated to fall in the range of 171 to 777 μA [43].

The power-off state of the environmental sensor, EEPROM memory and LoRa transceiver is attained by using dedicated power switches. Individual devices are powered on and off sequentially. First, the BME688 sensor performs a measurement. According to the technical documentation [44], a measurement may take up 10.8 s, and the current during measurement is approximately 3.9 mA. One page of data is then written to the 24CW1280 EEPROM. This operation requires 5 ms, and the current is 1 mA [45]. After EEPROM access, a wireless transmission via the LoRaWAN network is performed. The class A communications profile is applied, and communication opens with a transmission burst, which takes around 500 ms with a current of up to 48 mA. Two similar receive operations are then attempted, each consisting of a 250 ms delay (I = 2.1 mA) followed by a 250 ms receive window stage (I = 8.2 mA) [45].

4.6. Reference Solutions

To evaluate the proposed TEG-powered module that harvests energy according to an air and near-surface ΔT, three reference solutions were applied. A comparison of the individual solutions is summarized in Table 2.

Ikeda et al. (R1) developed a sensor prototype driven by a TEG (KELK Ltd. KTGM 199-2, Hiratsuka, Japan) that was able to harvest on average more than 100 μW at a ΔT of 2–35 °C between the air and underground soil at a depth of 30 cm. [27]. Huang et al. (R2) used a TEG (TG12-6-02 from Marlow Industries) to generate electricity at a ΔT of 0–26.5 °C at a depth of 0.3–3.0 m. The results showed the feasibility of powering wireless sensors with a power of 76–335 μW [28]. Wang et al. (R3) used a hybrid system (photovoltaics and a TG12-8 from Marlow Industries) to provide a stable WSN output at a ΔT of 0–25 °C and depth of 2.5 m. The TEG produced an average output power of 200–324 μW [29].

By evaluating and comparing reference solutions, the following pros and cons were found. The advantage of the R1 article is a research impact by realizing a battery-free sensor. The advantage of the R2 article is proof of TEG powering wireless sensors in remote areas by performing experimental monitoring of devices for 6 months. The advantage of the article R3 includes providing stable power to WSNs by using a hybrid energy harvesting system.

4.7. *Performance Evaluation*

The simulation was evaluated by several criteria and performance characteristics. At first, the simulation was evaluated in terms of successfully (complete) cycles, missed cycles and the ratio. A missed cycle is a period during which transmission was required but energy was insufficient. The ratio represents the percentage of successful cycles and total periods. The ratio can be calculated as follows.

$$\text{Ratio} = \frac{\text{Complete}}{\text{Complete} + \text{Missed}} \cdot 100\% \tag{11}$$

The next evaluation parameter is maximal delay (Max. delay). The maximum delay is the maximum time of system outage caused by insufficient energy in the harvesting module. It is calculate as a maximal time between consecutive missed cycles.

The simulation is also evaluated in terms of energy usage. The unused energy E_U is sum of energy when the supercapacitor is fully charged and the load does not utilize all produced energy. The unused energy is expressed as the ratio between the sum of unused energy and sum of produced energy.

The simulation was performed for various load properties and time to discharge (TTD), and overheads were calculated. TTD was calculated as follows:

$$E_{\text{store}} = \frac{1}{2}C_{\text{store}} \cdot V^2_{\text{store max}} - \frac{1}{2}C_{\text{store}} \cdot V^2_{\text{store min}}, \tag{12}$$

where E_{store} denotes usable energy, C_{store} is the capacity of the supercapacitor, $V_{\text{storemax}} = 5.25$ V denotes maximum C_{store} voltage and $V_{\text{storemin}} = 2.35$ V is the minimal C_{store} voltage:

$$\text{TTD} = \frac{E_{\text{store}}}{E_{\text{run}} + E_{\text{sleep}}} \cdot \Delta T, \tag{13}$$

where TTD is time to discharge parameter, E_{run} is required energy for one cycle, E_{sleep} is required energy for one cycle of sleep and ΔT is the simulation period.

Overheads are defined as ratio between sleep mode energy E_{sleep}, and total energy consumption and can be calculated as follows.

$$\text{Overheads} = \frac{E_{\text{sleep}}}{E_{\text{sleep}} + E_{\text{run}}} \cdot 100\% \tag{14}$$

5. Results

The experimental section described a simulation performed using historical air and soil temperature data. Figure 11 shows the air and soil temperatures at several depths over the course of a year. Variation in the soil temperature decreased with soil depth; therefore, the variation in difference between air temperature and temperature in deeper soil increased.

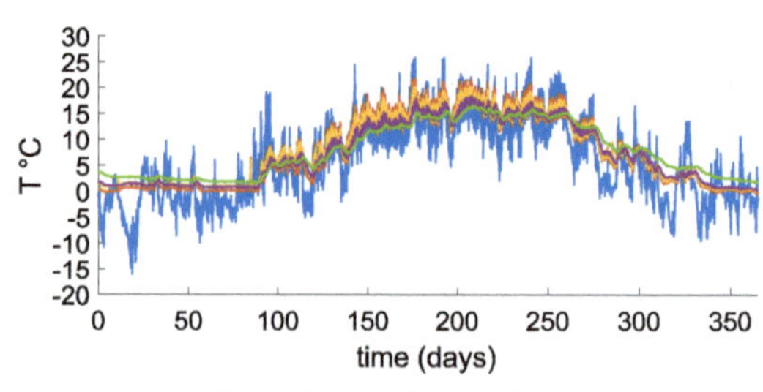

Figure 11. Input data: T is air temperature, T5 is soil temperature at 5 cm, T10 is soil temperature at 10 cm, T20 is soil temperature at 20 cm and T50 is soil temperature at 50 cm.

Soil temperature and air temperature are inputs for the sensor model to determine temperature difference (ΔT). The dataset of temperature differences was calculated as an output of the FEM simulation.

The active surface of the heat conductor is located at a depth of 50 cm. Figures 12a and 13a show the input temperatures acting on the active surface of the heat conductor and the heat sink's surface. Figures 12b and 13b show the difference between air and soil temperature, and the temperature difference on the TEG. Figure 12 charts measurements for the summer season: Air temperature oscillates around soil temperature, and the temperature difference decreases below the absolute value of one degree Celsius, resulting in the DC/DC converter not being able to function as required. Figure 13 charts measurements for the winter season: For most of the time, air temperature is either below or above soil temperature and provides a sufficient temperature difference. Air temperature varies with soil temperature during significant changes in the weather.

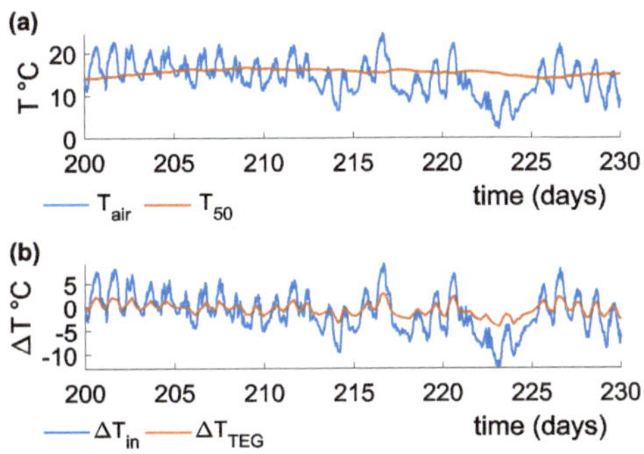

Figure 12. (**a**) Air and soil temperatures at a of depth 50 cm during summer, showing a large oscillation of air temperature around soil temperature; (**b**) temperature difference at the sensor input and temperature difference on the TEG.

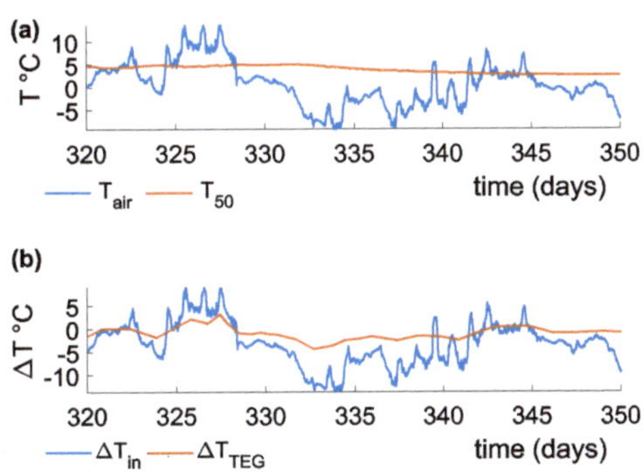

Figure 13. (**a**) Air and soil temperatures at a depth of 50 cm during winter. Air temperature differs from the soil temperature and creates a stable temperature difference; (**b**) temperature difference at the sensor input and temperature difference on the TEG.

Figure 14 shows the daily average temperature differences (absolute values) on the TEG and daily totals of harvested energy. The quantity of harvested energy is estimated by the TEG model and is highly dependent on the temperature difference between the cold and hot sides. Although the temperature difference between soil and air reaches up to 12 °C (Figure 11) the temperature difference on the surface of the TEG is not greater than 4 °C. Even so, the temperature difference on the TEG is mostly less than 4 °C, thereby allowing the DC/DC converter to function and harvest a small quantity of energy. The DC/DC converter is able to function as required when the input voltage is greater than 30 mV, which corresponds to a temperature difference of approximately 1 °C. Figure 14 also shows the quantity of harvested energy over the course of a day, the maximum quantity being around 60 J, although the median is only 4.82 J.

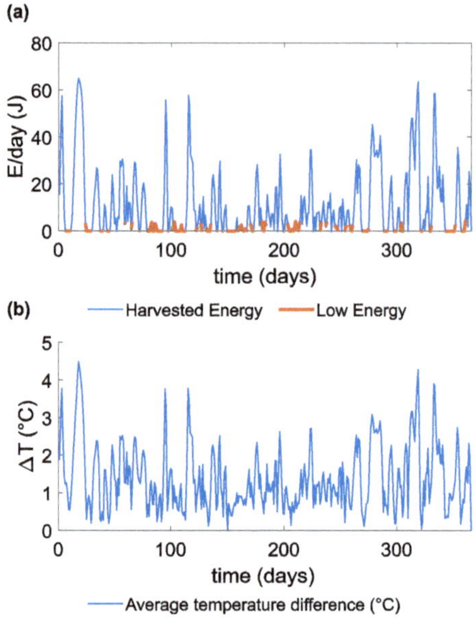

Figure 14. (**a**) Graph of daily harvested energy over the course of one year, where the quantity of harvested energy less than the median of harvested energy (4.82 J) is indicated as Low Energy; (**b**) graph of the average daily temperature difference on the TEG.

The simulation modeled the efficiency of DC/DC converter and also efficiency of charging and discharging os supercapacitor C_{STRORE}. The efficiency of the DC/DC converter is defined by a datasheet (see Figure 9). The charging efficiency calculation is based on power dissipation on R_{ESR}. Higher charging current causes lower charging efficiency. Assume the worst case, which is calculated as the maximum power from TEG used for supercapacitor charging (no energy is used by load). The supercapacitor is charged when V_{out} reaches 2.35 V, so charging efficiency in these conditions is at least 99.7%. The discharging efficiency is calculated for two cases: sleep and run mode. Assume that the supercapacitor volatage is approximately 2.35 V; thus, the output current is equal to the supercapacitor current (worst case). In this case, the efficiency is 99.9%. For the run mode, it is assumed that the average output current is 5.8 mA. For a minimal supercapacitor voltage of 2.35 V, the discharging efficiency is 93.1%, and for a maximal supercapacitor voltage of 5.25 V, the discharge efficiency is 98.5%.

Figure 15 shows the histogram of daily harvested energy over the course of one year. On more than 50% (186 from 365) of days, the DC/DC converter harvests less than 5 J of energy. Five joules of harvested energy is equivalent to an electrical current of 24.6 μA at a voltage of 2.35 V per day.

Table 7 shows the statistical parameters of daily harvested energy, daily average electrical power and calculated output current for the specified output voltage. The daily harvested energy at the 25th percentile was zero, indicating at least 91 days of the year when it was not possible to harvest any energy.

Harvested energy is used to power the load block, which consist of a microcontroller, a transmission module and a data collection module. The load block modes are controlled by the MCU, which switches between run and sleep modes. Implementation of this finite state machine allows several combinations of duty cycles. A simulation of the MCU duty cycle was executed for six cases. The load model measured and transmitted data simultaneously according to a defined period only when the power good pin was active. The simulation was performed for three period settings (10, 60 and 240 min) and two sleep current settings (1.65 μA and 15.3 μA).

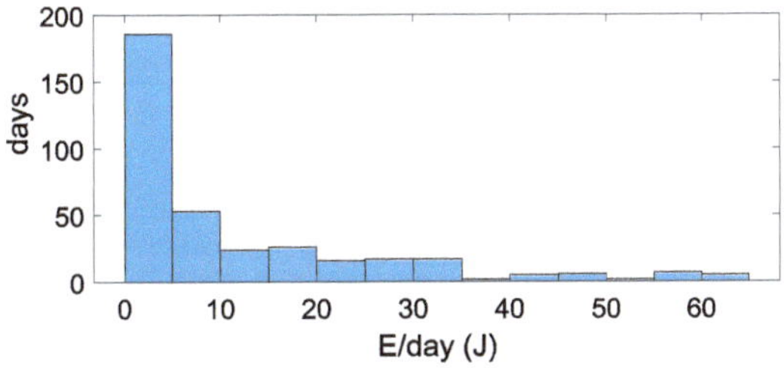

Figure 15. Histogram of daily harvested energy over the course of one year. On most days of the year, the quantity of energy harvested was between 0 and 5 J.

Table 7. Statistical parameters of daily harvested energy, daily average electrical power and calculated output current for the specified voltage.

	$\|\Delta T\|$	E (J/day)	P (μW)	I (μA) for $V_{OUT} = 2.35$ V
mean	1.40	11.15	129.04	54.91
25th	0.64	0.00	0.00	0.00
50th	1.23	4.82	55.78	23.74
75th	2.05	17.12	198.15	84.32
maximum	4.58	65.00	752.27	320.11

Table 8 documents the simulation results for complete and missed cycles over the course of one year. A complete cycle is an executed cycle, while a missed cycle is a cycle when run mode should be enabled but the available energy is insufficient. The ratio parameter represents the ratio of complete and total (complete + missed) cycles. The maximum delay is the maximum time of system outage caused by insufficient energy in the harvesting module. The unused energy parameter represents the ratio of unused energy and total harvested energy over the course of one year. Case 1 and Case 4 represent a short operation period; in these cases, the total number of complete or missed cycles reached the highest values. By contrast, Case 3 demonstrates the opposite, missing only 35 cycles over the course of the year. The EWSN functioned as required for 98.4% of the time, representing only 35 missed cycles out of a total 2190. In this case, the maximum delay was 2 days, representing 12 consecutive missed cycles. Other configurations experienced significantly larger maximum delays in the range of 7.3–10 days.

Table 8. Results of the simulated configurations. A missed cycle is a period during which transmission was required but energy was insufficient. The ratio represents the percentage of successful cycles and total periods. E_U is unused energy. TTD represents the time to discharge, and overheads represent consumption during sleep.

Case	I_{SLEEP} (μA)	Period (min)	Complete (-)	Missed (-)	Ratio (%)	Max. Delay (days)	E_U (%)	TTD (days)	Overheads (%)
1	1.65	10	18,751	33,810	35.7	10.0	17.3	0.42	1.2
2	1.65	60	6690	2070	76.4	7.5	66.7	2.37	6.7
3	1.65	240	2155	35	98.4	2.0	86.4	7.88	22.4
4	15.3	10	17,848	34,713	34.0	10.0	14.4	0.38	9.9
5	15.3	60	5588	3172	63.8	8.7	55.8	1.52	40.0
6	15.3	240	1723	467	78.7	7.3	69.5	2.76	72.8

The time to discharge (TTD) parameter represents operational time without input energy from the TEG and a fully charged supercapacitor. The configurations with a short period (one and four) did not store additional energy, and during periods of insufficient input energy, energy was fully drained from storage after approximately 0.4 days, and the device stalled. Cases with long periods required lower quantities of energy. Case 3 was able to operate up to 7.88 days without input energy. The disadvantage of configurations with a long operational mode is that a large quantity of energy is used for sleep. The overhead represents the ratio of sleep consumption for two operational modes and the total energy consumption for one operational mode and sleep. Case 1 demonstrated the lowest overhead. Case 1 (1.2% overhead) applied a short sleep period and small sleep current. Cases 2, 3 and 4 demonstrated acceptable overheads of less than 25%. Cases 5 and 6 demonstrated relatively high overheads. Case 6 consumed 72.8% of energy in sleep mode. Table 8 also lists figures for unused energy E_U. Case 4 attained the highest power consumption, using 85.6% of harvested energy but missing 66% of transmission periods. Case 3 produced the lowest power consumption, using only 13.6% of harvested energy and missing only 1.6% of cycles.

Boxplots of the times between the first and last missed cycles are shown in Figure 16. Cases 1 and 4 represent configurations with the shortest default operational period. This scenario uses energy immediately, and no stored energy is available for situations when a lack of input energy occurs. The boxplot outliers represent situations of long-term energy outages. Cases 1 and 4 both have similar statistical parameters since power is instantly dissipated as a result of power consumption having minimal effect during sleep mode.

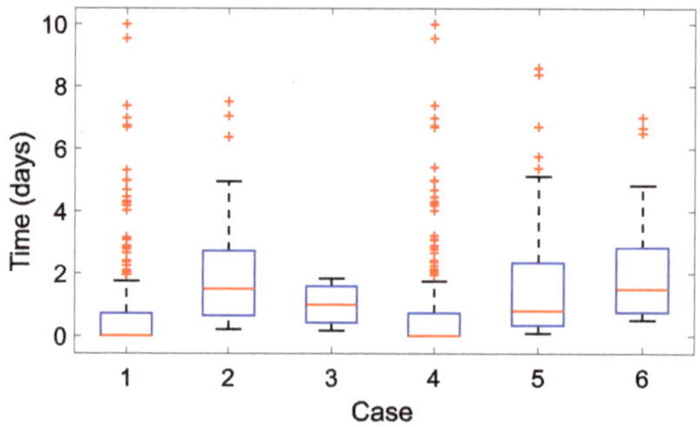

Figure 16. Boxplots of times between the first and last missed cycles in each configuration.

The results of the 60-minute interval in Case 2 indicate very low operational performance due to numerous outages and a median delay time of 1.5 days. Case 3 demonstrated the most suitable results. It is a very stable solution, with 50% of the times between the first and the last missed cycles falling in the range 0.42–1.58.

The scenarios with high sleep power consumption (Cases 5 and 6) produced results different from Cases 2 and 3, revealing a significant effect from sleep current. Case 5 demonstrated low performance operation, similarly to Case 2. In Case 6, sleep current had a major effect, resulting in low performance compared to the 60-minute configuration (Case 5).

Figures 17 and 18 chart the progress of the simulation. Figure 17 shows the EWSN configuration with a sleep current of 15.3 µA and 4-h period. Figure 17a displays the simulation results for the course of a year; Figure 17b provides a detailed view of a part of this period. The figures graph the simulated voltage parameters on the DC/DC converter's output. Voltage V_{STORE} reflects the current status of the supercapacitor, with a maximum value of 5.25 V. V_{OUT} and V_{LDO} are DC/DC outputs, and PGOOD is the output pin that indicates that a nominal voltage of V_{OUT} is present. The cycle curve illustrates an active operational mode.

It is interesting that the EWSN functions are suitable as long as V_{STORE} maintains 2.35–5.25 V. If V_{STORE} falls below the 2.35 V threshold, there will not be sufficient energy availble to supply the EWSN. This range also represents a safe range in which the supercapacitor is able to supply the EWSN when no energy is available from the input. The V_{STORE} voltage is limited to 5.25 V. If the supercapacitor attains this value, no more storage is available for incoming energy, and the EWSN will not use all the input.

The results in the graphs indicate that the EWSN is able to function without input energy when the supercapacitor is charged, i.e., without any additional input energy, the EWSN in Case 3 operated for up to 7.88 days (Figure 18), and the EWSN in Case 6 operated for up to 2.76 days (Figure 17).

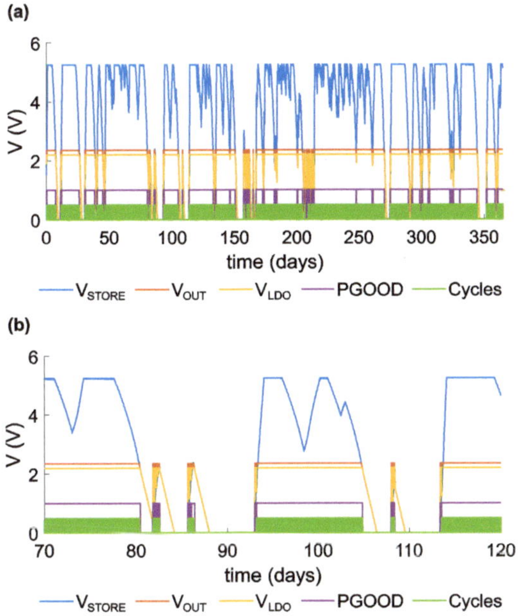

Figure 17. Graph of voltage levels; case I_{SLEEP} = 15.3 µA, T = 4 h. (**a**) One year; (**b**) detail of days 70–120.

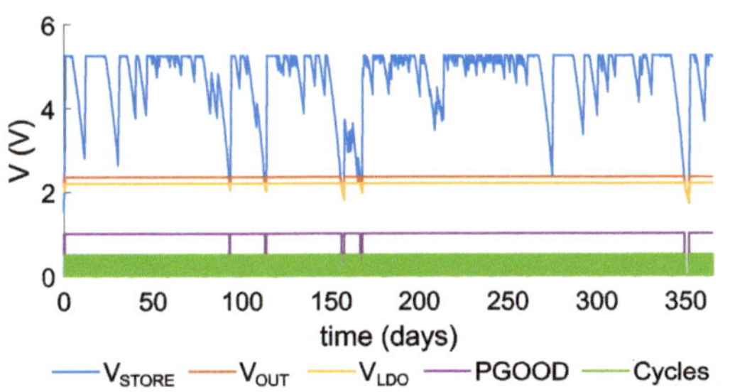

Figure 18. Graph of voltage levels; case I_{SLEEP} = 1.65 µA, T = 4 h.

6. Discussion

The results stimulate several interesting areas of discussion, one area being a comparison of the experimental results to reference solutions [27–29]. The results indicate a daily average electrical power in the range 0–320.11 µW, which is directly comparable to the power generated by the TEGs listed in Table 2. The power ranges of the reference solutions are larger because of the TEG harvesting conditions and especially because of a larger range in the input temperature difference. The reference articles also describe infield experiments that bring facts about output power changes according to deployment location.

Another notable feature is the sensor module's design, particularly the sensor's length and ability to be deployed in a specific environment. The sensor's length affects the temperature difference on the cold and hot sides of the TEG built into the module, corresponding to a quantity of generated energy. Ikeda et al. [27] concluded that the temperature difference between near-surface air and shallow underground soil at a depth of 0.3 m can be used to generate power. Jury et al. [35] indicated that at a depth of 0.5 m in soil, the daily temperature fluctuations are very small and that at a depth of 1 m, the temperature does not change over the course of a day. Applying these findings, we simulated the temperature data at a depth of 0.5 m. The module can be deployed at various locations in diverse climates. The temperature of soil at a depth of a few units of tenths of a metre is affected mainly by the longer annual cycle of changes in air temperature and not by the shorter daily cycles. The module can, therefore, be located in any geographical area in the world without significantly affecting the principle of harvesting energy from the TEG used to power an IoT device that monitors environmental parameters.

Management of available energy for the module also presented challenges. Energy harvested from a TEG is boosted by a step-up (boost) DC/DC converter designed specifically for thermoelectric energy harvesting. The DC/DC converter (LTC3109) used in the experiment was most effective when operating with small temperature differences (above 1 °C). At a temperature of 2 °C, the efficiency of the converter declined sharply. Future circuit designs by the manufacturer would require improvements to this drop in efficiency in the device. It is also worth considering the use of another type of DC/DC converter, which is equally suitable for harvesting thermoelectric energy when such an electronic component becomes available on the market. The stability of each simulated configuration with regard to sleep consumption and time to discharge is also an interesting question (i.e., operation with a minimum of outages). The results showed that the device is the most stable if the operating period is long (240 min) and the sleep current is low (set to 1.65 μA). A trade-off between sleep consumption and the operating period parameters may be required to minimize sleep consumption and energy wastage. The technology presents a broad area of application in smart energy management strategies based on machine learning.

Experiments have demonstrated and verified that TEGs can be used as a suitable power source for maintenance-free EWSN devices without batteries. EWSN devices are equipped with MCUs, measurement sensors and transmission modules such as LoRaWAN modules. These devices function in duty cycle modes but are inactive (sleep mode) most of the time. This configuration allows the device to harvest sufficient energy for measurement and data transmission. The results of the presented study show that the experimental device suffered delays caused by a temporary lack of energy in the energy harvesting subsystem. In the case of a fixed duty cycle, the delays were only directly dependent on the incoming energy. Again, the technology presents a broad area of applications in smart energy management strategies based on machine learning.

7. Conclusions and Future Work

The article presented a simulation of an environment-monitoring IoT device powered by a TEG which exploits the differences in air and soil temperature. The device included a sensor, a TEG, a DC/DC boost converter and a load. The sensor module was used to power a DC/DC converter from a TEG, while the load simulated data transmission, a control algorithm and data collection. The simulation represented a mathematical description of an EWSN and applied historical air and soil temperature data. The energy harvested from the TEG was used to power a load block controlled by an MCU to operate in two modes (sleep and active). The MCU operation cycle was simulated in six configurations with different settings for different periods (10, 60 and 240 min) and sleep currents (1.65 μA and 15.3 μA). The results showed that 0 to 65 J of energy could be harvested daily. The design of the EWSN load module included an LTC3109 DC/DC converter that operated under defined conditions. The device's capacitor stored energy when insufficient energy was produced by the TEG.

Future work with this technology can be applied in several areas. One interesting area is optimization of the sensor's construction itself. In the present configuration, the sensor's weight, dimensions and material costs, especially for copper, are not optimal. The aim of future studies may involve reducing the dimensions of the sensor while maintaining sufficient input energy from the TEG. Another area involves optimization of the EWSN hardware, for example, selection of a superior storage capacitor, improved DC/DC converter design and embedding all components on an electronic board (MCU, sensors, data storage, IoT communication modules, etc.). One very interesting area is in designing energy management strategies to optimize the stability of the EWSN node and to specify its behaviour. Broad scope exists in implementing various machine learning methods to manage energy consumption when sufficient energy is available or the storage of energy in a local supercapacitor when a lack of energy is indicated. In terms of practical implementation, the authors of the presented study are now testing the presented solution in field conditions and are aiming to publish the results of long-term testing (months or a year).

Author Contributions: Conceptualization, M.P.; methodology, T.P. and M.P.; software, J.K. and S.O.; validation, P.Z.; formal analysis, D.W.; investigation, A.P.; writing—original draft preparation, T.P., J.K., M.S., S.O. and M.P.; writing—review and editing, T.P., J.K. and M.P.; visualization, J.K.; supervision, M.P. All authors have read and agreed to the published version of the manuscript.

Funding: This work was supported by the European Regional Development Fund in Research Platform focused on Industry 4.0 and Robotics in the Ostrava project, CZ.02.1.01/0.0/0.0/17_049/0008425 within the Operational Programme Research, Development and Education. This work was also supported by project SP2021/29, "Development of algorithms and systems for control, measurement and safety applications VII" of Student Grant System, VSB-TU Ostrava. This work has received funding from the European Union's Horizon 2020 research and innovation programme under grant agreement N° 856670.

Institutional Review Board Statement: Not applicable.

Informed Consent Statement: Not applicable.

Conflicts of Interest: The authors declare no conflict of interest.

References

1. Ramya, R.; Saravanakumar, G.; Ravi, S. Energy harvesting in wireless sensor networks. In *Artificial Intelligence and Evolutionary Computations in Engineering Systems*; Springer: New Delhi, India, 2016; pp. 841–853.
2. Rokonuzzaman, M.; Mishu, M.K.; Amin, N.; Nadarajah, M.; Roy, R.B.; Rahman, K.S.; Buhari, A.M.; Binzaid, S.; Shakeri, M.; Pasupuleti, J. Self-Sustained Autonomous Wireless Sensor Network with Integrated Solar Photovoltaic System for Internet of Smart Home-Building (IoSHB) Applications. *Micromachines* **2021**, *12*, 653. [CrossRef]
3. Chandragandhi, S.; Udayakumar, E.; Srihari, K. A Novel Approach on Renewable Energy Harvesting Using Internet of Things (IoT). In *Smart Sensors for Industrial Internet of Things*; Springer: Berlin/Heidelberg, Germany, 2021; pp. 271–285.
4. Prauzek, M.; Kromer, P.; Rodway, J.; Musilek, P. Differential evolution of fuzzy controller for environmentally-powered wireless sensors. *Appl. Soft Comput. J.* **2016**, *48*, 193–206. [CrossRef]
5. Prauzek, M.; Konecny, J.; Borova, M.; Janosova, K.; Hlavica, J.; Musilek, P. Energy harvesting sources, storage devices and system topologies for environmental wireless sensor networks: A review. *Sensors* **2018**, *18*, 2446. [CrossRef] [PubMed]
6. Musilek, P.; Prauzek, M.; Krömer, P.; Rodway, J.; Bartoň, T. Intelligent energy management for environmental monitoring systems. In *Smart Sensors Networks*; Elsevier: Amsterdam, The Netherlands, 2017; pp. 67–94.
7. Li, Y.; Hamed, E.A.; Zhang, X.; Luna, D.; Lin, J.S.; Liang, X.; Lee, I. Feasibility of harvesting solar energy for self-powered environmental wireless sensor nodes. *Electronics* **2020**, *9*, 2058. [CrossRef]
8. Jiao, P.; Borchani, W.; Hasni, H.; Lajnef, N. Enhancement of quasi-static strain energy harvesters using non-uniform cross-section post-buckled beams. *Smart Mater. Struct.* **2017**, *26*, 085045. [CrossRef]
9. Shaikh, F.K.; Zeadally, S. Energy harvesting in wireless sensor networks: A comprehensive review. *Renew. Sustain. Energy Rev.* **2016**, *55*, 1041–1054. [CrossRef]
10. Shad, R.; Steingart, D.; Frechette, L.; Wright, P.; Rabaey, J. Power Sources for Wireless Sensor Networks. In *Wireless Sensor Networks*; Karl, H., Wolisz, A., Willig, A., Eds.; Springer: Berlin/Heidelberg, Germany, 2004; pp. 1–17.
11. Akhtar, F.; Rehmani, M.H. Energy replenishment using renewable and traditional energy resources for sustainable wireless sensor networks: A review. *Renew. Sustain. Energy Rev.* **2015**, *45*, 769–784. [CrossRef]
12. Zulkepli, N.; Yunas, J.; Mohamed, M.A.; Hamzah, A.A. Review of Thermoelectric Generators at Low Operating Temperatures: Working Principles and Materials. *Micromachines* **2021**, *12*, 734. [CrossRef]

13. Nguyen, N.Q.; Pochiraju, K.V. Behavior of thermoelectric generators exposed to transient heat sources. *Appl. Therm. Eng.* **2013**, *51*, 1–9. [CrossRef]
14. Jaziri, N.; Boughamoura, A.; Müller, J.; Mezghani, B.; Tounsi, F.; Ismail, M. A comprehensive review of Thermoelectric Generators: Technologies and common applications. *Energy Rep.* **2020**, *6*, 264–287. [CrossRef]
15. Cui, H.; Guan, Y.; Chen, H.; Deng, W. A Novel Advancing Signal Processing Method Based on Coupled Multi-Stable Stochastic Resonance for Fault Detection. *Appl. Sci.* **2021**, *11*, 5385. [CrossRef]
16. Pullwitt, S.; Kulau, U.; Hartung, R.; Wolf, L.C. A feasibility study on energy harvesting from soil temperature differences. In Proceedings of the 7th International Workshop on Real-World Embedded Wireless Systems and Networks, Shenzhen, China, 4 November 2018; pp. 1–6.
17. Sullivan, O.; Mukhopadhyay, S.; Kumar, S. Thermoelectric Generators Embedded in Microelectronic Chip. In *ASME International Mechanical Engineering Congress and Exposition*; American Society of Mechanical Engineers: Houston, TX, USA, 2012; Volume 45257, pp. 961–968.
18. Jin, T.; Gao, S.; Xia, H.; Ding, H. Reliability analysis for the fractional-order circuit system subject to the uncertain random fractional-order model with Caputo type. *J. Adv. Res.* **2021**, *32*, 15–26. [CrossRef] [PubMed]
19. Jin, T.; Xia, H. Lookback option pricing models based on the uncertain fractional-order differential equation with Caputo type. *J. Ambient. Intell. Humaniz. Comput.* **2021**, 1–14. [CrossRef]
20. Datta, U.; Dessouky, S.; Papagiannakis, A. Harvesting thermoelectric energy from asphalt pavements. *Transp. Res. Rec.* **2017**, *2628*, 12–22. [CrossRef]
21. Tahami, A.; Gholikhani, M.; Dessouky, S. A Novel Thermoelectric Approach to Energy Harvesting from Road Pavement. In *International Conference on Transportation and Development 2020*; American Society of Civil Engineers: Reston, VA, USA, 2020; pp. 174–181.
22. Lan, S.; Yang, Z.; Chen, R.; Stobart, R. A dynamic model for thermoelectric generator applied to vehicle waste heat recovery. *Appl. Energy* **2018**, *210*, 327–338. [CrossRef]
23. Wang, N.; Gao, C.; Ding, C.; Jia, H.Z.; Sui, G.R.; Gao, X.M. A thermal management system to reuse thermal waste released by high-power light-emitting diodes. *IEEE Trans. Electron Devices* **2019**, *66*, 4790–4797. [CrossRef]
24. Priya, V.; Rajendran, M.K.; Kansal, S.; Chowdary, G.; Dutta, A. A human body heat driven high throughput thermal energy harvesting single stage regulator for wearable biomedical IoT nodes. *IEEE Internet Things J.* **2018**, *5*, 4989–5001.
25. Praveena, B.; Sowmiya, A.; Logeshwari, P.; Kowshalya, G.; Sonia, B.; Kavipriya, R. Thermo electric generator module in driving the vehicle and monitoring using IoT. In Proceedings of the 2018 2nd International Conference on Inventive Systems and Control (ICISC), Coimbatore, India, 19–20 January 2018; IEEE: Coimbatore, India, 2018; pp. 822–825.
26. Seyoum, B.B.; Rossi, M.; Brunelli, D. Energy neutral wireless bolt for safety critical fastening. *Sensors* **2017**, *17*, 2211. [CrossRef]
27. Ikeda, N.; Shigeta, R.; Shiomi, J.; Kawahara, Y. Soil-Monitoring Sensor Powered by Temperature Difference between Air and Shallow Underground Soil. *Proc. ACM Interact. Mobile Wear. Ubiquitous Technol.* **2020**, *4*, 1–22. [CrossRef]
28. Huang, Y.; Xu, D.; Kan, J.; Li, W. Study on field experiments of forest soil thermoelectric power generation devices. *PLoS ONE* **2019**, *14*, e0221019. [CrossRef]
29. Wang, H.; Li, W.; Xu, D.; Kan, J. A hybrid microenergy storage system for power supply of forest wireless sensor nodes. *Electronics* **2019**, *8*, 1409. [CrossRef]
30. Yao, C.J.; Zhang, H.L.; Zhang, Q. Recent progress in thermoelectric materials based on conjugated polymers. *Polymers* **2019**, *11*, 107. [CrossRef] [PubMed]
31. da Rosa, A. Chapter 5—Thermoelectricity. In *Fundamentals of Renewable Energy Processes*, 3rd ed.; da Rosa, A., Ed.; Academic Press: Boston, MA, USA, 2013; pp. 149–212. [CrossRef]
32. Shen, B.; Hendry, R.; Cancheevaram, J.; Watkins, C.; Mantini, M.; Venkatasubramanian, R. DC-DC converter suitable for thermoelectric generator. In Proceedings of the ICT 2005, 24th International Conference on Thermoelectrics, Clemson, SC, USA, 19–23 June 2005; pp. 529–531. [CrossRef]
33. Richelli, A.; Colalongo, L.; Tonoli, S.; Kovacs-Vajna, Z.M. A 0.2–1.2 V DC/DC boost converter for power harvesting applications. *IEEE Trans. Power Electron.* **2009**, *24*, 1541–1546. [CrossRef]
34. Tran-Dinh, T.; Pham, H.M.; Dao, B.P.; Hoang-Thi, H.; Pham-Nguyen, L.; Lee, S.G.; Le, H.P. Low-Power PMIC with Two Hybrid Converters for TEG Application. In Proceedings of the 2020 IEEE Asia Pacific Conference on Circuits and Systems (APCCAS), Ha Long, Vietnam, 8–10 December 2020; IEEE: Ha Long, Vietnam, 2020; pp. 15–18.
35. Jury, W.A.; Horton, R. *Soil Physics*; John Wiley & Sons: Hoboken, NJ, USA, 2004.
36. Czech Hydrometeorological Institute. 2021. Available online: https://www.chmi.cz/ (accessed on 1 November 2021).
37. Pryor, R.W. *Multiphysics Modeling Using COMSOL®: A First Principles Approach*, 1st ed.; Jones & Bartlett Learning: Burlington, MA, USA, 2009.
38. Pryor, R. *MLI Physics Collection*; David Pallai Mercury Learning and Information: Dulles, VA, USA, 2018.
39. Demir, M.E.; Dincer, I. Development of a hybrid solar thermal system with TEG and PEM electrolyzer for hydrogen and power production. *Int. J. Hydrogen Energy* **2017**, *42*, 30044–30056. [CrossRef]
40. TEC1-12706—Thermoelectric Cooler, 05-03-2021. Available online: https://peltiermodules.com/peltier.datasheet/TEC1-12706.pdf (accessed on 1 November 2021).

41. LTC3109—Auto-Polarity, Ultralow Voltage Step-Up Converterand Power Manager, 21-02-2021. Available online: https://www.analog.com/en/products/ltc3109.html (accessed on 1 November 2021).
42. KW-5R5C105-R—Farnell, 20-05-2021. Available online: https://cz.farnell.com/eaton-bussmann-series/kw-5r5c105-r/cap-1f-5-5v-super-radial/dp/2302170 (accessed on 1 November 2021).
43. NXP KL25Z MCU, 16-01-2021. Available online: https://www.nxp.com/design/development-boards/freedom-development-boards/mcu-boards/freedom-development-platform-for-kinetis-kl14-kl15-kl24-kl25-mcus:FRDM-KL25Z (accessed on 1 November 2021).
44. Bosch BME688 Sensor, 22-07-2021. Available online: https://www.bosch-sensortec.com/products/environmental-sensors/gas-sensors/bme688/ (accessed on 1 November 2021).
45. 24CW1280 EEPROM, 15-04-2021. Available online: https://www.microchip.com/wwwproducts/en/24CW1280 (accessed on 1 November 2021).
46. Semtech SX1261 LoRa Transceiver, 09-08-2021. Available online: https://www.semtech.com/products/wireless-rf/lora-core/sx1261 (accessed on 1 November 2021).

sensors

MDPI

Article

Solar Energy Harvesting to Improve Capabilities of Wearable Devices

Alba Páez-Montoro *, Mario García-Valderas, Emilio Olías-Ruíz and Celia López-Ongil

Department of Electronic Technology, Universidad Carlos III de Madrid, 28911 Madrid, Spain; mgvalder@ing.uc3m.es (M.G.-V.); olias@ing.uc3m.es (E.O.-R.); celia@ing.uc3m.es (C.L.-O.)
* Correspondence: alpaezm@ing.uc3m.es

Abstract: The market of wearable devices has been growing over the past decades. Smart wearables are usually part of IoT (Internet of things) systems and include many functionalities such as physiological sensors, processing units and wireless communications, that are useful in fields like healthcare, activity tracking and sports, among others. The number of functions that wearables have are increasing all the time. This result in an increase in power consumption and more frequent recharges of the battery. A good option to solve this problem is using energy harvesting so that the energy available in the environment is used as a backup power source. In this paper, an energy harvesting system for solar energy with a flexible battery, a semi-flexible solar harvester module and a BLE (Bluetooth® Low Energy) microprocessor module is presented as a proof-of-concept for the future integration of solar energy harvesting in a real wearable smart device. The designed device was tested under different circumstances to estimate the increase in battery lifetime during common daily routines. For this purpose, a procedure for testing energy harvesting solutions, based on solar energy, in wearable devices has been proposed. The main result obtained is that the device could permanently work if the solar cells received a significant amount of direct sunlight for 6 h every day. Moreover, in real-life scenarios, the device was able to generate a minimum and a maximum power of 27.8 mW and 159.1 mW, respectively. For the wearable system selected, Bindi, the dynamic tests emulating daily routines has provided increases in the state of charge from 19% (winter cloudy days, 4 solar cells) to 53% (spring sunny days, 2 solar cells).

Keywords: energy harvesting; internet of things; physiological sensors; solar energy; wearables; wireless communication; wireless sensor network

Citation: Páez-Montoro, A.; García-Valderas, M.; Olías-Ruíz, E.; López-Ongil, C. Solar Energy Harvesting to Improve Capabilities of Wearable Devices. *Sensors* **2022**, *22*, 3950. https://doi.org/10.3390/s22103950

Academic Editor: Slim Naifar

Received: 28 April 2022
Accepted: 20 May 2022
Published: 23 May 2022

1. Introduction

Wearables are smart devices that are usually worn as watches, bands, clothes or even skin implants. In Figure 1, some examples are presented. These devices normally include one or more sensors, a power source, a processing unit and wireless communication modules. The most common applications of these smart devices are healthcare [1], activity tracking and sports [2], education [3], etc.

The wearable devices market has been growing over the past few years. According to [4], this market will reach 150 billion EUR by 2028. Recently, due to the COVID-19 outbreak, the use of wearable devices has also increased, since many of them include physiological sensors and are helpful for remote patient monitoring [5].

Wearables are usually part of the Internet of Things (IoT) systems. IoT provides a platform to connect and exchange information between multiple elements, from wearable sensors that collect information from the physical world to remote servers that store data and/or command different actions [6]. The combination of wearable devices with communication and computing capabilities results in a Cyber-Physical System (CPS) [7].

This paper is focused on testing the viability of including energy harvesting techniques in a smart bracelet, which is part of an IoT system called Bindi (presented in [8]). Bindi consists of a bracelet and a pendant that collect physiological and acoustic data and send them to a mobile phone application (app), which is connected to a remote server. Bindi has been made to protect women against gender-based violence by automatically detecting fear with the measurements from smart sensors and embedded machine learning algorithms.

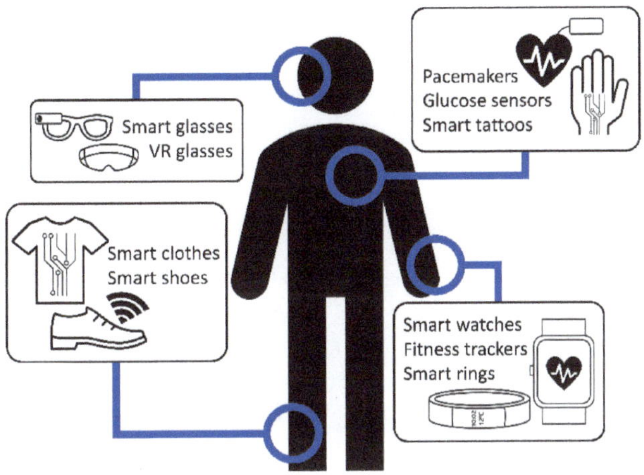

Figure 1. Examples of wearable devices.

According to [9], an IoT system can be divided into three main layers with different data computing capabilities: cloud, fog and edge. In Bindi, cloud computing is conceived as the remote server, fog computing as the mobile application and edge computing as the wearable devices. In Figure 2, the architecture of Bindi and the IoT layers for each component are shown.

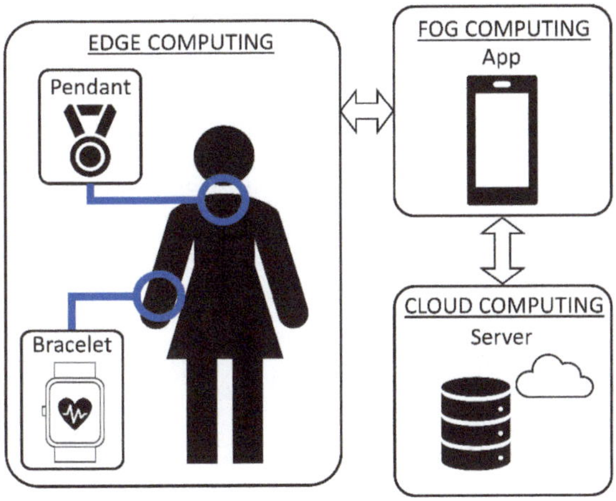

Figure 2. Bindi architecture as an IoT system.

The smart bracelet in Bindi has three physiological sensors: a heart rate sensor, an electrodermal activity sensor (EDA) and a temperature sensor. The sampling frequencies are 100 Hz, 10 Hz and 5 Hz, respectively. The data collected from them are processed (noise elimination filtering, artifacts removal, signal quality assessment, band frequencies selection, main feature extraction and machine learning algorithms), encrypted and sent wirelessly through a BLE (Bluetooth® Low Energy) protocol with a data rate of around 1 kBps. Even though the bracelet can automatically detect fear by using the embedded machine learning algorithms, ciphered raw data from the sensors are sent so that the physiological signals can be kept in the server and can be used as judicial evidence in gender-based violent crimes. The bracelet also has other functionalities, such as a panic button, an accelerometer, a buzzer or power management, that can be seen in Figure 3.

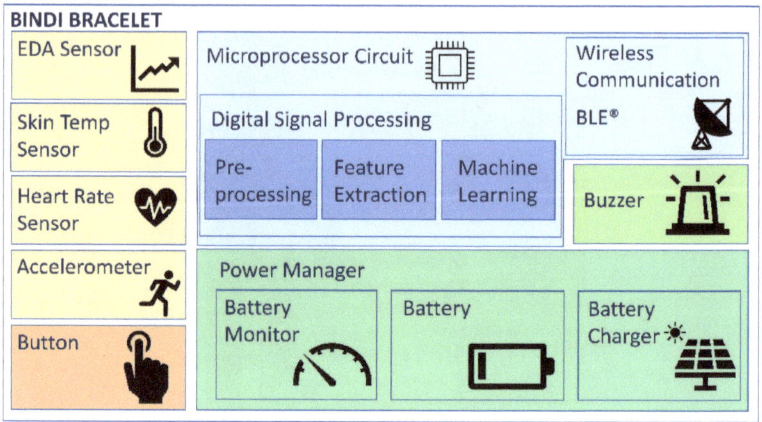

Figure 3. Bindi bracelet schematic.

The system includes a low power microprocessor, from Nordic Semiconductor® (nRF52840), which manages the sampling schedule for sensors, the I2C and SPI intranode communications, the tasks related to BLE wireless protocol including pairing, command reception and processing, as well as data continuous streaming and battery monitoring. The system implies a firmware of 512 kB of RAM and 64 kB of Flash memory, including the BLE stack. The average power consumption of the bracelet during its normal operation, three physiological sensors, accelerometer, battery monitoring and wireless data streaming, is 18.5 mW (5 mA, 3.7 V).

So many capabilities have a direct impact on the power consumption of the bracelet, and, in consequence, its battery life. The longer the battery life, the better the user experience, since the wearable device will not need to be charged as frequently, making it easier to use and more comfortable.

Wearable devices are powered by electrochemical Energy Storage Devices (ESDs), such as supercapacitors or rechargeable batteries, since these are efficient and clean technologies. Nowadays, rechargeable batteries lead the wearable market since they have a higher energy density than supercapacitors. Moreover, supercapacitors are traditionally larger and it is hard to integrate them within a wearable device. Even though a lot of research has been done into new experimental, flexible and small-sized supercapacitors, they are still inadequate for large scale, efficient fabrication [10].

Ideally, it would be more convenient to have wearable devices that can be totally powered with the energy available in their surroundings. Energy harvesting technologies are used to collect energy from the environment and use it to power a device. The most common energy harvesting sources are solar energy, thermal energy and mechanical energy. However, it is not possible to power a device only by using energy harvesting with the currently available technologies. Energy harvesters such as solar cells (which obtain power from the sunlight) or Thermoelectric Generators (TEGs) (which obtain power from temperature differences) have very low efficiencies and the energy they harvest is not always available (e.g., solar cells in the night). Hence, they are used as backup sources that work alongside rechargeable batteries. Using two or more energy sources in a hybrid energy harvesting system is also a good option, since one power source can back up the others when they are not available [11].

Even though solar energy is not always available, it has become the most used energy source for energy harvesting applications in wearables. Photovoltaic technologies have been developing since the discovery of the photovoltaic effect in 1839 by Becquerel [12]. There are three generations of solar cells, from the first generation of silicon solar cells to the second generation of III-V composites, to the third generation of perovskite, dye-sensitized, quantum dot and organic solar cells [13].

In summary, there must be a trade-off between wearable device capabilities and their consequent power consumption. The objective of this work is to design a solar energy harvesting system and test the viability of including solar cells in the smart bracelet of the Bindi system, to cope with its many capabilities. Solar cell performance depends on many factors, such as the operating conditions of the load (sensor measurements and connectivity), time, location and position of the solar cells, and it is difficult to theoretically estimate the generated power when they are included in a wearable device. Thus, the objective of this work is also to test the performance of the solar energy harvesting system included in a wearable device by using different daily life scenarios. Commercial devices will be used in this design, since they are more mature technologies and most energy harvesting wearable devices in the literature are experimental and do not consider large-scale production and fabrication costs [14]. Technology accessibility in the mid-term is a condition for Bindi design.

2. Related Work

Using energy harvesting not only increases the battery life of portable devices with the energy available in the environment, but it also reduces the ecological footprint compared to fossil fuels, the main power source used worldwide [15,16].

Solar energy is the most common energy source used in energy harvesting systems since it is possible to obtain tens or even hundreds of mW with relatively small-sized solar cells [17].

Solar cells are made of semiconductor junctions that can convert sunlight energy into electrical energy via the photovoltaic effect. Semiconductors have a characteristic energy level called a band gap. When incident photons have higher energy than this band gap, the solar cell will generate an electrical current.

Solar cells have a theoretical efficiency limit (called the Shockley–Queisser limit) at 33.7% on an ideal semiconductor with a band gap of 1.34 eV. This limit was calculated considering that all photons with energies higher than the band gap of the semiconductor are absorbed, but not all of them can be converted to electrical energy, since the band gap can change due to the temperature increase. Furthermore, only radiative recombination was considered. In real solar cells, there are more factors that affect the efficiency, so the actual limit is lower than the theoretical Shockley–Queisser limit (around 29% [18]). To overcome this intrinsic limit, exhaustive research into new materials and configurations must be done.

Nowadays, more than 90% of solar cells available in the market are silicon solar cells, due to their higher efficiency levels and their relatively low production costs [18,19]. Silicon solar cells can be divided into three groups depending on their internal structure: monocrystalline, polycrystalline and amorphous. Monocrystalline silicon solar cells have the highest efficiency, with a reported value of 26.7% [18]. On the other hand, amorphous silicon is mainly used to manufacture flexible solar cells, which are a good option for wearable devices in terms of comfort. However, their efficiency levels are still very low (around 12%).

Other semiconductor materials, such as GaAs, can also be used to manufacture solar cells. In [20], Alta Devices Inc. demonstrated an efficiency of 27.6% in a single junction GaAs thin-film solar cell. However, this semiconductor needs a good quality encapsulation when used in wearable devices due to the toxicity of the arsenic. On the other hand, it is possible to use more than one semiconductor junction (heterojunction solar cells), such as the one presented by LONGi Inc. with an efficiency of 26.3% [21]. Organic solar cells, like the one presented by Epishine [22], are also being studied, since they have potential for both roll-to-roll fabrication and good performance with indoor lighting. Recently, perovskite solar cells have achieved an efficiency value similar to silicon solar cells and it is estimated that the production costs would be halved. However, the stability of perovskite solar cells is still a challenge, since they suffer from fast degradation when exposed to typical ambient conditions [23,24].

Energy harvesting circuits usually follow the same structure. The input from the harvester goes to a conditioning circuit that optimizes the generated power, then the energy is stored in an ESD or is used to power an electronic load. Solar energy harvesting is no exception. The schematic shown in Figure 4 is a representation of the common structure for solar energy harvesting circuits. The input from the solar cell goes through a boost converter controlled by a Maximum Power Point Tracking (MPPT) algorithm. MPPT algorithms force the solar cells to work around their Maximum Power Point (MPP), which is normally at 80% of their open circuit voltage. The energy obtained from the solar cell is used to power a load or is stored in an ESD.

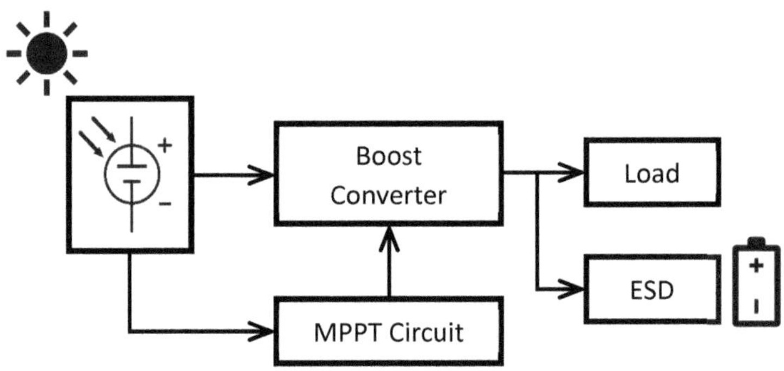

Figure 4. Schematic of a typical solar energy harvesting circuit.

As previously stated, in wearables, the load is typically an electronic system that includes one or more sensors, a processing unit and wireless communication devices. The ESD is usually a rechargeable battery whose chemical composition depends on the application.

In the 20th Century, NiCd rechargeable batteries were used to power portable devices, but cadmium is an expensive and toxic component, so they were mostly replaced by NiMH rechargeable batteries, which have a higher power density and a lower toxicity. Nowadays, most commercial wearable devices use Lithium-ion (Li-ion) or Lithium-ion polymer (LiPo) rechargeable batteries. Both of them have good values of power density and they can maintain a constant voltage (called nominal voltage) during almost all of the discharge, as seen in Figure 5 [25]. Moreover, it is possible to manufacture thin and flexible Li-ion and LiPo batteries. However, their main disadvantage is that they may explode when overcharged, so designers must be especially cautious and include battery protection circuits in their systems [26].

In this work, a solar energy harvesting system is designed to test the viability of including commercial monocrystalline silicon solar cells and a LiPo flexible battery in a wearable device with a high number of functionalities. Previous to this work, other energy harvesting applications in wearables have been studied from the literature so that the proposed system could be compared to other systems.

In [27] Toh et al. use a flexible solar cell with an MPPT circuit and a supercapacitor to power a wearable temperature sensor and a radio frequency transceiver. The wearable is able to work only with the energy harvested from the solar cells whenever it is not transmitting data from the temperature sensor. During the transmission, the average power consumption of the system is 3.91 μW and it needs to be powered from the supercapacitor. The device can work for 15 h since body temperature data are only sent once every minute.

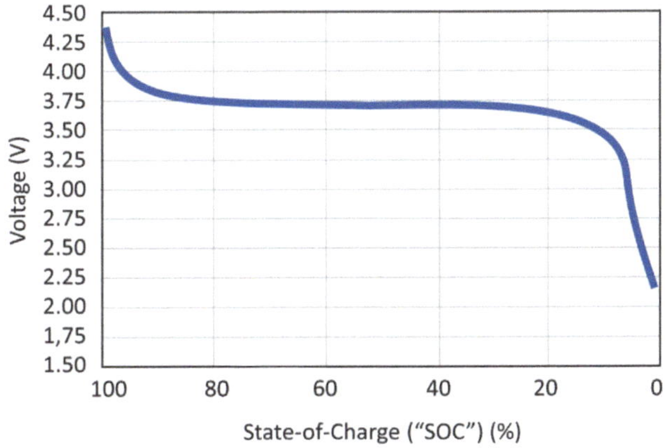

Figure 5. Typical Li-ion discharge curve [25].

In [28], Kim et al. use a TEG and an LiS battery to power a commercial glucose sensor with a power consumption of 64 µW. TEGs do not have high voltage levels when used in the human body due to the low temperature difference between the skin and the ambient. Thus, using an LiS battery is a good option, since its typical voltage is around 2 V, which is lower than the typical voltage in Li-ion and LiPo batteries (3.7 V). Moreover, LiS batteries have a higher energy density than Li-ion batteries. Nevertheless, they deteriorate faster, so they are not ready to replace commercial Li-ion batteries [29].

As for mechanical energy, in [30], Nastro et al. designed energy harvesters for wearables based on piezoelectric cantilevers. Piezoelectric materials generate power from mechanical energy, but they often need high frequency excitation. Since human movements are in the order of a few Hz, in this work, the authors use two and four piezoelectric cantilevers to improve the power generated by two different devices at low frequencies. They consist of an internal steel ball which moves when the user is moving and impacts on the piezoelectric cantilevers. They were able to generate a maximum instant power of 1.58 mW with an average power of 9.65 µW in 0.7 s when wearing the designed devices as a wristband and when moving the hand at a low frequency.

The average power values generated by the devices presented in [27,28,30] are in the order of µW, which is enough for wearables with only a few functionalities, such as the presented glucose sensor or the temperature sensor that sends one sample every minute. However, wearable smart devices include more functionalities, so these proposals would not be able to significantly impact on their power consumption.

On the other hand, a great amount of work has been done on the development of hybrid energy harvesting devices. The hybridization of two or more power sources can be done within the energy harvester itself or by using independent energy harvesters in one wearable device.

One hybrid system example is the jacket designed by Brogan et al. that includes 16 solar cells and 12 TEGs to charge a battery pack [31]. They obtained a maximum power output of 500 mW in a sunny day and outdoors. Note that the power generated from the solar cell was the main contribution to the total power, since power from the TEGs was on the order of the microwatts µW.

In [32], Magno et al. present a low power smart bracelet that combines various sensors with solar and thermal energy harvesting. Both the solar cells and TEGs are embedded in the band of the bracelet. They can generate up to 550 µW, with the solar cells indoor and up to 250 µW with the TEGs and a difference of 5 K.

In [33], Lemey et al. present a hybrid textile module that consists of a textile antenna integrated with a TEG and two solar cells to harvest solar energy in indoor and outdoor scenarios. When a user moves into a room with artificial light at 17.5 °C, the harvesters can generate up to 310.9 μW with a total system efficiency of 47.7%. Using the device in an outdoor scenario increases considerably the generated power up to 46 mW.

Another example is [34], where Liu et al. present a photo-thermo-electrochemical device. This device consists of a TEG covered with a photothermal polymer that improves solar light absorption and, thus, creates a higher temperature difference for the TEG to generate power. They can charge a supercapacitor over 250 mV. Even though the device has a good performance and the production costs were considered, wearable smart devices normally use rechargeable batteries, which work at a higher voltage than hundreds of mV.

Moreover, in [35], Ren et al. present a hybrid harvester for solar and mechanical energy. The device includes both a flexible organic solar cell and a triboelectric generator, which can generate power from the contact of two materials. The harvester is covered with a layer of groove-shaped micro/nanostructured haze thin film, which is a hydrophobic material that improves light trapping, triboelectric effect and encapsulation.

Another hybrid harvester is designed in [13] using organic solar cells and TEGs. Organic solar cells are usually not considered for commercial products, since their efficiency is lower than that of other solar cell technologies. However, the authors of this work increased the collected electrical energy by 46.3% for a temperature difference of 10 K.

After the analysis of the state-of-the-art energy harvesting systems, it was decided that commercial solar cells and a commercial flexible battery would be used. Even if some of the examples previously presented have good results, they are mainly experimental devices, meaning that production costs and long-term stability have not been considered. Furthermore, the energy harvesting systems presented in these studies are mainly integrated with low-power wearables that do not have a high number of functionalities, which is not the case in the study selected for this work, Bindi.

3. System Design

In this work, an electronic system has been designed as a proof-of-concept to test and validate a solar energy harvesting application proposal. Specifically, it has been made so that a microprocessor module with a power consumption similar to a smart wearable could be powered by a flexible LiPo battery and solar cells. Moreover, the device also offers the possibility of recharging the battery through a USB (Universal Serial Bus) charger in case a faster charge is needed. This system is intended to be included in a Bindi smart bracelet.

The schematic with the basic electronic components of the system is presented in Figure 6. They are described throughout the rest of this section.

The microprocessor module (μP Module in Figure 6) is the Aconno™ ACN52840, which is especially made for IoT applications. It includes the Nordic Semiconductor® nRF52840 SoC (System on Chip) and an on-board antenna for BLE communication, alongside other features. The SoC is the same as the one used in Bindi wearables. The module can be powered by a voltage between 1.7 V and 3.6 V. More information relating to the module can be found in [36].

Two power management Integrated Circuits (ICs) made by Texas Instruments® (Dallas, TX, USA) are used to recharge the battery and to power the microprocessor module. In Figure 6 they are presented as EH IC (for the Energy Harvesting power supply) and USB Charger IC (for the USB charging). They are never enabled at the same time, so a manual switch is used to select between them. Furthermore, both ICs include safety measures for battery charging such as overvoltage protection or charging current control.

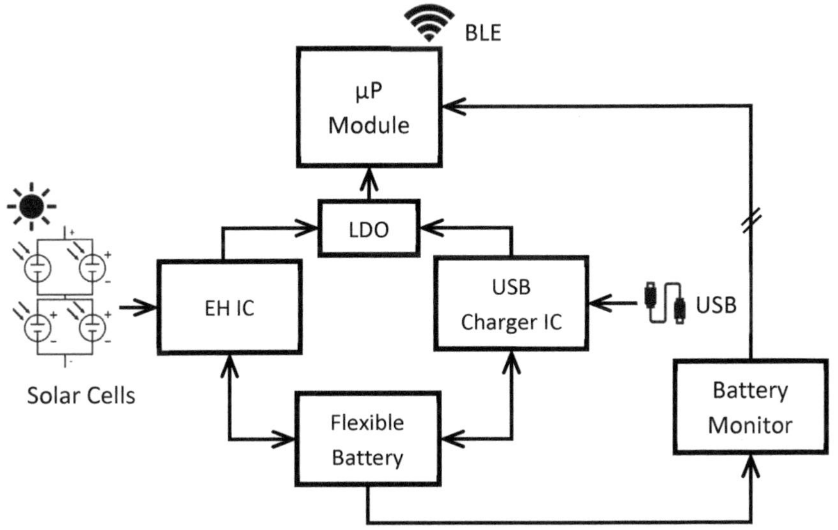

Figure 6. Schematic of the designed device.

The EH IC (BQ25570) is used to manage the power acquired from the solar cells to both charge the battery and power the microprocessor module. This IC includes an MPPT algorithm that can be configured depending on the energy source that is being used. It can normally charge an ESD with an input power as low as 5 µW connected to its boost converter. Moreover, the IC can power a load at a fixed output voltage value with its buck converter. The output voltage (VOUT) and other parameters such as a battery good indicator (VBAT_OK) and overvoltage threshold (VBAT_OV) can be configured with a resistor divider. In this case, VOUT is set to 1.7996 V, the VBAT_OK threshold was set to 2.823 V and 3.097 V (two different values due to charging/discharging hysteresis) and VBAT_OV was set to 4.308 V. If the power from the solar cells is high enough, the load is powered and the battery is recharged by them. If the power is too low, the load is powered directly by the battery. More information about this IC can be found in [37].

The USB Charger IC (BQ24232) is used to recharge the battery and power the load when an external Direct Current (DC) source is connected through a type-C USB connector. The output voltage when a DC source is connected is 4.4 V and cannot be changed. The maximum charging current is programmed with a resistor and it is set to 136 mA. In case there is no DC external source connected, the load is powered by the battery. More information about this IC can be found in [38].

In this design, the load (µP Module) is powered by 1.8 V. Since the USB Charger IC has an output voltage of 4.4 V, voltage regulation is needed. A Low-Dropout Voltage regulator (LDO) was used. The selected LDO is the same used in Bindi system: the Microchip Technology® MIC5365 has a fixed output voltage of 1.8 V. Although the EH IC does not need voltage regulation (since the output voltage is 1.7996 V), it was decided that its output would be also connected to the LDO input so that manually switching between the two ICs is easier. More information can be found in [39].

As for the battery, a Grepow® flexible LiPo battery with a capacity of 135 mAh and a nominal voltage of 3.7 V is used [40]. In addition to the USB Charger IC and the EH IC, the battery also includes a protection circuit to prevent overcharging. Even though LiPo batteries can be dangerous, they are leading the wearable market. Safer batteries are mostly experimental prototypes, which are not ready to be manufactured on a large scale. For these reasons, it was decided to use a LiPo battery in the proposed system, taking care of the overcharging protection and considering a good encapsulation design as future work. Furthermore, the selected battery is flexible, so it is a good choice for a wearable device.

A battery monitor is also needed to measure the battery current and the State-of-Charge (SOC). The Maxim Integrated^{TM} Fuel Gauge IC MAX17055 [41] has been selected since it is also used in the wearable devices in Bindi. This IC uses the Maxim ModelGauge^{TM} m5 EZ algorithm for battery measurements with tolerance against lithium batteries diversity. It uses an I2C interface to communicate with the microprocessor.

The selection of the solar cells used for this device is based on theoretical calculations. These calculations have been made considering the following:

- In the place where the experiments are made (Madrid, Spain), there are 8 h of light and 16 h of no light during a day, on average.
- During the 8 h of light time, the solar cells would fully power the device while recharging the battery for the remaining 16 h of no light.
- During the 16 h of no light, the device would be fully powered only by the battery.
- The average current drawn by the device is 5 mA, which is similar to the power consumption of the smart bracelet in Bindi, according to their battery monitor, when the bracelet is in continuous operation: receiving and processing data from sensors and wireless data streaming.
- The voltage provided by the battery is 3.7 V (nominal voltage).
- The typical efficiency of a solar cell is 25%.

Taking these into consideration, the energy required to power the device during the day and recharge the battery for 16 h of no light is the power of the device multiplied by the 24 h in a day. Considering that the power can be obtained as the supply voltage of a system multiplied by the drawn current:

$$E = V_{nom} \cdot I_{avg} \cdot t_{day}, \tag{1}$$

where E is the energy required by the device during a day, V_{nom} is the nominal value of the battery voltage, I_{avg} is the average current drawn by the device and t_{day} is the 24 h in a day.

Hence, the energy required would be 1598.4 J. However, this value of energy should be acquired by the solar cell during the 8 h of light, meaning that their power (energy divided by time) should be:

$$P_{sc} = \frac{E}{t_{light} \cdot 3600 \cdot \eta}, \tag{2}$$

where P_{sc} is the power given by the solar cell, E is the energy required by the device during a day, t_{light} is the hours of light during a day and η is the solar cell efficiency.

Therefore, the solar cell should be able to supply 222 mW. After obtaining this value, the ANYSOLAR Ltd. monocrystalline solar cells SM141K04LV were selected [42]. They have a maximum power of 123 mW and an efficiency of 25%. Four of these solar cells were connected in a series-parallel configuration, so that the total power is four times higher than the power of one solar cell. The configuration can be seen in the schematic of the device in Figure 6. Even though they are rigid solar cells, they are connected by soldered wire, so the result is a 60 × 45 mm for a semi-flexible module.

Using amorphous silicon flexible solar cells was also considered but, since their efficiency is lower than monocrystalline silicon solar cells, this consideration was discarded. The entire circuit (except for the battery and the solar cells) was integrated in a PCB (Printed Circuit Board). The PCB, the flexible battery and the solar cells module that form the complete device are presented in Figure 7. The dimensions of these components are 48 × 48 mm^2, 60 × 45 mm^2, 43 × 53 mm^2 for battery, solar cells module and PCB board, respectively.

Figure 7. Parts of the designed device. Flexible battery (upper left), PCB (lower left) and solar cells module (right). 1 EUR coin used as a size reference.

4. Testing Procedure

The testing procedure for wearable devices using energy harvesting from the solar panels considers the different conditions of sunlight and user's activity, which is related to the periods of solar exposure and wireless protocol connectivity. Two different types of test have been defined. The first set of tests was devoted to measuring the static conditions for the device, except for the sunlight received. The experiments consisted of placing the solar cells in specific positions for the entire discharge time to observe their performance and contribution to the battery life in different sunlight conditions. These tests were aimed mainly at measuring the capacity of the solar panels to recharge the battery under different sunlight conditions. The second set of tests was aimed at measuring the behaviour of the energy harvesting module under dynamic conditions. The whole system was worn by the user during her daily routines, while the state of charge of the battery was sent wirelessly to a smart phone. A 3D-printed flexible module has been prototyped to contain the designed PCB and the flexible battery to allow easiness and comfort of use.

It is difficult to state the standard dynamic conditions for testing the performance of an energy harvesting proposal. In this work, the authors have proposed a set of typical daily routines attending to the application of the wearable system selected, Bindi. The procedure for assessing the energy harvesting capacity may differ for another application, but the two types of tests proposed should be performed in different conditions.

In both tests, the processing unit of the designed device is loaded with a firmware that allows the battery monitor to measure the battery current in mA (milliamps) and SOC in % and send them to a mobile phone through a BLE wireless protocol every second.

4.1. First Set

In the first set of tests, the solar cells were manually laid so that the light would hit in different ways. A lux meter 5032C from MAVOLUX® was placed next to the solar cells in every test in this set to collect data relating to the illuminance, which is the luminous flux incident on a surface measured in lux (lx). This magnitude is related to the distance to the light source and the human perception of light, i.e., only visible wavelengths are measured [43].

Three experiments were proposed with three sunlight scenarios: No Sunlight (NSL), Indirect Sunlight (ISL) and Direct Sunlight (DSL). The season of the year (even month and day), the latitude and weather conditions should be registered, for the sake of rigorous comparison. In this work, the three experiments took place on different sunny days in December 2021, Madrid, Spain.

The NSL experiment consists of a total discharge cycle of the flexible battery, from 100% (battery fully charged) to 0% (battery fully discharged) with no solar cells connected. Some of the performed tests showed that there is no difference between covering the cells so they receive no light and removing them. This experiment serves as a reference to obtain the average current of the system and the battery lifetime without solar energy harvesting.

The ISL experiment follows the same methodology as the NSL experiment (a full discharge cycle) but, in this case, the solar cells were connected to the system and fixed to a window (from the inside), perpendicular to the ground and facing north.

The DSL experiment was similar to the ISL experiment but, this time, the solar cells were placed parallel to the ground so that they could receive sunlight more directly. The solar cells were placed on the eastern side of the building during the morning and on the western side during the afternoon, so that they could receive as much light as possible. Note that direct sunlight is a term that is typically used when incident sun rays are perpendicular to the surface of the solar cell. During this experiment, the solar cells are left under the sun for the whole day, meaning that the incident rays are not perpendicular to their surface the whole time. Even so, in this paper, we will refer to this type of scenario as Direct Sunlight.

4.2. Second Set

As for the second set of experiments, the designed device was tested as a wearable device by using a two-pieced, 3D-printed, flexible housing, as seen in Figure 8. The prototype is intended to be used as a bracelet, wearing each part of the housing on opposite sides of the wrist or the arm. The first part of the housing (marked as "1" in Figure 8) contains the solar cell semi-flexible module, while the second part of the housing ("2" in Figure 8) contains the PCB and the flexible battery. The user wore the prototype while performing four different Daily Routines (DR) in a city environment. This prototype can be worn easily, as the band is 45 mm in width and parts 1 and 2 are 70 × 47 mm^2 (a cellular phone is around 70 × 150 mm^2). Although this device is a prototype and not the final implementation, it has been designed as an accessory, used for street sports, running, jogging, bicycle, etc.

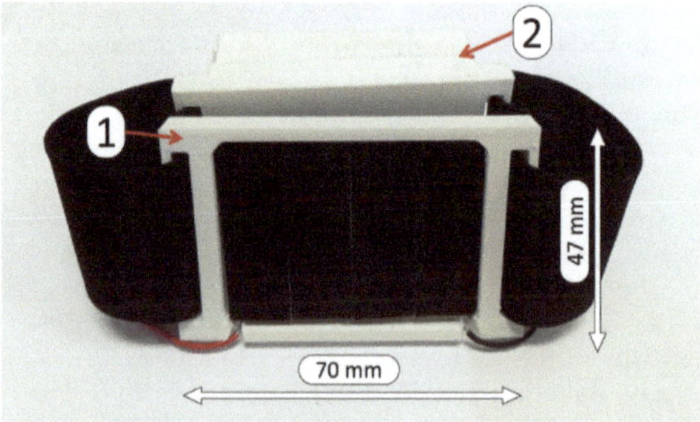

Figure 8. Wearable device prototype. 1: solar cells module housing. 2: PCB and flexible battery housing.

The first Daily Routine (DR1) consisted of wearing the device during two working days with sunny weather in February 2022, Madrid, Spain. The user stayed inside an office during the working day, stayed at home during free time and commuted by public transport. While indoors, the user tried to place the arm wearing the device next to a window when possible. On the other hand, the user tried to walk under the sun as much as possible when outdoors.

The second DR (DR2) consisted of wearing the device during the weekend with cloudy weather in February 2022, Madrid, Spain. The user wore the device at home during free time and outdoors while doing daily errands.

In the third and fourth DR (DR3 and DR4), only two solar cells connected in series were used, so that the dimension of the wearable could be reduced. However, this meant that the power generated by the semi-flexible solar cell module was halved.

In DR3, the device was worn by the user for two working days in April 2022, Madrid, Spain. The first day was sunny and the second day was cloudy. The user wore the device normally while walking outdoors and took the device off to place it under direct sunlight when in the office and stayed still when outside.

In DR4, the device was worn for two working days, both especially sunny, in May 2022, Madrid, Spain. In this case, the user always wore the device and did not purposedly take it off to place it near direct sunlight, but moved the wearing arm next to a window when possible.

4.3. Summary

In this subsection, the most important characteristics of every test and other specific details are summarized in Table 1: month and season, the weather during the experiment, the percentage of outdoor time, the number of solar cells used in every test and the position of the solar cell module during the experiment. The first set of experiments were run in 40°16′29.9″ N 3°54′26.9″ W location. The second set of experiments were run in 40°19′59.5″ N 3°45′59.3″ W (DR1, DR3 and DR4) and 40°25′0″ N, 3°42′12″ W (DR2).

Table 1. Summary of the characteristics of the tests.

Set	Test	Month (Season)	Weather	% Outdoor Time	Number of Solar Cells	Position of Solar Cells
1	NSL	December (Winter)	-	-	-	-
	ISL	December (Winter)	Sunny	-	4	Window
	DSL	December (Winter)	Sunny	-	4	Window
2	DR1	February (Winter)	Sunny	9	4	Wrist
	DR2	February (Winter)	Cloudy	25	4	Wrist
	DR3	April (Spring)	Sunny/cloudy	12	2	Wrist *
	DR4	May (Spring)	Sunny	12	2	Wrist

* The user purposedly took the device off and place it under the sun when possible.

5. Results

In this section, the results of the experiments from the first and second sets of tests are discussed.

5.1. First Set

After the three experiments, the data collected from the lux meter (illuminance) and from the battery monitor sent to the mobile phone wirelessly through BLE protocol (current and State-of-charge, SOC) were analyzed.

In Figure 9, the SOC vs. time of the NSL (No Sunlight), ISL (Indirect Sunlight) and DSL (Direct Sunlight) experiments are presented in green, blue and red, respectively. All curves represent a total discharge of the battery (from 100% to 0%). Thus, the battery lifetime increase can be compared. A vertical line is drawn at 24 h to compare the SOC after a whole day with the different sunlight scenarios. The increases in SOC in every curve are the results of solar cells recharging the battery. Please note that for DSL the exposition to sunlight was terminated after 55 h, leaving the battery to run out. If the experiment had been left working, the energy harvesting circuitry would have continued to recharge the battery. In the NSL scenario, the battery lasted 28.57 h with an average current of −4.2 mA (i.e., the designed device has a power consumption of around 15.5 mW, considering an average battery voltage of 3.7 V). Note that a negative current value means that the battery is discharging. The experiment started at 16:07 and ended at 20:42 the following day. The SOC was 17% 24 h after the start of the experiment.

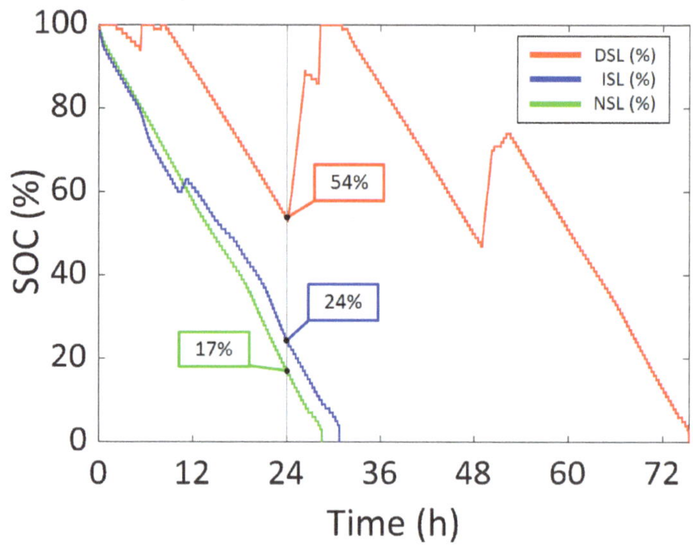

Figure 9. SOC vs. time of the first set of experiments.

As for the ISL experiment, in Figure 10, the battery current vs. time of both the ISL experiment and the NSL experiment are presented in blue and green, respectively, so that the increase in battery lifetime can be clearly seen. The illuminance vs. time during the ISL experiment is also added to the graph as a colour scale. The battery life reached 30.78 h in the ISL experiment. It started at 00:33 and ended at 7:20 the following day. The solar cells received a significant amount of sunlight for 1.61 h during the experiment, which is 1.25 h per day. It is considered that the solar cells were receiving a significant amount of sunlight whenever the battery current surpasses −2.1 mA (half the average current in the NSL experiment). The solar cells were able to recharge the flexible battery for 0.81 h, i.e., 0.63 h per day. This time was calculated considering that the battery is charging when the measured current is positive. The maximum battery current was 13.6 mA (50.3 mW, considering an average battery voltage of 3.7 V) with an illuminance of 32,000 lux. After 24 h from the start, the SOC was 24%.

As for the DSL experiment, in Figure 11, the battery current vs. time of the DSL and NSL experiments are presented in blue and green respectively, so that the battery lifetime increase can be compared. Moreover, the illuminance vs. time during the DSL experiment is represented as a colour scale to gain information relating to the amount of light hitting the solar cells module. In this case, the battery lasted 75.42 h. As previously stated, the solar cells were covered after 55 h because the system reached an SOC of 100% in the first two days, as seen in Figure 9. This means that the device could have been permanently working under the same sunlight conditions.

The experiment started at 9:00 and was concluded three days later at 12:43. During the 75.42 h of the experiment, there was a total of 19.00 h of a significant amount of sunlight and the battery was recharged for 13.01 h, i.e., 6.05 h of a significant amount of light and 4.14 h of recharging per day. This means that the system would be able to constantly work if it received significant direct sunlight for around 6 h every day. The highest battery current was 27.8 mA (102.9 mW, considering an average battery voltage of 3.7 V) with 40,870 lux. After 24 h, the SOC was 54%.

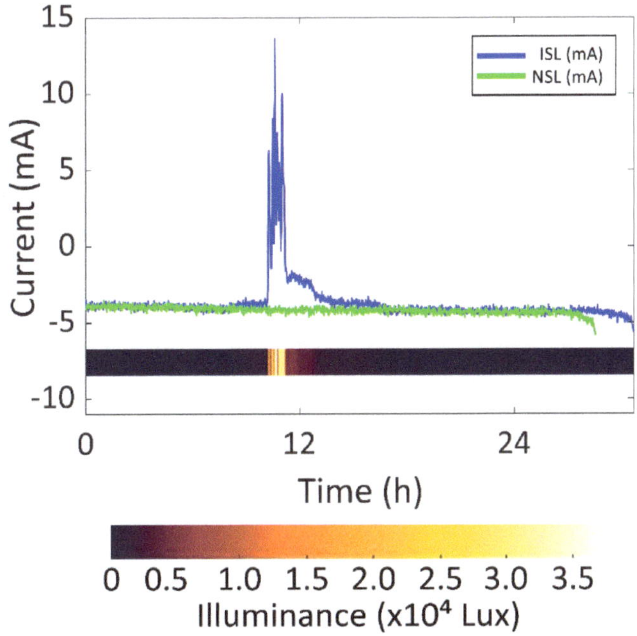

Figure 10. Battery current vs. time of ISL and NSL experiments.

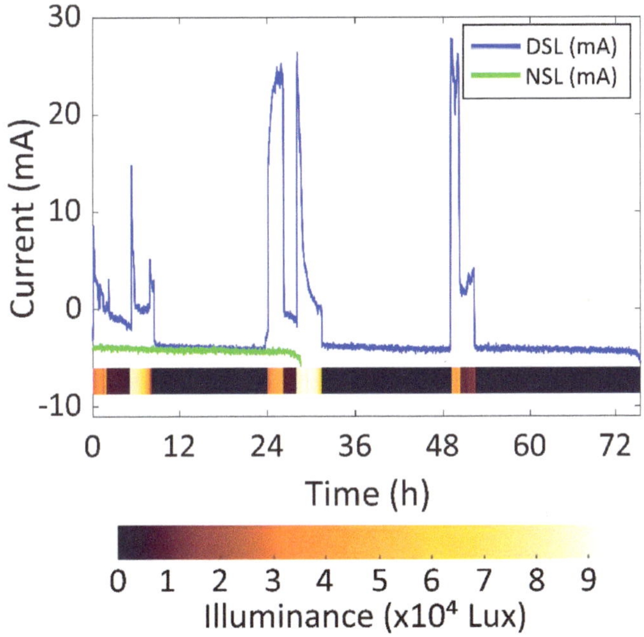

Figure 11. Battery current vs. time of DSL and NSL experiments.

5.2. Second Set

To examine the results of the three experiments from the second set (Daily Routines, DR), the data obtained from the mobile phone (battery current and SOC sent by BLE from the device) were analyzed.

In Figure 12, the SOC of the experiments for four Daily Routines (DR1, DR2, DR3 and DR4) are presented in red, blue, black and magenta (respectively), compared to the NSL experiment in green. Furthermore, a vertical line is drawn at 24 h to compare the SOC after a whole day with the different DRs. In DR1, there were increases in the SOC due to using the device outdoors and receiving enough sunlight to recharge the battery. In the DR2 experiment, there was no increase in the SOC because of the cloudy weather. The sunlight outdoors was not able to recharge the battery; however, it was able to decrease current consumption and increase the battery life compared to NSL. In the DR3 experiment, there were two increases in the SOC at the start due to the user purposedly placing the device under the sun while being outdoors and in the office. On the second day, the user repeated the same process, but the device was not able to recharge the battery due to the cloudy weather. In DR4, the first and third increases were due to placing the arm next to a window and the second peak was due to outdoor lighting.

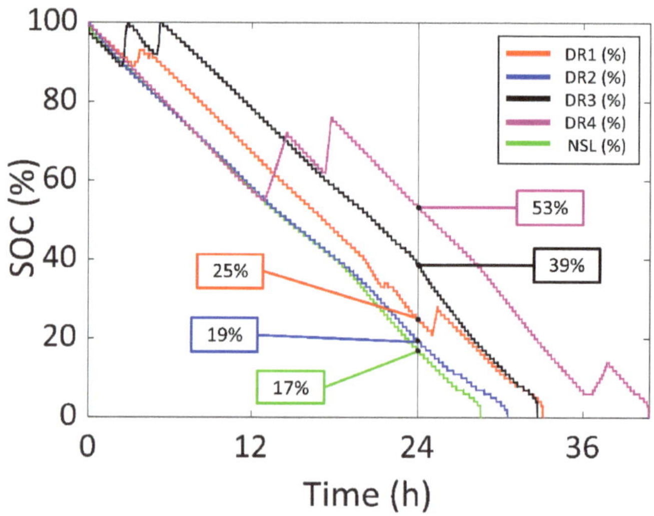

Figure 12. SOC vs. time of the second set experiments and NSL.

As for the first Daily Routine (DR1), the battery current vs. time of DR1 compared to the NSL experiment is shown in Figure 13 (DR1 in blue and NSL in green). The experiment started at 12:38 of day one and ended at 21:44 of day two (33.11 h). There was an increase of 4.54 h in the battery life (compared to NSL), with 1.37 h of a significant amount of sunlight (0.99 h per day). The battery was recharged for 0.93 h (0.68 h per day). Three main peaks can be appreciated in the blue curve. The first one was due to the outdoor sunlight received by the device when commuting from work to home. The highest battery current value was reached at this time (40 mA, 148.0 mW). The second peak (13 mA, 48.1 mW) was due to the user placing the wearing arm next to a window while being indoors (office). The third peak (37 mA, 136.9 mW) was due to staying outdoors during a lunch break. After 24 h, the SOC was 25%.

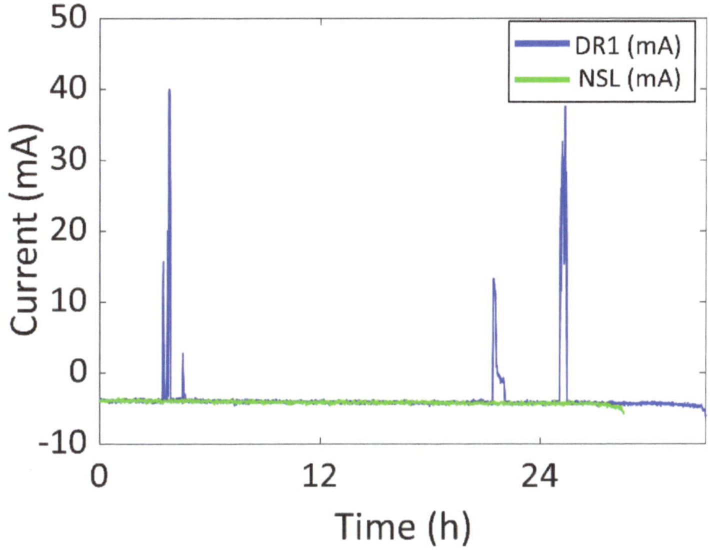

Figure 13. Battery current vs. time of DR1 and NSL experiments.

On the other hand, in Figure 14, the battery current vs. time of both DR2 and NSL experiments are presented in blue and green, respectively. The experiment started at 10:24 of day one and ended at 16:56 of day two (30.54 h). In the DR2 experiment there was an increase of 1.96 h in battery life, compared to NSL. There was a significant amount of sunlight for 1.14 h (0.89 h per day) and the battery was recharged for 0.06 h (0.05 h per day). Note that, even if the light is not enough to charge the battery, decreasing the current consumption also contributes to an increase in the battery life. Three main peaks can be seen in Figure 14. The highest battery current value is 7.5 mA (27.8 mW) due to the user purposedly placing the wearable directly facing the sun. The second and third peaks have a value of −2 mA and 0 mA, respectively. These are due to walking outdoors in a city environment. Note that, compared to DR1, the maximum values of the peaks are lower due to the cloudy weather. After 24 h, the SOC was 19%.

As for the third Daily Routine (DR3), the battery current vs. time of DR3 compared to the NSL experiment is shown in Figure 15 in blue and green, respectively. The experiment started at 11:30 of day one and ended at 20:13 of day two (32.72 h). There was an increase of 4.14 h in the battery life (compared to NSL), with 0.85 h of a significant amount of sunlight (0.63 h per day). The battery was recharged for 0.79 h (0.58 h per day). The two peaks in the blue curve (41 mA (151.7 mW) and 43 mA (159.1 mW), respectively) are due to the user purposefully taking off the device and placing it under direct sunlight. There is another increase in the battery current 22 h after the start of the experiment (−2.5 mA) due to the user taking the device off again and placing it under the sunlight. However, it was not enough to recharge the battery due to cloudy weather. After 24 h, the SOC was 39%.

In Figure 16, the battery current vs. time of DR4 compared to NSL is presented in blue and green respectively. The experiment started at 20:23 of day one and ended at 13:13 of day three, which is a total of 40.84 h, an increase of 12.26 h. There was 3.54 h of a significant amount of sunlight (2.08 h per day) and the battery was recharged for 3.31 h (1.95 h per day). The first and third peaks in the blue curve were due to placing the wearing arm next to a window during a meeting (13 mA (48.1 mW) and 12 mA (44.4 mW), respectively). The second peak was due to staying under outdoor sunlight after lunch (40 mA (148 mW)). After 24 h, the SOC is 53%.

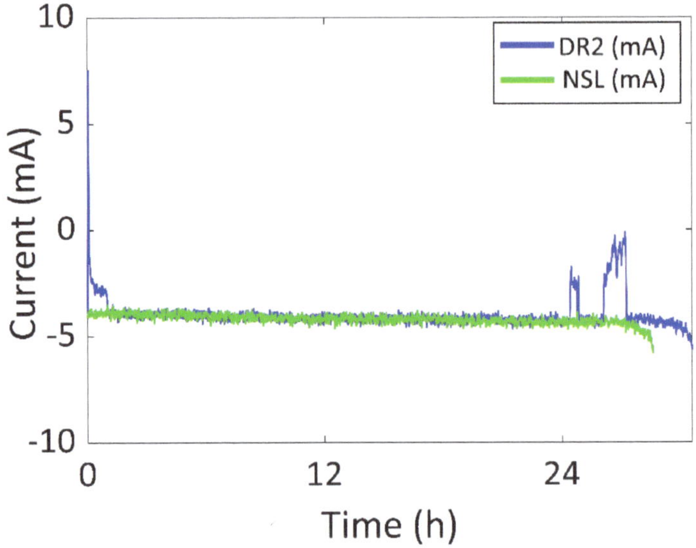

Figure 14. Battery current vs. time of DR2 and NSL experiments.

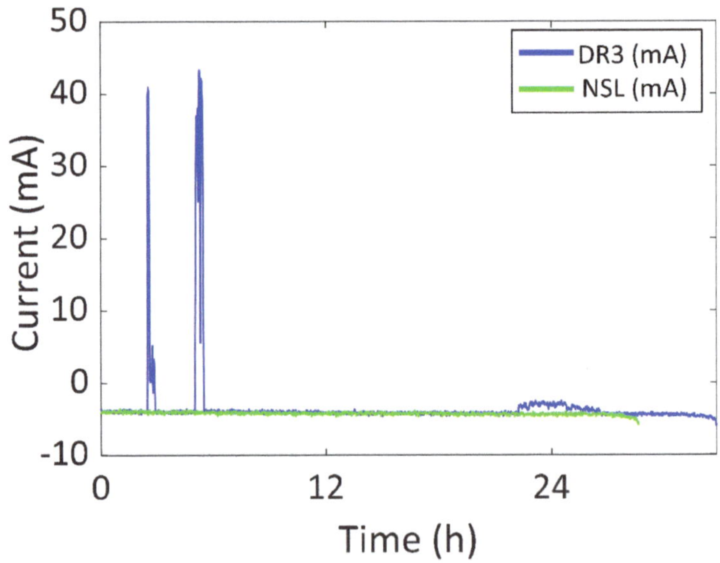

Figure 15. Battery current vs. time of DR3 and NSL experiments.

5.3. Discussion

In this subsection, the results obtained in the experiments are discussed and compared between them and with other studies in the literature.

In Table 2, the results obtained in this section for all the experiments are presented. These results include the total time in hours of each experiment, the SOC after 24 h, the number of hours with significant amount of sunlight every day (when current surpasses −2.1 mA) and the amount of charging hours due to enough sunlight every day (when current is higher than 0 mA).

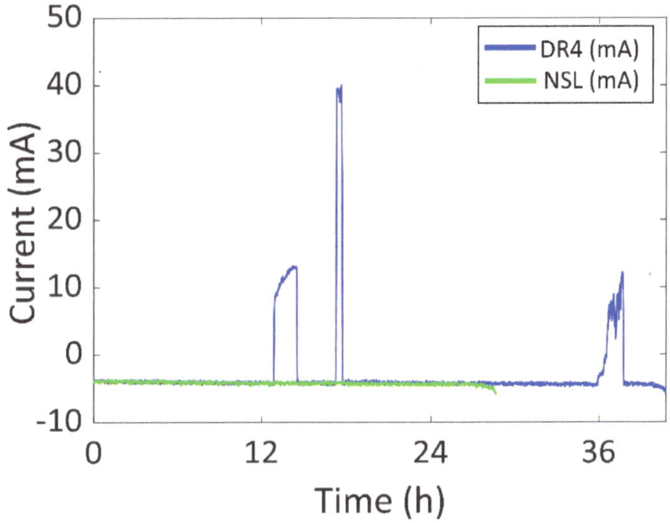

Figure 16. Battery current vs. time of DR4 and NSL experiments.

Table 2. Time results for all experiments.

	NSL	ISL	DSL	DR1	DR2	DR3	DR4
Total time (h)	28.58	30.78	75.42	33.11	30.54	32.72	40.84
SOC after a day (%)	17	24	54	25	19	38	53
Sunlight time per day (h/day)	-	1.25	6.05	0.99	0.89	0.63	3.54
Charging time per day (h/day)	-	0.63	4.14	0.68	0.05	0.58	3.31
Maximum power (mW)	-	50.3	102.9	148.0	27.8	159.1	148.0

These results may seem predictable (battery lifetime increases when adding an energy harvesting system to a wearable). However, the values cannot be easily deduced from the datasheets of the commercial components. It is very difficult to characterize the sunlight conditions of a wearable device due to the high number of variables that must be considered (e.g., user's movement, weather, solar cells position, country, season, wireless communication connectivity, etc.) For this reason, it was necessary to try to characterize the solar energy harvesting system included in a wearable device by using it in different real-life scenarios, such as the ones presented in this work.

Note that experiments DR3 and DR4 have similar or better results than DR1 and DR2, even when the number of solar cells was halved. DR3 and DR4 were performed in spring, so the incidence angle of sun rays was more favorable. Thus, we can conclude that the performance of the solar energy harvesting system is valid even for winter, and is improved in the other seasons.

Moreover, it is also proven that solar energy is the most efficient technology for a wearable device. The generated power in every test is in the order of the milliwatts, which is higher than most values presented in Section 2 (Related Work). For example, in [28,30], TEGs and piezoelectric devices were used for energy harvesting and achieved results in the order of the microwatts. TEGs need a high temperature difference to generate the required power, which is difficult to accomplish in terms of body temperature and environment temperature. It is possible to increase the generated power by using heat sinks, but they are normally too big to use in a wearable device. TEGs are commonly used in industrial environments, since the machinery and equipment can reach high temperatures compared to the ambient. As for piezoelectric devices, they usually need high frequencies to generate enough power, but human movements normally have low frequencies.

When compared to other solar energy harvesting systems in the literature, the device presented in this work is also proven to have a good performance. In [31], for example, the presented jacket with 16 solar cells and 12 TEGs can generate up to 500 mW (mostly by solar energy), but 16 solar cells are too many for the smart bracelet in Bindi. As for [32], a smart, multi-sensor bracelet, similar to the one in Bindi, is also presented. The device also includes the energy harvesters in its band, including the heat sinks of the TEGs, which could make the bracelet slightly uncomfortable. The average generated power is in the order of the hundreds of microwatts.

Therefore, energy harvesting for wearable devices in IoT systems through other technologies different to solar cells is not competitive. Even in the literature, the approaches proposed with the newest techniques and/or materials are not applicable to wearable devices, which require energies in the range of mW.

6. Conclusions

In this work, a device to test the viability of solar-powered wearable devices, including solar energy harvesting in smart wearables with a high number of functionalities, specifically the smart bracelet in Bindi system, has been presented. Furthermore, a procedure for testing the performance of the proposed system has been defined, including a flexible housing 3D-printed design, to easily test the prototype in real situations, since theoretical results are difficult to obtain due to the high number of variables that must be considered (e.g., weather, country, user's movement, etc.)

In real-life scenarios, the device was able to generate a minimum power of 27.8 mW and a maximum power of 159.1 mW.

The solar energy harvesting system was tested under different light conditions: NSL (No Sunlight), ISL (Indirect Sunlight) and DSL (Direct Sunlight). The SOC after 24 h was 3.18 times higher than NSL in the DSL scenario and 1.41 times higher in the ISL scenario. The maximum generated power was 50.5 mW in ISL and 102.9 mW in DSL. It was concluded that the device could permanently work if the solar cells received a significant amount of DSL for 6 h every day.

Furthermore, the device was also tested in four different Daily Routines (DR) so that the experiment was similar to common use. In DR1, the device was worn during working days in winter with sunny weather and the battery life was 1.16 times higher than NSL. The maximum generated power was 148.0 mW. As for DR2, the device was worn during the weekend in winter with cloudy weather and the battery life was 1.07 times higher than NSL. The maximum generated power was 27.8 mW. To reduce the dimensions of the prototype, two alternative DRs were included in this set (DR3 and DR4) using only two solar cells in the semi-flexible module. In DR3, the device was used during both a sunny and a cloudy working day in spring and the battery life was 1.14 higher than NSL. The maximum generated power was 159.1 mW. In DR4, the device was used for three working days in spring and the battery life was 1.43 times higher than NSL. The maximum generated power was 148.0 mW.

It can be concluded that, even when the number of solar cells is halved, the season has a very high influence on the results.

The dimensions of the presented proof-of-concept study are not optimal for a wearable device. However, it was proven that there is an important contribution to battery life even if only two solar cells are used. The second part of the prototype (PCB and flexible battery) can be easily reduced, since the PCB layout can be optimized and the flexible battery can adapt its shape around the wrist.

The users of the device know about the possibility of recharging the battery with solar energy, so they can deliberately place the device under a significant amount of light to charge it (e.g., sitting next to a window when indoors or changing the device position so that it receives more direct sunlight when outdoors).

On the other hand, commercial devices have been used in this design. Even if there has been a rapid increase in energy harvester development, most of them are experimental and they are usually hard to produce on a large scale because of the fabrication processes and costs.

The main conclusion of this work is that using solar energy harvesting with commercial solar cells and a commercial flexible LiPo battery as an ESD is a viable solution to increase battery life in wearables with intense data processing and wireless communications, although a rigorous characterization should be done depending on the latitude, weather, wireless connectivity and users' daily routines.

Author Contributions: Conceptualization, A.P.-M. and C.L.-O.; methodology, A.P.-M. and E.O.-R.; software, A.P.-M.; validation, A.P.-M., E.O.-R. and C.L.-O.; formal analysis, A.P.-M.; investigation, A.P.-M.; resources, M.G.-V.; data curation, A.P.-M.; writing—original draft preparation, A.P.-M.; writing—review and editing, A.P.-M., M.G.-V., C.L.-O. and E.O.-R.; visualization, A.P.-M.; supervision, M.G.-V., C.L.-O. and E.O.-R.; project administration, M.G.-V. and C.L.-O.; funding acquisition, M.G.-V. and C.L.-O. All authors have read and agreed to the published version of the manuscript.

Funding: This research was funded by the Department of Research and Innovation of Madrid Regional Authority, in the EMPATIA-CM research project (reference Y2018/TCS-5046). This work has been partially supported by the European Union—NextGenerationEU, with the SAPIENTIAE4BINDI project "Proof of Concept" 2021. (Ref: PDC2021-121071-I00/AEI/10.13039/501100011033). This work has been supported by the Madrid Government (Comunidad de Madrid-Spain) under the Multiannual Agreement with UC3M in the line of Excellence of University Professors (EPUC3M26), and in the context of the V PRICIT (Regional Programme of Research and Technological Innovation).

Institutional Review Board Statement: Not applicable.

Informed Consent Statement: Not applicable.

Data Availability Statement: Not applicable.

Acknowledgments: The authors acknowledge the technical help given by José Ángel Miranda.

Conflicts of Interest: The authors declare no conflict of interest.

References

1. E4 Wristband | Real-Time Physiological Signals | Wearable PPG, EDA, Temperature. Available online: https://www.empatica.com/research/e4 (accessed on 18 March 2022).
2. Fitbit. Available online: https://www.fitbit.com/global/es/home (accessed on 18 March 2022).
3. Bos, D.; Miller, S.; Bull, E. Using Virtual Reality (VR) for Teaching and Learning in Geography: Fieldwork, Analytical Skills, and Employability. *J. Geogr. High. Educ.* **2021**, 1–10. [CrossRef]
4. Ometov, A.; Shubina, V.; Klus, L.; Skibińska, J.; Saafi, S.; Pascacio, P.; Flueratoru, L.; Gaibor, D.Q.; Chukhno, N.; Chukhno, O.; et al. A Survey on Wearable Technology: History, State-of-the-Art and Current challenges. *Comput. Netw.* **2021**, *193*, 108074. [CrossRef]
5. Channa, A.; Popescu, N.; Skibinska, J.; Burget, R. The Rise of Wearable Devices during the COVID-19 Pandemic: A Systematic Review. *Sensors* **2021**, *21*, 5787. [CrossRef]
6. Haghi Kashani, M.; Madanipour, M.; Nikravan, M.; Asghari, P.; Mahdipour, E. A Systematic Review of IoT in Healthcare: Applications, Techniques, and Trends. *J. Netw. Comput. Appl.* **2021**, *192*, 103164. [CrossRef]
7. Jin, X.; Li, L.; Dang, F.; Chen, X.; Liu, Y. A Survey on Edge Computing for Wearable Technology. *Digit. Signal Process.* **2021**, *125*, 103146. [CrossRef]
8. Rituerto-González, E.; Miranda, J.A.; Canabal, M.F.; Lanza-Gutiérrez, J.M.; Peláez-Moreno, C.; López-Ongil, C. A Hybrid Data Fusion Architecture for BINDI: A Wearable Solution to Combat Gender-Based Violence. In Proceedings of the 10th International Conference the Multimedia Communications, Services and Security, Kraków, Poland, 8–9 October 2020.
9. Portilla, J.; Mujica, G.; Lee, J.S.; Riesgo, T. The Extreme Edge at the Bottom of the Internet of Things: A Review. *IEEE Sens. J.* **2019**, *19*, 3179–3190. [CrossRef]
10. Ren, M.; Di, J.; Chen, W. Recent Progress and Application Challenges of Wearable Supercapacitors. *Batter. Supercaps.* **2021**, *4*, 1279–1290. [CrossRef]
11. Khan, A.S.; Khan, F.U. A Survey of Wearable Energy Harvesting Systems. *Int. J. Energy Res.* **2022**, *46*, 2277–2329. [CrossRef]
12. Becquerel, E. On Electron Effects under the Influence of Solar Radiation. *Comptes Rendus De "L" Acad. Sci. Paris.* **1839**, *9*, 561.
13. Kim, Y.J.; Park, S.E.; Cho, B.J. A Wearable Organic Photovoltaic-Thermoelectric (OPV-TE) Hybrid Generator to Minimize the Open-Circuit Voltage Losses of OPV Module. *Nano Energy* **2022**, *93*, 106775. [CrossRef]
14. Xu, C.; Song, Y.; Han, M.; Zhang, H. Portable and Wearable Self-Powered Systems Based on Emerging Energy Harvesting Technology. *Microsyst Nanoeng.* **2021**, *7*, 1–14. [CrossRef] [PubMed]
15. Leonov, V. *Wearable Monitoring Systems*; Springer: Boston, MA, USA, 2011.

16. Selvan, K.V.; Mohamed Ali, M.S. Micro-Scale Energy Harvesting Devices: Review of Methodological Performances in the Last Decade. *Renew. Sustain. Energy Rev.* **2016**, *54*, 1035–1047. [CrossRef]
17. Bito, J.; Bahr, R.; Hester, J.G.; Nauroze, S.A.; Georgiadis, A.; Tentzeris, M.M. A Novel Solar and Electromagnetic Energy Harvesting System With a 3-D Printed Package for Energy Efficient Internet-of-Things Wireless Sensors. *IEEE Trans. Microw. Theory Tech.* **2017**, *65*, 1831–1842. [CrossRef]
18. Andreani, L.C.; Bozzola, A.; Kowalczewski, P.; Liscidini, M.; Redorici, L. Silicon Solar Cells: Toward the Efficiency Limits. *Adv. Phys.* **2019**, *4*, 1548305. [CrossRef]
19. Sharma, S.; Jain, K.K.; Sharma, A. Solar Cells: In Research and Applications—A Review. *Mater. Sci. Appl.* **2015**, *6*, 1145–1155. [CrossRef]
20. Kayes, B.M.; Nie, H.; Twist, R.; Spruytte, S.G.; Reinhardt, F.; Kizilyalli, I.C.; Higashi, G.S. 27.6% Conversion Efficiency, a New Record for Single-Junction Solar Cells under 1 Sun Illumination. In Proceedings of the 2011 37th IEEE Photovoltaic Specialists Conference, Seattle, WA, USA, 19–24 June 2011.
21. LONGi Solar. Available online: https://www.longi.com/en/news/7469// (accessed on 18 March 2022).
22. Epishine. Available online: https://www.epishine.com/product (accessed on 18 March 2022).
23. Rao, K.M.; Sangeetha, D.N.; Selvakumar, M.; Sudhakar, Y.N.; Mahesha, M.G. Review on Persistent Challenges of Perovskite Solar Cells' Stability. *Sol. Energy* **2021**, *218*, 469–491. [CrossRef]
24. Polman, A.; Knight, M.; Garnett, E.C.; Ehrler, B.; Sinke, W.C. Photovoltaic Materials: Present Efficiencies and Future Challenges. *Science* **2016**, *352*, 6283. [CrossRef]
25. Silicon Lightworks. Available online: https://siliconlightworks.com/li-ion-voltage (accessed on 18 March 2022).
26. Liang, Y.; Zhao, C.-Z.; Yuan, H.; Chen, Y.; Zhang, W.; Huang, J.-Q.; Yu, D.; Liu, Y.; Titirici, M.-M.; Chueh, Y.-L.; et al. A Review of Rechargeable Batteries for Portable Electronic Devices. *InfoMat.* **2019**, *1*, 6–32. [CrossRef]
27. Toh, W.Y.; Tan, Y.K.; Koh, W.S.; Siek, L. Autonomous Wearable Sensor Nodes With Flexible Energy Harvesting. *IEEE Sens. J.* **2014**, *14*, 2299–2306. [CrossRef]
28. Kim, J.; Khan, S.; Wu, P.; Park, S.; Park, H.; Yu, C.; Kim, W. Self-Charging Wearables for Continuous Health Monitoring. *Nano Energy.* **2021**, *79*, 105419. [CrossRef]
29. Chen, Y.; Wang, T.; Tian, H.; Su, D.; Zhang, Q.; Wang, G. Advances in Lithium–Sulfur Batteries: From Academic Research to Commercial Viability. *Adv. Mater.* **2021**, *33*, 3666. [CrossRef] [PubMed]
30. Nastro, A.; Pienazza, N.; Baù, M.; Aceti, P.; Rouvala, M.; Ardito, R.; Ferrari, M.; Corigliano, A.; Ferrari, V. Wearable Ball-Impact Piezoelectric Multi-Converters for Low-Frequency Energy Harvesting from Human Motion. *Sensors* **2022**, *22*, 772. [CrossRef] [PubMed]
31. Brogan, Q.; O'Connor, T.; Ha, D.S. Solar and Thermal Energy Harvesting with a Wearable Jacket. In Proceedings of the 2014 IEEE International Symposium on Circuits and Systems (ISCAS), Melbourne, Australia, 1–5 June 2014.
32. Magno, M.; Brunelli, D.; Sigrist, L.; Andri, R.; Cavigelli, L.; Gomez, A.; Benini, L. InfiniTime: Multi-sensor wearable bracelet with human body harvesting. *Sustain. Comput. Inform. Syst.* **2016**, *11*, 38–49. [CrossRef]
33. Lemey, S.; Declercq, F.; Rogier, H. Textile Antennas as Hybrid Energy-Harvesting Platforms. *Proc. IEEE* **2014**, *102*, 1833–1857. [CrossRef]
34. Liu, Y.; Zhang, S.; Beirne, S.; Kim, K.; Qin, C.; Du, Y.; Zhou, Y.; Cheng, Z.; Wallace, G.; Chen, J. Wearable Photo-Thermo-Electrochemical Cells (PTECs) Harvesting Solar Energy. *Macromol. Rapid Commun.* **2022**, *43*, 6. [CrossRef]
35. Ren, Z.; Zheng, Q.; Wang, H.; Guo, H.; Miao, L.; Wan, J.; Xu, C.; Cheng, S.; Zhang, H. Wearable and Self-Cleaning Hybrid Energy Harvesting System Based on Micro/Nanostructured Haze Film. *Nano Energy* **2020**, *67*, 104243. [CrossRef]
36. Aconno, ACN52840. Fully Integrated, Ultra-Low Power, Bluetooth 5 Module ACN52840 Datasheet. Available online: https://1897079276.rsc.cdn77.org/wp-content/uploads/2018/08/20191203_ACN52840_Datasheet_V1.3.pdf. (accessed on 21 April 2022).
37. Available online: https://www.ti.com/lit/ds/sym-780link/bq25570.pdf?ts=1650519248622&ref_url=https%253A%252F%252Fwww.ti.com%252Fproduct%252FBQ25570 (accessed on 21 April 2022).
38. bq25570 Nano Power Boost Charger and Buck Converter for Energy Harvester Powered Applications. Available online: https://www.ti.com/lit/ds/symlink/bq24232.pdf?ts=1650456025712&ref_url=https%253A%252F%252Fwww.ti.com%252Fproduct%252FBQ24232. (accessed on 21 April 2022).
39. bq2423x USB-Friendly Lithium-Ion Battery Charger and Power-Path Management IC. Available online: https://ww1.microchip.com/downloads/en/DeviceDoc/mic5365.pdf. (accessed on 21 April 2022).
40. MIC5365/6 High-Performance Single 150 mA LDO. Available online: https://www.grepow.com/page/ultra-thin-battery.html. (accessed on 21 April 2022).
41. Grepow UltraThin Battery. Available online: https://datasheets.maximintegrated.com/en/ds/MAX17055.pdf (accessed on 21 April 2022).
42. Available online: https://waf-e.dubudisk.com/anysolar.dubuplus.com/techsupport@anysolar.biz/O18Ae0B/DubuDisk/www/Gen3/SM141K04LV%20DATA%20SHEET%20202007.pdf. (accessed on 21 April 2022).
43. Amajama, J. Effect of Solar Illuminance (or Intensity) on Solar (Photovoltaic) Cell's Output and the Use of Converging Lenses and X or Gamma Rays to Enhance Output Performance. *Int. J. Eng. Res. Gen. Sci.* **2016**, *4*, 284–289.

sensors

MDPI

Article

Boosting Quantum Battery-Based IoT Gadgets via RF-Enabled Energy Harvesting †

Sumit Gautam [1,2,*], Sourabh Solanki [2,*], Shree Krishna Sharma [2], Symeon Chatzinotas [2] and Björn Ottersten [2]

1 Department of Electrical Engineering, Indian Institute of Technology, Indore 453552, India
2 Interdisciplinary Centre for Security, Reliability and Trust (SnT), University of Luxembourg, L-1855 Luxembourg, Luxembourg; shree.sharma@uni.lu (S.K.S.); symeon.chatzinotas@uni.lu (S.C.); bjorn.ottersten@uni.lu (B.O.)
* Correspondence: sumit.gautam@iiti.ac.in (S.G.); sourabh.solanki@uni.lu (S.S.)

† This paper is an extended version of our work in: Gautam, S.; Sharma, S.K.; Chatzinotas, S.; Ottersten, B. Modeling and Optimization of RF-Energy Harvesting-assisted Quantum Battery System. In Proceedings of the 2021 IEEE 93rd Vehicular Technology Conference: VTC2021-Spring, Helsinki, Finland, 25–28 April 2021.

Abstract: The search for a highly portable and efficient supply of energy to run small-scale wireless gadgets has captivated the human race for the past few years. As a part of this quest, the idea of realizing a Quantum battery (QB) seems promising. Like any other practically tractable system, the design of QBs also involve several critical challenges. The main problem in this context is to ensure a lossless environment pertaining to the closed-system design of the QB, which is extremely difficult to realize in practice. Herein, we model and optimize various aspects of a Radio-Frequency (RF) Energy Harvesting (EH)-assisted, QB-enabled Internet-of-Things (IoT) system. Several RF-EH modules (in the form of micro- or nano-meter-sized integrated circuits (ICs)) are placed in parallel at the IoT receiver device, and the overall correspondingly harvested energy helps the involved Quantum sources achieve the so-called quasi-stable state. Concretely, the Quantum sources absorb the energy of photons that are emitted by a photon-emitting device controlled by a micro-controller, which also manages the overall harvested energy from the RF-EH ICs. To investigate the considered framework, we first minimize the total transmit power under the constraints on overall harvested energy and the number of RF-EH ICs at the QB-enabled wireless IoT device. Next, we optimize the number of RF-EH ICs, subject to the constraints on total transmit power and overall harvested energy. Correspondingly, we obtain suitable analytical solutions to the above-mentioned problems, respectively, and also cross-validate them using a non-linear program solver. The effectiveness of the proposed technique is reported in the form of numerical results, which are both theoretical and simulations based, by taking a range of operating system parameters into account.

Keywords: 5G and beyond/6G wireless networks; greencom; IoT; quantum battery; RF-energy harvesting; transmit power optimization

Citation: Gautam, S.; Solanki, S.; Sharma, S.K.; Chatzinotas, S.; Ottersten, B. Boosting Quantum Battery-Based IoT Gadgets via RF-Enabled Energy Harvesting. *Sensors* **2022**, *22*, 5385. https://doi.org/10.3390/s22145385

Academic Editors: Slim Naifar, Olfa Kanoun and Carlo Trigona

Received: 30 May 2022
Accepted: 13 July 2022
Published: 19 July 2022

1. Introduction

1.1. Motivation

The world has consistently witnessed technological revolutions pertaining to Wireless Communications over the past few decades. Recent advancements in the field have narrowed down the gap between so-called binary information bits and the envisioned Quantum bits (Qubits). In this vein, the rapid transition to Quantum-based methods naturally demands an advanced framework that derives a technically synergistic relationship between the Quantum theories and Wireless-Communications-based techniques [1,2]. However, the main challenge in this context is our dependence on the energy sources, which is as strong as our dependence on the devices themselves [3–6].

Several works in the literature have focused on addressing these issues, and the interesting idea of a Quantum battery (QB) has been put forth by various researchers [7–10] in this context.

More recently, the conceptual developments around the use of a nano-meter-sized QB has shown immense promise [11]. The working principle of the nano-meter-sized QB is similar to the phenomenon of Luminescence. Herein, nanostructured solid-state of matter in the nano-meter-sized QB works on the principle that the energy of photons can be transferred to the electrons present at the core of the QB, thereby increasing their lifetime (i.e., being in the energized state) to prolonged periods. The theoretical feasibility of this possibility is often determined by a quasi-stable state, which is primarily achievable within a lossless closed system. It is noteworthy that this phenomenon remains the backbone of several closed-system types of QBs [12]. Considering the present practical systems, it is almost impossible to ensure the lossless criterion for any closed-system design. Given that the persistent losses are bound to occur in the QBs, it could still be possible to utilize complementary systems to compensate for these losses, such as the use of radio-frequency (RF)-based wireless energy-harvesting (EH) techniques.

Nikola Tesla suggested an EH concept of extracting energy through the air during the late 1890s and early 1900s [13,14]. Since then, this concept has been applied to a variety of sectors of research, including wireless devices [15], flying objects [16], vehicles [17], and so on. Varshney recently explored the potential of adopting RF-based simultaneous wireless information and power transfer (SWIPT), which combines a standard information receiver with an EH module [18]. In this regard, various SWIPT receiver topologies, including time-switching (TS), power-splitting (PS), or separated architecture (SA), were investigated in the literature [19–22]. As such, the RF-based EH module is made up of several components, such as an RF-antenna, matching circuit, polarized capacitor, and rectifier circuit, and is hence considered a passive device [23]. The saturation effect in the diode element frequently causes problems for the EH module. As a result of the saturation effect, the non-linear nature of EH operation is noticed. This impact can be mitigated by connecting numerous EH circuits in a parallel manner, resulting in a linear region for operation [24,25]. The RF-based EH is also seen as a potential next step in the development of green wireless communication systems [26].

1.2. Current Studies and Related Works

Several studies have looked into the possibility of EH using ambient sources, i.e., thermal, mechanical, solar, and other sources, to power the QBs [27]. QBs using RF-EH mechanisms, on the other hand, are still an untapped and unexplored area of research. Due to the large size of the EH module, creating a system that integrates a QB and EH modules is difficult in the current scenario. Furthermore, the authors believe that, in the future, EH modules could be reduced to micro- to nano-meter-sized integrated circuits (ICs), which could be used to somehow aid QBs. The QBs require external help to compensate for any losses that may occur as a result of internal operations while maintaining their stability. As a result, a unique concept aimed at modeling and optimizing a design using a group of RF-EH ICs stacked in parallel to help a QB appears promising. The proposed approach could be used in Internet-of-Things (IoT) systems due to the small size of RF-EH-aided QBs. Quantum communications and QBs may also be explored, or used in conjunction in the IoT systems.

Regarding the practical or real-life scenarios, the framework proposed in this paper, which is based on the EH-enabled QBs, may be utilized in sensors, biomolecular analysis, photovoltaic devices, photochemical reagents, fluorescence, and light-emitting diodes (LEDs) [28]. When it comes to EH-assisted QB cores, the management of energy becomes crucial. In this regard, the frameworks and subsequent discussions provided in [29,30] tackle the main blocks used to manage energy in IoT nodes, which can further be translated towards management of the harvested energy for assisting the QB systems. In [31], the authors discussed power management in ultrasound-based EH in implantable medical devices, where the work is more focused on systems to manage power provided by an alternating current (AC) source (such as rectifiers and rectenna), and solutions for direct current (DC)-type sources are discussed in [32,33].

Most approaches concerning operation of QBs consider two-level operational frameworks (qubits), where a charger system assists in their promotion from the ground to excited state [9,34,35]. The charger system can be either quantum based or a classical time-dependent operation, driving the qubits [36,37]. In this vein, the key features of QBs are characterized in terms of energy storage and its maximum charging time [38]. In general, the characterization metrics, such as voltage, current, power and lifetime of QBs, can vary based on their types and operations. Correspondingly, the complexity of the circuit needed to sustain the QB operations may also be variable accordingly (e.g., utilization of micromasers as QBs [39]).

1.3. Novelty and Contributions

In this invited paper, we present a novel and futuristic framework considering an IoT system comprised of an RF-based transmitter and an end user equipped with a QB assisted by several RF-EH ICs. The RF-signals emitted by the transmit device are captured by a number of RF-EH ICs, where the overall harvested energy is then managed by a micro-controller. The micro-controller is also responsible for controlling a photon-emitting device. The photons dispatched from the photon-emitting device carry energy for the power-hungry electrons present at the core of a QB. In order to model and design the considered system, we formulate an optimization problem to minimize the overall transmitted power, subject to constraints on the overall harvested energy at the IoT-user node, and the number of RF-EH ICs. We investigate the problem in two ways. First, we analyze the problem using theoretical rigour to obtain the closed form solutions for the intended parameters. Second, we directly solve the (main) formulated non-convex problem using an optimization solver. Unlike [40], we formulate and solve an additional optimization problem to minimize the total number of RF-EH ICs, under constraints on the total transmit power and overall harvested energy (please note that this paper is an extension of the authors' prior work, published as [40] (© 2021 IEEE)). Specifically, the main contributions and novelty of this work are summarized below.

1. We present a novel and futuristic model of an IoT-receiver device equipped with RF-EH ICs-assisted QB. In the proposed design, a QB is present at the core of the IoT-user device and is assisted via numerous RF-EH ICs.
2. In order to design and optimize this novel scheme, we formulate a couple of mathematical problems to minimize: (i) the overall transmitted power, subject to constraints on demanded harvested energy and number of RF-EH ICs; and (ii) the number of RF-EH ICs, subject to constraints on demanded harvested energy and total transmit power.
3. Correspondingly, we apply suitable convexification methods to find the optimal solutions to the above-mentioned problems. In this vein, we first obtain the closed-form solutions using theoretical analysis and then validate the results by directly solving the main (primal) problem via a non-convex optimization solver, sequentially, for both the problems.
4. We illustrate, with the help of numerical results, the benefits of the proposed design and related optimization of the considered framework. The outcomes motivate towards the realization of the proposed concept to a prospective real-life implementation.

The subsequent sections of this invited paper are organized as follows. Section 2 provides an introduction to the system model. The proposed RF-EH-assisted QB System design, and the problems corresponding to minimization of transmit power minimization and optimization of the number of RF-EH ICs at the IoT user, respectively, are presented in Section 3. Numerical results are shown and discussed in Section 4. Finally, concluding remarks are provided in Section 5.

2. System Model

We consider an IoT-based framework comprising a transmitter and an RF-EH ICs-assisted QB-enabled user, and discuss its possible prototyping and related challenges. Concerning the overall system design, we deliberate on some key aspects to provide adequate motivation for our proposed design. Acknowledging various possibilities of closed- and open-system models, several conceptualizations of QB designs have been proposed in the literature [7–10]. Owing to a wide range of evident challenges around the realization of open-system designs pertaining to the QBs [12], we therefore focus on the closed-system design of QB for our analysis.

Regarding the closed-system design, the notion of QBs broadly works on the principle of a photon being absorbed by an electron, which eventually reaches an excited state using the photon's energy. Consequently, the excitation condition of the electron may prevail for extremely long periods of time. Based on this principle, conceptual nano-meter-sized QBs [11] have been proposed recently, which are proficient in holding the electric charge for a prolonged time period. The corresponding illustration of such a mechanism is depicted as in Figure 1. The above-mentioned principle also shares a similarity with the Luminescence effect, wherein certain goods, e.g., plastic objects, dials of watches, etc., glow in a dark environment for a very long time following the process of photon absorption.

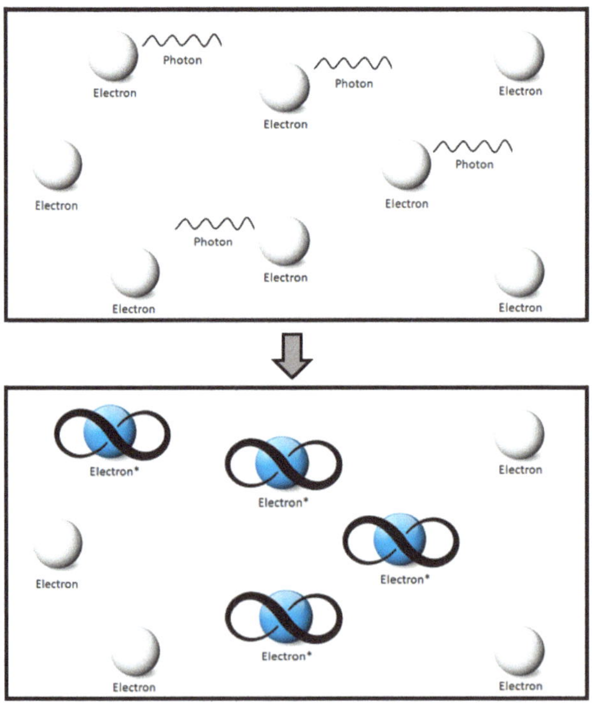

Figure 1. The phenomenon of the absorption of photons' energy by the electrons in nano-structured solid state of matter within a closed system, where they may theoretically stay in this condition for extremely long periods of time [40].

It is noteworthy that, from a theoretical point of view, the above-mentioned process may seem sound and viable. However, it is extremely challenging to realize in practice given the wide range of influencing parameters, such as heat losses, environmental effects, losses pertaining to the respective functionalities of the components, etc. This makes it difficult to ensure a completely lossless environment, thereby leading to intractability, considering the current state of the technology. In this direction, there is now a growing interest around the design of QBs that can leverage from the harvested energy via ambient free sources of power, such as thermal, mechanical, and solar sources, etc., [27].

Harvesting energy via electromagnetic (EM) waves is another interesting prospect that has intrigued humankind for several years. In this regard, RF-based signals have been utilized for the purpose of EH to meet the needs of the battery-constrained wireless devices. Correspondingly, RF-based EH is yet another potential technique, wherein a passive elements-based matching circuit, a rectifier circuit and a charge-holding capacitor element constitute a basic EH module. Considering the current state of the art, the existing EH modules are constrained by large sizes, which are nearly equal to or greater than the dimensions of common handheld gadgets. The authors, however, envision that the EH modules will be miniaturized in future and can be made available in the form of RF-EH ICs, as shown in Figure 2. This motivation is backed by the fact supporting the evolution of RF-based information receivers, where the large sizes of modules have been/were drastically scaled down over time.

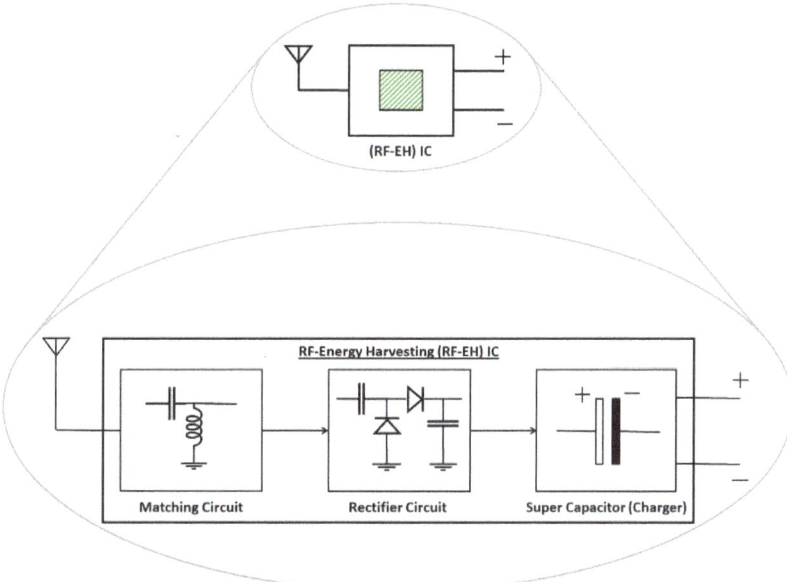

Figure 2. The miniaturization concept of RF-EH modules into micro- or nano-meter-sized ICs is depicted herein. As its working principle, an RF antenna absorbs the ambient (or specific band(s) of) RF signal and converts its energy into the equivalent DC type of output [40].

To this end, let us assume that the considered wireless RF-transmitter emits the signal $x(t)$ via a single antenna, which is intended for the RF-EH IC(s). Noticeably, the benefits of using different waveforms to improve the EH efficiency may also be explored and exploited [41]. More specifically, a variety of waveform designs have been investigated and presented in the literature. However, it is difficult to comment on the best waveform that improves EH significantly, since it depends on the type of employed EH circuit. As a generalized assumption, we consider zero mean and unit variance signal in this invited paper to ensure a theoretical tractability.

Let us consider and denote the overall transmit power as P_T. The received signal at the QB-equipped IoT user is then given by: $y(t) = h(t) * x(t) + n_R(t)$, where $h(t)$ is the complex channel gain coefficient for the wireless channel and $n_R(t)$ is the additive zero mean Gaussian noise at the corresponding IoT user's receiving antenna equipment, with a noise variance of σ_R^2. Correspondingly, in case a linear EH operation is considered at the RF-EH IC unit of the receiver, then the relevant expression in this context is given as: $\mathcal{E}^{\mathcal{L}} = \zeta_i(P_T|h|^2 + \sigma_R^2)$, where $0 < \zeta_i \leq 1$ is the energy conversion efficiency of the corresponding receiver. In practice, we note that $\mathcal{E}^{\mathcal{L}}$ is analytically valid for theoretical evaluations; however, its real-time implementation is dubious. Therefore, we adopt the sigmoidal-function-based non-linear EH model [42–44] to characterize the EH operation, defined as

$$\mathcal{E}^{\mathcal{N}} = \frac{\mathcal{E}'}{1-\phi} \cdot \left(\frac{1}{1 + e^{(-\alpha P_T |h|^2 + \alpha\beta)}} - \phi \right), \tag{1}$$

where $\phi \triangleq \frac{1}{1+\exp(\alpha\beta)}$, the constant $\mathcal{E}'$ is obtained by determining the maximum harvested energy on the saturation of the EH circuit, and α and β are specific to the capacitor and diode turn-on voltage metrics at the EH circuit. For further analysis, we consider normalized time slots to use the terms power and energy interchangeably.

3. Proposed Design of RF-EH-Assisted QB System

Based on the concepts discussed and developed in the previous section, we present herein the proposed novel and futuristic model of an IoT system comprising a transmitter and QB-enabled end user assisted by numerous RF-EH ICs. The basic schematic of the proposed RF-EH ICs-assisted QB framework is shown in Figure 3, where the QB is assumed to be present at the core of the IoT user. The complete battery system is comprised of multiple RF-EH ICs, micro-controllers, photon-emitting devices, and the QB at the core. The RF-EH ICs are placed all across the system, such that the transmitted RF signal's energy is extracted and then managed by the micro-controllers placed in different zones. The accumulated energy is then utilized by the micro-controllers to power the correspondingly connected photon-emitting devices, which thereby ensure the quasi-stability at the core of the QB.

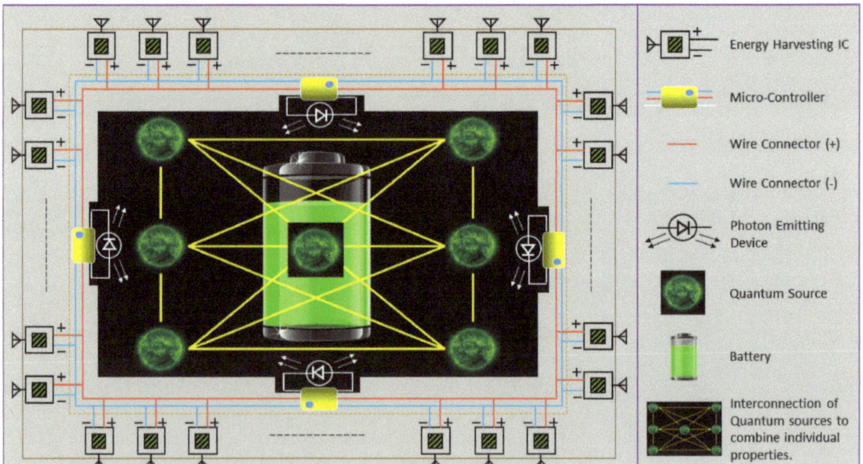

Figure 3. Prototype of the proposed novel and futuristic RF-EH ICs-assisted QB system [40].

As discussed before, ensuring a closed-system design in any QB is challenging, and hence, some losses are bound to occur in practice. However, these losses may be compensated by providing enough energy to the electrons present at the core of the QB. In this regard, the IoT-user device may contact the nearest transmitter, which may occur through regular RF-based Wireless Communications or via Quantum-based wireless communications systems using entanglement of Qubits [1,2]. The transmit source then uses RF signaling to transfer the demanded energy to the IoT user, where the energy is harvested with the help of multiple RF-EH ICs arranged all around the receive unit (it is explicit that the saturation effect of the diode prohibits a single unit of RF-EH IC from harvesting energy beyond a certain limit, and hence, multiple RF-EH ICs are assumed to be placed in parallel to overcome this issue). For our analysis, we consider that n number of RF-EH ICs are available, with a maximum cap of N. The selection of N is naturally decided according to the design requirements, e.g., the size of the device.

Upon the reception of the RF signal emitted by the transmit source, each RF-EH IC at the IoT user converts the signal energy using its miniaturized passive circuit. The management of the overall harvested energy is performed by a micro-controller. To emulate a lossless criterion corresponding to the closed-system design of the QB, the micro-controller is also responsible for the operation of an extremely low-intensity photon-emitting device (e.g., light-emitting diode (LED)), wherein the emitted photons facilitate the energy-transfer process at the core of the QB. The overall harvested energy demand at the IoT user is assumed to be $\xi = \eta + \kappa$. Herein, 'η' is the harvested energy demand for enabling the photon emission process. The additional 'κ' represents the harvested energy demand required for other consumption metrics, such as operation of the micro-controller, compensation for losses, and maintaining a suitable temperature within the closed system of the QB.

To proceed, we intend to model and optimize the involved metrics and parameters in the considered framework. Concretely, we first develop the formulation to minimize the overall transmitted power under limitations on the minimum harvested energy and the number of RF-EH ICs. Next, we formulate and solve another problem to optimize the number of RF-EH ICs at the IoT user, subject to constraints on the overall harvested energy demand and a maximum limitation on the transmit power. The following sections provide more insight into the considered problems and the obtained (proposed) solutions.

3.1. Transmit Power Minimization

In this section, we minimize the total transmit power, subject to the constraints on the overall energy harvested by the RF-EH ICs, and the number of (non-linear) EH ICs. The corresponding optimization problem can then be represented in its analytical form, as follows

$$(P1): \min_{n,P_T} \; P_T \tag{2}$$

$$\text{s.t.} \;\; (C1): n\mathcal{E}^{\mathcal{N}} \geq \xi, \tag{3}$$

$$(C2): n \leq N. \tag{4}$$

We find that the joint optimization of n and P_T in (3) makes $(P1)$ non-convex and intractable. However, an analytical solution based on a graphical/direct-evaluation method is possible, wherein one parameter may be relaxed/fixed to solve for the other, and vice versa. In this context, let us first observe that the objective in $(P1)$ is independent of n. Next, we note that $\mathcal{E}^{\mathcal{N}}$ is an increasing function of P_T. Moreover, note that the value of either n or P_T may be increased to satisfy the constraint in $(C1)$. Concretely, the optimized solution of P_T may be obtained by enforcing $n = N$, meaning that the equality in (4) must hold. In this vein, we leverage from this solution, i.e., $n = N$, to obtain/represent the the following (reduced) problem.

$$(P2): \min_{P_T} \; P_T \tag{5}$$

$$\text{s.t.} \;\; (C1): N\mathcal{E}^{\mathcal{N}} \geq \xi. \tag{6}$$

Note that $(P2)$ is of a convex form, which implies that it should be possible to obtain either a closed-form solution or that the problem may be solved directly via well-known convex optimization methods [45]. Let us consider the former case, wherein we seek a suitable closed-form solution for P_T. To proceed, we observe that an optimal solution is achievable by letting the equality in (6) hold. This notion is based on the fact that the overall harvested energy demand at the IoT user will be sufficed by utilizing a minimized transmit power, provided that all the considered N RF-EH ICs are active. Therefore, we obtain a valid solution of P_T based on the analysis mentioned above, expressed as follows (a detailed solution is provided in Appendix A; please refer to the same for more insights).

$$P_T = \frac{1}{\alpha|h|^2}\left(\alpha\beta - \ln\left(\frac{N\mathcal{E}' - \xi(1-\phi) - N\phi\mathcal{E}'}{\xi(1-\phi) + N\phi\mathcal{E}'}\right)\right). \tag{7}$$

Hereby, we provide Algorithm 1 for the minimization of transmit power.

Algorithm 1 Minimization of Transmit Power.

1: **Initialize:** n, $N = N_{\text{Max}}$, $P_T = P_{\text{Max}}$, ξ, α, and β.
2: **Generate Channel Coefficients:** h_i, $i = \{1, 2, \cdots, M_{\text{Max}}\}$, where M_{Max} is the upper limit for Monte Carlo experiment runs.
3: **For theoretical validation:** Compute P_T according to (7), keeping $n = N_{\text{Max}}$.
4: **For simulation validation:** Compute P_T based on optimization via an interior point algorithm using $fmincon(\cdot)$ function in MATLAB, $\forall h_i$
5: **To obtain different results (averaged over M_{Max}):** Keep varying values of N_{Max}, P_{Max}, and ξ.

Based on our analysis, it is clear that N plays a crucial role in the system design. An intuitive interpretation from the solution provided in (7) would be that the higher the number of RF-EH ICs in the device, the lower the transmitted power. As discussed before, the micro-controller can then manage the harvested energy to cater to the various needs to maintain the stability of the QB. In the following section, we present another optimization problem to optimize the number of RF-EH ICs, under the constraint on the overall harvested energy demand and the maximum transmit power limitation.

3.2. Optimization of the Number of RF-EH ICs

In this section, we optimize the total number of RF-EH ICs, subject to the constraints on the harvested energy by RF-EH ICs, and a limitation on the total transmit power. The corresponding optimization problem is represented in its mathematical form, as follows

$$(P3): \min_{n, P_T} \; n \tag{8}$$

$$\text{s.t.} \; (\text{C1}): n\mathcal{E}^{\mathcal{N}} \geq \xi, \tag{9}$$

$$(\text{C2}): P_T \leq P_{\text{Max}}. \tag{10}$$

We note that $(P3)$ is a non-convex problem, primarily due to the joint optimization of n and P_T in (9). Similar to before, an analytical solution based on a graphical/direct-evaluation method is possible. Correspondingly, it is observed that the objective in $(P3)$ is independent of P_T. In addition, we note that $\mathcal{E}^{\mathcal{N}}$ is an increasing function of P_T, while the presence of n in (C1) indicates that the constraint may be satisfied by either increasing its value or the value of P_T, as discussed in the previous section. Concretely, it becomes natural to choose the upper limit (value) of P_T, for the optimization of n as the objective of $(P3)$. This translates to an implication that the equality in (10) must hold. Therefore, by leveraging from one solution, i.e., $P_T = P_{\text{Max}}$, we reach the following (reduced) problem.

$$(P4): \min_{n} \; n \tag{11}$$

$$\text{s.t.} \; (\text{C1}): n \cdot \frac{\mathcal{E}'}{1-\phi} \cdot \left(\frac{1}{1 + e^{(-\alpha P_{\text{Max}}|h|^2 + \alpha\beta)}} - \phi\right) \geq \xi. \tag{12}$$

Upon obtaining the convexified form $(P4)$, we are now in a position to seek a closed-form solution or solve it via well-known convex optimization techniques [45]. To obtain a closed-form solution for n, we observe that an optimum solution is obtained when the equality in (12) holds. This is based on the fact that each RF-EH IC would be operational at the maximum possible limit (saturation regime). Hence, we turn to the following solution of n to address the saturation concern, based on the above-mentioned analysis (for a detailed solution, refer to Appendix B).

$$n = \frac{\xi(1-\phi)}{\mathcal{E}'\left(\frac{1}{1+e^{(-\alpha P_{\text{Max}}|h|^2+\alpha\beta)}} - \phi\right)}. \tag{13}$$

In the following, Algorithm 2 describes the steps to minimize the number of RF-EH ICs.

Algorithm 2 Minimization of Number of RF-EH ICs.

1: **Initialize:** n, $N = N_{\text{Max}}$, $P_T = P_{\text{Max}}$, ξ, α, and β.
2: **Generate Channel Coefficients:** h_i, $i = \{1, 2, \cdots, M_{\text{Max}}\}$, where M_{Max} is the upper limit for Monte Carlo experiment runs.
3: **For theoretical validation:** Compute n according to (13), keeping $P_T = P_{\text{Max}}$.
4: **For simulation validation:** Compute n based on optimization via interior point algorithm using $fmincon(\cdot)$ function in MATLAB, $\forall h_i$
5: **To obtain different results (averaged over M_{Max}):** Keep varying values of N_{Max}, P_{Max}, and ξ.

It is noteworthy that P_{Max} plays a crucial role in the system design. An intuitive interpretation from the solution provided in (13) would be that the higher the transmitted power of the system, the lower the number of RF-EH ICs. Based on the optimized number of RF-EH ICs, the micro-controller can then manage the harvested energy to cater to the various needs to maintain the stability of QB. In the subsequent section, we present some simulation results to test the effectiveness of the proposed design in terms of the obtained solutions.

4. Simulation Results

Herein, we evaluate the performance of the proposed framework in terms of the minimized transmit power and the optimized number of RF-EH ICs, as obtained by solving the concerned mathematical problems discussed in the previous section. Direct implementation of the closed-form expressions are performed with the help of MATLAB R2019b, and the relevant closed-form solutions are referred to as 'Theoretical'. Concerning the non-linear program solver, we make use of the *fmincon(·)* solver present in the optimization toolbox of MATLAB R2019b [46]. The corresponding outcomes are hereby described as 'Simulation' for further reference.

We assume an ITU-R outdoor framework (site-general model for propagation within street canyons) [47] to generate channel realizations with the path-loss exponent:

$$PL(D,F) = 10a\log_{10}(D) + b + 10c\log_{10}(F) + N(0,\sigma)\ \text{dB}, \tag{14}$$

where D is the 3D direct distance between the transmitting and receiving stations (m), F is the operating frequency (GHz); the coefficients a, b, and c are associated with the increase in the path loss with distance, the offset value of the path loss, and the increase in the path loss with frequency, respectively; and $N(0,\sigma)$ is a zero mean Gaussian random variable with a standard deviation σ (dB). The channel coefficients for the link between the transmitter and IoT user are generated accordingly. Specifically, we choose: F = 24 GHz; D is assumed to be 5 m (unless specified otherwise) for the IoT user with respect to the transmitter. a = 2.12, b = 29.2, c = 2.11 and σ = 5.06 dB (the value of F is chosen according to frequency range 2 (FR2) pertaining to the 5G-and-beyond systems [22]. Concerning the EH process via RF, the proposed method can be applied to other kind of sources and for different (more) scaled frequencies, such as frequency in the FM or AM bands or Bluetooth low energy (BLE) range (2.4 GHz) [48,49]). The constants corresponding to the non-linear EH circuit are chosen as $\mathcal{E}'$ = 2.8 mJ, α = 1500, and β = 0.0022 [42–44]. Concerning every experimental run, we consider an average of 500 random channel realizations. For proper distinction among the closed-form and simulations-based solutions (obtained via *fmincon(·)* solver), we refer to the former as theoretical and the latter as simulation results. Table 1 illustrates the parameter value selections pertaining to the simulation environment.

Table 1. Table for parameter value selections pertaining to the simulation environment.

Parameter	Value
a	2.12
b	29.2
c	2.11
σ	5.06 dB
$\mathcal{E}'$	2.8 mJ
α	1500
β	0.0022
F	24 GHz
D	5 m (variable)

In Figure 4, we illustrate the effect on the optimized transmit power values with the growing demand of harvested energy at the IoT user. Both the proposed theoretical and simulation-based solutions are compared herein. A non-linear increment is observed in the optimized transmit power with the escalating demand of harvested energy. On the other hand, when the number of RF-EH ICs is incremented steadily, we observe a considerable depletion in the corresponding values of optimized transmit power. The obtained results show that both theoretical and simulations-based solutions are more or less the same. Additionally, it is noted that the outcomes are in line with our intuitive analysis.

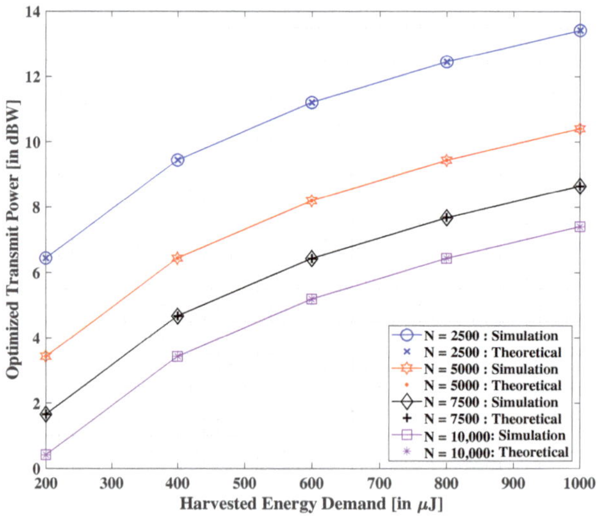

Figure 4. Minimized transmit power (in dBW) versus the demanded harvested energy for altering values of N [40].

Using Figure 5, we will now discuss the percentage increase in the transmit power for various transiting states concerning the number of RF-EH ICs. The transit states are defined as the jump from one value of N to another, which determines our experimental outcomes. For example, $N(10{,}000 \rightarrow 7500)$ indicates the outcome when $N = 10{,}000$ is transited to $N = 7500$ for a fixed value of harvested energy demand. The corresponding percentage increment is defined by: % increment $= \frac{\text{Final Value}-\text{Starting Value}}{|\text{Starting Value}|} \times 100$. We consider three cases, vis-á-vis (a) $\eta = 400$ μJ, (b) $\eta = 600$ μJ, and (c) $\eta = 800$ μJ, to investigate the percentage increment in the transmit power for various transiting states of N. The escalation in the percentage increment of transmit power from one state to another (in a successive manner) is noteworthy for each of the cases, as shown. The overall results imply the non-linear incremental relationship of the transmit power with the number of RF-EH ICs, for increasing values of harvested energy demands.

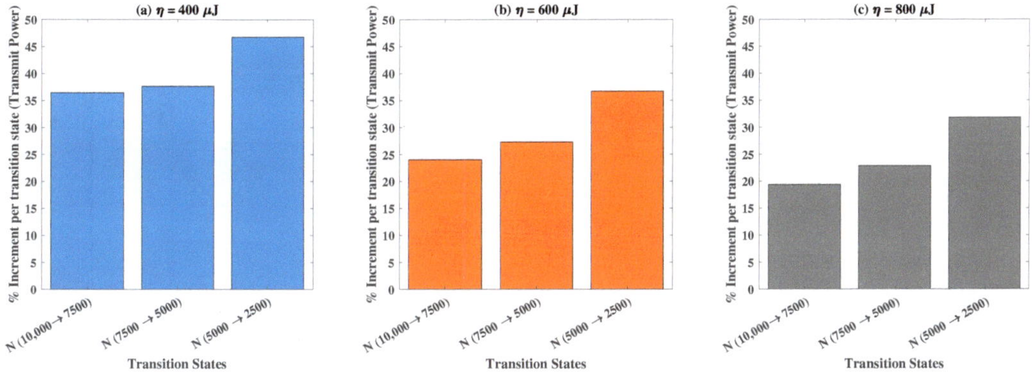

Figure 5. Percentage increment in transmit power versus different transition states pertaining to N for: (**a**) $\eta = 400\ \mu$J, (**b**) $\eta = 600\ \mu$J, and (**c**) $\eta = 800\ \mu$J.

Figure 6 depicts the results corresponding to the impact of increasing distance values between the transmit source and the end user on the optimized transmit power, while keeping $N = 5000$. As above, we compare the theoretical and simulation-based outcomes and find an identical nature concerning both the yielded solutions. As the distance between the transmitter and the IoT user increases, we see corresponding (significant) increments to the minimum transmit power values. It is additionally noteworthy that the above-mentioned effect amplifies with growing demands of harvested energy at a higher level. From the reported outcomes, the distance limitation pertaining to the EH mechanism is explicit.

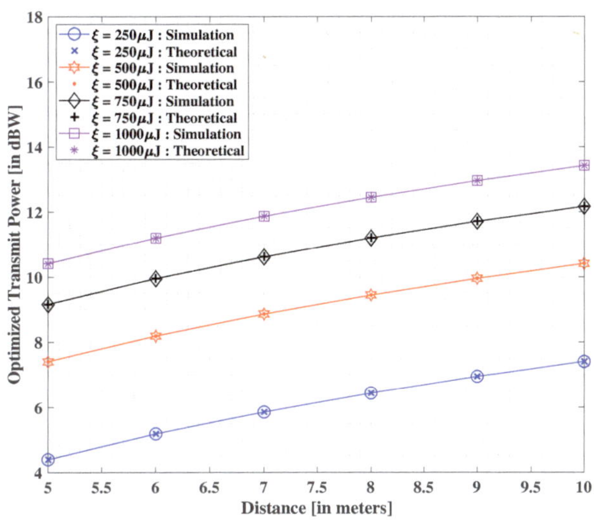

Figure 6. Minimized transmit power (in dBW) versus the demanded harvested energy, with $N = 5000$ [40].

In Figure 7, we present the bar plots of the percentage increase in the transmit power for various transiting states pertaining to the harvested energy demands. Herein, we let transit states represent the move from one value of ξ to another consecutive value, which regulates the experimental outcomes. Yet again, e.g., $\xi(250\ \mu\text{J} \rightarrow 500\ \mu\text{J})$, implies the transit of the outcomes from $\xi = 250\ \mu$J to $\xi = 500\ \mu$J. The corresponding percentage increment in the transmit power is computed according to the definition provided previously. We compare the three cases: (a) $D = 6$ m, (b) $D = 8$ m, and (c) $D = 10$ m, to find the percentage increment in the transmit power for various transiting states of ξ. Additionally, it is important to note the relative nature of percentages while examining two cases concerning D together. A depreciation in the percentage increment of the transmit powers is noted for the transits between the successive states for all cases. The relative outcomes indicate an incremental relationship of the transmit power with the harvested energy demands for appreciating values of D, while observing each transiting state individually.

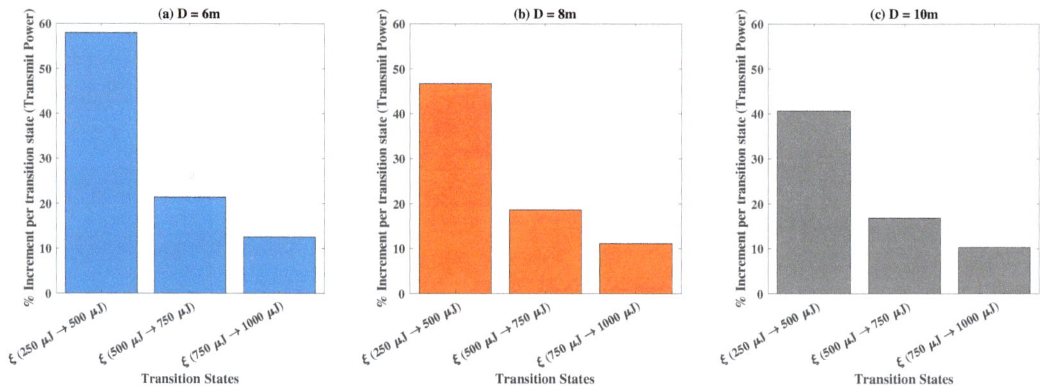

Figure 7. Percentage increment in transmit power versus different transition states pertaining to ξ for: (**a**) $D = 6$ m, (**b**) $D = 8$ m, and (**c**) $D = 10$ m.

The results shown in Figure 8 unveil the effect on the minimized transmit power values by increasing the number of RF-EH ICs, while assuming that $\xi = 500\ \mu$J. We find again that both the analytical and simulation results are identical in nature, thereby implying the optimality of the theoretical outcomes. Herein, we observe a swift decrement in the values of the optimized transmit power in contrast with the growing values of N. In addition, a significant impact on this pattern is seen when the distance between the transmit source and the IoT user is increased, wherein an increment in the minimized transmit power values is reported. Nonetheless, it is needless to conclude that the transmit power optimization will be significantly better when the RF-EH ICs present at the IoT user are placed very close to the transmitter. Moreover, the results imply that a large enough number of RF-EH ICs at the IoT user would further ensure that the transmit power is minimized by a considerable amount.

We depict in Figure 9 that the bar plots of the percentage increase in the transmit power for various transiting states pertaining to D. The move from one value of D to another (consecutive) value is defined as the transit state in this case. More specifically, the notation $D(2.5\text{ m} \rightarrow 5\text{ m})$ implies that the starting point of the experiment is the optimized transmit power at $D = 2.5$ m, the final value is obtained at $D = 5$ m, and so on. Again, the percentage increment in the transmit power is computed as per the definition stated above. Herein, we compare the results corresponding to: (a) $N = 2000$, (b) $N = 3000$, and (c) $N = 4000$ to find the percentage increment in the transmit power for various transiting states of D. Note that the percentage values are relative in nature for any two transition states or corresponding cases. This is derived from the results obtained in Figure 8, wherein the higher values of transmit power are seen to be at higher levels for $D > 2.5$ m, while the increment in power values corresponding to $D(5\text{ m} \to 7.5\text{ m})$ and $D(7.5\text{ m} \to 10\text{ m})$ is seen to be marginal in comparison to $D(2.5\text{ m} \to 5\text{ m})$. Therefore, the depreciation in the percentage increase in the transmit power per transit state for the three considered cases is justified. The overall relative results indicate an incremental non-linear relationship of the transmit power, with respect to D, for increasing values of N, while observing each transiting state individually.

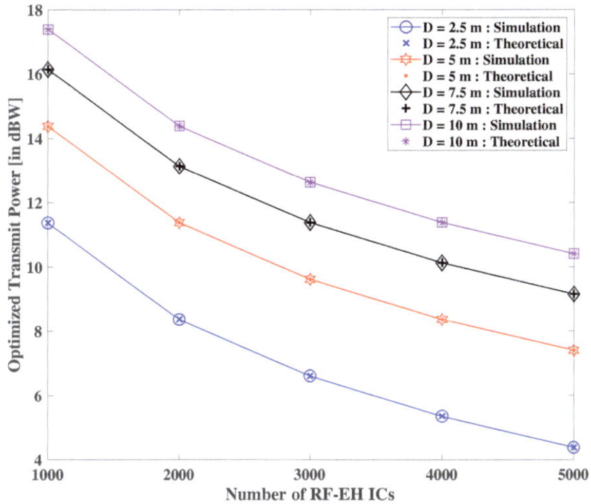

Figure 8. Minimized transmit power (in dBW) versus N for different value selections pertaining to the separation distance between the transmitter and IoT user [40].

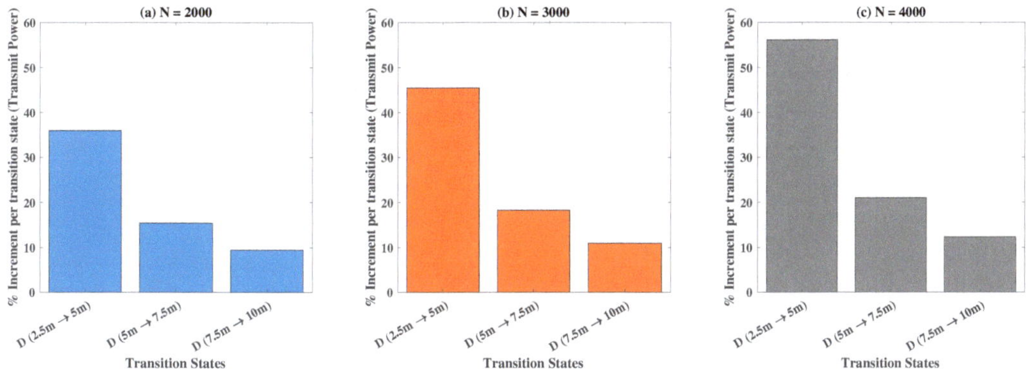

Figure 9. Percentage increment in transmit power versus different transition states pertaining to D for: (**a**) $N = 2000$, (**b**) $N = 3000$, and (**c**) $N = 4000$.

Next, in Figure 10, we illustrate the variation in the optimized number of RF-EH ICs with the increasing demand of harvested energy at the IoT user. Similar to before, we compare the similarities between the outcomes pertaining to the proposed theoretical and simulation-based solutions. Herein, we observe that the optimized number of RF-EH ICs increases non-linearly with growing demands of harvested energy. However, a considerable decrement in the values of the optimized number of RF-EH ICs is seen with the increasing values of total transmit power P_{Max}. We can note that both theoretical and simulation-based solutions yield similar results. It is noteworthy that these results seem to be in line with our intuitive analysis.

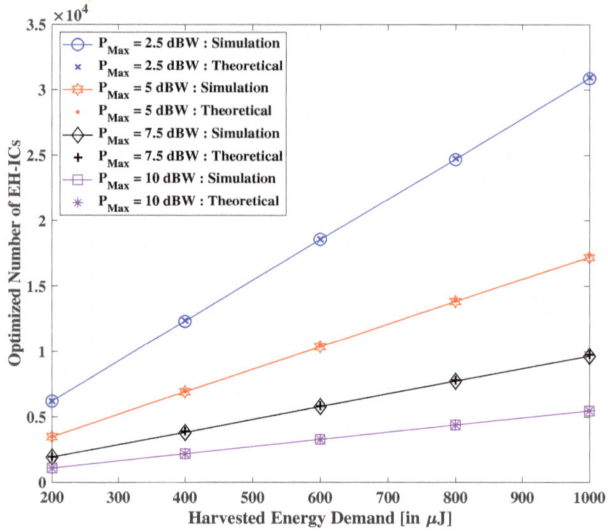

Figure 10. Optimized number of RF-EH ICs versus the overall demanded harvested energy demand for altering values of P_{Max}.

In Figure 11, we show the effect of increasing the distance between the transmit source and the IoT user equipped with the QB on the optimized number of RF-EH ICs, while assuming $P_{\text{Max}} = 10$ dBW. Herein, we draw a comparison between the theoretical and simulation-based solutions, wherein both the solutions are reported to be almost identical. We observe that the minimized number of RF-EH ICs escalates with increasing distance values between the transmitter and IoT user. However, this effect amplifies at a higher level with the growing demands of overall harvested energy. The clear impact of distance limitation may be inferred herein, as seen in the previous results.

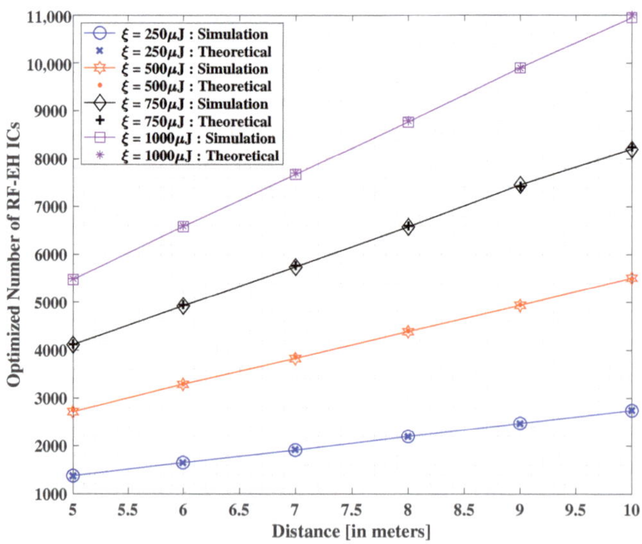

Figure 11. Optimized number of RF-EH ICs versus varying separation distances between the transmitter and end user, for different values of demanded harvested energy, where $P_{\text{Max}} = 10$ dBW.

In Figure 12, we depict the results showing the impact of increasing the transmit power values on the optimized number of RF-EH ICs, with an assumption of $\zeta = 500\ \mu$J. Both the theoretical and simulation-based solutions are found to be more or less identical in nature. We observe that the minimum optimized number of RF-EH ICs depreciates with the increasing values of P_{Max}. We additionally observe that the optimization yield corresponding to the transmit power intensifies with the increasing values of distance between the transmitter and the IoT user. This outcome suggest that the more closer the IoT user is to the transmitter, the better the optimization of the number of RF-EH ICs. Additionally, if the transmit power is large enough, the number of RF-EH ICs can be further minimized drastically. However, it is important to respect the maximum transmissible (allowable) power limitations deemed safe for living beings.

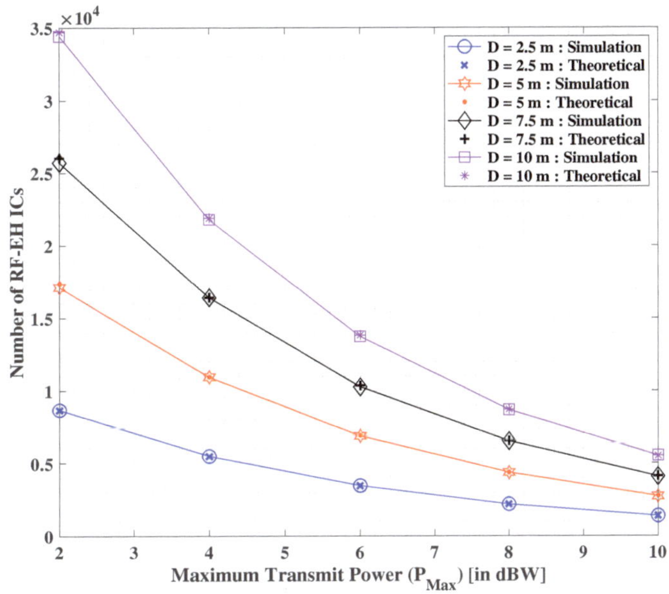

Figure 12. Optimized number of RF-EH ICs versus P_{Max} for altering selections of the separation distance between the transmitter and IoT user.

5. Conclusions

In this invited paper, we investigated a novel IoT system comprised of RF-EH ICs-assisted QB. A micro-controller was utilized to manage the overall harvested energy from several RF-EH ICs placed in parallel, and was also responsible for powering a photon-emitting device. The considered system was composed of a single transmitter and a single QB-enabled IoT user. In this vein, we firstly minimized the total transmitted power, subject to constraints on minimum number of RF-EH ICs, and harvested energy demand of the IoT user. Next, we optimized the number of RF-EH ICs subject to constraints on the total transmit power and harvested energy demands. For both the above-mentioned problems, we obtained suitable closed-form solutions and also cross-validated them using a non-convex optimization solver. Concerning future prospects, the outcomes from this work hold great promise in terms of both transmit power optimization while respecting the safety limits, and use of a minimal number of RF-EH ICs to maintain the stability of QB in a single IoT receiver unit. The proposed novel framework based on IoT systems equipped with RF-EH ICs-assisted QB may find applications in the domains where battery concerns are prevalent, specifically in compact devices with displays. Herein, the internal movement of electrons within a QB may be translated to a visual information via display, where this interesting idea can be taken up for future research. Some practical examples where RF-EH

ICs assisted QB systems could come in handy are: electric automotives, wireless gadgets, biomedical sensors, and many other limited battery-constrained devices. In future, we plan to extend this work to a multi-user and multi-carrier scenario for the RF-EH process to improve the efficiency of the QB system.

Appendix A. Closed-Form Solution for the Transmit Power Minimization Problem

The non-linear EH constraint at IoT user, as in (6), is represented as

$$N \cdot \frac{\mathcal{E}'}{1-\phi} \cdot \Big(\frac{1}{1+e^{(-\alpha P_T|h|^2+\alpha\beta)}} - \phi\Big) \geq \xi, \tag{A1}$$

where ξ is the harvested energy demand at the IoT user.

We rearrange the expression in (A1) as

$$\frac{1}{1+e^{(-\alpha P_T|h|^2+\alpha\beta)}} \geq \xi \cdot \left(\frac{1-\phi}{N\mathcal{E}'}\right) + \phi, \tag{A2}$$

and rewrite it as

$$1+e^{(-\alpha P_T|h|^2+\alpha\beta)} \leq \frac{N\mathcal{E}'}{\xi(1-\phi)+\phi N\mathcal{E}'}. \tag{A3}$$

The equation in (A3) implies that

$$e^{(-\alpha P_T|h|^2+\alpha\beta)} \leq \frac{N\mathcal{E}' - (\xi(1-\phi)+\phi N\mathcal{E}')}{\xi(1-\phi)+\phi N\mathcal{E}'}. \tag{A4}$$

Next, we take the natural logarithm on both sides of (A4)

$$-\alpha P_T|h|^2 + \alpha\beta \leq \ln\left(\frac{N\mathcal{E}' - (\xi(1-\phi)+\phi N\mathcal{E}')}{\xi(1-\phi)+\phi N\mathcal{E}'}\right). \tag{A5}$$

An equivalent constraint on the transmit power is obtained upon further simplification of (A5), represented by

$$P_T \geq \frac{1}{\alpha|h|^2}\Big(\alpha\beta - \ln\Big(\frac{N\mathcal{E}' - \xi(1-\phi) - N\phi\mathcal{E}'}{\xi(1-\phi)+N\phi\mathcal{E}'}\Big)\Big). \tag{A6}$$

From (A6), it is clear that in order to minimize the total transmit power, the equality in (A6) must hold. Therefore, we present the closed-form expression for P_T, as follows.

$$P_T = \frac{1}{\alpha|h|^2}\Big(\alpha\beta - \ln\Big(\frac{N\mathcal{E}' - \xi(1-\phi) - N\phi\mathcal{E}'}{\xi(1-\phi)+N\phi\mathcal{E}'}\Big)\Big). \tag{A7}$$

QED. ∎

Appendix B. Closed-Form Solution for the Optimal Number of RF-EH ICs

From (12), we have the following non-linear EH constraint at the IoT user

$$n \cdot \frac{\mathcal{E}'}{1-\phi} \cdot \left(\frac{1}{1+e^{(-\alpha P_{\text{Max}}|h|^2+\alpha\beta)}} - \phi\right) \geq \xi, \tag{A8}$$

where ξ is the harvested energy demand at the IoT user, as before.

Rearrangement of (A8) leads to

$$n \cdot \frac{\mathcal{E}'}{1-\phi} \geq \xi \cdot \left(\frac{1}{1+e^{(-\alpha P_{\text{Max}}|h|^2+\alpha\beta)}} - \phi\right)^{-1}, \tag{A9}$$

which can be further simplified as

$$n \geq \frac{\xi(1-\phi)}{\mathcal{E}'\left(\frac{1}{1+e^{(-\alpha P_{\text{Max}}|h|^2+\alpha\beta)}} - \phi\right)}. \quad \text{(A10)}$$

From (A10), it is clear that in order to minimize the total number of RF-EH ICs, the equality in (A10) must hold. Therefore, we reach to the closed-form expression for n, as follows.

$$n = \frac{\xi(1-\phi)}{\mathcal{E}'\left(\frac{1}{1+e^{(-\alpha P_{\text{Max}}|h|^2+\alpha\beta)}} - \phi\right)}. \quad \text{(A11)}$$

QED. ■

Author Contributions: Conceptualization, S.G. and S.K.S.; Data curation, S.G. and S.S.; Formal analysis, S.G.; Investigation, S.G.; Methodology, S.G. and S.S.; Resources, S.G. and S.K.S.; Supervision, S.K.S., S.C. and B.O.; Validation, S.G.; Writing—original draft, S.G.; Writing—review & editing, S.G., S.S., S.K.S., S.C. and B.O. All authors have read and agreed to the published version of the manuscript.

Funding: The research leading to these results has received funding from the Luxembourg National Research Fund (FNR), Luxembourg, under the FNR-FNRS bilateral InWIP-NET: Integrated Wireless Information and Power Networks (R-AGR-0700-10-X). The authors also acknowledge the joint research work carried out between the Indian Institute of Technology—Indore and the University of Luxembourg, enabled by their mutual Memorandum of Understanding (MoU).

Conflicts of Interest: The authors declare no conflicts of interest.

References

1. Salek, F.; Anshu, A.; Hsieh, M.; Jain, R.; Fonollosa, J.R. One-Shot Capacity Bounds on the Simultaneous Transmission of Classical and Quantum Information. *IEEE Trans. Inf. Theory* **2020**, *66*, 2141–2164. [CrossRef]
2. Cacciapuoti, A.S.; Caleffi, M.; Van Meter, R.; Hanzo, L. When Entanglement Meets Classical Communications: Quantum Teleportation for the Quantum Internet. *IEEE Trans. Commun.* **2020**, *68*, 3808–3833. [CrossRef]
3. Gautam, S.; Lagunas, E.; Bandi, A.; Chatzinotas, S.; Sharma, S.K.; Vu, T.X.; Kisseleff, S.; Otterst, B. Multigroup Multicast Precoding for Energy Optimization in SWIPT Systems With Heterogeneous Users. *IEEE Open J. Commun. Soc.* **2020**, *1*, 92–108. [CrossRef]
4. Gautam, S.; Lagunas, E.; Sharma, S.K.; Chatzinotas, S.; Ottersten, B.; Vandendorpe, L. Weighted Sum-SINR and Fairness Optimization for SWIPT-Multigroup Multicasting Systems With Heterogeneous Users. *IEEE Open J. Commun. Soc.* **2020**, *1*, 1470–1484. [CrossRef]
5. Gautam, S.; Vu, T.X.; Chatzinotas, S.; Ottersten, B. Cache-Aided Simultaneous Wireless Information and Power Transfer (SWIPT) With Relay Selection. *IEEE J. Sel. Areas Commun.* **2019**, *37*, 187–201. [CrossRef]
6. Ridddle. A Quantum Battery That Never Runs Out. 2020. Available online: https://www.youtube.com/watch?v=7fM4e0cx5zU (accessed on 20 May 2022).
7. Campaioli, F.; Pollock, F.A.; Binder, F.C.; Céleri, L.; Goold, J.; Vinjanampathy, S.; Modi, K. Enhancing the Charging Power of Quantum Batteries. *Phys. Rev. Lett.* **2017**, *118*, 150601. [CrossRef] [PubMed]
8. Andolina, G.M.; Keck, M.; Mari, A.; Campisi, M.; Giovannetti, V.; Polini, M. Extractable Work, the Role of Correlations, and Asymptotic Freedom in Quantum Batteries. *Phys. Rev. Lett.* **2019**, *122*, 047702. [CrossRef]
9. Andolina, G.M.; Farina, D.; Mari, A.; Pellegrini, V.; Giovannetti, V.; Polini, M. Charger-mediated energy transfer in exactly solvable models for quantum batteries. *Phys. Rev. B* **2018**, *98*, 205423. [CrossRef]
10. Quach, J.Q.; Munro, W.J. Using Dark States to Charge and Stabilize Open Quantum Batteries. *Phys. Rev. Appl.* **2020**, *14*, 024092. [CrossRef]
11. Liu, J.; Segal, D.; Hanna, G. Loss-Free Excitonic Quantum Battery. *J. Phys. Chem. C* **2019**, *123*, 18303–18314. [CrossRef]
12. Dou, F.Q.; Wang, Y.J.; Sun, J.A. Closed-loop three-level charged quantum battery. *arXiv* **2020**, arXiv:2004.09429.
13. Lumpkins, W. Nikola Tesla's Dream Realized: Wireless power energy harvesting. *IEEE Consum. Electron. Mag.* **2014**, *3*, 39–42. [CrossRef]
14. Valenta, C.R.; Durgin, G.D. Harvesting Wireless Power: Survey of Energy-Harvester Conversion Efficiency in Far-Field, Wireless Power Transfer Systems. *IEEE Microw. Mag.* **2014**, *15*, 108–120.
15. Ruan, T.; Chew, Z.J.; Zhu, M. Energy-Aware Approaches for Energy Harvesting Powered Wireless Sensor Nodes. *IEEE Sens. J.* **2017**, *17*, 2165–2173. [CrossRef]
16. Yang, Z.; Xu, W.; Shikh-Bahaei, M. Energy Efficient UAV Communication with Energy Harvesting. *IEEE Trans. Veh. Technol.* **2020**, *69*, 1913–1927. [CrossRef]

17. Wang, L.; Todaria, P.; Pandey, A.; O'Connor, J.; Chernow, B.; Zuo, L. An Electromagnetic Speed Bump Energy Harvester and Its Interactions With Vehicles. *IEEE/ASME Trans. Mechatron.* **2016**, *21*, 1985–1994. [CrossRef]
18. Varshney, L.R. Transporting information and energy simultaneously. In Proceedings of the 2008 IEEE International Symposium on Information Theory, Toronto, ON, Canada, 6–11 July 2008; pp. 1612–1616. [CrossRef]
19. Ding, Z.; Zhong, C.; Ng, W.K.D.; Peng, M.; Suraweera, H.A.; Schober, R.; Poor, H.V. Application of smart antenna technologies in simultaneous wireless information and power transfer. *IEEE Comm. Mag.* **2015**, *53*, 86–93. [CrossRef]
20. Gautam, S.; Lagunas, E.; Chatzinotas, S.; Ottersten, B. Sequential Resource Distribution Technique for Multi-User OFDM-SWIPT Based Cooperative Networks. In Proceedings of the 2018 IEEE Global Communications Conference (GLOBECOM), Abu Dhabi, United Arab Emirates, 9–13 December 2018; pp. 1–7. [CrossRef]
21. Gautam, S.; Lagunas, E.; Sharma, S.K.; Chatzinotas, S.; Ottersten, B. Relay selection strategies for SWIPT-enabled cooperative wireless systems. In Proceedings of the 2017 IEEE 28th Annual International Symposium on Personal, Indoor, and Mobile Radio Communications (PIMRC), Montreal, QC, Canada, 8–13 October 2017; pp. 1–7. [CrossRef]
22. Gautam, S.; Solanki, S.; Sharma, S.K.; Chatzinotas, S.; Ottersten, B. Hybrid Active-and-Passive Relaying Model for 6G-IoT Greencom Networks with SWIPT. *Sensors* **2021**, *21*, 6013. [CrossRef]
23. Perera, T.D.P.; Jayakody, D.N.K.; Sharma, S.K.; Chatzinotas, S.; Li, J. Simultaneous Wireless Information and Power Transfer (SWIPT): Recent Advances and Future Challenges. *IEEE Commun. Surv. Tutor.* **2018**, *20*, 264–302. [CrossRef]
24. Kang, J.; Kim, I.; Kim, D.I. Joint Tx Power Allocation and Rx Power Splitting for SWIPT System With Multiple Nonlinear Energy Harvesting Circuits. *IEEE Wirel. Commun. Lett.* **2019**, *8*, 53–56. [CrossRef]
25. Ma, G.; Xu, J.; Zeng, Y.; Moghadam, M.R.V. A Generic Receiver Architecture for MIMO Wireless Power Transfer With Nonlinear Energy Harvesting. *IEEE Signal Process. Lett.* **2019**, *26*, 312–316. [CrossRef]
26. Awan, A.Y.; Ali, M.; Naeem, M.; Qamar, F.; Sial, M.N. Joint Network Admission Control, Mode Assignment, and Power Allocation in Energy Harvesting Aided D2D Communication. *IEEE Trans. Ind. Inform.* **2019**, *16*, 1914–1923. [CrossRef]
27. Jordan, A.N.; Sothmann, B.; Sánchez, R.; Büttiker, M. Powerful and efficient energy harvester with resonant-tunneling quantum dots. *Phys. Rev. B* **2013**, *87*, 075312. [CrossRef]
28. Verma, S.; Arya, S. Chapter 15—Applications of quantum dots in batteries. In *Green Sustainable Process for Chemical and Environmental Engineering and Science*; Inamuddin, Boddula, R., Asiri, A.M., Rahman, M.M., Eds.; Elsevier: Amsterdam, The Netherlands, 2021; pp. 287–318. [CrossRef]
29. Ballo, A.; Grasso, A.D.; Palumbo, G. A Review of Charge Pump Topologies for the Power Management of IoT Nodes. *Electronics* **2019**, *8*, 480. [CrossRef]
30. Ballo, A.; Grasso, A.D.; Palumbo, G.; Tanzawa, T. Charge Pumps for Ultra-Low-Power Applications: Analysis, Design, and New Solutions. *IEEE Trans. Circuits Syst. II Express Briefs* **2021**, *68*, 2895–2901. [CrossRef]
31. Ballo, A.; Bottaro, M.; Grasso, A.D. A Review of Power Management Integrated Circuits for Ultrasound-Based Energy Harvesting in Implantable Medical Devices. *Appl. Sci.* **2021**, *11*, 2487. [CrossRef]
32. Ballo, A.; Grasso, A.D.; Giustolisi, G.; Palumbo, G. Optimized Charge Pump With Clock Booster for Reduced Rise Time or Silicon Area. *IEEE Trans. Circuits Syst. II Express Briefs* **2019**, *66*, 1977–1981. [CrossRef]
33. Ballo, A.; Grasso, A.D.; Palumbo, G. A Subthreshold Cross-Coupled Hybrid Charge Pump for 50-mV Cold-Start. *IEEE Access* **2020**, *8*, 188959–188969. [CrossRef]
34. Gemme, G.; Grossi, M.; Ferraro, D.; Vallecorsa, S.; Sassetti, M. IBM Quantum Platforms: A Quantum Battery Perspective. *Batteries* **2022**, *8*, 43. [CrossRef]
35. Qi, S.; Jing, J. Magnon-mediated quantum battery under systematic errors. *Phys. Rev. A* **2021**, *104*, 032606. doi: 10.1103/PhysRevA.104.032606. [CrossRef]
36. Crescente, A.; Carrega, M.; Sassetti, M.; Ferraro, D. Charging and energy fluctuations of a driven quantum battery. *New J. Phys.* **2020**, *22*, 063057. [CrossRef]
37. Zhang, Y.Y.; Yang, T.R.; Fu, L.; Wang, X. Powerful harmonic charging in a quantum battery. *Phys. Rev. E* **2019**, *99*, 052106. [CrossRef]
38. Binder, F.C.; Vinjanampathy, S.; Modi, K.; Goold, J. Quantacell: Powerful charging of quantum batteries. *New J. Phys.* **2015**, *17*, 075015. [CrossRef]
39. Shaghaghi, V.; Singh, V.; Benenti, G.; Rosa, D. Micromasers as Quantum Batteries. *arXiv* **2022**, arXiv:2204.09995.
40. Gautam, S.; Sharma, S.K.; Chatzinotas, S.; Ottersten, B. Modeling and Optimization of RF-Energy Harvesting-assisted Quantum Battery System. In Proceedings of the 2021 IEEE 93rd Vehicular Technology Conference (VTC2021-Spring), Helsinki, Finland, 25–28 April 2021; pp. 1–6. [CrossRef]
41. Clerckx, B.; Bayguzina, E. Waveform Design for Wireless Power Transfer. *IEEE Trans. Signal Process.* **2016**, *64*, 6313–6328. [CrossRef]
42. Guo, J.; Zhang, H.; Zhu, X. Theoretical analysis of RF-DC conversion efficiency for class-F rectifiers. *IEEE Trans. Microw. Theory Tech.* **2014**, *62*, 977–985. [CrossRef]
43. Boshkovska, E.; Ng, D.W.K.; Zlatanov, N.; Schober, R. Practical Non-Linear Energy Harvesting Model and Resource Allocation for SWIPT Systems. *IEEE Wirel. Commun. Lett.* **2015**, *19*, 2082–2085. [CrossRef]
44. Gautam, S.; Lagunas, E.; Chatzinotas, S.; Ottersten, B. Relay Selection and Resource Allocation for SWIPT in Multi-User OFDMA Systems. *IEEE Trans. Wirel. Commun.* **2019**, *18*, 2493–2508. [CrossRef]

45. Boyd, S.; Boyd, S.P.; Vandenberghe, L. *Convex Optimization*; Cambridge University Press: Cambridge, UK, 2004.
46. *MATLAB Optimization Toolbox (Fmincon)*; 9.7.0.1296695 (R2019b) Update 4; The MathWorks: Natick, MA, USA. Avaliable online: https://nl.mathworks.com/products/optimization.html (accessed on 20 May 2022).
47. Series, P. *Propagation Data and Prediction Methods for the Planning of Short-Range Outdoor Radiocommunication Systems and Radio Local Area Networks in the Frequency Range 300 MHz to 100 GHz*; Tech. Rep. ITU-R; ITU: Geneva, Switzerland, 2017.
48. Gautam, S.; Kumar, S.; Chatzinotas, S.; Ottersten, B. Experimental Evaluation of RF Waveform Designs for Wireless Power Transfer Using Software Defined Radio. *IEEE Access* **2021**, *9*, 132609–132622. [CrossRef]
49. Gautam, S.; Kumar, S.; Chatzinotas, S.; Ottersten, B. Experimental Comparison of RF Waveform Designs for Wireless Power Transmission. In Proceedings of the 2021 IEEE 26th International Workshop on Computer Aided Modeling and Design of Communication Links and Networks (CAMAD), Porto, Portugal, 25–27 October 2021; pp. 1–7. [CrossRef]

sensors

MDPI

Article

A New RF Energy Harvesting System Based on Two Architectures to Enhance the DC Output Voltage for WSN Feeding

Chemseddine Benkalfate [1,2,*], Achour Ouslimani [1], Abed-Elhak Kasbari [1] and Mohammed Feham [2]

1 Quartz Laboratory, Department of Electrical and Electronic Engineering, Ecole Nationale Supérieure de l'Electronique et de ses Applications, 95014 Cergy, France; achour.ouslimani@ensea.fr (A.O.); abed-elhak.kasbari@ensea.fr (A.-E.K.)

2 STIC Laboratory, Department of Telecommunications, Faculty of Technology, University Abou Bekr Belkaid, Tlemcen BP 230-13000, Algeria; mohammed.feham@univ-tlemcen.dz

* Correspondence: benkalfate.chemseddine@ensea.fr

Abstract: In this paper, a new RF Energy Harvesting (RF-EH) system for Wireless Sensor Network (WSN) feeding is proposed. It is based on two different monitored architectures using switch circuits controlled by the input powers. One architecture is more adapted to high input powers and the other to low input powers. The two different architectures and the system are designed and realized on Teflon glass substrate with a relative permittivity of 2.1 and thickness of 0.67 mm. They are tested separately as a function of the distance from the relay antenna. A new multiband antenna with a size of 40×30 mm^2 is used for both architectures and the system. The measured antenna gains are 2.7 dB, 2.9 dB, and 2.55 dB for the frequencies of 1.8 GHz, 2.1 GHz, and 2.66 GHz corresponding to the mobile communication networks, respectively. The rectifier consists of two Schottky diodes forming a full-wave rectifier and voltage doubler. The maximum measured RF-to-DC conversion efficiency is 71.5%. The proposed RF-EH system provides a maximum DC output voltage of 5.6 V and 3.15 V for an open and 2 kΩ resistance load, respectively.

Keywords: RF energy harvesting system; multi-bands antenna; patch antenna; rectifier; impedance matching circuit; rectenna; switch

Citation: Benkalfate, C.; Ouslimani, A.; Kasbari, A.-E.; Feham, M. A New RF Energy Harvesting System Based on Two Architectures to Enhance the DC Output Voltage for WSN Feeding. *Sensors* **2022**, *22*, 3576. https://doi.org/10.3390/s22093576

Academic Editors: Carlo Trigona, Olfa Kanoun and Slim Naifar

Received: 25 March 2022
Accepted: 3 May 2022
Published: 7 May 2022

1. Introduction

RF Energy Harvesting (RF-EH) applications are the focus of several researchers around the world. Several techniques have been proposed to ultimately solve the problem of battery life and ensure the energy autonomy of electronic equipment.

RF-EH systems consist of an antenna, impedance matching circuit, and a rectifier as presented in Figure 1.

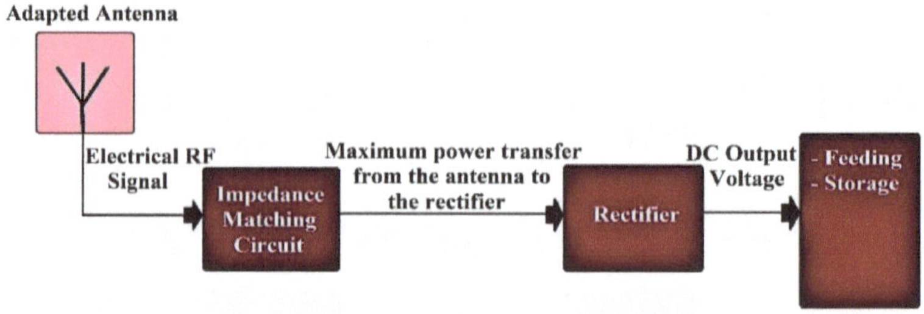

Figure 1. RF Energy Harvesting system blocks.

Several researchers propose many architectures to enhance the output voltages. Multi-stages Dickson voltage multiplier circuits based on Schottky diodes [1–3] or CMOS transistors [4,5] are proposed. The use of many active components decreases the efficiency of RF-EH systems due to the energy consumption and leakages. In [6–9], the authors propose a system in which the antenna is followed by a power divider circuit. Each output port of the divider is matched to a rectifier for a given frequency and all the DC output voltages of the rectifiers are then combined. This architecture, which divides the RF power collected by the antenna to several rectifiers, reduces the efficiency of the energy harvesting system. In addition, to maintain sufficient efficiency, this architecture requires a large antenna, which is contrary to the miniaturization.

In this paper, we propose a new RF-EH system based on two different architectures. One architecture (Figure 2a) is more adapted to low input powers and the other to high input powers (Figure 2b). The proposed system uses switches to select the more suitable architecture (Figure 2c) according to the input power. Both architectures and the system have been designed and realized on Teflon glass substrate hybrid technology and tested as a function of the distance from the mobile communication base station.

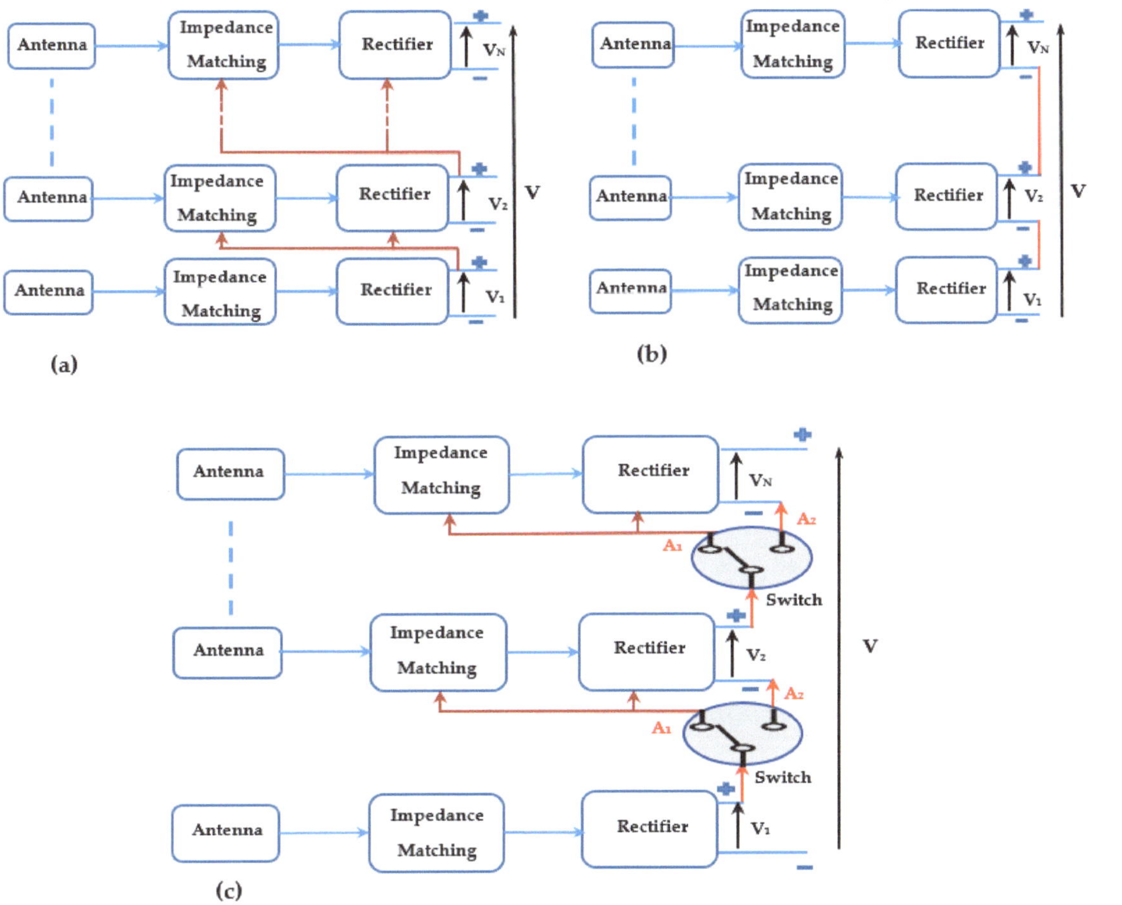

Figure 2. Block diagram of (**a**) first architecture (RF-EH1), (**b**) second architecture (RF-EH2), and (**c**) the proposed system RF-EH. A_1 and A_2 are the two positions corresponding to first and second architectures.

Generally speaking, the architectures consist in exploiting N rectennas (antenna + rectifier) that are matched to the N frequency bands that exist in the environment.

In this work, three rectennas are used to demonstrate the feasibility of the system. These rectennas are matched to the most powerful mobile communication network bands (GSM, UMTS, LTE-2.6) as presented in Figure 3.

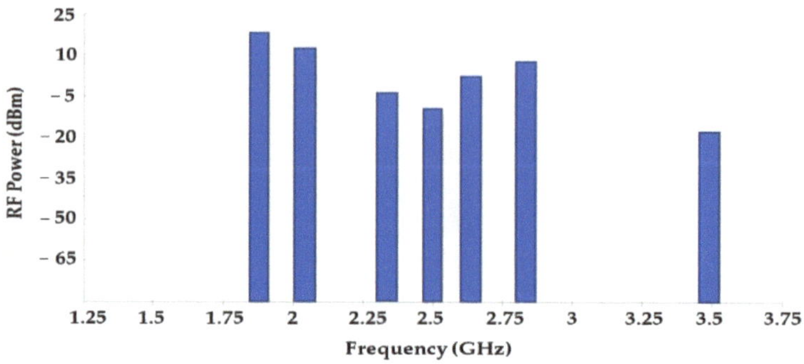

Figure 3. Measured ambient RF power at 22 m from mobile communication base station.

The first architecture, presented in Figure 2a, uses the DC output voltage V_1, provided by the rectenna-1 matched to the GSM network, to shift up the average value of the RF signal received by the rectenna-2, which is adapted to the UMTS network. The DC output voltage V_2, supplied by the rectenna-2, is used to increase the average value of the RF signal received by the rectenna-3 that is matched to the LTE-2.6 network.

The second architecture realizes a summation of the three DC output voltages provided by the three rectennas as depicted in Figure 2b.

The proposed system (Figure 2c) uses two switches to move from one architecture to another depending on the input power levels. A switching zone is observed in the input power interval of [−6–−5.5] dBm. For low input powers, the system switches to the first architecture and switches to the second one for high input powers.

The proposed system provides a maximum measured DC output voltage of 3.15 V at 22 m from the base station and a maximum efficiency of 71.5% at 75 m for a 2 kΩ resistance load. This system combines the advantages of both architectures.

For each rectenna, the proposed antenna is a new multi-band one covering all the mobile communications and wireless network bands (GSM-1800, UMTS, LTE-2.3, Wi-Fi, and LTE-2.6). The choice of this multi-band antenna results in a compact size antenna, compared to a single mono-frequency antenna that only covers one of these bands. The dimension of the proposed antenna is 40×30 mm^2. The rectifier is realized with two Schottky diodes (SMS7630-079LF) that form a full-wave rectifier and voltage doubler. This rectifier is matched to the antenna by using two LC resonators. Simulations were carried out using CST and ADS softwares.

Figure 4 shows that the effective distance between the relay antenna and the system under test is given as a function of the height of the mobile communication relay antenna and other parameters. The effective height can be defined as the difference between the height of the relay antenna and that of the device under test.

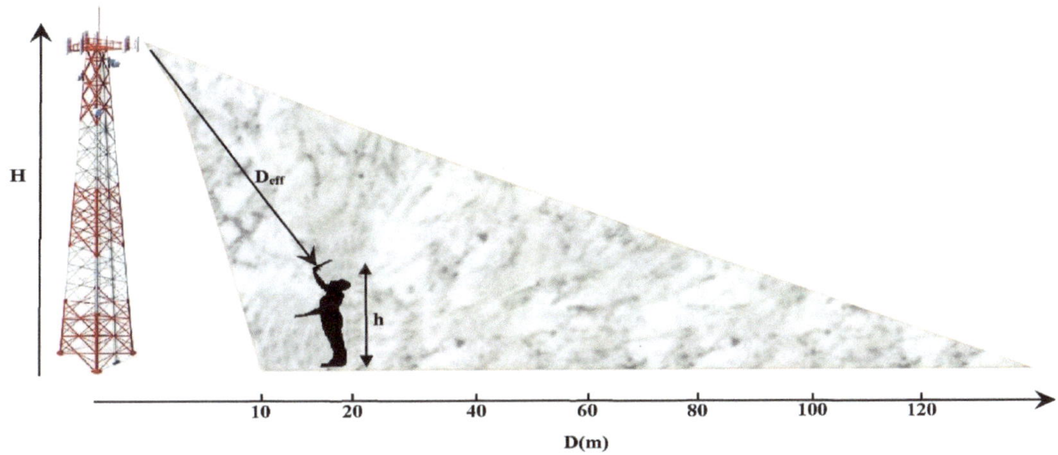

Figure 4. Geometric parameters of measurement field. D_{eff}: Effective distance.

The expression of the effective distance can be extracted by using Pythagorean law on the right triangle formed by the effective height and the horizontal distance between the relay antenna and the measuring point, as given by the Equation (2).

$$H_{eff} = H - h = 21.7 - 2 = 19.7\ m \tag{1}$$

$$D_{eff} = \sqrt{D^2 + H_{eff}{}^2} = \sqrt{D^2 + 388} \tag{2}$$

The coverage of this relay antenna starts from D = 10 m. Applying Equation (2), the effective distance is 22 m at this position. The measured received RF powers by the three rectennas (GSM-1800, UMTS, and LTE-2.6) are depicted in Table 1.

Table 1. Measured received powers as a function of distance from the mobile communication antenna relay for the three selected networks (GSM-1800, UMTS-2100 and LTE-2.6) in mW.

D_{eff} (m)	22	28	36	45	54	63	72	82	92	102	112	122
P(GSM)	28	7.2	1.8	0.6	0.35	0.25	0.17	0.12	0.09	0.07	0.05	0.043
P(UMTS)	18	2.8	0.8	0.35	0.22	0.17	0.13	0.097	0.08	0.05	0.032	0.028
P(LTE)	12.8	1.8	0.6	0.25	0.16	0.1	0.08	0.07	0.05	0.036	0.018	0.015

In the next paragraphs, we emphasize the antenna part, rectennas, both proposed architectures, and the proposed RF-EH system, respectively.

2. Stat of the Art on RF Energy Harvesting Systems

Many works on RF energy harvesting system architectures have been reported in the literature [10–24]. In these systems, different antenna and rectifier structures have been proposed to obtain sufficient RF output power with improved gain and efficiency. The rectifiers are connected to the antennas via impedance matching circuits. From these reported works, we can classify the RF-EH systems into five categories: the single-band, the broadband, the multi-band, the multi-port, and the reconfigurable RF-EH systems, which are highlighted in this section.

2.1. Single-Band RF-EH Systems

These systems are designed with a single band antenna with a full or half rectifier based on Schottky diodes [10–12]. In [10], the proposed system uses a single-band wire printed antenna realized by tow symmetric patches adapted to the 2.45 GHz frequency.

The rectifier consists of two Schottky diodes (HSMS-2852) connected to the antenna by a coplanar strip-line. The rectifier is matched to the antenna using an impedance matching circuit designed by microstrip lines and open stubs. This system provides a maximum RF-to-DC conversion efficiency of 50% for a −17.2 dBm input power and 1.4 kΩ load resistance. In [11], the authors present a system with a flexible and multilayered (3 layers) antenna matched to the 2.45 GHz frequency. It demonstrates a maximum radiation efficiency of 62%. The flexibility of this antenna allows it to reduce the size of the RF-EH system. The system uses a full wave rectifier designed by Schottky diodes (SMS7630) and an impedance matching circuit based on microstrip lines with open stubs. It provides a maximum RF-to-DC conversion efficiency of 65% for a 0 dBm input power. In [12], the RF-EH system is proposed to exploit the provided RF power from the GSM-900 network. The used antenna covers a frequency bandwidth of (823.7–1336.9) MHz. It demonstrates a radiation efficiency of 80.8% and maximum gains of 2.96 dB and 3.62 for frequencies of 900 MHz and 1227 MHz, respectively. The RF-EH system provides a maximum RF-to-DC conversion efficiency of 61.7% for a −5 dBm input power. The matching of the antenna to the rectifier is ensured using an LC resonator as an impedance matching network.

2.2. Broad-Band RF-EH Systems

These systems use a broadband antenna followed by a matched rectifier(s). The matched rectifier circuit can be either a single broadband one [13,14] or many narrowband rectifiers connected to the outputs of RF power divider [15]. In [13], a wide-band circular patch antenna is reported. It covers a frequency bandwidth of [0.8–3.6] GHz. To exploit the full frequency band available at the output of the antenna, the authors propose a broadband matched rectifier circuit ranging from 1 GHz to 3.5 GHz. This RF-EH system provides a maximum RF-to-DC conversion efficiency of 73.4% for a 3 dBm input power. In [14], the authors adopt the same technique, but for lower frequencies ranging from 0.45 GHz to 1 GHz. This system demonstrates a maximum RF-to-DC conversion efficiency of 77% for a −1 dBm input power. In [15], the authors propose an RF-EH system in which an antenna covering a band of [0.9–2.6] GHz is connected to the RF power divider with two outputs. Each output of the divider is then connected to a matched rectifier circuit to realize either a multi-output RF energy harvesting system or to sum the output voltages for improved system performances. The two rectifiers are matched to 900 MHz and 2.45 GHz. The overall DC voltage obtained by this system at 400 m from a base station (relay antenna) is 243 mV for a 4.7 kΩ load resistance.

2.3. Multi-Band RF-EH Systems

In these systems, the antenna is a multi-band one followed either by a multi-band rectifier or several single-band rectifiers that are matched to the antenna for different frequencies [16–19]. In [16], the authors present a 3D RF energy harvesting system with a multi-band antenna covering the frequencies 1.8 GHz, 2.1 GHz, 2.45 GHz, and 2.6 GHz. It demonstrates a maximum gain of 10 dBi at 2.45 GHz. This antenna is connected to four rectifiers using a four-output RF power divider. The rectifiers are designed by two SMS7630 Schottky diodes. The maximum measured RF-to-DC conversion efficiency for each rectifier are 27%, 26%, 25.5%, and 27.5% for the four frequencies of 1.8 GHz, 2.1 GHz, 2.45 GHz, and 2.6 GHz, respectively, and a −6 dBm input power. A maximum voltage of 484 mV is provided by the rectifier matched to the 2.6 GHz frequency for a 3.3 kΩ load resistance. In [17], the proposed RF energy harvesting system uses a multi-band antenna with 12 output ports that are matched to bands of [1.7–1.8] GHz and [2.1–2.7] GHz. Each output port is connected to a full-wave rectifier based on SMS7630 Schottky diodes and matched to the antenna by open stubs. This system demonstrates a maximum RF-to-DC conversion efficiency of 67% for a −3 dBm input power. The output voltages of each rectifier are summed to improve the system performance. This provides maximum DC output powers of 65 µW and 20 µW outside and inside the building (Outdoor/Indoor) for ambient power densities of -13 dBm/cm^2 and -40 dBm/cm^2, respectively. In [18], the authors propose a dual-band system in which the antenna is a dual band one covering the UHF [855–935] MHz and ISM 2.45 GHz bands. The maximum radiation efficiencies and gains of this antenna are 88% and 92% and 2 dBi and 3.8 dBi for the UHF and ISM bands, respectively. The antenna is connected to two rectifiers using a two output RF power divider. One rectifier is matched to the UHF band while the other is matched to the ISM band. The maximum voltage delivered by this system is 0.7 V for an input power of −12 dBm and a 10 kΩ load resistance. The maximum RF-to-DC conversion efficiency is 65% for an input power of −5 dBm. in [19], the authors propose a dual band system. It uses a dual band antenna based on many resonators that are connected together using diodes to route the currents provided by each resonator to a single point and ensure a combination of the RF signals. The two bands covered by this antenna are ISM 2.45 GHz and ISM 5.8 GHz. The antenna is connected to a dual band matched rectifier covering the same bands as the antenna. The maximum RF-to-DC conversion efficiency provided by this system is 70% for a 10 dBm input power. This efficiency is obtained using the rectifier designed by the HSMS2860 Schottky diode.

2.4. Multi-Port RF-EH Systems

These systems use RF power combiners with two or more inputs [20–22]. The used multi-port antenna could be either a single band, multi-band, or broadband one. Each antenna port is then connected to the combiner and the output of the combiner is connected to the rectifier. In [20], the authors propose a complex structure (3D structure) for an RF-EH system that consists of a multi-band/multi-port antenna (8 ports) where each pair of ports is then connected to a 2-way to 1-way combiner (two inputs, one output). The output of each combiner is then connected to a rectifier based on HSMS-2850 Schottky diodes. This RF-EH system covers the GSM-900, GSM-1800, 3G (2.1 GHz), Wi-Fi (2.45 GHz), and LTE (2.6 GHz) bands. For a −20 dBm input power and 2.1 kΩ load resistance, the authors register maximum RF-to-DC conversion efficiencies of 66.52%, 51.5%, 56.5%, 31.9%, and 30% for the 3G (2.12 GHz), GSM 900, GSM 1800, Wi-Fi, and LTE bands, respectively. For a lower input power (−35 dBm), the efficiencies drop to 45%, 41.2%, 38.1%, and 34% for the summed outputs of GSM 900 + GSM 1800, GSM 900 + GSM 1800 + 3G, GSM 900 + GSM 1800 + 3G + Wi-Fi and GSM 900 + GSM 1800, + 3G, + Wi-Fi, and + LTE, respectively. The overall DC voltage collected is 0.75 V for an input power of −27 dBm and a 2.1 kΩ load resistance. In [21], the proposed RF-EH system consists of a single band multi-pixel antenna with 4 output ports. Each output of this antenna is connected to a Schottky diode HSMS-2850 full wave rectifier adapted for the 1.8 GHz frequency, which is the center frequency of the band covered by the antenna. The impedance matching circuit is realized using microstrip lines and shorted stubs. For each port, the RF-to-DC conversion efficiency is 11.9%, 33.4%, and 51.3% for −30 dBm, −20 dBm, and −10 dBm input powers, respectively. Numerous studies have been made to select one combination that allows for the achievement of a maximum power of 11.2 μW. In [22], the authors propose a multi-port RF-EH system that uses a single-band/multi-layer Grid Array Antenna (GAA) with dual output ports matched to the 2.45 GHz frequency. It allows it to obtain a high sensitivity at low input powers and has a wide incidence angle range. Each output port is connected to a half-wave rectifier designed by an SMS-7630 Schottky diode. The RF-EH system consists of two balanced rectifiers whose output DC powers are efficiently combined. One rectifier is connected to a 3 kΩ load resistance and the other to a 6 kΩ load resistance. The output DC power is greater than 200 μW for an incidence angle range of [−30° to 30°]. The maximum obtained RF-to-DC conversion efficiency is 45% for an input power density of 1 $\mu W/cm^2$.

2.5. Reconfigurable RF-EH Systems

These systems are manufactured (designed) with a single band, a multiband, or a broadband antenna. The rectifier is matched to the antenna by using an impedance matching structure based on active components (Diodes or Transistors) that form a switch [23,24]. The system proposed in [23] is a reconfigurable adaptive system consisting of a single-band microstrip patch antenna and a reconfigurable rectifier circuit. To have better adaptation for variable input power, a reconfigurable adapter is deployed between the antenna and the rectifier that is flexibly adjusted. To miniaturize the circuit size, the matching section takes the shape of an L, where a PIN diode switch (SMP1345-SC-79) is in series to implement the reconfigurable matching section. Depending on the input power levels, the PIN switch states can be automatically monitored to disconnect/connect the matching stub from/to the rectifier circuit to achieve better matching performance at different power levels. The proposed system operates at 5.8 GHz for a wide input power range. In this design, a 150-pF pre-capacitor is used to protect the antenna from a reverse current. An HSMS 2850 Schottky diode with a low junction capacitance (C_{j0} = 0.18 pF) is connected in parallel to the rectifier circuit. Two radial microstrip stubs are used as a DC pass filter and serve to suppress high-order harmonics and smooth the output DC voltage waveform—in cooperation with the pre-capacitor. The measured results for a 220 Ω load resistance show that this RF-EH system achieves an RF-to-DC conversion efficiency of more than 50% for a wide input power range of 0 to 20 dBm, with a maximum efficiency of 68% at 9 dBm and a maximum DC output voltage of 3.6 V at a 20 dBm input power.

The authors in [24] present a frequency reconfigurable RF-EH system. It is designed by a broadband patch antenna and a half-wave rectifier based on an HSMS-2850 Schottky diode. The frequency bandwidth covered by the antenna is [3.5–6.5] GHz. For the frequencies of 5.2 GHz and 5.8 GHz, the measured gains are 4.45 dBi and 4.49 dBi, respectively. The measured radiation efficiencies are 94% and 90.1%, respectively, for both frequencies. The rectifier is matched to the antenna for both frequencies of 5.2 GHz and 5.8 GHz by a reconfigurable shorted stub. Depending on the DC output voltage for both frequencies, an automatic ON-OFF of an adaptive FET switching is performed, and the stub is connected/disconnected to/from the rectifier circuit for a good matching between the antenna and the rectifier. For a 6.5 dBm input power and 200 Ω load resistance, the maximum obtained RF-to-DC conversion efficiencies are 54.9% and 52.2% for the 5.2 GHz and 5.8 GHz frequencies, respectively.

Table 2 shows the substrates used to realize the RF-EH systems seen in this section.

Table 2. Substrates used in the design of the RF-EH systems.

Ref.	[10]	[11]	[12]	[13]	[14]	[15]	[16]	[17]	[18]	[19]	[20]	[21]	[22]	[23]	[24]
Substrates	RT/ Duroid 6002	Flet Polyester/ Woven Polyester	FR-4	Rogers 5880	Rogers 5880	FR-4	Roger 4003C	Rogers RT6002	Rogers 5880	Rogers 3210	FR-4	Rogers 4003C	FR-4	Rogers 4003C	Rogers 5880
ε_r	2.94	1.2/1.5	4.4	2.2	2.2	4.4	3.55	2.94	2.2	10.2	4.4	3.38	4.4	3.38	2.2

3. Antenna Design and Measurements

The proposed antenna is a meander line one. Meander forms can be considered as deformed wires that create resonance effects. These resonators can either resonate at the same frequency, which allows them to enhance the antenna performance at this frequency, or resonate at different frequencies in order to obtain a multi-band antenna. Three-dimensional Electromagnetic simulations under CST software are carried out to optimize the antenna performances and dimensions. The proposed antenna consists of 8 meanders of different widths and lengths where each meander resonates at its own frequency, resulting in 8 frequencies covered by this antenna. The antenna is realized on Teflon glass substrate (ε_r = 2.1, thickness = 0.67 mm). The size of this antenna is 40 × 30 mm^2. The input impedance is 50 Ω. Figure 5 shows the proposed antenna model. Table 3 gives all the antenna dimensions. Figure 6 depicts the simulated and measured S_{11} parameter of the antenna.

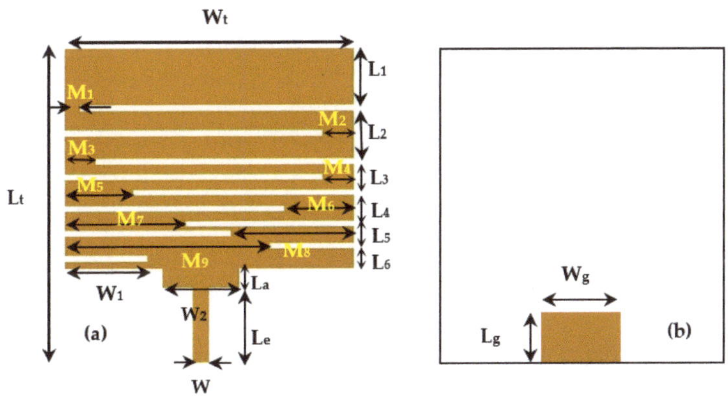

Figure 5. Proposed antenna structure, (**a**) Top side, (**b**) Bottom side.

Table 3. All antenna dimensions.

Dimensions	Values (mm)
L_t	40
W_t	30
W	2
L_a	3.5
L_e	13.5
L_1	6
L_2	6
L_3	3
L_4	3.5
L_5	2
L_6	2.5
M_1	1.2
M_2	2
M_3	3
M_4	2
M_5	7
M_6	6
M_7	13
M_8	12
M_9	8
W_1	9
W_2	9.5
W_g	5
L_g	9

The simulation and measurements are in good agreement. The measurements demonstrate that the antenna covers the frequencies of 1.8 GHz, 2.11 GHz, 2.3 GHz, 2.5 GHz, and the band from [2.6–3.6] GHz with S_{11} values of −18 dB, −21 dB, −18 dB, −13 dB, and −19 dB at the central frequency of the band, respectively. Figure 7 shows the realized antenna and measurement prototype.

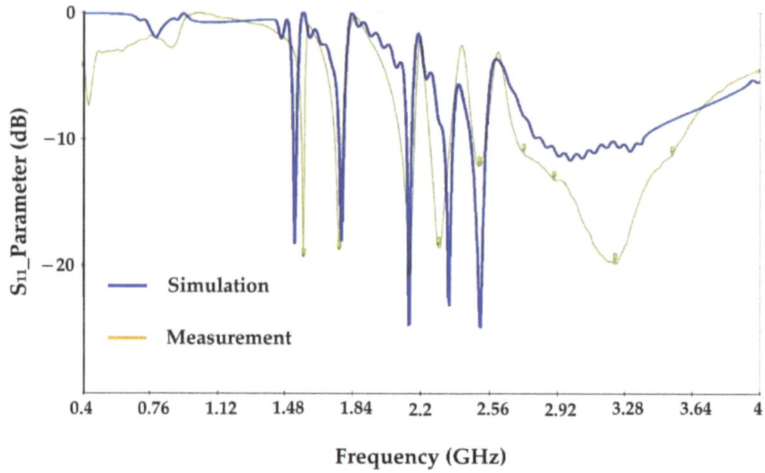

Figure 6. The simulated and measured S_{11} parameter of the proposed antenna.

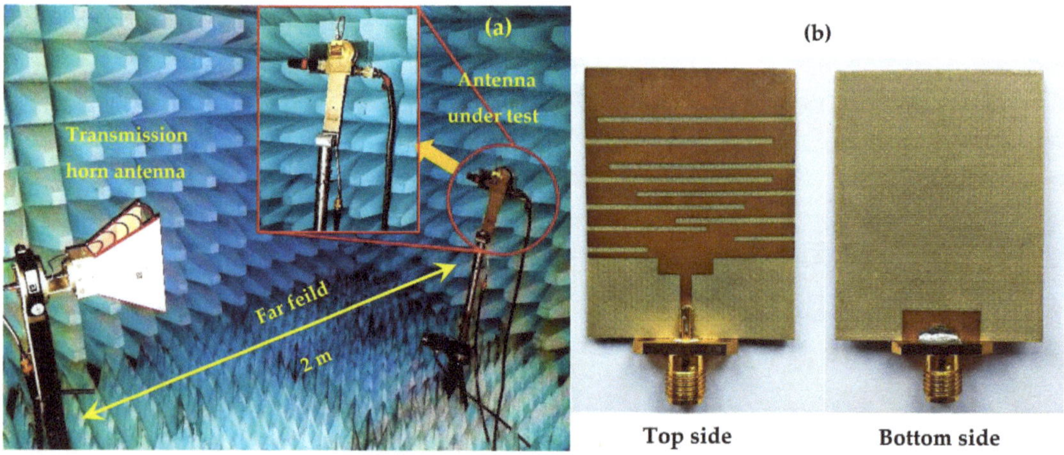

Figure 7. (**a**) Measurement prototype and (**b**) realized antenna.

Figure 8 depicts the simulated and measured radiation pattern versus gain in dB of the proposed antenna for the frequencies 1.8 GHz, 2.1 GHz, and 2.66 GHz, respectively.

It can be noted that the simulated and measured radiation pattern of this antenna are in good agreement and remain almost the same at the three frequencies of 1.8 GHz, 2.1 GHz, and 2.66 GHz with measured gain values of 2.7 dB, 2.9 dB, and 2.55 dB, respectively. The measured efficiencies are 70%, 65%, and 68.5% for 1.8 GHz, 2.1 GHz, and 2.66 GHz, respectively. All these results make this antenna suitable for RF-EH applications.

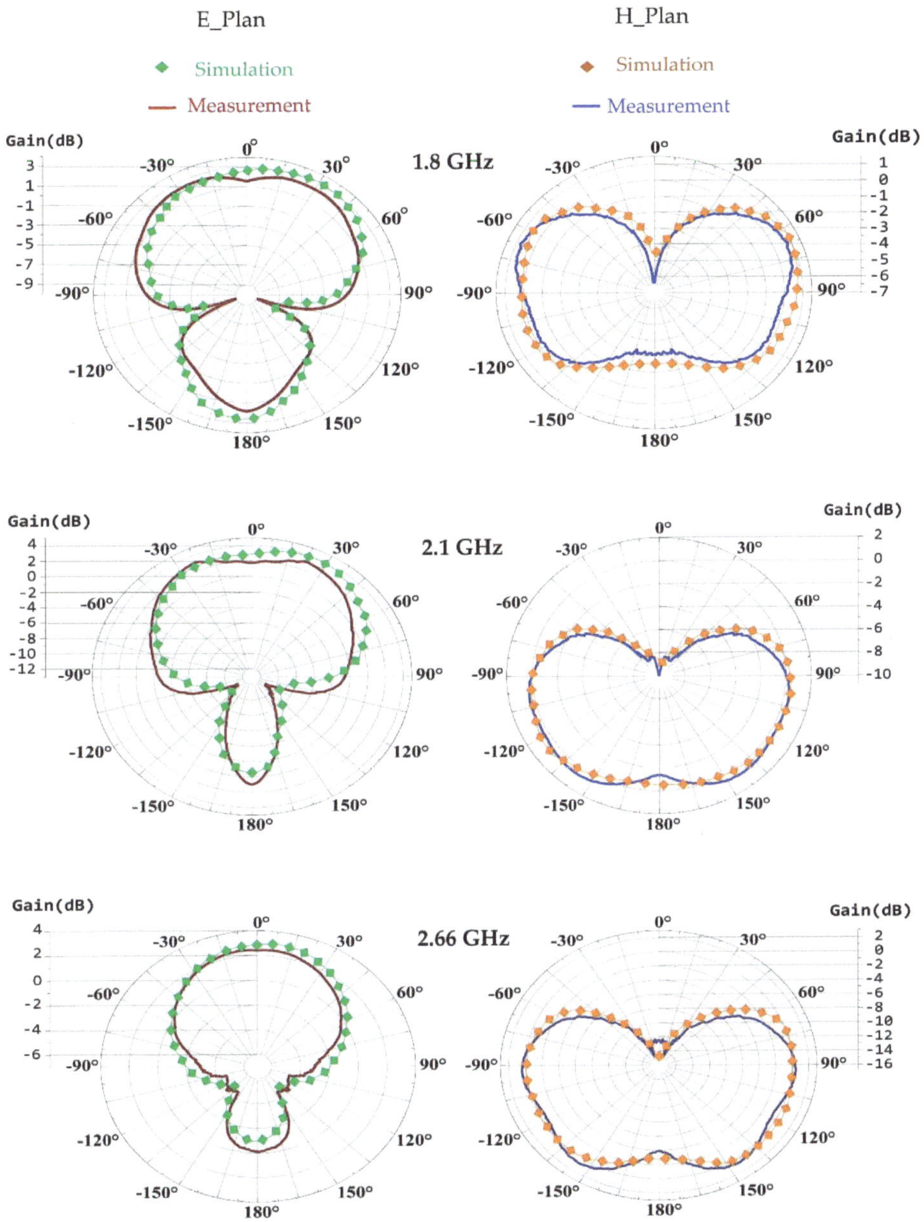

Figure 8. Simulated and measured radiation pattern for GSM, UMTS, and LTE bands, respectively.

4. Rectenna Design and Measurements

Figure 9 depicts the rectenna, which is the association of the antenna, rectifier, and impedance matching circuit in which inductors and capacitors are ideal. The rectifier is formed by two Schottky diodes and two capacitors forming a DC volage doubler. The impedance matching circuit is based on two L, C resonators (L_1, C_1 and L_2, C_2). Three rectennas are designed and realized for the GSM-1800, UMTS (3G), and LTE-2.6 bands.

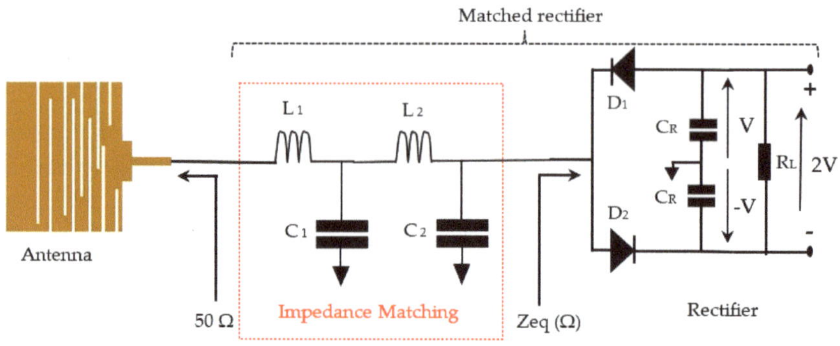

Figure 9. Rectenna circuit with ideal impedance matching, D_1, D_2: (Schottky diodes: SMS7630-079LF), C_R = 470 uF, R_L = 2 kΩ.

For each band, the structure of the rectenna remains the same, only the values of the electrical elements (L_1, C_1 and L_2, C_2) of the impedance matching circuit change with the frequency. Table 4 gives the values of the electrical components of the impedance matching circuit for 1.8 GHz, 2.11 GHz, and 2.66 GHz.

Table 4. Electrical ideal element values of the impedance matching circuits for different frequency bands.

Frequency	L_1 (nH)	L_2 (nH)	C_1 (pF)	C_2 (pF)
1.8 GHz	1.5	7.9	7.5	0.55
2.11 GHz	1.2	3.9	9.9	1.1
2.66 GHz	0.2	2.7	6.5	0.8

In practice, the capacitors and inductors are a complex component combining resistive, inductive, and capacitive phenomena as shown in Figure 10.

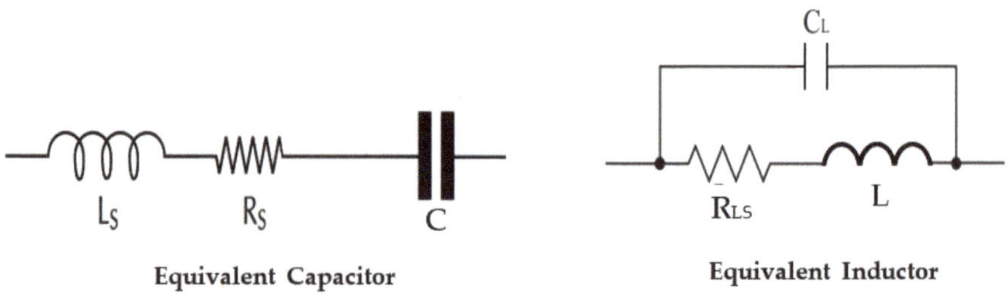

Figure 10. Equivalent circuit of capacitor and inductor. L_s: Series inductance, R_s: Electrodes and terminations resistance, C: Capacitance, L: Inductor, C_L: Parasitic capacitor, R_{LS}: Intern resistor.

The quality factors Q_L and Q_C values of each component are given in their datasheet. The other element values could be derived using the resonance condition and quality factor expressions for both components (capacitor and inductor), as is given in Equations (3)–(5). Figure 11 depicts the practical rectenna circuit with all the parasitic elements. Table 5 gives all the parasitic element values.

$$\omega_0{}^2 = \frac{1}{LC} \tag{3}$$

$$Q_L = \frac{L\omega_0}{R_{LS}} \tag{4}$$

$$Q_C = \frac{1}{R_S C\omega_0} \tag{5}$$

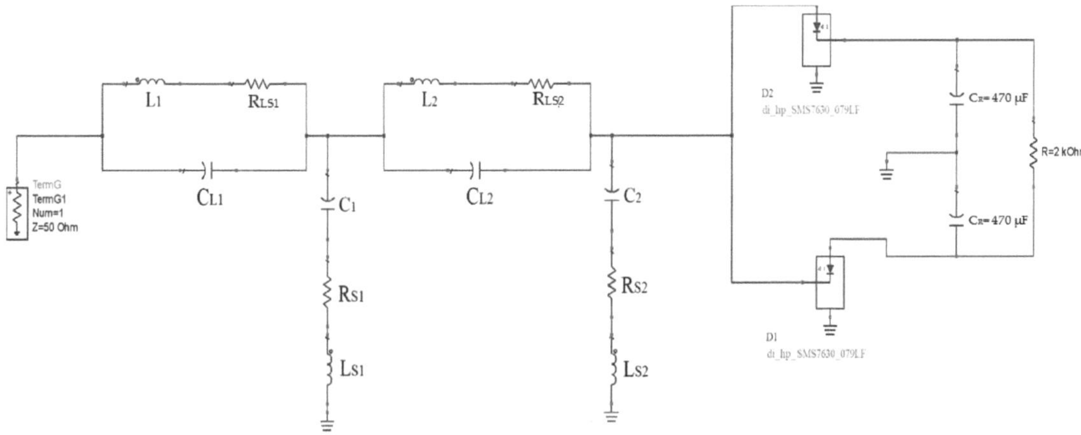

Figure 11. Practical rectenna circuit with all parasitic elements.

Table 5. Values of the parasitic elements of the used inductors and capacitors for the three frequency bands.

Frequency	L_{s1} (pH)	L_{s2} (pH)	C_{L1} (pF)	C_{L2} (pF)	R_{LS1} (mΩ)	R_{LS2} (mΩ)	R_{S1} (mΩ)	R_{s2} (mΩ)
1.8 GHz	0.25	7.9	1.8	0.55	60	52	96	82
2.11 GHz	0.17	5	1.98	0.6	78	50	80	70
2.66 GHz	0.1	0.1	14.8	0.5	150	150	5	50

Figure 12 depicts the simulated, co-simulated, and measured S_{11} parameters of the three matched rectifiers (rectifier + impedance matching) for the three bands of 1.8 GHz, 2.1 GHz, and 2.66 GHz. In the co-simulation part, we take into consideration the parasitic elements of the impedance matching circuit. The simulation and co-simulation are performed using ADS software for a −15 dBm input power. At −15 dBm, a good correlation between measurement and co-simulation is obtained. The matching level varies with the input power such that each time the power is increased, the S_{11} degrades with a small offset for the three resonance frequencies. In general, it can be observed that the measured S_{11} deeps of the prototype are at the desired frequency operation range for the GSM-1800, UMTS, and LTE bands, thereby allowing the rectenna to operate in all three bands for this order of input powers.

The realized rectenna prototype, which simulated and measured the DC output voltages and RF-to-DC conversion efficiencies for the three rectennas matched to the GSM-1800, UMTS, and LTE-2.6 bands, respectively, are depicted in Figure 13. The RF-to-DC conversion efficiency can be calculated by using Equation (6):

$$\eta(\%) = \frac{P_{out}}{P_{int}} \cdot 100 = \frac{V_{DC}^2}{R_L \cdot P_{int}} \cdot 100 \quad (6)$$

where P_{int}, P_{out}, R_L,V_{DC} are the input RF power, output DC power, resistance load, and the DC output voltage, respectively.

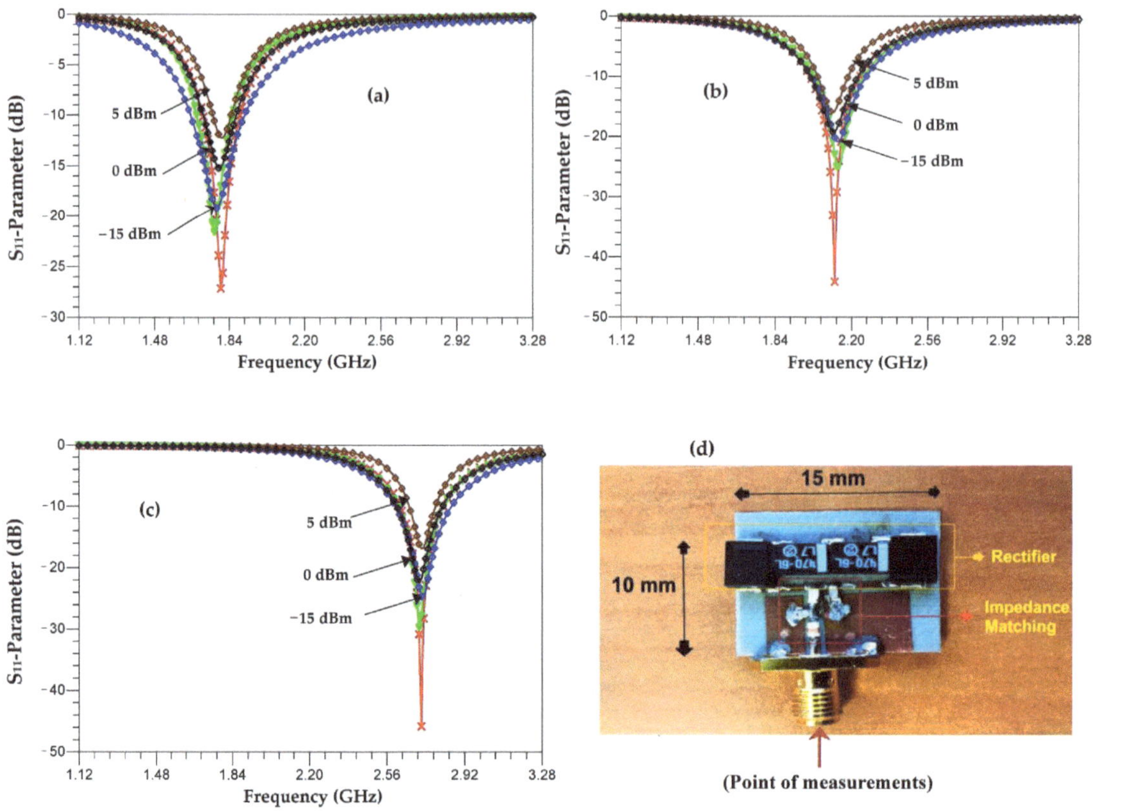

Figure 12. Simulated and measured S_{11} parameters of the three rectifiers for each frequency bands ((**a**) 1.8 GHz, 2.1 GHz (**b**) and 2.66 GHz (**c**)) and realized prototype (**d**). ◊◊◊ Measurement, × Simulation ▽ Co-simulation with parasitic elements.

Figure 13b presents efficiency as a function of the load resistance. It can be seen that the maximum efficiency is obtained for a load resistance of 2 kΩ.

Figure 13 shows a good agreement between the simulations and measurements. At an effective distance of 60 m from the relay antenna, the rectenna matched to the GSM-1800 network has a maximum simulated and measured RF-to-DC efficiency of 53% and 52%, respectively. At this distance, the measured harvested DC power is 180 µW. At a distance of 54 m, the rectenna adapted to the UMTS network has a maximum simulated and measured RF-to-DC conversion efficiency of 47.5% and 46.7%, respectively. At this position, the measured harvested DC power is 101 µW. The rectenna that is adapted to the LTE-2.6 network has a simulated and measured RF-to-DC conversion efficiency of 42% and 40.5%, respectively, at 54 m from the relay antenna. At this distance, this rectenna provides a harvested DC power of 64.8 µW.

At 122 m, the measured RF-to-DC conversion efficiencies are 37.7%, 17.9%, and 12%, and the measured DC output powers are 16.2 µW, 5 µW, and 0.18 µW for GSM-1800, UMTS, and LTE-2.6, respectively.

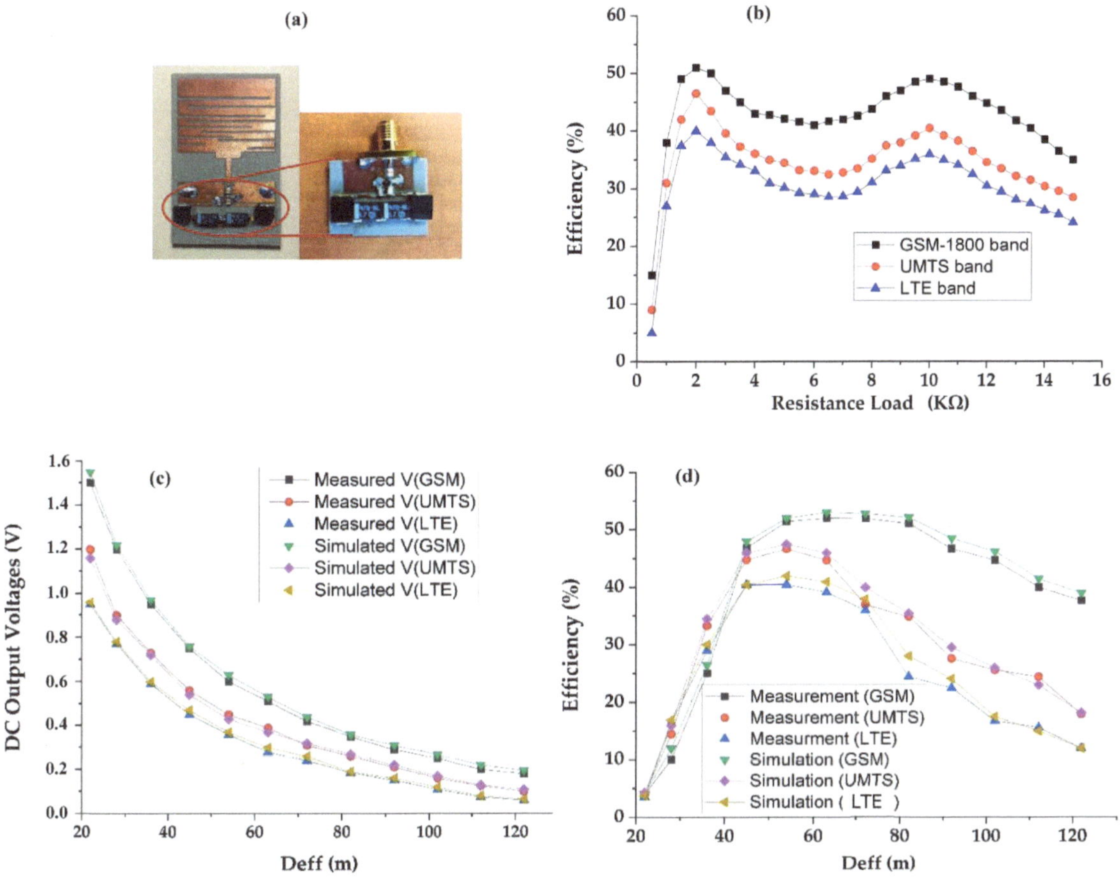

Figure 13. (**a**) Realized rectenna prototype, (**b**) Efficiency variation versus resistance load at 54 m, and (**c**,**d**) simulated and measured DC output voltages and RF-to-DC conversion efficiency, respectively, for GSM-1800, UMTS, and LTE-2.6 bands for 2 kΩ resistance load.

5. Proposed Architectures to Enhance the Performances of the RF Energy Harvesting Systems

In order to enhance the performance of the RF energy harvesting system in terms of DC output power as well as RF-to-DC efficiency even far from relay antenna (low input powers) two architectures have been designed, simulated, and tested.

The efficiency of the architectures are given by Equation (7):

$$\eta(\%) = \frac{P_{out}}{P_{GSM} + P_{UMTS} + P_{LTE}} \cdot 100 = \frac{V_{DC}^2}{R_L \cdot (P_{GSM} + P_{UMTS} + P_{LTE})} \cdot 100 \quad (7)$$

where P_{out}, R_L,V_{DC} are the input RF power, output DC power, resistance load, and the DC output voltage, respectively.

5.1. First Architecture (RF-EH$_1$)

This architecture consists of a superposition of the three rectennas studied previously. Due to the higher transmission power of the base station for the GSM-1800 network, the rectenna adapted to GSM-1800 must be placed in the first position to shift the RF signals of the following rectenna by a large DC component, which allows the diodes to operate in their stable state. In this case, the losses at the diodes become negligible for a certain input power range. Points S_1 and S_2 become the references for the second and third rectenna, respectively, as shown in Figure 14.

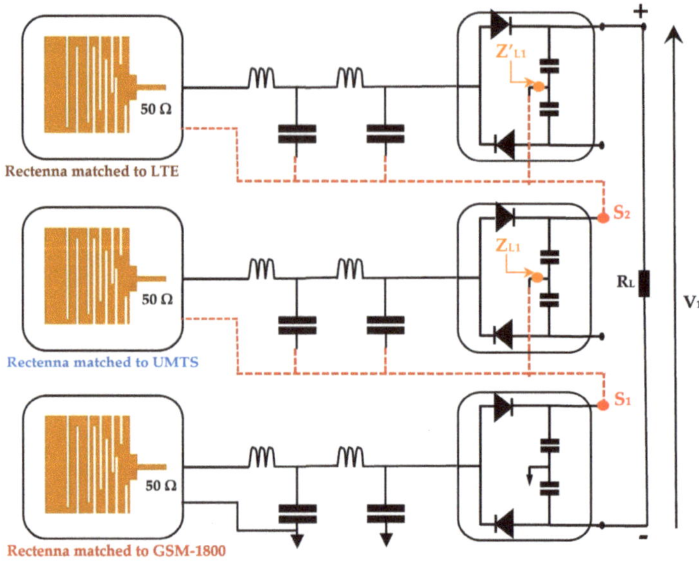

Figure 14. First architecture with three rectennas matched to GSM-1800, UMTS, and LTE-2.6. Z_{L1}, Z'_{L1} are the impedances at the selected points.

The impedances Z_{L1} and Z'_{L1} are characterized using ADS software (data items). The real parts are equal to 7.2 kΩ and 8.13 kΩ, respectively, with a negligeable imaginary one.

Figure 15 shows the simulated and measured DC output voltages as a function of the effective distance between the RF-EH_1 and the mobile communication relay antenna. Figure 16 presents the variation in the simulated and measured RF-to-DC conversion efficiency.

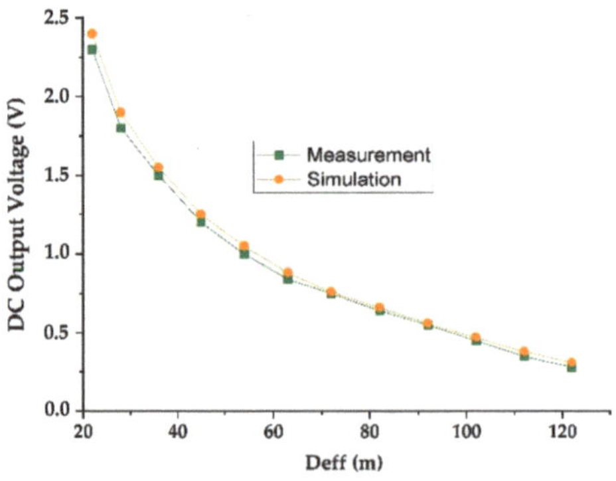

Figure 15. Simulated and measured DC output voltage for 2 kΩ resistance load.

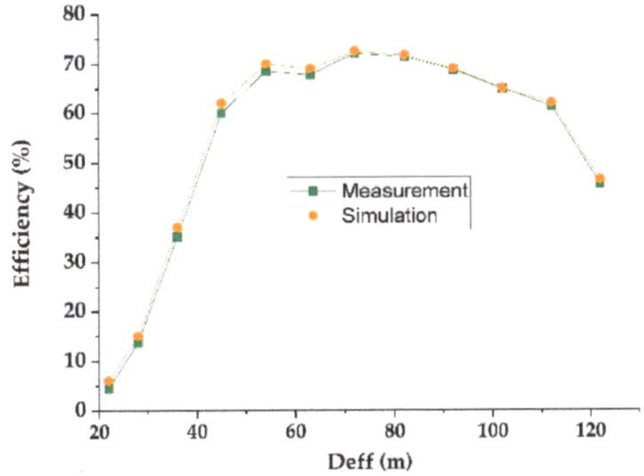

Figure 16. Simulated and measured RF-to-DC conversion efficiency for 2 kΩ resistance load.

The maximum simulated and measured efficiencies of 71.5% and 71% are achieved at 72 m from the mobile communication relay antenna. These efficiencies stay above 50% until 122 m. The measured DC power supplied by this architecture is 288.8 μW at the effective distance of 72 m and 39.2 μW at 122 m. These results demonstrate that this architecture is suitable for low input powers (far distances from relay antenna) and limited for high input powers (near to relay antenna).

Figure 17 shows the realized RF-EH_1 with a dimension of 64 × 55 mm^2.

Figure 17. (**a**) Realized RF-EH1 based on the first architecture, (**b**) Measurement prototype at 22 m from relay antenna.

5.2. Second Architecture (RF-EH_2)

Figure 18 presents the second architecture. It consists of summing the DC output voltages delivered by each rectenna (antenna + rectifier) [14–16,25]. This architecture presents an ideal summation when the system is an open load. The presence of the load degrades it's performance in terms of the DC output power.

The impedances Z_{L2} = 0.1 + j0.0015 and Z'_{L2} = 0.095 − j0.0084 are characterized using ADS software (data items). Figures 19 and 20 show the simulated and measured DC output voltages and the variation of the RF-to-DC conversion efficiency, respectively, as a function of the effective distance between the RF-EH_2 and the mobile communication base station (relay antenna).

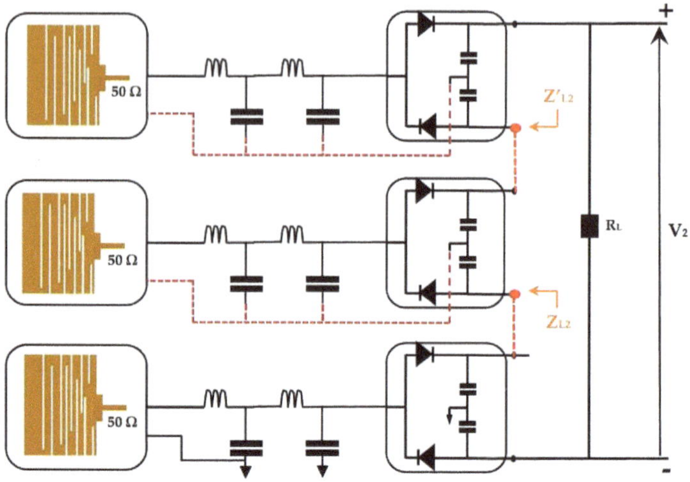

Figure 18. Second architecture with three rectennas matched to GSM-1800, UMTS, and LTE-2.6. Z_{L2}, Z'_{L2} are the impedances at the selected points.

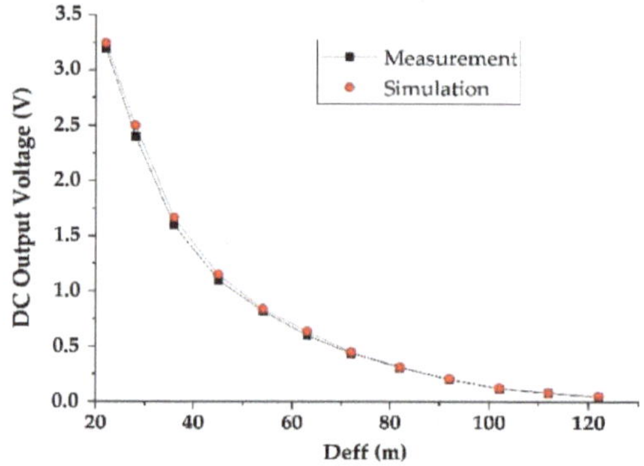

Figure 19. Measured and simulated DC output voltage as a function of the effective distance for 2 kΩ resistance load.

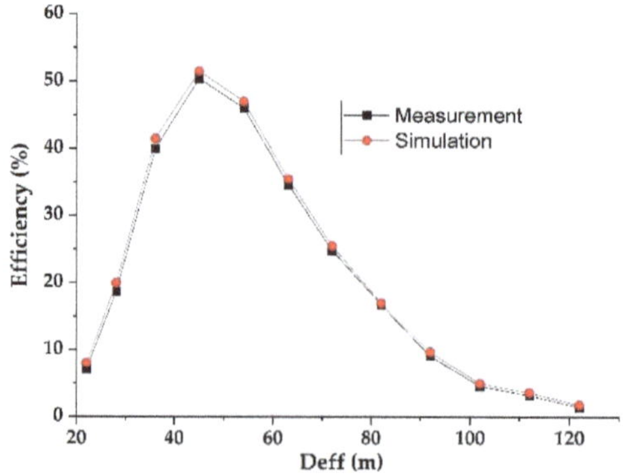

Figure 20. Measured and simulated efficiencies provided by the second architecture for 2 kΩ resistance load.

We note that the output voltage drops rapidly. At 92 m from the base station, the voltage becomes very low 0.2 V and 80 mV at 122 m. Maximum simulated and measured efficiencies of 52% and 50% respectively, are achieved at 45 m from the mobile communication relay antenna. This efficiency decreases quickly beyond 54 m. The DC power supplied by this architecture is 660 μW at the effective distance of 45 m and 3.2 μW at 122 m. The maximum measured DC power of 5.12 mW is achieved at the effective distance of 22 m from the relay antenna. These results demonstrate that this architecture is suitable for height input powers (near relay antenna) and limited for low input powers (far distances from relay antenna). Figure 21 shows the realized RF-EH_2 with a dimension of 64×55 mm^2.

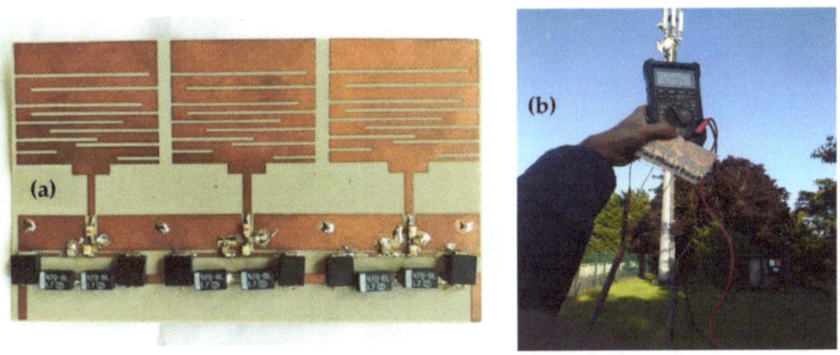

Figure 21. (**a**) Realized RF-EH_2, (**b**) Prototype of measurement at 22 m from relay antenna.

6. The Proposed System (RF-EH)

The RF-EH system consists of switching between the two architectures tested previously by using a switch based on two MOSFET transistors (IRFD_220) in order to realize a suitable system for the low and high input power levels. The switching is done according to the input power levels (close or far from relay antenna). The two outputs, Out_1 and Out_2, of the switch are connected to the high and low impedances Z_{L1} and Z_{L2}, respectively.

6.1. Design and Realization of Switch Circuit

Figure 22 depicts the switch circuit with the simulated output currents variation according to the DC input voltage (Gate voltage of the transistors).

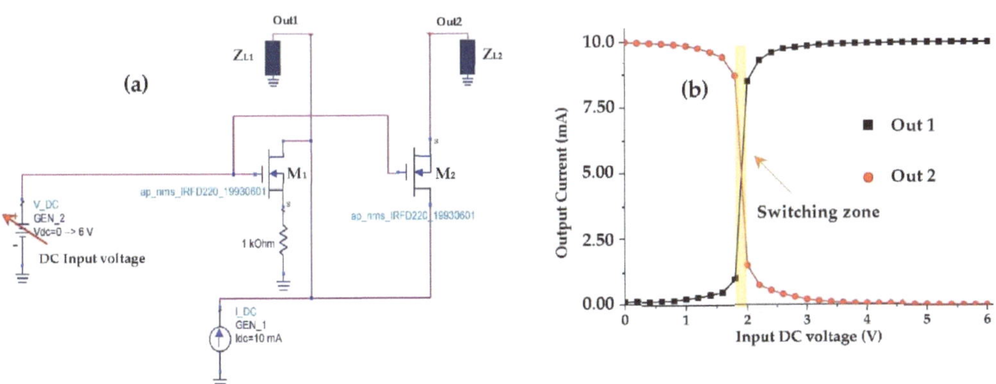

Figure 22. (**a**) Proposed switch circuit, (**b**) Simulated DC output currents (Out1, Out2). Z_{L1} = 7.2 kΩ, Z_{L2} = 0.1 Ω.

The variable voltage generator simulates the DC voltage supplied by the rectenna (far or near the base station). In the case of high input powers (near the base station), the output voltage is raised, and it is applied at the gate of the transistor. The gate current is null, which implies an infinite input impedance. The voltage delivered by the rectenna is exploited at open load, which makes it possible to reach higher voltages than that of the threshold quickly. In this case, the transistors M_1 and M_2 are conducting, which gives the equivalent circuit high input powers illustrated in Figure 23a. For low input powers, both transistors are blocked due to the lower DC voltage provided by the rectenna than the threshold one, which can be modelled by Figure 23b.

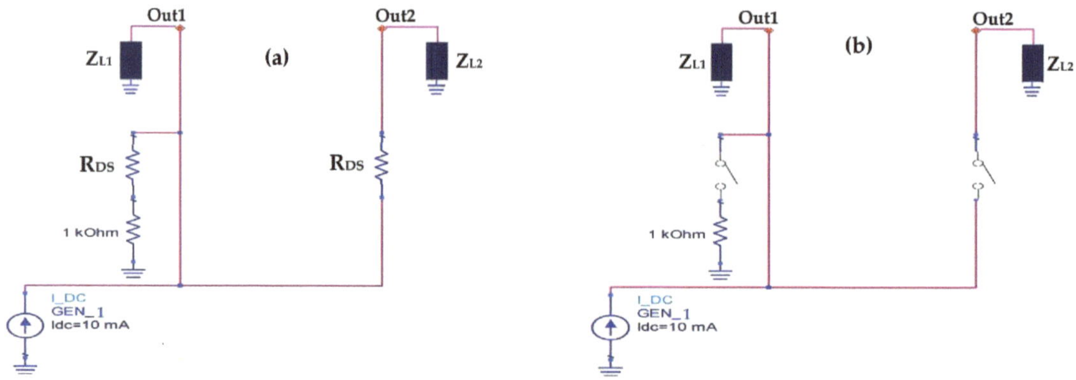

Figure 23. (**a**) Equivalent circuit for high input powers, (**b**) Equivalent circuit for low input powers.

For high input powers, both transistors are conducting for a difference of $V_G - Vs > V_{th}$, such that the resistance R_{DS} of the transistor IRFD_220 is equal to 0.8 Ω. The majority of the power provided by the source (GEN_1) passes to output 2 with a small ohmic loss due to the R_{DS} resistance. For output 1, the majority of the power supplied by the source (GEN_1) passes through the R_{DS} resistance to the ground with a small leakage to output 1, making this output isolated for high input power.

For low input powers, both transistors are blocked for a difference of $V_G - Vs < V_{th}$ and the R_{DS} resistance is very large (theoretically infinite). In this case, output 2 is completely attenuated and the majority of the power supplied by the source (GEN_1) goes to output 1. Table 6 summarizes the proposed switch states as a function of the input power levels.

Table 6. Output stats of the proposed switch according to the input power levels.

Input Power Levels	Low Input Power	High Input Power
Output 1	ON	OFF
Output 2	OFF	ON

The proposed switch is realized on Teflon glass substrate ($\varepsilon_r = 2.1$) with 0.67 mm of thickness. The total size of this switch is 15 × 15 mm^2.

Figures 24 and 25 present the realized switch with a measurement prototype and the measured results, respectively.

For $I_{gn} = 10$ mA and $V_{gn} = 1$ V, the switching zone is located in the DC input voltage range of [2–2.2] V. The width of the measured switching zone is slightly wider than the simulated one and low leakage currents are observed for both isolated outputs (Out1 and Out2) due to material imperfections. In general, the measurements and simulations are in good agreement.

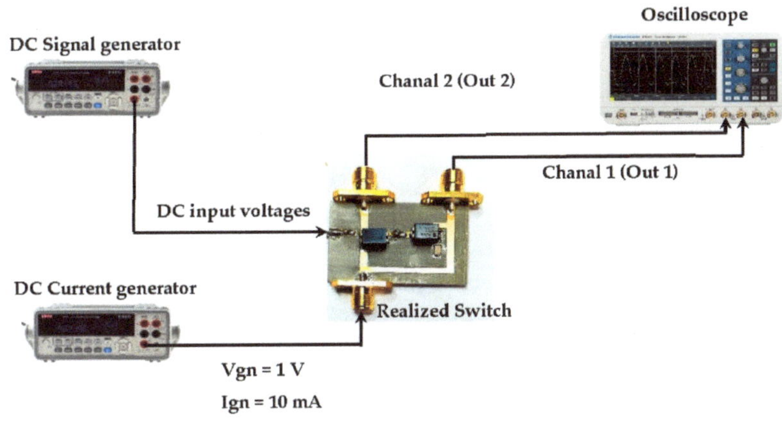

Figure 24. Measurement prototype of the output currents for the first and second outputs of the proposed switch.

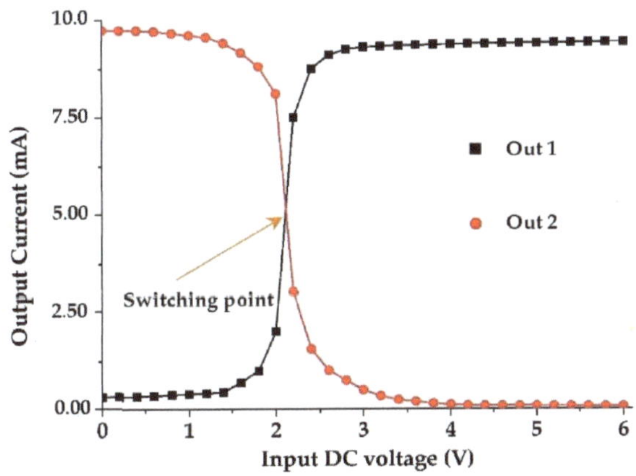

Figure 25. Measured DC output currents (Out1, Out2) of the proposed switch.

The next step is using this switch to realize the proposed RF-EH system, which is discussed in the following paragraph.

6.2. Proposed RF-EH System

Figure 26 presents the proposed RF-EH system that combines the two architectures studied previously using the presented switch.

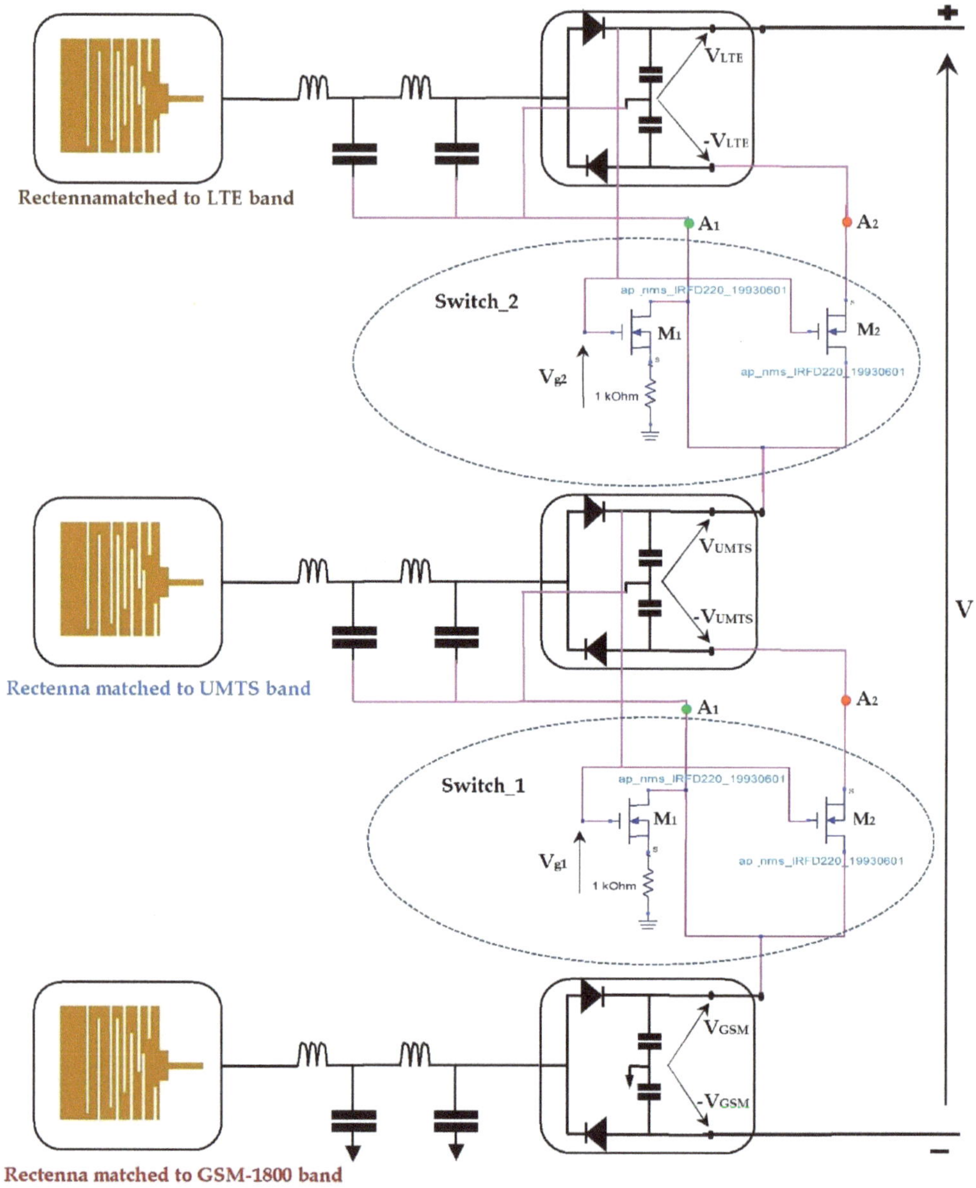

Figure 26. Proposed RF-EH system combining both previous architectures using the studied switch. A_1 and A_2 are the two positions corresponding to first and second architectures.

Due to the infinite gate impedance of the MOSFET, both transistors are controlled by the output voltages of the rectennas for open load (V_{g1} and V_{g2}). For high input powers (near relay antenna), the switches are pointing into A_2. Figure 27a gives the structure of the second architecture. For low input powers (far from relay antenna), the switches pointed into A_1, resulting in the first architecture, which is presented in Figure 27b.

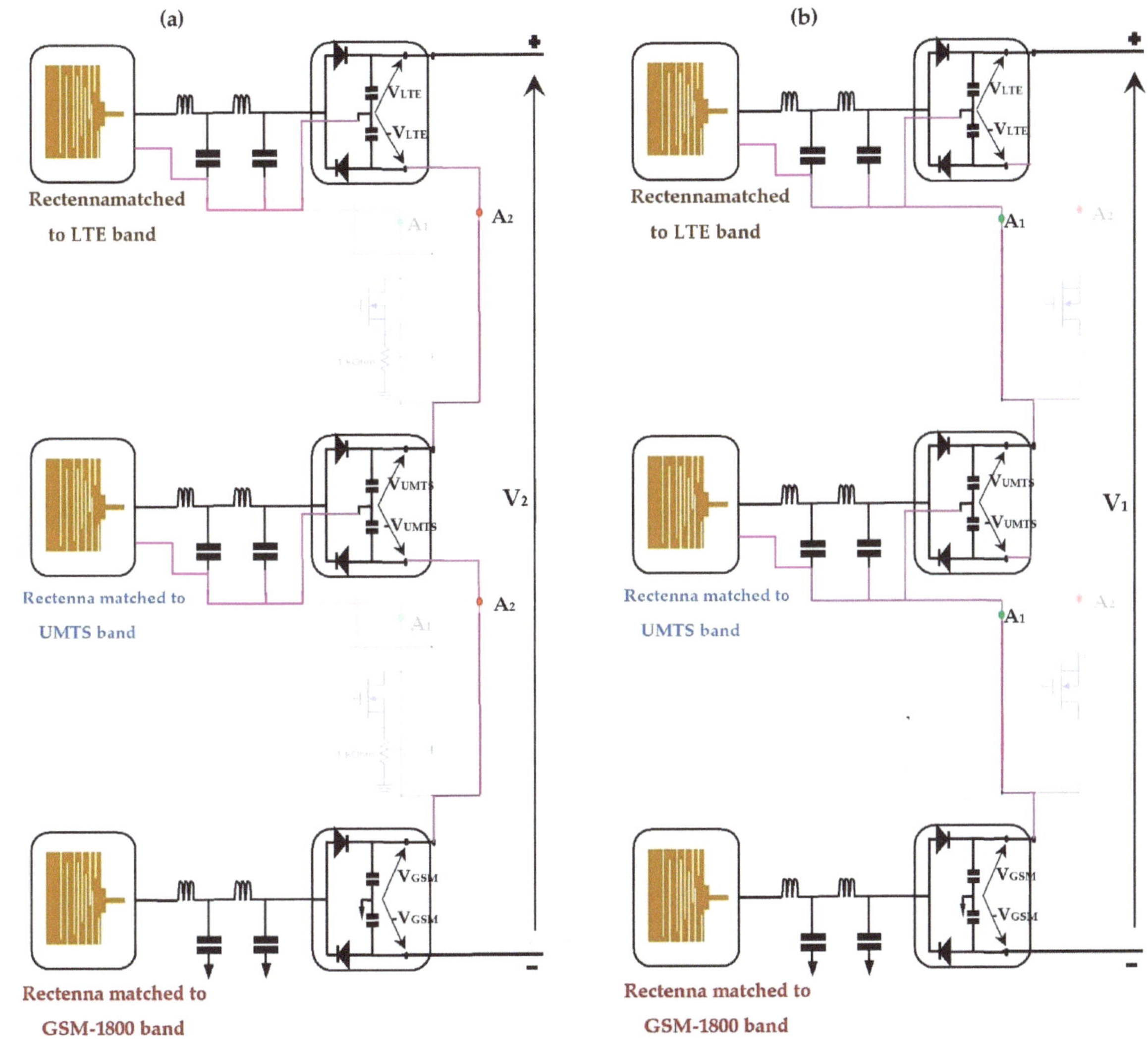

Figure 27. Switching between both previous architectures according to the input power levels. (**a**) for high input power levels, (**b**) for low input power levels.

Figure 28 depicts the realized RF-EH system, measurement prototype, measured DC output voltages, and measured RF-to-DC conversion efficiencies for a 2 kΩ resistance load. We can note that the RF-EH system provides a maximum DC voltage of 3.15 V at 22 m from the relay antenna, and a maximum RF-to-DC conversion efficiency of 71.5% at 75 m. The switching from one architecture to another is located at 40 m from the mobile communication base station. From 22 m to 40 m (high input powers), the RF-EH system points to the second architecture, which is more matched for high input power levels. A little difference between the voltages provided by the second architecture (3.2 V) and by the RF-EH system (3.15 V) is due to the transistor leakages for on-transistors (Drain-to-Source resistance R_{DS}). From 40 m to 122 m (low input powers), the RF-EH system points to the first architecture, which is more adapted to the low input power levels. Both voltages provided by the RF-EH system and the first architecture are almost the same due to the very low transistor leakages for off-transistors (infinite R_{DS}).

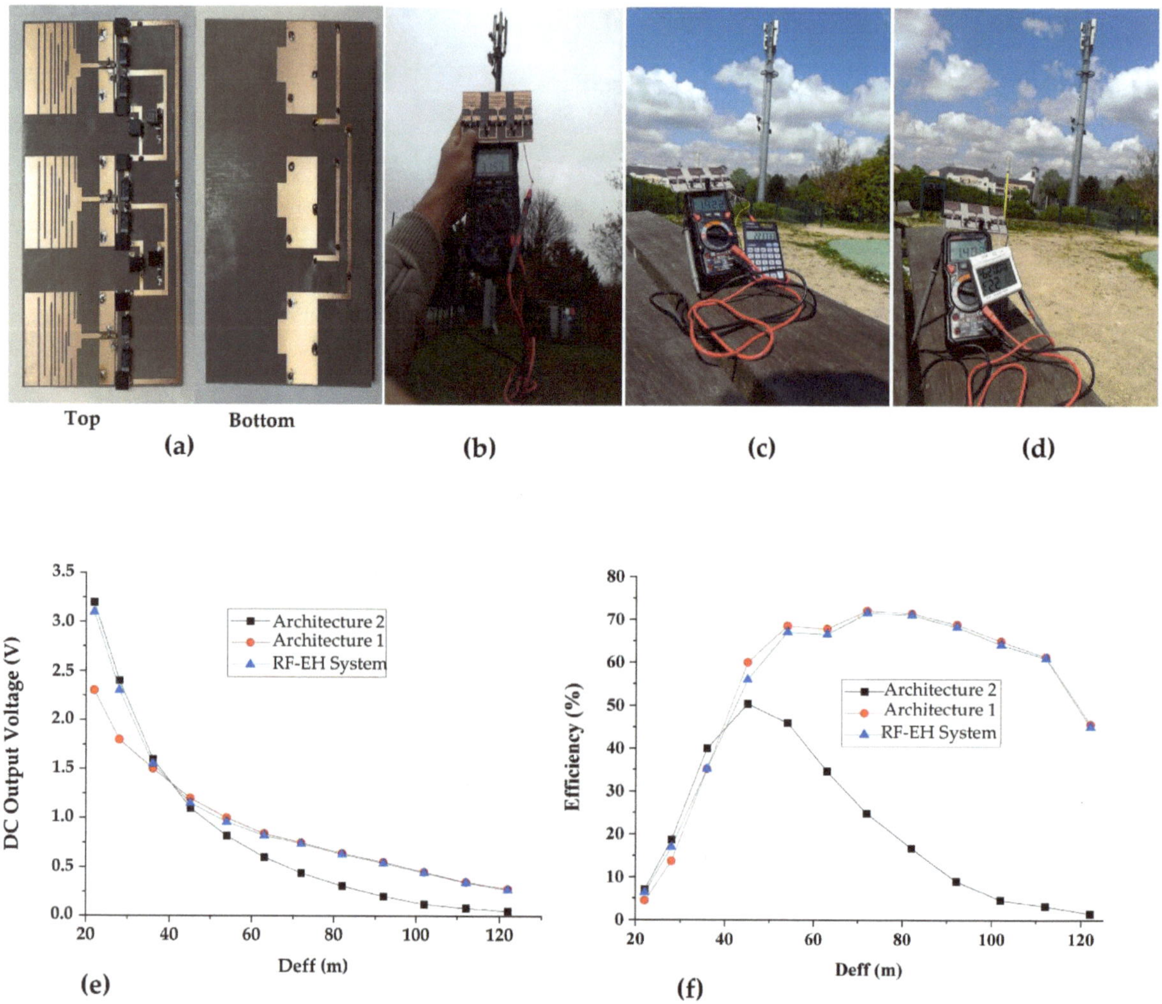

Figure 28. Measurements results: (**a**) realized proposed RF-EH system, (**b**) prototype of measurements, (**c**) Application of the proposed RF-EH system to feed digital calculate at 38 m from relay antenna, (**d**) Application of the proposed RF-EH system to feed temperature sensor at the same distance, (**e**) Measured DC output voltages, and (**f**) Measured RF-to-DC conversion efficiency.

At 122 m, the RF-EH system provides a DC voltage of 0.37 V and RF-to-DC conversion efficiency of 45%.

7. Discussion

The first architecture presented is very suitable for low input power and limited for high input power. This can be explained by the current flow at the link between the rectennas that form each architecture, as shown in Figure 29.

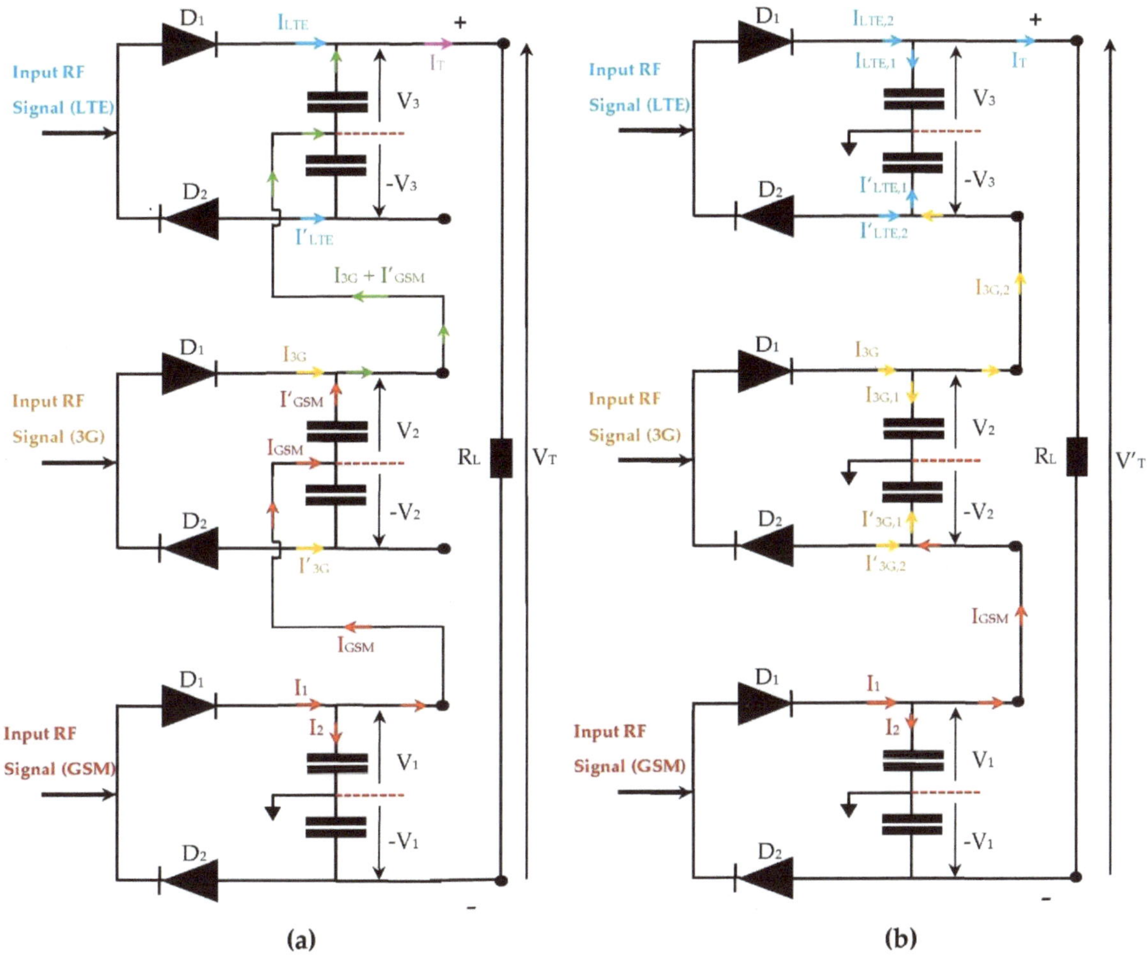

Figure 29. Current flow for both architectures. (**a**) First architecture, (**b**) Second architecture.

For the first architecture, the link between the rectennas adapted to the GSM and 3G networks is established by the current flow I_{GSM}, and the link between the rectennas adapted to the 3G and LTE networks is established by the current flow $I_{3G} + I'_{GSM}$. The RF transmission power of the relay antenna for the 3G network is relatively low compared to that of the GSM network. For low input powers, the current I_{3G} is considerably weakened, and the currents I_{GSM} and $I_{3G} + I'_{GSM} \approx I'_{GSM}$ remain sufficient to maintain the link between the rectennas up to 122 m from the relay antenna, which guarantees a good matching for low input powers.

For the second architecture, the link between the rectennas matched to the GSM and 3G networks is ensured by the current flow I_{GSM}, and the link between the rectennas adapted to the 3G and LTE networks is established by the current flow $I_{3G,2}$, which is a part of the current I_{3G} ($I_{3G,2} < I_{3G}$). For low input power, the current $I_{3G,2}$ is considerably attenuated. The current I_{GSM} is sufficient to maintain the link between the rectennas matched to the GSM and 3G networks up to 122 m from the relay antenna. The current $I_{3G,2}$ is completely attenuated, which prevents the maintenance of the link between the rectennas adapted to the 3G and LTE networks and degrades the performances of this architecture for low input powers.

For high input powers, the second architecture provides higher DC power than the first architecture. For an open load, the DC output voltages supplied by the first and the second architectures are $V_T = 2V_1 + V_2 + V_3$ and $V'_T = 2(V_1 + V_2 + V_3)$, respectively. These voltages degrade a little when the R_L load is present due to ohmic leakages.

The proposed switch placed between the rectennas in order to select the adapted architecture according to the input power levels is based on two MOSFET transistors. These transistors could be electrically modelized by the source-to-drain equivalent resistor R_{DS}, which varies in its function of the input power levels (On/Off Transistor stats) as presented in Figure 30.

$$Z_{eq} = \frac{(R_{DS} + 1k\Omega)(Z_{L2} + R_{DS})}{Z_{L2} + 2R_{DS} + 1k\Omega} \tag{8}$$

$$I_{Out1} = \frac{Z_{eq}}{Z_{L1} + Z_{eq}} \cdot I_{gn} \tag{9}$$

$$I_{Out2} = \frac{Z_{L1}(R_{DS} + 1k\Omega)}{(Z_{L2} + 2R_{DS} + 1k\Omega) \cdot (Z_{L1} + Z_{eq})} \cdot I_{gn} \tag{10}$$

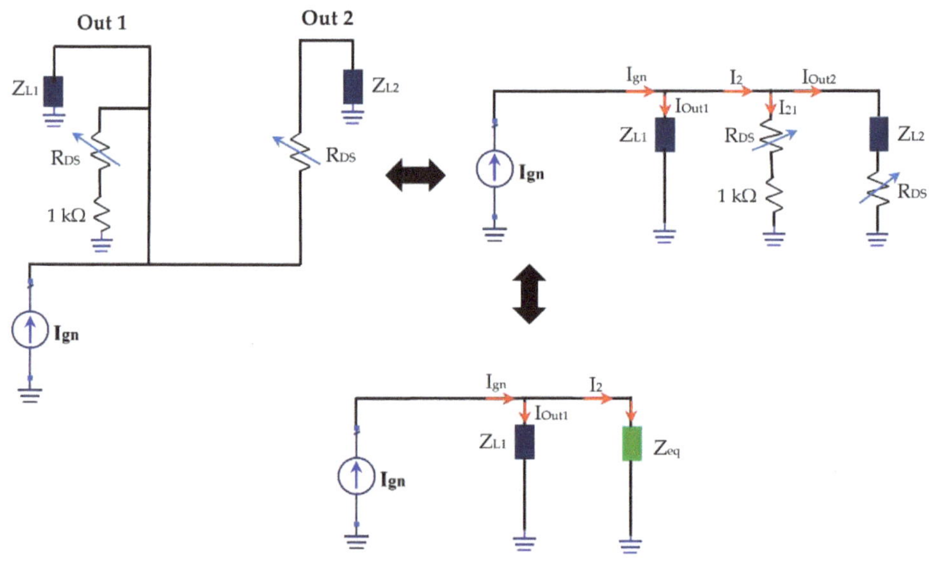

Figure 30. Electrical representation of the proposed switch.

For high input power levels, $R_{DS(on)} = 0.8\ \Omega$ (on-transistor), $Z_{L1} = 7.2\ k\Omega$, $Z_{L2} = 0.1\ \Omega$, and $Z_{eq} = 0.9\ \Omega$.

$$I_{out1} = 1.24 \cdot 10^{-4} \cdot I_{gn} \tag{11}$$

$$I_{out2} = 0.999\ I_{gn} \approx I_{gn} \tag{12}$$

For $I_{gn} = 10$ mA, $I_{out1} = 1.24\ \mu A \approx 0$ A, and $I_{out2} = 9.99$ mA $\approx I_{gn}$. It can be noted that the current I_{out1} is almost null (out1 isolated) and that the current I_{out2} is the majority, which implies that the switch points to output 2 for high input powers select the second architecture.

For low input power levels, $R_{DS(off)} \to \infty$ (off-transistor), $Z_{L1} = 7.2\ k\Omega$, $Z_{L2} = 0.1\ \Omega$, and $Z_{eq} \to \infty$.

$$\lim_{R_{DS};\, Z_{eq} \to \infty} I_{Out1} = I_{gn} \tag{13}$$

$$\lim_{R_{DS};Z_{eq}\to\infty} I_{Out2} = 0 \tag{14}$$

For I_{gn} = 10 mA, I_{out1} = 10 mA, and I_{out2} = 0 A (in real case I_{out2} is almost null due to leakage currents of the transistor MOSFET).

We can note that the current I_{out2} is null (out2 isolated) and that the current I_{out1} is the majority, which implies that the switch points to output 1 for low input powers select the first architecture.

In Table 7, a comparison between the proposed RF-EH system and other RF energy harvesting systems reported in the literature is presented.

Table 7. Comparison between proposed RF-EH system performances and other systems reported in the literature.

References	Input Power (dBm)	Freq. Band (GHz)	Max RF-DC Eff (%)	Maximum Output Voltage (V)	System Dimensions (mm^3)	Minimum Load (kΩ)
[10]	0	2.45	81	1 V at −3.5 dBm	125 × 80 × 0.25	1.4
[11]	−7	2.45	28.7	3 V at −7 dBm	500 × 64 × 3.3	4
[12]	−8	0.823–1.337	61.7	2 V at 0 dBm	120 × 120 × 0.52	10
[13]	3	0.9–3	73.4	1.2 V at 3 dBm	200 × 180 × 0.787	1
[14]	−1	0.45–0.9	75	1 V at −8 dBm	160 × 160 × 1.5	15
[15]	−10	0.91–2.55	68	0.243 V at −10 dBm	165 × 125 × 0.8	4.7
[16]	−6	1.4–2.95	27.5	0.484 V at −6 dBm	100 × 80 × 0.87	0.8
[17]	−3	1.7–1.8 and 2.1–2.7	67	1.1 V at −3 dBm	145 × 145 × 1.53	2
[18]	−5	0.866, 0.925 and 2.45	65	0.7 V Ambient	115 × 71 × 0.787	10
[19]	0	2.45 and 5.8	58	NA	64 × 64 × 1.24	0.7
[20]	−27	0.95, 1.74, 2.12, 2.42 and 2.90	66.5	0.797 V Ambient	220 × 220 × 30	2.11
[21]	−10	1.8	51.3	0.129 V Ambient	100 × 100 × 60	6
[22]	−3	2.45	45.3	0.782 V Ambient	337 × 226 × 10	3
[23]	10	5.8	68	1.1 V at 10 dBm	86 × 24 × 0.813	0.22
[24]	6.5	5.2 and 5.8	54.9	0.7 V at 6.5 dBm	87 × 90.9 × 0.787	0.2
[25]	5.8	0.9, 1.8, 2.1 and 2.45	84	0.9 V at −15 dBm	100 × 100 × NA	11
This Work	**10**	**1.8, 2.1 and 2.66**	**71.5%**	**3.1 V Ambient**	**65 × 64 × 0.67**	**2**

The proposed system that combines the advantages of both architectures offers a large feeding possibility of many sensors reported in many citations. There are five types of sensors often used: gas sensors, image sensors, pressure sensors, biomedical sensors, and temperature sensors.

In the following paragraph, we give different sensors cited in the most recent literature according to their DC power consumption.

7.1. Gas Sensors

In [26], the authors propose a Si FET-type gas sensor with a localized micro-heater capable of heating up to 124 °C with a power consumption of 4 mW. In [27], a high linearity detection circuit with a constant detection voltage for a resistive gas sensor is proposed. The maximum power consumption of this circuit is only 4.6 mW. The authors of [28] report a hexagonal gas sensor cell for multi-channel gas detection. Six sensors are integrated with a hexagonal micro-hotplate to make up the sensor cell. The average power consumption of the sensor cell is 3.03 mW when the sensor cell starts working. In [29], a novel metal oxide semiconductor (MOS) gas sensor based on a single cantilever is reported. The static power consumption is 2.96 mW. The authors of [30] propose the hydrogen gas detection property of Pt-AlGaN/GaN high electron mobility transistor (HEMT) sensors with a recessed gate structure. The power consumption of the sensor is 2.95 mW. Figure 31 presents the feeding abilities of both proposed architectures and the system compared to the minimum required power supply of these sensors.

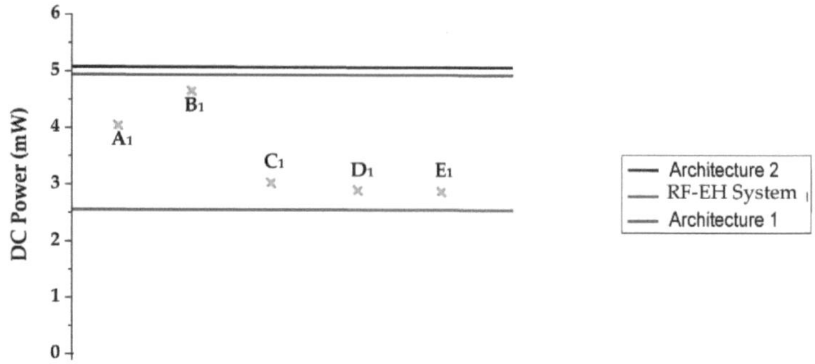

Figure 31. Feeding ability of the system and both architectures at 22 m from relay antenna compared to minimum required power supply of the gas sensors. The DC powers A_1, B_1, C_1, D_1 and E_1 correspond to the power supply of sensors given in the references [26–30], respectively.

As we can see, the RF-EH system as well as the second architecture could feed the cited sensors [26–30] at 22 m from relay antenna. Contrary to the first architecture, which provides a DC power lower than what theses sensors require to be turned on given that they are far from relay antenna (D > 30 m), so these sensors couldn't be powered.

7.2. Image Sensors

The authors of [31] propose a single-pixel sensor for near-field imaging based on the startup time of an oscillator. The average power consumption of this sensor is 2.7 mW. In [32], the authors report a counter structure for a single slope analog-to-digital converter that is parallel to a column (SS-ADC) in CMOS image sensors. The total power consumption is 2.25 mW for a 640 × 480 effective image resolution at 60 frame rates. In [33], the authors propose a mixed-signal perception chip, in which a 32 × 32 ADC-free image sensor and a BNN processing array are directly integrated into a standard 180 nm CMOS process. The entire processing system consumes only 1.8 mW. In [34], the authors report a CMOS image sensor (CIS) with column-parallel single-shot compressive sensing (CS) for always-on Internet-of-Things (IoT) application. A prototype VGA image sensor consumes only 0.7 mW at 45 frames/s. In [35], a CMOS image sensor with a programmable kernel for feature extraction is proposed. The power consumption of this sensor is 117 μW for 480 frames/s.

Figure 32 depicts the minimum required power supply of these sensors and the DC output power supplied by both architectures and the RF-EH system at different distances from the relay antenna.

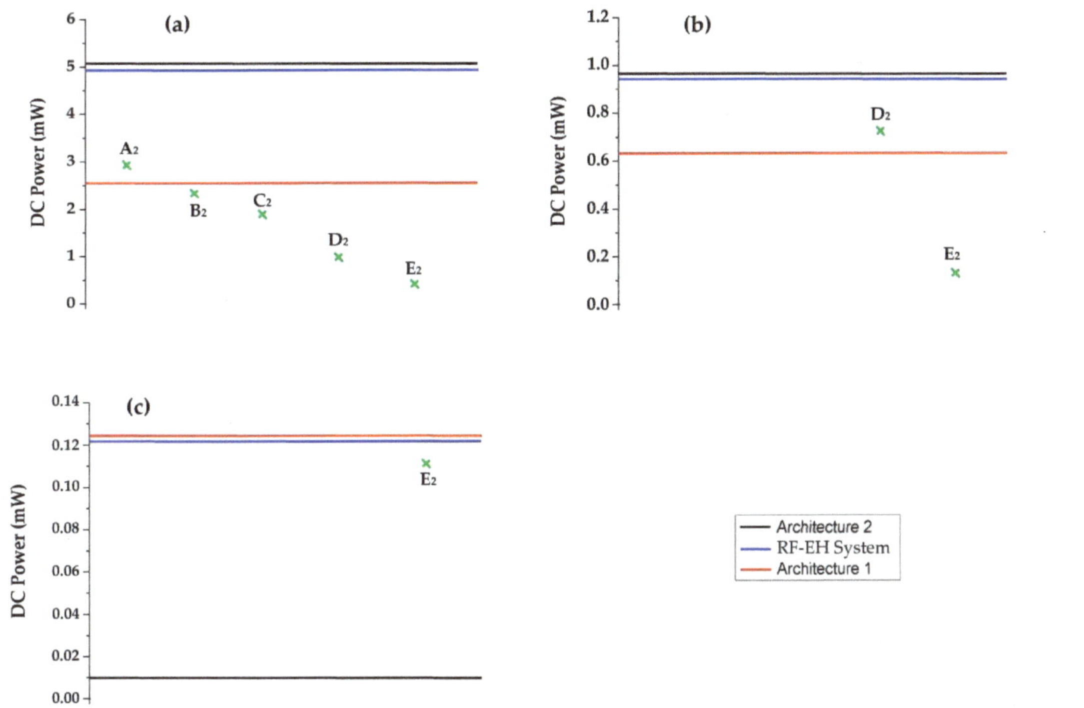

Figure 32. Feeding ability of the RF-EH system and architectures at (**a**): 22 m, (**b**): 40 m, and (**c**): 110 m from relay antenna compared to minimum required power supply of the image sensors. The DC powers A_2, B_2, C_2, D_2 and E_2 correspond to the power supply of sensors proposed in the references [31–35], respectively.

We can note that the RF-EH system and the second architecture could feed all sensors cited in [31–33] at 22 m form the relay antenna. The first architecture provides enough DC power to feed the sensors of [32,33], except for that of [31]. At 40 m, the proposed RF-EH system and the second architecture could feed the sensors of [34,35]. The first architecture supplies a sufficient DC power to feed only the sensor reported in [35]. At 110 m from the relay antenna, the second architecture provides insufficient DC power to feed any sensor. The proposed RF-EH system and the first architecture could feed the sensor cited in [35].

7.3. Pressure Sensors

In [36], a pressure sensor array with a low-power near-sensor CMOS chip for Human Gait Monitoring is proposed. The average power consumption of this sensor is 1.45 mW. The authors of [37] report an iontronic pressure sensor based on organic electrochemical transistors. The average power consumption of this sensor is 0.5 mW. In [38], a thin-film MOSFET-based pressure sensor is proposed. The power consumption of this sensor is 390 μW. In [39], the authors propose a sensor to detect pressure changes from 30 mmHg to 1000 mHg. It consumes a power of 72 μW.

Figure 33 shows the feeding abilities of the RF-EH system and both architectures according to the distance from relay antenna.

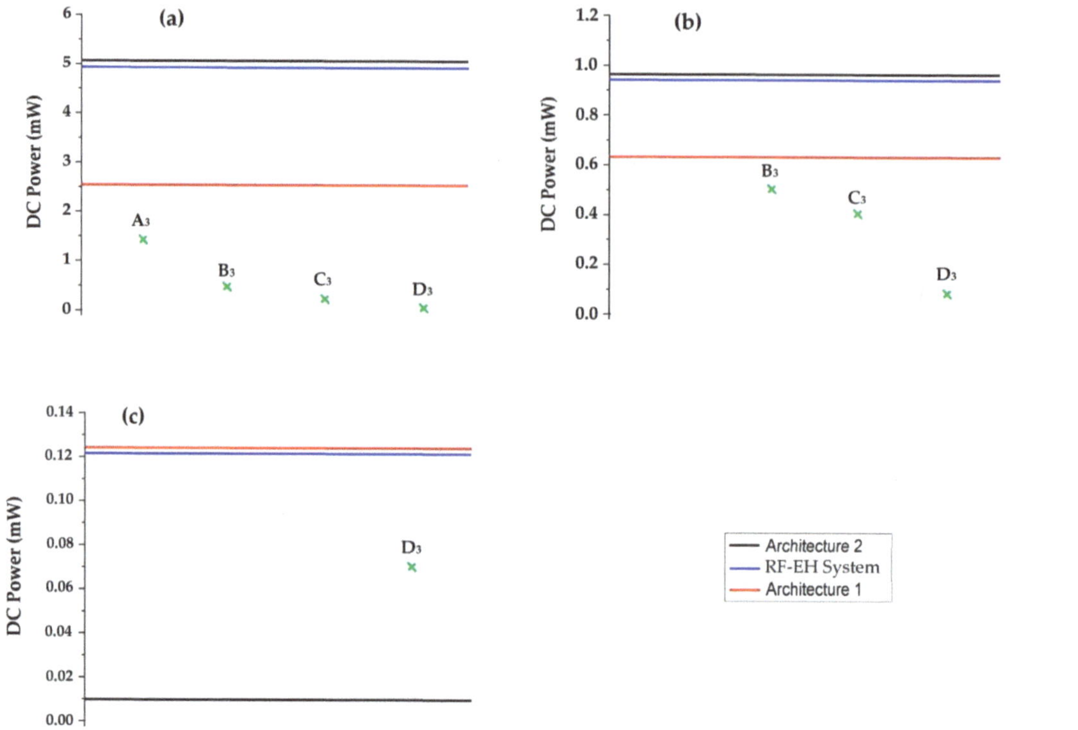

Figure 33. Feeding ability of the RF-EH system and architectures at (**a**): 22 m, (**b**): 40 m, and (**c**): 110 m from relay antenna compared to minimum required power supply of the pressure sensors. The DC powers A_3, B_3, C_3 and D_3 correspond to the power supply of sensors cited in the references [36–39], respectively.

As we can see, at 22 m from the relay antenna the RF-EH system, as well as both architectures, could feed all pressure sensors cited in [36–39]. At 40 m, the sensors reported in [37–39] could be fed. At 110 m, the RF-EH system and the first architecture could feed the sensor from [39]. The second architecture doesn't provide enough DC power.

7.4. Biomedical Sensors

The proposed system offers a large feeding possibility for many sensors reported in many citations. In [40], the authors propose a Photoplethysmography (PPG) sensory system for continuous health monitoring. The average power consumption of this sensor is 196 µW, including the Reference/Bias generation and analog-to-digital conversion section ADC. The same sensor topology is presented in [41] and [42] with a reduced power consumption level of 121 µW and 50.75 µW, respectively. The sensor proposed in [43] is a CMOS Reconfigurable Multi-Sensor SoC for Biomedical Applications. It consists of three blocks that consume the most energy (10-bit SAR-ADC block, OOK transmitter, Digital processor, and bias circuit and buffer). The total power consumed by this sensor is 942.9 µW. In [44], The authors propose an optical Sensor for Accurate Heartbeat Monitoring with a voltage range of [1.8–3.3] V for 60 µA, which gives a power consumption range of [108–198] µW. Figure 34 shows the feeding ability of both architectures and the RF-EH system in the function of the distance from the relay antenna.

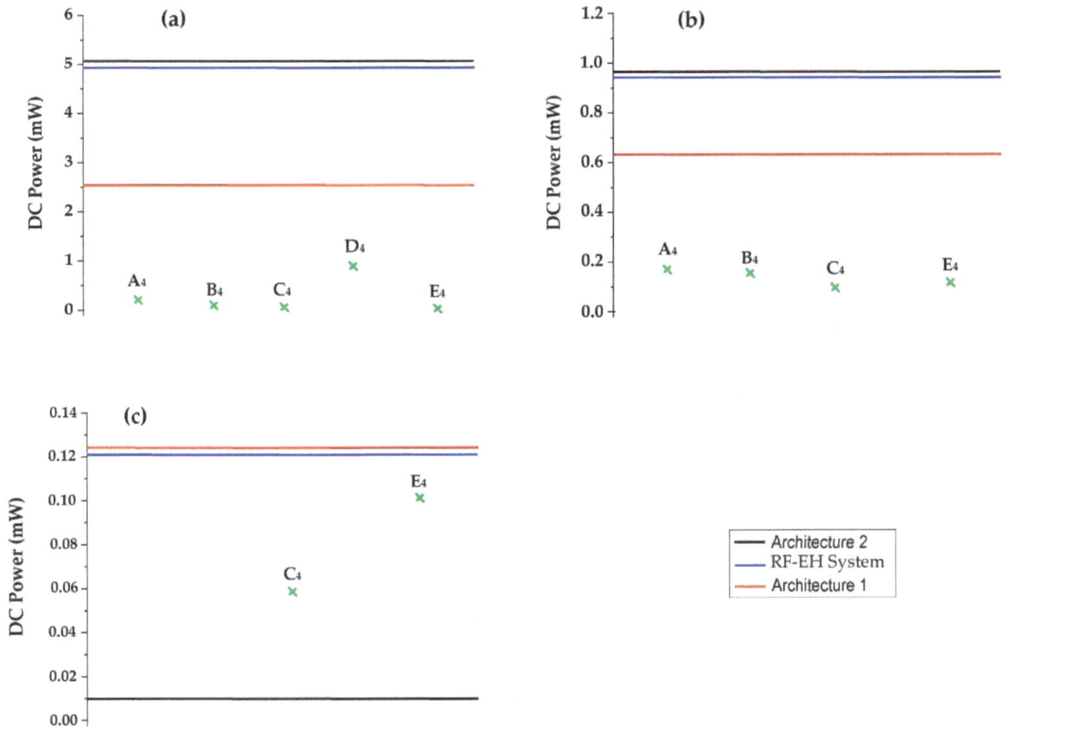

Figure 34. Feeding ability of the RF-EH system and architectures at (**a**): 22 m, (**b**): 40 m, and (**c**): 110 m from relay antenna compared to minimum required power supply of the biomedical sensors. The DC powers A_4, B_4, C_4, D_4 and E_4 correspond to the power supply of sensors presented in the references [40–44], respectively.

We can remark that all sensors could be powered by the RF-EH system and both architectures at 22 m from the relay antenna. At 40 m, the RF-EH system and the second architecture allow for all sensors feeding. The first architecture provides an insufficient DC power to feed the sensor reported in [43]. At 110 m, the RF-EH system and the first architecture could feed only the sensors from [42,44]. The second architecture couldn't feed any sensor at this distance.

7.5. Temperature Sensors

In [45], the authors report a spintronic/CMOS-based thermal sensors. This consists of a thermal aware system composed of hundreds of distributed thermal sensor nodes. This sensor system consumes a power of 11.9 µW during its on-state. In [46], an integrated pulse width modulated (PWM) temperature sensor is proposed. The power consumption of this sensor is 47.2 nW (for 27 °C) and 17.6 nW (for −20 °C). Both sensors could be fed by the RF-EH system as well as the two architectures within the distance range of [22–110] m from the relay antenna, as presented in Figure 35. At 122 m, the second architecture provides an almost null DC power that doesn't allow for the feeding of these sensors.

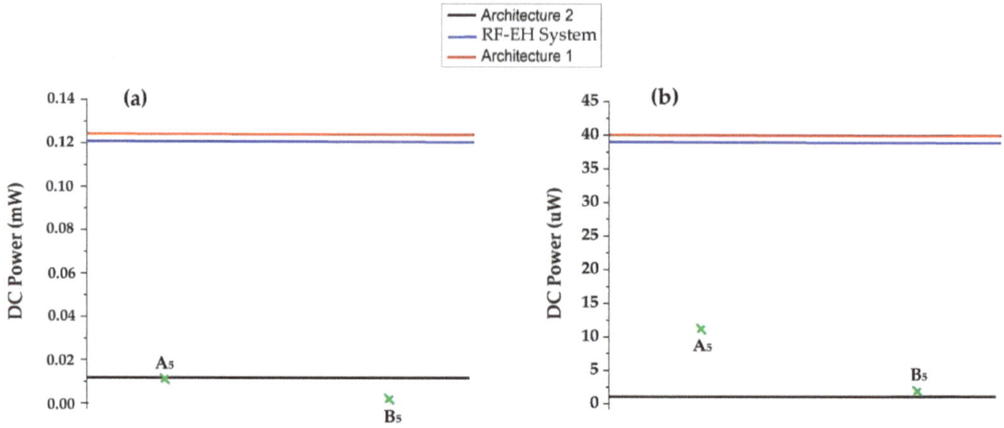

Figure 35. Feeding ability of the proposed system and architectures at (**a**): 110 m, (**b**): 122 m from relay antenna compared to minimum required power supply of the temperature sensors. The DC powers A_5 and B_5 correspond to the power supply of sensors reported in the references [45,46], respectively.

8. Application of RF-EH System to Feed WSN

8.1. Ground Effect on RF Energy Harvesting Process

The presence of an obstacle, especially when it is near the radiating element, can change a little of the overall radiation properties of the RF-EH antenna. In practice, the most common obstacle that is always present is the ground. Any energy provided from the radiating element (patch) directed towards the ground will be reflected. The amount of reflected energy and its direction are controlled by the geometry and the constituent parameters of the ground.

In general, the ground is a lossy medium ($\sigma \neq 0$) whose effective conductivity increases with frequency. Thus, it acts as a very good conductor for certain frequencies, depending mainly on its composition and moisture content. The Volumetric Water Content (VWC) parameter is the main factor affecting the response of an antenna. Thus, it broadens the bandwidth and modifies the radiation pattern by increasing the directivity and gain [47–49].

In [47,48], the presence of the ground (VWC) allows a bandwidth broadening from 8 GHz to a band of 10.5 GHz. In [49], the presence of the ground provides an advantage to the antenna by expanding the bandwidth from 3 GHz to 3.8 GHz. In our case, the energy harvesting process will not be affected by the presence of the ground because even if the bandwidth of the antenna is widened, the rectifiers remain suitable for the GSM-1800, UMTS, and LTE bands. A slight improvement in gain [50] would improve the received RF power, and therefore the DC output power.

8.2. Presentation of WSN System to Be Powered by the Proposed RF-EH System

Many applications of wireless sensor networks are limited by the insufficient battery life of the sensors. The power consumption of processors and microcontrollers is being significantly reduced due to new advances in micro-electronics. This gives the possibility for using RF energy harvesting sources to feed wireless sensor nodes [51].

Wireless sensor systems are designed mainly from several sensor nodes that are dispersed around several management stations called cluster heads (CH), which form clusters. These are located around a sink station. Each sensor node collects information from its environment and transmits it to its cluster head (CH), then each CH communicates with its sink station via wireless communication that allows users to access this information.

These sensor systems are deployed inside the cells of the mobile communication network managed by relay antennas as shown in Figure 36 [52,53].

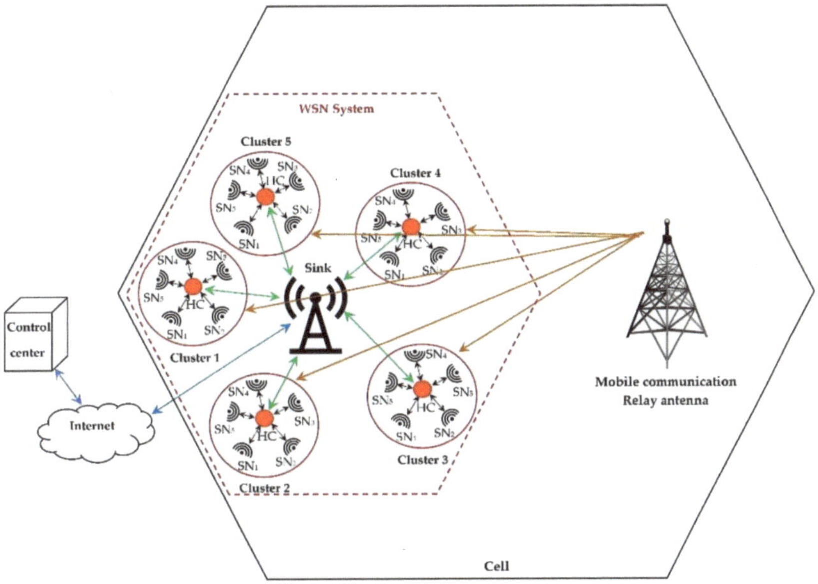

Figure 36. WSN system architecture within mobile communication cell.

The proposed RF-EH system mainly exploits the electromagnetic waves radiated by these relay antennas due to their high transmission power and the multitude of bands available to feed each sensor node. The RF-EH system can also exploit the electromagnetic waves coming from the sink stations as well as from the sensors themselves.

To properly study how the proposed RF-EH system self-powers its wireless sensor systems (RF-EHWSN), two cases can be discussed.

- WSN system close to the relay antenna.
- WSN system far from the relay antenna.

8.3. WSN System Close to the Relay Antenna

In this case, the distance between the relay antenna and the RF-EHWSN is small, which implies a high RF power received by the RF-EH system. This allows it to provide a high enough DC power to feed the sensor nodes. The output DC power could be exploited directly by using a DC voltage regulator connected between the output of the RF-EH system and the sensor feed point.

8.4. WSN System Far from the Relay Antenna

In this case, the distance between the relay antenna and the RF-EHWSNs is large, which greatly reduces the RF power received by the RF-EH system, and thus the output DC power. For this scenario, the direct use of the output DC power does not ensure a good self-powering of the sensor nodes. A super-capacitor placed between the RF-EH system and the sensors is recommended. This super-capacitor allows for the storage of the energy supplied by the RF-EH system while the sensors are in standby mode. The full charge of these super-capacitors takes a relatively large amount of time, which could impact the communication process between the sensor nodes. The addition of a power management block is essential to solve this problem. The power management circuit consists of a DC–DC converter for optimizing the energy transfer from the output of the RF-EH system to the accumulator, and a linear low-dropout (LDO) voltage regulator for generating a suitable sensor supply voltage [54].

8.5. WSN System Self-Powering

The exploitation of RF waves during an information exchange between sensor nodes could be used to self-power them. Indeed, in the case where the ZigBee protocol was adopted by the WSN system, for example, each RF-EHWS (RF energy harvesting of Wireless Sensor) retrieves the electromagnetic waves coming from the other transmitting sensors. The process is similar for all sensor nodes that exist in the WSN system, as shown in Figure 37.

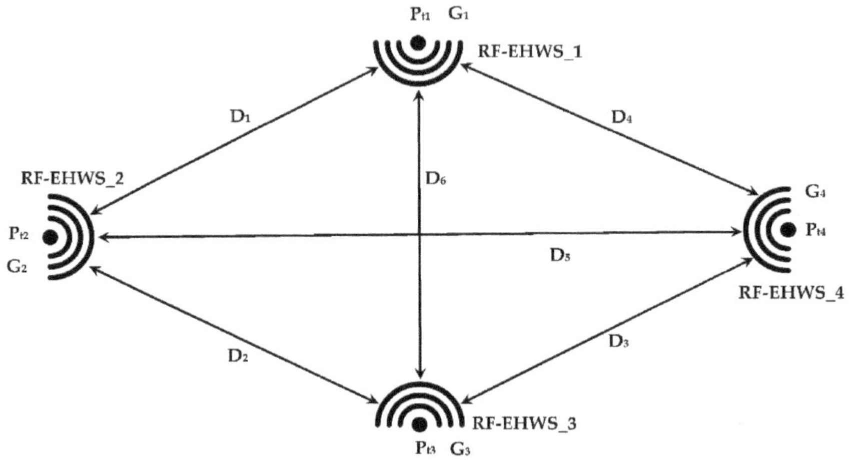

Figure 37. WSN system architecture for self-powering process.

P_{ti}, G_i, and D_i are the transmission power of each sensor, gain of sensor antenna, and the distances between sensors, respectively.

The received power by each RF-EHWS could be expressed by the following equations using the Friis formula:

$$P_r = G_t G_r \left(\frac{\lambda}{4\pi D}\right)^2 P_t \tag{15}$$

where P_r, G_t, G_r, P_t, D, and λ are the transmission power, transmission gain, reception gain, the distance between sensors, and the wavelength, respectively.

RF-EHWSN_1 receives power from the three other sensors, which gives Equations (16)–(19):

$$P_{r1} = G_{t2}G_{r1}\left(\frac{\lambda}{4\pi D_1}\right)^2 P_{t2} + G_{t3}G_{r1}\left(\frac{\lambda}{4\pi D_6}\right)^2 P_{t3} + G_{t4}G_{r1}\left(\frac{\lambda}{4\pi D_4}\right)^2 P_{t4} \tag{16}$$

$$P_{r2} = G_{t1}G_{r1}\left(\frac{\lambda}{4\pi D_1}\right)^2 P_{t1} + G_{t3}G_{r1}\left(\frac{\lambda}{4\pi D_2}\right)^2 P_{t3} + G_{t4}G_{r1}\left(\frac{\lambda}{4\pi D_5}\right)^2 P_{t4} \tag{17}$$

$$P_{r3} = G_{t1}G_{r1}\left(\frac{\lambda}{4\pi D_6}\right)^2 P_{t1} + G_{t2}G_{r1}\left(\frac{\lambda}{4\pi D_2}\right)^2 P_{t2} + G_{t4}G_{r1}\left(\frac{\lambda}{4\pi D_3}\right)^2 P_{t4} \tag{18}$$

$$P_{r4} = G_{t1}G_{r1}\left(\frac{\lambda}{4\pi D_4}\right)^2 P_{t1} + G_{t2}G_{r1}\left(\frac{\lambda}{4\pi D_5}\right)^2 P_{t2} + G_{t3}G_{r1}\left(\frac{\lambda}{4\pi D_3}\right)^2 P_{t3} \tag{19}$$

where P_{ri} are the received RF powers by RF-EHWS_i.

As we can see, each RF-RHWS receives three powers, this increases the output DC voltages provided by the RF-EH systems, which allows for the self-powering of WSNs. In order to use the proposed RF-EH system efficiently, it is necessary to match all the rectennas of the system to the communication frequency of the WSN.

8.6. Obstacle Facing the Application of RF Energy Harvesting in WSNs and Proposed Protocols

RF energy is abundantly available in the environment from various sources of electromagnetic radiation, which allows it to be used with WSNs. This type of energy effectively provides mobile self-power, which is consistent with the wireless functionality of the WSN.

The real challenge facing the application of RF energy harvesting in WSNs is the difficulty of dividing the sensor operating time efficiently between the charging process and data transmission times. For this purpose, several protocols have been proposed in the literature to solve this problem [55,56]. These protocols are applied to a microcontroller placed between the storage capacitor and the node sensor, as presented in Figure 38 [57].

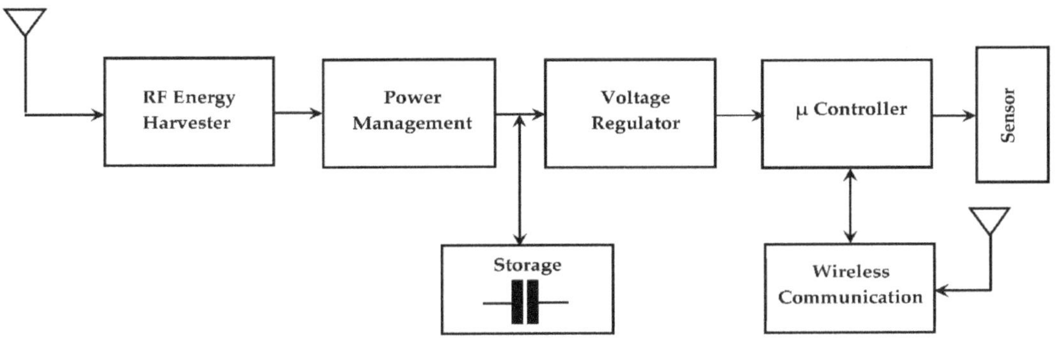

Figure 38. RF-EHWSN system complete architecture.

Two MAC (Medium Access Control) protocols take into account the RF energy harvesting applications: RF-AASP (Radio Frequency energy harvesting process based on active and adaptive sleep period) and RF-MAC (RF Medium Access Control). The RF-AASP determines the charging time of the super-capacitor based on the data traffic and the RF-MAC determines the charging time based on the sensor's participation in the data communication.

In [55], an RF-AASP algorithm is proposed to adaptively manage the sleep period of a sensor. The period varies according to the traffic model and the available energy of the sensors. A low-energy sensor checks the traffic model and the quality of service (QoS) satisfaction. Based on these results, the sensor adjusts parameters like Beacon Order (BO) and Super frame Order (SO). The two adjusted parameters determine the sleep period by the equation $t_{sleep} = 2BO - 2SO$. The sensor harvests energy during the sleep period to charge the supercapacitors. In [56], an RF-MAC algorithm is proposed with a new approach for RF energy harvesting. The energy harvesting is done with the RFE (Request-for-Energy) packet. A low energy sensor dispatches the RFE packet. Peripheral energy transmitters (for this protocol it is the Sink station) respond with CFE (Cleared-for-Energy) packets. When the sensor dispatches the ACK, the sink station emits energy. In this case, the charging time depends on the value of the Important Index (IDX), which indicates how the sensor participated in the data communication on the channel. This algorithm ensures continuous data communication when there is no energy demand. However, when there is a frequent demand for energy, starvation can occur. A sensor that sends data falls into starvation because of the higher priority of the energy demand.

9. Conclusions

In this paper, a new RF energy harvesting system is proposed in order to enhance the DC output voltage for low power WSN feeding. It is based on two architectures, one suitable for low input powers and the other for high input powers. Input power-controlled switching circuits are used to select the more suitable architecture. Both architectures and the system have been designed and realized on Teflon glass hybrid technology and experimentally characterized separately. The efficiency variation as a function of the load resistance yield were at a load of 2 kΩ for maximum efficiency. This value is used for all measurements. DC output voltages and efficiencies have been measured as a function of the distance from the mobile communication base station (relay antenna). The transfer function of the switches has been studied, simulated, and experimentally characterized in order to verify their functionality. The first architecture supplies a maximum DC output voltage of 2.3 V at 22 m from the base station and presents a maximum efficiency of 72% at 70 m. The second one provides a maximum DC output voltage of 3.2 V at 22 m from the base station and a maximum efficiency of 50% at 40 m. The proposed system provides a maximum DC output voltage of 3.15 V at 22 m from the relay antenna and a maximum efficiency of 71.5% at 70 m from the relay antenna. A DC voltage of 0.37 V is provided by the proposed RF-EH system at 122 m from the relay antenna with an RF-to-DC conversion efficiency of 43%.

Each rectenna that forms both architectures and the RF-EH system consists of a new multi-band antenna matched to a rectifier. The antenna covers all the mobile communications and wireless network bands (GSM-1800, UMTS, LTE-2.3, Wi-Fi, and LTE-2.6). The antenna has been characterized by simulating and measuring its S_{11} parameter and radiation pattern. The measured efficiencies of the antenna are 70%, 65%, and 68.5% for 1.8 GHz, 2.1 GHz, and 2.66 GHz, respectively.

Finally, the functioning of the architectures as well as that of the switching circuits has been explained and discussed. The DC output voltages provided by the architectures and by the RF-EH system have been applied to several types of sensors reported in the literature in order to ensure the operation of these sensors as a function of the distance from the relay antenna.

Author Contributions: Methodology, C.B., A.O., A.-E.K. and M.F.; Supervision, A.O., A.-E.K. and M.F.; Validation, A.O., A.-E.K. and M.F.; Writing—original draft, C.B., A.O., A.-E.K. and M.F.; Review & editing, C.B., A.O., A.-E.K. and M.F. All authors have read and agreed to the published version of the manuscript.

Funding: This research received no external funding.

Institutional Review Board Statement: Not applicable.

Informed Consent Statement: Not applicable.

Data Availability Statement: Not applicable.

Conflicts of Interest: The authors declare no conflict of interest.

References

1. Zhang, M.; Liu, X.; Guo, H.; Yang, X.; Xie, D. Development of high-efficiency rectification circuit for RF energy harvesting. In Proceedings of the 2015 Asia-Pacific Microwave Conference (APMC), Nanjing, China, 6–9 December 2015; pp. 1–3. [CrossRef]
2. Saraiva, H.M.; Borges, L.M.; Pinho, P.; Goncalves, R.; Chavez-Santiago, R.; Barroca, N.; Tavares, J.; Gouveia, P.T.; Carvalho, N.B.; Balasingham, I.; et al. Experimental Characterization of Wearable Antennas and Circuits for RF Energy Harvesting in WBANs. In Proceedings of the 2014 IEEE 79th Vehicular Technology Conference (VTC Spring), Seoul, Korea, 18–21 May 2014; pp. 1–5. [CrossRef]
3. Chaour, I.; Fakhfakh, A.; Kanoun, O. Enhanced Passive RF-DC Converter Circuit Efficiency for Low RF Energy Harvesting. *Sensors* **2017**, *17*, 546. [CrossRef]
4. Taghadosi, M.; Albasha, L.; Quadir, N.A.; Rahama, Y.A.; Qaddoumi, N. High Efficiency Energy Harvesters in 65nm CMOS Process for Autonomous IoT Sensor Applications. *IEEE Access* **2017**, *6*, 2397–2409. [CrossRef]
5. Zhang, L.; Cheng, X.; Deng, X. A modified Dickson's charge pump circuit with high output voltage and high pumping efficiency. *Analog Integr. Circuits Signal Process.* **2019**, *101*, 601–609. [CrossRef]
6. Okba, A.; Takacs, A.; Aubert, H.; Charlot, S.; Calmon, P.F. Multiband rectenna for microwave applications. *Comptes Rendus. Phys.* **2017**, *18*, 107–117. [CrossRef]
7. Tung, N. Multi-band ambient RF energy harvesting rectifier for autonomous Wireless Sensor networks. In Proceedings of the 2016 IEEE Region 10 Conference (TENCON), Singapore, 22–25 November 2016; pp. 3736–3739. [CrossRef]
8. Nimo, A.; Beckedahl, T.; Ostertag, T.; Reindl, L. Analysis of Passive RF-DC Power Rectification and Harvesting Wireless RF Energy for Micro watt Sensors. *AIMS Energy* **2015**, *3*, 184–200. [CrossRef]

9. Singh, N.; Kanaujia, B.K.; Beg, M.T.; Mainuddin; Khan, T.; Kumar, S. A dual polarized multiband rectenna for RF energy harvesting. *AEU-Int. J. Electron. Commun.* **2018**, *93*, 123–131. [CrossRef]
10. Sun, H.; Guo, Y.; He, M.; Zhong, Z. Design of a High-Efficiency 2.45-GHz Rectenna for Low-Input-Power Energy Harvesting. *IEEE Antennas Wirel. Propag. Lett.* **2012**, *11*, 929–932. [CrossRef]
11. Adami, S.-E.; Proynov, P.; Hilton, G.S.; Yang, G.; Zhang, C.; Zhu, D.; Li, Y.; Beeby, S.P.; Craddock, I.J.; Stark, B.H. A Flexible 2.45-GHz Power Harvesting Wristband with Net System Output From −24.3 dBm of RF Power. *IEEE Trans. Microw. Theory Tech.* **2018**, *66*, 380–395. [CrossRef]
12. Vishal, S.; Shankar, P.V.; Vivek, S. A Wideband Star-Shaped Rectenna for RF Energy Harvesting in GSM Band. In *Advances in Electromechanical Technologies*; Springer: Singapore, 2021; pp. 163–173. [CrossRef]
13. Moreno, J.M.; Vázquez, A.S.M.; Barragán, C.A.B.; González, J.M.V.; Rosas, J.C.A. Radio Frequency Energy Harvesting System Making Use of 180° Hybrid Couplers and Multiple Antennas to Improve the DC Output Voltage. *IEEE Lat. Am. Trans.* **2020**, *18*, 604–612. [CrossRef]
14. Song, C.; Huang, Y.; Zhou, J.; Carter, P. Improved Ultrawideband Rectennas Using Hybrid Resistance Compression Technique. *IEEE Trans. Antennas Propag.* **2017**, *65*, 2057–2062. [CrossRef]
15. Fakharian, M.M. A Wideband Rectenna Using High Gain Fractal Planar Monopole Antenna Array for RF Energy Scavenging. *Int. J. Antennas Propag.* **2020**, *2020*, 3489393. [CrossRef]
16. Nguyen, H.Q.; Le, M.T. Multiband Ambient RF Energy Harvester with High Gain Wideband Circularly Polarized Antenna toward Self-Powered Wireless Sensors. *Sensors* **2021**, *21*, 7411. [CrossRef] [PubMed]
17. Song, C.; Lu, P.; Shen, S. Highly Efficient Omnidirectional Integrated Multiband Wireless Energy Harvesters for Compact Sensor Nodes of Internet-of-Things. *IEEE Trans. Ind. Electron.* **2021**, *68*, 8128–8140. [CrossRef]
18. Quddious, A.; Zahid, S.; Tahir, F.A.; Antoniades, M.A.; Vryonides, P.; Nikolaou, S. Dual-Band Compact Rectenna for UHF and ISM Wireless Power Transfer Systems. *IEEE Trans. Antennas Propag.* **2021**, *69*, 2392–2397. [CrossRef]
19. Li, L.; Zhang, X.; Song, C.; Zhang, W.; Jia, T.; Huang, Y. Compact Dual-Band, Wide-Angle, Polarization- Angle -Independent Rectifying Metasurface for Ambient Energy Harvesting and Wireless Power Transfer. *IEEE Trans. Microw. Theory Tech.* **2021**, *69*, 1518–1528. [CrossRef]
20. Roy, S.; Tiang, R.J.-J.; Roslee, M.B.; Ahmed, M.T.; Mahmud, M.A.P. Quad-Band Multiport Rectenna for RF Energy Harvesting. *IEEE Access* **2021**, *9*, 77464–77481. [CrossRef]
21. Shen, S.; Zhang, Y.; Chiu, C.-Y.; Murch, R. Directional Multiport Ambient RF Energy-Harvesting System for the Internet of Things. *IEEE Internet Things J.* **2021**, *8*, 5850–5865. [CrossRef]
22. Hu, Y.-Y.; Sun, S.; Xu, H.; Sun, H. Grid-Array Rectenna with Wide Angle Coverage for Effectively Harvesting RF Energy of Low Power Density. *IEEE Trans. Microw. Theory Tech.* **2019**, *67*, 402–413. [CrossRef]
23. Lu, P.; Song, C.; Cheng, F.; Zhang, B.; Huang, K. A Self-Biased Adaptive Reconfigurable Rectenna for Microwave Power Transmission. *IEEE Trans. Power Electron.* **2020**, *35*, 7749–7754. [CrossRef]
24. Lu, P.; Huang, K.M.; Yang, Y.; Cheng, F.; Wu, L. Frequency-Reconfigurable Rectenna with an Adaptive Matching Stub for Microwave Power Transmission. *IEEE Antennas Wirel. Propag. Lett.* **2019**, *18*, 956–960. [CrossRef]
25. Kuhn, V.; Lahuec, C.; Seguin, F.; Person, C. A Multi-Band Stacked RF Energy Harvester with RF-to-DC Efficiency Up to 84%. *IEEE Trans. Microw. Theory Tech.* **2015**, *63*, 1768–1778. [CrossRef]
26. Jang, D.; Jung, G.; Jeong, Y.; Hong, Y.; Hong, S.; Shin, W.; Chang, K.S.; Jeong, C.B.; Park, B.-G.; Lee, J.-H. Efficient Integration of Si FET-type Gas Sensors and Barometric Pressure Sensors on the Same Substrate. In Proceedings of the 2019 IEEE International Electron Devices Meeting (IEDM), San Francisco, CA, USA, 7–11 December 2019; pp. 26.4.1–26.4.4. [CrossRef]
27. Li, C.; Chen, M.; Peng, S.; Qi, W.; Wang, N.; Rong, Q.; Chen, D.; Xu, L. A High Linearity Detection Circuit with Constant Detection Voltage for Resistive Gas Sensor. In Proceedings of the 2019 IEEE 4th Advanced Information Technology, Electronic and Automation Control Conference (IAEAC), Chengdu, China, 20–22 December 2019; pp. 1051–1058. [CrossRef]
28. Xie, D.; Liu, R.; Adedokun, G.; Xu, L.; Wu, F. A Novel Low Power Hexagonal Gas Sensor Cell for Multi-Channel Gas Detection. In Proceedings of the 2021 IEEE 34th International Conference on Micro Electro Mechanical Systems (MEMS), Gainesville, FL, USA, 25–29 January 2021; pp. 430–433. [CrossRef]
29. Xie, D.; Chen, D.; Peng, S.; Yang, Y.; Xu, L.; Wu, F. A Low Power Cantilever-Based Metal Oxide Semiconductor Gas Sensor. *IEEE Electron Device Lett.* **2019**, *40*, 1178–1181. [CrossRef]
30. Sokolovskij, R.; Zhang, J.; Zheng, H.; Li, W.; Jiang, Y.; Yang, G.; Yu, H.; Sarro, P.M.; Zhang, G. The Impact of Gate Recess on the H_2 Detection Properties of Pt-AlGaN/GaN HEMT Sensors. *IEEE Sens. J.* **2020**, *20*, 8947–8955. [CrossRef]
31. Fu, H.; Han, Z.; Li, Y.; Ma, K. A Single-Pixel Sensor for Near-field Imaging Based on Startup Time of Oscillator. *IEEE Microw. Wirel. Components Lett.* **2021**, *31*, 1323–1326. [CrossRef]
32. Park, K.; Yeom, S.; Kim, S.Y. Ultra-Low Power CMOS Image Sensor with Two-Step Logical Shift Algorithm-Based Correlated Double Sampling Scheme. *IEEE Trans. Circuits Syst. I Regul. Pap.* **2020**, *67*, 3718–3727. [CrossRef]
33. Liu, Z.; Liu, Z.; Ren, E.; Luo, L.; Wei, Q.; Wu, X.; Li, X.; Qiao, F.; Liu, X.J. A 1.8mW Perception Chip with Near-Sensor Processing Scheme for Low-Power AIoT Applications. In Proceedings of the 2019 IEEE Computer Society Annual Symposium on VLSI (ISVLSI), Miami, FL, USA, 15–17 July 2019; pp. 447–452. [CrossRef]
34. Park, C.; Zhao, W.; Park, I.; Sun, N.; Chae, Y. A 51-pJ/Pixel 33.7-dB PSNR 4× Compressive CMOS Image Sensor with Column-Parallel Single-Shot Compressive Sensing. *IEEE J. Solid-State Circuits* **2021**, *56*, 2503–2515. [CrossRef]

35. Hsu, T.-H.; Chen, Y.-R.; Liu, R.-S.; Lo, C.-C.; Tang, K.-T.; Chang, M.-F.; Hsieh, C.-C. A 0.5-V Real-Time Computational CMOS Image Sensor with Programmable Kernel for Feature Extraction. *IEEE J. Solid-State Circuits* **2020**, *56*, 1588–1596. [CrossRef]
36. Tian, X.; Liu, Z.; Guo, C.; Yang, J.; Chen, J.; Lyu, S.; Bi, H.; Qiao, F.; Wu, X.; Lu, Y. Pressure Sensor Array with Low-Power Near-Sensor CMOS Chip for Human Gait Monitoring. *IEEE Sens. Lett.* **2021**, *5*, 1–4. [CrossRef]
37. Wang, X.; Meng, X.; Zhu, Y.; Ling, H.; Chen, Y.; Li, Z.; Hartel, M.C.; Dokmeci, M.R.; Zhang, S.; Khademhosseini, A. A Sub-1-V, Microwatt Power-Consumption Iontronic Pressure Sensor Based on Organic Electrochemical Transistors. *IEEE Electron Device Lett.* **2021**, *42*, 46–49. [CrossRef]
38. Gardner, E.L.W.; de Luca, A.; Philippe, J.; Dragomirescu, D.; Udrea, F. Thin-Film MOSFET-Based Pressure Sensor. *IEEE Sens. Lett.* **2019**, *3*, 1–4. [CrossRef]
39. Zhang, C.; Gallichan, R.; Lapshev, S.; Budgett, D.M.; McCormick, D. A Low Power Capacitance to Frequency Converter in 180nm CMOS for an Implantable Capacitive Pressure Sensor. In Proceedings of the 2019 IEEE Biomedical Circuits and Systems Conference (BioCAS), Nara, Japan, 17–19 October 2019; pp. 1–4. [CrossRef]
40. Lin, Q.; Xu, J.; Song, S.; Breeschoten, A.; Konijnenburg, M.; Van Hoof, C.; Tavernier, F.; Van Helleputte, N. A 119dB Dynamic Range Charge Counting Light-to-Digital Converter for Wearable PPG/NIRS Monitoring Applications. *IEEE Trans. Biomed. Circuits Syst.* **2020**, *14*, 800–810. [CrossRef] [PubMed]
41. Song, S.; Konijnenburg, M.; Van Wegberg, R.; Xu, J.; Ha, H.; Sijbers, W.; Stanzione, S.; Biswas, D.; Breeschoten, A.; Vis, P.; et al. A 769 µW Battery-Powered Single-Chip SoC with BLE for Multi-Modal Vital Sign Monitoring Health Patches. *IEEE Trans. Biomed. Circuits Syst.* **2019**, *13*, 1506–1517. [CrossRef] [PubMed]
42. Lin, B.; Ma, Z.; Atef, M.; Ying, L.; Wang, G. Low-Power High-Sensitivity Photoplethysmography Sensor for Wearable Health Monitoring System. *IEEE Sens. J.* **2021**, *21*, 16141–16151. [CrossRef]
43. Huang, Y.-J.; Tzeng, T.-H.; Lin, T.-W.; Huang, C.-W.; Yen, P.-W.; Kuo, P.-H.; Lin, C.-T.; Lu, S.-S. A Self-Powered CMOS Reconfigurable Multi-Sensor SoC for Biomedical Applications. *IEEE J. Solid-State Circuits* **2014**, *49*, 851–866. [CrossRef]
44. Boukhayma, A.; Barison, A.; Haddad, S.; Caizzone, A. Earbud-Embedded Micro-Power mm-Sized Optical Sensor for Accurate Heart Beat Monitoring. *IEEE Sens. J.* **2021**, *21*, 19967–19977. [CrossRef]
45. Qoutb, A.G.; Friedman, E.G. Spintronic/CMOS-Based Thermal Sensors. In Proceedings of the 2020 IEEE International Symposium on Circuits and Systems (ISCAS), Seville, Spain, 12–14 October 2020; pp. 1–5. [CrossRef]
46. Azam, A.; Bai, Z.; Walling, J.S. An Ultra-Low Power CMOS Integrated Pulse-Width Modulated Temperature Sensor. *IEEE Sens. J.* **2021**, *21*, 1294–1304. [CrossRef]
47. Zemmour, H.; Baudoin, G.; Hamouda, C.; Diet, A.; Biancheri-Astier, M. Impact of soil on UWB buried antenna and communication link in IR-UWB WUSN applications. In Proceedings of the 2015 European Radar Conference (EuRAD), Paris, France, 9–11 September 2015; pp. 353–356. [CrossRef]
48. Zemmour, H.; Baudoin, G.; Diet, A. Soil Effects on the Underground-to-Aboveground Communication Link in Ultrawideband Wireless Underground Sensor Networks. *IEEE Antennas Wirel. Propag. Lett.* **2017**, *16*, 218–221. [CrossRef]
49. Alkhalifeh, K.; Hislop, G.; Ozdemir, N.A.; Craeye, C. Efficient MoM Simulation of 3-D Antennas in the Vicinity of the Ground. *IEEE Trans. Antennas Propag.* **2016**, *64*, 5335–5344. [CrossRef]
50. Han, D.; Polycarpou, A.C.; Balanis, C.A. Ground effects for VHF/HF antennas on helicopter airframes. *IEEE Trans. Antennas Propag.* **2001**, *49*, 402–412. [CrossRef]
51. Varghese, B.; John, N.E.; Sreelal, S.; Gopal, K. Design and Development of an RF Energy Harvesting Wireless Sensor Node (EH-WSN) for Aerospace Applications. *Procedia Comput. Sci.* **2016**, *93*, 230–237. [CrossRef]
52. Din, M.S.U.; Rehman, M.A.U.; Ullah, R.; Park, C.-W.; Kim, B.S. Towards Network Lifetime Enhancement of Resource Constrained IoT Devices in Heterogeneous Wireless Sensor Networks. *Sensors* **2020**, *20*, 4156. [CrossRef] [PubMed]
53. Liu, T.; Qu, X.; Tan, W.; Cheng, Y. An Energy Efficient Cooperative Communication Scheme in Ambient RF Powered Sensor Networks. *IEEE Access* **2019**, *7*, 86545–86554. [CrossRef]
54. Caselli, M.; Ronchi, M.; Boni, A. Power Management Circuits for Low-Power RF Energy Harvesters. *J. Low Power Electron. Appl.* **2020**, *10*, 29. [CrossRef]
55. Nguyen, T.D.; Khan, J.Y.; Ngo, D.T. Energy harvested roadside IEEE 802.15.4 wireless sensor networks for IoT applications. *Ad Hoc Netw.* **2017**, *56*, 109–121. [CrossRef]
56. Nguyen, T.D.; Khan, J.Y.; Ngo, D.T. An adaptive MAC protocol for RF energy harvesting wireless sensor networks. In Proceedings of the 59th IEEE Global Communications Conference, GLOBECOM, Washington, DC, USA, 4–8 December 2016; pp. 1–6.
57. Grossi, M. Energy Harvesting Strategies for Wireless Sensor Networks and Mobile Devices: A Review. *Electronics* **2021**, *10*, 661. [CrossRef]

sensors

MDPI

Article

A Highly Efficient RF-DC Converter for Energy Harvesting Applications Using a Threshold Voltage Cancellation Scheme

Muhammad Basim [1], Danial Khan [1], Qurat Ul Ain [1], Khuram Shehzad [2], Syed Adil Ali Shah [1], Byeong-Gi Jang [1], Young-Gun Pu [1], Joon-Mo Yoo [2], Joon-Tae Kim [3] and Kang-Yoon Lee [1,*]

[1] Department of Electrical and Computer Engineering, Sungkyunkwan University, Suwon 16419, Korea; basim@skku.edu (M.B.); danialkhan@skku.edu (D.K.); quratulain@skku.edu (Q.U.A.); adilshah@skku.edu (S.A.A.S.); seeys17@skku.edu (B.-G.J.); hara1015@skku.edu (Y.-G.P.)
[2] SKAiChips, Sungkyunkwan University, Suwon 16419, Korea; khuram1698@skku.edu (K.S.); fiance2@g.skku.edu (J.-M.Y.)
[3] Department of Electrical and Electronic Engineering, Konkuk University, Seoul 05029, Korea; jtkim@konkuk.ac.kr
* Correspondence: klee@skku.edu; Tel.: +82-31-299-4954

Abstract: In this paper, a self-threshold voltage (V_{th}) compensated Radio Frequency to Direct Current (RF-DC) converter operating at 900 MHz and 2.4 GHz is proposed for RF energy harvesting applications. The threshold voltage of the rectifying devices is compensated by the bias voltage generated by the auxiliary transistors and output DC voltage. The auxiliary transistors compensate the threshold voltage (V_{th}) of the PMOS rectifying device while the threshold voltage (V_{th}) of the NMOS rectifying device is compensated by the output DC voltage. The proposed RF-DC converter was implemented in 180 nm Complementary Metal-Oxide Semiconductor (CMOS) technology. The experimental results show that the proposed design achieves better performance at both 900 MHz and 2.4 GHz frequencies in terms of *PCE*, output voltage, sensitivity, and effective area. The peak power conversion efficiency (*PCE*) of 38.5% at −12 dBm across a 1 MΩ load for 900 MHz frequency was achieved. Similarly, for 2.4 GHz frequency, the proposed circuit achieves a peak *PCE* of 26.5% at −6 dBm across a 1 MΩ load. The proposed RF-DC converter circuit shows a sensitivity of −20 dBm across a 1 MΩ load and produces a 1 V output DC voltage.

Keywords: self-threshold voltage cancellation (STVC); RF energy harvesting; power conversion efficiency (*PCE*); CMOS technology; RF-DC converter

Citation: Basim, M.; Khan, D.; Ain, Q.U.; Shehzad, K.; Shah, S.A.A.; Jang, B.-G.; Pu, Y.-G.; Yoo, J.-M.; Kim, J.-T.; Lee, K.-Y. A Highly Efficient RF-DC Converter for Energy Harvesting Applications Using a Threshold Voltage Cancellation Scheme. *Sensors* **2022**, *22*, 2659. https://doi.org/10.3390/s22072659

Academic Editors: Slim Naifar, Olfa Kanoun and Carlo Trigona

Received: 28 February 2022
Accepted: 28 March 2022
Published: 30 March 2022

1. Introduction

In the past decade, the interest in energy harvesting for portable and wearable electronic devices, biomedical implanted devices, radio and frequency identification (RFID), and the Internet of Things (IoT) is increasing day by day [1–5]. However, in IoT, the near-field technique cannot scale well where wireless sensor nodes obtain power over wide indoor and outdoor environments [6]. Solar energy, RF energy, thermal energy, and vibration energy are some of the main sources for energy harvesting applications. The density of wireless devices rapidly increased in this decade. Moreover, mostly ultrahigh-frequency ISM bands are used for communication systems and harvest multiband RF energy simultaneously [7].

The concept of using RF signals as a source of power for wireless devices is quite appealing. The power associated with communication signals is unpredictable and often minimal; thus, it is difficult to use it for providing power supply to wireless electronic devices. To reduce the cost, it is necessary to integrate an RF energy harvesting system with a low-power system on a CMOS integrated chip. A far-field RF energy harvesting system is used to harvest energy from ambient RF energy sources, i.e., frequency modulation (FM) or dedicated RF sources and amplitude modulation (AM) radio transmission, cellular transmission signals, and television signals. A limited RF energy power source is transmitted for RF energy harvesting as specified by the Federal Communication Commission. This RF signal is strongly affected by a number of factors that degrade the performance of the signal, including weak signal strength and path loss. The available power source for collecting RF energy depends upon the path loss of the available RF signal in free space, which can be calculated using the following equation:

$$L_P = 10\ \log_{10}\left(\frac{4\Pi D}{\lambda}\right)^2 \tag{1}$$

where L_P is the path loss of the free space, D is the distance from the source, and λ is the wavelength of the signal. The equation shows that the free space path loss increases if the wavelength (λ) increases.

Figure 1 shows the block diagram of the proposed RF-DC architecture in which the RF-DC converter is a key component of the RF energy harvesting system that converts the incoming RF signal to DC voltage. The RF energy harvesting unit consists of an antenna, impedance matching network, radiofrequency to direct current (RF-DC) circuit, and a storage device. The antenna collects electromagnetic wave signals from the environment. The strength of the electromagnetic signal is very low and decreases rapidly as the distance from the antenna to the RF source increases. The input impedance of the RF-DC converter to a 50 Ω antenna is matched by the matching network. Moreover, it also maximizes the power transfer between them. Impedance mismatch may occur due to the varying input received from RF signals that reduce the power of the rectifier. For this purpose, a Pi-matching network is used in the proposed structure. Finally, the storage device stores the converted output DC voltage and provides for further use in wireless electronic devices. The performance of the system can be assessed based on the power conversion efficiency (*PCE*) of the RF-DC converter.

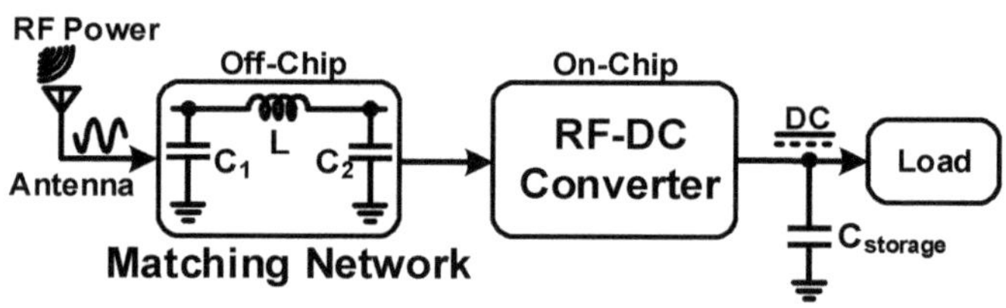

Figure 1. Block diagram of an RF energy harvester.

Several RF-DC converter architectures have been published which have enhanced the *PCE* for energy harvesting applications. RF energy harvesting can be adopted to operate IoT devices. RF energy harvesting is beneficial for distance flexibility. Since its output is unpredictable and small, it degrades the system's strength. A CMOS RF rectifier proposed in [8] is based on a self-threshold voltage cancellation scheme. In the circuitry, positive feedback is used to minimize the threshold voltage of the transistors, but the disadvantage of this mechanism is that *PCE* rapidly rises and falls in response to the input RF signal. Moreover, with an increase in input RF voltage level, the RF-DC converter's maximum output voltage saturates. A digital control loop and an on-chip capacitor array are implemented with an adaptive matching function [9]. The effect of the matching network must be considered to improve the power transfer to the rectifier. In the case of perfect matching, a pi-type matching network is used to provide a passive voltage enhancing factor. Higher efficiency can be attained to minimize the threshold voltage of MOSFETs in RF-DC circuitry. Threshold voltage causes loss in the RF-DC converter circuit, which is one of the major difficulties in design [10]. A control circuit and two differential sub-rectifiers are used in [11] for adaptive power harvesters. The rectifier is a switch between the serial and parallel modes by a control signal generated by the control circuit. The control circuit senses the output voltage of two sub-rectifiers. In reference [12], a class-E amplifier that is based on time-reversal duality theory is used as a class-E rectifier. The other methods use a voltage multiplier which makes them different from other RF rectifiers. A full-wave matching network and cross dipole antenna are proposed in [13] for a rectifier-booster regulator. This rectifier converts the RF signal to DC voltage and boosts it. A cross-connected differential rectifier is presented in [14,15] with a differential custom antenna. In [16,17], an optimum number of the rectifier is used to maintain high efficiency by using the maximum power point tracking (MPPT) technique over a wide input power range. The author in [18] reported a dual path adaptive control CMOS differential rectifier. The adaptive control circuit switches the rectifier between low- and high-power paths according to the input power level. The limitation in this research is more power loss due to the use of multiple stages of a cross-coupled rectifier. A 17-stage rectifier is presented in [19], which can deliver 1.2 V at 1 MΩ load with a self-compensated structure. A self-compensation scheme is presented in [20] in which an individual body biasing is provided to triple-well NMOS transistors. The triple-well NMOS transistors are used as a rectifying device, although triple-well NMOS transistors are not available in all CMOS technologies [21,22]. The design in [23] presents a differential cross-connected CMOS rectifier that minimizes the leakage current and compensates for the threshold voltage of the rectifying device. The design in [24,25] also proposed a self-threshold voltage cancellation scheme for RF energy harvesting applications. The referred papers also uses the threshold voltage cancellation scheme-based rectifier, which works for a low frequency of 402 MHz by using N-numbers of stages of rectifier with PMOS transistors. In [26], the author proposed a 10-stage cross-connected rectifier by using the threshold voltage compensation technique for the heavy load of 5 MΩ. A 900 MHz RF-DC converter is proposed in [27] with the aim to optimize the sensitivity and generate 50 µW output power. A multi-path energy harvesting architecture is proposed to maintain the high *PCE* of the system [28]. Similarly, in reference [29], the design uses a passive multi-stage RF to DC rectifier that is based on n-well technology. Moreover, the backward connection is used to generate the negative source bias in the Dickson charge multiplier.

The design of a highly efficient RF-DC converter using the threshold voltage cancellation scheme for energy harvesting applications is presented in this paper. This paper is organized as follows. Section 2 discusses the self-threshold voltage cancellation scheme. The proposed RF-DC converter is explained in Section 3. Section 4 presents the measurement results. Finally, Section 5 explains the conclusion of the paper.

2. Self V_{th} Cancellation Scheme

The V_{th} of the rectifying device plays a significant role in the performance and operation of the RF-DC converter for energy harvesting. For an RF-DC converter operation to rectify a low RF power to DC power, a low-threshold voltage rectifying device is required. Different technology-based approaches are proposed to reduce the threshold voltage of the devices. Some of the devices include SMS, HSMS, Schottky diodes, SOS, and floating gate transistors that usually lower the threshold voltage by storing the pre-charged voltage at the gate. The additional fabrication steps are the main drawback of the technology-based approach. Moreover, it also prevents the integration of RF energy harvesters in normal CMOS ICs. To reduce the threshold voltage, an active/passive circuit technique can also be used. The active technique uses an external power source that increases the cost and maintenance, while additional circuitry is needed to compensate for the threshold voltage in the passive technique. Most of the RF-DC converter designs focused on the reduction in threshold voltage while neglecting the increase in the reverse leakage current and power losses. Consequently, decreasing the threshold voltage may cause an increase in leakage current which has an adverse effect on the output DC voltage and *PCE* of an RF-DC converter. Considering the threshold voltage and reverse leakage, the overall *PCE* can be defined as:

$$PCE = \frac{P_{OUT} - P_{leakage}}{P_{IN}} \tag{2}$$

where P_{OUT} is the power delivered to the load when the transistors act as forward-biased, $P_{leakage}$ is the output leakage power when the transistors act as reverse-biased, while P_{IN} is the input power. The *PCE* of the proposed rectifier starts to increase if the input power is increased. To attain the highest *PCE*, several strategies for lowering the voltage drop across MOS transistors have been proposed. However, when the voltage drop across transistors is reduced, the reverse leakage current in the negative half cycle is increased. This behavior produces a decrease in the *PCE* because it introduces the energy loss stored in previous cycles. The ratio of the useable DC power given to the load resistor divided by the total input RF power delivered to the RF-DC converter is the actual *PCE* of the proposed rectifier. This *PCE* can be computed by the following equation:

$$\eta(\%) = \frac{P_{OUT}}{P_{IN}} = \frac{V^2{}_{OUT}}{R_L \times P_{IN}} \times 100 \tag{3}$$

where V_{OUT} is the output DC voltage across the load resistor R_L and P_{IN} is the input RF power from the ambient sources. Figure 2a shows the voltage doubler based on CMOS transistors by connecting the gate and drain terminal with each other. The PMOS transistor works in a one-half cycle while the NMOS transistor works in the other half cycle of the input. Figure 2b shows the diode-based voltage doubler which converts the AC input to DC output. If the voltage drop across each transistor reaches zero, the output voltage of the voltage doubler can be double that of the amplitude of the RF signal. Therefore, the main challenge is to minimize the voltage drop across the forward-biased transistors to maximize the power flow to the output and to minimize the reverse leakage current to avoid the loss of energy.

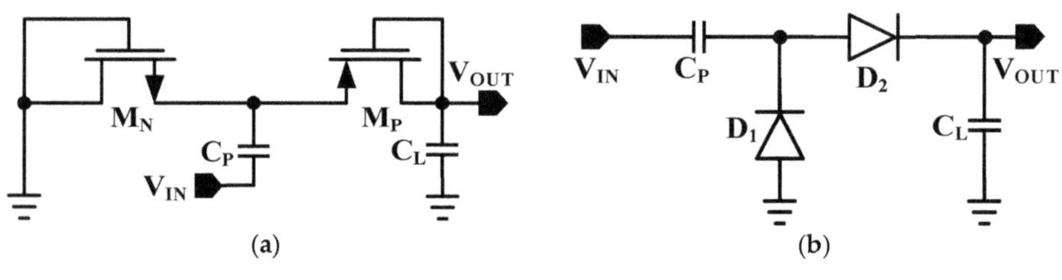

Figure 2. Conventional rectifier design. (**a**) CMOS-based voltage doubler. (**b**) Diode connected voltage doubler.

3. Proposed RF-DC Converter

Figure 3 shows the circuit architecture of the proposed RF-DC converter. The proposed RF-DC converter uses both positive and negative levels of incoming RF signal. The rectifier cancels the threshold voltage effect using the input RF voltage and the output DC voltage. The basic RF-DC converter is designed by using a PMOS transistor (M_{P1}) and an NMOS transistor (M_{N1}). The threshold voltage compensation circuitry in the proposed architecture is for the PMOS transistor M_{P1}. As from the circuitry, the gate terminal of the NMOS transistor M_{N1} is connected to the output of the rectifier to obtain the threshold compensation. C_P is a pumping capacitor while C_L is used as a battery for charge storing purposes, while R_L is the load resistor. Two NMOS transistors (M_{N2} and M_{N3}) and one PMOS (M_{P2}) are auxiliary transistors operating in the subthreshold region. These transistors provide an optimum gate to source compensation voltage to the main transistor M_{P1}. To produce this optimum gate to source compensation voltage, proper sizing of the auxiliary transistors is needed, while the leakage current can be minimized by a high impedance path to the ground. In practice, to obtain high output voltage, a transistor with very small resistance is needed. Therefore, the optimization should be carried out under the worst condition for RF energy harvesting applications. The leakage and parasitic loss increase as the size of the transistor increases. Similarly, by the selection of small transistors, an improper and undesirable transfer of charges occur. For this purpose, the width of the transistors M_{N1} and M_{P1} are chosen as 8 and 16 μm, respectively, while their channel lengths are chosen as a minimum. The sizes of the auxiliary transistors M_{P2}, M_{N2}, and M_{N3} are chosen as 1 μm/8 μm, 1 μm/4 μm, and 1 μm/2 μm, respectively. The capacitors' values of C_P and C_L are set as 400 fF and 1 pF. The *PCE* has been decreasing by further increasing the sizes of the transistors. The sizes of the transistors are selected to achieve high efficiency with a low input power range.

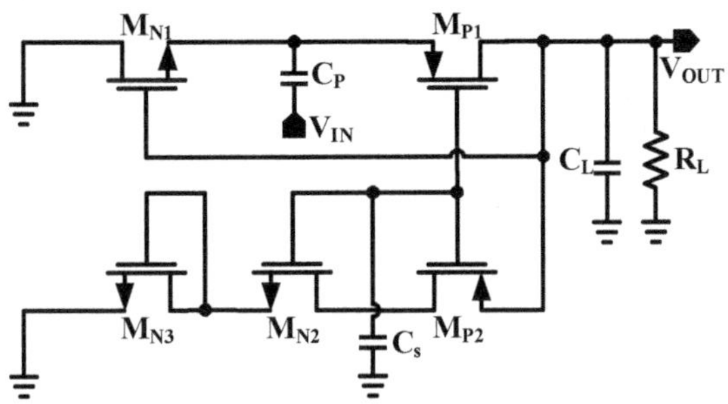

Figure 3. Circuit diagram of proposed RF-DC converter.

Figure 4 shows the operating principle of the proposed RF-DC converter. The RF-DC converter enters the charging phase (Figure 4a), as the negative input signal appears. In the negative phase, the NMOS transistor M_{N1} will first go into conducting mode, and then the capacitor C_P will begin to charge. The voltage produced across C_P can be found by applying Kirchhoff's voltage law (KVL) as:

$$V_{CP} = -V_{IN} + V_{dMN1} \tag{4}$$

where V_{IN} is the peak amplitude of the input RF voltage and V_{dN} is the voltage drop across the transistor M_{N1} due to the threshold voltage. By considering C_P as an ideal capacitor, the whole charge will be transferred to the C_L without any loss during the discharging phase. The first-order equivalent circuit of the second path during the discharging phase of C_P is depicted in Figure 4b to determine the voltage across C_{AUX}. Thus, by applying KVL, we can write:

$$V_{OUT} = V_{AUX} - V_{th} + I_d R_{ON} - V_{IN} \tag{5}$$

where V_{AUX} is a voltage that appears across the auxiliary capacitor C_{AUX}, V_{th} is the threshold voltage, and the R_{ON} is the ON resistance of the transistor M_{P1} while the drain current is represented by I_d. The I_d is also given by the equation:

$$I_d = \beta(V_{AUX} - V_{th})^2 \tag{6}$$

Principally, "β" is equal to ($\mu_n C_0/2L^2$) where μ_n is the mobility of the electron in the channel, the capacitance between the channel and gate of the transistor is C_0, and L is the length of the source to the drain channel. Several different properties affect the threshold voltage of the MOS structure. The presence of a threshold voltage, for example, can be a major constraint in circuits built to work with low-voltage batteries. As a result, substantial effort was directed toward developing MOS architectures with low V_{th} values. By solving Equation (4) for V_{CL}, we get

$$V_{AUX} = V_{th} - \frac{1}{\beta R} + \sqrt{\left(\frac{1}{\beta R}\right)^2 + 4(V_{OUT} + V_{IN})} \quad (7)$$

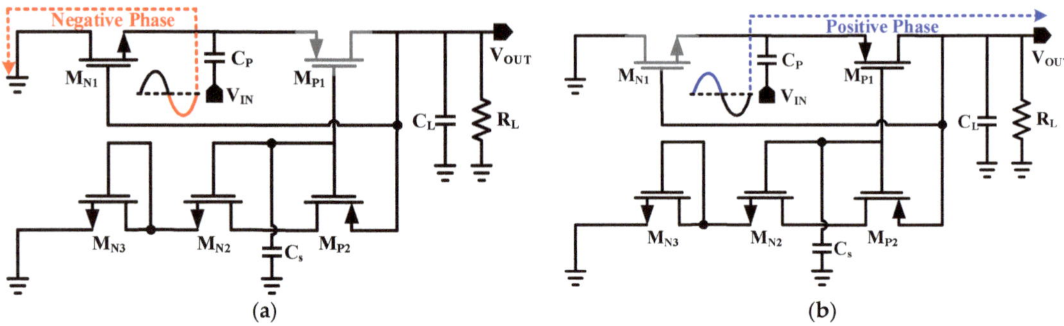

Figure 4. Operating principle of the RF-DC converter. (**a**) Charging (negative) phase. (**b**) Discharging (positive) phase.

Equation (7) shows that the voltage across C_{AUX} and V_{AUX} is developed using the contribution of both input and output voltages. The rectifier will enter the discharging phase as the positive RF signal appears. Figure 4b shows the equivalent circuit for the discharging phase of the rectifier. It was shown clearly in V_{CL} that the voltage is the DC biasing gate for the source voltage of the transistor M_{N1}. During this phase, the threshold voltage compensation in transistor M_{P1} occurs. Applying KVL on this discharging path and replacing Equation (3) will result as:

$$V_{OUT} = 2V_{IN} - V_{dMN1} - V_{dMP1} \quad (8)$$

where V_{dMP1} is ideally compensated by V_{AUX}, hence, we can also write Equation (7) as

$$V_{OUT} = 2V_{IN} - V_{dMN1} \quad (9)$$

As a result, it can be deduced that, ideally, M_{N1} will account for the majority of the loss in the rectified output DC voltage obtained by the proposed RF-DC converter.

4. Measurement Results

The proposed self-threshold voltage cancellation RF-DC converter is designed and implemented in standard 180 nm CMOS technology. Figure 5a shows the fabricated chip micrograph. The total active area of the fabricated chip is 160 × 120 μm excluding the pads. The fabricated chip is soldered and packaged on an FR4 PCB board. Figure 5b shows the measurement setup and the fabricated chip is measured with a single-tone sinusoidal signal of 900 MHz and 2.4 GHz generated by the signal generator (Agilent E4438C). The output DC voltage is measured using an oscilloscope and digital multi-meter. The impedance matching network is integrated off-chip between the 50 Ω signal generator and the fabricated chip. The impedance matching network enhances the incoming RF signal and transfers it to the chip. Reflection between the impedance matching circuit and the RF-DC converter, PCB trace losses, and impedance matching network losses caused by passive elements, are the factors that affect the overall performance of the RF-DC converter. The net input power can be calculated after excluding all these losses and transfers to the chip. The maximum output DC voltage and the *PCE* were measured for the resistive loads of 100 kΩ, 500 kΩ, and 1 MΩ.

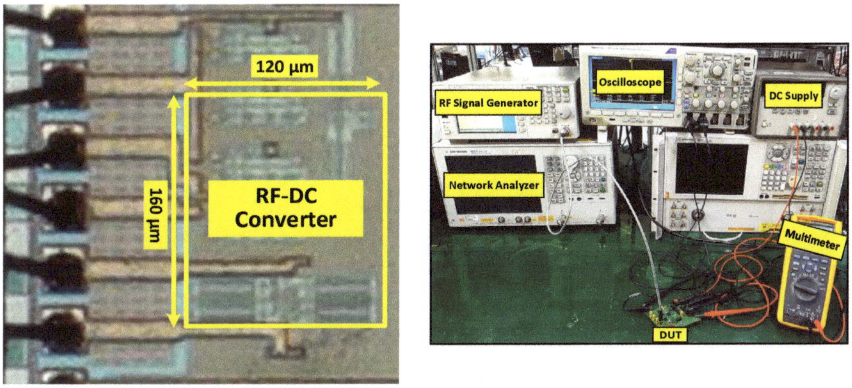

(**a**) (**b**)

Figure 5. (**a**) Chip microphotograph and (**b**) measurement setup of the proposed RF-DC converter architecture.

Figure 6 shows the measured |S11| parameter for the proposed RF-DC converter. The measured values of |S11| at 900 MHz and 2.4 GHz are −24.958 and −22.892 dB, respectively, for the 1 MΩ load which shows excellent matching.

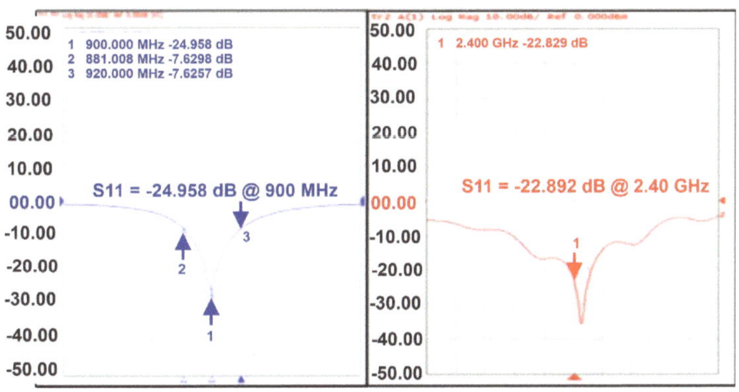

Figure 6. Measured |*S*11| for the RF-DC converter at a 1 MΩ load.

The performance of the proposed RF-DC converter can be measured by the output DC voltage and its efficiency. It achieves better efficiency and output voltage with very low input power. The input power is varied from −20 to −3 dBm. The output rectified DC voltage after simulation and measurement results are plotted across various resistive loads, as shown in Figure 7a. By increasing the load resistance, the output DC voltage of the proposed RF-DC circuit increases gradually. Similarly, Figure 7b shows the simulated and measured power conversion efficiency of the proposed RF-DC converter as a function of input power for different load resistances for the 900 MHz frequency. The proposed architecture achieves maximum simulated efficiency of 45% and measured efficiency of 42.5% with −7 dBm input power level and 100 KΩ of load resistance. However, the *PCE* starts decreasing by further increasing the input power level. For a load of 500 KΩ, the maximum simulated efficiency is 42.5% and the measured efficiency is 39% at −10 dBm of input power level.

Similarly, a simulated efficiency of 40% and measured efficiency of 38% is achieved at −12 dBm of the input power level at 1 MΩ load resistance. If the load resistance increases, the efficiency curve shifts towards the left side.

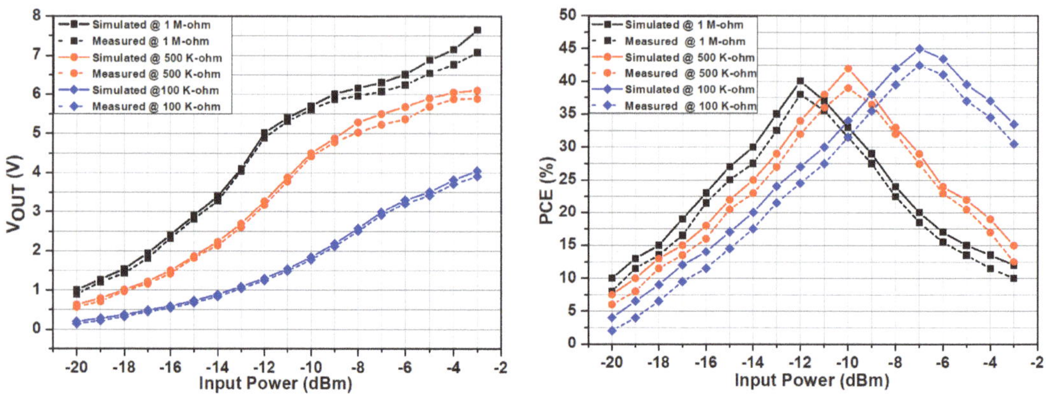

(**a**) (**b**)

Figure 7. Simulated and measured results of the RF-DC converter for different loads. (**a**) Output DC voltage and (**b**) *PCE*.

The performance of the proposed RF-DC converter was checked in all PVT variations (typical corner points, slow corner points, and fast corner points) and compared with the measurement results. The performance was first checked on simulation and post-simulation levels for PVT variations and then measurement level. The proposed RF-DC converter satisfied both the simulation and measurement results. Figure 8 shows simulated and measured results of the RF-DC converter at different frequencies, i.e., 900 MHz and 2.4 GHz. Figure 8a presents the simulated and measured DC voltage at both 900 MHz and 2.4 GHz frequencies. The recorded simulated and measured output DC voltage is 5 and 4.8 V, respectively, at −12 dBm input power level for 900 MHz frequency. An 8.38 V simulated and 8.1 V measured voltage is achieved for 2.4 GHz frequency at −6 dBm input power level. Similarly, for 2.4 GHz, the maximum simulated and measured efficiencies are 28% and 26.5% at −6 dBm input power level, as shown in Figure 8b. Moreover, the proposed architecture obtains high *PCE* for a wide input power range. The sensitivity of the proposed architecture to obtain a 1 V output DC voltage is −20 dBm for 900 MHz. Figure 9 shows the measured results including the output DC voltage and *PCE* of the proposed RF-DC converter including all losses for different frequencies with 1 MΩ load resistance for both 900 MHz and 2.4 GHz frequencies. Figure 9a shows the output measured voltage at both 900 MHz and 2.4 GHz frequencies.

The measured voltage including a loss at 900 MHz frequency is 4.24 V with a −12 dBm input power level. Similarly, a 7.3 V is achieved at a 2.4 GHz frequency with a −6 dBm input power level including all losses, respectively. As the output voltage is relatively high, therefore, the breakdown voltage of the transistors (MN1 and MP1) used in the main rectification chain is large enough to bear the voltage larger than 5 V. Figure 9b shows the *PCE* of the proposed RF-DC converter including all the losses at both 900 MHz and 2.4 GHz frequencies. The maximum *PCE* achieved at 900 MHz including losses is 37%. Similarly, a 25% efficiency is achieved at 2.4 GHz frequency including all the losses.

Table 1 shows the summary of the performance of the proposed architecture and compares it with the prior work. The proposed circuit shows decent performance and minimum active area. At −12 and −6 dBm input powers, the proposed RF-DC converter achieves maximum efficiency of 38.5% and 26% for the 1 MΩ load resistance. Moreover, the proposed RF-DC converter maintains more than 20% *PCE* from the −8 to −16 dBm input power range. It can be seen that the DC output voltage achieved through our work, i.e., 4.8 V for 900 MHz and 8.1 for V for 2.4 GHz is higher than the compared reference works. However, the circuit reported in [26] shows higher sensitivity than this work but the achieved *PCE* is much lower than our proposed architecture. Overall, our proposed circuit shows better performance and sensitivity than the previous architectures.

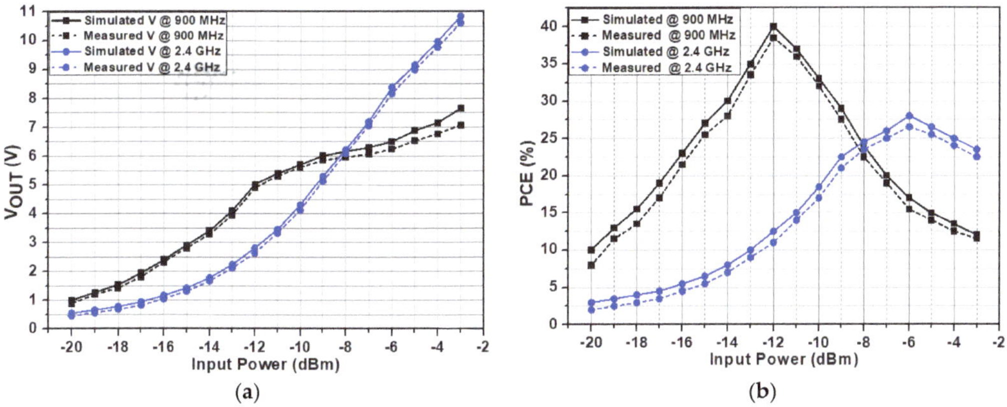

Figure 8. Simulated and measured results of the RF-DC converter at different frequencies for 1 MΩ. (**a**) Output DC voltage and (**b**) *PCE*.

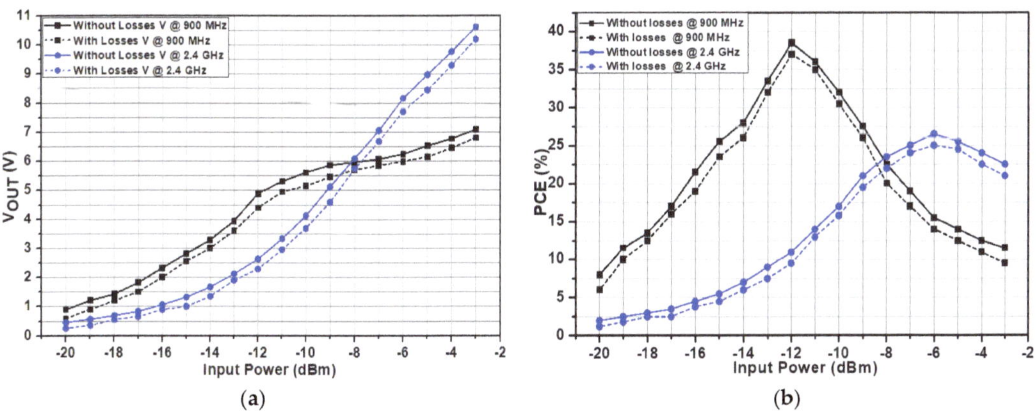

Figure 9. Measured results of the RF-DC converter at different frequencies for 1 MΩ with and without losses. (**a**) Output DC voltage and (**b**) *PCE*.

Table 1. Summary and performance comparison.

Parameters	This Work	[2]	[13]	[18]	[22]	[23]	[24]
		2015	**2019**	**2017**	**2017**	**2020**	**2014**
Technology (nm)	**180**	130	Diode-Based	65	180	180	130
Energy harvesting	**RF**	RF	RF	RF	RF	RF	RF
Frequency (GHz)	**0.9/2.4**	0.902~0.928	2.45	0.953	0.915	0.902	0.915
Load (MΩ)	**1**	1	0.2	0.147	1	0.2	1
Input power (dBm)	**−12/−6**	−15	13	−10	−2	−8	−16.8
DC output (V)	**4.8/8.1**	3.2	1.7	2.6	2.4	3.23	2.2
***PCE* (%)**	**38.5/26.5**	32	37.5	36.5	27	33	22.6
Effective Area	**0.19 mm^2**	-	0.74 mm^2	0.47 mm^2	-	0.105 mm^2	-
Sensitivity: 1 V for 1 MΩ	**−20 dBm**	−20.5 dBm	-	−17.5 dBm	14.8 dBm	−20.2 dBm	−21.6 dBm

5. Conclusions

This paper presented a self-threshold voltage (V_{th}) compensated RF-DC converter operating at 900 MHz and 2.4 GHz frequencies for RF energy harvesting applications. The proposed RF-DC converter was implemented in 180 nm CMOS technology. In the proposed RF-DC circuit, the auxiliary transistors and output DC voltage generated the bias voltage for the V_{th} compensation of the rectifying devices in the main rectification chain. The measurement results showed that the proposed circuit obtained a peak *PCE* of 38.5% at −12 dBm input power across a 1 MΩ load for 900 MHz frequency. Similarly, for 2.4 GHz, the proposed RF energy harvester showed a peak *PCE* of 26.5% at −6 dBm across a 1 MΩ load resistance.

Author Contributions: Conceptualization, M.B. and D.K.; methodology, M.B.; software, M.B., D.K., K.S. and Q.U.A.; validation, M.B., D.K., Q.U.A. and S.A.A.S.; formal analysis, M.B., D.K., Q.U.A. and B.-G.J.; investigation, M.B. and D.K.; resources, M.B.; data curation, M.B. and D.K.; writing—original draft preparation, M.B. and K.-Y.L.; writing—review and editing, M.B., D.K., Y.-G.P., J.-M.Y., J.-T.K. and K.-Y.L.; visualization, M.B., D.K., Q.U.A. and K.-Y.L.; supervision, Y.-G.P., J.-M.Y., J.-T.K. and K.-Y.L.; project administration, K.-Y.L. All authors have read and agreed to the published version of the manuscript.

Funding: There is no external funding for this research.

Institutional Review Board Statement: Not applicable.

Informed Consent Statement: Not applicable.

Data Availability Statement: Not applicable.

Acknowledgments: This work was supported by the Institute of Information & communications Technology Planning & Evaluation (IITP) grant funded by the Korea government (MSIT) (No.2020-0-00261, Development of low power/low delay/self-power suppliable RF simultaneous information and power transfer system and stretchable electronic epineurium for wireless nerve bypass implementation).

Conflicts of Interest: The authors declare no conflict of interest.

References

1. Stoopman, M.; Keyrouz, S.; Visser, J.H.; Philips, K.; Serdijn, A.W. Co-design of a CMOS rectifier and small loop antenna for highly sensitive RF energy harvesters. *IEEE J. Solid-State Circuits* **2014**, *49*, 622–634. [CrossRef]
2. Hameed, Z.; Moez, K. A 3.2 V–15 dBm adaptive threshold-voltage compensated RF energy harvester in 130 nm CMOS. *IEEE Trans. Circuits Syst. I Regul. Pap.* **2015**, *62*, 948–956. [CrossRef]
3. Nakamoto, H.; Yamazaki, D.; Yamamoto, T.; Kurata, H.; Yamada, S.; Mukaida, K.; Ninomiya, T.; Ohkawa, T.; Masui, S.; Gotoh, K. A Passive UHF RF Identification CMOS Tag IC Using Ferroelectric RAM in 0.35 µm CMOS Technology. *IEEE J. Solid-State Circuits* **2006**, *42*, 101–110. [CrossRef]
4. Kotani, K.; Sasaki, A.; Ito, T. High-efficiency differential-drive CMOS rectifier for UHF RFIDs. *IEEE J. Solid-State Circuits* **2009**, *44*, 3011–3018. [CrossRef]
5. Lazaro, O.; Rincón-Mora, G.A. A Nonresonant Self-Synchronizing Inductively Coupled 0.18m CMOS Power Receiver and Charger. *IEEE J. Emerg. Sel. Top. Power Electron.* **2014**, *3*, 261–271. [CrossRef]
6. Bito, J.; Hester, J.G.; Tentzeris, M.M. Ambient RF energy harvesting from a two-way talk radio for flexible wearable wireless sensor devices utilizing inkjet printing technologies. *IEEE Trans. Microw. Theory Tech.* **2015**, *63*, 4533–4543. [CrossRef]
7. Kotani, K.; Ito, T. High-efficiency CMOS rectifier circuit with self-Vth-cancellation and power regulation functions for UHF RFIDs. In Proceedings of the 2007 IEEE Asian Solid-State Circuits Conference, Jeju, Korea, 12–14 November 2007; pp. 119–122.
8. O'Driscoll, S.; Poon, A.S.; Meng, T.H. An mm-sized implantable power receiver with adaptive link compensation. In Proceedings of the 2009 IEEE International Solid-State Circuits Conference-Digest of Technical Papers, San Francisco, CA, USA, 8–12 February 2009; pp. 294–295.
9. Kuhn, V.; Lahuec, C.; Seguin, F.; Person, C. A multi-band stacked RF energy harvester with RF-to-DC efficiency up to 84%. *IEEE Trans. Microw. Theory Tech.* **2015**, *63*, 1768–1778. [CrossRef]
10. Mikeka, C.; Arai, H. Design issues in radio frequency energy harvesting system. In *Sustainable Energy Harvesting Technologies—Past, Present and Future*; IntechOpen: London, UK, 2011; pp. 235–256.
11. Li, C.J.; Lee, T.C. 2.4-GHz high-efficiency adaptive power. *IEEE Trans. Very Large Scale Integr. (VLSI) Syst.* **2013**, *22*, 434–438. [CrossRef]
12. Dehghani, S.; Johnson, T. A 2.4-GHz CMOS class-E synchronous rectifier. *IEEE Trans. Microw. Theory Tech.* **2016**, *64*, 1655–1666. [CrossRef]

13. Mansour, M.M.; Ichihashi, M.; Kanaya, H. Wide-band and efficiency-improved 0.18 um CMOS RF differential rectifier for wireless energy harvesting. In Proceedings of the 2019 IEEE Asia-Pacific Microwave Conference (APMC), Singapore, 10–13 December 2019; pp. 141–143.
14. Lee, D.; Kim, T.; Kim, S.; Byun, K.; Kwon, K. A CMOS rectifier with 72.3% RF-to-DC conversion efficiency employing a tunable impedance matching network for ambient RF energy harvesting. In Proceedings of the 2018 International SoC Design Conference (ISOCC), Daegu, Korea, 12–15 November 2018; pp. 259–260.
15. Li, X.; Lu, Y.; Huang, M.; Martins, R.P. A 2.4-GHz mid-field CMOS wireless power receiver achieving 46% maximum PCE and 163-mW output power. *IEEE Trans. Circuits Syst. II Express Briefs* **2019**, *67*, 360–364. [CrossRef]
16. Kim, S.Y.; Abbasizadeh, H.; Rikan, B.S.; Oh, S.J.; Jang, B.G.; Park, Y.J.; Khan, D.; Nga, T.T.K.; Kang, K.T.; Pu, Y.G.; et al. A-20 to 30 dBm input power range wireless power system with an MPPT-based reconfigurable 48% efficient RF energy harvester and 82% efficient A4WP wireless power receiver with open-loop delay compensation. *IEEE Trans. Power Electron.* **2018**, *34*, 6803–6817. [CrossRef]
17. Khan, D.; Abbasizadeh, H.; Kim, S.Y.; Khan, Z.H.N.; Shah, S.A.A.; Pu, Y.G.; Hwang, K.C.; Yang, Y.; Lee, M.; Lee, K.Y. A design of ambient RF energy harvester with a sensitivity of− 21 dBm and power efficiency of 39.3% using internal threshold voltage compensation. *Energies* **2018**, *11*, 1258. [CrossRef]
18. Yan, L.; Haojuan, D.; Mo, H.; Man, K.L.; Sai, W.S.; Seng, P.U.; Rui, P.M. A Wide Input Range Dual-Path CMOS Rectifier for RF Energy Harvesting. *IEEE Trans. Circuits Syst.* **2017**, *64*, 166–170.
19. Papotto, G.; Carrara, F.; Palmisano, G. A 90-nm CMOS threshold-compensated RF energy harvester. *IEEE J. Solid-State Circuits* **2011**, *46*, 1985–1997. [CrossRef]
20. Khan, D.; Oh, S.J.; Shehzad, K.; Verma, D.; Khan, Z.H.N.; Pu, Y.G.; Lee, M.; Hwang, K.C.; Yang, Y.; Lee, K.Y. A CMOS RF energy harvester with 47% peak efficiency using internal threshold voltage compensation. *IEEE Microw. Wirel. Compon. Lett.* **2019**, *29*, 415–417. [CrossRef]
21. Scorcioni, S.; Larcher, L.; Bertacchini, A. A reconfigurable differential CMOS RF energy scavenger with 60% peak efficiency and-21 dBm sensitivity. *IEEE Microw. Wirel. Compon. Lett.* **2013**, *23*, 155–157. [CrossRef]
22. Abouzied, M.A.; Ravichandran, K.; Sánchez-Sinencio, E. A fully integrated reconfigurable self-startup RF energy-harvesting system with storage capability. *IEEE J. Solid-State Circuits* **2017**, *52*, 704–719. [CrossRef]
23. Khan, D.; Oh, S.J.; Shehzad, K.; Basim, M.; Verma, D.; Pu, Y.G.; Lee, M.; Hwang, K.C.; Yang, Y.; Lee, K.Y. An efficient reconfigurable RF-DC converter with a wide input power range for RF energy harvesting. *IEEE Access* **2020**, *8*, 79310–79318. [CrossRef]
24. Hameed, Z.; Moez, K. Hybrid forward and a backward threshold-compensated RF-DC power converter for RF energy harvesting. *IEEE J. Emerg. Sel. Top. Circuits Syst.* **2014**, *4*, 335–343. [CrossRef]
25. Munir, A.A.-A.; Sami, R.A.-B. Hybrid Internal Vth Cancellation Rectifiers for RF Energy Harvesting. *IEEE Access* **2020**, *8*, 51976–51980.
26. Noghabaei, S.M.; Radin, R.L.; Savaria, Y.; Sawan, M. A High-Sensitivity Wide Input-Power-Range Ultra-Low-Power RF Energy Harvester for IoT Applications. *IEEE Trans. Circuits Syst. I Regul. Pap.* **2022**, *69*, 440–451. [CrossRef]
27. Hsieh, P.; Chiang, T. An RF energy harvester with 35.7% PCE at PIN of −15 dBm. In Proceedings of the Symposium on VLSI Circuits, Kyoto, Japan, 11–13 June 2013; pp. C224–C225.
28. Khan, D.; Oh, S.J.; Yeo, S.K.; Ryu, Y.; In, S.H.; Rad, R.E.; Ali, I.; Pu, Y.G.; Yoo, S.S.; Lee, M.; et al. A High-Efficient Wireless Power Receiver for Hybrid Energy-Harvesting Sources. *IEEE Trans. Power Electron.* **2021**, *36*, 11148–11162. [CrossRef]
29. Hameed, Z.; Moez, K. Fully-integrated passive threshold-compensated PMOS rectifier for RF energy harvesting. In Proceedings of the IEEE 56th International Midwest Symposium on Circuits and Systems (MWSCAS), Columbus, OH, USA, 4–7 August 2013; pp. 129–132.

sensors

MDPI

Article

Multiband Ambient RF Energy Harvester with High Gain Wideband Circularly Polarized Antenna toward Self-Powered Wireless Sensors

Hong Quang Nguyen and Minh Thuy Le *

Department of Instrumentation and Industrial Informatics, School of Electrical Engineering, Hanoi University of Science and Technology, 1 Đại Cồ Việt, Bách Khoa, Hai Bà Trưng, Hà Nội 10000, Vietnam; quang.nh174135@sis.hust.edu.vn

* Correspondence: thuy.leminh@hust.edu.vn

Abstract: In this work toward a sustainable operation of a self-powered wireless sensor, we investigated a multiband Wi-Fi/3G/4G/5G energy harvester based on a novel wideband circularly polarized antenna, a quadplexer, and rectifiers at four corresponding bands. This proposed antenna consisted of four sequentially rotated dual-dipoles, fed by a hybrid feeding network with equal amplitude and an incremental 90° phase delay. The feeding network was composed of three Wilkinson power dividers and Schiffman phase shifters. Based on the sequential rotation method, the antenna obtained a −10 dB reflection coefficient bandwidth of 71.2% from 1.4 GHz to 2.95 GHz and a 3 dB axial ratio (AR) bandwidth of 63.6%, from 1.5 GHz to 2.9 GHz. In addition, this antenna gain was higher than 6 dBi in a wide bandwidth from 1.65 GHz to 2.8 GHz, whereas the peak gain was 9.9 dBi. The quad-band rectifier yielded the maximum AC–DC conversion efficiency of 1.8 GHz and was 60% at −1 dBm input power, 2.1 GHz was 55% at 0 dBm, 2.45 GHz was 55% at −1 dBm, and 2.6 GHz was 54% at 0.5 dBm, respectively. The maximum RF–DC conversion efficiency using the wideband circularly polarized antenna was 27%, 26%, 25.5%, and 27.5% at −6 dBm of input power, respectively.

Keywords: circularly polarized antenna; Schiffman phase shifter; sequential rotation feeding network; ambient RF energy harvesting; self-powered wireless sensor

Citation: Nguyen, H.Q.; Le, M.T. Multiband Ambient RF Energy Harvester with High Gain Wideband Circularly Polarized Antenna toward Self-Powered Wireless Sensors. *Sensors* **2021**, *21*, 7411. https://doi.org/10.3390/s21217411

Academic Editors: Slim Naifar, Olfa Kanoun and Carlo Trigona

Received: 9 October 2021
Accepted: 3 November 2021
Published: 8 November 2021

1. Introduction

Wireless sensors have recently been used in a variety of applications, including smart homes, human health monitoring and personal medical care, animal surveillance, food safety, and infrastructure monitoring [1]. Finding a reliable energy source for long-term functioning of ubiquitous wireless sensors is a major concern. Batteries not only take a long time and money to replace or recharge, but they also pollute the environment. Therefore, energy harvesting is an effective solution to solve the above problems. The solutions are to harvest energy from triboelectricity, thermoelectricity, solar energy, or radio-frequency (RF) energy sources. Given the increasing demand for wireless equipment in all aspects of daily life and the continuity of RF energy, ambient RF energy harvesting could be a feasible and viable way for replacing batteries in wireless sensors.

GSM 900, GSM 1800/4G, UMTS 2100/3G, LTE/NR 2600/4G/5G, and Wi-Fi are popular RF sources in the ambient environment. However, as many surveys have indicated [2,3], the major challenge in ambient RF energy harvesting is their extremely low power density. In particular, the average power density of GSM 900, GSM 1800, and UMTS 2100 are 140 $\mu W/m^2$, 850 $\mu W/m^2$, and 1450 $\mu W/m^2$, respectively. The low RF power density limits the amount of RF power that antennas can receive, resulting in low RF-to-DC power conversion efficiency. To design a highly efficient rectenna, many solutions have been presented [4–15]. The first approach was to use a single-band antenna combined with a single-band rectifier [4,5]. Because the power density of a single frequency is low, a single-band rectenna is no longer suitable. In another approach to increase the efficiency and harvested power of rectennas, wideband antennas were used [6,7]. This choice allows the rectennas to harvest more bands of RF waves. However, in some earlier works, due to the small gain and linear polarization of the patch [6] and monopole antennas [7], the harvesters only reached 18% and 8% of conversion efficiency at −5 dBm, respectively.

A recent approach using multiband antennas was presented in [8–11]. Multiband energy harvesting can improve the amount of harvested energy. However, these works necessitate a huge load or achieve low conversion efficiency at a low-input power density. A dual-port patch rectenna in [8] required a high 5 kΩ load and large size. The conversion efficiency was smaller than 5% at −10 dBm in [9]. Circularly polarized (CP) rectennas have been used to increase the conversion efficiency because of their flexibility in electromagnetic transmission [12–15]. At 2.45 GHz, a 75.6% RF–DC conversion efficiency was observed for 12 dBm input power [12]. The antenna had a narrow 3 dB axial ratio (AR) bandwidth. In addition, when the input power was reduced below −10 dBm, it had low efficiency. A 2.45 GHz rectenna based on a slot antenna with a wider 3 dB AR bandwidth was presented in [13,14]. At −6 dBm, the efficiency can reach up to 59.5% and 59%, respectively. However, that rectenna can operate in just one single band; thus, it is difficult to supply self-powered wireless sensors due to the low power density of Wi-Fi. Therefore, the need for a circularly polarized, wide-band, high gain antenna to increase ambient RF energy harvesting efficiency and power has to be addressed.

Recently, circularly polarized antennas have been made by using single feed [16,17], dual feed [18–22], or a sequential phase feed (multiple-feeds) [23–27]. Using single feeds can reduce the design complexity, but they have narrow 3 dB AR bands [16] or low gain and radiation efficiency [17]. A wider bandwidth can be obtained using dual-band, dual-sense CP antennas or crossed dipoles using side parasitic elements [19,20]. However, those antennas require large sizes and suffer from complicated designs. High gain and wide AR bandwidth are achieved by using a wideband quadrature coupler [21] or branch-line coupler [22] feeding to dipoles. Complicated designs and large sizes are some limitations to using an RF harvester. Among different feeding networks, the sequential phase technique is one of the most efficient and widely used feeding networks to build wideband CP and high gain antennas. In [23], by using the sequential-phase feed network, the proposed antenna featured a wide impedance bandwidth (54.3%) as well as a wide AR bandwidth (42%). The antennas in [24,27] showed the peak gains of 9.8 dBi and 10.73 dBi, respectively. Due to the large difference in amplitude and phase excitation in a narrow range of frequency, they only showed 12% and 31% AR bandwidths. A wideband circularly polarized bowtie antenna array was presented in [25]. The antenna array's 3-dB AR bandwidth was increased by using the sequential rotation approach. A 2 × 2 CP planar antenna array was presented in [26]. The results showed that the 3 dB AR bandwidth was extended from 1.84 GHz to 2.51 GHz and the peak gain wad 5.7 dBi.

This study presents a wideband circularly polarized dual-dipole antenna utilizing a progressively rotated approach. As radiating elements, the array included 2 × 2 dual-dipoles and a broadband phase shifting network. Each element had two parallel haft-wave dipoles with one director to enhance the antenna gain. The dual-dipole had a simple structure and presents good impedance performance. A Wilkinson power divider was combined with a double Schiffman phase shifter to create the phase shifting network. The 3-dB AR bandwidth was increased by adopting this hybrid feed. A quadplexer was designed to separate the four frequency bands and transfer them to their corresponding rectifiers in the RF–DC system application. The following are the sections of this paper: The design of a dual-dipole element is shown in Section 2. Then, the array configuration and feeding network structure are presented to achieve a 90° phase shifter and equal amplitude between four elements. Then, the 2 × 2 antenna array performance is investigated. Section 3 shows the design of the quadplexer with four rectifiers and applies this proposed antenna in an ambient RF energy harvesting system. In the last section, the conclusions are presented.

2. Antenna Array Design

2.1. Wideband Dual-Dipole Antenna Element Design

To design the antenna array, we first designed the antenna element. A printed J-shaped balun and two identical parallel half-wavelength dipoles formed the proposed element antenna. To boost gain and front-to-back ratio, a half-wavelength rectangular director was placed in front of the dual-dipole, as shown in Figure 1a.

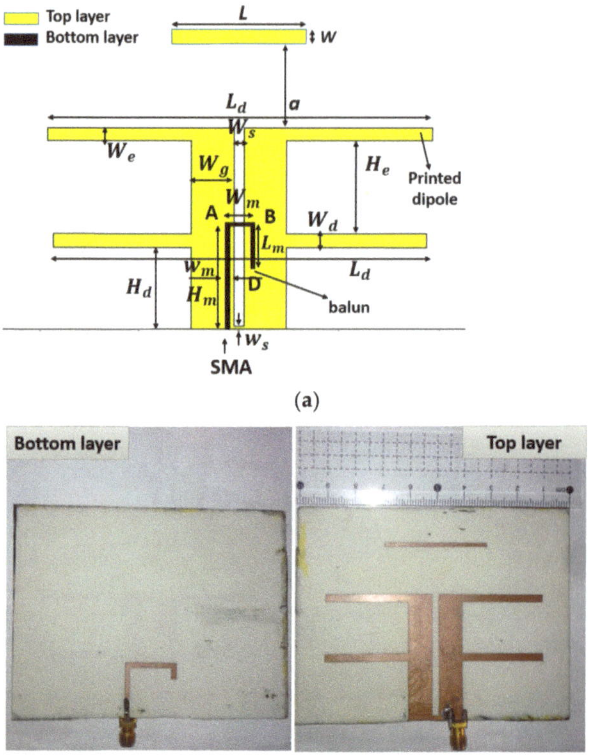

(**a**)

(**b**)

Figure 1. (**a**) The antenna element structure; (**b**) the antenna prototype.

The proposed element was designed on a 0.8 mm-thick Roger 4003C substrate with a dielectric constant of 3.55 and a loss tangent of 0.0027, as shown in Figure 1a. The manufactured prototype of the antenna with a dimension of 100 mm × 80 mm × 0.87 mm (0.74λ × 0.59λ × 0.006λ) is shown in Figure 1b. The top layer of the antenna consisted of three parts. The first part included two quarter-wave feeding lines with the length H_d and the width W_g, separated by an air gap of length $(H_d - w_s)$ and width W_s, the first haft-wave dipole of length L_d and width W_d. The second part, which was employed for expanding bandwidth, is the extending length H_e of the feed line connected to the second dipole with the length L_d and the width W_e. The third part was a director of length L and width w placed in parallel with the second printed dipole at a distance a. The bottom layer was an adjusted integrated J-shaped balun that is evaluated in [24,28]. The entire length of the J-shaped balun was $H_m + W_m + L_m$.

The first resonance at 1.85 GHz was created by optimizing the length of the two dipoles. Then, we increased the distance H_e between them, which created the second resonance as shown in Figure 2a. Then, to obtain the second resonance at 2.7 GHz, H_e was selected as 19 mm. The antenna impedance was mainly influenced by W_s, H_d, and H_m. As demonstrated in Figure 2b, the placement of the feeding point H_m had a significant effect on the input impedance from 1.8 GHz to 2.7 GHz. The impedance of the antenna element was well matched with 50 Ω of the SMA connector when H_m equals a quarter wavelength at 2.1 GHz. All of these dimensions were then optimized to obtain the desired bandwidth from 1.8 GHz to 2.6 GHz. The final antenna geometry after optimizing is shown in Table 1.

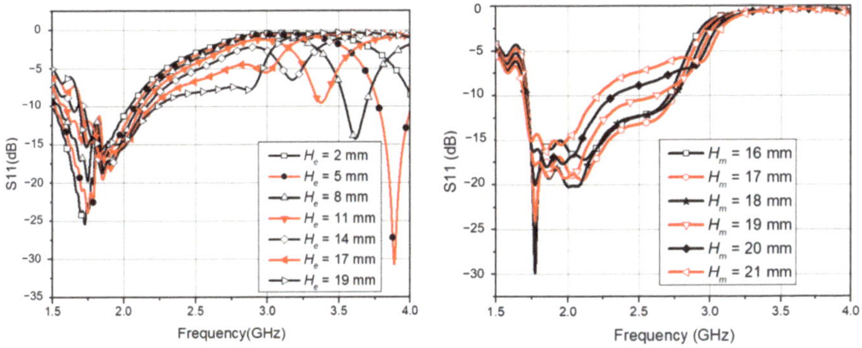

(**a**) (**b**)

Figure 2. (**a**) Effect of H_e on the reflection coefficient; (**b**) effects of H_m on the reflection coefficient.

Table 1. The dual-dipole antenna dimensions.

Para.	Val. (mm)	Para.	Val. (mm)	Para.	Val. (mm)
L_d	80	W_g	8.85	W_s	2.3
L_m	6	W_m	18	a	18
H_d	21.5	W_d	2.9	W_e	2.9
H_m	20	w_m	1.8	L	38
W_h	2	H_e	19	w	1.5

Figure 3a shows the simulated and measured reflection coefficient of the antenna element. In simulation, the frequency bands were 1.73–2.81 GHz (with a bandwidth of 1080 MHz) and in measurement, 1.575 GHz to 2.875 GHz, fully encompassing the GSM 1800/4G, UMTS 2100/3G, LTE/NR 2600/4G/5G, and Wi-Fi. Unfortunately, the reflection coefficient was measured in the environment, not in an anechoic chamber, so the 3G and 4G in the environment will cause errors in measurement and simulation. Furthermore, the solder joints between the SMA and the antenna generate losses. They cause a difference between simulated and measured values. Figure 3b presents the simulated and measured antenna gain over the operating frequency.

2.2. Hybrid Feeding Network

To construct a CP antenna array based on the proposed antenna element, a sequential phase rotated feeding network is required. The impedance matching and phase-adjusting duties were performed by the hybrid feeding network shown in this section. A sequential feeding strategy was used to achieve CP radiation in both the element orientation and phase distribution.

To obtain the required wideband phase differences between the elements, three Schiffman phase shifters and Wilkinson power dividers were used. Wilkinson power dividers help isolate the element antennas from each other, whereas the Schiffman phase shifters create a 90° phase difference between adjacent antennas and maintain this difference over a wide frequency spectrum [29]. First, ports 4 and 5 were separated from ports 2 and 3 with the first Wilkinson power divider. Then, each pair of ports was separated by another power divider. Each path of the power divider was incorporated with a half-wavelength and quarter-wavelength couple line to create the 90° phase difference. The proposed feeding network for the 2 × 2 array antenna is shown in Figure 4. The Wilkinson power dividers are designed to have an equal power division ratio in order to supply all four antennas with similar amplitude. The feeding network was designed on a RO4003C substrate of 0.8 mm thickness. As the input port, the feed point was connected to a SMA connector at the opposite side of the substrate.

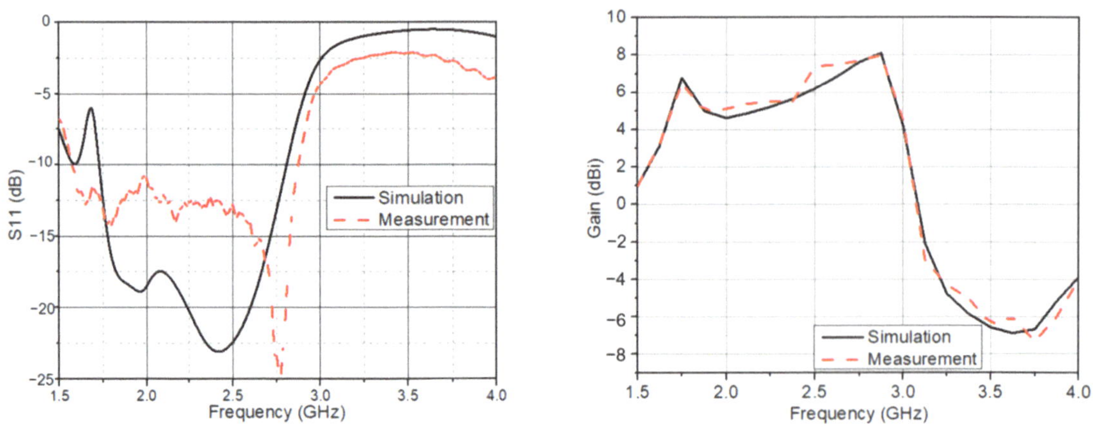

(a) (b)

Figure 3. (**a**) Simulated and measured reflection coefficient; (**b**) simulated and measured antenna gain.

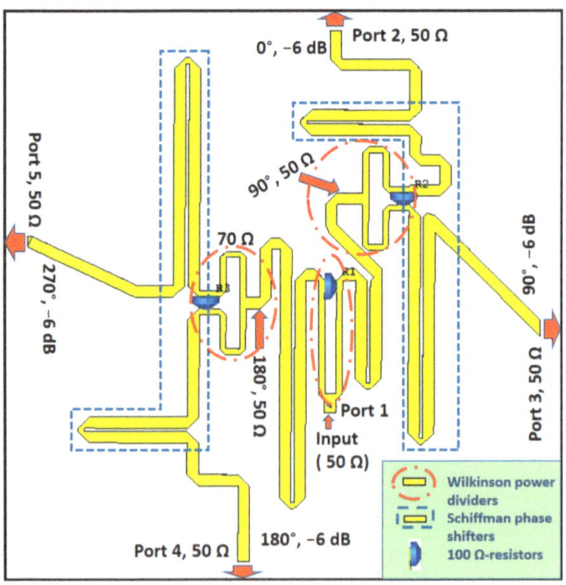

Figure 4. Schematic of hybrid feed network based on the Schiffman phase shifter.

The reflection coefficient of the feeding network is shown in Figure 5a. At 2.3 GHz, a minimum reflection coefficient of −26 dB was obtained. The excitation of four dual-dipole elements had an amplitude close to the predicted −7 dB as shown in Figure 5b. In addition, they were nearly equal to each other from 1.8 GHz to 3.0 GHz. The power was distributed evenly between the four ports by the feeding network; however, some losses in the isolation resistors and long transmission lines cause the output phases to shift. The achieved output phases for four elements are shown in Figure 5c. The phase delay between two inputs of four pairs port 2/port 3, port 3/port 4, and port 4/port 5 was roughly 90°. This phase difference was 107° of maximum at 1.8 GHz and 72° of minimum at 2.45 GHz. It interprets why the proposed antenna array has the CP at this band.

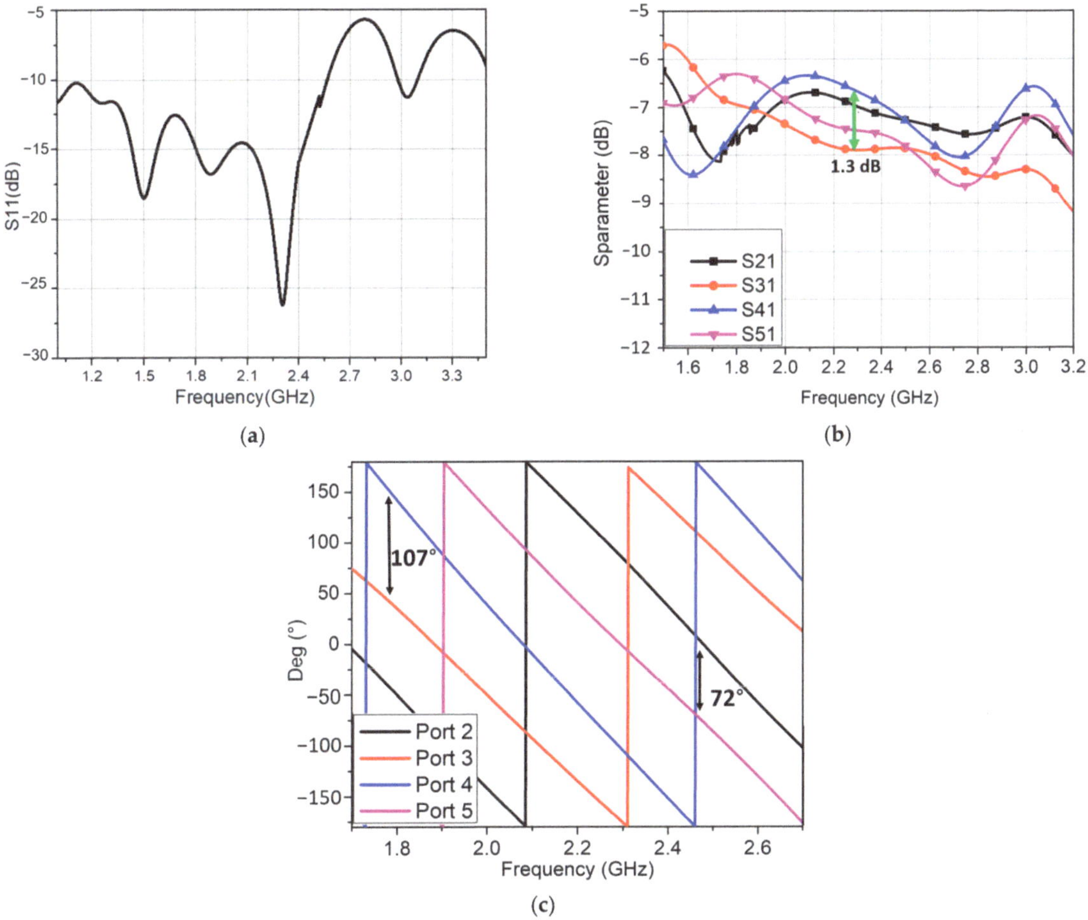

Figure 5. The feeding network performances. (**a**) Reflection coefficient; (**b**) amplitude distribution; (**c**) phase distribution.

2.3. Wideband Circularly Polarized Antenna

In this section, we attached the hybrid feeding network to four elements. This is illustrated in Figure 6a. It is worth noting that the elements were also oriented in a sequential rotation fashion. Figure 6b depicts an image of the manufactured 2 × 2 array with the sequential feeding network. The total size of the proposed antenna was 100 mm × 100 mm × 70 mm. The simulated and measured S_{11} are shown in Figure 7a. The proposed antenna had a wide bandwidth from 1.46 GHz to 2.63 GHz (57.3%) in simulation and 1.4 GHz to 2.95 GHz (71.2%) in measurement. In the 1.6–2.85 GHz frequency range, the antenna gain ranged from 6 dBi to 9.9 dBi. In addition, the peak gain was 9.9 dBi at 2.45 GHz as shown in Figure 7b. In Figure 8, with sequential rotation, the AR bandwidth of the array was smaller than 3 dB within the wide band of 1.5–2.9 GHz (63.6%).

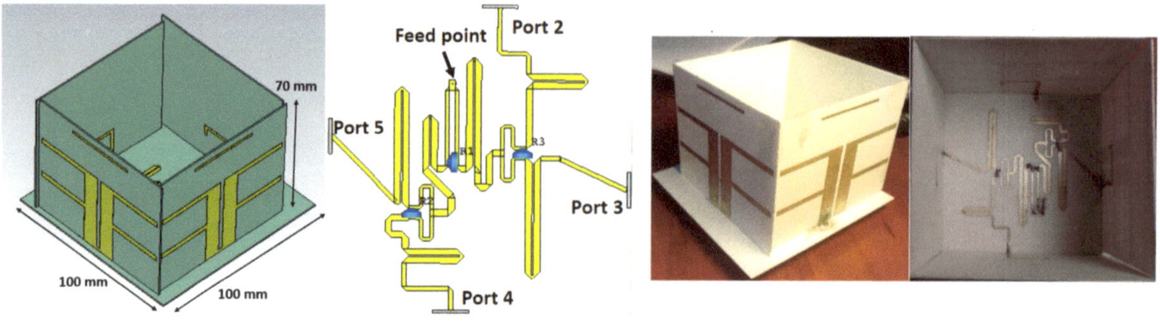

(a) (b)

Figure 6. (**a**) Proposed 3D antenna and 2D feeding network; (**b**) fabricated antenna.

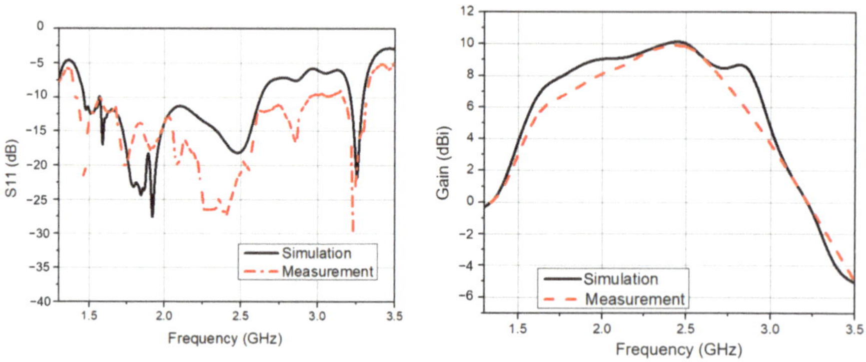

(a) (b)

Figure 7. (**a**) Simulated and measured reflection coefficient of the wideband and CP antenna; (**b**) simulated and measured gain of the wideband and CP antenna.

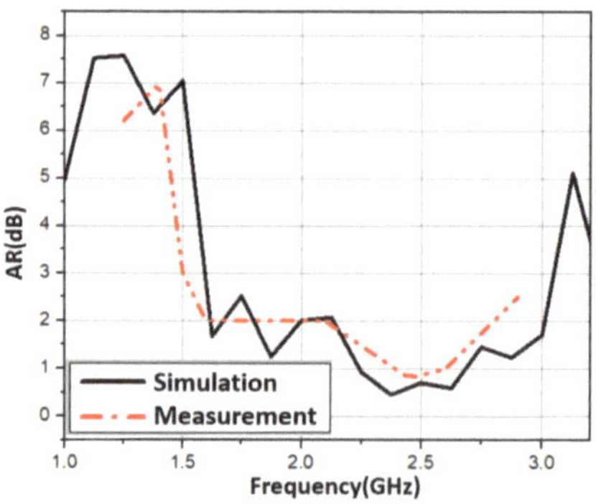

Figure 8. Simulated and measured axial ratio of the wideband and CP antenna.

As a result, by employing the sequential rotation method, the AR performance was significantly improved. The E-plane and H-plane radiation patterns of the array antenna at 1.8 GHz, 2.1 GHz, 2.45 GHz, and 2.6 GHz are shown in Figure 9. At those bands with half-power beamwidths (HPBW) of roughly 63.6°, 57.9°, 51.9°, and 47.5°, respectively, the antenna was more focused toward the director. The simulation and measurement were quite similar.

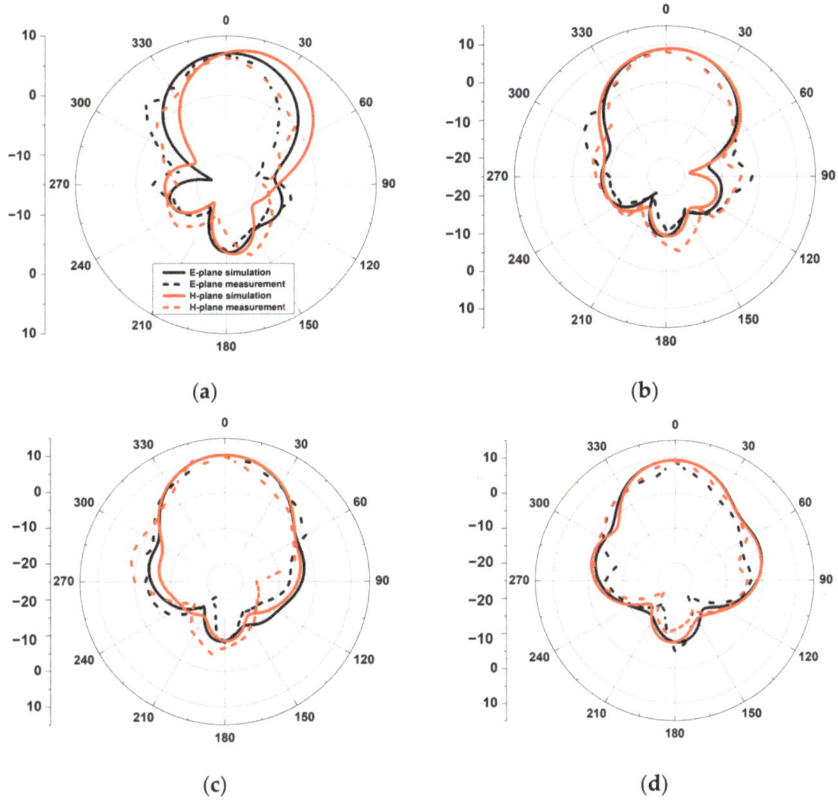

Figure 9. Measured and simulated radiation patterns of the proposed wideband and CP antenna at: (**a**) 1.8 GHz; (**b**) 2.1 GHz; (**c**) 2.45 GHz; (**d**) 2.6 GHz.

Table 2 shows a comparison of the proposed antenna and related works. Our antenna had the widest −10 dB reflection bandwidth and 3 dB AR bandwidth. In addition, it had an average size smaller than the antennas in [24–26,30,31], which had lower isolation, complicated design, and a fabricated process, but were bigger than the antenna in [23].

Table 2. Comparison with related works.

Ref.	Freq. f_0 (GHz)	BW (%)	3 dB-AR BW (%)	Dim. ($\lambda_0{}^3$)	Gain (dBi)
[23]	7	54.3	42.1	0.49 × 0.49 × 0.04	11.3
[24]	5.8	48.3	12	0.9 × 0.9 × 0.38	9.8
[25]	1.45	46	N/A	1.93 × 1.93 × 0.3	12.2
[26]	2.45	N/A	31	1.21 × 1.21 × 0.01	5.7
[30]	3.75	37.3	25.4	1.06 × 1.06 × 0.5	9.52
[31]	3.3	71	N/A	1.64 × 1.64 × 0.22	9.63
This work	**2.2**	**71.2**	**63.6**	**0.73 × 0.73 × 0.51**	**9.9**

3. Multiband Ambient RF Energy Harvester Using the Proposed Antenna

3.1. Quadplexer Design

The quadplexer consisted of four parallel channels made by parallel band-stop filters (BSFs) on top of a 0.8 mm thick Roger 4003C substrate as shown in Figure 10a. The theoretical background was based on the proposed triplexer design in [32]. Note that f_1 = 2.6 GHz, f_2 = 1.8 GHz, f_3 = 2.1 GHz, and f_4 = 2.45 GHz. Each one of the two middle channels consisted of three open stubs as BSFs corresponding to three rejected bands. The two outer channels used only two stubs each because their three reject bands were close enough to each other; two BSFs are enough to filter the other band. We bent them into an L-shape to reduce size and placed them in serial to each other. The spacings between the first and third stub were first designed to obtain the maximum transmittance within the allowed bands. To be more explicit, each distance was initially assigned as a quarter wavelength of the allowed frequency. The second stub was placed into the middle.

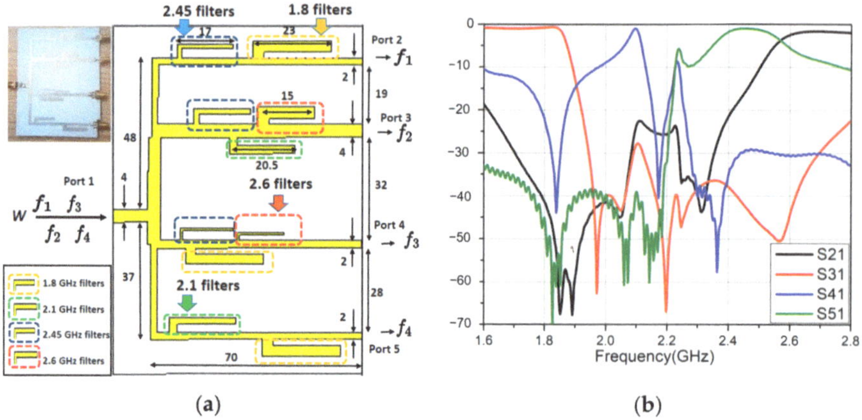

Figure 10. (**a**) The proposed quadplexer (dimension in mm); (**b**) simulated S-parameters of the quadplexer.

A 1–4 power divider connected the four channels. To make the design process easier, the power divider was constructed independently of four channels, with equivalent impedances replacing the four channels. Figure 10b shows the simulated S-parameters of the quadplexer. The efficiency at 1.8, 2.1, 2.45, and 2.6 GHz were 93%, 88%, 90%, and 75% in simulation, respectively. The high frequencies make this quadplexer suitable for multiband energy harvesting.

3.2. Rectifiers Design and Full Testing System

First, we designed four single-band rectifiers. Because of its high conversion efficiency at low power densities, the SMS7630 Schottky diode was used in the voltage doubler rectifier circuits. This diode is suitable for an input power level of around of −10 to 0 dBm. Figure 11 depicts the prototype and configuration of these rectifier circuits. The operating frequency was controlled by l_1, w_1, R_L, and R, and they were optimized in simulation. Four resistive loads of 800 Ω, 870 Ω, 1200 Ω, and 3300 Ω were used for the 1.8 GHz, 2.1 GHz, 2.45 GHz, and 2.6 GHz rectifiers, respectively. The capacitor was chosen as 470 pF. The DC filters were large radial stubs that reject the harmonics in order to improve conversion efficiency.

We proposed a quad-band rectenna working at 1.8 GHz, 2.1 GHz, 2.45 GHz, and 2.6 GHz, four common frequency of bands available in the environment, formed by the proposed wideband circularly polarized antenna, the quadplexer, and four single rectifiers. The quadplexer helps separate the four frequencies and transfer them to corresponding rectifiers. The rectifiers are single-banded, so they are easy to design and also likely to achieve good performance. The proposed quad-band rectenna is shown in Figure 12.

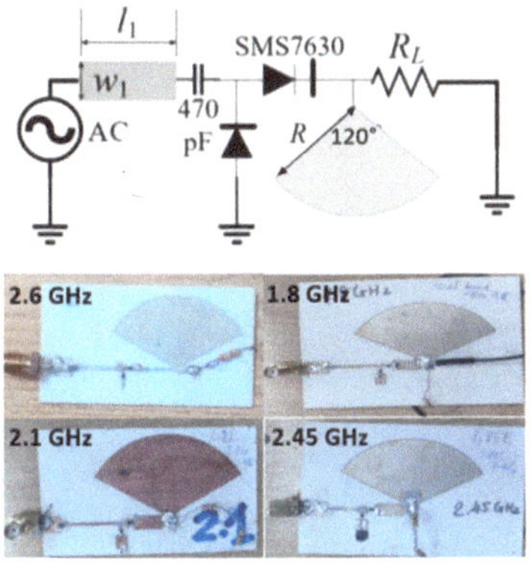

Figure 11. Schematic and fabricated rectifiers.

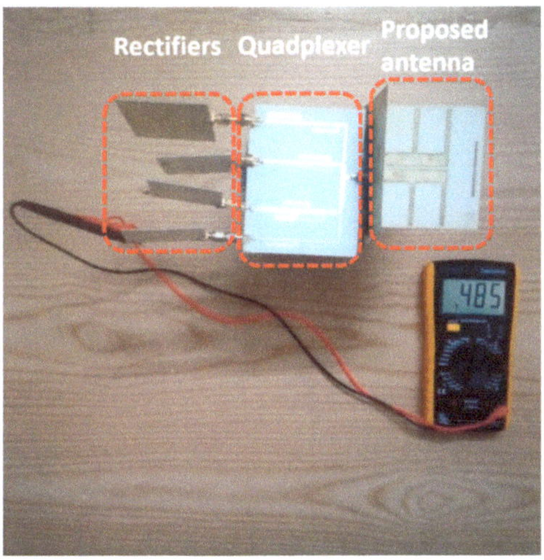

Figure 12. Quad-band rectenna system using the wideband and CP antenna with a maximum output voltage at 2.6 GHz.

We investigated the advantages of employing the 2×2 circular polarized array for RF energy harvesting in self-power devices. As shown in Figure 13a, the highest simulated AC–DC efficiency of the 1.8 GHz rectifier is 65% at −1 dBm input power, for the 2.1 GHz it is 64%, the 2.45 GHz is 61%, and the 2.6 GHz is 61%, respectively. The maximum measured AC–DC efficiency was 60%, 55%, 55%, and 54% at −1 dBm, 0 dBm, −1 dBm, and 0.5 dBm input power, respectively. The highest reported RF–DC output voltage was 1568 mV at 2 dBm, occurring with the 2.6 GHz rectifier. After connecting the rectifiers to the quadplexer, forming a quad-band rectifier, the AC –DC efficiency slightly decreased to 49%, 43.5%, 41%, and 38% for the 1.8, 2.1, 2.45, and 2.6 GHz bands at −0.5 dBm input power, as shown in Figure 13b.

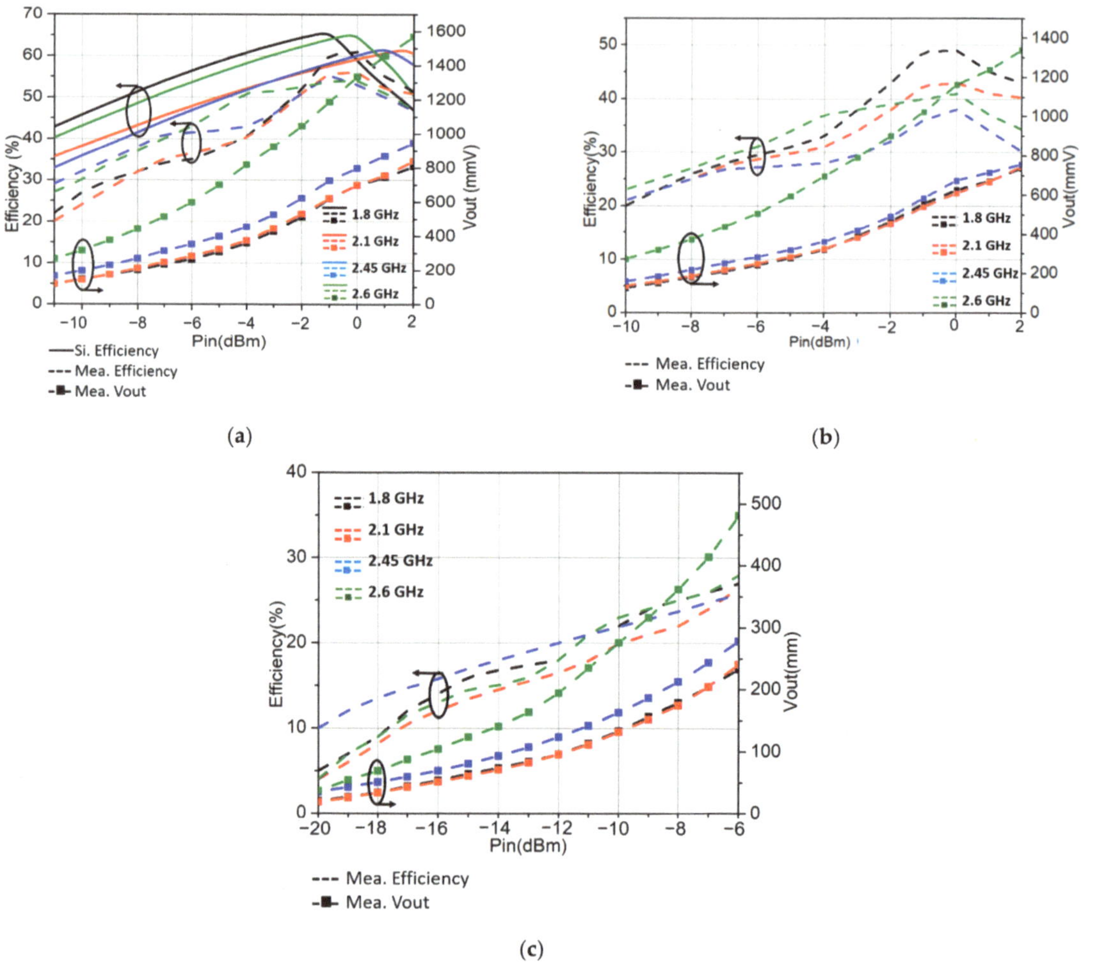

Figure 13. (**a**) Simulated and measured AC–DC efficiency of the rectifiers; (**b**) measured AC–DC efficiency of the quadplexer and rectifiers; (**c**) measured RF–DC efficiency.

In the full system testing, the proposed wideband and CP antenna were connected to the quadplexer and four rectifiers to form the multiband rectenna prototype via SMA connectors, as shown in Figure 12. A signal generator excited monotone incident waves via a reference antenna. The RF–DC efficiency was smaller than the AC–DC due to the loss on the connectors and quadplexer. The maximum RF–DC efficiency was 27%, 26%, 25.5%, and 27.5% at −6 dBm, respectively, as in Figure 13c. The maximum measured voltage was 484 mmV at 2.6 GHz, at nan input power level of −6 dBm. This result shows its potential for a self-powered wireless sensor when connecting with a power management circuit. The quad-band rectenna was also tested in the ambient environment, 70 m from the base station. The rectenna collected and harvested the 3G and 4G waves transmitted from the base station, along with the Wi-Fi signal in the surrounding houses into DC votltage. The maximum recorded voltage was 194.5 mV, corresponding to 47 µW power, as shown in Figure 14. When the rectenna is put closer to the base station, the amount of power gathered should be substantially higher. It should be noted that for practical application, the locations of the power sources are unknown. For locations very close to power sources, such as base stations, the harvested power can be high enough to support a wireless sensor, which consumes several milliwatts in full active mode. For our experiment, the distance from the rectenna to the power source was quite far away; therefore, the harvested power was not enough to supply any sensor. However, it was enough for low-power wireless sensors in sleep modes, which only need several microwatts [33,34].

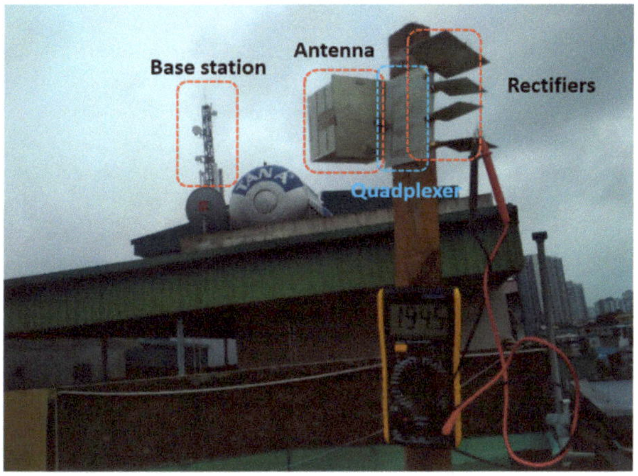

Figure 14. Experimentation of the proposed multiband rectenna.

Table 3 shows the maximum efficiencies over the power levels of this work and previous works. The proposed rectenna can harvest energy from the four bands of 3G/4G/5G/Wi-Fi with the conversion efficiency from 25.5% to 27.5% at −6 dBm. The conversion efficiency at −10 dBm of this work is higher than the triple-band rectenna in [35] and the quad-band rectenna in [11], thanks to the CP antenna. Although the efficiency can reach 52% at 3.5 dBm in [35], the conversion efficiency drops to 20% at −10 dBm, and requires a load of up to 14 kΩ. Although the linearly polarized antennas in [11,35] are easy to realize, they can only harvest in one polarization. The RF–DC conversion efficiency in [8] is higher than in our work because the rectifier was designed on a high-cost substrate RT/Duroid 5880. It also requires a very high load of 5 kΩ, leading to a very small output current. The RF–DC efficiency in [9] is the highest with an input power of 20 dBm, and it is more suitable for wireless power transfer as the incident RF power in the environment is usually low (less than −5 dBm). At −10 dBm, the efficiency is less than 5%. Although the dual-polarized rectennas in [8,9] harvest from many polarizations, one multiband rectifier is required for each polarization, leading to bulky systems. Our work already has the advantages of being suitable for energy harvesting systems for self-powered wireless sensors with low load and higher conversion efficiency at low power, less than −10 dBm.

Table 3. Comparison with previous works.

Ref.	Input Power	Freq. Band (GHz)	Antenna polarization	Max.Eff	Eff. (−10 dBm)	Min. Load
[8]	−35~−10 dBm	0.91, 1.85, 2.1	Dual-porlarized	40%, 33%, 25%	40%	5 KΩ
[9]	−10~30 dBm	4.75, 5.42, 5.76, 6.4, 6.9, 7.61	Dual-porlarized	84%	4%	3 KΩ
[11]	−25~−5 dBm	0.84, 1.86, 2.1, 2.45	Linear	30%, 22%, 33%, 16.5%	22%	N/A
[35]	−10~10 dBm	0.866, 1.841, 1.957	Linear	52%, 27%, 29%	20%	14 KΩ
This work	−20~−6 dBm	1.8, 2.1, 2.45, 2.6	CP	27%, 26%, 25.5%, 27.5%	23%	800 Ω

4. Conclusions

In this paper, a high gain wideband circularly polarized antenna using the sequential rotation feeding technique was presented. The dual-dipole element was investigated by cascading two dipoles, showing a bandwidth of 58.4% from 1.575–2.875 GHz. The proposed feeding network was created by using three Wilkinson power dividers and three Schiffman phase shifter units to maintain stable circular polarization over a wide bandwidth. The wide bandwidth element and the hybrid feeding network with a good performance helped the proposed antenna achieve a 71.2% bandwidth ($S_{11} \leq -10$ dB), 63.6% 3 dB AR bandwidth, and a 9.9 dBi peak gain. These results show that the proposed antenna is a good candidate for multiband energy harvesting systems. The antenna is applied as the receiving antenna in a multiband ambient 3G/4G/5G/Wi-Fi energy harvester, the measured RF–DC efficiency is 27% at 1.8 GHz, 26% at 2.1 GHz, 25.5% at 2.45 GHz, and 27.5% at 2.6 GHz under the input power of -6 dBm.

Author Contributions: Data curation, H.Q.N.; Investigation, M.T.L.; Methodology, M.T.L.; Supervision, M.T.L.; Validation, M.T.L.; Visualization, H.Q.N. and M.T.L.; Writing—original draft, H.Q.N.; Writing—review & editing, M.T.L. All authors have read and agreed to the published version of the manuscript.

Funding: This research is funded by Hanoi University of Science and Technology under projectnumber T2021-SAHEP-007.

Institutional Review Board Statement: Not applicable.

Informed Consent Statement: Not applicable.

Data Availability Statement: All data that support the findings of this study are included within the article.

Conflicts of Interest: The authors declare no conflict of interest.

References

1. Jian, L.; Okada, H.; Itoh, T.; Harada, T.; Maeda, R. Toward the World Smallest Wireless Sensor Nodes with Ultralow Power Consumption. *IEEE Sens. J.* **2014**, *14*, 2035–2041. [CrossRef]
2. Sun, H.; Guo, Y.; He, M.; Zhong, Z. A Dual-Band Rectenna Using Broadband Yagi Antenna Array for Ambient RF Power Harvesting. *IEEE Antennas Wirel. Propag. Lett.* **2013**, *12*, 918–921. [CrossRef]
3. Piñuela, M.; Mitcheson, P.D.; Lucyszyn, S. Ambient RF Energy Harvesting in Urban and Semi-Urban Environments. *IEEE Trans. Microw. Theory Tech.* **2013**, *61*, 2715–2726. [CrossRef]
4. Shi, Y.; Jing, J.; Fan, Y.; Yang, L.; Wang, M. Design of a novel compact and efficient rectenna for WIFI energy harvesting. *Prog. Electromagn. Res. C* **2018**, *83*, 57–70. [CrossRef]
5. Chuma, E.; Iano, Y.; Bravo-Roger, L.; De la Torre Rodríguez, L.; Sanchez-Soriano, M. Compact Rectenna Based on a Fractal Geometry with a High Conversion Energy Efficiency per Area. *IET Microw. Antennas Propag.* **2017**, *12*, 173–178. [CrossRef]
6. Chang, M.-C.; Weng, W.-C.; Chen, W.-H.; Li, T.-Y. A Wideband Planar Rectenna for WLAN Wireless Power Transmission. In Proceedings of the 2017 IEEE Wireless Power Transfer Conference (WPTC), Taipei, Taiwan, 10–12 May 2017; pp. 1–3.
7. Dardeer, O.M.A.; Elsadek, H.A.; Abdallah, E.A. Compact Broadband Rectenna for Harvesting RF Energy in WLAN and WiMAX Applications. In Proceedings of the 2019 International Conference on Innovative Trends in Computer Engineering (ITCE), Aswan, Egypt, 2–4 February 2019; pp. 292–296.
8. Shen, S.; Chiu, C.-Y.; Murch, R.D. A Dual-Port Triple-Band L-Probe Microstrip Patch Rectenna for Ambient RF Energy Harvesting. *IEEE Antennas Wirel. Propag. Lett.* **2017**, *16*, 3071–3074. [CrossRef]
9. Singh, N.; Kanaujia, B.; Beg, M.; Siddique, M.; Kumar, S.; Khan, T. A Dual Polarized Multiband Rectenna for RF Energy Harvesting. *AEU-Int. J. Electron. Commun.* **2018**, *93*, 123–131. [CrossRef]
10. Zahra, W.; Djerafi, T. Ambient RF Energy Harvesting for Dual-Band Frequencies below 6 GHz. In Proceedings of the 2018 IEEE Wireless Power Transfer Conference (WPTC), Montreal, QC, Canada, 3–7 June 2018; pp. 1–2.
11. Vu, H.S.; Nguyen, N.; Ha-Van, N.; Seo, C.; Thuy Le, M. Multiband Ambient RF Energy Harvesting for Autonomous IoT Devices. *IEEE Microw. Wirel. Compon. Lett.* **2020**, *30*, 1189–1192. [CrossRef]
12. Chen, Q.; Zhao, P.; Chen, S.; Wang, G.; Chen, X. A Dual-Frequency Circularly Polarized Rectenna for 2.45 and 5.8 GHz Wireless Power Transmission. In Proceedings of the 2018 IEEE Asia-Pacific Conference on Antennas and Propagation (APCAP), Auckland, New Zealand, 5–8 August 2018. [CrossRef]
13. Mansour, M.M.; Kanaya, H. Novel L-Slot Matching Circuit Integrated with Circularly Polarized Rectenna for Wireless Energy Harvesting. *Electronics* **2019**, *8*, 651. [CrossRef]

14. Shen, S.; Zhang, Y.; Chiu, C.-Y.; Murch, R. An Ambient RF Energy Harvesting System Where the Number of Antenna Ports Is Dependent on Frequency. *IEEE Trans. Microw. Theory Tech.* **2019**, *67*, 3821–3832. [CrossRef]
15. Zhang, L.; Zhang, W.; Chen, X. Design of a Compact Conformal Circularly Polarized Rectenna. In Proceedings of the 2019 International Workshop on Electromagnetics: Applications and Student Innovation Competition (iWEM), Qingdao, China, 18–20 September 2019; pp. 1–3.
16. Wang, X.-C.; Sun, L.; Lu, X.-L.; Liang, S.; Lu, W.-Z. Single-Feed Dual-Band Circularly Polarized Dielectric Resonator Antenna for CNSS Applications. *IEEE Trans. Antennas Propag.* **2017**, *65*, 4283–4287. [CrossRef]
17. Zhou, C.F.; Cheung, S.W. A Wideband CP Crossed Slot Antenna Using 1- $\lambda $ Resonant Mode With Single Feeding. *IEEE Trans. Antennas Propag.* **2017**, *65*, 4268–4273. [CrossRef]
18. Lee, C.S.; Fan, Y.; Ezzat, M. Single-Feed Circularly Polarized Microstrip Antenna With Bethe Holes on the Radiating Patch. *IEEE Trans. Antennas Propag.* **2020**, *68*, 4935–4938. [CrossRef]
19. Saini, R.K.; Dwari, S.; Mandal, M.K. CPW-Fed Dual-Band Dual-Sense Circularly Polarized Monopole Antenna. *IEEE Antennas Wirel. Propag. Lett.* **2017**, *16*, 2497–2500. [CrossRef]
20. Luo, Y.; Chu, Q.-X.; Bornemann, J. Enhancing Cross-Polarization Discrimination or Axial Ratio Beamwidth of Diagonally Dual or Circularly Polarized Base Station Antennas by Using Vertical Parasitic Elements. *IET Microw. Antennas Propag.* **2017**, *11*, 1190–1196. [CrossRef]
21. Sun, H.-H.; Zhu, H.; Ding, C.; Guo, Y.J. Wideband Planarized Dual-Linearly-Polarized Dipole Antenna and Its Integration for Dual-Circularly-Polarized Radiation. *IEEE Antennas Wirel. Propag. Lett.* **2018**, *17*, 2289–2293. [CrossRef]
22. Wen, L.; Gao, S.; Luo, Q.; Hu, W.; Sanz-Izquierdo, B. Design of a Wideband Dual-Feed Circularly Polarized Antenna for Different Axial Ratio Requirements. *IEEE Antennas Wirel. Propag. Lett.* **2021**, *20*, 88–92. [CrossRef]
23. Mohammadi-Asl, S.; Nourinia, J.; Ghobadi, C.; Majidzadeh, M. Wideband Compact Circularly Polarized Sequentially Rotated Array Antenna with Sequential-Phase Feed Network. *IEEE Antennas Wirel. Propag. Lett.* **2017**, *16*, 3176–3179. [CrossRef]
24. Bui, T.D.; Nguyen, Q.C.; Le, M.T. Novel Wideband Circularly Polarized Antenna for Wireless Applications. In Proceedings of the 2017 IEEE Asia Pacific Microwave Conference (APMC), Kuala Lumpur, Malaysia, 13–16 November 2017; pp. 430–433.
25. Zhang, Z.-Y.; Zuo, S.; Zhao, Y.; Ji, L.-Y.; Fu, G. Broadband Circularly Polarized Bowtie Antenna Array Using Sequentially Rotated Technique. *IEEE Access* **2018**, *6*, 12769–12774. [CrossRef]
26. Dardeer, O.M.A.; Elsadek, H.A.; Abdallah, E.A. 2 $\times$ 2 Circularly Polarized Antenna Array for RF Energy Harvesting in IoT System. In Proceedings of the 2018 IEEE Global Conference on Internet of Things (GCIoT), Alexandria, Egypt, 5–7 December 2018; pp. 1–6.
27. Cao, T.M.; Phuong, T.H.T.; Bui, T.D. Circularly Polarized Antenna Array Based on Hybrid Couplers for 5G Devices. *Bull. Electr. Eng. Inform.* **2021**, *10*, 1446–1454. [CrossRef]
28. Li, R.; Wu, T.; Pan, B.; Lim, K.; Laskar, J.; Tentzeris, M.M. Equivalent-Circuit Analysis of a Broadband Printed Dipole with Adjusted Integrated Balun and an Array for Base Station Applications. *IEEE Trans. Antennas Propag.* **2009**, *57*, 2180–2184. [CrossRef]
29. Quirarte, J.L.R.; Starski, J.P. Novel Schiffman Phase Shifters. *IEEE Trans. Microw. Theory Tech.* **1993**, *41*, 9–14. [CrossRef]
30. Cao, T.M.; Vu, H.S.; Bui, T.D.; Le, M.T. Left Hand and Right Hand Circularly Polarized Antenna for 5G Devices. In *Industrial Networks and Intelligent Systems*; Vo, N.-S., Hoang, V.-P., Vien, Q.-T., Eds.; Springer International Publishing: Cham, Switzerland, 2021; pp. 119–127.
31. Kim, S.; Kim, J.-I. A Circularly Polarized High-Gain Planar 2 $\times$ 2 Dipole-Array Antenna Fed by a 4-Way Gysel Power Divider for WLAN Applications. *IEEE Antennas Wirel. Propag. Lett.* **2019**, *18*, 1051–1055. [CrossRef]
32. Dinh, M.Q.; Le, M.T. Triplexer-Based Multiband Rectenna for RF Energy Harvesting from 3G/4G and Wi-Fi. *IEEE Microw. Wirel. Compon. Lett.* **2021**, *31*, 1094–1097. [CrossRef]
33. Nguyen, N.; Nguyen, Q.C.; Le, M.T. A Novel Autonomous Wireless Sensor Node for IoT Applications. *TELKOMNIKA Telecommun. Comput. Electron. Control* **2019**, *17*, 2389–2399. [CrossRef]
34. Nguyen, T.H.; Tran, M.N.; Le, Q.H.; Vu, T.A.; Nguyen, Q.C.; Nguyen, D.D.; Le, M.T. Smart Shoe Based on Battery-Free Bluetooth Low Energy Sensor. In *Industrial Networks and Intelligent Systems*; Vo, N.-S., Hoang, V.-P., Vien, Q.-T., Eds.; Springer International Publishing: Cham, Switzerland, 2021; pp. 156–166.
35. Boursianis, A.D.; Papadopoulou, M.S.; Koulouridis, S.; Rocca, P.; Georgiadis, A.; Tentzeris, M.M.; Goudos, S.K. Triple-Band Single-Layer Rectenna for Outdoor RF Energy Harvesting Applications. *Sensors* **2021**, *21*, 3460. [CrossRef] [PubMed]

sensors

MDPI

Article

Metamaterial-Integrated High-Gain Rectenna for RF Sensing and Energy Harvesting Applications †

Woosol Lee, Suk-il Choi, Hae-in Kim, Sunghyun Hwang, Saeyoung Jeon and Yong-Kyu Yoon *

Department of Electrical and Computer Engineering, University of Florida, Gainesville, FL 32611, USA; leewoosol@ufl.edu (W.L.); sukilchoi@ufl.edu (S.-i.C.); kimhaein@ufl.edu (H.-i.K.); sunghyun.hwang@ufl.edu (S.H.); s.jeon@ufl.edu (S.J.)

* Correspondence: ykyoon@ece.ufl.edu

† This paper is an extended version of our abstract presented in a conference: Lee, W.; Kim, H.; Hwang, S.; Jeon, S.; Cho, H.; Yoon, Y.-K. 3D integrated high-gain rectenna in package with metamaterial superstrates for high efficiency wireless power transfer applications. In Proceedings of the 2021 IEEE 71st Electronic Components and Technology Conference (ECTC), Virtual Meeting, 1 June–4 July 2021; pp. 1317–1322.

Abstract: This paper presents a metamaterial (MTM)-integrated high-gain rectenna for RF sensing and energy harvesting applications that operates at 2.45 GHz, an industry, science, medicine (ISM) band. The novel MTM superstrate approach with a three-layered integration method is firstly introduced for rectenna applications. The integrated rectenna consists of three layers, where the first layer is an MTM superstrate consisting of four-by-four MTM unit cell arrays, the second layer a patch antenna, and the third layer a rectifier circuit. By integrating the MTM superstrate on top of the patch antenna, the gain of the antenna is enhanced, owing to its beam focusing capability of the MTM superstrate. This induces the increase of the captured RF power at the rectifier input, resulting in high-output DC power and high entire end-to-end efficiency. A parametric analysis is performed in order to optimize the near-zero property of the MTM unit cell. In addition, the effects of the number of MTM unit cells on the performance of the integrated rectenna are studied. A prototype MTM-integrated rectenna, which is designed on an RO5880 substrate, is fabricated and characterized. The measured gain of the MTM-integrated rectenna is 11.87 dB. It shows a gain improvement of 6.12 dB compared to a counterpart patch antenna without an MTM superstrate and a maximum RF–DC conversion efficiency of 78.9% at an input RF power of 9 dBm. This results in the improvement of the RF–DC efficiency from 39.2% to 78.9% and the increase of the output DC power from 0.7 mW to 6.27 mW (a factor of 8.96 improvements). The demonstrated MTM-integrated rectenna has shown outstanding performance compared to other previously reported work. We emphasize that the demonstrated MTM-integrated rectenna has a low design complexity compared with other work, as the MTM superstrate layer is integrated on top of the simple patch antenna and rectifier circuit. In addition, the number of MTM units can be determined depending on applications. It is highly envisioned that the demonstrated MTM-integrated rectenna will provide new possibilities for practical energy harvesting applications with improved antenna gain and efficiency in various IoT environments.

Keywords: metamaterial; rectenna; RF energy harvesting; high-gain antenna; RF–DC conversion efficiency; rectifier

Citation: Lee, W.; Choi, S.-i.; Kim, H.-i.; Hwang, S.; Jeon, S.; Yoon, Y.-K. Metamaterial-Integrated High-Gain Rectenna for RF Sensing and Energy Harvesting Applications. *Sensors* **2021**, *21*, 6580. https://doi.org/10.3390/s21196580

Academic Editors: Slim Naifar, Olfa Kanoun and Carlo Trigona

Received: 31 August 2021
Accepted: 27 September 2021
Published: 1 October 2021

1. Introduction

With the development of wireless technologies, the ambient wireless power density is growing owing to the increasing number of electromagnetic (EM) power sources. Harvesting this radio frequency energy has gained a lot of attention in recent years. A rectenna (rectifier + antenna), which is the key component of the wireless RF energy harvesting, continues to be a highly expected solution to the power challenges related to ubiquitous sensor connectivity as envisioned in the Internet of Things (IoT). Rectennas with a high-gain antenna and a high-efficiency rectifier are highly required.

Recently, various rectenna designs have been proposed in the literature for RF energy harvesting, such as monopole rectennas [1,2], dipoles [3–5], circularly polarized (CP) antennas [6–8], dual-band rectennas [9,10], and a broadband rectenna [11]. In order to solve the limited power challenges, researchers have studied various techniques for improving the gain of the rectenna, such as reflecting surface [4,12], differential microstrip [13], differential feeding [14], antenna array [15,16], SIW cavity [17], air gap [9], and aperture coupling [9,18]. In [12], a 5.8 GHz high-gain dual-rhombic-loop rectenna was investigated to achieve an antenna gain of 10.7 dBi. By having a dual-rhombic-loop antenna structure with terminated gaps, left-hand circular polarization was achieved. The maximum RF–DC conversion efficiency was 80% at an input power of 20 dBm. This dual-rhombic-loop structure had an advantage of high gain and broadband properties, but the improved gain was still limited, and an integrated rectifier made the entire system large. In [13], a differential microstrip antenna with a gain of 8.5 dBi was introduced. The differential antenna was utilized in either center-grounded or differential configuration. The rectenna was designed to operate at the GSM900 band (890–960 MHz), with a peak RF–DC conversion efficiency of 65% at an input power of 2.19 dBm. However, a two-layered differential patch antenna with the additional ground layer made the design complexity very high. In [16], a monopole wideband fractal antenna array fed by a microstrip-to-slot-line transition feeding network was designed. As the feeding line was based on slot-line stubs and radial microstrip, the broadband microstrip-to-slot-line transitions were realized. An antenna gain of 8.5 dBi and RF–DC conversion efficiency of 76% at 5 dBm were obtained. This fractal antenna array had an advantage of broadband property (0.91 to 2.55 GHz), but the design process of the fractal antenna array increased the design complexity. In addition, the rectifier was connected to the antenna in a perpendicular direction, thereby increasing the overall dimension of the rectenna and limiting its practical applicability. In [17], a 24 GHz substrate integrated waveguide (SIW) cavity-backed antenna array was demonstrated for millimeter-Wave energy harvesting. By having the SIW cavity in the antenna, the gain was increased from 10.3 dBi to 12.6 dBi. The maximum RF–DC conversion efficiency was 42% at 12.55 dBm. Even though the gain of the demonstrated antenna was relatively high, the measured RF–DC conversion efficiency was too low, and the overall normalized device was comparably large (0.776 λ^3). Moreover, the SIW cavity made the fabrication and design process more difficult. In [9], a microstrip patch antenna with a right-angle triangular aperture slot was exhibited. By having an air gap between the patch antenna and feed line, the improved antenna gain was realized. An antenna gain of 7.82 dBi and an RF–DC conversion efficiency of 32.5% at 10 dBm were obtained, but the rectifier circuit was not integrated to the antenna, thereby increasing overall system size, as the rectifier was connected to the same layer of antenna. Overall, the improved antenna gains of the previously reported work were still limited, and the methods to enhance the gain highly increased the device complexity. Therefore, a fully integrated rectenna with high gain, low design complexity, and high RF–DC conversion efficiency is needed.

In the meantime, some researchers have introduced metamaterials (MTMs) for the antenna gain improvement as a form of MTM superstrates [19–22], where MTM superstrates help increase the antenna gain by utilizing the near-zero refraction characteristic of the MTM. MTMs are artificially designed materials that have uncommon EM properties, such as evanescent wave amplification, negative refraction, and near-zero refraction, thereby improving the antenna gain [23,24]. Figure 1 shows the principle of a gain improvement property of an MTM superstrate. The MTM superstrate focuses and amplifies the EM field at the boundary, owing to its near-zero refractive index, improving the gain of the antenna. By simply employing the MTM superstrate layer on top of the antenna, the gain improvement of the patch antenna can be achieved.

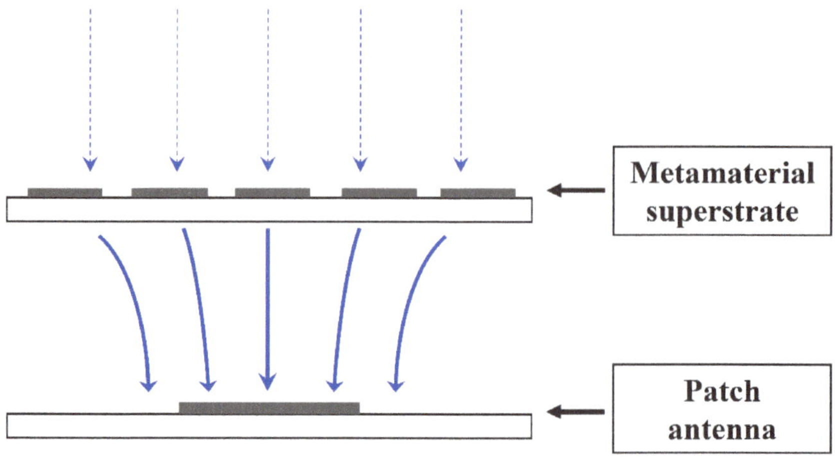

Figure 1. Principle of a gain improvement property of the MTM superstrate.

To the best of our knowledge, these novel properties of the MTM superstrate have not been fully investigated in the field of RF energy harvesting. In this work, the novel MTM superstrate approach with a three-layered integration method is firstly introduced for rectenna applications. The MTM-integrated high-gain rectenna, which consists of a first layer of MTM superstrate, second layer of a patch antenna, and third layer of rectifier circuit, is demonstrated. The integrated MTM superstrate enhances the gain of the antenna, inducing the increase of the captured RF power at the rectifier input. It will lead to high-output DC power and high entire efficiency. In addition, the rectifier circuit is incorporated with the backside of the patch, which is isolated from the patch antenna ground plane. While the cross-coupling between the rectifier and the MTM superstrate is suppressed, the rectenna system footprint remains the same. The demonstrated MTM-integrated rectenna solves the power challenges of the rectenna systems owing to its high-gain property, and it also reduces the design complexity by simply stacking the MTM superstrate layer on top of the patch antenna. In order to verify the effectiveness of the MTM-integrated rectenna, a prototype MTM-integrated rectenna is designed, fabricated, and characterized. The optimized near-zero property of the MTM unit cell is obtained using a parametrical analysis. In addition, the effects of the number of MTM unit cells on the performance of the integrated rectenna are studied. The conventional patch antenna is compared with the patch antenna integrated with the MTM superstrate to verify the focusing property of the MTM superstrate. Finally, the rectenna properties of the MTM-integrated rectenna are characterized and compared with those of the previously reported studies.

This paper is an extended work of the preliminary study [25]. Compared with [25], more results and discussions are reported, including the following: (1) the MTM-integrated rectenna is newly designed and fabricated on an RO5880 substrate to improve its rectenna performance, (2) parametric analysis is performed on the design of the MTM superstrate, (3) the studies on the effects of the number of MTM unit cells on the performance of the integrated rectenna is carried out, and (4) the detailed comparison of the MTM-integrated rectenna with previously reported rectenna systems is provided.

This paper is organized as follows. Section 2 discusses the design and analysis of the MTM integrated high-gain rectenna, and Section 3 presents the fabrication of the MTM-integrated rectenna. The simulated and measured results are presented in Section 4. Section 5 is the conclusion of this work.

2. Design and Analysis of the MTM Integrated High-Gain Rectenna

The schematic of the MTM integrated high-gain rectenna is presented in Figure 2. As shown in Figure 2a, the MTM-integrated rectenna consists of 1st layer of the MTM superstrate, 2nd layer of the patch antenna, and 3rd layer of the rectifier circuit. The MTM superstrate is stacked on the patch antenna with the separation s. The rectifier circuit is connected to the coaxial feeding point of the patch antenna through the interconnect.

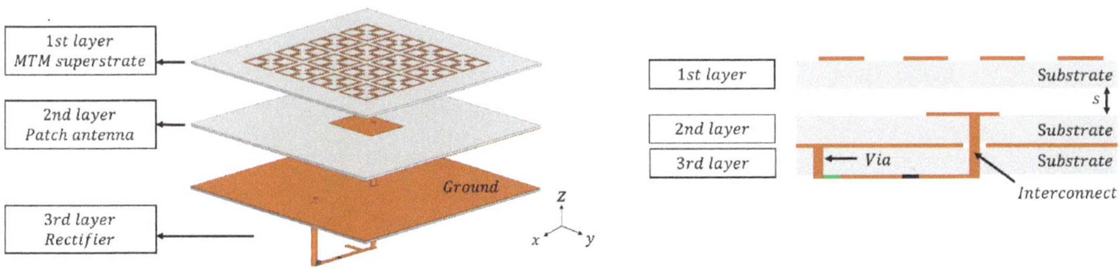

(**a**) (**b**)

Figure 2. Schematic of a 3D integrated rectenna: (**a**) perspective view, (**b**) side view, where $s = 10$ mm.

2.1. Design of the MTM Unit Cell

In this subsection, the MTM unit cell is designed to have the gain improvement property. In this work, a 4-leaf clover shaped MTM unit cell [21] is carefully designed to have a near-zero refractive index for 2.45 GHz, a popular industry, science, and medicine (ISM) band, where multiple applications such as Wi-Fi, Bluetooth, and ZigBee are used and its RF waves are abundant around us, and the unit cell is designed on an RO5880 substrate (ε = 2.2, δ = 0.0009) instead of FR4 (ε = 4.4, δ = 0.03) in order to minimize the substrate loss. To design the MTM unit cell that has the near-zero refractive index at 2.45 GHz, a parametric analysis is performed using High Frequency Structure Simulator (HFSS, Ansys Inc. Canonsburg, PA, USA). In the parametric analysis, the effective refractive index can be obtained from the simulation results by using the standard retrieval methods [26–28].

$$z = \pm\sqrt{\frac{(1+S_{11})^2 - S_{21}{}^2}{(1-S_{11})^2 - S_{21}{}^2}} \tag{1}$$

$$e^{ink_0 d} = \frac{S_{21}}{1 - S_{11}\frac{z-1}{z+1}} \tag{2}$$

$$n_{eff} = \frac{1}{k_0 d}\left\{\left[[\ln\left(e^{ink_0 d}\right)]'' + 2m\pi\right] - i[\ln\left(e^{ink_0 d}\right)]'\right\} \tag{3}$$

where S_{11} and S_{21} are the reflection and transmission coefficients; $(\cdot)'$ and $(\cdot)''$ denote the real part and imaginary part of the complex numbers, respectively; n_{eff} is the effective refractive index; k_0 is the wavenumber; d is the maximum thickness of the slab; z is the impedance; and m is the integer related to the branch index of n'.

Figure 3a shows the schematic of the 4-leaf clover shaped MTM unit cell, and Figure 3b shows the simulated effective refraction index of the MTM unit cell with the varied L values while keeping other variables same for simplicity. In this analysis, the near-zero range is defined as the range where the real values of the effective refractive index are equal to or less than 0.5 ($\eta_{eff}(re) \leq 0.5$) [25]. As shown in Figure 3b, the value of L is varied from 27 to 30 mm. When L increases, the near-zero range shifts to low frequencies, as depicted in Figure 3b. The L value is determined to be 28 mm, since the near-zero range of the MTM unit cell should be designed to include 2.45 GHz. As a result, the optimized near-zero range is shown to be from 2.4 to 2.54 GHz, which means the designed MTM changes the direction of the EM field by the boundary conditions to near-zero, thereby improving the gain of the antenna at the frequencies within this range.

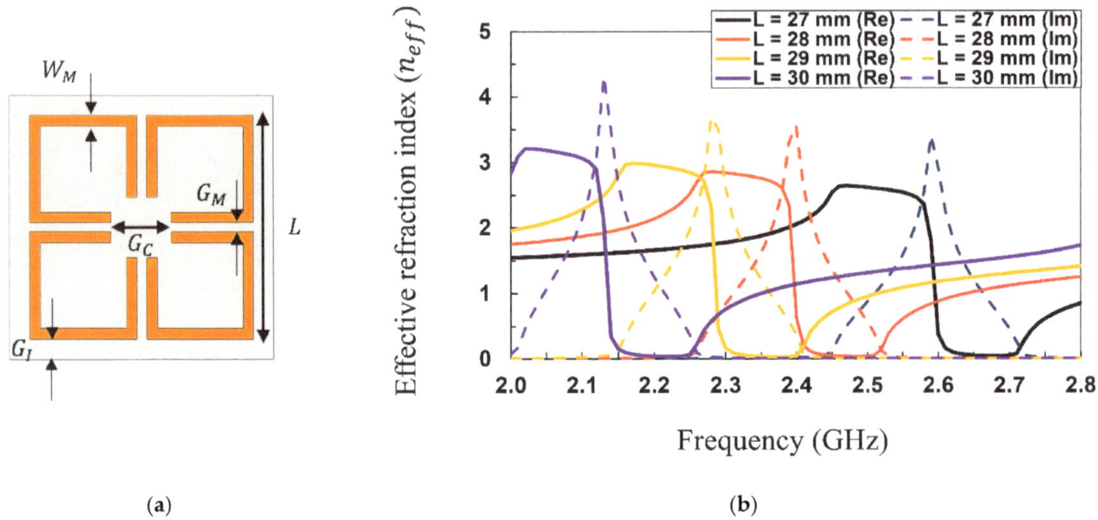

Figure 3. (**a**) Schematic of the 4-leaf clover shaped MTM unit cell; (**b**) Simulated effective refraction index of the MTM unit cell with the varied L values ($L = 28$ mm, $W_M = 1$ mm, $G_M = 1$ mm, $G_C = 6$ mm, $G_I = 2$ mm).

2.2. Design of the MTM Integrated Patch Antenna

In this subsection, the MTM integrated patch antenna is designed and simulated to verify the gain improvement property of the MTM superstrate. As shown in Figure 1, the MTM superstrate is stacked on the patch antenna. Firstly, the 2nd layer of the patch antenna is designed to be 47.8 mm × 39.9 mm for 2.45 GHz. Then, the MTM superstrate is placed over the patch antenna with a separation of 10 mm. In order to optimize the number of the MTM unit cells, the effects of the number of MTM unit cells on the performance of the integrated rectenna are investigated. We compare the radiation patterns of the patch antenna with a 2 × 2, 3 × 3, 4 × 4, and 5 × 5 MTM arrayed superstrate and without an MTM superstrate.

As shown in Figure 4, the gain of the patch antenna with a 4 × 4 MTM superstrate shows the highest peak gain, followed by the peak gain of the patch with a 5 × 5 MTM superstrate, and that with a 3 × 3 MTM superstrate, and that with a 2 × 2 MTM superstrate, and that without an MTM superstrate. The simulated peak gains of the patch antenna without and with an MTM superstrate are summarized in Table 1. As the number of the MTM unit cells increases, the peak gain tends to increase except for the case of the patch with the 5 × 5 MTM superstrate. Interestingly, the peak gain of the patch with the 5 × 5 MTM superstrate is slightly inferior to that with the 4 × 4 MTM superstrate. When the number of MTM unit cells in the array increases to 5 × 5, we expected the EM beam focusing coverage to be extended. However, since the beam focusing coverage of the 4 × 4 MTM superstrate is already enough to cover the size of the patch antenna, the peak gain improvement with the 5 × 5 MTM superstrate could be minute. Moreover, the simulation result indicates that the 5 × 5 MTM superstrate has increased loss compared with the 4 × 4 MTM superstrate, thereby decreasing the resultant peak gain of the antenna. Based on this analysis, it is concluded that the optimized number of the MTM unit cell in the array for the MTM superstrate is 4 × 4. Furthermore, it is worth noting that the number of MTM units could be determined depending on applications.

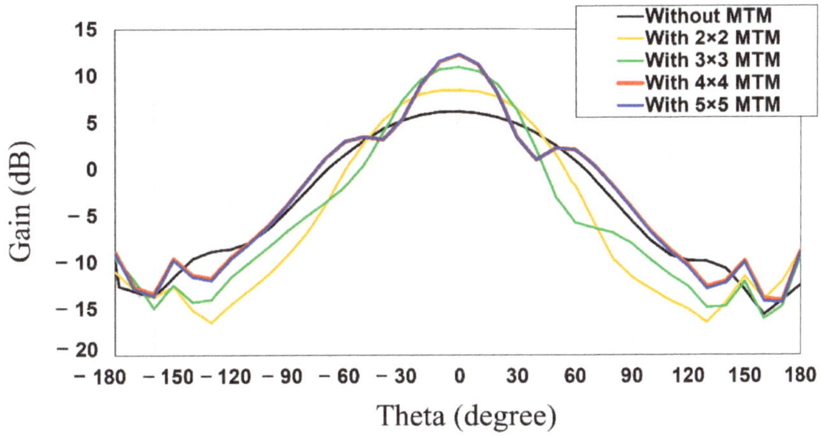

Figure 4. Simulated radiation patterns of the patch antenna with and without an MTM superstrate.

Table 1. Simulated peak gains of the patch antenna without and with an MTM superstrate.

Case	Peak Gain (dBi)	Improved Peak Gain (dB)
Patch only	6.15	-
Patch with 2 × 2 MTM superstrate	8.44	2.29
Patch with 3 × 3 MTM superstrate	10.93	4.78
Patch with 4 × 4 MTM superstrate	12.3	6.15
Patch with 5 × 5 MTM superstrate	12.25	6.1

2.3. Design of the Rectifier Circuit

Figure 5 shows the block diagram of the rectifier circuit. A designed half-wave rectifier includes the matching circuit, rectifying component, DC-pass filter, and load. In this work, the rectifier is designed to have maximum RF–DC conversion efficiency at an input power of 10 dBm in order to provide sufficient output power to the end IoT devices. For the performance of the rectifier circuit, we utilize an HSMS2860 Schottky diode, which has a maximum forward voltage of 0.35 V. The rectifier circuit also contains a low-pass filter (100 pF shunt capacitor) and a resistive load (1.1k ohms). The choice of the capacitor and resistor values needs to fulfil a number of requirements. Firstly, the values must be chosen considering its cut-off frequency as its time constant should be very much longer than the time interval between the successive peaks of the rectified waveform. The value of the capacitor does not significantly affect the RF–DC conversion efficiency as long as the aforementioned criteria is satisfied. However, as for the value of the resistor, the optimum load resistor varies with frequency and input power. Here, the value of the load resistor and capacitor are optimized in Advanced Design System (ADS) software under an input power of 10 dBm to maximize the RF–DC conversion efficiency at 2.45 GHz. By sweeping the load resistance from 0.5 to 5 k ohms and the capacitance from 50 to 200 pF during the circuit optimization, the optimal values are found to be 1.1 k ohms for a resistor and 100 pF for a capacitor so that the conversion efficiency is the highest. The calculated cut-off frequency of the low-pass filter using selected values of the capacitor and resistor becomes 1.45 MHz, which can block high frequency band effectively. In addition, in order to ensure the maximal power transfer to the diode, a matching circuit is carefully designed at the diode input port. The impedance matching circuit is a crucial and also difficult part of the rectifier design. Due to the nonlinearity of the rectifier, the input impedance of the rectifier varies with the frequency, input power level, etc. In this work, an open stub is utilized to match the input impedance at 2.45 GHz at an input power of 10 dBm. The rectifier circuit is matched to 50 ohms since the input impedance of the MTM integrated antenna is 50 ohms.

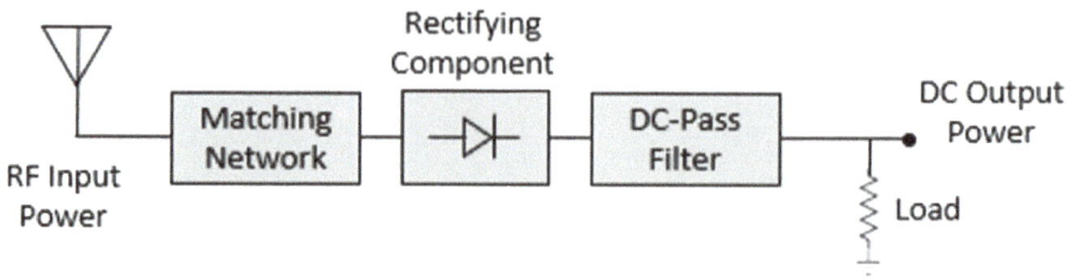

Figure 5. Block diagram of the rectifier circuit.

As shown in Figure 6, ADS has been utilized for the simulation of the rectifier circuit. In order to consider the nonlinearity of the Schottky diode, Harmonic Balance simulation was utilized. In addition, the HSMS2860 spice model was utilized to obtain accurate simulation results. The detailed simulation results of the rectifier circuit are further discussed with the measurement results in Section 4.

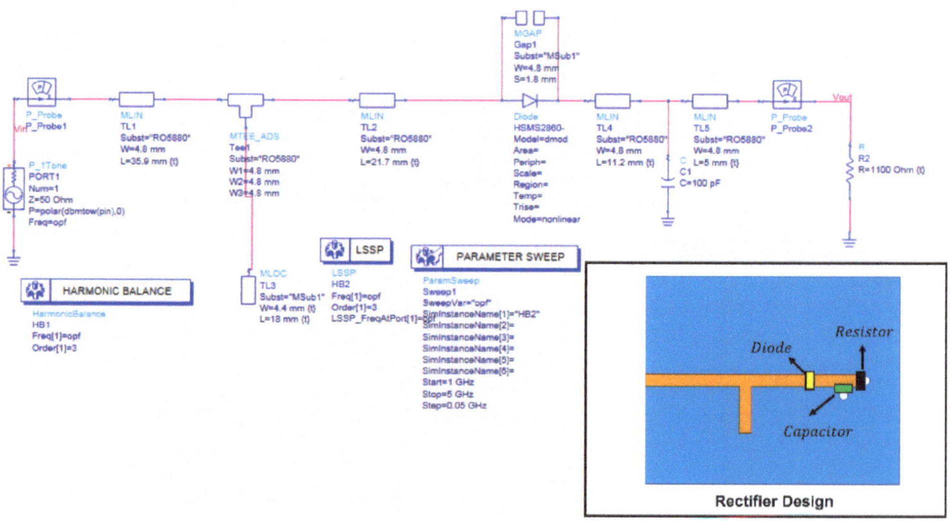

Figure 6. ADS schematic of the rectifier circuit.

3. Fabrication of the MTM Integrated High-Gain Rectenna

The prototype of the MTM integrated high-gain rectenna is fabricated on an RO5880 substrate. Each layer of the MTM-integrated rectenna is fabricated using a milling machine. Figure 7a–c shows the fabricated MTM superstrate, patch antenna, and rectifier circuit. Then, the patch antenna is fully integrated with the MTM superstrate and rectifier circuit, as shown in Figure 7d,e. For the fabrication of the rectifier circuit, The Schottky diode (HSMS 2860, Avago Technologies Inc. San Jose, CA, USA), 1.1k ohms resistor (ERA-6AEB112V, Panasonic Electronics Inc. Kadoma, Osaka, Japan), and 100 pF capacitor (GCM1885G2A101FA16D, Murata Electronics Inc. Nagokakyo, Kyoto, Japan) are selected.

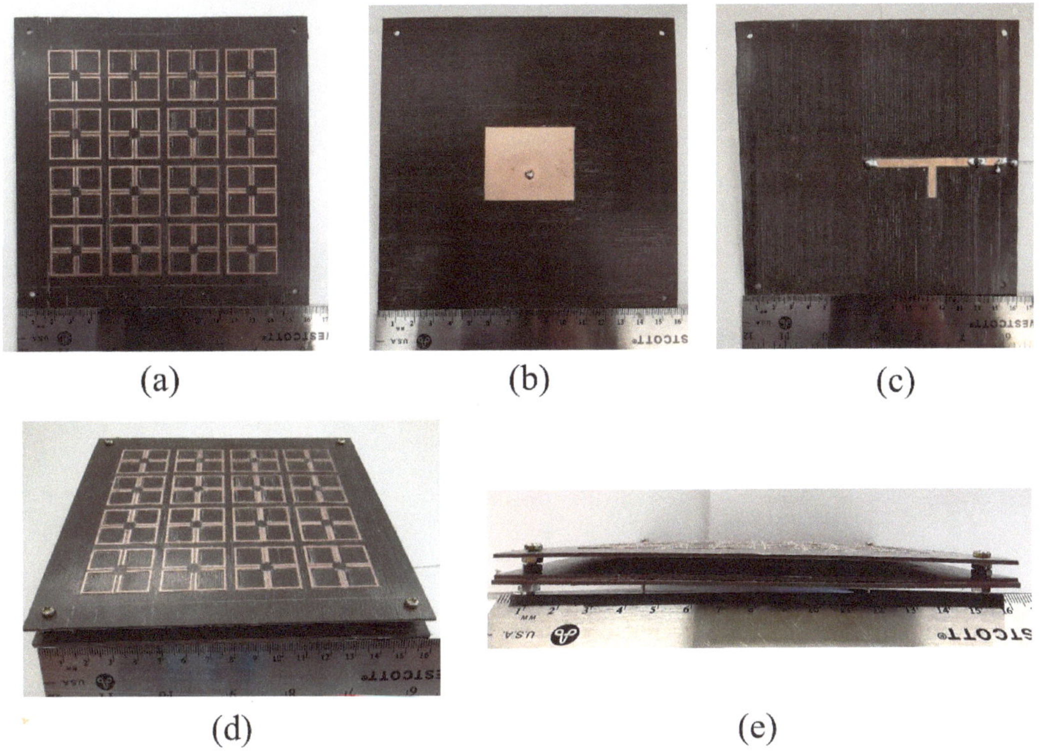

Figure 7. (a) Fabricated MTM superstrate, (b) fabricated patch antenna, (c) fabricated rectifier, (d) fabricated MTM-integrated rectenna, (e) side view of the fabricated MTM-integrated rectenna.

4. Simulated and Measured Results

In this section, we evaluate the performances of the MTM integrated high-gain rectenna and compare the measured results with the simulation results.

4.1. Simulated and Measured Results of the Patch Antenna with and without MTM Superstrate

In this subsection, the patch antenna with and without an MTM superstrate are simulated and measured, and their characteristics are compared. The return loss and peak gain of the antennas were measured using a vector network analyzer (HP E8361A, Agilent, Inc. Santa Clara, CA, USA) after standard one port short-open-load (SOL) calibration between 2.3 GHz and 2.6 GHz. Figure 8 shows the measured and simulated return losses of the patch with and without a 4×4 MTM superstrate. It is shown that the measured results are matched well with the simulated results. The measured resonant frequencies are a bit upshifted compared with the simulated ones, which are attributed to fabrication tolerance, i.e., the dimension of the fabricated devices is decreased during milling machine fabrication. In addition, the input impedances of the patch antenna with and without the MTM superstrate are simulated, as shown in the Figure 9. It is shown that, without MTM superstrate, the input resistance lies in the vicinity of 50 ohms with input reactance approximately near-zero at 2.46 GHz. With the MTM superstrate, the input resistance is shifted closer to ideal value of 50 ohms, and input reactance crossed zero at 2.45 GHz. Thus, the resonant frequency of the patch with the MTM superstrate is a bit downshifted compared with the one of the patch without MTM superstrate.

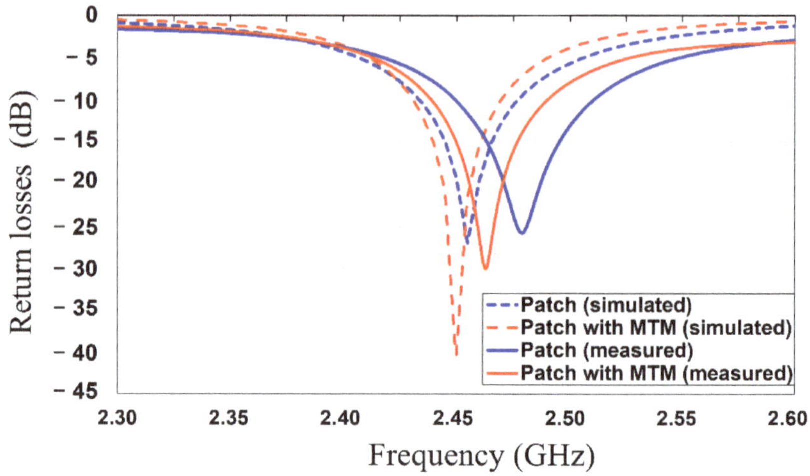

Figure 8. Simulated and measured return loss results of the patch antenna without and with MTM superstrate.

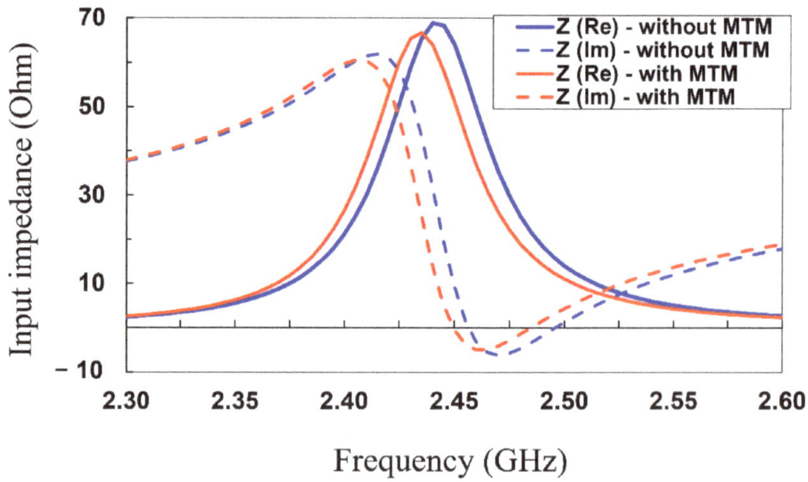

Figure 9. Simulated input impedance of the patch antenna without and with MTM superstrate.

In addition, Figure 10 exhibits the simulated and measured peak gain of the antenna without and with an MTM superstrate as a function of frequency. The measured peak gain of the patch antenna is greatly enhanced from 5.75 dBi to 11.87 dBi (6.12 dB improvement) when the MTM superstrate is integrated on top of the patch. It means that the patch antenna architecture with an MTM superstrate can transfer 4.09 times as much power as the patch antenna without an MTM superstrate. The simulated and measured results in this section prove that the implemented MTM superstrate can serve as the beam-focusing lens, and the MTM superstrate is very effective for enhancing the antenna gain. The measured and simulated return loss and peak gain results are summarized in Table 2.

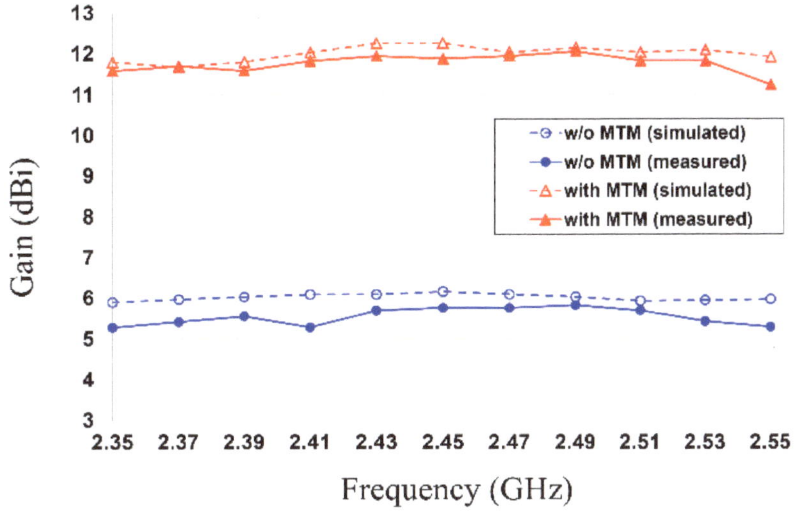

Figure 10. Simulated and measured peak gain of the patch antenna with and without MTM superstrate.

Table 2. Simulated and measured performances of the patch antenna with and without MTM superstrate.

Parameters	Patch Only		Patch with MTM	
	Simulated	**Measured**	**Simulated**	**Measured**
Return loss (dB)	−27.05	−25.88	−40.34	−30.13
Resonant frequency (GHz)	2.46	2.479	2.45	2.465
10-dB bandwidth (GHz)	0.039	0.044	0.051	0.061
Peak gain (dBi)	6.15	5.75	12.3	11.87

4.2. Simulated and Measured Results of the Rectifier Circuit

In this subsection, the third layer of the rectifier circuit is simulated, measured, and characterized. The ADS software was utilized for the simulation of the rectifier circuit. The measurement was performed using a vector network analyzer (HP E8361A, Agilent, Inc. Santa Clara, CA, USA), RF signal generator (HP 8648D, Agilent Inc. Santa Clara, CA, USA), and multimeter (Fluke 189). Firstly, the return loss of the rectifier circuit was simulated and measured using a vector network analyzer. Figure 11 shows that the measured results of the rectifier circuit exhibit a return loss of 22.4 dB at 2.47 GHz, which matches well with the simulated return loss of 33.6 dB at 2.45 GHz, which means the matching network of the rectifier matches well with the antenna.

Then, the RF–DC conversion efficiency of the rectifier circuit was simulated and measured. The RF–DC conversion efficiency of the rectifier can be obtained using the following Equation (4):

$$\eta_{RF-\mathrm{dc}} = \frac{P_{out}}{P_{in}} \times 100\ (\%) = \frac{{V_{DC}}^2}{P_{in} R_L} \times 100\ (\%), \quad (4)$$

where P_{out} is the output DC power, P_{in} is the input RF power to the rectifier circuit, R_L is the load resistance value, and V_{DC} is the output DC voltage. The simulated and measured RF–DC conversion efficiency and output DC power as a function of the input RF power levels to the rectifier circuit are depicted in Figure 12. The measured RF–DC conversion efficiencies are shown to be 2.69%, 40.71%, and 79.7% at −10 dBm, 0 dBm, and 10 dBm, respectively. A maximum RF–DC conversion efficiency of 79.96% is obtained at an input power of 10.5 dBm. The differences between simulated and measured results in Figure 10 can be attributed to: (1) The insertion loss of the SMA connector, which is connected between the RF signal generator and the rectifier circuit. This loss has not been taken into account in the ADS simulation. (2) The tolerance of the spice simulation model provided by the vendor of HSMS 2860 Schottky diode.

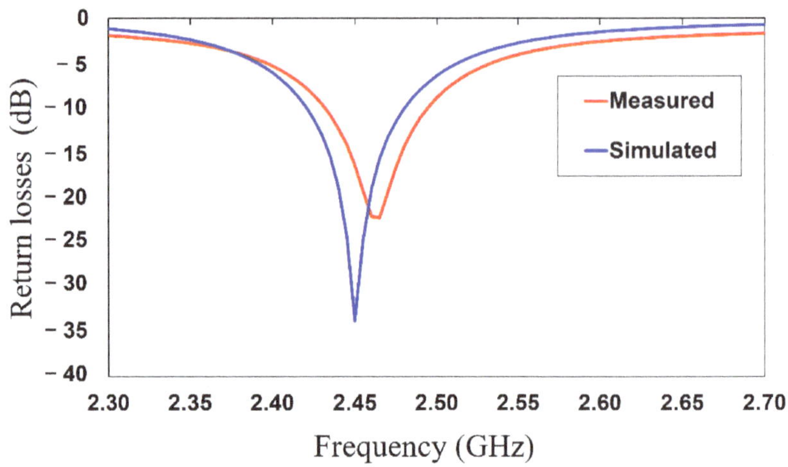

Figure 11. Simulated and measured return loss results of the rectifier circuit.

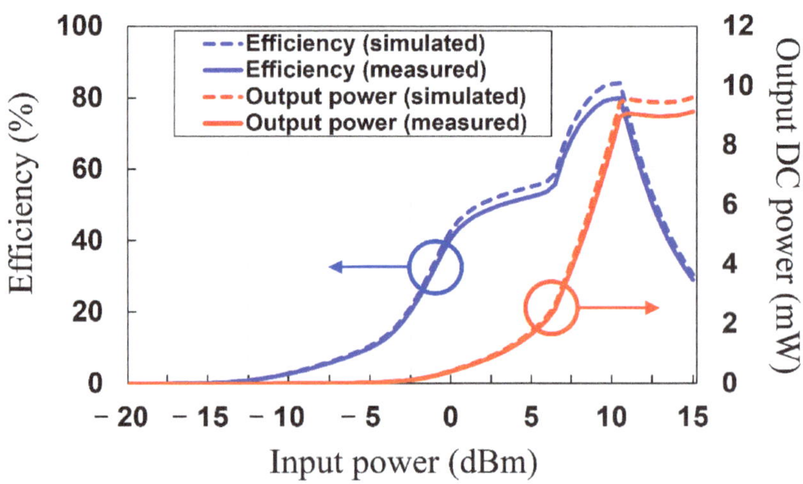

Figure 12. Simulated and measured return loss results of the rectifier circuit.

4.3. Measured Results of the MTM Integrated High-Gain Rectenna

Finally, the MTM integrated high-gain rectenna is fully integrated. Figure 13 shows the measurement setup of the MTM-integrated rectenna which consists of the RF signal generator, Tx horn antenna (JXTXLB-10180, A-INFO Inc. Irvine, CA, USA), Rx MTM-integrated rectenna, and multimeter. In the measurement, the MTM-integrated rectenna is placed at a distance of 40 cm away from the Tx antenna which satisfies the far-field condition at 2.45 GHz ($R \geq \frac{2D^2}{\lambda} = 26.8$ cm, where D is the largest dimension of the aperture of the antenna and λ is the free-space wavelength at the operating frequency). The RF–DC conversion efficiency of the rectenna can be calculated by:

$$\eta_{rectenna} = \frac{{V_L}^2}{P_r R_L} \times 100\ (\%) \quad (5)$$

where P_r is the RF power captured by the Rx integrated rectenna, R_L is load resistance value, and V_L is output voltage on the load. The RF power captured by the Rx integrated rectenna, P_r, can be calculated by the Friis transmission equation:

$$P_r = \left(\frac{\lambda}{4\pi r}\right)^2 G_t G_r P_t \quad (6)$$

where λ is the free-space wavelength at the operating frequency, r is the distance between Tx and Rx rectenna, G_t and G_r is the gain of the Tx and Rx antenna, respectively, and P_t is the Tx power. In the measurement, the values of G_t and G_r are 11 dB and 11.87 dB, respectively.

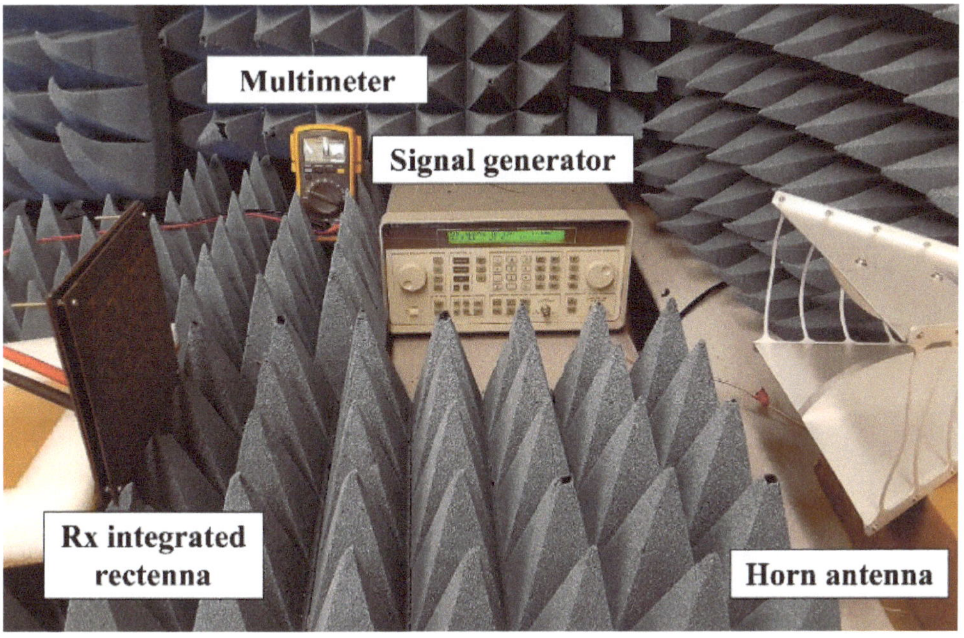

Figure 13. Measurement setup for the MTM-integrated rectenna.

As shown in Figure 14, the measured RF–DC conversion efficiency and output DC power of the MTM-integrated rectenna as a function of the RF input power can be achieved using the aforementioned analysis. It is shown that the measured RF–DC conversion efficiency of 2.67%, 33.84%, and 78.63 is obtained at −10 dBm, 0 dBm, and 10 dBm, respectively, with the maximum RF–DC conversion efficiency of 78.9% being at an input power of 9 dBm. When the input power of the rectenna without MTM is 2.88 dBm, the input power can be increased to 9 dBm by integrating MTM. This results in the improvement of the RF–DC efficiency from 39.2% to 78.9% and the increase of the output DC power from 0.7 mW to 6.27 mW (a factor of 8.96 improvements). Moreover, the RF–DC conversion efficiency versus the frequency at an input power of 10 dBm is plotted in Figure 15. It is shown that the measured RF–DC conversion efficiency is higher than 50% in the range of 2.38 to 2.53 GHz.

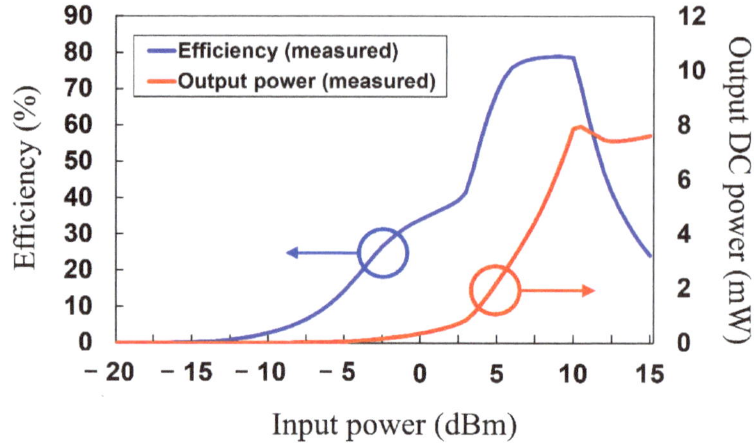

Figure 14. Measured RF–DC conversion efficiency of the MTM-integrated rectenna.

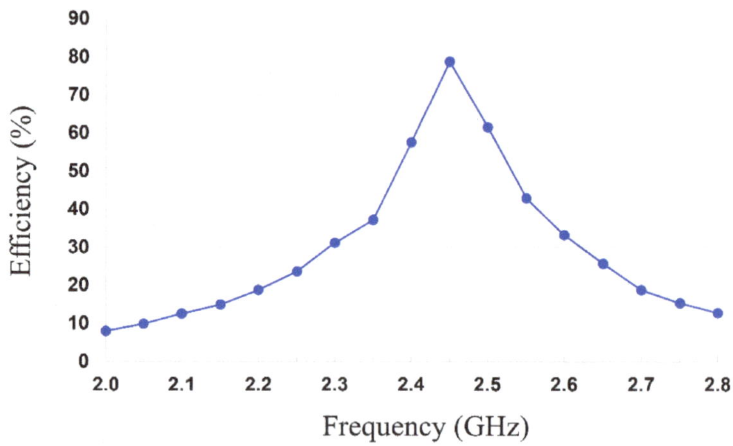

Figure 15. Measured RF–DC conversion efficiency versus frequency of the MTM-integrated rectenna.

4.4. Comparison

In this subsection, the MTM integrated high-gain rectenna is compared with previously reported high-gain rectennas. Firstly, the demonstrated MTM-integrated rectenna is compared with the preliminary study [25] in Table 3. The MTM-integrated rectenna shows overall improved performance in terms of peak gain, improved gain, and RF–DC conversion efficiency, which are attributed to the decreased substrate loss. The peak gains without/with MTM are improved by 0.97 dB and 1.17 dB, respectively. This enhancement can be interpreted as the substrate loss difference between FR4 and RO 5880. Furthermore, the demonstrated MTM-integrated rectenna shows improved RF–DC conversion efficiency compared with [25], which is also due to the reduced substrate loss of RO5880.

Table 3. Comparison of This Work with Preliminary Work.

Parameters	[25]	This Work
Substrate	FR4 (ε = 4.4, δ = 0.03)	RO5880 (ε = 2.2, δ = 0.0009)
Operating Frequency	2.45 GHz	2.45 GHz
Dimensions (mm^3)	150 × 150 × 14.71 (0.18 λ^3)	158 × 158 × 14.71 (0.199 λ^3)
Peak gain without MTM	4.78 dBi	5.75 dBi
Peak gain with MTM	10.7 dBi	11.87 dBi
Improved gain	5.92 dB	6.12 dB
Resistive load	1 kΩ	1.1 kΩ
Peak RF–DC conversion efficiency	63.5% @ 10.5 dBm input	78.9% @ 9 dBm input
DC power at peak efficiency	6.35 mW	6.27 mW

The demonstrated MTM-integrated rectenna is compared with the previously reported high-gain rectennas. Comparison parameters include the operating frequency; a technique that is used for the rectenna design; diode; antenna dimensions; RF input power/density (P_{in}/S); RF–DC conversion efficiency; design complexity/rectenna integration; and antenna gain. As shown in Table 4, the demonstrated MTM-integrated rectenna has shown outstanding performance compared to previously reported work. It is especially worth emphasizing that this is the first rectenna system utilizing the MTM superstrate integrated with the patch antenna and rectifier in a single device. Furthermore, the demonstrated work shows comparably high antenna gain and RF–DC conversion efficiency among selected work. Although the rectenna system demonstrated in [17] shows the highest gain, the rectenna exhibits a low RF–DC conversion efficiency of 42%, which leads to low end-to-end efficiency. In addition, the demonstrated MTM-integrated rectenna has a low design complexity compared with other work, as the MTM superstrate layer is integrated on top of the simple patch antenna and rectifier circuit.

Table 4. Comparison of This Work with Other High-Gain Rectennas.

Ref.	Operating Frequency (GHz)	Technique	Diode	Antenna Dimensions (mm^3)	P_{in} (dBm)/ S ($\mu W/cm^2$)	RF–DC Conversion Efficiency (%)	Design Complexity/Rectenna Integration	Antenna Gain (dBi)
[12]	5.8	Reflecting surface	MA40150-119	-	20/-	80	High/Yes	10.7
[4]	2.45	Reflecting surface	HSMS2852	110 × 90 × 20.6 * (0.113 λ^3)	-/1.95	80.03	Low/Yes	8.6
[13]	0.9	Differential patch	-	137 × 137 × 21.2 (0.01 λ^3)	2.19/-	65.3	High/No	8.5
[16]	0.95	Antenna array	HSMS285C	165 × 165 × 0.8 (0.18 λ^3)	5/-	76	High/No	8.5
[17]	24	SIW cavity	MA4E1317	55 × 55 × 0.5 * (0.776 λ^3)	12.55/-	42	High/Yes	12.6
[9]	2.4	Air gap, aperture coupling	HSMS286B	110 × 89 × 5.07 (0.026 λ^3)	10/-	32.52	Intermediate/No	7.82
[18]	5.78	Aperture coupling	HSMS2860	40 × 40 × 1.6 (0.026 λ^3)	13.98/-	63	High/No	7
This work	2.45	MTM superstrate	HSMS2560	158 × 158 × 14.71 * (0.199 λ^3)	9/-	78.9	Low/Yes	11.87

* Includes the rectifier dimension as the rectenna integrated in a single device.

5. Conclusions

In this work, the MTM integrated high-gain rectenna is demonstrated. The novel MTM superstrate approach with a three-layered integration method is firstly introduced for rectenna applications. The optimized near-zero property of the MTM unit cell is obtained using a parametric analysis. In addition, the effects of the number of MTM unit cells on the performance of the integrated rectenna are studied. By integrating the MTM on top of the patch antenna, the gain of the antenna is enhanced by 6.12 dB, resulting in the increase of incident RF power at the rectifier input, thereby inducing high-output DC power and high end-to-end efficiency. A measured RF–DC conversion efficiency of 2.67%, 33.84%, and 78.63 is achieved at −10 dBm, 0 dBm, and 10 dBm, respectively. A maximum RF–DC conversion efficiency of 78.9% is obtained at an input power of 9 dBm. When the input power of the rectenna without MTM is 2.88 dBm, the input power can be increased to 9 dBm by integrating MTM. This results in the improvement of the RF–DC efficiency from 39.2% to 78.9% and the increase of the output DC power from 0.7 mW to 6.27 mW (a factor of 8.96 improvements). The demonstrated MTM-integrated rectenna has shown outstanding performance compared to previously reported other work. It is especially worth emphasizing that this is the first rectenna system utilizing the MTM superstrate integrated with the patch antenna and rectifier in a single device. The demonstrated MTM-integrated rectenna has a low design complexity compared with other work, as the MTM superstrate layer is integrated on top of the simple patch antenna and rectifier circuit. In addition, the number of MTM units can be determined depending on applications. It is highly envisioned that the demonstrated MTM-integrated rectenna will provide new possibilities for practical energy harvesting applications with improved antenna gain and efficiency in various IoT environments.

Author Contributions: W.L.: conceptualization, software, methodology, data curation, validation, and writing—original draft preparation; S.-i.C.: data curation, validation, and writing—review and editing; H.-i.K.: software, and writing—review and editing; S.H.: data curation and visualization; S.J.: data curation and visualization; Y.-K.Y.: supervision, conceptualization, writing—review and editing, project administration, and funding acquisition. All authors have read and agreed to the published version of the manuscript.

Funding: This research was supported in part by 'Beyond Limit Academia Collaboration Program' funded by Samsung Electronics.

Institutional Review Board Statement: Not applicable.

Informed Consent Statement: Not applicable.

Data Availability Statement: Not applicable.

Acknowledgments: W. Lee is a recipient of the fellowship program by the Republic of Korea Army Headquarters. RF experiment/characterization was performed at the University of Florida.

Conflicts of Interest: The authors declare no conflict of interest.

References

1. Arrawatia, M.; Baghini, M.S.; Kumar, G. Broadband bent triangular omnidirectional antenna for RF energy harvesting. *IEEE Antennas Wirel. Propag. Lett.* **2016**, *15*, 36–39. [CrossRef]
2. Zakaria, Z.; Zainuddin, N.A.; Aziz, M.Z.A.A.; Husain, M.N.; Mutalib, M.A. Dual-band monopole antenna for energy harvesting system. In Proceedings of the IEEE Symposium on Wireless Technology and Applications (ISWTA), Kuching, Malaysia, 22–25 September 2013; pp. 225–229.
3. Lin, W.; Ziolkowski, R.W. Electrically small, single-substrate huygens dipole rectenna for ultracompact wireless power transfer applications. *IEEE Trans. Antennas Propag.* **2021**, *69*, 1130–1134. [CrossRef]
4. Sun, Y.; Guo, X.; He, M.; Zhong, Z. Design of a high-efficiency 2.45-GHz rectenna for low-input-power energy harvesting. *IEEE Antennas Wirel. Propag. Lett.* **2012**, *11*, 929–932.
5. Tu, W.H.; Hsu, S.H.; Chang, K. Compact 5.8-GHz rectenna using stepped-impedance dipole antenna. *IEEE Antennas Wirel. Propag. Lett.* **2017**, *6*, 282–284. [CrossRef]

6. Harouni, Z.; Cirio, L.; Osman, L.; Gharsallah, A.; Picon, O. A dual circularly polarized 2.45-GHz rectenna for wireless power transmission. *IEEE Antennas Wirel. Propag. Lett.* **2021**, *10*, 306–309. [CrossRef]
7. Song, C.; Huang, Y.; Zhou, J.; Zhang, J.; Yuan, S.; Carter, P. A high-efficiency broadband rectenna for ambient wireless energy harvesting. *IEEE Trans. Antennas Propag.* **2015**, *63*, 3486–3495. [CrossRef]
8. Song, C.; Huang, Y.; Carter, P.; Zhou, J.; Yuan, S.; Xu, Q.; Kod, M. A novel six-band dual CP rectenna using improved impedance matching technique for ambient RF energy harvesting. *IEEE Trans. Antennas Propag.* **2016**, *64*, 3160–3171. [CrossRef]
9. Hassan, N.; Zakaria, Z.; Sam, W.Y.; Hanapiah, I.N.M.; Mohamad, A.N.; Roslan, A.F.; Ahmad, B.H.; Ismail, M.K.; Aziz, M.Z.A.A. Design of dual-band microstrip patch antenna with rightangle triangular aperture slot for energy transfer application. *Int. J. RF Microw. Comput.-Aided Eng.* **2019**, *29*, e21666. [CrossRef]
10. Shi, Y.; Fan, Y.; Li, Y.; Yang, L.; Wang, M. An efficient broadband slotted rectenna for wireless power transfer at LTE band. *IEEE Trans. Antennas Propag.* **2019**, *67*, 814–822. [CrossRef]
11. Hagerty, J.A.; Helmbrecht, F.B.; McCalpin, W.H.; Zane, R. Popovic, Z.B. Recycling ambient microwave energy with broad-band rectenna arrays. *IEEE Trans. Microw. Theory Tech.* **2004**, *52*, 1014–1024. [CrossRef]
12. Strassner, B.; Chang, K. 5.8-GHz circularly polarized rectifying antenna for wireless microwave power transmission. *IEEE Trans. Microw. Theory Tech.* **2002**, *50*, 1870–1876. [CrossRef]
13. Arrawatia, M.; Baghini, M.S.; Kumar, G. Differential microstrip antenna for RF energy harvesting. *IEEE Trans. Antennas Propag.* **2015**, *63*, 1581–1588. [CrossRef]
14. Sun, H. An enhanced rectenna using differentially-fed rectifier for wireless power transmission. *IEEE Antennas Wirel. Propag. Lett.* **2016**, *15*, 32–35. [CrossRef]
15. Ren, Y.J.; Chang, K. 5.8 GHz circularly polarized dual-diode rectenna and rectenna array for microwave power transmission. *IEEE Trans. Microw. Theory Tech.* **2006**, *54*, 1495–1502.
16. Fakharian, M.M. A wideband rectenna using high gain fractal planar monopole antenna array for rf energy scavenging. *Int. J. Antennas Propag.* **2020**, *2020*, 3489323. [CrossRef]
17. Ladan, S.; Guntupalli, A.B.; Wu, K. A high-efficiency 24 GHz rectenna development towards millimeter-wave energy harvesting and wireless power transmission. *IEEE Trans. Circuits Syst. I Regul. Pap.* **2014**, *61*, 3358–3366. [CrossRef]
18. Yang, X.X.; Jiang, C.; Elsherbeni, A.Z.; Yang, F.; Wang, Y. A novel compact printed rectenna for data communication systems. *IEEE Trans. Antennas Propag.* **2013**, *61*, 2532–2539. [CrossRef]
19. Samantaray, D.; Bhattacharyya, S. A Gain-Enhanced Slotted Patch Antenna Using Metasurface as Superstrate Configuration. *IEEE Trans. Antennas Propag.* **2020**, *68*, 6548–6556. [CrossRef]
20. Singh, A.K.; Abegaonkar, M.P.; Koul, S.K. High-gain and high-aperture-efficiency cavity resonator antenna using metamaterial superstrate. *IEEE Antennas Wirel. Propag. Lett.* **2017**, *16*, 2388–2391. [CrossRef]
21. Kim, C.; Lee, K.; Lee, S.; Kim, K.T.; Yoon, Y. A surface micromachined high directivity GPS patch antenna with a four-leaf clover shape metamaterial slab. In Proceedings of the 2012 IEEE 62nd Electronic Components and Technology Conference (ECTC), San Diego, CA, USA, 29 May–1 June 2012; pp. 942–947.
22. Lee, W.; Kim, H.; Yoon, Y.-K. Metamaterial-inspired dual-function loop antenna for wireless power transfer and wireless communications. In Proceedings of the 2020 IEEE 70th Electronic Components and Technology Conference (ECTC), Orlando, FL, USA, 3–30 June 2020; pp. 1351–1357.
23. Padilla, W.J.; Basov, D.N.; Smith, D.R. Negative refractive index metamaterials. *Mater. Today* **2006**, *9*, 28–35. [CrossRef]
24. Lee, W.; Yoon, Y.-K. Wireless power transfer systems using metamaterials: A review. *IEEE Access* **2020**, *8*, 147930–147947. [CrossRef]
25. Lee, W.; Kim, H.; Hwang, S.; Jeon, S.; Cho, H.; Yoon, Y.-K. 3D integrated high gain rectenna in package with metamaterial superstrates for high efficiency wireless power transfer applications. In Proceedings of the 2021 IEEE 71st Electronic Components and Technology Conference (ECTC), Virtual Meeting, 1 June–4 July 2021; pp. 1317–1322.
26. Smith, D.; Schultz, S.; Markos, P.; Soukoulis, C. Determination of effective permittivity and permeability of metamaterials from reflection and transmission coefficients. *Phys. Rev. B* **2002**, *65*, 195104. [CrossRef]
27. Smith, D.; Vier, D.; Koschny, T.; Soukoulis, C. Electromagnetic parameter retrieval from inhomogeneous metamaterials. *Phys. Rev. E* **2005**, *71*, 036617. [CrossRef]
28. Chen, X.; Grzegorczyk, T.M.; Wu, B.-I.; Pacheco, J., Jr.; Kong, J.A. Robust method to retrieve the constitutive effective parameters of metamaterials. *Phys. Rev. E* **2004**, *70*, 016608. [CrossRef]

sensors

Article

Multiple Concurrent Slotframe Scheduling for Wireless Power Transfer-Enabled Wireless Sensor Networks

Sol-Bee Lee [1], Sam Nguyen-Xuan [2], Jung-Hyok Kwon [3] and Eui-Jik Kim [1,*]

1 Division of Software, Hallym University, 1 Hallymdaehak-gil, Chuncheon 24252, Gangwon-do, Korea; thfqla3535@hallym.ac.kr
2 Department of Information Technology, Posts and Telecoms Institute of Technology at HCM City Campus, Ho Chi Minh City 71007, Vietnam; samnx@ptithcm.edu.vn
3 Smart Computing Laboratory, Hallym University, 1 Hallymdaehak-gil, Chuncheon 24252, Gangwon-do, Korea; jhkwon@hallym.ac.kr
* Correspondence: ejkim32@hallym.ac.kr

Abstract: This paper presents a multiple concurrent slotframe scheduling (MCSS) protocol for wireless power transfer (WPT)-enabled wireless sensor networks. The MCSS supports a cluster-tree network topology composed of heterogeneous devices, including hybrid access points (HAPs) serving as power transmitting units and sensor nodes serving as power receiving units as well as various types of traffic, such as power, data, and control messages (CMs). To this end, MCSS defines three types of time-slotted channel hopping (TSCH) concurrent slotframes: the CM slotframe, HAP slotframe, and WPT slotframe. These slotframes are used for CM traffic, inter-cluster traffic, and intra-cluster traffic, respectively. In MCSS, the length of each TSCH concurrent slotframe is set to be mutually prime to minimize the overlap between cells allocated in the slotframes, and its transmission priority is determined according to the characteristics of transmitted traffic. In addition, MCSS determines the WPT slotframe length, considering the minimum number of power and data cells required for energy harvesting and data transmission of sensor nodes and the number of overprovisioned cells needed to compensate for overlap between cells. The simulation results demonstrated that MCSS outperforms the legacy TSCH medium access control protocol and TSCH multiple slotframe scheduling (TMSS) for the average end-to-end delay, aggregate throughput, and average harvested energy.

Keywords: energy harvesting; multiple concurrent slotframes; slotframe length determination; wireless power transfer; wireless sensor network; IEEE 802.15.4 TSCH

Citation: Lee, S.-B.; Nguyen-Xuan, S.; Kwon, J.-H.; Kim, E.-J. Multiple Concurrent Slotframe Scheduling for Wireless Power Transfer-Enabled Wireless Sensor Networks. *Sensors* **2022**, *22*, 4520. https://doi.org/10.3390/s22124520

Academic Editors: Slim Naifar, Olfa Kanoun and Carlo Trigona

Received: 29 April 2022
Accepted: 12 June 2022
Published: 15 June 2022

1. Introduction

The wireless sensor network (WSN) is an essential technology to provide wireless connectivity between devices in Internet of Things (IoT) systems, serving as a key enabler for implementing various IoT applications, such as smart grid management, process automation, and monitoring [1–5]. In WSNs, sensor nodes require frequent maintenance, such as recharging or replacing the battery, and their communication is often limited when the remaining energy is low because they are generally battery-powered [6,7]. Recently, wireless power transfer (WPT) technology that harvests energy from radio frequency (RF) signals has been widely applied to WSNs; thus, WPT-enabled WSNs that transmit both power and data using an RF signal have the advantage of reducing the battery dependency of sensor nodes and extending the network lifespan [8,9].

Unlike the traditional WSNs, the WPT-enabled WSN includes a hybrid access point (HAP), a device that acts as a power transmitting unit (PTU) to supply power to sensor nodes that act as a power receiving unit (PRU), and includes the existing sensor nodes [10,11].

In WPT-enabled WSN, these heterogeneous devices (i.e., HAPs and sensor nodes) constitute a cluster-tree network, where each cluster includes one HAP serving as a cluster head (CH) and multiple sensor nodes serving as a cluster member [12,13]. Thus, sensor nodes harvest energy from adjacent HAPs and transmit data to the CH. The HAPs collect data from sensor nodes and forward them to the sink. In such a WPT-enabled WSN structure, for timely energy harvesting and data exchange between heterogeneous devices, comprehensively designing a medium access control (MAC) layer protocol considering a network topology and various traffic types (i.e., control messages (CMs), intra-cluster traffic, and inter-cluster traffic) is a challenging issue [14–16].

In the literature, many studies have tried to develop the MAC layer protocol for WPT-enabled WSNs. In [17–19], the authors proposed carrier sense multiple access with collision avoidance (CSMA/CA)-based MAC protocols, in which HAPs and sensor nodes competitively occupy the channel to perform WPT and data transmission. However, they cause unnecessary energy consumption of sensor nodes and nondeterministic power and data transmission using a random backoff period and retransmission mechanism.

The authors in [20,21] proposed a hybrid MAC protocol that uses a superframe structure, including CSMA/CA and time-division multiple access (TDMA) periods. The use of the CSMA/CA period still causes unnecessary energy consumption of the sensor nodes, while their work combines the benefits of both approaches.

In [22–24], the authors proposed TDMA-based harvest-then-transmit protocols, in which energy harvesting and data transmission are performed in a contention-free manner. They allowed sensor nodes to perform energy harvesting and data transmission through frame scheduling, preventing unnecessary energy consumption due to collisions or idle listening. However, the existing TDMA-based harvest-then-transmit protocols may waste resources in WPT-enabled WSNs because they use a fixed-length frame regardless of network topology and traffic type [25].

Time-slotted channel hopping (TSCH) specified in the IEEE 802.15.4-2015 standard [26] can address the problems of the existing MAC protocols by allowing the use of concurrent slotframes. Specifically, TSCH simultaneously uses one or more slotframes with different lengths to efficiently support various communication schedules that vary depending on the network topology and traffic type. Each slotframe consists of multiple cells expressed as a pair with a timeslot and channel offset, and the length of each slotframe is determined according to the number of timeslots. Thus far, several studies have proposed TSCH scheduling schemes using multiple concurrent slotframes to support the network topology and traffic type.

In [27–29], the authors employed three types of slotframes. Each slotframe is used for the enhanced beacon (EB), CM, and data. The length of each slotframe is empirically determined, considering the amount of traffic that varies depending on the network topology. However, TSCH scheduling schemes are specialized only for data delivery and use a slotframe structure in which one cell for data transmission and one cell for data reception are allocated within a slotframe based on the sensor node ID. Therefore, their inflexible slotframe structure is unsuitable to support the WPT operation required by the sensor nodes.

In [30], the authors increased the link capacity using a resource allocation algorithm based on the network topology information to support WPT and data transmission in a WSN. However, in [30], a homogeneous network environment composed only of sensor nodes was considered; thus, it is unsuitable for WPT and data transmission operations between heterogeneous devices in WPT-enabled WSNs, including both PTU and PRU.

In [31], the authors proposed a TSCH-based MAC protocol to support energy harvesting and data transmission of sensor devices with different transmission periods. In [31], the sensor devices perform cell allocation using a concurrent slotframe with the same length as their transmission period, causing frequent overlap between cells, and the computational load for solving this increased.

This paper proposes a multiple concurrent slotframe scheduling (MCSS) protocol, supporting a cluster-tree network topology and various traffic types in WPT-enabled WSNs. To this end, MCSS uses three types of TSCH concurrent slotframes: the CM slotframe, HAP slotframe, and WPT slotframe. Each slotframe is dedicated to transmitting the CM traffic, inter-cluster traffic, and intra-cluster traffic. To mitigate the influence of overlapped schedules (i.e., overlapped cells) among slotframes on the network performance, MCSS sets the length of each TSCH concurrent slotframe to be mutually prime and determines the transmission priority of slotframes considering the characteristics of the transmitted traffic. In addition, MCSS determines the WPT slotframe length for each cluster using the numbers of required power and data cells, including the number of overprovisioned cells. The contributions of this work are summarized as follows:

- First, multiple TSCH concurrent slotframes with different lengths are used according to the three traffic types (i.e., CMs traffic, inter-cluster traffic, and intra-cluster traffic) so that MCSS can support a cluster-tree network comprising heterogeneous devices (i.e., HAPs and sensor nodes), which is a general form of the WPT-enabled WSN.
- Second, the WPT slotframe length may be set differently for each cluster based on the amount of intra-cluster traffic, reducing the number of empty cells within the WPT slotframe and improving the channel utilization for energy harvesting and data transmission of sensor nodes.
- Finally, the lengths of TSCH concurrent slotframes are set to be mutually prime to each other, and their transmission priorities are differentiated, thereby, reducing the influence of overlapped cells among the concurrent slotframes.

To verify the superiority of MCSS, we performed an experimental simulation and compared the performance of MCSS with that of the legacy TSCH MAC protocol with a single slotframe. The results demonstrated that MCSS achieved 70.63% and 70.93% shorter end-to-end delay, 17.89% and 18.80% higher aggregate throughput, and 111.96% and 116.28% higher average harvested energy compared with the legacy TSCH MAC and TSCH multiple slotframe scheduling (TMSS), respectively.

The remainder of this paper is organized as follows. Section 2 presents a system model for MCSS. We describe the design of MCSS in detail in Section 3. We provide the simulation configuration and results in Section 4. Finally, Section 5 concludes this paper.

2. System Model

In this section, we introduce a system architecture of MCSS in which heterogeneous devices (i.e., HAPs and sensor nodes) constitute a cluster-tree network. Then, we describe the overview of TSCH used as the basic MAC scheme.

2.1. System Architecture

Figure 1 illustrates a system architecture of MCSS, consisting of multiple HAPs and sensor nodes. In MCSS, the cluster composed of one HAP and sensor nodes is considered a HAP cluster (i.e., intra-cluster). The HAP is equipped with directional and omnidirectional antennas and acts as a CH in the HAP cluster. It transfers power to adjacent sensor nodes using beamforming technology that steers the beam in a specific direction.

In addition, the HAP transmits the data collected from its cluster members (i.e., sensor nodes) to the neighboring HAP (i.e., parent HAP) so that the collected data can be sent to the root. As HAPs are always connected to a power source, they do not need to harvest energy from any other HAPs; only data are exchanged between HAPs. The sensor node is considered a battery-powered device with a single omnidirectional antenna, harvesting energy from the HAP (i.e., its CH), and transmitting data to the HAP.

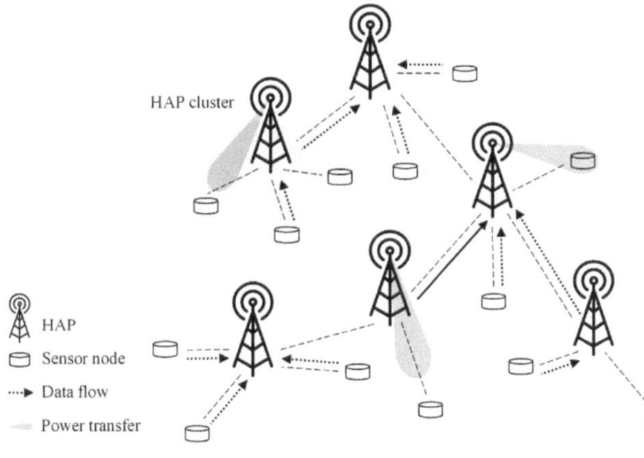

Figure 1. System architecture of MCSS.

In MCSS, the amount of energy consumed by a sensor node to transmit a data packet ($E_{Tx,timeslot}$) is derived based on the energy consumption model in [32] as follows.

$$E_{Tx,timeslot} = \left(P_{sleep}L_{TsTxOffset} + P_{tx}L_{data} + P_{idle}L_{TsRxAckDelay} + P_{rx}L_{Ack}\right)/L_{timeslot}, \quad (1)$$

where P_{sleep}, P_{tx}, P_{idle}, and P_{rx} represent the amount of energy consumed when a sensor node is in sleep, transmission (Tx), idle, and reception (Rx) states during a timeslot, respectively. In addition, $L_{TsTxOffset}$ and $L_{TsRxAckDelay}$ indicate the time intervals for the sensor node to prepare data packet transmission and for the sensor node to wait to receive an acknowledgment (*Ack*) from the HAP, respectively. Further, L_{data}, L_{Ack}, and $L_{timeslot}$ indicate the time it takes to transmit a data packet, an *Ack*, and the timeslot length, respectively. The HAP supplies power to the sensor nodes based on the amount of energy consumed by each sensor node derived by Equation (1), enabling the data transmission of sensor nodes and preventing their energy depletion.

In MCSS, an IPv6 routing protocol for low-power and lossy networks (RPL) is used for routing between HAPs and sensor nodes [33]. The RPL provides path diversity by organizing a destination-oriented directed acyclic graph (DODAG) according to a routing metric called the *Rank*. The *Rank* refers to the routing distance from a sensor node to the root and is determined by the objective function (OF). The OF of each sensor node computes the *Rank* of each sensor node based on the expected transmission count.

For DODAG formation, each sensor node in RPL broadcasts the DODAG information object (DIO) message containing its *Rank* to its neighbors. Upon receiving the DIO messages, the sensor node selects its parent among the neighbor HAPs considering the *Rank* included in the DIO messages. In MCSS, only HAPs can be selected as the parent of HAPs and sensor nodes. When the parent selection is completed, the sensor node transmits the DODAG advertisement object message toward the root. The root can recognize the new destination information and participation of the sensor node through the received DODAG advertisement object message.

2.2. Time-Slotted Channel Hopping

Further, TSCH is a MAC protocol for low-power and lossy networks based on multichannel and time-slotted access standardized in IEEE 802.15.4-2015 [26]. In addition, TSCH supports reliability and deterministic communication between nodes by using a slotframe structure. The slotframe consists of multiple cells and repeats over time. Each cell is identified by the timeslot and channel offset. The cell is long enough to exchange a maximum-size data frame and its *Ack* frame, typically 10 ms long.

The cell can be classified as a dedicated or shared cell. The dedicated cell guarantees transmission between a pair of nodes. In contrast, the shared cell allows multiple nodes to access the same channel simultaneously. In each cell, the nodes communicate on the channel determined by channel hopping as follows:

$$Channel = channelHoppingSequence[(ASN + offset_{ch})\% \, n_{availableCh}], \quad (2)$$

where the *channelHoppingSequence* represents a list of channels to be hopped; *ASN* denotes the absolute slot number, indicating the total number of timeslots that increase every timeslot since the network started; $offset_{ch}$ refers to the channel offset; and $n_{availableCh}$ indicates the number of available channels in the TSCH network.

In the TSCH network, all nodes are globally synchronized to the network through the EB. The EB includes information, such as the slotframe structure, *ASN*, and *channelHoppingSequence*. After the TSCH network formation is completed, cell allocation is performed. The IPv6 over the TSCH mode of the IEEE 802.15.4e (6TiSCH) operation sublayer (6top) protocol (6P) defined by the IETF 6TiSCH working group provides 6P transaction for TSCH cell allocation [34].

The 6P transaction is a mechanism that enables neighboring nodes to add, delete, and relocate cells to their TSCH schedule. In particular, the 6P transaction can be composed of two or three steps. In a two-step 6P transaction, a candidate cell list consists of cells that can be allocated by the node sending a 6P request message. However, in a three-step 6P transaction, the list contains cells that can be allocated by the node receiving the 6P request message.

Figure 2 depicts an example of two-step and three-step 6P transactions. In Figure 2a, node A transmits a 6P ADD request message to node B to allocate the required cells. The number of cells to be added (i.e., NumCells), and the list of candidate cells (i.e., CellList) determined by node A are included in the 6P ADD request message. Node B compares the CellList of the received 6P ADD request message to its TSCH schedule to identify which cells are available and selects two available cells. The list of selected cells becomes a CellList to be included in the 6P response message.

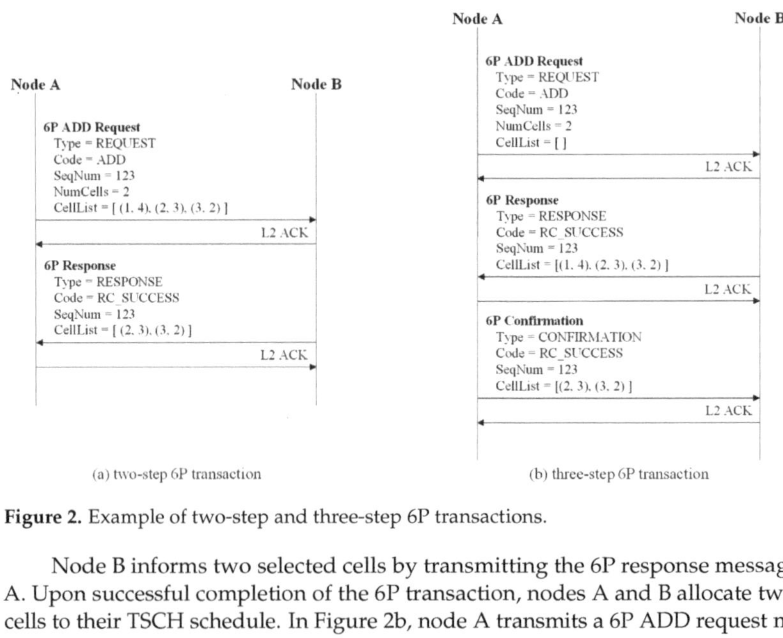

Figure 2. Example of two-step and three-step 6P transactions.

Node B informs two selected cells by transmitting the 6P response message to node A. Upon successful completion of the 6P transaction, nodes A and B allocate two selected cells to their TSCH schedule. In Figure 2b, node A transmits a 6P ADD request message to node B, including an empty CellList. When node B receives the 6P ADD request message, node B composes the CellList, including NumCells or more candidate cells available in its TSCH schedule, and transmits the 6P response message to node A.

Node A constructs a CellList by selecting only available candidate cells from the CellList of the 6P response message and transmits a 6P confirmation message containing the CellList to Node B. As a result, the 6P transaction is completed, and nodes A and B allocate the selected cells to their TSCH schedules. In MCSS, we used the three-step 6P transaction to allow each HAP, which knows the schedule (i.e., the results of cell allocation) for all sensor nodes included in the HAP cluster, to construct a candidate cell list.

3. Design of MCSS

To support the cluster-tree topology and various types of traffic in WPT-enabled WSN, the MCSS defines multiple TSCH concurrent slotframes, allocates power cells and data cells, and determines the transmission priorities. In the following subsections, we describe the operations of MCSS in detail.

3.1. Multiple Concurrent Slotframes

The MCSS uses three slotframes to handle heterogeneous traffic types (i.e., CM traffic, inter-cluster traffic, and intra-cluster traffic) in the WPT-enabled WSN, as follows:

1. The CM slotframe is used to send and receive EB frames, which notify the MCSS schedule of the HAP cluster. It is also used to exchange other CMs, such as RPL messages and 6P messages. In the CM slotframe, one shared cell (TxRxS) capable of both transmission and reception is allocated, through which all CMs are exchanged between neighbors. The length of the CM slotframe is set in consideration of the EB transmission period.
2. The HAP slotframe is used to deliver data traffic (i.e., inter-cluster traffic) between HAPs, for which the HAP allocates at least one data transmission cell (DTx) to transmit data traffic to its parent. In addition, multiple data reception cells (DRxs) can be allocated to the HAP slotframe. The number of data reception cells equals the number of child HAPs of the HAP. The HAP slotframe length is set shorter than other concurrent slotframes to minimize the delivery latency of data traffic.
3. The WPT slotframe supports the transmission of power and data traffic (i.e., intra-cluster traffic) within the HAP cluster. In the WPT slotframe, multiple power transmission cells (PTxs) and data reception cells (DRxs) can be allocated by the HAP and sensor nodes through the three-step 6P transaction described in Section 3.2. The length of the WPT slotframe can be set differently for each HAP cluster and is determined by the HAP.

The MCSS establishes communication schedules by projecting the multiple concurrent slotframes onto a single plane. This operation may cause overlap between cells allocated to different slotframes. To reduce the overlapped cells, MCSS sets the number of timeslots within each slotframe to be a different prime number, making the length of the slotframes mutually prime. In addition, the transmission priority for each slotframe is set to the following priority order considering the traffic characteristics: CM slotframe > HAP slotframe > WPT slotframe.

Accordingly, if overlap occurs between cells, MCSS selects the cells allocated to a slotframe with the highest transmission priority. The WPT slotframe has the lowest transmission priority among slotframes. To compensate for the overlapped cells of the WPT slotframe, MCSS determines the length of the WPT slotframe for each HAP cluster using the minimum number of power and data cells and the number of overprovisioned cells. The length determination for the WPT slotframe is described in Section 3.2 in detail.

Figure 3a presents an example of a WPT-enabled WSN topology consisting of three HAPs (HAP 0, 1, and 2) and two sensor nodes (sensor node 0 and 1). Figure 3b–d represent the CM, HAP, and WPT slotframes, and their lengths are 19, 5, and 11, respectively. In addition, HAP 1 can broadcast its EB frame in timeslots 0 and 19 (CM TxRxS) within the CM slotframes.

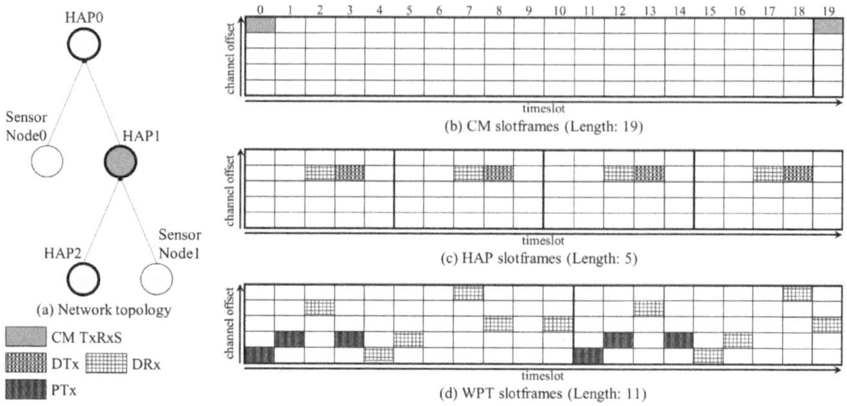

Figure 3. Example of multiple concurrent slotframes in MCSS.

In addition, it can exchange the RPL and 6P messages with its parent (i.e., HAP 0) or child nodes (i.e., HAP 2 and sensor node 1) in timeslots 0 and 19. Moreover, HAP 1 receives data packets from HAP 2 in timeslots 2, 7, 12, and 17 (DRxs) and transmits data packets to HAP 0 in timeslots 3, 8, 13, and 18 (DTxs) within the HAP slotframes. In the WPT slotframe, HAP 1 transmits power to sensor node 1 using power transmission cells (PTxs) and receives data packets from sensor node 1 using data reception cells (DRxs).

Figure 4 illustrates the MCSS schedule of HAP 1 in Figure 3a. In the figure, multiple concurrent slotframes of HAP 1 are projected onto one plane, and as a result, only one cell is allocated to each timeslot. Specifically, the data reception cell (DRx) of the HAP slotframe and data reception cell (DRx) of the WPT slotframe overlap in the cell (1 and 2) of the MCSS schedule. In this case, the data reception cell (DRx) of the HAP slotframe is allocated to the MCSS schedule of HAP 1 because the HAP slotframe has a higher priority than the WPT slotframe. At timeslots 0, 2, 3, 7, 8, 12, 13, 18, and 19, two or more cells with different channel offsets are scheduled; however, only one cell is allocated to the MCSS schedule according to the priority of concurrent slotframes.

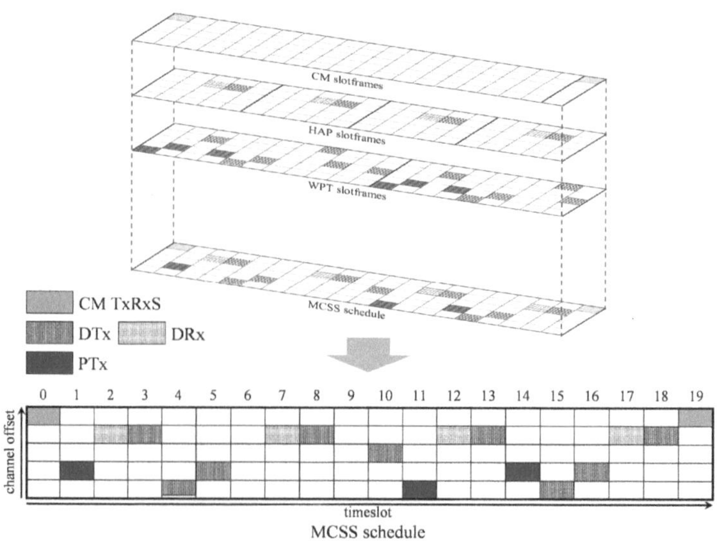

Figure 4. Example of an MCSS schedule.

3.2. Length Determination and Cell Allocation for WPT Slotframe

In WPT-enabled WSNs, the amount of intra-cluster traffic for each HAP cluster depends on the number of cluster members. Therefore, MCSS determines the WPT slotframe length differently for each HAP cluster, considering the minimum number of power and data cells required for energy harvesting and data transmission of cluster members and the number of overlapped cells among slotframes. To allocate power and data cells to the WPT slotframe, MCSS uses the three-step 6P transaction specified in the 6TiSCH standard [34]. However, unlike the existing standard, MCSS uses the modified 6P messages to accommodate all information for power cell allocation and data cell allocation.

To determine the WPT slotframe length, the HAP broadcasts the EB frame, including the length of the initial WPT slotframe. We assume that the initial WPT slotframe length is the same as the default slotframe length in the minimal 6TiSCH configuration [35]. Then, the HAP waits until it receives hello messages from all cluster members in the HAP cluster. The hello message includes the minimum number of data cells, which should be allocated to the initial WPT slotframe to accommodate the data traffic transmitted from the cluster member. Upon receiving the message from all cluster members, the HAP calculates the minimum number of power cells using the minimum number of data cells. The minimum number of power cells required for the i-th cluster member ($n_{minPC,i}$) is calculated as follows:

$$n_{minPC,i} = n_{minDC,i} \left\lceil \frac{E_{Tx,timeslot}}{E_{Rx,timeslot,i}} \right\rceil \quad (0 < i \leq n_{node,j}, 0 < j \leq M), \tag{3}$$

where i refers to the index of the cluster member in the HAP cluster, j denotes the index of the HAP, and $n_{node,j}$ is the number of cluster members for the j-th HAP cluster. Moreover, M represents the total number of HAPs in the network, $n_{minDC,i}$ denotes the minimum number of data cells required for the i-th cluster member, and $E_{Rx,timeslot,i}$ refers to the energy received by the i-th cluster member from the HAP (i.e., its parent) during one timeslot. Finally, $E_{Rx,timeslot,i}$ is calculated as follows:

$$E_{Rx,timeslot,i} = P_{Rx,i} L_{timeslot} \quad (0 < i \leq n_{node,j}, 0 < j \leq M), \tag{4}$$

where $P_{Rx,i}$ represents the power that the i-th cluster member receives per second. In addition, $P_{Rx,i}$ is calculated as follows:

$$P_{Rx,i} = \eta P_{Tx} |h|^2 = \eta P_{Tx} / (1 + |d_i|^{\alpha}) \quad (0 < i \leq n_{node,j}, 0 < j \leq M), \tag{5}$$

where η refers to the energy harvesting efficiency factor, P_{Tx} represents the transmission power of the HAP, h denotes the harvested power gain from the HAP, d_i indicates the distance between the i-th cluster member and its parent, and α refers to the path-loss parameter. Moreover, d_i is calculated as follows.

$$d_i = 10^{\frac{-rssi_{avg,i} + A}{10\alpha}} \quad (0 < i \leq n_{node,j}, 0 < j \leq M) \tag{6}$$

where A refers to the received signal strength of the cluster member at a distance of 1 m from its parent, and $rssi_{avg,i}$ denotes the average received signal strength indicator (RSSI) of data packets received from the parent of the i-th cluster member.

Then, the HAP calculates the numbers of overprovisioned power and data cells, which refer to power and data cells that must be allocated to compensate for the overlap between cells. The allocation of overprovisioned cells prevents power and data traffic delays caused by a difference in priority between slotframes. The numbers of overprovisioned power and data cells for the i-th cluster member ($n_{overPC,i}$ and $n_{overDC,i}$) are calculated by considering the ratio of $n_{minPC,i}$ and $n_{minDC,i}$ to the sum of the minimum numbers of the power and data cells for all cluster members in the j-th HAP cluster, expressed as follows:

$$n_{overPC,i} = round\left(\frac{n_{overcell,j}n_{minPC,i}}{\sum\limits_{i=1}^{n_{node,j}}(n_{minPC,i}+n_{minDC,i})}\right) \ (0 < i \leq n_{node,j}, 0 < j \leq M), \quad (7)$$

$$n_{overDC,i} = round\left(\frac{n_{overcell,j}n_{minDC,i}}{\sum\limits_{i=1}^{n_{node,j}}(n_{minPC,i}+n_{minDC,i})}\right) \ (0 < i \leq n_{node,j}, 0 < j \leq M), \quad (8)$$

where $n_{overcell,j}$ represents the total number of overprovisioned cells in the j-th HAP cluster. Further, $n_{overcell,j}$ is determined based on the least common multiple (LCM) of multiple concurrent slotframes of the j-th HAP cluster ($l_{LCM,j}$), calculated as follows:

$$n_{overcell,j} = \left\lfloor \frac{l_{WPT,j}n_{totalcell,j}}{l_{LCM,j}n_{node,j}} \right\rfloor \ (0 < j \leq M), \quad (9)$$

where $l_{WPT,j}$ represents the initial WPT slotframe length for the j-th HAP cluster. In addition, $n_{totalcell,j}$ denotes the total number of cells expected to be allocated to the MCSS schedule for the j-th HAP cluster during $l_{LCM,j}$ timeslots, calculated as follows:

$$n_{totalcell,j} = n_{CMcell,j} + n_{HAPcell,j} - n_{dupcell,CM\&HAP,j} \ (0 < j \leq M) \quad (10)$$

where $n_{CMcell,j}$ and $n_{HAPcell,j}$ are the respective numbers of cells expected to be independently allocated to the CM and HAP slotframes of the j-th HAP cluster during $l_{LCM,j}$ timeslots. Additionally, $n_{dupcell,CM\&HAP,j}$ is the number of cells expected to be redundantly allocated between the CM and HAP slotframes in the j-th HAP cluster during $l_{LCM,j}$ timeslots. Moreover, $n_{CMcell,j}$ and $n_{HAPcell,j}$ in Equation (10) are calculated as follows:

$$n_{CMcell,j} = \frac{l_{LCM,j}}{L_{CM}} \ (0 < j \leq M), \quad (11)$$

$$n_{HAPcell,j} = \frac{l_{LCM,j}n_{HAP,j}}{L_{HAP}} \ (0 < j \leq M), \quad (12)$$

where L_{CM} and L_{HAP} denote the length of CM and HAP slotframes, respectively. The term $n_{HAP,j}$ refers to the total number of HAPs in the j-th HAP cluster. In addition, $n_{dupcell,CM\&HAP,j}$ in Equation (10) is determined based on the relationship between repeated cycles of multiple concurrent slotframes for the same period, calculated as follows:

$$n_{dupcell,CM\&HAP,j} = \frac{l_{LCM,j}n_{HAP,j}}{LCM(L_{CM}, L_{HAP})} \ (0 < j \leq M) \quad (13)$$

where $LCM(L_{CM}, L_{HAP})$ represents the LCM of the lengths of the CM and HAP slotframes. Furthermore, $LCM(x_1, x_2, \cdots, x_w)$ refers to the LCM of x_1, x_2, ..., x_w, where w is the number of multiple concurrent slotframes.

Consequently, the numbers of required power and data cells for the i-th cluster member in the HAP cluster ($n_{reqPC,i}$ and $n_{reqDC,i}$) are calculated as follows:

$$n_{reqPC,i} = n_{minPC,i} + n_{overPC,i} \ (0 < i \leq n_{node,j}, 0 < j \leq M), \quad (14)$$

$$n_{reqDC,i} = n_{minDC,i} + n_{overDC,i} \ (0 < i \leq n_{node,j}, 0 < j \leq M). \quad (15)$$

For the WPT slotframe length determination, each HAP maintains a HAP cluster table (HCT) containing information about the cluster members. Specifically, the table entry of the HCT for the j-th HAP (i.e., the j-th HCT) includes the ID list ($\mathbf{ID}_{(j)}$) and numbers of required power and data cell lists ($\mathbf{N}_{reqPC(j)}$ and $\mathbf{N}_{reqDC(j)}$) of its cluster members of the j-th HAP. Finally, $\mathbf{ID}_{(j)}$, $\mathbf{N}_{reqPC(j)}$, and $\mathbf{N}_{reqDC(j)}$ are represented as follows:

$$\mathbf{ID}_{(j)} = \left[id_{(1)}, id_{(2)}, \cdots, id_{(i)}\right], \ (0 < i \leq n_{node,j}, 0 < j \leq M), \tag{16}$$

$$\mathbf{N}_{reqPC(j)} = \left[n_{reqPC,1}, n_{reqPC,2}, \cdots, n_{reqPC,i}\right], \ (0 < i \leq n_{node,j}, 0 < j \leq M), \text{ and} \tag{17}$$

$$\mathbf{N}_{reqDC(j)} = \left[n_{reqDC,1}, n_{reqDC,2}, \cdots, n_{reqDC,i}\right], \ (0 < i \leq n_{node,j}, 0 < j \leq M), \tag{18}$$

where $id_{(i)}$ indicates ID of the i-th cluster member in the j-th HAP cluster.

Algorithm 1 presents the procedure of the WPT slotframe length determination for each HAP. The HAP initializes variables (i.e., $l_{minWPT,j}$ and $primeNumFlag$), and $l_{minWPT,j}$ represents the minimum length required to determine the length of the WPT slotframe for the j-th HAP cluster, which is the sum of $n_{reqPC,i}$ and $n_{reqDC,i}$ for all cluster members in the HCT. In addition, $primeNumFlag$ determines whether the HAP locates the smallest prime number greater than $l_{minWPT,j}$. Then, the HAP calculates $l_{minWPT,j}$ by adding $n_{reqPC,i}$ and $n_{reqDC,i}$ for all cluster members included in $\mathbf{ID}_{(j)}$. Moreover, $l_{minWPT,j}$ of the j-th HAP, $l_{minWPT,j}$, is calculated as follows:

$$l_{\text{min}WPT,j} = \sum_{k=0}^{n_{node,j}-1} \left(n_{reqPC,k} + n_{reqDC,k}\right), \ 0 \leq k \leq n_{node,j} - 1. \tag{19}$$

If $l_{minWPT,j}$ is not a prime number, the j-th HAP increases $l_{minWPT,j}$ by increments of 1 to determine the smallest prime number greater than $l_{minWPT,j}$. Finally, the HAP sets $l_{WPT,j}$ to the determined $l_{minWPT,j}$. After the length determination for the WPT slotframe is completed, the HAP broadcasts the EB frame to notify its cluster members of $l_{WPT,j}$.

Algorithm 1. WPT slotframe length determination

INITIALIZE $primeNumFlag$ to TRUE, $l_{minWPT,j}$ to 0
FOR $id_{(i)}, i, i \in \left[0, n_{node,j}\right]$
  $l_{minWPT,j} \leftarrow l_{minWPT,j} + n_{reqPC,i} + n_{reqDC,i}$
ENDFOR
WHILE $primeNumFlag$
 FOR each iteration, $k, j \in \left[2, \text{math.sqrt}(l_{minWPT,j}) + 1\right]$
  IF $l_{minWPT,j} \,\%\, k == 0$
   $primeNumFlag \leftarrow$ TRUE
   $l_{minWPT,j} = l_{minWPT,j} + 1$
   Break
  ELSE
   $primeNumFlag \leftarrow$ FALSE
  ENDIF
 ENDFOR
ENDWHILE
$l_{WPT,j} \leftarrow l_{minWPT,j}$
RETURN $l_{WPT,j}$

Upon receiving the EB frame, including the numbers of required power and data cells ($n_{reqPC,i}$ and $n_{reqDC,i}$), the cluster member initiates the three-step 6P transaction to allocate power and data cells to the WPT slotframe. To this end, the i-th cluster member transmits the modified 6P ADD request, including $n_{reqPC,i}$ and $n_{reqDC,i}$ to its parent along with the empty PowerCellList and empty DataCellList. When the HAP receives the modified 6P ADD request from the i-th cluster member in the HAP cluster, it configures the PowerCellList and DataCellList for the i-th cluster member ($\mathbf{PC}_{(i)}$ and $\mathbf{DC}_{(i)}$) by considering its WPT slotframe schedule. In addition, $\mathbf{PC}_{(i)}$ and $\mathbf{DC}_{(i)}$ are represented as follows:

$$\mathbf{PC}_{(i)} = \left[pc_{(i,1)}, pc_{(i,2)}, \cdots, pc_{(i,k)}\right], \ (0 < i \leq n_{node,j}, 0 < j \leq M, 0 \leq k \leq l_{WPT,j}), \tag{20}$$

$$\mathbf{DC}_{(i)} = \left[dc_{(i,1)}, dc_{(i,2)}, \cdots, dc_{(i,l)}\right], \ (0 < i \leq n_{node,j}, 0 < j \leq M, 0 \leq l \leq l_{WPT,j}), \tag{21}$$

where $pc_{(i,k)}$ and $dc_{(i,l)}$ represent cells (i.e., timeslot and channel offset) for energy harvesting and data transmission, respectively. Moreover, k and l refer to the cell index in $\mathbf{PC}_{(i)}$ and $\mathbf{DC}_{(i)}$. Note that $\mathbf{PC}_{(i)}$ and $\mathbf{DC}_{(i)}$ consist of at least $n_{reqPC,i}$ and $n_{reqDC,i}$. When the configuration of $\mathbf{PC}_{(i)}$ and $\mathbf{DC}_{(i)}$ is completed, the HAP transmits the modified 6P ADD response to the i-th cluster member.

Upon receiving the modified 6P ADD response, the i-th cluster member selects the required power and data cells from $\mathbf{PC}_{(i)}$ and $\mathbf{DC}_{(i)}$. Then, the i-th cluster member transmits the modified 6P ADD confirmation to its parent, including the selected required power and data cells. Finally, HAP adds the selected required power and data cells to its WPT slotframe schedule. Likewise, the i-th cluster member adds them to its WPT slotframe schedule. Thus, the three-step 6P transaction is completed. Figure 5 presents an example of the three-step 6P transaction in MCSS.

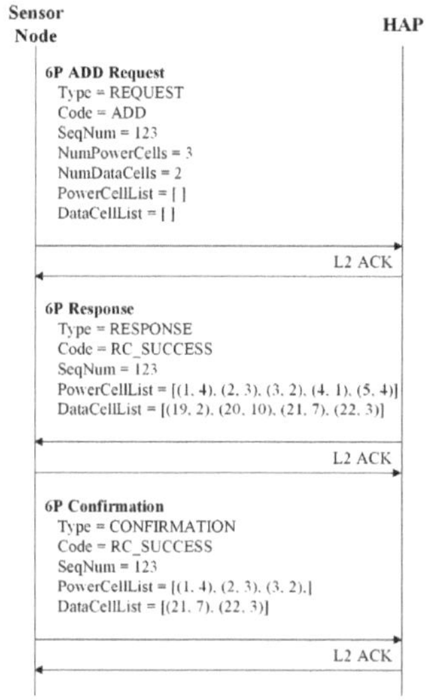

Figure 5. Example of the three-step 6P transaction.

4. Performance Evaluation

The experimental simulation was conducted under the IEEE 802.15.4 physical (PHY)/MAC layer environment using the MATLAB simulator to evaluate the performance of MCSS. The simulation results of MCSS were compared to those of the legacy TSCH MAC [26] with a single slotframe and TMSS [31] to verify the superiority of MCSS. We describe the simulation configuration in the following subsections and discuss the simulation results in detail.

4.1. Simulation Configuration

In the simulation, we considered a cluster-tree network topology consisting of five HAPs and multiple sensor nodes, in which one HAP is designated as the root, and the other HAPs maintain the same number of sensor nodes as the cluster members and operate as their CHs. The number of cluster members varies from one to 10 to investigate the performance of MCSS as the number of sensor nodes changes. Moreover, we assumed that each HAP is randomly placed within the communication range of one of the other adjacent HAPs, and each sensor node is randomly placed within a distance of 2 m from its CH.

To investigate the effect of setting the initial structure of the TSCH concurrent slotframe suitable for the network size and topology, we considered a static cluster-tree network environment without changes in the number of sensor nodes and traffic within the cluster. We further assumed that all sensor nodes maintain the same residual energy of 140 mA at the start of the simulation. The timeslot length was set to 10 ms, and we assumed that the legacy TSCH MAC uses a single slotframe with a fixed length of 100 timeslots.

For MCSS, the lengths of the CM slotframe and HAP slotframe were set to 331 and five timeslots, respectively. The CM slotframe length was set by referring to the default setting in the Contiki implementation [36], which is a representative TSCH+RPL network implementation, including the exchange of CMs. The HAP slotframe length was empirically set to a prime number that maximizes the aggregate throughput through experiments.

The WPT slotframe length was determined by Algorithm 1, and its maximum value is limited to 101 timeslots, the nearest prime number to the length of a single slotframe of the legacy TSCH MAC (i.e., 100). If no limitation exists on the WPT slotframe length, then the number of sensor nodes that can be accommodated in the network is almost the same; thus, no meaningful change in the network performance can be obtained.

TMSS uses an EB slotframe for transmitting EB frames and CMs and a TMSS slotframe for energy harvesting and the data transmission of sensor nodes. Thus, in TMSS, the EB slotframe length was set to 331 timeslots, the same as the CM slotframe of MCSS, and the TMSS slotframe length was set to 100 timeslots considering the transmission period of data packets.

Moreover, each sensor node was assumed to transmit a certain number of data packets per second to its CH, set to one, two, and four. The data packet size and data rate were set to 127 bytes and 250 Kbps, respectively. The simulation parameters, such as the Tx power (P_{Tx}) for each HAP, path-loss parameter (α), and energy harvesting efficiency factor (η), were set to 100 mW, 2.7, and 0.65, respectively, to calculate the number of power cells. In addition, the energy consumed when a sensor node is in *Tx* (P_{tx}), *Rx* (P_{rx}), idle (P_{idle}), and sleep (P_{sleep}) states during one timeslot was set to 20.98, 17.96, 0.001, and 0.001 mA, respectively.

The performance of MCSS in terms of average end-to-end delay, aggregate throughput, and average harvested energy was compared to those of the legacy TSCH MAC and TMSS. In addition, the effect of WPT on the aggregate throughput was investigated. The simulation was iterated 100 times. The detailed parameters are listed in Table 1.

Table 1. Simulation parameters.

Parameter	Value	Parameter	Value
PHY/MAC	IEEE 802.15.4	d	0–2 m
Number of sensor nodes	4–40	Data packets transmitted per second	1, 2, 4
Packet size	127 bytes	P_{tx}	20.98 mA
Data rate	250 Kbps	P_{rx}	17.96 mA
Timeslot length	10 ms	P_{idle}	0.001 mA
CM slotframe length	331	P_{sleep}	0.001 mA
HAP slotframe length	5	P_{Tx}	100 mW
WPT slotframe length	2–101	α	2.7
TSCH slotframe length	100	η	0.65

4.2. Simulation Results

In MCSS, the WPT slotframe length is determined according to the numbers of required power and data cells, among which the number of overprovisioned cells depends on the length of the CM slotframe and HAP slotframe. In the simulation, we first set the CM slotframe length shorter than the EB transmission period to ensure that the EB frame is transmitted at least once within the EB transmission period. Then, we experimentally investigated the HAP slotframe length that maximizes the aggregate throughput. In this experiment, the HAP slotframe length varied from three to 20.

Figure 6a–c illustrates the variations in the aggregate throughput of MCSS according to the change of the HAP slotframe length and the number of sensor nodes when the numbers of data packets transmitted per second are one, two, and four, respectively. Commonly in each figure, the aggregate throughput of MCSS increases as the HAP slotframe length decreases from 23 to five and decreases again when it changes from five to three. The shorter the HAP slotframe length, the more frequently data packets can be exchanged between HAPs.

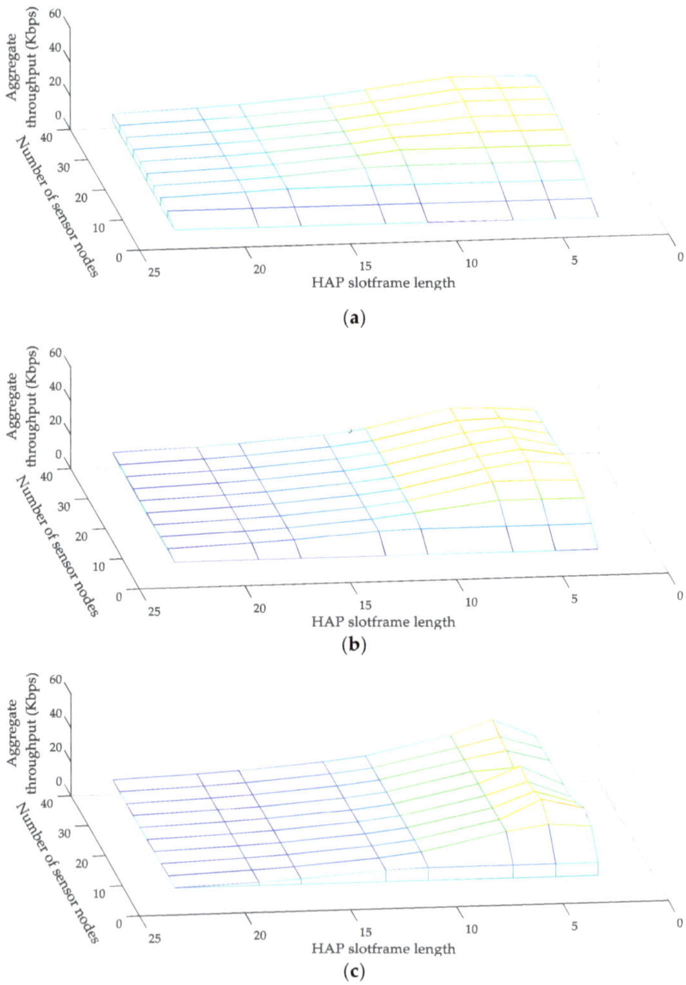

Figure 6. Aggregate throughput of MCSS: (**a**) one, (**b**) two, and (**c**) four packets/s.

However, the priority of a cell in the HAP slotframe is always higher than that of a cell in the WPT slotframe. Accordingly, when the HAP slotframe length is very short (i.e., three), the number of data packets transmitted from the sensor node to the CH in the WPT slotframe decreases, and as a result, the aggregate throughput decreases. In the figure, the color represents the level of the aggregate throughput; the brighter the color, the higher the aggregate throughput.

As a result, when the HAP slotframe length is five, the overall aggregate throughput is the highest. Therefore, we set the HAP slotframe length to five in later experiments. Figure 7 illustrates the variations in the average end-to-end delay for various numbers of data packets transmitted per second. The average end-to-end delay indicates the average time taken by a data packet to be transmitted from the sensor node to the root. MCSS exhibits a lower average end-to-end delay than the legacy TSCH MAC and TMSS, regardless of the number of data packets transmitted per second. In MCSS, each HAP transmits data packets received from its cluster members to the root using the HAP slotframe, which is shorter than the TSCH slotframe and TMSS slotframe.

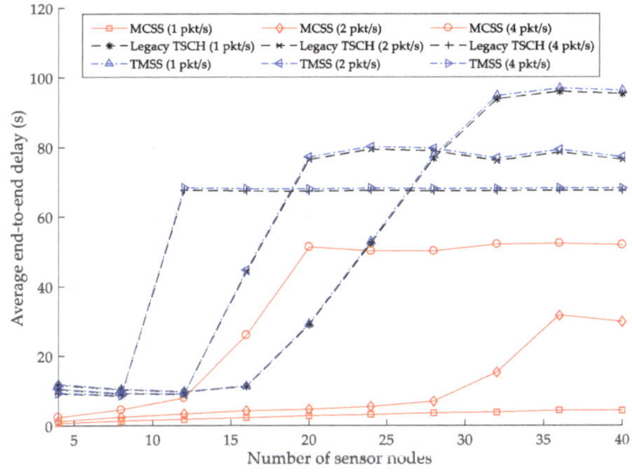

Figure 7. Average end-to-end delay.

In MCSS, when the number of data packets transmitted per second is one, the average end-to-end delay does not increase significantly as the number of sensor nodes increases. This outcome is because, in MCSS, the HAP slotframe can sufficiently accommodate data traffic using the HAP slotframe, and only data transmission delays due to the overlapped cells occur. When the numbers of data packets transmitted per second are two and four, the average end-to-end delay increases until the numbers of sensor nodes are 36 and 20, respectively. This result is because, even if the number of data packets transmitted from the sensor nodes to HAPs (i.e., intra-cluster traffic) increases, the number of cells exchanging data packets between HAPs does not increase.

In such a case, if the numbers of sensor nodes are greater than 36 and 20, all cells in the WPT slotframe are occupied. Additional data cells cannot be allocated to the WPT slotframe, and thus the end-to-end delay is no longer increased. In the legacy TSCH MAC, data cells are randomly allocated to a TSCH slotframe of fixed length. Accordingly, the HAP may wait for the next TSCH slotframe to transmit because, in the TSCH slotframe, the data cells used for inter-cluster traffic transmission may be allocated to an earlier timeslot compared with the data cells used for intra-cluster traffic transmission. The average end-to-end delay of TMSS is similar to that of the legacy TSCH MAC but slightly longer.

The lengths of the TMSS slotframe and TSCH slotframe are equal to 100 timeslots; thus, in both cases, the same number of power and data cells for inter-cluster traffic and intra-cluster traffic transmissions are allocated in one slotframe. However, in TMSS, due to the use of a separate EB slotframe, overlap between cells may cause the delay of data packets. Quantitatively, when the numbers of data packets transmitted per second are one, two, and four, the average end-to-end delays of MCSS are 94.09% and 94.15%, 80.40% and 80.60%, and 37.41% and 38.03% shorter than those of the legacy TSCH MAC and TMSS, respectively.

Figure 8a–c illustrates the variations in the WPT slotframe length for each HAP according to the change in the number of cluster members for one, two, and four data packets transmitted per second, respectively. Commonly in each figure, the WPT slotframe length of HAP 1 equals two because HAP 1 is the root and does not have any cluster members. When the number of data packets transmitted per second is one, the WPT slotframe length of HAPs (except HAP 1) increases as the number of cluster members increases because the WPT slotframe length is determined to accommodate the increased power and data traffic in the HAP cluster. In contrast, when the number of data packets transmitted per second is two or four, it increases and is limited to 101, the maximum value.

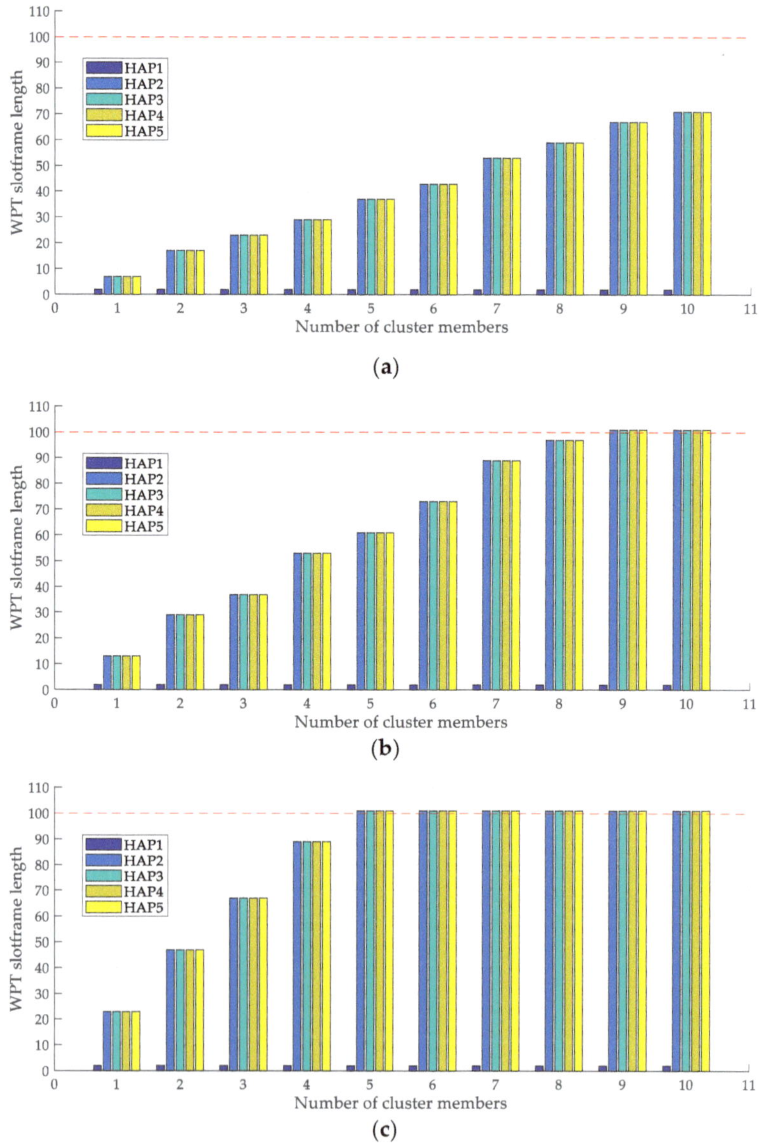

Figure 8. WPT slotframe lengths of HAPs: (**a**) one, (**b**) two, and (**c**) four packets/s.

Figure 9 displays the change in the aggregate throughput as the number of data packets transmitted per second varies. In MCSS, only the data and power cells are allocated to the WPT slotframe; however, in the existing TSCH, all types of cells are allocated to one TSCH slotframe. Similarly, in TMSS, all power and data cells except for EB frames and CMs are allocated to the same TMSS slotframe. In other words, the WPT slotframe of the MCSS can accommodate more data cells of sensor nodes than the slotframe of the legacy TSCH MAC.

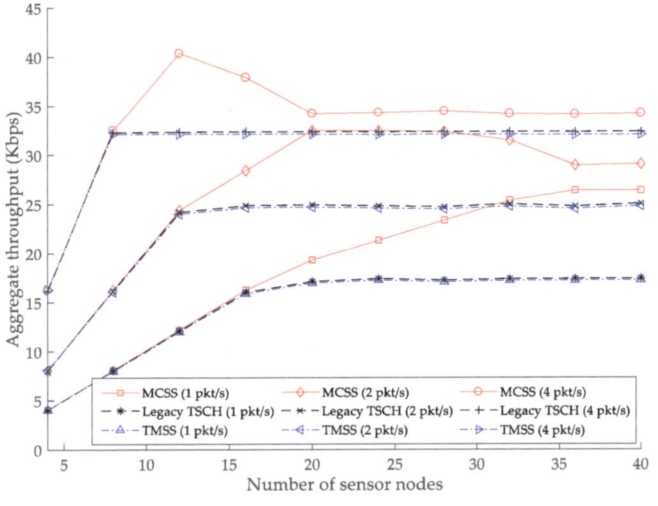

Figure 9. Aggregate throughput.

Therefore, in the figure, MCSS obtains a higher aggregate throughput than the legacy TSCH MAC and TMSS when the number of sensor nodes is larger than a certain number and the number of data packets transmitted per second is fixed. If the number of sensor nodes is less than a certain number, the aggregate throughput of all protocols is the same because their slotframe accommodates all network traffic. Quantitatively, when one, two, and four data packets are transmitted per second, the aggregate throughputs of MCSS are 26.82% and 26.60%, 18.63% and 18.37%, and 8.21% and 7.87% higher than those of the legacy TSCH MAC and TMSS, respectively.

Figure 10 depicts the average harvested energy for varying numbers of data packets transmitted per second. When the number of data packets transmitted per second is one, MCSS exhibits higher average harvested energy than legacy TSCH MAC and TMSS regardless of the number of sensor nodes because the WPT slotframe length is always shorter than the slotframe lengths of legacy TSCH MAC and TMSS.

When the number of packets transmitted per second is two or four, and if the number of sensor nodes is greater than 24 or 12, then the average harvested energy of the MCSS is lower than those of the legacy TSCH MAC and TMSS, as illustrated in Figure 8. Additionally, MCSS can accommodate power and data traffic for a larger number of sensor nodes than the legacy TSCH MAC and TMSS by using multiple concurrent slotframes. However, such multiple concurrent slotframes causes overlap between cells, thus, reducing the average energy harvested by individual sensor nodes in MCSS compared to the legacy TSCH MAC and TMSS.

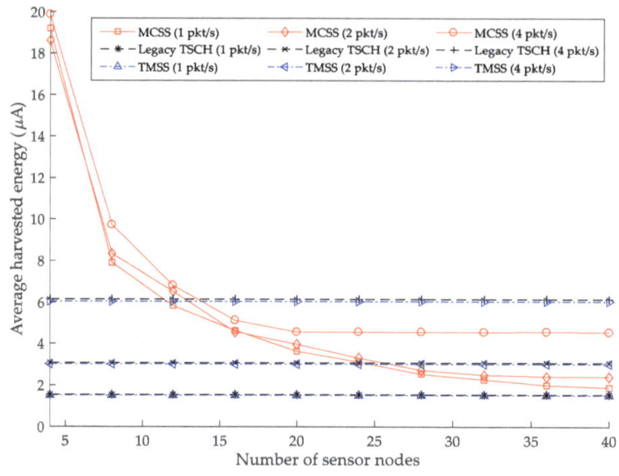

Figure 10. Average harvested energy.

When the number of data packets transmitted per second is one, the average harvested energy of MCSS decreases as the number of sensor nodes increases because an increase in the number of sensor nodes causes a decrease in the number of power cells for each sensor node allocated to the WPT slotframe. If the number of data packets transmitted per second is two or four, the average harvested energy of MCSS decreases only until it reaches a certain level. The reason is that the number of power cells successfully allocated to the WPT slotframe does not increase when the number of sensor nodes exceeds a certain number.

In contrast, the legacy TSCH MAC and TMSS maintain a constant average harvested energy that is almost similar to each other regardless of the number of sensor nodes because the number of power cells for each sensor node allocated to the slotframe is the same. In TMSS, the use of a separate EB slotframe enables energy harvesting of more sensor nodes compared to the legacy TSCH MAC. However, TMSS exhibits a slightly lower average harvested energy than the legacy TSCH MAC due to the overlap between cells. When the numbers of data packets transmitted per second are one, two, and four, the average harvested energies of MCSS are 244.30% and 244.28%, 79.68% and 79.68%, and 11.90% and 11.90% higher than those of the legacy TSCH MAC and TMSS, respectively.

To investigate the effect of WPT on the performance of MCSS, we compared the aggregate throughput for the case where the HAP transmits power to the sensor nodes (i.e., MCSS with WPT) and the case where it does not (i.e., MCSS without WPT). In this experiment, the initial residual energy of sensor nodes was randomly set between 70 and 140 mA, and the simulation time increased from 50 to 500 min in 50 min increments. Figure 11a,b illustrates the variations in the aggregate throughput of MCSS for varying simulation times when the numbers of sensor nodes are 12 and 36, respectively.

In the case of MCSS with WPT, the aggregate throughput is maintained as constant regardless of the simulation time because sensor nodes perform energy harvesting so that the energy-depleted dead nodes do not occur. On the other hand, as the simulation time increases, the aggregate throughput in the case of MCSS without WPT remains constant until a certain simulation time, and then starts to decrease. In the part in which the aggregate throughput is maintained constant, the dead nodes do not occur because the amount of energy consumed by the sensor nodes is less than their remaining energy. In contrast, in the opposite case, as the simulation time increases, the number of dead nodes increases, and, as a result, the aggregate throughput decreases.

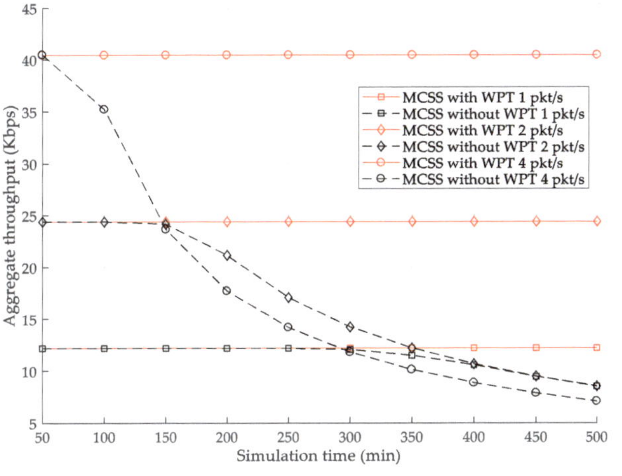

(a)

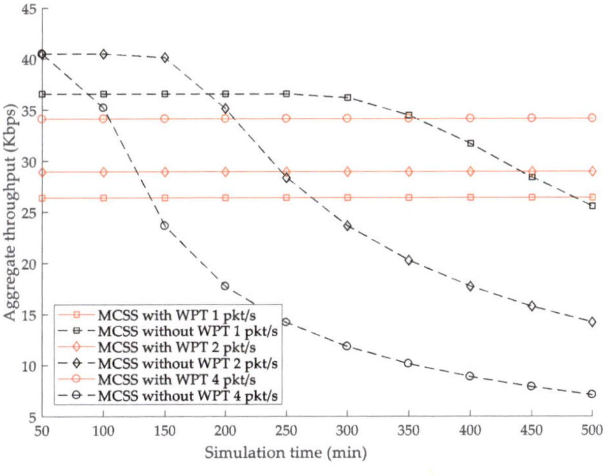

(b)

Figure 11. Effect of WPT on aggregate throughput: (**a**) 12 and (**b**) 36 sensor nodes.

In Figure 11a, in both cases, all sensor nodes successfully allocate the required data cells in the WPT slotframe; thus, the aggregate throughput remains the same until a certain simulation time. On the other hand, in Figure 11b, in the case of MCSS with WPT, there are sensor nodes that cannot allocate their own data cells due to the increase of the number of power cells allocated in the WPT slotframe; thus, there is a part in which the aggregate throughput in the case of MCSS without WPT is higher than that in the case of MCSS with WPT. However, in the case of MCSS without WPT, the number of dead nodes increases as the simulation time increases. Accordingly, after a certain simulation time, the aggregate throughput of MCSS without WPT becomes lower than that of MCSS with WPT.

5. Conclusions

This paper presented an MCSS protocol designed to support a cluster-tree network topology and various traffic types in WPT-enabled WSNs. In addition, MCSS defines three types of TSCH concurrent slotframes: CM slotframes, HAP slotframes, and WPT slotframes, to schedule the CM traffic, inter-cluster traffic, and intra-cluster traffic for WPT-enabled

WSNs, respectively. To address the problems caused by overlapped cells, MCSS sets each slotframe to a different length and transmission priority.

In particular, the WPT slotframe length for each cluster was determined based on the total number of required power and data cells, including the calculated number of overprovisioned cells. The experimental simulation was conducted in an environment of varying network sizes. The performances of MCSS, legacy TSCH MAC, and TMSS were compared in terms of the average end-to-end delay, aggregate throughput, and average harvested energy to demonstrate the superiority of MCSS.

Author Contributions: Conceptualization, S.-B.L. and E.-J.K.; methodology, S.-B.L., S.N.-X., J.-H.K., and E.-J.K.; software, S.-B.L., J.-H.K., and E.-J.K.; validation, S.-B.L., S.N.-X., and E.-J.K.; formal analysis, S.-B.L. and E.-J.K.; investigation, J.-H.K. and E.-J.K.; resources, S.N.-X. and E.-J.K.; data curation, J.-H.K.; writing—original draft preparation, S.-B.L., S.N.-X., J.-H.K., and E.-J.K.; writing—review and editing, S.-B.L., S.N.-X., J.-H.K., and E.-J.K.; visualization, S.-B.L.; supervision, E.-J.K.; project administration, E.-J.K.; funding acquisition, E.-J.K. All authors have read and agreed to the published version of the manuscript.

Funding: This research was supported by Hallym University Research Fund, 2022 (HRF-202201-007).

Institutional Review Board Statement: Not applicable.

Informed Consent Statement: Not applicable.

Data Availability Statement: No new data were created or analyzed in this study. Data sharing is not applicable to this article.

Conflicts of Interest: The authors declare no conflict of interest.

Abbreviations

The following abbreviations are used in this manuscript:

6P	6top Protocol
6TiSCH	IPv6 over the TSCH mode of the IEEE 802.15.4e
6top	6TiSCH Operation Sublayer
Ack	Acknowledgment
CH	Cluster Head
CM	Control Message
CSMA/CA	Carrier Sense Multiple Access with Collision Avoidance
DIO	DODAG Information Object
DODAG	Destination-Oriented Directed Acyclic Graph
EB	Enhanced Beacon
HAP	Hybrid Access Point
IoT	Internet of Things
LCM	Least Common Multiple
MAC	Medium Access Control
MCSS	Multiple Concurrent Slotframe Scheduling
OF	Objective Function
PHY	Physical
PRU	Power Receiving Unit
PTU	Power Transmitting Unit
RF	Radio Frequency
RPL	Routing Protocol for Low-power and Lossy Networks
RSSI	Received Signal Strength Indicator
Rx	Reception
TDMA	Time-Division Multiple Access
TMSS	TSCH Multiple Slotframe Scheduling
TSCH	Time-Slotted Channel Hopping
Tx	Transmission
WPT	Wireless Power Transfer
WSN	Wireless Sensor Network

References

1. Chen, S.; Xu, H.; Liu, D.; Hu, B.; Wang, H. A vision of IoT: Applications, challenges, and opportunities with china perspective. *IEEE Int. Things J.* **2014**, *1*, 349–359. [CrossRef]
2. Majid, M.; Habib, S.; Javed, A.R.; Rizwan, M.; Srivastava, G.; Gadekallu, T.R.; Lin, J.C.-W. Applications of Wireless Sensor Networks and Internet of Things Frameworks in the Industry Revolution 4.0: A Systematic Literature Review. *Sensors* **2022**, *22*, 2087. [CrossRef] [PubMed]
3. Pundir, S.; Wazid, M.; Singh, D.P.; Das, A.K.; Rodrigues, J.J.P.C.; Park, Y. Intrusion Detection Protocols in Wireless Sensor Networks Integrated to the Internet of Things Deployment: Survey and Future Challenges. *IEEE Access* **2020**, *8*, 3343–3363. [CrossRef]
4. Ramezani, P.; Jamalipour, A. Toward the evolution of wireless powered communication networks for the future Internet of Things. *IEEE Netw.* **2017**, *31*, 62–69. [CrossRef]
5. López, O.L.A.; Alves, H.; Souza, R.D.; Montejo-Sánchez, S.; Fernández, E.M.G.; Latva-Aho, M. Massive wireless energy transfer: Enabling sustainable IoT towards 6G era. *IEEE Int. Things J.* **2021**, *8*, 8816–8835. [CrossRef]
6. Pereira, F.; Correia, R.; Pinho, P.; Lopes, S.I.; Carvalho, N.B. Challenges in Resource-Constrained IoT Devices: Energy and Communication as Critical Success Factors for Future IoT Deployment. *Sensors* **2020**, *20*, 6420. [CrossRef]
7. Mishu, M.K.; Rokonuzzaman, M.; Pasupuleti, J.; Shakeri, M.; Rahman, K.S.; Hamid, F.A.; Tiong, S.K.; Amin, N. Prospective Efficient Ambient Energy Harvesting Sources for IoT-Equipped Sensor Applications. *Electronics* **2020**, *9*, 1345. [CrossRef]
8. Han, B.; Ran, F.; Li, J.; Yan, L.; Shen, H.; Li, A. A Novel Adaptive Cluster Based Routing Protocol for Energy-Harvesting Wireless Sensor Networks. *Sensors* **2022**, *22*, 1564. [CrossRef]
9. Choi, H.-H.; Lee, J.-R. Energy-Neutral Operation Based on Simultaneous Wireless Information and Power Transfer for Wireless Powered Sensor Networks. *Energies* **2019**, *12*, 3823. [CrossRef]
10. Lee, S.-B.; Kwon, J.-H.; Kim, E.-J. Residual Energy Estimation-Based MAC Protocol for Wireless Powered Sensor Networks. *Sensors* **2021**, *21*, 7617. [CrossRef]
11. Lu, X.; Niyato, D.; Wang, P.; Kim, D.I.; Han, Z. Wireless charger networking for mobile devices: Fundamentals, standards, and applications. *IEEE Wirel. Commun.* **2015**, *22*, 126–135. [CrossRef]
12. Bi, S.; Zeng, Y.; Zhang, R. Wireless powered communication networks: An overview. *IEEE Wirel. Commun.* **2016**, *23*, 10–18. [CrossRef]
13. Maeng, J.; Dahouda, M.K.; Joe, I. Optimal Power Allocation with Sectored Cells for Sum-Throughput Maximization in Wireless-Powered Communication Networks Based on Hybrid SDMA/NOMA. *Electronics* **2022**, *11*, 844. [CrossRef]
14. Sadeq, A.S.; Hassan, R.; Sallehudin, H.; Aman, A.H.M.; Ibrahim, A.H. Conceptual Framework for Future WSN-MAC Protocol to Achieve Energy Consumption Enhancement. *Sensors* **2022**, *22*, 2129. [CrossRef] [PubMed]
15. Shim, K.; Nguyen, T.-V.; An, B. Exploiting Opportunistic Scheduling Schemes and WPT-Based Multi-Hop Transmissions to Improve Physical Layer Security in Wireless Sensor Networks. *Sensors* **2019**, *19*, 5456. [CrossRef]
16. Li, K.; Ni, W.; Duan, L.; Abolhasan, M.; Niu, J. Wireless Power Transfer and Data Collection in Wireless Sensor Networks. *IEEE Trans. Veh. Technol.* **2018**, *67*, 2686–2697. [CrossRef]
17. Naderi, M.Y.; Nintanavongsa, P.; Chowdhury, K.R. RF-MAC: A Medium Access Control Protocol for Re-Chargeable Sensor Networks Powered by Wireless Energy Harvesting. *IEEE Trans. Wirel. Commun.* **2014**, *13*, 3926–3937. [CrossRef]
18. Ha, T.; Kim, J.; Chung, J.-M. HE-MAC: Harvest-then-transmit based modified EDCF MAC protocol for wireless powered sensor networks. *IEEE Trans. Wirel. Commun.* **2018**, *17*, 3–16. [CrossRef]
19. Kim, T.; Park, J.; Kim, J.; Noh, J.; Cho, S. REACH: An Efficient MAC Protocol for RF Energy Harvesting in Wireless Sensor Network. *Wirel. Commun. Mob. Comput.* **2017**, *2017*, 6438726. [CrossRef]
20. Kim, Y.-S.; Kwon, J.-H.; Lim, Y.; Kim, E.-J.; Kim, D.; Kim, Y.S. Hybrid Medium Access Control for Time-switching Simultaneous Wireless Information and Power Transfer. *Sens. Mater.* **2019**, *31*, 3549–3558. [CrossRef]
21. Cho, S.; Lee, K.; Kang, B.; Joe, I. A hybrid MAC protocol for optimal channel allocation in large-scale wireless powered communication networks. *EURASIP J. Wirel. Commun. Netw.* **2018**, *2018*, 9. [CrossRef]
22. Ju, H.; Zhang, R. Throughput maximization in wireless powered communication networks. *IEEE Trans. Wirel. Commun.* **2014**, *13*, 418–428. [CrossRef]
23. Xie, L.; Xu, J.; Zhang, R. Throughput maximization for UAV-enabled wireless powered communication networks. *IEEE Int. Things J.* **2018**, *6*, 1690–1703. [CrossRef]
24. Niyato, D.; Wang, P.; Kim, D.I. Performance Analysis and Optimization of TDMA Network With Wireless Energy Transfer. *IEEE Trans. Wirel. Commun.* **2014**, *13*, 4205–4219. [CrossRef]
25. Tavakoli, R.; Nabi, M.; Basten, T.; Goossens, K. Hybrid Timeslot Design for IEEE 802.15.4 TSCH to Support Heterogeneous WSNs. In Proceedings of the 2018 IEEE 29th Annual International Symposium on Personal, Indoor and Mobile Radio Communications (PIMRC), Bologna, Italy, 9–12 September 2018; pp. 1–7. [CrossRef]
26. *IEEE Std 802.15.4-2015*; IEEE Standard for Low-Rate Wireless Networks. IEEE: Piscataway, NJ, USA, 2016; pp. 1–709.
27. Duquennoy, S.; Al Nahas, B.; Landsiedel, O.; Watteyne, T. Orchestra: Robust mesh networks through autonomously scheduled TSCH. In Proceedings of the 13th ACM Conference on Embedded Networked Sensor Systems, Seoul, Korea, 1–4 November 2015; pp. 337–350. [CrossRef]

28. Jeong, S.; Paek, J.; Kim, H.S.; Bahk, S. TESLA: Traffic-aware Elastic Slotframe Adjustment in TSCH Networks. *IEEE Access* **2019**, *7*, 130468–130483. [CrossRef]
29. Kim, S.; Kim, H.; Kim, C. ALICE: Autonomous Link-based Cell Scheduling for TSCH. In Proceedings of the 2019 18th ACM/IEEE International Conference on Information Processing in Sensor Networks (IPSN), Montreal, QC, Canada, 16–18 April 2019; pp. 121–132. [CrossRef]
30. He, S.; Tang, Y.; Li, Z.; Li, F.; Xie, K.; Kim, H.-j.; Kim, G.-j. Interference-Aware Routing for Difficult Wireless Sensor Network Environment with SWIPT. *Sensors* **2019**, *19*, 3978. [CrossRef] [PubMed]
31. Kim, D.; Kwon, J.-H.; Kim, E.-J. TSCH Multiple Slotframe Scheduling for Ensuring Timeliness in TS-SWIPT-Enabled IoT Networks. *Electronics* **2021**, *10*, 48. [CrossRef]
32. Vilajosana, X.; Wang, Q.; Chraim, F.; Watteyne, T.; Chang, T.; Pister, K.S. A realistic energy consumption model for TSCH networks. *IEEE Sens. J.* **2013**, *14*, 482–489. [CrossRef]
33. Alexander, R.; Brandt, A.; Vasseur, J.; Hui, J.; Pister, K.; Thubert, P.; Levis, P.; Struik, R.; Kelsey, R.; Winter, T. RPL: IPv6 Routing Protocol for Low-Power and Lossy Networks. RFC 6550. 2012. Available online: https://www.rfc-editor.org/info/rfc6550 (accessed on 29 April 2022).
34. Wang, Q.; Vilajosana, X.; Watteyne, T. 6TiSCH Operation Sublayer (6top) Protocol (6P). RFC 8480. 2018. Available online: https://www.rfc-editor.org/info/rfc8480 (accessed on 29 April 2022).
35. Vilajosana, X.; Pister, K.; Watteyne, T. Minimal IPv6 over the TSCH Mode of IEEE 802.15.4e (6TiSCH) Configuration. RFC 8180. 2017. Available online: https://www.rfc-editor.org/info/rfc8180 (accessed on 29 April 2022).
36. Contiki-NG: The OS for Next Generation IoT Devices. Available online: https://github.com/contiki-ng/contiki-ng (accessed on 19 May 2022).

sensors

MDPI

Article

Energy-Efficient Optimal Power Allocation for SWIPT Based IoT-Enabled Smart Meter

Zaki Masood [1], Ardiansyah [2] and Yonghoon Choi [1,*]

[1] Department of Electrical Engineering, Chonnam National University, Gwangju 61186, Korea; 186204@jnu.ac.kr
[2] Department of Artificial Intelligence Convergence, Chonnam National University, Gwangju 61186, Korea; ardi@ejnu.net
* Correspondence: yh.choi@jnu.ac.kr

Abstract: This paper presents an internet of things (IoTs) enabled smart meter with energy-efficient simultaneous wireless information and power transfer (SWIPT) for the wireless powered smart grid communication network. The SWIPT technique with energy harvesting (EH) is an attractive solution for prolonging the battery life of ultra-low power devices. The motivation for energy efficiency (EE) maximization is to increase the efficient use of energy and improve the battery life of the IoT devices embedded in smart meter. In the system model, the smart meter is equipped with an IoT device, which implements the SWIPT technique in power splitting (PS) mode. This paper aims at the EE maximization and considers the orthogonal frequency division multiplexing distributed antenna system (OFDM-DAS) for the smart meters in the downlink with IoT enabled PS-SWIPT system. The EE maximization is a nonlinear and non-convex optimization problem. We propose an optimal power allocation algorithm for the non-convex EE maximization problem by the Lagrange method and proportional fairness to optimal power allocation among smart meters. The proposed algorithm shows a clear advantage, where total power consumption is considered in the EE maximization with energy constraints. Furthermore, EE vs. spectral efficiency (SE) tradeoff is investigated. The results of our algorithm reveal that EE improves with EH requirements.

Keywords: distributed antenna system; energy efficiency; energy harvesting; internet of things; smart grid; wireless power transfer

Citation: Masood, Z.; Ardiansyah; Choi, Y. Energy-Efficient Optimal Power Allocation for SWIPT Based IoT-Enabled Smart Meter. *Sensors* **2021**, *21*, 7857. https://doi.org/10.3390/s21237857

Academic Editors: Carlo Trigona, Olfa Kanoun and Slim Naifar

Received: 27 October 2021
Accepted: 19 November 2021
Published: 25 November 2021

1. Introduction

In recent years, there has been growing interest in smart grid technology to build a green and energy-efficient smart city. The smart grid is transforming the way of conventional power generation and renewable resources to distributed energy between utilities and consumers. Various solutions have been proposed for the efficient use of energy and for the reduced operational cost of a smart city [1–5]. Within the next few years, the smart energy meter is likely to become an important component in smart grid technology. In general terms, smart meters can be described as a way to obtain power consumption in more detail than conventional power meters. Smart meters allow two-way communications between the meter and the grid network, which reduce cost and increase reliability. The smart meters also coordinate bidirectional electricity trading in the smart grid. This bidirectional communication provides a smart meter to collect information and energy at the same time. Multiple approaches have been suggested for smart meters to be more intelligent and reliable [6–10]. Much work on the potential of smart grid communication has been investigated, among them also power-line communication (PLC) [5,11–13]. PLC carries data on a conductor simultaneously, which is used in power distribution to the consumers. A major draw back of using PLC is the current attenuation due to the high voltage lines. Another disadvantage is the noise generated by the power line which is far more than telephone lines. On the other hand, advanced metering infrastructure (AMI) system collects and analyzes energy consumption on-demand [14]. It forwards the collected data back to the utility for serval purposes such as load forecasting, monitoring, and billing.

The internet of things (IoT) is an open network to transport data over the internet without demanding human to computer interaction. Many studies have been published on IoT, which provides a platform to control smart appliance and sensor devices [15–21]. Bedi et al. [15] highlight the evolution of IoT in transforming electric power and energy systems (EPESs) into secure, reliable, and intelligent EPESs. The real-time co-simulation of demand response (DR) policies in the smart grids network is investigated in [18]. The report in [18] exhibits the internet-connected intelligent devices located at customer premises and the smart grid to collect energy information and send commands. Lee et al. [20] propose a new method to secure data transport in a cellular network from a cellular-connected IoT device to a host. It has been suggested that the use of IoT is reliable for the future smart grid communication [21]. This approach of [21] seems to be reliable due to the remote monitoring and efficient control of the power flow among connected devices in an electric grid. In general, the IoT infrastructure requires smart devices, which can harvest energy from the different sources to wireless nodes. The wireless energy transfer (WET) is one of the possible solutions, which can transfer energy to prolong the battery life of the IoT device. Many attempts [22–24] have been made with the purpose of wireless power transfer (WPT), which comprises inductive coupling and electromagnetic (EM) radiations. Furthermore, radio frequency (RF) signals carry energy, which can be used in wireless powered communication networks (WPCNs) [25], where wireless nodes can utilize the RF energy into the power. Therefore, wireless energy harvesting (EH) can be a promising solution, which converts the received RF into energy. As a result, EH can be useful to increase the battery life for the IoT device. In recent years [26–29], there has been considerable interest in simultaneous wireless information and power transfer (SWIPT), where the user equipment (UE) can harvest energy and transfer information at the same time.

This approach is well suited for energy constraint relay systems to improve energy efficiency (EE) and information transfer simultaneously. In the last few years, much more information on SWIPT has become available, where power splitting (PS) and time switching (TS) modes are investigated. In the PS mode, the SWIPT technique performs two functions: one for EH and the other for information decoding (ID). Chae et al. [30] report a new scheme for PS-SWIPT based IoT sensor networks, which reduces the transmit power under EH constraints. Additionally, PS-SWIPT operation has been widely studied with various multiple-input multiple-output (MIMO) channels and distributed antenna system (DAS) [31–33]. In [34], optimal energy cooperation policy has been investigated, which focuses on the advantage of using DAS in a smart grid system. The DAS spatially divides antenna nodes connected to a centralized processor via a physical medium and delivers wireless connectivity to a fixed area. It can be applied to indoors or outdoors in any communication network, where antenna elevations at or below the clutter level, and port connections compact. In [35], the downlink orthogonal frequency division multiple (OFDM) SWIPT system with multiple IoT devices is investigated. Specifically, the resource allocation problem to maximize the secrecy rate for OFDM access (OFDMA), and time division multiple access (TDMA) systems are investigated, in a full-duplex network. More recent evidence [36] reports the EE maximization in the DAS-based IoT network with SWIPT technology adopting an optimal power allocation scheme. Ariffin et al. [37] investigate real-time energy trading strategy, where beamforming technique in a downlink green cloud radio access network (C-RAN) with SWIPT is adopted. Furthermore, the authors of [38] perform multiuser resource allocation for OFDMA system with SWIPT in the PS mode.

Most studies have only focused on EE with SWIPT in conventional cellular communication networks. Various papers have been presented to solve this issue. In previous work [39], we consider the DAS with PS-SWIPT system, which maximizes EE by optimal power allocation for an IoT device, the solution based on the Lagrange method and KKT conditions. In contrast to the previous work, in this paper we consider multiple IoT-enabled smart meters and propose an optimal power allocation algorithm for a OFDM-DAS system with PS-SWIPT. This paper aims EE maximization in the OFDM-DAS system for smart grid, where the smart meters can harvest energy and transfer information with an IoT enabled SWIPT technology. Thus, this approach provides the energy-efficient power allocation scheme in a smart grid and improves the performance of each smart meter, which is equipped with SWIPT function. This method is chosen because it is one of the feasible ways to implement SWIPT in wireless powered smart grid communication networks. The main contributions of this paper are summarized as follows:

1. This research deals with multiple IoT-enabled smart meters, where the non-convex EE maximization problem is transformed into a subtractive form and the solution is based on proportional fairness.
2. This paper takes a new look at the EE objective function and considers three constraints, i.e., EH constraint, PS ratio at the energy harvester, and DAS transmit power.
3. An optimal power allocation algorithm is proposed for the non-convex EE maximization problem by adopting nonlinear fractional programming and the Lagrangian method.

We believe that our work presented in this paper will promote EH technologies in wireless powered smart grid communication networks.

This paper is organized as follows: Section 2 gives a brief overview of IoT-based smart grid system model. In Section 3, we analyze a non-convex EE maximization problem. A new methodology is outlined in Section 3 and a power allocation algorithm is proposed. Simulation results are presented in Section 4, followed by conclusion in Section 5.

2. System Model

As shown in Figure 1, wireless smart grid communication model can be classified in two layers: the smart grid layer and the communication layer. The renewable energy resources such as solar and wind are the self-generated power sources for the grid to satisfy the commercial, industrial and residential consumer load demands. In the smart grid layer, electric power flows from the distributed energy resources (DER) to the consumers. The DER generates electric power via conventional power plants or renewable sources, which are utilized to transport electricity to the electric load at the customer. On the other hand, in the communication layer, the consumer load is monitored by the smart meter. The essential part of the system in communication layer is a smart meter, which measures the power consumption of the home appliances and other devices. Here, it should be noted that in DAS all distributed antenna (DA) ports are connected via physical link to a central processor (CP), which performs DA port selection, DA power allocation, and subcarrier assignments. The useful data from the consumer devices are transferred transfer to the smart meter which enables IoT to exchange information via the Internet. Let us consider K smart meters and received signal at the kth smart meter is given by:

$$Y_k = \mathbf{h} X_k + z, \tag{1}$$

where $\mathbf{h}$ is the complex channel vector between DA port and smart meter, X_k is the transmit signal from the DA port, and z indicates the additive Gaussian noise $\mathcal{CN}(0, \sigma^2)$ at the destination. In the system model, it is assumed that each smart meter is equipped with an IoT device, which implements SWIPT functions in the PS mode. Throughout this paper, we will use the term PS-SWIPT to refer to SWIPT operation in PS mode at the smart meter. For the PS-SWIPT function, ρ_k denotes the splitting ratio. The information decodes with a part at the receiver is ρ_k, where the rest part $1 - \rho_k$ for the EH part. Consequently, the two split signals at the destination for EH and ID, respectively.

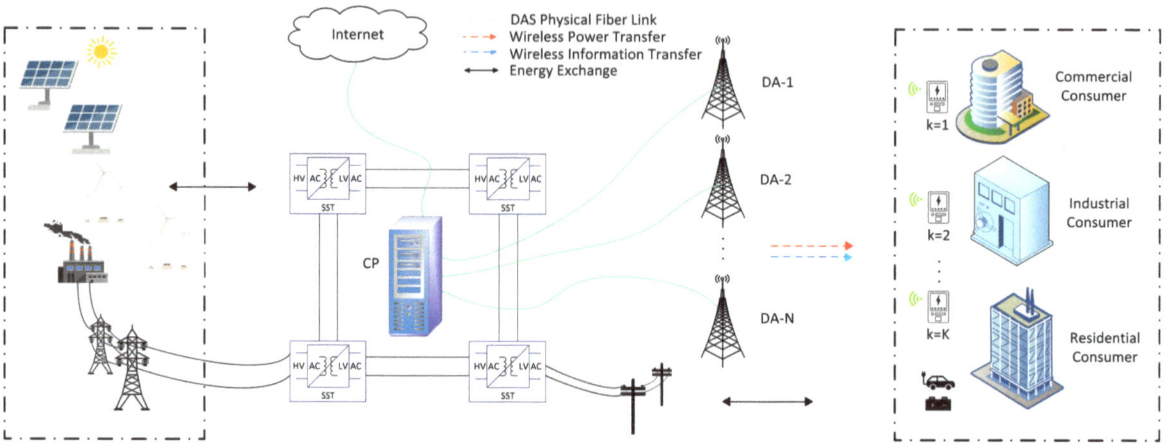

Figure 1. Structure of OFDM-DAS based IoT enabled PS-SWIPT in smart grid communication network.

Figure 2 shows the building area network (BAN), where a local user intercommunicates own smart appliances, electric vehicle (EV) control, battery storage, and solar panel. Moreover, building energy management systems (BEMS) allow the user to monitor and automatically control the use of energy [40]. The BAN uses wireless technology based on Zigbee standards or home wiring over PLC. It can be used to remotely monitor or control the connected devices mentioned above. Furthermore, this management system connects a customer's building network to the internet via WiFi connection. The smart meter interacts with the utility energy management system (UEMS) over the wireless communication network. We consider a downlink, OFDM-DAS system, where each DA port transmits data to the smart meter on multiple carrier frequencies. The bandwidth spectrum of this network is uniformly distributed into M subcarriers.

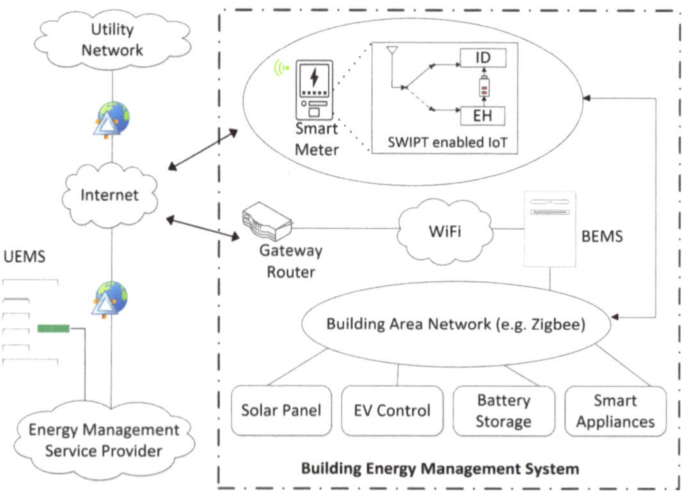

Figure 2. BEMS with IoT enabled PS-SWIPT, where smart meter in smart grid communication network exchange information between utility and BAN.

For information transmission among utility and smart grid, subcarriers can be allocated to each IoT device within smart meter. Moreover, each subcarrier allows only one user in the selected transmission time from the CP unit. Therefore, transmit power p_{mnk} from DA port n, on subcarrier m, to the smart meter k is written as (2), where P_{tot}^{max} denotes the total maximum transmit power. It is worth noting that each subcarrier is allocated to only a single IoT device in (3).

$$\sum_{k=1}^{K}\sum_{n=1}^{N}\sum_{m=1}^{M} p_{mnk} \leq P_{tot}^{max}, \tag{2}$$

$$p_{mnk} \cdot p_{mnk'} = 0, \ \forall m, n, k, \ k' = \{1, 2, \ldots, K\}, k' \neq k, \tag{3}$$

$$\sum_{k=1}^{K}\sum_{m=1}^{M} p_{mnk} \leq P_{n}^{max}, \tag{4}$$

where $m = 1, 2, \ldots, M$ and $n = 1, 2, \ldots, N$, which represents the m-th subcarrier and n-th DA port, respectively, and $k = 1, 2, \ldots, K$ is the smart meters equipped IoT enabled PS-SWIPT function. The maximum transmit power at the n-th DA port is P_n^{max} in (4).

The acquisition of perfect CSI is available between DA port and IoT device, although it is practically challenging, however, there are techniques to obtain CSI for IoT devices [41]. It is assumed that the channel state information (CSI) is available between the DA port and IoT devices [42,43]. h_{mnk} represents channel power gain between the IoT device and DA port. In the smart grid network, we consider only the downlink case for the coded information rate at the IoT device. The OFDM transfers the coded information on multiple subcarrier frequencies, which results in resilient to interference, and multipath effects. Additionally, it satisfies the spectral efficiency (SE) over the communication channel. It is well known that the information transmission rate R_k of the received RF signal for ID receiver is given as

$$R_k(\rho_k, p_{mnk}) = \frac{1}{M}\sum_{n=1}^{N}\sum_{m=1}^{M} \log_2\left(1 + \frac{\rho_k h_{mnk} p_{mnk}}{\sigma^2}\right), \tag{5}$$

where σ^2 is the Gaussian noise variance for the channel between IoT device and DA port, and M represents the total number of OFDM subcarriers.

It is important to note that, although the non-linear EH model is more practical [44], there are two reasons that we add the linear model, particularly into the PS-SWIPT system. The first reason is that the IoT-enabled smart meter has a low power regime for the SWIPT. Thus, it can be closely approximated by the linear EH model [45,46]. The second reason is that non-linear model is piecewise linear, where EH is easily trackable. Thus, for the EH part of the energy harvester, the harvested energy at the each IoT device can be written as:

$$E_k(\rho_k, p_{mnk}) = \xi(1 - \rho_k)\sum_{n=1}^{N}\sum_{m=1}^{M} h_{mnk}\sum_{k'=1}^{K} p_{mnk'}. \tag{6}$$

Therefore, the received power at the k-th device is $\sum_{n=1}^{N}\sum_{m=1}^{M} h_{mnk}\sum_{k'=1}^{K} p_{mnk'}$, where smart meter are equipped with IoT enabled PS-SWIPT function. In the relationship of EH, ξ is the energy conversion efficiency.

The RF-EH is considered as the dominant source in wireless green communication technology, as the demand for IoT and mobile base stations is increasing. Furthermore, from (6), it can be determined that an IoT device is able to decode information for its channel, and harvests energy from remaining channels. The efficiency of conversion adopting RF-EH is normally low, but energy is collected enough to recharge micropower devices such as IoT or remote sensors.

3. Problem Formulation

In the classical approach, the total power consumption is defined as (7), where μ is the reciprocal of the power amplifier drain efficiency, P_c is the power conversion which represents power consumption into the circuit at the receiver. Thus, received power at the harvester can be written as:

$$P_{con} = \mu \sum_{k=1}^{K} \sum_{n=1}^{N} \sum_{m=1}^{M} p_{mnk} + P_c. \tag{7}$$

Let us take advantage of utility function to model the modulation schemes in a power allocation problem. In our system model, IoT device with pre-specified application utility function can ensure optimal performance for a DAS system which outperforms regular schedulers. Unlike the conventional model, the power consumption in the SWIPT system can be compensated by the harvested power. This method is chosen because it is one of the most possible ways to include harvested energy into the system [47].

$$P_{total}(\rho_k, p_{mnk}) = \mu \sum_{k=1}^{K} \sum_{n=1}^{N} \sum_{m=1}^{M} p_{mnk} + P_c - E_k(\rho_k, p_{mnk}). \tag{8}$$

Generally, EE can be defined as the ratio of total achievable transmission rate and the total power consumption. Therefore, EE of IoT enabled PS-SWIPT system can be written as:

$$\eta_{EE} = \frac{\frac{1}{M}\sum_{k=1}^{K}\sum_{n=1}^{N}\sum_{m=1}^{M}\log_2\left(1+\frac{\rho_k h_{mnk} p_{mnk}}{\sigma^2}\right)}{\mu\sum_{k=1}^{K}\sum_{n=1}^{N}\sum_{m=1}^{M} p_{mnk} + P_c - \xi(1-\rho_k)\sum_{k=1}^{K}\sum_{n=1}^{N}\sum_{m=1}^{M} h_{mnk}\cdot\sum_{k'=1}^{K} p_{mnk'}}, \tag{9}$$

$$(P1): \quad \max_{p_{mnk}, \rho_k} \quad \eta_{EE} \tag{10}$$

$$\text{subject to} \quad E_k \geq E_{min}, \tag{11}$$

$$\sum_{k=1}^{K} \sum_{m=1}^{M} p_{mnk} \leq P_n^{max}, \ \forall m,n, \tag{12}$$

$$p_{mnk} \geq 0, \ \forall m, n, \tag{13}$$

$$p_{mnk} \cdot p_{mnk'} = 0, \ \forall m,n,k \neq k', \tag{14}$$

$$0 < \rho_k \leq 1. \tag{15}$$

In the problem formulation, we consider the transmit power at the DA port, E_{min}, and PS ratio constraints (11)–(15). The constraint in (11) is the EH constraint, which limits the harvested power at the harvester with minimum EH requirements. The constraint in (12) is the transmit power constraint, which limits the transmit power at the DA port. This limits the p_{mnk} to the peak transmit power P_n^{max} at the DA port. The constraints (14) and (15) express the subcarrier allocation in the OFDM-DAS system, and limit the PS-SWIPT ratio for EH and ID function, respectively.

Since OFDM divides a channel into multiple subcarriers, let θ_{mnk} denote subcarrier assignment indicator for each IoT device. Thus, only one user is allowed on the same subcarrier for the proposed optimal power allocation. θ_{mnk} can have a value of 0 or 1. In other words, $\theta_{mnk} = 1$ implies that subcarrier is allocated to the k-th user, otherwise $\theta_{mnk} = 0$. Therefore, (P1) is modified with indicator constraint (19) and (20) as:

$$(P2): \quad \max_{p_{mnk}, \rho_k} \quad \eta_{EE} \tag{16}$$

$$\text{subject to} \quad E_k \geq E_{min}, \tag{17}$$

$$\sum_{k=1}^{K}\sum_{m=1}^{M} p_{mnk} \leq P_n^{max}, \ \forall m,n, \tag{18}$$

$$\sum_{k=1}^{K}\sum_{n=1}^{N} \theta_{mnk} = 1, \ \forall m, \tag{19}$$

$$\theta_{mnk} \in \{0,1\}, \ \forall m,n,k. \tag{20}$$

Next, we calculate the second order condition (SOC) of the information transmission rate of (5) and total power of (8) with respect to p_{mnk}. According to (21) and (22), R_k is concave and P_{total} is convex with respect to p_{mnk}. Therefore EE objective function with respect to p_{mnk} is quasiconcave function.

$$\frac{\partial^2 R_k(\rho_k, p_{mnk})}{\partial p_{mnk}^2} = -\frac{1}{Mln2}.\frac{\rho_k^2 h_{mnk}^2}{(\sigma^2+\rho_k h_{mnk} p_{mnk})^2} < 0, \tag{21}$$

$$\frac{\partial^2 P_{total}(\rho_k, p_{mnk})}{\partial p_{mnk}^2} = 0. \tag{22}$$

In the case of (P2), the EE maximization problem is a quasi-concave for the power allocation variable P_n^{max} [47]. Therefore, EE maximization problem (P2) with respect to constraints is a non-convex optimization problem. Next, we will propose an optimal power allocation scheme for the smart grid enabled PS-SWIPT in OFDMA-DAS system. We adopt proportional fairness to solve the EE maximization problem. The key benefit of this method is to achieve optimal power allocation while satisfying the minimum EH requirements. In addition to (P2), another constraint is expressed in (25), where θ_{mni} and θ_{mnj} are the set of predetermined values for proportional rate constraint between the number of smart meters on DA port. Therefore, (P2) is transformed to a new optimization problem in its equivalent subtractive form, which is formulated as (P3). Similar to [48], we adopt subcarrier allocation for the DAS-OFDM system, where Ω_n is the set of a subcarrier for a transmission rate of the i-th smart meter on DA port. It is important to note that EE objective function (P2) and (P3) are equivalent if and only if $F(\omega^*) = 0$ and $f(\omega^*) = \mathbf{p}^*$ [49].

$$\begin{aligned}(P3): \max_{p_{mnk},\rho_k} & \sum_{k=1}^{K}\sum_{m\in\Omega_n} \frac{1}{M}\log_2\left(1+\frac{\rho_k h_{mnk} p_{mnk}}{\sigma^2}\right) \\ & -\omega.\left[\mu\sum_{k=1}^{K}\sum_{m\in\Omega_n} p_{mnk} + P_c - \xi(1-\rho_k)\sum_{k=1}^{K}\sum_{n=1}^{N}\sum_{m=1}^{M} h_{mnk}\sum_{k'=1}^{K} p_{mnk'}\right]\end{aligned} \tag{23}$$

subject to Equations (17) to (20), (24)

$$\frac{R_i}{R_j} = \frac{\sum_{m=1}^{M}\theta_{mni}}{\sum_{m=1}^{M}\theta_{mnj}}, \forall i,j = \{1,2,\ldots,K\}, i \neq j. \tag{25}$$

The function EE is a multivariable function and subject to constraints (17)–(20). Lagrangian function for EE maximization problem (P3) is given in (26), where $\lambda_{1,k}$, λ_2 and

$\lambda_{3,k}$ are the Lagrange multipliers for the constraints, respectively, $\forall\ k$. The Lagrangian function for the EE objective function can be written as:

$$\begin{aligned}\Lambda(p_{mnk},\rho_k,\lambda_{1,k},\lambda_2,\lambda_{3,k}) &= \sum_{k=1}^{K}\sum_{m\in\Omega_n}\frac{1}{M}\log_2\left(1+\frac{\rho_k h_{mnk}p_{mnk}}{\sigma^2}\right) - \omega.\left[\mu\sum_{k=1}^{K}\sum_{m\in\Omega_n} p_{mnk} + P_c\right.\\ &\left.-\xi(1-\rho_k)\sum_{k=1}^{K}\sum_{n=1}^{N}\sum_{m=1}^{M} h_{mnk}\sum_{k'=1}^{K} p_{mnk'}\right] + \sum_{k=1}^{K}\lambda_{1,k}(E_k - E_{min}) +\\ &\lambda_2\left(P_{m,n}^{max} - \sum_{k=1}^{K}\sum_{m\in\Omega_n} p_{mnk}\right) + \sum_{k=2}^{K}\lambda_{3,k}\left[\frac{1}{M}\sum_{m\in\Omega_n}\log_2\left(1+\frac{\rho_1 h_{mn1}p_{mn1}}{\sigma^2}\right) -\right.\\ &\left.\frac{\sum_{m=1}^{M}\theta_{mni}}{\sum_{m=1}^{M}\theta_{mnj}}\frac{1}{M}\sum_{m\in\Omega_n}\log_2\left(1+\frac{\rho_k h_{mnk}p_{mnk}}{\sigma^2}\right)\right].\end{aligned} \tag{26}$$

Next, the optimal power allocation for the EE objective function can be obtained by partial derivatives of the Lagrange function with respect to p_{mnk}. The first order partial derivative condition for p_{mnk} with Lagrange function can be written as:

$$\begin{aligned}\frac{\partial(p_{mnk},\rho_k,\lambda_{1,k},\lambda_2,\lambda_{3,k})}{\partial p_{mn1}} &= -\omega\mu + \xi(1-\rho_1)\, h_{mn1}\left(\sum_{k'=1}^{K}\lambda_{1,k} + \omega\right) - \lambda_2\\ &+ \frac{\rho_1 h_{mn1}}{(\sigma^2+\rho_1 h_{mn1}p_{mn1})}\left(\frac{1}{Mln2} + \frac{1}{Mln2}\sum_{k=2}^{K}\lambda_{3,k}\right),\end{aligned} \tag{27}$$

$$\begin{aligned}\frac{\partial(p_{mnk},\rho_k,\lambda_{1,k},\lambda_2,\lambda_{3,k})}{\partial p_{mnk}} &= -\omega\mu + \xi(1-\rho_k)h_{mnk}\left(\sum_{k=1}^{K}\lambda_{1,k} + \omega\right) - \lambda_2 +\\ &\frac{\rho_k h_{mnk}}{(\sigma^2+\rho_k h_{mnk}p_{mnk})}\cdot\left(\frac{1}{Mln2} - \frac{\sum_{m=1}^{M}\theta_{mni}\lambda_{3,k}}{\sum_{m=1}^{M}\theta_{mnj}Mln2}\right).\end{aligned} \tag{28}$$

Thus, from (27) and (28), we set the partial derivative equal to 0 and obtain optimal power as:

$$\tau_1^* = \frac{1}{\left[\omega\mu - \xi(1-\rho_1)h_{mn1}(\omega + \sum_{k=1}^{K}\lambda_{1,k}) + \lambda_2\right]}\cdot\left(\frac{1}{Mln2} + \frac{1}{Mln2}\sum_{k=2}^{K}\lambda_{3,k}\right) - \frac{\sigma^2}{\rho_1 h_{mn1}} \tag{29}$$

$$\tau_2^* = \frac{1}{\left[\omega\mu - \xi(1-\rho_k)h_{mnk}(1 + \sum_{k=1}^{K}\lambda_{1,k}) + \lambda_2\right]}\cdot\left(\frac{1}{Mln2} - \frac{\sum_{m=1}^{M}\theta_{mni}\lambda_{3,k}}{\sum_{m=1}^{M}\theta_{mnj}Mln2}\right) - \frac{\sigma^2}{\rho_k h_{mnk}} \tag{30}$$

where $\lambda_{1,k}^{(i+1)}, \lambda_2^{(i+1)}$ and $\lambda_{3,k}^{(i+1)}$ update the Lagrange multipliers. In order to obtain the optimal solution, we use the gradient method to update the Lagrange multipliers. Therefore, $\alpha_1^{(i)}$, $\alpha_2^{(i)}$ and $\alpha_3^{(i)}$ are the step size of the multipliers. Thus, multipliers can be written as:

$$\lambda_{1,k}^{(i+1)} = \left[\lambda_{1,k}^{(i)} + \alpha_1^{(i)}(E_{min} - E_k)\right]^+, \tag{31}$$

$$\lambda_2^{(i+1)} = \left[\lambda_2^{(i+1)} + \alpha_2^{(i)}\left(P_{m,n}^{max} - \sum_{k=1}^{K}\sum_{m\in\Omega_n} p_{mnk}\right)\right]^+, \tag{32}$$

$$\lambda_{3,k}^{(i+1)} = \lambda_{3,k}^{(i+1)} + \alpha_3^{(i)}\left[\frac{1}{M}\sum_{m\in\Omega_n}\log_2\left(1+\frac{\rho_1 h_{mn1}p_{mn1}}{\sigma^2}\right) - \frac{\sum_{m=1}^{M}\theta_{mni}}{\sum_{m=1}^{M}\theta_{mnj}}\frac{1}{M}\sum_{m\in\Omega_n}\log_2\left(1+\frac{\rho_k h_{mnk}p_{mnk}}{\sigma^2}\right)\right]^+. \tag{33}$$

Finally, optimal power allocation for the EE maximization problem can be written as:

$$p^*_{mn1} = min\{\tau^*_1, P^{max}_n\}, \quad (34)$$

$$p^*_{mnk} = min\{\tau^*_2, P^{max}_n\}. \quad (35)$$

In order to obtain the optimal solution for PS splitting ratio, similarly, we solve the partial derivatives for PS ratio, which can be written as:

$$\begin{aligned}\frac{\partial\Lambda(p_{mnk},\rho_k,\lambda_{1,k},\lambda_{2,1},\lambda_{3,k})}{\partial\rho_1} =& -\xi\left(\omega+\sum_{k=1}^{K}\lambda_{1,k}\right)h_{mn1}P_{mn1'}+ \\ & \left(\frac{1}{Mln2}+\frac{1}{Mln2}\sum_{k=2}^{K}\lambda_{3,k}\right)\left(\frac{h_{mn1}p_{mn1}}{\sigma^2+\rho_k h_{mn1}p_{mn1}}\right),\end{aligned} \quad (36)$$

$$\begin{aligned}\frac{\partial\Lambda(p_{mnk},\rho_k,\lambda_{1,k},\lambda_{2,1},\lambda_{3,k})}{\partial\rho_k} =& -\xi\left(\omega+\sum_{k=1}^{K}\lambda_{1,k}\right)\sum_{m\in\Omega_n}h_{mnk}P_{mnk'}+ \\ & \left(\frac{1}{Mln2}+\frac{\theta_k}{\theta_{k'}}\frac{1}{Mln2}\sum_{k=2}^{K}\lambda_{3,k}\right)\left(\frac{h_{mnk}p_{mnk}}{\sigma^2+\rho_k h_{mnk}p_{mnk}}\right),\end{aligned} \quad (37)$$

$$\rho^*_1 = \frac{\left(\frac{1}{Mln2}+\frac{1}{Mln2}\sum_{k=2}^{K}\lambda_{3,k}\right)}{\sigma^2+h_{mn1}p_{mn1}}\cdot\frac{h_{mn1}p_{mn1}}{\xi\left(\omega+\sum_{k=1}^{K}\lambda_{1,k}\right)h_{mn1}P_{mn1'}}, \quad (38)$$

$$\rho^*_k = \frac{\left(\frac{1}{Mln2}+\frac{\theta_{mni}}{\theta_{mnj}}\frac{\sum_{k=2}^{K}\lambda_{3,k}}{Mln2}\right)}{\sigma^2+h_{mnk}p_{mnk}}\cdot\frac{h_{mnk}p_{mnk}}{\xi\left(\omega+\sum_{k=1}^{K}\lambda_{1,k}\right)\sum_{m\in\Omega_n}h_{mnk}P_{mnk'}}. \quad (39)$$

First, subcarrier is assigned to each DA port, and later remaining subcarriers are assigned in such a way that it maximizes the overall SE of the OFDM-DAS. Similar to [50], we adopt the subcarrier allocation to find the optimal ratio for θ_{mni} and θ_{mnj} constraints in (25), with rate constraint ϕ_k in Table 1. The optimal power allocation and PS-SWIPT ratio for EE maximization are summarized as Algorithm 1. The convergence of the proposed algorithm can be observed from the number of iterations in the inner and outer loops. Each iteration step in the inner loop finds an optimal transmit power for the given parameters in Table 2. Since (26) is a non-convex function, the closed-form solution is computationally challenging. Therefore, after $\lambda_{1,k}, \lambda_2$ and $\lambda_{3,k}$ is set to fixed in the inner loop, optimal solution for p^*_{mnk} can be obtained. Next, we adopt the gradient method to compute the Lagrange multipliers, which are small step sizes $\alpha^{(i)}_1$, $\alpha^{(i)}_2$ and $\alpha^{(i)}_3$ for updating [48].

Table 1. Fairness Rate Constraint.

Fairness index k	0	1	2	3	4
$\phi_1 = \phi_2 = \phi_3 = \phi_4 = \phi_5$	2^0	2^1	2^3	2^4	2^5
$\phi_6 = \phi_7 = \ldots = \phi_k$	1	1	1	1	1

Table 2. Simulation Paramters.

Parameter	Value
Number of DA ports (N)	5
Number of IoT enabled smart meter devices (K)	15
Number of subcarriers (M)	64
Noise power (σ_z^2)	−104 dBm
Path loss exponent (α)	3.7
Circuit power consumption (P_c)	5 W
Shadow fading standard deviation	8 dB
Radius of the cell (R)	1000 m
Maximum transmit power (P_{max}^n)	30 dBm
Number of channel realization	10^4

Algorithm 1: Optimal transmit power for EE.

Initialization: $\eta_{EE} = 0, \omega = 0.01, \zeta = 0.0005, i = 0$
Set channel gain $h_{mnk}, \ \forall m, n, k$
for n = 1 : N **do**
while$\left(F(\omega) \geq \zeta \text{ and } \left|\lambda_{1,k}^{(i+1)} - \lambda_{1,k}^{(i)}\right| < \zeta \text{ and } \left|\lambda_2^{(i+2)} - \lambda_2^{(i)}\right| < \zeta \text{ and } \left|\lambda_{3,k}^{(i+3)} - \lambda_{3,k}^{(i)}\right| < \zeta\right)$
  for k = 1 : K **do**
    for m = 1 : M **do**
      if $k = 1$ **then,**
        Solve τ_1^*, obtain solution p_{mn1}^* (29),
      else
        Solve τ_2^*, obtain solution p_{mnk}^* (30).
    end for
  end for
Update $\lambda_{1,k}^{(i+1)}, \lambda_2^{(i+1)}$ and $\lambda_{3,k}^{(i+1)}$ for $i = i + 1$ by using (31), (32) and (33), respectively.
end while
Compute PS ratio ρ_1^* and ρ_k^*
By using $p_{mn1}^*, p_{mnk}^*, \rho_1^*$ and ρ_k^*, calculate optimal value of objective function η_{EE}^*
end for

4. Numerical Results

In this section, simulation results are provided to validate the performance of the proposed algorithm for EE maximization. The results demonstrate the effectiveness and convergence of the proposed power allocation algorithm. In the simulations, we set the total number of DA ports and smart meter enabled IoT devices $N = 5$ and $K = 15$, respectively. The number of subcarriers that confirms our finding set to $M = 64$.

Commonly, DAS is implemented as a cell, therefore in the simulation, it is considered that the DAS has a single cell structure in the smart grid network. Thus, main CP is fixed at the center and the number of DA ports is uniformly distributed in a cell. We assume that the DA ports are uniformly distributed in the smart grid communication network and located with polar coordinates $(\sqrt{\frac{3}{7}}R, \frac{2\pi(n-1)}{N})$, where $R = 1000$ m is the radius of the cell. For the numerical analysis, simulation parameters are considered according to Table 2. Therefore, for the $N = 5$ DA ports, the location of n-th DA port can be determined in the OFDM-DAS based IoT-enabled smart grid network. The drain efficiency of the power amplifier μ is set to 0.38, whereas the energy conversion efficiency ξ at the IoT device is 0.6. The log-distance path loss model is adopted, and the path loss exponent is 3.7 dB. For PS-SWIPT operation, the splitting ratio ρ_k at each IoT device ranges between 0 and 1. The shadow fading adopted as 8 dB, and noise power is set to −104 dBm.

For results comparisons, existing work shows EE maximization for DAS-SWIPT and OFDM-SWIPT, which have been published in [38,51–53]. A comparative table of these existing articles is listed in Table 3. First, we investigate the performance of EE and SE versus the number of iteration for the proposed power allocation algorithm. The relationship can be observed against the proportional fairness index ϕ_k.

Table 3. Recent Power Allocation Schemes.

Related work	OFDM	DAS	SWIPT
Xu et al. [38]	✓	-	✓
Xu et al. [51]	✓	✓	-
Zhou et al. [52]	✓	-	✓
Yu et al. [53]	✓	✓	-
Our paper	✓	✓	✓

In Figures 3 and 4, we plot SE and EE versus the number of iterations, with fairness index ϕ_k, according to rate constraint in Table 1. The fairness index controls the subcarrier allocation to meet constraints requirements. The minimum EH requirement E_{min} is set to 2 mW. The static circuit power P_c is assumed to be 5 W, for a fixed circuit power and we can conclude that EE is nonincreasing for the number of smart meters. For the convergence, the step sizes (α_1, α_2, and α_3) are set to 0.005, which update Lagrange multipliers. For each DA port, the maximum transmit power is set to 30 dBm. It can be observed that EE converges with a few number of iterations. After four iterations, the optimal power maximizes EE and achieves a steady value. This method gives a fairness strategy to control power allocation among PS-SWIPT based smart meters.

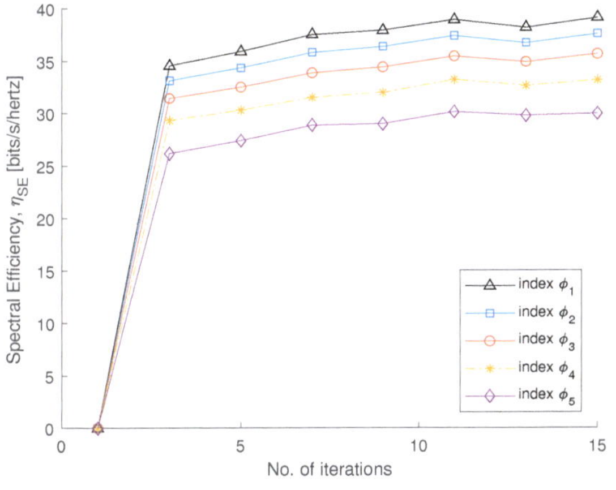

Figure 3. SE vs. number of iterations.

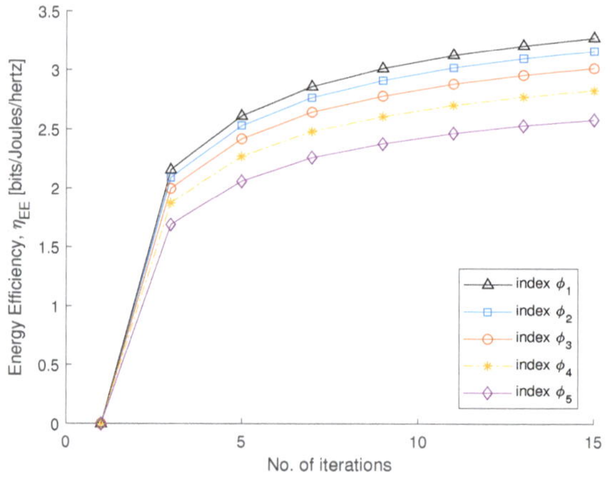

Figure 4. EE vs. number of iterations.

In the next simulations, we compare the EE and SE performance for different power constraints with fairness index ϕ_k. In Figure 5, we plot SE with respect to the maximum transmit power constraint. The SE is obtained by the optimal power allocation, which rises dramatically in the maximum allowed transmit power constraint. Initially, no significant difference is found in the case of the small transmit power, whereas after 15 dBm it increases distinctly to achieve the convergence. The result shows that SE of the proposed algorithm converges with an increase in transmit power to meet EH requirements. In Figure 6, the proposed EE maximization algorithm under different maximum transmit power constraint is evaluated. Similar behavior is also reported in [51], where the EE maximization problem approached without SWIPT. It can be seen from Figure 6 that the EE increases sharply for small transmit power constraints and steadily increases for high. Thus, a balance between EE and maximum transmit power constraint can be achieved for the higher transmit power constraint.

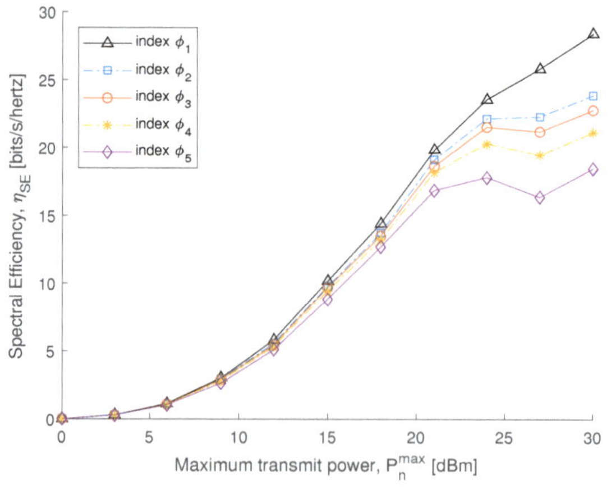

Figure 5. SE vs. transmit power.

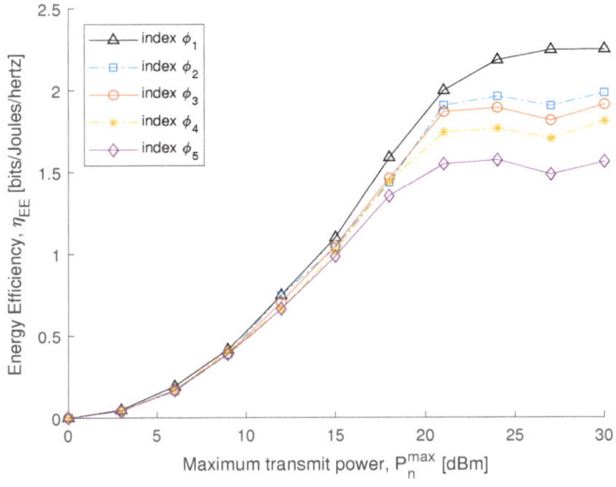

Figure 6. EE vs. transmit power.

Figure 7 illustrates the SE and EE tradeoff for PS-SWIPT with different rate constraints in Table 1. It shows that the EE increases and approaches a certain peak for a fixed P_c and E_{min} acquires no better energy-efficient transmission that is closed to this asymptote.

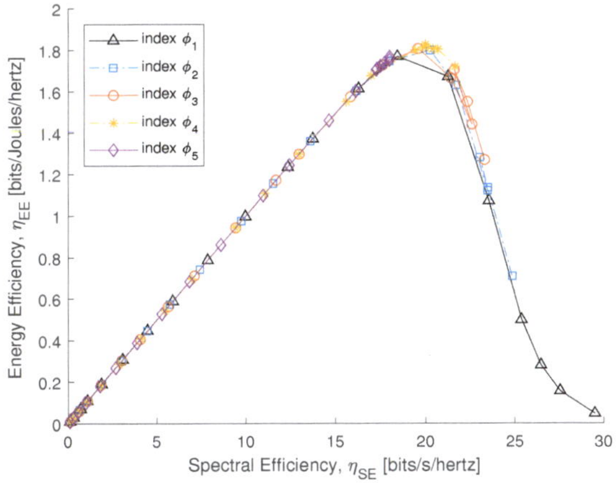

Figure 7. SE vs. EE.

To further understand the benefits of the proposed algorithm, in Figures 8 and 9, we compare the SE and EE performance of the proposed power allocation with the number of IoT-enabled smart meters. Figures 8 and 9 represent the SE and EE for the PS-SWIPT, as the number of IoT-enabled devices increases with different rate constraint index. These results confirm that sufficient transmit power is available at the DA port if there is an increase in IoT-enabled smart meters.

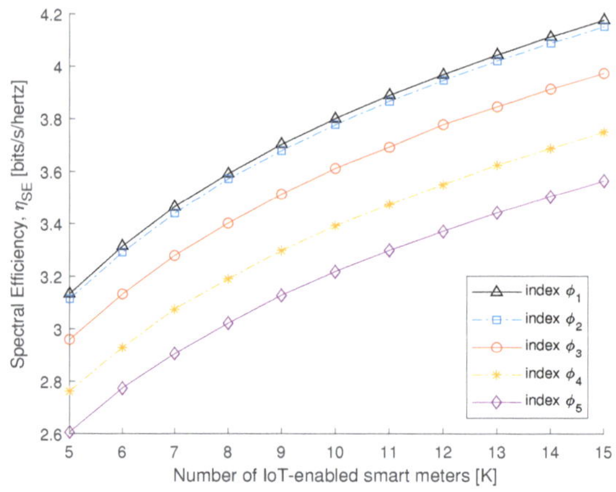

Figure 8. SE vs. number of smart meters.

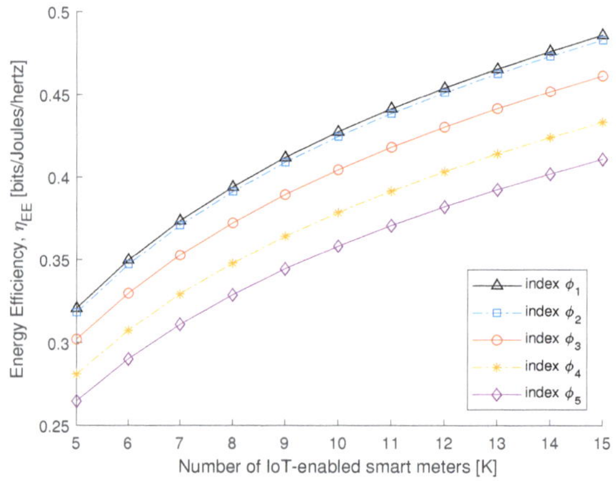

Figure 9. EE vs. number of smart meters.

5. Conclusions

This paper addressed the EE maximization problem for IoT-enabled smart meters in DAS-OFDM system. We formulated the EE maximization problem and proposed an algorithm for the optimal transmit power and PS-SWIPT ratio. The algorithm found an optimal solution, which maximizes the EE in the DAS-OFDM system with proportional fairness among IoT-enabled smart meters. The performance of the proposed algorithm demonstrated that the EE converges towards an optimal solution with minimum EH requirements. Furthermore, the numerical results showed that our proposed algorithm achieved the optimal solution with a few iterations. It also confirmed that as the number of IoT-enabled smart meters increases, the EE increases under minimum EH requirements. Therefore, our research will be constructive in solving the difficulty of EE maximization with IoT-enabled PS-SWIPT in the wireless powered smart grid communication networks. Hence, further studies on the current topic are required towards energy-efficient solutions in the uplink case to establish green communication technology for smart grid networks.

Author Contributions: Z.M., the main research idea and manuscript preparation, software, validation; A.A., software, manuscript preparation; Y.C., the main research idea and manuscript preparation. All authors have read and agreed to the published version of the manuscript.

Funding: This research was supported by Basic Science Research Program through the National Research Foundation of Korea (NRF) funded by the Ministry of Education (NRF-2019R1I1A3A01060631).

Institutional Review Board Statement: Not Applicable.

Informed Consent Statement: Not Applicable.

Data Availability Statement: Not Applicable.

Conflicts of Interest: The authors declare no conflict of interest.

References

1. Morello, R.; Mukhopadhyay, S.C.; Liu, Z.; Slomovitz, D.; Samantaray, S.R. Advances on sensing technologies for smart cities and power grids: A review. *IEEE Sens. J.* **2017**, *17*, 7596–7610. [CrossRef]
2. Ye, F.; Qian, Y.; Hu, R.Q.; Das, S.K. Reliable energy-efficient uplink transmission for neighborhood area networks in smart grid. *IEEE Trans. Smart Grid* **2015**, *6*, 2179–2188. [CrossRef]
3. Mohamed, N.; Al-Jaroodi, J.; Jawhar, I.; Lazarova-Molnar, S.; Mahmoud, S. SmartCityWare: A service-oriented middleware for cloud and fog enabled smart city services. *IEEE Access* **2017**, *5*, 17576–17588. [CrossRef]
4. Wen, G.; Yu, X.; Liu, Z.W.; Yu, W. Adaptive consensus-based robust strategy for economic dispatch of smart grids subject to communication uncertainties. *IEEE Trans. Ind. Inform.* **2017**, *14*, 2484–2496. [CrossRef]
5. Li, B.; Kisacikoglu, M.C.; Liu, C.; Singh, N.; Erol-Kantarci, M. Big data analytics for electric vehicle integration in green smart cities. *IEEE Commun. Mag.* **2017**, *55*, 19–25. [CrossRef]
6. Hussain, S.S.; Tak, A.; Ustun, T.S.; Ali, I. Communication modeling of solar home system and smart meter in smart grids. *IEEE Access* **2018**, *6*, 16985–16996. [CrossRef]
7. Worighi, I.; Maach, A.; Hafid, A.; Hegazy, O.; Van Mierlo, J. Integrating renewable energy in smart grid system: Architecture, virtualization and analysis. *Sustain. Energy Grids Netw.* **2019**, *18*, 100226. [CrossRef]
8. Sun, Q.; Li, H.; Ma, Z.; Wang, C.; Campillo, J.; Zhang, Q.; Wallin, F.; Guo, J. A comprehensive review of smart energy meters in intelligent energy networks. *IEEE Internet Things J.* **2015**, *3*, 464–479. [CrossRef]
9. Siryani, J.; Tanju, B.; Eveleigh, T.J. A machine learning decision-support system improves the internet of things smart meter operations. *IEEE Internet Things J.* **2017**, *4*, 1056–1066. [CrossRef]
10. Aalamifar, F.; Lampe, L. Optimized WiMAX profile configuration for smart grid communications. *IEEE Trans. Smart Grid* **2016**, *8*, 2723–2732. [CrossRef]
11. Rabie, K.M.; Adebisi, B.; Gacanin, H.; Yarkan, S. Energy-per-bit performance analysis of relay-assisted power line communication systems. *IEEE Trans. Green Commun. Netw.* **2018**, *2*, 360–368. [CrossRef]
12. Sharma, D.; Dubey, A.; Mishra, S.; Mallik, R.K. A frequency control strategy using power line communication in a smart microgrid. *IEEE Access* **2019**, *7*, 21712–21721. [CrossRef]
13. Papadopoulos, T.A.; Kaloudas, C.G.; Chrysochos, A.I.; Papagiannis, G.K. Application of narrowband power-line communication in medium-voltage smart distribution grids. *IEEE Trans. Power Del.* **2013**, *28*, 981–988. [CrossRef]
14. Ke, C.H.; Hsieh, S.Y.; Lin, T.C.; Ho, T.H. Efficiency network construction of advanced metering infrastructure using Zigbee. *IEEE Trans. Mobile Comput.* **2018**, *18*, 801–813. [CrossRef]
15. Bedi, G.; Venayagamoorthy, G.K.; Singh, R.; Brooks, R.R.; Wang, K.C. Review of Internet of Things (IoT) in electric power and energy systems. *IEEE Internet Things J.* **2018**, *5*, 847–870. [CrossRef]
16. Andrawes, A.; Nordin, R.; Abdullah, N.F. Energy-efficient downlink for non-orthogonal multiple access with SWIPT under constrained throughput. *Energies* **2020**, *13*, 107. [CrossRef]
17. Zafari, F.; Papapanagiotou, I.; Christidis, K. Microlocation for internet-of-things-equipped smart buildings. *IEEE Internet Things J.* **2015**, *3*, 96–112. [CrossRef]
18. Barbierato, L.; Estebsari, A.; Pons, E.; Pau, M.; Salassa, F.; Ghirardi, M.; Patti, E. A distributed IoT infrastructure to test and deploy real-time demand response in smart grids. *IEEE Internet Things J.* **2018**, *6*, 1136–1146. [CrossRef]
19. Montori, F.; Bedogni, L.; Bononi, L. A collaborative internet of things architecture for smart cities and environmental monitoring. *IEEE Internet Things J.* **2017**, *5*, 592–605. [CrossRef]
20. Lee, C.; Fumagalli, A. Internet of things security-multilayered method for end to end data communications over cellular networks. In Proceedings of the 2019 IEEE 5th World Forum on Internet of Things (WF-IoT), Limerick, Ireland, 15–18 April 2019; IEEE: Piscataway, NJ, USA, 2019; pp. 24–28.
21. Morello, R.; De Capua, C.; Fulco, G.; Mukhopadhyay, S.C. A smart power meter to monitor energy flow in smart grids: The role of advanced sensing and IoT in the electric grid of the future. *IEEE Sens. J.* **2017**, *17*, 7828–7837. [CrossRef]
22. Clerckx, B.; Bayguzina, E. Low-complexity adaptive multisine waveform design for wireless power transfer. *IEEE Antennas Wirel. Propag. Lett.* **2017**, *16*, 2207–2210. [CrossRef]
23. Lee, J.; Lee, K.; Cho, D.H. Stability improvement of transmission efficiency based on a relay resonator in a wireless power transfer system. *IEEE Trans. Power Electron.* **2017**, *32*, 3297–3300. [CrossRef]
24. Zhou, W.; Sandeep, S.; Wu, P.; Yang, P.; Yu, W.; Huang, S.Y. A wideband strongly coupled magnetic resonance wireless power transfer system and its circuit analysis. *IEEE Microw. Wirel. Compon. Lett.* **2018**, *28*, 1152–1154. [CrossRef]

25. Hu, S.; Chen, X.; Ni, W.; Wang, X.; Hossain, E. Modeling and analysis of energy harvesting and smart grid-powered wireless communication networks: A contemporary survey. *IEEE Trans. Green Commun. Netw.* **2020**, *4*, 461–496. [CrossRef]
26. Perera, T.D.P.; Jayakody, D.N.K.; Sharma, S.K.; Chatzinotas, S.; Li, J. Simultaneous wireless information and power transfer (SWIPT): Recent advances and future challenges. *IEEE Commun. Surv. Tuts.* **2017**, *20*, 264–302. [CrossRef]
27. Shi, L.; Ye, Y.; Hu, R.Q.; Zhang, H. Energy efficiency maximization for SWIPT enabled two-way DF relaying. *IEEE Signal Process. Lett.* **2019**, *26*, 755–759. [CrossRef]
28. Jalali, J.; Khalili, A. Optimal resource allocation for MC-NOMA in SWIPT-enabled networks. *IEEE Commun. Lett.* **2020**, *24*, 2250–2254. [CrossRef]
29. Rajaram, A.; Khan, R.; Tharranetharan, S.; Jayakody, D.N.K.; Dinis, R.; Panic, S. Novel SWIPT schemes for 5G wireless networks. *Sensors* **2019**, *19*, 1169. [CrossRef]
30. Chae, S.H.; Jeong, C.; Lim, S.H. Simultaneous wireless information and power transfer for Internet of Things sensor networks. *IEEE Internet Things J.* **2018**, *5*, 2829–2843. [CrossRef]
31. Zhu, Z.; Huang, S.; Chu, Z.; Zhou, F.; Zhang, D.; Lee, I. Robust designs of beamforming and power splitting for distributed antenna systems with wireless energy harvesting. *IEEE Syst. J.* **2018**, *13*, 30–41. [CrossRef]
32. Li, J.; Xiong, K.; Cao, J.; Yang, X.; Liu, T. Energy efficiency in RF energy harvesting-powered distributed antenna systems for the internet of things. *Sensors* **2020**, *20*, 4631. [CrossRef]
33. Zhu, Z.; Chu, Z.; Zhou, F.; Niu, H.; Wang, Z.; Lee, I. Secure beamforming designs for secrecy MIMO SWIPT systems. *IEEE Wirel. Commun. Lett.* **2017**, *7*, 424–427. [CrossRef]
34. Yuan, F.; Jin, S.; Wong, K.K.; Zhao, J.; Zhu, H. Wireless information and power transfer design for energy cooperation distributed antenna systems. *IEEE Access* **2017**, *5*, 8094–8105. [CrossRef]
35. Xu, D.; Zhu, H. Secure Transmission for SWIPT IoT Systems With Full-Duplex IoT Devices. *IEEE Internet Things J.* **2019**, *6*, 10915–10933. [CrossRef]
36. Yu, X.; Chu, J.; Yu, K.; Teng, T.; Li, N. Energy-Efficiency Optimization for IoT-Distributed Antenna Systems With SWIPT Over Composite Fading Channels. *IEEE Internet Things J.* **2019**, *7*, 197–207. [CrossRef]
37. Ariffin, W.N.S.F.W.; Zhang, X.; Nakhai, M.R. Sparse beamforming for real-time resource management and energy trading in green C-RAN. *IEEE Trans. Smart Grid* **2016**, *8*, 2022–2031. [CrossRef]
38. Xu, D.; Zhu, H. Outage minimized resource allocation for multiuser OFDM systems with SWIPT. *IEEE Access* **2019**, *7*, 79714–79725. [CrossRef]
39. Masood, Z.; Park, H.; Jang, H.S.; Yoo, S.; Jung, S.P.; Choi, Y. Optimal Power Allocation for Maximizing Energy Efficiency in DAS-Based IoT Network. *IEEE Syst. J.* **2021**, *15*, 2342–2348. [CrossRef]
40. Nugraha, G.D.; Musa, A.; Cho, J.; Park, K.; Choi, D. Lambda-based data processing architecture for two-level load forecasting in residential buildings. *Energies* **2018**, *11*, 772. [CrossRef]
41. Xu, Y.; Li, G.; Yang, Y.; Liu, M.; Gui, G. Robust resource allocation and power splitting in SWIPT enabled heterogeneous networks: A robust minimax approach. *IEEE Internet Things J.* **2019**, *6*, 10799–10811. [CrossRef]
42. Tham, M.L.; Chien, S.F.; Holtby, D.W.; Alimov, S. Energy-efficient power allocation for distributed antenna systems with proportional fairness. *IEEE Trans. Green Commun. Netw.* **2017**, *1*, 145–157. [CrossRef]
43. Xu, J.; Zhang, R. Energy beamforming with one-bit feedback. *IEEE Trans. Signal Process.* **2014**, *62*, 5370–5381. [CrossRef]
44. Ye, Y.; Shi, L.; Chu, X.; Lu, G. Throughput fairness guarantee in wireless powered backscatter communications with HTT. *IEEE Wirel. Commun. Lett.* **2020**, *10*, 449–453. [CrossRef]
45. Tang, J.; Luo, J.; Liu, M.; So, D.K.; Alsusa, E.; Chen, G.; Wong, K.K.; Chambers, J.A. Energy efficiency optimization for NOMA with SWIPT. *IEEE J. Sel. Topics Signal Process.* **2019**, *13*, 452–466. [CrossRef]
46. Di, X.; Xiong, K.; Fan, P.; Yang, H.C. Simultaneous wireless information and power transfer in cooperative relay networks with rateless codes. *IEEE Trans. Veh. Technol.* **2016**, *66*, 2981–2996. [CrossRef]
47. Yu, H.; Guo, S.; Yang, Y.; Ji, L.; Yang, Y. Secrecy energy efficiency optimization for downlink two-user OFDMA networks with SWIPT. *IEEE Syst. J.* **2017**, *13*, 324–335. [CrossRef]
48. He, C.; Sheng, B.; Zhu, P.; You, X.; Li, G.Y. Energy-and spectral-efficiency tradeoff for distributed antenna systems with proportional fairness. *IEEE J. Sel. Areas Commun.* **2013**, *31*, 894–902. [CrossRef]
49. Dinkelbach, W. On nonlinear fractional programming. *Manage. Sci.* **1967**, *13*, 492–498. [CrossRef]
50. Wang, S.; Shi, W.; Wang, C. Energy-efficient resource management in OFDM-based cognitive radio networks under channel uncertainty. *IEEE Trans. Commun.* **2015**, *63*, 3092–3102. [CrossRef]
51. Xu, W.; Yu, X.; Leung, S.H.; Chu, J. Energy-Efficient Power Allocation Scheme for Distributed MISO System With OFDM Over Frequency-Selective Fading Channels. *IEEE Access* **2018**, *6*, 51217–51226. [CrossRef]
52. Zhou, X.; Zhang, R.; Ho, C.K. Wireless information and power transfer in multiuser OFDM systems. *IEEE Trans. Wirel. Commun.* **2014**, *13*, 2282–2294. [CrossRef]
53. Yu, X.; Xu, W.; Leung, S.H.; Dang, X.; Teng, T. Power allocation schemes for DAS with OFDM under per-antenna power constraint. *IEEE Trans. Veh. Technol.* **2018**, *67*, 10102–10106. [CrossRef]

sensors

MDPI

Article

A Comprehensive Study of Event Detection in WPCN Networks with Noisy Measurements

Michael Koutsioumpos [1,*], Evangelos Zervas [2], Efstathios Hadjiefthymiades [1] and Lazaros Merakos [1]

1 Department of Informatics and Telecommunications, National and Kapodistrian University of Athens, Panepistimioupolis, Ilissia, 15772 Athens, Greece; shadj@di.uoa.gr (E.H.); merakos@di.uoa.gr (L.M.)
2 Department of Electrical and Electronics Engineering, University of West Attica, 12244 Egaleo, Greece; ezervas@uniwa.gr
* Correspondence: mkouts@di.uoa.gr; Tel.: +30-6976-44-0726

Abstract: Various aspects of the detection of events in wireless powered communication networks (WPCN) are studied and analyzed under the assumption of highly noisy sensor measurements. In WPCN, networks sensor nodes' stored energy is a scarce resource and must be treated sparingly. Frequent false alarm detections force superfluous transmissions, thus depleting nodes' energy storage. This has an adverse effect on the probability of successful transmission of the information message and its delay in case of a true positive detection. In this work, the detection problem is approached using an optimal stopping framework, where the involved likelihoods are highly unstable due to the noisy measurements. A classical AR filter is adopted in order to smooth the posterior likelihoods prior to their usage in the detection phase and its performance is contrasted to that of a novel Beta Particle Filter smoother. The effects of the smoothing filters on the achieved false alarm rate and detection delay are examined using numerical and simulation results. Moreover, the assessment of the detection process takes into account critical WPCN parameters, such as the charging efficiency and the location of the sensors, thus aiding the system design.

Keywords: WPCN; optimal stopping; mmWave sensor network

Citation: Koutsioumpos, M.; Zervas, E.; Hadjiefthymiades, E.; Merakos, L. A Comprehensive Study of Event Detection in WPCN Networks with Noisy Measurements. *Sensors* **2022**, *22*, 2163. https://doi.org/10.3390/s22062163

Academic Editors: Slim Naifar, Olfa Kanoun and Carlo Trigona

Received: 29 January 2022
Accepted: 7 March 2022
Published: 10 March 2022

1. Introduction

In recent years, technological advancements have paved the way for applications of wireless sensor networks in many aspects of daily life, such as environmental monitoring, agricultural monitoring, weather forecasting, and fire detection [1,2]. A wireless sensor network consists of battery-powered sensor devices dedicated to collecting information by continuously monitoring the physical environment and detecting critical events of interest. In case of a critical event, the sensor must be capable of immediately detecting and informing the network of the current situation.

A known problem for sensor nodes powered by batteries is that once the battery of a sensor is low charged or exhausted, the device becomes unavailable. Consequently, parts of the monitored area, with time, become unsupervised, affecting the performance of the entire network. It is conceivable that a network of battery-powered sensors is likely to degrade with time and special provisions should be made for its uninterrupted operation. Using wireless power transfer (WPT) technologies in such networks offers an important additional tool for extending the network lifetime by periodically charging the sensor batteries. Millimeter-wave radiofrequency technology supports high gains and is well fitted to dense access networks. As a result, using a millimeter-band antenna that conveys energy and information via the downlink is an excellent choice for maintaining the sensor network's performance.

We focus on an event detection application using a wireless powered communication network (WPCN) under the assumption of extremely noisy measurements. To this end, we consider multiple sensor nodes scattered randomly in the monitored area that are capable of harvesting energy wirelessly. An access point (AP), equipped with a mmWave antenna, orchestrates the information exchange with the nodes and the charging process. Each sensor takes measurements from the environment and tries to detect a possible change in the distribution of the monitored parameter. In case of an event, the sensor transmits the available information at the time to the AP with as much energy as possible. As sensor measurements may be noisy due to the nature of the monitored phenomenon or the quality of the sensors themselves, the detection process may result in an extremely large number of false alarms. The sensor nodes treat each alarm as a true event and try to pass relative information to the AP. This in turn exhausts frequently the node's stored energy since increasing the probability of successful message reception demands transmissions using all the available power. Harvesting energy from scratch regularly (especially with a small charging efficiency index) leads to partially charged nodes at any time, thus jeopardizing the functionality of the system. The aim of this work is to reduce the number of false alarms, which deplete the node energy (because of multiple unnecessary transmissions), and at the same time reduce the delay of detecting the critical event. We base the solution under investigation on Shiryayev's theory of optimal stopping and specifically on the theory of the optimization problem of disruption [3,4].

Showing that the measurements of the posterior probability of the event occurring before the event may be quite unstable, we propose a solution to remedy the problem. A novel smoothing technique based on particle filtering principles is proposed and compared with the classic AR filtering technique.

Specifically, the main contributions of this paper are the following:

- We derive analytical expressions for the average harvested energy per slot and the probability of successful information reception for a node.
- We define the optimal stopping problem and show that the node has to postpone its transmission, at least until the accumulated energy satisfies a specific criterion.
- We propose two solutions to overcome the posterior probability measurement variability problem. The first relies on the use of an AR filter, whereas the second uses a novel technique based on the particle filter theory.
- We assess the performance of the proposed solutions through simulations.

We organize the rest of this paper as follows. Section 2 summarizes the existing related research work. In Section 3, we present the system model and provide analytical expressions of the average harvesting energy per time slot and the probability of successful information reception. In Section 4, we define and solve the optimization problem and introduce two posterior smoothing techniques, a classical AR filter and a second that relies on particle filter principles. Simulation results are presented in Section 5. Finally, Section 6 contains our conclusions.

2. Related Work

The feasibility of a wireless power sensor network powered by a mmWave MIMO antenna is present in [5,6]. The authors present the benefits in terms of energy harvesting efficiency. Optimal section selection algorithms are introduced and analyzed to maximize the power coverage of the network. The work in [7] considers the performance of the network in terms of energy harvest efficiency and the ability of the sensor network to successfully detect environmental changes. The authors propose a solution that balances the two criteria. The authors in [8] introduce a reporting scheme that allows sensor nodes to inform the AP about their energy levels. Based on these reports, efficient energy-beamforming strategies are devised so that, acting proactively, all sensors are sufficiently charged most of the time and are capable of successfully transmitting their information upon detection of an event.

Event detection is one of the most important tasks of sensor networks, with a wide range of applications like intrusion detection, environmental monitoring, fire detection, and outlier detection in sensor networks [9–11]. Event detection most of the time is expressed as a change in the distribution of a monitored parameter from the sensor network.

Observations are assumed to be taken sequentially in time, and the objective is to detect the change as quickly as possible subject to the probability of false alarm. The problem of event detection in a multiple sensors network has been studied in various contexts. One area of interest is how the event is perceived from the sensor network. Works in [12,13] assume that only one sensor senses the change in the environment and its identity is unknown. An extension in [14] assumes that an unknown subset of affected sensors detects the event at the same time. The case where the sensors perceive the change at different times is studied in [15,16].

Another aspect of multiple sensor detection focuses on where the event information from the sensors is located for decision making. For the decentralized version, each sensor based on his measurements decides if a change in the monitored environment consists of an alarm [4,17]. On the contrary, for the centralized scenario [18], event information is available at several wireless sensors in the network, which transmit their information to a fusion center. The fusion center is responsible to raise an alarm as soon as possible subject to a false alarm rate constraint. The work in [19] considers the scenario where all sensors communicate their information to a fusion center. The fusion center is responsible for the decision process by monitoring the posterior probability of the received measurements. Additionally, the fusion center, in order to minimize the energy consumption of the sensors, chooses each time which sensors to keep enabled. Sensors are collocated and events happen at a random time following a geometric distribution. The proposed algorithm accomplishes to detect the intrusion (event) as early as possible using a minimal number of observations subject to a false alarm probability. One disadvantage of the method, as is concluded from the simulation results, is the computational complexity of the algorithm that depends on the number of active sensors. The authors in [20] propose a two-stage detection process in a sensor network used for fire detection. At first, each sensor based on its measurements monitors for any differences in parameter distribution. By defining a soft threshold and using a modified CUMSUM algorithm to sequentially detect the change, the sensor upon detection sends all the available information to a fusion center. At the second step, the fusion center, based on the measurements received by detecting nodes, applies a hard threshold to raise a fire alarm.

There are two common approaches to solve a change detection problem. Firstly, the minimax method [21], where the objective is to minimize the worst-case delay subject to a lower bound in the meantime between false alarms. Secondly, the Bayesian approach [3], where the time of the change (event) is assumed to be a random variable with known distribution and the goal is to minimize the detection delay subject to the probability of false alarm. Minimax optimal solution minimizes the cost under the worst-case change-time distribution, which is generally unknown. Therefore, we follow the Bayesian approach which uses prior information about the distribution of the change time. This prior distribution can be estimated using historic data. An additional constraint has been incorporated in the optimization process to accommodate the probability of successful information transmission.

3. System Model and Analysis

We consider a mmWave WPCN with sensor nodes scattered randomly in a circular area of radius ρ. The sensor nodes monitor the area for possible events, and upon detection of an event, they report it to the AP.

3.1. Network Topology

Sensor nodes are wirelessly charged by an Access Point (AP) centered at the origin of the monitored area. A uniform linear array of M antennas at the AP is used to switch the radiation beam along S sectors. Switching among sectors can be performed in a circular fashion or randomly. Figure 1 depicts the topology of the system. For this work, no blockage effect is considered but the channel between the AP and a sensor node is modeled as a fast-fading Rayleigh channel, independent for each sensor node.

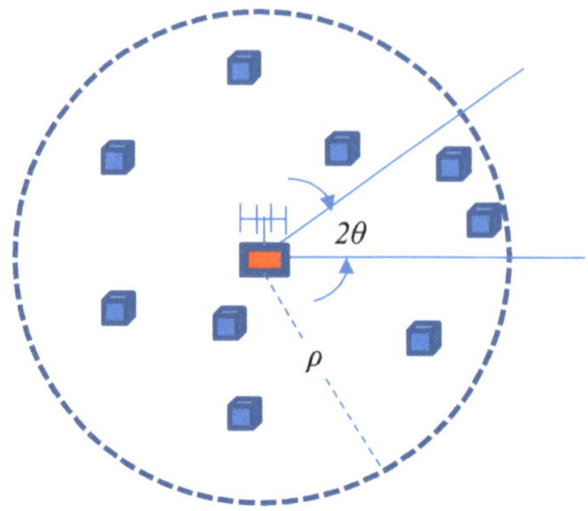

Figure 1. Network topology with sensor nodes depicted with boxes.

Time is slotted and the duration of each slot, T_S, is taken equal to the residence time of the AP beam to each sector. For the time slot per se we consider two formats. In the simplest one, each slot is comprised of two fields: the harvesting period T_H and the information period T_I, as it is shown in Figure 2a.

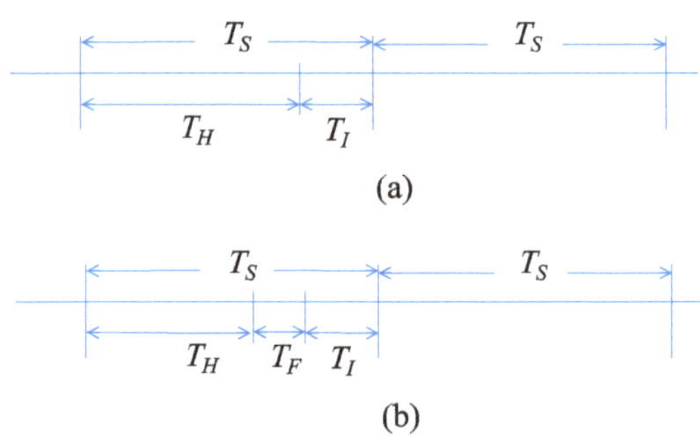

Figure 2. System time is slotted with slot duration T_s equal to the residence time of the charging beam to each sector. (**a**) During each slot nodes in the served sector charge for a period T_H and transmit (if necessary) for a period T_I. (**b**) An alternative slot format allows sensor nodes to exchange information for fusion or message relaying using period T_F.

During the energy harvesting period (T_H), sensors located in the served sector harvest energy and charge their battery. This process is discussed in the next subsection. Upon detection of an event, the node will transmit its identity (ID) and proper information (measurements) regarding the event during the information period (T_I). To maximize the probability of successful reception of the information message by the AP, the sensor nodes use all their stored energy for transmission. Since, neighboring nodes can simultaneously detect the event, to avoid collisions of multiple transmissions at the AP, an orthogonal multiple access scheme, like FDMA, is adopted. The justification for such a choice is the abundance of bandwidth in the mmWave band and the low rate of information transmission.

The second time slot format was proposed in [8] to facilitate sensors' feedback to the AP. This was achieved by introducing an additional field, called the reporting period, during which the sensors report their energy levels to the AP. The same time slot format, which is depicted in Figure 2b, can be used to accommodate sensor fusion. In this case, each slot is divided into three periods: the energy harvesting period (T_H), the fusion period (T_F), and the information period (T_I). T_F is introduced to allow neighboring sensor nodes to exchange information such as their measurements, for fusion purposes, or their energy levels if more sophisticated energy-conserving schemes are desirable. Although several comments in this work refer to the second time slot format, an in-depth exploitation of its advantages is left for future work.

3.2. Energy Harvesting Phase

The amount of harvested energy by a node depends on the received RF power, denoted by P^{RF}, and the exposure time to radiation. For this work, we adopt the nonlinear model presented in [22], according to which

$$\mathcal{E} = \frac{t\gamma\mathcal{E}_b(1-\exp(-\zeta_1 P^{\text{RF}}))}{1+\exp(-\zeta_1(P^{\text{RF}}-\zeta_2)} \tag{1}$$

The constants ζ_1, ζ_2 are circuit implementation dependent, $\mathcal{E}_b$ denotes the batteries' maximum energy storage, and the critical parameter $\gamma \in [0,1]$ models the efficiency of the charging process. The harvested energy is an increasing function of the received power, $\mathcal{E} = h(P^{\text{RF}})$, and therefore:

$$P\{\mathcal{E} \leq \epsilon\} = P\{P^{\text{RF}} \leq h^{-1}(\epsilon)\} \tag{2}$$

Solving Equation (1) for P^{RF}, we obtain

$$P^{\text{RF}} = -\frac{1}{\zeta_1}\ln\left(\frac{t\gamma\mathcal{E}_b - \epsilon}{t\gamma\mathcal{E}_b + \epsilon\exp(\zeta_1\zeta_2)}\right) = \eta(t,\epsilon) \tag{3}$$

Thus,

$$\mathcal{E} \leq \epsilon \iff P^{\text{RF}} \leq \eta(t,\epsilon) \tag{4}$$

For the received power P^{RF}, we use the model in [23]. For a node located at distance r from the AP and at a normalized angle ϕ from the direction of the serving beam, the received power is

$$P^{\text{RF}} = \frac{P_0|g|^2 F_M(\phi)}{1+r^a} \tag{5}$$

where P_0 is the BS transmit power, α is the path loss exponent, and g is the complex channel gain between the sensor node and the AP, modeled as complex Gaussian with zero mean and unit variance, i.e., $g \sim CN(0,1)$. The function $F_M(\phi)$ represents the Fejér kernel of order M, that is

$$F_M(\phi) = \frac{1}{M}\frac{\sin^2\left(\frac{\pi M\phi}{2}\right)}{\sin^2\left(\frac{\pi\phi}{2}\right)} \tag{6}$$

Therefore, for a harvesting period T_H, we obtain

$$\begin{aligned} F_{\mathcal{E}}(\epsilon) = P\{\mathcal{E} \leq \epsilon\} &= P\{P^{\text{RF}} \leq \eta(T_H,\epsilon)\} = P\left\{|g|^2 \leq \frac{\eta(T_H,\epsilon)(1+r^a)}{P_0F_M(\phi)}\right\} \\ &= 1-\exp\left(-\eta(T_H,\epsilon)\frac{1+r^a}{P_0F_M(\phi)}\right) \end{aligned} \tag{7}$$

where the equality in the second line of equations is due to the fact that $|g|^2$ is exponentially distributed. For a given r and ϕ the average harvested energy per time slot can be found by

$$\mathcal{E}_{av,s}(r,\phi) = \int_0^{\infty}(1-F_{\mathcal{E}}(\epsilon))d\epsilon = \int_0^{\gamma T_H\mathcal{E}_b}(1-F_{\mathcal{E}}(\epsilon))d\epsilon \tag{8}$$

Figure 3 shows the average harvested energy per slot for the parameter set used in the simulation section (see Table 1) and for three different values of the parameter γ. Note that for large distances from the AP, the harvested energy for nodes located at the edge of the beam (dotted lines, $\phi = 12°$) is considerably lower than the energy harvested by nodes located at the axial direction of the beam (solid lines, $\phi = 0°$). If a sensor node is aware of its location, i.e., explicit deployment of nodes, it can anticipate the number of time slots needed to reach a specific energy level and thus act appropriately in a case of an event. For example, for $\gamma = 0.03$ and a node located at the axial direction of the beam, 50 time slots are required at most for the node to become fully charged.

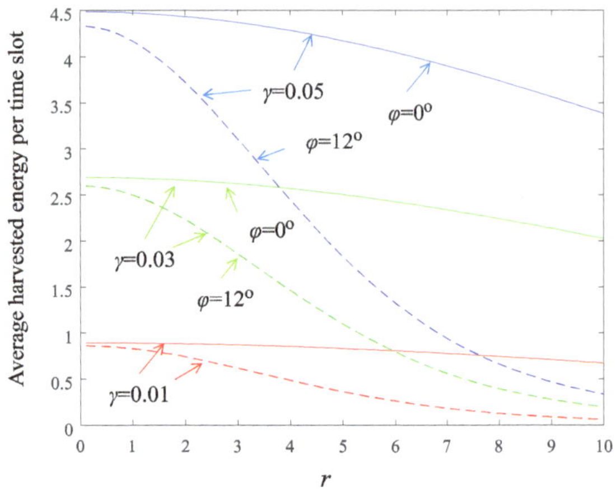

Figure 3. Average harvested energy $\mathcal{E}_{av,s}(r,\phi)$ per time slot. Solid lines: $\phi = 0°$, dotted lines: $\phi = 12°$. Blue color corresponds to $\gamma = 0.05$, green color to $\gamma = 0.03$, and red color to $\gamma = 0.01$.

Table 1. Parameters used in simulations.

Parameter	Value
$\zeta_1, \zeta_2, \mathcal{E}_b$	1500, 0.0022, 100
ρ, θ, α, M	10 m, $\pi/15$, 2, 8
P_0, σ^2, λ	−10 db, −30 db, 1
D_I, W'	1024 bits, 1 KHz

If the location (r,ϕ) of the sensor node is unknown, we have to average over the spatial distribution of distances and angles in the serving sector to estimate the average harvested energy per time slot. In this case

$$\mathcal{E}_{av,s} = \frac{1}{\theta\rho^2}\int_{-\theta}^{\theta}\int_{0}^{\rho}\int_{0}^{\gamma T_H \mathcal{E}_b} \exp\left(-\eta(T_H,\epsilon)\frac{(1+r^a)}{P_0 F_M(\phi)}\right) r dr d\phi d\epsilon \tag{9}$$

The average harvested energy per slot, $\mathcal{E}_{av,s}$, is plotted vs. γ in Figure 4 for three values of the path loss exponent α. As it is expected, as α increases the amount of energy harvested by a sensor node at each time slot is reduced. The differences are magnified for larger values of the parameter γ.

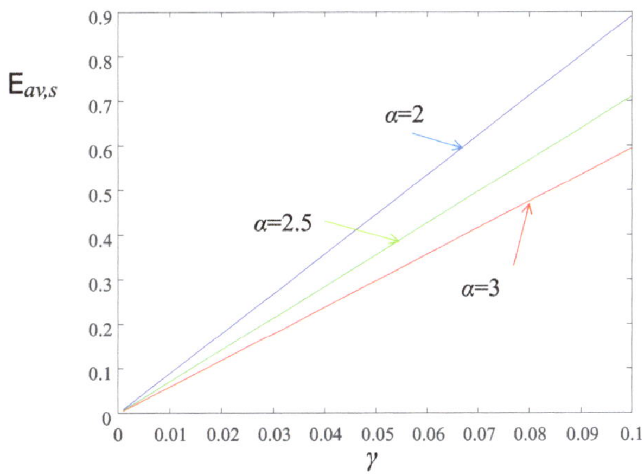

Figure 4. Average harvested energy $\mathcal{E}_{av,s}$ vs. γ for various values of the path loss exponent α. Blue color corresponds to $\alpha = 2$, green color to $\alpha = 2.5$, and red color to $\alpha = 3$.

3.3. Information Transmission Phase

Upon detection of an event, the sensing node awaits until the AP switches the beam towards its direction, charges for a period T_H and then transmits during the period T_I.

If after charging the node's battery is at energy level $\mathcal{E}$, the transmission power is

$$P = \frac{\mathcal{E}}{T_I} \tag{10}$$

The signal-to-noise ratio (SNR) at the AP from the transmitted node located at (r, ϕ) is

$$\text{SNR}(r, \phi) = \frac{P|g|^2 F_M(\phi)}{(1 + r^a)\sigma^2} \tag{11}$$

where σ^2 is the noise power. Therefore,

$$\begin{aligned} P\{\text{SNR}(r,\phi) \le x \mid P\} &= P\left\{ \frac{P|g|^2 F_M(\phi)}{(1+r^a)\sigma^2} \le x \mid P \right\} \\ &= P\left\{ |g|^2 \le \frac{x(1+r^a)\sigma^2}{PF_M(\phi)} \mid P \right\} \\ &= 1 - \exp\left(-x \frac{(1+r^a)\sigma^2}{PF_M(\phi)} \right) \end{aligned} \tag{12}$$

Successful reception of the information message is possible if

$$R_I < W' \log_2(1 + \text{SNR}(r, \phi)) \tag{13}$$

where R_I is the information rate and W' the bandwidth allocated to a node. The information rate R_I equals D_I/T_I, where D_I is the amount of data (bits) that the node sends to the AP and T_I is the transmission time interval ($T_I = T_S - T_H$). Using orthogonal signaling we assume that the available bandwidth W is equally divided to all nodes, and therefore

$$W' = \frac{W}{S_u} \tag{14}$$

with S_u an upper bound on the total number of nodes. Solving Equation (13) for SNR(r,ϕ) and using Equation (12), we obtain

$$\begin{aligned} p_s(r,\phi) &= P\{\text{successful information reception for a node located at } (r,\phi)\} \\ &= \exp\left(-(2^{R_I S_u/W}-1)\frac{(1+r^a)\sigma^2}{PF_M(\phi)}\right) \end{aligned} \quad (15)$$

Again, if the location of the sensor node is unknown we integrate over the spatial distribution of the nodes to find the average probability of successful information reception.

$$p_s = E_\Phi[p_s(r,\phi)] = \frac{1}{\theta\rho^2}\int_{-\theta}^{\theta}\int_0^{\rho}\exp\left(-(2^{R_I S_u/W}-1)\frac{(1+r^a)\sigma^2}{PF_M(\phi)}\right) r dr d\phi \quad (16)$$

Figure 5 depicts the probability of successful reception of messages vs. the available energy level for two nodes located at the edge of the network ($r = 10$). The first node is located at the axial direction of the beam (blue color) whereas the second node is located at the boundary of the sector (red color). The values of the parameters used are given in Table 1. Moreover, in Figure 5 the average probability of successful reception p_s (Equation (16)) is plotted with green solid line.

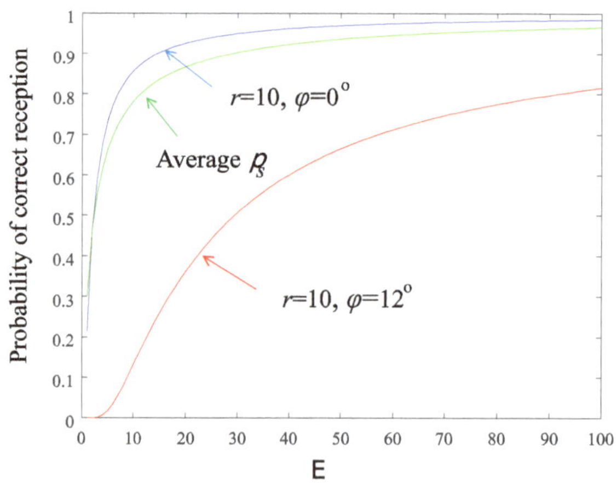

Figure 5. Probability of successful information reception. Green color: the average p_s, blue color: node located at $(10, 0°)$, red color: node located at $(10, 12°)$.

3.4. Sensing Model

A very simple sensing model is adopted for this work. A sensor node senses all events within a distance r_s from its location. The situation is shown schematically in Figure 6 where nodes #1 and #2 detect the event while nodes #3 and #4 fail to do so. More sophisticated models, like the one in [24], can be used but their added value for this work is limited.

A distribution $f_0(\cdot)$ is used to model sensor measurements prior to the event. The same distribution, but with different parameters, say $f_1(\cdot)$, is used to model measurements after the event. For the rest of this paper, $f_0(\cdot)$ and $f_1(\cdot)$ are assumed to be Gaussian with means 0 and 1 respectively and standard deviation equal to 0.4. The large value of the standard deviation, compared to the difference of the means, indicates extremely noisy measurements.

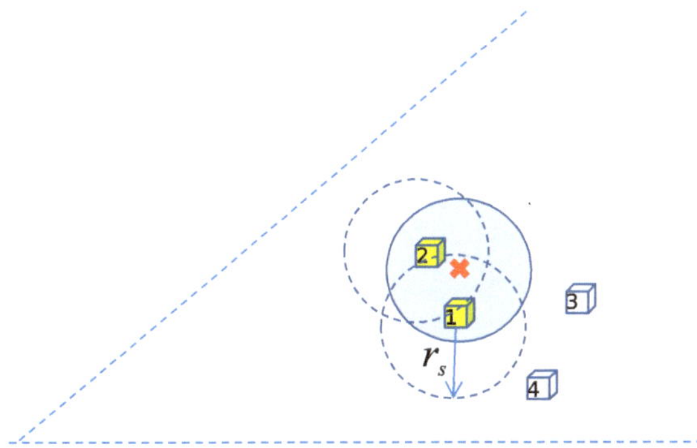

Figure 6. Nodes within a distance r_s from the event (like nodes #1 and #2) can sense it.

4. Detection of Events and Data Fusion

A well-known optimal stopping technique due to Shiryayev [4] is used for event detection and described in Section 4.1. As the involved posterior probabilities are highly noisy, a Beta Particle Filter is introduced in Section 4.2 to smooth them. Finally, data fusion principles that could improve the system's performance are presented in Section 4.3.

4.1. Event Detection

In the problem considered we assume for simplicity that the time at which an event occurs, $\mathcal{T}$, follows the geometric distribution with probability of success ϱ. The parameter ϱ is assumed known, for example from historical data. We furthermore denote the prior probability that an event happened before starting time by π. That is

$$P\{\mathcal{T}=0\} = \pi \tag{17}$$

$$P\{\mathcal{T}=n\} = (1-\pi)(1-\varrho)^{n-1}\varrho \tag{18}$$

The time instant n represents the nth period of beamforming the tagged sensor node. Let τ be the time at which the sensor node decides to transmit its information (measurements) to the BS. There are three risk functions associated with τ. First is the risk function $J_1(\tau) = P\{\tau < \mathcal{T}\}$ which is interpreted as the probability of false alarm. Second is the risk function $J_2(\tau) = E[\max(\tau - \mathcal{T}, 0)]$ which is the average delay of transmission after occurrence of the event. Third is the risk function $J_3(\tau) = P\{\text{unsuccessful transmission at } \tau\}$. Minimizing the aforementioned risks has adversary effects on the selection of τ. For example, to maximize the probability of a successful transmission we have to postpone it until the sensor node is sufficiently charged to overcome bad channel conditions or even bad location conditions (distant node or edge beamforming node). This postponement imposes in turn a large delay on information reporting to the BS. If the nodes act alone, that is they do not relay information by other nodes, then it is crucial to transmit with as much power as possible in order to increase the probability the successful information reception by the BS. Thus, we formulate the problem as

$$\begin{aligned} &\min P\{\tau < \mathcal{T}\} + \lambda_d E[\max(\tau - \mathcal{T}, 0)] \\ &\text{s.t.c.} \quad P_{ui} < \psi \end{aligned} \tag{19}$$

where P_{ui} denotes the probability of unsuccessful information reception and λ_d is a weight factor balancing the importance between false alarm rate and detection delay. Let us first focus on the constraint and reveal the role of τ in it. Successful reception of the information message is possible if Equation (13) is satisfied. Using Equation (15) and defining for notation simplicity

$$Q(r,\phi) = \frac{(2^{R_I/W'} - 1)(1 + r^a)\sigma^2}{F_M(\phi)} \tag{20}$$

we obtain

$$P_{ui} < \psi \Leftrightarrow 1 - \exp\left\{-\frac{Q(r,\phi)}{P^\tau}\right\} < \psi \Leftrightarrow P^\tau > -\frac{Q(r,\phi)}{\ln(1-\psi)} \tag{21}$$

where P^τ is the transmission power at time instant τ. Moreover, P^τ is equal to $\mathcal{E}^\tau / T_I$, where $\mathcal{E}^\tau$ is the total energy harvested by the node up to time τ and T_I is the signaling period. Thus, the constraint of the optimization problem (19) imposes a lower bound on the selected time τ' to transmit. The node has to postpone its transmission at least until the accumulated stored energy satisfies

$$\mathcal{E}^{\tau'} > -\frac{T_I Q(r,\phi)}{\ln(1-\psi)} \tag{22}$$

Inequality (22) presupposes knowledge of the location of the node. If this is not possible, the threshold on the right side of (22) may be replaced by an average over the possible locations of the sensor nodes, that is

$$\mathcal{E}^{\tau'} > \frac{1}{\theta\rho^2}\int_{-\theta}^{\theta}\int_0^{\rho} -\frac{T_I Q(r,\phi)}{\ln(1-\psi)} r dr d\theta \tag{23}$$

We turn now to the minimization of the objective function in (19). To this end we define the posteriori probability of the event occurring before time n

$$\pi_n = P\{\mathcal{T} \le n | \mathcal{F}_n\}, \qquad \pi_0 = \pi \tag{24}$$

where $\mathcal{F}_n = \sigma\{X_1, X_2, \ldots, X_n\}$ is the σ-algebra characterizing the information before time n. Sensor measurements $X_1, X_2, \ldots$ are assumed independent having pdf $f_0(\cdot)$ prior to the event and pdf $f_1(\cdot)$ after occurrence of the event. Using Bayes' formula

$$\pi_{n+1} = \frac{(\pi_n + (1-\pi_n)\varrho) f_1(X_{n+1})}{(\pi_n + (1-\pi_n)\varrho) f_1(X_{n+1}) + (1-\pi_n)(1-\varrho) f_0(X_{n+1})} \tag{25}$$

Noting that

$$P\{\tau < \mathcal{T}\} = E[1 - \pi_\tau] \tag{26}$$

and following Shiryayev, the objective function in the optimization problem (19) can be transformed into the more convenient form

$$\begin{aligned} J(\tau) &= P\{\tau < \mathcal{T}\} + \lambda_d E[\max(\tau - \mathcal{T}, 0)] \\ &= E\left[1 - \pi_\tau + \lambda_d \sum_{k=0}^{\tau-1} \pi_k\right] \end{aligned} \tag{27}$$

For a given prior π, we seek an optimal stopping time $\tau^\star$ that satisfies $J^\star(\tau^\star) = \min J(\tau)$. It turns out that the optimal stopping time is independent of π and can be found by a simple thresholding rule

$$\tau^\star = \min\{n \ge 0 : \pi_n \ge \Gamma\} \tag{28}$$

Indeed, according to the results of Shiryayev (Theorem 2.23) the minimum risk function satisfies the equations

$$J^\star(\pi) = \min\{1 - \pi, \lambda_d \pi + E[J^\star(\pi_1)]\} \tag{29}$$

$$= \lim_n Q^n(1 - \pi) \tag{30}$$

where π_1 is the posterior probability given by (25) (one step forward after the current prior π) and the expectation is over the pdf

$$f(x) = (\pi_n + (1-\pi_n)\varrho)f_1(x) + (1-\pi_n)(1-\varrho)f_0(x) \tag{31}$$

The operator $Q(\cdot)$ is defined as

$$Q(h(\pi)) = \min\{h(\pi), \lambda_d\pi + E[h(\pi_1)]\} \tag{32}$$

The $Q(\cdot)$ function is concave as the pointwise minimum of concave functions. Repeating this argument, $Q^n(1-\pi)$ is concave and therefore $J^\star(\pi)$ is concave. The optimal stopping time $\tau^\star$ is given by

$$\tau^\star = \min\{n \geq 0 : J^\star(\pi_n) = 1 - \pi_n\} \tag{33}$$

and due to the concavity of the risk function it assumes the form given by (19). The threshold Γ can be found using the equation

$$1 - \Gamma = \lambda_d\Gamma + J^\star(\Gamma) \tag{34}$$

Figure 7 depicts $J^\star(\pi)$ for three values of λ_d. As λ_d decreases more emphasis is given in minimizing the false alarm rate and therefore the threshold Γ drifts to larger values.

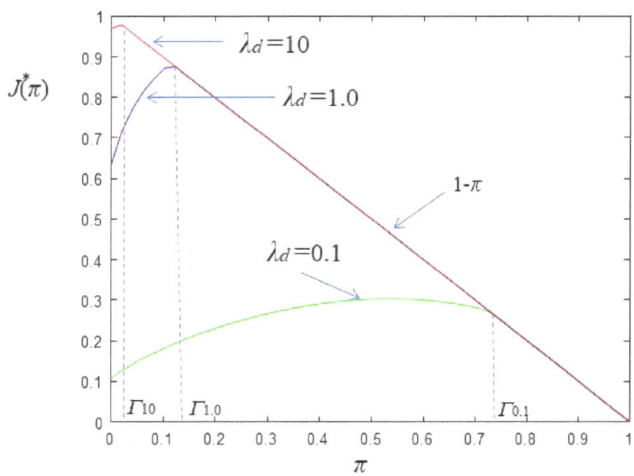

Figure 7. $J^\star(\pi)$ and related thresholds.

It is clear at this point that the choice of τ (the time at which the sensor node decides to transmit its information) is affected by two independent criteria. The first one is the energy criterion (22) which depends on the location of the sensor node, the probability of successful transmission $1-\psi$, and the charging parameter γ. The second criterion, say false-alarm/delay criterion, is described by (28) and it depends, among other things, on λ_d and the "frequency" parameter ϱ. Combining these two criteria into one is a matter of the nature of the events and the system's parameters. For example, for false-alarm and successful reporting critical events, both criteria should be satisfied, so we can combine them as

$$\tau_o = \max\{\tau', \tau^\star\} \tag{35}$$

whereas for delay critical events we can simply set

$$\tau_o = \tau^\star \tag{36}$$

which means that the sensor-node transmits as soon as possible regardless of its accumulated energy. Moreover, the values of the parameters of the system play a major role in the decision process, as it is demonstrated next. Consider the aforementioned network topology with $\rho = 10$ m, $S = 15$, sensing area of radius $r_s = 1$ m, and the parameter ϱ set to 0.01. This means that events happen every 100 system time slots on average, which is quite high for rare events. If the events are uniformly distributed in the network area, then a specific sensor node will be "hit" with probability $\varrho r_s^2/\rho^2 = 0.0001$, that is every 10,000 time slot on average. Setting the successful transmission constraint to $1 - \psi = 0.9$, then approximately $\mathcal{E}_{av,s} = 30$ (eyeball Figure 5). Using the curve of the smallest γ (=0.01) in Figure 3 and for the worst case of $r = 10$, $\phi = 12°$, we find that $30/0.1 = 300$ time slots are needed for the node to charge at the value of 30 energy units starting from zero conditions. Since the tagged sensor is charged every $S = 15$ time slot (round-robin beamforming), the number of system time slots that are needed for the sensor to reach its target energy level is $300 \times 15 = 3500$. Thus the event, with high probability, will find the node sufficiently charged for transmission. In cases like this, the constraint of the optimization problem (15) can be neglected.

4.2. *Smoothing the Posterior Probability*

Measurements of the posterior probability of the event occurring before time n, as they are described by Equation (25) may be highly unstable. Thus, for example, a low valued threshold as Γ_{10} or $\Gamma_{1.0}$ in Figure 7 may be crossed several times before the occurrence of an event. This in turn will impose unnecessary sensor node transmissions which will deplete its energy reserve. Therefore, if the energy harvesting process has a low rate compared to ϱ (the events' occurrence rate), it is highly probable that at the time of an event the sensor node will be uncharged. This implies a large delay in information reporting. To remedy this problem we resort to posterior probability smoothing and we test two techniques. The first technique is straightforward and relies on the use of an AR filter whereas the second technique is a novel one and relies on particle filter principles.

4.2.1. AR Smoothing

For simplicity we consider an AR filter of order one to smooth the posterior probability π_n. Thus,

$$\tilde{\pi}_n = \alpha\tilde{\pi}_{n-1} + (1-\alpha)\pi_n \tag{37}$$

where $\alpha \in (0,1)$ determines the degree of smoothing. Using a value α, we have essentially an average over the last $1/(1-\alpha)$ values of the posterior probability.

Figure 8 shows a trace of the posterior probability π_n (Equation (25), dotted black line) generated using $\varrho = 0.01$, $f_0 \sim \mathcal{N}(0, 0.4^2)$ and $f_1 \sim \mathcal{N}(1, 0.4^2)$. The time of the event was taken at $\mathcal{T} = 100$. As it is observed there are large false spikes both prior to the event and after. The red solid line depicts $\tilde{\pi}_n$ for $\alpha = 0.9$. Although there is some sort of smoothing there are substantial high values of the posterior probability prior to the event which may lead to several false detections. Decreasing α improves smoothing but the slope of the curve after the event decreases. This means that an extra delay may be introduced before detecting the event.

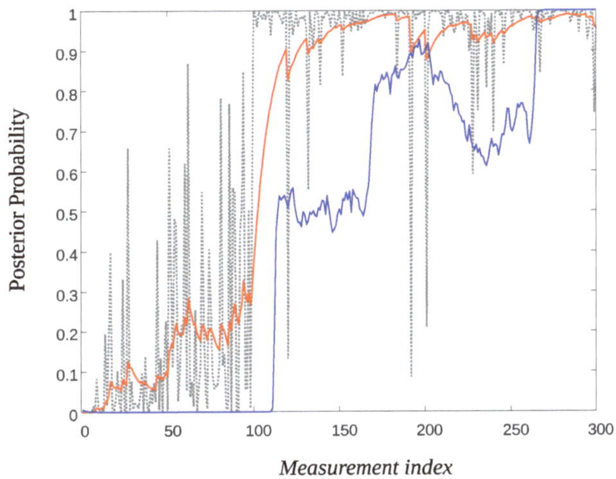

Figure 8. Smoothed posterior probabilities: Original π_n (dotted-black), AR filter $\hat{\pi}_n$ for $\alpha = 0.9$ (solid-red), and Beta Particle Filter $\hat{\pi}_n$ (solid-blue).

4.2.2. Beta Particle Filter Smoothing

The second technique relies on the use of particle filtering and specifically the sampling-importance-resampling (SIR) method. In Bayesian filtering, such as Kalman or particle filtering, estimating the system state $\mathbf{x}_n$ based on observations up to the time n, $\mathbf{y}_{1:n}$, involves two steps. A prediction step to compute the prior $p(\mathbf{x}_n|\mathbf{y}_{1:n-1})$, followed by a filtering step to compute the posterior $p(\mathbf{x}_n|\mathbf{y}_{1:n})$. Thus

$$\text{Prediction:} \quad p(\mathbf{x}_n|\mathbf{y}_{1:n-1}) = \int p(\mathbf{x}_n|\mathbf{x}_{n-1})p(\mathbf{x}_{n-1}|\mathbf{y}_{1:n-1})d\mathbf{x}_{n-1} \tag{38}$$

$$\text{Filtering:} \quad p(\mathbf{x}_n|\mathbf{y}_{1:n}) \propto \int p(\mathbf{y}_n|\mathbf{x}_n)p(\mathbf{x}_n|\mathbf{y}_{1:n-1}) \tag{39}$$

In the proposed smoothing technique, the posterior probabilities π_n, given by Equation (25), play the role of the observations $\mathbf{y}_n$ while the smoothed values $\hat{\pi}_n$ will represent the unknown internal system state $\mathbf{x}_n$. In both the prediction step and filtering step we use the Beta distribution to model the state transition probability distribution $p(\mathbf{x}_n|\mathbf{x}_{n-1}) = p(\hat{\pi}_n|\hat{\pi}_{n-1})$ and the likelihood $p(\mathbf{y}_n|\mathbf{x}_n) = p(\pi_n|\hat{\pi}_n)$. The Beta distribution is given by

$$\text{Beta}_x(a,b) = \frac{\Gamma(a+b)}{\Gamma(a)\Gamma(b)}x^{a-1}(1-x)^{b-1} \tag{40}$$

with the shape parameters a and b controlling the peak and the sharpness of the peak and $\Gamma(\cdot)$ is the Gamma function. Since the filtered value $\hat{\pi}_n$ has to be close to $\hat{\pi}_{n-1}$ we model the state transition probability as

$$p(\hat{\pi}_n|\hat{\pi}_{n-1}) \sim \text{Beta}_{\hat{\pi}_n}(\nu_1\hat{\pi}_{n-1}, \nu_1(1-\hat{\pi}_{n-1})) \tag{41}$$

with the parameter ν_1 taking a large value, to guarantee $a > 1$ and $b > 1$. Similarly, we model the likelihood as

$$p(\pi_n|\hat{\pi}_n) \sim \text{Beta}_{\pi_n}(\nu_2\hat{\pi}_n + 1, \nu_2(1-\hat{\pi}_n) + 1) \tag{42}$$

with the parameter ν_2 having a relatively small value. The bias term (+1) was introduced to avoid numerical problems when π_n assumes small values. The Beta Particle Filtering algorithm is presented in Algorithm 1.

Algorithm 1 The Beta Particle Filter smoothing algorithm.

Initialization
- Select the number of particle streams K
- Generate K samples for the initial state $\hat{\pi}_0 = \pi$
  For $k = 1, \ldots, K$
  - draw $\hat{\pi}_0^k \sim \text{Beta}_{\hat{\pi}_0}(\nu_1 \pi, \nu_1(1-\pi))$
  - set $w_0^k = 1/K$ % set initial weights

  end for

Main Loop
For $n = 1, 2, \ldots$
  For $k = 1, 2, \ldots, K$
  - draw $\hat{\pi}_n^k \sim \text{Beta}_{\hat{\pi}_n}(\nu_1 \hat{\pi}_{n-1}^k, \nu_1(1-\hat{\pi}_{n-1}^k))$
  - $w_n^k = w_{n-1}^k \text{Beta}_{\pi_n}(\nu_2 \hat{\pi}_n^k + 1, \nu_2(1-\hat{\pi}_n^k) + 1)$

  end for
  $W_n^k = w_n^k / \sum_k w_n^k$ % normalize weights
  $K_{\text{eff}} = 1/\sum_k (W_n^k)^2$
  If $K_{\text{eff}} < K_T$ % K_T was taken equal to $0.85K$
  - Resample $\{\hat{\pi}_n^k, W_n^k\}_{k=1}^K$ to obtain $\{\bar{\pi}_n^k, 1/K\}_{k=1}^K$
  - Set $\hat{\pi}_n^k = \bar{\pi}_n^k$, $W_n^k = 1/K$

  end if
- Estimate $\hat{\pi}_n = \sum_k \hat{\pi}_n^k W_n^k$

end for

The solid-blue line in Figure 8 depicts the smoothed value $\hat{\pi}_n$ using the Beta Particle Filter method. The plot was obtained using $K = 200$ particle streams and the parameters $(\nu_1, \nu_2) = (500, 1)$. As it is observed, prior to the event the smoothed posterior probability stays close to zero but there is a substantial delay after the event to reach the true value one.

4.3. Fusion of the Sensor Measurements

From the discussion in the previous subsection, it becomes clear that more reliable estimates of the posterior probability are needed before the sensor nodes decide to transmit their information. This may be achieved by fusing the measurements of the sensors that cover the same area. To this end, sensor nodes utilize the second time-slot format and exchange measurement information during T_F prior to any transmission.

Figure 9 shows the smoothed posterior probabilities of three sensors and as it is observed false alarm probabilities (prior to the event and for the AR filtering technique), have been reduced by a factor 2 to 3 compared to that of Figure 8. Note that if m sensors are involved in the fusion process, the posterior probabilities (Equation (25)) are given by

$$\pi_{n+1} = \frac{(\pi_n + (1-\pi_n)\rho) \prod_{i=1}^m f_1^{(i)}(X_{n+1}^{(i)})}{(\pi_n + (1-\pi_n)\rho) \prod_{i=1}^m f_1^{(i)}(X_{n+1}^{(i)}) + (1-\pi_n)(1-\rho) \prod_{i=1}^m f_0^{(i)}(X_{n+1}^{(i)})} \tag{43}$$

However, there is a price to pay for the advantages of fusion. Covering an area of interest by more than one sensor node is not always feasible and it depends mainly on the type of sensors used. For example, an optical sensor has a much larger sensing radius compared to a temperature sensor but nevertheless, line-of-sight blocking limits the size of the monitored area. Let us assume that the distribution of the sensor nodes is a homogeneous Poisson Point Process (PPP) with intensity λ. If it is desirable for the event to be detected by at least two sensors with a probability higher than α_c, then

$$P\{\#\text{sensors} \geq 2\} \geq \alpha_c \Rightarrow 1 - \left(e^{-\lambda A_s} + e^{-\lambda A_s} \lambda A_s\right) \geq \alpha_c \tag{44}$$

with $A_s = \pi r_s^2$. Setting $r_s = 1$ m and $\alpha_s = 0.5$ results in 168 nodes on average that have to be deployed over an area of radius 10 m. If the sensing radius is reduced to 0.5 m then the number of nodes has to be quadrupled. Moreover, using an extra time period T_F in the time slot format in order for the sensors to exchange their measurements, reduces either the harvesting period T_H, or the transmission period T_I, or both. One may neglect T_F and resort to centralized information fusion by letting all the sensors sensing an event independently transmit their measurements to the AP.

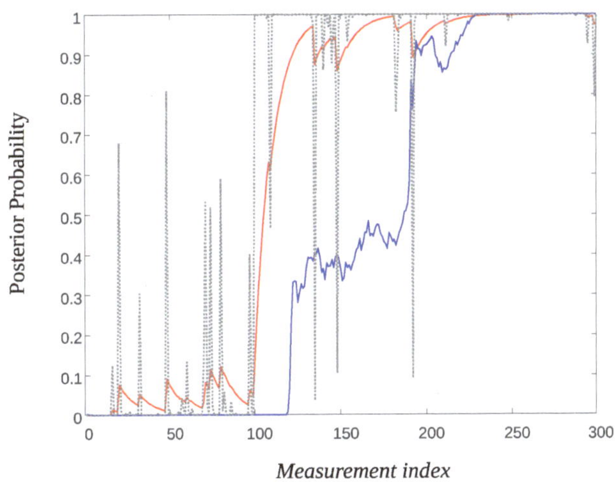

Figure 9. Data fusion of three sensors for smoothing the posterior probabilities: Original π_n (dotted-black), AR filter $\tilde{\pi}_n$ for $\alpha = 0.9$ (solid-red) and Beta Particle Filter $\hat{\pi}_n$ (solid-blue).

There are also certain advantages to introducing the time period T_F. During this period sensor nodes may exchange information, beyond measurements for data fusion, such as their energy reserves. In this case, more sophisticated transmission strategies may be explored. For example, among two or more sensor nodes that detect an event, only the node with the higher energy level transmits the information to the AP. The rest of the nodes delay their transmission either for extra charging purposes or to increase time diversity. Since for neighboring nodes the channel gains are highly correlated, postponing the transmission for a period greater than the time coherence of the channel may prevent nodes transmitting under the same (possibly bad) channel conditions.

5. Simulation Results

In this section, some simulation results are presented that corroborate the analysis of the previous section. The values of the system parameters are given in Table 1 taken from [6] and the simulation platform was MATLAB.

In our simulation model, a base station lies at the center of a cell of radius 10 m and a linear antenna array of M = 8 elements is used to energy beamform 15 sectors in a round-robin fashion. The location of the event is selected randomly in the monitored area and the event interarrival time follows the geometric distribution. Channel gains are modeled as independent random variables following an exponential distribution. We assume fast varying channels in the sense that the gains change at every time slot.

We focus on the behavior of two sensor nodes located at the edge of the network, that is at a distance $r = 10$ m from the AP. The first node lies along the direction of the beam pattern ($\phi = 0°$), whereas the second sensor node lies at $\phi = 12°$, which is at the edge of the beam pattern. Sensor information transmission and charging are possible only when the AP beamforms towards the tagged sector. However, the nodes continuously sense the environment for possible events and run the detection algorithm. The simulation horizon is large enough to ensure the creation of several events in the neighborhood of the sensors.

We present two sets of simulations. In the first set we illustrate the effects of smoothing, charging efficiency γ, and event probability ϱ on the false alarm rate and the accumulated node energy prior to information transmission. In the second set, we deal with the effects of the aforementioned parameters on the information reporting delay.

5.1. Charging and False Alarm Rate

Figure 10 shows the average accumulated energy for various values of the threshold parameter Γ (Criterion 36). The averages were obtained using 200 runs, the charging efficiency was set to $\gamma = 0.01$ and AR smoothing of the posterior probabilities was used. A few observations are immediate. First of all, the node at the edge of the beam pattern fails to reach high energy levels as it is with the node along the main direction of the beam. A remedy to this problem is to offset the direction of the beam (by half of the beam width) before starting a new charging cycle. Secondly, the energy storage from one value onwards remains constant. Values of the threshold Γ greater than 0.3 do not affect the charge percentage of the sensors. For small values of Γ, the threshold is exceeded several times forcing the sensor to transmit and therefore to deplete regularly its energy storage. Figure 11 repeats the experiment with a larger value of the charging efficiency parameter γ. In this case, sensors charge at a tenfold rate and manage to recover from false transmissions.

The effect of smoothing the posterior probabilities is more obvious for small values of the threshold Γ. As it is observed from Figures 10 and 11, high smoothing ($\alpha = 0.9$) prevents, to some extent, false alarms and moderates unnecessary threshold crossings.

Figure 12 compares the performance of the Beta Particle Filter smoother with that of the AR filter using $\alpha = 0.9$. The threshold Γ was set equal to 0.15 and the accumulated energy prior to information transmission is plotted vs. the charging efficiency γ. Note that for this experiment $\varrho = 1$, which means a high rate of events. One event occurs in the monitored network area every one time slot on average. It is conceivable from the figure that BPF outperforms AR smoothing by almost 10%.

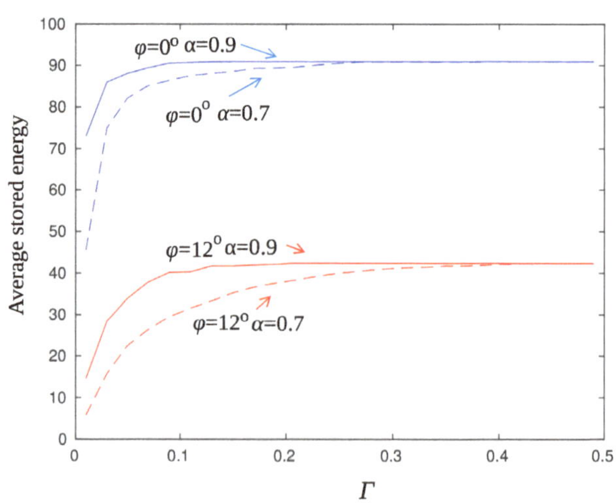

Figure 10. AR filter—average stored energy for $\gamma = 0.01$, $\varrho = 0.1$.

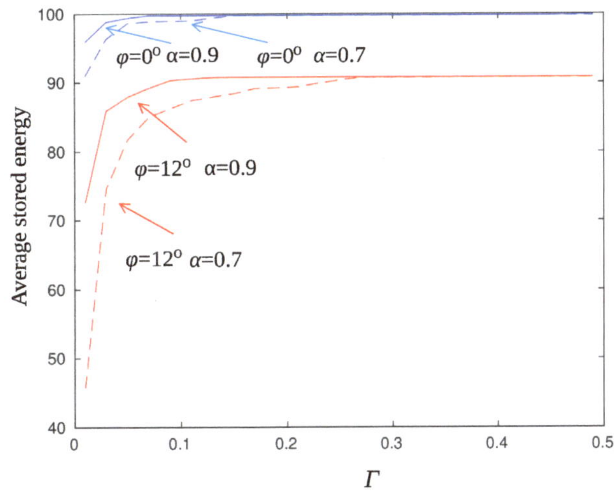

Figure 11. AR filter—average stored energy for $\gamma = 0.1$, $\varrho = 0.1$.

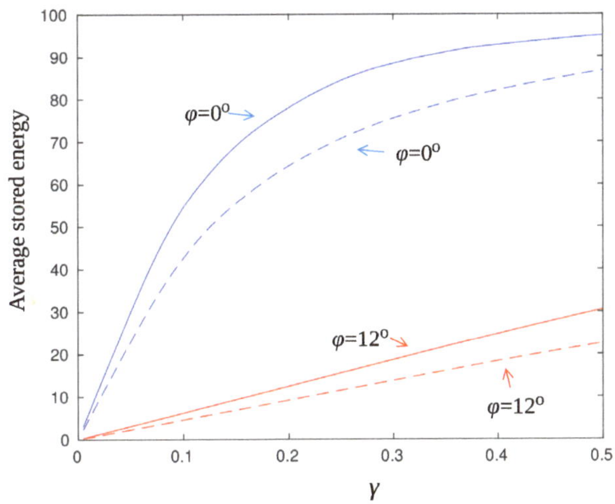

Figure 12. Average stored energy for $\Gamma = 0.15$, $\varrho = 1$, $\alpha = 0.9$, and $r = 10$ m. Solid line: Particle filter method, dashed line: AR filtering.

Figure 13 compares the performance of the smoothing filters under detection criteria 35 and 36. Black lines correspond to the case of a sensor that lies on the main direction of the beam pattern $\phi = 0°$. The target probability of unsuccessful transmission was set equal to $\psi = 0.05$. There are two observations that reveal the superiority of the BPF against AR smoothing regarding the treatment of false alarms. Firstly, using BPF the sensor node reaches higher values of accumulated energy that are almost independent on the threshold parameter Γ. Secondly, Criterion 35, that is postponement of information transmission until both energy and detection threshold constraints are satisfied, does not have a major impact on the accumulated energy. This is justified by the fact that the BPF smoother is not "fooled" by temporary ripples of the posterior probabilities and therefore the sensor node does not discharge for unnecessary transmissions. Red lines correspond to the case of a sensor that lies at the edge of the beam pattern $\phi = 12°$. Although the BPF smoother has steadily better performance than the AR filter, the differences are not that large. An edge sensor is hard to charge and therefore it is difficult to recover after false alarms. In cases like this, it is preferable to postpone transmissions until the sensor is sufficiently charged in order to guarantee the successful transmission of the information message.

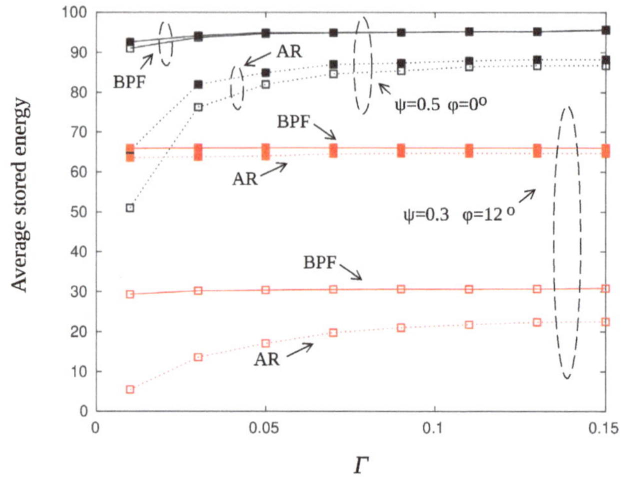

Figure 13. Average stored energy for particle filter (solid line) and AR filter (dotted line), with parameters $\gamma = 0.5$, $\varrho = 1$, $r = 10$ m , $\psi = 0.05$, $\phi = 0°$ (black lines), and $\psi = 0.3$, $\phi = 12°$ (red lines). Filled squares mark both threshold and energy criteria (Cr. 35), squares mark only threshold criteria (Cr. 36).

5.2. Charging and Detection Delay

Figure 14 plots the detection delay, measured in time slots, for a sensor that charges well ($\phi = 0°$ and high values of the charging efficiency parameter γ. An immediate observation is that the use of BPF results in larger delays compared to that of AR filter by almost 25–30 time slots. This is in agreement with the results of Figures 8 and 9, where it is demonstrated that the response of the BPF to the changes of the posterior likelihood is delayed. Another, rather obvious, observation is that smaller values of threshold parameter Γ produces smaller delays. Indeed, comparing the circle marked curves, which correspond to $\Gamma = 0.3$, to the square marked curves, which correspond to $\Gamma = 0.15$, we note a difference in the detection delay of 10 time slots for the BPF and 4 time slots for the AR filter. The criterion selection, i.e., criterion 35 or 36, does not have a major impact on the detection delay for nodes that charge easily. As it is observed from Figure 14 criterion 35 (filled square and circle markers) provides the same performance with criterion 36 (empty markers). However, this is not the case for sensor nodes that are difficult to charge, as is shown next. Figure 15, depicts the detection delay performance for a node that lies at the edge ($\phi = 12°$) of the charging beam. Unless the charging efficiency γ is high enough, detection delays reach large values if the node has to postpone its transmission (criterion 35, filled markers). Thus, in such cases alternative solutions must be sought. For example, one may take off edge sensors from their unfavorable position by beamforming using an angle offset or by letting sensor nodes relay the information of their neighbors. In the latter scheme, nodes transmit using less power since the objective is to reach a neighbor node that is well charged and capable to relay the information to the AP.

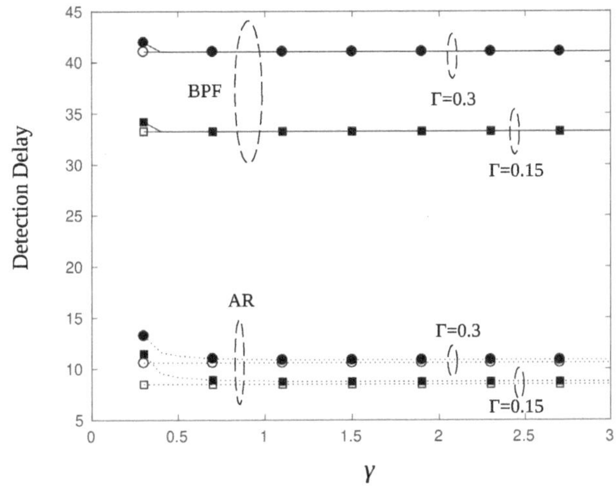

Figure 14. Detection delay for particle filter (solid line) and AR filter (dotted line), with parameters $\phi = 0°$, $\varrho = 1$, $\psi = 0.05$, $r = 10$ m, $\Gamma = 0.15$ (square), and $\Gamma = 0.3$ (circle). Filled squares mark both threshold and energy criteria (Cr. 35), squares mark only threshold criteria (Cr. 36).

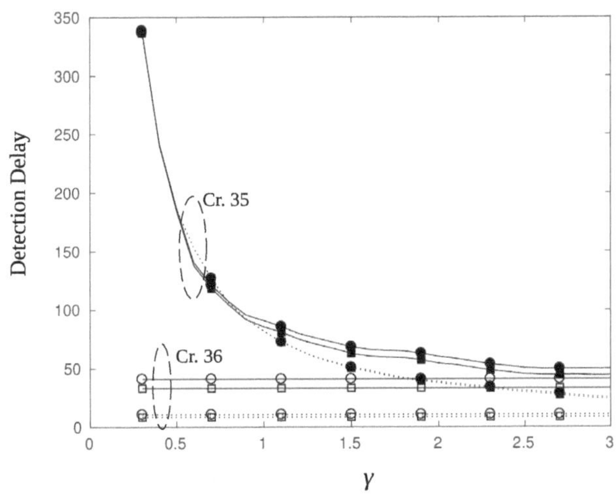

Figure 15. Detection delay for particle filter (solid line) and AR filter (dotted line), with parameters $\phi = 12°$, $\varrho = 1$, $\psi = 0.3$, $r = 10$ m, $\Gamma = 0.15$ (square), and $\Gamma = 0.3$ (circle). Filled squares mark both threshold and energy criteria (Cr. 35), squares mark only threshold criteria (Cr. 36).

Figure 16 summarizes some comparative results for the case of a sensor that charges well ($\phi = 0°$). The left column of figures corresponds to criterion 36 whereas the right column corresponds to criterion 35. The first row of figures is for AR filtering ($\alpha = 0.9$) and the second row is for Beta Particle Filter smoothing. The average number of false alarms is depicted with bars, whereas the detection delay is depicted with filled diamond marks. As it is observed, for very small values of the threshold parameter Γ, the Beta Particle Filter smoothing results in considerably smaller values (tenfold reduction) of false alarms compared to AR filtering. However, the situation is reversed regarding the detection delay. Using criterion 35, although the detection delay is not affected, the false alarm rates are reduced by a factor of three (or more) for small values of the parameter Γ.

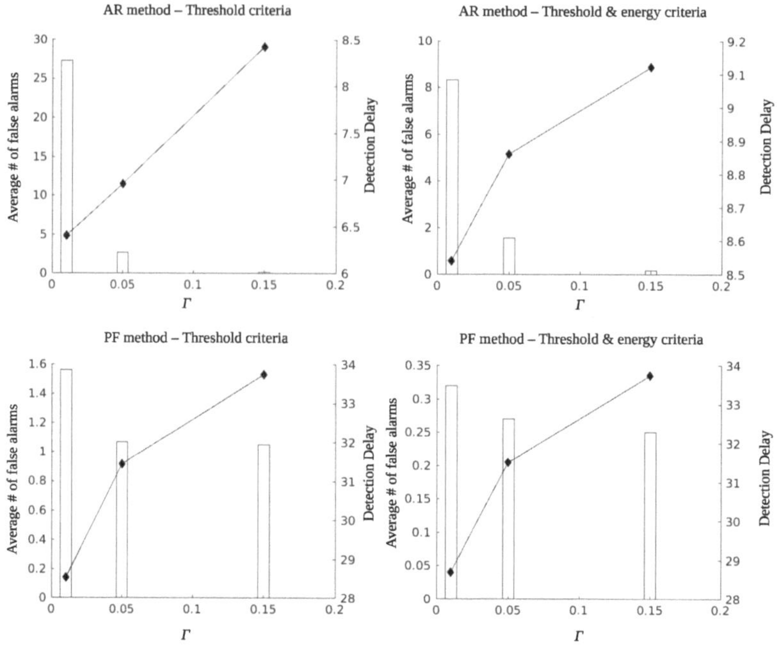

Figure 16. Detection delay—average number of false alarms before the event, with parameters $\phi = 0°$, $\varrho = 1$, $\psi = 0.05$, $r = 10$ m, $\gamma = 0.5$, and $\alpha = 0.9$.

6. Discussion

A thorough study was conducted on the performance aspects of a WPCN network used to detect events under the assumption of noisy sensor measurements. Detecting changes in the distribution of environmental parameters was based on the Optimal Stopping Theory by Shiryaev. The unstable nature of the involved likelihoods imposes smoothing on the posterior estimates and to this end, a Beta Particle Filter was proposed and tested against a classical AR filter. Analytical and simulation results show that performance indices, such as false alarm rate, detection delay, and probability of successful transmission, rely heavily on the charging efficiency and the location of the sensor nodes. Future work is oriented to incorporating mechanisms, such as data/information fusion and relaying of information, that alleviates the problem of partly charged sensor nodes, in an effort to increase the reliability of the system.

Author Contributions: Conceptualization, M.K. and E.Z.; methodology, L.M.; software, M.K.; validation, E.H. and E.Z.; formal analysis, M.K. and E.Z.; writing—original draft preparation, M.K.; writing—review and editing, E.H.; supervision, L.M. All authors have read and agreed to the published version of the manuscript.

Funding: This research received no external funding.

Institutional Review Board Statement: Not applicable.

Informed Consent Statement: Not applicable.

Data Availability Statement: Data set available on request to corresponding authors.

Conflicts of Interest: The authors declare no conflict of interest.

References

1. Djidi, N.E.H.; Gautier, M.; Courtay, A.; Berder, O.; Magno, M. How can wake-up radio reduce lora downlink latency for energy harvesting sensor nodes? *Sensors* **2021**, *21*, 733. [CrossRef] [PubMed]
2. Ahmad, I.; Hee, L.M.; Abdelrhman, A.M.; Imam, S.A.; Leong, M.S. Scopes, challenges and approaches of energy harvesting for wireless sensor nodes in machine condition monitoring systems: A review. *Measurement* **2021**, *183*, 109856. [CrossRef]
3. Shiryaev, A.N. On optimum methods in quickest detection problems. *Theory Probab. Its Appl.* **1963**, *8*, 22–46. [CrossRef]
4. Shiryaev, A.N. *Optimal Stopping Rules*; Springer Science & Business Media: Berlin/Heidelberg, Germany, 2007; Volume 8.
5. Guo, J.; Zhou, X.; Durrani, S. Wireless power transfer via mmWave power beacons with directional beamforming. *IEEE Wirel. Commun. Lett.* **2018**, *8*, 17–20. [CrossRef]
6. Psomas, C.; Krikidis, I. Energy beamforming in wireless powered mmWave sensor networks. *IEEE J. Sel. Areas Commun.* **2018**, *37*, 424–438. [CrossRef]
7. Hong, Y.W.P.; Hsu, T.C.; Chennakesavula, P. Wireless power transfer for distributed estimation in wireless passive sensor networks. *IEEE Trans. Signal Process.* **2016**, *64*, 5382–5395. [CrossRef]
8. Koutsioumpos, M.; Zervas, E.; Hadjiefthymiades, E.; Merakos, L. Monitoring for Rare Events in a Wireless Powered Communication mmWave Sensor Network. *Sensors* **2020**, *20*, 3341. [CrossRef] [PubMed]
9. Godala, S.; Vaddella, R.P.V. A study on intrusion detection system in wireless sensor networks. *Int. J. Commun. Netw. Inf. Secur.* **2020**, *12*, 127–141.
10. Kandris, D.; Nakas, C.; Vomvas, D.; Koulouras, G. Applications of wireless sensor networks: An up-to-date survey. *Appl. Syst. Innov.* **2020**, *3*, 14. [CrossRef]
11. Safaei, M.; Asadi, S.; Driss, M.; Boulila, W.; Alsaeedi, A.; Chizari, H.; Abdullah, R.; Safaei, M. A systematic literature review on outlier detection in wireless sensor networks. *Symmetry* **2020**, *12*, 328. [CrossRef]
12. Tartakovsky, A.G.; Rozovskii, B.L.; Blažek, R.B.; Kim, H. Detection of intrusions in information systems by sequential change-point methods. *Stat. Methodol.* **2006**, *3*, 252–293. [CrossRef]
13. Tartakovsky, A.G.; Veeravalli, V.V. Change-point detection in multichannel and distributed systems. *Appl. Seq. Methodol. Real-World Examples Data Anal.* **2004**, *173*, 339–370.
14. Xie, Y.; Siegmund, D. Sequential multi-sensor change-point detection. In Proceedings of the 2013 Information Theory and Applications Workshop (ITA), San Diego, CA, USA, 10–15 February 2013; pp. 1–20.
15. Raghavan, V.; Veeravalli, V.V. Quickest change detection of a Markov process across a sensor array. *IEEE Trans. Inf. Theory* **2010**, *56*, 1961–1981. [CrossRef]
16. Hadjiliadis, O.; Zhang, H.; Poor, H.V. One shot schemes for decentralized quickest change detection. *IEEE Trans. Inf. Theory* **2009**, *55*, 3346–3359. [CrossRef]
17. Moustakides, G.V. Optimal stopping times for detecting changes in distributions. *Ann. Stat.* **1986**, *14*, 1379–1387. [CrossRef]
18. Veeravalli, V.V. Decentralized quickest change detection. *IEEE Trans. Inf. Theory* **2001**, *47*, 1657–1665. [CrossRef]
19. Premkumar, K.; Kumar, A. Optimal sleep-wake scheduling for quickest intrusion detection using wireless sensor networks. In Proceedings of the IEEE INFOCOM 2008-The 27th Conference on Computer Communications, Phoenix, AZ, USA, 13–18 April 2008; pp. 1400–1408.
20. Zervas, E.; Mpimpoudis, A.; Anagnostopoulos, C.; Sekkas, O.; Hadjiefthymiades, S. Multisensor data fusion for fire detection. *Inf. Fusion* **2011**, *12*, 150–159. [CrossRef]
21. Lorden, G. Procedures for reacting to a change in distribution. *Ann. Math. Stat.* **1971**, *42*, 1897–1908. [CrossRef]
22. Boshkovska, E.; Ng, D.W.K.; Zlatanov, N.; Schober, R. Practical non-linear energy harvesting model and resource allocation for SWIPT systems. *IEEE Commun. Lett.* **2015**, *19*, 2082–2085. [CrossRef]
23. Ding, Z.; Fan, P.; Poor, H.V. Random beamforming in millimeter-wave NOMA networks. *IEEE Access* **2017**, *5*, 7667–7681. [CrossRef]
24. Manatakis, D.V.; Manolakos, E.S. Estimating the spatiotemporal evolution characteristics of diffusive hazards using wireless sensor networks. *IEEE Trans. Parallel Distrib. Syst.* **2014**, *26*, 2444–2458. [CrossRef]

sensors

MDPI

Article

Downlink Performance Modeling and Evaluation of Batteryless Low Power BLE Node

Ashish Kumar Sultania [1], Carmen Delgado [2], Chris Blondia [1] and Jeroen Famaey [1,*]

[1] IDLab-Department of Computer Science, University of Antwerp-imec, 2000 Antwerp, Belgium; ashishkumar.sultania@uantwerpen.be (A.K.S.); chris.blondia@uantwerpen.be (C.B.)
[2] i2CAT Foundation, 08034 Barcelona, Spain; carmen.delgado@i2cat.net
* Correspondence: jeroen.famaey@uantwerpen.be; Tel.: +32-4726-449-37

Abstract: Deploying low maintenance and long-life systems is an important requirement of emerging commercial Internet of Things (IoT) solutions. Such systems can be envisioned in which the connected devices are powered by energy harvested from ambient sources and stored in long-lifetime capacitors rather than short-lived and polluting batteries. However, due to the unpredictable nature of ambient energy harvesting, such batteryless IoT devices might not always have enough energy to initiate communication. The Bluetooth Low Energy (BLE) specification defines support for Low Power Nodes (LPNs) using the friendship feature, where the LPN is associated with a neighbouring friend node (FN). The LPN can receive downlink (DL) data and remain connected to the network via the FN that buffers the LPN's incoming packets while allowing the LPN to save energy by sleeping or turning itself off. This novel BLE feature makes the LPN highly suitable to support the connection of batteryless ambiently-powered IoT devices. While the LPN can decide when to transmit uplink (UL) packets and does not depend on the FN to receive downlink (DL) data, the LPN needs to poll its FN to receive the buffered packets. However, the DL packet latency increases with this process due to the buffering time at the FN. Therefore, in this work, we present an analytical model to characterize the performance as a function of DL data latency and packet delivery ratio (PDR) of a batteryless LPN powered by different harvesting powers and capacitor sizes. This would help to optimally choose the correct configuration of the batteryless LPN for its network deployment. We also compare the analytical model and simulation results, showing consistency with an average error of 2.23% for DL data latency and 0.09% for the PDR.

Keywords: batteryless; Bluetooth Low Energy; energy harvesting; IoT; low power node

Citation: Sultania, A.K.; Delgado, C.; Blondia C.; Famaey, J. Downlink Performance Modeling and Evaluation of Batteryless Low Power BLE Node. *Sensors* **2022**, *22*, 2841. https://doi.org/10.3390/s22082841

Academic Editors: Carlo Trigona, Olfa Kanoun and Slim Naifar

Received: 31 January 2022
Accepted: 5 April 2022
Published: 7 April 2022

1. Introduction

The emerging Internet of Things (IoT) concept aims to wirelessly connect billions of devices to the internet. Usually, these devices are powered by batteries. However, replacing batteries for such a large number of devices increases the cost of system maintenance, as it requires manual intervention. Battery life is also reduced due to over-charging and over-discharging, as well as environmental conditions (e.g., temperature fluctuations) [1]. Moreover, disposing of batteries is also a serious environmental concern. Therefore, considering the downside of battery usage and the requirement of low maintenance 'deploy and forget' devices for the IoT, batteries can be replaced by easily recyclable and cheap capacitors combined with energy harvesting to obtain and store energy in a more sustainable way [2,3]. Such devices can achieve a long-term multi-decade active lifetime.

Bluetooth Low Energy (BLE) has become one of the most important technologies for the IoT thanks to its features (i.e., extremely low power consumption, low cost, and mesh topology support). It supports many use-cases, such as asset tracking, sensing, monitoring, control, and automation, with thousands of interconnected devices [4]. However, most existing BLE-based IoT solutions consider battery-powered devices. Meanwhile, designing a batteryless solution is important for long-term sustainability (i.e., no adverse impact on the environment), but it is also a challenging task. Batteryless devices usually charge their capacitor by harvesting ambient energy, which is often uncontrollable and unpredictable, allowing them to only work for a short amount of time. These conditions can result in intermittent behaviour where the device can experience multiple power failures. The duration of the power failure cycle depends on the harvesting power, capacitor size, and operating voltage. Moreover, for a specific harvesting power, the time to charge a capacitor increases with its capacitance [5]. Therefore, choosing the optimal capacitor size based on the minimum available harvesting power and required operations to achieve the desired performance is important.

The BLE specification [6] already provides a friendship feature to aid Low Power Nodes (LPNs) to save energy, keeping themselves in SLEEP or OFF mode for most of the time. The LPN awakes only (or turns on) to transmit or receive data packets. For sending uplink (UL) data packets, the LPN can broadcast them at any time. On the other hand, the LPN can start a poll process (request/response) to receive the downlink (DL) packets that are temporarily buffered at the friend node (FN). The polling happens periodically to receive any incoming data packets. This can reduce the overall energy consumption at the cost of increased DL data latency.

In battery-powered nodes (non-harvesting), long-term energy consumption is a key factor as it affects the total lifetime of the device. This factor is less important in batteryless devices powered by ambient energy as they can perpetually recharge themselves. Instead, short-term energy consumption bursts are more important as they may drain the capacitor and result in a power failure. In this article, we focus on analyzing the achievable DL latency and packet delivery ratio (PDR) using a batteryless BLE LPN supported by different harvesting powers and capacitor sizes. We model the system utilizing a discrete-time Markov Chain and compare the results with our developed simulator [7]. The simulator needs to simulate extended periods of time and be repeated many times to average out stochastic variability, whereas a model can immediately provide the results for a specific use case at steady state. This allows the model, compared to a full-fledged simulator, to be integrated into a real-time optimizer in the node that can help to easily reconfigure the BLE configuration parameters based on current conditions. The model can be used to theoretically perform a feasibility analysis of the friendship feature for bidirectional battery-less communication and to deduce its limitations. Moreover, it would aid in designing battery-less BLE devices by calculating its optimal hardware and software configuration for IoT use-cases that require downlink or bidirectional communication. Specifically, our model can be used to optimally derive the hardware configuration in terms of capacitor size. The software configuration parameters, in terms of ReceiveDelay, ReceiveWindow, Tx power level, and Poll interval of the batteryless LPN can be determined based on the use-case requirement in terms of data rate, DL latency, and PDR considering the neighboring FN queue size and minimum harvesting power at the network site.

The paper is organized as follows: Section 2 summarizes related work; Section 3 gives an overview of the BLE friendship feature; Section 4 explains the batteryless LPN system; and Section 5 presents the analytical model. Subsequently, Section 6 presents all the evaluation results and Section 7 shows the usage of model to derive optimal parameters by considering a logistic management IoT use-case. Section 8 draws the main conclusions. Finally, Section 9 briefly discusses limitations of the model and potential avenues for future work.

2. Related Work

There are several batteryless BLE prototypes designed by researchers. Silva et al. [8] presented a prototype where an implantable batteryless glucose monitor node was capable of transmitting data to a mobile phone when supplied power wirelessly using a Radio Frequency (RF) energy source. However, their design did not consider the Bluetooth mesh friendship feature and sent only UL data. Liu et al. recently proposed another RF-power based prototype for indoor BLE beaconing [9]. They tested the device for both UL and DL data. However, they did not analyze the effect of the capacitor size nor the performance in terms of DL latency. Fraternali et al. [10] designed a batteryless BLE sensor node based on ambient light energy harvesting. They presented a predictive algorithm that adapted to the changing ambient conditions to maximize the node lifetime and application quality of service. However, their focus was also only on the UL data. Nilsson et al. also conducted indoor and outdoor experiments to evaluate power consumption, range, throughput, and UL latency of a Bluetooth mesh network [11]. In summary, most of the research on batteryless BLE nodes is performed considering UL data without considering the friendship feature or DL communications.

There are two papers that considered the friendship feature. However, they considered battery-powered LPNs rather than batteryless LPNs. The first paper is by Alvarez et al. [12], who performed an in-depth analysis of security on the friendship concept. They demonstrated that a denial-of-friendship attack could reduce the LPN's battery life by 70-fold. The second paper is by Darroudi et al. [13]. They presented an analytical model to study the energy performance parameters of a battery-operated BLE LPN. These parameters include current consumption, lifetime, and energy consumed per delivered bit for different PollTimeout and ReceiveWindow values. They calculated the evaluation parameters based on the average value of time and current consumption for all the radio states and for fixed UL data rate interval, in contrast to DL traffic, which does not require the polling feature or friend queue.

In contrast, our work is focused on the DL latency and packet delivery metrics. We focus on a batteryless LPN harvesting energy from ambient sources and not on battery power LPNs. As such, we do not consider current consumption as the evaluation metric, but rather a constraint that should be satisfied. Moreover, we focus on DL packet arrivals, which are first queued at the FN, and so our model studies the queueing performance based on the Markov chain with Poisson distributed data arrivals which provides discrete random distribution of data. In real-world applications, the data arrivals are also random, and so it is used to represent the message arrivals in the BLE network. The model provides the opportunity to analyze the batteryless LPN powered by different capacitor sizes receiving different harvested powers for various Bluetooth mesh network parameters, such as friend queue size, packet size, or receive delay. The model is also compared to our developed simulator as presented in [7].

3. BLE Friendship Feature

The Bluetooth mesh network is based on a publish-subscribe model for data communication. It supports up to 32,767 devices with a maximum of 128 hops [14]. The devices need to have addresses and security information to become part of the mesh network to send or receive data. This is enabled by a provisioner, and the provisioned device is called anode. Any node can support four additional optional features (namely, relay, proxy, friend, and low power [6]), as shown in Figure 1.

The relay node supports receiving and re-transmitting data packets to all the nodes in its connection range. It helps to extend the range of the entire network. The proxy node helps the non-mesh-supported BLE devices to communicate via the mesh network. To enable both these features, the node needs to listen continuously for advertising packets and, therefore, should be connected to a continuous power source [15]. To support energy-constraint nodes, BLE introduces the LPN, which can offload the task of listening to its incoming channel to a FN that can temporarily buffer any incoming data packets. For example, in Figure 1, LPN 'A' is associated with FN 'X' for the DL data. However, it does not need help to transmit UL data. The LPN can wake up any time to send UL data, assuming that at least one of its neighbouring nodes actively listens to the incoming channels. With the help of its FN, the LPN can save much of its energy by turning off its components and keeping itself in SLEEP mode for most of the time. However, before switching to the SLEEP or OFF state, the LPN needs to first establish a friendship with one of its single-hop neighbouring nodes. This neighbour needs to support the optional friendship feature.

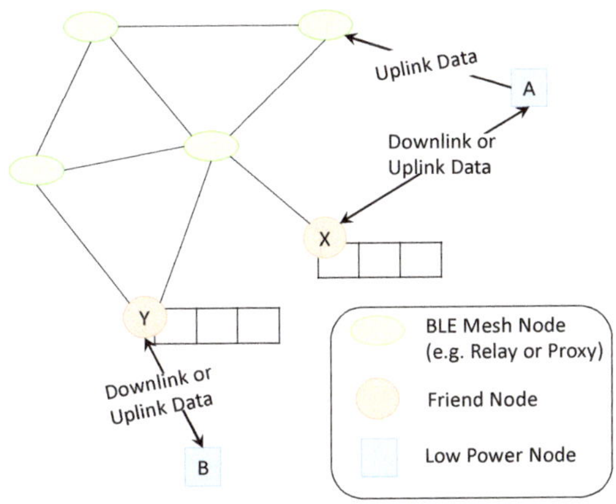

Figure 1. BLE mesh network.

3.1. Friendship Establishment

Figure 2 illustrates the messages exchanged during the friendship establishment procedure. The same Friend Request is received by one or more neighbouring friend feature enabled nodes. This message is 28 bytes of a network Protocol Data Unit (PDU), which consists of Criteria, ReceiveDelay (RD), PollTimeout (PT), PreviousAddress, NumElements, and NumReqCounter [6].

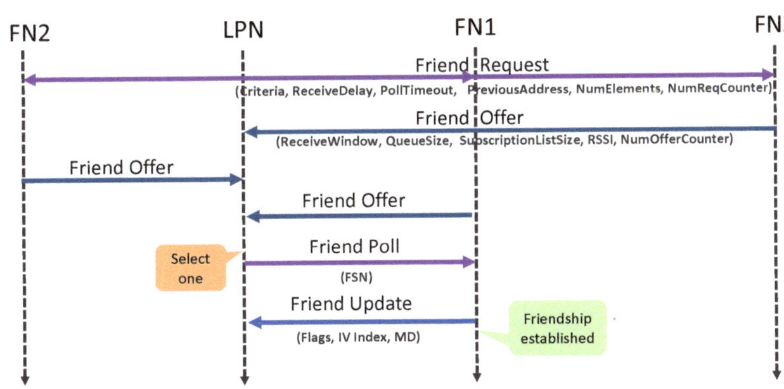

Figure 2. Friendship establishment procedure of BLE.

The Criteria is the minimum requirement in terms of the Received Signal Strength Indicator (RSSI), ReceiveWindow (RW), and supported queue size that an FN should satisfy. The RD is the waiting time of the LPN before it should expect to receive a poll response from the FN. PT is used to ensure that the FN does not keep the friendship context indefinitely and terminates the relationship if it does not receive any poll before the PT timer expires (since the last poll). Lastly, the RW is the maximum time that the LPN should listen to receive a response from the FN after sending a poll request. These timers can be seen in Figure 3. The RD and RW can be configured with a maximum value of 255 ms and a PT with a maximum of 96 h. PreviousAddress defines the address of its previous FN, NumElements defines its number of addressable elements (as messages in the BLE mesh network are exchanged from the elements), and NumReqCounter defines the number of sent requests required to calculate the Friendship Security Material.

The FNs that satisfy the initial criteria defined by the LPN will offer the friendship by sending a Friend Offer (FO). This response message has a network PDU of 24 bytes, which consists of RW, supported friend queue (FQ) size, size of subscription list, calculated

RSSI, and number of FO sent by this FN. Based on all the received FOs, the LPN selects the best among them, considering the Criteria satisfactions. To confirm this selection, the LPN acknowledges the selected FN with a Friend Poll (FP) message of 19 bytes, with the Friend Sequence Number (FSN) equal to zero. Next, the FN finalizes the friendship procedure by sending a Friend Update (FU) message of 24 bytes that contains the flags, More Data (MD), indicating if the queue is empty or not, and another flag that indicates the update of the initialization vector (IV) index (which is a shared network resource) or security key. It also contains the current IV index value known by the FN. Once the friendship is established, the LPN also notifies the FN with all the subscribed group list addresses so that the FN knows which messages it should buffer and forward to the LPN.

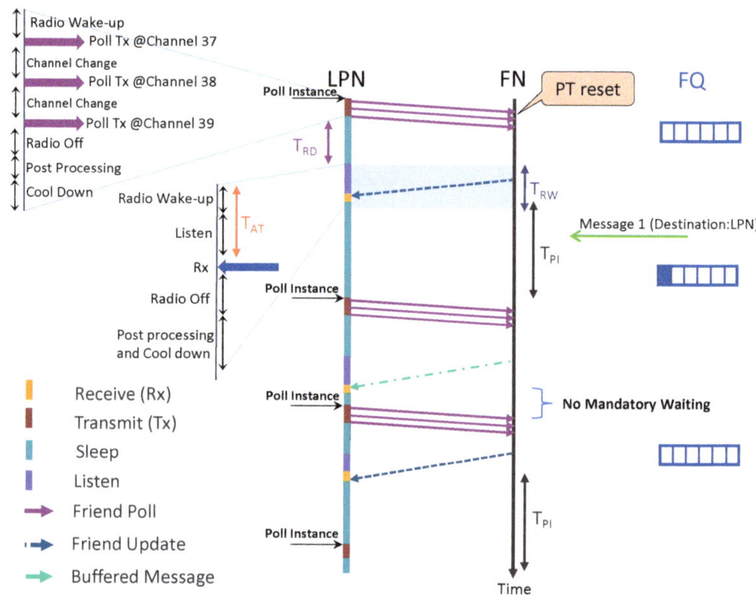

Figure 3. Sequence diagram of the messages exchanged between the LPN and FN to receive DL packets.

3.2. *Receiving Downlink Data*

As the friendship is established, the LPN can now switch to SLEEP mode, and all the incoming DL packets of the LPN will be buffered in the FQ. The process is shown in Figure 3.

The LPN can retrieve them from the FN by periodically waking up and sending an FP message in all three broadcast channels (37 to 39). On reception of the FP, the FN gets a time window of RW (T_{RW}) seconds, starting (T_{RD}) seconds after receiving the FP to send a response [16]. We also define an arrival time (AT or T_{AT}), which denotes the total listening time until the LPN starts receiving the message. While sending the FP, the LPN needs to toggle the FSN value to indicate the acknowledgement of the previous message. If the FN receives the same FSN field value as the last FP message received from the FN, it responds with the same message that it has previously sent, unless the FQ head is updated. This can be updated by any new incoming packet(s) whenever it arrives at a full FQ. In this case, the oldest packet (FQ head) would be discarded (dequeued) and the newly arrived packet would be added in the queue tail (queued). This strategy is called 'push-out'. Therefore, if the previously sent message has been discarded (on the arrival of new packets), then the oldest entry in the FQ shall be sent. If the FQ is empty, an FU is sent with MD equal to zero. To avoid receiving frequent, continuous FU messages, a timer poll interval timer (T_{PI}) can be considered. On receiving an FU message, an LPN needs to wait until the PI expires to send the next FP message. As we consider a batteryless LPN, it may happen that it is unable to send the FP exactly at the expiration of PI due to energy constraints. The friendship can also be terminated if the LPN does not receive any response from the FN after sending the FP for a certain amount of attempts. Moreover, BLE mesh provides infrastructure-less connectivity without any overheads of time synchronization [17]. As the nodes (except LPNs) are assumed to be always listening, there is no need to synchronize their listening timeslots. Moreover, the LPN always initiates communication with the FN itself (i.e., receiver-initiated communication); therefore, the LPN does not need to synchronize explicitly for receiving data either [16]. Timing of the RW is done relative to the FP message sent by the LPN to its FN.

4. Batteryless BLE LPN

Let the LPN have an already established friendship with a neighbouring FN. To receive DL packets, the radio of an LPN can be in 15 different states, as shown in Figure 3 (radio wake-up to post-processing and cool down). The voltage (energy) of a capacitor as a function of time, supporting various states of the radio as used in the remainder of this section, was first introduced by Delgado et al. [18].

4.1. Capacitor Voltage

Equation (1) provides the voltage on the capacitor after the device operates at a load current of $I_L(X)$ (amperes) in state X for a time Δt, harvesting from a constant voltage source, providing a power equal to P_h (watts).

$$V(t+\Delta t) = \frac{V_{max} \cdot P_h}{P_h + V_{max} I_L(X)} \left(1 - e^{\frac{-\Delta t \cdot (P_h + V_{max} I_L(X))}{V_{max}^2 \cdot C}}\right) + V(t) \cdot e^{\frac{-\Delta t \cdot (P_h + V_{max} I_L(X))}{V_{max}^2 \cdot C}}, \quad (1)$$

where, V_{max} (volt) is the maximum capacitor voltage for capacitance C (farad).

As the capacitor has a small energy density and due to the unpredictable availability of harvested energy, it can result in intermittent on-off behaviour of the batteryless LPN, as shown in Figure 4. When the capacitor voltage drops below the minimum operating voltage (V_{cutoff}), the LPN transitions to the OFF state. If the LPN is in the OFF state, it transitions to the SLEEP state only when achieving a predefined turn-on voltage V_{turnon}.

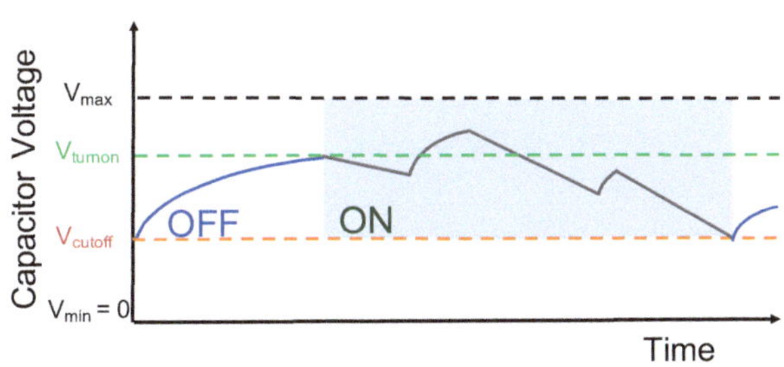

Figure 4. Intermittent behaviour of a batteryless LPN harvesting continuously.

However, in our model, an energy-aware LPN is considered that tries to avoid the OFF state. It can use an ultra-low-power comparator, with power consumption in the order of pico-watts [19], which can determine if a defined voltage level in the capacitor has been reached. This voltage could be set to the minimum voltage level required to execute a task (such as poll, receive data) without going below V_{cutoff}. We refer to that approach as an energy-aware device.

As stated before, the friendship can be terminated if no response is received either after sending $\dot{N}$ FPs or if the time between two consecutive FPs is larger than T_{PT}. To avoid this, the P_h is considered such that it allows the capacitor to recharge itself from V_{min} to a threshold voltage level that is sufficient to perform a poll request-response cycle in less than T_{PT}. The time to send the FP when the batteryless LPN needs to have a required threshold voltage of V_{th}^{FP} to start the FP and the remaining voltage of the capacitor after an FP is calculated in Section 4.2.

4.2. Time to Next Poll

The time at which the next FP can take place depends on whether the FQ is empty or not and the available energy after the previous poll.

4.2.1. FQ Is Not Empty

The voltage of the capacitor after all the polling stages can be calculated according to Equation (1), starting from an initial voltage (V_0) and considering 'radio wake-up' as an initial state. Let V_f be the final voltage after all the polling radio stages (as shown in Figure 3) of the previous poll are executed. Therefore, there are two scenarios:

- $V_f \geq V_{th}^{FP}$: The next FP can be sent immediately, and the time between two FPs (T_{FP}) is the sum of all the state timings as defined in Table 1 and is calculated as Equation (2). The capacitor voltage at the start of the next poll instance will be V_f.

$$T_{FP} = \sum_{i=AllStates} T_i \tag{2}$$

- $V_f < V_{th}^{FP}$: As V_f is less than the required voltage to poll, the device needs to switch to SLEEP mode to charge the capacitor and avoid its voltage from reaching V_{cutoff}. Let the node sleep for the time T_{sleep} which can be deduced using Equation (1). Therefore, the time between two FPs has an extra wait time of T_{sleep} in addition to all the FP state's timings and is calculated in Equation (3).

$$T_{FP} = \sum_{i=AllStates} T_i + T_{sleep} \tag{3}$$

The capacitor voltage at the start of the next poll instance will be V_{th}^{FP} (after the capacitor is charged while the LPN is in SLEEP mode).

4.2.2. FQ Is Empty

If the FQ is empty, the LPN needs to wait for the time PI (T_{PI}) to poll, as it received an FU after the previous FP message. Therefore, in this case, the voltage V_f would be calculated after a sleep time of T_{PI}. Now again, there are two cases.

- $V_f \geq V_{th}^{FP}$: The time to poll has an extra delay of T_{PI} in addition to all the FP state's timings.

$$T_{FP} = \sum_{i=AllStates} T_i + T_{PI} \tag{4}$$

The capacitor voltage at the start of the next poll instance will be V_f.

- $V_f < V_{th}^{FP}$: Similar to above, the node needs to switch to SLEEP mode to provide enough time to charge the capacitor. Therefore, the time between two FPs has an additional time T_{sleep} and T_{PI} in addition to all the FP state's timings.

$$T_{FP} = \sum_{i=AllStates} T_i + T_{PI} + T_{sleep} \tag{5}$$

The capacitor voltage at the start of the next poll instance is V_{th}^{FP}.

The threshold voltage of the FP (V_{th}^{FP}) can be calculated by taking the inverse of the function, as mentioned in Equation (1) calculating V(t), knowing the value of $V(t + \Delta t)$ for all the states of the poll, considering the initial voltage as V_{cutoff}, and with starting state 'post-processing and cool down'.

Table 1. State sequence for a poll request and response of a LPN.

Identifier	Radio State	Time [µs] of the LPN [mA]	Current Cons.	Comments
1	Radio wake-up	2940	0.587	Prepare for poll transmission
2	TX (at 4 dBm)	471	9.09	Tx advertisement packet on channel 37
3	Change channel	29	7.78	Change radio frequency
4	TX (at 4 dBm)	380	9.15	Tx advertisement packet on channel 38
5	Change channel	29	7.57	Change again the radio frequency
6	TX (at 4 dBm)	384	9.16	Tx advertisement packet on channel 39
7	Radio off	34	6.60	Main radio in OFF state
8	Post processing	420	2.17	
9	Cool down	20,820	0.00653	Prepare to switch to the SLEEP state
10	Sleep	RD	0.00896	Node in SLEEP mode
11	Wake-up prescan	1910	1.13	Wake-up for Rx
12	Listen	AT	8.68	Actively Listen for incoming DL packet
13	Scan message	1160 (or 900)	9.64	Rx 24 B FQ data (or 22 B FU)
14	Radio off	389	4.69	Main radio in OFF state
15	Post-processing and Cool down	23,410	0.00619	Set up the sleep timer for the next state and switch to SLEEP state

5. System Model

This section presents the analytical model of the system described in Section 4. The DL latency, FQ occupancy, and PDR are calculated. All the symbols used in this section are summarized in Appendix A.

5.1. The Queueing Model

We call the time interval between two FPs, when the FQ at the start of the interval is not empty, a 'service'. The time interval between two FPs, during which an FU is transmitted from the FN to the LPN (i.e., the FQ at the start of the interval is empty), is called a 'vacation'. The system can be modelled as a finite capacity queue, where the server takes repeated vacations. The duration of a service and a vacation depends on the LPN's capacitor voltage. Let the packets arrive at the FN according to a Poisson distribution with data arrival rate ˘, and the FN temporarily buffers them in a buffer (FQ) of capacity N. The vector $\mathbf{U}$ represents the discretized voltages that a capacitor can take between 1 and V_{max}. Hence, the duration of a service when the voltage equals $i \in \mathbf{U}$ can be denoted by S_i with distribution function $\mathbb{S}_i(t)$ and Laplace Transform (LST) $\mathbb{S}_i^*(\nu)$. Similarly, for the vacation, W_i, $\mathbb{W}_i(t)$ and $\mathbb{W}_i^*(\nu)$ can represent the duration, the distribution function, and its LST, respectively. The duration of a service consists of all the radio states during polling and possibly followed by a sleep period to reach the threshold index V_{th}^{FP}. The radio states during the polling can be assumed to have a constant duration for a fixed size of the DL packet. Hence, S_i can be computed considering FQ is not empty and W_i considering it is empty, as discussed in Section 4.2. If at the start of a service the voltage is equal to i, then the voltage at the next poll instance is denoted by $\mathbb{V}_i^S$ and calculated using Equation (1) for all the polling radio states. Similarly, if the voltage is equal to i at the start of a vacation, its values at the next inspection instant denoted by $\mathbb{V}_i^W$ also include the SLEEP state for T_{PI}.

The probability $\mathbb{P}_{k,i}^S$ and $\mathbb{P}_{k,i}^W$ that k packets arrive during the service time and the vacation time respectively, starting from $\mathbf{U}$, equal to i, can be calculated as Equation (6).

$$\begin{aligned} \mathbb{P}_{k,i}^S &= \int_0^\infty \frac{(\lambda t)^k}{k!} e^{-\lambda t} d\mathbb{S}_i(t) \\ \mathbb{P}_{k,i}^W &= \int_0^\infty \frac{(\lambda t)^k}{k!} e^{-\lambda t} d\mathbb{W}_i(t) \end{aligned} \quad (6)$$

The column vectors of the averages of the service times and the vacation times are given by Equation (7).

$$\begin{aligned}\overline{\mathbf{E}[S]} &= (\mathbf{E}[S_1], \mathbf{E}[S_2], ..., \mathbf{E}[S_{V_{max}}])' \\ \overline{\mathbf{E}[W]} &= (\mathbf{E}[W_1], \mathbf{E}[W_2], ..., \mathbf{E}[W_{V_{max}}])'\end{aligned} \tag{7}$$

5.2. Embedded Markov Chain

The system is observed at the poll instants (t_n, n = 1, 2, ..., ∞). We refer to these poll instants as *inspection instants*. Let Q_n be the number of packets in the FQ and V_n be the available capacitor voltage at time instant t_n. It should be noted that V_n will never take values below $V_{startpoll}$, which is the minimum voltage available at a poll instant that allows all functions of the radio to receive a packet. Hence, the stochastic process (Q_n, V_n) at t_n is a discrete-time finite Markov Chain with state-space size equal to $(N+1) \cdot (V_{max} - V_{startpoll} + 1)$. The limiting probability distribution function of this Markov Chain is denoted by $\overline{p} = (\overline{p}_0, \overline{p}_1, ..., \overline{p}_N)$. Let $(\overline{p}_k)_i$ denote the joint probability distribution of the voltage and the queue length, as shown in Equation (8).

$$\begin{aligned}(\overline{p}_k)_i &= \lim_{n\to\infty} Prob\Big(Q_n = k, V_n = i\Big), \\ &\quad \text{where, } 0 \le k \le N, i = V_{startpoll}, ...V_{max}\end{aligned} \tag{8}$$

The transition matrix of this Markov Chain is given by Equation (9).

$$\mathbf{P} = \begin{bmatrix} B_0 & B_1 & B_2 \dots & B_{N-2} & B_{N-1} & \sum_{n=N}^{\infty} B_n \\ A_0 & A_1 & A_2 \dots & A_{N-2} & \sum_{n=N-1}^{\infty} A_n & 0 \\ 0 & A_0 & A_1 \dots & A_{N-3} & \sum_{n=N-2}^{\infty} A_n & 0 \\ \vdots & \vdots & \vdots \; \vdots & \vdots & \vdots & \vdots \\ 0 & 0 & 0 \dots & A_0 & \sum_{n=1}^{\infty} A_n & 0 \\ 0 & 0 & 0 \dots & 0 & \sum_{n=0}^{\infty} A_n & 0 \end{bmatrix} \tag{9}$$

where, B_n represents the transitions starting from an empty system and ending up with n packets in the FQ at the next inspection point. Whereas, A_n represents the transitions starting from a non-empty system. The $(V_{max} - V_{startpoll} + 1) \cdot (V_{max} - V_{startpoll} + 1)$ size-matrices A_n are defined by Equation (10).

$$(A_n)_{i,j} = \begin{cases} \mathbb{P}^{\mathbb{S}}_{n,i}, & \text{for i} = [V_{startpoll}, V_{max}], \text{j} = \mathbb{V}^{S}_{i} \\ 0, & \text{elsewhere} \end{cases} \tag{10}$$

Similarly, B_n can also be calculated as Equation (11).

$$(B_n)_{i,j} = \begin{cases} \mathbb{P}^{\mathbb{W}}_{n,i}, & \text{for i} = [V_{startpoll}, V_{max}], \text{j} = \mathbb{V}^{W}_{i} \\ 0, & \text{elsewhere} \end{cases} \tag{11}$$

In the numerical examples, for a large value of k, $\mathbf{P}^k$ has positive real entries. Therefore, the matrix $\mathbf{P}$ is regular and has a unique normalized eigenvector $\overline{p}$ corresponding to the eigenvalue 1. Hence, the probability distribution satisfies Equation (12).

$$\begin{aligned}&\overline{p} = \overline{p} \cdot P \\ &\overline{p} \cdot \overline{e} = 1, \\ &\text{where, } \overline{e} \text{ is the (N+1)} \cdot (V_{max} - V_{startpoll} + 1) \text{ unit vector.}\end{aligned} \tag{12}$$

The steady-state probability of the system occupancy at arbitrary time instants has been derived in [20]. Let T_{FP} be the average time between two consecutive inspection instants given by Equation (13).

$$T_{FP} = [1 - \overline{p}_0 \cdot \overline{e}] \cdot \mathbf{E}[S] + \overline{p}_0 \cdot \overline{e} \cdot \mathbf{E}[W], \tag{13}$$

where

$$\mathbf{E}[S] = \begin{cases} \frac{\sum_{n=1}^{N} \overline{p}_n \cdot \overline{\mathbf{E}[S]}}{1 - \overline{p_0} \cdot \overline{e}}, & if\ \overline{p_0} \cdot \overline{e} \neq 1 \\ 0, & if\ \overline{p_0} \cdot \overline{e} = 1\ (system\ is\ always\ empty) \end{cases} \tag{14}$$

and

$$\mathbf{E}[W] = \begin{cases} \frac{\overline{p_0} \cdot \overline{\mathbf{E}[W]}}{\overline{p_0} \cdot \overline{e}}, & if\ \overline{p_0} \cdot \overline{e} \neq 0 \\ 0, & if\ \overline{p_0} \cdot \overline{e} = 0\ (system\ is\ never\ empty) \end{cases} \tag{15}$$

The queue occupancy distribution, i.e., the probability that there are n packets in the FQ, is then given by Equation (16) [20].

$$O_n = \frac{1}{\lambda \cdot T_{FP}} \cdot \left[\overline{p}_n \cdot \overline{e} - \overline{p}_0 \cdot B_n \cdot \overline{e}\right], \qquad \text{where, n = [0, N−1]}$$
$$\text{And} \tag{16}$$
$$O_N = 1 - \frac{1}{\lambda \cdot T_{FP}} \cdot \left[1 - \overline{p}_0 \cdot \overline{e}\right].$$

Therefore, the total percentage of packet loss is equal to $100 \cdot O_N$.

5.3. Performance Metrics

In this section, the metrics, average FQ occupancy, and average DL latency, are computed.

- Average FQ occupancy: The FN uses a push-out strategy to manage its FQ, which means when a packet arrives at a full FQ, the oldest packet in the FQ would be dropped. However, the number of packets in the *push-out* system is the same even if the arriving packet is dropped (drop-tail). Since the probability that an arriving packet finds a full FQ is equal to O_N, the average system occupancy in a 'drop-tail' system is given by Equation (17).

$$\mathbf{O}_{DT} = \sum_{n=1}^{N} n \cdot O_n . \tag{17}$$

Now, the average system occupancy in a 'push-out' system ($\mathbf{O}_{PO}$) is equal to the $\mathbf{O}_{DT}$.

- Average DL latency: The actual arrival rate in a drop-tail system is given by $(1 - O_N) \cdot \lambda$ and, applying Little's well-known formula [21], the average response time in a drop-tail system is given by Equation (18).

$$\mathbf{T}_{DT} = \frac{\mathbf{O}_{DT}}{(1 - O_N) \cdot \lambda} . \tag{18}$$

However, the waiting time of a packet in the system using push-out is quite different from the waiting time in a system using drop-tail. In a drop-tail system, the packets behind a tagged packet do not impact the delay of that packet. However, in a push-out system, packets arriving after the tagged packet may, in case they arrive at a full queue, push out the first packet in the queue and hence influence the delay experienced by the tagged packet. The waiting time in the push-out system consists of two parts: firstly, the time interval between the packet arrival instant (at FQ) and the next inspection instant (i.e., until the end of the vacation if it arrives during a vacation or until the end of a service if it arrives during a service); secondly, the time interval between this inspection instant and the start of the transmission of the packet.

- Between arrival instant and the next inspection instant: There are two scenarios, either the packet arrives during the vacation period or during the service period.
    * An arbitrary time instant falls in a vacation: Let $P_n^{\omega}(t)$ be the probability that an arbitrary time instant falls in a vacation period, that there are n packets (maximum N) present in the system at time t, and that the remaining vacation time is T_v that satisfies $t \leq T_v < t + dt$. The LST of this probability has already been determined by Blondia [20]. Hence, the inverse LST can provide us the value of $P_n^{\omega}(t)$ and $P_N^{\omega}(t)$, as defined in Equation (19).

$$P_n^{\omega}(t) = \frac{1}{\lambda \cdot T_{FP}} \cdot \left[\overline{p}_0 \cdot \overline{e} \cdot \mu_{\mathbf{E}[W]}(t) \cdot (-1)^{n+1} \cdot e^{\lambda \cdot (t - \mathbf{E}[W])} \cdot \frac{(t - \mathbf{E}[W])^n}{n!} - \sum_{l=0}^{n} (\overline{p}_0 \cdot B_l \cdot \overline{e} \cdot \lambda^{n-l+1} \cdot (-1)^{n-l+1} \cdot e^{\lambda \cdot t} \cdot \frac{t^{n-l}}{(n-l)!}) \right]. \quad (19)$$

and

$$P_N^{\omega}(t) = \frac{\overline{p}_0 \cdot \overline{e}}{T_{FP}} \cdot \left[1 - \mu_{\mathbf{E}[W]}(t)\right] - \sum_{n=0}^{N-1} P_n^{\omega}(t),$$

where, n = $[0, N-1]$ and $\mu_{\mathbf{E}[W]}(t)$ is a Heaviside function that outputs (0) if $t \leq \mathbf{E}[W]$ and (1) if $t > \mathbf{E}[W]$.

$P_{n,k}^{\omega}(t)$ represents the probability that an arbitrary time instant falls in a vacation period, that there are n packets (maximum N) present in the system at time t, and there are k packet arrivals (maximum $N-1$) during the remaining vacation time T_v. This probability is given by Equation (20).

$$P_{n,k}^{\omega}(t) = e^{-\lambda \cdot t} \frac{(\lambda \cdot t)^k}{k!} \cdot P_n^{\omega}(t). \quad (20)$$

The probability that at an arbitrary time instant there are n packets (maximum $N-1$) in the system and that, during the remaining time of the vacation k, packets (maximum $N-1$) have arrived is given in Equation (21).

$$P_{n,k}^{\Omega} = \int_0^{\infty} P_{n,k}^{\omega}(t)dt = \frac{1}{\lambda T_{FP}} \sum_{l=0}^{n} \left(\overline{p}_0 \cdot B_l \cdot \overline{e}\right) \cdot \left[\frac{(\lambda \cdot \mathbf{E}[W])^{n-l+k+1} \cdot (-1)^{n-l}}{(n-l)! \cdot k! \cdot (n-l+k+1)} \right]$$

And (21)

$$P_{N,k}^{\Omega} = \frac{\overline{p}_0 \cdot \overline{e}}{\lambda T_{FP}} \cdot \left[1 - e^{-\lambda \cdot \mathbf{E}[W]} \cdot \sum_{i=0}^{k} \frac{(\lambda \cdot \mathbf{E}[W])^i}{(i)!} \right] - \sum_{n=0}^{N-1} P_{n,k}^{\Omega}.$$

The average waiting time until the next inspection instant of a packet that arrives at an arbitrary time instant in a vacation period is then given by Equation (22).

$$T_{n,k}^{\Omega} = \int_0^{\infty} t \cdot P_{n,k}^{\omega}(t)dt = \frac{1}{\lambda T_{FP}} \sum_{l=0}^{n} \left(\overline{p}_0 \cdot B_l \cdot \overline{e}\right) \cdot \left[\frac{(\lambda \cdot \mathbf{E}[W])^{n-l+k+1} \cdot (-1)^{n-l} \cdot \mathbf{E}[W]}{(n-l)! \cdot k! \cdot (n-l+k+2)}\right] \quad (22)$$

And

$$T_{N,k}^{\Omega} = \frac{\overline{p}_0 \cdot \overline{e} \cdot (k+1)}{\lambda^2 \cdot T_{FP}} \cdot \left[1 - e^{-\lambda \cdot \mathbf{E}[W]} \cdot \sum_{i=0}^{k+1} \frac{(\lambda \cdot \mathbf{E}[W])^i}{(i)!}\right] - \sum_{n=0}^{N-1} T_{n,k}^{\Omega}.$$

* An arbitrary time instant falls in a service: Let $P_n^{\pi}(t)$ be the probability that an arbitrary time instant falls in a service period, that there are n packets (maximum N) present in the system at time t, and that the remaining service time T_s satisfies $t \leq T_s < t + dt$. Based on the Laplace transform, [20] $P_n^{\pi}(t)$ can be calculated as Equation (23).

$$P_n^{\pi}(t) = \frac{1}{\lambda \cdot T_{FP}} \cdot \left[\sum_{l=1}^{n} \left(\overline{p}_1 \cdot \overline{e} \cdot \lambda^{n-l} \cdot \mu_{\mathbf{E}[S]}(t) \cdot (-1)^{n-l+1} \cdot e^{\lambda \cdot (t - \mathbf{E}[S])} \cdot \frac{(t-\mathbf{E}[S])^{n-l}}{(n-l)!}\right) - \sum_{l=1}^{n} \left(\overline{p}_{l-1} - \overline{p}_0 \cdot B_{l-1} \cdot \overline{e} \cdot \lambda^{n-l} \cdot (-1)^{n-l+1} \cdot e^{-\lambda \cdot t} \cdot \frac{t^{n-l}}{(n-l)!}\right)\right],$$

and (23)

$$P_N^{\pi}(t) = \frac{\sum_{l=1}^{N-1} \overline{p}_1 \cdot \overline{e}}{T_{FP}} \cdot \left[1 - \mu_{\mathbf{E}[S]}(t)\right] - \sum_{n=1}^{N-1} P_n^{\pi}(t).$$

where, n = [0, ..., N−1].

The probability that a packet arrives in a service period and n packets are present in the FQ at time t and k packets arrive in the remaining service time is given by Equation (24).

$$P_{n,k}^{\pi}(t) = e^{-\lambda t} \frac{(\lambda t)^k}{k!} \cdot P_n^{\pi}(t) \quad (24)$$

where, n = [0,N] and k = [0,N-1].

Similar to Equation (21), the probability ($P_{n,k}^{\Pi}$) that, at an arbitrary time instant, there are n packets in the system and that during the remaining time of the service k packets have arrived can be calculated as Equation (25).

$$P_{n,k}^{\Pi} = \int_0^{\infty} P_{n,k}^{\pi}(t)dt = \frac{1}{\lambda T_{FP}} \sum_{l=1}^{n} \left(\overline{p}_{l-1} - \overline{p}_0 \cdot B_{l-1} \cdot \overline{e}\right) \cdot \left[\frac{(\lambda \cdot \mathbf{E}[S])^{n-l+k+1} \cdot (-1)^{n-l}}{(n-l)! \cdot k! \cdot (n-l+k+1)}\right]. \quad (25)$$

and

$$P_{N,k}^{\Pi} = \frac{\sum_{l=1}^{N} \overline{p}_l \cdot \overline{e}}{\lambda T_{FP}} \cdot \left[1 - e^{-\lambda \cdot \mathbf{E}[S]} \cdot \sum_{i=0}^{k} \frac{(\lambda \cdot \mathbf{E}[S])^i}{(i)!}\right] - \sum_{n=1}^{N-1} P_{n,k}^{\pi}.$$

The average waiting time of a packet that arrives at an arbitrary time instant in a service period is then given by Equation (26).

$$T_{n,k}^{\Pi} = \int_0^{\infty} t \cdot P_{n,k}^{\pi}(t)dt$$
$$= \frac{\sum_{l=1}^{N}(\overline{p}_l \cdot \overline{e})}{\lambda T_{FP}} \cdot \left[\frac{(\lambda \cdot \mathbf{E}[S])^{n-l+k+1} \cdot (-1)^{n-l} \cdot \mathbf{E}[S]}{(n-l)! \cdot k! \cdot (n-l+k+2)}\right]. \tag{26}$$

and

$$T_{N,k}^{\pi} = \frac{\sum_{l=1}^{N} \overline{p}_l \cdot \overline{e} \cdot (k+1)}{\lambda^2 \cdot T_{FP}} \cdot \left[1 - e^{-\lambda \cdot \mathbf{E}[S]} \cdot \sum_{i=0}^{k+1} \frac{(\lambda \cdot \mathbf{E}[S])^i}{(i)!}\right] - \sum_{n=0}^{N-1} T_{n,k}^{\pi}.$$

- Between this inspection instant and the start of the transmission of the packet: Let $T_{n,k}^{s}$ be the mean waiting time of a packet in the FQ at the *end* of a service that will be served and that has n packets (maximum $N-2$) ahead and k packets (maximum $N-n-2$) behind is given by Equation (27).

$$\begin{aligned}
T_{0,k}^{s} &= 0, \qquad \text{for } k = [0, N-2] \\
T_{1,k}^{s} &= \sum_{j=0}^{N-k-2} \left(T_{0,k+j}^{s} + \mathbf{E}[S] \cdot \frac{(\lambda \cdot \mathbf{E}[S])^j}{j!} \cdot e^{-\lambda \cdot \mathbf{E}[S]}\right), \qquad \text{for } k = [0, N-3] \\
T_{n,k}^{s} &= \sum_{j=0}^{N-n-k-1} \left(T_{n-1,k+j}^{s} + \mathbf{E}[S] \cdot \frac{(\lambda \cdot \mathbf{E}[S])^j}{j!} \cdot e^{-\lambda \cdot \mathbf{E}[S]}\right) + \\
&\sum_{j=N-n-k}^{N-k-2} \left(T_{N-k-2-j,k+j}^{s} + \mathbf{E}[S] \cdot \frac{(\lambda \cdot \mathbf{E}[S])^j}{j!} \cdot e^{-\lambda \cdot \mathbf{E}[S]}\right), \text{ for } k = [0, N-n-2] \\
&\qquad \text{and } n = [2, N-2].
\end{aligned} \tag{27}$$

Similar to Equation (27), the mean waiting time of a packet in the FQ at the end of a vacation period $T_{n,k}^{v}$ that will be served and has n packets ahead (maximum $N-1$) and k packets (maximum $N-n-1$) behind can be calculated.

The total waiting time of a packet arriving at an arbitrary instant at the FN or average DL latency, served eventually in a system using the push-out strategy, is given by Equation (28).

$$\begin{aligned}
\mathbf{T}_{PO} = &\sum_{k=0}^{N-1}\left(T_{0,k}^{\Omega} + P_{0,k}^{\Omega} \cdot T_{0,k}^{v}\right) + \sum_{n=1}^{N-1}\left[\sum_{k=0}^{N-n-1}\left(T_{n,k}^{\Omega} + P_{n,k}^{\Omega} \cdot T_{n,k}^{v}\right) + \right. \\
&\left.\sum_{k=N-n}^{N-1}\left(T_{N,k}^{\Omega} + P_{N,k}^{\Omega} \cdot T_{N-k-1,k}^{v}\right)\right] + \sum_{k=0}^{N-2}\left(T_{1,k}^{\Pi} + P_{1,k}^{\Pi} \cdot T_{0,k}^{s}\right) + \\
&\sum_{n=2}^{N-1}\left[\sum_{k=0}^{N-n-1}\left(T_{n,k}^{\Pi} + P_{n,k}^{\Pi} \cdot T_{n-1,k}^{s}\right) + \sum_{k=N-n}^{N-2}\left(T_{N,k}^{\Pi} + P_{N,k}^{\Pi} \cdot T_{N-k-2,k}^{s}\right)\right]
\end{aligned} \tag{28}$$

6. Evaluation

This section presents the evaluation of the DL latency and PDR results using the analytical model and the simulator. Firstly, the description of the simulation setup is presented. This is followed by comparing the results of our analytical model and the simulator. The model is implemented in Matlab R2020b, while the simulation program is written in Python 3.9.6. Thereafter, these results are analyzed in detail based on parameters such as harvesting power, capacitor size, data arrival rate, and poll interval timer of a batteryless BLE LPN.

Note that the execution time of the model in Matlab scales with the size of the transition matrix. Therefore, if it is known that, at a specified harvesting rate, the capacitor would be unable to recharge itself up to V_{max} and the matrix size can be reduced by assuming a lower value of V_{max}. This behaviour can be easily perceivable for low harvesting rates in combination with large capacitor sizes.

6.1. Simulation Setup

We implemented an event-based simulator to imitate the friendship communication mechanism of the batteryless LPNs as proposed in [7]. The simulator follows an event-based packet scheduling and is capable of reproducing the BLE radio activities, such as sending FP messages and FQ buffered data packets. It also uses the same capacitor model to imitate the capacitor behaviour, as mentioned in Section 4. Each experiment of the simulator is run until a total of 25,000 packets have been generated according to a Poisson arrival process. To request and receive a buffered data packet, the LPN follows the sequence of states listed in Table 1, in the order from top to bottom. The table also mentions the execution time and the current consumption of the corresponding states of the LPNs. These time and current consumption values can be used as Δt and $I_L(X)$ to calculate the capacitor's voltage at any time using Equation (1). The time and the current consumption were measured at 1.8V supply voltage on the Nordic nRF52840 BLE devkit [22] using the Nordic Power Profiler Kit II [23] while running the developed LPN application. All the states of the FP and receiving message mentioned in Table 1 can be seen in Figure 5.

The parameters listed in Table 2 are used to evaluate the model. As such, the values of V_{cutoff} and V_{max} are taken considering one of the configurations of a representative off-the-shelf power management unit [24]. The following performance metrics are considered in the comparison:

- Packet delivery ratio (PDR): The ratio of packets successfully received by the LPN compared to the total reaching the FQ.
- Downlink packet latency: The average latency to receive a packet by the LPN from the time it arrives at the FQ.

Table 2. Simulation parameters.

Parameters	Symbol	Values/Range
Poisson packet arrival rate	μ_p	[1, 10, 30, 60, 120] s
Harvesting power	P_h	[0.1,0.15,0.25,0.3,0.35, 0.2,0.4,0.6,0.8,1] mW
Capacitor size	C	[0.1,0.2,0.3,0.4,0.5,1] mF
Poll interval	T_{PI}	[1, 10, 20, 40, 60, 80, 100, 150, 200, 250, 300, 400, 500, 600, 800, 1000] s
Turn-off voltage	V_{cutoff}	2.8 V
Max operating voltage	V_{max}	4.5 V
Friend Queue size	N	[8, 16] packets
Receive Delay	T_{RD}	255 ms
Arrival time (max up to T_{RW})	T_{AT}	0 ms
Tx power level		0 dBm

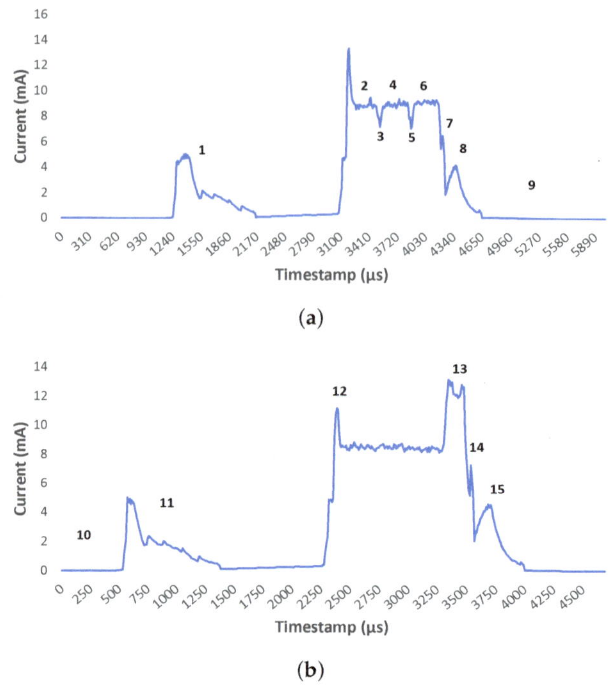

Figure 5. Current consumption of nRF52840 with the state number as defined in Table 1. (**a**) Current consumption during FP; (**b**) Current consumption during Scan.

6.2. Model Validation

The model is validated using the simulation and the emulator. We ran a total of 5640 test cases to compare the results of the analytical model and the simulation. The test cases encompass all possible parameter combinations listed in Table 2. Whereas, using an emulator, a hardware with a lack of automation, we manually ran 25 tests.

6.2.1. Using Simulation

The harvesting power is considered for harvesting techniques, which are low indoor light or human body temperature, that generate up to 1 mW. To avoid longer latency delay, the poll interval is considered only up to 10 times the data rate. These test cases only analyze DL data. The cumulative distribution shown in Figure 6 presents the relative deviation in PDR and DL latency between the simulation and the model results. We also simulate the device for the large power (more than 1 mW), generating harvesting techniques such as industrial temperature, direct sunlight, or mechanical movement. However, the result obtained with such harvesting techniques would result in 100% PDR with the lowest possible DL latency. Therefore, we exclude these test cases from the result analysis.

There is a mean difference of 2.23% and 0.099% in DL latency and PDR, respectively, found in a run of all the test cases. A high relative difference in PDR value ($\geq$2%) is observed when the PI is set to a relatively larger value than the data arrival rate. For these cases, the relative difference in latency is around 13%. Moreover, the large relative deviation in latency value ($\geq$15%) is also observed for the low P_h when the PDR is less than 100%. However, the deviation in latency is higher for the lowest FQ sizes (8 or less). For such cases, the deviation in PDR is only 0.58%.

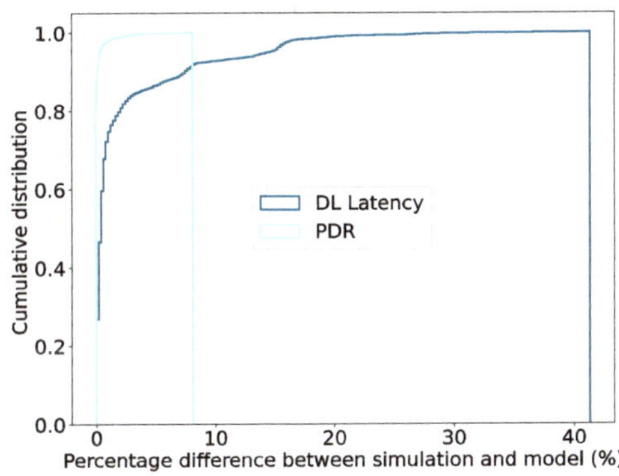

Figure 6. Cumulative distribution of difference in model and simulation results for both the DL latency and PDR metrics.

There are a total of 92.73% and 100% test cases for DL latency and PDR, respectively, with a percentage difference less than 10%. By analyzing the results for the PDR and DL latency of the model and simulation, it is evident that the model's accuracy is high. However, solving the model is computationally complex for large FQ sizes (32 or more), and the relative mean difference in the DL latency test cases for the lowest FQ size (2 packets) is higher compared to larger queue sizes.

6.2.2. Using Emulator

We use an energy-harvesting emulator called Environment Emulator (EnvEmu) [25], which is a stripped-down version of the TMote Sky. It is connected to the Nordic nRF52840 development board that runs the BLE LPN application, as shown in Figure 7. The EnvEmu is powered by USB and provides a variable input voltage between 1.5 and 3.3 V to Nordic BLE LPN via its VDD-GND pin. The output voltage is based on the emulation of a virtual capacitor and energy harvesting source, which can both be dynamically configured in the EnvEmu via simple commands. An instruction manual for the EnvEmu using Linux can be found in [26]. Two more Nordic nRF52832 development boards powered by USB are also part of the BLE mesh network, of which one acts as an FN and another generates DL packets. The FN temporarily accumulates the incoming DL data from the mesh node, which the batteryless LPN polls.

Figure 7. EnvEmu setup.

Figure 8 shows the comparison of the model and emulator for different FQs. It can be observed that the results are similar. The mean difference in the result of the emulator and model is 1.22% and 0.3% in PDR and DL latency, respectively.

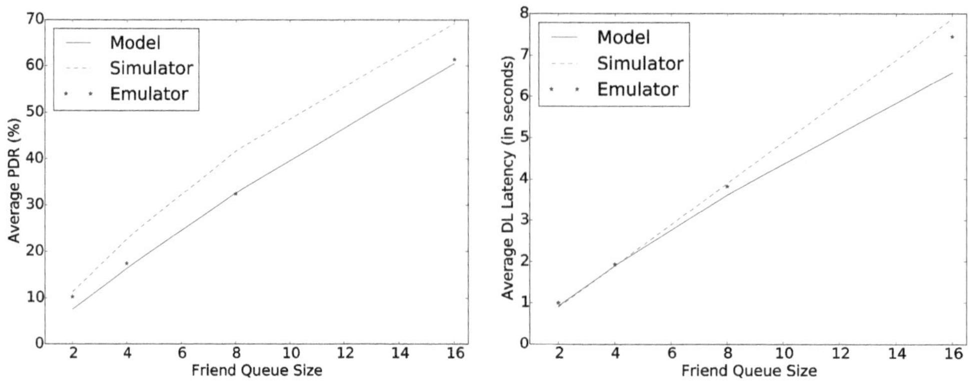

(a) (b)

Figure 8. Comparing emulator, simulator, and model for different FQ at μ_p = 1 s, T_{PI} = 30 s, C = 1 mF, P_h = 0.1 mW, T_{AT} = 5 ms. (**a**) Average packet delivery ratio; (**b**) DL latency.

6.3. Result Analysis Based on Analytical Model and Simulation

Figures 9–11 plot the results of the analytical model and the simulation, showing the effect of all the dependent parameters. It is clear that the analytical model and simulation follow the same behaviour, and the results match closely in all the plotted graphs. It can be observed from Figure 9 that, as the harvesting power increases, the average PDR and DL latency improve until they reach their corresponding maximum achievable values.

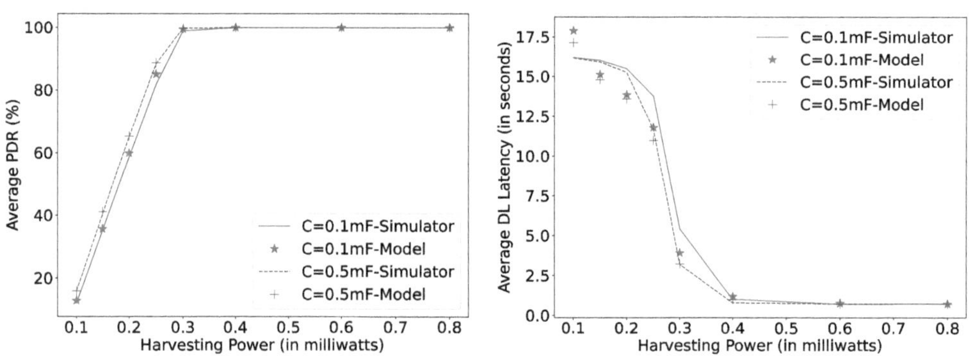

(a) (b)

Figure 9. Comparing simulator and model for different harvesting powers at μ_p = 1 s, T_{PI} = 1 s, N = 16, T_{AT} = 0 ms. (**a**) Average packet delivery ratio; (**b**) DL latency.

The improvement in the PDR and DL latency accelerate faster for low harvesting power until the PDR crosses 80% (0.25 mW in Figure 9a,b). Further, an increase in the harvesting power improves the DL latency at a higher rate than the PDR because, at such harvesting power, the capacitor can recharge faster again to V_{th}^{FP} to receive packets. As soon as the PDR and DL latency reach their corresponding best achievable values, an increase in the harvesting power does not affect the results because the capacitor voltage never drops below V_{th}^{FP}, and therefore, the node does not need to go into SLEEP mode to recharge its capacitor. For an arrival rate of 1 s, this harvesting power point is 0.6 mW, as shown in Figure 9. With the decrease in arrival rate, the capacitor gets more time to recharge itself. Therefore, for an arrival rate of 10 s, the mentioned harvesting power point value reduces to 0.3 mW (not shown in Figures). The size of the capacitor also plays a vital role in achieving the desired performance. As a larger capacitor can store more energy, the PDR for the 0.5 mF capacitor is higher than for 0.1 mF. However, the increase of the capacitor size also affects its charging time. Therefore, the performance does not improve much with the increase in the capacitor size after a certain optimal capacitor size for each specific harvesting power. There is an optimal capacitor size for each set of parameters. It can be observed from Figure 10b that, at a harvesting power of 0.1 mW, the optimal capacitor size, where the DL latency is lowest, is 0.35 mF, and for 0.2 mW of harvested power, it is 0.45 mF. An increase in the capacitor size higher than 0.5 mF shows reduced growth in the PDR (harvesting power 0.1 and 0.2 mW). Additionally, when the PDR is 100%, the increase in the capacitor size would improve the DL latency (yellow line at P_h 0.4 mW), but not substantially.

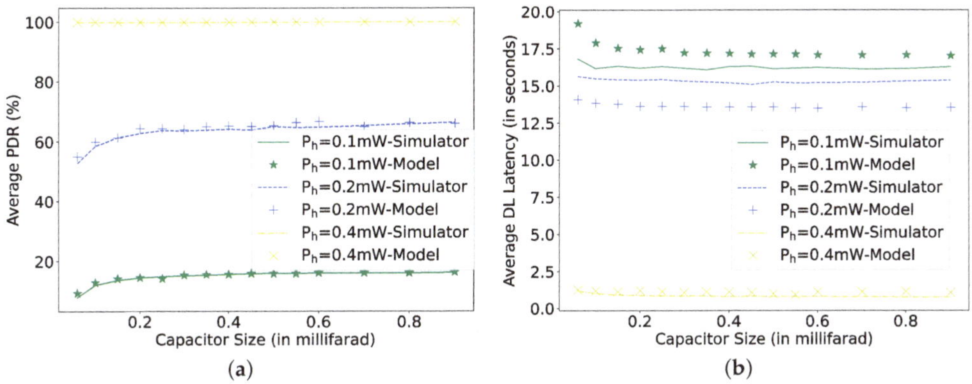

Figure 10. Comparing simulator and model for different capacitor sizes and harvesting power at μ_p = 1 s, T_{PI} = 1 s, N = 16, T_{AT} = 0 ms. (**a**) Average packet delivery ratio; (**b**) DL latency.

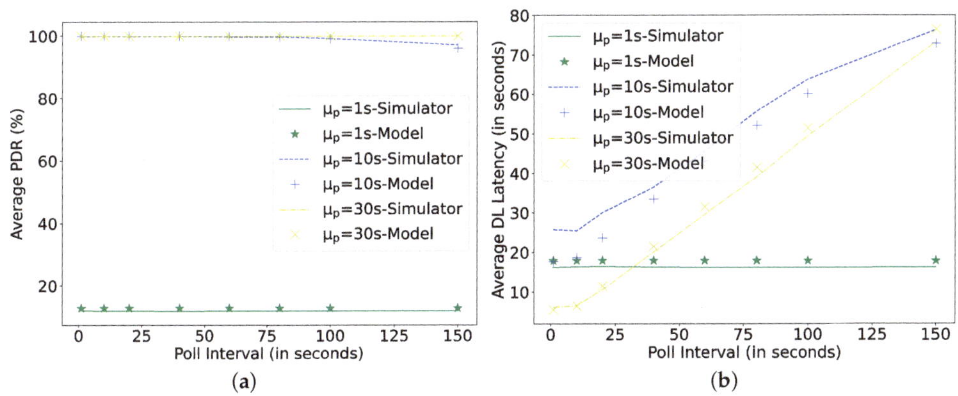

Figure 11. Comparing simulator and model for different Poll Intervals at C = 0.1 mF, P_h = 0.1 mW, N = 16, T_{AT} = 0 ms. (**a**) Average packet delivery ratio; (**b**) DL latency.

Figure 11 shows that, for a fixed harvesting power (0.1 mW), with the decrease in Poisson arrival rate for downlink packets, the PDR can be improved to 100%. This is because the capacitor gets ample time to charge itself before the subsequent transmission. However, the achieved latency can increase with the increase in the poll interval timer as the node has to wait for this PI once it receives the FU. Therefore, the higher the PI, the longer the arriving packets wait in the FQ to be polled. As such, when the data arrival rate is 10 s, some packets are also being dropped for large PI (150 s). Although frequent polling consumes additional energy, the considered harvesting power (0.1 mW) is sufficient to support a PI as low as 1 s without negatively affecting PDR. Moreover, the DL latency at the arrival rate of 1 s is not affected by the increase in PI, because there are always packets waiting in the FQ to be polled, and the LPN never receives an FU message.

If the size of the FQ decreases, then the number of dropped packets increases for higher PI. This can be observed in Figure 12. The smaller the FQ, the lower the PDR. Another interesting parameter of the LPN is the AT, which also affects the power consumption and DL latency, as the LPN needs to actively listen during the configured time. With the increase in AT, the LPN wastes energy in idle listening on the incoming channels.

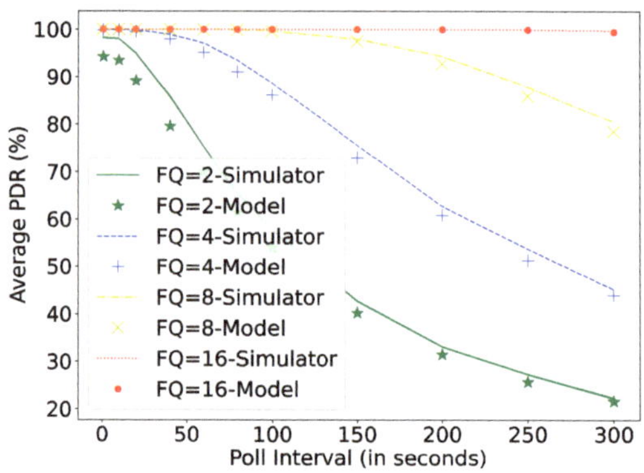

Figure 12. Comparing simulator and model for different poll intervals and FQ at C = 0.1 mF, P_h = 0.1 mW, μ_p = 30 s, T_{AT} = 0 ms.

Figure 13 shows its effect on DL latency and PDR. It can be observed that a slight increase in AT can drastically affect the PDR and, therefore, the DL latency. However, when the data arrival rate is low, the LPN has sufficient time to recharge its capacitor to maintain a PDR of 100%, but the increase in AT affects the DL latency exponentially. As seen for the Poisson arrival rate of 30 s, the latency increases 60-fold from 0.65 s at 0 s AT to 38.43 s at AT of 255 ms. At a high Poisson arrival rate (i.e., 1 s), the impact of the AT is higher because the LPN has less time to charge its capacitor and needs to receive more packets compared to lower Poisson data arrival rates. Increasing the AT further affects the performance marginally. Therefore, AT should be configured as small as possible to achieve high LPN performance. However, a small AT value provides less time to the FN to send a response to the LPN, which in turn affects the scalability (i.e., it causes missed response opportunities when multiple LPNs expect a response simultaneously).

Furthermore, from Figure 14, it can be observed that an LPN having a capacitor of 0.1 mF and a harvesting power of 0.1 mW can receive almost all the data packets when the arrival rate is 10 s for poll interval timers lower than 60 s. Further, decreasing the data arrival rate (from 7 s to 30 s) would improve the latency because fewer packets are entering into the FQ and the capacitor gets more time to recharge itself. Similarly, it improves the PDR when the data arrival rate decreases from 1 s to 7 s. However, the DL latency increases in this interval because with the decrease in data interval fewer packets are entered into the FQ and thus wait longer to be polled.

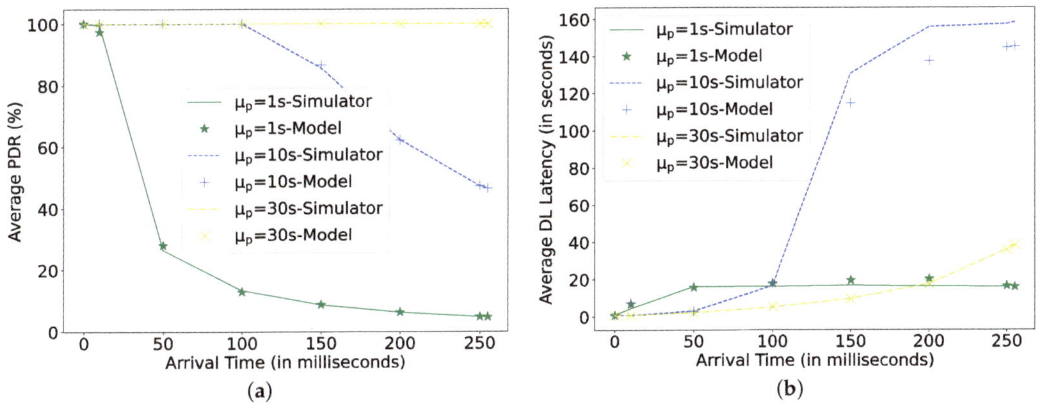

Figure 13. Comparing simulator and model for different arrival times (AT) at C = 4 mF, P_h = 1 mW, N = 16, T_{PI} = 1 s. (**a**) Average packet delivery ratio; (**b**) DL latency.

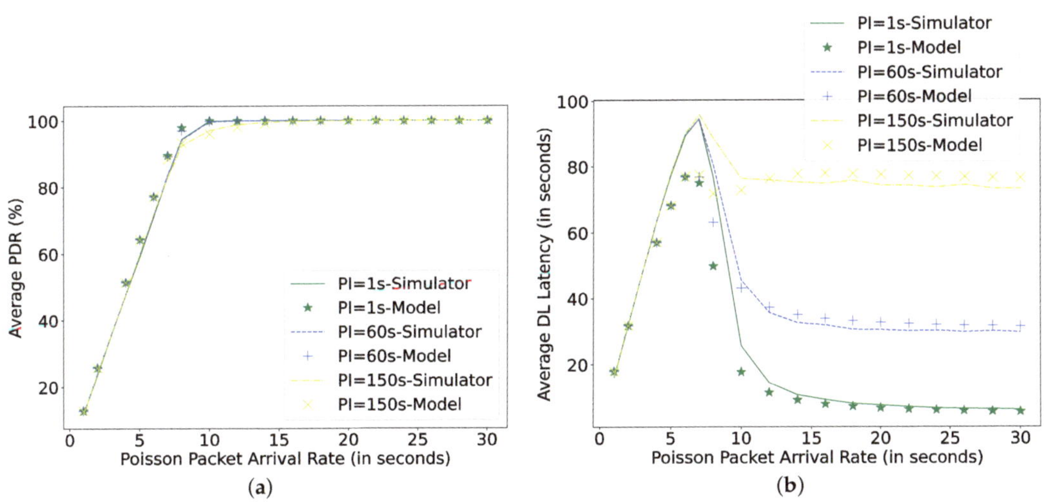

Figure 14. Comparing simulator and model for different packet arrival rate at C = 0.1 mF, P_h = 0.1 mW, N = 16. (**a**) Average packet delivery ratio; (**b**) DL latency.

7. Use-Case: Logistic Management

This section presents the use of the proposed model to derive optimal parameters for a batteryless BLE node, considering the IoT use-case of the logistics process. Logistics is one of the major industries adopted by IoT at scale. IoT applications can facilitate asset monitoring in warehouses or shops. Instead of checking the assets manually, they can be connected via batteryless BLE devices that allow effortless tracking of each item. This batteryless device can listen to incoming queries to respond about its availability or transmit sensor data. Consider a use-case where a server queries for the asset availability every 30 s and the batteryless node can harvest 100 μW. The model can be used to find the optimal BLE batteryless LPN configuration, including the capacitor size and poll interval. Assuming a friend queue size of eight packets and an arrival time of 5 min, the model provides the result as shown in Figure 15. It can be observed that the PDR is 100% for lower PI values and the capacitor which are greater than equal to 0.5 mF. However, DL latency is lower for a capacitor of 5 mF or more and does not vary much. Therefore, considering the charging and discharging time of a capacitor, we should select the lowest possible capacitor, which, for this use-case, can be considered as 5 mF. Additionally, the DL latency is lowest (18.27 s) at a PI equal to 20 s.

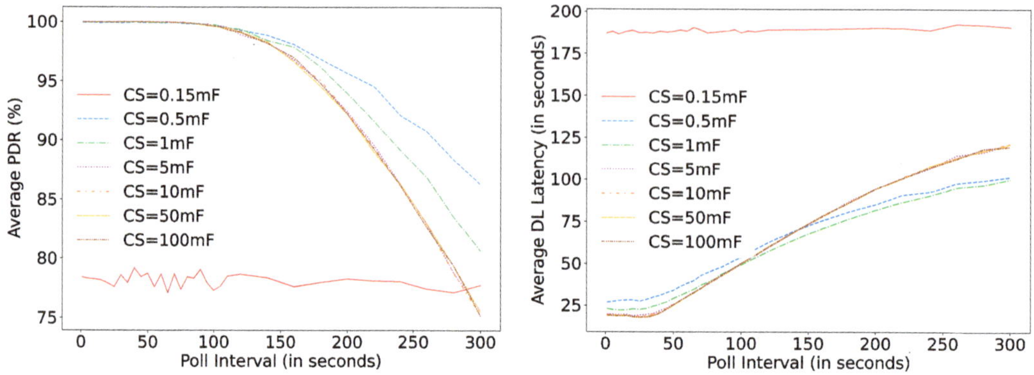

(a) (b)

Figure 15. Asset monitoring at DR = 30 s, P_h = 0.1 mW, N = 8, T_{AT} = 5 ms. (**a**) Average packet delivery ratio; (**b**) DL latency.

Moreover, if the packet arrival rate from the server changes, the model can be used to determine the new expected performance of this batteryless LPN that harvests 100 µW and can be configured with a capacitor of 5 mF and polling at 20 s. The result is shown in Figure 16. It can be observed that, at a high arrival rate (1 and 10 s), many packets drop, whereas, lowering the packet arrival rate (60 and 600 s) improves the DL latency up to 55% without affecting the PDR.

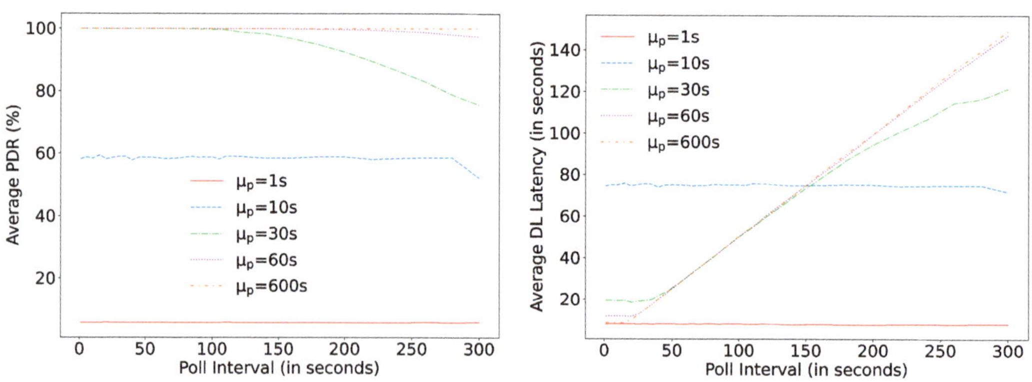

(a) (b)

Figure 16. Asset monitoring at C = 5 mF, P_h = 0.1 mW, N = 8, T_{AT} = 5 ms for different packet arrival rate. (**a**) Average packet delivery ratio; (**b**) DL latency.

8. Conclusions

In this paper, we presented an analytical model to calculate the DL latency and PDR for a batteryless BLE LPN operating on power harvested from natural or artificial indoor light. The model's accuracy was compared with Python-based simulation results for various parameters (i.e., capacitor size, harvesting power, data arrival rate, poll interval timer, arrival time, and friend queue size). The results derived from these two methods showed an average deviation of 2.23% in the results of DL latency and 0.09% in PDR. The graphs of different parameters showed diverse behaviour for DL latency and PDR, whereas the model and simulation show the same pattern.

The results showed that a harvesting power of 0.6 mW can maintains a DL packet inter-arrival rate as low as 1 s, and at 0.3 mW, the LPN can receive all the packets arriving at a rate of 10 s to an FQ of size 16. The smallest optimal capacitor size should be chosen depending on the available minimum harvesting power so that it can be recharged rapidly. The poll interval timer affects the DL latency linearly if the PDR is 100%. Therefore, it is important to choose the lowest possible poll interval timer that can be supported by the minimum harvesting power. If the harvesting power varies over time, the poll interval timer should be dynamically adapted. The external factors, such as Friend Queue size, also affect the performance of the batteryless LPN. A small queue size increases the packet loss. In our experiments, a queue size of 16 packets was enough to maintain 100% PDR, for polling intervals as high as 300 s at a downlink packet arrival rate of 30 s. Moreover, the AT parameter should be configured to be as small as possible so that the LPN does not lose energy in idle listening on the incoming channels. However, a very low AT may negatively affect scalability. For a harvesting power of 1mW, an AT up to 100 ms could be supported without a decrease in PDR.

9. Discussion and Future Work

There are several future research directions. In our model, we have considered a power management system that supplies variable voltage directly from the capacitor. However, more advanced power management solutions exist that equalize the supply voltage and only provide the device with power if the capacitor voltage is within certain minimum and maximum bounds [27]. Such solutions offer more consistent and stable performance, as then the power consumption for each radio state would be stable with a stable voltage supply. Most of the existing works in the batteryless domain do not consider this [28–31]. Therefore, in the future, we will explore adapting the batteryless circuit model for such power management solutions. However, in this case, an additional circuit would be required to read the capacitor voltage. CusTARD uses an Analog-to-digital (ADC) pin to read the capacitor voltage [32]. The ADC pin can read values in a range that is usually lower than the capacitor voltage. Therefore, a voltage divider would be required, which increases the power consumption of the device (several tens of μW) and therefore also affects the performance in terms of PDR and DL data latency for different harvesting powers. This could be incorporated into our model by changing Equation (1) and adding the current consumption of the additional circuit to the load current. Other modules used for voltage reading are Torpor Logic and Power switch [33], which can consume up to 2 μW rather than the assumed ultra-low-power comparator that consumes power in the order of pico-watts. As such, we need to perform measurements using real hardware to incorporate this power consumption into the model, as the hardware circuitry to enable measuring the capacitor voltage can impact the results.

The harvesting architecture considered is harvest-store-use. However, there is another architecture 'harvest-use' that eliminates the use of capacitors for energy storage [34]. This architecture can reduce the device cost and efficiency. A prototype of such an architecture is presented by Lee et al. [35]. Our model can be extended to a storage-less architecture. In this case, the performance is mainly governed by the harvesting power without a capacitor. Therefore, the stochastic process would consider the number of packets in the FQ and the harvesting source power. A certain minimum harvesting power is then required before and during execution of each task. Therefore, we need another additional circuitry, such as a sense resistor to measure the power of the harvesting source, which further increases the power demand of the batteryless device. Therefore, the requirements and performance of this architecture would be different.

Additionally, we aim to extend the analytical model to consider a variable harvesting power, in contrast to the currently considered constant harvesting power. This would provide a realistic harvesting model to evaluate harvesting sources. such as solar or wind. The model would consider only the current amount of harvested energy. Such solutions are realized by Zhou et al. [36], Patil et al. [37], and Kumar et al. [38] for investigating wireless sensor networks with energy harvesting. An approach to achieve this is by considering multiple energy chunks that are harvested during different Poisson distributed time slots, where the harvesting power during a slot remains constant [37]. As such, our existing model can be relatively easily extended to such a scenario by applying it to different harvesting power inputs for each variable-length time slot.

Finally, the increased CPU utilization due to servicing multiple LPNs may slow down the LPN's response and thus affect the DL latency. Therefore, another aspect is to study the interactions and effects of multiple LPNs associated with one FN. To date, there are no works that study this aspect. However, research has been done to study the effect on discovery latency due to the number of scanners [39] or due to interference in the presence of many BLE nodes [40]. With an increasing number of BLE devices, delays of device discovery show exponential growth. Due to interference, the latency can increase up to 400% in a network of 30 devices. This implies that severe contention can occur among multiple BLE devices while considering bi-directional communication using a batteryless device.

Author Contributions: Conceptualization, A.K.S., C.D, C.B. and J.F.; methodology, A.K.S. and C.B.; software, A.K.S. and C.B.; validation, A.K.S. and C.D.; formal analysis, A.K.S., C.B. and J.F.; investigation, A.K.S. and C.B.; resources, A.K.S., C.D. and C.B.; data curation, A.K.S. and C.B.; writing—original draft preparation, A.K.S. and C.B.; writing—review and editing, C.D. and J.F.; visualization, C.B. and J.F.; supervision, C.B. and J.F.; project administration, J.F.; funding acquisition, J.F. All authors have read and agreed to the published version of the manuscript.

Funding: This research was funded in part by the ICON project BLUESS, the University of Antwerp IOF funded project COMBAT (Time-Sensitive Computing on batteryless IoT Devices), the Flemish FWO SBO S001521N IoBaLeT(Sustainable Internet of batteryless Things) project, and the CERCA Programme from the Generalitat de Catalunya.

Institutional Review Board Statement: Not applicable.

Informed Consent Statement: Not applicable.

Data Availability Statement: Not applicable.

Conflicts of Interest: The authors declare no conflict of interest.

Appendix A. Parameters Used in the Model

Table A1 summarizes all the important parameters' symbol with their definition.

Table A1. Parameters Used in the Markov Chain Model.

Symbol	Meaning
λ	Poisson data arrival rate
N	FQ size
$\mathbf{U}$	Voltage Vector
S_i	Duration of service when $\mathbf{U}$ = i
W_i	Duration of vacation when $\mathbf{U}$ = i
$\mathbb{V}_i^S$	Voltage at next poll instance when $\mathbf{U}$ = i at the start of a service
$\mathbb{V}_i^W$	Voltage at next poll instance when $\mathbf{U}$ = i at the start of a vacation
$\mathbf{E}[S_i]$	Average duration of S_i
$\mathbf{E}[W_i]$	Average duration of W_i
$\overline{p}$	Probability distribution vector of Markov Chain

Table A1. *Cont.*

Symbol	Meaning
$\overline{p}_i$	Probability distribution of the voltage equals i
$(\overline{p}_k)_i$	Probability distribution of the voltage equals i if the queue length is k
B_i	Transitions starting from an empty system and ending up with i packets in FQ at the next FP
A_i	Transitions starting from a non-empty system and ending up with i packets in FQ at the next FP
O_n	System occupancy when n number of packets are present in FQ
$\mathbf{O}_{DT}$	Average system occupancy in drop-tail system
$\mathbf{O}_{PO}$	Average system occupancy in push-out system
$\mathbb{P}^S_{k,i}$	Probability that k packets arrive at service time starting from $\mathbf{U}$ = i
$\mathbb{P}^W_{k,i}$	Probability that k packets arrive at vacation time starting from $\mathbf{U}$ = i
$P^{\omega}_n(t)$	Probability that the packet arrives in a vacation period and n packets are present in the FQ at time t
$P^{\omega}_{n,k}(t)$	Probability that there are packet arrives in a vacation period and n packets are present in the FQ at time t and k packets arrive in the remaining vacation time
$P^{\Omega}_{n,k}$	Probability that the packet arrives in a vacation period and n packets are present in the FQ and k packets arrive in the remaining vacation time
$P^{\pi}_n(t)$	Probability that the packet arrives in a service period and n packets are present in the FQ at time t
$P^{\pi}_{n,k}(t)$	Probability that the packet arrives in a service period and n packets are present in the FQ at time t and k packets arrive in the remaining service time
$P^{\Pi}_{n,k}$	Probability that the packet arrives in a service period and n packets are present in the FQ and k packets arrive in the remaining service time
$\mu_{\mathbf{E}[W]}(t)$	Heaviside function for $\mathbf{E}[W]$
$\mu_{\mathbf{E}[S]}(t)$	Heaviside function for $\mathbf{E}[S]$
T_{FP}	Average Time between two consecutive FPs
$\mathbf{T}_{DT}$	DL latency in drop-tail system
$\mathbf{T}_{PO}$	DL latency in push-out system
$T^{\Omega}_{n,k}$	Average waiting time of a packet that arrives in a vacation period
$T^{\Pi}_{n,k}$	Average waiting time of a packet that arrives in a service period
$T^{s}_{n,k}$	Mean waiting time of a future served packet in the FQ at the end of a service period that has n packets ahead and k packets behind it
$T^{v}_{n,k}$	Mean waiting time of a future served packet in the FQ at the end of a vacation period that has n packets ahead and k packets behind it

References

1. Buchmann, I. *Batteries in a Portable World: A Handbook on Rechargeable Batteries for Non-Engineers*; Cadex Electronics Inc.: Richmond, BC, Canada, 2001.
2. Batteries vs. Ultracapacitors? The Answer Is Both. Available online: https://batteryindustry.tech/batteries-vs-ultracapacitors-the-answer-is-both/ (accessed on 22 January 2022).
3. Wang, Z.L.; Lin, L.; Chen, J.; Niu, S.; Zi, Y. *Triboelectric Nanogenerators Green Energy and Technology*; Springer: Berlin/Heidelberg, Germany, 2016.
4. Enterprise Bluetooth Applications, Rigado. Available online: https://www.rigado.com/enterprise-bluetooth-applications/ (accessed on 22 January 2022).
5. RC Charging Circuit, Electronics Tutorials. Available online: https://www.electronics-tutorials.ws/rc/rc_1.html (accessed on 22 January 2022).
6. Mesh Profile Bluetooth Specifications. Available online: https://www.bluetooth.com/specifications/mesh-specifications/ (accessed on 22 January 2022).
7. Sultania, A.K.; Delgado, C.; Famaey, J. Enabling Low-Latency Bluetooth Low Energy on Energy Harvesting Batteryless Devices Using Wake-Up Radios. *Sensors* **2020**, *20*, 5196. [CrossRef] [PubMed]
8. Silva, S.; Martins, H.; Valente, A.; Soares, S. A Bluetooth Approach to Diabetes Sensing on Ambient Assisted Living systems. *Procedia Comput. Sci.* **2012**, *14*, 181–188. [CrossRef]

9. Liu, Q.; IJntema, W.; Drif, A.; Pawełczak, P.; Zuniga M.; Yıldırım, K. S. Perpetual Bluetooth Communications for the IoT. *IEEE Sens. J.* **2021**, *21*, 829–837. [CrossRef]
10. Fraternali, F.; Balaji, B.; Agarwal, Y.; Benini, L.; Gupta, R. Pible: Battery-free mote for perpetual indoor ble applications: Demo abstract. In Proceedings of the 5th Conference on Systems for Built Environments (BuildSys '18), Shenzen, China, 7–8 November 2018; pp. 168–171.
11. Nilsson, M.; Deknache, H. Investigation of Bluetooth Mesh and Long Range for IoT Wearables. Ph.D. Thesis, Malmo Universitet/Teknik och Samhälle, Malmö, Sweden, 2018.
12. Álvarez, F.; Almon, L.; Hahn, A.; Hollick, M. Toxic Friends in Your Network: Breaking the Bluetooth Mesh Friendship Concept. In Proceedings of the 5th ACM Workshop on Security Standardisation Research Workshop (SSR'19), London, UK, 11 November 2019; pp. 1–12.
13. Darroudi, S.; Caldera-Sànchez, R.; Gomez, C. Bluetooth Mesh Energy Consumption: A Model. *Sensors* **2019**, *19*, 1238. [CrossRef] [PubMed]
14. Álvarez, F.; Almon, L.; Radtki, H.; Hollick, M. Bluemergency: Mediating Post-disaster Communication Systems using the Internet of Things and Bluetooth Mesh. In Proceedings of the 9th IEEE Global Humanitarian Technology Conference, Seattle, WA, USA, 17–20 October 2019.
15. Designing with Bluetooth Mesh: Nodes and Feature Types. Available online: https://www.embedded.com/designing-with-bluetooth-mesh-nodes-and-feature-types/ (accessed on 22 January 2022).
16. Hortelano, D.; Olivares, T.; Carmen Ruiz, M. Reducing the energy consumption of the friendship mechanism in Bluetooth mesh. *Comput. Netw.* **2021**, 195, 108172. [CrossRef]
17. Aijaz, A. Infrastructure-less Wireless Connectivity for Mobile Robotic Systems in Logistics: Why Bluetooth Mesh Networking is Important? *arXiv* **2021**, arXiv:2107.05563.
18. Delgado, C.; Sanz, J.M.; Famaey, J. On the Feasibility of Batteryless LoRaWAN Communications using Energy Harvesting. In Proceedings of the IEEE Global Communications Conference (GLOBECOM), Waikoloa, HI, USA, 9–13 December 2019; pp. 1–6.
19. Gawhare, S.B.; Gaikwad, A. Area and Power efficient High Speed Voltage Comparator. In Proceedings of the International Conference on Automatic Control and Dynamic Optimization Techniques, Pune, India, 9–10 September 2016; pp. 198–201.
20. Blondia, C. A queueing model for a wireless sensor node using energy harvesting. *Telecommun. Syst.* **2021**, 77, 335–349. [CrossRef]
21. Stochastic, Karl Sigman. Available online: http://www.columbia.edu/~ks20/stochastic-I/stochastic-I-LL.pdf (accessed on 4 January 2022).
22. nRF52840 DK, Nordic Semiconductor. Available online: https://www.nordicsemi.com/Software-and-tools/Development-Kits/nRF52840-DK (accessed on 22 January 2022).
23. Power Profiler Kit II, Nordic Semiconductor. Available online: https://www.nordicsemi.com/Software-and-tools/Development-Tools/Power-Profiler-Kit-2 (accessed on 22 January 2022).
24. User Guide Evaluation Board for AEM10941, Epeas. Available online: https://e-peas.com/wp-content/uploads/2020/06/epeas_Solar_Energy_Harvesting_User_Guide_AEM10941.pdf (accessed on 22 January 2022).
25. w-iLab.1 Hardware. Available online: https://doc.ilabt.imec.be/ilabt/wilab/hardware.html (accessed on 4 March 2022).
26. Using the Sensor Nodes. Available online: https://doc.ilabt.imec.be/ilabt/wilab/tutorials/sensor.html (accessed on 21 March 2022).
27. Dekimpe, R.; Xu, P.; Schramme, M.; Gerard, P.; Flandre, D.; Bol, D. A batteryless ble smart sensor for room occupancy tracking supplied by 2.45 ghz wireless power transfer. *Integration* **2019**, *67*, 8–18. [CrossRef]
28. Gleonec, P.; Ardouin, J.; Gautier, M.; Berder, O. Poster: Multi-source Energy Harvesting for IoT nodes. In Proceedings of the Conference on Design and Architectures for Signal and Image Processing (DASIP), Rennes, France, 12–14 October 2016.
29. Ahn, J. I.; Kim, D. ; Ha, R.; Cha, H. State-of-Charge Estimation of Supercapacitors in Transiently-powered Sensor Nodes. *IEEE Trans. -Comput.-Aided Des. Integr. Circuits Syst.* **2021**, *41*, 225–237. [CrossRef]
30. Gleonec, P.; Ardouin, J.; Gautier, M.; Berder, O. Architecture exploration of multi-source energy harvester for IoT nodes. In Proceedings of the IEEE Online Conference on Green Communications (OnlineGreenComm), Piscataway, NJ, USA, 14 November–17 December 2016; pp. 27–32.
31. Naresh, B.; Singh, V.K.; Sharma, V.K. Flexible Hybrid Energy Harvesting System to Power Wearable Electronics. In Proceedings of the Fourth International Conference on Advances in Electrical, Electronics, Information, Communication and Bio-Informatics (AEEICB), Chennai, India, 27–28 February 2018; pp. 1–5.
32. Hester, J.; Tobias, N.; Rahmati, A.; Sitanayah, L.; Holcomb, D.; Fu, K.; Burleson, W. P.; Sorber, J. Persistent Clocks for Batteryless Sensing Devices. *ACM Trans. Embed. Comput. Syst.* **2016**, *15*, 4. [CrossRef]
33. Anagnostou, P.; Gomez, A.; Hager, P.A.; Fatemi, H.; Gyvez, J.P.; Thiele, L.; Benini, L. Torpor: A Power-Aware HW Scheduler for Energy Harvesting IoT SoCs. In Proceedings of the 28th International Symposium on Power and Timing Modeling, Optimization and Simulation (PATMOS), Platja d'Aro, Spain, 2–4 July 2018; pp. 54–61.
34. Elahi, H.; Munir, K.; Eugeni, M.; Atek, S.; Gaudenzi, P. Energy Harvesting towards Self-Powered IoT Devices. *Energies* **2020**, *13*, 5528. [CrossRef]
35. Lee, H.G.; Chang, N. Powering the IoT: Storage-less and converter-less energy harvesting. In Proceedings of the 20th Asia and South Pacific Design Automation Conference, Chiba, Japan, 19–22 January 2015; pp. 124–129.

36. Zhou, H.; Jiang, T.; Gong, C.; Zhou, Y. Optimal Estimation in Wireless Sensor Networks With Energy Harvesting. *IEEE Trans. Veh. Technol.* **2016**, *65*, 9386–9396. [CrossRef]
37. Patil, K.; Fiems, D. The value of information in energy harvesting sensor networks. *Oper. Res. Lett.* **2018**, *46*, 362–366. [CrossRef]
38. Kumar, A.; Saad, W. On the tradeoff between energy harvesting and caching in wireless networks. In Proceedings of the IEEE International Conference on Communication Workshop (ICCW), London, UK, 8–12 June 2015; pp. 1976–1981.
39. Cho, K.; Park, W.; Hong, M.; Park, G.; Cho, W.; Seo, J.; Han, K. Analysis of Latency Performance of Bluetooth Low Energy (BLE) Networks. *Sensors* **2014**, *15*, 59–78. [CrossRef] [PubMed]
40. Treurniet, J.J.; Sarkar, C.; Prasad, R.V.; De Boer, W. Energy Consumption and Latency in BLE Devices under Mutual Interference: An Experimental Study. In Proceedings of the 3rd International Conference on Future Internet of Things and Cloud, Rome, Italy, 24–26 August 2015; pp. 333–340.

sensors

MDPI

Article

Light-Based IoT: Developing a Full-Duplex Energy Autonomous IoT Node Using Printed Electronics Technology

Malalgodage Amila Nilantha Perera [1,*], Marcos Katz [1], Juha Häkkinen [2] and Roshan Godaliyadda [3]

[1] Centre for Wireless Communications, University of Oulu, 90570 Oulu, Finland; marcos.katz@oulu.fi
[2] Circuits and Systems Research Unit, University of Oulu, 90570 Oulu, Finland; juha.hakkinen@oulu.fi
[3] Department of Electrical and Electronic Engineering, University of Peradeniya, Kandy 20400, Sri Lanka; roshangodd@ee.pdn.ac.lk
* Correspondence: malalgodage.perera@oulu.fi

Abstract: The light-based Internet of things (LIoT) concept defines nodes that exploit light to (a) power up their operation by harvesting light energy and (b) provide full-duplex wireless connectivity. In this paper, we explore the LIoT concept by designing, implementing, and evaluating the communication and energy harvesting performance of a LIoT node. The use of components based on printed electronics (PE) technology is adopted in the implementation, supporting the vision of future fully printed LIoT nodes. In fact, we envision that as PE technology develops, energy-autonomous LIoT nodes will be entirely printed, resulting in cost-efficient, flexible and highly sustainable connectivity solutions that can be attached to the surface of virtually any object. However, the use of PE technology poses additional challenges to the task, as the performance of these components is typically considerably poorer than that of conventional components. In the study, printed photovoltaic cells, printed OLEDs (organic light-emitting diodes) as well as printed displays are used in the node implementation. The dual-mode operation of the proposed LIoT node is demonstrated, and its communication performance in downlink and uplink directions is evaluated. In addition, the energy harvesting system's behaviour is studied and evaluated under different illumination scenarios and based on the results, a novel self-operating limitation aware algorithm for LIoT nodes is proposed.

Keywords: LIoT; VLC; energy-autonomous; printed electronics; zero-energy; IoT

Citation: Perera, M.A.N.; Katz, M.; Häkkinen, J.; Godaliyadda, R. Light-Based IoT: Developing a Full-Duplex Energy Autonomous IoT Node Using Printed Electronics Technology. *Sensors* **2021**, *21*, 8024. https://doi.org/10.3390/s21238024

Academic Editors: Slim Naifar, Olfa Kanoun, Carlo Trigona and Fabio Viola

Received: 19 October 2021
Accepted: 28 November 2021
Published: 1 December 2021

1. Introduction

The Internet of things (IoT) paradigm aims at providing wireless connectivity to virtually any object, such as machines, specialised sensors and actuators, home and office appliances, and everyday items. The number of connected IoT devices is rapidly growing, and it is estimated a several-fold increase in the number of these devices by the end of this decade from the current 10 billion (2021). Providing wireless connectivity to a massive number of IoT nodes needs to be performed in an efficient manner, particularly considering the use of key resources such as energy, spectrum and network infrastructure. Moreover, the environmental impact of such a massive connectivity should be minimised. Thus, IoT technology should be designed not only to be an effective communication system but also a sustainable one. Clearly, the design of both network infrastructure and IoT nodes needs to be carefully considered to fulfil these goals.

To this end, we consider in this paper the concept of light-based IoT (LIoT) as a sustainable IoT approach to provide wireless connectivity. LIoT exploits the zero-energy node concept, where an energy-autonomous node harvests from the environment the energy needed to operate. Light is exploited in the LIoT concept not only to power up the node but also to transfer information to and from the IoT node. Similar approaches have been studied in [1]. In general, all these approaches exploit the lighting infrastructure to also provide wireless connectivity, and they have the well-known advantages of optical wireless communications, such as the available extensive unlicensed spectrum, inherent security, and interference-free operation. A crucial aspect of the LIoT concept studied here is that its implementation in a sustainable manner is investigated. For that purpose, printed electronics (PE) technology is used in the implementation of the concept. In the future, we envision that fully printed IoT nodes will be implemented on a thin substrate (e.g., paper, plastic. etc.), resulting in a very low cost, highly flexible and sustainable connectivity solution that could be attached to any object. As such, LIoT is promising not only for conventional IoT applications but also for novel use cases, including logistics, industrial, healthcare and others. The prospect of producing a small form factor layer containing a complete self-powered IoT node with some functionalities is truly appealing, and in this paper, we discuss the design, implementation and performance evaluation of such a LIoT node using PE technology. Given the current state of PE technology, it is not possible today to implement a fully printed LIoT node, a task which remains our long-term ultimate goal. However, in this research, we make partial use of PE, employing several critical components using printed technology, including photovoltaic (PV) cells, light-emitting diodes (LEDs) and displays. The fact that the performance of PE components is typically poorer than that of conventional electronic components is one of the key challenges of LIoT. Particularly, since energy harvesting is vital for LIoT, the fact that the efficiency of printed PV cells is considerably smaller than that of typical cells requires careful system design, as it will be discussed in this paper. Moreover, it is shown that efficient energy harvesting, storing and managing are essential tasks that need to be mastered in order to successfully make LIoT a reality. The paper discusses the detailed design and implementation of a full-duplex LIoT node using PE technology, taking into account the design constraints posed by that technology.

The main contributions of this work can be summarised as follows. Initially, this paper introduces and discusses the unique advantages, the applicability of the Light-based IoT concept, and the potential of implementing LIoT using printed electronics. Moreover, the paper proposes using methods and essential technologies such as DC-DC converters, energy harvesting protocols, MPPT and modulations to improve the dual-mode operation (energy harvesting and communication) of LIoT utilising indoor illumination. By selecting suitable technologies and methods from the above mentioned, a LIoT based smart label prototype was implemented as a case study. Based on the prototype's performance, the practicality of the proposed framework, effect of using directed and non-directed radiation patterned optical links were analysed. In addition, a novel self-operating limitation aware algorithm for LIoT was proposed allowing the node to decide its operating mode (e.g., active/sleeping) based on the balance between its current energy capabilities and the task energy requirements.

There are numerous research studies that have been carried out to create energy-autonomous IoTs using renewable energy sources. According to the survey conducted in [2], energy sources such as ambient light, wind, vibration, thermal, motion energy and radiofrequency can be used for IoT energy harvesting purposes. In addition, according to the study, ambient light can be considered as the most efficient renewable source with the highest efficiency of consistent energy conversion capabilities compared to other sources.

Therefore, there is an extensive range of possible applications for ambient light energy harvesting based IoT designs. In [3], authors propose energy-autonomous radio frequency (RF) based IoT proof of concept with a solar energy harvesting approach. The energy harvesting system proposed in the design consists of traditional Si rigid photovoltaics and lithium-ion batteries capable of catering to the power requirement of the Wi-Fi, sensors, and microcontroller units. The node is designed to harvest energy during the daytime and save the required energy in the battery. In order to keep the node in a low power consuming state during idle periods, the use of duty cycle schedule-based sleep and wakeup mechanism was exploited in the work. Moreover, the authors propose a microcontroller-based adaptive duty cycling schedule that changes according to the system's solar irradiance and battery levels. In [4], the authors propose an energy-autonomous IoT design using the visible light backscatter approach. The node is initially powered up by photovoltaic energy harvesting and expected to communicate using visible light during the operation. The node's uplink is activated by an electronically activated liquid crystal display (LCD) shutter which is designed to expose the retroreflector according to the uplink data modulation. The IoT node is operating passively; hence, no energy is stored during any part of the operation.

The designs from the above-discussed works limit the practicality of integrating these with many modern applications. Using traditional electronics and high energy density, chemical reaction-based batteries for implementation could introduce limitations on IoT applications requiring compact form factor and chemical hazardous-free designs. Even though the nodes are expected to operate maintenance-free for a long lifetime, the limited number of battery cycles can constrain the product's life span at some point. Approaches such as retroreflectors and RF antenna-based communication can be less advantageous when the designs are required to fulfil modern internode communication and power-sharing without adding additional hardware/circuitries. In addition, a study in [5] shows that long term Wi-Fi radiation exposure can cause minor health risks even if they are within the standard guideline exposure range. Therefore, it is essential to address the modern IoT requirements and features such as internode data /power-sharing, small compact designs with flexibility characteristics, low manufacturing costs, minimum electromagnetic radiation exposure to users and environmental impact at the design stage of the IoT technology.

In the presence of these considerations, this paper is organised as follows. Section 2 introduces in detail the LIoT concept and discusses some interesting applications. In Section 3, a brief overview of key technologies for sustainable implementation are introduced. Section 4 discusses requirements and identifies research challenges. The design of a LIoT node supporting uplink and downlink communications is described in Section 5, while in Section 6 performance of the implemented LIoT node is assessed. Finally, Section 7 concludes the paper.

2. LIoT Concept

The Light-based IoT concept proposes a novel design for IoT which enables light to be used to energise and to provide wireless connectivity to an IoT node. This concept follows the nature of "expose and connect", simply describing the operating principle: when a node is exposed to light, it will be connected to the Internet [1]. According to the LIoT concept, indoor photovoltaic energy harvesting technology is utilised for the node's power requirement, making the nodes entirely energy-autonomous. At the same time, VLC is used to wirelessly connect with the network. As an advantage of this technology, the indoor lighting infrastructure used for indoor illumination can be simultaneously used for IoT networking purposes. In addition, inter-node communication, localisation awareness and power-sharing technologies such as optical wireless power transfer (OWPT) can be possibly integrated with the concept-based nodes, making them comparative to the traditional IoTs. The structure of LIoT node is depicted in Figure 1.

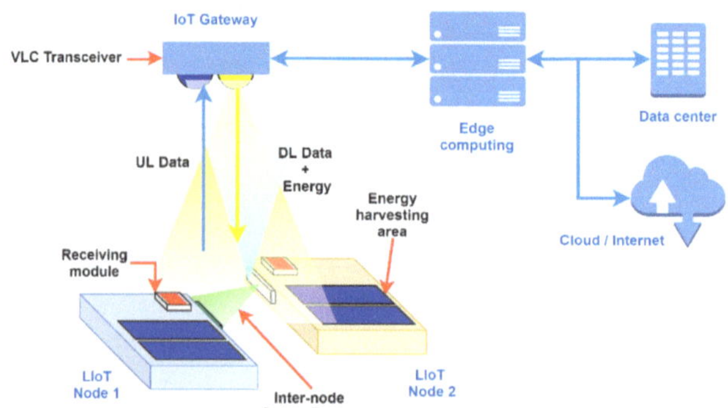

Figure 1. The light-based IoT concept and architecture: example of two networked LIoT nodes.

The dual-purpose transmitter is expected to transmit modulated data to the LIoT nodes creating a visible light communication link while providing flicker-free eye-friendly indoor illumination. The LIoT node is designed to receive the optical signal through the optical channel while harvesting the energy from the incident irradiation on the integrated PV cells. Depending on the optical signal detecting performance, a sensitive photo sensor or energy harvesting integrated PV cells can perform the receiving task.

Similarly to the operation of the transmitter, the visible light transmitter integrated on the node can be used for both power-sharing and communication with neighbouring nodes. In that way, this technology can be used to communicate and power up wireless sensor networks simultaneously and cooperatively. In addition, integration of location awareness on the node based on the received optical signal could further improve inter-node communication and power-sharing efficiency. Awareness of the proximity of neighboring nodes can be used to improve internode communication and power transferring capability of the wireless sensor network.

There are several unique benefits that can be attained from applying this technology to future IoT. Since VLC is used for communication, all unique advantages of VLC, such as license-free spectrum with unlimited reusability, robustness to electromagnetic interference and inherent physical security, are valid for the LIoT nodes. In addition, with the battery-free energy-autonomous design, the nodes implemented with LIoT technology are expected to operate without any regular maintenance. These promising characteristics will pave the way towards future green IoT communications. Moreover, the LIoT concept lends itself to the development of sustainable IoT solutions, particularly when implemented with novel technologies such as printed electronics and others, as will be discussed in the paper.

Applications

LIoT can be applied in most of the conventional IoT scenarios. In addition, there are some applications that particularly match LIoT characteristics. A few compelling applications of LIoT technology are briefly discussed next to highlight its potential. LIoT can improve the quality of the logistics sector. With the use of sustainable implementation technology such as printed or biodegradable electronics, very low-cost smart labels or container-based IoT nodes can be made to monitor and sense the environment and to display relevant real-time information. Since LIoT nodes are battery-free, the risk of fire hazards and quality degradation over time is minimal. The transceivers can be directly integrated with the warehouse/storage areas existing lighting infrastructure without costly modifications. Easily integrable sensors such as temperature, humidity, and colour can be used to monitor each product condition in real-time while supporting novel adaptive storage climate controlling systems. The low power flexible displays can be used to display real-time updating information such as price, discounts, condition-based expiry dates or even on product advertisement platforms. The self-localisation and proximity awareness of the nodes can be utilised to improve the operations of management and supply services. In addition, LIoT nodes can operate harmlessly even in electromagnetic sensitive areas such as hospitals, aviation, or manufacturing plants. Low-cost implementation (with existing lighting infrastructure), long time reusability, electromagnetic interference- and exposure-free robust operation, ability of supporting large range of sensor applications and maintenance-free uninterrupted operation can be highlighted as some of the key advantages that can be obtained from integrating LIoT technology.

The duplex LIoT design can be used to replace the traditional design of the day-to-day devices such as headphones, hearing aids and museum guides. Since VLC has the ability to reach higher data rates, traditional radio frequency technologies used for portable devices such as Bluetooth can be replaced with LIoT technology. In this way, users can use these products wherever LIoT friendly indoor illumination is available.

The underwater VLC (UVLC) is a promising technology which has already proven to go beyond acoustic signal based underwater communication [6,7]. With UVLC friendly modulations, more robust communications links can be made with underwater applications [8]. Since recharging or replacing the batteries of battery-based IoTs is a tedious and challenging process, moving to the energy harvesting design approach is considered a more feasible solution in the long run. In [9] authors propose a hybrid optical and acoustic communication technique with an energy harvesting approach. Similarly, by combining UVLC technology with energy-autonomous LIoT, underwater IoT quality of service can be further improved. In contrast, the LIoT designs' less energy-autonomous nature becomes more advantageous than traditional designs when it comes to underwater IoT operations.

3. Printing Electronics for a Lower Environmental Impact

Recent years have seen a lot of interest in printed electronics, a type of electronics created using conventional printing techniques. The key advantages of PE include low-cost manufacturing with high throughput, compatibility with system fabrication on flexible or stretchable substrates utilising inorganic, organic, and bioinspired materials, and relative ease of integration of complete systems or hybrid systems on a single sheet of plastics or even paper. Contrary to this, silicon-based microfabrication processes such as photolithography and vacuum-based methods (e.g., sputtering, chemical vapour deposition) tend to be complex and costly. A variety of applications have been enabled by printing functional inks containing soluble or dispersed materials on PE, such as transparent conductive films, flexible energy harvesting, thin-film transistors, electroluminescent devices, and wearable sensors [10].

3.1. *Manufacturing Techniques*

Printing electronics and sensors traditionally involves bringing pre-patterned parts of a circuit in contact with flexible substrates and transferring functional inks or solutions onto them. The two major approaches for developing printing systems are contact and non-contact printing. Contact printing involves bringing patterned structures with inked surfaces into physical contact with the substrate. Non-contact processes deploy the solution through openings or nozzles arranged in a pre-programmed pattern, while structures are defined based on the movement of the stage (substrate holder). Contact-based printing technologies include gravure printing, gravure-offset printing, flexographic printing, and R2R printing. Inkjet printing, screen-printing, and slot-die coating are among the non-contact printing methods. In the non-contact printing field, the advantages are several. These include simplicity, affordability, speed, adaptability to fabrication processes, low material waste, high precision of patterns, and easy control by adjusting just a few process parameters. The newly emerging polymeric stamp-based printing methods such as nanoimprint, micro-contact printing, and transfer printing have also generated significant interest, especially for monocrystalline semiconductor-based flexible electronics.

According to [11], printing technologies such as gravure and slot-die can achieve 50–200 μm high print resolution while printing at 0.02–12 μm and 0.15–60 μm thickness levels respectively. On the other hand, flexographic and screen printing technologies are more capable of achieving beyond 100 m/min printing speeds. Technologies such as Inkjet and transfer printing have the unique advantage of low material wastage compared to other methods.

3.2. *Materials*

In order to build (all) printed systems, both functional inks suitable for different printing techniques and flexible substrates described next are needed.

3.2.1. Functional Inks

Typical applications for conductors include wiring, interconnects, electrodes, antennas, touch buttons, etc. In general, copper, gold, and aluminium are the most common materials used. Silver is the most widely used printed conductor material, and there are many commercial silver inks available for numerous printing processes. Printed electronics use both nanoparticle and microparticle silver inks. Compared to microparticle silver inks, nanoparticle inks have typically better conductivity but are also usually more expensive. The much higher resistivity of both nano and microparticle inks compared to solid metal conductors is perhaps the biggest difference when designing printed vs. printed circuit board-based electronics, especially in high current applications or in applications requiring low resistance levels (e.g., low-noise circuits) [12].

Typical examples of dielectric/insulating materials are glass (inorganic), ceramics (inorganic), plastics (organic) and paper (organic). Insulators are widely used in electronics, and they are crucial elements of widely used components such as thin-film transistors, sensors and interconnections. Low-cost organic dielectric materials that are available in large quantities and can be dissolved in various solvents and solutions can be printed easily as compared to their inorganic counterparts [12].

In printed electronics, polymers, polymer composites, and small molecule-based semiconductors are very important. Semiconductive materials are widely used in transistors, solar cells, light-emitting diodes and sensors. Similar to the conducting materials discussed above for printing technologies, organic/inorganic semiconducting materials are also used for printable sensors and electronics. Inorganic materials have superior properties in terms of performance and stability, while solution-processable organic semiconductors are attractive due to low-cost processing in an ambient environment and flexibility [12].

3.2.2. Substrate Materials

There are many substrate materials, such as plastics, paper, textiles, glass, and metals, each with its own pros and cons. Most of the time, the substrates for printed electronics need to be cheap and flexible, so papers and polymer films (PET, PEN) are commonly used. The advantages of paper as a material over polymer films are its recyclable nature, its lower cost, and its lower thermal expansion. Papers are limited in their use by their roughness and absorption, but different coatings improve this situation. When it comes to incorporating functionality into clothing, textiles also offer an intriguing alternative. In Roll-2-Roll manufacturing, dimensional stability and mechanical properties of the substrate are also important. In cases where greater thermal and chemical stability is needed, more expensive plastics can be used, such as polyimide, PETS, and PEEK. Glass substrates continue to provide excellent barrier properties, but even flex glass varieties pose some challenges in R2R processing [12].

However, these very thin and flexible substrates tend to have thermal conductivity issues when they are used for applications with high power dissipating profiles. The low thickness and low thermal conductivity of these substrate materials can cause poor heat spreading performance while, factors such as excessive thermal expansion and low glass transition temperature can change the physical and chemical properties of the substrate materials under the heat [13].

3.3. Printed Components for Light-Based IoT

Thinking about the flexible and thin realisation of LIoT systems, we will now look at some of the main components of a LIoT node in their printed form—namely, organic photovoltaics (OPVs/solar cells), batteries and light-emitting devices. Printed gas, temperature and pressure, etc., sensors, which undoubtedly are at the heart of both traditional IoT and current trends in printed electronics, will not be considered in detail here, since the topic in this paper is more leaning towards the communication and energy harvesting parts of the system, thus ignoring the specific sensory function of the node.

3.3.1. Solar Cells for Indoor Energy Harvesting

Billions of IoT devices are expected to be installed over the coming years, with almost half of them connected inside buildings. Currently, the use of batteries to power these devices places significant constraints on this development. Energy harvesting has the potential to solve these hardware issues, providing greater reliability and operational lifetimes in wireless sensor networks.

Photovoltaic energy harvesting, which is the conversion of light into electrical energy, is the most common method of harvesting energy and is now a well-established technology. The amount of energy harvested depends on the intensity and spectral content of the light falling on the surface of the solar cell, the incident angle of the light, and the size, sensitivity, temperature, and type of solar cells used [14].

PV cell technologies such as Dye sensitised solar cell (DSSC), organic photovoltaics (OPV), copper indium gallium selenide solar cells (CIGS) and perovskite can be manufactured by using printing technologies [15]. According to the solar cell efficiency tables in [16], the currently achieved maximum power conversion efficiency of these PV technologies can be summarised as in Table 1.

Table 1. Power conversion efficiency under global AM1.5 spectrum (1000 W/m^2) at 25 °C.

PV Type	Power Conversion Efficiency (PCE)
CIGS (cell) (Cd-free)	23.35 $\pm$ 0.5
Perovskite (cell)	22.6 $\pm$ 0.6
DSSC (cell)	11.9 $\pm$ 0.4
OPV	15.2 $\pm$ 0.2

Recent work using a new high bandgap organic semiconductor called (IO–4Cl), demonstrates organic solar cells with a PCE of 26.1% measured under the indoor illumination conditions, simulated by a 2700 K LED lamp at 1000 lux [17]. In comparison with the competing photovoltaic technologies, such as traditional c-Si and newly emerging perovskite solar cells, these new OPV devices are the most compatible for indoor applications. In fact, the OPV materials have great tunability of their absorption spectra, thus, it is very easy to match them with the majority of light sources currently used for indoor lighting, which are fluorescent lamps, incandescent lamps, and white light-emitting diode (LED) lamps [18].

3.3.2. Light-Emitting Devices

The most prominent alternatives for light signal production in an all-printed LIoT node (i.e., ignoring solid-state LEDS/micro-LEDs and quantum dots) are the alternating current powder electroluminescent (ACPEL) device, the organic light-emitting diode (OLED) and the light-emitting electrochemical cell (LEC) thanks to thin flexible and printable layers. For example, ACPEL devices can be screen printed establishing economic, simple manufacturing. Their high AC-driving voltage makes them well-suited for energy harvesting applications. However, OLEDs are much more well-suited for low-power LIoT thanks to their low DC-driving voltages. The high sensitivity to oxygen and water of the OLED materials, however, necessitates a high barrier encapsulation. Unlike OLEDs, LECs realise a balanced charge injection using mobile ions in their active layer, broadening the choice in electrodes, but the improvement in lifetime is limited because of incomplete knowledge of the working mechanisms [19].

3.4. *Printed Electronics for a Sustainable Future*

The consequences of the digital transformation and consumer behaviour result in massive amounts of e-waste produced globally. For instance, in 2019, approximately 53.6 Mt of e-waste was generated, and it is increasing at an alarming rate of 2 Mt per year. It is common for e-waste to end up in waste dumps, despite the declared percentages of recycling (20–25%); this has negative impacts on the environment and human health [20].

Organic materials often used in printed electronics are uniquely suited to produce electronics that can be sustainable and biodegradable. Many organic materials have been shown to be biodegradable, safe, and non-toxic, including compounds of natural origin. Additionally, the unique features of such organic materials suggest they will be useful in biofunctional electronics, demonstrating functions that would be inaccessible for traditional inorganic compounds. Such materials may lead to fully biodegradable and even biocompatible and metabolisable electronics for many low-cost applications. For example, numerous materials with a bio-origin have been identified as suitable substrates for the fabrication of organic electronics. Materials such as paper, silk and gelatin enable several functionalities: low-cost, non-toxicity, biodegradability, and often biocompatibility and bioresorbability. Based on these findings, recent research has shown that 'green' organic electronics can be very practical and is likely to have a positive impact in the future as it uses biomaterials to make the recycling of e-waste more efficient [21].

For printed electronics to be environmentally friendly, a comprehensive assessment during the device lifetime, known as life cycle analysis (LCA), is required. A life cycle assessment of printed electronics provides a quantitative measure of the environmental impacts at each stage, from production to end-of-life. Researchers have conducted several studies that cover the life cycle stages of printed products and systems. The environmental assessment of printed antennas, as well as assessments of toxic emissions and ageing of polymer films, have been investigated by several research groups. In addition, reports have been published on LCA and recycling options for photovoltaics and the energy demand and reusability of biodegradable and recycled printed LED foils. Similar studies have been conducted for OFETs, carbon-based inks on thermoplastic substrates, and recyclable sensors and TFTs composed of fully carbon-based inks and cellulose-based substrates. While there are limitations, the advancement in printed electronics shows there is a possibility of producing electronic devices with a low carbon footprint [20].

4. Requirements and Challenges

In this section, some requirements and challenges for the development of LIoT are identified and discussed. In addition, possible technology alterations according to the design requirements are suggested in the text.

4.1. Eye-Friendly Data Transmission

When using indoor luminaries for illumination and communication purposes, it is essential to maintain eye-friendly, flicker-free and colour consistent indoor illumination while transmitting data. The IEEE 802.15.7 standard for short-range optical communications [22], mainly highlights flicker mitigation and dimming support requirements for short-range optical communications. According to standards, dimming support describes a VLC system that supports changing the perceived brightness of the lights according to the user's requirement. Moreover, the brightness changing intervals should not exceed the maximum flickering time period (MFTP), which expresses the maximum time period the human eye cannot perceive brightness fluctuations. In [23], MFTP is approximated as a time period within 5 milliseconds. Unregulated flickering VLC transmission could trigger illnesses such as photosensitive epilepsy in sensitive people [24]. The flickering rate of the transmission relies on the modulation scheme used for data transmission. According to [25], modulation schemes such as on-off keying (OOK), colour shift keying (CSK), pulse position modulation (PPM) and orthogonal frequency-division multiplexing (OFDM) are some of the compatible modulation schemes with VLC. However, in [26], authors mention that multicolour mixed white LEDs used for CSK modulation-based VLC systems are more prominent to phenomena such as colour shifting caused by unequal ageing between multicolour sources, which could affect the performance in the long term.

Moreover, in VLC systems, which use RC filter-based pre-equalisers at the transmitter, modulation schemes such as OOK with more power spectral density (PSD) at the DC level are more prominent to cause baseline wandering phenomena (BLW) at the receiver [27]. This could randomly degrade the BER performance of the VLC link based on the transmitting bit sequence. The modulation schemes derived from PPM tend to be immune to BLW as they have less PSD at the DC level.

4.2. Harvesting and Storing Energy at Low Voltage Efficiently

In order to design low powered, energy-autonomous IoT nodes, it is important to select the suitable energy harvesting usage protocol for the application's requirement. There are three main types of energy harvesting usage protocols, namely Harvest-Use (HU), Harvest-Store-Use (HSU) and Harvest-Use-Store (HUS). As the name suggests, there is no energy store in HU, while there is an energy store step in HSU and HUS [28]. Even though the energy storage-less, simple HU protocol is more supportive of the concept of "expose and connect" based LIoT node designs, energy storage-based protocols have the ability to cater for the power requirement of most currently available microcontrollers and sensors. Dynamically varying power requirements for different functionalities of the node can be smoothly rectified by using energy storage. The structure of HSU can be described as in Figure 2.

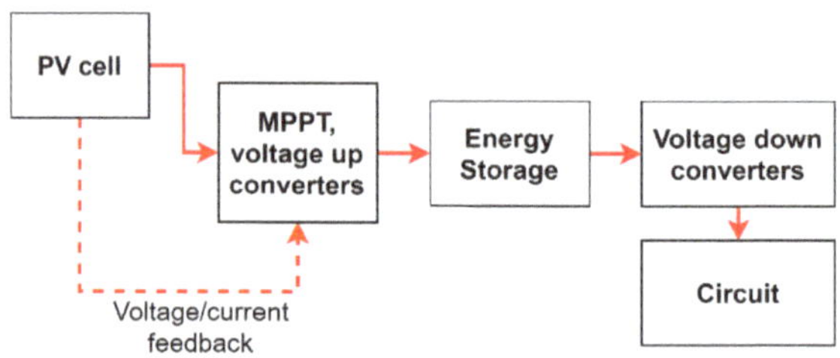

Figure 2. Harvest-Store-Use energy harvesting structure.

For energy-autonomous design, it is vital to obtain the maximum possible energy from harvesters and store them with minimum energy losses and leakages during the operation. Typically, these designs consist of energy harvesting units such as photovoltaics, piezoelectrics or thermoelectric modules, and energy-storing units such as capacitors, supercapacitors, or batteries. When it comes to photovoltaics-based energy harvesting, varying incident illumination on PV cells could result in fluctuations at the storing voltages. Generally, a charge pump circuit or boost converter circuit could increase the final storing voltage at storage in situations such as this. Typically charge pump circuits double or multiply the input voltage by using capacitors, while boost converters use inductors for voltage increase. According to [29,30], charge pump designs are more suitable for steady input, more compact, the on-chip solution required designs, and boost converters are more suitable for high efficiency required designs with fewer space constraints due to the external inductors. Therefore, a boost converter could be the best solution for energy-autonomous designs due to their high efficiency compared to the charge pump approach. Equation (1) expresses the connection between PV terminal voltage and the DC-DC converter voltage output:

$$V_{out} = \frac{1}{1-D} V_{PV}, \tag{1}$$

where, V_{out} = output voltage of the converter, V_{PV} = voltage at PV terminals, D = duty ratio.

Since PV cells are nonlinear, their efficiency can vary due to the level of irradiance and ambient temperature. Therefore, operating them at the maximum power point (MPP) is necessary to achieve maximum efficiency. For this task, the maximum power point tracking (MPPT) method can be used. In this method, the voltage at the terminal of the PV cell is regulated to keep it closer to the MPP corresponding voltage (Vmpp) of the PV cell at the moment. To determine the maximum power point, controllers such as P&O, incremental conductance, fuzzy logic, and artificial neural networks can be used [31–35]. Based on the controller's output signal, MPP is achieved by regulating the connection between the voltage up-converter and PV cell terminals [36,37].

On the other hand, the required surface area for energy harvesting can be minimised by using high-performance PV cells with optimal conditions. The number of PV cells and their series or parallel connectivity configuration can be selected based on the voltage or current requirement of the application. Since the human eye is less sensitive to wavelengths corresponding to the blue colour region of the spectrum, studies in [38] show that blue colour light has to emit more radiant power to achieve a specified lux level, compared to red and green colours. The comparatively more radiant power of the blue colour leads to more energy generation at the PV cells compared to other colours. Therefore, it is beneficial to use luminaires that emit wavelengths of the blue region when designing VLC transmitters for energy harvesting purposes in the indoor environment. According to [39], even though it is an electroluminescence process, Wien law can be used to estimate colour temperature based on the required wavelength region. Using the Wien equation below (2), both eye and energy harvesting friendly suitable light types can be determined.

$$\lambda_{max} = \frac{hc}{4.965kT} \tag{2}$$

where, h = Planck's constant, c = velocity of light, k = Boltzmann's constant, T = color temperature.

When designing a low-powered IoT node, selection of energy storage type should be made according to the application's nature. Generally, batteries and supercapacitors are used for these kinds of energy-storing requirements. Among these, batteries are capable of having more energy density, while capacitors have more power density. The high energy density storage devices are more capable of providing steady energy for a long period of time, while high power density storage devices can quickly get energy in or out from the device [40]. Therefore, to power up the "expose and connect" concept based LIoT, it is important to minimise the storage device's initial cold start charge time and charge quickly to starting voltage. By using high power density storage such as supercapacitors, this requirement can be fulfilled. The voltage response for charging a supercapacitor can be expressed as the below (3).

$$V_c(t) = V_{in}(1 - e^{-\frac{t}{\tau}}), \tag{3}$$

where, τ = time constant defined as

$$T = R * C, \tag{4}$$

where, C = capacity, R = series resistance.

$$C = \frac{Q}{V}, \tag{5}$$

where, C = capacity, V = voltage.

According to (3), when time is equal to one time constant, the capacitor is charged up to approximately 63% of its final voltage. Similarly, the capacitor gets discharged to 37% of the final charged voltage during a one-time constant equal period. On the other hand, the equation explains that capacity value is directly proportional to the accumulated charges at a given voltage. Therefore, it is essential to keep the optimum resistance and capacity values when selecting a capacitor [41]. The equal series resistance (ESR) of the capacitor can be identified by the ESR rating value. By using a low ESR rated capacitor with a sufficient capacity rating for the required application, the time constant can be minimised.

In addition, according to (5), it is important to keep sufficient voltage at the storage unit to store more charges. However, when supplying the energy to a connected load such as a microcontroller or sensor, the voltage level needs to be reduced. Typically for this purpose, a low dropout regulator (LDO) or buck converter can be used. According to [42], buck converters have more efficiency than LDOs when the load draws milliamperes rated current. However, LDOs are more suitable for compact in-chip designs as buck converters require space-consuming external inductors for their operation. Therefore, a suitable converter needs to be chosen based on the requirement of the application. For applications that involve low power-autonomous designs, using high-efficiency buck converter designs would be beneficial.

4.3. Low Energy Consumption Optimisation

In order to preserve the harvested energy while the IoT node is not actively operating, a sleep state can be introduced to the low power node design. In this way, the node is put into a low power consuming intermediate status, which can be woken up with external or internal interrupts. Generally, the duty cycling method can be effectively used for this purpose. In this method, nodes periodically alternate between the sleep and wake up modes. In addition, duty cycling can be mainly classified as synchronous and asynchronous approaches. In duty cycling, nodes self-wake up periodically so that the transmitting device has to wait till the scheduled node wakes up time to communicate. However, the transmission delay between the states' alternative period is considered the main drawback of this method. On the other hand, the wake-up radio (WuR) approach can be used for on-demand wake-up purposes without causing any delay. Typically, this approach uses a low power consuming simple modulated secondary receiver unit for the node's sleep interruption. In [43], the authors describe an energy-autonomous VLC based WuR receiver with a pattern correlator. In this way, the WuR receiver can filter out the dedicated wake-up signal, among many other wake-up signals. Since a secondary receiver unit is active throughout the sleep state, this method has less power efficiency compared to duty cycling [44]. Therefore, considering the trade-off between transmission delay and less power efficiency, these methods need to be integrated with the node designs.

4.4. Early-Stage of Sustainable Electronic Technologies

In contrast to traditional electronics, sustainable electronics technologies are still early in their development process. Hence, finding specific active and passive components for developing an entire prototype using printed or biodegradable electronics appears to be a true challenge at this time. However, according to [45–47], essential components for LIoT nodes, such as printed photodiodes, OLEDs, indoor PV cells, displays, and temperature sensors, are currently available as stable components for the integrations. Therefore, it is clear that an early-stage LIoT prototype development process can be initiated using traditional and sustainable electronics-based components. With the development of the technologies in future, these traditional electronics can be replaced with their alternative counterparts, and a fully printed LIoT node will be implemented.

5. Design of a Duplex LIoT Node

In order to demonstrate the concept of LIoT, we have designed and implemented a working prototype. The design is a LIoT-based smart product label, which contains display and temperature monitoring sensors. When the node is exposed to the transmitter's illumination, the product label is expected to update its information display while transmitting the temperature variation of the wrapped product. The design of the node can be illustrated as in Figure 3.

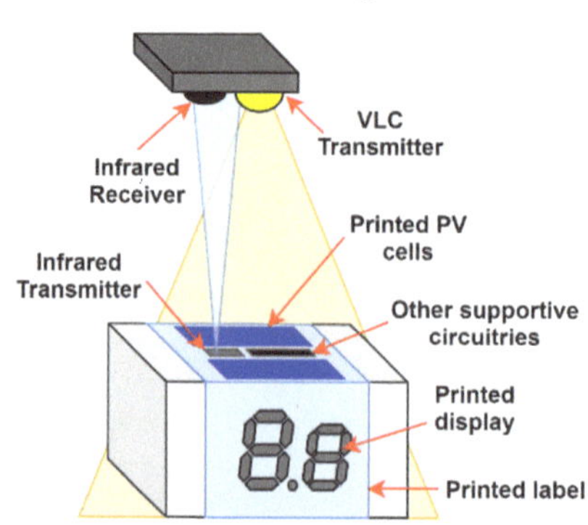

Figure 3. LIoT-based product smart label design.

The smart label was designed to work as a duplex communication system with both receiving (i.e., downlink) and transmitting (i.e., uplink) capabilities. In addition, the system is designed to support multiple users using the same transmitter illumination infrastructure (multi-user operation). In this prototype, visible light and infrared links are used as the downlink and the uplink, respectively. During the idle period of time, the node is configured to go to the energy-saving sleep state, and whenever the downlink is operating, the node is designed to wake up and receive data.

5.1. Transceiver—Illumination System

The overall structure of the transceiver-illumination system is illustrated in Figure 4.

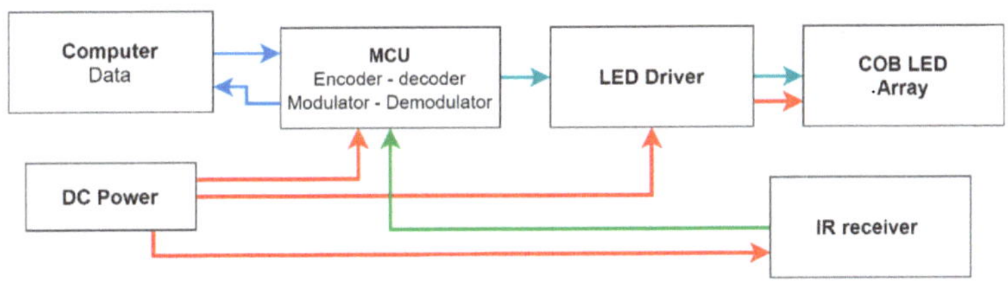

Figure 4. Structure of the transceiver-illumination system.

The input data from the main computer is initially transferred to the microcontroller unit (MCU) by using serial communication. From the MCU, the data is encoded and modulated according to the selected modulation. After that, the modulated data is converted to an electrical signal, and it is fed to the LED driver to control the transmitting lights according to the signal. Inside the LED driver, the electrical signal is levelled up to provide sufficient voltage and current level required by the selected lights. On the other hand, the transmitted optical signals from nodes are received from the receiver at the transceivers surface. The generated electrical signals based on the received signals are transferred to the MCU for decode and demodulation.

In order to select the modulation scheme of the prototype, factors such as VLC compatibility and robustness against capacitive component-specific phenomena such as BLW are considered. In the implemented prototype, pulse position modulation (PPM) was used as the digital modulation scheme for the optical transmitting signal. VLC friendliness and the modulated data frame is designed to facilitate 48 bits, containing address bits and payload bits for each targeted LIoT node. Moreover, additional on-off keying (OOK) based wake-up bits are added in front of the data frame to wake up the node from sleeping mode. In addition, to minimise the interferences from external sources, the transmitting signal was modulated using a 38 kHz carrier signal. In the final design, an Atmega328P based Arduino Uno development board was used as the MCU of the transceiver prototype. A TSOP 1838 IR sensor is used as the IR receiver of the transceiver. A captured waveform of a data frame is illustrated in Figure 5.

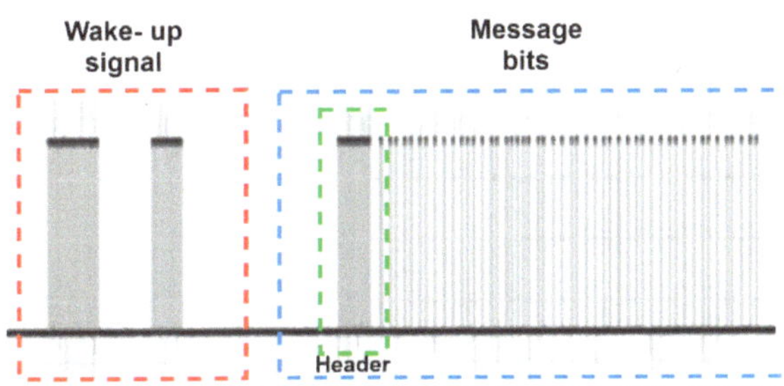

Figure 5. Oscilloscope capture of a designed data frame.

The LED driver of the transceiver was designed to provide the required DC bias voltage for lights while detecting the MCU's TTL level voltage signals and scale them to create better modulation depth. Moreover, the required DC bias voltage, and factors such as rise fall time, drain current limitation, and compatibility with TTL level signal was considered for the design of the LED driver. Based on the above factors, an IRF520 N channel metal–oxide–semiconductor field-effect transistor (MOSFET) was used as the switching unit for the LED driver. The design diagram of the LED driver is illustrated in Figure 6.

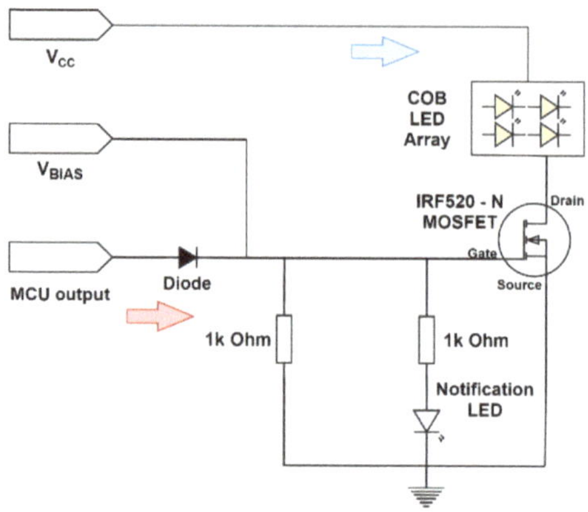

Figure 6. LED driver structure.

As the light-emitting sources, indoor illumination and energy harvesting friendly 5700 K—cool white chip on board (COB) type LED sources were used. In order to mitigate the perceivable flickering effects and enhance the energy harvesting performance, an idle transmitting mode was introduced to the system. In previous work, we have shown that idle transmit mode can improve the energy harvesting mode compared to relying on bias voltage [48]. In the idle transmitting mode, when the transmitter is not transmitting any data signals, unmodulated long square pulses are fed to the LED driver and transmitted through the lights. In this way, voltage fluctuation on PV cells can be minimised between the data transmitting and non-transmitting interval.

5.2. LIoT Node

The structure of the prototype node is depicted in Figure 7. The node consists of main segments such as transceiver system, energy harvesting system, temperature sensor, MCU and output printed display.

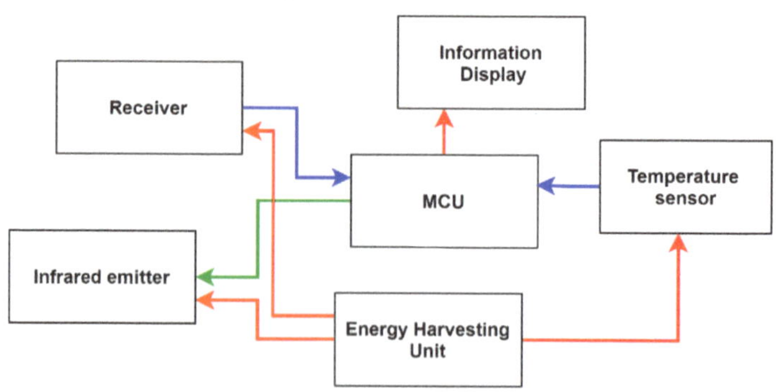

Figure 7. The structure of the prototype LIoT node.

The designed algorithm for the LIoT smart label is described in Figure 8. Initially, the node receives the optical signal and checks the request correspondence for the node's main functions. Sensing and transmitting the temperature and managing the info display are the two main functions of the designed prototype.

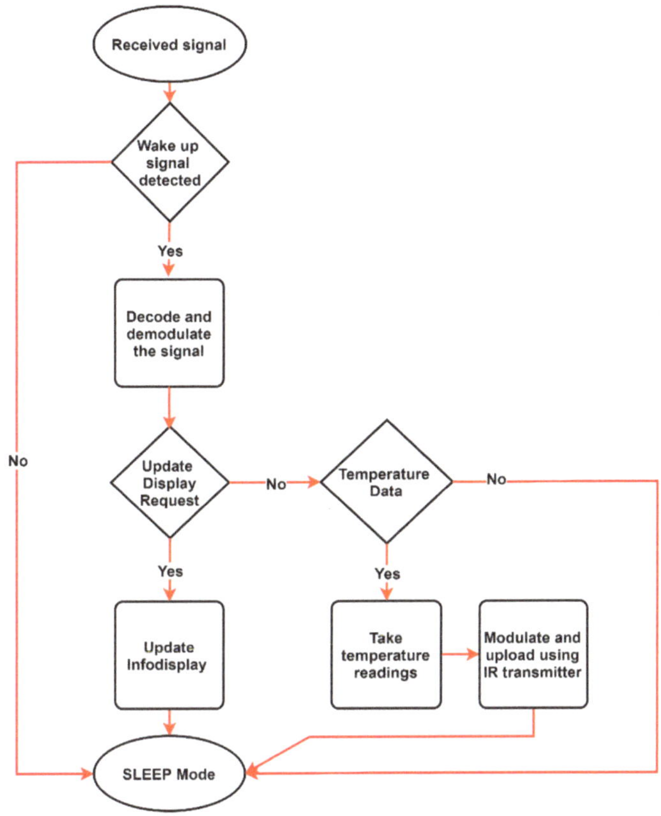

Figure 8. Designed algorithm for the LIoT smart label.

5.2.1. Transceiver System

The transceiver system consists of a visible light receiver and infrared emitter to facilitate downlink and uplink functions. In this design, an on-demand wake up capable WuR inspired wake-up approach was used for energy conservation. Since the VLC receiving sensors have better power efficiencies compared to RF receiving antennas, the same receiving sensor is used for wake-up signal reception in the implementation. This has mitigated the need for a separate receiver when using the WuR approach, thus, improving the overall power efficiency, the common deficiency of the approach. As WuR approach already has less transmission delay compared to its duty cycle approach-based counterpart, this has resulted in a solution that contained the best of both approaches. The visible light receiver was designed to receive the downlink data signal and extract the 38 kHz modulated signal in order to demodulate and decode. In this circuit, initially, the photodiode receives the optical signal and creates a corresponding current signal. After that, a trans-impedance amplifier (TIA) converts the current signal to a voltage–electrical signal. Then the electrical signal is amplified by using an automatic gain control amplifier (AGC). By analysing the header, the AGC adjusts the gain level and provides the waveform for the bandpass filter with 38 kHz centre frequency. In this way, the bandpass filter filters out the external noise components from the converted electrical signal and passes only the message electrical signal for processing. During AGC processing, the electrical signal is modified to TTL voltage levels to be detected at MCU's analogue to digital converter. The MCU algorithm initially waits for the wake-up interruption and then identifies the data signal measuring the header time length. Thereafter, MCU is configured to demodulate and decode the message signal and proceed according to it. For the implementation, a 3.3 V voltage compatible ATmega328P based Arduino mini development board was used for the MCU. TIA, AGC and bandpass filter all on-chip integrated VSOP 38338 integrated chip (IC) with 525 nm PIN diode used for the receiver prototype. The structure of the receiving system is given in Figure 9.

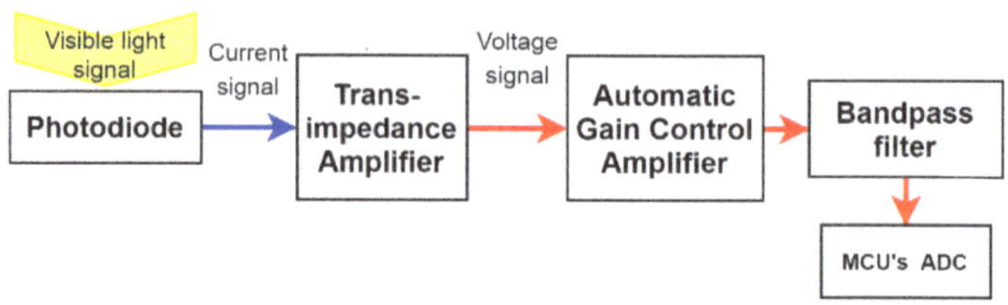

Figure 9. The structure of the receiving system.

In order to minimise the uplink-downlink optical channel interferences and avoid naked eye perceivable flicker, which could be caused by a visible light emitter, an IR emitter was used as the uplink for the node. The uplink transmitting unit was designed to transmit temperature sensor readings using infrared (IR) light. According to the design, the voltage variations provided by the 3.3 v compatible TMP36GT9Z temperature sensor are fed into the MCU for temperature mapping. Thereafter, the processed temperature information is encoded and modulated for uplink transmission. Similar to the downlink, uplink data are also modulated using a 36 kHz carrier signal in order to mitigate the noise effects. As the information display of the smart label prototype, a flexible printed electrochromic seven segment display was used. The bistable nature of the electrochromic display helps to hold the display information during node sleep state, which is advantageous for a low power product label design. Since the display is 3.3 v compatible, it was directly connected to the digital output pins of the MCU unit.

5.2.2. Energy Harvesting Unit

In order to achieve energy-autonomous, battery-less design, the energy harvest unit (EHU) of the prototype is designed to keep minimum energy storage on the device while providing a continuous power supply for the node to operate. The overall design of the EHU followed the type of HSU energy harvesting and usage protocol to facilitate the fluctuating power requirement of the commercial MCU and sensor nodes. The designed EHU consists of two PV cells, a power management IC (PMIC) with a built-in boost converter, prismatic capacitor and buck converter to support the energy harvest process. The structure of the designed EHU is illustrated in Figure 10.

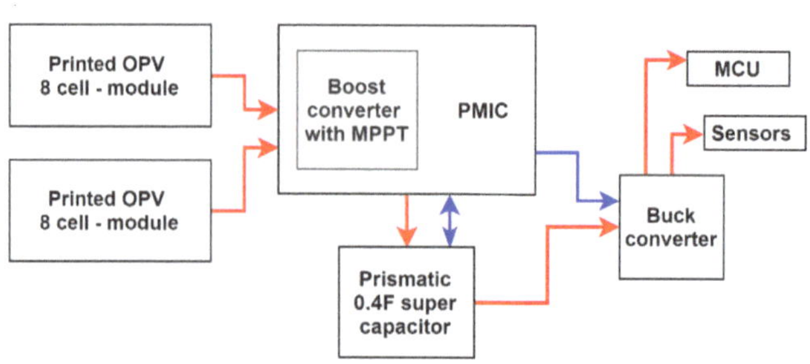

Figure 10. The structure of the designed EHU.

In the EHU, two parallelly connected Epshine LEH printed organic PV 8 cell modules were used to harvest sufficient energy for the requirement. Then, with the use of a boost converter and the MPPT tracking feature integrated with the e-peas AEM 10941 PMIC, the input voltage further improved while keeping PV cells at the maximum efficiency region. A low ESR 0.4F supercapacitor manufactured CAP -XX was used as the harvested energy storage of the node. According to the configured algorithm, whenever storing voltage reaches 3.9 v, PMIC turns on the buck converter to step down the storing voltage to 3.3 v voltage. For this, a Texas TPS 62740 step-down buck DCDC converter was used for the implementation. When the storing voltage reaches 3.3 v at the capacitor, the PMIC turns off the buck converter and detaches the load to speed up the charging process. The designed energy harvesting system's turn on-off capacitor voltages can be tabulated as in Table 2.

Table 2. Turn on-off voltages of the designed EHU.

State	Capacitor Voltage (Vcp)
Turn off	3.3 v
Turn on	3.9 v

During this turn on-off period of the window, the node can be fully functional and perform its designed task. The state transition and its dependable parameters are illustrated in Figure 11a. The final view of the designed node is shown in Figure 11b.

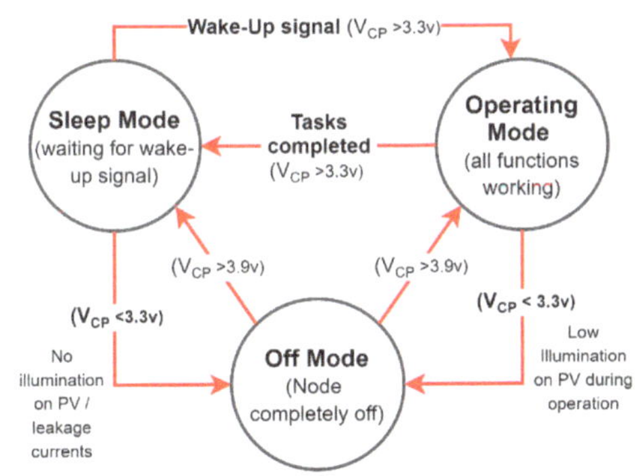

(a)

(b)

Figure 11. (**a**) State transition diagram and conditional parameters; (**b**) LIoT product label prototype using printed components.

6. Performance Evaluation

In this section, key performance experiments and results are described. The main functions of the LIoT concept are experimented with and evaluated here. Based on the experiment results, some hardware components selection were made for the prototype implementation. Since the prototype of LIoT smart label does not require a higher data rate for its functionality, approximately 1.4 kbps (max) of low data rate was used for the communication.

6.1. Uplink—IR LED (5 mm) vs. Printed OLED

In order to compare the performance of a traditional IR uplink (Lambertian radiation pattern) with VLC and printed OLED emitter-based uplink (non—Lambertian radiation pattern) under the LIoT smart label usage scenarios (on market shelves, inside refrigerators) the following experiment was carried out. The performances of the two types of links under aligned and non-aligend transmitter-receiver alignment scenarios were tested during the experiment. To simulate a possible LIoT label-based product displacement during an actual application scenario, a 15 cm horizontal displacement was introduced. For the implementation, a 5 mm IR LED and 4 cm × 7 cm printed white OLED was used as the transmitting sources. A monolithic photodiode and on-chip trans-impedance amplifier integrated OPT 101 analogue sensors were used as the receiver for both setups. The setup arrangement is depicted in Figure 12a. In this experiment, the IR and printed OLED transmitters were initially placed directly facing the photodiode sensor, and the SNR was measured. The transmitters were placed 0.20 m away from the receiver sensor, and minimum external illumination conditions (below 20 lux) were maintained during the experiment. Then, the receiver sensor was moved 15 cm horizontally away from the previous position making the transmitter and receiver nonaligned, and the experiment was repeated. The results are plotted in Figure 12b.

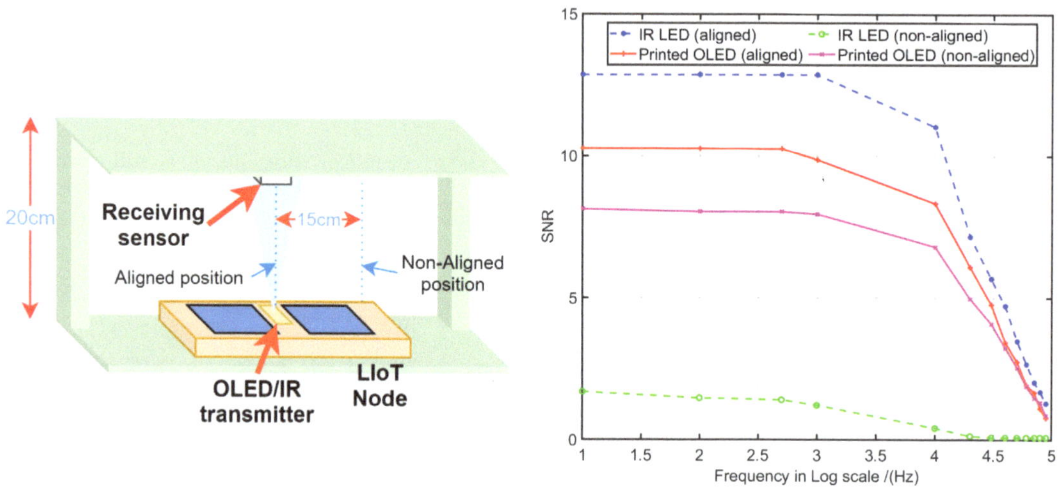

Figure 12. (**a**) Experiment setup arrangement; (**b**) IR vs. printed OLED under different link alignments.

The results suggest that a typical 5 mm IR LED based uplink has a better frequency response than the printed OLED when both transmitters and receiver sensors are placed directly. However, during the non-direct experiment, the results showed that the signal emitted by the planar lighting source (printed OLED and sensor act as non-directional link) has a better frequency response than directional lighting source (5 mm IR led and sensor act as a hybrid link) when they are not spatially aligned. The Lambertian radiation pattern emitted by the 5 mm IR led causes more concentrated radiation on the receiver in the aligned link scenario, while the non-directional, wider emitting angles of the OLED results in better performance when links were not aligned.

Therefore, according to the results, printed OLED with higher emitting angles would be a good choice for more dynamic, close-ranged LIoT applications. The large irradiation pattern caused by the high emitting pattern could provide stability compared to the more concentrated radiation pattern. On the other hand, the concentrated directional IR link can be advantageous for stationary applications such as smart labels which are considered in the implementation. The high SNR levels of the IR link could result in more robust communication links under noisy environments.

6.2. Frequency Response of Printed PV Cells

In order to determine the feasibility of using commercially available printed PV cells for LIoT data receiving purposes, the frequency responses of different printed PV cell types were measured in this the experiment. To create a similar environment as in the previously discussed experiment, the PV cells were placed 0.2 m below the transmitter, and square pulses were transmitted as the signal. Indoor optimised Perovskite PV cells and both indoor and outdoor optimised organic printed PV cells were used in this measurement setup. Since each printed PV cell panel had a different active area, the normalised voltage per square centimetre of the area was considered for the SNR calculation. The details of used PV cells are given in Table 3. The obtained results are plotted in Figure 13.

Table 3. Commercially available printed PV cells used for the experiment.

Printed PV Type	Manufacturer	Active Area (cm^2)
OPV (indoor)	Epshine	35
OPV (outdoor)	Infinity PV	300
Perovskite	VTT	95

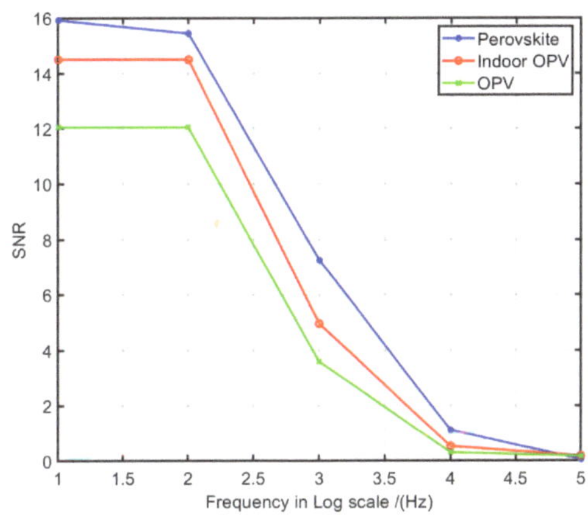

Figure 13. Frequency response of the PE based perovskite and OPV technologies.

According to the results, both perovskite and organic printed PV cells show low SNR performance when the frequency increases. Therefore, commercially available printed PV technologies such as OPV and perovskite have low-frequency characteristics which are not capable of supporting high data rate applications for LIoT. The designed receiver circuit for the prototype node was not able to demodulate and decode the extracted data signals from the PV cells. Hence, a photodiode based receiving circuit was added to the prototype for high data rate receiving purposes.

6.3. *Effect of MPPT and Boost Converter on Indoor Energy Harvesting*

The feasibility of using dynamic outdoor sunlight energy harvest optimising technologies for the static transmitter based LIoT application scenarios were tested in this experiment. In order to simulate the receiving levels with dynamic movements of LIoTs under the fixed transmitter, the illumination of the transmitter was varied with 120 s fixed interval during the experiment. The MPPT and boost converter's effect on the LIoT node's energy harvesting process was analysed during this experiment. The MPPT-based and boost converter-equipped AEM 10941 PMIC and non-actively energy harvesting contributing LTC3588 PMIC, were used for this experiment. To store the harvested energy, a cap-xx high-performance 0.4 F supercapacitor was used. In the experiment, the illumination levels were changed with a fixed interval of 120 s, and the voltage between the capacitor was measured. The experiment was repeated for both types of PMICs. The obtained results are plotted in Figure 14.

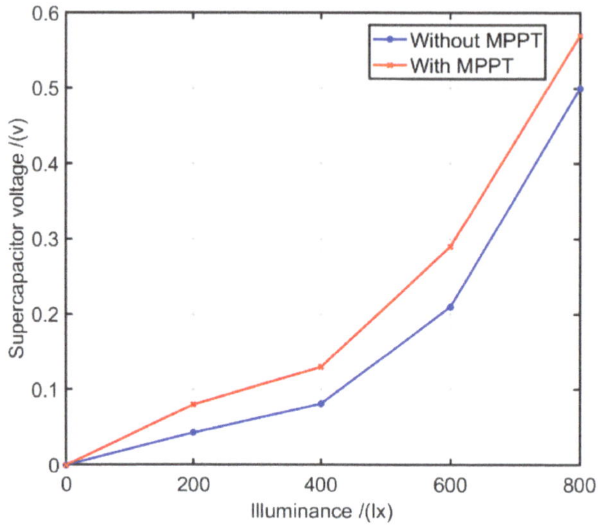

Figure 14. Capacitor voltage variation with and without MPPT.

The results suggest that under low lux levels, MPPT and boost converter managed to store more energy compared to other approaches while keeping PV panels at optimal efficiency. However, with the increasing illuminance levels, the gradients of the graphs show that the advantage of MPPT is negligible compared to low illuminance levels. This can be explained by the fact that higher illuminance on the PV cell naturally drives the cell closer to the MPP so that the effect of the algorithm based MPPT controller is less dominant under these conditions.

Therefore, using MPPT and boost converter setups will be more viable for LIoT applications designed to operate under dynamic light-receiving conditions. In addition, since MPPT and boost converter equipped PMICs have a limited input voltage range (typically 0–5 v), extra precautions need to be taken to avoid overvoltage generating scenarios. For the static LIoT applications, fixing the ideal illumination levels at the designing stage could achieve maximum energy harvesting at the PV cells. In this way, static LIoT applications can avoid additional PV optimising circuitries and follow a less complex energy harvesting design approach.

6.4. Performance Analysis of the Designed Energy Harvesting Unit

In order to assess the performance of the designed energy harvesting system, the following experiment was carried out. A 5 mm IR LED was used as an uplink emitter for calculating the transmitting current requirement. The max ratings of the drawing currents for each operating mode are listed in Table 4.

Table 4. Maximum ratings for different operating modes of the node.

Operating Mode	Maximum Drawing Current
Sleep	<1 mA
Receiving/Displaying	3 mA
Transmitting	25 mA

In order to create a fairly high current consumption for the evaluation, the node was programmed to operate in transmitting and receiving mode in equal operating periods without falling to sleep mode. Initially, while the node is completely turned off (zero consumption), the charging time from Vcp 3.3 v to 3.9 v was measured (described in

Table 2). Even though the PMIC has the ability to charge the Vcp up to 4.5 v under no-load conditions, to determine the minimum operational time values, a turn-on value of 3.9 v is considered as the maximum charging Vcp throughout the experiment. Then, as soon as the node starts to turn on when Vcp reaches to 3.9 v, the discharging time from 3.9 v to 3.3 v under the given illuminance was measured. The experiment was repeated by varying the illuminance on the two parallelly connected printed organic PV cells during the experiment. The received results are plotted in Figure 15a.

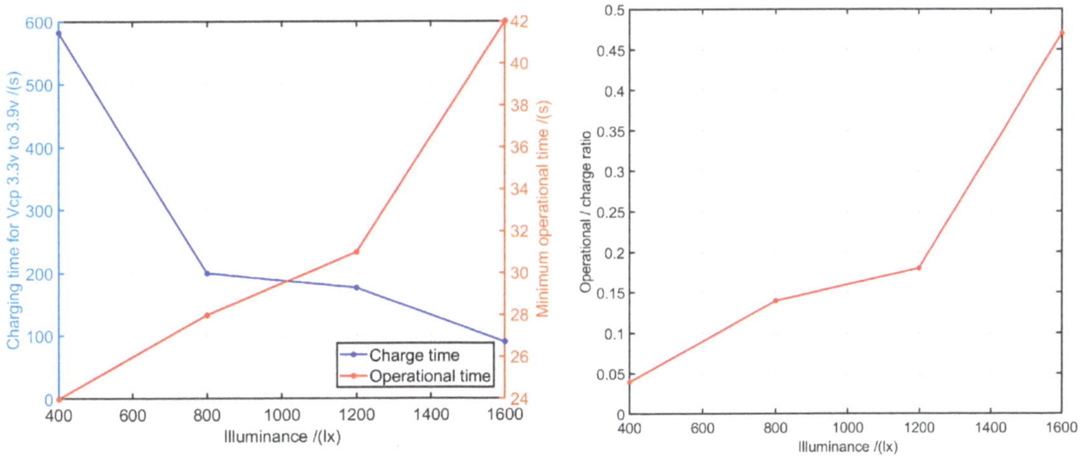

(**a**) (**b**)

Figure 15. (**a**) Charge time and operational time variation for different illuminance; (**b**) illumination vs. operational-charge ratio variation.

According to the results, the node's energy harvesting system works at maximum performance when the illuminance levels are relatively higher. Even though the energy harvesting and storing system manages to provide only around 30 s of minimum operational time for 800 to 1200 lux illuminance levels, it is fully sufficient for the functionality of the LIoT smart label application. As previously mentioned, by keeping the node at a sleeping state (no load) throughout the charging time, it can achieve a maximum capacitor voltage of 4.5 v. Hence, the operational time can be further increased without any additional effort.

By using the values obtained from the previous experiment, the operations-to-charge time ratio curve was determined. The equation used to determine the operation-to-charge time ratios given below.

$$operation - to - charge\ time\ ratio = \frac{Minimum\ operational\ time\ per\ given\ illumination\ level}{Charging\ time\ for\ Vcp\ to\ \text{off to on}state\ voltage\ threshold}, \quad (6)$$

The obtained curve for the prototype node is illustrated in Figure 15b.

Using these curves, the node will be self-aware about its limitations under the given illumination condition and can accordingly decide its own operation and sleep times by coordinating with the transmitter. In this way, LIoT nodes can be designed to operate as a "smart LIoT node" which is aware of its current (situational) limitations and has the ability to adapt accordingly to execute required tasks. The proposed algorithm for smart LIoT is illustrated in Figure 16.

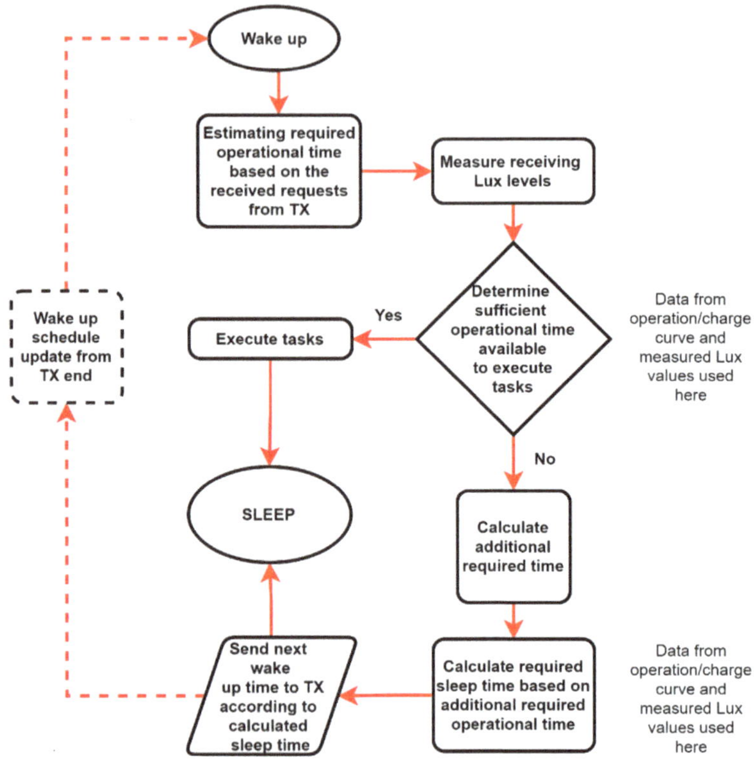

Figure 16. Proposed algorithm for Smart LIoT.

7. Discussion

According to the study of Lambertian vs. non-Lambertian optical emitters usability for LIoT, the results indicate that lighting sources with small emitting angles can achieve better SNR rates for stationary LIoT applications, while sources with large emitting angles are more suitable for dynamic LIoT applications, where mobility is expected. The frequency response comparison of two organic and Perovskite printed PV technologies showed that printed PVs have a slower frequency response for LIoT signal receiving purposes. Furthermore, the experiment on the applicability of MPPT and DC-DC boost converter-based energy harvesting for LIoT highlighted that MPPT and boost converter could improve the energy harvesting efficiency in low light conditions. The operation-to-charge time curve derived from the EHU's performance analysis results can be used to implement a novel limitation awareness smart algorithm for the LIoT. In this way, the smart LIoTs will execute tasks adoptively under the varying power constraint scenarios.

8. Conclusions

A feasibility study and implementation work of the energy-autonomous LIoT node has been presented in this paper. In this work, a LIoT based smart label prototype was implemented as a case study. Based on the performance and results of the prototype, the practicality of the proposed LIoT framework was analysed. The use of visible light for the LIoT nodes' communication enables indoor energy harvesting through the same infrastructure while adding unique VLC advantages over radio communications. In order to support the idea of the sustainable green communication concept, the usability of sustainable implementation technologies for the implementation was considered throughout the work. The printed technologies-based components used for the implementation, such as display, OLED and PV cells, worked as expected with other conventional components and sensors, despite their lower performance. Technologies such as visible light communication and self-energy harvesting are getting attention to cater to the requirement of sustainable zero energy nodes, while massive wireless energy transfer is expected to be a key technology in the new 6G IoT. As for the future development of this work, inter-nodes power-sharing approach will be developed and implemented. In addition, the smart LIoT concept can be further improved considering adaptive transmitter illumination level adjustments and internode transmission optimisation with proximity information. With the development of thin electronics technologies integrable on any surface, such as PE, it will be possible to integrate different functionalities on any available surface. These integrated nodes, sensors, actuators and others, will unlock new boundaryless "living surfaces" technology that does not require any hands-on devices in the future.

Author Contributions: Conceptualisation, M.K.; software, M.A.N.P.; writing—original draft preparation, M.A.N.P.; writing—review and editing, M.K., J.H., R.G.; supervision, M.K., J.H., R.G. All authors have read and agreed to the published version of the manuscript.

Funding: This work was supported by the Academy of Finland—6G Flagship.

Institutional Review Board Statement: Not applicable.

Informed Consent Statement: Not applicable.

Data Availability Statement: Not applicable.

Conflicts of Interest: The authors declare no conflict of interest.

References

1. Katz, M.; O'Brien, D. Exploiting Novel Concepts for Visible Light Communications: From Light-Based IoT to Living Surfaces. *Optik* **2019**, *195*, 163176. [CrossRef]
2. Garg, N.; Garg, R. Energy Harvesting in IoT Devices: A Survey. In Proceedings of the 2017 International Conference on Intelligent Sustainable Systems (ICISS), Palladam, India, 7–8 December 2017; pp. 127–131. [CrossRef]
3. Vitoriano, F.A.; Gomes, G.P.; de Lima Monteiro, D.W. Towards a Wireless Energy Autonomous IoT Module with a Self-Adaptable Data Diet. In Proceedings of the 2021 5th International Symposium on Instrumentation Systems, Circuits and Transducers (INSCIT), Campinas, Brazil, 23 August 2021; pp. 1–6.
4. Xu, X.; Shen, Y.; Yang, J.; Xu, C.; Shen, G.; Chen, G.; Ni, Y. PassiveVLC: Enabling Practical Visible Light Backscatter Communication for Battery-Free IoT Applications. In Proceedings of the 23rd Annual International Conference on Mobile Computing and Networking, Salt Lake City, UT, USA; 4 October 2017; pp. 180–192.
5. Dongus, S.; Jalilian, H.; Schürmann, D.; Röösli, M. Health Effects of WiFi Radiation: A Review Based on Systematic Quality Evaluation. *Crit. Rev. Environ. Sci. Technol.* **2021**, 1–20. [CrossRef]
6. Kao, C.-C.; Lin, Y.-S.; Wu, G.-D.; Huang, C.-J. A Comprehensive Study on the Internet of Underwater Things: Applications, Challenges, and Channel Models. *Sensors* **2017**, *17*, 1477. [CrossRef] [PubMed]
7. Wang, F.; Liu, Y.; Jiang, F. Nan Chi High Speed Underwater Visible Light Communication System Based on LED Employing Maximum Ratio Combination with Multi-PIN Reception. *Opt. Commun.* **2018**, *425*, 106–112. [CrossRef]
8. Chi, N.; Shi, M. Advanced modulation formats for underwater visible light communications. *Chin. Opt. Lett.* **2018**, *16*, 120603.
9. Saeed, N.; Celik, A.; Al-Naffouri, T.Y.; Alouini, M.-S. Energy Harvesting Hybrid Acoustic-Optical Underwater Wireless Sensor Networks Localization. *Sensors* **2017**, *18*, 51. [CrossRef]
10. Huang, Q.; Zhu, Y. Printing Conductive Nanomaterials for Flexible and Stretchable Electronics: A Review of Materials, Processes, and Applications. *Adv. Mater. Technol.* **2019**, *4*, 1800546. [CrossRef]
11. Khan, S.; Lorenzelli, L.; Dahiya, R.S. Technologies for Printing Sensors and Electronics Over Large Flexible Substrates: A Review. *IEEE Sens. J.* **2015**, *15*, 3164–3185. [CrossRef]
12. PrintoCent. *PrintoCent Introduction to Printed Intelligence: Handbook for Technology Training and Coaching*, 2nd ed.; Rousu, S., Särkijärvi, K., Prescott, D., Eds.; PrintoCent: Oulu, Finland, 2019; ISBN 978-951-597-179-1.
13. Vishwakarma, V.; Waghela, C.; Jain, A. Measurement of Out-of-Plane Thermal Conductivity of Substrates for Flexible Electronics and Displays. *Microelectron. Eng.* **2015**, *142*, 36–39. [CrossRef]
14. Mathews, I.; Kantareddy, S.N.; Buonassisi, T.; Peters, I.M. Technology and Market Perspective for Indoor Photovoltaic Cells. *Joule* **2019**, *3*, 1415–1426. [CrossRef]
15. Demirci Sankir, N.; Sankir, M. (Eds.) *Printable Solar Cells*; John Wiley & Sons, Inc.: Hoboken, NJ, USA, 2017; ISBN 978-1-119-28373-7.
16. Green, M.A.; Dunlop, E.D.; Hohl-Ebinger, J.; Yoshita, M.; Kopidakis, N.; Hao, X. Solar Cell Efficiency Tables (Version 58). *Prog. Photovolt. Res. Appl.* **2021**, *29*, 657–667. [CrossRef]

17. Cui, Y.; Wang, Y.; Bergqvist, J.; Yao, H.; Xu, Y.; Gao, B.; Yang, C.; Zhang, S.; Inganäs, O.; Gao, F.; et al. Wide-Gap Non-Fullerene Acceptor Enabling High-Performance Organic Photovoltaic Cells for Indoor Applications. *Nat. Energy* **2019**, *4*, 768–775. [CrossRef]
18. Saeed, M.A.; Kim, S.H.; Kim, H.; Liang, J.; Woo, H.Y.; Kim, T.G.; Yan, H.; Shim, J.W. Indoor Organic Photovoltaics: Optimal Cell Design Principles with Synergistic Parasitic Resistance and Optical Modulation Effect. *Adv. Energy Mater.* **2021**, *11*, 2003103. [CrossRef]
19. Verboven, I.; Deferme, W. Printing of Flexible Light Emitting Devices: A Review on Different Technologies and Devices, Printing Technologies and State-of-the-Art Applications and Future Prospects. *Prog. Mater. Sci.* **2021**, *118*, 100760. [CrossRef]
20. Wiklund, J.; Karakoç, A.; Palko, T.; Yiğitler, H.; Ruttik, K.; Jäntti, R.; Paltakari, J. A Review on Printed Electronics: Fabrication Methods, Inks, Substrates, Applications and Environmental Impacts. *JMMP* **2021**, *5*, 89. [CrossRef]
21. Irimia-Vladu, M.; Głowacki, E.D.; Voss, G.; Bauer, S.; Sariciftci, N.S. Green and Biodegradable Electronics. *Mater. Today* **2012**, *15*, 340–346. [CrossRef]
22. IEEE Standards Association. *IEEE Standard for Local and Metropolitan Area Networks—Part 15.7: Short-Range Wireless Optical Communication Using Visible Light*; IEEE: New York, NY, USA, 2011. [CrossRef]
23. Berman, S.M.; Greenhouse, D.S.; Bailey, I.L.; Clear, R.D.; Raasch, T.W. Human Electroretinogram Responses to Video Displays, Fluorescent Lighting, and Other High Frequency Sources. *Optom. Vis. Sci.* **1991**, *68*, 645–662. [CrossRef]
24. Wilkins, A.; Veitch, J.; Lehman, B. LED Lighting Flicker and Potential Health Concerns: IEEE Standard PAR1789 Update. In Proceedings of the 2010 IEEE Energy Conversion Congress and Exposition, Atlanta, GA, USA, 12–16 September 2010; pp. 171–178. [CrossRef]
25. Aliaberi, A.; Sofotasios, P.C.; Muhaidat, S. Modulation Schemes for Visible Light Communications. In Proceedings of the 2019 International Conference on Advanced Communication Technologies and Networking (CommNet), Rabat, Morocco, 12–14 April 2019; pp. 1–10. [CrossRef]
26. Protzman, J.B.; Houser, K.W. LEDs for General Illumination: The State of the Science. *LEUKOS* **2006**, *3*, 121–142. [CrossRef]
27. Haigh, P.A.; Ghassemlooy, Z.; Le Minh, H.; Rajbhandari, S.; Arca, F.; Tedde, S.F.; Hayden, O.; Papakonstantinou, I. Exploiting Equalization Techniques for Improving Data Rates in Organic Optoelectronic Devices for Visible Light Communications. *J. Lightwave Technol.* **2012**, *30*, 3081–3088. [CrossRef]
28. Tony, A.; Hiryanto, L. A Review on Energy Harvesting and Storage for Rechargeable Wireless Sensor Networks. *IOP Conf. Ser. Mater. Sci. Eng.* **2019**, *508*, 012120. [CrossRef]
29. Zhao, L.; Qian, J. DC-DC Power Conversions and System Design Considerations for Battery Operated System. Available online: https://www.ti.com/download/trng/docs/seminar/Topic%206%20-%20DC-DC%20Power%20Conversion%20and%20System%20Design%20Considerations%20for%20Battery%20Operated%20System.pdf (accessed on 28 November 2021).
30. Saif, M.B.; Xu, Y.; Hofmann, K. Comparison of Charge Pump and Boost Circuits for High Voltage & Low Power Voltage Conversions. In Proceedings of the 2016 International Symposium ELMAR, Zadar, Croatia, 12–14 September 2016; pp. 67–71.
31. Bianconi, E.; Calvente, J.; Giral, R.; Mamarelis, E.; Petrone, G.; Ramos-Paja, C.A.; Spagnuolo, G.; Vitelli, M. Perturb and Observe MPPT Algorithm with a Current Controller Based on the Sliding Mode. *Int. J. Electr. Power Energy Syst.* **2013**, *44*, 346–356. [CrossRef]
32. Loukriz, A.; Haddadi, M.; Messalti, S. Simulation and Experimental Design of a New Advanced Variable Step Size Incremental Conductance MPPT Algorithm for PV Systems. *ISA Trans.* **2016**, *62*, 30–38. [CrossRef] [PubMed]
33. Algarín, C.R.; Fuentes, R.L.; Castro, A.O. Implementation of a Cost-Effective Fuzzy MPPT Controller on the Arduino Board. *Int. J. Smart Sens. Intell. Syst.* **2018**, *11*, 1–10. [CrossRef]
34. Aouchiche, N.; Ait Cheikh, M.S.; Becherif, M.; Ebrahim, M.A.; Hadjarab, A. Fuzzy Logic Approach Based Mppt for the Dynamic Performance Improvement for PV Systems. In Proceedings of the 2017 5th International Conference on Electrical Engineering—Boumerdes (ICEE-B), Boumerdes, Algeria, 29–31 October 2017; pp. 1–7.
35. Thilakarathna, D.B.; Dissanayake, D.R.R.K.; Dissanayake, D.M.K.V.B.; Ekanayake, M.P.B.; Godaliyadda, G.M.R.I.; Ekanayake, J.B. A Comparison between MPPT Algorithm and DC-DC Convertor Configurations for PV Systems. In Proceedings of the 2021 IEEE 10th International Conference on Information and Automation for Sustainability, Negambo, Sri Lanka, 11–13 August 2021.
36. AEM10941 Solar Harvesting | Photovoltaic Energy Harvesting. Available online: https://e-peas.com/product/aem10941/ (accessed on 17 October 2021).
37. Wu, T.; Arefin, M.S.; Redoute, J.-M.; Yuce, M. A Solar Energy Harvester with an Improved MPPT Circuit for Wearable IoT Applications. In Proceedings of the 11th International Conference on Body Area Networks, Turin, Italy, 15–16 December 2017. [CrossRef]
38. Li, Y.; Grabham, N.J.; Beeby, S.P.; Tudor, M.J. The Effect of the Type of Illumination on the Energy Harvesting Performance of Solar Cells. *Sol. Energy* **2015**, *111*, 21–29. [CrossRef]
39. Neupane, R. Study on the Validity of Wien's Displacement Law on Tungsten Bulb. *J. Nep. Phys. Soc.* **2020**, *6*, 85–96. [CrossRef]
40. Energy Density vs Power Density—Energy Education. Available online: https://energyeducation.ca/encyclopedia/Energy_density_vs_power_density (accessed on 17 October 2021).
41. Significance of the RC Time Constant, or Tau, for Super Capacitors—Technical Support Knowledge Center Open. Available online: https://edadocs.software.keysight.com/kkbopen/significance-of-the-rc-time-constant-or-tau-for-super-capacitors-598781345.html (accessed on 17 October 2021).

42. Kilani, D.; Mohammad, B.; Saleh, H.; Ismail, M. LDO Regulator versus Switched Inductor DC-DC Converter. In Proceedings of the 2014 21st IEEE International Conference on Electronics, Circuits and Systems (ICECS), Marseille, France, 7–10 December 2014; pp. 638–641.
43. Ramos, J.S.; Demirkol, I.; Paradells, J.; Vossing, D.; Gad, K.M.; Kasemann, M. Towards Energy-Autonomous Wake-up Receiver Using Visible Light Communication. In Proceedings of the 2016 13th IEEE Annual Consumer Communications & Networking Conference (CCNC), Las Vegas, NV, USA, 6–13 January 2016; pp. 544–549.
44. Kozłowski, A.; Sosnowski, J. Energy Efficiency Trade-Off between Duty-Cycling and Wake-Up Radio Techniques in IoT Networks. *Wirel. Pers. Commun.* **2019**, *107*, 1951–1971. [CrossRef]
45. Strobel, N.; Droseros, N.; Köntges, W.; Seiberlich, M.; Pietsch, M.; Schlisske, S.; Lindheimer, F.; Schröder, R.R.; Lemmer, U.; Pfannmöller, M.; et al. Color-Selective Printed Organic Photodiodes for Filterless Multichannel Visible Light Communication. *Adv. Mater.* **2020**, *32*, 1908258. [CrossRef]
46. Kang, J.; Koo, Y.; Ha, J.; Lee, C. 41-2: Invited Paper: Recent Developments in Inkjet-printed OLEDs for High Resolution, Large Area Applications. *SID Symp. Dig. Tech. Pap.* **2020**, *51*, 591–594. [CrossRef]
47. Wang, Y.-F.; Sekine, T.; Takeda, Y.; Yokosawa, K.; Matsui, H.; Kumaki, D.; Shiba, T.; Nishikawa, T.; Tokito, S. Fully Printed PEDOT:PSS-Based Temperature Sensor with High Humidity Stability for Wireless Healthcare Monitoring. *Sci. Rep.* **2020**, *10*, 2467. [CrossRef]
48. Perera, A.; Katz, M.; Godaliyadda, R.; Hakkinen, J.; Strommer, E. Light-Based Internet of Things: Implementation of an Optically Connected Energy-Autonomous Node. In Proceedings of the 2021 IEEE Wireless Communications and Networking Conference (WCNC), Nanjing, China, 29 March 2021; pp. 1–7. [CrossRef]

sensors

MDPI

Article

Energy Autonomous Wireless Sensor Nodes for Freight Train Braking Systems Monitoring

Federico Zanelli *, Marco Mauri, Francesco Castelli-Dezza, Edoardo Sabbioni, Davide Tarsitano and Nicola Debattisti

Department of Mechanical Engineering, Politecnico di Milano, 20156 Milan, Italy; marco.mauri@polimi.it (M.M.); francesco.castellidezza@polimi.it (F.C.-D.); edoardo.sabbioni@polimi.it (E.S.); davide.tarsitano@polimi.it (D.T.); nicola.debattisti@polimi.it (N.D.)

* Correspondence: federico.zanelli@polimi.it; Tel.: +39-0223998407

Abstract: Nowadays, railway freight transportation is becoming more and more crucial since it represents the best alternative to road transport in terms of sustainability, pollution, and impact on the environment and on public health. Upgrading the potentiality of this kind of transportation, it would be possible to avoid delays in goods deliveries due to road accidents, traffic jams, and other situation occurring on roads. A key factor in this framework is therefore represented by monitoring and maintenance of the train components. Implementing a real time monitoring of the main components and a predictive maintenance approach, it would be possible to avoid unexpected breakdowns and consequently unavailability of wagons for unscheduled repair activities. As highlighted in recent statistical analysis, one of the elements more critical in case of failure is represented by the brake system. In this view, a real time monitoring of pressure values in some specific points of the system would provide significant information on its health status. In addition, since the braking actions are related to the load present on the convoy, thanks to this kind of monitoring, it would be possible to appreciate the different behavior of the system in case of loaded and unloaded trains. This paper presented an innovative wireless monitoring system to perform brake system diagnostics. A low-power system architecture, in terms of energy harvesting and wireless communication, was developed due to the difficulty in applying a wired monitoring system to a freight convoy. The developed system allows acquiring brake pressure data in critical points in order to verify the correct behavior of the brake system. Experimental results collected during a five-month field test were provided to validate the approach.

Keywords: wireless sensor node; MEMS; energy harvesting; pressure measurements; brake system; freight transportation; monitoring; predictive maintenance

Citation: Zanelli, F.; Mauri, M.; Castelli-Dezza, F.; Sabbioni, E.; Tarsitano, D.; Debattisti, N. Energy Autonomous Wireless Sensor Nodes for Freight Train Braking Systems Monitoring. *Sensors* **2022**, *22*, 1876. https://doi.org/10.3390/s22051876

Academic Editors: Carlo Trigona, Olfa Kanoun and Slim Naifar

Received: 28 January 2022
Accepted: 21 February 2022
Published: 27 February 2022

1. Introduction

As highlighted in recent EU-funded research projects, the performances of railway freight transportation have increased in the last few years and the projection is that the trend will continue in the future because it represents the best alternative from the sustainability point of view with respect to road freight transportation, which instead has a huge impact on the environment [1–4].

However, rail freight transport is still affected by issues that limit its usage and efficiency. An example of this is represented by derailments that, although decreasing in number in the last few years, still occur on tracks around the world, leading to a decline in confidence in this transportation mode [5].

The growing idea in the field is therefore to improve the reliability and performances of these vehicles by implementing smart and innovative technologies useful to carry out continuous monitoring of the main crucial components and to apply predictive maintenance logistics [6]. In this way, it could be possible to avoid expensive maintenance interventions and higher costs due to the unavailability of the wagon for operations. However, limited data have been collected so far on these vehicles for predictive maintenance purposes because of lack of appropriate instrumentation. Monitoring systems suitable to the purpose must in fact be characterized by low costs and easy installation on the wagon, by impacting the re-approval procedure as little as possible. Moreover, due to the lack of any power sources on board freight wagons, the use of energy harvesting techniques is necessary to supply the installed instrumentation [7].

A recurrent failure with a high impact on reliability and service is related to the braking system in which, if a malfunction occurs, it results in complexity in finding the cause of the fault within the system. In addition to vehicle failures, a possible common consequence of the malfunctioning of the braking system is the onset of wear on the rolling surface (facets). The presence of facets implies a damaging action of the traveling wagon on the infrastructure itself [8,9].

For these reasons, the braking system plays a fundamental role in traffic safety, and it has a great influence on the occurrence of anomalies on other components of the mechanical system (such as running gear, traction systems, etc.). This, therefore, translates into greater costs in case of breakdown and in the difficulty of identifying anomalies during routine maintenance operations on this type of vehicle. The problem has been recently widely studied more from the numerical point of view through the development of models able to reproduce the air brake system behavior [10,11] than from the experimental side, where few examples of experimental tests on the field can be found [12]. Clearly, the possibility of monitoring the braking system would represent a smart solution since it is present almost on all the wagons and would allow the recording of the braking actions according to the load and the line traveled.

The air brake system of a freight train has a very similar structure among the various models present on the market. Every freight wagon is equipped with a brake main pipe, which connects the locomotive to the wagons. Downstream of the brake main pipe, it is connected the control valve which manages the auxiliary tank, the control tank, and the brake cylinder. When the driver activates the brake system, a pressure variation occurs in the main brake pipe because it is put in communication with the atmosphere or with the main tank.

Another fundamental component of the brake system is represented by the weighing valve mounted on every bogie. The weighing valve allows regulation of the pressure of the brake cylinder by supplying a pressure value proportional to the load carried on the wagon. The pressure generated in the brake cylinder is almost proportional to the pressure drop in the main brake pipe and the performance characteristic depending on the transfer curve of the control valve. The control valve's main task is to compare the pressure value in the main brake pipe with the control tank one. If a pressure drop in the main brake pipe is observed, the control valve put in communication the auxiliary tank with the brake cylinder and the braking phase is started. Moreover, the control valve has the role to guarantee the braking and release time according the UIC540 standard, which are different in case the braking mode is set to P (passenger train) or G (freight train) [13]. It is therefore evident that the monitoring of pressure trends in some crucial point of the braking system allows the identification of eventual malfunctioning of its main components and a smart solution for the monitoring of the braking system of freight wagons, as was proposed by [13].

In [13], the temperature of brake blocks and the pressure variations was measured in some key points of the braking plant, and the significance of the collected data and the efficiency of the developed monitoring systems was demonstrated. However, the main issues of this monitoring apparatus were represented by the presence of wired sensors and by the limited autonomy due to the lack of any energy harvesters. The previous problems underline that to improve system reliability it is necessary to install a smart monitoring system on wagons to perform predictive maintenance of the braking system. This kind of apparatus should take advantage of wireless communication due to the unfeasibility of wirings on this kind of vehicles and of energy harvesting since no power supply is nowadays present on freight wagons. Moreover, measurements should be performed in peculiar points of the system useful to identify possible failures of the main components, without undermining the safety of the braking plant.

Regarding the wireless communication, it has successfully employed on several kinds of monitoring activities on different mechanical and civil structures [14–21], while its use on rail vehicles is not common due to the presence of a huge quantity of iron that can affect the quality of transmission. The durability of this kind of sensor is also a concern because this monitoring activity surely represents a harsh scenario in terms of environmental actions. An example can be represented by the ballast lifting from the track bed that can impact instrumentation installed on the bogie [22].

The need for a low-energy consumption of the system is a priority in this context, and it can be reached both by adopting ultra-low-power electronics component and optimizing the acquisition cycles. The introduction of a low-consumption status when the acquisition is not running is of paramount importance to save unneeded power.

In the end, many kinds of energy harvesters have been tested and implemented in recent years to be used in monitoring systems with the aim to recharge batteries when an energy excess is produced [23–27]. Among them, solar energy is seen as the most reliable and widely available in outdoor scenarios [28,29]. It can be found in literature that the implementation of photovoltaic panels used to harvest solar energy on wireless sensor nodes resulted in successful application of vibration monitoring of a freight wagon [30].

For these reasons, an innovative system composed by wireless sensor nodes endowed of pressure transducers was developed to be used for the monitoring of pressure variation inside the braking system of a freight wagon. Sensor nodes were designed to be energetically autonomous thanks to the choice of low-power electronic components and to the development of an optimized state machine able to minimize energy consumptions. The developed system was then installed on a freight wagon for a 5-month field test. Sensor nodes were connected to crucial points of the braking plant in order to perform pressure measurements useful for continuous monitoring of the system.

The paper is organized as follows. Section 2 describes the design of the wireless sensor node, both from the hardware and software point of view. A laboratory calibration of the developed device is shown in Section 3, while Section 4 is dedicated to outline the field test and the arrangement of the experimental set-up. In the end, in Section 5, we propose an analysis of the experimental data collected with a focus on the consistency of the acquired data, on the energy harvester performance, and on the overall efficiency of the monitoring system. Finally, it is shown how the collected data can be used to pursue a predictive maintenance approach able to identify possible faults in the brake system.

2. Sensor Description

The developed pressure wireless sensor node was designed to acquire brake system pressures over time and to perform some preliminary calculations in order to minimize the transfer of data reducing the power consumption. Indeed, the power supply represents a critical issue in this application due to the lack of train power line on freight convoys and the complexity of installing cables on existing wagon. For this last reason, wireless communication is also an essential element to overcome these limits, assuring a proper data transfer while keeping a low power consumption.

In this application, the sensor nodes were developed starting from the design of vibration sensor nodes described in [31]. The power source of the sensor nodes is essentially provided by a Lithium Polymer (Li-Po) battery and by a mini photovoltaic (PV) panel able to recharge the battery when sufficiently illuminated.

The wireless communication was based on Bluetooth Low Energy 5.0 (BLE) stack, which allows to ensure a good communication range (up to 200 m in line of sight) with very low power consumption (10–500 mW) [32,33] (Figure 1).

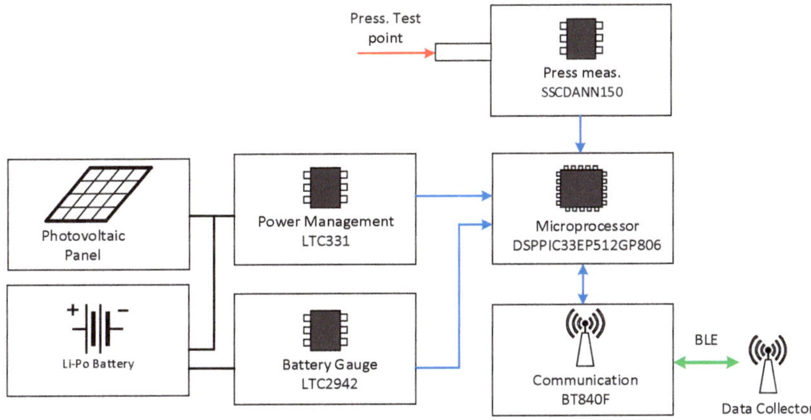

Figure 1. Hardware architecture.

2.1. Hardware

From the hardware side, the main innovation with respect to the sensor architecture described in [31] is represented by the implementation of a Micro Electro Mechanical Systems (MEMS) pressure transducer, namely the Honeywell SSCDANN150PAAA3. Among the different models present on the market, this device was chosen for various reasons. The selection of this model was carried out by taking into account the requirements of the monitoring activity and the features of the developed electronic board.

Firstly, the pressure range typical of this application is 0–8 Bar less than the full-scale range of this sensor (150 psi approximately 10,34 Bar). Another feature of this model is that it is an absolute pressure transducer, translating into a more robust solution for this context useful in obtaining reliable diagnostic data.

The single-axial barbed port of the sensor has the dimensions suitable to host a 4-mm-diameter plastic tube, which allows a safe connection to the wagon test points. Moreover, the I2C analog interface and the 3.3 V supply voltage are essential features for the transducer to be hosted on the developed sensor node. A crucial aspect of the implemented transducer is represented by the current consumption, as already pointed out previously. The chosen analog version of the pressure transducer has a current consumption of 2.1 mA while the serial interface version absorbs approximately 3.1 mA. In the end, another interesting feature is constituted by the power supply. In this case, the transducer is power supplied directly by a digital pin of the microcontroller. Therefore, in this way, the sensor is supplied only when a measure is needed. The measurement procedure takes some milliseconds as is shown in the following, while during the remaining time the current consumption of the pressure transducer is zero.

The main features of the chosen transducer are summed up in Table 1.

Table 1. Pressure sensor characteristics.

SSCDANN150PAAA3 Characteristics	
Sensor Type	Absolute
Pressure Range	0–150 Psi
Interface	Analog (0.33 V~2.97 V)
Resolution	12 bit (internal ADC)
Operating Temperature	0–50 °C
Supply Voltage	3.3 V
Current consumption	2.1 mA

The board's main power source comes from the solar energy harvested by means of a mini-PV panel, whose main features are reported in Table 2. This energy harvester was chosen to optimize the ratio between compact dimensions (suitable to be host on the top of the sensor board) and output power necessary to recharge the Li-Po battery. The PV panel is connected to the LTC3331 component that performs on board power management. This component implements a Buck-Boost DC/DC converter with an Energy Harvesting Battery Charger, which manages the battery charge using solar energy providing a stabilized voltage supply for the microprocessor and for all on board sensors. The LTC3331 chip does not allow a Maximum Power Point Tracking (MPPT), hence it has been used with the internal hysteretic controller with a proper selection of the voltage thresholds. In order to continuously monitor the sensor node battery status, a LTC2942 battery gauge was inserted. This IC (Integrated Circuit) perform the estimation of the battery State of Charge (SoC) and measures the battery voltage and temperature.

Table 2. Photovoltaic panel characteristics.

PV SEEED 313070004 Characteristics	
Material	Monocrystalline Sylicon
Typical peak power	0.55 W
Open circuit voltage power	8.2 V
Voltage at peak power	5.5 V
Current at peak power	100 mA
Length	70 mm
Width	55 mm
Weight	17 g
Efficiency	17%

The realized electronic board prototypes are enclosed in 3D printed boxes realized in CPE (Chlorinate Polyethylene), whose characteristics are suitable for the harsh railway environment. The final dimensions of the sensor node box are 91 × 70 × 61 mm, with a weight of 445 g (inclusive of insulating gel filling the box). The hardware functional scheme of the sensor is shown in Figure 1 and a photo of the realized sensor node prototype is shown in Figure 2.

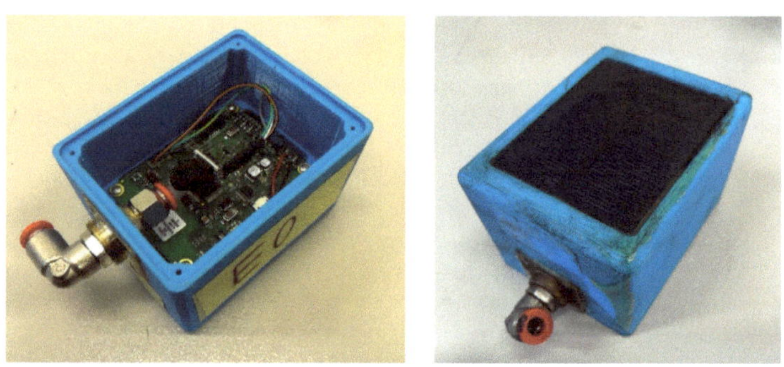

(**a**) (**b**)

Figure 2. (**a**) Internal view of the sensor node; (**b**) external view of the sensor node showing the PV panel installed on the top of the box.

2.2. *Software*

The sensor node firmware, implemented on the DSPIC33EP512 microcontroller and on the BT840F module, was written with a special attention to the power consumption minimization, which is a crucial aspect of the design. Despite the presence of the mini photovoltaic panel, it is important to save energy due to the difficulty in obtaining a good sun exposure when the sensor is mounted on wagon, or when the PV panels are shielded for any reason during travel (e.g., in long tunnels or underground sections of the railway) in order to acquire data continuously. The software architecture and functionalities are represented in Figure 3.

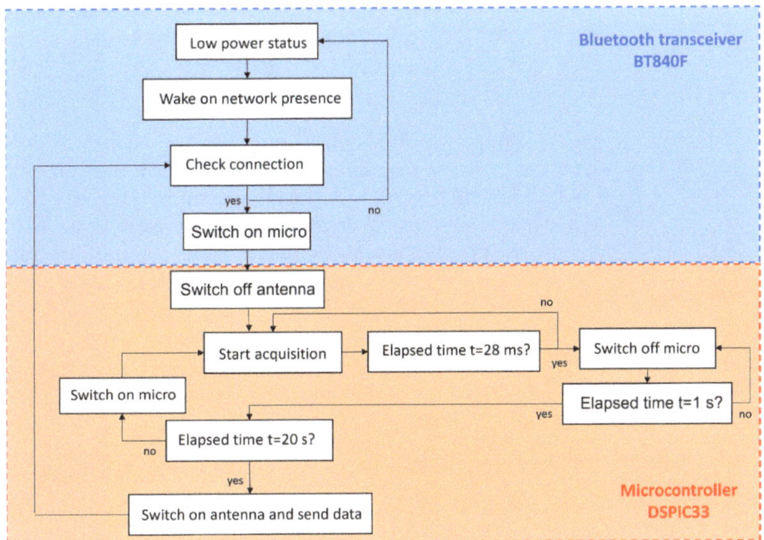

Figure 3. Software architecture and functionalities. The software is implemented on two microcontrollers: BT840F (blue part) and DSPIC33 (red part).

According to Figure 3, the acquisition software is implemented on two microcontrollers: the BT840F (which controls the communication) and the DSPIC33 (which controls the acquisition). The acquisition procedure starts when the sensor node detects the presence of the data collector gateway on the same BLE sub-network. In order to simplify the connection between the sensor nodes and the data collector, the sensor node was not paired with the data collector but just starts to communicate when the gateway subnetwork is discovered. When this occurs, the BT840F turns off the antenna and wakes up the DSPIC33. The DSPIC33 starts to acquire pressure data continuously and a pressure measure is obtained every second (1 Hz) as a mean over 4 pressure samples at 200 Hz. This mean measures are stored in a buffer for a duration of 20 s. These acquisition parameters (1 Hz sampling frequency and 20 s of acquisition) were selected according to the transient pressure dynamics in the brake system [10,11]. The acquisition phase of a pressure measure lasts approximately 28 ms each second (4×5 ms = 20 ms for the acquisition phase +8 ms for other operations) and the microcontroller goes to sleep after this period and it automatically wakes up every second in order to collect the next sample. Data transmission of the entire buffer will occur before the end of 20 s and it takes 3 s approximately. In this way, the antenna is always turned off and it is switched on only when data have to be sent to the gateway. This procedure allows a huge saving in energy consumption since the transmission phase is the most power consuming. An estimation of sensor node power consumption could be obtained using the absorbed currents in Table 3 and using (1).

Table 3. Current consumption.

	Pressure Sensor (I_p)	Micro (I_μ)	Sleep Phase (I_s)	Commun. (I_a)	Total (I_{cyle})
Mean curr. (mA)	1.7	7.24	0.69	8	3.72
Time (s)	20	0.020 every 1 s	0.98 every 1 s	3	20

The proposed sensor nodes were equipped with a 2000 mAh Li-Po battery, so according to the estimated absorbed current, every node is able to work continuously more than two months (67 days) without any inflow current coming from the PV panel considering an active time of 8 h a day.

$$I_{mean} = \frac{I_p \cdot t_p + I_\mu \cdot t_\mu + I_s \cdot t_s + I_a \cdot t_a}{T_{cycle}}. \tag{1}$$

Data transmitted by sensor nodes to the gateway every 20 s are constituted basically by the pressure measurements and the time when the acquisition has been performed, which is fundamental to correlate data during the postprocessing phase. In addition, sensor nodes transmit useful diagnostic data concerning the battery voltage, the environmental temperature, and an index of the quality communication between them and the data collector (RSSI value).

3. Sensor Calibration

A preliminary calibration of the pressure sensors was carried out in laboratory, taking advantage of the available instrumentation, in order to evaluate the sensor nodes performance and to characterize them. The instrument used to produce a pressure reference was a WIKA CPC6050 and the pressure range adopted as reference was −90–90 mBar with 10 mBar steps. The sensor node was connected to the output port of the reference instrument through a pneumatic tube, while measurement data were read and stored by means of serial communication with a laptop (Figure 4).

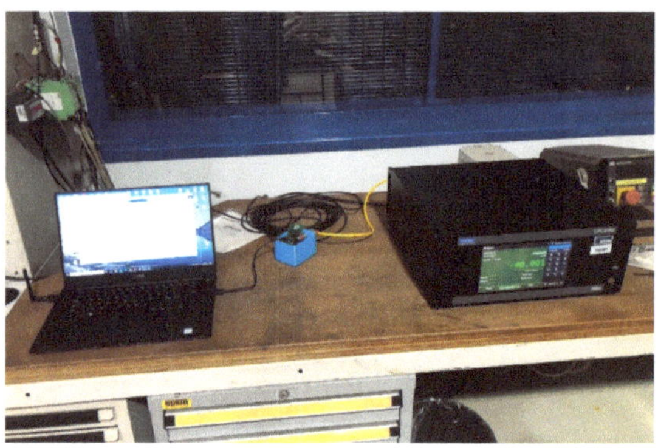

Figure 4. Calibration test bench.

The adopted calibration procedure involves the following steps. Firstly, a pressure difference Δp was imposed using the reference instrumentation. The reference instrumentation pressure P_{WIKA} (which includes a very accurate measurement of the atmospheric/barometric pressure) and the sensor node pressure P_{sens} were collected. This operation was repeated 19 times using a 10 mBar step in the −90–90 mBar pressure range. In the end, the offset P_{off} characterizing the sensor node measurement was computed according to (2):

$$p_{off} = \frac{\sum_{k=1}^{19} P_{WIKA}(\Delta p)_k - P_{sens}(\Delta p)_k}{19}. \tag{2}$$

The corrected sensor node pressure measurement p_s was then obtained through (3)

$$p_s(\Delta p) = p_{sens}(\Delta p) + p_{off}. \tag{3}$$

In Figure 5, an example of calibration data is shown.

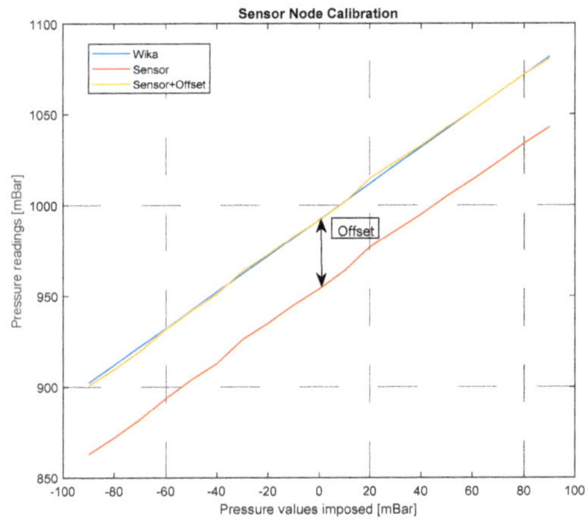

Figure 5. Example of calibration experimental data.

4. Field Campaign and Experimental Set-Up Description

The developed sensor nodes were tested during a 5-month field campaign (from 11 November 2020 to 20 April 2021) where it was possible to install them on a T3000e freight wagon, made available by Mercitalia Intermodal SpA (Figure 6). This kind of wagon is employed in very long travels for intermodal transportation of both trailers and containers. This situation represents clearly the perfect workbench for the braking system monitoring. The track sections and number of travels run by the instrumented wagon during the field test are summed up in Table 4.

Figure 6. Mercitalia Intermodal T3000e freight wagon.

As in the case of other railway vehicles, each T3000e wagon is endowed of some test points placed in suitable position of the braking system that are used during maintenance interventions to perform tests on the braking system efficiency. It seemed the best choice to take advantage of these test points to mount the sensor nodes on the wagon.

Table 4. Travels during field test.

Track Section	Estimated Distance (km)	Number of Travels
Milano-München	610	2
München-Milano	610	2
Verona-München	450	2
München-Verona	450	2
Verona-Rotterdam	1120	9
Rotterdam-Verona	1120	9
Total	24,400	26

The T3000e is composed of two semi-wagons. Three sensor nodes were mounted on the first semi-wagon on the three points of interest of the braking system, namely one on the Main Pipe (MBP), one on the braking cylinder (BC), and the last on the weighing valve (WV). The monitoring of the braking system in these points was chosen accordingly to the diagnostic need, allowing to verify if a wheel is braked when it should not have been. In addition, other diagnostic information concerning wheel flattening, wear of pads, and axle box overheating could be collected. During this experimental campaign commercial, wired sensors were installed on the same test points in the adjacent semi-wagon in order to use the collected data as benchmark. The wired pressure sensor characteristics are listed in Table 5. A scheme of the instrumented wagon is shown in Figure 7.

Table 5. Pressure sensor characteristics.

Wired Pressure Sensors Characteristics	
Sensor Type	Relative
Pressure Range	0–10 Bar
Precision	0.1%
Interface	CANOpen
Operating Temperature	−40–125 °C
Supply Voltage	8–32 V

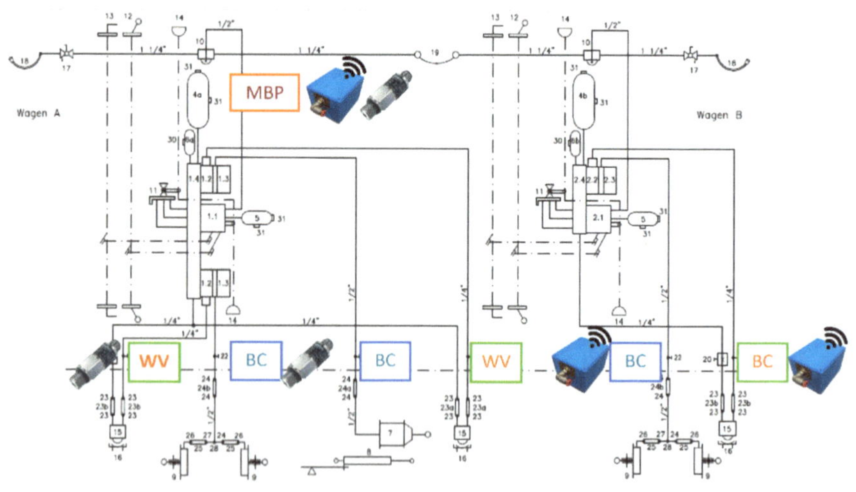

Figure 7. Position of sensor nodes on the Mercitalia Intermodal T3000e wagon.

4.1. Test Points and Sensor Node Connections

As already pointed out, it was chosen to take advantage of test points already present on-board the wagon to carry out pressure measurements. Usually during maintenance operations, the instrumentation is connected through tests points by means of the universal connector described in Appendices A.3 of UIC 543-1. Therefore, the idea was to replicate those connectors in order to link safely sensor nodes to test point during the field tests. An example of the test point for the weighing valve and the realized connection is shown in Figure 8.

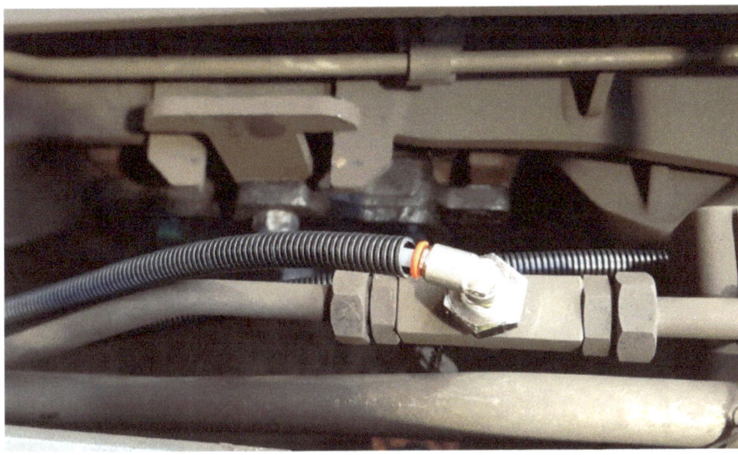

Figure 8. Mechanical connections.

The three sensor nodes were positioned on the external side of the wagon to guarantee a good sun exposure of the photovoltaic panels. Due to this external position, the sensor nodes were connected with pressure test points through the use of the previous described connectors and of plastic tubes appropriately shielded to avoid any failure related to debris impact. Sensor nodes connected to the braking cylinder (BC) and the weighing valve (WV) are visible in Figure 9a, while the one used to perform measurements on the main brake pipe (MBP) is shown in Figure 9b.

(a) (b)

Figure 9. Position of sensor nodes on Mercitalia Intermodal wagon (Figure 7). (**a**) Node BC and WV (**b**) Node MBP.

4.2. Data Collector and Gateway

Sensor nodes transmit their data wirelessly to a gateway data collector placed on the wagon (white box of Figure 10). The data collector also implements a 4G/LTE gateway to transfer the data on the cloud. This system consists of an industrial pc, a BLE master board to communicate with sensor nodes, a 4G/LTE modem, a GPS receiver, and a custom electronic board developed to manage the power supply of the entire system.

Figure 10. Data collector box on Mercitalia Intermodal wagon.

The gateway data collector's function is to receive data from the wireless nodes through a dedicated BLE board, store them locally, and send them remotely. The gateway is fed through a DC/DC converter and a Lead-Acid battery. This battery is recharged by an axle-box synchronous generator when the train overpasses 10 km/h. The developed custom electronic board was used to implement an energy saving logic based on the speed values acquired from the axle-box generator through CAN communication. Going into detail, the electronic board turns on the industrial PC and the devices connected (BLE board, 4G/LTE modem, ...) when a wagon movement is detected while it turns them off when the train does not move for enough time (approximately 5 min). In addition, to save energy and increase the system efficiency, this technique allows the gathering of only useful data when the train is in motion, limiting the amount of data stored on the PC and which is transmitted remotely. The power manager board, in addition, acquires and transfers to the PC via serial communication, and diagnostic data regarding the main status and the speed values come from the axle box.

The acquisition software is started automatically when the PC is active, and after the collection of the pressure measurements, it associates them with the GPS information in order to achieve a complete fleet position monitoring. A scheme of the described system is represented in Figure 11.

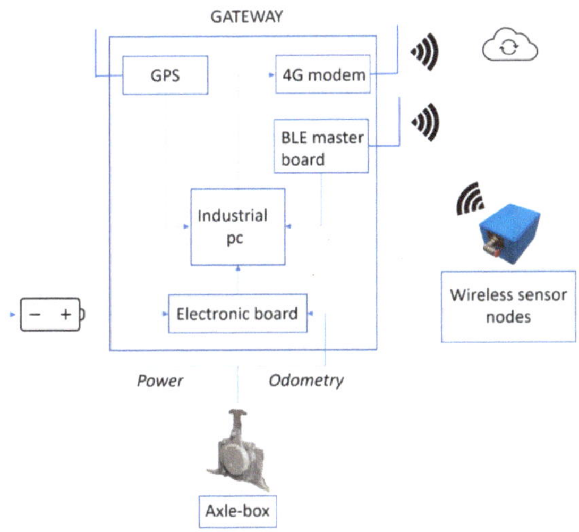

Figure 11. Data collector architecture.

5. Experimental Results and Data Analysis

During the field campaign, the instrumented wagon traveled mainly on the Verona-Rotterdam track section. This railway track nowadays represents a very important route crossing Europe for goods transportation (Figure 12).

Figure 12. Example of GPS track data during field experiments.

The field campaign allowed us to test the functioning of the monitoring system in a real case scenario and to collect a good amount of experimental data. Data acquired by the industrial wired sensors were used to validate the measurements performed by the developed wireless sensor nodes. In addition, during the post processing phase, the speed measurement and GPS signals acquired were taken advantage to superimpose the data carried out on different travels, verifying the measurements' repeatability. Thanks to this possibility, data comparison and superposition results are much more accurate. The availability of this information, moreover, made it possible to correlate the occurring of a braking event with the position of the wagon on the railway track. In the end, an analysis on the possibility of using the collected data to perform a predictive maintenance of the braking plant was carried out. Energy harvester performances during the field campaign were deeply analyzed in paragraph 5.2.

5.1. Validation of the Pressure Measurements Acquired by Sensor Nodes

In order to acquire a validation of the developed measurement devices, the first step was the comparison between data acquired by industrial sensors and the ones coming from the wireless sensor nodes. A good coherence between the two datasets can be observed looking at Figure 13, referring to the track section Verona-Rotterdam. In addition, it is possible to appreciate the braking events correlation in terms of pressure in the braking cylinder and train speed trends. As soon as the pressure in the braking cylinder increased, the train speed began to decrease as one would expect. The difference in values between measurements related to the weighing valve of the two systems can be explained by the fact that the two semi-wagons monitored were loaded differently, which often happens.

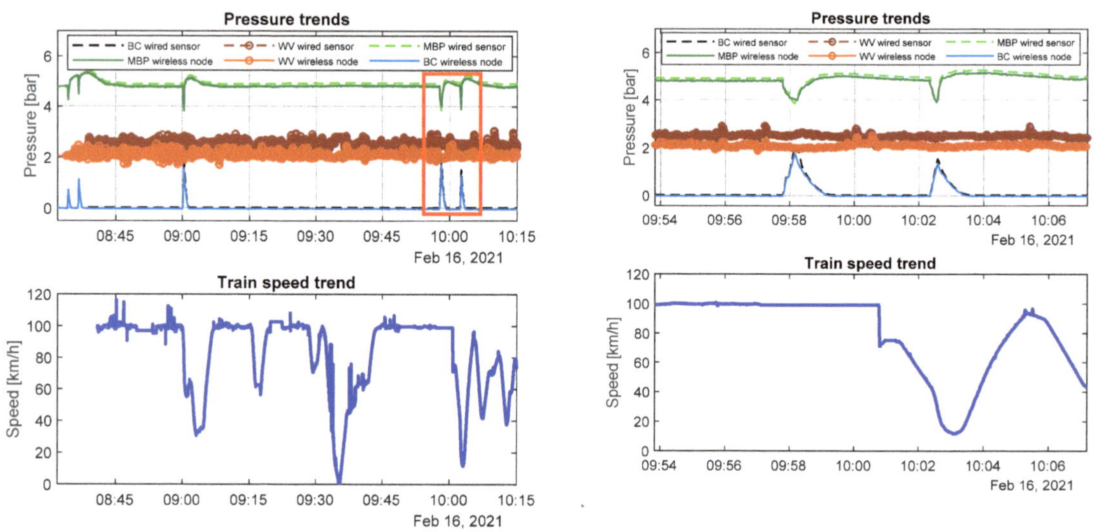

(a) (b)

Figure 13. Example of comparison of pressure data. (**a**) Pressure data during travel; (**b**) example of a braking event.

To complete the validation procedure, different travels on the same track section (Verona-Rotterdam) were superimposed to verify the measurements repeatability (Figure 14). Data acquired during different travels were realigned considering the speed measurement and GPS information and making the superimposition procedure straightforward. As can be seen, although braking phases can occurr in different situations from time to time, there was a good coherence between braking events performed in different travels, and also speed trends were quite repeatable from travel to travel. This analysis could also be useful to assess the different driving styles used by machinists in various travels on the same track section.

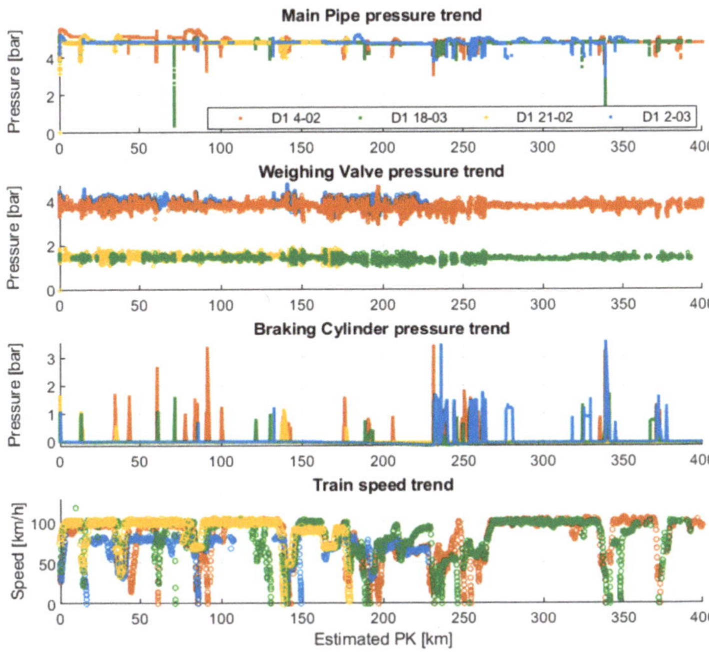

Figure 14. Example of collected pressure data during four different travels (4 February, 21 February, 2 March, 18 March) on the Verona-Rotterdam track.

5.2. Analysis of the Energy Harvester Performances

An aspect of paramount importance to be carefully analyzed during field tests on wireless sensor nodes is the efficiency of the implemented energy harvesting system. This aspect represents a key feature of the developed device since it allows us to make sensor nodes almost independent from an energetic point of view by any wired power supply. Laboratory tests are surely useful to check, in a preliminary stage, that the developed board works properly on the electrical side. However, it is nearly impossible to reproduce, in a confined environment, all the possible threats that can undermine the efficiency of the harvester (cloudy weather, tunnels, dust, debris impact, presence of few solar hours, temperature variation). Only field tests can therefore assure that the chosen harvester is suitable for the monitoring purposes of the specific application (in terms of scenario and duty cycle). Some data collected during the field tests are analyzed in the following.

In most cases, the balance between the inflow energy coming from the PV panel and the very low consumption offered by the implemented acquisition logic and by the use of ultra-low-power electronic components allows sensor nodes to perform continuous measurements without discharging batteries. In Figure 15, the battery voltage time history of the three sensor nodes during a travel on the Verona-Rotterdam track section is represented as example of the performance. The data were acquired after 4 months of experimentation, and so it is also possible to see the aging effect on batteries.

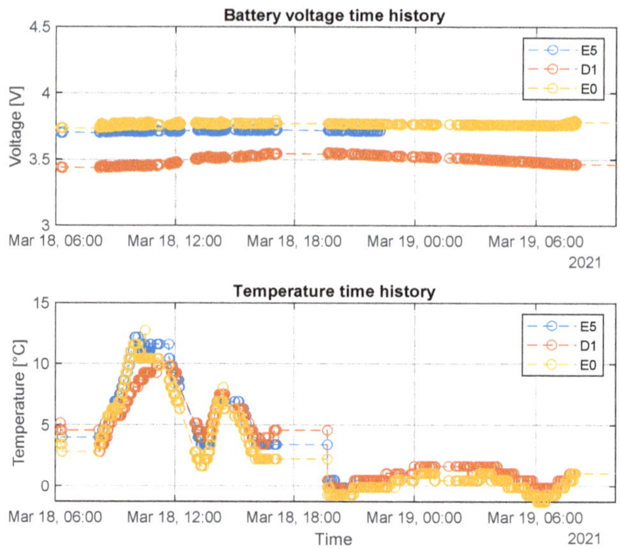

Figure 15. Example of battery voltage and temperature data.

In particular, the voltage trend of the three sensors kept almost constant during several hours of working activities, witnessing the efficiency of the power consumption logic implemented and of the energy harvester employed. The blank voids are due to the fact that, as already pointed out, when the train stops for a certain amount of time (that happens in case of certain crossings or stations), the whole system is put to sleep to save energy. However, some limitations of the energy harvester employed came out for this application. An example of this situation is visible in Figure 15.

The battery voltage curve of sensor nodes E5 and E0 (blue and red curve respectively) showed a steep decreasing trend during night hours, reaching values under the 3V threshold. This condition represents almost the limit for sensor nodes to stay awake and active for acquisition purposes. The reason for this situation can be found in some limitations of the implemented energy harvesters for the specific application. In particular, the main cause could be that, from the experience collected in the field test, freight trains travel very often at night, which translates in the system operating clearly with no sunny conditions. If the train is made to stop in railway depots, the PV panels could be hidden by adjacent convoys or by the structure of the depot itself. Therefore, when the instrumented convoy begins a new journey, the sensor nodes might not be sufficiently charged to run acquisitions, at least initially especially after some months of installation. Other factors that, resulting from the experimental activity, can undermine the PV panels' efficiency are the presence of very long tunnels on the traveled track sections and the presence of dust covering the external layer of panels. Anyway, as can be appreciated from Figure 16, as soon as the sunlight hits the PV panels, the developed electronics were able to recharge the batteries in an efficient way, allowing the sensor nodes to perform their acquisition task with aged batteries.

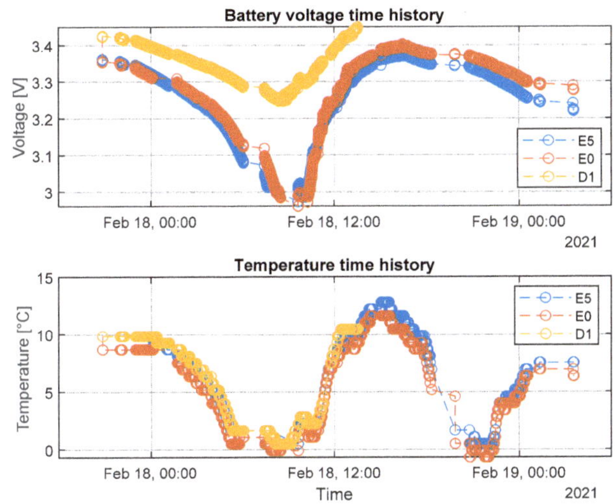

Figure 16. Example of battery voltage and temperature data.

5.3. Diagnostic Analysis

The collected data could be used for predictive maintenance of the braking systems. In particular, the pressure measures could be used to describe the state of health of the brake system. A complete pressure time-history related to a travel carried out on the Verona-Rotterdam is shown in Figure 17a. As can be observed from the pressure curves of main pipe and braking cylinder, several braking events occurred during the travel. The focus was then moved on to single braking (Figure 17b). It can be easily noted that the sampling frequency adopted for acquisition is enough to describe the braking pressure dynamics with good accuracy the.

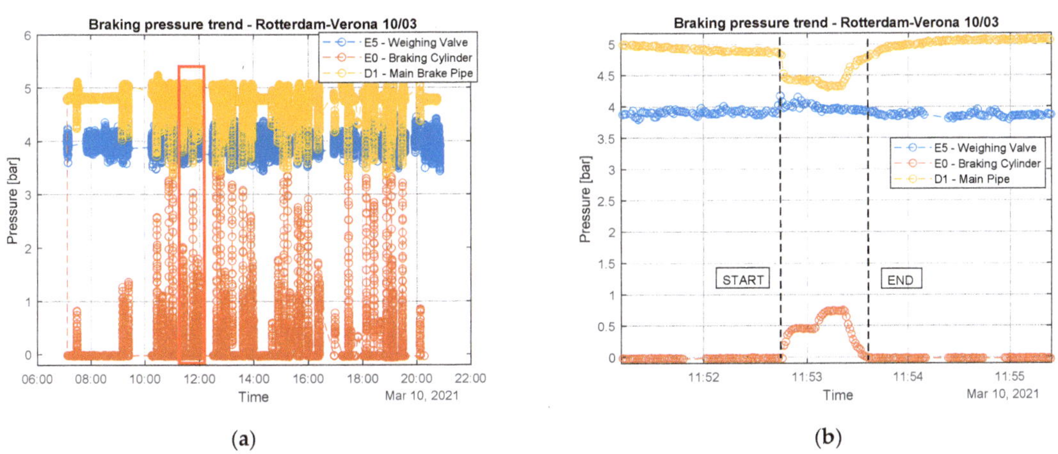

Figure 17. Pressure measures during a wagon travel (**a**) Pressure history (**b**) Pressure variation during a braking event.

Analyzing Figure 17a, it is possible to carry out some considerations regarding possible applications of a predictive maintenance approach [1]. Some approaches mainly based on machine learning [34,35] are already available in the literature, however the innovative solution proposed in this paper was based on the smart measurement of pressure values in some peculiar points of the systems to identify possible faults in the main components.

This way it could be possible to run simple and efficient algorithm directly on the on-board gateway in real time to perform these diagnostic activities. Some examples of this approach are proposed in the following.

Referring to Figure 17b, when the braking phase begins, a decreasing in pressure values at the main pipe was observed and there was a contemporary pressure increase in the braking cylinder. If this does not happen, it would mean that a fault on the monitored braking cylinder may have happened, and that bogie is not able to brake. Moreover, faulty conditions of the brake cylinder can be identified through the analysis of the duration of the braking phase, since its lasting in time is due to the type of braking mode selected before the travel (usually the selection is taken between freight mode and passenger mode though a specific mechanical leverage). If the time duration of the braking phase in not coherent with the selected mode (and prescribed by the UIC540 standard), some problems occurred.

Regarding the magnitudes of pressure, the one at braking cylinder is a function of the wagon load and it is managed by the weighing valve. Clearly, the higher the loading, the higher the brake force requested. This higher braking request requires higher pressure values at the braking cylinder. Therefore, monitoring the weighing valve pressure and the brake cylinder pressure, it is possible to understand if there is a malfunctioning of the weighing valve component. In addition, the monitoring of the weighing valve pressure can be used to distinguish the loaded or unloaded wagon condition. To appreciate the effectiveness of this method, two journeys with similar train speed profiles were taken into account. Looking at Figure 18, it is possible to underline that the braking phases of the unloaded wagon on travel 18th of March were less severe than the ones performed by the loaded wagon during travel 4th of February. In conclusion, it could be possible to run a diagnostic algorithm based on the previous considerations in real time on the on-board PC in order to perform continuous monitoring of the system useful for a predictive maintenance approach.

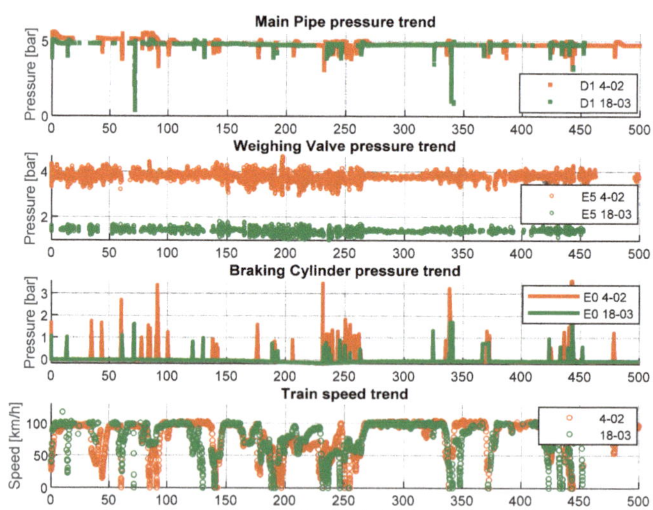

Figure 18. Comparison of braking events measures on different brake system points during two-wagon travel. Red lines refer to 4 February's journey (loaded wagon), and green lines refer to 18 March's journey (non-loaded wagon).

6. Conclusions

The development of an innovative monitoring system composed of wireless sensor nodes aimed at performing pressure measurements on freight train braking systems was proposed. The choice of the electronic components and the implementation of a smart firmware were carried out to minimize the power consumption. Thanks to this software implementation and to the adoption of a suitable energy harvester, the developed sensor nodes result as autonomous from an energetic point of view.

The designed system was installed on a freight wagon for a 5-month field test. The set-up was arranged to perform measurements in specific points of the plant, realizing safe mechanical connection to the existing vehicle brake system test points. During the field test, the wagon traveled more than 24,000 km on important track sections across Europe and the correct functioning of the system was verified.

In the post-processing phase, an initial focus was devoted to check the efficiency of the implemented energy harvester in recharging sensor nodes batteries and to verify the consistency of pressure measurements through the comparison with wired commercial sensors. The use of the acquired data for diagnostic purposes of the brake system was then validated through a deep analysis of the experimental data. It was demonstrated how a predictive maintenance approach can be employed taking advantage of the monitoring of the pressure dynamics in some peculiar points of the plant. Through the development of specific algorithms to be run on-board, the system makes it possible to identify the presence of faults in the braking system, reducing the cost and time related to unexpected breakdowns and increasing the reliability of this transportation system.

Author Contributions: Conceptualization, M.M., F.C.-D., E.S. and D.T.; methodology, F.Z., M.M., F.C.-D. and E.S.; software, F.Z., M.M., F.C.-D. and N.D.; validation, F.Z., M.M., F.C.-D. and E.S.; formal analysis, F.Z., M.M. and N.D.; investigation, F.Z., M.M. and N.D.; resources, F.C.-D., E.S. and D.T.; data curation, F.Z. and M.M.; writing—original draft preparation, F.Z.; writing—review and editing, F.Z. and M.M.; visualization, F.Z. and M.M.; supervision, F.C.-D., E.S. and D.T.; project administration, F.C.-D., E.S. and D.T.; funding acquisition, F.C.-D., E.S. and D.T. All authors have read and agreed to the published version of the manuscript.

Funding: This research received no external funding.

Institutional Review Board Statement: Not applicable.

Informed Consent Statement: Not applicable.

Data Availability Statement: Not applicable.

Acknowledgments: This work is part of the "Upgrade Carro Intelligente" project of the Joint Research Center Transportation, established by Fondazione Politecnico di Milano, carried out in collaboration with Mercitalia Intermodal (Polo Mercitalia—Gruppo FS Italiane).

Conflicts of Interest: The authors declare no conflict of interest.

References

1. Ulianov, C.; Hyde, P. Innovative Monitoring and Predictive Maintenance Solutions on Lightweight Wagon. 2017. 114ff. Available online: http://d-rail-project.eu/IMG/pdf/DR-D1-1-F1-Summary_Report_and_Database_of_derailment_incidents_20120405.pdf (accessed on 20 January 2022).
2. Budak, A.; Sarvari, P.A. Profit margin prediction in sustainable road freight transportation using machine learning. *J. Clean. Prod.* **2021**, *314*, 127990. [CrossRef]
3. De Campos, R.S.; Simon, A.T.; De Campos Martins, F. Assessing the impacts of road freight transport on sustainability: A case study in the sugar-energy sector. *J. Clean. Prod.* **2019**, *220*, 995–1004. [CrossRef]
4. Li, L.; Zhang, X. Reducing CO_2 emissions through pricing, planning, and subsidizing rail freight. *Transp. Res. D Transp. Environ.* **2020**, *87*, 102483. [CrossRef]
5. Robinson, P.M.; Scott, P.; Lafaix, B.; Kozyr, G.; Zarembski, A.; Vasic, G.; Franklin, F.; Gilmartin, B.; Schoebel, A.; Ripke, B. Development of the Future Rail Freight System to Reduce the Occurrences and Impact of Derailment- D-RAIL—Summary Report and Database of Derailments Incidents. 2011. Available online: https://projects.shift2rail.org/download.aspx?id=0a2c0725-8392-42b8-8a96-3cfb5361953f (accessed on 20 January 2022).
6. Bernal, E.; Spiryagin, M.; Cole, C. Onboard Condition Monitoring Sensors, Systems and Techniques for Freight Railway Vehicles: A Review. *IEEE Sens. J.* **2019**, *19*, 4–24. [CrossRef]
7. Gao, M.; Cong, J.; Xiao, J.; He, Q.; Li, S.; Wang, Y.; Yao, Y.; Chen, R.; Wang, P. Dynamic modeling and experimental investigation of self-powered sensor nodes for freight rail transport. *Appl. Energy* **2020**, *257*, 113969. [CrossRef]

8. Lan, Q.; Dhanasekar, M.; Handoko, Y.A. Wear damage of out-of-round wheels in rail wagons under braking. *Eng. Fail. Anal.* **2019**, *102*, 170–186. [CrossRef]
9. Sui, S.; Wang, K.; Ling, L.; Chen, Z. Effect of wheel diameter difference on tread wear of freight wagons. *Eng. Fail. Anal.* **2021**, *127*, 105501. [CrossRef]
10. Wu, Q.; Cole, C.; Spiryagin, M.; Chang, C.; Wei, W.; Ursulyak, L.; Shvets, A.; Murtaza, M.A.; Mirza, I.M.; Zhelieznov, K.; et al. Freight train air brake models. *Int. J. Rail Transp.* **2021**, 1–49. [CrossRef]
11. Natarajan, S.V.; Subramanian, S.C.; Darbha, S.; Rajagopal, K.R. A model of the relay valve used in an air brake system. *Nonlinear Anal. Hybrid Syst.* **2007**, *1*, 430–442. [CrossRef]
12. Grudén, M.; Hinnemo, M.; Dancila, D.; Zherdev, F.; Edvinsson, N.; Brunberg, K.; Andersson, L.; Byström, R.; Rydberg, A. Field operational testing for safety improvement of freight trains using wireless monitoring by sensor network. *IET Wirel. Sens. Syst.* **2014**, *4*, 54–60. [CrossRef]
13. Aimar, M.; Somà, A. Study and results of an onboard brake monitoring system for freight wagons. *Proc. Inst. Mech. Eng. F J. Rail Rapid Transit* **2018**, *232*, 1277–1294. [CrossRef]
14. Alves, M.M.; Pirmez, L.; Rossetto, S.; Delicato, F.C.; de Farias, C.M.; Pires, P.F.; dos Santos, I.L.; Zomaya, A.Y. Damage prediction for wind turbines using wireless sensor and actuator networks. *J. Netw. Comput. Appl.* **2017**, *80*, 123–140. [CrossRef]
15. Federici, F.; Alesii, R.; Colarieti, A.; Faccio, M.; Graziosi, F.; Gattulli, V.; Potenza, F. Design of wireless sensor nodes for structural health monitoring applications. *Procedia Eng.* **2014**, *87*, 1298–1301. [CrossRef]
16. Chew, Z.J.; Ruan, T.; Zhu, M. Strain Energy Harvesting Powered Wireless Sensor Node for Aircraft Structural Health Monitoring. *Procedia Eng.* **2016**, *168*, 1717–1720. [CrossRef]
17. Chae, M.J.; Yoo, H.S.; Kim, J.Y.; Cho, M.Y. Development of a wireless sensor network system for suspension bridge health monitoring. *Autom. Constr.* **2012**, *21*, 237–252. [CrossRef]
18. Nåvik, P.; Rønnquist, A.; Stichel, S. A wireless railway catenary structural monitoring system: Full-scale case study. *Case Stud. Struct. Eng.* **2016**, *6*, 22–30. [CrossRef]
19. Fedele, R.; Merenda, M. An iot system for social distancing and emergency management in smart cities using multi-sensor data. *Algorithms* **2020**, *13*, 254. [CrossRef]
20. Sofi, A.; Jane Regita, J.; Rane, B.; Lau, H.H. Structural health monitoring using wireless smart sensor network—An overview. *Mech. Syst. Signal Process.* **2022**, *163*, 108113. [CrossRef]
21. Diana, G.; Tarsitano, D.; Mauri, M.; Castelli Dezza, F.; Zanelli, F.; Manenti, A.; Ripamonti, F. A wireless monitoring system to identify wind induced vibrations in HV transmission lines. In Proceedings of the 3rd SEERC Conference, Vienna, Austria, 29 November–2 December 2021.
22. Zhu, J.Y.; Hu, Z.W. Flow between the train underbody and trackbed around the bogie area and its impact on ballast flight. *J. Wind Eng. Ind. Aerodyn.* **2017**, *166*, 20–28. [CrossRef]
23. Jin, L.; Deng, W.; Su, Y.; Xu, Z.; Meng, H.; Wang, B.; Zhang, H.; Zhang, B.; Zhang, L.; Xiao, X.; et al. Self-powered wireless smart sensor based on maglev porous nanogenerator for train monitoring system. *Nano Energy* **2017**, *38*, 185–192. [CrossRef]
24. Zuo, J.; Dong, L.; Ding, J.; Wang, X.; Diao, P.; Yu, J. Design and validation of a self-powered device for wireless electronically controlled pneumatic brake and onboard monitoring in freight wagons. *Energy Convers. Manag.* **2021**, *239*, 114229. [CrossRef]
25. Wang, P.; Wang, Y.F.; Gao, M.Y.; Wang, Y. Energy harvesting of track-borne transducers by train-induced wind. *J. Vibroeng.* **2017**, *19*, 1624–1640. [CrossRef]
26. Ahmad, I.; Hee, L.M.; Abdelrhman, A.M.; Imam, S.A.; Leong, M.S. Scopes, challenges and approaches of energy harvesting for wireless sensor nodes in machine condition monitoring systems: A review. *Measurement* **2021**, *183*, 109856. [CrossRef]
27. Wang, Z.; Wang, W.; Gu, F.; Wang, C.; Zhang, Q.; Feng, G.; Ball, A.D. On-rotor electromagnetic energy harvester for powering a wireless condition monitoring system on bogie frames. *Energy Convers. Manag.* **2021**, *243*, 114413. [CrossRef]
28. Miller, T.I.; Spencer, B.F., Jr.; Li, J.; Jo, H. *Solar Energy Harvesting and Software Enhancements for Autonomous Wireless Smart Sensor Networks*; Report No. NSEL-022; NSEL Report Series; Newmark Structural Engineering Laboratory: Urbana, IL, USA, 2010; pp. 1–63.
29. Ibrahim, R.; Chung, T.D.; Hassan, S.M.; Bingi, K.; Salahuddin, S.K.B. Solar Energy Harvester for Industrial Wireless Sensor Nodes. *Procedia Comput. Sci.* **2017**, *105*, 111–118. [CrossRef]
30. Cii, S.; Tomasini, G.; Bacci, M.L.; Tarsitano, D. Solar Wireless Sensor Nodes for Condition Monitoring of Freight Trains. *IEEE Trans. Intell. Transp. Syst.* **2020**, 1–13. [CrossRef]
31. Zanelli, F.; Castelli-Dezza, F.; Tarsitano, D.; Mauri, M.; Bacci, M.L.; Diana, G. Design and field validation of a low power wireless sensor node for structural health monitoring. *Sensors* **2021**, *21*, 1050. [CrossRef] [PubMed]
32. Yang, J.; Poellabauer, C.; Mitra, P.; Neubecker, C. Beyond beaconing: Emerging applications and challenges of BLE. *Ad Hoc Netw.* **2020**, *97*, 102015. [CrossRef]
33. Liu, L.S.; Ni, J.Q.; Zhao, R.Q.; Shen, M.X.; He, C.L.; Lu, M.Z. Design and test of a low-power acceleration sensor with Bluetooth Low Energy on ear tags for sow behaviour monitoring. *Biosyst. Eng.* **2018**, *176*, 162–171. [CrossRef]
34. Appoh, F.; Yunusa-Kaltungo, A.; Kumar Sinha, J. Hybrid adaptive model to optimise components replacement strategy: A case study of railway brake blocks failure analysis. *Eng. Fail. Anal.* **2021**, *127*, 105539. [CrossRef]
35. Wang, Q.; Gao, T.; Tang, H.; Wang, Y.; Chen, Z.; Wang, J.; Wang, P.; He, Q. A feature engineering framework for online fault diagnosis of freight train air brakes. *Measurement* **2021**, *182*, 109672. [CrossRef]

sensors

MDPI

Article

Residual Energy Estimation-Based MAC Protocol for Wireless Powered Sensor Networks

Sol-Bee Lee, Jung-Hyok Kwon and Eui-Jik Kim *

School of Software, Hallym University, 1 Hallymdaehak-gil, Chuncheon 24252, Gangwon-do, Korea; thfqla3535@hallym.ac.kr (S.-B.L.); jhkwon@hallym.ac.kr (J.-H.K.)
* Correspondence: ejkim32@hallym.ac.kr; Tel.: +82-33-248-2333

Abstract: This paper presents a residual energy estimation-based medium access control (REE-MAC) protocol for wireless powered sensor networks (WPSNs) composed of a central coordinator and multiple sensor devices. REE-MAC aims to reduce overhead due to control messages for scheduling the energy harvesting operation of sensor devices and provide fairness for data transmission opportunities to sensor devices. REE-MAC uses two types of superframes that operate simultaneously in different frequency bands: the wireless energy transfer (WET) superframe and wireless information transfer (WIT) superframe. At the beginning of each superframe, the coordinator estimates the change in the residual energy of individual sensor devices caused by their energy consumption and energy harvesting during the previous superframe. It then determines the devices' charging priorities, based on which it allocates dedicated power slots (DPSs) within the WET superframe. The simulation results demonstrated that REE-MAC exhibits superior performance for the harvested energy, average freezing time, and fairness to existing representative WPSN MAC protocols.

Keywords: energy harvesting; Internet of Things; out-of-band approach; residual energy estimation; wireless power transfer; wireless powered sensor network

Citation: Lee, S.-B.; Kwon, J.-H.; Kim, E.-J. Residual Energy Estimation-Based MAC Protocol for Wireless Powered Sensor Networks. *Sensors* **2021**, *21*, 7617. https://doi.org/10.3390/s21227617

Academic Editors: Slim Naifar, Olfa Kanoun and Carlo Trigona

Received: 19 October 2021
Accepted: 13 November 2021
Published: 16 November 2021

1. Introduction

Recent advances in radio frequency (RF)-based wireless energy transfer (WET) techniques enable battery-powered sensor devices to receive energy remotely without time and space constraints on ambient resources such as solar, thermal, wind, and vibration, enabling perpetual operations. Thus, wireless sensor networks (WSNs) with RF-based WET—wireless powered sensor networks (WPSNs)—are considered one of the most promising technologies for a sustainable Internet of Things [1–9]. In the WPSN, a power station wirelessly transfers energy to sensor devices that use the harvested energy to transmit their collected information to a fusion center [10–12]. The power station and fusion center may be included in one device or separated into different devices.

Although such WPSNs are expected to have a potentially infinite network life, they cannot always prevent short-term disconnections due to temporary energy shortages of some sensor devices. Such short-term disconnections cause an imbalance in transmission opportunities between sensor devices, resulting in an unfairness problem for WPSNs [13–16]. Moreover, in contrast to the traditional WSNs, in the WPSN, when sensor devices operate, wireless information transfer (WIT) and WET must be jointly considered. Consequently, it is necessary to design an appropriate medium access control (MAC) protocol for WPSNs.

Many studies have been conducted to design an efficient MAC protocol for WPSNs. In [17–19], the authors proposed a MAC protocol based on carrier-sense multiple access with collision avoidance (CSMA/CA) for WPSNs. Sensor devices access the channel competitively to conduct energy harvesting and data transmission; their channel access priorities are probabilistically differentiated by the backoff duration and inter-frame space (IFS), determined by the remaining energy. In [20–22], the authors proposed the time-division multiple access (TDMA)-based MAC protocol, in which a central coordinator allocates time resources for energy harvesting and data transmission considering the remaining energy of the sensor devices and the energy consumption required for data transmission. The sensor devices harvest energy in the dedicated time slots and transmit data.

Cho et al., used both TDMA and CSMA/CA methods to support energy harvesting and data transmission for two types of traffic patterns: periodic and non-periodic [23]. In [23], a coordinator allocates the dedicated TDMA time slots for energy harvesting and data transmission within a superframe to sensor devices that generate periodic traffic requiring transmission reliability on-demand. In contrast, energy harvesting and data transmission for non-periodic traffic are performed in the rest of the superframe using the CSMA/CA scheme. The studies in [17–23] commonly suffer from network performance degradation due to a lack of bandwidth resources from performing both WET and WIT operations within the same frequency band.

In [24–26], the authors used an out-of-band WET approach where sensor devices perform WET and WIT operations independently in different frequency bands, mitigating the effect of bandwidth limitation on WIT operation and improving network performance. However, their research suffers from high overhead due to the exchange of numerous control messages required to schedule WET and WIT operations in separate frequency bands.

Furthermore, the WET scheduling method used in [24–26] depends on simple criteria such as the distance to the sensor device and the energy required to transmit the data packet, increasing the difference in the residual energy between the sensor devices in the network. This difference causes an imbalance in transmission opportunities between sensor devices, resulting in an unfairness problem for network performance. This unfairness problem also applies to the in-band WET approach of [17–23].

This paper proposes a residual energy estimation-based MAC (REE-MAC) protocol, with two advantages for WPSNs composed of a central coordinator and multiple sensor devices. First, REE-MAC increases the residual energy of individual sensor devices by reducing overhead due to control messages for scheduling the energy harvesting operation of sensor devices. The coordinator numerically estimates the residual energy of individual sensor devices rather than exchanging numerous control messages. Second, REE-MAC establishes fairness among the data transmission opportunities for sensor devices. The coordinator allocates WET slots within the superframe to the sensor device by comprehensively considering the distance, harvested energy, and consumed energy for individual sensor devices. Accordingly, the residual energy of the sensor devices in the network is maintained at a similar level.

To this end, REE-MAC uses two types of superframes that operate simultaneously in different frequency bands: WET superframe and WIT superframe. In the WET superframe, a power transmitting unit (PTU) serving as a central coordinator supplies power to power receiving units (PRUs) (i.e., sensor devices) using the TDMA scheme. In the WIT superframe, multiple PRUs compete to transmit data packets to the PTU using CSMA/CA. At the beginning of each superframe, the PTU estimates the residual energy of individual PRUs changed due to their energy consumption and energy harvesting during the previous superframe. The PTU then determines the PRUs' charging priorities, based on which it allocates dedicated power slots (DPSs) within the WET superframe.

We performed an experimental simulation to verify the superiority of REE-MAC over FF-WPT [25] and HE-MAC [19], which are the representative MAC protocols for WPSNs of out-of-band and in-band WET approaches, respectively. The results demonstrated that REE-MAC achieves 18.08% and 145.60% higher average harvested energy, 81.03% and 64.21% shorter average freezing time, and 100.49% and 135.56% higher fairness than FF-WPT and HE-MAC, respectively.

The rest of this paper is organized as follows. In Section 2, we present a system model for REE-MAC. In Section 3, the detailed operation of REE-MAC is described. The simulation configuration and results are presented in Section 4. Finally, Section 5 concludes this paper.

2. System Model

Figure 1 illustrates the system architecture of the considered WPSN, which consists of a single PTU and multiple PRUs. The PRUs are deployed within the transmission range of the PTU. At the request of the PRU, the PTU either transfers power to the PRU or receives data from the PRU, for which the PTU and PRU are equipped with two types of transceivers: a power transceiver (P-Tx/Rx) for WET and a data transceiver (D-Tx/Rx) for WIT. The P-Tx/Rx and D-Tx/Rx in a device operate independently in different frequency bands and interact using internal signaling—enabling the WET and WIT of REE-MAC to be performed simultaneously.

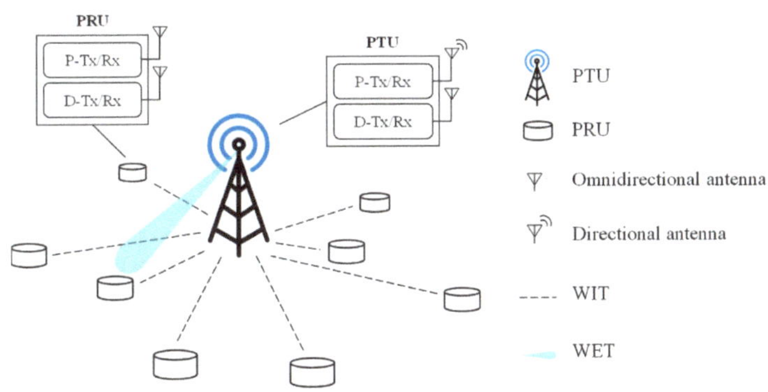

Figure 1. System architecture of REE-MAC.

Furthermore, the P-Tx/Rx and D-Tx/Rx of PTU always keep their radio on, while for the PRU, only its D-Tx/Rx is always on. The PRU keeps its P-Tx/Rx turned on to receive a beacon and power only in WET slots allocated to it and turns it off in other WET slots to prevent unnecessary energy consumption due to idle listening. Both the PTU and PRU have two antennas. The PTU has a directional antenna with a fixed beamwidth for WET and an omnidirectional antenna for WIT. We consider an adaptive array smart antennas as a directional antenna, which adjusts the beam direction by adapting the phase distribution of its antenna array elements. The PTU can transfer power to multiple PRUs using such a directional antenna. The PRU has two omnidirectional antennas for WET and WIT, respectively.

In REE-MAC, for determining when and how long to transfer power to individual PRUs, the PTU estimates the power that each PRU can receive per unit time, derived by Equation (1).

$$P_{r,i} = P_t G_t G_r \left(\frac{\lambda}{4\pi}\right)^2 d_i^{-\alpha}, \tag{1}$$

where $P_{r,i}$ is the power that the i-th PRU receives per second and P_t is the transmission power of the PRU. G_t and G_r are the antenna gain of the PTU and the PRU, respectively. We consider a flat-top radiation pattern as the directional antenna model of the PTU [27]; thus, G_t is the same as $2\pi/\theta$, where θ is the beamwidth of the directional antenna of the PTU. In contrast, because the PRU receives power from the PTU using its omnidirectional antenna, G_r is equal to one [28]. λ, d_i, and α refer to the wavelength, distance between the PTU and i-th PRU, and path loss exponent, respectively.

3. Design of REE-MAC

Figure 2 illustrates a superframe structure of the REE-MAC. The REE-MAC maintains a dual superframe structure—WET and WIT superframes—operated in separated frequency bands. In the WET superframe, the PTU transfers power to the PRUs according to their dedicated schedule using the TDMA scheme. The PRUs use the harvested energy to exchange control messages and data packets using contention-based CSMA/CA in the

WIT superframe. The WET superframe is divided into multiple equally sized DPSs, each allocated to an individual PRU for its exclusive power reception.

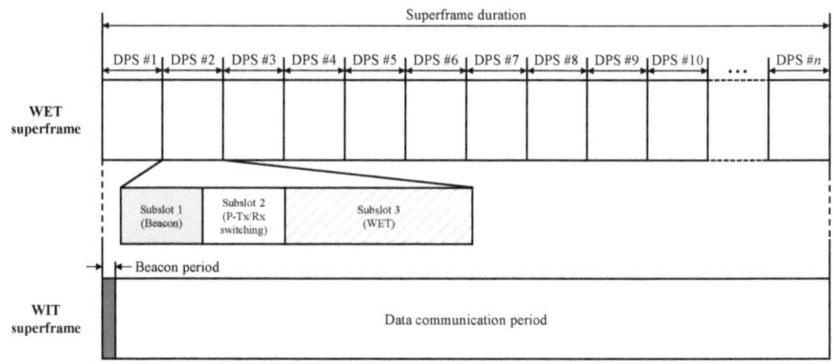

Figure 2. WET and WIT superframe structure.

The DPS consists of three subslots: a beacon subslot, a P-Tx/Rx switching subslot, and a WET subslot. In the beacon subslot, the PTU or PRU beacon is broadcasted. The PTU beacon is transmitted in the beacon subslot of the first DPS in the WET superframe, through which the PTU informs the PRUs of the allocation result for the DPSs in the WET superframe. In the beacon subslot of the remaining DPSs in the WET superframe, the PRUs request power transfer by transmitting a PRU beacon to the PTU. In the P-Tx/Rx switching subslot, the PTU obtains the direction of the PRU from the received PRU beacon, and the PRU switches the mode of its P-Tx/Rx from Tx to Rx for harvesting the energy.

The direction of the PRU indicates the phase of the PRU beacon signal transmitted from the PRU, through which the PTU can change the beam direction of its antenna toward the corresponding PRU to transfer power to it. In the WET subslot, the PTU transfers power to the allocated PRU. The WIT superframe consists of a beacon period and a data communication period. In the beacon period in the WIT superframe, the PTU broadcasts a beacon including the WIT superframe parameters through D-Tx/Rx. The PTU and PRUs compete freely with each other for channel access using the CSMA/CA scheme.

Figure 3 illustrates the overall operation of the REE-MAC. The solid and dashed lines represent the operation of the devices in WET and WIT superframes, respectively. These two superframes are expressed to overlap each other because the WET and WIT operations are performed simultaneously in different frequency bands. At the beginning of the WIT superframe, the PTU first broadcasts a beacon using D-Tx/Rx. Upon receiving the beacon from the PTU, the PRU transmits energy harvesting information, such as the node identifier (ID), residual energy, maximum battery capacity, and distance from the PTU, to the PTU.

This transmission of energy harvesting information for each PRU is performed only once throughout the entire network lifetime. The PTU and PRUs then exchange control messages and data packets using CSMA/CA. At the beginning of the WET superframe, the PTU performs the DPS allocation, which consists of three operations: calculating the number of DPSs required for individual PRUs to be fully charged, calculating the number of DPSs to be allocated to individual PRUs, and determining the charging priority for individual PRUs.

After the DPS allocation is completed, the PTU broadcasts the PTU beacon to notify PRUs of the scheduled WET superframe structure in the beacon subslot of the first DPS in the WET superframe. Then, in the beacon subslot in the subsequent DPSs allocated to individual PRUs, the PRU sends its PRU beacon to the PTU. With the PRU beacon, the PTU acquires the direction of the PRU and transfers power to the corresponding PRU in the WET subslot.

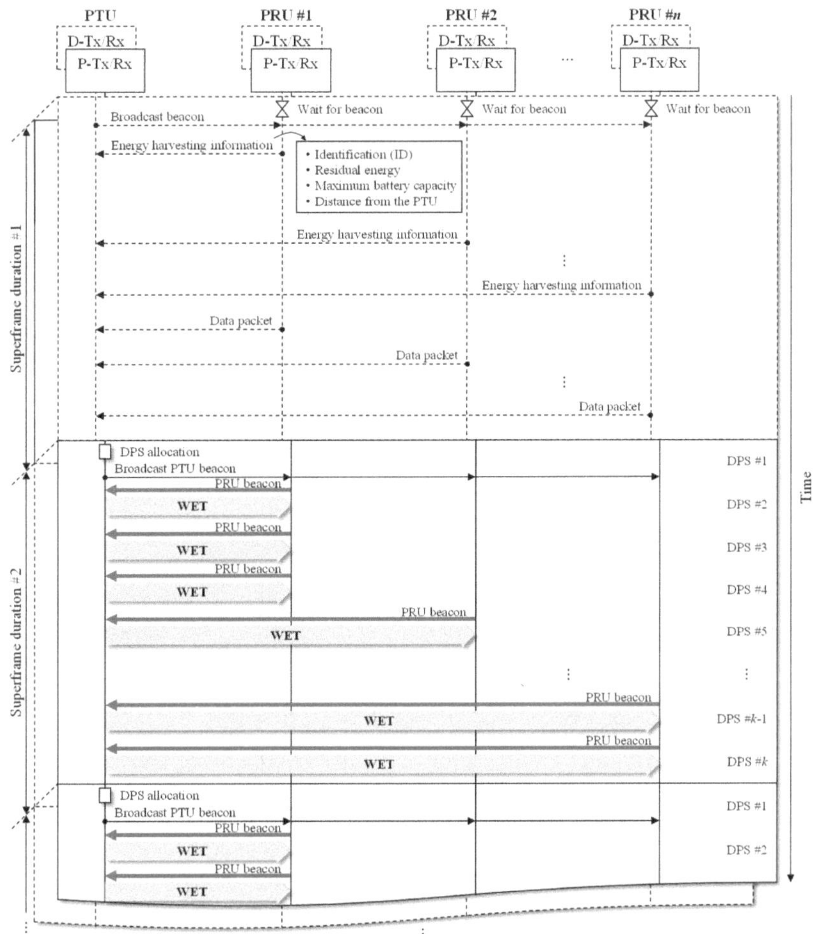

Figure 3. Overall operation of REE-MAC.

In estimating the residual energy of individual PRUs, the PTU first calculates the number of beacon, successful, collided, and idle slots in the previous WIT superframe. The beacon slot indicates a slot used by the PTU and PRU to transmit and receive the beacon. The successful slot indicates a slot used for successful transmission between the PTU and PRU. The collided slot indicates a slot in which collision occurs due to the data packets simultaneously transmitted from two or more PRUs. The idle slot is one in which no PRUs have transmitted the data packets due to the random backoff. It is assumed that n PRUs in the network always have the data packets to transmit in a saturation condition. The PTU maintains the node ID of PRUs (i.e., **ID**), the distance from itself to each PRU obtained from the received PRU beacons (i.e., **D**), the neighbor's node ID of each PRU (i.e., $\mathbf{Nbr}_{(i)}$), and the number of data packets received from each PRU during the previous superframe (i.e., **Npkt**), which are represented by the matrices, as in Equation (2).

$$
\begin{aligned}
\mathbf{ID} &= \left[id_{(1)},\ id_{(2)},\ \cdots,id_{(i)},\ \cdots,\ id_{(n)}\right],\ \ 0 < i \leq n \\
\mathbf{D} &= \left[d_{(1)},\ d_{(2)},\ \cdots,\ d_{(i)},\ \cdots,d_{(n)}\right],\ \ 0 < i \leq n \\
\mathbf{Nbr}_{(i)} &= \left[id_{(i,1)},\ id_{(i,2)},\ \cdots,\ id_{(i,l)}\right],\ \ 0 < i \leq n,\ \ 0 \leq l \leq (n-1) \\
\mathbf{Npkt} &= \left[npkt_{(1)},\ npkt_{(2)},\ \cdots,npkt_{(i)},\cdots,\ npkt_{(n)}\right],\ \ 0 < i \leq n,
\end{aligned}
\tag{2}
$$

where $id_{(i)}$ is the node ID for the i-th PRU, n is the number of PRUs in the network, $d_{(i)}$ is the distance from the PTU to the i-th PRU, $id_{(i,l)}$ is the node ID of the l-th neighbor PRU of the i-th PRU, and $npkt_{(i)}$ is the number of data packets received from the i-th PRU during the previous WIT superframe.

The number of beacon and successful slots in the previous WIT superframe can be calculated using L_{beacon} and **Npkt**, where L_{beacon} is the length of the beacon subslot of DPS. We calculate the number of collided and idle slots by considering the collision and idle probabilities in the remaining slots except for beacon and successful slots in the previous WIT superframe as p_{col} and p_{idle}, respectively [29]. p_{col} is the probability that two or more PRUs transmit the data packet in a randomly chosen slot excluding beacon and successful slots in the previous WIT superframe, as defined by Equation (3).

$$
p_{col} = \sum_{k=2}^{n} \binom{n}{k} \tau^k (1-\tau)^{n-k}, \tag{3}
$$

where τ is the probability that the PRU transmits a data packet in a randomly chosen slot excluding beacon and successful slots in the previous WIT superframe. p_{idle} is the probability that no PRU transmits a data packet in a randomly chosen slot, excluding beacon and successful slots in the previous WIT superframe, as defined by Equation (4).

$$
p_{idle} = 1 - p_{col} \tag{4}
$$

The total number of slots in the previous WIT superframe, $n_{totalSlots}$, can then be calculated by Equation (5).

$$
n_{totalSlots} = L_{dataSF} / L_{BP}, \tag{5}
$$

where L_{BP} is the slot length, which is the same as a unit backoff period. The numbers of beacon, successful, collided, and idle slots in the previous WIT superframe, $n_{beaconSlots}$, $n_{succSlots}$, $n_{colSlots}$, and $n_{idleSlots}$, can be calculated by Equations (6)–(9), respectively.

$$
n_{beaconSlots} = L_{beacon} / L_{BP} \tag{6}
$$

$$
n_{succSlots} = \left(L_{succ} \sum_{i=1}^{n} n_{pkt,i} \right) / L_{BP} \tag{7}
$$

$$
n_{colSlots} = p_{col}(n_{totalSlots} - n_{beaconSlots} - n_{succSlots}) \tag{8}
$$

$$
n_{idleSlots} = p_{idle}(n_{totalSlots} - n_{beaconSlots} - n_{succSlots}), \tag{9}
$$

where L_{succ} is the time taken for successful transmission between the PTU and PRUs. Figure 4 illustrates the timing diagrams for the successful transmission and collision. L_{data}, L_{SIFS}, L_{ACK}, and L_{DIFS} are the length of a data packet, a short inter-frame space (SIFS), an acknowledgment (ACK), and a distributed inter-frame space (DIFS), respectively.

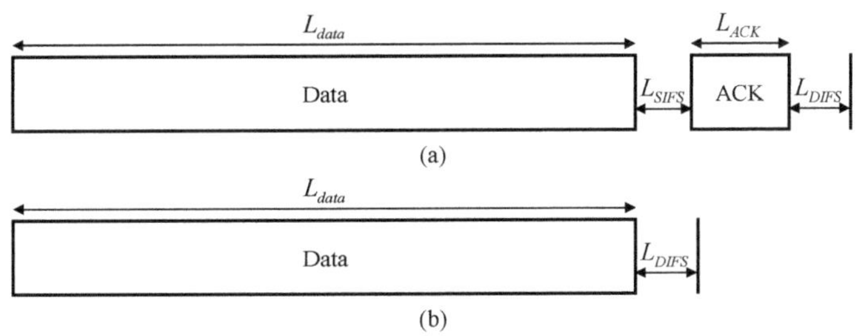

Figure 4. Timing diagram: (**a**) successful transmission, (**b**) collision.

The PTU then calculates $E_{beacon,i}$, $E_{succ,i}$, $E_{col,i}$, and $E_{idle,i}$ which are the amounts of energy consumed by the i-th PRU in the beacon, successful, collided, and idle slots in the previous WIT superframe, respectively, considering the timing diagrams. $E_{beacon,i}$ is the energy consumed by the i-th PRU to receive the beacon, as defined by Equation (10).

$$E_{beacon,i} = n_{beaconSlots} E_{rx} L_{BP}, \tag{10}$$

where E_{rx} is the energy consumed per second by a PRU when its D-Tx/Rx is in the Rx state. $E_{succ,i}$ is the energy consumed by the i-th PRU during successful slots in the previous WIT superframe, as defined by Equation (11).

$$E_{succ,i} = E_{succTx,i} + E_{succRx,i} + E_{succIdle,i}, \tag{11}$$

where $E_{succTx,i}$ is the energy consumed by the successful transmission of the i-th PRU when its D-Tx/Rx is in the Tx state. $E_{succRx,i}$ is the energy consumed by the successful transmission of neighbors of the i-th PRU when its D-Tx/Rx is in the Rx state. Specifically, $E_{succRx,i}$ is the energy consumption of the i-th PRU in successful slots when it overhears the data packets transmitted by other PRUs in the previous WIT superframe.

Furthermore, $E_{succIdle,i}$ is the energy consumed by the successful transmission of the PRUs not adjacent to the i-th PRU when the D-Tx/Rx is in an idle state. $E_{succTx,i}$ can be calculated by Equation (12).

$$E_{succTx,i} = n_{pkt,i}(E_{tx} L_{data} + E_{rx} L_{ACK} + E_{idle}(L_{SIFS} + L_{DIFS})), \tag{12}$$

where E_{tx} and E_{idle} are the energy consumed per second by the PRU when its D-Tx/Rx is in the Tx and idle states, respectively. $E_{succRx,i}$ can be calculated by Equation (13).

$$E_{succRx,i} = n_{succNbr,i}(E_{rx} L_{data} + E_{idle} L_{ACK} + E_{idle}(L_{SIFS} + L_{DIFS})), \tag{13}$$

where $n_{succNbr,i}$ is the total number of data packets successfully transmitted to the PTU by neighbor PRUs of the i-th PRU. $E_{succIdle,i}$ can be calculated by Equation (14).

$$E_{succIdle,i} = \left(n_{succSlots} - \left(n_{pkt,i} L_{data}\right) - \left(n_{succNbr,i} L_{data}\right)\right) E_{idle} L_{BP} \tag{14}$$

$E_{col,i}$ is the energy consumed by the i-th PRU during collided slots in the previous WIT superframe, as defined by Equation (15).

$$E_{col,i} = E_{colTx,i} + E_{colRx,i} + E_{colIdle,i}, \tag{15}$$

where $E_{colTx,i}$ is the energy consumed by the data packet transmission of the i-th PRU when one or more other PRUs transmits the data packet simultaneously. $E_{colRx,i}$ is the energy consumed by the i-th PRU due to the collision caused by two or more PRUs including at least one neighbor. Specifically, $E_{colRx,i}$ is the energy consumption of the i-th PRU when it overhears the data packets transmitted by two or more PRUs, including at least one neighbor, simultaneously. $E_{colIdle,i}$ is the energy consumed by the i-th PRU when two or more PRUs, excluding itself and its neighbors, transmit the data packet simultaneously. In this case, two or more PRUs outside the transmission range of the i-th PRU transmit the data packet simultaneously. $E_{colTx,i}$, $E_{colRx,i}$, and $E_{colIdle,i}$ can be calculated by Equations (16)–(18), respectively.

$$E_{colTx,i} = n_{colSlots}(E_{tx}L_{data} + E_{idle}L_{DIFS})\left(\tau\sum_{k=1}^{n-1}\binom{n-1}{k}\tau^k(1-\tau)^{n-k-1}\right), \quad 1 \le k \le (n-1) \tag{16}$$

$$E_{colRx,i} = \left(n_{colSlots}(E_{rx}L_{data} + E_{idle}L_{DIFS})(1-\tau)\sum_{k=2}^{n-1}\binom{n-1}{k}\tau^k(1-\tau)^{n-k-1}\right) - E_{colIdle,i}, \quad 2 \le k \le (n-1) \tag{17}$$

$$E_{colIdle,i} = n_{colSlots}E_{idle}\left((1-\tau)(1-\tau)^{n_{nbr,i}}\sum_{k=2}^{n-n_{nbr,i}-1}\binom{n-n_{nbr,i}-1}{k}\tau^k\right), \quad 2 \le k \le (n-n_{nbr,i}-1) \tag{18}$$

Finally, $E_{idle,i}$ is the energy consumed by the i-th PRU during the idle slots in the previous WIT superframe when its D-Tx/Rx is in an idle state, as defined by Equation (19).

$$E_{idle,i} = n_{idleSlots}E_{idle}L_{BP} \tag{19}$$

For estimating the residual energy of the i-th PRU, the PTU must calculate not only the energy consumed by the i-th PRU but also the energy harvested by the i-th PRU (i.e., $E_{rx,i}$) in the previous WET superframe, as defined by Equation (20).

$$E_{rx,i} = \sum_{k=1}^{n_{DPS,i}} E_{DPS,i}, \tag{20}$$

where $n_{DPS,i}$ is the number of DPSs allocated to the i-th PRU in the previous WET superframe. $E_{DPS,i}$ is the energy harvested by the i-th PRU during one DPS, as defined by Equation (21).

$$E_{DPS,i} = \eta P_{r,i}L_{WET}, \tag{21}$$

where η is the energy harvesting efficiency of the PRU, $P_{r,i}$ is the power received per second of the i-th PRU (refer to Equation (1) in Section 2), and L_{WET} is the length of the WET subslot.

Consequently, the residual energy of the i-th PRU (i.e., $E_{res,i}$) can be represented by Equation (22).

$$E_{res,i} = E_{res,i} - E_{beacon,i} - E_{succ,i} - E_{col,i} - E_{idle,i} + E_{rx,i} \tag{22}$$

Based on the knowledge of the residual energy estimation for individual PRUs, the PTU performs the DPS allocation at the beginning of every WET superframe. Algorithm 1 presents the DPS allocation procedure, which consists of three operations: (1) the calculation of the number of DPSs required for individual PRUs to be fully charged, (2) the calculation of the number of DPSs to be allocated to individual PRUs, and (3) the determination of charging priority for individual PRUs.

Algorithm 1. DPS allocation

1: **INITIALIZE** $\mathbf{N}_{DPS}$ to [],$\mathbf{N}_{sortedDPS}$ to [],$\mathbf{I}_{DPS}$ to [],$\mathbf{I}_{startDPS}$ to [],S_{DPS} to 0, $n_{reqDPS,i}$ to 0, cnt to 2, n_{avaDPS} to $n_{totalDPS} - 1$, $E_{\max,i}$ to 1
2: Calculation of the number of DPSs required for/* individual PRUs to be fully charged */
3: **FOR** each PRU, i, $i \in [1, n]$
4:     $n_{reqDPS,i} \leftarrow (E_{\max,i} - E_{res,i})/E_{DPS,i}$
5:     $S_{DPS} \leftarrow S_{DPS} + n_{reqDPS,i}$
6: **ENDFOR**
7: /* Calculation of the number of DPSs to be allocated to individual PRUs */
8: **FOR** each PRU, i, $i \in [1, n]$
9:     $n_{DPS,i} \leftarrow round(n_{avaDPS}(n_{reqDPS,i}/S_{DPS}))$
10:     $\mathbf{N}_{DPS}[i] \leftarrow n_{DPS,i}$
11: **ENDFOR**
12: /* Determination of the charging priority for individual PRUs */
13: $[\mathbf{N}_{sortedDPS}, \mathbf{I}_{DPS}] \leftarrow sort(\mathbf{N}_{DPS}, 'descend')$
14: **FOR** each PRU, i, $i \in [1, n]$
15:     $\mathbf{I}_{startDPS}[\mathbf{I}_{DPS}[i]] \leftarrow cnt$
16:     $cnt \leftarrow cnt + \mathbf{N}_{sortedDPS}[i]$
17: **ENDFOR**
18: **RETURN** $\mathbf{N}_{DPS}$ and $\mathbf{I}_{startDPS}$

In the algorithm, the PTU initializes the attributes and variables (i.e., $\mathbf{I}_{DPS}$, $\mathbf{I}_{startDPS}$, $\mathbf{N}_{DPS}$, $\mathbf{N}_{sortedDPS}$, S_{DPS}, $n_{reqDPS,i}$, cnt, n_{avaDPS}, and $E_{\max,i}$), where $\mathbf{I}_{DPS}$ and $\mathbf{I}_{startDPS}$ are the attributes to store the indices of the PRUs, $\mathbf{N}_{DPS}$ and $\mathbf{N}_{sortedDPS}$ are the attributes to track the number of DPSs allocated to them, and S_{DPS}, $n_{reqDPS,i}$, cnt, n_{avaDPS}, and $E_{\max,i}$ are the variables for counting the number of DPSs. $\mathbf{I}_{DPS}$ includes the indices of elements in $\mathbf{N}_{DPS}$ before $\mathbf{N}_{DPS}$ is sorted in descending order as $\mathbf{N}_{sortedDPS}$. $\mathbf{I}_{startDPS}$ includes the starting indices of DPSs allocated to individual PRU in the WET superframe. $\mathbf{N}_{DPS}$ includes the numbers of DPSs allocated to PRUs. $\mathbf{N}_{sortedDPS}$ is $\mathbf{N}_{DPS}$ sorted in descending order according to the values of elements included in $\mathbf{N}_{DPS}$. $\mathbf{I}_{DPS}$, $\mathbf{I}_{startDPS}$, $\mathbf{N}_{DPS}$, and $\mathbf{N}_{sortedDPS}$ are represented by one-dimensional arrays. S_{DPS} is the sum of the number of DPSs required by PRUs, $n_{reqDPS,i}$, which is the number of DPSs required for the i-th PRU to be fully charged from the current residual energy to the maximum battery capacity. cnt is a counter value used to calculate the starting index of DPSs allocated to individual PRU in the WET superframe. n_{avaDPS} is the number of available DPSs in the WET superframe excluding the first DPS and DPSs already allocated to PRUs, initialized to $n_{totalDPS} - 1$. $n_{totalDPS}$ is the total number of DPSs in the WET superframe.

In the first operation, the PTU obtains the number of DPSs required for individual PRUs to be fully charged (i.e., $n_{reqDPS,i}$), calculated using the energy required for each PRU's battery to be charged to its maximum battery capacity (i.e., $E_{\max,i} - E_{res,i}$) and the energy harvested during one DPS (i.e., $E_{DPS,i}$) (line 4). The PTU then calculates the sum of the number of DPSs required by all PRUs (i.e., S_{DPS}) (line 5).

In the second operation, the PTU obtains the number of DPSs to be allocated to individual PRUs (i.e., $n_{DPS,i}$) according to the ratio of the number of DPSs required by each PRU to the sum of the number of DPSs required by all PRUs (i.e., $n_{reqDPS,i}/S_{DPS}$) (line 9). Accordingly, the PTU maintains the list of the number of DPSs to be allocated to individual PRUs (i.e., $\mathbf{N}_{DPS}[i]$) (line 10).

Finally, the PTU determines the charging priority for individual PRUs by sorting $\mathbf{N}_{DPS}[i]$ in descending order according to the number of DPSs (line 13). The starting index of the DPSs in the WET superframe allocated to each PRU is calculated according to the charging priority for individual PRUs (line 15). Consequently, the PRU with low residual energy can perform the energy harvesting ahead of other PRUs.

After completing the DPS allocation algorithm, the PTU includes the number of DPSs (i.e., $\mathbf{N}_{DPS}$) and the starting index (i.e., $\mathbf{I}_{startDPS}$) of DPSs allocated to PRUs in the PTU beacon and broadcasts the PTU beacon. Based on the results of DPS scheduling, the PTU transfers power to the PRUs, and the PRUs perform energy harvesting.

4. Performance Evaluation

We evaluated the performance of REE-MAC using experimental simulations with the MATLAB simulator. The simulation results were compared with those of FF-WPT [25] and HE-MAC [19]. FF-WPT is an out-of-band solution that transfers power to devices using a different frequency band separated from that used to transmit data packets. In contrast, HE-MAC is an in-band solution that performs both power transfer and data transmission within the same frequency band. In the following subsections, we present in detail the simulation setup and configuration and the simulation results.

4.1. Simulation Configuration

In the simulation, we considered a WPSN consisting of one PTU and multiple PRUs. We assumed that each PRU is randomly deployed within the communication range of the PTU, set to 4 m. We further assumed that the PRU always has the data packets to transmit to the PTU. In the simulation, the number of PRUs varies from 2 to 20. The performance of REE-MAC was compared with those of FF-WPT [25] and HE-MAC [19] in terms of average harvested energy, average consumed energy, average freezing time, residual energy distribution, throughput distribution, and fairness indices for residual energy and throughput.

We investigated fairness by considering the freezing state, which indicates a state of the device in which the PRU cannot transmit the data packets due to a lack of residual energy. In the simulation, the PRU enters the freezing state when the residual energy is less than a pre-configured energy threshold. Then, if it exceeds a pre-configured active threshold, it is released from the freezing state. The active threshold indicates the minimum residual energy required for the PRU to resume the data packet transmission in the freezing state. It is set larger than the energy threshold to prevent the PRUs from entering the freezing state again immediately after they are released from the freezing state.

In FF-WPT, the WET operation is performed according to TDMA-based round-robin scheduling, while the WIT operation is performed using the CSMA/CA scheme. In FF-WPT, the interval required for exchanging control messages was set to 300 μs. Furthermore, the number of energy frames of each PRU is determined considering packet size and harvested energy per second. In HE-MAC, the arbitration inter-frame spaces (AIFSs) for PTU and PRU (i.e., $AIFS_{PTU}$ and $AIFS_{PRU}$) were set to 50 and 70 μs, respectively. The simulation was iterated 50 times. The detailed simulation parameters are listed in Table 1.

Table 1. Simulation parameters.

Parameter	Value	Parameter	Value
Number of PRUs	2–20	d	0–4 m
Data rate	2 Mbps	Maximum battery capacity	1 mJ
WET superframe	1 s	Initial residual energy	0.6 mJ
WIT superframe	1 s	Active threshold	0.6 mJ
DPS length	10 ms	Energy threshold	0.1 mJ
Number of DPSs	100	CW_{min}	31
Beacon subslot	40 µs	CW_{max}	1023
P-Tx/Rx switching subslot	10 µs	G_t	12
WET subslot	9950 µs	G_r	1
RTS	20 bytes	P_{tx}	31.47 mA
CTS	14 bytes	P_{rx}	26.94 mA
ACK	14 bytes	P_{idle}	0.00156 mA
SIFS	10 µs	P_t	3000 mW
DIFS	50 µs	α	2.7
Beacon	15 bytes	η	0.85
Packet size	100–200 bytes	Slot length	20 µs

4.2. Simulation Results

Figure 5a,b illustrate the variations in the average harvested energy for 100- and 200-byte packets, respectively. The average harvested energy decreases as the number of PRUs in the network increases due to decreased energy harvested by individual PRUs per WET superframe. In both figures, REE-MAC exhibits a higher average harvested energy compared with both FF-WPT and HE-MAC. Regarding the DPS allocation operation, REE-MAC requires a shorter time overhead than FF-WPT. In REE-MAC, the PTU performs DPS allocation by estimating the residual energy for individual PRUs and informs the PRUs of the number of allocated DPSs through a PTU beacon at the beginning of the WET superframe. Therefore, REE-MAC entails only time overhead for sending one PTU beacon.

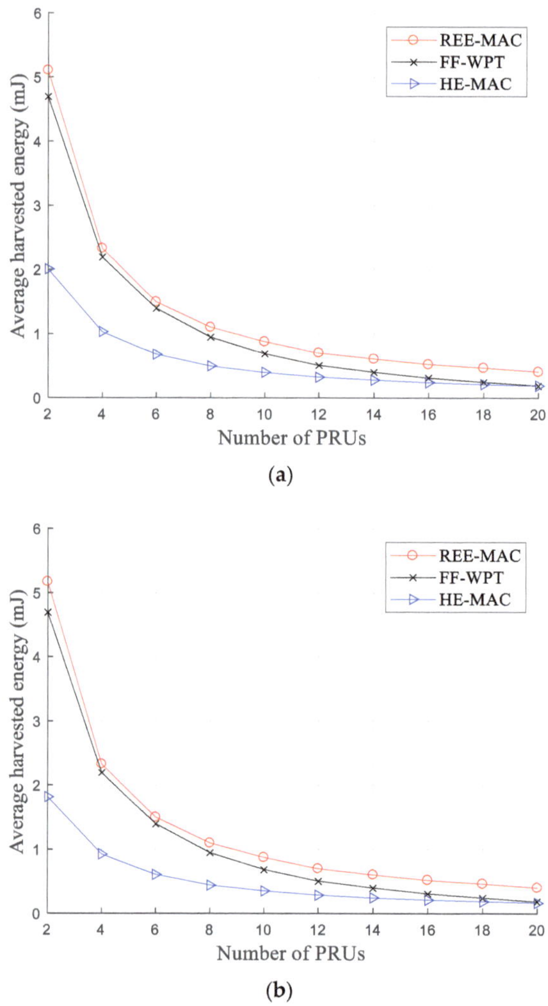

Figure 5. Average harvested energy: (**a**) packet size of 100 bytes; (**b**) packet size of 200 bytes.

In contrast, FF-WPT incurs a long time overhead because, in FF-WPT, the PTU needs to exchange control messages with all PRUs to obtain the distance between itself and each PRU. Furthermore, HE-MAC exhibits the lowest average harvested energy because, in HE-MAC, the exchange of control messages, the transmission of data packets, and energy harvesting are all performed in the same frequency band.

As presented in Table 2, the average harvested energy of REE-MAC and FF-WPT are almost identical, even if the packet size increases from 100 to 200 bytes, because the packet size does not affect the average harvested energy of PRUs in out-of-band approaches. However, in HE-MAC, the average harvested energy decreases slightly when the packet size increases from 100 to 200 bytes because the proportion of the energy harvesting time within the WET superframe is reduced. Quantitatively, the average harvested energy of REE-MAC is 17.79% and 18.38% higher than that of FF-WPT when the PRUs transmit the 100- and 200-byte packets, respectively. Moreover, the average harvested energy of REE-MAC is 132.15% and 159.04% higher than that of HE-MAC when the PRUs transmit the 100- and 200-byte packets, respectively.

Table 2. Average harvested energy (mJ).

Packet Size	Protocol	Number of PRUs									
		2	4	6	8	10	12	14	16	18	20
100 bytes	REE-MAC	2.65	1.56	1.03	0.76	0.61	0.52	0.48	0.41	0.36	0.31
	FF-WPT	2.57	1.47	0.95	0.67	0.51	0.41	0.36	0.30	0.25	0.21
	HE-MAC	1.31	0.72	0.48	0.35	0.28	0.24	0.20	0.18	0.16	0.14
200 bytes	REE-MAC	2.65	1.56	1.03	0.76	0.61	0.52	0.46	0.41	0.36	0.31
	FF-WPT	2.57	1.47	0.95	0.67	0.51	0.41	0.36	0.30	0.25	0.21
	HE-MAC	1.20	0.65	0.44	0.32	0.25	0.21	0.18	0.16	0.14	0.13

Figure 6a,b illustrate the variations in the average consumed energy for 100- and 200-byte packets, respectively. In REE-MAC, the average consumed energy decreases as the number of PRUs increases because the number of data packets transmitted by PRUs gradually decreases due to the increase in collisions and backoff delay. Consequently, the number of transmissions of PRUs is reduced, reducing the energy consumed by PRUs. In FF-WPT, as in REE-MAC, the average consumed energy tends to decrease overall as the number of PRUs increases.

Furthermore, FF-WPT exhibits an average consumed energy similar to REE-MAC. However, in some sections (i.e., when the number of PRUs is 14 to 18), the average consumed energy of FF-WPT slightly increases. As the number of PRUs increases, the energy each PRU can harvest decreases, and accordingly, the number of PRUs entering the freezing state increases. Therefore, a relatively small number of PRUs transmit data packets to the PTU, and the average consumed energy can increase due to the reduced contention level.

HE-MAC exhibits lower average consumed energy compared with both REE-MAC and FF-WPT. In HE-MAC, the PRUs have relatively few transmission opportunities for data packets because both WET and WIT operations are performed within the same frequency band. Therefore, a small number of data packet transmissions reduces the energy consumed by the PRU. Quantitatively, when the PRUs transmit 100- and 200-byte packets, the average consumed energy of REE-MAC is 7.79% and 8.29% higher than that of FF-WPT, respectively. Moreover, it is 43.78% and 43.74% higher compared with HE-MAC, respectively.

Figure 7a,b illustrate the variations in the average freezing time for 100- and 200-byte packets, respectively. The freezing time is the time the PRU is in the freezing state. As the number of PRUs increases, the energy harvested by each PRU decreases, and accordingly, the average freezing time of PRUs is highly likely to increase. REE-MAC exhibits a shorter average freezing time than both FF-WPT and HE-MAC. In REE-MAC, the DPS allocation is performed considering the residual energy of individual PRUs. Accordingly, in REE-MAC, PRUs with less residual energy are allocated more DPSs within the superframe. Therefore, all PRUs in the network maintain similar residual energy, and the time they are in the freezing state is relatively short.

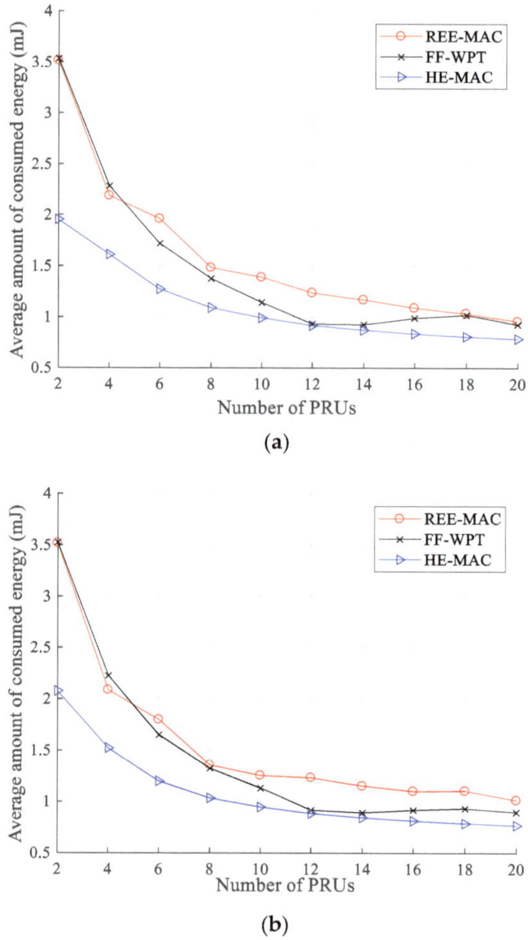

Figure 6. Average consumed energy: (**a**) packet size of 100 bytes; (**b**) packet size of 200 bytes.

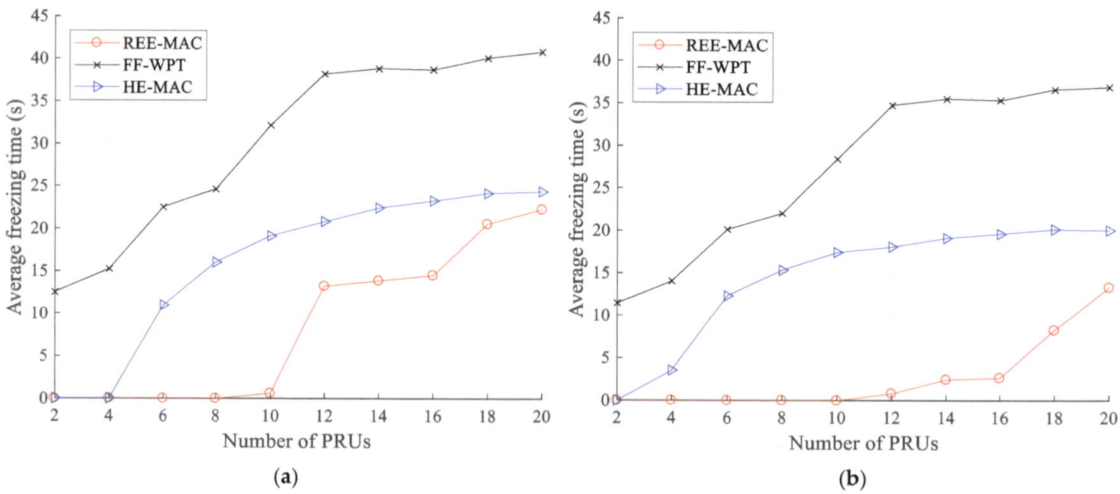

Figure 7. Average freezing time: (**a**) packet size of 100 bytes; (**b**) packet size of 200 bytes.

When the sizes of data packets transmitted by PRUs are 100 and 200 bytes, the average freezing time of REE-MAC is almost zero until the numbers of PRUs are 10 and 12, respectively. However, if the numbers of PRUs exceed 10 and 12, the average freezing time of REE-MAC increases as the number of PRUs increases. FF-WPT consistently exhibits a longer average freezing time than both REE-MAC and HE-MAC because, in FF-WPT, the PTU transfers power to PRUs according to the distance without considering the residual energy of PRUs. Furthermore, individual PRUs harvest smaller energy due to the exchange of control messages required for WET operation. Accordingly, in FF-WPT, PRUs enter the freezing state more frequently.

When the number of PRUs is more than 12, the average freezing time of FF-WPT increases gradually. When the number of PRUs is increased to more than 12, the PRUs enter the freezing state more quickly due to the decrease in harvested energy. Therefore, it takes longer for the residual energy of PRUs in the freezing state to reach the active threshold.

In HE-MAC, the PRU occupying the channel uses the harvest-then-transmit scheme. Therefore, the PRU first harvests the energy required to transmit the data packet and then uses it to transmit the data packet. Other PRUs maintain an idle state to minimize energy consumption. Consequently, the average freezing time of HE-MAC is shorter than that of FF-WPT. Quantitatively, when the PRUs transmit 100- and 200-byte packets, the average freezing time of REE-MAC is 72.03% and 90.04% shorter than that of FF-WPT, respectively. It is also 47.26% and 81.15% shorter than that of HE-MAC.

Figure 8a,b illustrate the variations in the residual energy distribution of individual PRUs for 100- and 200-byte packets, respectively. Both figures indicate the residual energy distribution of individual PRUs in a specific round of the experimental simulation. In REE-MAC and FF-WPT, when the packet size increases, the residual energy of individual PRUs decreases due to an increase in the consumed energy.

In HE-MAC, the number of dead PRUs increases. A dead PRU indicates a PRU with a residual energy of zero. In REE-MAC, the difference between the residual energy of individual PRUs is slight compared with both FF-WPT and HE-MAC. REE-MAC enables PRUs to maintain similar residual energy through the DPS allocation considering the residual energy. In FF-WPT, the fluctuation in the residual energy distribution of individual PRUs is larger than that of REE-MAC because the DPSs are allocated considering only the distance between the PTU and individual PRUs. In HE-MAC, the residual energy of individual PRUs is almost zero except when the number of PRUs is 2 because the PRUs consume additional energy in addition to data packet transmission, and the residual energy of the PRUs gradually decreases. Consequently, in HE-MAC, more dead PRUs occur compared with REE-MAC and FF-WPT.

Figure 9a,b illustrate the variations in the throughput distribution of individual PRUs for 100- and 200-byte packets, respectively. Both figures indicate the throughput distribution of individual PRUs in a specific round of the experimental simulation. The fluctuation of the throughput distribution increases as the number of PRUs in the network increases. Moreover, when the packet size increases, the throughput of individual PRUs increases due to the decrease in the backoff delay. REE-MAC exhibits a constant throughput distribution regardless of the number of PRUs, compared with FF-WPT and HE-MAC. In REE-MAC, all PRUs in the network maintain similar throughput performance.

From the results of Figures 7 and 8, in REE-MAC, the PRUs maintain the shortest freezing time, on average, and dead PRUs with zero residual energy rarely occur. Therefore, compared with FF-WPT and HE-MAC, the PRUs using REE-MAC can have relatively even transmission opportunities. However, in FF-WPT and HE-MAC, the transmission opportunity is biased toward some PRUs as the number of PRUs in the network increases. Accordingly, in FF-WPT and HE-MAC, the throughput distribution for specific PRUs becomes severely concentrated.

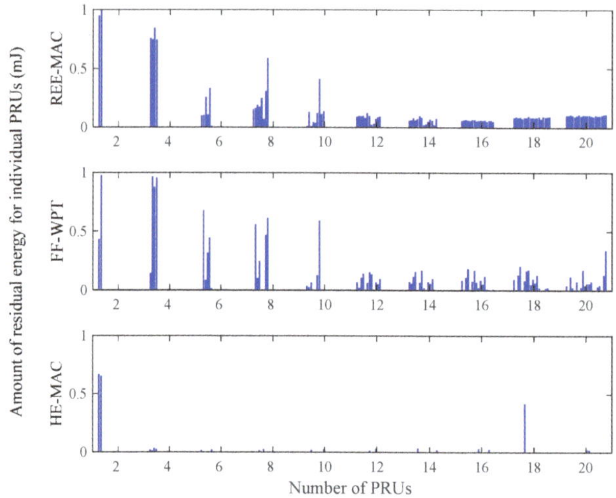

(**a**)

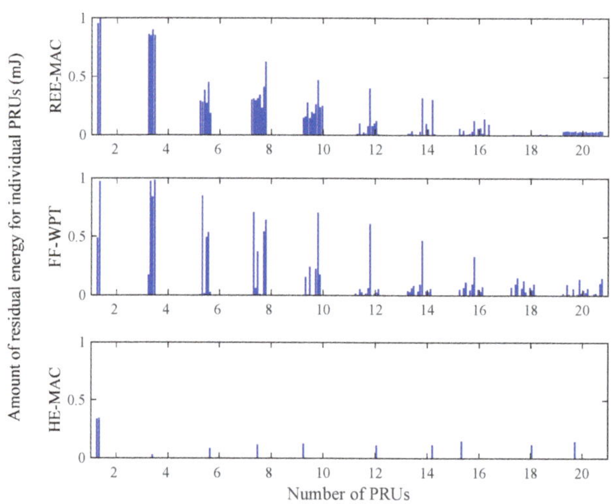

(**b**)

Figure 8. Residual energy distribution of individual PRUs: (**a**) packet size of 100 bytes; (**b**) packet size of 200 bytes.

Figure 10a,b illustrate the fairness index for residual energy for 100- and 200-byte packets, respectively. The fairness index for residual energy (F_{res}) can be calculated by Equation (23) [30].

$$F_{res} = \frac{\left(\sum_{i}^{n} x_i^2\right)}{n \cdot \sum_{i=1}^{n} x_i^2}, \tag{23}$$

where n is the number of PRUs and x_i is the fairness parameter, which represents the residual energy of the i-th PRU.

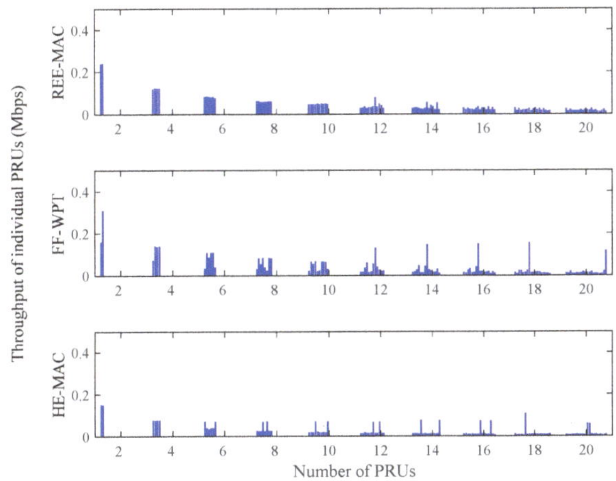

(a)

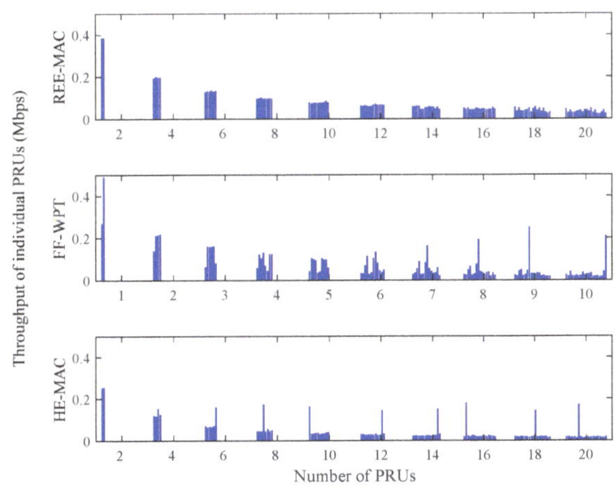

(b)

Figure 9. Throughput distribution of individual PRUs: (**a**) packet size of 100 bytes; (**b**) packet size of 200 bytes.

In REE-MAC, the fairness index for residual energy decreases and then increases again as the number of PRUs increases. For 100- and 200-byte packets, when the number of PRUs is 10 and 14 or less, respectively, the fairness index for residual energy of REE-MAC decreases as the number of PRUs increases. As the number of PRUs in the network increases, the difference in the residual energy between individual PRUs increases due to a decrease in the energy harvested by individual PRUs and the difference in the energy consumed by individual PRUs (refer to Figure 8).

In contrast, when the numbers of PRUs are larger than 10 and 14, respectively, the fairness index for residual energy of REE-MAC increases as the number of PRUs increases. As the number of PRUs increases, the energy harvested by individual PRUs decreases, and the number of PRUs in the freezing state increases accordingly. PRUs in the freezing state perform only WET operations until their residual energy reaches the active threshold. Therefore, as the residual energy of many PRUs in the freezing state approaches the active threshold, the fairness index for residual energy increases.

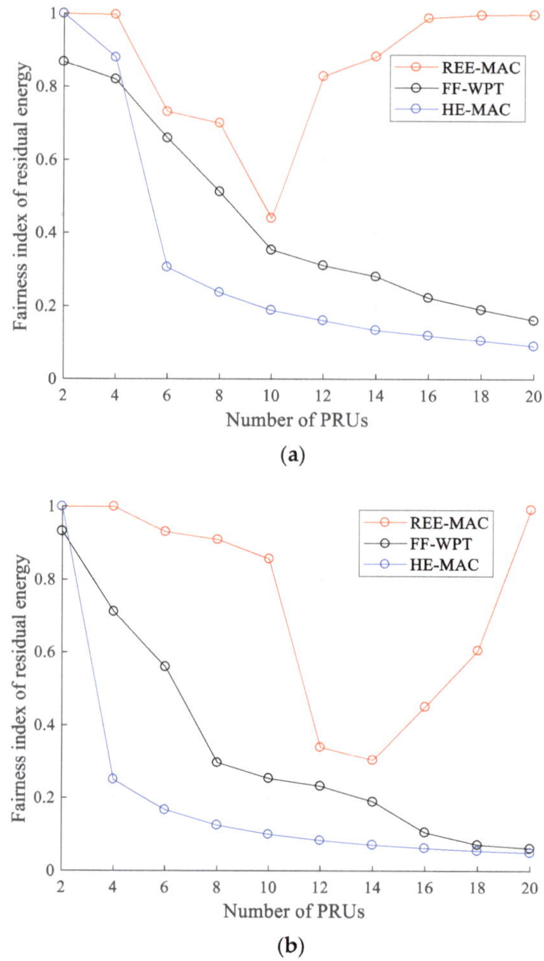

Figure 10. Fairness index for residual energy: (**a**) packet size of 100 bytes; (**b**) packet size of 200 bytes.

However, in FF-WPT and HE-MAC, the fairness index for residual energy decreases as the number of PRUs in the network increases. As depicted in Figure 7, the gap in transmission opportunities between PRUs increases as the average freezing time of PRUs increases. This gap increases the difference between the energy consumed by individual PRUs. Therefore, the fairness index for residual energy gradually decreases as the difference between the residual energy of individual PRUs increases. Quantitatively, when the PRUs transmit 100- and 200-byte packets, the fairness index for residual energy of REE-MAC is 95.34% and 116.23% higher than that of FF-WPT, respectively. It is also 165.88% and 276.00% higher compared with that of HE-MAC, respectively.

Figure 11a,b illustrate the fairness index for throughput for 100- and 200-byte packets, respectively. The fairness index for throughput (F_{th}) can be calculated by Equation (24).

$$F_{th} = \frac{\left(\sum_{i}^{n} y_i^2\right)}{n \cdot \sum_{i=1}^{n} y_i^2}, \tag{24}$$

where n is the number of PRUs and y_i is the fairness parameter indicating the number of data packets that the i-th PRU transmits to the PTU.

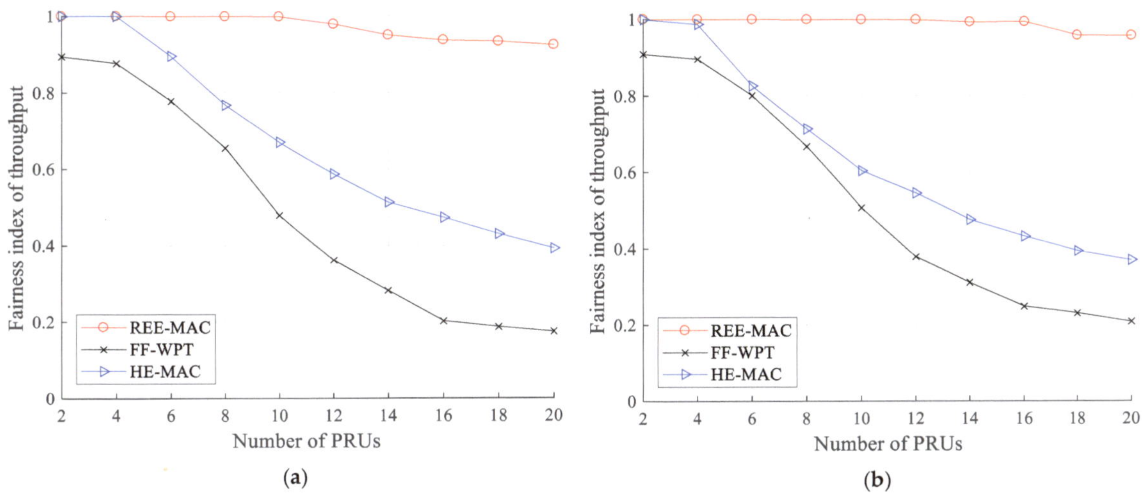

Figure 11. Fairness index for throughput: (**a**) packet size of 100 bytes; (**b**) packet size of 200 bytes.

In each case of 100- and 200-byte packets, REE-MAC maintains the fairness index for throughputs of 0.924 and 0.956 or higher, respectively, regardless of the number of PRUs. As the number of PRUs increases, the energy harvested by individual PRUs decreases and the difference in energy consumed by individual PRUs increases. Nevertheless, in REE-MAC, the residual energy of PRUs is maintained similar to each other due to DPS allocation considering the residual energy of individual PRUs. Accordingly, individual PRUs achieve a high fairness index for throughput by having a similar transmission opportunity.

In contrast, FF-WPT exhibits a lower fairness index for throughput compared with both REE-MAC and HE-MAC. From the results in Figure 7, in FF-WPT, PRUs maintain the longest freezing time on average. The difference in throughput performance between PRUs in FF-WPT becomes significant because the increase of the freezing time causes the transmission opportunities to be biased to some PRUs. The fairness index for throughput of FF-WPT decreases as the number of PRUs increases regardless of the packet size because the average freezing time of PRUs increases due to the decrease in the energy harvested by individual PRUs.

HE-MAC exhibits a higher fairness index for throughput than FF-WPT regardless of the packet size. In HE-MAC, PRUs that are not in the freezing state maintain similar residual energy through the harvest-then-transmit scheme, and thus they have a relatively similar transmission opportunity. Quantitatively, when the PRUs transmit 100- and 200-byte packets, the fairness index for throughput of REE-MAC is 98.58% and 91.80% higher than that of FF-WPT, respectively. It is also 44.46% and 55.91% higher compared with HE-MAC, respectively.

5. Conclusions

This paper presents the REE-MAC protocol for WPSNs, which aims to reduce overhead due to control messages for scheduling the WET operation and provide fairness for data transmission opportunities to the sensor devices. REE-MAC achieves low overhead by numerically estimating the residual energy of individual PRUs without exchanging control messages. Furthermore, in REE-MAC, the PTU allocates the DPSs inversely proportional to the residual energy of individual PRUs, so that all PRUs in the network maintain similar residual energy. Thereby, it minimizes the energy depletion of some PRUs and provides individual PRUs with a fair data transmission opportunity. At the beginning of each superframe, the PTU calculates the consumed and harvested energy of individual PRUs and then estimates their residual energy. It then performs the DPS allocation based on the results of the residual energy estimation.

We conducted an experimental simulation to evaluate the performance of REE-MAC under the environment of changing network size and packet size. The results demonstrate that REE-MAC uses the residual energy estimation to reduce unnecessary waste of bandwidth due to the exchange of control messages, increasing the energy harvested by individual PRUs. Moreover, REE-MAC prevents the DPSs from biased allocation to some PRUs, reducing the freezing time of the PRUs due to lack of energy.

These operations of REE-MAC give similar transmission opportunities to PRUs in the network, ensuring higher fairness compared with FF-WPT and HE-MAC in terms of residual energy and throughput. On average, REE-MAC achieves 18.08% and 145.60% higher energy harvested, 81.03% and 64.21% shorter average freezing time, 105.79% and 220.94% higher fairness index for residual energy, and 95.19% and 50.18% higher fairness index for throughput, compared with FF-WPT and HE-MAC, respectively.

Author Contributions: Conceptualization, S.-B.L. and E.-J.K.; methodology, S.-B.L., J.-H.K. and E.-J.K.; software, S.-B.L., J.-H.K. and E.-J.K.; validation, S.-B.L. and E.-J.K.; formal analysis, S.-B.L. and E.-J.K.; investigation, J.-H.K. and E.-J.K.; resources, E.-J.K.; data curation, J.-H.K.; writing—original draft preparation, S.-B.L., J.-H.K. and E.-J.K.; writing—review and editing, S.-B.L., J.-H.K. and E.-J.K.; visualization, S.-B.L.; supervision, E.-J.K.; project administration, E.-J.K.; funding acquisition, J.-H.K. and E.-J.K. All authors have read and agreed to the published version of the manuscript.

Funding: This research was supported by Basic Science Research Program through the National Research Foundation of Korea (NRF) funded by the Ministry of Education (NRF-2020R1I1A3052733 and NRF-2019R1I1A1A01059787). This work was also supported by the NRF grant funded by the Korea government (MSIT) (NRF-2021R1C1C2095696).

Institutional Review Board Statement: Not applicable.

Informed Consent Statement: Not applicable.

Data Availability Statement: No new data were created or analyzed in this study. Data sharing is not applicable to this article.

Conflicts of Interest: The authors declare no conflict of interest.

References

1. Choi, K.W.; Ginting, L.; Rosyady, P.A.; Aziz, A.A.; Kim, D.I. Wireless-powered sensor networks: How to realize. *IEEE Trans. Wirel. Commun.* **2017**, *16*, 221–234. [CrossRef]
2. Perera, T.D.P.; Jayakody, D.N.K.; Sharma, S.K.; Chatzinotas, S.; Li, J. Simultaneous wireless information and power transfer (SWIPT): Recent advances and future challenges. *IEEE Commun. Surv. Tutor.* **2018**, *20*, 264–302. [CrossRef]
3. Setiawan, D.; Aziz, A.A.; Kim, D.I.; Choi, K.W. Experiment, modeling, and analysis of wireless-powered sensor network for energy neutral power management. *IEEE Syst. J.* **2018**, *12*, 3381–3392. [CrossRef]
4. Bi, S.; Ho, C.K.; Zhang, R. Wireless powered communication: Opportunities and challenges. *IEEE Commun. Mag.* **2015**, *53*, 117–125. [CrossRef]
5. Ramezani, P.; Jamalipour, A. Toward the evolution of wireless powered communication networks for the future Internet of Things. *IEEE Netw.* **2017**, *31*, 62–69. [CrossRef]
6. Du, R.; Shokri-Ghadikolaei, H.; Fischione, C. Wirelessly-powered sensor networks: Power allocation for channel estimation and energy beamforming. *IEEE Trans. Wirel. Commun.* **2020**, *19*, 2987–3002. [CrossRef]
7. Du, R.; Xiao, M.; Fischione, C. Optimal node deployment and energy provision for wirelessly powered sensor networks. *IEEE J. Sel. Areas Commun.* **2019**, *37*, 407–423. [CrossRef]
8. López, O.L.A.; Alves, H.; Souza, R.D.; Montejo-Sánchez, S.; Fernández, E.M.G.; Latva-Aho, M. Massive wireless energy transfer: Enabling sustainable IoT towards 6G era. *IEEE Internet Things J.* **2021**, *8*, 8816–8835. [CrossRef]
9. Clerckx, B.; Huang, K.; Varshney, L.R.; Ulukus, S.; Alouini, M.-S. Wireless power transfer for future networks: Signal processing machine learning computing and sensing. *IEEE J. Sel. Top. Signal Process.* **2021**, *15*, 1060–1094. [CrossRef]
10. Chu, Z.; Zhou, F.; Zhu, Z.; Hu, R.Q.; Xiao, P. Wireless powered sensor networks for Internet of Things: Maximum throughput and optimal power allocation. *IEEE Internet Things J.* **2018**, *5*, 310–321. [CrossRef]

11. Mai, V.V.; Shin, W.-Y.; Ishibashi, K. Wireless power transfer for distributed estimation in sensor networks. *IEEE J. Sel. Top. Signal Process.* **2017**, *11*, 549–562. [CrossRef]
12. Hong, Y.-W.P.; Hsu, T.-C.; Chennakesavula, P. Wireless power transfer for distributed estimation in wireless passive sensor networks. *IEEE Trans. Signal Process.* **2016**, *64*, 5382–5395. [CrossRef]
13. Shi, H.; Prasad, R.V.; Onur, E.; Niemegeers, I.G.M.M. Fairness in wireless networks: Issues measures and challenges. *IEEE Commun. Surv. Tutor.* **2014**, *16*, 5–24. [CrossRef]
14. Lu, X.; Wang, P.; Niyato, D.; Kim, D.I.; Han, Z. Wireless networks with RF energy harvesting: A contemporary survey. *IEEE Commun. Surv. Tutor.* **2015**, *17*, 757–789. [CrossRef]
15. Lei, M.; Zhang, X.; Ding, H.; Yu, B. Fairness-aware resource allocation in multi-hop wireless powered communication networks with user cooperation. *Sensors* **2018**, *18*, 1890. [CrossRef]
16. Zhu, Z.; Peng, J.; Gu, X.; Li, H.; Liu, K.; Zhou, Z.; Liu, W. Fair resource allocation for system throughput maximization in mobile edge computing. *IEEE Access* **2018**, *6*, 5332–5340. [CrossRef]
17. Naderi, M.Y.; Nintanavongsa, P.; Chowdhury, K.R. RF-MAC: A medium access control protocol for re-chargeable sensor networks powered by wireless energy harvesting. *IEEE Trans. Wirel. Commun.* **2014**, *13*, 3926–3937. [CrossRef]
18. Iqbal, A.; Kim, Y.; Lee, T.-J. Access mechanism in wireless powered communication networks with harvesting access point. *IEEE Access* **2018**, *6*, 37556–37567. [CrossRef]
19. Ha, T.; Kim, J.; Chung, J.-M. HE-MAC: Harvest-then-transmit based modified EDCF MAC protocol for wireless powered sensor networks. *IEEE Trans. Wirel. Commun.* **2018**, *17*, 3–16. [CrossRef]
20. Ju, H.; Zhang, R. Throughput maximization in wireless powered communication networks. *IEEE Trans. Wirel. Commun.* **2014**, *13*, 418–428. [CrossRef]
21. Hadzi-Velkov, Z.; Nikoloska, I.; Chingoska, H.; Zlatanov, N. Opportunistic scheduling in wireless powered communication networks. *IEEE Trans. Wirel. Commun.* **2017**, *16*, 4106–4119. [CrossRef]
22. Yao, Q.; Huang, A.; Shan, H.; Quek, T.Q.S.; Wang, W. Delay-aware wireless powered communication networks—Energy balancing and optimization. *IEEE Trans. Wirel. Commun.* **2016**, *15*, 5272–5286. [CrossRef]
23. Cho, S.; Lee, K.; Kang, B.J.; Joe, I. A hybrid MAC protocol for optimal channel allocation in large-scale wireless powered communication networks. *EURASIP J. Wirel. Commun. Netw.* **2018**, *2018*, 1–13. [CrossRef]
24. Iannello, F.; Simeone, O.; Spagnolini, U. Medium access control protocols for wireless sensor networks with energy harvesting. *IEEE Trans. Commun.* **2012**, *60*, 1381–1389. [CrossRef]
25. Xia, M.; Aissa, S. On the efficiency of far-field wireless power transfer. *IEEE Trans. Signal Process.* **2015**, *63*, 2835–2847. [CrossRef]
26. Nobar, S.K.; Niya, J.M.; Tazehkand, B.M. Performance analysis of cognitive wireless powered communication networks under unsaturated traffic condition. *IEEE Trans. Green Commun. Netw.* **2020**, *4*, 819–831. [CrossRef]
27. Kwon, J.-H.; Kim, E.-J. Asymmetric directional multicast for capillary Machine-to-Machine using mmWave communications. *Sensors* **2016**, *16*, 515. [CrossRef]
28. Cai, L.X.; Cai, L.; Shen, X.; Mark, J.W. REX: A randomized exclusive region based scheduling scheme for mmWave WPANs with directional antenna. *IEEE Trans. Wirel. Commun.* **2010**, *9*, 113–121. [CrossRef]
29. Bianchi, G. Performance analysis of the IEEE 802.11 distributed coordination function. *IEEE J. Sel. Areas Commun.* **2000**, *18*, 535–547. [CrossRef]
30. Babu, A.V.; Jacob, L. Fairness analysis of IEEE 802.11 multirate wireless LANs. *IEEE Trans. Veh. Technol.* **2007**, *56*, 3073–3088. [CrossRef]

sensors

MDPI

Article

The Smart Meter Challenge: Feasibility of Autonomous Indoor IoT Devices Depending on Its Energy Harvesting Source and IoT Wireless Technology

Edgar Saavedra *, Laura Mascaraque, Gonzalo Calderon, Guillermo del Campo and Asuncion Santamaria

CeDInt-UPM, Universidad Politécnica de Madrid, Campus de Montegancedo, Pozuelo de Alarcón, 28223 Madrid, Spain; lmascaraque@cedint.upm.es (L.M.); gcalderon@cedint.upm.es (G.C.); gcampo@cedint.upm.es (G.d.C.); asun.santamaria@upm.es (A.S.)
* Correspondence: e.saavedra@upm.es

Abstract: Most smart meters are connected and powered by the electric mains, requiring the service interruption and qualified personnel for their installation. Wireless technologies and energy harvesting techniques have been proved as alternatives for communications and power supply, respectively. In this work, we analyse the energy consumption of the most used IoT wireless technologies nowadays: Sigfox, LoRaWAN, NB-IoT, Wi-Fi, BLE. Smart meters' energy consumption accounts for metering, standby and communication processes. Experimental measurements show that communication consumption may vary upon the specific characteristics of each wireless communication technology—payload, connection establishment, transmission time. Results show that the selection of a specific technology will depend on the application requirements (message payload, metering period) and location constraints (communication range, infrastructure availability). Besides, we compare the performance of the most suitable energy harvesting (EH) techniques for smart meters: photovoltaic (PV), radiofrequency (RF) and magnetic induction (MIEH). Thus, EH technique selection will depend on the availability of each source at the smart meter's location. The most appropriate combination of IoT wireless technology and EH technique must be selected accordingly to the very use case requirements and constraints.

Keywords: smart meter; Internet of Things; energy harvesting; energy efficiency; LPWAN; Sigfox; LoRaWAN; NB-IoT; Wi-Fi; BLE; IoT; IIoT

Citation: Saavedra, E.; Mascaraque, L.; Calderon, G.; del Campo, G.; Santamaria, A. The Smart Meter Challenge: Feasibility of Autonomous Indoor IoT Devices Depending on Its Energy Harvesting Source and IoT Wireless Technology. *Sensors* **2021**, *21*, 7433. https://doi.org/10.3390/s21227433

Academic Editors: Carlo Trigona, Olfa Kanoun and Slim Naifar

Received: 20 October 2021
Accepted: 6 November 2021
Published: 9 November 2021

1. Introduction

In the last few years, the number of IoT devices has grown significantly, so does the spending in IoT-related fields, which is expected to increase by 24% in 2021 [1]. Moreover, the global IoT market will be worth more than 1200 billion ($\times 10^9$) euros by 2027 [2]. It is estimated that the IoT was born in 2008, with the things-to-people ratio growing from 0.08 in 2003 to 1.84 in 2010 [3], and leading to almost 4.5 in 2021, with 35 billion devices connected [4], currently adding more than 120 new devices per second.

IoT devices may be used for a wide range of applications: healthcare, ambient conditions monitoring, workout tracking, infrastructures monitoring, failure prevention, smart metering, etc. The latter is the one we focus on in this work. Following our research presented in [5], in which an autonomous Sigfox smart meter was presented, in this paper, we analyse the feasibility of indoor IoT nodes. Specifically, we focus on smart meters with different energy harvesting (EH) sources and wireless technologies—although it might be extrapolated to any kind of indoor sensor.

Regarding the smart meter field, most of them are directly connected to the mains—such as those installed by electric companies—[6–8], so an intrusion in the electrical panel is required for installation, which leads to the requirement of a specialist and a temporary interruption in power supply. All these cons increase costs and prevent new deployments [9,10].

Considering the great growth in the IoT industry, the ability to keep smart meters autonomous, adapting their working conditions by their scenario necessities is crucial to encourage companies—and people—to adopt smart meters. Thus, we are about to compare the main trendy wireless communication technologies workable in smart meters—Sigfox, LoRaWAN, NB-IoT, Wi-Fi, BLE—for an extensive variety of scenarios. At the same time, we compare the most adequate EH sources for indoor wireless sensors: photovoltaic (PV), radiofrequency (RF); and specifically for smart meters: magnetic induction (MIEH). Based on metering application requisites and smart meter location constraints, this work aims to help smart meters' developers and adopters with a roadmap to select the most suitable wireless communication technologies and energy harvesting techniques.

The rest of the paper is organised as follows: Section 2 describes the smart meter under consideration for this review, characterising both its consumption and working principle. Section 3 assesses the actual electrical consumption for each wireless technology in the spotlight. In Section 4, we sum up the existing literature to provide data on EH generations for every technique. Finally, in Section 5, the feasibility of different combinations between IoT technologies and EH sources are discussed, supported by a brief conclusion in Section 6.

2. Targeted Smart Meter

The smart meter used in the development of this work is inspired by the one we developed in [5], except for the EH Subsystem (see Figure 1), which is removed and characterised accordingly as EH source in Section 4. In this section, we will briefly explain the main characteristics of the smart meter.

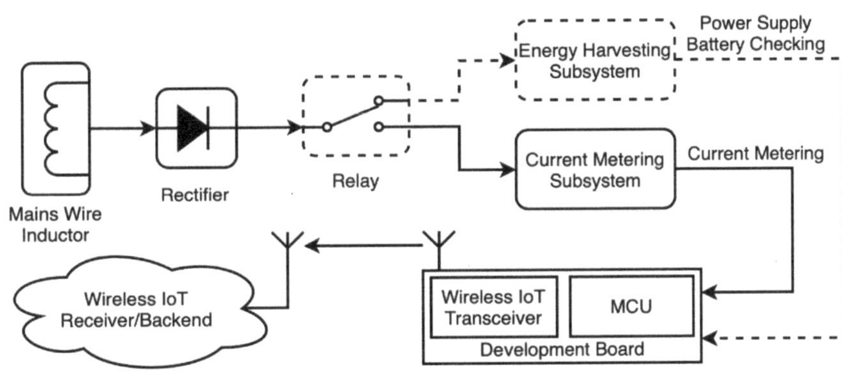

Figure 1. General block-diagram of the smart meter under consideration.

2.1. Working Principle

The working principle of the smart meter is that it is event-driven, with a variable metering period (*A* minutes) depending on the working conditions. Current measurements are stored in a buffer (sized to *N* measurements) and then sent when the buffer is full—see Figure 2. One measurement is 1 byte in size, as it is encoded in tenths on amperes (dA) with 1 dA (0.1 A) precision, which would allow measurements between 0 and 255 dA (0–25.5 A).

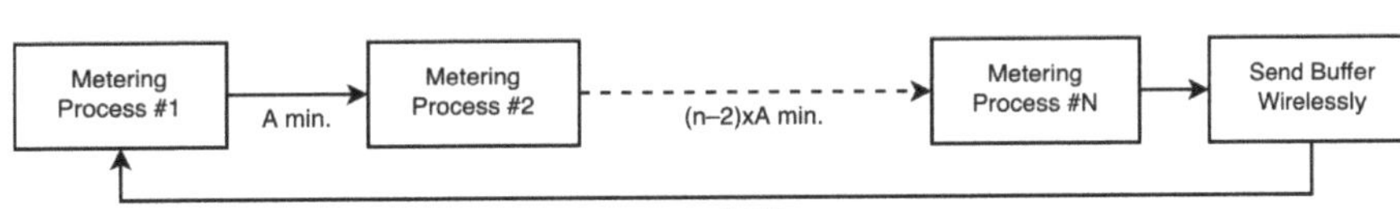

Figure 2. Graphical representation of the main behaviour of the smart meter.

This buffer is sent to the corresponding wireless receiver depending on the wireless technology that we are testing. It may be a base station that sends the data to a backend

(Sigfox, NB-IoT), an ad-hoc wireless gateway/host (LoRaWAN, Wi-Fi), or a receiver device scanning messages (BLE).

As well as the metering period may be changed based on the scenario conditions, so is the buffer. In the original design, the buffer was stated to 12 measurements due to Sigfox constrains [5], and the metering period was 15 min first, with latter tests of 5-min periods.

In this work, as other wireless technologies are used, the buffer size and the period may vary and adapt to the characteristics of each wireless technology.

2.2. Development Board

The development used for the smart meter is a FiPy [11]. This board provides Sigfox, LoRaWAN, NB-IoT, Wi-Fi and BLE in the same device, thus making the comparison between wireless technologies easier, as the basal consumption of the board is the same with no regard to the wireless technology under study. The microcontroller (MCU) is an Xtensa dual-core 32-bit LX6, which is programmed in MicroPython.

Figure 3 shows the basic layout for the five different wireless technologies embedded into the development board. Being the smart meter workable across different technologies makes the server ubiquitous and interoperable between smart metering instances—be they any of the aforementioned wireless technologies.

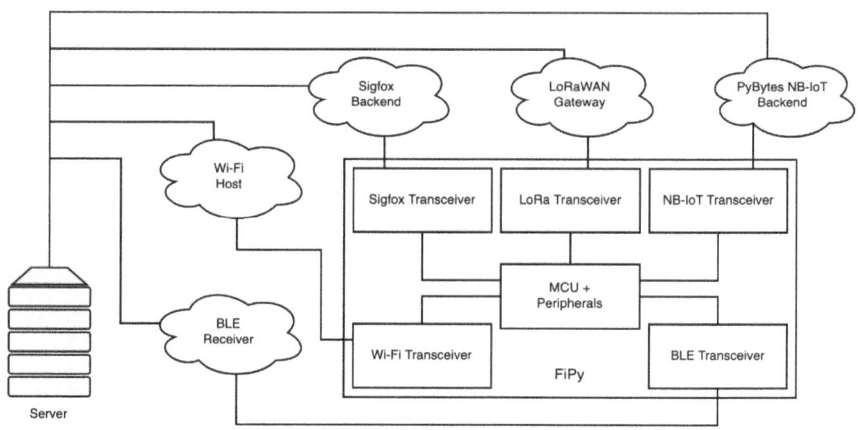

Figure 3. Layout for the five different wireless technologies under consideration.

2.3. Current Metering

The metering range is 0–25 A as it is an adequate range for most domestic scenarios. However, this range is easily adjusted by making small changes in the current metering subsystem [5], and it is not crucial in the object of this paper.

The Metering Subsystem is made as one can see in Figure 4. It is based on a voltage divider to convert the incoming current-proportional signal from the current probe into an ADC-measurable voltage signal. As depicted in Figure 2, the signal outcoming the current probe is first rectified. This allows a simpler circuit design for both current metering and energy harvesting (MIEH).

Resistors were adjusted accordingly to set the proper metering range within the ADC input range.

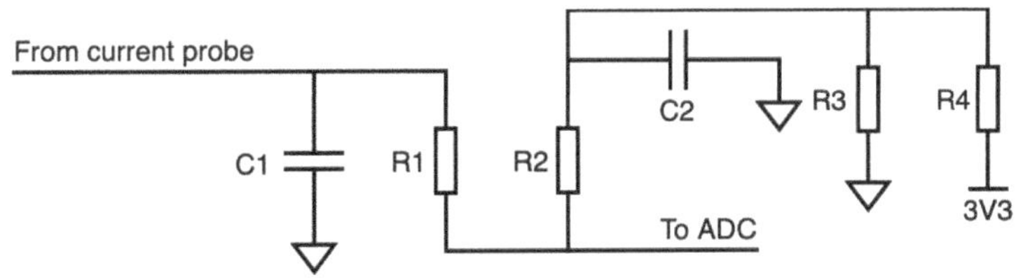

Figure 4. Conceptual schematic of the current metering subsystem.

2.4. Metering Consumption

Considering the functioning principle described in Section 2.1 hereof, the specific metering—data communication apart—consumption may be: (1) that coming from every metering process, and (2) stand-by time between metering processes:

1. The consumption due to the metering process will be the same indistinctively to the wireless technology used, as no wireless transceiver is turned on for current metering and the device is always the same: 44.16 mA during 2.55 s, i.e., 112.77 mAs—see Figure A1.
2. The stand-by consumption is a constant current value that is demanded during the deep-sleep period of the device, i.e., when the device is waiting for the timer to expire, and another current measurement is about to be done—this time is A minutes in Figure 2: 16.28 μA during A min, i.e., $A \times 1.29$ mAs. Datasheet [11] claims a deep-sleep current consumption of 25 μA at 3.3 V, fact that agrees with our 16.28 μA at 5 V.

These two different consumptions will compute for the energy consumption of the very smart meter. Measurements taken with the aid of a Keysight/Agilent 34410A multimeter [12] at 5 V.

In this manner, we split apart the consumption due to the wireless IoT technology—which is evaluated in Section 3 and will vary across protocols. The consumption of the wireless transmission happens eventually when the measurements buffer is full—corresponding to #N in Figure 2.

3. IoT Wireless Technologies Characterisation

In this section, the five IoT wireless technologies under study are evaluated. As the main objective of this work is to evaluate the feasibility of different wireless technologies according to their energy consumption, current consumption measurements are taken for each technology: Sigfox, LoRaWAN, NB-IoT, Wi-Fi and BLE.

We have selected these technologies as they tend to be the most used, supported, and well-known IoT wireless technologies nowadays. Furthermore, they represent a wide range of characteristics: some of them need to deploy a dedicated infrastructure (LoRaWAN, Wi-Fi, BLE) and others provide their own network for receiving messages (Sigfox, NB-IoT); Wi-Fi and BLE provide a short range whilst the others provide a very long range; they differ as well in data rate, payload, or the necessity of a handshake to send messages. These facts make the energy consumption comparison tougher as their behaviour is not equivalent. This will be discussed in Section 5.

For consumption portrayal, message payloads from 12 bytes (Sigfox limitation) up to 768 bytes will be considered—double-fold increments. Furthermore, the consumption of the handshake or communication establishment for those technologies requiring it is considered apart from that of the transmission itself.

Then, we will evaluate the choice of using larger messages with the technologies supporting that. Several protocols provide very high data rates, which may lead to an increase in energy efficiency provided that larger messages are used, as they will deliver fast, avoiding headers and eventual handshakes. Table 1 [13–15] summarises some critical points for the wireless technologies under consideration—for the transceivers embedded in our development board.

Table 1. Summary for some critical characteristics of the IoT Wireless technologies.

Technology	Data Rate	Payload [1]	Handshake
Sigfox	100 bps	12 B	No
LoRaWAN	5470 bps	222 B	Yes
NB-IoT	250 kbps	768+ B	Yes
Wi-Fi	16 Mbps	768+ B	Yes
BLE	1 Mbps	25 B	No

[1] User-available payload per message/packet.

Depending on the message size, a different number of transmissions will be needed, due to the maximum payload allowed by each technology for a single message. Hence, the consumptions required for each technology would be determined by the consumption of the connection establishment plus X times that of the message transmission, depending on the number of messages required—see Table 2.

Table 2. Number of transmissions needed for each wireless technology for different payloads.

Technology	24-Byte	48-Byte	96-Byte	192-Byte	384-Byte	768-Byte
Sigfox	2	4	8	16	32	64
LoRaWAN	1	1	1	1	2	4
NB-IoT	1	1	1	1	1	1
Wi-Fi	1	1	1	1	1	1
BLE	1	2	4	8	16	31

For current consumption characterisation, the equipment we used is a Nordic Semiconductor Power Profiler Kit II [16], which provides current sample rates of up to 100 kS/s (thousand-samples per second) with a resolution of up to 200 nA. Every current measurement is made with just the development board attached to the Power Profiler, using its embedded power source and high-precision ammeter, and only the essential peripherals turned on (see Figure 5).

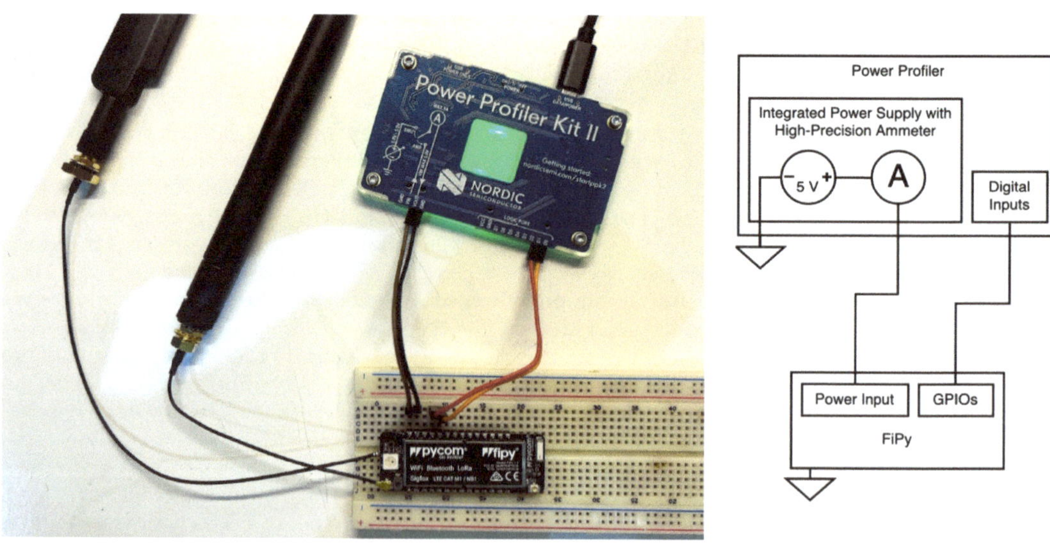

(a) (b)

Figure 5. Current measurement setup: (**a**) Real picture of the Power Profiler Kit II measuring current consumption of the development board; (**b**) Block-diagram of the current measurement setup.

So as to evaluate the actual accuracy of the Power Profiler, we set a test bed as described in Appendix B, Figure A11. The set was excited with a 100 Hz sinusoidal wave, characterised at the same time by the Power Profiler and a Keysight/Agilent 34410A multimeter—device taken as reference. Readings from the multimeter claim:

- I_{DC} = 92.825 mA
- I_{AC} = 20.539 mA_{RMS}

Considering that we are working with a sinusoidal wave, its peak value—which is that happened at 100 Hz—can be calculated as in Equation (1):

$$I_{100\text{ Hz}} = I_{AC_{RMS}} \times \sqrt{2} = 20.539 \times \sqrt{2} = 29.05\text{ mA} \quad (1)$$

The signal measured by the Power Profiler for 10 periods is depicted in Figure A12, as well as its Fast Fourier Transform (FFT)—performed with the aid of Matlab. In the FFT plot, we can see that there are only noticeable current values for 0 Hz (DC) and 100 Hz. These values are, with their relative errors (ε):

- $I_{0\text{ Hz}}$ = 90.61 mA; ($\varepsilon_{0\text{ Hz}}$ = 2.4%)
- $I_{100\text{ Hz}}$ = 28.71 mA; ($\varepsilon_{100\text{ Hz}}$ = 1.3%)

Besides, we extracted the value of the signal in 1000 same-phase sampling points (0 radians, which corresponds to a sine wave's nulls). For this set of values, the standard deviation was reckoned, happening to be 0.2492 mA for a mean value of 90.61 mA (0.3%).

Table 3 sums up the consumption for every technology characterised, which can be seen in detail in the following subsections and Appendix A. For the consumption characterisation, we measured the current demand for each wireless technology synchronously with the FiPy's firmware. This was made by triggering a digital signal during the transmission process. The consumption appearing in Table 3 corresponds to an average of a cluster of 100 current measurements for each technology. Range for different measurements is less than 5%, which is an acceptable tolerance value for characterisation taking into account the accuracy of the Power Profiler.

Table 3. Summary of energy consumptions for IoT wireless technologies (12-byte messages).

Technology	MTC (mA)	ETC (mA)	Establishment Consumption (mAs)	Transmission Consumption (mAs)	Total Consumption (mAs)
Sigfox [1]	<59.4	53.1	0	804.80	804.80
LoRaWAN	<59.4	62.5	256.73	6.33	263.06
NB-IoT	<277.2	141.2	4341.9	8.88	4350.8
Wi-Fi	118.8	109.9	198.46	3.40	201.86
BLE	85.8	95.9	51.98	1.01	52.99

[1] Sigfox's ETC is not calculated with the average consumption of the transmission process (Appendix A), but with the average current demand of one of the three transmissions occurred in Sigfox (90.1 mA), so as to not include the period between transmissions in the calculus.

In this table, we also depict the current demand of the development board's transceivers for each technology according to the manufacturer (Wi-Fi/BLE [17], LoRa/Sigfox [18], NB-IoT [11]), converted to 5 V (MTC). The current demands provided by the manufacturers do not always comply with our specific working characteristics, reason why they are shown as an upper limit—nevertheless, these data are useful to check that measured consumptions are in a reasonable margin within those of the manufacturer:

- Sigfox/LoRa transceiver's current demand is specified for +17 dBm, but not for +14 dBm, which is our case.
- NB-IoT transceiver's current demand is only specified for a possible maximum 1.5 W power, but actual power profile would depend upon network conditions.

The idle consumption of the board was also characterised and happened to be around 37 mA. This current must be subtracted to that measured for the whole board when transmitting (Appendix A) in order to isolate the specific experimental consumption of the transceiver's transmission (ETC), since idle current demand is due to the development board's MCU and peripherals before wireless transceivers are activated. However, this is not the case for Wi-Fi and BLE as the transceiver is embedded into the MCU, and the consumption data provided by the manufacturer take the whole board into consideration.

The consumption coming from connection establishment (or just modem initialisation) is also depicted in Table 3. This consumption is relevant, as the device will need to do this process when waking up from deep sleep for the technologies requiring it.

In the upcoming subsections, each technology will be deeply analysed. First, the consumption for the initial 12-byte-payload message is explained; then, a table shows the measured consumptions for larger payload sizes for each technology. Detailed consumption figures are attached in Appendix A.

3.1. Sigfox

The Sigfox 12-byte transmission process lasts 10.6 s with an average current consumption of 75.93 mA, which leads to an energy consumption of 804.80 mAs. Sigfox does not need a handshake for the connection to be established, so this energy is the only needed when sending a message.

In the consumption plot (Figure A2), it is clearly noticeable the three transmissions in three different carriers that Sigfox performs to ensure delivery: the three big waves that last about two seconds, which is the time on air of a 12-byte Sigfox message.

For larger payloads, check Table 4.

Table 4. Summary of energy consumptions (mAs) for Sigfox.

Sigfox Payload (Bytes)	12	24	48	96	192	384	768
Establishment consumption	0	0	0	0	0	0	0
Number of transmissions	1	2	4	8	16	32	64
Transmission consumption	804.80	804.80	804.80	804.80	804.80	804.80	804.80
Total consumption	804.80	1609.6	3219.2	6438.4	12,876.8	25,753.6	51,507.2

3.2. LoRaWAN

LoRaWAN allows different configurations regarding spreading factor (SF) and bandwidth. In this work, we assume the most used in the LoRa field is the proper one to use, with corresponds to using an SF7 with 125 kHz. Other SF could be used if messages were lost, but this configuration works well for us whilst providing the fastest data rates, which means less time on air, thus less energy spent.

With this set, the LoRaWAN 12-byte transmission process lasts for 63.63 ms, with an average consumption of 99.51 mA, which leads to an energy consumption of 6.33 mAs for the LoRaWAN process (see Figure A3). However, it is important to note that after transmission, the LoRa transceiver may keep activated for a while to be able to receive messages. We will not consider that phenomenon here as we truly think that it is not necessary for message transmission—the main purpose of this work is comparing only the uplink channel.

LoRaWAN needs connection establishment before sending messages. When the device is put in a deep sleep mode, the connection needs to be done again as the device loses its previous state. The consumption for the connection establishment can be seen in Figure A4. This process lasts 5.2 s with an average consumption of 49.42 mA, which leads to an energy consumption of 256.73 mAs.

Thus, the full energy consumption for a LoRaWAN 12-byte message is 263.06 mAs—considering both the establishment and the transmission. For larger payloads, check Table 5.

Table 5. Summary of energy consumptions (mAs) for LoRaWAN.

LoRaWAN Payload (Bytes)	12	24	48	96	192	384	768
Establishment consumption	256.73	256.73	256.73	256.73	256.73	256.73	256.73
Number of transmissions	1	1	1	1	1	2	4
Transmission consumption	6.33	9.34	12.75	20.79	36.81	36.81	36.81
Total consumption	263.06	266.07	269.48	277.52	293.54	330.35	403.97

3.3. NB-IoT

The NB-IoT 12-byte transmission process lasts for 49.8 ms, with an average current of 178.2 mA, which leads to an energy consumption of 8.88 mAs (see Figure A5). The consumption for NB-IoT does not increase much as message payload increases—as measured in Table 6. This wireless technology provides very high data rates which makes the time spent not definitely increase when we use small byte-sized increases.

Table 6. Summary of energy consumptions (mAs) for NB-IoT.

NB-IoT Payload (Bytes)	12	24	48	96	192	384	768
Establishment consumption	4341.9	4341.9	4341.9	4341.9	4341.9	4341.9	4341.9
Number of transmissions	1	1	1	1	1	1	1
Transmission consumption	8.88	8.88	8.88	9.13	9.37	9.87	11.15
Total consumption	4350.8	4350.8	4350.8	4351.0	4351.3	4351.8	4353.1

NB-IoT takes a long time to establish the connection. The modem first needs to be initialised, then attach to the network, then connect to the service. The time needed for attaching and connecting to the network can vary depending on the NB-IoT network conditions (Vodafone Spain), so will the energy consumption. An average process is shown in Figure A6, with an energy consumption of 4342 mAs.

We can see that the consumption coming from the transmission is negligible considering that from the connection establishment (8.88 mAs vs. 4341.9 mAs), with a total of 4350.8 mAs for a 12-byte message. For larger payloads, check Table 6.

3.4. Wi-Fi

The Wi-Fi transmission process lasts for 30.98 ms, with an average current of 109.0 mA, which leads to an energy consumption of 3.4 mAs. The Wi-Fi consumption does not depend on the packet size, as it is meant for larger messages. The time it takes for the transmission to be made is not consistent, however, as it depends on the current network traffic and network characteristics. For illustration, let us consider an average transmission as in Figure A7.

Wi-Fi needs to establish a connection before sending messages. The modem needs to be reinitialised every time the device wakes from a sleep mode as well. This time will vary depending on real working conditions as the Wi-Fi connection time is not only dependant on the IoT device. In Figure A8, the consumption for modem initialisation and connection establishment can be seen.

Thus, the whole consumption required to send a Wi-Fi message is that coming from the establishment and the transmission per se, which is 201.86 mAs. For larger payloads, check Table 7.

Table 7. Summary of energy consumptions (mAs) for Wi-Fi.

Wi-Fi Payload (Bytes)	12	24	48	96	192	384	768
Establishment consumption	198.46	198.46	198.46	198.46	198.46	198.46	198.46
Number of transmissions	1	1	1	1	1	1	1
Transmission consumption	3.40	3.40	3.40	3.40	3.40	3.40	3.40
Total consumption	201.86	201.86	201.86	201.86	201.86	201.86	201.86

3.5. BLE

For the case of BLE, the FiPy board still provides a limited support of the protocol. Only advertisements are supported, and no security can be implemented. Thus, we implement transmissions over BLE as advertisements that are received by another BLE device.

Moreover, the size of message does not matter as the driver always fill the user datagram with dumb bytes. We have checked in the lab that the time and current waveform—i.e., the energy consumed—for the advertisement transmission is the same either with one user byte or the maximum 25 user bytes.

In Figure A9, the current consumption for the BLE transmission process (12-byte message) can be seen.

Since the BLE advertisement is set until the transmission finishes, the time spent is 10.52 ms, with an average consumption of 95.99 mA, which leads to an energy consumption of 1.01 mAs for the transmission of a 12-byte BLE message.

BLE does not need a connection establishment per se, but when the device wakes up from a sleep mode, the modem needs to be reinitialised. This consumption (see Figure A10) must be taken into consideration every time the device sends a message: 51.98 mAs.

Thus, the full energy consumption for a 12-byte BLE transmission—considering both the modem initialisation and the message transmission—gets to be 52.99 mAs. For larger payloads, check Table 8.

Table 8. Summary of energy consumptions (mAs) for BLE.

BLE Payload (Bytes)	12	24	48	96	192	384	768
Establishment consumption	51.98	51.98	51.98	51.98	51.98	51.98	51.98
Number of transmissions	1	1	2	4	8	16	31
Transmission consumption	1.01	1.01	1.01	1.01	1.01	1.01	1.01
Total consumption	52.99	52.99	54.00	56.02	60.06	68.14	83.29

4. Energy Harvesting Generation

The global increase in IoT nodes, and especially indoor ones, has led to the search for new ways of providing power supply to them. IoT devices are usually wireless, which means they need a battery (or other storage element) to operate. This battery needs to be replaced occasionally, which increase costs—in fact, the cost of replacing the battery can be greater than the IoT device itself. For exemplification, the device explained in Section 2 hereof has an average consumption of about 1 mW [5]. This device was wisely designed to be very power efficient, so it can run autonomously by means of MIEH.

Energy harvesting technology provides a green, carbon-free, sustainable, and virtually infinite power supply to wireless devices, obtaining the available energy from the environment to reduce—or even eliminate—the need for storage elements and wired power supply. Some of the most relevant EH methods for IoT devices are ambient radiation (RF), photovoltaic, piezoelectric, magnetic induction, vibration, pyroelectric and thermoelectric [5].

This paper is focused on the development and possibilities of what we consider the three main sources for indoor smart meters: photovoltaic (transforming light radiation into electrical current), radiofrequency (using the already-present electromagnetic waves) and magnetic induction (exploiting the changing magnetic fields that occurred in AC wires).

The nature of these three technologies entails a complex comparative study. In the literature, one can find different results for a bunch of designs, configurations and sets. Power generations depend widely on the real characteristics of the ambient energy availability, device's design, system's requisites, and implementation. However, it is crucial to compare these techniques with one another. Hence, for this work, we are using data from similar use cases found in the literature—comparing them by magnitude order instead of concrete values as they tend to be specific for each case. Be they:

- Photovoltaic: tens of milliwatts (10–20 mW)
- Radiofrequency: tenths of milliwatts (0.1–0.2 mW)
- Magnetic induction: milliwatts (1–2 mW)

4.1. *Photovoltaic*

Photovoltaic cells are used to power a multitude of sensors and other IoT systems. In [19], Xicai Yue et al. developed a PV EH model for a CO_2 sensor, which produces an output of 4.2 V and a pulse current of 100 mA for 600 ms. They also determined that the best average storage efficiency is obtained at 200 lux. This is remarkable because, if a storage element is added to the PV harvester, the device could achieve its self-sustainability. Abhiman Hande et al. depicted in [20] the use of several ultracapacitors as energy storage devices. This leads to a compact and robust system with fewer solar panels in a series-parallel combination.

The values of the indoor photovoltaic technique depend mainly on two criteria: light and material. Regarding the nature of light it usually would be artificial for inside use cases; in terms of material, the efficiency of new generation cells compared to commercial silicon is more than twice [21]. Nonetheless, it is important to note that real-world working conditions may drop theoretical efficiencies of PV panels to half [22].

Indoor lighting conditions are more complex than the exterior ones, since spaces are smaller, and the light levels are lower and intermittent. Indoor PV cells have a small size of around 30 cm^2 [23]. A polymer-based silicon cell has a conversion power efficiency of 9% for F12 fluorescent lamps (100 lux). The light resource comes from natural outdoor light and artificial light from luminaires—fluorescent, LED. Solar cells require an irradiance of around 7 mW/m^2 to work properly.

On the one hand, the non-constant presence of people reduces the daylight hours of the surroundings. On the other hand, average consumption is much lower than active consumption, saving energy and enabling this technology. The most common outdoor solar cells are made of silicon, a cheap material highly studied that produces 0.1 mW/lm. However, for indoor conditions where size requirements are tougher, researchers are studying new materials with larger bandgap energy, higher efficiency and smaller dark currents [24].

In [25], Lin Xie et al. use an organic photovoltaic cell for high intensity indoor environments, such as hospitals, with a theoretical efficiency of 60%. In the same way, the interest in gallium cells for indoor EH is also growing thanks to the lack of transparency losses, leading to a 40% energy conversion [22]. A small GaAs cell that is one-third of an office paper can power devices of up to 10 mW [26].

PV could be mixed with other technologies. For example, in [27] Yen Kheng Tan et al. create a hybrid of solar photovoltaic and thermal energy. The goal of this system is to extend the lifetime of the wireless sensor nodes. For a solar irradiance of 1010 lux and a thermal gradient of 10 K, an average of 621 W is obtained—three times more than just thermal.

Indoor and outdoor uses can also be combined, for example for smart building applications. This is described in [28], where a PV EH generates an average of 0.5 W, getting to be a maximum of 2.3 W with 130 lux.

4.2. Radiofrequency

Radio signals, mobile phones or Wi-Fi are some of the examples that originate these electromagnetic waves which are available almost everywhere. In addition, the fluctuations it presents are usually caused by human factor, not by weather conditions.

It is essential to consider the frequency, the modulation, and the power transmission to properly design the circuit of indoor devices. The key element is known as rectenna. Its main components are a rectifier and an antenna. In order to optimise energy conversion, a DC-DC boost converter can be added [29]. Many rectennas have been created, achieving different efficiencies depending on the design characteristics.

The study of radiofrequency EH is related to the dBm value of the incident signal and the conversion efficiency. For –20 dBm, equivalent to 10 μW, the average efficiency may be around 10%, while if it is goes up to 0 dBm (1 mW), the efficiency could increase up to 50% [30].

In [31], a paper substrate rectenna designed for an input power of −20 dBm over LTE bands presented a range of 5–16% efficiency; while the six-band dual rectenna presented in [32] showed a range of efficiency between 37.7% and 41.4% with −15 dBm input power. In [33], A. Eid et al. presented an ultra-compact and flexible rectenna that can operate at the 2.4 GHz ISM band reaching up to 40% efficiency.

For a 2.4 GHz Wi-Fi Multi-SSID router and power of 0 dBm, the energy harvester presented in [30] has an efficiency of 50.18%, due to the power management unit which minimise losses. Using the same circuit design, it is possible to operate on several ISM band or cellular RF as well [34].

Huaguang He et al. depicted in [35] a 2×4 rectenna array for the range of 895–925 MHz and 1.6–2.65 GHz. They obtained an output pulse voltage of 3.3 V and 33 mA for 100 ms, with an efficiency of 33% for 1800 MHz, 41.5% for 900 MHz and 55% for 2.4 GHz. However, it is important to note that the use of rectenna arrays at indoor environments might be inappropriate if the multipath effect is not eliminated since RF energy harvesting is sensitive to the angle of incidence, so adding more rectennas to the harvester is not a straightforward solution.

The current challenge is to improve the conversion capacity and energy harvesting of the rectennas. There are several possibilities, such as broadband antennas or antenna arrays. Nevertheless, the first requires a matching network for each frequency band and the second ones add complexity.

4.3. Magnetic Induction

The third most relevant energy harvesting technique for indoor IoT smart meters or similar applications is magnetic induction. The basic requirement is to have a coil, either near or surrounding the wires carrying alternating current so that the change of inducted magnetic fields excite the coil, thus generating an electromotive force. This means that the electrical panel or distribution box must be opened, but no electrical disruption is needed as electrical circuits are not modified. This EH technique was studied and characterised in detail in our previous work "Smart Metering for Challenging Scenarios: A Low-Cost, Self-Powered and Non-Intrusive IoT Device". In this paper [5], a metering IoT device was power supplied by the electromotive force generated inside the coil, providing around 1 mW of power for an average household mains line.

When sensors or devices require current measurement, as in this case, MIEH is the optimal choice since it uses the same physical principle as current probes. In this way, the whole system is more compact, simpler, and cost-efficient. This appreciation of the use of magnetic induction as EH has hardly been developed and even less applying wireless IoT communications. The effective mains current energy harvesting range of this device is similar to that at home, corresponding to a current between 2.2 and 6.4 A. Using IEEE 802.15.4 as wireless technology, Danilo Porcarelli et al. achieved in [36] a device is self-powered with a load of 300 W for at least 60 s of measurement.

5. Discussion

In this section, we assess the main challenge for autonomous indoor smart meters: power balance. We address the trade-off between working conditions, energy harvesting generation and wireless technology energy consumption.

The five wireless technologies in the scope are compared with different working conditions for the smart meter—metering period and buffer size—, at the same time they are evaluated to check if they comply with the available power generated by the EH techniques.

5.1. Wireless Technology Selection

After having characterised the five wireless technologies, we can now analyse the energy consumption each of them would have, and therefore discuss which one would be better depending on the power available.

This work was planned so that every wireless technology runs in the same development board. This allows us to not be dependent on the consumption of the metering process per se, being the same for all technologies. This was characterised in Section 2 hereof, and it is not worth considering for the wireless technologies comparison.

Figure 6 depicts a quantitative representation of the consumption needed for each technology—it is plotted out on a non-linear scale to enhance visualisation, since for the technologies requiring establishment, the vast majority of their consumption comes from that very process.

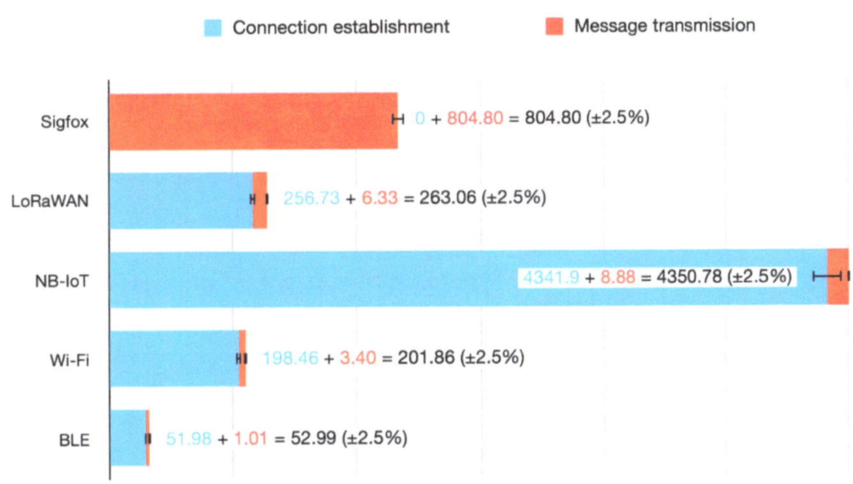

Figure 6. Quantitative representation of the wireless technologies' consumption. A confidence interval of 5% is depicted as the set of measurements for every technology were within this range.

These consumptions are those from Table 3, where they were characterised for 12-byte messages as it is the Sigfox limitation. However, for some technologies, increasing the message payload does not proportionally increase the message transmission consumption—as one could see in Section 3, Tables 4–8.

This way, we could reduce the number of messages—thus lowering the number of connection establishments—, and therefore decrease the whole energy consumption. For instance, with LoRaWAN in our configuration (SF7, 125 kHz), we can send messages of up to 222 bytes. For the case of BLE, our board lets us send messages of up to 25 bytes. For the case of both Wi-Fi and NB-IoT, this limit is much higher and above 1024 bytes—768 bytes is the limit considered for comparison.

Table 9 shows the full consumption for each wireless technology and every message size. Moreover, in Figure 7 (logarithmic scale) we can see the evolution on the consumption required by each technology as the payload increases from 12 to 768 bytes.

Table 9. Summary of energy consumptions (mAs) for all message sizes for all technologies.

Technology	12-Byte	24-Byte	48-Byte	96-Byte	192-Byte	384-Byte	768-Byte
Sigfox	804.80	1609.6	3219.2	6438.4	12,876.8	25,753.6	51,507.2
LoRaWAN	263.06	266.07	269.48	277.52	293.54	330.35	403.97
NB-IoT	4350.8	4350.8	4350.8	4351.0	4351.3	4351.8	4353.1
Wi-Fi	201.86	201.86	201.86	201.86	201.86	201.86	201.86
BLE	52.99	52.99	54.00	56.02	60.06	68.14	83.29

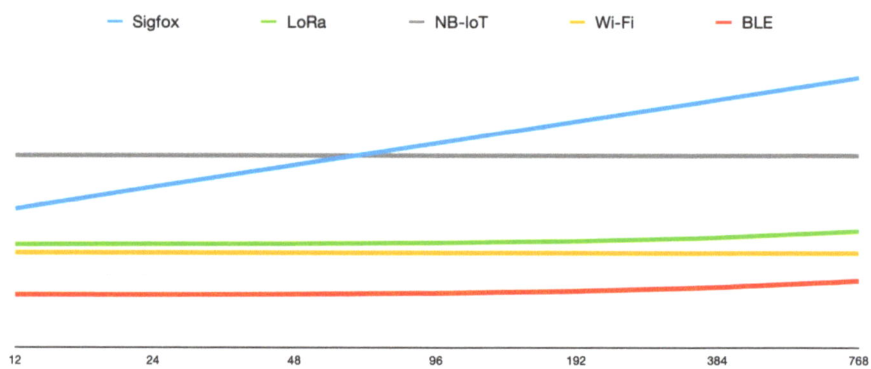

Figure 7. Plot of the wireless technologies' consumption in logarithmic scale for different message sizes.

Apart from the features of the communications—payload, communication establishment, number of transmissions—the characteristics of every use case will also determine the best suitable wireless technology for the application: what range should the device support, is there a strict real-time requirement for the data, can a public infrastructure be used, etc.

5.2. Energy Harvesting Technique Selection

For feasibility characterisation, not only the raw power available from the energy harvesting sources is important, but also their availability. For some EH sources, their availability may vary importantly whilst for others may be steadier, more predictable. For instance, indoor PV might be very different along time for the same space, as it will depend on the external weather, several people switching lights, and even the state of blinds. On the other hand, RF will be steadier along time as ambient radiation tends to be constant in the long term—it really only changes provided that new wireless technologies are deployed. MIEH may be also fairly predictable and constant, as the energy consumption of a building tends to be the same by periods—working days/weekends, working hours/nights, and holidays.

In order to avoid the lack of power supply, it could be possible to combine two types of EH techniques, such as PV and MIEH. However, this combination makes the device more complicated as two harvesting circuits would be required. Furthermore, the solar panel should be located outside the electrical panel to receive light, which adds complexity to the whole system. The same manner, if RF were used, the rectenna would also need to be outside the electrical panel to receive stronger signals.

The use of one EH source over another will depend on the specific use case: where will the device be located, will there be enough mains consumption to exploit MIEH, will there be ambient light to install a PV harvester, will there be much data to be transmitted, etc.

Nonetheless, for smart meters the best approach gets to be MIEH: no elements need to be placed outside the electrical panel, and the same physical device acts as current probe—crucial for metering—and harvester, which simplifies the device, hence providing an easy, fast way of deploying ubiquitous smart meters.

Although adding more harvesters or using better ones could solve the power balance problem, this would also make the device more expensive—and complex if several harvesters are combined. Since this work is focused on smart meters, MIEH is the most appropriate solution in terms of scalability, simplicity and ease of installation; even if some trade-offs are faced using this technique, such as reducing metering period.

5.3. Constraints for Energy Consumption of Smart Meters

We can reckon the average power demand for every case with Equation (2). The result (Υ) would be in mAs, which is directly converted to mW multiplying by the voltage supply of the characterisation (5 V).

This calculus let us write Table 10, where all possible combinations are shown, properly depicting the average consumption of the device. This is the consumption that must be counter-balanced by the EH source in order to make the device autonomous.

Table 10. Smart meter average consumption (mW) depending on buffer size—i.e., message payload—, metering period and wireless technology.

Buffer Size (Bytes)	Metering Period (min.)	Sigfox	LoRaWAN	NB-IoT	Wi-Fi	BLE
12	1	15.07	11.31	39.69	10.89	9.85
	5	3.08	2.33	8.00	2.24	2.03
	10	*1.58*	*1.20*	4.04	*1.16*	*1.06*
	15	*1.08*	*0.83*	2.72	*0.80*	*0.73*
24	1	15.07	10.40	24.59	10.18	9.66
	5	3.08	2.15	4.98	2.10	2.00
	10	*1.58*	*1.11*	2.53	*1.09*	*1.04*
	15	*1.08*	*0.77*	*1.72*	*0.75*	*0.72*
48	1	15.07	9.95	17.03	9.83	9.57
	5	3.08	2.05	3.47	2.03	*1.98*
	10	*1.58*	*1.07*	*1.78*	*1.06*	*1.03*
	15	*1.08*	*0.74*	*1.21*	*0.73*	*0.71*
96	1	15.07	9.72	13.26	9.65	9.53
	5	3.08	2.01	2.72	*1.99*	*1.97*
	10	*1.58*	*1.05*	*1.40*	*1.04*	*1.03*
	15	*1.08*	*0.72*	*0.96*	*0.72*	*0.71*
192	1	15.07	9.60	11.37	9.57	9.50
	5	3.08	*1.99*	2.34	*1.98*	*1.97*
	10	*1.58*	*1.03*	*1.21*	*1.03*	*1.02*
	15	*1.08*	*0.71*	*0.83*	*0.71*	*0.71*

Table 10. Cont.

Buffer Size (Bytes)	Metering Period (min.)	Sigfox	LoRaWAN	NB-IoT	Wi-Fi	BLE
384	1	15.07	9.55	11.42	9.50	9.49
	5	3.08	*1.98*	2.15	*1.97*	*1.96*
	10	*1.58*	*1.03*	*1.12*	*1.03*	*1.02*
	15	*1.08*	*0.71*	*0.77*	*0.71*	*0.70*
768	1	15.07	9.52	9.95	9.50	9.49
	5	3.08	*1.97*	2.06	*1.97*	*1.96*
	10	*1.58*	*1.03*	*1.07*	*1.02*	*1.02*
	15	*1.08*	*0.71*	*0.74*	*0.71*	*0.71*

Considering EH generations as in Section 4 hereof, each table cell text is marked indicating the EH techniques that would comply to make the device autonomous—we consider a qualitative magnitude order, as claimed in Section 4:

- Circled: RF complies, so do MIEH and PV (never happens)
- Italics: MIEH complies, so does PV
- Underline: only PV complies
- Grey text: no EH technique would comply

$$\Psi = \frac{\Sigma \times \Gamma + \Phi + \Sigma \times \mathrm{T} \times \Theta}{\Sigma \times \mathrm{T}} \tag{2}$$

where:

- Σ is the buffer size (bytes)
- Γ is the consumption of a standard metering process (mAs)
- Φ is the consumption of a transmission for the designated technology (mAs)
- T is the metering period (s)
- Θ is the stand-by consumption (mA)

The optimal combination of energy harvesting technique and IoT wireless technology will depend on functional requisites (metering period), communication requirements (buffer size) and location constraints (EH source availability).

Enlarging the stand-by periods of event-driven IoT nodes enhances power requirements, making them lower. However, this means losing time-precision in measurements because fewer samples are taken. This phenomenon could even lead to missing information about short events, such as the consumption peak that happened when heating something in a microwave oven, which may last for 3–5 min, but might be missed if the metering period is set to a higher value.

Nevertheless, enlarging the measurements buffer does not always enhance energy consumption. In the case of Sigfox, there is no difference at all since Sigfox's consumption increases linearly as the buffer size increases, due to both its 12-byte payload limitation and the fact that it does not need any establishment consumption (see Table 4). The technology with the largest reduction in consumption by buffer enlargement is NB-IoT, mainly because it allows big payloads but requires a big establishment consumption; however, the transmission consumption per se is nearly the same with no regard to the message size (see Table 6). The others (LoRaWAN, Wi-Fi, BLE) present a restraint enhancement that, in most cases, will not be worth it as the data will update less frequently.

Moreover, as fewer measurements are taken—stand-by period extended, buffer size is increased, fewer wireless transmissions—the predominant consumption tends to be that coming from the smart meter per se, thus making wireless technologies consumption differences less noticeable in the overall average consumption of the IoT device. Notice that the consumption coming from the stand-by period and the metering process is the same regardless of the wireless IoT technology.

It is important to note that, with the development board and wireless technologies used for this work, radiofrequency energy harvesting cannot be used since the amount of energy that can be harvested is lower than the necessary to meet the autonomous working conditions.

6. Conclusions

Feasibility of autonomous indoor IoT smart meters is mainly restricted by the limitations on the ambient energy available to harvest. Although IoT devices have achieved very low power consumptions in the recent years, wireless technologies still require a considerably large amount of power to perform. With the aim of facilitating the design and development of autonomous wireless smart meters, we have performed a comparative analysis of the energy consumption of the trendiest IoT wireless communication technologies (Sigfox, LoRaWAN, NB-IoT, Wi-Fi, BLE), and the most appropriate EH techniques for smart meter applications (PV, RF, MIEH). This analysis will help smart meter developers, metering solutions providers and smart metering adopters in the selection of the most suitable combination of wireless communication technology and energy harvesting technique—based on their use case characteristics and location constraints.

Regarding IoT wireless technologies, it has been shown that the most energy efficient happens to be BLE, followed by LoRaWAN and Wi-Fi, finally tailing NB-IoT and Sigfox. However, their specific consumption will be determined by the metering period and buffer size. Customarily, BLE will be the best choice if communication range is not a limiting factor; otherwise, it will be LoRaWAN. However, if a commercial managed receiving network is wished, NB-IoT or Sigfox will be the suitable ones depending on data constrains.

Concerning EH techniques, the most powerful is PV, followed by MIEH, and finally RF. Yet, considering the smart meter field, MIEH gets to be the most appropriate since it takes advantage of the metering component (current sensor) as energy harvester, reducing device complexity and costs. Besides, smart meters are usually located inside electrical panels or distribution boxes—where no light is present.

Results prove that selection of wireless technology and energy harvesting technique will depend on the desirable application requisites—metering period, communication period—and the location constraints—EH source availability.

Author Contributions: Conceptualization, methodology, validation, E.S. and G.d.C.; software, hardware, laboratory work, E.S. and L.M.; formal analysis, resources, visualization, data curation, E.S.; investigation, E.S., L.M. and G.C.; writing—original draft preparation, E.S. and L.M.; writing—review and editing, supervision, G.d.C.; project administration, funding acquisition, A.S. All authors have read and agreed to the published version of the manuscript.

Funding: This work is part of the CHIST-ERA research project "ABIDI: Context-aware and Veracious Big Data Analytics for Industrial IoT" and is funded by the State Research Agency (AEI) from Ministerio de Ciencia e Innovación (MICINN) of Spain, grant number PCI2019-103762.

Institutional Review Board Statement: Not applicable.

Informed Consent Statement: Not applicable.

Data Availability Statement: Not applicable.

Conflicts of Interest: The authors declare no conflict of interest. The funders had no role in the design of the study; in the collection, analyses, or interpretation of data; in the writing of the manuscript, or in the decision to publish the results.

Appendix A

This appendix contains current consumption graphs for every wireless technology characterised and the smart metering process.

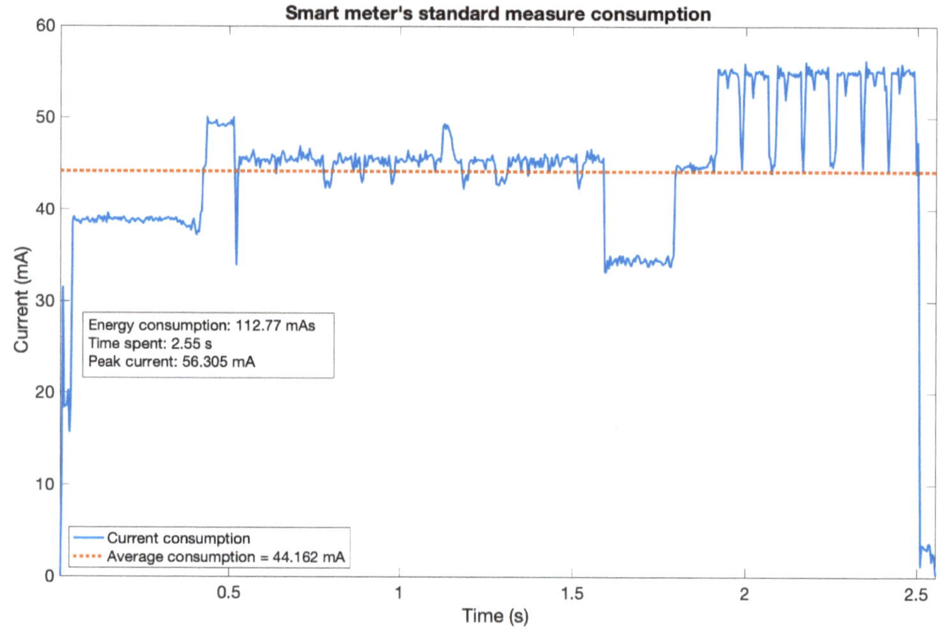

Figure A1. Current consumption detail of the smart meter's metering process.

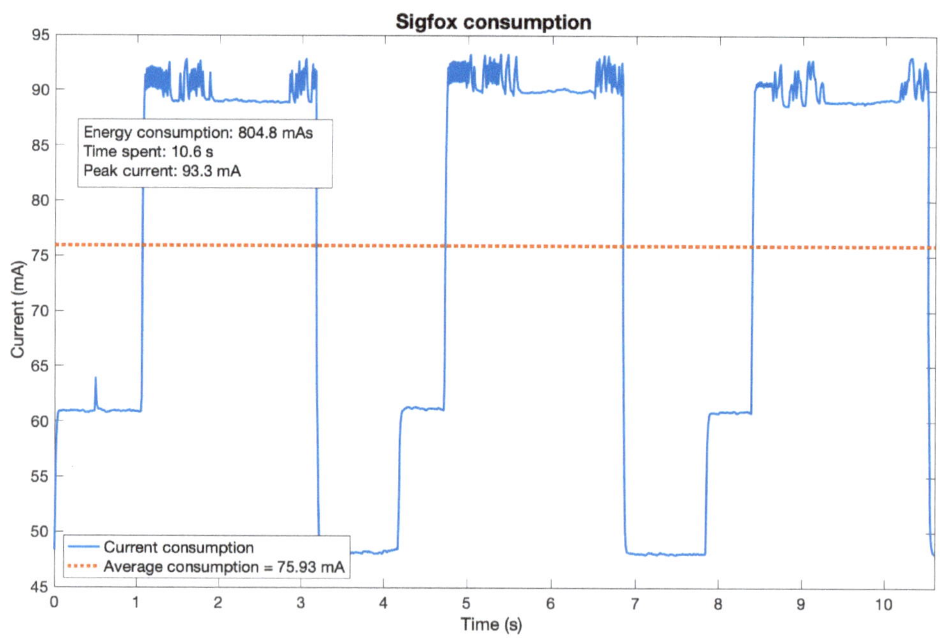

Figure A2. Current consumption detail for the Sigfox transmission. Data acquired at 100 s/s (samples per second).

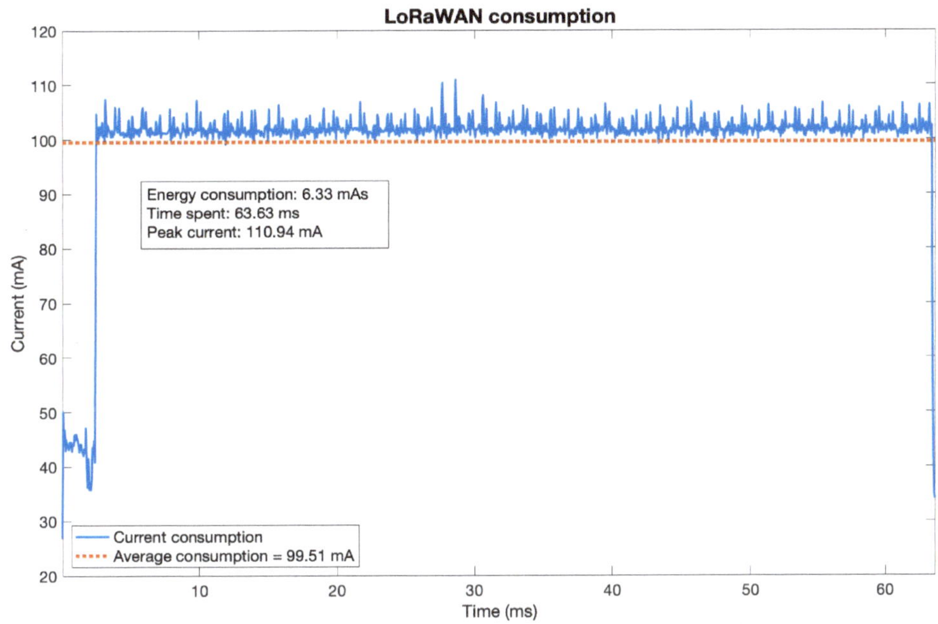

Figure A3. Current consumption detail for the LoRaWAN transmission. Data acquired at 100 kS/s, resampled at 1/6 for plotting.

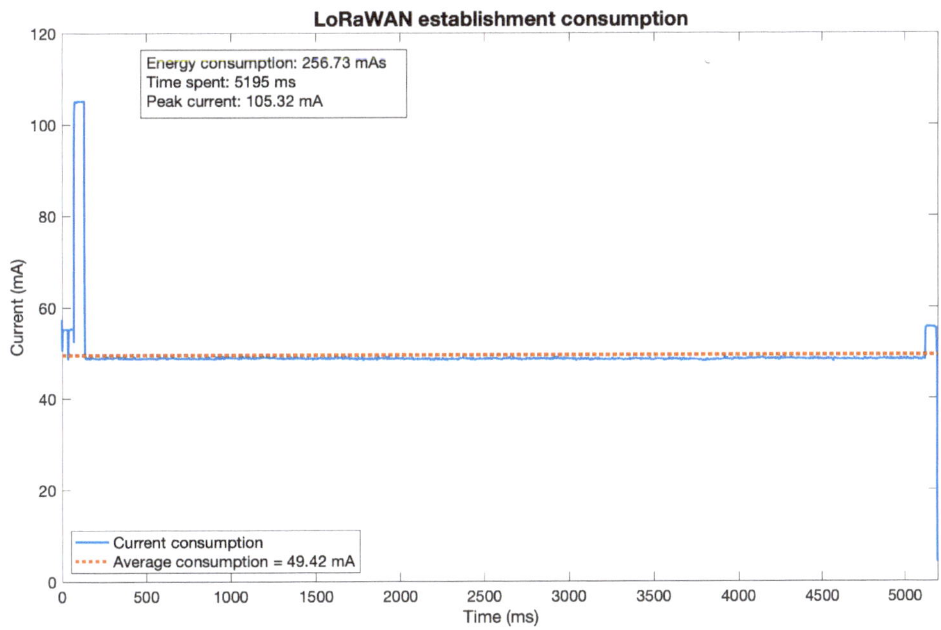

Figure A4. Current consumption detail for the LoRaWAN connection. Data acquired at 1 kS/s.

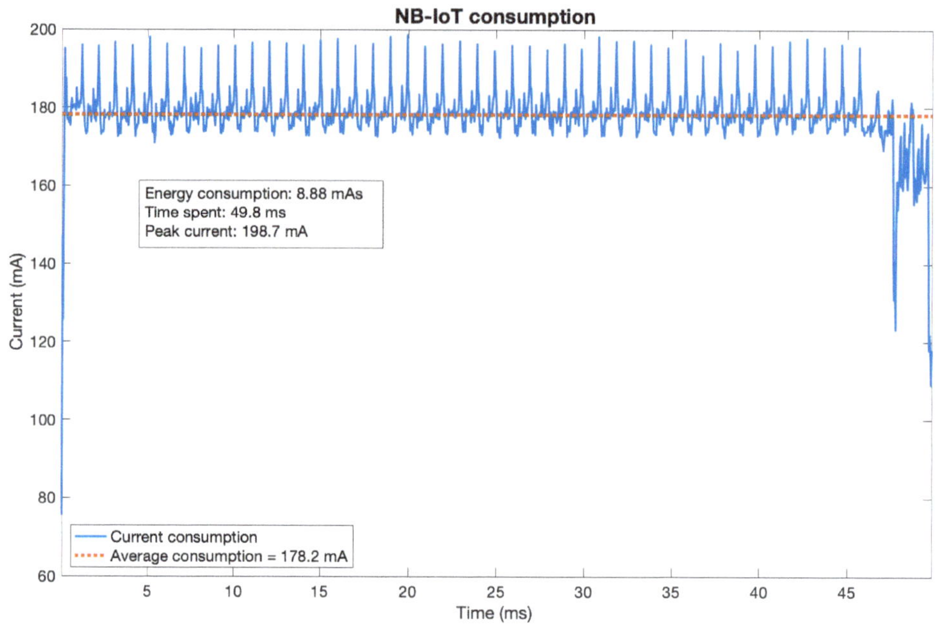

Figure A5. Current consumption detail for the NB-IoT transmission. Data acquired at 100 kS/s, resampled at 1/4 for plotting.

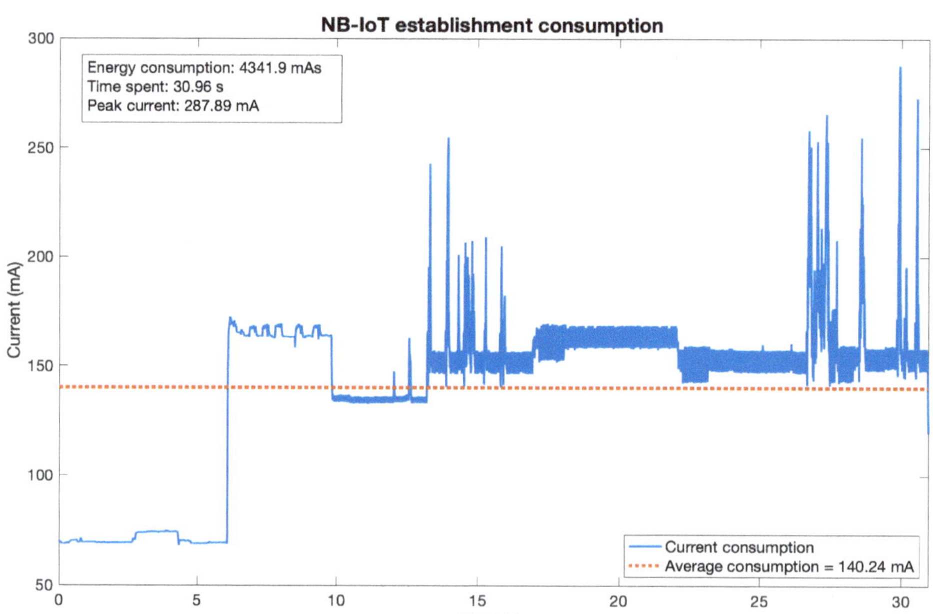

Figure A6. Current consumption for the NB-IoT connection establishment. Data acquired at 100 S/s.

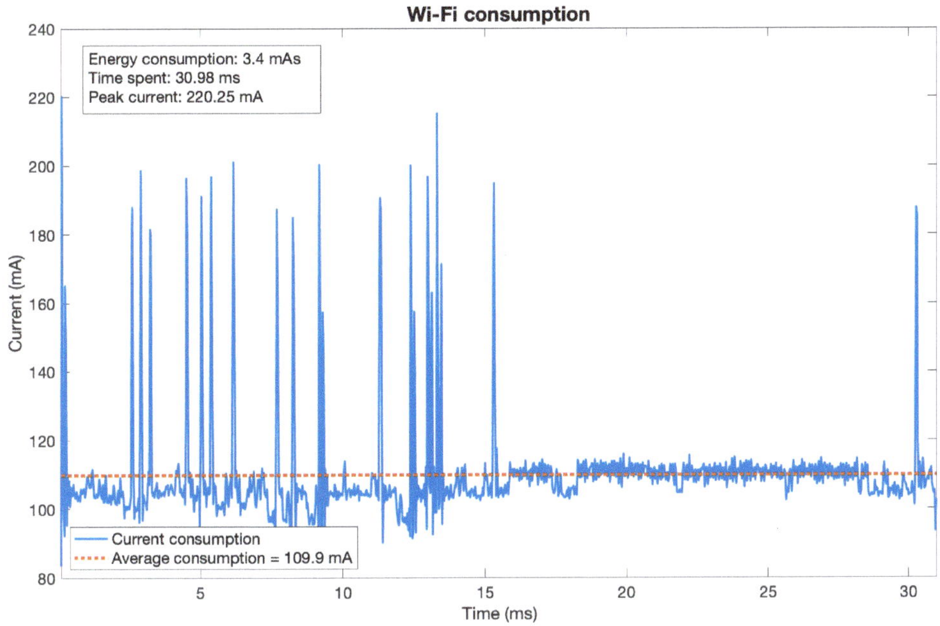

Figure A7. Current consumption detail for the Wi-Fi transmission. Data acquired at 100 kS/s, resampled at 1/3 for plotting.

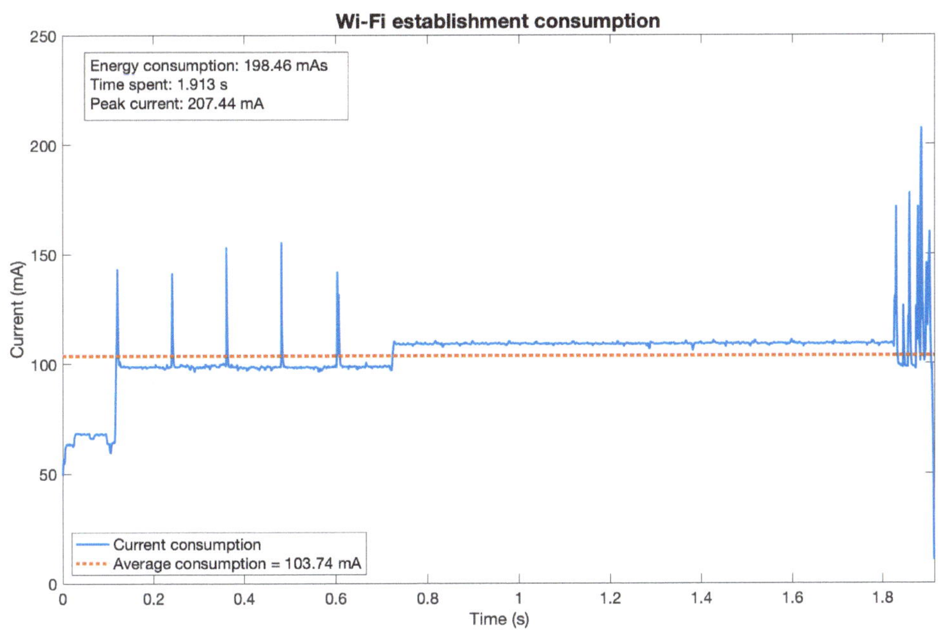

Figure A8. Current consumption detail for the Wi-Fi connection establishment. Data acquired at 1 kS/s.

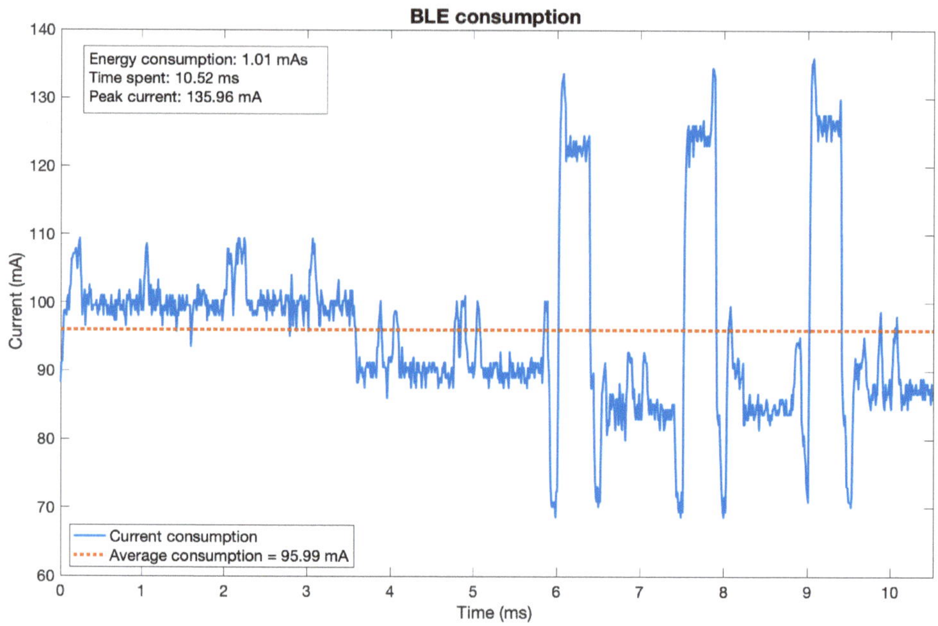

Figure A9. Current consumption detail for the BLE transmission. Data acquired at 100 kS/s.

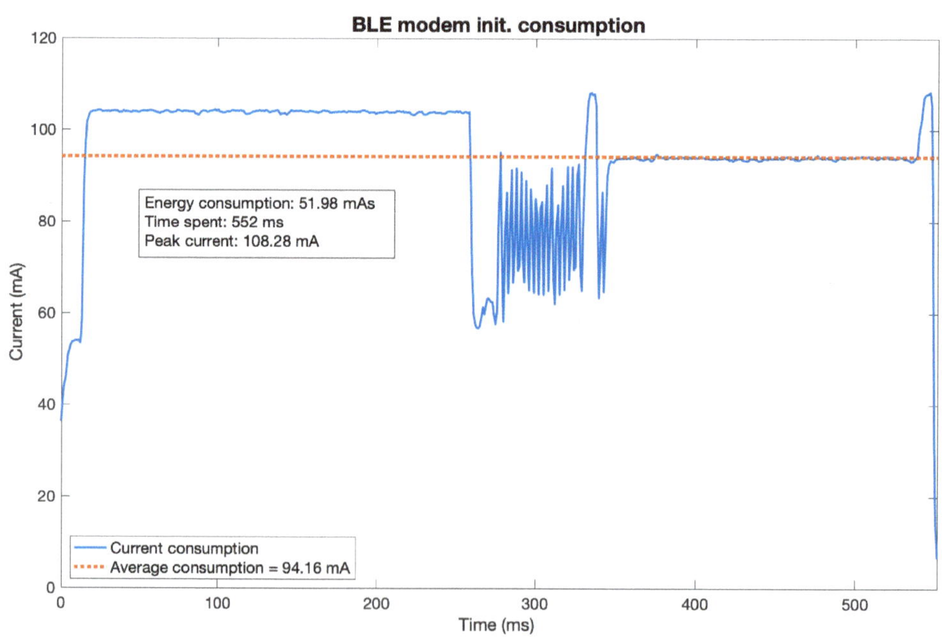

Figure A10. Current consumption for the BLE modem initialisation. Data acquired at 1 kS/s.

Appendix B

This appendix contains information about the process used to determine the Power Profiler's metering accuracy.

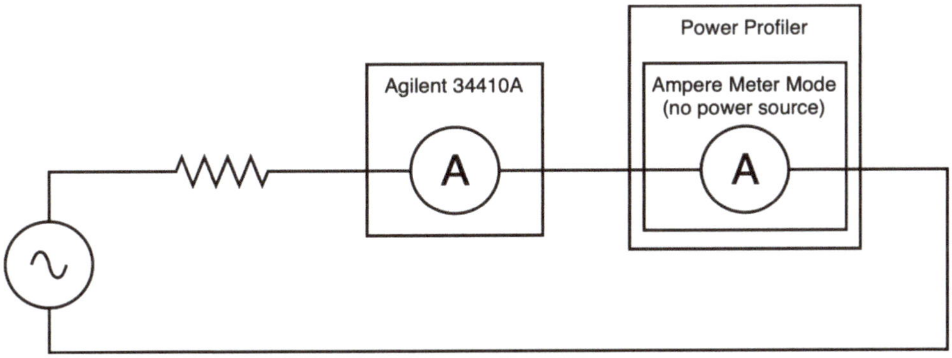

Figure A11. Setup used to determine the Power Profiler's accuracy.

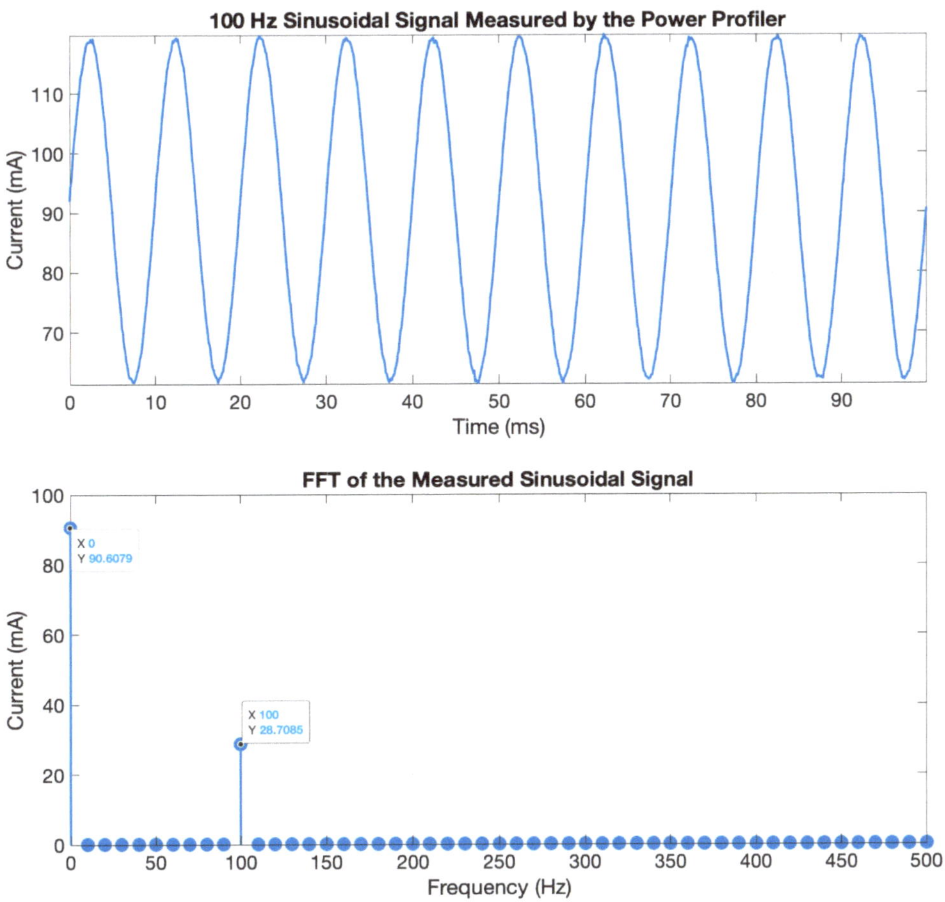

Figure A12. 100 Hz sinusoidal signal used to determine the Power Profiler's accuracy and its FFT.

References

1. Wegner, P. Global IoT Spending to Grow 24% in 2021, Led by Investments in IoT Software and IoT Security. 2021. Available online: https://iot-analytics.com/2021-global-iot-spending-grow-24-percent/ (accessed on 8 November 2021).

2. Fotrune Business Insights Global IoT Market to Be Worth USD 1463.19 Billion by 2027 at 24.9% CAGR; Demand for Real-Time Insights to Spur Growth. 2021. Available online: https://www.globenewswire.com/en/news-release/2021/04/08/2206579/0/en/Global-IoT-Market-to-be-Worth-USD-1-463-19-Billion-by-2027-at-24-9-CAGR-Demand-for-Real-time-Insights-to-Spur-Growth-says-Fortune-Business-Insights.html (accessed on 8 November 2021).
3. Evans, D. The Internet of Things. How the Next Evolution of the Internet Is Changing Everything. 2021. Available online: https://www.cisco.com/c/dam/en_us/about/ac79/docs/innov/IoT_IBSG_0411FINAL.pdf (accessed on 8 November 2021).
4. Maayan, G.D. The IoT Rundown for 2020: Stats, Risks, and Solutions. 2020. Available online: https://securitytoday.com/Articles/2020/01/13/The-IoT-Rundown-for-2020.aspx (accessed on 8 November 2021).
5. Saavedra, E.; Del Campo, G.; Santamaria, A. Smart Metering for Challenging Scenarios: A Low-Cost, Self-Powered and Non-Intrusive IoT Device. *Sensors* **2020**, *20*, 7133. [CrossRef] [PubMed]
6. Sendin, A.; Simon, J.; Urrutia, I.; Berganza, I. PLC deployment and architecture for Smart Grid applications in Iberdrola. In Proceedings of the 18th IEEE International Symposium on Power Line Communications and Its Applications, Glasgow, UK, 30 March–2 April 2014; pp. 173–178.
7. Sendin, A.; Berganza, I.; Arzuaga, A.; Pulkkinen, A.; Kim, I.H. Performance results from 100,000+ PRIME smart meters deployment in Spain. In Proceedings of the 2012 IEEE Third International Conference on Smart Grid Communications (SmartGridComm), Tainan, Taiwan, 5–8 November 2012; pp. 145–150.
8. Reinhardt, A.; Burkhardt, D.; Mogre, P.S.; Zaheer, M.; Steinmetz, R. SmartMeter.KOM: A Low-Cost Wireless Sensor for Distributed Power Metering. In Proceedings of the 2011 IEEE 36th Conference on Local Computer Networks, Bonn, Germany, 4–7 October 2011; pp. 1032–1039.
9. Woo, C.-K.; Ho, T.; Shiu, A.; Cheng, Y.-S.; Horowitz, I.; Wang, J. Residential outage cost estimation: Hong Kong. *Energy Policy* **2014**, *72*, 204–210. [CrossRef]
10. Praktiknjo, A.; Hähnel, A.; Erdmann, G. Assessing energy supply security: Outage costs in private households. *Energy Policy* **2011**, *39*, 7825–7833. [CrossRef]
11. Pycom FiPy Specsheets. Available online: https://docs.pycom.io/gitbook/assets/specsheets/Pycom_002_Specsheets_FiPy_v2.pdf (accessed on 8 November 2021).
12. Keysight Keysight 34410A Multimeter Datasheet. Available online: https://www.keysight.com/Us/En/Assets/7018-01326/Data-Sheets/5989-3738.Pdf (accessed on 8 November 2021).
13. Deutsche Telekom IoT NB-IoT, LoRaWAN, Sigfox: An up-to-Date Comparison. 2021. Available online: https://iot.telekom.com/resource/blob/data/492968/e396f72b831b0602724ef71056af5045/mobile-iot-network-comparison-nb-iot-lorawan-sigfox.pdf (accessed on 8 November 2021).
14. del Campo, G.; Gomez, I.; Cañada, G.; Piovano, L.; Santamaria, A. Guidelines and criteria for selecting the optimal low-power wide-area network technology. In *LPWAN Technologies for IoT and M2M Applications*; Elsevier: Amsterdam, The Netherlands, 2020; pp. 281–305. ISBN 978-0-12-818880-4.
15. Technology | Sigfox. Available online: https://www.sigfox.com/en/what-sigfox/technology (accessed on 8 November 2021).
16. Nordic Semiconductor Power Profiler Kit II Specifications. Available online: https://infocenter.nordicsemi.com/index.jsp?topic=%2Fug_ppk2%2FUG%2Fppk%2FPPK_user_guide_Intro.html (accessed on 8 November 2021).
17. Espressif ESP32 Series Datasheet. Available online: https://www.espressif.com/sites/default/files/documentation/esp32_datasheet_en.pdf (accessed on 8 November 2021).
18. Semtech Semtech SX1272 Datasheet. Available online: https://semtech.my.salesforce.com/sfc/p/#E0000000JelG/a/440000001NCE/v_VBhk1IolDgxwwnOpcS_vTFxPfSEPQbuneK3mWsXlU (accessed on 8 November 2021).
19. Yue, X.; Kauer, M.; Bellanger, M.; Beard, O.; Brownlow, M.; Gibson, D.; Clark, C.; MacGregor, C.; Song, S. Development of an Indoor Photovoltaic Energy Harvesting Module for Autonomous Sensors in Building Air Quality Applications. *IEEE Internet Things J.* **2017**, *4*, 2092–2103. [CrossRef]
20. Hande, A.; Polk, T.; Walker, W.; Bhatia, D. Indoor solar energy harvesting for sensor network router nodes. *Microprocess. Microsystems* **2007**, *31*, 420–432. [CrossRef]
21. Batista, D.; Oliveira, J.; Paulino, N.; Carvalho, C.; Farinhas, J.; Charas, A.; Dos Santos, P.M. Combined Organic Photovoltaic Cells and Ultra Low Power CMOS Circuit for Indoor Light Energy Harvesting. *Sensors* **2019**, *19*, 1803. [CrossRef] [PubMed]
22. Teran, A.S.; Moon, E.; Lim, W.; Kim, G.; Lee, I.; Blaauw, D.; Phillips, J. Energy Harvesting for GaAs Photovoltaics Under Low-Flux Indoor Lighting Conditions. *IEEE Trans. Electron Devices* **2016**, *63*, 2820–2825. [CrossRef] [PubMed]
23. Foti, M.; Tringali, C.; Battaglia, A.; Sparta, N.; Lombardo, S.; Gerardi, C. Efficient flexible thin film silicon module on plastics for indoor energy harvesting. *Sol. Energy Mater. Sol. Cells* **2014**, *130*, 490–494. [CrossRef]
24. Teran, A.S.; Wong, J.; Lim, W.; Kim, G.; Lee, Y.; Blaauw, D.; Phillips, J. AlGaAs Photovoltaics for Indoor Energy Harvesting in mm-Scale Wireless Sensor Nodes. *IEEE Trans. Electron Devices* **2015**, *62*, 2170–2175. [CrossRef]
25. Xie, L.; Song, W.; Ge, J.; Tang, B.; Zhang, X.; Wu, T.; Ge, Z. Recent progress of organic photovoltaics for indoor energy harvesting. *Nano Energy* **2021**, *82*, 105770. [CrossRef]
26. Politi, B.; Parola, S.; Gademer, A.; Pegart, D.; Piquemil, M.; Foucaran, A.; Camara, N. Practical PV energy harvesting under real indoor lighting conditions. *Sol. Energy* **2021**, *224*, 3–9. [CrossRef]
27. Tan, Y.K.; Panda, S. Energy Harvesting From Hybrid Indoor Ambient Light and Thermal Energy Sources for Enhanced Performance of Wireless Sensor Nodes. *IEEE Trans. Ind. Electron.* **2011**, *58*, 4424–4435. [CrossRef]

28. Rokonuzzaman, M.; Mishu, M.; Amin, N.; Nadarajah, M.; Roy, R.; Rahman, K.; Buhari, A.; Binzaid, S.; Shakeri, M.; Pasupuleti, J. Self-Sustained Autonomous Wireless Sensor Network with Integrated Solar Photovoltaic System for Internet of Smart Home-Building (IoSHB) Applications. *Micromachines* **2021**, *12*, 653. [CrossRef]
29. Visser, H.J.; Vullers, R.J.M. RF Energy Harvesting and Transport for Wireless Sensor Network Applications: Principles and Requirements. *Proc. IEEE* **2013**, *101*, 1410–1423. [CrossRef]
30. Angulo, F.; Navarro, L.; Quintero M, C.Q.; Pardo, M. A Simple WiFi Harvester with a Switching-Based Power Management Scheme to Collect Energy from Ordinary Routers. *Electronics* **2021**, *10*, 1191. [CrossRef]
31. Palazzi, V.; Hester, J.; Bito, J.; Alimenti, F.; Kalialakis, C.; Collado, A.; Mezzanotte, P.; Georgiadis, A.; Roselli, L.; Tentzeris, M.M. A Novel Ultra-Lightweight Multiband Rectenna on Paper for RF Energy Harvesting in the Next Generation LTE Bands. *IEEE Trans. Microw. Theory Tech.* **2017**, *66*, 366–379. [CrossRef]
32. Haboubi, W.; Takhedmit, H.; Luk, J.-D.L.S.; Adami, S.-E.; Allard, B.; Costa, F.; Vollaire, C.; Picon, O.; Cirio, L. An efficient dual-circularly polarized rectenna for RF energy harvesting in the 2.45 GHZ ISM Band. *Prog. Electromagn. Res.* **2014**, *148*, 31–39. [CrossRef]
33. Eid, A.; Hester, J.; Nauroze, A.; Lin, T.-H.; Costantine, J.; Tawk, Y.; Ramadan, A.H.; Tentzeris, M. A Flexible Compact Rectenna for 2.40 Hz ISM Energy Harvesting Applications. In Proceedings of the 2018 IEEE International Symposium on Antennas and Propagation & USNC/URSI National Radio Science Meeting, Boston, MA, USA, 8–13 July 2018; pp. 1887–1888.
34. Muncuk, U.; Alemdar, K.; Sarode, J.D.; Chowdhury, K.R. Multiband Ambient RF Energy Harvesting Circuit Design for Enabling Batteryless Sensors and IoT. *IEEE Internet Things J.* **2018**, *5*, 2700–2714. [CrossRef]
35. He, H.; Li, T. Design of an indoor RF Energy Harvesting Module using Dual-Band Rectenna Array for IoT Applications. In Proceedings of the 2019 IEEE 4th Advanced Information Technology, Electronic and Automation Control Conference (IAEAC), Chengdu, China, 20–22 December 2019; Volume 1, pp. 1407–1410.
36. Porcarelli, D.; Brunelli, D.; Benini, L. Clamp-and-measure forever: A MOSFET-based circuit for energy harvesting and measurement targeted for power meters. In Proceedings of the 5th IEEE International Workshop on Advances in Sensors and Interfaces IWASI, Bari, Italy, 13–14 June 2013; pp. 205–210.

sensors

MDPI

Article

Minimum Length Scheduling for Multi-Cell Full Duplex Wireless Powered Communication Networks

Muhammad Shahid Iqbal [1,*,†], Yalcin Sadi [2,†] and Sinem Coleri [3,†]

1 Department of Electrical Engineering, National University of Technology, Islamabad 44000, Pakistan
2 Department of Electrical and Electronics Engineering, Kadir Has University, Istanbul 34083, Turkey; yalcin.sadi@khas.edu.tr
3 Department of Electrical and Electronics Engineering, Koc University, Istanbul 34450, Turkey; scoleri@ku.edu.tr
* Correspondence: miqbal16@ku.edu.tr or shahid.iqbal@nutech.edu.pk
† These authors contributed equally to this work.

Abstract: Wireless powered communication networks (WPCNs) will be a major enabler of massive machine type communications (MTCs), which is a major service domain for 5G and beyond systems. These MTC networks will be deployed by using low-power transceivers and a very limited set of transmission configurations. We investigate a novel minimum length scheduling problem for multi-cell full-duplex wireless powered communication networks to determine the optimal power control and scheduling for constant rate transmission model. The formulated optimization problem is combinatorial in nature and, thus, difficult to solve for the global optimum. As a solution strategy, first, we decompose the problem into the power control problem (PCP) and scheduling problem. For the PCP, we propose the optimal polynomial time algorithm based on the evaluation of Perron–Frobenius conditions. For the scheduling problem, we propose a heuristic algorithm that aims to maximize the number of concurrently transmitting users by maximizing the allowable interference on each user without violating the signal-to-noise-ratio (SNR) requirements. Through extensive simulations, we demonstrate a 50% reduction in the schedule length by using the proposed algorithm in comparison to unscheduled concurrent transmissions.

Keywords: wireless powered communication networks; energy harvesting; full-duplex; power control; scheduling; multi-cell network

Citation: Iqbal, M.S.; Sadi, Y.; Coleri, S. Minimum Length Scheduling for Multi-Cell Full Duplex Wireless Powered Communication Networks. *Sensors* **2021**, *21*, 6599. https://doi.org/10.3390/s21196599

Academic Editors: Slim Naifar, Olfa Kanoun and Carlo Trigona

Received: 1 September 2021
Accepted: 21 September 2021
Published: 2 October 2021

1. Introduction

According to the recent Ericsson mobility report, 24.6 billion sensor nodes are expected to be installed by 2025 [1]. The majority of the applications in real time monitoring and industrial automation have strict delay requirements [2]. Therefore, increasing the lifetime of this massive number of sensor nodes while satisfying their latency requirements is a major challenge for future wireless networks. Moreover, to fulfill these diverse requirements, this massive installation will be deployed by using low power transceivers with energy harvesting capability, simple receiver circuitry, and intelligent medium access protocols [3].

For perpetual energy, due to full control of the energy transfer, small form factor, long range and no alignment requirements, the radio frequency (RF) energy harvesting (EH) is a promising technique for the wireless energy transfer for IoT and MTC type applications. In the literature, RF-EH has been investigated with two models named the simultaneous wireless information and power transfer (SWIPT) and wireless powered communication network (WPCN). In SWIPT, the same signal is used for simultaneous energy and information transfer.

The authors in [4,5] analyzed the trade off between energy efficiency and information transmission capacity for a single user using additive white Gaussian noise and a flat fading channel, respectively. The multi-user SWIPT was investigated for data buffer constrained sum throughput maximization [6] and energy efficiency maximization under the energy causality constraint [7].

On the other hand, WPCN consists of a hybrid access point (HAP), which is responsible for dedicated energy transmission in the downlink and a set of users that harvest this energy to transmit back their information to the HAP in the uplink. The authors in [8] introduced the first protocol for the sum throughput maximization of a WPCN, in which the frame length is divided into two non-overlapping phases dedicated for energy and information transfer. In this half-duplex model, as energy and information take place in different phases, the information transmission order of the users is not important, thus, no scheduling algorithm is required.

To increase the energy harvesting rate of such networks, the usage of multiple energy transmitters is investigated in [9–11]. Particularly, [9] studied the time allocation and load balancing for the throughput maximization, [10] provided the implementation of a test-bed to maximize the amount of harvested energy, and [11] investigated a wireless body area network with the objective of maximizing the sum throughput. Furthermore, to increase the network coverage, multi-cell WPCN, in which multiple HAPs are used for both energy transmission and information reception, were investigated for minimum throughput maximization [12] and throughput maximization through beam-forming [13].

However, these objectives fail to provide any delay guarantee, which is necessary for the time critical networks. Only [14] considered the minimum length scheduling problem. All of these problems investigated for multi-cell network consider half-duplex model, which results in an under utilization of the time and energy resources due to non-overlapping energy and information transmission phases.

1.1. Related Works

Due to the recent advances in self-interference cancellation techniques for wireless networks [15–18] and the related practical implementations [19,20], full-duplex has become a potential transceiving technique for 5G and beyond networks. In full-duplex, the uplink and downlink transmissions take place simultaneously using the same frequency reducing the spectrum requirements to half, i.e., either the spectral efficiency or the total number of users within a cell zone can be increased by a factor of two. From a practical implementation point of view, [21] presented an experimental setup for low complexity full-duplex design by employing the passive suppression with a patch antenna and active self-interference cancellation by using estimation and reconstruction of the interfering signal.

The authors in [22] implemented a prototype for full-duplex operation by implementing a wide-band RF canceller and an adaptive non-linear digital canceller. In the context of full-duplex networks, [23] categorized all the users into lethargic users, i.e., users with low energy levels and energetic users, i.e., the users with sufficient energy stored in their batteries. The authors proposed a medium access protocol named FarMac, which aims to maximize the energy transferred to the lethargic users and information received from the energetic users.

The authors in [24] investigated the minimum length scheduling problem for a single cell continuous rate transmission model. To solve the problem, the authors used a penalty function defined as the difference between the actual transmission time and the minimum transmission time of a user at a particular scheduling decision time. Since the penalty function is purely characterized for a single user, the solutions are not applicable for concurrent transmissions. The authors in [25] solved the sum throughput maximization problem for a single cell continuous rate model.

The continuous rate model is hard to realize practically for such low energy harvesting networks. To address this issue and to simplify the receiver structure, The authors in [26–28] studied an on–off transmission scheme for a single cell network with the objective to derive the outage probability, minimize the schedule length, and maximize the sum throughput, respectively. In an on–off transmission scheme, the users either transmit information using the maximum allowed transmit power or remain silent if they do not not have enough power to achieve this power level throughout their transmission.

The researchers in [3,29,30] further extended these formulations to investigate the minimum length scheduling problem and sum throughput maximization problem for a discrete rate model, where any finite set of discrete transmission rate values are supported. In [3], the authors defined the minimum length scheduling (MLS) slot, which is defined as the slot corresponding to the minimum transmission completion time while starting anytime after the scheduling decision time.

The scheduling problem is solved based on the allocation of the MLS slots to the users. The MLS slot is purely defined for a single user without considering the concurrent transmissions and inter-cell interference. Since the MLS slot specifically focuses on the first time instant at which transmission is feasible for a specific user; it is not generalizable to multi-cell networks in which the feasibility of the transmission of a particular user depends on the other concurrently transmitting users. All of these studies consider a single cell network, which limits their wide deployment due to low energy harvesting rate, limited applications and impractical model for wide coverage, which does not consider the inter-cell interference.

The major challenge for multi-cell WPCN is to determine the group of users that will transmit simultaneously. For a single cell scenario, it is easy to come up with some certain characteristics that can help to determine the optimal solutions. However, for a multi-cell scenario, the per user characteristic based solutions, such as penalty or MLS slot-based solutions, are not applicable since the transmit powers, transmission rates, interference, and energy harvesting characteristics are now dependent on a set of concurrently transmitting users rather than the characteristics of a single user.

In [3,24], the authors proposed solutions relying on the characteristics of the user without considering the other users; however, in the multi-cell scenario, we need to consider all the simultaneously transmitting users. To the best of the authors' knowledge, this is the first study to investigate the multi-cell full-duplex WPCN while considering the self-interference and inter-cell interference for a simultaneous transmission of the users.

1.2. Contributions

The goal of this paper is to determine the optimal power control and scheduling for the constant transmission rate model with the objective of minimizing the schedule length for a realistic multi-cell full-duplex WPCN system. In full-duplex, all the users can harvest energy during the information transmission of other users. Therefore, the users with lower energy level can be scheduled later so that they harvest more energy and can transmit their information in shorter time. Moreover, in multi-cell WPCN, the users transmitting information simultaneously interfere with each other, which necessitates an intelligent grouping mechanism to minimize the interference among the concurrently transmitting users. The original contributions of this paper are listed below:

- We propose an optimization framework for the minimization of the schedule length in a full-duplex multi-cell WPCN for the first time in the literature. The framework considers the concurrent transmissions of the users, and incorporates the non-linear energy harvesting model for a full-duplex WPCN, in which the HAPs and users both operate in full-duplex mode.
- We formulate a mixed-integer non-linear optimization problem to minimize the schedule length. The formulated optimization problem is non-convex and generally hard to solve for the global optimal solution. As a solution strategy, we perform decomposition of the optimization problem into the power control problem and scheduling problem.
- For the power control problem, we propose an optimal polynomial time algorithm based on the evaluation of the Perron–Frobenius conditions.
- For the scheduling problem, we propose a heuristic algorithm based on the maximization of the number of concurrently transmitting users within a transmission slot by maximizing the allowable interference on each user without violating their signal-to-noise-ratio (SNR) requirements.

1.3. Organization of the Paper

The rest of the paper is organized as follows. Section 2 describes the system model and assumptions. Section 3 presents the mathematical formulation of the minimum length scheduling problem. Section 4 provides the problem formulation and solution for the power control problem. The scheduling algorithm is given in Section 5. Section 6 provides the performance evaluation of the proposed scheduling scheme. Finally, our concluding remarks are given in Section 7.

2. System Model and Assumptions

The system model and related assumptions are described as follows:

- We consider a multi-cell WPCN, which consists of K HAPs and N users denoted by $\mathcal{N}$ as shown in Figure 1. All the HAPs and users are equipped with a full-duplex antenna. The HAPs are connected to a stable power line and transmit at a constant power P_h continuously. On the other hand, users do not have any external power supply and can only harvest energy from the HAPs. The harvested energy is stored in a rechargeable battery with initial level B_n for user n, where $n \in \mathcal{N}$, and capacity B_{max}. The users can harvest energy from all the HAPs but only transmit information to a single HAP to which they are connected to. The set of users connected to HAP k is denoted by L^k for $k \in \{1, \cdots, K\}$ such that $\bigcup_{k=1}^{K} L^k = \mathcal{N}$. The important symbols used in this paper and their definitions are given in Table 1.
- We consider the time division multiple access (TDMA) protocol within each cell for the uplink information transmission. The TDMA is more power efficient since it allows the users to keep their transmitter circuitry inactive until their particular allocated time slot. One of the HAP works as the central HAP, i.e., central gateway and network manager in WirelessHART or system manager in International Society of Automation (ISA) 100.11a [31]. During the initialization phase, all the HAPs collect the channel state information (CSI) from the users and share it with the central HAP. The central HAP then runs the algorithm and shares the resulting resource allocation, scheduling decisions, and synchronization information by using a beacon transmitted to all the HAPs.
Since the focus of this study is to design a resource allocation and scheduling algorithm, the mechanisms for synchronization are out of scope of this paper and can be found in [32]. The total time in which the system remains operational is partitioned into frames, and each frame is further divided into M variable length non-overlapping time slots. In each time slot, a subset of users from different cells transmits their information simultaneously to their respective HAP. No intra-cell interference exists except the self interference at the HAP due to full duplex operational mode. However, simultaneous transmitting users from different cells may create interference to each other.
- The uplink information transmission and downlink energy transfer channels are assumed to be different. The uplink channel gain from user n to HAP k is denoted by g_{nk}, and the downlink channel gain from HAP k to user n is denoted by h_{nk}. We assume that all the channels are quasi-static, i.e., the channel gains remain the same in the current frame and can vary independently in the next frame [12,33–35]. We further assume that the channel state information is perfectly known at the HAP [8,35,36].
- We assume that user n has a traffic demand D_n bits to be transmitted in each frame.
- We assume that each user allocated to the transmission slot m transmits information at constant rate r if the SNR of user n is above a fixed threshold γ_n as

$$\frac{P_n g_{nn}}{N_o W + W \sum_{j \neq n} P_j g_{jn} + \beta P_h} \geq \gamma_n, \quad n \in \mathcal{N} \tag{1}$$

where P_n is the transmit power of user n, W is the channel bandwidth, N_o is the noise density, βP_h represents the interference created by the energy radiation by the HAPs including self-interference and interference from the other HAPs, and the term

$W\sum_{j\neq n} P_j g_{jn}$ is the interference at the HAP to which user n is connected from other users that are concurrently transmitting information in slot m.

- We consider a realistic non-linear energy harvesting model [37], in which the energy harvesting rate for user n is

$$C_n = P_s[\Psi_n - \Omega_n][1 - \Omega_n]^{-1} \tag{2}$$

where $\Omega_n = (1 + e^{a_n b_n})^{-1}$ is a constant to guarantee zero-input zero-output response; P_s is the maximum harvested power during saturation; and $\Psi_n = (1 + e^{-a_n(P_n^T - b_n)})^{-1}$ is the logistic function related to user n. a and b are the positive constants related to the non-linear charging rate with respect to the input power and turn-on threshold, respectively. For a given energy harvesting circuit, the parameters P_s, a and b are determined by curve fitting.

- We assume that the users can harvest energy from all the HAPs, and the total power received by user n is given as follows:

$$P_n^T = \sum_{k=1}^{K} h_{nk} P_h \tag{3}$$

Table 1. Symbols and notations.

Symbol	Definition
N	Total number of users
K	Total number of HAPs
M	Total number of time slots
g_{nk}	Uplink channel gain from user n to HAP k
h_{nk}	Downlink channel gain from HAP k to user n
τ^m	The length of transmission slot m
x_n^m	Transmission rate of user n in slot m
z_n^m	Time slot allocation variable
D_n	Data requirement of user n
r^i	Transmission rate of user i
γ_n	Signal-to-noise ratio threshold for user n
P_h	Transmit power of HAP
βP_h	Power of Self-interference at HAP
P_n	Transmit power of user n
I_{ij}	Interference of user i on HAP j
P_{max}	Maximum transmit power of a user
C_n	Energy harvesting rate of user n
B_n	Initial battery level of user n
I_n^{max}	Maximum allowed interference level for user n
$\mathbf{S}$	Set of concurrently transmitting users
W	Bandwidth
N_0	Noise density
P_n^T	Total received power by user n
E_n	Energy of user n
α	Path loss exponent

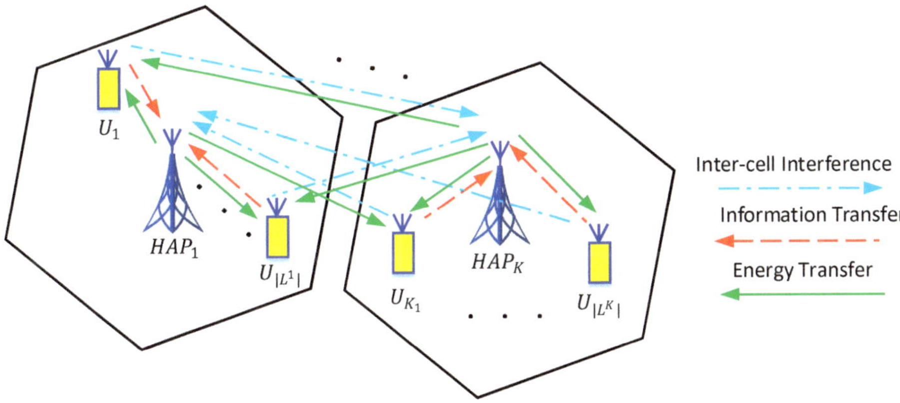

Figure 1. Multi-cell wireless powered communication network architecture.

3. Minimum Length Scheduling Problem Formulation

The joint optimization of the power control and scheduling for concurrently transmitting users with the objective of minimizing the schedule length with the traffic demand, energy causality, and user transmit power constraints is formulated and denoted by $\mathcal{MC}-\mathcal{MLSP}$. The formulation of the constant transmission rate model is as follows:

$\mathcal{MC}-\mathcal{MLSP}$:

minimize

$$\sum_{m=1}^{M} \tau^m \tag{4}$$

subject to

$$\sum_{m=1}^{M} z_n^m = 1, \;\; n \in \mathcal{N} \tag{5}$$

$$\sum_{n \in L^k} z_n^m \leq 1, \;\; m \in \{1, \cdots, M\}, k \in \{1, \cdots, K\}, \tag{6}$$

$$x_n^m \tau^m \geq D_n \;\; n \in \mathcal{N}, \;\; m \in \{1, \cdots, M\}, \tag{7}$$

$$x_n^m = z_n^m r \;\; n \in \mathcal{N}, \;\; m \in \{1, \cdots, M\}, \tag{8}$$

$$\frac{P_n g_{nn}}{N_o W + W \sum_{j \neq n} P_j g_{jn} + \beta P_h} \geq \gamma_n, \;\; n \in \mathcal{N}, \tag{9}$$

$$B_n + \sum_{j=1}^{m} C_n \tau^j - z_n^m P_n \tau^m \geq 0, \;\; n \in \mathcal{N}, \;\; m \in \{1, \cdots, M\} \tag{10}$$

$$P_n \leq z_n^m P_{max}, \;\; n \in \mathcal{N}, \;\; m \in \{1, \cdots, M\}, \tag{11}$$

variables

$$P_n \geq 0, \; \tau^m \geq 0, \; z_n^m \in \{0, 1\}, \; n \in \mathcal{N}, \; m \in \{1, \cdots, M\}. \tag{12}$$

The variables of the problem are P_n, the transmit power of user n; τ^m, the length of transmission slot m; and z_n^m, a binary variable, which takes a value 1 if user n is allocated to slot m and 0 otherwise. x_n^m denotes the transmission rate of user n in time slot m and r is the constant rate level.

The objective of the optimization problem is to minimize the schedule length as given by Equation (4). Equation (5) states that each user should be allocated to only one time slot. Equations (6) represents that, at most, one user can be allocated to a slot for the same HAP. Equations (7) and (8) together represent the traffic demand of users. Equation (9) represents the condition for SNR to achieve a constant rate. Equation (10) gives the energy causality constraint: The total amount of available energy, including both the initial energy and the energy harvested until and during the transmission of a user, should be greater than or equal to the energy consumed during its transmission. Equation (11) represents the maximum transmit power constraint for the users.

The optimization problem formulated in Equations (4)–(12) includes both continuous variables, which are P_n and τ^m, and integer (binary) variables, such as z_n^m. Furthermore, it is a non-linear problem due to the multiplication of two variables in energy causality constraint given in Equation (10). Hence, this is a Mixed Integer Non-Linear Programming (MINLP) problem, which are, in general, difficult to solve for optimal solutions in polynomial time [7].

Solution Framework

As the optimization problem is a Mixed Integer Non-Linear Programming (MINLP) problem, finding the global optimal solution requires exponential time algorithms, which are intractable even for medium size networks [38]. To overcome this intractability, we decompose the optimization problem into the optimal power control problem and scheduling problem as detailed below:

- For a given set of concurrently transmitting users, we formulate the optimization problem to determine the minimum transmission slot length and the corresponding transmit power vector while considering the maximum transmit power constraint, traffic demand, and energy causality of the users. We first show that the power control problem is a feasibility problem, and then, by using the Perron–Frobenius condition, we find the optimal power vector if the problem is feasible.
- Determining the power vector and transmission slot length for a given set of users reduces the problem to the optimization of the concurrently transmitting users within a transmission slot. For the scheduling problem, we exploit the affordable interference levels of all the users and group them based on these interference levels while considering the maximum transmit power and energy causality of the users.

4. Power Control Problem

In this section, we determine the optimal power control and time slot length for a given set of concurrently transmitting users from different cells with the objective of minimizing the time slot length. For simplification, we remove the superscript m in Section 3, which represents the time slot index. The optimal power control problem is formulated as follows:
minimize

$$t \tag{13}$$

subject to

$$\frac{D_n}{x_n} \leq t, \;\; n \in \mathbf{S}, \tag{14}$$

$$B_n + C_n t - P_n t \geq 0, \;\; n \in \mathbf{S}, \tag{15}$$

$$P_n \leq P_{max}, \;\; n \in \mathbf{S}, \tag{16}$$

$$x_n = r, \;\; n \in \mathbf{S}, \tag{17}$$

$$\frac{P_n g_{nn}}{N_o W + W \sum_{j \neq n} P_j g_{jn} + \beta P_h} \geq \gamma_n, \;\; n \in \mathbf{S} \tag{18}$$

variables

$$P_n \geq 0, \; t > 0, n \in \mathbf{S}. \tag{19}$$

The objective of this problem is to determine the optimal time slot length for a given set of concurrently transmitting users in **S**. The constraints in Equations (14)–(18) represent the traffic demand, energy causality, maximum transmit power, and constant transmission rate constraint, respectively. Note that B_n refers to the energy available at the start of the transmission slot, and we used the same notation for simplicity.

This optimization problem can be reduced to a feasibility problem. If a feasible power vector exists, then the optimal time slot length is $t = \max_{n \in \mathbf{S}} D_n/r$. In the following, we present the feasibility of the concurrent transmission of the users in set $\mathbf{S}$. For the feasibility of concurrent transmission, there should exist a set of transmit power levels $P_n, \forall n \in \mathbf{S}$, such that the SNR constraint given in Equation (18) is satisfied for each user in $\mathbf{S}$ without violating the maximum transmit power and energy causality constraints given in Equations (15) and (16).

At any decision time t_{dec}, the existence of such a set of transmit power levels that satisfy the P_{max} constraint can be determined by using Perron–Frobenius conditions described as follows: Let $\mathbf{G_S}$ be a $|\mathbf{S}| \times |\mathbf{S}|$ relative channel gain matrix, in which $a_{ij} = g_{ji}/g_{ii}$ represents the entry of the ith row and jth column for $i \neq j$ and $a_{ij} = 0$ for $i = j$. Let $\mathbf{D_S}$ represent a $|\mathbf{S}| \times |\mathbf{S}|$ diagonal matrix with the ith entry equal to γ_i.

Let $\boldsymbol{\sigma}_\mathbf{S}$ be a $|\mathbf{S}| \times 1$ normalized noise power vector with the ith entry equal to $\gamma_i W N_0/g_{ii}$. Then, the Perron–Frobenius condition states that there exists a power allocation vector that satisfies the P_{max} constraint if and only if the largest real eigenvalue of $\mathbf{D_S G_S}$ is less than 1 and every element of the component-wise minimum power vector $(\mathbf{I} - \mathbf{D_S G_S})^{-1}{}_\mathbf{S}$ is less than or equal to P_{max}, i.e., $(\mathbf{I} - \mathbf{D_S G_S})^{-1} \boldsymbol{\sigma}_\mathbf{S} \preceq \mathbf{P_{max}}$, where $\mathbf{P_{max}}$ is a vector with all entries equal to P_{max}.

For a feasible solution, the users must also satisfy the energy causality constraint, therefore, if the transmit powers also follow the energy causality constraint, the solution is feasible at the time of the decision.

Lemma 1. *If* $\mathbf{P^{min}} = (\mathbf{I} - \mathbf{D_S G_S})^{-1} \boldsymbol{\sigma}_\mathbf{S}$ *is an infeasible solution to problem in Equations (13)–(19), then, there is no feasible solution.*

Proof. Let $\mathbf{P^{min}} = \{P_1^{min}, \cdots, P_{|\mathbf{S}|}^{min}\}$ denote the minimum power vector for a set $\mathbf{S}$. Any other power vector $\mathbf{P}'$ satisfying the SNR constraint is component-wise greater than $\mathbf{P^{min}}$, i.e., $\mathbf{P}' \succ \mathbf{P^{min}}$. Then, if $\mathbf{P^{min}}$ is infeasible, there exists either $P_k^{min} > P_{max}$ for any particular user $k \in \mathbf{S}$ or required energy E_j is greater than the available energy, i.e., $E_j = P_j^{min} D_j/r > E_j^{available}$ for any particular use j. Then, if $\mathbf{P^{min}}$ is infeasible, there is no feasible solution for problem in Equations (13)–(19), since either P_k' will be greater than P_{max} or E_j' will be still greater than $E_j^{available}$, making $\mathbf{P}'$ also infeasible. □

The following theorem states that a feasible $\mathbf{P^{min}}$ vector is an optimal solution to the optimization problem in Equations (13)–(19).

Theorem 1. *Let* $\mathbf{P^{min}} = (\mathbf{I} - \mathbf{D_S G_S})^{-1} \boldsymbol{\sigma}_\mathbf{S}$ *denote the minimum feasible power vector obtained from Perron–Frobenius conditions. Then, if* $\mathbf{P^{min}}$ *is feasible, i.e.,* $\mathbf{P^{min}} \preceq \mathbf{P_{max}}$*, and it satisfies the energy causality constraint of each user, then,* $\mathbf{P^{min}}$ *is an optimal solution to optimization problem in Equations (13)–(19).*

Proof. Any feasible power vector gives the same solution to problem in Equations (13)–(19), i.e., $t = \max_{n \in \mathbf{S}} D_n/r$. Thus, if $\mathbf{P^{min}}$ is feasible, it is an optimal solution. □

Since the computational complexity of evaluating Perron–Frobenius conditions for $|\mathbf{S}|$ users is $\mathcal{O}(|\mathbf{S}|^3)$, the computational complexity of solving the power control problem for the constant transmission rate model is $\mathcal{O}(|\mathbf{S}|^3)$.

5. Scheduling

In Section 4, we solved the power control problem optimally for a known set of users transmitting information simultaneously. For all the users transmitting concurrently, the information transmission may be too long due to the extensive interference caused to each other. Therefore, the schedule length can further decrease by intelligently grouping the users into a set of subsets for concurrent transmission. However, determining the optimal subsets of the users transmitting concurrently and their schedule requires an exponential computational effort. Let $\nu = \{\nu_j : 1 \leq j \leq |\nu|\}$ denote the set of all possible partitions of the set of users $\mathcal{N} = \{1, \cdots, N\}$, where a partition of set $\mathcal{N}$ is defined as a set of nonempty mutually exhaustive and mutually exclusive subsets of set $\mathcal{N}$, e.g., for the set of 10 users, $\{\{1,5,8\},\{4,7,9,10\},\{2,3,6\}\}$ is one possible partition of set $\mathcal{N}$.

The total number of such partitions for a set of N users is $|\nu| = \sum_{j=0}^{N-1} \binom{N-1}{j} B_j$, where B_j is a Bell number with $B_0 = 1$ and $B_j = \sum_{z=0}^{j-1} \binom{j-1}{z} B_z$. For a given partition of $\mathcal{N}$, each subset of users transmits information concurrently within a time slot. Therefore, for a specific partition ν_j with ζ_j number of subsets, the total number of possible transmission sequences are $\zeta_j!$, i.e., for the above mentioned example $\zeta_j = 3$ resulting in six possible transmission sequences.

Then, total number of possible transmission sequences for a set of users $\mathcal{N}$ is $\sum_{j=1}^{|\nu|} \zeta_j!$. This exponential computational complexity of the scheduling problem makes the problem intractable even for medium size networks. Hence, scalable and fast solutions are required. In the following, we present the scheduling problems.

In this section, we present a heuristic scheduling algorithm, which aims at maximizing the number of users within each concurrently transmitting set by maximizing the interference on each user without violating the energy causality, SNR requirements, and maximum transmit power constraint.

Let I_n^{max} denote the maximum allowed interference level from other users that a particular user n can afford, i.e., any interference level greater than I_n^{max} results in an infeasible transmission scenario for user n. This maximum allowed interference can be determined by using Equation (9) as follows:

$$I_n^{max} = \frac{1}{W}\left[\frac{P_n g_{nn}}{\gamma_n} - N_0 W - \beta P_h\right], \tag{20}$$

It is important to note that the maximum allowed interference level of any user is an increasing function of the transmit power of that user. Hence, increasing the transmit power allows the user to accommodate more interference and possibly the allocation of more simultaneously transmitting users in a slot.

Furthermore, for a user n, the earliest scheduling time ts_n is defined as the first time instant when it can afford transmission at constant rate individually and is given by $ts_n = \max(0, P_n^{min} D_n / C_n r - B_n / C_n - D_n / r)$, where $P_n^{min} = \gamma_n (N_0 W + \beta P_h) / g_{nn}$, is the minimum power required to achieve the SNR γ_n without any interference, which is derived from Equation (18) for the case when the user transmits individually. In the following analysis, we assume that $P_n^{min} \leq P_{max}$ exists, otherwise, the problem is infeasible. The addition of any user in set $\mathbf{S}$ results in a delay in the earliest time instant ts_n due to increased interference.

The scheduling algorithm should determine the subsets of users such that the maximum number of users transmit their information simultaneously with the interference level close to the I_n^{max} for each user n. Therefore, at any decision time t_{dec}, first, we determine a set of users that can afford the constant rate at t_{dec}, i.e., the users for which $ts_n \leq t_{dec}$. Then, for this set of users, we evaluate the maximum power that they can afford, i.e., $P_n = max(P_{max}, E_n(t_{dec})/(D_n/r))$ and the maximum level of interference they can afford from other users in a concurrent transmission, i.e., $I_n^{max} = \frac{1}{W}(P_n g_{nn}/\gamma_n - N_0 W - \beta P_h)$.

Then, CRSA evaluates the pairwise interference of the feasible users and groups the users from different cells such that the number of concurrently transmitting users within each time slot is maximized without violating the maximum allowed interference level.

The Constant Rate Scheduling Algorithm (CRSA) is given in Algorithm 1 and described in detail next. Let $\mathbf{S} = \{\mathbf{S^1}, \cdots, \mathbf{S^M}\}$ be a set of subsets of users concurrently transmitting in time slots $\mathbf{t}^* = \{t_1^*, \cdots, t_M^*\}$ by using powers $\mathbf{P}^* = \{\mathbf{P}_1^*, \cdots, \mathbf{P_M}\}$. The mth entry of $\mathbf{S}$ denoted by $\mathbf{S^m}$ contains a subset of users that transmit simultaneously in the mth time slot with duration $t_m^* = \max_{s \in \mathbf{S^m}} D_s/r$ and corresponding power vector $\mathbf{P_m^*}$.

Algorithm 1 Constant Rate Scheduling Algorithm (CRSA)

Input: A set of users $\mathcal{N}$
Output: Simultaneously transmitting user set in each slot $\mathbf{S}$, transmission time for each slot $\mathbf{t}^*$, transmit power vector $\mathbf{P}^*$

$m \leftarrow 1, t_{dec} \leftarrow 0, \mathbf{P} \leftarrow \mathbf{0}, \mathbf{I^{max}} \leftarrow \mathbf{0}, \mathbf{I} \leftarrow \mathbf{0}$
determine ts_n, for all $n \in \mathcal{N}$
while $\mathcal{N} \neq \emptyset$ **do**
  $\mathcal{F} \leftarrow \emptyset, \mathbf{S^m} \leftarrow \emptyset$
  if $t_{dec} < \min_{n \in \mathcal{N}} ts_n$ **then**
    $t_{dec} = \min_{n \in \mathcal{N}} ts_n$
  end if
  for $n \in \mathcal{N}$ **do**
    if $ts_n \leq t_{dec}$ **then**
      $\mathcal{F} \leftarrow \mathcal{F} + n$
      $E_n(t_{dec}) = B_n + C_n(t_{dec} + D_n/r)$
      $P_n \leftarrow min(P_{max}, E_n(t_{dec})/(D_n/r))$
      $I_n^{max} \leftarrow \frac{1}{W}\left[\frac{P_n g_{nn}}{\gamma_n} - N_0 W - \beta P_h\right]$
    end if
  end for
  evaluate interference I_{ns} of user n to s, for all $n, s \in \mathcal{F}$
  $u \leftarrow arg \max_{j \in \mathcal{F}} I_j^{max}$
  $I_s \leftarrow 0$, for all $s \in \mathcal{F}$
  $\mathbf{S^m} \leftarrow \mathbf{S^m} + \{u\}$
  for $k = 1 : K$ **do**
    if $u \notin L_k$ **then**
      $\mathcal{F}^k \leftarrow L_k \cap \mathcal{F}$
      sort $\mathcal{F}^k$ in descending order of I_n^{max}
      for $v \in \mathcal{F}^k$ **do**
        if $\sum_{j \in \mathbf{S^m}} I_{jv} \leq I_v^{max}$ **then**
          if $I_s^{max} - I_s \geq I_{vs}, \quad \forall s \in \mathbf{S^m}$ **then**
            $I_v \leftarrow \sum_{j \in \mathbf{S^m}} I_{jv}$
            $I_s = I_s + I_{vs}, \quad \forall s \in \mathbf{S^m}$
            $\mathbf{S^m} \leftarrow \mathbf{S^m} + \mathcal{F}^k(v)$
            break,
          end if
        else
          break,
        end if
      end for
    end if
  end for
  $t_m^* \leftarrow \max_{s \in \mathbf{S^m}} D_s/r,$
  $\mathcal{N} \leftarrow \mathcal{N} - \mathbf{S^m}$
  $t_{dec} = t_{dec} + t_m^*$
  $m \leftarrow m + 1$
end while

Let $\mathbf{I^{max}}$ be a $N \times 1$ maximum affordable interference vector, in which j-th entry represents the maximum allowed interference level from other users on user j. Let $\mathbf{I}$ be a $N \times K$ pairwise interference matrix, in which I_{ij} is the entry of ith row and jth column representing the interference caused by user i on the HAP j.

The input of CRSA is a set of users denoted by $\mathcal{N}$, whereas its outputs include a set $\mathbf{S}$, and corresponding time and power allocation vectors denoted by $\mathbf{t}^*$ and $\mathbf{P}^*$, respectively. The algorithm starts by initializing the transmission slot number m to 1, decision time t_{dec} to 0, power vector $\mathbf{P}$, maximum affordable interference $\mathbf{I^{max}}$, and interference matrix $\mathbf{I}$ to $\mathbf{0}$ (Line 1). Then, it determines the earliest starting time ts_n for every user $n, n \in \mathcal{N}$ (Line 2).

In each iteration, the algorithm determines the set $\mathbf{S^m}$ as well as their powers. The $CRSA$ first updates the decision time by setting it equal to the minimum of earliest starting times of the users, since no user has enough energy for transmission earlier than (Lines 5–7). Then, it checks the feasibility of each unallocated user based on their earliest scheduling time (Line 9). For the set of feasible users at time t_{dec}, denoted by $\mathcal{F}$, $CRSA$ evaluates the amount of energy each user needs to complete its transmission, the maximum transmit power, and the maximum interference each user can afford (Lines 11–13).

Once the set of feasible users $\mathcal{F}$ and their parameters are determined, the algorithm groups the users in set $\mathcal{F}$ with similar interference characteristics for concurrent transmission. For this purpose, the algorithm evaluates the pairwise interference that each user in $\mathcal{F}$ causes to others while transmitting simultaneously (Line 16). $CRSA$ picks the user that can afford the maximum interference from other users, since this allows the addition of more users in the set of simultaneously transmitting users (Line 17).

Then, it initializes the interference that is created on a certain user by concurrent transmissions I_s to 0 and adds the user that can afford maximum interference to the set $\mathbf{S^m}$ for transmission in the current time slot (Lines 18–19). The algorithm attempts to group the most suitable user from each cell iteratively (Lines 20–37). For this, $CRSA$ visits each cell one by one and terminates the search when either a feasible user is found or no user can be grouped from the current cell.

For each cell, $CRSA$ first sorts all the feasible users within the cell based on their I_n^{max} values (Line 23). Then, it checks the compatibility for each user for possible concurrent transmission until it finds a suitable user from current cell. For a user v to be compatible for concurrent transmission in the set S^m, the total interference caused by all the users in set $\mathbf{S^m}$ must be less than I_v^{max}, i.e., $\sum_{j \in \mathbf{S^m}} I_{jv} \leq I_v^{max}$ (Line 25), and the interference caused by user v must be less than the affordable interference margin of the existing users in set $\mathbf{S^m}$, i.e., $I_s^{max} - I_s \geq I_{vs}, \quad \forall s \in \mathbf{S^m}$ (Line 26).

If the first condition is not valid for the user with maximum I_n^{max} value in the current cell, this implies that no other user from current cell can afford this much interference, and hence the algorithm moves to the next cell without picking any user from this cell (Line 33). If the first condition is true but the user is causing excessive interference to the users in set S^m, the algorithm continues searching among the remaining feasible users within current cell for simultaneous transmission (Lines 24–32).

If both of these conditions are true, the algorithm updates the interference levels of set S^m, adds this user to set S^m and moves to the next cell (Lines 27–29). Once all the cells are visited, $CRSA$ evaluates the transmission time slot length for S^m, updates the set of unallocated users, the decision time, and the slot number (Lines 38–41). The algorithm terminates when all the users are allocated (Line 3).

Note that in half-duplex, the transmission frame is divided into two non-overlapping phases named the energy harvesting phase, in which all users harvest energy, and the information transmission phase, in which the users transmit their information by using the harvested energy. Due to fixed and equal energy harvesting duration for all users, the scheduling is not that important, i.e., the order of transmissions in the information transmission phase in not relevant since the users are not harvesting energy during the information transmission time of other users.

On the other hand, in full-duplex, since the users can harvest energy during the transmission of previously allocated users, this allows us to reduce the schedule length by intelligently scheduling the users. If we eliminate the impact of self-interference, the proposed algorithm can be used for a half duplex system with minor modifications as follows. The users can be grouped by exploiting the maximum affordable interference from other users as performed in CRSA algorithm, and the energy harvesting duration τ_0^m can be determined for each group by using the half-duplex assumption in which no energy is harvested during the information transmission phase. Then, the overall energy harvesting duration will be $\max \tau_0^m$ for $m \in \{1, 2, \cdots, M\}$.

6. Simulation Results

The goal of this section is to evaluate the performance of the proposed scheduling algorithm in comparison to the multi-cell transmission with no scheduling and a previously proposed penalty-based scheduling algorithm for a single cell network. The multi-cell transmission with no scheduling, denoted by MCNS, randomly selects a single user from each cell for the concurrent transmission by using the constant rate within a transmission slot. The previously proposed penalty-based scheduling algorithm in [24], denoted by MPA, aims to minimize the schedule length for a single cell full-duplex WPCN by scheduling the users in increasing order of their penalties.

The penalty function of user i is defined as the difference between the actual transmission time and the minimum possible transmission time at any decision time, where the minimum possible transmission time of user i is the time corresponding to maximum transmit power, i.e., $P_i = P_{max}$. MPA uses the continuous rate model, in which the Shannon channel capacity formula for an additive white Gaussian noise (AWGN) channel is used to calculate the transmission rate of the users as a function of the SNR. For our simulations, MPA picks the user with the minimum penalty among all the users within all cells and allocates it to the current time slot without allowing concurrent transmissions.

Simulation results are obtained by averaging 1000 independent random network realizations. The cells are uniformly distributed within a circle with radius 100 m, and the users are uniformly distributed within the coverage of each cell with radius 10 m. The attenuation of the links considering large-scale statistics are determined by using the path loss model given by

$$PL(d) = PL(d_0) + 10\alpha log_{10}\left(\frac{d}{d_0}\right) + Z, \tag{21}$$

where $PL(d)$ is the path loss at distance d, d_0 is the reference distance, α is the path loss exponent, and Z is a zero-mean Gaussian random variable with standard deviation σ. The small-scale fading is modeled using Rayleigh fading with the scale parameter Ω set to the mean power level obtained from the large-scale path loss model. The parameters used in the simulations are $D_n = 100$ bits for $n \in \mathcal{N}$ and the constant rate $r = 50$ Kbps. The constant rate and corresponding SNR are linked to each other by the Shannon capacity formulation; $W = 1$ MHz; $d_0 = 1$ m; $PL(d_0) = 30$ dB; $\alpha = 2.7$, $\sigma = 4$ [8,24]. The self interference coefficient β is -70 dBm. The initial battery levels of the users are 10^{-9} J, $P_{max} = 1$ mW and $P_h = 1$ W for the simulations, unless otherwise stated [24].

Figure 2 shows the schedule length of the algorithms for different HAP transmit powers in a network of 10 cells and 50 users. The schedule length decreases as the HAP transmit power increases, since higher HAP power allows the users to harvest more energy, which allows users to reach the constant rate earlier and, hence, a lower waiting time. For lower values of HAP power, the schedule length of CRSA is 50% less than MCNS, since CRSA considers the concurrent transmission of users while considering their energy harvesting rates and interference levels.

On the other hand, MCNS randomly selects users for concurrent transmission without considering the energy harvesting rates and interference characteristics, which may lead to grouping the users with very low energy harvesting rates or severe interference levels, resulting in longer slot lengths. The MPA has a higher schedule length than CRSA, although the MPA uses the continuous rate model, which has a natural advantage over the constant transmission rate model; however, CRSA still performs better. Since CRSA considers the concurrent transmissions while MPA only schedules a single user in each time slot, i.e., no concurrent transmissions. It is important to note that CRSA gives a lower schedule length in comparison to MPA due to the proper scheduling of the users for concurrent transmission.

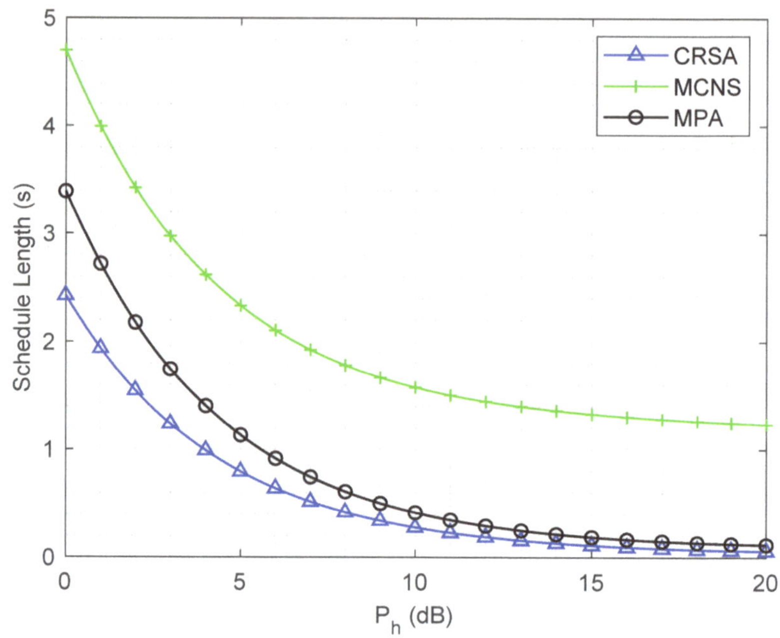

Figure 2. Schedule length of the algorithms for different HAP transmit powers in a network of 10 cells and 50 users.

Figure 3 shows the performance of the algorithms for different number of cells in a network with five users in each cell. The schedule length increases as the number of cells increases in the network, since a higher number of cells results in a higher traffic load, which requires more transmission combinations due to the different interference and energy harvesting rates of the users. For a single cell network, MPA and CRSA almost perform similar. This is due to the fact that there are no concurrent transmissions in a single cell network.

The schedule length of CRSA is half of MCNS and 30% lower than MPA for larger networks due to the intelligent grouping of the users for concurrent transmission. For the MCNS, the steep increase in the schedule length is due to the fact that MCNS randomly selects one user from each cell for simultaneous transmission, which may cause severe interference to other users. For MCNS and MPA algorithms, the addition of a new cell results in an almost constant increase. This is due to the fact that the addition of a new cell increases the number of users in the network and, hence, interference for the concurrent transmission.

For CRSA, the addition of a new cell results in less increase in the schedule length compared to other algorithms. This is because the CRSA attempts to allocate more users within a transmission slot; therefore, the user from a new cell is mostly accommodated within the same transmission slot without any increase in the transmission slot length.

Figure 4 illustrates the impact of network size on the schedule length for the algorithms. The schedule length increases as the number of users increases within each cell, since this increase requires more transmission slots to complete the transmission of all the users. The schedule length of MCNS increases almost linearly as the number of users increases in a cell. On the other hand, the effect of each new user on the CRSA algorithm diminishes, since a higher number of users within each cell increases the probability of finding a more suitable user for the concurrent transmission, resulting in smaller schedule length.

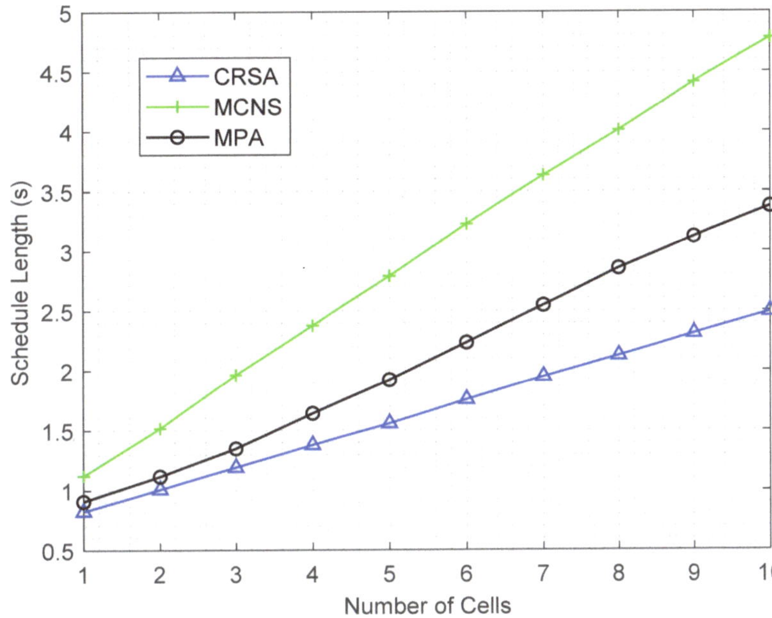

Figure 3. Schedule length of the algorithms for different number of HAPs with five users in each HAP.

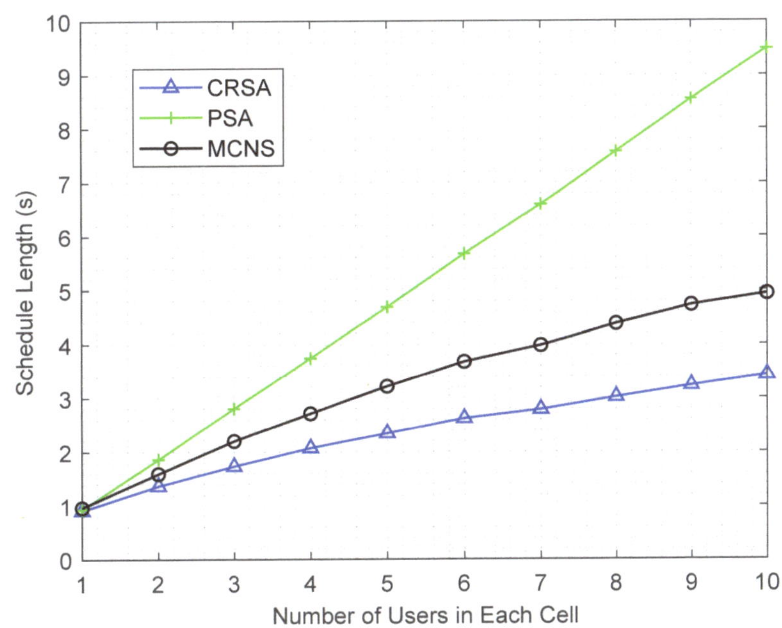

Figure 4. Schedule length of the algorithms for different number of users in each cell in a network of 10 HAPs.

Figure 5 illustrates the schedule length for different values of self-interference coefficient β. For lower values of β, i.e., -90 to -70 dBm, the schedule length is constant due to the low impact of self-interference. Since the self-interference is very small as compared to noise value; therefore, the SNR values do not change too much to affect the required energy and transmission time. However, as the β increases, the schedule length increases drastically due to the dominating effect of self-interference as compared to the noise power. Due to this reason, the SNR drops quickly resulting in a lower probability of concurrent transmissions. It is important to note that, due to intelligent scheduling of the users, the schedule length of CRSA is almost half of the MCNS for higher values of β.

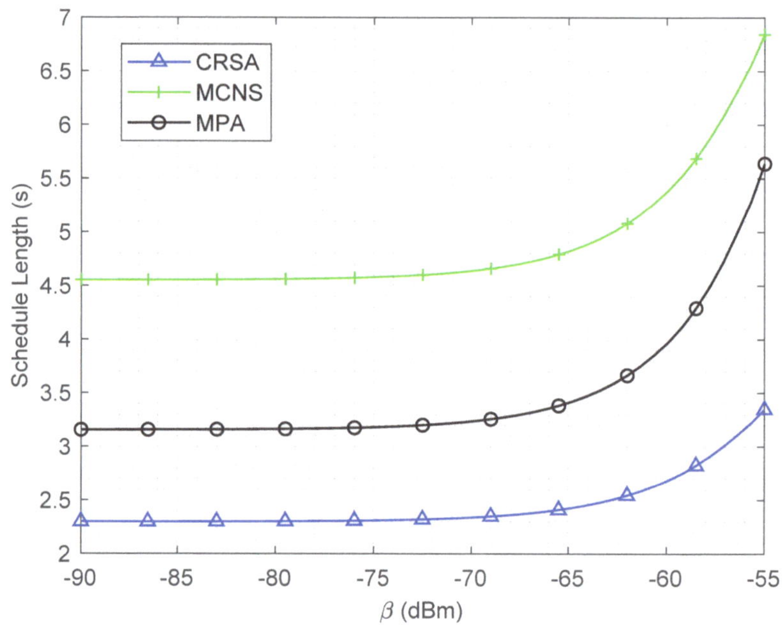

Figure 5. Schedule length of the algorithms for different interference levels in a network of 10 cells and 50 users.

7. Conclusions

In this paper, we formulated a mixed integer programming problem for the minimum length scheduling problem to determine the power control and scheduling, which is difficult to solve for the global optimal solution. As a solution strategy, we decomposed the optimization problem into the power control and scheduling problems. First, we solved the power control problem based on the evaluation of Perron–Frobenius conditions. Then, the proposed optimal power control solution was used to determine the optimal transmission time for a subset of users that will be scheduled by the scheduling algorithm.

For the scheduling problem, we proposed a heuristic algorithm based on the maximization of the number of concurrently transmitting users by maximizing the allowed interference on each user. Through extensive simulations, we demonstrated that the proposed solutions outperformed the conventional successive transmission and concurrent transmission of randomly selected users for different HAP transmit powers, network densities, and network size.

Author Contributions: Conceptualization, M.S.I., Y.S., and S.C.; methodology, Y.S.; software, M.S.I.; validation, M.S.I.; formal analysis, M.S.I., Y.S., and S.C.; original draft preparation, M.S.I.; writing, review and editing, Y.S. and S.C.; supervision, S.C.; project administration, S.C.; funding acquisition, S.C. All authors have read and agreed to the published version of the manuscript.

Funding: This research is funded by the Scientific and Technological Research Council of Turkey Grant #117E241.

Institutional Review Board Statement: Not applicable.

Informed Consent Statement: Not applicable.

Data Availability Statement: Not applicable.

Conflicts of Interest: The funders had no role in the design of the study; in the collection, analyses, or interpretation of data; in the writing of the manuscript, or in the decision to publish the results.

References

1. Jejdling, F. *The Ericsson Mobility Report*; Ericsson: Stockholm, Sweden, 2020.
2. Blanckenstein, J.; Klaue, J.; Karl, H. A Survey of Low-Power Transceivers and Their Applications. *IEEE Circuits Syst. Mag.* **2015**, *15*, 6–17. [CrossRef]
3. Iqbal, M.S.; Sadi, Y.; Coleri, S. Minimum Length Scheduling for Discrete-Rate Full-Duplex Wireless Powered Communication Networks. *IEEE Trans. Wirel. Commun.* **2021**, 1. [CrossRef]
4. Zhou, X.; Zhang, R.; Ho, C.K. Wireless Information and Power Transfer: Architecture Design and Rate-Energy Tradeoff. *IEEE Trans. Commun.* **2013**, *61*, 4754–4767. [CrossRef]
5. Liu, L.; Zhang, R.; Chua, K.C. Wireless Information and Power Transfer: A Dynamic Power Splitting Approach. *IEEE Trans. Commun.* **2013**, *61*, 3990–4001. [CrossRef]
6. Li, Z.; Jiang, Y.; Gao, Y.; Sang, L.; Yang, D. On Buffer-Constrained Throughput of a Wireless-Powered Communication System. *IEEE J. Sel. Areas Commun.* **2019**, *37*, 283–297. [CrossRef]
7. Gupta, A.; Singh, K.; Sellathurai, M. Time-Switching EH-Based Joint Relay Selection and Resource Allocation Algorithms for Multi-User Multi-Carrier AF Relay Networks. *IEEE Trans. Green Commun. Netw.* **2019**, *3*, 505–522. [CrossRef]
8. Ju, H.; Zhang, R. Throughput Maximization in Wireless Powered Communication Networks. *IEEE Trans. Wirel. Commun.* **2014**, *13*, 418–428. [CrossRef]
9. Guo, C.; Liao, B.; Huang, L. Time Allocation and Load Balancing in Multi-Cell Wireless Powered Communication Networks. *IEEE Access* **2016**, *4*, 7795–7805. [CrossRef]
10. Kwan, J.C.; Fapojuwo, A.O. Radio Frequency Energy Harvesting and Data Rate Optimization in Wireless Information and Power Transfer Sensor Networks. *IEEE Sensors J.* **2017**, *17*, 4862–4874. [CrossRef]
11. Shen, S.; Qian, J.; Cheng, D.; Yang, K.; Zhang, G. A Sum-Utility Maximization Approach for Fairness Resource Allocation in Wireless Powered Body Area Networks. *IEEE Access* **2019**, *7*, 20014–20022. [CrossRef]
12. He, C.; Liang, J.; Qian, G.; Guo, C.; Feng, D. Optimal Time Allocation in Multi-Cell Wireless Powered Communication Networks. *IEEE Access* **2019**, *7*, 26519–26526. [CrossRef]
13. Kwan, J.C.; Fapojuwo, A.O. Performance Optimization of a Multi-Source, Multi-Sensor Beamforming Wireless Powered Communication Network With Backscatter. *IEEE Sensors J.* **2019**, *19*, 10898–1090. [CrossRef]
14. Salik, E.D.; Gurur Onalan, A.; Coleri, S. Minimum Length Scheduling for Multi-Cell Wireless Powered Communication Networks. In Proceedings of the 2020 IEEE 31st Annual International Symposium on Personal, Indoor and Mobile Radio Communications, London, UK, 31 August–3 September 2020; pp. 1–6. [CrossRef]
15. Sabharwal, A.; Schniter, P.; Guo, D.; Bliss, D.W.; Rangarajan, S.; Wichman, R. In-band Full-duplex Wireless: Challenges and Opportunities. *IEEE J. Sel. Areas Commun.* **2014**, *32*, 1637–1652. [CrossRef]
16. Liu, G.; Yu, F.R.; Ji, H.; Leung, V.C.M.; Li, X. In-band Full-duplex Relaying: A survey, Research Issues and Challenges. *IEEE Commun. Surv. Tutor.* **2015**, *17*, 500–524. [CrossRef]
17. Sultan, R.; Seddik, K.G.; Han, Z.; Aazhang, B. Joint Transmitter-Receiver Optimization and Self-Interference Suppression in Full-Duplex MIMO Systems. *IEEE Trans. Veh. Technol.* **2021**, *70*, 6913–6929. [CrossRef]
18. Singh, K.; Wang, K.; Biswas, S.; Ding, Z.; Khan, F.A.; Ratnarajah, T. Resource Optimization in Full Duplex Non-Orthogonal Multiple Access Systems. *IEEE Trans. Wirel. Commun.* **2019**, *18*, 4312–4325. [CrossRef]
19. Bharadia, D.; McMilin, E.; Katti, S. Full Duplex Radios. *ACM SIGCOMM Comput. Commun. Rev.* **2013**, *43*, 375–386. [CrossRef]
20. Duarte, M.; Sabharwal, A. Full-duplex Wireless Communications using Off-the-shelf Radios: Feasibility and First Results. In Proceedings of the Asilomar Conference on Signals, Systems and Computers, Pacific Grove, CA, USA, 7–10 November, 2010; pp. 1558–1562.
21. Amjad, M.S.; Nawaz, H.; Ozsoy, K.; Gurbuz, O.; Tekin, I. A Low-Complexity Full-Duplex Radio Implementation With a Single Antenna. *IEEE Trans. Veh. Technol.* **2018**, *67*, 2206–2218. [CrossRef]
22. Korpi, D.; Tamminen, J.; Turunen, M.; Huusari, T.; Choi, Y.S.; Anttila, L.; Talwar, S.; Valkama, M. Full-duplex mobile device: pushing the limits. *IEEE Commun. Mag.* **2016**, *54*, 80–87. [CrossRef]
23. Shao, C.; Roh, H.; Kim, T.; Lee, W. Multisource wireless energy harvesting-based medium access control for rechargeable sensors. *IEEE Trans. Consum. Electron.* **2016**, *62*, 119–127. [CrossRef]
24. Iqbal, M.S.; Sadi, Y.; Coleri, S. Minimum Length Scheduling for Full Duplex Time-Critical Wireless Powered Communication Networks. *IEEE Trans. Wirel. Commun.* **2020**, *19*, 5993–6006. [CrossRef]

25. Iqbal, M.S.; Sadi, Y.; Coleri, S. Throughput Maximization for Full Duplex Wireless Powered Communication Networks. In Proceedings of the IEEE International Conference on Communications (ICC), Dublin, Ireland, 7–11 June 2020; pp. 1–6.
26. Morsi, R.; Michalopoulos, D.S.; Schober, R. Performance Analysis of Near-Optimal Energy Buffer Aided Wireless Powered Communication. *IEEE Trans. Wirel. Commun.* **2018**, *17*, 863–881. [CrossRef]
27. Morsi, R.; Michalopoulos, D.S.; Schober, R. On-off transmission policy for wireless powered communication with energy storage. In Proceedings of the 2014 48th Asilomar Conference on Signals, Systems and Computers, Pacific Grove, CA, USA, 2–5 November 2014; pp. 1676–1682.
28. Iqbal, M.S.; Sadi, Y.; Coleri, S. Optimal On-Off Transmission Schemes for Full Duplex Wireless Powered Communication Networks. In Proceedings of the IEEE International Black Sea Conference on Communications and Networking (BlackSeaCom), Odessa, Ukraine, 26–29 May 2020; pp. 1–6.
29. Iqbal, M.S.; Sadi, Y.; Coleri Ergen, S. Minimum Length Scheduling for Discrete Rate Based Full Duplex Wireless Powered Communication Networks. In *Ad-Hoc, Mobile, and Wireless Networks*; Springer International Publishing: Berlin/Heidelberg, Germany, 2019; pp. 343–354.
30. Iqbal, M.S.; Sadi, Y.; Coleri, S. Throughput maximization in Discrete Rate Based Full Duplex Wireless Powered Communication Networks. *Internet Technol. Lett.* **2020**, *4*, e206. [CrossRef]
31. HART Communication Foundation. *HART Communication Application Guide*; HART Commun. Foundation: Austin, TX, USA, 2013.
32. Coleri Ergen, S.; Varaiya, P. PEDAMACS: Power efficient and delay aware medium access protocol for sensor networks. *IEEE Trans. Mob. Comput.* **2006**, *5*, 920–930. [CrossRef]
33. Xu, J.; Liu, L.; Zhang, R. Multiuser MISO Beamforming for Simultaneous Wireless Information and Power Transfer. *IEEE Trans. Signal Process.* **2014**, *62*, 4798–4810. [CrossRef]
34. Chen, H.; Li, Y.; Rebelatto, J.L.; Uchoa-Filho, B.F.; Vucetic, B. Harvest-then-cooperate: Wireless-powered Cooperative Communications. *IEEE Trans. Signal Process.* **2015**, *63*, 1700–1711. [CrossRef]
35. Wu, Q.; Tao, M.; Ng, D.W.K.; Chen, W.; Schober, R. Energy-Efficient Resource Allocation for Wireless Powered Communication Networks. *IEEE Trans. Wirel. Commun.* **2016**, *15*, 2312–2327. [CrossRef]
36. Kang, X.; Ho, C.K.; Sun, S. Full-Duplex Wireless-Powered Communication Network With Energy Causality. *IEEE Trans. Wirel. Commun.* **2015**, *14*, 5539–5551. [CrossRef]
37. Boshkovska, E.; Ng, D.W.K.; Zlatanov, N.; Schober, R. Practical Non-Linear Energy Harvesting Model and Resource Allocation for SWIPT Systems. *IEEE Commun. Lett.* **2015**, *19*, 2082–2085. [CrossRef]
38. Boyd, S.; Vandenberghe, L. *Convex Optimization*; Cambridge University Press: New York, NY, USA, 2004.

sensors

MDPI

Article

Hybrid Active-and-Passive Relaying Model for 6G-IoT Greencom Networks with SWIPT

Sumit Gautam [1,2,*], Sourabh Solanki [1,*], Shree Krishna Sharma [1], Symeon Chatzinotas [1] and Björn Ottersten [1]

1 Interdisciplinary Centre for Security, Reliability and Trust (SnT), University of Luxembourg, L-1855 Luxembourg, Luxembourg; shree.sharma@uni.lu (S.K.S.); symeon.chatzinotas@uni.lu (S.C.); bjorn.ottersten@uni.lu (B.O.)
2 Huawei Technologies Sweden AB, Vädursgatan 5, 412 50 Göteborg, Sweden
* Correspondence: sumit.gautam@uni.lu (S.G.); sourabh.solanki@uni.lu (S.S.)

Abstract: In order to support a massive number of resource-constrained Internet-of-Things (IoT) devices and machine-type devices, it is crucial to design a future beyond 5G/6G wireless networks in an energy-efficient manner while incorporating suitable network coverage expansion methodologies. To this end, this paper proposes a novel two-hop hybrid active-and-passive relaying scheme to facilitate simultaneous wireless information and power transfer (SWIPT) considering both time-switching (TS) and power-splitting (PS) receiver architectures, while dynamically modelling the involved dual-hop time-period (TP) metric. An optimization problem is formulated to jointly optimize the throughput, harvested energy, and transmit power of a SWIPT-enabled system with the proposed hybrid scheme. In this regard, we provide two distinct ways to obtain suitable solutions based on the Lagrange dual technique and Dinkelbach method assisted convex programming, respectively, where both the approaches yield an appreciable solution within polynomial computational time. The experimental results are obtained by directly solving the primal problem using a non-linear optimizer. Our numerical results in terms of weighted utility function show the superior performance of the proposed hybrid scheme over passive repeater-only and active relay-only schemes, while also depicting their individual performance benefits over the corresponding benchmark SWIPT systems with the fixed-TP.

Keywords: 5G and beyond/6G wireless networks; Greencom; IoT; passive repeater; relaying systems; SWIPT

Citation: Gautam, S.; Solanki, S.; Sharma, S.K.; Chatzinotas, S.; Ottersten, B. Hybrid Active-and-Passive Relaying Model for 6G-IoT Greencom Networks with SWIPT. *Sensors* **2021**, *21*, 6013. https://doi.org/10.3390/s21186013

Academic Editors: Slim Naifar, Olfa Kanoun and Carlo Trigona

Received: 28 July 2021
Accepted: 4 September 2021
Published: 8 September 2021

1. Introduction

1.1. General Motivation

The world witnesses the introduction of a novel generation of wireless communications approximately every ten years. With the current speedy deployment of hardware for the fifth-generation (5G) of mobile wireless systems in 2021, we are now at a juncture to anticipate what lies ahead for the constitution of the sixth generation (6G) mobile cellular systems [1]. The incoming crucial updates to the hardware equipment for 5G may require dense infrastructure deployment [2]. In future, it will become important to leverage the 5G architecture for internet-of-things (IoT) systems, along with the introduction of some methods to increase the coverage area as well as considering energy harvesting (EH) frameworks while embracing the concept of Green Communications (Greencom)-IoT systems.

1.2. Background to the Considered Topics

With the wide deployment of 5G systems, it is becoming crucial to understand the main notion behind the working of such technology. The 5G systems may be implemented in low-band, mid-band or high-band millimeter-wave (mmWave from 24 GHz up to 100 GHz are expected to be used for 5G) [3]. The air interface defined by the 3rd

Generation Partnership Project (3GPP) for 5G is known as New Radio (NR), and the specification is subdivided into two frequency bands, frequency range-1 (FR1) (below 6 GHz) and frequency range-2 (FR2) (in the mmWave regime) [4], each with different capabilities [5]. Considering the low-band, 5G uses a frequency range similar to 4G cellphones, i.e., 600–850 MHz. However, the 5G is able to provide download speeds little higher than 4G, i.e., 30–250 megabits per second (Mbit/s) [6]. The range and coverage area of the low-band 5G cell towers area similar to the ones in 4G. Regarding the mid-band 5G, microwaves of 2.5–3.7 GHz are used which allow speeds of 100–900 Mbit/s, with each cell tower providing service up to several kilometers in radius. This level of service is the most widely deployed within many metropolitan areas in 2020. The high-band 5G uses frequencies of 25–39 GHz, near the bottom of the mmWave band. In this vein, there is also a possibility that higher frequencies will be used in the future, e.g., for the 6G systems. Using this scheme often achieves download speeds in the gigabit-per-second (Gbit/s) range, which is comparable to cable internet. However, mmWave bands have a more limited range, requiring many small cells [7].

Since the proposed 5G setting will be more prominent in the coming future, it cannot be denied that the future 6G systems will leverage a lot of the hardware architectures and techniques from the 5G framework. In this regard, deployment of several small-cells in various places on the planet will become cumbersome and hence some supporting techniques need to be sought to enhance the network coverage. Cooperative communications, which works on the principle of introducing repeater(s) or relay(s) between the transmit source and end-user, has been established as a promising technique to address the limited coverage concerns. Herein, repeater(s) or relay(s) assist in forwarding the transmit signal towards the end-user by preserving the signal properties and power, which can enable enhanced coverage. In this vein, the cooperative mechanism may be performed via single or multiple signal hops [8,9].

As the technological advancements in the field of wireless communications continue to amaze the mankind in several ways, a major hurdle in this progress is our dependence on the energy sources, which are as strong as on the gadgets themselves [10]. An early EH concept of getting energy from/through the air to power all equipment, was proposed by Nikola Tesla in the late 1890s to early 1900s [11,12]. Ever since then, this concept has found its applications in several fields of research including vehicles [13], wireless gadgets [14], flying objects [15], etc. More recently, several researchers have proposed using RF signals for facilitating simultaneous wireless transmission of information and energy , which involves the traditional information receiver unit along with an EH module. This idea of joint information and energy transmission is likely to play significant role in emerging technologies [16–18]. Applications like wearable devices, wireless sensor nodes etc., are expected to adopt the proposed wireless recharging alternative more rapidly. With the increasing demand for energy to operate the devices, it has almost become a necessity to adopt such techniques to compensate for rapidly draining battery sources [19–21].

In order to investigate the performance above-mentioned scenarios, optimization of system parameters such as, e.g., data throughput, harvested energy, or transmit power, come in handy. Herein, the mathematical formulation of the optimization problem corresponding to the wireless scenario is considered under a certain set of constraints. In case the formulated optimization problem is convex, it can be solved directly via simple methods, such as gradient descent algorithm [22]. For the deterministic and continuous constrained optimization under non-linear programming, obtaining a direct solution is cumbersome [23]. Herein, suitable transformations or approximations may be employed to convexify the primal problem in order to obtain a suitable solution [24,25]. Alternatively, a Lagrange dual method based solution could be a possibility, which operates on the dual regime corresponding to the main problem. It is noteworthy that the dual problem is always convex and hence a suitable solution can be sought [22,26]. However, in this case, it is important to check the duality gap between the primal solution and the one obtained via dual method. Extremely less or zero duality gap indicates that the optimal solution has been achieved, while a large duality gap indicates a sub-optimal solution is obtained [27]. On the other hand, an optimization problem with fractional objective is extremely challenging to solve. In this context, Dinkelback algorithm based solution is found to be most effective [28,29]. In the context of this paper, we leverage some steps from the above-mentioned techniques to solve the formulated problem, which will be presented in the later sections. To facilitate the reading process, we have summarized the list of abbreviations in Table 1. In the following, we discuss some related works to the framework considered in this paper.

Table 1. List of Abbreviations.

Acronym	Full-Form
3GPP	3rd Generation Partnership Project
5G	Fifth Generation Mobile Wireless Communications
6G	Sixth Generation Mobile Wireless Communications
AF	Amplify-and-Forward
AGN	Active Relay Device-based Green Communication Network
AR	Active Relay
AWGN	Additive White Gaussian Noise
CSI	Channel State Information
DCCP	Disciplined Convex-Concave Programming
DF	Decode-and-Forward
EH	Energy Harvesting
FR1	Frequency Range-1 for 5G
FR2	Frequency Range-2 for 5G
Greencom	Green Communications
HAP	Hybrid Active-and-Passive
ID	Information Decoding
IoT	Internet-of-Things
IRS	Intelligent Reflecting Surface
KKT	Karush-Kuhn-Tucker
MRC	Maximum Ratio Combining
NR	New Radio
PGN	Passive Repeater Device-based Green Communication Network
PR	Passive Repeater
PS	Power Splitting
QoS	Quality-of-Service
RIS	Reconfigurable Intelligent Surface
SNR	Signal-to-Noise-Ratio
SWIPT	Simultaneous Wireless Information and Power Transfer
TP	Time Period
TS	Time Switching
TSF	Time-Splitting Factor

1.3. Related Works

To enhance the network coverage of wireless communications systems, certain well-known methods such as passive repeater (PR) and/or relaying have shown immense potential [30]. A PR is a battery-less device used to reflect the transmit signal (with encoded data) towards the desired user, where it may be utilized to energize the IoT devices as well as to enable the deployment of battery-free devices in future networks [31]. Noticeably, traditional backscattering may be considered as a special form of PR. In case of an active relay (AR), the transmitted signal from the source is received at the relay device and re-transmitted to the end-user after either a direct amplification according to the amplify-and-forward (AF) protocol, or a decoding or quantization process before its re-transmission as a re-encoded signal with message suppression, according to the decode-and-forward (DF) protocol [32,33]. Regarding the facilitation of EH at the end-user, the simultaneous wireless information and power transfer (SWIPT) technique seems promising [34,35]. In this regard, the power-splitting (PS) [36] and time-switching (TS) [37] schemes have been widely considered to deploy SWIPT receivers in practice.

Considering the latest trends in relaying methodologies pertaining to the recent advancements in electromagnetic materials [38–41], the notion of Reconfigurable Intelligent Surface (RIS) [42] or Intelligent Reflecting Surface (IRS) [43] holds immense potential. Concretely, an RIS is a two-dimensional surface that interacts with the incoming electromagnetic waves using several small reflecting elements that are electronically tuned [44]. More specifically, a transmitted signal intended for an end-user may be more efficiently sent over a wireless medium, where the channel could now be controlled via an RIS such that a metal plate is rotated or bent to redirect the corresponding incident wave [45]. The meta-materials considered herein are expected to inherit some extra-ordinary electromagnetic features that are unnatural [46], such as, adverse refraction [47], flawless assimilation/absorption, and atypical reflection/scattering [48]. To ensure the possibilities discussed above, an ideal RIS must be comprised of large number of elements having controllable properties [49], wherein any alteration to the reflection coefficient (e.g., phase shift or re-modulation) of the elements can control the beamforming of the incident waves.

However, the above prospect of realizing RIS seems promising, there are still several challenges associated while considering its practical implementation aspect [50]. In this vein, three major concerns include: (i) Difficulty in providing continuous and uninterrupted power-supply to the RIS elements given their active nature; (ii) continuous switching methods to adapt the RIS elements according to the reflection needs, which require additional computational resources and energy; and (iii) management of thermal noise at the very small RIS elements (given their designs are more suitable for high frequency operations) becomes an issue, primarily due to continuous frictional operations and power supply. Besides, it is extremely difficult to ensure proper supply of energy where power resources are scarce. Therefore, it becomes natural to seek the PR alternatives as other complementary methods to the RIS. In this manner, the benefits of employing the reflectors may be harnessed using the passive means in the instances where the deployment of active-RIS becomes challenging [51].

Recent developments in the area of PR and AR type of schemes have focused on important aspects such as cooperation enhancement, however, with the consideration of a linear EH model [52,53]. In [54], the authors presented an interesting framework for wireless energy beamforming in the reflector-and-relay based communications system, wherein a non-linear EH mechanism is considered at the intermediary device but not at the end-user. Additionally, most related works focus on the maximization of either the signal-to-noise ratio (SNR) [55], or the throughput [56]. However, it is important to optimize the joint metrics of data throughput, harvested energy and transmit power of the SWIPT-enabled 6G systems. On one hand, it is noteworthy that the reflector-only mechanism may not be suitable for dense deployment scenarios [31,57]. Whereas on the other hand, the power-dependence of relays is critical in the relaying-only scenarios, wherein a power-failure may lead to the collapse of the entire wireless system without any backup. In this regard, the hybrid PR and AR based scheme may ensure a sustainable wireless ecosystem where both devices may supplement or complement each other accordingly. Moreover, in the context of the hybrid framework, a comparative study between the PS and TS SWIPT schemes with non-linear EH model along with the design of dynamic dual-hop time-period (TP) has not yet been investigated in the literature so far.

1.4. Our Contributions

In this invited paper, we present a dual-hop system wherein a transmit source transfers both information and energy to a SWIPT-enabled end-user employing either PS or TS architecture. We investigate three distinct Greencom systems incorporating a PR node, an AR node, and a novel hybrid active-and-passive relaying node, respectively; which either reflects and/or re-transmits the information-and-energy signal to the end-user, in addition to the weak direct link. The novelty and main contributions of this work are listed as follows.

(i) In contrast to the existing reflector-only and relaying-only techniques, we propose a novel hybrid active-and-passive relaying scheme to facilitate SWIPT to a PS- or TS-enabled end-user, along with the dynamic designing of dual-hop TP, under certain receive processing assumptions.
(ii) We formulate a novel problem (incorporating the three systems) to maximize the weighted utility function comprising data throughput, harvested energy and transmit power, subjected to some quality-of-service (QoS) constraints. Unlike the other works that assume equal time-intervals in the two hops (also considered herein as the benchmarks), we present a framework to dynamically design the TP for the dual-hop link, along with the computation of the PS or TS factors.
(iii) In order to solve the aforementioned problems, we present two distinct methods based on the Lagrange dual technique and Dinkelbach method assisted convex programming, respectively, where both the approaches yield an appreciable solution within polynomial computational-time.
(iv) The effectiveness of the proposed hybrid active-and-passive relaying scheme is shown over the reflector-only and relay-only schemes for both PS and TS SWIPT schemes via numerical analysis, with individual benefits shown over their respective benchmark designs having a fixed TP.

1.5. Further Organization of the Paper

The remainder of this paper is organized as follows. Section 2 presents a description of the system model and Section 3 provides an analysis of the considered Greencom networks. Section 4 focuses on the joint optimization of data throughput, harvested energy and transmit power. Section 5 presents the simulation results. Finally, Section 6 concludes the paper.

2. System Model

We consider three wireless Greencom network scenarios for investigation, with a single transmit source, a single reflection and/or re-transmission device (i.e., either PR or AR or both), and a single user at the receiving end, as represented in Figure 1. The transmit source communicates with the end-user via two half-duplex communication links within a single time-slot. In particular, besides the conventional direct link, we assume the availability of a couple of intermediary links to assist in the delivery of the desired transmit signal. The end-user is assumed to be capable of performing both information decoding (ID) and energy harvesting (EH) simultaneously.

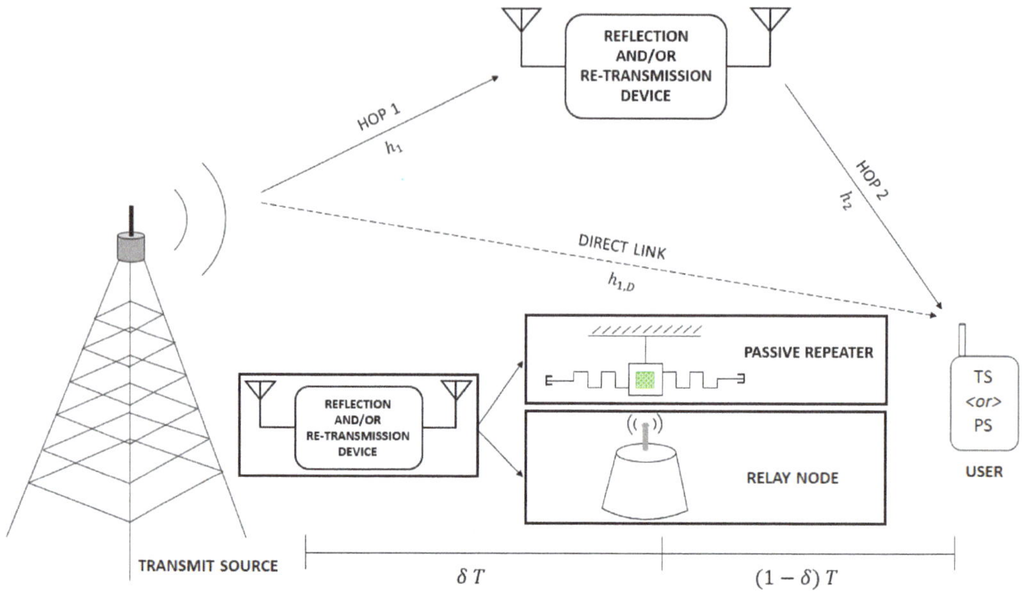

Figure 1. Dual hop system with a PR and/or AR device.

Let us define $P_{\mathcal{T}}$ as the transmit power at the source, $\mathcal{B}$ as the overall bandwidth, and T as the TP for further analysis. In each system, the direct link is assumed to be active throughout the TP, with a pre-defined channel coefficient of $h_{1,D}$. For analytical convenience, we assume that the channel state information (CSI) is known to the intended devices. The transmit source emits the signal intended for the corresponding SWIPT-enabled end-user. The process is carried out using a direct link through the whole TP, with additional dual-hop assistance via a reflecting and/or re-transmitting device over the same TP. More specifically, the source transmits a symbol $s \in \mathbb{C}$, which is received by the destination and the reflecting and/or re-transmitting device. The propagation path of the signal is sufficiently small due to the short communication distance. In this vein, the destination node is considered to utilize adequate delay compensation methods to synchronize and coherently process the received signals from the second hop and the direct link. Without loss of generality, we assume $\mathbb{E}[|s|^2] = 1$, with operation $\mathbb{E}[\cdot]$ denoting the expectation value. The signal received by the end-user via direct link is defined as: $y_{1,D} = \sqrt{P_{\mathcal{T}}}\, h_{1,D}\, s + n_{u_a}$, where n_{u_a} is the additive white Gaussian noise (AWGN) at the end-user which is an independent and identically distributed (i.i.d.) complex Gaussian random variable with zero mean and variance $\sigma^2_{n_{u_a}}$.

Concerning the SWIPT-enabled receiver device, two different SWIPT schemes namely PS or TS, respectively, may be adopted. For the PS scheme, an optimal fraction of the received signal is provided to the information decoder and the remaining part is provided to the energy harvester. In this case, the PS ratio is denoted by ρ, where $0 \leq \rho \leq 1$. Alternatively, when the receiver incorporates the TS scheme, we define a time switching ratio, τ, where $0 \leq \tau \leq 1$. In particular, for the first fraction of time period, all the received signal power is used for harvesting energy, whereas during the remaining time, information decoding from the received signal takes place.

In order to estimate the amount of harvested energy at the end-user, we adopt the sigmoidal function based non-linear EH model [27,58], defined as

$$\mathcal{E}(x) = \frac{\mathcal{E}'}{1-\phi} \cdot \Big(\frac{1}{1+e^{(-\alpha x+\alpha\beta)}} - \phi \Big), \tag{1}$$

where x is the input power at the energy harvesting module, $\phi \triangleq \frac{1}{1+\exp(\alpha\beta)}$, the constant $\mathcal{E}'$ is obtained by determining the maximum harvested energy on the saturation of the energy harvesting circuit, and α and β are specific to the capacitor and diode turn-on voltage metrics at the EH circuit. Practically, a standard curve-fitting tool based on analytical data may be used to decide the appropriate values of $\mathcal{E}'$, α, and β. A comparison between the linear and non-linear EH models is depicted in Figure 2, where the energy conversion efficiency for linear EH module is assumed to be 0.75, $\mathcal{E}'$ = 2.8 mJ, α = 1500, and β = 0.0022 (for the non-linear EH module) [27,58]. The non-linearity introduced due to the diode and capacitor elements is observed at lower input power at the EH module while a constant EH operation is seen for higher values of input power which implies the saturation at diode element of the EH module. On the other hand, the linear EH model increases constantly based on the increasing input power at the EH module, without considering the saturation points of the diode element [58]. This is due to the assumption of a constant energy conversion efficiency at the EH module.

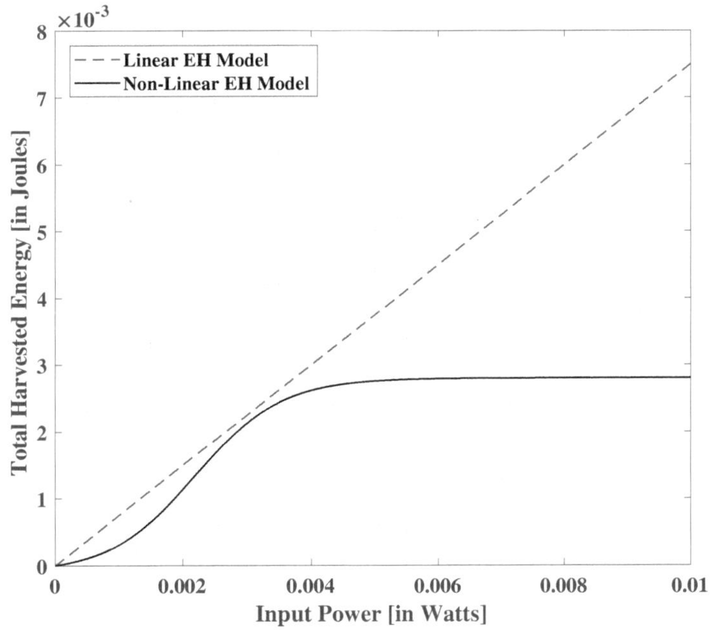

Figure 2. Comparison between the energy extraction capabilities of linear and non-linear energy harvesting models.

The harvested energy using the PS and TS schemes are respectively given by

$$\mathcal{E}^{\mathrm{PS}}(x,\rho) = \frac{\mathcal{E}'}{1-\phi} \cdot \left(\frac{1}{1+e^{(-\alpha\rho x+\alpha\beta)}} - \phi \right), \tag{2}$$

$$\mathcal{E}^{\mathrm{TS}}(x,\tau) = \tau \cdot \frac{\mathcal{E}'}{1-\phi} \cdot \left(\frac{1}{1+e^{(-\alpha x+\alpha\beta)}} - \phi \right). \tag{3}$$

The energy harvested at the end-user using the PS and TS schemes via direct link signal is given by $\mathcal{E}_{1,D}^{\mathrm{PS}}(\mathbb{E}[|y_{1,D}|^2],\rho)$ and $\mathcal{E}_{1,D}^{\mathrm{TS}}(\mathbb{E}[|y_{1,D}|^2],\tau)$, computed according to (2) and (3), respectively. Further, we assume that the normalized time slots use the terms power and energy interchangeably.

The effective signal-to-noise ratio (SNR) via direct link as seen at the ID branch of the end-user, respectively for PS and TS schemes, is given by

$$\Upsilon_{1,D}^{\text{PS}} = \frac{(1-\rho)\, P_{\mathcal{T}}\, |h_{1,D}|^2}{(1-\rho)\, \sigma_{n_{ua}}^2 + \sigma_{n_{uID}}^2}, \text{ and } \Upsilon_{1,D}^{\text{TS}} = \frac{P_{\mathcal{T}}\, |h_{1,D}|^2}{\sigma_{n_{ua}}^2 + \sigma_{n_{uID}}^2}, \tag{4}$$

where $n_{u_{ID}} \in \mathbb{CN}(0,\, \sigma_{n_{u_{ID}}}^2)$ is the noise introduced by the baseband processing circuit. The effective throughput achieved at the end-user, combining the PS and TS schemes is given by

$$R_{1,D} = \begin{cases} R_{1,D}^{\text{PS}} = \mathcal{B}\log_2\left(1 + \Upsilon_{1,D}^{\text{PS}}\right) \\ R_{1,D}^{\text{TS}} = (1-\tau)\mathcal{B}\log_2\left(1 + \Upsilon_{1,D}^{\text{TS}}\right) \end{cases} \tag{5}$$

3. Analysis of Greencom Network Scenarios

In this section, we present the discussion on the considered Greencom network scenarios in detail. To proceed, we define the time-splitting factor (TSF) of TP as δ, where $0 < \delta < 1$, to assist the dual-hop mechanism. It is evident that significant benefits may be obtained by employing dynamic TSF in contrast to the equal-time period-based splitting in dual-hop cooperative systems [59,60]. In this vein, it is noteworthy that the phenomenon of dynamic TSF modelling in the context of AF relaying primarily translates to a better power optimization and scheduling mechanism at the transmitter, while ensuring an improved management of the network resources. Specifically, we assume that the first-hop spans over δT of the TP while in the second hop, the remainder of the TP, i.e., $(1-\delta)T$, is allocated for successful reflection and/or re-transmission of the intended signal from the PR and/or AR device(s) to the end-user. Additionally, we assume that the direct and indirect links are managed via different supervised signalling methods (e.g., use of separate directional antenna elements). More specifically, the transmission via direct link is assumed to be active for the whole time duration T, while the transmission through the indirect link remains active only for the stipulated δT duration. Following this assumption, it is needless to mention that the interference between the signal via (weaker) direct and (stronger) indirect links will be negligible, and thus this possible interference aspect is discarded in the further analysis.

3.1. *Traditional Passive Repeater or Active Relay-Based Systems*

In this case, we assume the availability of either reflection-antenna equipped PR device as the reflector node, or a single antenna based AR node as a re-transmission device, on the second hop. Further, we refer the two above-mentioned systems as PR device-based Greencom network (PGN) and AR device-based Greencom network (AGN), respectively. The channel coefficient in the first phase is defined as h_1.We represent the signal seen at the reflecting/re-transmitting device as: $r_1 = \sqrt{P_{\mathcal{T}}}\, h_1\, s + n_1$, where n_1 is the AWGN introduced due to the reflector/relay, which is an i.i.d. complex Gaussian random variable with zero mean and variance $\sigma_{n_1}^2$. In this context, we express $(r_1, h_1, n_1) := \{(r_{1,P}, h_{1,P}, n_P) : \text{For PGN}; (r_{1,R}, h_{1,R}, n_R) : \text{For AGN}\}$.

The SNR and the effective data throughput expressions are respectively defined as

$$\Upsilon_1 = \frac{P_{\mathcal{T}}\, |h_1|^2}{\sigma_{n_1}^2}, \text{ and } R_1 = \delta\, \mathcal{B}\, \log_2(1 + \Upsilon_1), \tag{6}$$

where $(\Upsilon_1, R_1, \sigma_{n_1}^2) := \{(\Upsilon_{1,P}, R_{1,P}, \sigma_{n_P}^2) : \text{For PGN}; (\Upsilon_{1,R}, R_{1,R}, \sigma_{n_R}^2) : \text{For AGN}\}$.

In the second hop, the channel coefficient is defined as h_2. In case of PGN, the intended signal from the PR is reflected to the end-user with the corresponding efficiency coefficient of $\sqrt{\eta}$, where $0 < \sqrt{\eta} \leq 1$. Whereas in the case of AGN, the AR re-transmits the signal after scaling it by a complex amplification coefficient w. It is noteworthy that in the case of AF protocol, the amplification coefficient is a complex number which incorporates the amplitude as well as the direction/phase components. However, we only consider the constraint on amplitude component and ignore the directional/phase part in this work for simplifying the further analysis. In order to ensure feasibility of the AGN system, we impose an upper bound on the total relay power defined by: $0 < |w|^2 \leq P_R$, where $P_R = \frac{P^\star - P_\mathcal{T}}{P_\mathcal{T}|h_{1,R}|^2 + \sigma^2_{n_R}}$ is the maximum overall available power at the relay, with the transmitter-relay system bounded by an overall power of $P^\star$, such that $P^\star > \max(P_\mathcal{T}, P_R)$. To proceed, we define the following metric:

$$\Xi = \begin{cases} \sqrt{\eta} : \text{PR's Reflection Coefficient in PGN} \\ w \ : \text{AR's Amplification Coefficient in AGN} \end{cases} \tag{7}$$

The received signal as seen at the end-user via reflection or re-transmission from the PR or AR device, respectively, is represented as $y_2 = \Xi\, h_2\, r_1 + n_{u_a}$, where $(y_2, h_2, r_1) := \{(y_{2,P}, h_{2,P}, r_{1,P}) : \text{For PGN};\ (y_{2,R}, h_{2,R}, r_{1,R}) : \text{For AGN}\}$.

We define the SNR obtained at the ID block of the end-user via second hop link, according to the PS and TS schemes, as

$$\Upsilon_2^{\text{PS}} = \frac{|\Xi|^2\,(1-\rho)\,P_\mathcal{T}\,|h_1|^2\,|h_2|^2}{(1-\rho)\,(|\Xi|^2\,|h_1|^2\,\sigma^2_{n_1} + \sigma^2_{n_{u_a}}) + \sigma^2_{n_{u_{ID}}}}, \tag{8}$$

$$\Upsilon_2^{\text{TS}} = \frac{|\Xi|^2\,P_\mathcal{T}\,|h_1|^2\,|h_2|^2}{|\Xi|^2\,|h_1|^2\,\sigma^2_{n_1} + \sigma^2_{n_{u_a}} + \sigma^2_{n_{u_{ID}}}}, \tag{9}$$

where $(\Upsilon_2^{\text{PS}}, \Upsilon_2^{\text{TS}}, h_1, h_2, \sigma^2_{n_1}) := \{(\Upsilon_{2,P}^{\text{PS}}, \Upsilon_{2,P}^{\text{TS}}, h_{1,P}, h_{2,P}, \sigma^2_{n_P}) : \text{For PGN};\ (\Upsilon_{2,R}^{\text{PS}}, \Upsilon_{2,R}^{\text{TS}}, h_{1,R}, h_{2,R}, \sigma^2_{n_R}) : \text{For AGN}\}$. The effective throughput achieved at the end-user, incorporating the respective PS and TS schemes is defined as

$$R_2 = \begin{cases} R_2^{\text{PS}} = (1-\delta)\mathcal{B}\log_2\left(1+\Upsilon_2^{\text{PS}}\right) \\ R_2^{\text{TS}} = (1-\tau)(1-\delta)\mathcal{B}\log_2\left(1+\Upsilon_2^{\text{TS}}\right) \end{cases}, \tag{10}$$

where $(R_2, R_2^{\text{PS}}, \Upsilon_2^{\text{PS}}, R_2^{\text{TS}}, \Upsilon_2^{\text{TS}}) := \{(R_{2,P}, R_{2,P}^{\text{PS}}, \Upsilon_{2,P}^{\text{PS}}, R_{2,P}^{\text{TS}}, \Upsilon_{2,P}^{\text{TS}}) : \text{For PGN};\ (R_{2,R}, R_{2,R}^{\text{PS}}, \Upsilon_{2,R}^{\text{PS}}, R_{2,R}^{\text{TS}}, \Upsilon_{2,R}^{\text{TS}}) : \text{For AGN}\}$.

The PS and TS expressions for harvested energy are denoted as $\mathcal{E}_2^{\text{PS}}(\mathbb{E}[|y_2|^2], \rho)$ and $\mathcal{E}_2^{\text{TS}}(\mathbb{E}[|y_2|^2], \tau)$, computed according to (2) and (3), respectively, where $(\mathcal{E}_2^{\text{PS}}, \mathcal{E}_2^{\text{TS}}, y_2) := \{(\mathcal{E}_{2,P}^{\text{PS}}, \mathcal{E}_{2,P}^{\text{TS}}, y_{2,P}) : \text{For PGN};\ (\mathcal{E}_{2,R}^{\text{PS}}, \mathcal{E}_{2,R}^{\text{TS}}, y_{2,R}) : \text{For AGN}\}$.

At the end-user, the overall data throughput achieved incorporating the yield from the second hop and direct link, for respective PS and TS schemes, is defined as

$$R_U = \begin{cases} R_U^{\text{PS}} = R_{1,D}^{\text{PS}} + R_2^{\text{PS}} \\ R_U^{\text{TS}} = R_{1,D}^{\text{TS}} + R_2^{\text{TS}} \end{cases}, \tag{11}$$

where $(R_U, R_U^{\text{PS}}, R_U^{\text{TS}}) := \{(R_{U,P}, R_{U,P}^{\text{PS}}, R_{U,P}^{\text{TS}}) : \text{For PGN};\ (R_{U,R}, R_{U,R}^{\text{PS}}, R_{U,R}^{\text{TS}}) : \text{For AGN}\}$. Note that due to the involvement of the time related metrics, the expressions in (11) are indicative of the overall data throughput and not the actual rate. Hence, it is possible to combine the data throughputs from the two links following a proper synchronization process at the receiver. This is not to be confused with the case of overall rate (traditionally measured in bits per seconds), wherein it is necessary to perform the SNR combining methods, such as, e.g., maximum ratio combining (MRC) [61].

The effective harvested energy expression incorporating both PS and TS schemes is:

$$E_U = \begin{cases} E_U^{\mathrm{PS}} = \mathcal{E}_{1,D}^{\mathrm{PS}}(\mathbb{E}[|y_{1,D}|^2], \rho) + (1-\delta)\mathcal{E}_2^{\mathrm{PS}}(\mathbb{E}[|y_2|^2], \rho) \\ E_U^{\mathrm{TS}} = \mathcal{E}_{1,D}^{\mathrm{TS}}(\mathbb{E}[|y_{1,D}|^2], \tau) + (1-\delta)\mathcal{E}_2^{\mathrm{TS}}(\mathbb{E}[|y_2|^2], \tau) \end{cases} \tag{12}$$

where $(E_U, E_U^{\mathrm{PS}}, E_U^{\mathrm{TS}}) := \{(E_{U,P}, E_{U,P}^{\mathrm{PS}}, E_{U,P}^{\mathrm{TS}}) : \text{For PGN}; (E_{U,R}, E_{U,R}^{\mathrm{PS}}, E_{U,R}^{\mathrm{TS}}) : \text{For AGN}\}$.

3.2. Proposed Hybrid Active-and-Passive Relaying Scheme

We present herein the proposed hybrid active-and-passive (HAP) relaying scheme in Greencom networks to facilitate SWIPT to the end-user over the second hop. In this case, the HAP relaying node receives the transmit signal on both the PR and AR system equipped with single antenna each, over the above-defined channel coefficients of $h_{1,P}$ and $h_{1,R}$, respectively. The dynamic TSF (ratio) definition is same as defined before. The signals, SNR and throughput expressions seen at the respective PR and the AR nodes are similar to the ones defined in the previous sections. Note that the PR and AR devices in the proposed HAP scheme may be co-located or not. However, for analytic convenience related to the dynamic TSF design, we assume a co-located hybrid PR and AR set-up.

Over the second hop, the PR device uses the coefficient $\sqrt{\eta}$ to reflect the main transmit signal while the AR employs the AF-protocol to boost the transmit signal with the help of w and re-transmit the same to the end-user over the second hop. For the remainder of TP, i.e., $(1-\delta)$T, the re-directed signals from the PR and AR experiences channel coefficients of $h_{2,P}$ and $h_{2,R}$, respectively. The corresponding signals, SNR and throughput expressions are similar as defined before. The overall effective throughput and the harvested energy expressions achieved at the end-user, incorporating the PS and TS schemes are respectively given by

$$R_{U,PR} = \begin{cases} R_{U,PR}^{\mathrm{PS}} = R_{1,D}^{\mathrm{PS}} + R_{2,P}^{\mathrm{PS}} + R_{2,R}^{\mathrm{PS}} \\ R_{U,PR}^{\mathrm{TS}} = R_{1,D}^{\mathrm{TS}} + R_{2,P}^{\mathrm{TS}} + R_{2,R}^{\mathrm{TS}} \end{cases},$$

$$E_{U,PR} = \begin{cases} E_{U,PR}^{\mathrm{PS}} = \mathcal{E}_{1,D}^{\mathrm{PS}}(\mathbb{E}[|y_{1,D}|^2], \rho) \\ \qquad +(1-\delta)\mathcal{E}_{2,P}^{\mathrm{PS}}(\mathbb{E}[|y_{2,P}|^2], \rho) \\ \qquad +(1-\delta)\mathcal{E}_{2,R}^{\mathrm{PS}}(\mathbb{E}[|y_{2,R}|^2], \rho) \\ E_{U,PR}^{\mathrm{TS}} = \mathcal{E}_{1,D}^{\mathrm{TS}}(\mathbb{E}[|y_{1,D}|^2], \tau) \\ \qquad +(1-\delta)\mathcal{E}_{2,P}^{\mathrm{TS}}(\mathbb{E}[|y_{2,P}|^2], \tau) \\ \qquad +(1-\delta)\mathcal{E}_{2,R}^{\mathrm{TS}}(\mathbb{E}[|y_{2,R}|^2], \tau) \end{cases}. \tag{13}$$

Considering a large number of parameters introduced in this work, we have provided the list of parameters and operators in Table 2. Additionally, the list of Greek symbols and their corresponding definitions are summarized in Table 3.

Table 2. List of Parameters and Operators—Definitions.

Notation	Definition
a	Coefficient associated with increase of path loss with distance
$\mathcal{B}$	Overall bandwidth
b	Coefficient associated with the offset value of path-loss
c	Coefficient associated with increase of path loss with frequency
d	Distance between the transmitting and receiving stations (in meters)
$\mathcal{E}'$	Maximum harvested energy obtained on the saturation of the EH circuit
$\mathcal{E}(x)$	Non-linear EH (sigmoidal) expression with input power x
$\mathcal{E}^{\mathrm{PS}}(x,\rho)$	Non-linear EH expression for PS scheme
$\mathcal{E}^{\mathrm{TS}}(x,\tau)$	Non-linear EH expression for TS scheme
$E_{\mathcal{U}}^{\mathrm{PS}}$	Harvested energy at the end-user with PS scheme in the PGN/AGN scenario
$E_{\mathcal{U}}^{\mathrm{TS}}$	Harvested energy at the end-user with TS scheme in the PGN/AGN scenario
$E_{\mathcal{U}}$	Parameter to refer to $E_{\mathcal{U}}^{\mathrm{PS}}$ or $E_{\mathcal{U}}^{\mathrm{TS}}$ according to the chosen scheme
$E_{\mathcal{U},PR}^{\mathrm{PS}}$	Overall harvested energy at the end-user with PS scheme in the HAP scenario
$E_{\mathcal{U},PR}^{\mathrm{TS}}$	Overall harvested energy at the end-user with TS scheme in the HAP scenario
$E_{\mathcal{U},PR}$	Parameter to refer to $E_{\mathcal{U},PR}^{\mathrm{PS}}$ or $E_{\mathcal{U},PR}^{\mathrm{TS}}$ according to the chosen scheme
$\mathbb{E}[\cdot]$	Expectation value operator
f	Operational frequency of the system
$h_{1,D}$	Channel coefficient for the direct link between transmit source and end-user
$h_{1,P}$	Channel coefficient for the first phase of indirect link pertaining to PR device
$h_{1,R}$	Channel coefficient for the first phase of indirect link pertaining to AR device
h_1	Parameter to refer to $h_{1,P}$ or $h_{1,R}$ according to the chosen scheme
$h_{2,P}$	Channel coefficient for the second phase of indirect link pertaining to PR device
$h_{2,R}$	Channel coefficient for the second phase of indirect link pertaining to AR device
h_2	Parameter to refer to $h_{2,P}$ or $h_{2,R}$ according to the chosen scheme
$J(\cdot)$	Lagrange Dual Function
$\mathcal{L}(\cdot)$	Lagrangian function operator
n_P	AWGN introduced at the PR device
n_R	AWGN introduced at the AR device
n_1	Parameter to refer to n_P or n_R according to the chosen scheme
n_{u_a}	AWGN introduced due to the antenna element of the end-user
$n_{u_{ID}}$	The noise introduced by the baseband processing circuit at the end-user
P_T	Transmit power at the source
P_R	Maximum overall power available at the relay
$P^{\star}$	Overall power limitation for the transmitter-relay system
$R_{\mathcal{U}}^{\mathrm{PS}}$	Data throughput at the end-user with PS scheme in the PGN/AGN scenario
$R_{\mathcal{U}}^{\mathrm{TS}}$	Data throughput at the end-user with TS scheme in the PGN/AGN scenario
$R_{\mathcal{U}}$	Parameter to refer to $R_{\mathcal{U}}^{\mathrm{PS}}$ or $R_{\mathcal{U}}^{\mathrm{TS}}$ according to the chosen scheme
$R_{\mathcal{U},PR}^{\mathrm{PS}}$	Overall data throughput at the end-user with PS scheme in the HAP scenario
$R_{\mathcal{U},PR}^{\mathrm{TS}}$	Overall data throughput at the end-user with TS scheme in the HAP scenario
$R_{\mathcal{U},PR}$	Parameter to refer to $R_{\mathcal{U},PR}^{\mathrm{PS}}$ or $R_{\mathcal{U},PR}^{\mathrm{TS}}$ according to the chosen scheme
$R_{1,D}^{\mathrm{PS}}$	Data throughput at the end-user with PS scheme via direct link
$R_{1,D}^{\mathrm{TS}}$	Data throughput at the end-user with TS scheme via direct link
R_1	Parameter to refer to $R_{1,D}^{\mathrm{PS}}$ or $R_{1,D}^{\mathrm{TS}}$ according to the chosen scheme
$r_{1,P}$	Transmit signal seen at the PR device
$r_{1,R}$	Transmit signal seen at the AR device
r_1	Parameter to refer to $r_{1,P}$ or $r_{1,R}$ according to the chosen scheme
s	Symbol tranmsitted from the source
T	Time period
w	Complex amplification coefficient of the AR device
$y_{1,D}$	Signal received by the end-user via direct link
$y_{2,P}$	Signal received by the end-user via PR device over the second phase
$y_{2,R}$	Signal received by the end-user via AR device over the second phase
y_2	Parameter to refer to $y_{2,P}$ or $y_{2,R}$ according to the chosen scheme

Table 3. List of Greek Symbols and Corresponding Definitions.

Symbol	Definition
(α, β)	Constants corresponding to the capacitor and diode turn-on voltage at EH circuit
χ	Mean corresponding to the confidence interval formula
δ	Time splitting factor corresponding to TP
$\bar{\epsilon}$	Threshold limit corresponding to Algorithm 1
$\sqrt{\eta}$	Reflection efficiency coefficient of the PR device
γ	Z-value corresponding to the confidence interval formula
κ	Parameter to compute the intermediary fraction during the Dinkelbach process
$\mathbf{\Lambda}$	Vector corresponding to the Lagrange dual variables: $(\lambda_1, \lambda_2, \lambda_3)$
ν	The number of observations corresponding to the confidence interval formula
ω_E	Weighing coefficient corresponding to the harvested energy
ω_P	Weighing coefficient corresponding to the transmit power
ω_R	Weighing coefficient corresponding to the data throughput
ρ	Power splitting ratio
σ	Standard deviation
$\sigma_{n_{ua}^2}$	Noise variance corresponding to n_{u_a}
$\sigma_{n_{uID}^2}$	Noise variance corresponding to $n_{u_{ID}}$
$\sigma_{n_P^2}$	Noise variance corresponding to n_P
$\sigma_{n_R^2}$	Noise variance corresponding to n_R
$\sigma_{n_1^2}$	Parameter to refer to $\sigma_{n_P^2}$ or $\sigma_{n_R^2}$ according to the chosen scheme
τ	Time-switching ratio
Θ	Metric to represent τ or ρ interchangeably
$\Upsilon_{1,D}^{PS}$	SNR obtained at the end-user with PS scheme via direct link
$\Upsilon_{1,D}^{TS}$	SNR obtained at the end-user with TS scheme via direct link
$\Upsilon_{1,P}$	SNR estimated at the PR via first hop indirect link
$\Upsilon_{1,R}$	SNR estimated at the AR via first hop indirect link
Υ_1	Parameter to refer to $\Upsilon_{1,P}$ or $\Upsilon_{1,R}$ according to the chosen scheme
$\Upsilon_{2,P}^{PS}$	SNR obtained at the end-user with PS scheme via second hop PR device link
$\Upsilon_{2,P}^{TS}$	SNR obtained at the end-user with TS scheme via second hop PR device link
$\Upsilon_{2,R}^{PS}$	SNR obtained at the end-user with PS scheme via second hop AR device link
$\Upsilon_{2,R}^{TS}$	SNR obtained at the end-user with TS scheme via second hop AR device link
Υ_2^{PS}	Parameter to refer to $\Upsilon_{2,P}^{PS}$ or $\Upsilon_{2,R}^{PS}$ according to the chosen scheme
Υ_2^{TS}	Parameter to refer to $\Upsilon_{2,P}^{TS}$ or $\Upsilon_{2,R}^{TS}$ according to the chosen scheme
ς	Minimum demanded data throughput
ξ	Minimum demanded harvested energy
Ξ	Metric combining reflection/amplification coefficients of PR/AR, respectively

4. Problem Formulation and Solution

Herein, we present the problem to jointly optimize the data throughput, harvested energy and transmit power, subjected to QoS constraints for each considered schemes (viz., PGN, AGN, and HAP). Before proceeding to the main hypothesis, we introduce some variables to simplify the problem representation.

4.1. Variable Definitions to Assist the Problem Formulation

In this section, we define some additional parameters corresponding to the overall transmit power, data throughput and harvested energy expressions obtained at the end-user, the dynamic TSF ratios, the relay amplification factor and the TS and PS ratios; to refer to the aforementioned frameworks of PGN, AGN, and HAP altogether. In this context, the variables are sequentially defined in Table 4.

Table 4. Parameter definitions to assist problem formulation.

System \ Parameter	P	R_1	R_2	R_U	E_U
PGN	P_T	$R_{1,P}$	$R_{2,P}$	$R_{U,P}$	$E_{U,P}$
AGN	$P_T + P_R$	$R_{1,R}$	$R_{2,R}$	$R_{U,R}$	$E_{U,R}$
HAP	$P_T + P_R$	$\begin{bmatrix} R_{1,P} \\ R_{1,R} \end{bmatrix}$	$\begin{bmatrix} R_{2,P} \\ R_{2,R} \end{bmatrix}$	$R_{U,PR}$	$E_{U,PR}$

We additionally consider another parameter to assist the PS and TS ratio metrics, defined as follows.

$$\theta := \{\rho : \text{For PS Scheme}; \ \tau : \text{For TS Scheme}\}. \tag{14}$$

It is clear that the newly defined parameters will adopt values according to the type of scenario chosen. Thus, we mathematically formulate the optimization problem in the following while incorporating the parameters defined in Table 4 and (14).

4.2. Optimization Problem with Weighted Utility Function

We formulate the problem with weighted utility function to jointly optimize the data throughput, harvested energy and total transmit power of the source, while ensuring that the demanded throughput and the harvested power at the destination node are both individually above the given thresholds. The overall optimization problem incorporating the PGN, AGN and HAP systems is analytically represented as

$$(P1): \max_{P_T, \delta, \theta, w} \quad \frac{\omega_R R_U + \omega_E E_U}{\omega_P P_T} \tag{15}$$

$$\text{subject to}: \quad (C1) \ : R_1 \geq R_2, \ \ (C2) \ : R_U \geq \varsigma, \tag{16}$$

$$(C3) \ : E_U \geq \xi, \quad (C4) \ : 0 \leq \delta \leq 1, \tag{17}$$

$$(C5) \ : 0 \leq \theta \leq 1, \ (C6) \ : 0 < P \leq P^{\star}, \tag{18}$$

$$(C7)^{*} : 0 \leq |w|^2 \leq P_R, \tag{19}$$

where ς and ξ are respectively the minimum demanded throughput and harvested energy by the destination SWIPT-capable user, ω_R, ω_E and ω_P are the constant weighing factors corresponding to the data throughput, harvested energy and the transmit power, respectively. We use θ to interchangeably refer to the PS or TS splitting factor ρ or τ, respectively. In order to ensure feasibility during the second hop transmission, we make use of $(C1)$ which essentially helps in designing δ and is also an indicator metric used at the transmitter for better resource management. It is noteworthy that $(C1)$ has a completely different meaning and purpose in this context and is not to be confused with a similar constraint used in the case of regenerative relaying (DF). The constraints $(C2)$ and $(C3)$ are the minimum thresholds on the overall data rate and overall harvested energy, respectively. The upper and lower bounds of the dynamic TSF fraction and the SWIPT splitting factors are represented in $(C4)$ and $(C5)$ respectively. The constraint with maximum transmit power budget of the system is incorporated using $(C6)$. Note that the variable w and $(C7)^*$ are considered only during the analysis of AGN and HAP systems. Focusing on the objective function, it is important to note the aspect of rate-energy (R-E) trade-off in SWIPT systems [34], which translates to either increase in data throughput at low EH demands, or vice-versa. Following this trade-off, the optimizer would intend to seek a balance point between the two metrics, viz., data throughput and harvested energy, such that both the metrics are maximized according to the stringent demands indicated in the constraints. In the following, we analyse $(P1)$ and seek a suitable solution.

4.3. Proposed Solutions to the Above-Mentioned Problem

In this section, we present a couple of approaches to solve the problem in $(P1)$. First, we employ a method based on the Lagrange dual approach to seek a close to optimal solution, without taking the computational complexity into consideration. The second technique is based on the alternating parameter based optimization, assisted by the Dinkelbach method. The nature of the latter process opens up the possibility of obtaining a sub-optimal solution within polynomial time. In this regard, the proposed solutions are discussed below.

4.3.1. Method to Seek an Asymptotically Optimal Solution

The problem $(P1)$ is difficult to solve since it has a non-convex form for both TS and PS schemes. However, it is clear from our optimization formulation that the dual method may be employed to seek an asymptotically optimal solution. Therefore, in order to obtain an analytical solution for the joint objective optimization problem in $(P1)$, we assume $\mathcal{D}$ as the set of all possibilities of $P_\mathcal{T}, \delta, \theta$ and w which satisfy the constraints (C4)–(C7). Then the corresponding Lagrange dual function is written as

$$J(\mathbf{\Lambda}) \triangleq \max_{\{P_\mathcal{T},\delta,\theta,w\}\in\mathcal{D}} \mathcal{L}(P_\mathcal{T},\delta,\theta,w;\mathbf{\Lambda}), \tag{20}$$

where the Lagrangian is expressed as

$$\mathcal{L}(P_\mathcal{T},\delta,\theta,w;\mathbf{\Lambda}) = F(P_\mathcal{T},\delta,\theta,w) - \lambda_1 G(P_\mathcal{T},\delta,\theta,w) \\ -\lambda_2 H(P_\mathcal{T},\delta,\theta,w) - \lambda_3 I(P_\mathcal{T},\delta,\theta,w), \tag{21}$$

with $\mathbf{\Lambda} = (\lambda_1, \lambda_2, \lambda_3) \succeq 0$ denoting the vector of the dual variables associated with the constraints (C1)–(C3), respectively, while the related functions are as defined in Table 5. The dual optimization problem is hence represented as

$$(P2) : \min_{\mathbf{\Lambda}} \quad J(\mathbf{\Lambda}) \tag{22}$$

$$\text{subject to}: \quad \mathbf{\Lambda} \succeq 0. \tag{23}$$

Table 5. Definitions to comprise the Lagrangian expression.

$F(P_\mathcal{T},\delta,\theta,w)$	$G(P_\mathcal{T},\delta,\theta,w)$	$H(P_\mathcal{T},\delta,\theta,w)$	$I(P_\mathcal{T},\delta,\theta,w)$
$\frac{\omega_R R_\mathcal{U}+\omega_E E_\mathcal{U}}{\omega_P P_\mathcal{T}}$	$R_2 - R_1 \leq 0$	$\varsigma - R_\mathcal{U} \leq 0$	$\xi - E_\mathcal{U} \leq 0$

Since it is explicit that a dual problem is always convex by definition [22], the gradient or subgradient-based methods can be used to minimize $J(\mathbf{\Lambda})$ with guaranteed convergence. In this regard, an asymptotic technique based on the block-coordinate descent approach may be employed [27]. Herein, an alternating optimization of the involved parameters is performed in an iterative manner, until the convergence of the weighted utility function value. The intermediary closed-form expressions for the involved parameters in the stage-based optimization may be obtained using the Karush-Kuhn-Tucker (KKT) conditions [22]. Even though a solution could be obtained within polynomial time, the computation process of the proposed method is still complex and cumbersome, as inferred from the analysis in [27]. Therefore, we omit the corresponding derivations for brevity, and seek an even simpler method to solve $(P1)$ as discussed in the succeeding section.

4.3.2. Dinkelback Method Assisted Convex Programming

As discussed before, the problem $(P1)$ involves a non-linear fractional programming, which is non-convex and difficult to solve directly. In such instances of non-convex non-linear fractional optimization, Dinkelbach method comes in handy to address this challenge [62]. Therefore, we transform the fractional form objective function into a difference of numerator and quasi-denominator form, using the following proposition.

Proposition 1. *The maximum achievable objective can be obtained using the following transformation, provided that*

$$\max_{P_{\mathcal{T}},\delta,\theta,w,\kappa^\star} \quad F_N(P_{\mathcal{T}},\delta,\theta,w) - \kappa^\star F_D(P_{\mathcal{T}},\delta,\theta,w)$$

$$= F_N(P^\star_{\mathcal{T}},\delta^\star,\theta^\star,w^\star) - \kappa^\star F_D(P^\star_{\mathcal{T}},\delta^\star,\theta^\star,w^\star) = 0, \tag{24}$$

for $F_N(P_{\mathcal{T}},\delta,\theta,w) \geq 0$ *and* $F_D(P_{\mathcal{T}},\delta,\theta,w) \geq 0$, *where*

$$F_N(P_{\mathcal{T}},\delta,\theta,w) = \omega_R R_U(P_{\mathcal{T}},\delta,\theta,w) + \omega_E E_U(P_{\mathcal{T}},\delta,\theta,w), \tag{25}$$

$$F_D(P_{\mathcal{T}},\delta,\theta,w) = \omega_P P_{\mathcal{T}}, \tag{26}$$

and

$$\kappa^\star = \frac{F_N(P^\star_{\mathcal{T}},\delta^\star,\theta^\star,w^\star)}{F_D(P^\star_{\mathcal{T}},\delta^\star,\theta^\star,w^\star)}. \tag{27}$$

Proof. Please refer to [62] for a proof of Proposition 1. □

It is clear that the adequate and compulsory conditions are provided by Proposition 1 for developing suitable strategy to optimize the intended parameters. Specifically, it should be possible to transform the primal problem in $(P1)$ having a fractional form-objective function into an equivalent optimization problem with a difference of numerator and quasi-denominator form-objective function (e.g., $F_N(P_{\mathcal{T}},\delta,\theta,w) - \kappa^\star F_D(P_{\mathcal{T}},\delta,\theta,w)$). Consequently, the solutions for both the primal and transformed problems will be equivalent [28].

Herein, it is noteworthy that an optimal solution could be obtained if the equality in (24) holds [62]. The equality condition may then be correspondingly used to validate the optimality of the solution. Therefore, instead of directly addressing the problem with a fractional-form objective, we transform it to another equivalent optimization problem with subtractive-form objective function whilst meeting the conditions in Proposition 1, whose solution is summarized in Algorithm 1.

In order to understand the working principle of Algorithm 1, we first note that the pivotal stage for the proposed Dinkelbach method-based solution is to develop an intermediate parameters' optimization policy in order to solve the following fixed κ optimization problem (steps 2–6 in Algorithm 1),

$$
\begin{aligned}
(P3):\ \max_{P_{\mathcal{T}},\delta,\theta,w} \quad & (\omega_R R_U + \omega_E E_U) - \kappa^{\star}(\omega_P P_{\mathcal{T}}) && (28)\\
\text{subject to}: \quad & (C1): R_1 \geq R_2, \quad (C2): R_U \geq \varsigma, && (29)\\
& (C3): E_U \geq \xi, \quad (C4): 0 \leq \delta \leq 1, && (30)\\
& (C5): 0 \leq \theta \leq 1, \ (C6): 0 < P \leq P^{\star}, && (31)\\
& (C7)^{*}: 0 \leq |w|^2 \leq P_R. && (32)
\end{aligned}
$$

Algorithm 1 Dinkelbach-assisted Alternating Parameter Optimization

1: **Initialize**: $\kappa = 0$, any feasible δ, θ, w, and $\bar{\epsilon}$: Threshold limit;
2: **REPEAT** (for a given κ, iteration: n)
3: Solve (28)–(32) to obtain $P_{\mathcal{T}}$, for given δ, θ, w;
4: Utilizing $P_{\mathcal{T}}$, solve (28)–(32) to obtain δ, for given θ, w;
5: Using $P_{\mathcal{T}}$ and δ , solve (28)–(32) to obtain θ, for given w;
6: Finally compute w by solving (28)–(32) via $P_{\mathcal{T}}$, δ, θ;
7: **IF** $(\omega_R R_U + \omega_E E_U) - \kappa(\omega_P P_{\mathcal{T}}) \leq \bar{\epsilon}$
8: Convergence = **TRUE**;
9: **RETURN** $\{P^{\star}_{\mathcal{T}}, \delta^{\star}, \theta^{\star}, w^{\star}\} = \{P_{\mathcal{T}}, \delta, \theta, w\}$, $\kappa^{\star} = \frac{(\omega_R R_U + \omega_E E_U)}{\omega_P P_{\mathcal{T}}}$;
10: **ELSE**
11: Set $\kappa = \frac{(\omega_R R_U + \omega_E E_U)}{\omega_P P_{\mathcal{T}}}$ and $n = n + 1$;
12: Convergence = **FALSE**;
13: **END IF**
14: **UNTIL** Convergence = **TRUE**.

Since the persisting non-convexity of problem in $(P3)$ causes a hindrance in seeking an optimal solution directly via joint optimization of the intended variable, it is natural to explore other relevant possibilities for obtaining a suitable solution. In this vein, we note that the problem can be made more tractable by optimizing a single parameter while keeping the other parameters fixed since the convexity of the problem is more probable. Such an alternating parameter based optimization technique has been found effective for solving various complicated optimization problems [63,64]. Herein, the optimality of the solution is, however, compromised. Next, we observe that following an alternating parameter based technique, each corresponding sub-problem belong to the category of disciplined convex-concave programming (DCCP) [65]. The consistently occurring convex-concave nature in $(C1)$ essentially contributes towards making $(P3)$ a DCCP. Therefore, suitable solutions could be obtained using standard optimization methods, e.g., the DCCP package in CVXPY [66]. Correspondingly, the problem (28)–(32) can be solved using the proposed Dinkelbach-assisted Alternating Parameter Optimization, summarized in Algorithm 1.

Regarding the convergence of the proposed Dinkelbach-assisted Alternating Parameter Optimization based solution, we initially prove that the auxiliary variable κ appreciates in each iteration. Following this, we prove that the value of κ converges to the optimal $\kappa^{\star}$ for sufficiently large number of iterations, such that the optimality condition in Proposition 1 is satisfied. More specifically, $F_N(P_{\mathcal{T}}, \delta, \theta, w) - \kappa^{\star} F_D(P_{\mathcal{T}}, \delta, \theta, w) = \widehat{F}(\kappa^{\star}) = 0$. Let us consider $\{P^{\star}_{\mathcal{T}}, \delta^{\star}, \theta^{\star}, w^{\star}\}$ as the optimized policies in the nth iteration. Suppose $\kappa(n) \neq \kappa^{\star}$ and $\kappa(n+1) \neq \kappa^{\star}$ represent the relevant values in the nth and $(n+1)$th iterations, respectively. We note from [62] that $\widehat{F}(\kappa(n)) > 0$ and $\widehat{F}(\kappa(n+1)) > 0$ hold. On the other hand, in the proposed algorithm, $\kappa(n+1) = \frac{F_N(P_{\mathcal{T}}, \delta, \theta, w)}{F_D(P_{\mathcal{T}}, \delta, \theta, w)}$. Thus, we can express $\kappa(n)$ as

$$
\begin{aligned}
\widehat{F}(\kappa(n+1)) &= F_N(P_{\mathcal{T}}, \delta, \theta, w) - \kappa(n) F_D(P_{\mathcal{T}}, \delta, \theta, w)\\
&= F_D(P_{\mathcal{T}}, \delta, \theta, w)\big(\kappa(n+1) - \kappa(n)\big). \qquad (33)
\end{aligned}
$$

Since $F_D(P_\mathcal{T}, \delta, \theta, w) = \omega_P P_\mathcal{T} > 0$, we have $\kappa(n+1) > \kappa(n)$. Therefore, we can show that as long as the number of iterations is large enough, $\widehat{F}(\kappa(n))$ will eventually approach zero and satisfy the optimality condition as stated in Proposition 1.

5. Numerical Results

In this section, we present a comparative study among the HAP, AGN and PGN systems. The three cases are analyzed using the MATLAB R2019b, with optimization performed via fmincon($\cdot$) solver present in the optimization toolbox [67], where the solutions are obtained with the help of interior point algorithm [22].

5.1. Simulation Set-Up

We assume an ITU-R outdoor framework (site-general model for propagation within street canyons) [68] to generate channel realizations with the path-loss exponent:

$$PL(d, f) = 10a\log_{10}(d) + b + 10c\log_{10}(f) + N(0, \sigma)\ \text{dB}, \tag{34}$$

where d is distance between the transmitting and receiving stations (m), f is the operating frequency (GHz), the coefficients a, b, and c are associated with the increase of the path loss with distance, the offset value of the path loss, and the increase of the path loss with frequency, respectively, and $N(0, \sigma)$ is a zero mean Gaussian random variable with a standard deviation σ (dB). The channels coefficients for the direct link and the dual-hop links are generated accordingly. Specifically, we choose: f = 24 GHz, d is randomly chosen between 8 m and 10 m in case of PR/AR device and randomly between 12 m and 15 m for the end-user with respect to the transmitter [69], a = 2.12, b = 29.2, c = 2.11 and σ = 5.06 dB. The simulation results presented in this section assume an overall bandwidth of $B = 50$ MHz. We set $\sigma^2_{n_B} = \sigma^2_{n_R} = \sigma^2_{n_{ua}} = -150$ dBW/Hz, and $\sigma^2_{n_{u_{ID}}} = -110$ dBW/Hz. The PR's reflection efficiency coefficient is chosen as η = 0.78 (corresponding to 1.1 dB loss [70]). The constants corresponding to the non-linear EH circuit are chosen as $\mathcal{E}'$ = 2.8 mJ, α = 1500, and β = 0.0022 [27,58]. The harvested energy demand is assumed to be $\xi = -70$ dBW, and the weighing factors are chosen as ω_R = 1 units/Gbit, ω_E = 1 units/nJ, and ω_P = 1 units/W. We use two test cases for the data demands of ς = 1.00 Gbits and ς = 1.05 Gbits. As a benchmark scheme, we assume equal time distribution for both the hops in all the cases, i.e., fixed δ = 0.5. Furthermore, an average of 1000 random channel realizations is presented for each experiment.

5.2. Experimental Findings and Analysis

In Figure 3a,b, we present the performance analyses of HAP, AGN and PGN in terms of the weighted utility function for the PS and TS schemes, respectively, versus the maximum limitation on the total transmit power. We observe that all the considered schemes can provide appreciable gains over their respective benchmark methods. The weighted utility function values for both PS and TS increases with growing overall maximum transmit power budget ($P^\star$) of the system. On the other hand, the performance is adversely affected when the data demand is increased from ς = 1.00 Gbits to ς = 1.05 Gbits, which is due to the joint optimization in the weighted utility function. Particularly, this effect is due to the well-known rate-energy trade-off incurred because of the joint maximization of data and harvested energy along with the minimization of total transmit power. In the case of PS, we notice improvements of 10% and 12.5% for HAP over the AGN and PGN scenarios, respectively for the data demand of ς = 1.00 Gbits, and improvements of 2% and 4% respectively for the data demand of ς = 1.05 Gbits, with an average decrement of 53% for a transition from former to the latter demand. Concerning the TS, we observe gains of 25% and 28% for HAP over the AGN and PGN scenarios, respectively for the data demand of ς = 1.00 Gbits, and improvements of 11.5% and 12.5%, respectively for the data demand of ς = 1.05 Gbits, with an average overall decrement of 80.5% for a transition from ς = 1.00 Gbits to ς = 1.05 Gbits. Clear benefits for HAP are observed over the AGN and PGN counterparts, with PS scheme outperforming the TS.

We now investigate the system performance from the perspective of so-called R-E trade-off in SWIPT systems [34]. Correspondingly, we show in Figure 4 the effect on harvested energy with varying data demand. It is noteworthy that the considered analysis may be easily carried out by keeping $\omega_R = 0$ in $(P1)$. In this vein, it is also important to mention that besides the use of ω_R and ω_E as the cost functions for data and energy metrics, respectively; their binary forms may be utilized to study the corresponding entities independently. Herein, we observe the (expected) decreasing nature of the harvested energy with increasing data demands for all the cases. The HAP system is found to outperform the AGN and PGN as anticipated, whereas the benefits of employing dynamic TSF are clearly seen over the fixed TSF (δ). We additionally note that the performance of the PS mechanism is way superior than the TS counterpart.

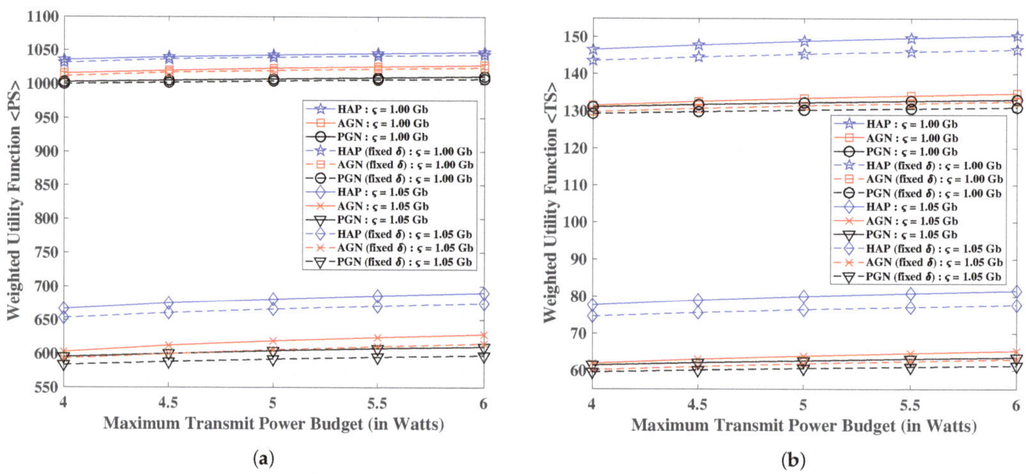

Figure 3. Performance analysis of the weighted utility function for (**a**) PS scheme, and (**b**) TS scheme; versus the maximum limitation on the total transmit power.

In Figure 5a,b, we show the effects on the weighted utility functions corresponding to the PS and TS, respectively, with varying TSF (i.e., fixed δ at an instance). Intuitively, when the TSF is small, the second hop reflection/re-transmission is expected to happen for significantly large amount of time with the considered TP. This would naturally enable better collection and interpretation of the energy and data, respectively, at the end-user following the signal synchronization process (as assumed). In other case, when the TSF is large, the second hop reflection/re-transmission would suffer due to a highly compact TP, which affects the overall throughput and hence the output at the end-user. We note from Figure 5a,b that the above-mentioned intuitive discussion is validated, wherein the depreciating values of the weighted utility function are observed with the increasing values of δ. On the other hand, the HAP system is found to outperform the AGN and PGN counterparts with PS providing significant gains over the TS scheme, thereby following in-line with the previously reported results.

In order to allow fair comparison among the different analyzed approaches, we calculate confidence intervals for the results and present the corresponding error bars in Figure 6a,b, pertaining to the PS and TS schemes, respectively. To compute the confidence interval, we use the formula: $\chi \pm \gamma \frac{\sigma}{\sqrt{\nu}}$, where χ is the mean, γ is the Z-value from the Table 6, σ is the standard deviation, and ν is the number of observations [71]. To proceed, we consider $\nu = 1000$ observations and compute the weighted utility functions corresponding to the PS and TS schemes for different values of the maximum transmit power budget ($P^{\star}$) and keeping $\varsigma = 1.00$ Gb. Therefore, it is clear that for 1000 instances the weighted utility functions are calculated per single consideration of $P^{\star}$. Correspondingly, the values of χ and σ can be computed. For our experiment, we assume $\gamma = 1.960$ pertaining to a confidence level of 90% to obtain the error bars shown in Figure 6a,b, for the PS and TS schemes, respectively. The HAP systems in both the cases are seen to outperform the AGN and PGN systems for PS and TS schemes. The fairness measure in terms of error bars is also explicit in this regard.

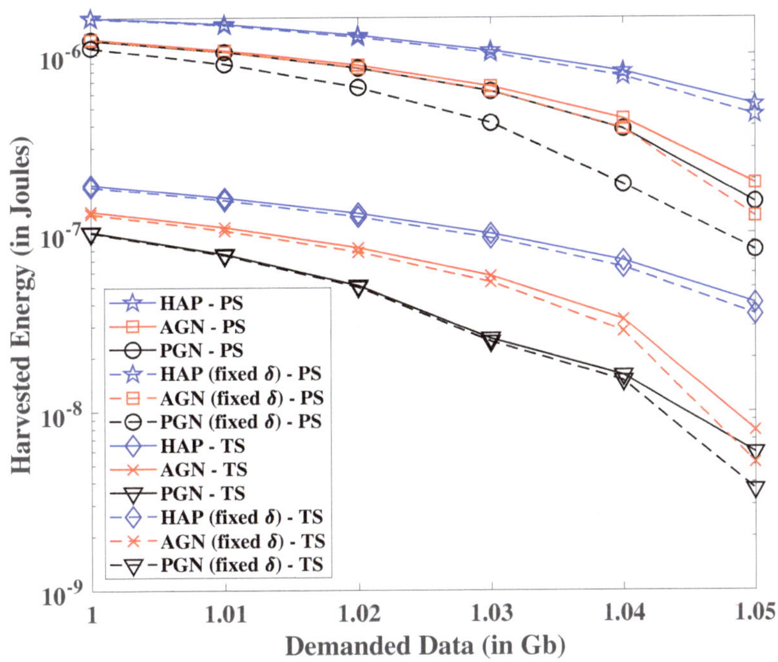

Figure 4. Analysis of harvested energy for PS and TS schemes versus the demanded data (R-E trade-off).

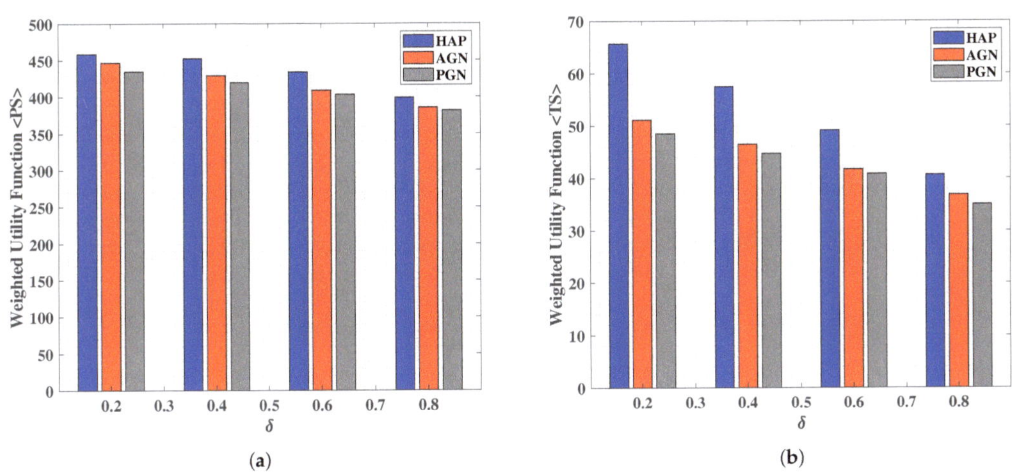

Figure 5. Performance analysis of the weighted utility function for (**a**) PS scheme, and (**b**) TS scheme; versus the variation in the (fixed) TSF (δ), while comparing HAP, AGN and PGN.

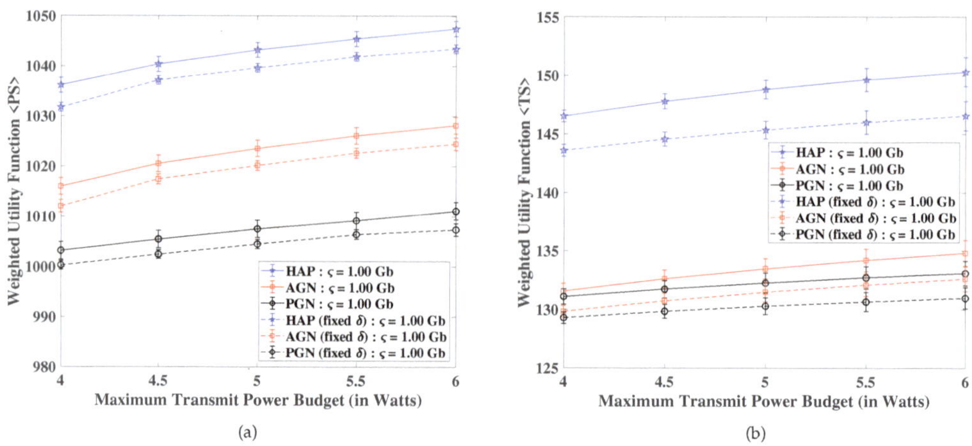

Figure 6. Analysis of harvested energy for (**a**) PS scheme, and (**b**) TS scheme; versus the demanded data (R-E trade-off) with error bars pertaining to a confidence level of 95%.

Table 6. Determination of Z-value according to the percentage of confidence level.

Percentage	80%	85%	90%	95%	99%	99.5%	99.9%
Z-value	1.282	1.440	1.645	1.960	2.576	2.807	3.291

5.3. General Outcomes and Trailing Discussion

From the results reported in the previous section and following an intuitive analysis, one important obvious observation is the superior performance of the HAP over the AGN and PGN systems. The significant benefits observed via the PS scheme over the TS are also explicit. The proposed framework shows that the existing AGN and PGN type of standalone systems may be fused together leading to the possibility of reaping maximum benefits within the system. In contrast, the RIS systems have evolved as the promising solutions, however, there are still considerable number of challenges involved in relation to their practical deployment. In this context, the primary concerns revolve around the ways to achieve flexible reflections and powering/switching mechanisms. Additionally, the practical implementation of dynamic TSF-enabled relaying may be challenging to achieve in practice [72], however, we cannot completely rule out this possibility on this type of AF-based cooperative system deployment. Alternatively, the benefits of equal-TSF based systems have also been addressed herein, which might be easier to deploy from the practical viewpoint. Nonetheless, it is needless to mention that an intermediary possibility of HAP comes in handy where an alternative deployment to RIS may be put into practice while the relevant active components/mechanisms are under development. The proposed methodology does not only seem to be a cost and energy efficient approach, but also serves as a promising solution for harnessing the maximum gains from the transmit signal power (within the considered wireless communications system).

6. Conclusions

In this invited paper, we investigated a hybrid active-and-passive (HAP) relaying scheme to not only improve the network coverage, but also to facilitate SWIPT in a 5G and beyond/6G Greencom environment, where the end-users are capable of decoding information and harvesting energy simultaneously, according to the PS and TS schemes while using the non-linear EH device for the latter. We formulated an optimization problem to jointly address the respective demands of data throughput, harvested energy and total transmit power, under some QoS constraints. In this context, we provided several polynomial computation-time based solutions which employ the Lagrange dual and Dinkelbach method assisted convex programming, respectively. We showed significant benefits of adopting HAP over the AGN and PGN counterpart systems with the help of numerical results, where the PS scheme was found to outperform the TS concerning SWIPT at the end-user. Considering significant benefits of the proposed HAP system, this work may be extended into several interesting directions in the future. Some extension possibilities include the investigation with distinct placement of the reflection and re-transmission elements, the incorporation of multiple antennas at the concerned devices, and the consideration of multiple HAP nodes and multiple users whose demands are addressed with the help of a multi-carrier framework. The examination of a cooperative system with DF protocol-enabled relays instead of AF (as considered herein) is also another interesting research prospect. Furthermore, the possibility of precoding/beamforming could also be considered to enhance the performance of the multiple antenna-enabled HAP systems.

Author Contributions: Conceptualization: S.G. and S.K.S.; methodology: S.G., S.S., S.K.S. and S.C.; software: S.G.; validation: S.G. and S.S.; formal analysis: S.G.; investigation: S.G.; resources: S.G., S.S. and S.K.S.; data curation: S.G.; writing-original draft preparation: S.G., S.S. and S.K.S.; writing—review and editing: S.G., S.S., S.K.S., S.C. and B.O.; visualization: S.G.; supervision: S.K.S., S.C. and B.O.; project administration: S.K.S., S.C. and B.O.; funding acquisition: S.K.S., S.C. and B.O. All authors have read and agreed to the published version of the manuscript.

Funding: The research leading to these results has received funding from the Luxembourg National Research Fund (FNR), Luxembourg, under the FNR-FNRS bilateral—InWIP-NET: Integrated Wireless Information and Power Networks (R-AGR-0700-10-X).

Data Availability Statement: Not applicable.

Conflicts of Interest: The authors declare no conflict of interest.

References

1. Viswanathan, H.; Mogensen, P.E. Communications in the 6G Era. *IEEE Access* **2020**, *8*, 57063–57074. [CrossRef]
2. Rappaport, T.S.; Xing, Y.; MacCartney, G.R.; Molisch, A.F.; Mellios, E.; Zhang, J. Overview of Millimeter Wave Communications for Fifth-Generation (5G) Wireless Networks-with a Focus on Propagation Models. *IEEE Trans. Antennas Propag.* **2017**, *65*, 6213–6230. [CrossRef]
3. TELCOMA-Global. 5G Technology, Vision, Performance, RAN and Core Network. *Telecom Daily Post*, 1 August 2021.
4. Sharetechnote. 5G/NR - FR/Operating Bandwidth. Available online: sharetechnote.com (accessed on 30 August 2021).
5. Haksuki, S. Fixed Wireless Communications at 60 GHz Unique Oxygen Absorption Properties. Available online: rfglobalnet.com (accessed on 30 August 2021).
6. Horwitz, J. The Definitive Guide to 5G Low, Mid, and High Band Speeds. *VentureBeat Online Magazine*, 10 December 2019.
7. Davies, D. Small Cells—Big in 5G. *Nokia*, 20 May 2019.
8. Kumbhani, B.; Kshetrimayum, R.S. Source and Relay Transmit Antenna Selection in Two Hop Cooperative Communication Systems Over κ-μ Fading Channels. In Proceedings of the 2016 Twenty Second National Conference on Communication (NCC), Guwahati India, 4–6 March 2016; pp. 1–5. [CrossRef]
9. Arti, M.K.; Bhatnagar, M.R. Performance Analysis of Hop-by-Hop Beamforming and Combining in DF MIMO Relay System over Nakagami-m Fading Channels. *IEEE Commun. Lett.* **2013**, *17*, 2080–2083. [CrossRef]
10. Gautam, S.; Sharma, S.K.; Chatzinotas, S.; Ottersten, B. Modeling and Optimization of RF-Energy Harvesting-assisted Quantum Battery System. In Proceedings of the 2021 IEEE 93rd Vehicular Technology Conference (VTC2021-Spring), Helsinki, Finland, 25–28 April 2021; pp. 1–6. [CrossRef]
11. Lumpkins, W. Nikola Tesla's Dream Realized: Wireless Power Energy Harvesting. *IEEE Consum. Electron. Mag.* **2014**, 3, 39–42. [CrossRef]
12. Valenta, C.R.; Durgin, G.D. Harvesting Wireless Power: Survey of Energy-Harvester Conversion Efficiency in Far-Field, Wireless Power Transfer Systems. *IEEE Microw. Mag.* **2014**, *15*, 108–120. [CrossRef]
13. Wang, L.; Todaria, P.; Pandey, A.; O'Connor, J.; Chernow, B.; Zuo, L. An Electromagnetic Speed Bump Energy Harvester and its Interactions with Vehicles. *IEEE/ASME Trans. Mechatron.* **2016**, *21*, 1985–1994. [CrossRef]
14. Ruan, T.; Chew, Z.J.; Zhu, M. Energy-Aware Approaches for Energy Harvesting Powered Wireless Sensor Nodes. *IEEE Sens. J.* **2017**, *17*, 2165–2173. [CrossRef]
15. Yang, Z.; Xu, W.; Shikh-Bahaei, M. Energy Efficient UAV Communication with Energy Harvesting. *IEEE Trans. Veh. Technol.* **2020**, *69*, 1913–1927. [CrossRef]
16. Huang, K.; Larsson, E. Simultaneous Information and Power Transfer for Broadband Wireless Systems. *IEEE Trans. Signal Process.* **2013**, *61*, 5972–5986. [CrossRef]

17. Gautam, S.; Ubaidulla, P. Simultaneous Transmission of Information and RF Energy in Multicarrier Systems. In Proceedings of the 2016 23rd International Conference on Telecommunications (ICT), Thessaloniki, Greece, 16–18 May 2016; pp. 1–5. [CrossRef]
18. Zhou, X.; Zhang, R.; Ho, C.K. Wireless Information and Power Transfer in Multiuser OFDM Systems. *IEEE Trans. Wirel. Commun.* **2014**, *13*, 2282–2294. [CrossRef]
19. Gautam, S.; Ubaidulla, P. Relay Selection and Transceiver Design for Joint Wireless Information and Energy Transfer in Cooperative Networks. In Proceedings of the 2017 IEEE 85th Vehicular Technology Conference (VTC Spring), Sydney, Australia, 4–7 June 2017; pp. 1–5. [CrossRef]
20. Gautam, S.; Krivochiza, J.; Haqiqatnejad, A.; Chatzinotas, S.; Ottersten, B. Boosting SWIPT via Symbol-Level Precoding. In Proceedings of the 2020 IEEE 21st International Workshop on Signal Processing Advances in Wireless Communications (SPAWC), Atlanta, GA, USA, 26–29 May 2020; pp. 1–5. [CrossRef]
21. Gautam, S.; Lagunas, E.; Vuppala, S.; Chatzinotas, S.; Ottersten, B. Pricing Perspective for SWIPT in OFDM-based Multi-User Wireless Cooperative Systems. In Proceedings of the 2019 IEEE Wireless Communications and Networking Conference (WCNC), Marrakech, Morocco, 15–18 April 2019; pp. 1–7. [CrossRef]
22. Boyd, S.; Boyd, S.P.; Vandenberghe, L. *Convex Optimization*; Cambridge University Press: Cambridge, UK, 2004.
23. Neos-Guide. Optimization Taxonomy. 2020. Available online: https://neos-guide.org/content/optimization-taxonomy (accessed on 30 August 2021).
24. Gautam, S.; Lagunas, E.; Kisseleff, S.; Chatzinotas, S.; Ottersten, B. Successive Convex Approximation for Transmit Power Minimization in SWIPT-Multicast Systems. In Proceedings of the ICC 2020—2020 IEEE International Conference on Communications (ICC), Dublin, Ireland, 7–11 June 2020; pp. 1–7. [CrossRef]
25. Gautam, S.; Lagunas, E.; Chatzinotas, S.; Ottersten, B. Feasible Point Pursuit and Successive Convex Approximation for Transmit Power Minimization in SWIPT-Multigroup Multicasting Systems. *IEEE Trans. Green Commun. Netw.* **2021**, *5*, 884–894. [CrossRef]
26. Gautam, S.; Lagunas, E.; Chatzinotas, S.; Ottersten, B. Wireless Multi-group Multicast Precoding with Selective RF Energy Harvesting. In Proceedings of the 2019 27th European Signal Processing Conference (EUSIPCO), A Coruña, Spain, 2–6 September 2019; pp. 1–5. [CrossRef]
27. Gautam, S.; Lagunas, E.; Chatzinotas, S.; Ottersten, B. Relay Selection and Resource Allocation for SWIPT in Multi-User OFDMA Systems. *IEEE Trans. Wirel. Commun.* **2019**, *18*, 2493–2508. [CrossRef]
28. Shen, K.; Yu, W. Fractional Programming for Communication Systems—Part I: Power Control and Beamforming. *IEEE Trans. Signal Process.* **2018**, *66*, 2616–2630. [CrossRef]
29. Gautam, S.; Chatzinotas, S.; Ottersten, B. Energy Efficiency Optimization Technique for SWIPT-Enabled Multi-Group Multicasting Systems with Heterogeneous Users. In Proceedings of the ICASSP 2021—2021 IEEE International Conference on Acoustics, Speech and Signal Processing (ICASSP), Toronto, ON, Canada, 6–11 June 2021; pp. 4880–4884. [CrossRef]
30. Yang, G.; Zhang, Q.; Liang, Y.C. Cooperative Ambient Backscatter Communications for Green Internet-of-Things. *IEEE Internet Things J.* **2018**, *5*, 1116–1130. [CrossRef]
31. Hiranandani, S.; Mohadikar, S.; Khawaja, W.; Ozdemir, O.; Guvenc, I.; Matolak, D. Effect of Passive Reflectors on the Coverage of IEEE 802.11ad mmWave Systems. In Proceedings of the 2018 IEEE 88th Vehicular Technology Conference (VTC-Fall), Chicago, IL, USA, 27–30 August 2018; pp. 1–6. [CrossRef]
32. Gautam, S.; Lagunas, E.; Chatzinotas, S.; Ottersten, B. Sequential Resource Distribution Technique for Multi-User OFDM-SWIPT Based Cooperative Networks. In Proceedings of the 2018 IEEE Global Communications Conference (GLOBECOM), Abu Dhabi, United Arab Emirates, 9–13 December 2018; pp. 1–7. [CrossRef]
33. Gautam, S.; Lagunas, E.; Sharma, S.K.; Chatzinotas, S.; Ottersten, B. Relay Selection Strategies for SWIPT-Enabled Cooperative Wireless Systems. In Proceedings of the 2017 IEEE 28th Annual International Symposium on Personal, Indoor, and Mobile Radio Communications (PIMRC), Montreal, QC, Canada, 8–13 October 2017; pp. 1–7. [CrossRef]
34. Zhou, X.; Zhang, R.; Ho, C.K. Wireless Information and Power Transfer: Architecture Design and Rate-Energy Tradeoff. *IEEE Trans. Commun.* **2013**, *61*, 4754–4767. [CrossRef]
35. Andrawes, A.; Nordin, R.; Abdullah, N.F. Energy-Efficient Downlink for Non-Orthogonal Multiple Access with SWIPT under Constrained Throughput. *Energies* **2019**, *13*, 107. [CrossRef]
36. Singh, U.; Solanki, S.; Gurjar, D.S.; Upadhyay, P.K.; da Costa, D.B. Wireless Power Transfer in Two-Way AF Relaying with Maximal-Ratio Combining under Nakagami-*m* Fading. In Proceedings of the 2018 14th International Wireless Communications Mobile Computing Conference (IWCMC), Limassol, Cyprus, 25–29 June 2018; pp. 169–173. [CrossRef]
37. Solanki, S.; Upadhyay, P.K.; Da Costa, D.B.; Ding, H.; Moualeu, J.M. Performance Analysis of Piece-Wise Linear Model of Energy Harvesting-Based Multiuser Overlay Spectrum Sharing Networks. *IEEE Open J. Commun. Soc.* **2020**, *1*, 1820–1836. [CrossRef]
38. Liaskos, C.; Nie, S.; Tsioliaridou, A.; Pitsillides, A.; Ioannidis, S.; Akyildiz, I. A New Wireless Communication Paradigm through Software-Controlled Metasurfaces. *IEEE Commun. Mag.* **2018**, *56*, 162–169. [CrossRef]
39. Di Renzo, M.; Zappone, A.; Debbah, M.; Alouini, M.S.; Yuen, C.; de Rosny, J.; Tretyakov, S. Smart Radio Environments Empowered by Reconfigurable Intelligent Surfaces: How It Works, State of Research, and The Road Ahead. *IEEE J. Sel. Areas Commun.* **2020**, *38*, 2450–2525. [CrossRef]
40. Björnson, E.; Özdogan, O.; Larsson, E.G. Reconfigurable Intelligent Surfaces: Three Myths and Two Critical Questions. *IEEE Commun. Mag.* **2020**, *58*, 90–96. [CrossRef]

41. Wu, Q.; Zhang, S.; Zheng, B.; You, C.; Zhang, R. Intelligent Reflecting Surface-Aided Wireless Communications: A Tutorial. *IEEE Trans. Commun.* **2021**, *69*, 3313–3351. [CrossRef]
42. Huang, C.; Zappone, A.; Alexandropoulos, G.C.; Debbah, M.; Yuen, C. Reconfigurable Intelligent Surfaces for Energy Efficiency in Wireless Communication. *IEEE Trans. Wirel. Commun.* **2019**, *18*, 4157–4170. [CrossRef]
43. Wu, Q.; Zhang, R. Intelligent Reflecting Surface Enhanced Wireless Network via Joint Active and Passive Beamforming. *IEEE Trans. Wirel. Commun.* **2019**, *18*, 5394–5409. [CrossRef]
44. Pei, X.; Yin, H.; Tan, L.; Cao, L.; Li, Z.; Wang, K.; Zhang, K.; Björnson, E. RIS-Aided Wireless Communications: Prototyping, Adaptive Beamforming, and Indoor/Outdoor Field Trials. *arXiv* **2021**, arXiv:2103.00534.
45. Björnson, E.; Wymeersch, H.; Matthiesen, B.; Popovski, P.; Sanguinetti, L.; de Carvalho, E. Reconfigurable Intelligent Surfaces: A Signal Processing Perspective with Wireless Applications. *arXiv* **2021**, arXiv:2102.00742.
46. Cui, T.J.; Liu, S.; Zhang, L. Information Metamaterials and Metasurfaces. *J. Mater. Chem. C* **2017**, *5*, 3644–3668. [CrossRef]
47. Shelby, R.A.; Smith, D.R.; Schultz, S. Experimental Verification of a Negative Index of Refraction. *Science* **2001**, *292*, 77–79. [CrossRef]
48. Liang, Y.C.; Long, R.; Zhang, Q.; Chen, J.; Cheng, H.V.; Guo, H. Large Intelligent Surface/Antennas (LISA): Making Reflective Radios Smart. *J. Commun. Inf. Netw.* **2019**, *4*, 40–50.
49. Tsilipakos, O.; Tasolamprou, A.C.; Pitilakis, A.; Liu, F. Wang, X.; Mirmoosa, M.S. Tzarouchis, D.C. Abadal, S.; Taghvaee, H.; Liaskos, C.; et al. Toward Intelligent Metasurfaces: The Progress from Globally Tunable Metasurfaces to Software-Defined Metasurfaces with an Embedded Network of Controllers. *Adv. Opt. Mater.* **2020**, *8*, 2000783. [CrossRef]
50. Liu, H.; Zhang, J.; Wu, Q.; Jin, Y.; Chen, Y.; Ai, B. RIS-Aided Next-Generation High-Speed Train Communications: Challenges, Solutions, and Future Directions. *arXiv* **2021**, arXiv:2103.09484.
51. Gautam, S.; Sharma, S.K.; Tran, D.H.; Chatzinotas, S.; Ottersten, B. Hybrid Active-and-Passive Relaying model for SWIPT-enabled 6G-IoT Greencom Networks. *Preprints* **2020**. [CrossRef]
52. Lyu, B.; Yang, Z.; Guo, H.; Tian, F.; Gui, G. Relay Cooperation Enhanced Backscatter Communication for Internet-of-Things. *IEEE Internet Things J.* **2019**, *6*, 2860–2871. [CrossRef]
53. Rajaram, A.; Khan, R.; Tharranetharan, S.; Jayakody, D.N.K.; Dinis, R.; Panic, S. Novel SWIPT Schemes for 5G Wireless Networks. *Sensors* **2019**, *19*, 1169. [CrossRef] [PubMed]
54. Gong, S.; Huang, X.; Xu, J.; Liu, W.; Wang, P.; Niyato, D. Backscatter Relay Communications Powered by Wireless Energy Beamforming. *IEEE Trans. Commun.* **2018**, *66*, 3187–3200. [CrossRef]
55. Gong, S.; Zou, Y.; Hoang, D.T.; Xu, J.; Cheng, W.; Niyato, D. Capitalizing Backscatter-Aided Hybrid Relay Communications with Wireless Energy Harvesting. *IEEE Internet Things J.* **2020**, *7*, 8709–8721. [CrossRef]
56. Gong, S.; Gao, L.; Xu, J.; Guo, Y.; Hoang, D.T.; Niyato, D. Exploiting Backscatter-Aided Relay Communications with Hybrid Access Model in Device-to-Device Networks. *IEEE Trans. Cogn. Commun. Netw.* **2019**, *5*, 835–848. [CrossRef]
57. Van Huynh, N.; Hoang, D.T.; Lu, X.; Niyato, D.; Wang, P.; Kim, D.I. Ambient Backscatter Communications: A Contemporary Survey. *IEEE Commun. Surveys Tuts* **2018**, *20*, 2889–2922. [CrossRef]
58. Guo, J.; Zhang, H.; Zhu, X. Theoretical Analysis of RF-DC Conversion Efficiency for Class-F Rectifiers. *IEEE Trans. Microw. Theory Technol.* **2014**, *62*, 977–985. [CrossRef]
59. Gautam, S.; Vu, T.X.; Chatzinotas, S.; Ottersten, B. Cache-Aided Simultaneous Wireless Information and Power Transfer (SWIPT) With Relay Selection. *IEEE J. Sel. Areas Commun.* **2019**, *37*, 187–201. [CrossRef]
60. Gautam, S.; Vu, T.X.; Chatzinotas, S.; Ottersten, B. Joint Wireless Information and Energy Transfer in Cache-Assisted Relaying Systems. In Proceedings of the 2018 IEEE Wireless Communications and Networking Conference (WCNC), Barcelona, Spain, 15–18 April 2018; pp. 1–6. [CrossRef]
61. Kahn, L.R. Ratio squarer. *Proc. Inst. Radio Eng.* **1954**, *42*, 1704.
62. Dinkelbach, W. On Nonlinear Fractional Programming. *Manag. Sci.* **1967**, *13*, 492–498. [CrossRef]
63. Gautam, S.; Lagunas, E.; Sharma, S.K.; Chatzinotas, S.; Ottersten, B.; Vandendorpe, L. Weighted Sum-SINR and Fairness Optimization for SWIPT-Multigroup Multicasting Systems with Heterogeneous Users. *IEEE Open J. Commun. Soc.* **2020**, *1*, 1470–1484. [CrossRef]
64. Gautam, S.; Lagunas, E.; Bandi, A.; Chatzinotas, S.; Sharma, S.K.; Vu, T.X.; Kisseleff, S.; Ottersten, B. Multigroup Multicast Precoding for Energy Optimization in SWIPT Systems with Heterogeneous Users. *IEEE Open J. Commun. Soc.* **2020**, *1*, 92–108. [CrossRef]
65. Shen, X.; Diamond, S.; Gu, Y.; Boyd, S. Disciplined Convex-Concave Programming. In Proceedings of the 2016 IEEE 55th Conference on Decision and Control (CDC), Vegas, NV, USA, 12–14 December 2016; pp. 1009–1014.
66. Diamond, S.; Boyd, S. CVXPY: A Python-Embedded Modeling Language for Convex Optimization. *J. Mach. Learn. Res.* **2016**, *17*, 2909–2913.
67. *MATLAB Optimization Toolbox (Fmincon)*; 9.7.0.1296695 (R2019b) Update 4; The MathWorks: Natick, MA, USA, 2019.
68. Series, P. Propagation Data and Prediction Methods for the Planning of Short-Range Outdoor Radiocommunication Systems and Radio Local Area Networks in the Frequency Range 300 MHz to 100 GHz. Tech. Rep., ITU, Tech. Rep. ITU-R 2017. Available online: https://www.itu.int/dms_pubrec/itu-r/rec/p/R-REC-P.1411-9-201706-S!!PDF-E.pdf (accessed on 4 June 2021).
69. Wikipedia. 5G. Available online: https://en.wikipedia.org/wiki/5G (accessed on 30 August 2021).

70. Kim, S.H.; Kim, D.I. Hybrid Backscatter Communication for Wireless-Powered Heterogeneous Networks. *IEEE Trans. Wirel. Commun.* **2017**, *16*, 6557–6570. [CrossRef]
71. Mathsisfun. Confidence Interval Calculator. Available online: https://www.mathsisfun.com/data/confidence-interval-calculator.html (accessed on 30 August 2021).
72. Mo, Z.; Su, W.; Matyjas, J.D. Amplify and Forward Relaying Protocol Design with Optimum Power and Time Allocation. In Proceedings of the MILCOM 2016—2016 IEEE Military Communications Conference, Baltimore, MD, USA, 1–3 November 2016; pp. 412–417. [CrossRef]

sensors

MDPI

Communication

Distributed Transmit Power Control for Energy-Efficient Wireless-Powered Secure Communications

Kisong Lee

Department of Information and Communication Engineering, Dongguk University, Seoul 04620, Korea; kslee851105@gmail.com; Tel.: +82-2-2260-3233

Abstract: In this study, we consider energy-efficient wireless-powered secure communications, in which N sets of transmitter, receiver, and energy harvesting (EH) nodes exist; each EH node is allowed only to harvest energy from the transmitted signals but is not to permitted to decode information. To maximize the sum secrecy energy efficiency (SEE) of the node sets while ensuring minimum EH requirement for each EH node, we propose a distributed transmit power control algorithm using a dual method, where each transmitter adjusts its transmit power iteratively until convergence without sharing information with the other node sets. Through simulations under various environments, we show that the proposed scheme surpasses conventional schemes in terms of the sum SEE and has significantly reduced computation time compared with the optimal scheme, which suggests the effectiveness and applicability of the proposed distributed method.

Keywords: secure communication; energy harvesting; secrecy energy efficiency; transmit power control; distributed algorithm

Citation: Lee, K. Distributed Transmit Power Control for Energy-Efficient Wireless-Powered Secure Communications. *Sensors* **2021**, *21*, 5861. https://doi.org/10.3390/s21175861

Academic Editors: Carlo Trigona, Olfa Kanoun and Slim Naifar

Received: 27 July 2021
Accepted: 27 August 2021
Published: 31 August 2021

1. Introduction

With the rapid growth of mobile traffic and smart devices, next-generation wireless communications are expected to have increased demands for high data rates, wide coverage, and high levels of security [1,2]. At the same time, the energy consumption would inevitably increase as the functions of mobile devices become more diverse. This may cause inconvenience to users as the battery may need to be recharged or replaced more often to extend the lifetime of the network. Accordingly, several methods, such as energy-efficient protocols [3], resource allocation [4,5], and multi-antenna techniques [6,7], have been proposed to improve the energy efficiencies of networks.

In addition to energy-efficient use of resources, energy harvesting (EH) technology has emerged to solve the problem of energy shortages in wireless nodes [8–12]. In particular, the methods to design EH systems were proposed in [8,9], and the potential of EH technology as a promising means to enable self-sustainable operations of wireless nodes was discussed in [10–12]. Recent investigations on energy efficiency optimization have been also reported for wireless-powered networks, in which EH-enabled nodes exist [13–17]. In [13], joint optimization of user scheduling and power allocation was studied to maximize energy efficiency considering the characteristics of EH. The authors of [14] investigated a proportional fair energy efficiency method that considers energy efficiency and user fairness simultaneously. In [15,16], resource allocation strategies were proposed to improve the energy efficiency of wireless-powered cognitive radio networks (CRNs). Moreover, the authors of [17] proposed the max–min antenna selection scheme for EH-based two-way relaying and analyzed the diversity gain.

Network diversification also causes increased concerns regarding security issues between different networks because there is increased risk of eavesdropping by unlicensed users when secret keys are shared between legitimate users [1,2]. In this context, a number of studies have been conducted on physical layer security (PLS) to ensure secure communications without relying on secret keys, such as cooperative relaying [18,19] and jamming signal transmission [20]. Regarding PLS and EH simultaneously, optimal policies for EH-enabled relays [21,22] and an optimal resource allocation considering transmit power and EH ratio [23] have been proposed for wireless-powered secure communications (WPSCs). The optimal policy of a friendly jammer capable of EH has also been studied to help secure communication between licensed nodes [24].

Unlike the aforementioned studies that focus only on energy efficiency [3–7,13–17] or PLS [18–24], secrecy energy efficiency (SEE), which is defined as the ratio of the secrecy rate to dissipated power, was suggested to achieve a balance between energy efficiency and secure communication [25–30]. For example, resource allocations including transmit power and beamforming vectors were optimized to maximize the SEE in CRNs [25,27] and multi-antenna systems [26,28], respectively. Furthermore, the problem of SEE optimization was investigated in multi-antenna and multi-user systems with confidentiality and reliability constraints [29], and the optimal control parameters including first- and second-layer power splitting ratios, beamforming vectors, and artificial noise covariance matrix were derived to maximize the SEE in multi-antenna wireless-powered networks [30]. However, only simple scenarios without co-channel interference were considered in [25,26], and high computational complexity and information sharing among the nodes are needed to resolve the problems noted in [27–30]. Therefore, a distributed algorithm with low complexity that can be operated in real systems is required for energy-efficient WPSCs with co-channel interference. The literature survey is summarized in Table 1.

Table 1. Literature Survey.

Literature	Metric	Research Issue	Limitation
[3]	EE	Protocol	Did not consider WPSCs.
[4,5]	EE	Resource allocation	Did not consider WPSCs.
[6,7]	EE	Multi-antenna tech.	Did not consider WPSCs.
[8,9]	EH capability	EH system design	Did not consider WPSCs.
[10–12]	Possibility of EH	Applicability of EH to netw.	Did not consider WPSCs.
[13]	EE	Resource allocation	Did not consider PLS.
[14]	Proportional fair EE	Resource allocation	Did not consider PLS.
[15,16]	EE	Resource allocation	Did not consider PLS.
[17]	EH power	Antenna selection	Did not consider PLS.
[18,19]	Secure commun.	Cooperative relaying	Did not consider EH.
[20]	Secure commun.	Jamming signal trans.	Did not consider EH.
[21,22]	Secure commun.	Optimal relaying policy	Did not consider EH.
[23]	Secure commun.	Resource allocation	Did not consider EH.
[24]	Secure commun.	Policy of a friendly jammer	Did not consider EH.
[25,26]	SEE	Resource allocation	Simple scenarios w/o interf.
[27–30]	SEE	Resource allocation	Centralized approach.

In this study, we investigate a distributed transmit power control (TPC) for energy-efficient WPCSs. The main contributions can be summarized as follows.

- We consider WPSCs where the EH nodes harvest energy from the transmitted signals but are not allowed to decode private information shared between the transmitter and receiver (Tx-Rx) pairs. Given this scenario, we formulate an optimization problem to find the optimal transmit powers of the Txs to maximize the sum SEE while guaranteeing that the amount of energy collected from each EH node is greater than a minimum required value.

- To solve this non-convex problem with low complexity, we propose a distributed TPC algorithm using dual decomposition, where each Tx determines its transmit power iteratively until convergence without sharing any information with the other node sets.
- Through performance evaluations under various environments, we show that the proposed scheme achieves a higher sum SEE than conventional schemes and remarkably reduced computation time compared with the optimal scheme.

The remainder of this paper is organized as follows. In Section 2, we provide the problem statement for the considered system model, and the distributed TPC algorithm for energy-efficient WPSCs is proposed in Section 3. In Section 4, the performance evaluations are shown through extensive simulations, and the conclusions are presented in Section 5.

2. System Model and Problem Statement

Figure 1 shows the schematic of the WPSCs, where N number of Tx-Rx pairs use the same frequency band for data transmissions while N number of EH nodes are permitted to only collect energy from the signals sent by the Txs; all nodes are equipped with single antenna. Since the EH nodes are not allowed to interpret the information shared between their corresponding Tx-Rx pairs, they are called untrusted nodes, and each Tx is required to adjust its transmit power to maintain information confidentiality while supplying sufficient energy to guarantee the minimum EH requirement for its EH node. The EH node associated with each Tx-Rx pair is assumed to be predetermined, and the set of these nodes is denoted as $\mathbb{N}$, i.e., $|\mathbb{N}| = N$. The channel gain between Tx i and Rx j is denoted as $h_{i,j}$ and that between Tx i and EH node j is denoted as $g_{i,j}$, which are assumed to follow a discrete time block-fading model.

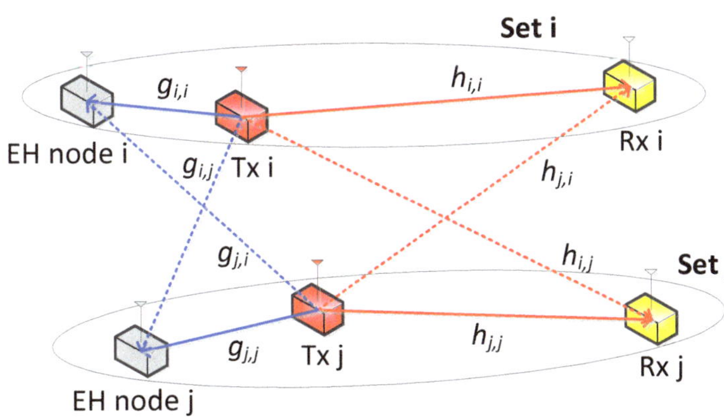

Figure 1. System model of WPSCs showing only two node sets for brevity.

Then, the signal received at Rx i is represented by

$$y_i = \sqrt{p_i} h_{i,i} x_i + \sum_{k \in \mathbb{N} \setminus \{i\}} \sqrt{p_k} h_{k,i} x_k + z_i, \tag{1}$$

where x_i denotes the normalized data symbol transmitted by Tx i with transmit power p_i, and $z_i \sim \mathcal{CN}(0, \sigma^2)$ indicates the noise at Rx i.

From (1), the achievable spectral efficiency (SE) is obtained as

$$r_i = \log_2\left(1 + \frac{p_i |h_{i,i}|^2}{\sigma^2 + \sum_{k \in \mathbb{N} \setminus \{i\}} p_k |h_{k,i}|^2}\right). \tag{2}$$

At the same time, the signal received at EH node i is expressed as

$$y_i^e = \sqrt{p_i} g_{i,i} x_i + \sum_{k \in \mathbb{N} \setminus \{i\}} \sqrt{p_k} g_{k,i} x_k + z_i^e, \tag{3}$$

where $z_i^e \sim \mathcal{CN}(0, \sigma^2)$. Considering that each EH node can harvest energy not only from the signal sent by Tx i but also from the signals sent by other Txs, the total harvested energy at EH node i is given by

$$e_i = \sum_{j \in \mathbb{N}} \zeta_i p_j |g_{j,i}|^2, \tag{4}$$

where ζ_i is the energy conversion efficiency. On the other hand, if EH node i overhears the signal sent by Tx i instead of harvesting energy, its achievable SE is represented by

$$r_i^e = \log_2 \left(1 + \frac{p_i |g_{i,i}|^2}{\sigma^2 + \sum_{k \in \mathbb{N} \setminus \{i\}} p_k |g_{k,i}|^2} \right). \tag{5}$$

From (2) and (5), the secrecy rate of node set i can be defined as the rate difference between the legitimate and eavesdropping links [31] as follows:

$$r_i^s = [r_i - r_i^e]^+, \tag{6}$$

where $[\cdot]^+ = \max(0, \cdot)$.

Moreover, the total power consumption at node set i can be obtained as

$$p_i^{CE} = p_C + p_i - e_i, \tag{7}$$

where p_C is the constant energy consumed by the circuits of each node set.

From (6) and (7), the SEE of node set i can be defined as the ratio of secrecy rate to total power consumption (bits/Hz/Joule), which can be formulated as follows:

$$\eta_i^s = \frac{r_i^s}{p_i^{CE}}. \tag{8}$$

It should be noted that this metric indicates how efficiently energy can be used to transmit secret information.

Based on these equations, we develop the optimization problem to find the optimal transmit powers of the Txs to maximize the sum SEE while guaranteeing the minimum EH requirement, $E_{\min}$, for each EH node, as follows.

$$\begin{aligned} \max_{0 \preceq \vec{p}} \quad & \sum_{i \in \mathbb{N}} \eta_i^s \\ s.t. \quad & e_i \geq E_{\min}, \ i \in \mathbb{N} \\ & p_i \leq P_{\max}, \ i \in \mathbb{N}, \end{aligned} \tag{9}$$

where $\vec{p} = \{p_1, p_2, \cdots, p_N\}$ and $P_{\max}$ is the maximum transmit power allowed for each Tx. The problem in (9) is non-convex because of the fractional objective function and interference term; hence, the optimal solution of $\vec{p}$ cannot be derived in a closed-form. The optimal solutions can be numerically found by brute-force search, where each p_i is quantized with M equally spaced values and all possible combinations are evaluated to find the optimal value. However, the channel state information (CSI) of all wireless links must be available to implement this method, and a high computational complexity of $O(M^N)$ is incurred, which increases exponentially with the number of node sets.

3. Distributed Transmit Power Control Algorithm

In this section, we present the distributed TPC algorithm, which can be operated with low complexity without sharing any information with the other node sets.

We decompose the original problem in (9) into N subproblems, which are then solved independently with low computational complexity [32]. In the subproblem, each Tx finds the transmit power to maximize its own SEE while ensuring the minimum EH requirement, which is formulated as follows:

$$\begin{aligned} \max_{0\leq p_i} \quad & \eta_i^{s} \\ s.t. \quad & \text{C1}: \; e_i \geq E_{\min} \\ & \text{C2}: \; p_i \leq P_{\max}. \end{aligned} \tag{10}$$

With defining $x_i = \frac{r_i^s}{p_i^{\text{CE}}}$, the objective function in (10) is converted from a fractional form to an equivalent subtractive form, $r_i^s - x_i p_i^{\text{CE}}$, using nonlinear fractional programming [33]. Accordingly, the subproblem in (10) is reformulated as

$$\begin{aligned} \max_{0\leq p_i} \quad & r_i^s - x_i p_i^{\text{CE}} \\ s.t. \quad & \text{C1 and C2}. \end{aligned} \tag{11}$$

To derive the transmit power of each Tx using the dual method, we denote the Lagrangian function of (11) as follows.

$$\mathcal{L}(p_i, \lambda_i, \mu_i) = r_i^s - x_i p_i^{\text{CE}} + \lambda_i(e_i - E_{\min}) + \mu_i(P_{\max} - p_i), \tag{12}$$

where $\lambda_i \geq 0$ and $\mu_i \geq 0$ are the respective Lagrange multipliers of each constraint of (11).

The dual objective is then defined as

$$\mathcal{G}(\lambda_i, \mu_i) = \max_{\vec{p}\succeq 0} \; \mathcal{L}(p_i, \lambda_i, \mu_i). \tag{13}$$

Using (13), the dual problem can be formulated as

$$\min_{0\leq\lambda_i,\, 0\leq\mu_i} \; \mathcal{G}(\lambda_i, \mu_i). \tag{14}$$

To find the suboptimal value of p_i, we build the Karush-Kuhn-Tucker (KKT) conditions with the complementary slackness, as follows.

$$\frac{\partial \mathcal{L}(p_i, \lambda_i, \mu_i)}{\partial p_i} = \frac{\partial r_i^s}{\partial p_i} - x_i \frac{\partial p_i^{\text{CE}}}{\partial p_i} + \lambda_i \frac{\partial e_i}{\partial p_i} - \mu_i = 0, \tag{15}$$

$$\lambda_i(e_i - E_{\min}) = 0, \tag{16}$$

$$\mu_i(P_{\max} - p_i) = 0, \tag{17}$$

$$0 \leq p_i \leq P_{\max}, \;\; 0 \leq \lambda_i, \;\; 0 \leq \mu_i. \tag{18}$$

Then, the transmit power that satisfies the KKT conditions in (15)–(18) can be derived as follows.

$$p_i = \left[\frac{1}{\ln 2 \left(x_i (1 - \zeta_i |g_{i,i}|^2) + \mu_i - \lambda_i \zeta_i |g_{i,i}|^2 \right) + t_i^{[s]}} - \frac{\Psi_i}{|h_{i,i}|^2} \right]^+, \tag{19}$$

where $\Psi_i = \sigma^2 + \sum_{j \in \mathbb{N} \setminus \{i\}} p_j |h_{j,i}|^2$ and $t_i^{[s]}$ is defined as

$$t_i^{[s]} = \frac{|g_{i,i}|^2}{\sigma^2 + \sum_{l \in \mathbb{N}} p_l |g_{l,i}|^2}. \tag{20}$$

In (19), Rx i can easily calculate Ψ_i by subtracting the signal power received from Tx i from the total received signal power. In addition, EH node i can measure $t_i^{[s]}$ readily because the denominator of (20) is the total received signal power at EH node i. Therefore, Tx i can determine its transmit power as shown in (19) by receiving information on Ψ_i from Rx i and information on $t_i^{[s]}$ and $g_{i,i}$ from EH node i, respectively. It should be noted that although the EH node is the untrusted node, it should send information on $t_i^{[s]}$ and $g_{i,i}$ to Tx i to receive enough energy to meet the EH requirement for operation. Moreover, Tx i does not need to share any information with the other node sets for calculating p_i, thereby allowing the proposed algorithm to operate in a distributed manner.

Moreover, the Lagrange multipliers are updated using a gradient algorithm as follows.

$$\begin{aligned} \lambda_i^{q+1} &= \left[\lambda_i^q - \nu_1(e_i - E_{\min})\right]^+, \\ \mu_i^{q+1} &= \left[\mu_i^q - \nu_2(P_{\max} - p_i)\right]^+, \end{aligned} \tag{21}$$

where ν_1 and ν_2 are sufficiently small step sizes for the update.

The operations of the proposed algorithm are described in Algorithm 1, where $\vec{o} = \{o_1, o_2, \cdots, o_N\}$. Specifically, each Tx initializes the transmit power and Lagrange multipliers randomly and calculates the SEE with the initialized transmit power. Next, the Txs determine the transmit powers according to (19) and update the Lagrange multipliers according to (21) iteratively until the transmit powers converge. The Txs also update the SEE and total consumed power with the converged transmit power to assess the convergence of the SEE. This process is repeated until the SEE converges.

Algorithm 1 Distributed transmit power control algorithm

Initialize $\vec{p}$, $\vec{\lambda}$, and $\vec{\mu}$ randomly
repeat
  Set $\vec{x} = \vec{r}^s / \vec{p}^{CE}$
  repeat
    $\vec{p}_{old} \leftarrow \vec{p}$
    for $i = 1$ to N
      Compute p_i according to (19)
      Update λ_i and μ_i according to (21)
    end for
    $\vec{p} = \{p_1, p_2, \cdots, p_N\}$
  until $\|\vec{p} - \vec{p}_{old}\| < \epsilon$
  Update $\vec{r}^s$ and $\vec{p}^{CE}$ with $\vec{p}$
until $\|\vec{r}^s - \vec{x}\vec{p}^{CE}\| < \delta$

Given that ϵ^{-2} iterations are needed to ensure that the norm of the gradient is less than ϵ in the worst-case scenario [34], the number of iterations required for convergence of the inner loop is ϵ^{-2}. Moreover, T denotes the number of iterations required for convergence of the outer loop [35], and the computational complexity of the proposed algorithm is $O(TN^2\epsilon^{-2})$, where N^2 is the number of computations required to calculate $\vec{p}$.

4. Simulation Results and Discussion

For the performance evaluations, the following system parameters are used as default unless stated otherwise: $N = 3$, $P_{\max} = p_C = 30$ dBm, $\sigma^2 = -100$ dBm, $E_{\min} = -10$ dBm, and $\eta_i = 0.5$ for $i \in \mathbb{N}$. All nodes are distributed randomly over an area of 50 m $\times$ 50 m, in which the maximum distance of each signal link and EH link in the same node set is 15 m. Because EH circuits have low power sensitivity, e.g., $-10\sim-30$ dBm for energy harvesters while $-60\sim-80$ dBm for information receivers [12,36], the small size area is appropriate for wireless-powered networks. A simplified path loss model with a path loss exponent of 2.7 is considered for urban areas [21]. Moreover, Rayleigh fading is considered for the signal links to reflect the non-line-of-sight (nLoS) characteristics while Rician fading with a K-factor of 6 is considered for the EH links to reflect the line-of-sight (LoS) characteristics [12]. The following five schemes are considered for performance evaluation in terms of the sum SEE, which can be mathematically written as $\mathbb{E}\left[\left(\prod_{i\in\mathbb{N}} \mathbb{1}_{e_i \geq E_{\min}}\right) \cdot \sum_{i\in\mathbb{N}} \eta_i^s\right]$. Note that a penalty is assigned by setting the sum SEE to zero when the minimum EH constraint is violated; therefore, the effects of EH violation are inherent in the sum SEE calculations.

- Optimal scheme: With the knowledge of the CSI of all wireless links, the near-optimal performance can be found using a brute-force search with $M = 100$; however, the performance for $N \geq 5$ is found using a divide and conquer algorithm because of the extremely high computational complexity of the brute-force search.
- Proposed scheme: The transmit powers of the Txs are determined using the proposed algorithm given in Algorithm 1.
- r^s max. scheme: The transmit powers of the Txs are determined to maximize the sum secrecy rate, $\sum_{i\in\mathbb{N}} r_i^s$, which is found from the divide and conquer algorithm.
- Equally reduced power (ERP) scheme [37]: All Txs use the same transmit power that maximizes the sum SEE while meeting the minimum EH constraint, and the optimal value of the transmit power is found by one-dimensional exhaustive search.
- EH max. scheme: The Txs use their maximum transmit powers to maximize the total harvested energies of the EH nodes, $\sum_{i\in\mathbb{N}} e_i$.
- Rand scheme: The transmit powers of the Txs are determined randomly.

Figure 2 depicts the transmit power of each Tx and the sum SEE against the number of iterations, which shows the convergence of the proposed scheme. Each Tx adjusts the transmit power to maximize its own SEE, which affects the SEEs of the other node sets. Although the transmit power update at each Tx influences the other Txs, the transmit powers of all Txs converge to stationary points with iteration progression; finally, the sum SEE also converges to a value of 8.85 within 70 iterations.

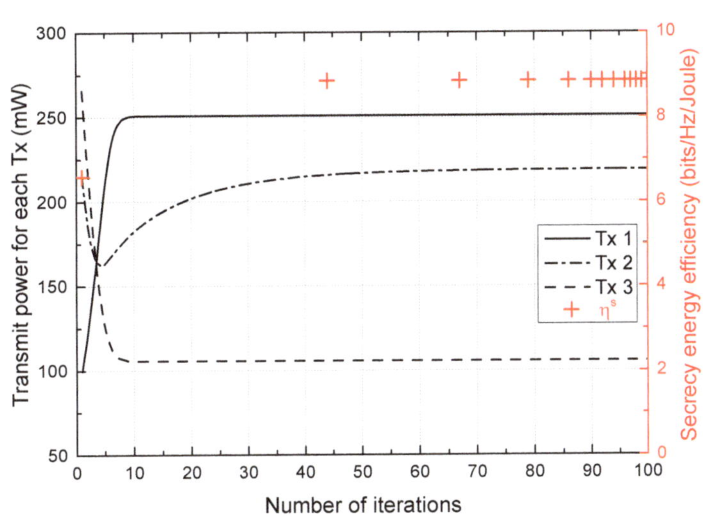

Figure 2. Convergence of the proposed scheme.

Figure 3 depicts the sum SEE versus maximum transmit power ($P_{\max}$). In the optimal, proposed, and ERP schemes, the sum SEE increases with $P_{\max}$ when $P_{\max} < 27$ dBm but converges to a stationary point when $P_{max} \geq 27$ dBm. This indicates that there is an optimal transmit power at which the maximum SEE is achieved, i.e., the use of extra transmit power beyond this value reduces the sum SEE. Therefore, these schemes do not use transmit powers of more than 27 dBm even though P_{max} increases over 27 dBm. On the other hand, the r^s max. and EH max. schemes use more transmit power to maximize the sum secrecy rate and total harvested energy, respectively, as P_{max} increases. However, the excessive use of transmit power exceeding 27 dBm causes inefficient energy consumption rather than improvement of the secrecy rate, so the sum SEE degrades seriously when P_{max} is greater than 27 dBm. The sum SEE of the rand scheme also decreases as P_{max} increases because it uses more transmit power without the adaptive TPC strategy.

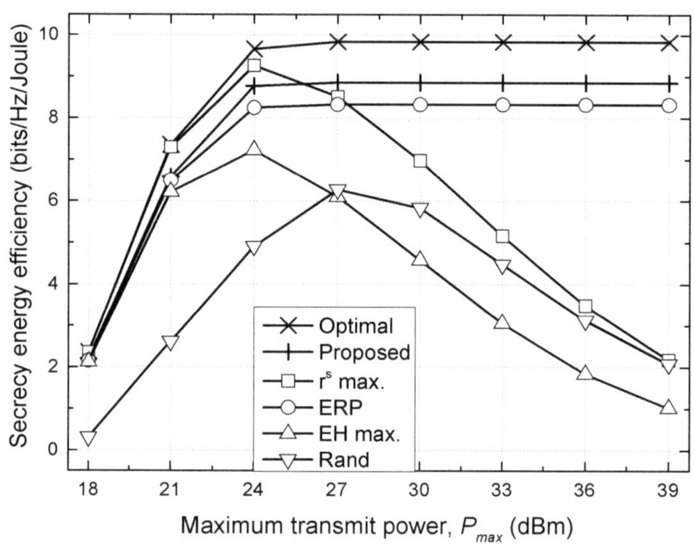

Figure 3. Sum secrecy energy efficiency vs. maximum transmit power (P_{max}) for different schemes.

Figures 4 and 5 depict the sum SEE versus required harvested energy (E_{min}) and energy conversion efficiency (η), respectively. In an environment where it is difficult to satisfy the EH requirements, i.e., larger E_{min} and smaller η, the Txs should use the additional transmit power inefficiently to satisfy the EH requirements. For some channel realizations, e.g., deep fading, the EH requirements cannot be inherently guaranteed at extremely large E_{min} or small η. Then, this causes serious degradation in the sum SEE because it is set to zero when the minimum EH constraint is violated. Hence, the sum SEEs of all schemes decrease as E_{min} increases or η decreases, especially when $E_{min} \geq -5$ dBm and $\eta \leq 0.3$. However, we can see that the proposed scheme shows a trend similar to the optimal scheme and outperforms the conventional schemes. In particular, the performance gain of the proposed scheme compared to the conventional schemes is greater in an environment that favorably satisfies the EH requirements, where the adaptive TPC strategy can operate more effectively.

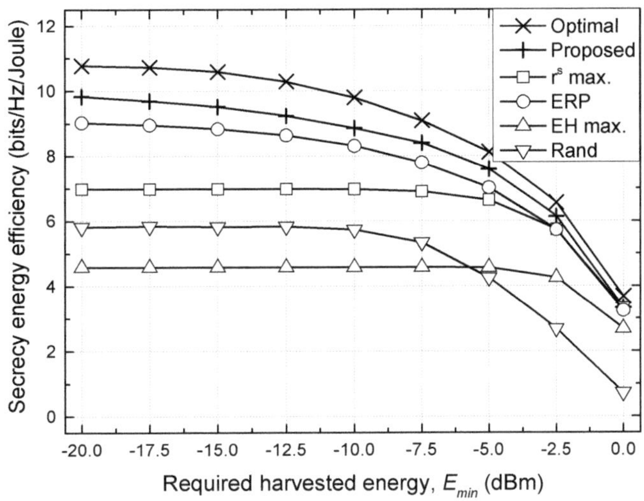

Figure 4. Sum secrecy energy efficiency vs. required harvested energy (E_{min}) for different schemes.

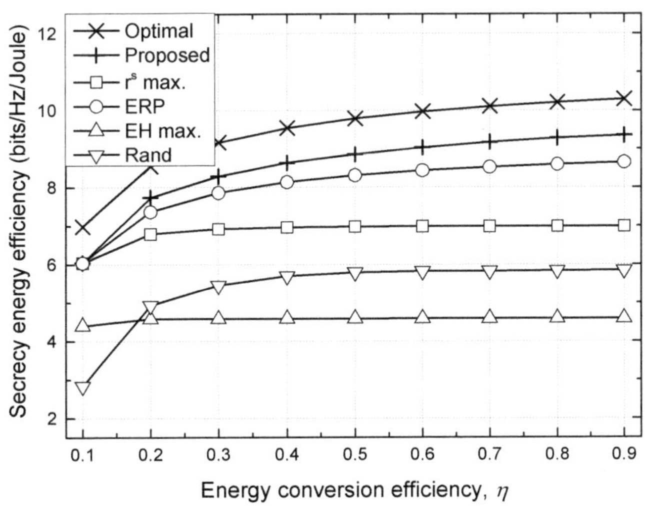

Figure 5. Sum secrecy energy efficiency vs. energy conversion efficiency (η) for different schemes.

Figure 6 depicts the cumulative distribution function (CDF) versus sum SEE. It is observed that the CDF of the proposed scheme is closest to that of the optimal scheme, thus confirming the effectiveness of the proposed TPC strategy. Given that high values of SEE are more distributed in the CDFs of the optimal and proposed schemes, rather than those of the conventional schemes, we note that the effective TPC is important for improving the SEE.

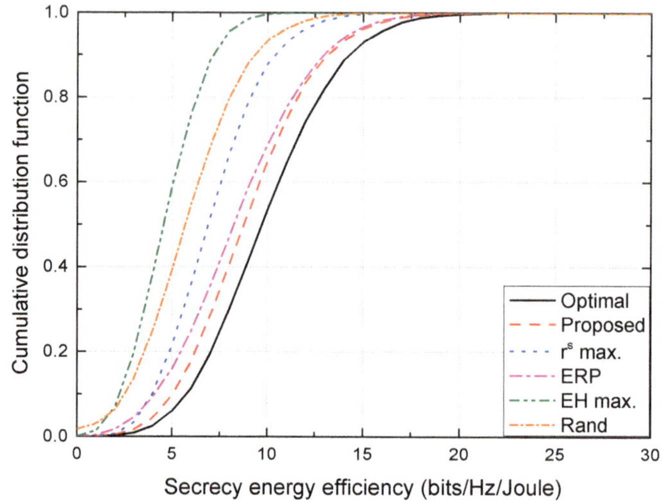

Figure 6. Cumulative distribution function vs. sum secrecy energy efficiency for different schemes.

Figure 7 depicts the average SEE for each node set ($\frac{\sum_{i\in\mathbb{N}} \eta_i^s}{N}$) and the computation time against number of node sets (N). It should be noted that the computation time of the EH max. and random schemes are omitted in subfigure (b) because they do not perform any computations to determine the transmit powers. As N increases, the Txs experience severe interference with each other. As a result, the improvement in the secrecy rate is insignificant compared to the transmit power used, which in turn deteriorates the average SEE with increasing N. However, the proposed scheme achieves a higher SEE than the conventional schemes by coping with the interference and reduces the computation time significantly compared with the optimal and r^s max. schemes. Although the ERP scheme can achieve the lowest computation time for a large number of N because the optimal transmit power can be found by one-dimensional search even as N increases, the proposed scheme achieves higher sum SEE than the ERP scheme, which verifies the effectiveness of the proposed distributed method in term of energy efficiency and secure communication.

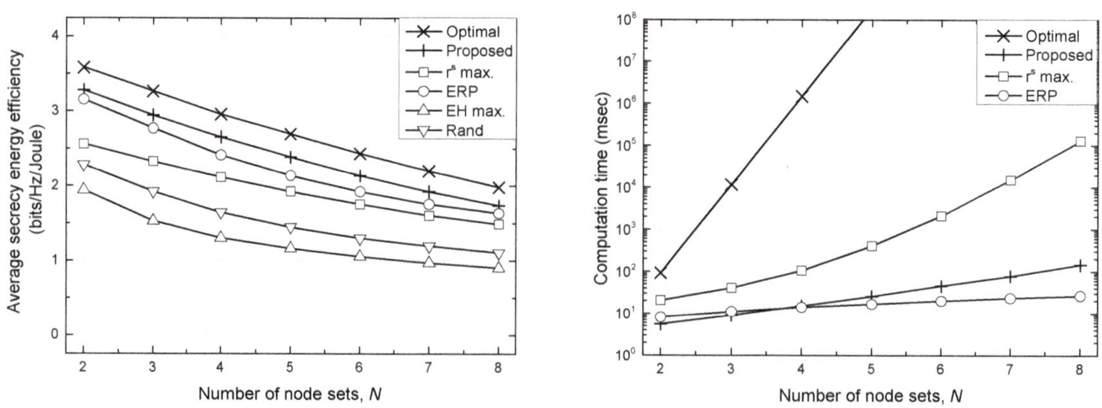

(a) (b)

Figure 7. Performance comparison against number of node sets (N) for different schemes. (**a**) Average secrecy energy efficiency per each node set vs. N. (**b**) Computation time vs. N.

Thus, the simulation results show that the proposed scheme has a slight difference in performance from the optimal scheme of about 10% because of its distributed nature; however, it is confirmed that the proposed scheme improves the sum SEE using the effective TPC strategy compared with the conventional schemes while significantly reducing computation time.

5. Conclusions

This study involves investigation of a distributed TPC algorithm for energy-efficient WPSCs, in which the transmit powers of the Txs are optimized to maximize their own SEEs while guaranteeing the minimum EH requirements for the corresponding EH nodes. Specifically, we analytically derived the closed-form equation for the transmit power, and proposed an iterative algorithm using a dual method, that can be operated in a distributed manner without sharing information with the other node sets. The simulation results confirm that the proposed scheme can achieve a higher sum SEE than existing schemes by adjusting the transmit power with respect to secrecy rate and energy efficiency; moreover, the computation time is significantly reduced compared with the optimal scheme. It is expected that our solution will be used to solve the complex TPC problems of WPSCs in a simple distributed manner. Interesting topics for future work include a deep learning-based distributed TPC algorithm for improving the performance of WPSCs.

Funding: This work was supported by the National Research Foundation of Korea (NRF) grant funded by the Korea government (MSIT) (No. 2021R1A2C4002024).

Institutional Review Board Statement: Not applicable.

Informed Consent Statement: Not applicable.

Data Availability Statement: Not applicable.

Conflicts of Interest: The authors declare no conflict of interest.

References

1. Frustaci, M.; Pace, P.; Aloi, G.; Fortino, G. Evaluating critical security issues of the IoT world: Present and future challenges. *IEEE Internet Things J.* **2018**, *5*, 2483–2495. [CrossRef]
2. Gui, G.; Liu, M.; Tang, F.; Kato, N.; Adachi, F. 6G: Opening new horizons for integration of comfort, security and intelligence. *IEEE Wirel. Commun.* **2020**, *27*, 126–132. [CrossRef]
3. Li, S.; Kim, J.G.; Han, D.H.; Lee, K.S. A survey of energy-efficient communication protocols with QoS guarantees in wireless multimedia sensor networks. *Sensors* **2019**, *19*, 199. [CrossRef]
4. Hoang, T.D.; Le, L.B.; Le-Ngoc, T. Energy-efficient resource allocation for D2D communications in cellular networks. *IEEE Trans. Veh. Technol.* **2016**, *65*, 6972–6986. [CrossRef]
5. Shi, J.; Yu, W.; Ni, Q.; Liang, W.; Li, Z.; Xiao, P. Energy efficient resource allocation in hybrid non-orthogonal multiple access systems. *IEEE Trans. Commun.* **2019**, *67*, 3496–3511. [CrossRef]
6. He, S.; Huang, Y.; Wang, J.; Yang, L.; Hong, W. Joint antenna selection and energy-efficient beamforming design. *IEEE Signal Process. Lett.* **2016**, *23*, 1165–1169. [CrossRef]
7. Lee, K. Distributed joint optimization of beamforming and power allocation for maximizing the energy efficiency of cognitive heterogeneous networks. *Sensors* **2021**, *21*, 3186. [CrossRef] [PubMed]
8. Raghunathan, V.; Chou, P.H. Design and power management of energy harvesting embedded systems. In Proceedings of the 2006 International Symposium on Low Power Electronics and Design (ISLPED), Tegernsee, Germany, 4–6 October 2006; pp. 369–374.
9. Lu, C.; Raghunathan, V.; Roy, K. Efficient design of micro-scale energy harvesting systems. *IEEE J. Emerg. Sel. Top. Circuits Syst.* **2011**, *1*, 254–266. [CrossRef]
10. Prauzek, M.; Konecny, J.; Borova, M.; Janosova, K.; Hlavica, J.; Musilek, P. Energy harvesting sources, storage devices and system topologies for environmental wireless sensor networks: A review. *Sensors* **2018**, *18*, 2446. [CrossRef]
11. Pinuela, M.; Mitcheson, P.; Lucyszyn, S. Ambient RF energy harvesting in urban and semi-urban environments. *IEEE Trans. Microw. Theory Techn.* **2013**, *61*, 2715–2726. [CrossRef]
12. Lu, X.; Wang, P.; Niyato, D.; Kim, D.I.; Han, Z. Wireless networks with RF energy harvesting: A contemporary survey. *IEEE Commun. Surv. Tut.* **2015**, *17*, 757–789. [CrossRef]
13. Lee, K.; Hong, J. Energy-efficient resource allocation for simultaneous information and energy transfer with imperfect channel estimation. *IEEE Trans. Veh. Technol.* **2016**, *65*, 2775–2780. [CrossRef]

14. Chung, B.C.; Lee, K.; Cho, D. Proportional fair energy-efficient resource allocation in energy-harvesting-based wireless networks. *IEEE Syst. J.* **2018**, *12*, 2106–2116. [CrossRef]
15. Lee, K.; Yoon, C.; Jo, O.; Lee, W. Joint optimization of spectrum sensing and transmit power in energy harvesting-based cognitive radio networks. *IEEE Access* **2018**, *6*, 30653–30662. [CrossRef]
16. Shahini, A.; Kiani, A.; Ansari, N. Energy efficient resource allocation in EH-enabled CR networks for IoT. *IEEE Internet Things J.* **2019**, *6*, 3186–3193. [CrossRef]
17. Noor-A-Rahim, M.; Khyam, M.O.; Guan, Y.L. Energy harvesting two-way relaying with antenna selection scheme. *IET Commun.* **2019**, *13*, 198–204. [CrossRef]
18. Chu, S. Secrecy analysis of modify-and-forward relaying with relay selection. *IEEE Trans. Veh. Technol.* **2019**, *68*, 1796–1809. [CrossRef]
19. Atapattu, S.; Ross, N.; Jing, Y.; He, Y.; Evans, J.S. Physical-layer security in full-duplex multi-hop multi-user wireless network with relay selection. *IEEE Trans. Wirel. Commun.* **2019**, *18*, 1216–1232. [CrossRef]
20. Park, K.; Wang, T.; Alouini, M. On the jamming power allocation for secure amplify-and-forward relaying via cooperative jamming. *IEEE J. Sel. Areas Commun.* **2013**, *31*, 1741–1750. [CrossRef]
21. Kalamkar, S.S.; Banerjee, A. Secure communication via a wireless energy harvesting untrusted relay. *IEEE Trans. Veh. Technol.* **2017**, *66*, 2199–2213. [CrossRef]
22. Lee, K.; Bang, J.; Choi, H. Secrecy outage minimization for wireless-powered relay networks with destination-assisted cooperative jamming. *IEEE Internet Things J.* **2021**, *8*, 1467–1476. [CrossRef]
23. Xu, D.; Zhu, H. Secure transmission for SWIPT IoT systems with full-duplex IoT devices. *IEEE Internet Things J.* **2019**, *6*, 10915–10933. [CrossRef]
24. Liu, W.; Zhou, X.; Durrani, S.; Popovski, P. Secure communication with a wireless-powered friendly jammer. *IEEE Trans. Wirel. Commun.* **2016**, *15*, 401–415. [CrossRef]
25. Ouyang, J.; Lin, M.; Zou, Y.; Zhu, W.; Massicotte, D. Secrecy energy efficiency maximization in cognitive radio networks. *IEEE Access* **2017**, *5*, 2641–2650. [CrossRef]
26. Wang, G.; Meng, C.; Heng, W.; Chen, X. Secrecy energy efficiency optimization in AN-aided distributed antenna systems with energy harvesting. *IEEE Access* **2018**, *6*, 32830–32838. [CrossRef]
27. Jiang, Y.; Zou, Y.; Ouyang, J.; Zhu, J. Secrecy energy efficiency optimization for artificial noise aided physical-layer security in OFDM-based cognitive radio networks. *IEEE Trans. Veh. Technol.* **2018**, *67*, 11858–11872. [CrossRef]
28. Jiang, Y.; Zou, Y.; Ouyang, J.; Zhu, J. Beamforming aided secrecy energy efficiency maximization in heterogeneous cellular networks. *IEEE Trans. Veh. Technol.* **2021**, *70*, 2576–2589. [CrossRef]
29. Zappone, A.; Lin, P.; Jorswieck, E.A. Secrecy energy efficiency for MIMO single- and multi-cell downlink transmission with confidential messages. *IEEE Trans. Inf. Forensics Secur.* **2019**, *14*, 2059–2073. [CrossRef]
30. Lu, Y.; Xiong, K.; Fan, P.; Ding, Z.; Zhong, Z.; Letaief, K.B. Secrecy energy efficiency in multi-antenna SWIPT networks with dual-layer PS receivers. *IEEE Trans. Wirel. Commun.* **2020**, *19*, 4290–4306. [CrossRef]
31. Wyner, A.D. The wire-tap channel. *Bell Syst. Tech. J.* **1975**, *54*, 1355–1387. [CrossRef]
32. Yu, W.; Lui, R. Dual methods for nonconvex spectrum optimization of multicarrier systems. *IEEE Trans. Commun.* **2006**, *54*, 1310–1321. [CrossRef]
33. Dinkelbach, W. On nonlinear fractional programming. *Manag. Sci.* **1967**, *13*, 492–498. [CrossRef]
34. Fliege, J.; Vaz, A.I.F.; Vicente, L.N. Complexity of gradient descent for multiobjective optimization. *Optim. Methods Softw.* **2018**, *34*, 949–959. [CrossRef]
35. Venturino, L.; Prasad, N.; Wang, X. Coordinated scheduling and power allocation in downlink multicell OFDMA networks. *IEEE Trans. Veh. Technol.* **2009**, *58*, 2835–2848. [CrossRef]
36. Stoopman, M.; Keyrouz, S.; Visser, H.J.; Philips, K.; Serdijn, W.A. Co-design of a CMOS rectifier and small loop antenna for highly sensitive RF energy harvesters. *IEEE J. Solid-State Circuits* **2014**, *49*, 622–634. [CrossRef]
37. Lee, W.; Ban, T.; Jung, B.C. Distributed transmit power optimization for device-to-device communications underlying cellular networks. *IEEE Access* **2019**, *7*, 87617–87633. [CrossRef]

sensors

MDPI

Review

Towards the Functional Ageing of Electrically Conductive and Sensing Textiles: A Review

Christian Biermaier, Thomas Bechtold and Tung Pham *

Research Institute of Textile Chemistry and Textile Physics, Faculty of Chemistry and Pharmacy, University of Innsbruck, 6850 Dornbirn, Austria; Christian.Biermaier@uibk.ac.at (C.B.); Thomas.Bechtold@uibk.ac.at (T.B.)
* Correspondence: Tung.Pham@uibk.ac.at

Abstract: Electronic textiles (e-textiles) have become more and more important in daily life and attracted increased attention of the scientific community over the last decade. This interdisciplinary field of interest ranges from material science, over chemistry, physics, electrical engineering, information technology to textile design. Numerous applications can already be found in sports, safety, healthcare, etc. Throughout the life of service, e-textiles undergo several exposures, e.g., mechanical stress, chemical corrosion, etc., that cause aging and functional losses in the materials. The review provides a broad and critical overview on the functional ageing of electronic textiles on different levels from fibres to fabrics. The main objective is to review possible aging mechanisms and elaborate the effect of aging on (electrical) performances of e-textiles. The review also provides an overview on different laboratory methods for the investigation on accelerated functional ageing. Finally, we try to build a model of cumulative fatigue damage theory for modelling the change of e-textile properties in their lifetime.

Keywords: e-textiles; conductive textiles; functional ageing; durability; endurance; fatigue lifetime; textile sensors; flexible sensors; accelerated ageing

Citation: Biermaier, C.; Bechtold, T.; Pham, T. Towards the Functional Ageing of Electrically Conductive and Sensing Textiles: A Review. *Sensors* **2021**, *21*, 5944. https://doi.org/10.3390/s21175944

Academic Editors: Carlo Trigona, Olfa Kanoun and Slim Naifar

Received: 29 July 2021
Accepted: 31 August 2021
Published: 4 September 2021

1. Introduction

Electronic textiles or e-textiles become more and more important in daily life and attract increased attention of the scientific community over the last decade. This interdisciplinary field of interest ranges from material science, over chemistry, physics, electrical engineering, information technology to textile design. Typically, e-textiles are defined as fabrics or cloths having sensors, actuators, power sources and data processors as integrated parts. All these parts are electrically connected by conductive lines [1].

Often, e-textiles are associated with the term "smart textiles". According to common understanding, smart textiles are, however, defined as "materials that sense and respond to environmental stimuli" [2,3]. The term "environmental stimuli" can be understood as chemical, electrical, thermal, mechanical, or other exposures. Smart textiles can also be divided into "passive smart textiles", e.g., humidity, temperature, pressure, gas sensors and "active smart textiles" which adapt their behaviour to their environmental exposure [3]. Apart from the typical functionality as a connector for signal transport and energy supply, conductive structures can also be utilised as a sensor and actuator areas in modern smart textiles, e.g., as temperature sensors, [4] pressure sensors [5] and piezo elements for energy harvesting [6].

In any case, electrical conductivity is an essential part of e-textiles, which is the focus of the review. In recent decades, many techniques have been developed to introduce conductive lines and patterns into textiles [7]. As the most important success criterion, the conductivity is often reported. However, the evolution of the functionality over usage time—where worsening is observed in most cases—is not yet well understood. Beside chronological age, the term ageing is used as an abstract term for a change on the material properties during usage or storage.

The classical ageing of textile can be categorised into physical ageing, photochemical degradation, thermal degradation, chemical attack and mechanical stress. Physical ageing is recognised as the re-crystallisation of the semi-crystalline polymers of the textile during usage, leading to a higher crystallinity degree and higher rigidity. Photochemical degradation results from light exposure and leads to the breakage of chemical bonds and possibly to the formation of new bonds. Thermal degradation is, depending on the temperature and exposure time, a kind of physical ageing, or a change in the chemical structure of the polymer considering oxidation and decomposition. Chemical attacks can be observed by an acidic or basic environment, depending on the polymer structure. Mechanical ageing, also referred to as mechanical stress, which results from movement of the textile, is divided into instantaneous response from elastic behaviour and a delayed response due to viscous behaviour [8].

With respect to e-textiles, the ageing phenomena consist of both substrate ageing and the ageing of electrical functionalities. While substrate ageing refers to changes in the material performance, e.g., mechanical strength, chemical inertness, etc., of the fabrics or fibres, functional ageing concerns losses in sensor behaviour, conductivity, capacitance, etc. It is worth noticing that both ageing mechanisms are strongly connected to each other. As an example, worsening in flexibility of a fabric as a result of fibre polymer degradation due to UV irradiation can directly impact the electrical property of the coated conductive paste under bending/stretching conditions. Therefore, it is not always possible to investigate the two ageing processes separately.

As e-textiles increasingly find applications in sport, medical and safety areas, there is a pressing need for ensuring and improving the reliability and durability of the products, in particular when it comes to sensitive applications such as in healthcare and life-saving. Thus, scientific knowledge in ageing behaviour is crucial.

To provide the readers with an overview on different basic mechanisms of functional ageing of e-textiles, in this review, we limit our focus to the changes and damages of conductive functionality as a basic feature of e-textiles under usage conditions and exclude the discussion on the ageing of other electronic devices. Furthermore, different approaches in modelling of functional ageing are discussed.

2. Conductive Textiles

In order to better understand the mechanism of functional ageing, especially the electrical conductivity of e-textiles, this section provides an overview of the basic possibilities in integrating electrical conductivity into fibre and textile structures as the integration technology also impacts the functional ageing of e-textiles.

Common techniques to obtain conductivity on yarns or textiles are, for example, metal deposition [9], embroidery of metal yarns into fabrics [10] and the application of metal-containing pastes and inks [11]. These conductive pastes and inks consist of a conductive filler distributed in a printable polymeric matrix. Often, also graphite and carbon black are used in conductive inks [12,13] as well as graphene [14] or types of carbon nanotubes (CNT) [15].

Another concept is the usage of intrinsically conductive polymers (ICP), discovered by A. J. Heeger, MacDiarmid and Shirakawa, achieving the Nobel Prize in 2000. Typically, conductive polymers consist of two main constituents: an extended π-electron system and a doping agent, e.g., an acid for protonation in polyaniline (PANI). Chemical doping by charge transfer or electrochemical doping are also possible [16]. Among a variety of conductive polymers, mainly polyethylenedioxythiophene (PEDOT:PSS) [17], polypyrrol (PPy) [18] and polyaniline (PANI) playing a role for conductive textiles [19].

The conductive material can be applied to a fibre or yarn for later integration into fabrics or the formation of conductive structures can be realised directly on fabric surfaces.

2.1. Conductive Yarns

There are many ways to introduce conductivity into textile yarns. One of the most common techniques is the utilisation of conductive polymers or carbon derivatives as conductive material in yarns.

The introduction of conductivity can be implemented in the melt mixing process. Here, the conductive polymer along with additives is mixed with thermoplastic polymers such as polyolefins and melt spun into conductive fibres. Typical examples are shown by Ikkalaa et al. and Yang et al. for PANI in combination with LDPE. This method is simple and the main advantage is the good distribution of conductive material across the entire fibre cross section [20,21]. Another example is compounding a small amount (7%) of multi-walled carbon nanotubes (MWCNTs) in polyamide 6 and for spinning conductive PA 6 fibres [22].

To maintain the bulk properties of fibres, conductive polymers can also be applied by coating on the surface of fibres or yarns. An example is given for the deposition of PANI on Nomex fibres by immersing the fibres into a solution and triggering an active polymerisation reaction [23]. The polymerisation of a conductive polymer may also take place on the surface of impregnated threads [24]. Furthermore, dip coating of fibres into a dissolved conductive polymer, such as commercially available polyaniline emeraldine salt in xylole, can also be applied [22]. The coating of yarns and fibres with carbon derivatives is also available in many varieties. For example, a bovine serum albumin pre-treated polyamide 6 yarn was coated with graphene oxide (GO) and consecutively reduced (i.e., RGO formed) with hydrogen iodide (HI) [14]. Additionally, the pre-impregnation of fibres with polymers (e.g., PVA) can be utilised to enhance the adsorption of carbon nanotubes (CNTs) [25].

Apart from melt mixing and coating, it is also possible to utilise a solvent spinning process for conductive polymers. A mixture of 2-acrylamido-2-methyl-1-propanesulfonic acid and PANI in dichloro acetic acid can be forced through a spinneret under nitrogen pressure into an acetone coagulation bath [26]. A similar technique was used for PEDOT:PSS dispersion in water that was spun into an acetone bath through a single hole spinneret [27].

Taking silver (Ag) and copper (Cu) as the most efficient conductive materials, the deposition of Ag and Cu on textiles is a fast-growing field. Silver can, for example, be deposited on a substrate by physical vapour deposition [28]. Silver-coated yarns are commercially available in different materials and strengths for the application in e-textiles [29].

The deposition of copper is mainly performed on fabrics. However, there are also some techniques for the coating of yarns with copper, e.g., by electrochemical deposition. This technique requires a conductive pre-coating step, e.g., of carbon nanotubes (CNT) [30].

Besides the electrochemical plating, the electroless copper deposition (ECD) plays an important role in metal deposition. Here, a reducing agent, e.g., formaldehyde, is responsible for the reduction in Cu^{2+} ions to copper metal under basic pH conditions, coming along with the development of hydrogen gas. The reaction is metal catalysed; thus, metal seeding is needed before starting the main deposition reaction. More detailed information can be found in literature reviews on electroless copper deposition [31]. Examples for ECD on yarns are given for copper seeding from copper sulphate with reduction by sodium borohydride on a high performance yarn of poly(p-phenylene-2,6-benzobisoxazole) (PBO) or the use of platinum particles as a catalyst on polypyrrole attached to aramid. Reduction agents here were glyoxylic acid or formaldehyde [32,33].

Another way of using metals as conductor in yarns is to manufacture the yarn from metal fibres, especially stainless steel [34,35] or copper [36] in a plied yarn of filaments or staple fibres, which are commercially available. Furthermore, metal fibres/threads can also be combined with polymers to form ply yarns, known as metal polymer hybrid yarns or metal composite yarns. Stainless steel staple fibre blended with polyester can be used in textile strain sensors [37]. Some recent studies on different blends of stainless steel yarn and viscose or PS show the effect of metal content and number of twists on the yarn conductivity. The higher the metal content, the lower the electrical resistance is.

Additionally, the moisture affinity of the polymer fibre influences the resistance of the ply yarn. The fibre with the higher absorption capacity led to about 60% of resistance [38].

The combination of conductive and non-conductive parts in a fibre or yarn can be realised in the form of a core–sheath construction. Here, the two components are distributed spatially as the inner part and an outer layer. Often, the outer layer acts as protection layer of the inner part, e.g., polypropylene-coated conductive yarns [39], metal wires with a twisted carbon nanotube web [40] or a PVDF sheath around a PDOT:PSS core [41].

2.2. Integration of Conductive Yarns in Textiles

To manufacture e-textiles, conductive yarns need to be integrated into fabrics. Common methods to integrate a conductive yarn in textiles include knitting and weaving. This means, the conductive yarn becomes a part of the nonconductive yarn network.

A conductive yarn can be integrated into a fabric by weft knitting, which means, to exchange at least one course of nonconductive yarn with a conductive yarn. This formed a strain sensor for an elastomeric lycra fabric with a silver-plated polyamide yarn [42]. The multiple contacts of neighboured yarn loops led to a low resistance in the relaxed state, whereas, during extension, these contacts became lost and the resistance increased. This process is part of the modelling of the resistance of knitted strain sensors by dividing each loop into different micro resistors [43]. The performance of a strain sensor is expressed by the gauge factor (GF), which is defined by ratio of resistance change and elongation. In the shown example, the gauge factor depended on the composition of conductive yarn. Silver-plated yarns show a strong, but steel–polyester hybrid yarns show a weak gauge behaviour [44].

Weaving also provides the possibility of integrating conductive fibres into fabrics. The woven textiles also show a gauge resistance behaviour, which is influenced by the weft density. In an example of a steel–polyester hybrid yarn, woven with a blend of polyester and cotton, the gauge sensitivity decreased with increasing density [45].

Models for the prediction of the resistance behaviour of woven conductive fabrics use micro-resistors and consider the crossings of warp and weft yarns as a critical factor. Models can be reduced to a unit cell of two warp and two weft yarns that consists of 16 different micro resistors [46], working as a textile resistor network [47]. Further models are shown by Beex et al. [48] or Zhang et al. [49].

Apart from the direct integration of conductive threads by knitting and weaving, conductive yarns can also be attached on a textile using embroidery. There are stitching embroidery and tailored fibre placement (TFP) machines [50]. The stitching embroidery is performed by using a conductive yarn as upper or lower yarn and an insulating yarn as counter-part. The flexibility of this method is shown by embroidering different patterns, e.g., spirals that function as electrical inductors or antennas [51].

Tailored fibre placement (TFP) (c.f. illustration Figure 1) puts less mechanical stress onto the placed fibre or yarn during its production and is known for placing carbon rovings on textiles. A fibre or bundle of fibres or a yarn is placed on a textile with subsequent zigzag embroidery on top [52]. The manufacturing of different patterns is very easy with TFP. An example is shown for a two dimensional strain sensor manufactured by placing a silver-coated polyamide yarn on a polyester based stretchable fabric [53,54].

2.3. Conductive Textiles by Surface Coating and Deposition

Another method of producing conductive structures on fabrics is surface coating with a layer of conductive material. Fabrics are similarly treated as yarns to gain conductivity.

Among others, dip coating is a very simple way of applying conductive materials to fabrics. In an example, a cotton textile was immersed into a colloidal solution of single-walled carbon nanotubes dissolved in 1,2-dichlorobenzene. The conductive textile showed a gauge and piezo-resistive behaviour [56]. Graphene oxide has good adsorption properties on polyethylene terephthalate. The graphene oxide layer needs to be reduced with, e.g., stannous ions after the adsorption to gain conductivity and crosslinking [57].

This technique is also referred to as dyeing [58]. Examples of dip coating that include conductive polymers are the direct application of PEDOT:PSS dispersed in water onto fabrics [59] or in situ synthesis of polythiophene compounds in the presence of a textile substrate [60].

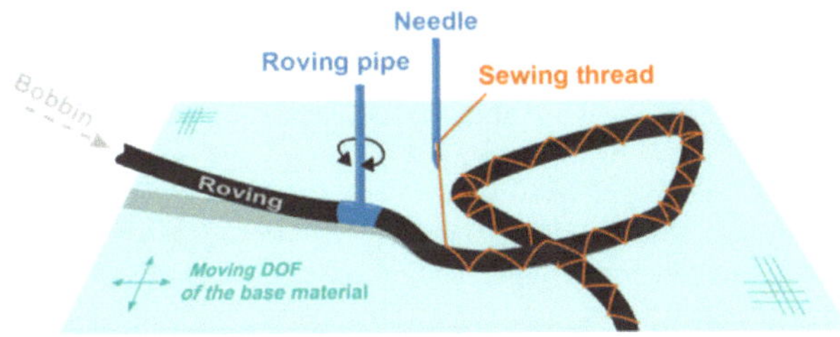

Figure 1. Sketch of tailored fibre placement in technical application [55], reprinted by Creative Commons Attribution (CC BY) license MDPI.

Considered as state of the art, the application of conductive inks and printing conductive lines on flexible substrates have been known since the earlier 1990s. In an example from 1994, a carbon-containing polymer was screen-printed onto a polymer foil for manufacturing a field effect transistor [61]. The panoply of conductive inks includes colloidal solutions and suspensions of conducting materials such as derivatives of carbon or conductive polymers. Solvents are water or organic liquids and contain additives such as a polyethene polyvinylacetate copolymer as a viscosity and stability agent. In an example, a spraying technique was used as a simple way of printing. A textile was sprayed on both sides with conductive ink. As the textile itself was utilised as an insulation layer, a flexible electrical capacitator was formed [62].

A more advanced technique is inkjet printing. Several factors influence the quality of inkjet printing. The surface tension of aqueous solutions is influenced by surfactants and the surface tension of the substrate can be influenced by different activation strategies, e.g., UV radiation, ozone or plasma [63]. An example for a conductive polymer-containing ink is an aqueous dispersion of PEDOT:PSS stabilised with DMSO and ethylene glycol for printing on a PET fabric to deliver the electrical energy for electroluminescence [64].

In general, inks for screen-printing need a higher viscosity than for inkjet printing. Because of the higher viscosity, it is possible to load the ink formulation with a higher conductive solid content. For example, an ink with a content of 60% silver flakes in a vinyl resin was used for screen printing on a textile fabric. The conductive layer was additionally coated with a protective polymer layer on top [65].

Printing methods can also be combined with other deposition techniques. Localised electroless copper deposition becomes possible by pre-printing the metallic catalytic seeds, mainly silver, onto polymeric substrates [66].

Apart from coating and printing, the application of metal deposition techniques obtains more attention with regard to the formation of conductive structures on textile surfaces. It is worth noticing that metal deposition in this context refers to the formation of a metallic layer on the atomic level such as sputtering and evaporation [28,67,68], galvanisation [69] and electroless deposition [70].

Copper and silver are dominating the metal deposition on fabrics. Normally, electroless copper deposition always needs a conductive pre-activation, for example, using formaldehyde as a reducing agent similar to as shown for yarns with silver nanoparticles as a catalyst [70]. Alternatively, copper can also be deposited by the reduction in copper ions with hydrosulphite. Subsequently, the layer is strengthened by electroplating in a copper sulphate solution [71]. Additionally, conductive polymers, e.g., PANI can be used as a pre-activation layer for a later copper deposition step [72].

3. Accelerated Functional Ageing

In order to investigate the ageing behaviour of functionalities and to be able to predict the usage lifetime and safety of e-textiles, accelerated functional ageing tests are often applied. Depending on the usage conditions, different types of accelerated functional ageing are discussed in this section: mechanical ageing, chemical ageing, ageing by washing and laundry and electrical ageing. The review also includes some examples of ageing behaviour of the conductive structure on non-textile substrates, e.g., flexible films, in order to highlight the general damage mechanism under usage conditions.

3.1. Mechanical Ageing

Functional ageing because of mechanical strain is one of the most important and most widely investigated ageing categories. Derived from common use, there is stretching, bending and abrasion. The mechanical ageing tests consist of cyclic treatments at a defined intensity. The damage does not occur immediately but is cumulative. Therefore, it is necessary to follow its emergence in order to investigate the causes.

3.1.1. Stretching

Stretching is possibly the most common applied mechanical stress to investigate the aging behaviour of e-textiles. However, in order to understand the ageing behaviour of e-textiles, their non-ageing behaviour (c.f. Section 5) needs to be understood as well. Both effects need to be considered separately in order to be able to explain the behaviour at the aged state. The gauge factor describes the relationship of elongation and electrical resistance (c.f. previous section). An increasing electrical resistance during elongation [73] is due to conductive particles or fractions of a cracked conductive layer moving away from each other [74]. The effect can be observed for conductive coated yarns or fibres by microscopy with an accompanied resistance measurement. An example was reported by Li et al. for a polyurethane core with a silver/graphene particle sheath and silicone protection coating [75]. Further studies used SEM imaging for the observation of the changes in conductive coating layers on textiles [76,77]. An example of the crack formation of the conductive layer and resulting increases can be seen in Figure 2 for PPy-coated Lycra fibres [77].

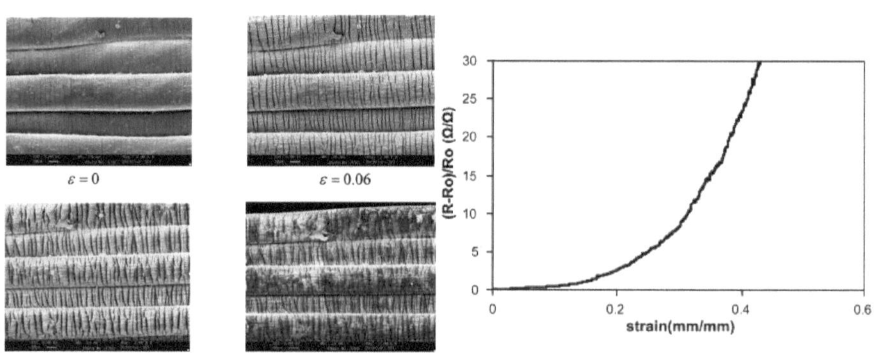

Figure 2. SEM images of crack formation (**left**) and electrical resistance (**right**) during stretching of PPy-coated Lycra fibres; reprinted from [77] with permission from Elsevier.

The role of the silver content in gauge behaviour was shown for screen-printed patterns of silver-coated polystyrene particles on a PDMS matrix. The resistance increased with elongation at a low silver content. A rising silver content led to a decreasing gauge factor. Hu et al. explained this observation by the Poisson effect that causes lateral contraction. This means, there is a reduction in the diameter of yarns or fibres while stretching. Hence, the conducting particles are moved closer together [78]. If the formation of new contacts predominates, the electrical resistance decreases during elongation. A descriptive example was published by Sadequi et al. by using "puffy threads" that were composed of polybutylene terephthalate dip-coated with a polydimethylsiloxane/carbon ink. Microscopic imaging during elongation showed decrimping and forming contacts of the fibres followed by a decrease in the electrical resistance [79]. Similar observations were reported for an electroless copper-plated cellulose fabric [9] (c.f. previous section).

The fabric structure plays a role in the non-ageing behaviour as well. Knitted textiles show three stages of stretching, dependent on strain: (1) slippage of the loops that causes variation in the contact points, (2) deformation of the yarn loops and (3) stretching of the yarns [80].

The methods of accelerated stretch ageing are simple and effective and, thus, favoured for simple durability evaluations. Frequently, cyclic stretch tests are aborted after a certain number of cycles, showing no or low damage. Examples for aborted tests are presented for carbon derivative-filled thermoplastic polyurethanes. They underwent 2000 test cycles of 10% or 20% of stretching with no ageing related change of the electrical resistance [81]. A carbon polydimethylsiloxane blend was stretched for 30 cycles at different strain rates without changes in conductivity [82]. Other examples are given for stretchable heat fibres consisting of a polyethylene core, copper nanowires and a silicone rubber protection coating [83]. However, also weft knitted and carbonised modal fabrics showed durability without ageing for 1000 cycles stretched up to 50% [84]. The aborted experiments show durability only for the performed cycle counts. However, as no analytical lifetime investigation is possible, it prevents comparisons with other conductive textile inventions.

A full break investigation requires the occurrence of damage. This is exemplarily presented for silver-sputtered plasma pre-activated PET fibres. Cyclic elongation to 5% led to an increasing mean electrical resistance and a wider distribution of the resistance values, i.e., a higher gauge factor, with rising cycle numbers. After 3500 cycles, first signs of material fatigue occurred. The treatment was stopped at 5000 cycles. A further failure investigation included SEM imaging, showing cracks and waves on the silver surface that were not present in the initial state [85].

Cyclic stretch durability testing is a possibility to compare different manufacturing methods. Koshi et al. investigated the influence of the permeation depth of a stretchable silver ink onto the ageing behaviour. Inks with two different viscosities were screen printed onto plain woven cotton fabrics. The printed conductive lines contained a gap with a resistor mounted with either an electrically conductive adhesive or a soldering paste. The different samples underwent cyclic elongation to 5%. The resistance of the e-textile was measured in situ. Beside the oscillation due to the gauge behaviour, the electrical resistance increased with an increasing cycle number up to a factor of 100. The e-textile with deep ink permeation and soldered contacting showed the best durability. SEM imaging illustrated the cracks and especially their propagation next to the contacts. Young's modulus of the mounting material is important for this type of connection geometry, to ensure the dissipation of the physical forces at the connection. Additionally, shearing forces were applied on the overhanging resistors. The soldered connections again showed the best durability [86].

The influence of relative textile direction and the dimension of printing patterns on functional aging was also investigated by Koshi et al. They printed a silver containing ink onto plain woven cotton fabrics parallel to the textile yarns and at a 45° angle between the warp and weft yarns. The width of the conductive lines was of either one or three yarns. All samples underwent cyclic elongation to 5% while measuring the electrical resistance with the four-probe method. The samples printed at a 45° angle fatigued faster because of the force distribution at the yarn crossings during stretching. The broader conductive lines survived longer due to the effect of the redundancy of pathways. Despite cracks in the coating, the current still had the possibility to take an alternative way around the damage. With the breakage of thin conductive lines, the electrical current is completely interrupted. SEM imaging after the first cycle and after the 100th cycle of stretching proved the explanations. They also showed an increase in number and width of the cracks for higher cycle numbers. The observations showed a high increase in resistance for the first stretch cycle after manufacturing. This is explained by yarn friction that hinders the displaced yarns from returning to the original state [87].

Another group investigated the fatigue behaviour of different meandering shapes using a double track approach. Before the realisation of material experiments, the behaviour of patterns was predicted by a finite element simulation. Figure 3 shows three investigated patterns—"elliptical" shaped, "U" shaped and "horseshoe" shaped. The simulation showed the distribution of the mechanical stress during stretching. For the material implementation, nickel and gold layers were encapsulated by PDMS films with the aid of a photoresist in the depicted patterns. The simulation as well as the cyclic load application experiments showed the best durability for the horseshoe-shaped patterns. The latter showed the best distribution of mechanical forces during elongation. A sample version with multi-track patterns consisting of four parallel thin tracks instead of one wide conductive line, showed still a higher durability. Despite the film model substrate, the effect of patterns on the mechanical ageing can be clearly demonstrated by both modelling and experimentally [88,89].

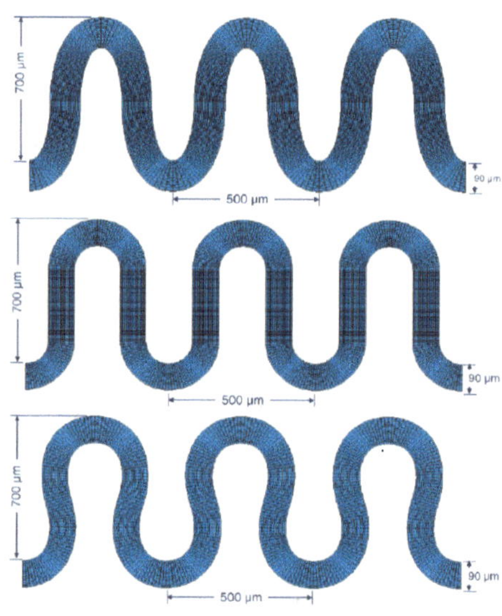

Figure 3. Investigated meandering patterns by Gonzales et al., elliptical, U-shaped and horseshoe-shaped; reprinted from [89] with permission from Elsevier.

Taking a similar approach on film substrate, Bossuyt et al. applied horseshoe-like patterns on different flexible substrates, e.g., copper foil on PU or PDMS, or silver-containing paste on nonwoven textiles. The maximum tensile load was determined ahead of durability tests. Then, the samples underwent cyclic load experiments with a statistical fatigue evaluation. Different elongation values were applied, i.e., 2.5%, 5% and 7.5% for the copper foil on PU and additionally 4% and 10% for the copper foil on PDMS. The lifetime was defined as 63.2% of the cycle number of total failure. The course of the resistance during cyclic elongation showed a sharp transition from functioning to failure, i.e., a "sudden fatigue". The numbers of cycles to failure were used for a statistical evaluation on the basis of Weibull functions. The evaluation demonstrates a cumulative damage behaviour for each level of strain (c.f. Section 4). Statistics revealed the assignment of the fatigue function to the Manson–Coffin relation for lower fatigue cycle numbers, i.e., high strain. Here, the deformation of the sample is plastic. Higher fatigue cycle numbers, i.e., lower strain, are assigned to the Basquin relation belonging to the elastic behaviour. A combination of both parts is stated as:

$$\frac{\Delta\varepsilon}{2} = \varepsilon_f(2N_f)^c + \frac{\sigma_f}{E}\left(2N_f\right)^b \tag{1}$$

with $\Delta\varepsilon$ as the total strain range in a cycle, N_f as the mean number of cycles to failure (=cyclic lifetime), c as the strain exponent, ε_f the as failure strain, σ_f as the failure stress and b as the stress exponent. The exponents and coefficients are material properties. The statistical results showed very good modelling of the fatigue behaviour and the potential for predicting the lifetime of e-textiles under defined conditions. Optical microscope imaging showed cracks of the conductive lines as material damage [90].

For application on a textile substrate, Isaia et al. knitted a blend of 20% Inox steel staple yarn and 80% PES to form a textile specimen, forming a kind of horseshoe-shaped structure. The specimens were elongated to 16% for 250 cycles with an in situ measurement of the electrical resistance by the four probe method. The course of the electrical resistance showed an increasing mean resistance in combination with an increasing gauge factor, both leading to saturation. This change was attributed to the shift of the contact points of the loops (c.f. resistor model). Microscope imaging in combination with a loop length measurement revealed a clear increase in the loop length comparing the initial and final state. As no material damage was found, the observations were identified as preconditioning. The preconditioning showed stability of the electrical resistance values with restarting the experiment, i.e., the preconditioning was permanent [91].

Other possibilities for enhancing the stretch durability of conductive textiles are pre-stretching during manufacturing and micro-patterning. On a PDMS film as a flexible model substrate, Jones et al. successfully reduced the aging affect by stretching the substrate during the manufacturing process to 25% elongation with a subsequent deposition of a chromium and a gold layer. For testing, samples were either stretched once to 100% or in slow cycles to 15%. The electrical resistance values showed a higher stability, i.e., lower value fluctuation, for the pre-stretched samples in comparison with the unstretched samples [92].

Another example of pre-conditioning was investigated by Graz et al. on PDMS film coated with a 50 nm gold layer. The samples underwent cyclic elongation to 20% with an in situ measurement of the electrical resistance. Initially, the resistance increased with elongation and decreased with relaxation. For higher cycle numbers, the peak shape of the resistance changed to a double peak, i.e., with a drop of resistance in its middle for high elongation. The cyclic elongation caused the formation of cracks and, thus, gold-layer islands. These islands were gaining distance while stretching. Lateral contraction led to new contacting and a lower resistance at high elongation. The breakage of the gold layer into micro gold islands stopped at a certain island size. The electrical properties then remained stable for a further elongation treatment. Taking this stability into consideration, the initial distress cycles can be applied as a pre-conditioning method to minimise later ageing during a possible application [93].

The possible recovery of conductivity after cyclic stretching can be performed by heating in certain polymer/conductive material blends. Inoue, Itabashi and Tada found this special effect for silver-containing polyurethane pastes with an 80% Ag content that were printed on textiles. The textiles underwent 10% elongation for 5000 cycles and showed an increase in resistance from 10^1 to 10^3 Ω together with an increasing gauge factor. Afterwards, the textiles were annealed at 100 °C for 10 min and the resistance showed a nearly full recovery, in the case of having no cracks in the conductive coating. Furthermore, the recovery process was repeatable. This kind of functional ageing is called reversible or recoverable functional ageing. As a possible mechanism of the observed regeneration, a rearrangement of the polymer structure was assumed [94–97].

3.1.2. Bending

Cyclic bending plays a similarly prominent role as cyclic stretching. The variety of bending methods is very broad. The bending angle and bending radius are two of the main parameters and low bending radii lead to sharp bending. Bending experiments sometimes also contain additional elongation. Bending, per se, is stretching at a gradient through the textile plane. In the inner part of the bending angle, the stretching factor becomes negative and the textile is compressed. Komolafe et al. used the basic beam bending theory for the calculation of the ideal position in a multilayer system to obtain the stretch factor to 0. The authors printed conductive tracks onto fabrics and encapsulated with polyurethane. The calculated thickness of the encapsulation shifted the conductive track into the calculated neutral axis [98]. Basically, the principle of the beam bending theory is based on the stress distribution over the sample cross-section during bending, i.e., one sample side is loaded with tension while the other side is loaded with compression, with the basic assumption of a homogeneous material. The application of the theory on textile substrate, however, has limitations, for example, because of the different elastic moduli of fabrics in tension and compression. Furthermore, conductive coatings also influence the very different movement behaviours of yarns and fibres in a textile structure (e.g., knitting vs. weaving), leading to changes in the elastic behaviour of the fabric substrates.

Guo et al. investigated accelerated aging by bending for Cu, Ag, Au and Ni electroless-coated PET and PI film substrates. Although the substrate does not represent a typical textile structure, it is a good way to separately investigate the ageing and damaging behaviour of the coated conductive layers. The polymers were activated by plasma and 2-(methacryloyloxy)ethyl-trimethylammoniumchloride (METAC) with the subsequent printing of a $(NH_3)_2PdCl_2$ catalyst ink. The specimen underwent bending tests with an angle of 180° and different radii. The samples were turned to stretch or compress the metal coating. The performance of 4000 cycles showed no resistance change for most samples [99]. In another work of Heo et al., silver nanowires were drop-coated on a PES-Spandex textile composite with PDMS to create a stretch sensor. No aging affect was observed after 2000 bending cycles with an angle from 0° to 150° [100]. These examples on polymer films demonstrated the use of the accelerated aging test method to prove the reliability of conductive coatings within a certain strain range. As no ageing occurred, the reliability is only valid for the performed low cycle numbers and no further damage or statistical evaluation is possible.

Stempien et al. printed a silver ink with a valve jet print head onto polyacrylnitrile, polyethyleneterephthalat, basalt, cotton, cotton/PET blend, aramid, wool and polypropylene (nonwoven) followed by a curing step at 130 °C. The samples were attached to a moveable and a fixed clamp and bent for ±30° out of plane for 50 k cycles. The edge of the moveable clamp functioned as a sharp bending line. The resistance was measured in situ by contacting the clamps. Rearrangements in the yarn system increased the electrical resistance for low cycle numbers and crack formation increased the electrical resistance and gauge factor for higher cycle numbers. The beginning of a significant increase in the electrical resistance was defined as the fatigue of lifetime. It ranged from 800 cycles for the polyacrylonitrile substrate to 50 cycles for the PET substrate [101].

De Kok et al. performed three related mechanical tests, i.e., stretching, bending and dynamic flex testing with different kinds of samples. For a start, a 1 µm Ag-coated nylon 66 single yarn was stretched cyclically. Increasing the cycle numbers led to an increasing electrical resistance and an increasing gauge factor. The increasing resistance was explained by the ongoing delamination of the silver coating, i.e., a loss of conductive material. For the aged state, this also meant interruptions of the conductive layers on the fibres and, by this, an interruption of the percolation. The increasing gauge factor was explained by the lateral contraction of the yarns while stretching, moving the fibres in the yarns together. The newly formed contacts lowered the resistance. This phenomenon is more effective for the aged state, as gaps in conductive layers of single fibres are bypassed by parts of functioning neighboured fibres. The stretching experiment was repeated with PU-coated filaments of the Ag-coated yarns. The additional coating prevented the formation of new contacts by lateral contraction. The electrical resistance increased with elongation contrary to the decrease for the unprotected samples.

The conductive yarn was applied on PET nonwoven fabrics by stitching embroidery and by TFP. For bending tests, the specimens were attached to springs at both ends. With a speed of 38/min a rod of 10 mm diameter deflected the middle of the sample to 80 mm below original height with a force of 20–25 N. In this process, the textile was elongated by about 0.5%. A total of 5000 cycles were performed with a 10 min interruption after the 500, 1000, 2000 and 4000 cycles. The course of the resistance showed an increasing mean resistance together with an increasing gauge factor. Both led into a saturation level. The 10 min interruptions showed an instable short-term recovery of the resistance. Among the sample manufacturing variety, TFP showed a better ageing performance than stitching embroidery.

Dynamical flex testing is a derivative of bending tests. The bending angle is fixed at 90°. The specimen is pulled over a corner, consisting of an insulated roll of 20 mm diameter. By this, bending effects the whole textile. A pulling force of 12.5 N was applied. One whole cycle comprised rolling back and forth. Within cycling, the experiments were paused with a regeneration time and initiated again. For flex testing, a silver coated nylon yarn was embroidered into different patterns. No correlation between the shape of the patterns and functional ageing was found. The behaviour of the electrical resistance in dynamical flex testing showed a similar trend as the bending ageing. After the experiments, the yarns were extracted from the textile for evaluation. Stitching embroidery exposed the yarns to the roller. This led to more damage on the inside than on the outside. The TFP textile showed less damage for two reasons. The yarn was mechanically less strained during fabrication and the yarn was placed on the outside of the roller during testing [50].

De Vries et al. applied the deflection method described above for the e-textile contacting investigation. For sample preparation, an LED was mounted on a siloxane substrate on a plain woven cotton fabric with a silver-coated copper wire power supply. A second fabric was attached on top with a cut-out for the LED. In a second version, a hard polycarbonate shell was attached on the LED before covering with the second textile. In the third version, a third textile layer was embedded between the hard shell and the power supply. The specimens were deflected with a polymer rod with gaps not to touch the LEDs. The electrical current was measured in situ. The samples fatigued with a sudden drop in the current. The cycle number until fatigue was used in a statistical lifetime evaluation with the help of Weibull distribution for the determination of the unreliability *F(t)*:

$$F(t) = 1 - e^{[-\frac{t}{\eta}]^{\beta}} \tag{2}$$

The shape parameter β is defined by the type of failure, i.e., early failure, random failure or wear out failure. η as the scale parameter is the number of cycles to failure. The location of breakage was analysed by X-ray imaging. The first sample showed the shortest fatigue lifetime with a breakage at the soldered connection next to the LED. The encapsulation stiffened the textile and shifted the point of inflection and breakage from the soldering connection to the edge of the encapsulation. The third version with the textile inlay showed the best fatigue behaviour, as the hard edge of the polycarbonate capsule was softened by the additional textile layer [102].

The previous described procedure was conducted in a variation. ELEKRISOLA litz wires were woven into a polyester fabric with floats for contacting LEDs with a conductive adhesive. The deflection method was applied with a frequency of 40 /min and different magnitudes of deflection. The breakage of the wires led to a sudden drop in the current. The bending radius was estimated as the function of deflection, bending angle and geometry. By this, the connection of the number of cycles to failure with the bending radius was drawn. The Manson–Coffin relation provided a mathematical evaluation for the high cycle fatigue problematics with the possibility of the calculation of the remaining lifetime [103].

With a decrease in the radius and an increase in the bending angle, bending becomes folding [99,104]. Thus, folding can be described as a "severe form of bending" [105]. A simple folding method was conducted by Cai et al. for cysteine-functionalised cotton fabrics with adsorbed Ag nanoparticles and a hydrogel protective coating. Folding was performed by hand, with an additional 100 g weight placed on the fold line. This method was cycled 80 times. The electrical resistance increased until reaching a saturation value. The experiment showed the positive effect of the protective hydrogel coating onto the electrical lifetime [106].

The hand-folding method was also applied in a variation by Zahid et al. on a plain woven cotton fabric, spray-coated with PEDOT:PSS and graphene nanoplatelets. The specimen was folded 180° and a 0.5 kg metal weight was placed for 1 min. After bending to 0° and relaxing for 1 min the electrical resistance was measured. The folding was repeated for 20 cycles showing an increase in the electrical resistance. Small cracks in the folding line were observed with an optical microscope [105].

An example for the mechanisation of folding is shown by Lim et al. An impacting machine with a crank applied a pressure of 15 kgf/cm^2 onto a folded fabric at its minimum height excursion. A textile capacity pressure sensor was prepared from a PEDOT-coated nylon/polyester microfibre fabric. The sensor was folded for 1000 cycles and tested with a reference weight in between [107].

Lee et al. used a folding endurance tester. The specimen was attached to the machine at three points. The first attachment was at the static part of the machine, the second in the middle of a rotating disk and the third at the edge of the rotating disk. The disk rotated back and forth to cover nearly 360°. A conductive silver ink was printed on PU-coated cotton fabrics and encapsulated. The treatment with the MIT folding endurance tester showed a clear ageing advantage for the encapsulated samples [108].

Stretching and bending showed in some parts similarities. As mentioned, bending contains some kind of stretching, mostly with a gradient. Hence, some conductive textiles show the similar resulting damage formation. The elongation in combination with different Young's modulus leads to at least breakages of the conductive layers.

3.1.3. Abrasion

Abrasion is the third important part of the mechanical stress testing on e-textiles. The test methods are simple in execution and mainly used to prove the durability of inventions, often accompanied with other accelerated ageing methods. Examples for simplified abrasion experiments on conductive textiles are given in [109–113].

Stempien et al. investigated the effect of the abrasion direction onto printed conductive lines in dimensions of a 14 cm length and 3 mm width on different substrates. The rubbing finger of an AATCC crock meter for testing colour fastness was prepared with a cotton cloth as the abradant. The textile was abraded for 2000 cycles in the printing direction and crosswise. The abradant was replaced every 200 rubs. The electrical resistance was measured in situ. The lengthwise abrasion caused a faster fatigue due to a higher loss of material. Additionally, the integration of the conductive ink into the substrate material played a role for the rubbing durability of the conductive lines [101].

The influence of plasma pre-treatment on the abrasion stability of the coated textile was investigated by Montarsolo et al. They utilised radio frequency-induced low-pressure plasma to activate a woven PET fabric at two different power stages. Additionally, reaction times and the process gas (oxygen and argon) were varied. The pre-activated fabrics were coated with polypyrrole. Martindale abrasion testing according to EN ISO 12947 was performed with samples of a 15 cm diameter and a pressure of 9 kPa for 10,000 cycles. Warp and weft yarns were examined separately. Despite their different absolute electrical resistance values, the same ageing behaviour was observed for both directions. The resistance graphs showed a fast increase for low cycle numbers, leading into saturation. The plasma-treated samples outperformed the untreated ones. Samples activated under harsher conditions showed the best abrasion durability for both fabric directions (Figure 4) [114].

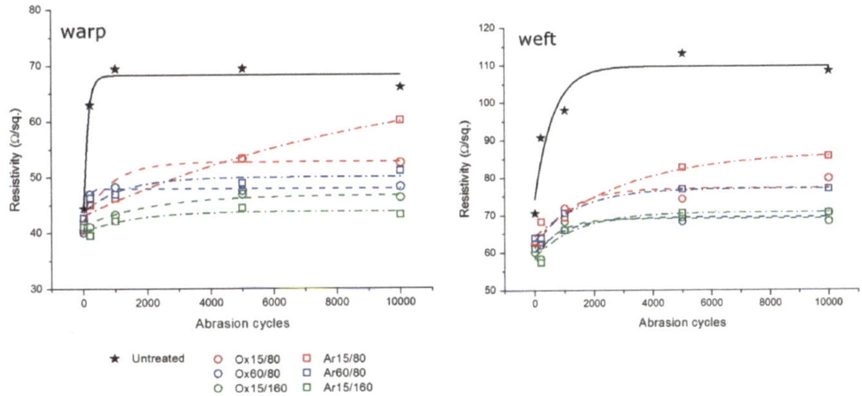

Figure 4. Behaviour of the electrical resistance in warp direction (**left**) and weft direction (**right**) during abrasion of PPy-coated polyester (PET) fabrics [114]; reprinted with permission from Wiley.

Further examples for the application of Martindale abrasion testing on e-textiles are described in [115–117].

An increase in the abrasion harshness due to the textile–sandpaper friction was performed by Bogan et al. The researchers tested a woven nylon fabric with a different content of twisted enamelled copper yarns in the warp direction. The Martindale method with a load of 9 kPa and standard worsted wool abradant did not cause damage over 50 k cycles. A Wyzenbeek apparatus was used for the heavier-duty stress. An 800-grit sandpaper functioned as the abradant and was applied onto the tensed sample. The electrical resistance was measured in situ. The course of the resistance showed sudden fatigue, i.e., an interruption of the conductivity. A lower number of conductive yarns per track and a lower content of non-conductive yarn in the twisted yarns showed an earlier lifetime fatigue. The failure of the conductive lines appeared together with the failure of the textile substrate [118].

3.2. Chemical Ageing

Chemical ageing is considered when exposing e-textiles to chemical environments, e.g., water and aqueous solutions of detergents or bleaching agents during the wash/laundry process. However, chemical ageing also comprises the influence of oxygen from air or the attack of acids, bases or solvents. The influence of the wash/laundry process on the functional aging will be discussed later in Section 3.3 as the process comprises both mechanical and chemical ageing.

However, the storage on the air atmosphere already affects oxygen-sensitive materials such as conductive polymers. For storage stability testing, Maity et al. coated a nonwoven textile substrate with polypyrrole. Storage at 25 °C and 65% RH for 18 weeks showed an increase in electrical resistance of approximately 10%. The ageing was attributed to de-doping of the conductive polymer because of oxidation [119]. Moss et al. [120] and Tabaciarova et al. [121] investigated the oxidation reaction of polypyrrole by means of XPS and IR spectroscopy. Additional carbonyl, aldehyde and hydroxyl functional groups were found on the polymer. The low activation energy of free radical mechanisms led to the reaction with oxygen from air. A high oxidation sensitivity during storage on air was reported for PANI, showing an increase from about 500 Ω/5 cm to a saturation level at 3000 Ω/5 cm [22]. Conductivity loss because of oxidation is also reported for PEDOT:PSS [122].

Relative humidity and temperature define the storage climate. Varesano et al. varied both to find out about their impact onto the defunctionalisation of polypyrrole-coated PET fibres. Whereas the change of the relative humidity had no or only a low impact, an increased storage temperature up to 80 °C yielded in a fast gain of electrical resistance.

In counter, storage at −28 °C did not significantly slow down the ageing in comparison to ambient conditions [123]. A protective coating was able to prevent oxidation by air, e.g., shown for liquid latex [124,125].

Storage ageing on air also impacts copper particles deposited on functionalised yarns. Liu et al. showed a decrease in electrical conductance of 90% within 7 days attributed to the oxidation of the metal [126]. They also showed a limitation in the increase in the electrical resistance onto the first 48 h after manufacturing, with a remaining constant saturation value over two weeks [127]. Hwang et al. printed ink containing copper nanoparticles or copper nanoparticles with additionally multi-walled carbon nanotubes onto a polyimide foil using the doctor blade method. The ink was flashlight sintered by xenon light. The samples were stored on air at 25 °C and 30% RH over a period of 30 days. The electrical resistance increased over time due to the oxidation of copper to copper(I)-oxide that was found by an XRD analysis. The ageing was accelerated by increasing the temperature to 85 °C and relative humidity to 85% over 14 days, showing a faster gain of electrical resistance [128].

In order to investigate the influence of acid, base and salt on ageing, Ali et al. immersed electrochemical copper-deposited fabrics in 0.3 M hydrogen chloride, 0.3 M sodium hydroxide or 50 g/L sodium chloride for 24 h. All solutions increased the electrical resistance with hydrochloric acid having the highest impact with 50% of increase. The impact of the sodium hydroxide solution onto the electrical resistance was lower, followed by the sodium chloride solution [71].

To test the protection efficiency against chemical ageing Jia et al. coated a tricot-knitted textile with silver nano-wires and a polyurethane protective layer. Samples of the conductive textiles were immersed in sulphuric acid at pH = 1 at 25 °C for 20 h. Unprotected samples nearly tripled their resistance and decreased the EMI shielding about one third in the range of 8 to 12.5 GHz, while the PU-protected samples showed no changes (Figure 5) [129]. An alternative for polyurethane protection is polydimethylsiloxane [130].

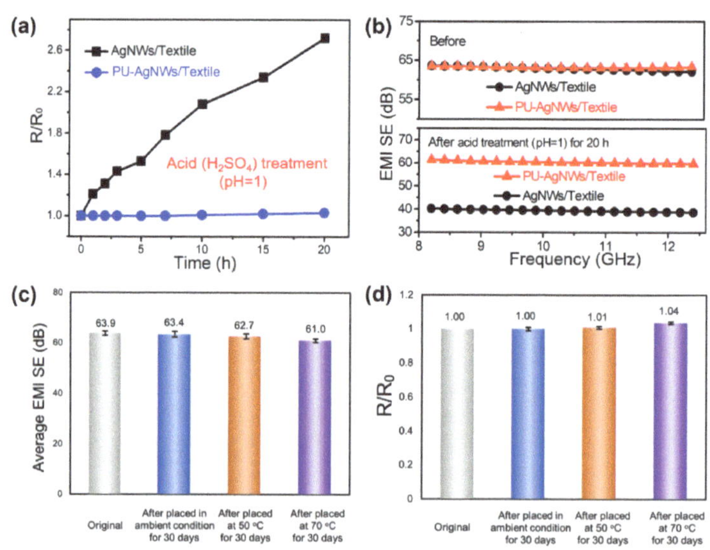

Figure 5. Effect of chemical ageing on the electrical resistance of silver nanowire/polyurethane-coated fabric: (**a**) relative resistance (R/R0), (**b**) electromagnetic interference (EMI) shielding, (**c**) EMI shielding before and after placed at different conditions, (**d**) R/R0 before and after placed at different conditions [129]; reprinted with permission from Wiley.

Erdogan et al. tested the impact of acids and bases onto polythiophene and polyaniline-coated PET yarns. The samples were immersed into hydrogen chloride or ammonia solutions of pH 5 to 11 at RT. After 1 h, they were rinsed and dried at 50 °C. The electrical resistance of the polyaniline coating strongly increased, especially for higher pH values. A mixture of polyaniline and polythiophene showed the lowest electrical resistance for pH 9 and below. The values ranged in the area of 10^6 Ω, as the conductive polymers were not doped. The resistance changes were explained by protonation phenomena in the pi-system [131].

Considering the exposure of wearables to body sweat, Priniotakis et al. manufactured woven, nonwoven and knitted ECG electrodes from stainless steel fibres. The electrodes were attached to a PTFE membrane with pores of 5 μm, simulating the skin. A mounted PVC tube contained artificial sweat. The electrochemical impedance between the electrodes was recorded. For short-term experiments, differently concentrated sodium chloride solutions showed a reduction in the electrical resistance. The artificial sweat for long-term experiments contained 50 g/L NaCl, 1 g/L urea and 0.5 g/L of other salts. The pH value of 5.8 was adjusted by NaOH and HCl. In the first 100 h the resistance increased very fast from 250 Ω to about 400 Ω, leading into a saturation for the next 400 h [132].

Another important aspect of chemical ageing is the influence of organic solvents, for example on conductive polymer coatings. Gosh et al. immersed a PEDOT:PSS-coated cotton fabric for a not closer stated time into boiling water, ethanol, toluene and tetrahydrofuran. Afterwards, the EMI shielding was tested in a range of 8 to 12.5 GHz. Boiling water lowered the shielding values slightly. The other solvents did not show a change compared to the initial state [133].

Han et al. immersed polypyrrole-coated aramid nano-fibrils into acetone, tetrahydrofuran, ethanol, dimethylformamide, glycol and glycerol. The electrical resistance was measured after 24 h and 175 h. The solvents showed a low impact. The highest effects occurred for acetone after 24 h and for ethanol after 175 h which increased the electrical resistance with a factor of about 2.5 of the initial state [134].

Chemical ageing has by far not been investigated as exhaustively as mechanical ageing, especially considering the long-term effects of the impact of water or sweat being important for wearables. However, chemical ageing plays a big role in wash ageing too, as there is an impact of water, detergents and additives.

3.3. Ageing by Washing and Laundry

Washability of conductive textiles is a striking challenge. It contains several single ageing compartments. Water, detergents and additives cause chemical ageing, and motion causes bending and abrasion. Several approaches and test methods have been developed to simulate the durability of e-textiles during the washing process.

In simple methods, conductive textiles or sensor specimens are immersed and stirred in water or detergent solutions. The results are difficult to transfer to machine washing [135]. As an example, gold-coated knitted polyester fabrics were stirred for 1 h in laundry detergent and dried at 60 °C overnight. The testing was stopped after a few cycles with no ageing effect [136]. Another example for beaker wash testing is shown for a PEDOT:PSS coating. The testing performed three cycles [137]. Additionally, the handwashing of copper-deposited samples [138] or the usage of an oscillating dyeing machine for a reduced graphene oxide-coated fabric [139] were described. All these methods, as described, do not achieve the harshness of a domestic machine wash.

In a different approach, Schwarz et al. coated activated para-aramid yarns by copper deposition and applied the wash durability testing ISO 6330:2000 standard for an ageing study. A domestic washing machine was loaded with the samples and 2 kg of cotton fabrics as dummies. The load underwent 25 washing cycles at 40 °C for 40 min each. The electrical resistance was measured by the four-probe method after 5, 10 and 25 cycles. The values increased, but the fabrics were still considered as functioning. SEM imaging showed damaging and the delamination of the conductive layers [33,140].

An example on damaging of conductive coatings and changes in resistance is given by Schwarz et al. [140] in Figure 6 for gold-coated aramid yarns after several washing steps.

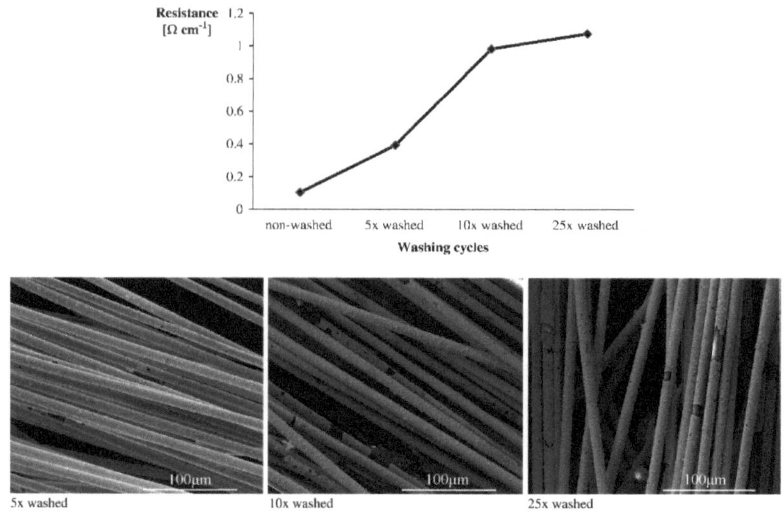

Figure 6. Increase in electrical resistance (**top**) and damage of coated conductive layer (**bottom**) after several washing cycles of gold-coated aramid yarns [140] reprinted with permission from Elsevier.

A further example for the application of the ISO 6330 standard is given for ECG electrodes composed of different PEDOT:PSS-coated fabrics. The electrical resistance was measured after each washing cycle. Despite an increase in the resistance, samples showed durability for 50 washing cycles [141]. In a similar simplified wash durability test, samples underwent cyclic washing and damages were visualised by imaging afterwards [109,142]. Additionally, standardisations from the American Association of Textile Chemists and Colorists (AATCC) were used for wash durability testing [143,144].

Considering washing/laundry as a combination of chemical and mechanical ageing, both processes contribute to the ageing of e-textiles. The effect of washing and its simulations onto silver-coated yarns was extensively investigated by the working group of Vladan Koncar. Zaman et al. used commercial and self-made textile ECG electrodes for the investigation in both ageing compartments. The commercial electrodes consisted of knitted polyester fabrics with a copper, copper/nickel or silver-coating. The self-made electrodes consisted of embroidered silver-coated yarns [145]. The resistance was determined by a four-probe measurement. Wash durability tests were performed according to the ISO 6330 standard domestic washing. The programmes "silk" and "express" were used with the main difference in the rotation speed of 15 rpm and 38.5 rpm. The spinning intensity was 400 rpm and the runtime 35 min at 40 °C. The amount of detergent was 20 g for a concentration of 1.25 g/L for a 2 kg load. In total, 50 wash cycles were carried out. The experiments were repeated four to six times.

The simulation of the mechanical part of wash ageing contained abrasion and pilling tests. The abrasion experiments were performed on a Martindale apparatus with a standard wool abradant, 9 kPa pressure and 10k back and forth rubs. For the pilling tests, the electrodes were embroidered to cotton fabrics and attached to tubes. The tubes were treated at ambient conditions at 60 rpm for 10k cycles in an orbiter pilling and snagging tester. The simulation of the chemical ageing consisted of a motionless immersion into solutions with a detergent concentration of 1–1.25 g/L. The time periods were varied from 30 min to 72 h at a temperature of 40 °C.

The silver-coated electrodes showed no ageing for 50 wash cycles for the "silk-wash". The other samples showed a slight increase in the electrical resistance. In the "express-wash", the silver electrodes were stable as well. The copper and nickel coated samples fatigued within the first 10 cycles. Frequency scans from 0 to 100 Hz proofed the functionality of the surviving samples. SEM imaging documented the damage as the delamination of the metal coatings. In comparison, the silver-coated self-made electrodes also showed wash ageing stability.

The Martindale abrasion testing over 10k cycles showed no gain of resistance for all commercial samples. The friction forces were too weak to delaminate the metals from the polymer. The self-made electrodes showed a slight increase in resistance. In the pilling tests, they showed no ageing as well. The immersing tests showed no impact on the electrical resistance of the silver-coated electrode. The copper and nickel-plated fabrics showed a small increase in the electrical resistance. It was concluded that the copper and nickel layer were oxidised during immersion and, thus, their adhesion to the polymer surface decreased [146].

A similar approach for the chemical part of wash ageing according to ISO 6330 was performed by Ismar et al. for commercially available silver-coated polyamide yarns. The proceeding delamination of the conductive coating was demonstrated by increasing the IR absorption of the polymer substrates [147].

Zaman et al. also performed a further approach for the simulation of the mechanical part of wash ageing. It was performed similar as in the previously described simulation with Martindale abrasion and pilling tests. Two and three-ply yarns were stitch embroidered on plain woven cotton fabrics with different stitch lengths in straight or zigzag lines and as single or three-line conductive path. The yarns were contacted with snap buttons in combination with silver paste. The machine wash treatment showed the best durability for the three-line embroidered sample. Among the ply yarns, the silver-coated yarn plied with two non-conductive yarns showed the slowest increase in resistance. Harsher conditions of "express-washing" were reflected in a faster gain of electrical resistance. The abrasion showed a resistance gain for all samples. The authors established correlations between the course of the resistance of machine-washing and abrasion testing. It was estimated that 1000 abrasion cycles could substitute eight or nine "silk-washes" and five or six "express-washes". Similar results were obtained for the pilling box test. About 1000 cycles were meant to substitute four "silk-washes" or three "express-washes". Non-fitting correlations, i.e., for a three-ply conductive yarn, were not taken into account for modelling. Microscopic imaging revealed a delamination of the silver coating as material damage [148].

Atakan et al. recombined the simulation of the mechanical ageing and chemical ageing in wash durability testing. Silver-coated polyamide yarns were subjected to Martindale abrasion in the dry state, soaked with water or soaked with an aqueous detergent solution. The addition of water to abrasion testing accelerated the delamination of the silver layer. The ISO 6630 conform addition of detergents increased the ageing even more, especially for commercially available detergents. The commercially available detergents contained particles that increased friction. The authors concluded that soaked abrasion is a suitable simulation for predicting the machine-washing lifetime of e-textiles [149].

Ankhili et al. tested the wash durability of protections and embroidery techniques against EN ISO 6330 conform domestic machine-washing. Silver-coated polyamide yarns were embroidered in different patterns. Beside convoluted zigzag lines that were crossing each other, two different types of protection were tested. The embroidered conductive lines were either coated with thermoplastic polyurethane (TPU) in a hot-press or covered with protective embroidery with a non-conductive polyester yarn. After 50 cycles of washing at 40 °C followed by 24 h of drying on air, the samples were investigated for change in electrical resistance, SEM-imaging and signal to noise ratio (SNR) in the frequency band 0 to 100 Hz. The unprotected zigzag-embroidered samples showed early fatigue. SEM imaging revealed the delamination of the silver layers as the cause of fatigue for the unprotected samples. The TPU-protected samples and the embroidery-protected samples survived the performed washing cycles with a slight increase in the electrical resistance. The embroidery-protected samples showed the best wash durability. This was led back to a better drying performance after washing. The penetration of water into the TPU-coated samples caused chemical ageing with a prevention of a fast evaporation due to the TPU coating. The electrodes surviving the machine wash testing showed no quality loss for the ECG signals measurement [145].

For bright clothes, bleaching agents with oxidising properties can be applied. Gaubert et al. knitted silver-coated nylon yarns together with cotton yarns into a double jersey fabric. The samples were washed with a bleaching agent containing detergent and with a bleaching agent-free detergent. A total of 30 washing cycles was performed in a domestic washing machine at 30 °C in the "delicate" programme. The samples were examined by electrical resistance, SEM imaging and colour lightness with a spectrophotometer. The lightness of the bleaching agent additive-washed samples decreased with increasing the cycle number. This was due to the formation of dark-appearing silver oxides. The samples washed with detergents only showed a much smaller decrease in lightness. The electrical resistance of the bleaching agent-washed samples was several magnitudes higher than for the reference. SEM imaging and EDS measurements showed the delamination of the silver layer. It was assumed that the formation of silver oxide decreased the adhesion forces on the polymer surface [150]. Similar results were obtained in a study using sodium percarbonate solutions in machine washing with and without detergent. The combination of oxidative chemical ageing and wash ageing led to faster fatigue, rather than each alone [151].

In order to increase the mechanical intensity, a steel-ball-aided wash ageing was investigated by Karim et al. with reduced graphene oxide-coated woven cotton fabrics. Machine wash ageing was simulated by the standard BS EN ISO 105 C06 A1S. The conductive textile was immersed into an aqueous solution of detergent with a concentration of 4 g/L. In total, 10 steel balls were added. The vessel was turned at a temperature of 40 °C for 30 min. The experiment was performed for 10 times and after each cycle, the sheet resistance was measured. It showed an increase with a high slope at the beginning of the treatment [152].

Apart from conductivity, Toivonen et al. investigated the effect of washing onto the performance of textile antennas. The antennas were manufactured by the embroidery of silver-plated yarns onto cotton fabrics. The water compatibility of the antenna was tested by soaking and drying afterwards. Every four minutes during drying, the read range was measured from 700 to 1200 MHz. The read range level was reduced in the wet state due to the high dielectric constant of water. Additionally, the maxima were dulled and shifted to lower frequencies. During drying, the performance of the antenna recovered to its original state. Three different machine washing programmes were performed: 40 °C, low spinning and a low amount of detergent, 60 °C, middle spinning and a middle amount of detergent and 95 °C with high spinning and a high amount of detergent. Both latter wash programmes led to a fast reduction in the transmitted power for different frequencies. The authors concluded that a protective coating was needed for the maintenance of the antenna performance over several washings [153].

A detailed comparison of wash durability testing methods was performed by Rotzler et al. Here, the different standards from, e.g., ISO, DIN and AATCC were discussed with respect to their suitability to e-textiles. The comparison also included different interpretations of the norms and, as a conclusion, the authors clearly addressed the need of test standards for e-textiles [154].

3.4. Electrical Ageing

Electrical ageing takes place due to the electrical operation of a device. It only plays a minor role in the endurance testing of e-textiles. Nagaraju et al. coated a commercially available copper-plated PET fabric with nickel–cobalt hydroxide nanosheets. In order to test the electrical current changes, galvanic or galvanostatic charging and discharging was applied. The sheets were cycled with a current density of 10 A/g in a voltage range of 0 to 0.5 V in a 1 M KOH solution. In total, 2000 cycles were performed and the specific capacity was investigated. It decreased from 1400 F/g to 1200 F/g, which was considered as a good stability [155].

A further example for an electrical testing method is given by Zhao et al. for super capacitators. The capacitators were manufactured form-reduced graphene oxide or polypyrrole blended with reduced graphene oxide-coated nylon fabrics, attached to nickel foam and separated by a filter paper. The samples were immersed into a 1 M Li_2SO_4 solution and cyclovoltammetry was performed in a range from 0 to 0.8 V. The capacitators containing only the reduced graphene oxide showed a stable performance over 2000 voltammetric cycles at about 90% of relative capacitance retention at 50% elongation and about 80% capacitance retention at 0% elongation. The coating of the reduced graphene oxide/polypyrrole blend showed a continuous decrease at both elongation states in the course of 1000 voltammetric cycles to 80% of its initial capacitance (Figure 7) [156].

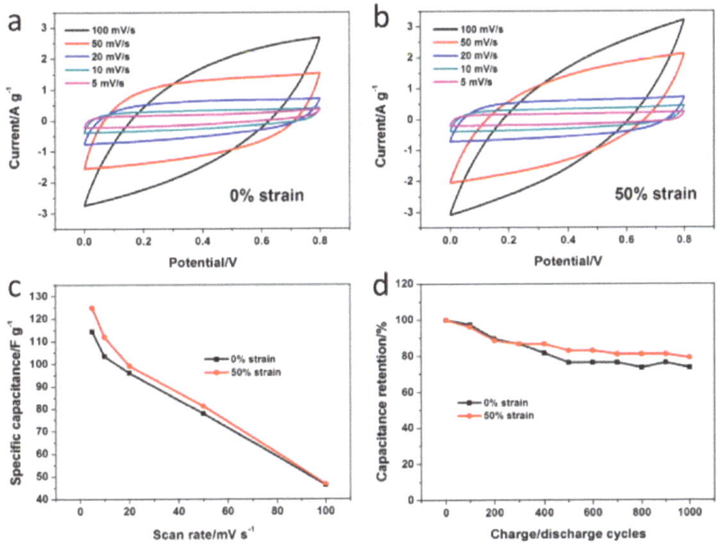

Figure 7. Depiction of voltammetric (CV) curves of graphene oxide/PPy-coated Lycra fabric for supercapacitor application: (**a**) 0% strain (**b**) 50% strain, (**c**) specific capacitances of the at 0% strain and (**d**) capacitance retention of the supercapacitor at 0% and 50% strain. [156]; reprinted with permission from Elsevier.

Another example is spray-coated fabrics with activated carbon ink subjected to voltammetric cycling from 0.8 to −0.8 V over 15,000 cycles. The relative capacitance showed an increase of about 5% for the first 4000 cycles and a slight decrease for the last 2500 cycles [62].

Hana et al. manufactured a copper oxide resistive switch. It consisted of a copper wire woven with a copper oxide-coated copper wire and a platinum spacer between the wire crossings. The switch persisted in a high resistance state until a certain voltage was applied. It then turned into a low resistance state and the current passed. It was used for data storage at a high voltage and data read out at a lower voltage. The presented textile was stable for 1000 readings and 30 switches but the experiments were not continued until ageing or failure [157].

The electrical ageing of conductive textiles was not investigated in depth and only plays a minor role in the literature on e-textiles. This is possibly due to small ageing effects compared to the strong forces of mechanical strains or reactive chemicals. It has not been clarified if electrical ageing is a real cause or just an indicator for other effects such as oxidation.

3.5. Challenges in Accelerated Functional Ageing

Overall, activities on mechanical ageing took a big share of research in the accelerated functional ageing of e-textiles. This included major typical loading types in textile application: stretching, bending and abrasion. There is a big difference in mechanical aging behaviour, depending on which technology is applied to the integration of conductive structures into textiles. While embroidery, e.g., TFP with a defined pattern or knitting technique may give a very good stretching behaviour of fabrics, these techniques show limitation in miniaturisation, i.e., generating conductive structures at a fibre level. Metal deposition, on the contrary, may provide a very thin, flexible conductive structure on fibre and textile surfaces. It would, however, need good protection coatings against abrasion and pre-conditioning for a better stretching behaviour. This makes the efforts in modelling more difficult, also when a taking substrate structure (knitted, woven) into consideration.

The biggest challenge in chemical ageing and ageing against washing and laundry is the combination of damages of both conductive structures and textile substrates. Furthermore, washing and laundry also consist of mechanical stresses, so losses in functionality during laundry for example need to be considered as cumulative results of different ageing processes.

Electrical ageing is the least investigated subject in the literature compared to other ageing processes. However, when considering possible electrochemical corrosion during (electrical) operation, this type of ageing deserves higher attention and needs to be investigated with a much higher focus in the future.

At practical application conditions of e-textiles, it is very often, that functional ageing is a combination of all mentioned ageing processes. It remains a big challenge to separately investigate each mechanism independent from each other. Thus, a helpful approach would be the cumulative damage modelling to obtain a combined view on the functional ageing behaviour of e-textiles.

4. Modelling of Functional Ageing

To provide a better understanding and forecasting the functional aging process, a mathematical modelling approach is applied. As conductivity is one of the dominant features of e-textiles, electrical resistance or conductance are often used as measures to build models to describe the functionalities (Section 4.1). In another approach, fatigue damage modelling is applied, considering ageing as a cumulative measure over the usage lifetime (Section 4.2).

4.1. Resistance Modelling of Conductive Textiles

Theoretical modelling of the electrical resistance in conductive textiles is difficult because of the textile structure and the frequently occurring gauge behaviour. However, there are some efforts in modelling of conductive textiles, e.g., the resistance of e-textiles is modelled as a resistor network (c.f. Section 2). Further examples for modelling of the non-ageing behaviour can be found in the following literature [48,49,78,80,158–161].

The formation of conductivity in heterogeneous material systems is explained by the percolation theory. The resistance is dependent from the volume fraction of the conductive material in relation to the non-conductive material. Therefore, recording and modelling of a percolation curve is a useful tool in manufacturing conductive fibres, yarns and textiles. This helps to find a fitting content of conductive material within the insulating substrate material and to compare manufacturing methods [162]. An important parameter is the critical volume fraction, which is also called percolation threshold. At this content of conductive material, a conductive path is formed and the resistance drops by some orders of magnitude within a small increase in the conductive material content. The percolation theory in combination with Poisson's law was used by Hasan et al. to model the course of the electrical resistance during the elongation of a silver particle-coated polyether ether ketone fibre. The change in volume during stretching caused a change in the volume fraction of the silver particles, followed by a change of the electrical resistance [163]. Often, the percolation theory is also drawn for the explanation of evolving damage. A loss of material lowers the surficial concentration of conductive material towards the percolation threshold. This additionally clarifies that not only the concentration, but also distribution is important for percolation.

Empirical electrical resistance modelling of conductive textiles requires the comparison of the resistance at an application condition to the resistance at defined conditions. In order to investigate the influence of a single condition onto the resistance, only this single parameter needs to be changed. An example is shown by Xue et al. for PPy-coated polyamide 6 and lycra fibres. Herein, temperature, moisture, elongation and number of stretch cycles were changed separately, all influencing the electrical resistance. Consequently, ageing caused by both, the strain rate and cycle numbers, was included into the empirical mathematical modelling of the ongoing resistance of the tested yarn [164].

4.2. Cumulative Fatigue Damage Modelling

In Section 3.1.1 on stretch ageing, the cumulative character of mechanical damage and a possible lifetime prediction of certain e-textiles were already brought up. On a more general level, cumulative fatigue damage modelling has been established for some material failure problems. The cumulative fatigue damage model describes the changes of materials performances from the initial state to complete failure. Herein, damage is the sum of finite small partial damages that are due to a repetitive cyclic stress treatment. The prediction of the remaining lifetime of an object is a goal of the cumulative fatigue damage theory. The simplest linear cumulative fatigue damage model for a material system includes two different levels of stress intensity. In an earlier stage, such a model was developed by Palmgren [165] and Miner [166]. It is described by the mathematical expression (3).

$$D = F(n, r, f,\ T, M, \ldots) = \sum_{i=1}^{m} \Delta D_i = \sum_{i=1}^{m} \frac{n_i}{N_i} = 1 \qquad (3)$$

when the cycle number n_i reaches the maximum number of cycles to failure N_i, the damage parameter D becomes one and the object is broken. This means $D = 0$ for $n_i = 0$ and $D = 1$ for $n_i = N_i$. The damage function F is dependent from the cycle number n, the intensity r, the frequency f and other factors such as temperature T or moisture M. Under lab and modelling conditions, frequency and ambient influences are kept constant. Thus, the damage is only dependent from n_i. The ratio n_i/N_i is also called the fractional lifetime. The damages of different applied stress levels are summed up to obtain the full damage D. The chronological sequence of the different stress levels does not play a role [167]. Fatigue damage modelling does not have to be linear and occurs in a wide variety [166]. For example, the Marco–Starkey ansatz in Formula (4) shows a combination of two stress levels in a test system. The empirical exponent α is introduced to describe the course of the damage accumulation [168].

$$D = \sum_{i=1}^{m} d_i = \left(\frac{n_1}{N_1}\right)^{\alpha} + \left(\frac{n_2}{2}\right)^{\alpha} = 1 \qquad (4)$$

Cumulative fatigue damage models are often used in lifetime measurements and calculations for composite materials, e.g., reinforced textiles [169].

5. Current Practices, Challenges and Development Trends

Despite a high number of publications on the functional aging of e-textiles, there are still many challenges, e.g., in harmonised test methods, the classification of ageing behaviours. In the following sections, current practises, challenges and recent development trends are discussed.

5.1. Variation in Test Methods

The functional ageing of e-textiles poses several questions that are related to the choice of material, manufacturing process and testing methods. There are questions on the damage effects and their recording or closer investigation methods but also on the statistical evaluation.

Basically, there are two main reasons for the execution of accelerated functional ageing experiments. The first one is the inducement of material damage with a subsequent investigation of its basic mechanisms for the revelation of its causes. The results can be used for statistical modelling and/or for the further enhancement of the e-textile properties. This has especially been performed for stretch-, bending- and wash ageing. It also includes the development of new test setups, the reconstruction of available systems or the application of standards.

The second goal of accelerated ageing experiments is proofing the durability of new inventions and developments, which covers the majority of the reported literature. Here, the ageing experiments only play a minor role. However, it also provides the possibility for a deeper investigation on a material and statistical scale, leading to an enhancements of the product.

Nonetheless, many functional ageing investigations are strongly minimised and simplified. We distinguish three main problems for these minimised supporting ageing experiments. One is the selection of the ageing category, that is sometimes focused on only one method, despite the availability of other possible tests. Another crucial point is the premature stop of accelerated ageing experiments at low treatment cycle numbers before any ageing occurs. The number of cycles sometimes seems to be chosen randomly. This kind of aborted testing only shows the durability for the performed cycles. However, a lifetime investigation would be better suitable for a fair comparison and for the estimation of lifetime under application conditions. The lack of ageing in some cases may also be derived from the low strain level intensity. Soft ageing methods do not generate material damage and by choosing the right conditions, e-textiles will survive any cycle numbers. This hinders simple comparability with other e-textiles.

As the research subject of accelerated aging is rather new and there are no standardised test methods, a fair judgment of the performance of the functionality as function over time is rather difficult.

We suggest considering the manufacturing process of a conductive textile for deciding on the test methods, apart from the textile substrate, conductive material and protection method. Additionally, the contacts to electrical devices and the intended application of e-textile play a role by defining the suitable test regime.

A good possibility to cope with several problems is the application of standards for classical textiles onto e-textiles. The EN ISO 12947 standardised Martindale abrasion testing works for testing all kind of coatings but also yarn-integrated textiles. Despite several attempts of the simulation of wash ageing, we consider wash testing according to ISO 6330 standard testing with domestic washing machines as the best solution to demonstrate washing durability.

5.2. Ageing Classification and Non-Ageing Behaviour

The ageing classes are mainly derived from the experimental testing setup. An example is mechanical ageing with stretching, bending or abrasion as the subcategory. The material damage mechanism does not play a role for the classification, e.g., "delamination-ageing" or "layer-breakage-ageing". The same is valid for wash ageing. Although comprising mechanical and chemical ageing, it exists as its own category. This is due to its experimental setup that contains only special segments of mechanical and chemical ageing treatments. The separation in mechanical and chemical aging is, however, reasonable and desired for a deeper investigation of the basic mechanisms. For simple durability testing the term wash ageing is more suitable, as, e.g., the number of machine wash cycles to failure is a magnitude for durability.

Chemical ageing is rather complex as it is due to the exposure to reactive substances, e.g., oxygen, acids or bases, but also leads to a change in the chemical composition of the conductive layer in e-textiles. The electrical ageing does not play a big role for e-textiles. It comprises damages due to the electrical operation that seems to play a small role in their extent compared to the other ageing mechanisms.

While the functional ageing was mainly due to material damaging, we summarised all non-permanent kinds of behaviour in the expression "non-ageing behaviour". Non-ageing behaviour is a direct response to an external influence that lasts for the time of impact with a subsequent full recovery. In contrast to the non-ageing behaviour, the reversible or recoverable ageing is permanent after an external influence but recovery is possible with the application of a further external effort.

5.3. Types of Failure

Durability testing of e-textiles is very often performed until failure occurs. There are several possibilities being considered for the indication of failure. On a material level, there is crack formation, delamination, connection breakage and micro-island formation of thin metal layers. These damages are directly detected by microscopic imaging or indirectly by, e.g., measurement of the intensity of IR spectra of the textile polymer or metal content. On an electrical level, capacitance, EMI-shielding and especially electrical conductivity, i.e., electrical resistance, are important and mostly drawn for the indication of functionality and failure.

Considering the electrical resistance, two main types of failure for conductive lines were found. Regarding the course of the electrical resistance during testing, there is sudden fatigue and stepwise fatigue. For sudden fatigue, the resistance/voltage/current signal of the object remains on a constant level during an accelerated ageing experiment until it changes strongly in a fast step. This is caused by the interruption of the conductive line due to the full breakage and was found, e.g., for wires and contact points. The constant remaining signal over a long time period is an advantage of this behaviour. However, on the other side, the remaining lifetime can only be estimated on a statistical basis, not from the measurement of electrical parameters.

For stepwise fatigue, the change of the electrical resistance is a slow but continuous process. It may start with the first cycle of an ageing test and show a certain course with elapsing time or cycle number. The average electrical resistance increases, often with a high slope at the beginning, but then leads into a saturation value. For this behaviour, it needs to be tested, in what way the changing electrical resistance influences a transmitted signal. On the other hand, a fixed course of the electrical resistance provides the possibility of modelling and making the gain of damage predictable just by measuring the electrical resistance. A further and frequent problem of the stepwise fatigue-type failure is an increasing variation of the values, due to a rising gauge factor. The stepwise fatigue does not have a clear point of failure. The end of lifetime needs to be determined ahead in dependence of the resistance level and the gauge factor. Causes for stepwise fatigue are mainly the dissolution of the conductor, delamination of coatings, formation of cracks in coatings and conductive layer breakages.

5.4. Cumulative Fatigue Damage Modelling for e-Textiles

Modelling the electrical resistance in dependence of a resistor network for the complex motion of textiles is a theoretical exercise. In the course of this review, an example for modelling the electrical resistance for stepwise fatigue was presented by Xue et al. among stretch ageing [166]. The statistical based lifetime modelling for sudden fatigue-type of failure was shown for bending and stretch ageing by Bossuyt and others [90,103]. Cumulative fatigue damage modelling was performed on different stress levels for e-textiles using the Basquin and Manson–Coffin relations.

These modelling approaches refer to stretching or bending. However, also abrasion, washing, oxidation or chemical attacks impact conductive textiles. Under realistic conditions, even several types of ageing may influence the quality of a textile during its lifetime. To summarize all effects, we propose an extension of the cumulative fatigue damage formula to all ageing categories. A hypothetical formula could sum up all damages of an e-textile during its lifetime. An example following Palmgren, Miner (1) and others could be an "all-in" approach (5):

$$D_{total} = \sum_{i=1}^{m} D_{mechanical} + \sum_{i=1}^{m} D_{wash} + \sum_{i=1}^{m} D_{chemical} + \sum_{i=1}^{m} D_{electrical} = 1 \quad (5)$$

$$\sum_{i=1}^{m} D_{mechanical} = \sum_{i=1}^{m} D_{stretch} + \sum_{i=1}^{m} D_{bend/fold} + \sum_{i=1}^{m} D_{abrasion} \quad (6)$$

$$\sum_{i=1}^{m} D_{wash} = F\left(T, speed, c_{detergent}, t, m\right) \quad (7)$$

$$\sum_{i=1}^{m} D_{chemical} = \sum_{i=1}^{m} D_{pH} + \sum_{i=1}^{m} D_{solvent} + \sum_{i=1}^{m} D_{salt} + \sum_{i=1}^{m} D_{oxidation} \quad (8)$$

In this proposal, the damage parameter $D_{ageing\ test}$ becomes 1, if the performed number of cycles of a treatment reaches the maximum number of cycles to failure, as in the original model. This could be stretching cycles, bending cycles, washing cycles, immersion cycles or immersion time (of maximum immersion time). As the cumulative damage value reaches one the e-textile is not functional any more. The total cumulative damage D_{total} is a sum of all different kinds of ageing, i.e., mechanical ageing, wash ageing, chemical ageing and electrical ageing. The mechanical damage (6) may be divided into damage from stretching, folding and abrasion. The example of the washing damage shows a function dependent from temperature T, the spinning speed, the concentration of detergent c, washing time t and washing cycles m (7). The chemical damage as damage from different pH values, solvents, salts and oxidation (8).

This summarising formula shows the complexity of impacts on e-textiles, especially when going into detail. It could also be suitable for a simplified description of the performed lifetime investigation and the behaviour of the textile. It is meant to give a more realistic view on the functional ageing of e-textiles because of the multiple impacts.

Considering the functional ageing of e-textiles as an accumulative event, new questions arise and more research is needed, e.g., concerning the interference of ageing methods or damage causes. Are there cross reactions? What role does the sequence of ageing tests play? What role does possible reversible ageing play? These questions need to be answered within each ageing category. However, the influence of the separated ageing mechanisms onto each other and on the overall ageing behaviour of e-textiles needs to be clarified as well.

6. Conclusions

The functional ageing of e-textiles is a very complex topic due to the wide variety of material systems and the multiple possible constrains and impacts. On the other hand, this research field is of high importance when considering the rapidly increased applications of e-textile, particularity in the sectors of healthcare and lifeguarding with a very high demand in durability and reliability.

The biggest remaining challenge is the combined character of the ageing behaviour as different mechanisms influence each other and cannot easily be separated for detailed investigations. From the material point of view, fibre chemistry with its specific polymer degradation behaviour and substrate structure (woven, knitted, embroidered) can have influences on the ageing of the conductive functionality. Further, different integration technologies (conductive yarns, TFP, coating and deposition) also show great influence on the functional ageing behaviour, in particular under the mechanical loading of e-textiles. On the other hand, different functional ageing processes (mechanical, chemical and electrical) also often take place simultaneously under typical textile usage conditions. Thus, research on functional ageing of e-textiles requires an interdisciplinarity approach where strong collaboration between different disciplines is required, e.g., material science, textile and polymer chemistry, electrical engineering and testing.

For obtaining a better understanding of the functional ageing behaviour and to predict the usage lifetime and safety of e-textiles, accelerated functional ageing tests are proven to be a suitable tool. Often, it is also utilised to differentiate and highlight new developments of e-textiles. Here, conductivity and electrical resistance, respectively, are the most important and most used as indicators for functional ageing. Some e-textiles also need the electrical capacitance or antenna parameters for characterisation. For the simulation of damages, a wide variety of treatment methods is available in the literature. While typical action modes in mechanical ageing such as stretching, bending and abrasion could be realised separately to investigate their influence in details (Section 3.1), challenges remain in chemical ageing and ageing against washing and laundry. Chemical ageing often combines damages of both conductive structures and textile substrates and mechanical stresses are also involved in washing and laundry; thus, making a separation of different ageing mechanisms difficult (Sections 3.2 and 3.3).

With regard to the visualisation of damages caused by ageing, scanning electron microscopy, laser microscopy, optical microscopy or X-ray imaging illustrate surficial processes for ageing and non-ageing behaviour. These show two main general mechanisms for the emergence of damage. Firstly, there is the formation of cracks and breakages in conductive layers that lead to a temporary decrease or interruption of electrical currents. There is also the breakage of connections and wires that directly interrupts conductive lines. Secondly, there is the loss of conductive material, which can be due to the delamination of conductive layers, dissolution of conductive material or transformation into a non-conductive material. Hence, also the determination of the metal content is an indicator for functional ageing.

At practical application conditions of e-textiles, it is very often that functional ageing is a combination of all mentioned ageing processes. It remains a big challenge to separately investigate each mechanism independent from each other. Thus, a helpful approach would be the cumulative damage modelling to obtain a combined view on the functional ageing behaviour of e-textiles.

Furthermore, as there are very few standardised ageing testing methods for conductive textiles, e-textiles and smart textiles available, we suggest considering the material system, manufacturing process, mounted devices and planned application to perform lifetime testing using several methods. For wash ageing, the use of the EN ISO 6630 for e-textile testing is quite convincing and appropriate. The same is valid for standardised Martindale abrasion testing according to EN ISO 12947. A similar recommendation is given in [154].

Author Contributions: Literature research, C.B.; detailed concept and discussion, C.B.; general concept, C.B. and T.P.; writing, C.B. and T.P.; methodology, T.B. and T.P.; supervision, T.P.; project administration and funding acquisition, T.B. and T.P. All authors have read and agreed to the published version of the manuscript.

Funding: This research was funded by the Austrian Funding Agency FFG, grant nrs. 874746 (Smart Tex Aging) and 882502 (TCCV2).

Institutional Review Board Statement: Not applicable.

Informed Consent Statement: Not applicable.

Data Availability Statement: Not applicable.

Conflicts of Interest: The authors declare no conflict of interest.

Nomenclature

Accelerated ageing	Experiment in a lab that simulates a possible impact. The expected time period of usage is shortened: often by a cyclic treatment with a higher frequency that in application or with higher temperature or general harsher conditions.
Conductivity	(Here short for electrical conductivity.) The inverse value of the electrical resistance, dimension for quality of a conductor.
Course (knitting)	Direction perpendicular to the direction of knitting.
Electrical resistance	Ratio of voltage and current in an electrical conductor, dimension for quality of a conductor or insulator.
Elongation	Relative extension, mostly in %.
EMI shielding	Protection from electromagnetic interferences, absorption of electromagnetic fields.
Encapsulation	Soft or rigid protection coating.
Galvanostatic charging	A device that keeps a certain current constant in an electrolytic cell.
Gauge factor GF	$\Delta R/\varepsilon$; electrical resistance change per elongation.
Non-ageing behaviour	Material behaviour that is not permanent or directly connected with the formation of damage; details in text.
Percolation theory	The randomized occupation of a grid that leads to clusters; here applied for the formation of conductive paths in blends of conductive and non-conductive material.
Ply yarn	Twisted threads that form a yarn.
Voltammetric cycling	Electrical measurement method for recording a current/voltage diagram while varying the voltage.
Wale (knitting)	Direction of knitting.
Warp yarn	Yarn in direction of weaving.
Weft yarn	Yarn perpendicular to direction of weaving.

References

1. Ehrmann, G.; Ehrmann, A. Electronic Textiles. *Encyclopedia* **2021**, *1*, 115–130. [CrossRef]
2. Cherenack, K.; van Pieterson, L. Smart textiles: Challenges and opportunities. *J. Appl. Phys.* **2012**, *112*, 91301. [CrossRef]
3. Van Langenhove, L.; Hertleer, C. Smart clothing: A new life. *Int. J. Cloth. Sci. Tech.* **2004**, *16*, 63–72. [CrossRef]
4. Landsiedel, J.; Root, W.; Aguiló-Aguayo, N.; Duelli, H.; Bechtold, T.; Pham, T. Multi-Point Flexible Temperature Sensor Array and Thermoelectric Generator Made from Copper-Coated Textiles. *Sensors* **2021**, *21*, 3742. [CrossRef] [PubMed]
5. Pizarro, F.; Villavicencio, P.; Yunge, D.; Rodríguez, M.; Hermosilla, G.; Leiva, A. Easy-to-Build Textile Pressure Sensor. *Sensors* **2018**, *18*, 1190. [CrossRef] [PubMed]
6. Matsouka, D.; Vassiliadis, S.; Bayramol, D.V. Piezoelectric textile fibres for wearable energy harvesting systems. *Mater. Res. Express* **2018**, *5*, 65508. [CrossRef]
7. Pham, T.; Bechtold, T. Chapter 9—Conductive Fibres. In *Handbook of Fibrous Materials*, 1st ed.; Hu, J., Kumar, B., Lu, J., Eds.; Wiley-VCH Verlag: Weinheim, Germany, 2020; pp. 237–265.
8. Bresee, R.R. General Effects of Ageing on Textiles. *J. Artif. Intell. Conscious.* **1986**, *25*, 39. [CrossRef]
9. Root, W.; Wright, T.; Caven, B.; Bechtold, T.; Pham, T. Flexible Textile Strain Sensor Based on Copper-Coated Lyocell Type Cellulose Fabric. *Polymers* **2019**, *11*, 784. [CrossRef]
10. Mecnika, V.; Scheulen, K.; Anderson, C.; Horr, M.; Breckenfelder, C. Joining technologies for electronic textiles. In *Electronic Textiles*; Woodhead Publishing: Sawston, UK, 2015; pp. 133–153. [CrossRef]
11. Merilampi, S.; Laine-Ma, T.; Ruuskanen, P. The characterization of electrically conductive silver ink patterns on flexible sub-strates. *Microelectron. Reliab.* **2009**, *49*, 782–790. [CrossRef]
12. Sebok, E.B.; Taylor, R.L. Carbon Blacks. In *Encyclopedia of Materials: Science and Technology*; Elsevier: Amsterdam, The Netherlands, 2001.
13. Bessonova, A.; Samiul, M.K.; Gartseeva, H.; Bailey, M.J.A. Highly reproducible printable graphite strain gauges for flexible devices. *Sens. Actuators A* **2014**, *206*, 75–80. [CrossRef]
14. Yun, Y.J.; Hong, W.G.; Kim, W.-J.; Jun, Y.; Kim, B.H. A Novel Method for Applying Reduced Graphene Oxide Directly to Electronic Textiles from Yarns to Fabrics. *Adv. Mater.* **2013**, *25*, 5701–5705. [CrossRef]
15. Jiang, K.; Li, Q.; Fan, S. Spinning continuous carbon nanotube yarns. *Nature* **2002**, *419*, 801. [CrossRef]
16. Heeger, A.J. Semiconducting and Metallic Polymers: The Fourth Generation of Polymeric Materials. *J. Phys. Chem. B* **2001**, *105*, 8475–8491. [CrossRef]
17. Invernale, M.A.; Ding, Y.; Sotzing, G.A. All-Organic Electrochromic Spandex. *ACS Appl. Mater. Interfaces* **2010**, *2*, 296–300. [CrossRef]

18. Boutrois, J.; Jolly, R.; Pétrescu, C. Process of polypyrrole deposit on textile. Product characteristics and applications. *Synth. Met.* **1997**, *85*, 1405–1406. [CrossRef]
19. MacDiarmid, A.G. Synthetische Metalle: Eine neue Rolle für organische Polymere (Nobel-Vortrag). *Angew. Chem.* **2001**, *113*, 2649–2659. [CrossRef]
20. Ikkala, O.; Laakso, J.; Väkiparta, K.; Virtanen, E.; Ruohonen, H.; Järvinen, H.; Taka, T.; Passiniemi, P.; Österholm, J.-E.; Cao, Y.; et al. Counter-ion induced processibility of polyaniline: Conducting melt processible polymer blends. *Synth. Met.* **1995**, *69*, 97–100. [CrossRef]
21. Yang, J.P.; Rannou, P.; Planès, J.; Pron, A.; Nechtschein, M. Preparation of low density polyethylene-based polyaniline conducting polymer composites with low percolation threshold via extrusion. *Synth. Met.* **1998**, *93*, 169–173. [CrossRef]
22. Devaux, E.; Koncar, V.; Kim, B.; Campagne, C.; Roux, C.; Rochery, M.; Saihi, D. Processing and characterization of conductive yarns by coating or bulk treatment for smart textile applications. *Trans. Inst. Meas. Control.* **2007**, *29*, 355–376. [CrossRef]
23. Kim, S.H.; Seong, J.H.; Oh, K.W. Effect of dopant mixture on the conductivity and thermal stability of polyaniline/nomex conductive fabric. *J. Appl. Polym. Sci.* **2002**, *83*, 2245–2254. [CrossRef]
24. Tan, S.N.; Ge, H. Investigation into vapour-phase formation of polypyrrole. *Polymer* **1996**, *37*, 965–968. [CrossRef]
25. Xue, P.; Park, K.; Tao, X.; Chen, W.; Cheng, X. Electrically conductive yarns based on PVA/carbon nanotubes. *Compos. Struct.* **2007**, *78*, 271–277. [CrossRef]
26. Pomfret, S.; Adams, P.; Comfort, N.; Monkman, A. Advances in processing routes for conductive polyaniline fibres. *Synth. Met.* **1999**, *101*, 724–725. [CrossRef]
27. Okuzaki, H.; Ishihara, M. Spinning and Characterization of Conducting Microfibers. *Macromol. Rapid Commun.* **2003**, *24*, 261–264. [CrossRef]
28. Hegemann, D.; Amberg, M.; Ritter, A.; Heuberger, M. Recent developments in Ag metallised textiles using plasma sputtering. *Mater. Technol.* **2009**, *24*, 41–45. [CrossRef]
29. Ismar, E.; Zaman, S.U.; Bahadir, S.; Kalaoglu, F.; Koncar, V. Seam Strength and Washability of Silver Coated Polyamide Yarns. *IOP Conf. Ser. Mater. Sci. Eng.* **2018**, *460*, 12053. [CrossRef]
30. Hannula, P.-M.; Peltonen, A.; Aromaa, J.; Janas, D.; Lundström, M.; Wilson, B.; Koziol, K.; Forsén, O. Carbon nanotube-copper composites by electrodeposition on carbon nanotube fibers. *Carbon* **2016**, *107*, 281–287. [CrossRef]
31. Gosh, S. Electroless copper deposition: A critical review. *Thin Solid Films* **2019**, *669*, 641–658. [CrossRef]
32. Geng, L.; Zhu, P.; Guo, R.; Lai, X.; Xiao, H.; Ren, E.; Zhou, M. Fabrication and characterization of electrically conductive copper coated poly(p-phenylene-2,6-benzobisoxazole) yarn. *Mater. Technol.* **2019**, *35*, 767–776. [CrossRef]
33. Schwarz, A.; Hakuzimana, J.; Westbroek, P.; De Mey, G.; Priniotakis, G.; Nyokong, T.; Van Langenhove, L. A study on the morphology of thin copper films on para-aramid yarns and their influence on the yarn's electro-conductive and mechanical properties. *Text. Res. J.* **2012**, *82*, 1587–1596. [CrossRef]
34. Roh, J.-S.; Chi, Y.-S.; Kang, T.J. Wearable textile antennas. *Int. J. Fash. Des. Technol. Educ.* **2010**, *3*, 135–153. [CrossRef]
35. Yang, B.; Tao, X.; Cai, J.; Yu, T. Strain sensing behavior of textile structures made of stainless steel continuous filament yarns under uni-axial tensile loading. *Int. Conf. Smart Mater. Nanotechnol. Eng.* **2007**, *6423*, 780624. [CrossRef]
36. Acti, T.; Chauraya, A.; Zhang, S.; Whittow, W.G.; Seager, R.; Vardaxoglou, J.C.; Dias, T. Embroidered Wire Dipole Antennas Using Novel Copper Yarns. *IEEE Antennas Wirel. Propag. Lett.* **2015**, *14*, 638–641. [CrossRef]
37. Guo, L.; Berglin, L.; Mattila, H. Improvement of electro-mechanical properties of strain sensors made of elastic-conductive hybrid yarns. *Text. Res. J.* **2012**, *82*, 1937–1947. [CrossRef]
38. Shahzad, A.; Ali, Z.; Ali, U.; Khaliq, Z.; Zubair, M.; Kim, I.S.; Hussain, T.; Khan, M.Q.; Rasheed, A.; Qadir, M.B. Development and characterization of conductive ring spun hybrid yarns. *J. Text. Inst.* **2019**, *110*, 141–150. [CrossRef]
39. Alagirusamy, R.; Eichhoff, J.; Gries, T.; Jockenhoevel, S. Coating of conductive yarns for electro-textile applications. *J. Text. Inst.* **2013**, *104*, 270–277. [CrossRef]
40. Zhang, D.; Miao, M.; Niu, H.; Wei, Z. Core-Spun Carbon Nanotube Yarn Supercapacitors for Wearable Electronic Textiles. *ACS Nano* **2014**, *8*, 4571–4579. [CrossRef] [PubMed]
41. Kerfeldt, M.A.; Lund, A.; Walkenström, P. Textile sensing glove with piezoelectric PVDF fibers and printed electrodes of PEDOT:PSS. *Text. Res. J.* **2015**, *85*, 1789–1799. [CrossRef]
42. Atalay, O.; Kennon, W.R. Knitted Strain Sensors: Impact of Design Parameters on Sensing Properties. *Sensors* **2014**, *14*, 4712–4730. [CrossRef]
43. Atalay, O.; Kennon, W.R.; Demirok, E. Weft-Knitted Strain Sensor for Monitoring Respiratory Rate and Its Electro-Mechanical Modeling. *IEEE Sens. J.* **2015**, *15*, 110–122. [CrossRef]
44. Atalay, O.; Tuncay, A.; Husain, M.D.; Kennon, W.R. Comparative study of the weft-knitted strain sensors. *J. Ind. Text.* **2016**, *46*, 1212–1240. [CrossRef]
45. Shanbeh, M.; Emadi, M. Effect of weft density and percentage of stainless steel fiber content of weft yarn on electrical prop-erties of woven fabric strain sensors. *J. Text. Inst.* **2016**, *107*, 958–966.
46. Gunnarsson, E.; Karlsteen, M.; Berglin, L.; Stray, J. A novel technique for direct measurements of contact resistance between interlaced conductive yarns in a plain weave. *Text. Res. J.* **2015**, *85*, 499–511. [CrossRef]
47. Banaszczyk, J.; De Mey, G.; Schwarz, A.; Van Langenhove, L. Current Distribution Modelling in Electroconductive Textiles. In Proceedings of the 14th International Conference of Mixed Design, Ciechocinek, Poland, 21–23 June 2007; pp. 418–423. [CrossRef]

48. Beex, L.; Verberne, C.; Peerlings, R.R. Experimental identification of a lattice model for woven fabrics: Application to electronic textile. *Compos. Part A Appl. Sci. Manuf.* **2013**, *48*, 82–92. [CrossRef]
49. Zhang, X.; Zhong, Y. An improved theoretical model of the resistive network for woven structured electronic textile. *J. Text. Inst.* **2020**, *111*, 235–248. [CrossRef]
50. De Kok, M.; de Vries, H.; Pacheco, K.; van Heck, G. Failure modes of conducting yarns in electronic-textile applications. *Text. Res. J.* **2015**, *85*, 1749–1760. [CrossRef]
51. Roh, J.-S.; Chi, Y.-S.; Lee, J.-H.; Nam, S.; Kang, T.J. Characterization of embroidered inductors. *Smart Mater. Struct.* **2010**, *19*. [CrossRef]
52. Mattheij, P.; Gliesche, K.; Feltin, D. Talored Fibre Placement—Mechanical Properties and Applications. *J. Reinf. Plast. Compos.* **1998**, *17*, 774–786. [CrossRef]
53. Vena, A.; Koski, K.; Moradi, E.; Babar, A.A.; Sydänheimo, L.; Ukkonen, L.; Tentzeris, M.M. An Embroidered Two-Dimensional Chipless Strain Sensor for Wireless Structural Deformation Monitoring. *IEEE Sens. J.* **2013**, *13*, 4627–4637. [CrossRef]
54. Mecnika, V.; Hoerr, M.; Krieviņš, I.; Jockenhoevel, S.; Gries, T. Technical Embroidery for Smart Textiles: Review. *Mater. Sci. Text. Cloth. Technol.* **2014**, *9*, 56–63. [CrossRef]
55. Sippach, T.; Dahy, H.; Uhlig, K.; Grisin, B.; Carosella, S.; Middendorf, P. Structural Optimization through Biomimetic-Inspired Material-Specific Application of Plant-Based Natural Fiber-Reinforced Polymer Composites (NFRP) for Future Sustainable Lightweight Architecture. *Polymers* **2020**, *12*, 3048. [CrossRef] [PubMed]
56. Kang, T.J.; Choi, A.; Kim, D.-H.; Jin, K.; Seo, D.K.; Jeong, D.H.; Hong, S.-H.; Park, Y.W.; Kim, Y.H. Electromechanical properties of CNT-coated cotton yarn for electronic textile applications. *Smart Mater. Struct.* **2011**, *20*, 15004. [CrossRef]
57. Babaahmadi, V.; Montazera, M.; Gaob, W. Surface modification of PET fabric through in-situ reduction and crosslinking of graphene oxide: Towards developing durable conductive fabric coatings. *Colloids Surfaces A* **2018**, *545*, 16–25. [CrossRef]
58. Panhuis, M.I.H.; Wu, J.; Ashraf, S.A.; Wallace, G. Conducting textiles from single-walled carbon nanotubes. *Synth. Met.* **2007**, *157*, 358–362. [CrossRef]
59. Ding, Y.; Invernale, M.A.; Sotzing, G.A. Conductivity Trends of PEDOT-PSS Impregnated Fabric and the Effect of Conductivity on Electrochromic Textile. *ACS Appl. Mater. Interfaces* **2010**, *2*, 1588–1593. [CrossRef]
60. Opwis, K.; Knittel, D.; Gutmann, J.S. Oxidative in situ deposition of conductive PEDOT:PTSA on textile substrates and their application as textile heating element. *Synth. Met.* **2012**, *162*, 1912–1918. [CrossRef]
61. Garnier, F.; Hajlaoui, R.; Yassar, A.; Srivastava, P. All-Polymer Field-Effect Transistor Realized by Printing Techniques. *Science* **1994**, *265*, 1684–1686. [CrossRef]
62. Yong, S.; Owen, J.; Beeby, S. Solid-State Supercapacitor Fabricated in a Single Woven Textile Layer for E-Textiles Applications. *Adv. Eng. Mater.* **2018**, *20*, 860. [CrossRef]
63. Kim, T.; Song, H.; Ha, J.; Kim, S.; Kim, D.; Chung, S.; Lee, J.; Hong, Y. Inkjet-printed stretchable single-walled carbon nanotube electrodes with excellent mechanical properties. *Appl. Phys. Lett.* **2014**, *104*, 113103. [CrossRef]
64. Hu, B.; Li, D.; Manandharm, P.; Fan, Q.; Kasilingam, D.; Calvert, P. CNT/conducting polymer composite conductors impart high flexibility to textile electroluminescent devices. *J. Mater. Chem.* **2012**, *22*, 1598–1605. [CrossRef]
65. Paul, G.; Torah, R.; Beeby, S.; Tudor, J. The development of screen printed conductive networks on textiles for biopotential monitoring applications. *Sens. Actuators A Phys.* **2014**, *206*, 35–41. [CrossRef]
66. Wang, Y.; Wang, Y.; Chen, J.-J.; Guo, H.; Liang, K.; Marcus, K.; Peng, Q.-L.; Zhang, J.; Feng, Z.-S. A facile process combined with inkjet printing, surface modification and electroless deposition to fabricate adhesion-enhanced copper patterns on flexible polymer substrates for functional flexible electronics. *Electrochim. Acta* **2016**, *218*, 24–31. [CrossRef]
67. Tan, X.-Q.; Liu, J.-Y.; Niu, J.-R.; Liu, J.-Y.; Tian, J.-Y. Recent Progress in Magnetron Sputtering Technology Used on Fabrics. *Materials* **2018**, *11*, 1953. [CrossRef] [PubMed]
68. Shahidi, S.; Moazzenchi, B.; Ghoranneviss, M. A review-application of physical vapor deposition (PVD) and related methods in the textile industry. *Eur. Phys. J. Appl. Phys.* **2015**, *71*, 31302. [CrossRef]
69. Kettlgruber, G.; Kaltenbrunner, M.; Siket, C.M.; Moser, R.; Graz, I.M.; Schwödiauer, R. Siegfried Bauer, Intrinsically stretchable and rechargeable batteries for self-powered stretchable electronics. *J. Mater. Chem. A* **2013**, *1*, 5505–5508. [CrossRef]
70. Root, W.; Aguiló-Aguayo, N.; Pham, T.; Bechtold, T. Conductive layers through electroless deposition of copper on woven cellulose lyocell fabrics. *Surf. Coat. Technol.* **2018**, *348*, 13–21. [CrossRef]
71. Ali, A.; Baheti, V.; Javaid, M.U.; Militky, J. Enhancement in ageing and functional properties of copper-coated fabrics by subsequent electroplating. *Appl. Phys. A* **2018**, *124*, 651. [CrossRef]
72. Zhao, Y.; Cai, Z.; Fu, X.; Song, B.; Zhu, H. Electrochemical deposition and characterization of copper crystals on polyaniline/poly(ethylene terephthalate) conductive textiles. *Synth. Met.* **2013**, *175*, 1–8. [CrossRef]
73. Kost, J.; Narkis, M.; Foux, A. Resistivity behavior of carbon-black-filled silicone rubber in cyclic loading experiments. *J. Appl. Polym. Sci.* **1984**, *29*, 3937–3946. [CrossRef]
74. Pramanik, P.K.; Khastagir, D.; Saha, T.N. Effect of extensional strain on the resistivity of electrically conductive nitrile-rubber composites filled with carbon filler. *J. Mater. Sci.* **1993**, *28*, 3539–3546. [CrossRef]
75. Li, X.; Hu, H.; Hua, T.; Xu, B.; Jiang, S. Wearable strain sensing textile based on one-dimensional stretchable and weavable yarn sensors. *Nano Res.* **2018**, *11*, 5799–5811. [CrossRef]

76. Xue, P.; Tao, X.; Tsang, H. In situ SEM studies on strain sensing mechanisms of PPy-coated electrically conducting fabrics. *Appl. Surf. Sci.* **2007**, *253*, 3387–3392. [CrossRef]
77. Wang, J.; Xue, P.; Tao, X. Strain sensing behavior of electrically conductive fibers under large deformation. *Mater. Sci. Eng. A* **2011**, *528*, 2863–2869. [CrossRef]
78. Hu, Y.; Zhao, T.; Zhu, P.; Zhang, Y.; Liang, X.; Sun, R.; Wong, C.-P. A low-cost, printable, and stretchable strain sensor based on highly conductive elastic composites with tunable sensitivity for human motion monitoring. *Nano Res.* **2018**, *11*, 1938–1955. [CrossRef]
79. Sadeqi, A.; Nejad, H.R.; Alaimo, F.; Yun, H.; Punjiya, M.; Sonkusale, S.R. Washable Smart Threads for Strain Sensing Fabrics. *IEEE Sens. J.* **2018**, *18*, 1. [CrossRef]
80. Wang, J.; Xue, P.; Tao, X.; Yu, T. Strain Sensing Behavior and Its Mechanisms of Electrically Conductive PPy-Coated Fabric. *Adv. Eng. Mater.* **2014**, *16*, 565–570. [CrossRef]
81. Zheng, Y.; Li, Y.; Dai, K.; Liu, M.; Zhou, K.; Zheng, G.; Liu, C.; Shen, C. Conductive thermoplastic polyurethane composites with tunable piezoresistivity by modulating the filler dimensionality for flexible strain sensors. *Compos. Part A Appl. Sci. Manuf.* **2017**, *101*, 41–49. [CrossRef]
82. Kong, J.-H.; Jang, N.-S.; Kim, S.-H.; Kim, J.-M. Simple and rapid micropatterning of conductive carbon composites and its application to elastic strain sensors. *Carbon* **2014**, *77*, 199–207. [CrossRef]
83. Cheng, Y.; Zhang, H.; Wang, R.; Wang, X.; Zhai, H.; Wang, T.; Jin, Q.; Sun, J. Highly Stretchable and Conductive Copper Nanowire Based Fibers with Hierarchical Structure for Wearable Heaters. *ACS Appl. Mater. Interfaces* **2016**, *8*, 32925–32933. [CrossRef]
84. Wang, C.; Zhang, M.; Xia, K.; Gong, X.; Wang, H.; Yin, Z.; Guan, B.; Zhang, Y. Intrinsically Stretchable and Conductive Textile by a Scalable Process for Elastic Wearable Electronics. *ACS Appl. Mater. Interfaces* **2017**, *9*, 13331–13338. [CrossRef]
85. Amberg, M.; Grieder, K.; Barbadoro, P.; Heuberger, M.; Hegemann, D. Electromechanical Behavior of Nanoscale Silver Coatings on PET Fibers. *Plasma Process. Polym.* **2008**, *5*, 874–880. [CrossRef]
86. Koshi, T.; Nomura, K.-I.; Yoshida, M. Electronic Component Mounting for Durable E-Textiles: Direct Soldering of Components onto Textile-Based Deeply Permeated Conductive Patterns. *Micromachines* **2020**, *11*, 209. [CrossRef]
87. Koshi, T.; Nomura, K.-I.; Yoshida, M. Requirements for Durability Improvement of Conductive Patterns Permeated in Textiles under Cyclic Tensile Deformation. *Micromachines* **2019**, *10*, 721. [CrossRef]
88. Brosteaux, D.; Axisa, F.; Gonzalez, M.; Vanfleteren, J. Design and Fabrication of Elastic Interconnections for Stretchable Electronic Circuits. *IEEE Electron Device Lett.* **2007**, *28*, 552–554. [CrossRef]
89. Gonzalez, M.; Axisa, F.; Bulcke, M.V.; Brosteaux, D.; Vandevelde, B.; Vanfleteren, J. Design of metal interconnects for stretchable electronic circuits. *Microelectron. Reliab.* **2008**, *48*, 825–832. [CrossRef]
90. Bossuyt, F.; Guenther, J.; Löher, T.; Seckel, M.; Sterken, T.; de Vries, J. Cyclic endurance reliability of stretchable electronic substrates. *Microelectron. Reliab.* **2011**, *51*, 628–635. [CrossRef]
91. Isaia, C.; McNally, D.; McMaster, S.A.; Branson, D.T. Effect of mechanical preconditioning on the electrical properties of knitted conductive textiles during cyclic loading. *Text. Res. J.* **2019**, *89*, 445–460. [CrossRef]
92. Jones, J.; Lacour, S.P.; Wagner, S.; Suo, Z. Stretchable wavy metal interconnects. *J. Vac. Sci. Technol. A* **2004**, *22*, 1723–1725. [CrossRef]
93. Graz, I.M.; Cotton, D.P.J.; Lacour, S.P. Extended cyclic uniaxial loading of stretchable gold thin-films on elastomeric substrates. *Appl. Phys. Lett.* **2009**, *94*, 71902. [CrossRef]
94. Inoue, M.; Itabashi, Y.; Tada, Y. Development of Bimodal Electrically Conductive Pastes with Ag Micro- and Nano-fillers for Printing Stretchable E-textile Systems. In Proceedings of the European Microelectronics Packaging Conference, Friedrichshafen, Germany, 14–16 September 2015.
95. Inoue, M.; Itabashi, Y.; Tada, Y. Recoverable fatigue behaviour of stretchable conductive adhesives with Ag loading. *Electron. Syst.-Integr. Conf.* **2014**. [CrossRef]
96. Itabashi, Y.; Inoue, M.; Tada, Y. Effect of filler morphology on fatigue properties of stretchable wires printed with Ag pastes. *ICEP* **2014**, 752–755. [CrossRef]
97. Inoue, M.; Tada, Y.; Itabashi, Y.; Tokumaru, T. Deterioration and recovery of electrical conductivity during fatigue testing of stretchable wires printed onto fabrics using Ag-loaded electrically conductive pastes. *MRS Proc.* **2014**, *1628*. [CrossRef]
98. Komolafe, O.; Torah, R.N.A.; Tudor, M.J.; Beeby, S.P.; Tudor, J. Modelling and experimental validation of the effect of the elastic properties of fabrics on the durability of screen printed e-textiles. *Smart Mater. Struct.* **2018**, *27*, 75046. [CrossRef]
99. Guo, R.; Yu, Y.; Xie, Z.; Liu, X.; Zhou, X.; Gao, Y.; Liu, Z.; Zhou, F.; Yang, Y.; Zheng, Z. Matrix-Assisted Catalytic Printing for the Fabrication of Multiscale, Flexible, Foldable, and Stretchable Metal Conductors. *Adv. Mater.* **2013**, *25*, 3343–3350. [CrossRef]
100. Heo, J.S.; Shishavan, H.H.; Soleymanpour, R.; Kim, J.; Kim, I. Textile-Based Stretchable and Flexible Glove Sensor for Monitoring Upper Extremity Prosthesis Functions. *IEEE Sens. J.* **2020**, *20*, 1754–1760. [CrossRef]
101. Stempien, Z.; Rybicki, E.; Patykowska, A.; Rybicki, T.; Szynkowska, M.I. Shape-programmed inkjet-printed silver electro-conductive layers on textile surfaces. *J. Ind. Text.* **2018**, *47*, 1321–1341. [CrossRef]
102. De Vries, H.; Cherenack, K.H.; Schuurbiers, R.; van Os, K. Failure Modes in Textile Interconnect Lines. *IEEE Electron. Dev. Lett.* **2012**, *33*, 10. [CrossRef]
103. De Vries, H.; Cherenack, K.H. Endurance behavior of conductive yarns. *Microelectron. Reliab.* **2014**, *54*, 327–330. [CrossRef]

104. Yu, Y.; Yan, C.; Zheng, Z. Polymer-Assisted Metal Deposition (PAMD): A Full-Solution Strategy for Flexible, Stretchable, Compressible, and Wearable Metal Conductors. *Adv. Mater.* **2014**, *26*, 5508–5516. [CrossRef]
105. Zahid, M.; Papadopoulou, E.L.; Athanassiou, A.; Bayer, I.S. Strain-responsive mercerized conductive cotton fabrics based on PEDOT:PSS/graphene. *Mater. Des.* **2017**, *135*, 213–222. [CrossRef]
106. Cai, D.; Zhou, J.; Duan, P.; Luo, G.; Zhang, Y.; Fu, F.; Liu, X. A hierarchical structure of l-cysteine/Ag NPs/hydrogel for conductive cotton fabrics with high stability against mechanical deformation. *Cellulose* **2018**, *25*, 7355–7367. [CrossRef]
107. Lim, S.J.; Bae, J.H.; Han, J.H.; Jang, S.J.; Oh, H.J.; Lee, W.; Kim, S.H.; Ko, J.H. Foldable and washable fully textile-based pressure sensor. *Smart Mater. Struct.* **2020**, *29*, 55010. [CrossRef]
108. Lee, W.J.; Park, J.Y.; Nam, H.J.; Choa, S.-H. The development of a highly stretchable, durable, and printable textile electrode. *Text. Res. J.* **2019**, *89*, 4104–4113. [CrossRef]
109. Kazani, I.; Hertleer, C.; De Mey, G.; Schwarz, A.; Guxho, G.; Van Langenhove, L. Electrical Conductive Textiles Obtained by Screen Printing. *Fibres Text. East. Eur.* **2012**, *20*, 57–63.
110. Wang, H.; Zhou, H.; Gestos, A.; Fang, J.; Niu, H.; Ding, J.; Lin, T. Robust, electro-conductive, self-healing superamphiphobic fabric prepared by one-step vapour-phase polymerisation of poly(3,4-ethylenedioxythiophene) in the presence of fluorinated decyl polyhedral oligomeric silsesquioxane and fluorinated alkyl silane. *Soft Matter* **2013**, *9*, 277–282. [CrossRef]
111. Boschi, A.; Arosio, C.; Cucchi, I.; Bertini, F.; Catellani, M.; Freddi, G. Properties and performance of polypyrrole (PPy)-coated silk fibers. *Fibers Polym.* **2008**, *9*, 698–707. [CrossRef]
112. Kerfeldt, M.A.; Straat, M.; Walkenström, P. Influence of coating parameters on textile and electrical properties of a poly(3,4-ethylenedioxythiophene):poly(styrene sulfonate)/polyurethane-coated textile. *Text. Res. J.* **2013**, *83*, 2164–2176. [CrossRef]
113. Malm, V.; Seoane, F.; Nierstrasz, V. Characterisation of Electrical and Stiffness Properties of Conductive Textile Coatings with Metal Flake-Shaped Fillers. *Materials* **2019**, *12*, 3537. [CrossRef]
114. Montarsolo, A.; Varesano, A.; Mossotti, R.; Rombaldoni, F.; Periolatto, M.; Mazzuchetti, G.; Tonin, C. Enhanced Adhesion of Conductive Coating on Plasma-Treated Polyester Fabric: A Study on the Ageing Effect. *J. Appl. Polym. Sci.* **2012**, *126*, 1385–1393. [CrossRef]
115. Garg, S.; Hurren, C.; Kaynak, A. Improvement of adhesion of conductive polypyrrole coating on wool and polyester fabrics using atmospheric plasma treatment. *Synth. Met.* **2007**, *157*, 41–47. [CrossRef]
116. Lin, T.; Wang, L.; Wang, X.; Kaynak, A. Polymerising pyrrole on polyester textiles and controlling the conductivity through coating thickness. *Thin Solid Films* **2005**, *479*, 77–82. [CrossRef]
117. Varesano, A.; Tonin, C.; Ferrero, F.; Stringhetta, M. Thermal stability and flame resistance of polypyrrole-coated PET fibres. *J. Therm. Anal. Calorim.* **2008**, *94*, 559–565. [CrossRef]
118. Bogan, K.; Seyam, A.-F.M.; Slade, J. Evaluation of the Electrical Integrity of E-Textiles Subjected to Abrasion. *J. Text. Appar. Technol. Manag.* **2019**, *11*, 1.
119. Maity, S.; Chatterjee, A.; Singh, B.; Singh, A.P. Polypyrrole based electro-conductive textiles for heat generation. *J. Text. Inst.* **2014**, *105*, 887–893. [CrossRef]
120. Moss, B.K.; Burford, R.P. Thermal ageing studies of conducting polypyrroles. *Polym. Int.* **1991**, *26*, 225–231. [CrossRef]
121. Tabačiarová, J.; Mičušík, M.; Fedorko, P.; Omastová, M. Study of polypyrrole aging by XPS, FTIR and conductivity measurements. *Polym. Degrad. Stab.* **2015**, *120*, 392–401. [CrossRef]
122. Jerkovic, I.; Koncar, V.; Grancaric, A.M. New Textile Sensors for In Situ Structural Health Monitoring of Textile Reinforced Thermoplastic Composites Based on the Conductive Poly(3,4-ethylenedioxythiophene)-poly(styrenesulfonate) Polymer Complex. *Sensors* **2017**, *17*, 2297. [CrossRef]
123. Varesano, A.; Ferrer, A.I.; Tonin, C. Electrical performance and stability of polypyrrole coated PET fibres. *e-Polymers* **2007**, *7*, 248–261. [CrossRef]
124. Gong, S.; Lai, D.; Wang, Y.; Yap, L.W.; Si, K.J.; Shi, Q.; Jason, N.N.; Sridhar, T.; Uddin, H.; Cheng, W. Tattoolike Polyaniline Microparticle-Doped Gold Nanowire Patches as Highly Durable Wearable Sensors. *ACS Appl. Mater. Interfaces* **2015**, *7*, 19700–19708. [CrossRef]
125. Lee, S.; Park, C.H. Electric heated cotton fabrics with durable conductivity and self-cleaning properties. *RSC Adv.* **2018**, *8*, 31008–31018. [CrossRef]
126. Liu, X.; Chang, H.; Li, Y.; Huck, W.T.S.; Zheng, Z. Polyelectrolyte-Bridged Metal/Cotton Hierarchical Structures for Highly Durable Conductive Yarns. *ACS Appl. Mater. Interfaces* **2010**, *2*, 529–535. [CrossRef]
127. Liu, S.; Hu, M.; Yang, J. A facile way of fabricating a flexible and conductive cotton fabric. *J. Mater. Chem. C* **2016**, *4*, 1320–1325. [CrossRef]
128. Hwang, H.-J.; Joo, S.-J.; Kim, H.-S. Copper Nanoparticle/Multiwalled Carbon Nanotube Composite Films with High Electrical Conductivity and Fatigue Resistance Fabricated via Flash Light Sintering. *ACS Appl. Mater. Interfaces* **2015**, *7*, 25413–25423. [CrossRef]
129. Jia, L.-C.; Ding, K.-Q.; Ma, R.-J.; Wang, H.-L.; Sun, W.-J.; Yan, D.-X.; Li, B.; Li, Z.-M. Highly Conductive and Machine-Washable Textiles for Efficient Electromagnetic Interference Shielding. *Adv. Mater. Technol.* **2019**, *4*, 201800503. [CrossRef]
130. Luo, J.; Wang, L.; Huang, X.; Li, B.; Guo, Z.; Song, X.; Lin, L.; Tang, L.-C.; Xue, H.; Gao, J. Mechanically Durable, Highly Conductive, and Anticorrosive Composite Fabrics with Excellent Self-Cleaning Performance for High-Efficiency Electromagnetic Interference Shielding. *ACS Appl. Mater. Interfaces* **2019**, *11*, 10883–10894. [CrossRef]

131. Erdoğan, M.K.; Karakışla, M.; Saçak, M. Conductive polyaniline-polythiophene/poly(ethylene terephthalate) composite fiber: Effects of pH and washing processes on surface resistivity. *J. Appl. Polym. Sci.* **2015**, *132*, 41979. [CrossRef]
132. Priniotakis, G.; Westbroek, P.; Van Langenhove, L.; Hertleer, C. Electrochemical impedance spectroscopy as an objective method for characterization of textile electrodes. *Trans. Inst. Meas. Control.* **2007**, *29*, 271–281. [CrossRef]
133. Ghosh, S.; Ganguly, S.; Remanan, S.; Das, N.C. Fabrication and investigation of 3D tuned PEG/PEDOT: PSS treated conductive and durable cotton fabric for superior electrical conductivity and flexible electromagnetic interference shielding. *Compos. Sci. Technol.* **2019**, *181*, 107682. [CrossRef]
134. Han, X.; Lv, L.; Yu, D.; Wu, X.; Li, C. Conductive Core–Shell Aramid Nanofibrils: Compromising Conductivity with Mechanical Robustness for Organic Wearable Sensing. *ACS Appl. Mater. Interfaces* **2019**, *11*, 3466–3473. [CrossRef]
135. Tang, Z.; Yao, D.; Hu, S.; Du, D.; Shao, W.; Tang, B.; Fan, J.-M.; Tang, X.-G.; Gao, J. Highly conductive, washable and super-hydrophobic wearable carbonnanotubes e-textile for vacuum pressure sensors. *Sens. Actuators A* **2020**, *303*, 111710. [CrossRef]
136. Wu, Y.; Mechael, S.S.; Chen, Y.; Carmichael, T.B. Solution Deposition of Conformal Gold Coatings on Knitted Fabric for E-Textiles and Electroluminescent Clothing. *Adv. Mater. Technol.* **2018**, *3*, 201700292. [CrossRef]
137. Guo, Y.; Otley, M.T.; Li, M.; Zhang, X.; Sinha, S.K.; Treich, G.M.; Sotzing, G.A. PEDOT:PSS "Wires" Printed on Textile for Wearable Electronics. *ACS Appl. Mater. Interfaces* **2016**, *8*, 26998–27005. [CrossRef]
138. Liu, X.; Guo, R.; Shi, Y.; Deng, L.; Li, Y. Durable, Washable, and Flexible Conductive PET Fabrics Designed by Fiber Interfacial Molecular Engineering. *Macromol. Mater. Eng.* **2016**, *301*, 1383–1389. [CrossRef]
139. Ren, J.; Wang, C.; Zhang, X.; Carey, T.; Chen, K.; Yin, Y.; Torrisi, F. Environmentally-friendly conductive cotton fabric as flexible strain sensor based on hot press reduced graphene oxide. *Carbon* **2017**, *111*, 622–630. [CrossRef]
140. Schwarz, A.; Hakuzimana, J.; Kaczynska, A.; Banaszczyk, J.; Westbroek, P.; McAdams, E.; Moody, G.; Chronis, Y.; Priniotakis, G.; De Mey, G.; et al. Gold coated para-aramid yarns through electroless deposition. *Surf. Coat. Technol.* **2010**, *204*, 1412–1418. [CrossRef]
141. Ankhili, A.; Tao, X.; Cochrane, C.; Coulon, D.; Koncar, V. Washable and Reliable Textile Electrodes Embedded into Underwear Fabric for Electrocardiography (ECG) Monitoring. *Materials* **2018**, *11*, 256. [CrossRef] [PubMed]
142. Varnaitė, S.; Katunskis, J. Influence of Washing on the Electric Charge Decay of Fabrics with Conductive Yarns. *Fibres Text. East. Eur.* **2009**, *17*, 69–75.
143. Zhao, Z.; Yan, C.; Liu, Z.; Fu, X.; Peng, L.-M.; Hu, Y.; Zheng, Z. Machine-Washable Textile Triboelectric Nanogenerators for Effective Human Respiratory Monitoring through Loom Weaving of Metallic Yarns. *Adv. Mater.* **2016**, *28*, 10267–10274. [CrossRef]
144. Lee, J.C.; Liu, W.; Lo, C.; Chen, C.-C. Laundering Reliability of Electrically Conductive Fabrics for E-Textile Applications. In Proceedings of the 69th Electronic Components and Technology Conference (ECTC), Las Vegas, NV, USA, 28–31 May 2019; pp. 1826–1832. [CrossRef]
145. Ankhili, A.; Zaman, S.U.; Tao, X.; Cochrane, C.; Koncar, V.; Coulon, D. How to Connect Conductive Flexible Textile Tracks to Skin Electrocardiography Electrodes and Protect Them Against Washing. *IEEE Sens. J.* **2019**, *19*, 11995–12002. [CrossRef]
146. Zaman, S.U.; Tao, X.; Cochrane, C.; Koncar, V. Understanding the Washing Damage to Textile ECG Dry Skin Electrodes, Embroidered and Fabric-Based; set up of Equivalent Laboratory Tests. *Sensors* **2020**, *20*, 1272. [CrossRef]
147. Ismar, E.; Zaman, S.U.; Tao, X.; Cochrane, C.; Koncar, V. Effect of Water and Chemical Stresses on the Silver Coated Polyamide Yarns. *Fibers Polym.* **2019**, *20*, 2604–2610. [CrossRef]
148. Zaman, S.U.; Tao, X.; Cochrane, C.; Koncar, V. Launderability of Conductive Polymer Yarns Used for Connections of E-textile Modules: Mechanical Stresses. *Fibers Polym.* **2019**, *20*, 2355–2366. [CrossRef]
149. Atakan, R.; Tufan, H.A.; Zaman, S.U.; Cochrane, C.; Bahadir, S.K.; Koncar, V.; Kalaoglu, F. Protocol to assess the quality of transmission lines within smart textile structures. *Measurement* **2020**, *152*, 107194. [CrossRef]
150. Gaubert, V.; Gidik, H.; Bodart, N.; Koncar, V. Towards Washable Smart Textiles: Investigating the Impact of Oxidizing Bleaching Agents on Silver-Plated Textile Electrodes. In Proceedings of the 19th World Textile Conference on Textiles at the Crossroads—AUTEX 2019, Ghent, Belgium, 11–15 June 2019.
151. Gaubert, V.; Gidik, H.; Bodart, N.; Koncar, V. Investigating the Impact of Washing Cycles on Silver-Plated Textile Electrodes: A Complete Study. *Sensors* **2020**, *20*, 1739. [CrossRef]
152. Karim, N.; Afroj, S.; Tan, S.; He, P.; Fernando, A.; Carr, C.; Novoselov, K. Scalable Production of Graphene-Based Wearable E-Textiles. *ACS Nano* **2017**, *11*, 12266–12275. [CrossRef]
153. Toivonen, M.; Björninen, T.; Sydänheimo, L.; Ukkonen, L.; Rahmat-Samii, Y. Impact of Moisture and Washing on the Performance of Embroidered UHF RFID Tags. *IEEE Antennas Wirel. Propag. Lett.* **2013**, *12*, 1590–1593. [CrossRef]
154. Rotzler, S.; von Krshiwoblozki, M.; Schneider-Ramelow, M. Washability of e-textiles: Current testing practices and the need for standardization. *Text. Res. J.* **2021**. [CrossRef]
155. Nagaraju, G.; Raju, G.S.R.; Ko, Y.H.; Yu, J.S. Hierarchical Ni–Co layered double hydroxide nanosheets entrapped on conductive textile fibers: A cost-effective and flexible electrode for high-performance pseudocapacitors. *Nanoscale* **2016**, *8*, 812–825. [CrossRef] [PubMed]
156. Zhao, C.; Shu, K.; Wang, C.; Gambhir, S.; Wallace, G. Reduced graphene oxide and polypyrrole/reduced graphene oxide composite coated stretchable fabric electrodes for supercapacitor application. *Electrochim. Acta* **2015**, *172*, 12–19. [CrossRef]
157. Hana, J.-W.; Meyyappan, M. Copper oxide resistive switching memory for e-textile. *AIP Adv.* **2011**, *1*, 32162. [CrossRef]

158. Vojtech, L.; Neruda, M. Modelling of Surface and Bulk Resistance for Wearable Textile Antenna Design. *Przegląd Elektrotechniczny* **2013**, *89*, 2.
159. Li, L.; Liu, S.; Ding, F.; Hua, T.; Au, W.M.; Wong, K.-S. Electromechanical analysis of length-related resistance and contact resistance of conductive knitted fabrics. *Text. Res. J.* **2012**, *82*, 2062–2070. [CrossRef]
160. Liu, S.; Tong, J.; Yang, C.; Li, L. Smart E-textile: Resistance properties of conductive knitted fabric—Single pique. *Text. Res. J.* **2016**, *87*, 1669–1684. [CrossRef]
161. Tokarska, M. Characterization of electro-conductive textile materials by its biaxial anisotropy coefficient and resistivity. *J. Mater. Sci. Mater. Electron.* **2019**, *30*, 4093–4103. [CrossRef]
162. Li, Y.; Wang, S.; Zhang, Y.; Zhang, Y. Electrical properties and morphology of polypropylene/epoxy/glass fiber composites filled with carbon black. *J. Appl. Pol.* **2005**, *98*, 1142–1149. [CrossRef]
163. Hasan, M.; Cherif, C.; Foisal, A.; Onggar, T.; Hund, R.; Nocke, A. Development of conductive coated Polyether ether ketone (PEEK) filament for structural health monitoring of composites. *Compos. Sci. Technol.* **2013**, *88*, 76–83. [CrossRef]
164. Xue, P.; Tao, X.; Kwok, K.W.; Leung, M.; Yu, T. Electromechanical Behavior of Fibers Coated with an Electrically Conductive Polymer. *Text. Res. J.* **2004**, *74*, 929–936. [CrossRef]
165. Palmgreen, A. Die Lebensdauer von Kugellagern. *Z. Ver. Dtsch. Ing.* **1924**, *68*, 339–341.
166. Miner, M.A. Cumulative Damage in Fatigue. *J. Appl. Mech.* **1945**, *12*, A159–A164. [CrossRef]
167. Hwang, W.; Han, K. Cumulative Damage Models and Multi-Stress Fatigue Life Prediction. *J. Compos. Mater.* **1986**, *20*, 125–153. [CrossRef]
168. Marco, S.M.; Starkey, W.L. concept of fatigue damage. *Trans. ASME* **1954**, *76*, 627–632.
169. Found, M.S.; Quaresimin, M. Two-stage fatigue loading of woven carbon fibre reinforced laminates. *Fatigue Fract. Eng. Mater. Struct.* **2003**, *26*, 17–26. [CrossRef]

MDPI AG
Grosspeteranlage 5
4052 Basel
Switzerland
Tel.: +41 61 683 77 34

Sensors Editorial Office
E-mail: sensors@mdpi.com
www.mdpi.com/journal/sensors

www.ingramcontent.com/pod-product-compliance
Lightning Source LLC
Chambersburg PA
CBHW042016260926
51850CB00020B/3423
9783725825097